AF182571

FIGHTING AT THE LEGAL BOUNDARIES

Fighting at the Legal Boundaries

CONTROLLING THE USE OF FORCE IN CONTEMPORARY CONFLICT

Kenneth Watkin

OXFORD
UNIVERSITY PRESS

OXFORD
UNIVERSITY PRESS

Oxford University Press is a department of the University of Oxford. It furthers the University's objective of excellence in research, scholarship, and education by publishing worldwide. Oxford is a registered trademark of Oxford University Press in the UK and certain other countries.

Published in the United States of America by Oxford University Press
198 Madison Avenue, New York, NY 10016, United States of America.

© Oxford University Press 2016

All rights reserved. No part of this publication may be reproduced, stored in a retrieval system, or transmitted, in any form or by any means, without the prior permission in writing of Oxford University Press, or as expressly permitted by law, by license, or under terms agreed with the appropriate reproduction rights organization. Inquiries concerning reproduction outside the scope of the above should be sent to the Rights Department, Oxford University Press, at the address above.

You must not circulate this work in any other form
and you must impose this same condition on any acquirer.

Library of Congress Cataloging-in-Publication Data

Names: Watkin, Kenneth, author.
Title: Fighting at the legal boundaries : controlling the use of force in
contemporary conflict / Kenneth Watkin.
Description: New York : Oxford University Press, 2016. | Includes
bibliographical references and index.
Identifiers: LCCN 2015037858 | ISBN 9780190457976 ((hardback) : alk. paper)
Subjects: LCSH: Aggression (International law) | War (International law) |
Self-defense (International law) | Combatants and noncombatants
(International law) | Non-state actors (International relations) |
Preemptive attack (Military science)
Classification: LCC KZ6396 .W38 2016 | DDC 341.5—dc23 LC record available at
http://lccn.loc.gov/2015037858

Note to Readers

This publication is designed to provide accurate and authoritative information in regard to the subject matter covered. It is based upon sources believed to be accurate and reliable and is intended to be current as of the time it was written. It is sold with the understanding that the publisher is not engaged in rendering legal, accounting, or other professional services. If legal advice or other expert assistance is required, the services of a competent professional person should be sought. Also, to confirm that the information has not been affected or changed by recent developments, traditional legal research techniques should be used, including checking primary sources where appropriate.

(Based on the Declaration of Principles jointly adopted by a Committee of the American Bar Association and a Committee of Publishers and Associations.)

You may order this or any other Oxford University Press publication by visiting the Oxford University Press website at www.oup.com

To my wife, Maureen, and daughters, Jessica, Allison, and Emily.
You are everything to me.

Contents

Preface xiii
Abbreviations xix

PART ONE | INTRODUCTION
1. An Outline of the Challenges 3
1.1. The Security Challenge 3
1.2. The Legal Challenge 10
1.3. The Operational Challenge: The Emerging "Law of 9/11" 20
1.4. The Outline 24

PART TWO | THE INTERACTION BETWEEN NORMATIVE FRAMEWORKS
2. Controlling State Involvement in Conflict 33
2.1. Introduction 33
2.2. A Focus on the State 34
2.3. The Destructive Power of the State 37
2.3.1. Reintroduction of Just War Theory 37
2.3.2. Inter-State Conflict: Limiting the Recourse to War 40
2.4. Noninternational Armed Conflict 43
2.4.1. Domestic Authority, International Law, and Non-State Actors 43
2.4.2. Internal Conflicts with an "International Character" 44
2.4.3. Conflict Evolution 46
2.4.4. Cross-Border Attacks by Non-State Actors 48
2.4.5. An Interpretive Struggle 52

3. Applying the Self-Defense Principles During Armed Conflict 55
3.1. Introduction 55
3.2. The Continuing Application of Self-Defense Principles 56
3.3. An "Overarching" Application 58
3.4. A More "Limited" Theory 63
3.5. Additional Restraining Mechanisms 70

3.6. The Self-Defense Principles: A Strategic or Tactical Issue? 72
3.6.1. Self-Defense and the Levels of War 72
3.6.2. Targeting: A Matter of Strategy and Tactics 78
3.6.3. Applying Self-Defense Principles Across the Armed Conflict Spectrum 83
3.7. Less Certain Boundaries 88

4. States, "Proper Authority," and Conflict 91
4.1. Introduction 91
4.2. Humanitarian Law and "Proper Authority" 92
4.3. Impact of the Proper Authority Principle 97
4.4. The State and Governance 99
4.4.1. Public, Private, and Mixed War 99
4.4.2. Governance 101
4.4.3. A Reluctant Application of Humanitarian Law 104
4.4.4. A Policy Solution for Legal Uncertainty 109
4.4.5. Small Wars and "Ungoverned Spaces" 112
4.5. The State, Governance, and the Intersection of Laws 119

5. The Humanitarian Law and Human Rights Law Interface 121
5.1. Introduction 121
5.2. Exclusionary Arguments and Operational Reality 122
5.2.1. The "Special Law" Debate 122
5.2.2. The Missing Discussion: Similarities and Overlap 131
5.3. A More Intimate History 134
5.3.1. Prior to the Mid-Twentieth Century 134
5.3.2. In the Aftermath of World War II 139
5.3.2.1. Neglect 139
5.3.2.2. Forced Integration 141
5.3.2.3. Divergence and Growing Reconciliation 143
5.4. Overlap and Convergence 152

PART THREE | THE THREAT, THE STATE RESPONSE, AND LEGAL UNCERTAINTY
6. Contemporary Threats: Insurgency and Terrorism 159
6.1. Introduction 159
6.2. Insurgency 160
6.2.1. Criteria for an Insurgency 160
6.2.2. Insurgent Organization 162
6.2.3. Purpose of the Conflict 166
6.3. Insurgency and Crime 168
6.3.1. The Nature of the Threat 168
6.3.2. Challenging the State 173
6.3.3. Other Criminal Activity 179

6.4. Terrorism and Terrorists 180
6.4.1. Terrorism 180
6.4.2. The "Terrorists" 190
6.4.3. Terrorism and Insurgency: Al Qaeda and an Islamic State 197
6.5. Counterinsurgency and Counterterrorism: An Integrated Response 208

7. Counterinsurgency and Converging Norms 215
7.1. Introduction 215
7.2. Counterinsurgency in the Twenty-First Century 216
7.3. The Primacy of Law Enforcement 223
7.4. Control and Operational Choices 228
7.4.1. The Nature of Control 228
7.4.2. The *"Targeted Killing" Case* and Control 231
7.5. The IED and Suicide Bombing Threat: Arrest or Kill? 237
7.6. Military Advantage and Counterinsurgency 244
7.6.1. Insurgents and Target "Value" 244
7.6.2. Strategic Advantage, Proportionality, and Counterinsurgency 254
7.6.2.1. Strategic Military Advantage and Civilian Casualties 254
7.6.2.2. Proportionality and Counterinsurgency 258
7.7. Overlapping Normative Frameworks 261

8. Counterterrorism and the "Away Game" 267
8.1. Introduction 267
8.2. A Lack of Consensus 269
8.3. Counterterrorism and "Direct Action" 271
8.4. New Technology/New Terminology: Placing Drones in Context 277
8.4.1. A Unique Weapon? 277
8.4.2. Drones: Special Terms and High-Profile Attacks 278
8.4.3. The "Ordinariness" of Drones 280
8.4.4. Controlling Drone Use 285
8.5. Legal Frameworks for Responding to Transnational Terrorism 287
8.5.1. The Restricted "Law Enforcement" Theory 287
8.5.2. The Permissive "Conduct of Hostilities" Approach 300
8.5.3. The "Self-Defense" Option 311
8.6. Seeking a Solution 318
8.7. An Emerging Broader "Unwilling or Unable" Justification 322

9. Non-State Actors and Armed Conflict 327
9.1. Introduction 327
9.2. Inter-State Conflict 330
9.3. Categorizing Conflict with Non-State Actors 335
9.3.1. Non-State Actors Controlled by a State 335
9.3.2. Harboring and Active Support 337

9.3.3. Entering the Territory of Another State: International Armed Conflict 339
9.3.4. The Israeli Supreme Court and Cross-Border Terrorism 341
9.3.4.1. Targeted Killing, Occupation, and "Conflicts of an International Character" 341
9.3.4.2. A Unique Security Threat 343
9.3.4.3. A Contrast: The American Approach 345
9.3.4.4. The Practical Impact 347
9.3.4.5. Unlawful Combatants 348
9.3.5. Extraterritorial Law Enforcement 349
9.3.6. Transnational Armed Conflict 352
9.3.7. Noninternational Armed Conflict and the Proper Authority 356
9.3.7.1. Not of an International Character: *Hamdan v. Rumsfeld* 356
9.3.7.2. A Concentration of Power and Responsibility 358
9.3.7.3. Borders or Barriers? 359
9.3.7.4. Acting in Place of the Territorial State 361
9.4. Determining the Existence of a Noninternational Armed Conflict 363
9.4.1. Conflict Criteria and the Intensity of the Violence 363
9.4.2. Group Organization 370
9.4.2.1. Varied Organizational Types 370
9.4.2.2. The Kosovo Liberation Army 372
9.4.3. The "Totality of the Circumstances" and Other Criteria 375
9.5. Addressing the Challenge 378

PART FOUR | APPLYING FORCE ACROSS THE CONFLICT SPECTRUM
10. Self-Defense and the Protection of Nationals 383
10.1. Introduction 383
10.2. Self-Defense and the Extraterritorial Protection of Nationals 385
10.2.1. A State Responsibility 385
10.2.2. Taking Action Across the Conflict Spectrum 395
10.3. Protecting Nationals and Hostage Rescue 399
10.3.1. Rescues During Armed Conflict 399
10.3.2. "Iconic" Counterterrorism: A Thunderbolt Strikes Entebbe 405
10.3.3. Modern Counterterrorism: Sierra Leone 412
10.3.4. Short-Term Conflicts 415
10.3.5. Rescues and Law Enforcement 421
10.3.5.1. State-Conducted Hostage Rescue: Columbia 421
10.3.5.2. Hostage Rescue and Foreign Involvement 423
10.3.5.3. States Acting Unilaterally to Rescue Their Own Nationals: Somalia 426
10.3.5.4. Law Enforcement and the Protection of Nationals 431
10.4. Converging Tactics: Special Forces, Law Enforcement, and Hostilities 433

11. Law Enforcement and "Self-Defense" 443
11.1. Introduction 443

11.2. Distinguishing Law Enforcement 445
11.2.1. The Terminology Challenge 445
11.2.2. Self-Defense: Common Roots and Different Goals 447
11.2.3. Law Enforcement and Its Unique Role 449
11.2.4. Imminence and the Nature of the Threat 454
11.2.5. Mission Accomplishment 458
11.3. Doctrinal Confusion: State Self-Defense and Law Enforcement 464
11.3.1. Rules of Engagement Doctrine 464
11.3.2. United Nations Operations and "Robust Peacekeeping" 467
11.3.3. United States Rules of Engagement and "Self-Defense" 474
11.4. The Way Ahead 479

12. The Narrow Operational and Normative Gap 483
12.1. Introduction 483
12.2. Narrowing the Gap 484
12.2.1. The Military Task: Applying Force of First and Last Resort 484
12.2.2. Force and Hostage Rescue 492
12.2.2.1. Unique Challenges and Responses 492
12.2.2.2. The Japanese Embassy, Lima, Peru, 1997 501
12.3. Explosives: The Game Changer 510
12.3.1. A Ubiquitous Weapon 510
12.3.2. Suicide Missions 515
12.3.3. Confronting the Explosive Threat 518
12.3.4. Strictly Necessary, Proportionate, and Overwhelming 521
12.4. A Unique Challenge 532

13. Limits of Law Enforcement and Human Rights Law 535
13.1. Introduction 535
13.2. The Shoot to Kill Controversy 536
13.2.1. The "Troubles" 536
13.2.2. Shooting to Kill 544
13.3. The Militarization of the Police 550
13.4. Human Rights Overreach 552
13.5. The Shoot-Down of Hijacked Aircraft 559
13.6. Limits and the Need for Clarity 564

PART FIVE | THE WAY AHEAD
14. A Holistic Solution 569
14.1. Introduction 569
14.2. From "Conventional" to "Hybrid" to "Holistic" 571
14.3. Responding to Transnational Attacks and Threats to Nationals 575
14.3.1. Armed Conflict or Law Enforcement? 575
14.3.2. Commencement of the Armed Conflict 584
14.3.3. Separating Armed Conflict from Law Enforcement 587

14.4. Applying a Human Rights–Based Law Enforcement Approach 588
14.4.1. As a Matter of Law 588
14.4.2. A Policy Choice by States 592
14.4.3. Discretion Exercised by the "Chain of Command" 595
14.4.4. A Law Enforcement Response to Attacks at Home and Abroad 597
14.4.5. Applying Law and Policy 600
14.5. Confronting Transnational Violence: A Holistic Approach 604
14.6. Law Enforcement and the Search for "Normalcy" 610
14.7. Appendix: Confronting Transnational Violence: A Holistic Approach 617

15. Preparing for Twenty-First-Century Conflict 619
15.1 Introduction 619
15.2. Legal Advisors and Twenty-First-Century Warfare 620

BIBLIOGRAPHY 633
TABLE OF CASES 661
INDEX 667

Preface

ON SEPTEMBER 11, 2001, I was the Deputy Judge Advocate General/Operations for the Canadian Forces. My experience, and that of many other State legal advisors, was the events of that tragic day represented a milestone marking a paradigm shift in the urgency, scope, and type of international law issues that had to be addressed. In Ottawa, after dealing with initial questions regarding the potential shoot-down of hijacked civilian aircraft over North America, the attention of the small group of extremely dedicated and talented military lawyers that I was privileged to lead quickly turned to an exceptionally broad array of legal questions. The focus was on confronting the menace of transnational terrorism overseas and protecting our country at home.

The range of legal issues was daunting and involved many of the matters that are the subject of this book. They included the use of force in national self-defense; the classification of armed conflict; the deployment of Special Forces and conventional military units to Afghanistan; Rules of Engagement; the status, treatment, and transfer of detainees; targeting, and "targeted killing"; the use of drones; the conduct of counterinsurgency and counterterrorism operations, particularly the challenges inherent in fighting among a civilian population; force protection; the wearing of uniforms and combatant status; maritime visit and search operations; the Proliferation Security Initiative, conduct after capture training; interrogation, intelligence operations; defense of nationals and hostage rescue; "friendly fire" investigations; as well as criminal investigations and other inquiries. Other matters included the legal review of weapons, means, or methods of warfare; Status of Forces and basing arrangements; Coalition operations and interoperability with Allies and international partners; and working with other government departments to develop a "whole of government" approach to operations.

At home legal responses had to be provided concerning North American Air Defense (NORAD); domestic land and maritime security; security at international meetings/events (including during G8, Organization of American States, and the 2010 Olympics); law enforcement assistance to other government departments; base and unit security; and the development of post-9/11 national antiterrorism legislation. Added to these issues were numerous questions related to piracy, counterdrug operations, other international law enforcement operations, the continuing role of advising military commanders on the maintenance of discipline ("the habit of obedience"), and responding to domestic court challenges regarding operational matters (i.e., transfer of detainees, use of force).

What was unique about the majority of these legal challenges was their association with transnational violence by non-State actors. These threats did not fit within the classic categories of inter-State conflict, internal State violence, or international law enforcement. A common refrain within our office was that each day felt like having to write an international law exam with little or no time to prepare. It was obvious that the legal issues confronting States were not directly covered by the treaty law governing inter-State warfare, with which both practitioners and academics were most comfortable. Anyone providing legal advice regarding operations was confronted with the need to search for legal sources that could be used to provide coherent, workable guidance to military commanders. For example, issues like the status of members of the Taliban armed forces and Al Qaeda led our officers to look at older academic articles like Professor/Judge Richard Baxter's "So-Called 'Unprivileged Belligerency': Spies, Guerrillas, and Saboteurs" for guidance on how to deal with these participants on the modern battlefield.[1] After some fifty years terms from another generation, such as "unprivileged belligerent" and "unlawful combatant," quickly entered the legal lexicon.

This was not the only area where Richard Baxter's work was a harbinger of the legal issues that need to be addressed in twenty-first-century conflict. In the late 1970s, as the international community started to turn its attention toward tackling the regulation of the "small wars" occurring after World War II, he had noted: "at the core of the law of war has always been a concern for human rights of those enemy military and civilian personnel who become the victims of war."[2] It is the requirement to consider extraterritorial application of human rights, not only in the context of their incorporation into international humanitarian law but also human rights law (international and domestic) in its own right that has become one of the most significant, disputed, and enduring legal issues arising in the post-9/11 period.

The lack of treaty law governing the conduct of hostilities in noninternational armed conflict and the unique challenges presented by transnational jihadist terrorist groups also resulted in greater engagement with the academic legal community. I was a beneficiary of a forward thinking decision by the Canadian Forces to send legal officers back to school to study contemporary law issues. As a Visiting Fellow at the Harvard Human Rights

---

1 Richard R. Baxter, *So-called "Unprivileged Belligerency": Spies, Guerrillas, and Saboteurs*, *in* HUMANIZING THE LAWS OF WAR: SELECTED WRITINGS OF RICHARD BAXTER 37 (Detlev F. Vagts et al. eds., 2013) (originally published in 1951 BRIT. Y.B. INT'L L. 323).

2 Richard Baxter, *Human Rights in War*, *in id.* 341, at 346 (the text first appeared in 31 BULLETIN OF THE AMERICAN ACADEMY OF ARTS AND SCIENCES 4–13 (1977).

Program, I had a unique opportunity to study international human rights law and engage with academics who themselves were beginning to address many of the same legal issues. It was in that forum, based on work that I did regarding the role of human rights in controlling the use of force during armed conflict,[3] where the idea of writing a book on the broader use of force issues challenging the international law and military practitioners was formed.

In some ways the military seemed at least partially prepared to address the transnational terrorist threat in the sense that State armed forces had to deal with complex nontraditional conflict following the end of the Cold War. This included difficult peacekeeping and peace enforcement operations in the former Yugoslavia, Somalia, Rwanda, and East Timor, which were captured under the doctrinal heading of Operations Other Than War. However, both practitioners and the academic community struggled with the post-9/11 "new normal" as they were forced to counter transnational "terrorist" groups that were not just criminals. It quickly became evident there was little appetite among States, academics, and human rights advocates to develop new treaty law. Instead, the mantra was to ensure the existing treaties were properly applied. Unfortunately, for practitioners this approach was decidedly unhelpful in addressing the existing and emerging operational challenges.

There was also a legal conflict brewing at the strategic level between human rights and humanitarian law advocates. The atmosphere was a defensive one with law of war advocates arguing there was no, or at least little, room for human rights law to be applied, and human rights law advocates, sometimes buoyed by the decisions of regional human rights tribunals, seeking to exclusively use human rights to govern the conflict. Adding to this discord were international lawyers, long schooled in a traditional emphasis on inter-State warfare, who resisted any extension of the law governing State self-defense to non-State actor conflict. Overall the discussion seemed to be less about exploring the interpretive boundaries of these various bodies of the law, and more about excluding the operation of one body of law or another.

Since then the international legal community has increasingly been forced to address the challenges presented by this complex security environment. The threat of inter-State conflict remains, but it is the "small wars" that will remain the dominant security challenge of this era. Advances have been made in the form of international working groups preparing guidance and manuals on legal issues such as direct participation in hostilities, air and missile warfare, occupation, detention, the use of force, and cyber conflict. This effort has largely been led by the venerable International Committee of the Red Cross, and although productive there remains significant areas of legal disagreement. Various legal theories were developed on a myriad of issues, however, what has been often missing is a consensus that is helpful to practitioners. What was becoming evident from an operational perspective as 9/11 turned into a "long war" was that the various bodies of law impacting on the State use of force: the law governing the recourse to war, international humanitarian law, human rights law (international and domestic), and international criminal law had to be applied concurrently, and seamlessly in order to properly govern military operations. Practitioners use the term "operational law" to explain this holistic approach. The result was that resolving the

[3] Kenneth Watkin, *Controlling the Use of Force: A Role for Human Rights Norms in Contemporary Armed Conflict*, 98 Am. J. Int'l L. 1 (2004).

interaction between the bodies of law applicable to armed conflict has become a particularly important task.

The purpose of this book is to explore the relationship between the bodies of law applicable to the use of force in contemporary conflict. It is a discussion about their common roots and history; the unique role the State plays in maintaining internal security; how these areas of law interface and overlap; and ways that they can be applied concurrently. In other words how they fit together, rather than concentrating on their differences. The book canvases the nature of the non-State actor threat (terrorists, insurgents, and transnational criminal gangs); the unique challenges posed by counterinsurgency and counterterrorism; the use of drones and conduct of hostage rescue; the military involvement in law enforcement; the State response to "one-off" transnational terrorist attacks; and the narrowing normative gap that occurs between these bodies of law as States counter groups fighting among the civilian population. The requirement to privilege the application of human rights–based law enforcement, as a matter of law or policy, will be seen to be an essential part of contemporary security operations. However, law enforcement will also be shown to have its limits, resulting in a continuing requirement to engage in hostilities where appropriate to counter the non-State actor threat. The reality is that State security forces have been required to fight at the boundaries of the established areas of law. What must not occur is for those boundaries to become barriers to the maintenance of international security and the protection of civilian populations.

The research and writing of this book has been a fascinating journey for which I had considerable help along the way. I was certainly inspired by the many young men and women of the Canadian Forces Office of the Judge Advocate General who were frequently deployed individually to the other side of the world to provide advice that literally had life-or-death consequences. Their professionalism, sense of service, dedication to defending their country, and clear commitment to protecting the rights of humans in the broadest sense was, and remains, exceptional. It was an honor to serve as the Judge Advocate General from 2006 to 2010, and to lead and work with them. I would also like to thank Professor Dennis Mandsager as head of the International Law Department of the U.S. Naval War College for the opportunity from 2011 to 2012 to be the Charles H. Stockton Professor of International Law. In that capacity I was given the time, resources, and freedom to do a considerable amount of the thinking and research that ultimately worked its way into this book.

One of the particular advantages of practicing international law is that your professional colleagues and good friends are often from other countries. In this regard I have truly been fortunate to be able to count Professor Geoffrey Corn as an exceptional friend and colleague. He stands out as a talented academic who has worked tirelessly to address some of the most challenging international legal issues of this century, while always keeping his eye on the operational needs of military commanders. In this regard he provides a unique perspective as both a theorist and practitioner. As I am certain his other "international" friends can attest, Geoff is an exceptionally generous colleague in terms of his time and advice, and he has been unflagging in his encouragement of my writing this book. I also remain grateful for the support provided by Harvey Yarosky, QC, Lieutenant-Colonel Tammy Tremblay, Tom Black, Hoshea Gottlieb and Moran Yahav.

Writing a book can be a daunting task and I would sincerely like to thank the production and editorial team at Oxford University Press for making it such a pleasant experience.

In particular, Blake Ratcliffe for agreeing to take on this project and for his steady hand in guiding this book from proposal to completion. David Lipp and Alden Domizio have been very helpful and responsive during the production stage, as has Arun Kumar Vasu of Newgen Knowledge Works Pvt Ltd. I am especially grateful for the careful copyediting and useful comments provided by Mary Rosewood. They have all been a pleasure to work with.

Finally, and most importantly, there is my family. I have, throughout this endeavor, and indeed throughout all the other activities in which I have embarked over thirty years of marriage, been blessed to have the most supportive and loving spouse. Maureen has withstood my endless hours spent in the study; many physical absences at international conferences, meetings, and trips to "exotic" locales; and not infrequent departures into a world of thought, with the grace and understanding so characteristic of a former service spouse. She has been unflagging in her support and so typically, without a word of complaint, read far too many drafts of this manuscript. I could not have done it without her. I am also blessed with three daughters, Jessica, Allison, and Emily, who are not only a constant source of pride but have also contributed directly to an underlying motivation to want in some small way to try and make this world a better place to live—ultimately, to help humanize and reduce the violence in the conflicts in which we find ourselves all too often engaged. That can only be done through respect for the rule of law and, importantly, caring equally for our fellow human beings regardless of their background or where they live.

Kenneth Watkin
Kingston Ontario Canada
December 18, 2015

Abbreviations

ANCOP	Afghan Civil Order Police
ANP	Afghan National Police
AP I or Additional Protocol I	Protocol Additional to the Geneva Conventions of 12 August 1949, and Relating to the Protection of Victims of International Armed Conflicts, *opened for signature* Dec. 12, 1977, 1125 UNTS 3
AP II or Additional Protocol II	Protocol Additional to the Geneva Conventions of 12 August 1949, and Relating to the Protection of Victims of Non-International Armed Conflicts, opened for signature Dec. 12, 1977, Art. 6(5), 1125 U.N.T.S. 609
AQAM	Al Qaeda and Associated Movements
AQAP	Al Qaeda in the Arabian Peninsula
AQI	Al Qaeda in Iraq, also known as the Al Qaeda in the Land of Two Rivers, or Al Qaeda in Mesopotamia
AQIM	Al Qaeda in the Islamic Maghreb
AQIS	Al Qaeda in the Indian Subcontinent
ATS	Maharashtra State Anti-Terrorism Squad, an Indian police unit
AUC	United Self-Defense Groups of Columbia
BBIED	Body-Borne Improvised Explosive Device
CANSOFCOM	Canadian Special Operations Forces Command
CAT	Campaign Against Terrorism
COIN	Counterinsurgency
CPL	Civilian Protection Law

DTO	Drug Trafficking Organization
DUF	Directives on the Use of Force
EFPs	Enhanced Formed Projectiles
ETA	Euskadi Ta Askatasuna, a Basque separatist organization in Spain
European Convention on Human Rights	Convention for the Protection of Human Rights and Fundamental Freedoms, 213 U.N.T.S. 222, *entered into force* Sept. 3, 1953, *as amended by* Protocols Nos. 3, 5, 8, and 11 *which entered into force* on Sept. 21, 1970, Dec. 20, 1971, Jan. 1, 1990, and Nov. 1, 1998 *respectively*
EU	European Union
FARC	Revolutionary Armed Forces of Columbia
FATA	Federally Administered Tribal Areas
FBI	Federal Bureau of Investigation
FLCS	Front de Libération de la Côte de Somalie
FSB	Federal Security Bureau (Russia)
GAULA	Unified Action Groups for Personal Liberty, A Columbian hostage rescue unit
GC I, or First Geneva Convention	Geneva Convention (I) for the Amelioration of the Condition of the Wounded and Sick in Armed Forces in the Field, 12 August 1949, 75 U.N.T.S. 31
GC II, or Second Geneva Convention	Geneva Convention (II) for the Amelioration of the Condition of Wounded, Sick and Shipwrecked Members of Armed Forces at Sea, 12 August 1949, 75 U.N.T.S. 85
GC III, or Third Geneva Convention	Geneva Convention (III) Relative to the Treatment of Prisoners of War, 12 August 1949, 75 U.N.T.S. 135
GC IV, or Fourth Geneva Convention	Geneva Convention (IV) Relative to the Protection of Civilian Persons in Time of War, 12 August 1949, 75 U.N.T.S. 287
GIGN	Groupe d'Intervention de la Gendarmerie Nationale, a French hostage rescue unit
GSG-9	Border Protection Group 9, a German hostage rescue unit
GSPC	Group for Preaching and Combat
GWOT	Global War on Terrorism
1907 Hague Land Warfare Regulations	Regulations Respecting the Laws and Customs of War on Land, annex to Hague Convention Respecting the Laws and Customs of War on Land, October 18, 1907, 36 Stat. 2277, 1 Bevans 631
HMSU	Royal Ulster Constabulary Headquarters Mobile Support Unit
HRT	Hostage Rescue Team, the United States Federal Bureau of Investigation
ICC	International Criminal Court
ICJ	International Court of Justice
ICRC	International Committee of the Red Cross
ICTY	International Criminal Tribunal for the former Yugoslavia

IDF	Israel Defence Forces
IED	Improvised Explosive Device
IHL	International Humanitarian Law
IHRL	International Human Rights Law
INLA	Irish National Liberation Army
IRA	Irish Republican Army
ISAF	International Security Assistance Force
ISIS	Islamic State in Iraq and al Sham
ISOI	Islamic State of Iraq
KLA	Kosovo Liberation Army
LeT	Lashkar-e-Tabia
LIFG	Libyan Islamic Fighting Group
LOAC	Law of Armed Conflict
LOIAC	Law of International Armed Conflict
LRM	Least Restrictive Means
LTTE	Liberation Tigers of Tamil Eelam
LVF	Loyalist Volunteer Force
MARCO	Marine Commando, Indian marines
MDI	Markaz al-Dawa wal-Irshad
MICG	Moroccan Islamic Combatant Group
MOOTW	Military Operations Other Than War
MRTA	Túpac Amaru Revolutionary Movement
MVD	Ministry of the Interior (Russia)
NSG	National Security Guard, an Indian joint military/police hostage rescue unit
NEO	Noncombatant Evacuation Operation
NORAD	North American Air Defense
NWFP	North-West Frontier Province
PFLP	Popular Front for the Liberation of Palestine
PID	Positive Identification
PIRA	Provisional Irish Republican Army
PLO	Palestine Liberation Organization
POMLET	Police Operational Mentoring and Liaison Team
RAID	Research Assistance Intervention Dissuasion
ROE	Rules of Engagement
1998 Rome Statute	Rome Statute of the International Criminal Court 1998, 2187 U.N.T.S. 90, entry into force July 1, 2002
RUC	Royal Ulster Constabulary
SAF	Special Action Force, a Philippines National Police unit.
SAS	Special Air Service, a United Kingdom Special Forces unit
SBS	Special Boat Service, a United Kingdom maritime Special Forces unit
SCO	Specialist Firearms Command, London Metropolitan Police

SEALs	"SEa, Air Land" teams, United States Navy special operations forces unit
SM	Suicide Missions
SROE	Standing Rules of Engagement
SRUF	Standing Rules for the Use of Force
SWAT	Special Weapons and Tactics
TBI	Trans-Border Institute
TTP	Tehrek-e-Taliban Pakistan
TTPs	Tactics, Techniques and Procedures
UDA	Ulster Defence Force
UFF	Ulster Freedom Fighters
UN Basic Principles on the Use of Force	Adopted by the Eighth United Nations Congress on the Prevention of Crime and the Treatment of Offenders, Havana, Cuba, Aug. 27 to Sept. 7, 1990, Basic Principles on the Use of Force and Firearms by Law Enforcement Officials
UVF	Ulster Volunteer Force
VBIEDs	Vehicle-Borne Improvised Explosive Devices

I Introduction

> To say a military tactic is legal, or even effective, is not to say it is wise or moral in every instance. For the same human progress that gives us the technology to strike half a world away also demands the discipline to constrain that power—or risk abusing it.[1]
>
> PRESIDENT BARAK OBAMA, 2013

1 An Outline of the Challenges

1.1. The Security Challenge

The international law governing armed conflict is at a crossroads. A well-established framework of laws primarily designed to control the resort to and conduct of inter-State conflict finds itself confronted with violence in the twenty-first century that is of a significantly different nature than in previous centuries. Instead of regulating wars between States, international law is mainly being asked to control violence, which is often and somewhat misleadingly referred to as "conflicts not of an international character."[2] Historically, such conflict has been viewed as being internal to States and, as a result, predominately fell within the purview of domestic law. Instead, contemporary security threats encompass an exceptionally broad range of violence that can have international impact. Those threats include insurgencies involving violence bordering on the level of inter-State conflict, terrorism, and threats from criminal gangs that transcend national borders. This chapter introduces the security, legal, and operational challenges presented by contemporary conflict. It also provides a chapter outline of the issues addressed in this book related to security forces having to fight at the boundaries of the various bodies of law applicable to such conflict.

Numerous contemporary noninternational conflicts, such as those that have occurred in Afghanistan and Iraq, are familiar to traditional interpreters of international law, at least to the extent they represent State armed forces fighting organized insurgent armed groups.

---

[1] *Obama's Speech on Drone Policy*, N.Y. TIMES, May 23, 2013 [hereinafter the *Drone Policy*], http://www.nytimes.com/2013/05/24/us/politics/transcript-of-obamas-speech-on-drone-policy.html.

[2] 1949 Geneva Conventions, common art. 3.

However, for a variety of reasons including religious-based ideology, increased globalization, and technological advances, many non-State participants simply do not respect the borders upon which the State-focused international law system is based.[3] This lack of the acceptance of and respect for international borders is reflected in the works of the Al Qaeda[4] writer, Abu Mus'Ab al-Suri, who uses the Arabic word for the Levant, *al-Sham*, in order to avoid reference to "Syria, Lebanon, Jordan, Israel, the Palestinian Territories, and the Sinai . . . ."[5] In this regard Al Qaeda is but one of a number of Salafi jihadist–based groups[6] seeking to restore a caliphate, which would be a global Islamic state.[7] This transnational approach is evident in the emergence of the self-styled "Islamic State," which had previously adopted the title of Islamic State in Iraq and al Sham (ISIS) to reflect its broader geographic reach.[8]

It has been suggested that "[t]he final stage of Al Qaeda's Islamic war is the gradual widening of the global battlefield to the point where Al Qaeda becomes less relevant and the Islamic caliphate begins to take control."[9] Even when an argument is made that a number of contemporary "noninternational" conflicts, such as those in Afghanistan, Yemen, Somalia, and Mali, may not appear to physically extend beyond the borders of the country, the violence can be strategically linked to a common jihadist cause that is transnational in scope.[10] In reality a major form of State versus non-State actor conflict has been elevated to a different transnational plane. The security reality is that the attacks by these jihadist groups, as well as many of the resulting State counterterrorist operations, transcend national borders.[11]

The issue for many international lawyers is whether such cross-border violence takes place in the form of an armed conflict or is a matter more properly subjected to law enforcement. As President Obama confirmed in a speech on May 23, 2013, the United States views itself, after over a decade of conflict, to still be engaged in a war with Al Qaeda, the Taliban, and their associated forces.[12] The crossing of borders during these evolving conflicts is not one

---

[3] Jim Lacey ed., A Terrorist's Call to Global Jihad: Deciphering Abu Musab Al-Suri's Islamic Jihad Manifesto 51 (2008) (where Al-Suri, one of Al Qaeda's leading theorists stated: "Islam pays no heed to the borders the Crusaders drew up between our countries and nationalities, citizenship regimes, and passports they invented.").

[4] "Al Qaeda" is used as the preferred spelling for that organization. Other spellings include al-Qa'ida, al Qaeda, al Qaida, and al-Qaeda. These spellings are reflected in quotations from other source material referred to in this book.

[5] Michael W. S. Ryan, Decoding Al-Qaeda's Strategy: The Deep Battle Against America 204–5 (2013).

[6] Mark E. Stout et al., Terrorist Perspectives Project: Strategic and Operational Views of Al Qaida and Associated Movements 5–7 (2008). *See also* Jessica S. Stern & J. M. Berger, ISIS: The State of Terror 268 (2015).

[7] Stout et al., *supra* note 6, at 19.

[8] Bruce Riedel, *ISIS Bids for Global Jihad Leadership with Mosul Attack*, Al-Monitor (June 11, 2014), http://www.al-monitor.com/pulse/originals/2014/06/isis-mosul-jihad.html. (ISIS appears to be commonly interpreted to mean Islamic State in Iraq and Syria. The use of the term al Sham suggests a much broader geographic goal for its caliphate). *See also ISIS*, Mapping Military Organizations, Stanford University, *available at* http://web.stanford.edu/group/mappingmilitants/cgi-bin/groups/view/1 (last visited May 25, 2015) (where it is indicated the "[t]he variation in translation stems from the word al-Sham, which refers to an area spanning Southern Turkey through Syria to Egypt, which can be translated as 'Greater Syria' or 'the Levant.'").

[9] Max G. Manwaring, Gangs Pseudo-Militaries and Other Modern Mercenaries 104 (2010) and Abdel Bari Atwan, After Bin Laden: Al Qaeda, The Next Generation 17 (2012). *See also* Ryan, *supra* note 5, at 39.

[10] Atwan, *supra* note 9, at 18–19.

[11] National Strategy for Counterterrorism, The White House 3 (June 2011), http://www.whitehouse.gov/sites/default/files/counterterrorism_strategy.pdf (where it is indicated "[t]he preeminent security threat to the United States continues to be from al-Qa'ida and its affiliates and adherents.").

[12] *Drone Policy*, *supra* note 1.

sided. As American General Stanley McChrystal has noted regarding the United States' networked response to terrorism: "[a]t one point, we were in 27 countries simultaneously."[13] The transnational nature of the conflict is evident in the frequently controversial use by the United States of unmanned aerial vehicles or "drones"[14] to strike at non-State opponents in the territory of other States.[15]

This jihadist-based conflict has been described as a form of global insurgency that "seeks to transform the entire Islamic world and remake its relationship with the rest of the globe."[16] The response required to confront this global threat has been identified as one "that is about years, even decades, rather than months."[17] However, unlike internal conflicts, there is no centralized form of "government" to coordinate the security response to this insurgent threat. This has placed considerable strain on an international system of legal governance, which is far more decentralized than that found within the States themselves. A unique aspect of this conceptually broader form of insurgency is its linkage to transnational terrorism, which plays a strategic role "as a form of long-range power projection against outsiders who support the government they are fighting."[18] This strategic use of terrorism poses a significantly different kind of threat than the international terrorism of the 1970s and 1980s. The international community had previously viewed terrorism mainly as criminal activity amenable to a law enforcement response.

The security challenges of the twenty-first century are not limited to the radical jihadist threat. As Steven Metz has identified, the nature of insurgency is changing. In this regard, "[t]wentieth century insurgency sought to eject the state from space it controlled (usually physical territory). Contemporary insurgency is a competition for uncontrolled spaces."[19] Reflecting both State and systemic weakness, "[i]t co-exists with many others, most importantly the rise of militias, powerful criminal gangs and syndicates, informal economies, the collapse of state services, humanitarian crises or disasters, crises of identity,

---

[13] *Generation Kill: A Conversation with Stanley McChrystal*, 92 Foreign Aff. 2, 2 (Mar./Apr., 2013).

[14] The term "drones" is used in this book to simplify the discussion regarding unmanned aircraft. For other terminology, *see* Commentary on the HPCR Manual on International Law Applicable to Air and Missile Warfare, Program on Humanitarian Policy and Conflict Research at Harvard University, Rule 1 (dd), at 54 and Rule 1(ee), at 55 (2010) [hereinafter the Air and Missile Warfare Manual Commentary], http://ihlresearch.org/amw/Commentary%20on%20the%20HPCR%20Manual.pdf (where a distinction is drawn between "unmanned aerial vehicles" and "unmanned combat aerial vehicles" with the latter carrying or controlling a weapon.). *See also* Michael N. Schmitt, *Unmanned Combat Aircraft Systems and International Humanitarian Law: Simplifying the Oft Benighted Debate*, 30 B.U. Int'l L.J. 595, 595 n.2 (2012) [for use of the term unmanned combat aircraft systems (UCAS). However, it is also noted "[t]he aircraft itself is also known as an unmanned aircraft, drone or unmanned aerial vehicle."].

[15] Elizabeth Bumiller, *A Day Job Waiting for a Kill Shot a World Away*, N.Y. Times, July 29, 2012, http://www.nytimes.com/2012/07/30/us/drone-pilots-waiting-for-a-kill-shot-7000-miles-away.html?pagewanted=all, and Matt J. Martin & Charles W. Sasser, Predator Remote-Control Air War Over Iraq and Afghanistan: A Pilot's Story 30 (2010).

[16] David Kilcullen, Counterinsurgency 185 (2010).

[17] Luke Harding, Andrew Sparrow, & Kim Willsher, *David Cameron Calls for "lobal ¶ Response" after Algerian Hostage Crisis*, The Guardian, Jan. 21, 2013, http://www.guardian.co.uk/world/2013/jan/20/david-cameron-algerian-hostage-crisis (quoting UK Prime Minister David Cameron).

[18] Steven Metz, Rethinking Insurgency, Strategic Studies Institute, U.S. Army War College 8 (2007), http://www.strategicstudiesinstitute.army.mil/pdffiles/PUB790.pdf.

[19] *Id.* at 11.

and transnational terrorism."[20] There is a synergy between criminal activity and contemporary insurgency. In this regard, "where insurgency takes root, organized crime will be pervasive" since they "sprout from common roots: ineffective governance, systemic weakness and pathology, and a culture or tradition of clandestine activity."[21] Transnational crime is also intimately linked to globalization. This has "emboldened ethnic-specific criminal organizations to set up operations far beyond their home operating zones."[22] Examples of these transnational actors include Mexican drug cartels; organized crime in many newly independent Soviet states; criminal groups in Iraq;[23] and even gangs operating among the Jamaican diaspora in Toronto, New York, and London.[24]

The contemporary security threat can also be extended to other actors such as gangs, agitators, militias, and other mercenary organizations "operating as state and non-state surrogates in the murky shadows of the contemporary asymmetric, irregular, and political global security arena."[25] As has been noted, "[g]angs and Iraqi insurgents, militias, and other non-state groups share common origins based on tribalism, and therefore, it is expected that they will exhibit similar structures and behaviors."[26] Transnational crime provides "a stateless milieu in which terrorism can flourish without worries of strong state intervention."[27] It has been observed that in Afghanistan "the overlap between crime, insurgency, tribal or business patronage networks, and governing itself is often so extensive that the same individuals play roles in multiple hierarchies."[28] The result has been "a type of contemporary irregular internal and/or external conflict that is playing an increasingly sophisticated role in creating multi-dimensional threats to stability, security, and effective national sovereignty all around the world today."[29]

A particular challenge for international law is how to deal with conflicts that are fundamentally "criminal insurgencies."[30] In some instances these insurgencies are conducted by gangs for which "*[d]rug trafficking organization* is no longer a sufficient term for them; they are a *criminal paramilitary complex*."[31] There often is little difference in terms of organization between such paramilitary gangs and insurgent groups.[32] Taken together, these new threats have been described as "*non*state" (e.g., gangs, insurgents, drug traffickers,

---

20 *Id.* at 12.

21 *Id.* at 29. *See also* PAUL J. SMITH, THE TERRORISM AHEAD: CONFRONTING TRANSNATIONAL VIOLENCE IN THE TWENTY-FIRST CENTURY 186 (2008).

22 SMITH, *supra* note 21, at 186.

23 *Id.*

24 DAVID KILCULLEN, OUT OF THE MOUNTAINS: THE COMING OF AGE OF THE URBAN GUERRILLA 101 (2013).

25 MANWARING, *supra* note 9, at 3.

26 John P. Sullivan & Robert J. Bunker, *Iraq & The Americas: 3 Gen Gangs Lessons and Prospects*, *in* JOHN P. SULLIVAN & ROBERT J. BUNKER, MEXICO'S CRIMINAL INSURGENCY 1 (2012).

27 SMITH, *supra* note 21, at 186.

28 KILCULLEN, *supra* note 24, at 128.

29 MANWARING, *supra* note 9, at 4.

30 IOAN GRILLO, EL NARCO: INSIDE MEXICO'S CRIMINAL INSURGENCY 206 (2011) (where reliance is placed on Steven Metz's 1993 reference to "commercial" or "criminal" insurgencies in an essay called *The Future of Insurgency*.).

31 *Id.* at 261. *See also* Steven Dudley, *The Emergence of Mexico's Small Armies*, INSIGHT CRIME: ORGANIZED CRIME IN THE AMERICAS (Nov. 26, 2012), http://www.insightcrime.org/investigations/part-i-the-emergence-of-mexicos-small-armies.

32 THE U.S. ARMY, MARINE CORPS, COUNTERINSURGENCY FIELD MANUAL ¶ 6–98, at 237 (2007) [hereinafter THE COUNTERINSURGENCY MANUAL] ("most insurgent groups are more similar to organized crime in their organizational structure and relations with the populace than they are to military units.").

transnational criminal organizations, terrorists, warlords), where conflict "thrive[s] in 'ungoverned or weakly governed space' between or within various host countries," and *intra*state, "which tends to involve direct and indirect conflict between state and nonstate actors."[33] Such conflicts challenge not only traditional notions of what constitutes armed conflict but also how amenable the resulting violence is to a law enforcement response. As a result, "[t]he power to deal with these kinds of situations is no longer hard combat firepower or even the more benign police power."[34] These situations of insecurity challenge traditional notions of the dividing line between armed conflict and ordinary law enforcement.

The complexity of the contemporary security environment may require a significant change in the conceptualization of armed conflict (i.e., war) and its associated security threats. Hugh Strachan has stressed the advantage of adopting a unitary vision of war, rather than applying the more traditional binary approach that divides conflict into regular versus irregular warfare (or, in contemporary terms, inter-State war versus noninternational conflict-insurgency/counterinsurgency/terrorism).[35] In effect, "war is war,"[36] and the fundamental nature of war remains the same regardless of how it is categorized.[37] This approach seeks to avoid the challenge of trying to equip military forces to fight radically different conceptualizations of conflict, including engaging in stabilization operations. It is a model that presents considerable advantages including, from the point of view of a historian, the avoidance of "the impact of short-term issues from swamping a sense of perspective on long-term continuities."[38]

Military doctrinal thinking has considerable distance to travel before such a fundamental change in thinking takes hold. In the interim, much of the discussion, including from a legal perspective, continues to center on how to reconcile conventional inter-State conflict with its seemingly unconventional noninternational counterpart. However, it is also evident to some analysts that war is morphing "into seemingly unfamiliar forms that combine regular and irregular forces on the same battlefields."[39] A particular challenge for many traditionalists is that the division of war into separate regular and irregular categories does not adequately describe the full scope of conflict. Even inter-State war has long had elements of both types of conflict that is captured by the contemporary use of the term "hybrid warfare."[40] This multifaceted character of warfare can be seen in the Al Qaeda doctrinal embrace of classic guerrilla theory, which envisages the development of more conventional forces in addition to its

---

[33] Manwaring, *supra* note 9, at 131.

[34] *Id.* at 132. *See also* Kilcullen, *supra* note 24, at 112 (where it is suggested that constabulary, gendarmerie, carabinieri, or coast guard forces may be more effective in dealing with irregular actors).

[35] Hew Strachan, The Direction of War 208 (2013). *See also id.* at 231, and Christopher Kolenda, The Counterinsurgency Challenge: A Parable of Leadership and Decision Making in Modern Conflict xiv (2012) (where insurgency/counterinsurgency is treated as a separate form of conflict).

[36] Strachan, *supra* note 35, at 207–8 (a phrase attributed to Général Vincent Desportes, former director of the French armed forces' doctrine center and head of the Collège Interarmées de Défense).

[37] Peter R. Mansoor, *Introduction*, *in* Hybrid Warfare: Fighting Complex Opponents from the Ancient World to the Present 1 (Williamson Murry & Peter R. Mansoor eds., 2012) ("as Clausewitz stated nearly two centuries ago, although war changes its characteristics in various circumstances, in whatever way it manifests itself, war is still war.").

[38] Strachan, *supra* note 35, at 209.

[39] Mansoor, *supra* note 37, at 1.

[40] *Id.* at 1–17 (for a discussion of "hybrid warfare").

guerrilla elements.[41] It is also evident in the State deployment of both conventional and Special Forces units[42] to counter this Salafi jihadist–based threat. It has been suggested, "American military forces must possess a wide range of means to combat *hybrid* opponents, from conventional power to counterinsurgency and counterterrorism capabilities."[43]

This application of hybrid terminology is not unique to the categorization of warfare. It is also increasingly used to describe organized armed groups that amalgamate conventional and guerrilla capabilities.[44] The term "hybrid" has also been used in relation to other groups that blend foreign and local fighters.[45] It can be applied in the legal realm regarding the contemporary merger of the humanitarian law and human rights normative approaches. In this context, humanitarian law–based targeting rules are applied by commanders when considering the killing of terrorists, while at the same time assessing the utility of applying human rights–based law enforcement norms to carry out their "arrest."[46] In this respect, the "kill or capture" decision, which was prevalent in the initial post-9/11 operations, has now migrated to one of "arrest or kill." It is no coincidence that the privileging of an arrest is placed ahead of killing an enemy in many operational contexts.

Importantly, the term "hybrid war" is itself underinclusive to the extent it neither overtly recognizes the prevalence and impact of criminal activity in contemporary conflict nor reflects the law enforcement responsibilities of military forces. Contemporary conflict shines a particularly bright spotlight on this aspect of military operations: the requirement to assist in the policing of populations. However, it is an approach that faces considerable doctrinal challenges within State military forces, as it will ultimately require a greater acceptance of a constabulary role. As has been suggested, "[a]ll armies worth their salt fear the threat they will become a gendarmerie."[47] As much as many military forces may dislike constabulary duties, it is clearly a part of an occupier's role in conventional inter-State conflict and a key aspect of counterinsurgency and counterterrorism operations in other contexts. The failure to recognize this fact can have tragic security and human consequences, as the Coalition forces discovered in Iraq in the aftermath of the 2003 invasion. The reality is that participation in war requires a capacity to perform conventional operations; conduct

---

[41] Norman Cigar, Al-Qa'da's Doctrine for Insurgency 12 (2009), and Ryan, *supra* note 5, at 230 (where he notes "[l]ike Naji, al-Qurashi, al-Muqrin, and other Al Qaeda strategists, al-Suri subscribes to a classic three-stage guerrilla war strategy loosely following the Maoist model.").

[42] Kevin Maurer, Gentlemen Bastards: On the Ground with America's Elite Special Forces 15–16 (2012) (where it is noted that in the U.S. military the "Special Forces" (e.g., the Green Berets) are different from special operations units (i.e., SEALs, Rangers, Pararescue, combat controllers). However, in this book the term "Special Forces" is applied generically to such units, although reference is periodically made to special operations units). *See also* Sean Naylor, Relentless Strike: The Secret History of Joint Special Operations Command 43–44 (2015) (where it is indicated U.S. special units can be further divided into "tiers" with Tier 1 units such as Army "Delta" unit and the SEAL "Team 6" also being referred to as special mission units).

[43] Mansoor, *supra* note 37, at 16 (emphasis added). *See also NATO to Counter "Hybrid Warfare" from Russia*, BBC News (May 14, 2015), http://www.bbc.com/news/world-europe-32741688 (where hybrid warfare is suggested "combines different types of threats, including conventional, subversion and cyber").

[44] Anthony H. Cordesman, George Sullivan, & William D. Sullivan, Lessons of the 2006 Israeli-Hezbollah War 80 (2007) (for a discussion of Hezbollah's military structure).

[45] David Kilcullen, The Accidental Guerrilla: Fighting Small Wars in the Midst of a Big One 38 (2009).

[46] Stanley McChrystal, My Share of the Task: A Memoire 229 (2013) (where it is indicated that the role of the Special Forces team dispatched to the site of the aerial strike on the Al Qaeda in Iraq leader al-Zaqwari was to "arrest anyone who escaped the blast").

[47] Strachan, *supra* note 35, at 208.

counterinsurgency operations, or otherwise fight guerrilla wars; and assist in maintaining law and order in respect of a civilian population.

In effect, the doctrinal divisions within inter-State and non-State actor conflict have a triumvirate quality, rather than simply a hybrid one. "War is war," but it involves conventional conflict, guerrilla warfare, and law enforcement tasks. The main difference between "regular" and "irregular" conflict (i.e., international and noninternational armed conflict) is not their binary categorization, but rather the inverted emphasis placed on these three areas of operations. For inter-State conflict, emphasis is normally placed on conventional operations. In contrast, irregular warfare focuses on a counterinsurgency or counterterrorism approach, which privileges the primacy of policing. However, this does not mean that non-State conflict cannot involve conventional warfare, or inter-State conflict will not require military forces to use force associated with law enforcement. Military forces must be prepared to conduct operations and use force in all three areas, just as police forces are often being asked to expand their operations beyond ordinary law enforcement.[48] Added to the security challenge are transnational criminal groups that are pushing the bounds of criminal activity toward an insurgency status; the large-scale involvement of insurgent and terrorist groups in a wide variety of illicit enterprises (i.e., kidnapping, drug trafficking, smuggling); and criminal gangs operating from territories that are inadequately governed, or on the high seas.

Further, to the extent international law governs internal conflict, it has been premised upon the existence of a strong and effective State in order to police its territory. While often domestic in origin, many of these conflicts have international consequences. This occurs not only in terms of the violence and crime spilling across borders but also in respect of the risks posed to non-nationals within the country. The violence linked to insurgency in spaces not effectively governed by States, such as the taking of foreign hostages, can prompt a State to cross geographic boundaries and, in effect, "enforce" the law in place of its failed or failing territorial counterpart. This can raise questions as to whether the State acting to defend its interests and nationals is participating in armed conflict or policing, and what laws govern such activity.

Finally, in terms of the ability of the existing international legal regime to address contemporary security challenges, perhaps the most significant issue is that insurgencies, global or otherwise, and conflict other than inter-State ones, represent the predominate form of warfare in the twenty-first century. In this changing security landscape the number of inter-State conflicts has significantly diminished. In this respect, it has been assessed that the ratio of internal to external war was two to one before 1945, however this increased to nearly five to one after that time.[49] Further, even when there has been inter-State conflict, such as in Afghanistan in 2001–2002 and Iraq in 2003–2004, these conflicts have been followed by lengthy and violent periods of insurgent violence. A fundamental question then arises regarding the ability of the international legal regime to control conflict as it turns away from the traditional path of inter-State conflict and takes the fork in the road toward these evolving types of conflict and violence.

---

[48] Jugdep S. Chima, *The Punjab Police and Counterinsurgency against Sikh Militants in India: The Successful Convergence of Interests, Identities and Institutions*, *in* POLICING INSURGENCIES: COPS AS COUNTERINSURGENTS 258, 274–75 (C. Christine Fair & Sumit Ganguly eds., 2014) (for reference to the creation of the 4,000-man Punjab Police commandos, which "eventually replaced the army at the cutting-edge of anti-insurgent operations in 1991 and 1992. . . .").

[49] JACK S. LEVY & WILLIAM R. THOMPSON, CAUSES OF WAR 12 (2010).

1.2. The Legal Challenge

There are significant challenges facing international law concerning its ability to regulate not only contemporary conflict but also explain the operation of that law to the nonlegal community. Before providing an outline of the substantive challenges facing the international law, it will be helpful to address the issue of terminology. The international law dealing with war often uses Latin terms—*jus ad bellum, jus in bello, lex specialis derogate lex generali* (the *lex specialis*), *lex lata*, and *lex ferenda*—which can be extremely confusing to those not immersed in the intricacies of this discipline. It is also ironic that a theoretical discipline which seeks to communicate fundamental aspects of how society is organized and should act places such heavy reliance on terms framed in a language that has largely lost contemporary relevance. To a certain extent this use of Latin highlights the degree to which this body of international law remains very technical in nature. It also reflects a theoretical construct significantly removed from the practical security reality facing the international community. For practitioners it can be challenging to translate these phrases into concepts that are meaningful to nonlawyers. In this book the use of these somewhat arcane technical terms will be avoided as much as possible, although it will not be possible in all circumstances, particularly when they are used in quotes or in a historical context.

Instead, the *jus ad bellum*, defined as "the right of making war," will be referred to as the law governing the recourse to war.[50] This is the body of law that establishes when it is lawful for a State to use force in self-defense in response to an attack. Since the enactment of the UN Charter, the most common reference to this body of law has been in terms of the law governing the exercise of State or national self-defense. The reference to State self-defense is also used to distinguish it from the exercise of individual self-defense and defense of others, which is integral to the use of force under human rights–based law enforcement norms. State self-defense is a narrower aspect of the more general law governing State recourse to war. What is often overlooked is that as a theoretical construct the law governing the recourse to war involves consideration of seven principles: just cause, right or proper authority, right intention, proportionality of ends, last resort, reasonable hope of success and the aim of peace.[51]

The continued relevance of these just war principles is highlighted in a 2001 report, *The Responsibility to Protect*, which assessed "the question of when, if ever, it is appropriate for states to take coercive—and in particular military—action, against another state for the purpose of protecting people at risk in that other state."[52] That report identified just cause, right intention, last resort, proportionate means, and reasonable prospects of success and acting under the direction of the right or proper authority as the principles for military intervention at the dawn of the twenty-first century.[53] In that context it was reported "[t]here is no better or more appropriate body than the United Nations Security Council

---

[50] Bryan A. Garner, Garner's Dictionary of Legal Usage, http://www.oxforddictionaries.com/secondary/garner_dict_legal_usage/jus-in-bello-jus-ad-bellum (last visited May 26, 2015).

[51] *Id.*

[52] The Responsibility to Protect: Report of the International Commission on Intervention and State Sovereignty, ICSS, Dec. 2001, http://responsibilitytoprotect.org/ICISS%20Report.pdf.

[53] *Id.* at XII.

to authorize military intervention for human protection purposes."[54] This reliance on what are fundamentally just war principles highlights a broader reach for the body of law governing the recourse to war than simply the exercise of State self-defense. With respect to this body of law, the "proper authority" principle will be seen to have an important contemporary impact on the status of detained persons and legitimacy of participants in noninternational armed conflict.

The *jus in bello*, which relates to "the corpus of the laws and customs of war," is a separate specialized body of law governing the actual conduct of hostilities.[55] It is more commonly referred to as humanitarian law, or the law of armed conflict, although historically it has also been referred to as the law of war. The "special" nature of humanitarian law is itself reflected in the term *lex specialis*, which has been interpreted by some States to exclude the application of another fundamental body of law during armed conflict: human rights law.[56] While the "special law" issue is the subject of further analysis in Chapter 5, the idea that one body of law completely excludes another highlights a particular challenge facing the international legal community when dealing with contemporary conflict.[57] These bodies of law have until recently largely been considered and applied separately from one another. As will be discussed, practical reality is forcing change.

The terms *lex lata* and *lex ferenda* respectively refer to the law as it is[58] and the law as it should be.[59] Simply determining what the law is (i.e., gaining broad agreement) can be a challenging endeavor. Much of the discussion, even post-9/11, has focused on simply reaching agreement on the existing law. As the drafters of the 2009 *Air and Missile Warfare Manual Commentary* noted in respect of their work:

> The goal is rather to present a methodical restatement of existing international law on air and missile warfare, based on the general practice of States accepted as law (*opinio juris*) and treaties in force. No attempt has been made to be innovative or to come up with a *lex ferenda* (however desirable this may appear to be): the sole aim has been to systematically capture in the text the *lex lata* as it is.[60]

Assessments of existing law are helpful. They are aided, particularly in respect of inter-State conflict, by reference to international treaties, such as the 1949 Geneva Conventions, which have enjoyed widespread State ratification. There is also extensive State practice upon which to base customary international law.

---

[54] *Id.* ¶ 3.A., at XII.

[55] GARNER, *supra* note 50.

[56] Andrea Gioa, *The Role of the European Court of Human Rights in Monitoring Compliance with Humanitarian Law in Armed Conflict*, *in* INTERNATIONAL HUMANITARIAN LAW AND INTERNATIONAL HUMAN RIGHTS LAW 201, 213 (Orna Ben-Naftali ed., 2011) (where such a view is attributed to the United States and Israel and is considered to be "radical").

[57] *Fragmentation of International Law: Difficulties Arising from the Diversification and Expansion of International Law*, Rep. of the Int'l L. Comm'n., 58th session, May 1–June 9, July 3–Aug. 11 2006, UN Doc. A/CN.4/L.702 ¶ 14(7), at 9 (July 18, 2006), http://legal.un.org/ilc/documentation/english/a_cn4_l682.pdf (where the term "special law" is used instead of *lex specialis*).

[58] BLACK'S LAW DICTIONARY 519 (Bryan A. Garner ed., 10th ed. 2009) ("Existing Law").

[59] *Id.* ("A principle that might be applied . . . .").

[60] AIR AND MISSILE WARFARE MANUAL COMMENTARY, *supra* note 14, at 2.

However, the same cannot be said for noninternational armed conflict, and the types of transnational terrorist attacks that dominate the contemporary security dialogue. While it is essential there is an understanding of what the law is, too heavy a reliance on treaties or the search for broad consensus on established law can leave the analysis locked into the conflicts of the past century. In the existing security environment, agreeing on what the applicable law is will be seen as only the starting point of the discussion regarding how the law should be applied to counter new and innovative security threats. Unfortunately, there is little consensus within the international law community on some crucial legal issues (e.g., the categorization of transnational conflict with non-State actors). In the post-9/11 security environment military practitioners are confronted far more often with questions demanding analysis of what the law "should be," rather than what it is "agreed" to be. The issues that often have to be addressed do not usually comfortably fit into the various international law theories that dominated much of the legal discussion up to the turn of the twenty-first century. As a result, this book will look not only at existing law but also consider what normative rules or law should be applied and how.

The historical focus on the State has also meant that traditionally there has been an overwhelming emphasis on inter-State conflict. This has occurred even though we live in a world where for some time significant levels of insecurity have primarily originated from within States. This concentration on the State and exclusionary applications of law can be seen in most contemporary legal discourse regarding the two bodies of law that are historically linked to just war theory: the law governing the recourse to war and humanitarian law. The separate application of these areas of law is suggested to be as a foundational principle of the international law governing warfare. The "equal application principle" considers the law governing the conduct of hostilities (i.e., humanitarian law) to apply equally to all participants regardless of the justness of their cause.[61]

The impact of the law governing the recourse to war on the body of law regulating hostilities is rarely discussed, although it is now being recognized that conflicts within States and action against terrorism "have always raised difficult challenges in relation to the application—let alone equal application—of the laws of war."[62] It is not that there has been no interaction between these two bodies of law. It has been noted that "historically and thematically, the *jus in bello* [humanitarian law] concerns have taken a second place to the *ad bellum* [recourse to war] criteria, since only after a decision is made that a particular use of force is just can one turn to the question of how to fight justly."[63] However, such a sequential approach belies a more complex relationship. The principles governing the exercise of State self-defense have been suggested to continue to apply during a resulting armed conflict (see Chapter 3, section 3.2, *The Continuing Application of Self-Defense Principles*). To the extent

---

[61] Adam Roberts, *The Equal Application of the Laws of War: A Principle Under Pressure*, 90 INT'L REV. RED CROSS 931, 932 (2008). *See also* Marco Sassoli, *Ius ad Bellum and Jus in Bello—The Separation between the Legality of the Use of Force and Humanitarian Rules to be Respected in Warfare: Crucial or Outdated?*, *in* INTERNATIONAL LAW AND ARMED CONFLICT: EXPLORING THE FAULTLINES 241, 247–48 (Michael Schmitt & Jelena Pejic eds., 2007) (for a discussion of the principle of equality of belligerents).

[62] Roberts, *supra* note 61, at 934. *See also* Eyal Benvenisti, *Rethinking the Divide Between Jus ad Bellum and Jus in Bello in Warfare Against Nonstate Actors*, 34 THE YALE J. INT'L L. 541 (2009).

[63] JAMES TURNER JOHNSON, MORALITY AND CONTEMPORARY WARFARE 36 (1999).

it is accepted that the right of State self-defense can be exercised in response to non-State actor attacks, that body of law can have an impact on the resulting hostilities.

Further, notwithstanding the important and laudatory goal of maintaining the "equal application principle," the law governing the conduct of hostilities cannot always be considered in a manner discrete from that controlling the State recourse to war. One of the challenges in assessing the impact of these principles has been that much of the dialogue concerning the interface between the legal framework that primarily applies in the run-up to war and that applying during war has concentrated on the "just cause" principle. For example, in the aftermath of World War II attempts to have belligerent status extended to civilians who use force to defend their country against "illegal aggressions" was rejected on the basis that extending rights on the basis of the justness of a cause was inconsistent with the "doctrine of belligerent equality."[64]

Another frequently overlooked issue is the role of the State as the "proper" or lawful authority to maintain peace and security. This just war principle impacts directly on both the recourse to war and the conduct of hostilities. The importance of fighting for the proper authority is evident in terms of attaining the protected status of prisoner of war. While the humanitarian law is said to apply equally, it is also widely accepted to have a unique application to a privileged class of participants: those who fight for States. On this issue, the equal application of the law is mainly in respect of State-authorized participants. The favoring of State-authorized participants has a significant discriminatory effect against their non-State counterparts. That effect permeates much of the law governing hostilities. Non-State participants are considered to be, at best, unprivileged belligerents,[65] and more often criminals, marauders, spies, bandits, rebels, or "terrorists," who are subject to criminal incarceration and prosecution.[66]

By effectively placing non-State actors "outside" the law, efforts to have all participants in warfare comply with humanitarian law rules can be weakened. This challenge is intensified in respect of noninternational armed conflict where non-State actors have little chance of gaining legitimate status without being victorious, although there also remains the possibility of being granted some form of amnesty at the termination of a conflict.[67] Nor is the asymmetric application of the law always restricted to issues related to the resort to, or

---

[64] GEOFFREY BEST, WAR AND LAW SINCE 1945, at 129 (2002) (1994).

[65] Richard R. Baxter, *So called "Unprivileged Belligerency": Spies, Guerrillas, and Saboteurs*, *in* HUMANIZING THE LAWS OF WAR: SELECTED WRITINGS OF RICHARD BAXTER 37, 42 (Detlev F. Vagts et al. eds., 2013) (originally appeared in1951 BRIT. Y.B. INT'L L. 328) (unprivileged belligerents are defined as "persons who are not entitled to treatment either as peaceful civilians or as prisoners of war by reason of the fact that they have engaged in hostile conduct without meeting the qualifications established by Article 4 of the Geneva Prisoners of War Convention of 1949"). *See also* THE MANUAL OF THE LAW OF ARMED CONFLICT, United Kingdom Ministry of Defence ¶ 11.4 at 279–80 (2004).

[66] HUGO GROTIUS, 3 THE RIGHTS OF WAR AND PEACE, 1246 (Richard Tuck ed., Liberty Fund Inc. 2005) (1646) ("the rest are called pilfering thieves or robbers."), EMER DE VATTEL, THE LAW OF NATIONS 507 (Knud Haakonssen ed., 2008) (1797) (where reference is made to banditti, buccaneers, pirates, and corsairs), Francis Lieber, *Guerrilla Parties Considered with Reference to the Law and Usages of War*, *in* RICHARD SHELLY HARTIGAN, LIEBER'S CODE AND THE LAW OF WAR 31, 34–37 (1983) (for a nineteenth-century reference to categories of participants such as freebooters, marauders, war rebels, conspirators, robbers, spies, and brigands who remained subject to prosecution), and more recently MAX BOOT, INVISIBLE ARMIES: AN EPIC HISTORY OF GUERRILLA WARFARE FROM ANCIENT TIMES TO THE PRESENT xxiii (2013) ["To further muddy the issue, their enemies usually try to band all 'guerrillas' or 'insurgents' (fairly laudatory titles) as 'terrorists' or other terms of abuse such as 'criminals,' 'bandits,' 'traitors,' or 'dead-enders.'"].

[67] AP II, Art. 6(5).

conduct of, hostilities. Since human rights law usually indicates it is States and not their non-State opponents who have responsibility for any breaches of that law, a powerful argument is provided as to why that body of law is limited in its ability to be extended to regulate contemporary warfare. Further, analysis of the recourse to war-related exercise of State self-defense also demonstrates a clear State bias, even though most contemporary conflict does not take place exclusively between States.

The bias in favor of States and their fights with one another has had a significant effect on the legal regime governing conflict. To the extent inter-State conflict will continue, there is a rich body of treaty and customary international law governing the recourse to and conduct of hostilities. However, far less attention has been paid to regulating internal conflicts, or transnational warfare with non-State actors. Notwithstanding efforts through "judge made" law[68] to clarify the degree to which humanitarian law applies to internal conflicts and the existence of guidance like that found in the 2005 International Review of the Red Cross, *Customary International Humanitarian Law Study*,[69] the full extent to which that law governs such conflict remains uncertain. This has led to considerable discord regarding interpretations of what the governing law is and how it applies to evolving conflicts.

Various bodies of law that impact on armed conflict are depicted in Figure 1.1. Applicable Legal Regimes

FIGURE 1.1 Applicable Legal Regimes.

The theoretical preference for treating these areas of law in an exclusionary fashion presents considerable challenges for practitioners attempting to apply the law across the full range of conflict. As has been noted, conventional wisdom prefers humanitarian law to be

[68] *See, e.g.*, Prosecutor v. Tadić, Case No. IT-94-1-A, Decision On The Defence Motion For Interlocutory Appeal On Jurisdiction, ¶¶ 96-127 (Int'l Crim. Trib. for the Former Yugoslavia, 2 October 1995), http://www.icty.org/x/cases/tadic/acdec/en/51002.htm.

[69] Customary International Humanitarian Law Study (Jean-Marie Henckaerts & Louise Doswald-Beck eds., 2005).

applied in isolation from the law governing the resort to war by States. Further, exclusionary approaches are also often applied in respect of the interaction of international human rights law and humanitarian law, with proponents of each body of law, at times, effectively denying the application of the other in the midst of armed conflict. However, even with respect to human rights, the international version of this body of law is often considered in isolation of its domestic counterpart. Despite the historical development of domestic human rights, its international counterpart is often seen as rising independently from the ashes of World War II virtually divorced from the principles and foundational documents that preceded it.

Perhaps the one body of law that can most readily be seen as crossing legal boundaries is international criminal law. Its genesis is found in "both *humanitarian law and human rights law*, as well as *national criminal law*."[70] Consideration of this integration must be tempered with the acknowledgment there is a tension between the application of the international accountability framework and its domestic counterpart. One of the reasons international tribunals have become empowered is that "domestic judicial authorities may be, and in fact are, loath to prosecute state agents or to institute criminal proceedings against private individuals, which might eventually involve state organs."[71] Within this relationship is yet another sign of overlap between normative frameworks. This can be seen in the "quasi-criminal" jurisdiction that the American Court of Human Rights has displayed in making its oversight of the "national prosecution of gross, state-sponsored crimes a centerpiece of its regional agenda."[72]

As a relatively new body of law, international criminal law is indicative of a trend that forms a central thesis of this book: that contemporary warfare can be most effectively regulated by a more integrated application of all the various bodies of law that impact on armed conflict. However, that trend is a nascent one, and exclusionary approaches toward interpreting the law continue to exert a powerful influence on how security threats are viewed and ultimately controlled. Those charged with assessing how the law is to be applied in the contemporary security environment are constantly confronted with limits or boundaries restricting in what way or even if a particular body of law is to be applied. These boundaries are twofold. First, there is the outer limit of established treaty and customary law that is mostly focused on international armed conflict. The result is that body of law is not specifically designed to apply to most contemporary hostilities. Second, there are the internal boundaries that are frequently applied between the various legal disciplines regardless of how conflict is characterized.

An emphasis on legal boundaries can lead to formal, and sometimes rigid, approaches to applying each body of law that does not always appear to work well in practice. It leads to the security personnel tasked with maintaining order having to fight literally at the limits of these legal boundaries. This can especially occur during the myriad of lower intensity operations such as counterinsurgency, counterterrorism, noncombatant evacuation, international hostage rescue, and peace support operations. In those situations operations are often conducted in a grey zone where international humanitarian law, international human

---

[70] Antonio Cassese, International Criminal Law 6 (2008).

[71] *Id.* at 435.

[72] Alexandra Huneeus, *International Criminal Law by Other Means: The Quasi-Criminal Jurisdiction of the Human Rights Courts*, 107 Am. J. Int'l L. 1, 1 (2013).

rights law, and domestic law intersect and often overlap. This does not mean that the law is fundamentally flawed or incapable of regulating contemporary conflict. In assessing the effectiveness of any legal framework there is a requirement to accept a degree of artificiality, which is an inherent part of any regime attempting to impose law and order on human activity. This may be particularly so in respect of activity that is as chaotic and destructive as warfare. That reality can lead to frustration with both the law and lawyers. This is reflected in Michael Walzer's observation that "[t]he lawyers have created a paper world, which fails at crucial points to correspond to the world the rest of us still live in."[73]

Efforts to provide structure and order can also lead to analytical frameworks that overly restrict the degree to which a particular body of law can be applied to a specific fact situation. Fundamental questions need to be asked as to whether the boundaries of the law are properly drawn, or are being interpreted in too rigid a fashion. Particular challenges have arisen as a result of the tendency to deny the common roots of bodies of law such as humanitarian law and human rights law or to accept there is an overlap in application of these various bodies of law. Another challenging issue for anyone endeavoring to apply international law to contemporary operations is the prevalence of conflicting interpretations of the law, which often dominate the legal dialogue. In part, this is a result of the absence of judicial rulings on many of the outstanding legal issues. Notwithstanding the existence of the International Court of Justice, the development of international criminal tribunals and courts over the past twenty years, the repeated foray of human rights tribunals into litigation relating to armed conflict, as well as periodic domestic court decisions, there is a dearth of jurisprudence regarding many controversial legal issues. While the prevalence of these unresolved and sometimes conflicting theories may lead to a vibrant debate in an academic context, the absence of decisions can make the application international law to complex and often-changing security situations very challenging for State legal advisors.[74]

It might be tempting to conclude the problem is simply that twenty-first-century warfare cannot be regulated by what is effectively twentieth-century law. However, the existing legal framework has been very successful in addressing conflict, albeit primarily between States. There has been a growing recognition that the significant drop in inter-State conflict following World War II is not solely impacted by the traditional theories of the existence of a nuclear umbrella[75] or suggestions of a "democratic peace."[76] The decline also appears attributable to the existence of the security framework limiting international conflict;[77] a humanitarian revolution, of which both human rights and humanitarian law are part;[78] and "the stability of international borders, which historically has been a major issue of contention

---

[73] Michael Walzer, Just and Unjust Wars xix (3rd ed., 1977).

[74] Michael N. Schmitt, *Narrowing the International Law Divide: The Drone Debate Matures*, Yale J. Int'l L. Online 2 (2014), http://www.yjil.org/docs/pub/o-39-schmitt-narrowing-the-international-law-divide.pdf.

[75] Levy & Thompson, *supra* note 49, at 12 (for reference to scholars tracing the decline of inter-State warfare to the development of nuclear weapons), and Martin Van Creveld, The Rise and Decline of the State 337 (1999).

[76] Steven Pinker, The Better Angels of Our Nature: Why Violence Has Declined 278–84 (2011).

[77] Joshua S. Goldstein, Winning the War on War 322 (2011) ("This does not always keep wars from breaking out anyway but has been successful in limiting violence in interstate wars."). *See also* Pinker, *supra* note 76, at 251.

[78] Pinker, *supra* note 76, at 188 ("despotism and major war, which had cast their shadow on humanity since the beginning of civilization, began to show cracks. The philosophy of Enlightenment humanism that united these developments got a toehold in the West and bided its time until more violent ideologies tragically ran their course.").

and cause of war."[79] Perhaps the greatest triumph of international law in the past century was the development of the security framework of the UN Charter controlling the recourse to war. As John Keegan noted, the 1991 action to remove Saddam Hussein from Kuwait following his "illegal sequestration of territory" represented "the first genuine triumph of just war mentality since Grotius had defined its guiding principles at the height of the Thirty Years' War in the seventeenth century."[80] Even though this principle has been challenged by Russia with its annexation of the Crimea and the support provided to paramilitary forces in eastern Ukraine, respect for national boundaries remains an essential component of the maintenance of international security and peace.[81]

However, it may be that the very success of this legal framework serves to undermine the application of these bodies of law to new and emerging conflicts. The laws governing the State exercise of self-defense and even international human rights have developed in a particular historical context. They are designed to address specific State centric challenges. As has been noted, for much of the twentieth century the focus by international lawyers was on inter-State conflict. Such a unitary focus carries with it a danger that the governing legal regime has come to be too closely aligned with such conflict. As a result, it may be that the interpretation of international law has become too tradition bound or path dependent.[82] Contemporary conflict exposes the need to look beyond inter-State warfare and focus on the lower-level violence that threatens the international order.

Adding to the debate about the ability of contemporary international law to regulate conflict is the schism that can exist between theory and practice. Such tensions are not new, as can be seen in the remarks by Quincy Wright in 1953 regarding the tension that existed between relying on "the scientific method of inducing law from observed behavior rather than the juristic method of deducing what behavior ought to be from the law."[83] As Tom Ruys has perceptively noted in respect of the exercise of self-defense:

> [i]nternational lawyers, on the one hand seem to focus on those large-scale interventions that generate important shockwaves within the international community.... The individual military commander, on the other hand, is more interested in knowing when he is permitted to open fire against incoming aircraft or a trespassing border patrol.[84]

He suggests that different mindsets develop, "one inspired by the [UN] Charter provisions as interpreted by the ICJ [International Court of Justice], the other by national rules

[79] GOLDSTEIN, *supra* note 77, at 278, and PINKER, *supra* note 76, at 258 ("One paradoxical contributor to the Long Peace was the freezing of national borders. The United Nations initiated a norm that existing states and their borders were sacrosanct.").

[80] JOHN KEEGAN, A HISTORY OF WARFARE 384 (1993).

[81] *Ukraine: Putin Signs Crimea Annexation*, BBC NEWS (Mar. 21, 2014), http://www.bbc.com/news/world-europe-26686949.

[82] Paul A. David, *Path Dependence—A Foundational Concept for Historical Social Science*, CLIOMETRICA — 1 J. HIST. ECONOMICS & ECONOMETRIC HIST 91 (2007).

[83] Quincy Wright, *The Outlawry of War and the Law of War*, 47 AM. J. INT'L L. 365, 368 (1953).

[84] TOM RUYS, ARMED ATTACK AND ARTICLE 51 OF THE UN CHARTER: EVOLUTIONS IN CUSTOMARY LAW AND PRACTICE 179–80 (2010). *See also* Daniel Bethlehem, *Principles Relevant to the Scope of a State's Right as Self-Defense Against an Imminent or Actual Armed Attack by Nonstate Actors*, 106 AM. J. INT'L L. 770, 774 (2012).

of engagement [ROE]. . . ."[85] However, this dichotomy goes much deeper than questions about opening fire against other State armed forces, or what military ROE should be. A portion of the international community has exhibited considerable reluctance to accept that the post–World War II concept of State self-defense could be extended to conflicts between States and non-State actors. Instead they remain the criminal acts of "terrorists."[86] The lack of consensus on such fundamental international law issues has directly contributed to a badly fractured and largely unresolved debate regarding what law governs counterterrorism operations involving the use of drones. It is a divide that must be bridged if international law is to have an effective role in addressing contemporary security threats. Practice must be informed by theory, and theory by practice.

There also appears to be an avoidance of substantively addressing broader questions relating to lower scale conflict involving non-State actors, including acting in defense of nationals on the high seas or in States where there is no functioning government. This extends to operations such as hostage rescue, the protection of nationals, noncombatant evacuations, counterinsurgency, counterpiracy, peace enforcement, and international law enforcement. It is these operations that take up a significant amount of the time, resources, and energy of military commanders and their legal advisors—not the potential of war between States, which is the focus of so much theoretical debate. The reluctance of international lawyers to deal with these lower intensity threats has often left State legal advisors looking for policy solutions to address the lack of consensus on what law applies in these complex operational settings. In effect, there is a gap that must be bridged. This bridging must be done in a manner that ensures the theory of international law properly addresses the practical security challenges facing States.

What is required is an assessment of why the boundaries and interpretational limits of these areas of law have developed in such a manner that the international legal regime appears to be so challenged in regulating contemporary conflict. In this regard there often seems to be a significant disconnect between the theory of how the legal regime works and the practice regarding the degree to which the various bodies of law actually apply and interact with one another. The way in which the law is theoretically interpreted at the strategic level is often quite different from its tactical application. This is perhaps best represented in the traditional approach followed by the United States until 2014 that favored the application of international humanitarian law over human rights law during armed conflict.[87] The common use of the term *lex specialis* (special law) by international lawyers to discuss this principle signals a complex and perhaps far too esoteric debate. This approach, combined with the view that human rights treaty law does not have extraterritorial application,

---

85 Ruys, *supra* note 84, at 180.

86 *See, e.g.*, Albrecht Randelzhofer, *Article 51*, *in* The Charter of the United Nations: A Commentary ¶ 34, at 802 (Bruno Simma et. al. eds., 2002) ("Acts of terrorism committed by private groups or organizations as such are not armed attacks in the meaning of Art. 51 of the UN Charter. But if large scale acts of terrorism of private groups *are attributable to a State*, they are an armed attack in the sense of Art. 51.") (emphasis added).

87 Operational Law Handbook, ¶ V.A.1., at 54 (Lieutenant-Commander David Lee, ed., 2015), http://www.loc.gov/rr/frd/Military_Law/pdf/operational-law-handbook_2015.pdf ("Traditionally, IHRL [international human rights law] and the LOAC have been viewed as separate systems of protection, where one wholly displaces the other. The displacement view is an all-or-nothing approach that results in either IHRL or LOAC setting the rules that govern the armed conflict at issue. This view applies IHRL and LOAC to distinct situations and relationships. The United States embraced this view until very recently.").

seemingly would preclude the human rights law from applying in much of contemporary armed conflict.[88] It is also an approach that does not appear to take into account the impact of customary international human rights law.

In stark contrast to the suggested theories of exclusion is the existence of an entire chapter on human rights in the 2015 U.S. *Operational Law Handbook*.[89] That chapter contains specific references to obligations for military personnel arising from the universal application of customary human rights law.[90] This indicates that at a practical level the experience gained from military operations in complex insurgencies such as Afghanistan and Iraq demands the application of both bodies of law. The requirement to simultaneously apply human rights norms related to law enforcement, as well as humanitarian law rules applicable to combat, is not an uncommon experience for military forces and their legal advisors. Therefore, while theoretical debates about exclusion and the position taken by some governments indicates a strategic-level battle is occurring for the legal "high ground," practitioners at the operational and tactical levels must get on with applying the most appropriate body of law. The challenge for the international community is how to bridge the apparent gap between the strategic dialogue and tactical reality.

Considerable strain has been placed on traditional interpretations of international law because of the shift in focus away from States fighting one another to combating "other" armed groups. New approaches have included human rights law–based challenges to the idea that humanitarian law exclusively governs military operations in noninternational armed conflict, or should even apply at all.[91] It has also been suggested that the body of international law governing internal armed conflict includes not only humanitarian law and human rights law, but also international criminal law.[92] Similarly, it is increasingly clear that the law governing the recourse to war and that controlling the conduct of hostilities cannot always be assessed completely divorced from one another as is commonly asserted in respect of international conflicts.[93] For a variety of reasons, including a trend by domestic/regional courts and tribunals to consider the extraterritorial application of domestic law, such noninternational law is gaining an enhanced profile.[94] This is in addition to the role that body of law inherently performs in controlling traditional internal conflicts. As a

---

[88] Michael J. Dennis, *Application of Human Rights Treaties Extraterritorially in Times of Armed Conflict and Military Occupation*, 99 Am. J. Int'l .L. 119 (2005), and Noam Lubell, Extraterritorial Use of Force Against Non-state Actors 197–98 (2010) (for a discussion of the U.S. position).

[89] Operational Law Handbook, *supra* note 87.

[90] *Id.* ch. 3, at 45–55.

[91] *See, e.g.*, William Abresch, *A Human Rights Law of Internal Armed Conflict: The European Court of Human Rights in Chechnya*, 16 Eur. J. Int'l L. 741, 747–48 (2005) (where it is suggested human rights law should govern the use of force in internal conflict).

[92] Sandesh Sivakumaran, *Re-evisaging the International Law of Internal Armed Conflict*, 22 Eur. J. Int'l L. 219, 232 (2011).

[93] Jasmine Moussa, *Can Jus Ad Bellum Override Jus In Bello? Reaffirming the Separation of the Two Bodies of Law*, 90 Int'l Rev. Red Cross 963, 987 (2008) ("Perhaps the most dangerous threat to the principle of separation of jus ad bellum and jus in bello arises in the context of what is now known as asymmetric warfare, or conflict between a state and non-state actor.").

[94] *See, e.g.*, Munaf v. Geren, 553 U.S. 674 (2008) (where the U.S. Supreme Court ruled "that the habeas statute extends to American citizens held overseas by American forces operating subject to an American chain of command, even when those forces are acting as part of a multinational coalition," although in those circumstances habeas corpus provided the petitioners with no relief).

result, the law governing operations is increasingly being applied by practitioners in a more holistic fashion, necessitating consideration as to whether these various bodies of law "are intimately interconnected and explicable only by reference to the whole."[95] It is an approach that must also seek to match legal theory with operational practice in order to better determine what the boundaries of those bodies of law are, and should be.

1.3. The Operational Challenge: The Emerging "Law of 9/11"

The degree to which a paradigm shift is occurring regarding the use of force is clearly reflected in President Barak Obama's May 23, 2013, *Drone Policy* speech in which he outlined the legal and policy basis for the use of drones as part of the ongoing U.S. campaign against terrorism. In that speech Obama stated that the war with Al Qaeda was lawful under both domestic and international law, with the high threshold of the constitutional protection a U.S. citizen enjoys against being targeted and killed without due process being extended "to all potential terrorist targets regardless of whether they are American citizens."[96] It is a threshold that "respects the dignity of every human life."[97] The day before the president's speech, Attorney General Eric Holder informed Congress of a new U.S. policy that "institutionalizes the Administration's exacting standards and processes for reviewing and approving operations to capture or use lethal force against terrorist targets outside the United States and areas of active hostilities."[98] In practical terms, this approach meant considering capture before killing a suspect and, if such action is not feasible, only using lethal force "when a terrorist target poses a continuing, imminent threat to Americans...."[99]

This broad and more holistic application of legal norms during armed conflict had already been telegraphed in remarks made by Harold Koh, the U.S. Department of State legal advisor, at the March 2010 annual meeting of the American Society of International Law.[100] In outlining the underlying legal rationale for the conduct of operations by the United States in armed conflict Koh made reference to a unique "law of 9/11." He also stressed that the nation, which is a major protagonist in these conflicts, was committed to "complying with all applicable law, including the laws of war, in all aspects of these ongoing armed conflicts."[101] The bodies of law referred to included the domestic constitutional law of the United States and that governing the exercise of the inherent right of self-defense.

---

[95] Oxford Dictionaries, http://www.oxforddictionaries.com/definition/american_english/holistic (last visited Dec. 5, 2014) (holistic is defined as "[c]haracterized by comprehension of the parts of something as intimately interconnected and explicable only by reference to the whole.").

[96] *Drone Policy, supra* note 1.

[97] *Id.*

[98] *Letter of Attorney General Eric Holder to The Honorable Patrick Leahy*, May 22, 2013, http://www.justice.gov/slideshow/AG-letter-5-22-13.pdf.

[99] *Id. See also* Robert Chesney, *Text of Attorney General's National Speech*, Lawfare (Mar. 5, 2012, 4:35 PM) *available at* http://www.lawfareblog.com/2012/03/text-of-the-attorney-generals-national-security-speech/ (where Attorney General Holder had already made clear that U.S. constitutional protections were to be applied outside the United States in respect of citizens being considered for targeted killing).

[100] Harold Hongju Koh, *The Obama Administration and International Law* (Mar. 25, 2010), http://www.state.gov/s/l/releases/remarks/139119.htm.

[101] *Id.*

The terminology used for matters referred to in that speech (e.g., detention, targeting and the prosecution of terrorist suspects[102]) could have also been addressed in terms of international humanitarian law, human rights law (i.e., right to life, fair trial, arbitrary detention), international criminal law, and the law governing the recourse to war. The reference to both international and domestic law reflects a broader consideration of the full spectrum of law impacting on contemporary conflict.

The president's *Drone Policy*, as well as Koh's presentation, echoed a dialogue that has dominated much of international discussion over the past decade concerning the nature of the interface between humanitarian law and human rights law in armed conflict.[103] It also highlighted the degree to which questions about what law governs the use of force against non-State actors have arisen in the context of exercise of the right of self-defense. Koh noted some have argued that the use of lethal force, in what is considered by the United States to be an armed conflict, "fails to provide adequate process and thus constitutes *unlawful extrajudicial killing*."[104] "Extrajudicial killing" is, of course, a human rights law term. This broader application of human rights concepts is not a uniquely American issue. The applicability of human rights norms or law is also represented in decisions of the European Court of Human Rights[105] and the Israeli Supreme Court,[106] as well as other domestic courts.[107] Further, discussions concerning the applicable law have not been limited to the use of force during armed conflict. They have also extended to questions regarding what rules govern the exercise of self-defense outside the territory of an established conflict[108] and the holding of detainees.[109] These simultaneous references to bodies of law, which are often considered to operate independently of one another, is significant in that it reflects a trend toward the more "holistic" approach. Due to the nature of contemporary conflict, legal practitioners, courts, and theorists are increasingly being required to consider the law governing military operations in its broadest sense. As has been noted, such operations can contain elements of conventional warfare, counterinsurgency, counterterrorism, and law enforcement.

---

[102] *Id.*

[103] *See, e.g.*, International Humanitarian Law and International Human Rights Law (Orna Ben-Naftali ed., 2011), and René Provost, International Human Rights and Humanitarian Law (2002).

[104] Koh, *supra* note 100.

[105] *See, e.g.*, McCann v. United Kingdom, App. No. 18984/91, Eur. Ct. H.R., Judgment (1995), http://hudoc.echr.coe.int/sites/eng/pages/search.aspx?i=001-57943.

[106] Public Committee Against Torture in Israel v. Israel, Israel Supreme Court [16 December 2006], 46 ILM 375, ¶40, at 393–94 (2007).

[107] *See, e.g.*, Nuhanović v. The State of the Netherlands [Court of Appeal in The Hague] Case no. 200.020.174/01, ¶ 6.35 (July 2011) (Neth.), *available at* http://uitspraken.rechtspraak.nl/inziendocument?id=ECLI:NL:GHSGR:2011:BR5388, and The State of the Netherlands v. Nuhanović [Supreme Court of the Netherlands] Case No. 12/03324 ¶¶ 3.17.2-3.17.3 at 33–34 (Sep. 16, 2013), *available at* http://www.asser.nl/upload/documents/20130909T125927-Supreme%20Court%20Nuhanovic%20ENG.pdf.

[108] *See, e.g.*, Kenneth Anderson, *Targeted Killing and Drone Warfare: How We Came to Debate Whether There Is a "Legal Geography of War," in* Future Challenges in National Security and Law 8 (Peter Berkowitz ed., 2011), http://www.futurechallengesessays.com, Jordan J. Paust, *Self-Defense Targetings of Non-State Actors and Permissibility of U.S. Use of Drones in Pakistan*, 19 J. Transnat'l L. & Pol'y 237 (2010), Mary Ellen O'Connell, *The Choice of Law Against Terrorism*, 4 Nat'l Sec. L. & Pol'y 343 (2010), and Michael N. Schmitt, *Drone Attacks Under the Jus Ad Bellum and Jus In Bello: Clearing the "Fog of Law,"* 13 Y.B. Int'l Hum. L. 311 (2010).

[109] John B. Bellinger & Vijay M. Padmanabhan, *Detention Operations in Contemporary Conflicts: Four Challenges for The Geneva Conventions and Other Existing Law*, 105 Am. J. Int'l L. 201 (2011), and Ryan Goodman, *The Detention of Civilians in Armed Conflict*, 103 Am. J. Int'l L. 48 (2009).

Perhaps most tellingly is that military law practitioners, operating where "the rubber meets the road," are in many respects leading this change. This is reflected most graphically in the use of the doctrinal term "operational law," which incorporates the wide range of international and domestic laws impacting on military operations.[110] However, it can also be seen in the approach taken by the UN Special Rapporteur on extrajudicial, summary, or arbitrary executions in a 2013 report on the use of drones, where it is noted "a holistic approach is needed in order to protect the right to life."[111] The idea that a broader application of law is required in contemporary operations is also reflected in Terry Gill and Dieter Fleck's excellent book, *The Handbook of the International Law of Military Operations.*[112] In this *Handbook* it is suggested that the international law of military operations is a branch of public international law, albeit one supplemented at some points by national constitutions, laws, and regulations.[113] Importantly, it is also a body of law that is influenced "by rules and practices which are not of a legal nature, but which are part of the policy of States and international organizations."[114] This is because for practitioners it is not enough to know the law. They must also sort out how it should be applied in a field where many of these bodies of law are operating simultaneously.

In effect, the international legal community finds itself in a period of transformation in assessing how law regulates conflict in the twenty-first century. As has been noted, there are two countervailing pressures. The trend, post-9/11, has seen a move toward greater integration of established bodies of law. However, the resistance toward such change, based on formalistic divisions of bodies of law in inter-State conflict, acts as a weighty anchor on this emerging effort. After twenty-eight years of practicing military law and providing operational legal advice in respect of an exceptionally broad range of operations, it is the conclusion of this author that the issue the international community needs to address is not

---

110 DEPARTMENT OF DEFENSE LAW OF WAR MANUAL, Office of the General Counsel ¶ 1.6.1, at 20 (June 2015), http://www.dod.mil/dodgc/images/law_war_manual15.pdf, where it is stated:

> The law of war is an important part of, but not the entirety of, operational law. Operational law consists of that body of domestic, foreign, and international law that specifically pertains to the activities of military forces across the entire conflict spectrum. Operational law includes diverse legal disciplines, such as military justice, administrative and civil law, legal assistance, claims, procurement law, national security law, fiscal law, and the law of war.

See LEGAL SUPPORT TO THE OPERATIONAL ARMY, FM-04, Headquarters Department of the Army ¶ 5-14, at 5–3 (Jan. 2012), http://armypubs.army.mil/doctrine/DR_pubs/dr_a/pdf/fm1_04.pdf, and the *Special Committee on the Canadian Mission in Afghanistan, Meeting No. 14*, 40th Parliament 2nd Session, Evidence No. 14 (Nov. 4, 2009), http://www.parl.gc.ca/content/hoc/Committee/402/AFGH/Evidence/EV4209808/AFGHEV14-E.PDF (where operational law was defined by this author in evidence before Parliament as "the domestic and international law applicable to all domestic and international Canadian Forces operations."). *See also* Terry Gill & Dieter Fleck, *Concept and Sources of the International Law of Military Operations, in* THE HANDBOOK OF THE INTERNATIONAL LAW OF MILITARY OPERATIONS 3 (Terry Gill & Dieter Fleck eds., 2010) [hereinafter HANDBOOK OF INTERNATIONAL OPERATIONAL LAW] (where it is stated that military operational law "is a term in general use in a growing number of armed forces, but it is relatively new, only having come into use in the past two decades to describe the various bodies of national and international law which are applicable to and regulate the planning and conduct of military operations.").

111 *Report of the Special Rapporteur on extrajudicial, summary or arbitrary executions*, UN Doc. AA/68/382, ¶ 24, at 6 (Sept. 13, 2013) (by Christof Heyns), http://www.un.org/en/ga/search/view_doc.asp?symbol=A/68/382.

112 HANDBOOK OF INTERNATIONAL OPERATIONAL LAW, *supra* note 110.

113 *Id.* at 8–10.

114 *Id.* at 10.

whether a more holistic approach is needed to regulate contemporary conflict. That should be a given. Rather the important issue is how extensively a change in approach needs to be implemented in order to ensure an effective and more comprehensive normative framework is adopted to limit violence and to protect civilian populations. At the heart of this discussion is the predominance that human rights law has claimed in countering twenty-first-century insurgent and terrorist threats. This requires consideration of how that body of law interacts with other applicable normative frameworks.

Not only does this broader concept of "operational law" need to be applied to contemporary security operations, but the manner in which it is interpreted must also be nestled within the necessary "holistic, long-term, strategic political and civil-military effort to address the legitimate and meaningful preservation (security) of a state and its society."[115] The numerous and diverse international security threats confronting the modern State, which occur below the level of traditional inter-State conflict, demand a more holistic approach. However, it must be an approach that addresses the fact these threats are not always amenable to a traditional military response. This will inevitably create a tension for those comfortable with the status quo that was primarily designed to address inter-State warfare in the twentieth century.

Such tension has already arisen between advocates of opposing camps, with the views of the traditionalists increasingly having to be reconciled with those of the integrators. Occurring in the context of warfare, this tension is perhaps reflective of Clausewitz's observation that "[a]ction in war is like movement in a resistive element . . . it is difficult for normal efforts to receive to achieve even moderate results."[116] The result may be that in order to effectively regulate what are proving to be novel and dangerous security threats, extraordinary efforts and new approaches will be needed to address the concerns of both camps. This assessment of the unfolding legal struggle suggests the real challenge for the broader international community will not be in considering whether the integration of legal regimes is necessary. Rather it will be to consider what policy choices need to be made in privileging one regime over another as the overlap between the various bodies of law is acknowledged and dealt with. Suggesting by rote that one body of law (i.e., humanitarian law) is special and therefore displaces the other (i.e., human rights law) will not provide the answer. The body of law that is applied must match the operational environment, as well as meet the ultimate operational and strategic goals.

If the international operating environment primarily involves the conduct of counterinsurgencies and counterterrorism, the principles relating to the conduct of law enforcement should take on an increasingly important place in the priorities of security forces. If victory is to be claimed in the struggle against twenty-first-century security threats, it will likely be based on a degree of "normalcy" being attained. For our societies such normalcy is represented in the default approach of maintaining order through law enforcement. Of necessity, due to the transnational nature of the insurgent and terrorist threat, operations will increasingly involve "policing" on an international scale. However, key to the practical employment of a law enforcement approach must be a realistic assessment of its limits

[115] MANWARING, *supra* note 9, at 150.

[116] CARL VON CLAUSEWITZ, ON WAR 120 (Michael Howard & Peter Paret eds. & trans., 1989).

regarding the use of force. Asking human rights law to do too much is fraught with danger given the influence such use may have in its wider application in peacetime. The result is that the requirement to conduct hostilities pursuant to humanitarian law during an armed conflict involving terrorists or insurgents will not disappear in the foreseeable future.

1.4. The Outline

It has been suggested that complex postmodern threats "do not fit neat categories and well-defined security fields."[117] They also do not fit comfortably within the boundaries of a legal framework largely developed to control inter-State conflict. This book looks specifically at the challenge of applying the "neat categories" of existing international law to contemporary conflicts. The analysis looks at the interaction between the applicable bodies of law by exploring whether their boundaries are improperly drawn or are being interpreted in too rigid a fashion. Emphasis is placed on the disconnect that can occur between theory and practice regarding how these legal regimes are applied and interact with one another. In looking at case studies, it will be established that the threat posed by insurgents, terrorists, and transnational criminal gangs often occurs not only at the point these bodies of law interact but also in situations where there is significant overlap. In this regard, the exercise of the long-standing right of States to defend nationals, including the conduct of operations such as hostage rescue, can involve the application of human rights–based law enforcement norms to counter threats transcending the conflict spectrum.

Particular attention is paid to the law enforcement framework governing the use of force. The trend has been to expand its application to the point where those norms, including the privileging of capture over killing, have in certain circumstances become the favored approach by military commanders. The utility of this approach ultimately depends upon the scope for the use of force that the human rights–based legal framework permits. It will be established that this paradigm is considerably more robust than often thought when using force in personal self-defense, in defense of others, or for mission accomplishment. However, there are also limits in the ability of a human rights–based framework to address contemporary threats without changing its fundamentally restrictive approach toward the use of force. Given the increasing State reliance on law enforcement, ascertaining the limits of that body of law is of particular importance in determining the optimal response to transnational security threats.

The book proceeds in five parts. The first part, "Introduction," which consists solely of this chapter, has set out the security, legal, and operational challenges of contemporary conflict. It ends by providing the following outline of the book chapters. Part II, "The Interaction Between Normative Frameworks," which include Chapters 2 to 5, focuses on the interaction between the law governing the recourse to war by States, the international law governing the conduct of hostilities (i.e., humanitarian law, or the law of armed conflict), and human rights, including an analysis of the historical influences that shaped their application as

[117] Robert J. Bunker, *The Mexican Cartel Debate: As Viewed Through Five Divergent Fields of Security Studies*, *in* Sullivan & Bunker, *supra* note 26, at 118.

separate bodies of law. Chapter 2, "Controlling State Involvement in Conflict," looks at the law governing the State recourse to war. After discussing the twentieth-century emphasis on inter-State conflict, which introduced the UN Charter–based security framework and the separation of the law governing the recourse to war from that controlling the conduct of hostilities, the analysis shifts to noninternational armed conflict.

The analysis of noninternational armed conflict identifies the traditional view of those conflicts being domestic in nature and the law governing the recourse to war having no application to such conflict. However, it also considers: the historical recognition that some can be international in character; the impact of contemporary conflicts transforming from international to noninternational ones; an increasing recognition that the law governing State self-defense can apply to responses against transnational non-State actor attacks; and the interpretive struggle that international lawyers have encountered in dealing with this new reality. Ultimately, it is shown there is a growing acceptance that the right to self-defense by States has applicability to non-State actor conflicts. Chapter 3, "Applying Self-Defense Principles During Armed Conflict," explores the interface between the governing self-defense legal principles (i.e., necessity, proportionality, imminence, and immediacy) and humanitarian law, which controls the use of force in resulting operations. Ultimately, this chapter establishes that the boundary between the law governing the exercise of State self-defense and that controlling operations is far more complex and nuanced than traditional theories appear to suggest. The former body of law is argued to have a strategic influence on the use of force by States, rather than exerting control over the actual conduct of operations at the tactical level.

Chapter 4, "States, 'Proper Authority,' and Conflict," looks at the significant impact that the just war principle of "proper authority" has on the international law governing armed conflict. It establishes that the special status accorded to States, as the authority to maintain order and participate in conflict, also introduces and mandates the consideration of human rights norms during non-State actor conflict. The criminalization of groups acting outside State authority permeates both international and noninternational armed conflict. This chapter discusses the linkage between State governance and human rights with a particular focus on States having been reluctant to have humanitarian law regulate internal hostilities. This leaves domestic law and therefore human rights law as frequently being the normative framework of choice for the resolution of non-State actor challenges to that governance role. Next, the approach adopted by some States of applying humanitarian law on international operations as a matter of policy because of the uncertainty attached to the application of humanitarian law in noninternational armed conflict is explored. Finally, this chapter looks at "ungoverned spaces" since this is where non-State actors particularly thrive and as a result often present the greatest threat to recognized States. As will be noted, these "ungoverned spaces" encompass a wide range of situations from the complete collapse of the territorial State to circumstances where a non-State actor may exercise near total control and carry out many of the functions of governance.

Chapter 5, "The Humanitarian Law and Human Rights Law Interface," addresses an issue that has dominated much of the post-9/11 legal dialogue regarding the use of force against non-State actors. The analysis highlights the often exclusionary interpretation applied to the two foundational bodies of law regulating the use of force in modern society. It sets out the history of that relationship, their common roots, as well as the impact that the periods of neglect, forced integration, divergence, and now growing reconciliation have had on the interface between humanitarian law and human rights law. This chapter highlights that masked behind the exclusionary arguments is the reality that human rights norms have always been an integral part of humanitarian law. Military forces have long had to apply human rights–based law enforcement norms during armed conflict. As States have increasingly had to address threats posed by organized armed groups operating among the people, often in the ungoverned spaces of the world, they are placed in a position of considering not only the overlap of these two bodies of law but also their increasing convergence in application.

Consisting of Chapters 6 to 9, Part III, "The Threat, the State Response, and Legal Uncertainty," analyzes the threats of insurgency and terrorism. This includes exploring their link to criminal activity and the phenomenon of transnational criminal organizations. Chapter 6, "Contemporary Threats: Insurgency and Terrorism," is primarily descriptive, discussing the insurgency and terrorist threats that have captured the attention of States and that will do so for some time. This chapter sets the scene for an analysis of how the law is and should be applied to address these threats. The unique attributes of each of these overlapping security threats will be explored, including a discussion of the emerging phenomenon of "criminal insurgency." The impact of criminal groups taking advantage of the general insecurity arising from insurgency, as well as the involvement of insurgent and terrorist organizations in such activity, will be discussed. An important issue that is explored is the political purpose of non-State actor conflict. This analysis examines "interference" with State governance as a possible threshold for when a criminal insurgency may evolve to one that is not amenable to being addressed solely by law enforcement means. The chapter argues that the increasingly integrated threat posed by insurgents and terrorists requires State responses capable of addressing these exceptionally broad threats. This means security personnel often have to conduct operations in an environment requiring consideration of both human rights and humanitarian law–based norms when using force.

Chapter 7, "Counterinsurgency and Converging Norms," discusses the unique challenges that arise during counterinsurgency operations. This begins with an assessment of the continuing relevance of counterinsurgency operations following the substantial withdrawal of Western States from Iraq and Afghanistan. The analysis of counterinsurgency includes exploring the adoption of the "police primacy" principle, which emphasizes a law enforcement approach and the necessity to resolve the overlapping application of humanitarian law and human rights law. The discussion then turns to the importance of being able to maintain sufficient physical control over an area of operations when assessing if a law enforcement approach is feasible. The next issue addressed is the challenge of countering the threat posed by improvised explosive device (IED) and suicide bombing cells during counterinsurgency operations. The group nature of those threats and factors affecting the responses available to State security forces will be considered. The focus then switches to targeting and the impact that the "value" of a target can have on operational decision making particularly during a counterinsurgency. Causing collateral civilian casualties and damage can particularly affect strategic assessments of the military advantage to be gained from an attack, as well as the application of the humanitarian law precaution against causing excessive collateral civilian casualties or damage. Finally, since States generally default to a law enforcement model domestically even when confronted with significant security threats, the option is explored of favoring that approach when conducting extraterritorial counterinsurgency operations.

Chapter 8, "Counterterrorism and the 'Away Game,'" looks at two methods of conducting counterterrorism operations: Special Forces raids and the use of drones. These capabilities provide unique operational choices for commanders when dealing with insurgent and terrorist forces operating in "ungoverned spaces." This chapter analyzes the unique capabilities Special Forces units provide—not only in their use to conduct missions to kill insurgents and terrorists but also in considering the capture of those persons, conducting hostage rescue missions, and countering violent transnational and other criminal organizations. Particular attention is also paid to the controversy that has developed concerning the use of drones. This includes assessing the degree to which they offer a new form of warfare and looks at the requirement to situate their counterterrorist use within a broader operational context.

The Chapter 8 analysis surveys three legal theories that have been suggested to apply to drone strikes: the "restricted" law enforcement approach, the permissive "conduct of hostilities" approach, and what is termed the "self-defense" option. Each approach will be demonstrated to have strengths and weaknesses. This will be done through the exploration of various legal issues such as the "geography of war," the reliance on neutrality law principles to justify intervention within another State, and the dangers of overreach in extending conduct of hostilities-based norms beyond situations of armed conflict. As becomes evident, this badly fractured and unresolved debate over what legal regime should govern the use of drones highlights the degree to which the international legal community has struggled with regulating use of force during operations occurring in the post-9/11 period. The chapter will demonstrate that the solution is trending toward one that more fully incorporates human rights–based law enforcement norms, either as a result of legal interpretation or policy into the rules governing drone strikes.

The final chapter of this part, Chapter 9, "Non-State Actors and Armed Conflict," also focuses on the legal uncertainty surrounding non-State actor conflict. In doing so attention is paid to two issues. First, to assessing how transnational violence is characterized across the conflict spectrum. This includes non-State actor attacks linked to inter-State conflict; their categorization as a conflict having an international character, as "extraterritorial law enforcement" and transnational armed conflict; and their classification as a noninternational armed conflict. The chapter identifies the *Hamdan v. Rumsfeld*[118] interpretation of "noninternational armed conflict" as the post-9/11 approach that resonates best with contemporary transnational threats, particularly because of its focus on status: State (i.e., the proper authority) versus non-State actors, rather than limitations based on geographic borders (i.e., internal conflicts). Second, there is the issue of the threshold for determining the existence of a noninternational armed conflict. In applying a "totality of the circumstances" approach and other recognized criteria, which are focused on the practical challenges facing security forces, the threshold will be seen to be lower than it is sometimes suggested to be the case. It will be argued the lack of theoretical consensus concerning the categorization of conflict has resulted in a tendency for practitioners to concentrate on whether an "armed

---

118 Hamdan v. Rumsfeld, 45 I.L.M. 1130, (2006) (U.S. Supreme Court 2006).

conflict" is in existence, rather than engage in the highly technical, confusing, and often divisive debate regarding conflict categorization.

Part IV, "Applying Force Across the Conflict Spectrum," includes Chapters 10 to 13. This part addresses the conduct of operations against non-State actors that span the conflict spectrum from inter-State warfare to international law enforcement. Chapter 10, "Self-Defense and the Protection of Nationals," highlights that notwithstanding the controversy post-9/11 concerning the exercise of the right to self-defense in response to attacks by non-State actors, the question of whether States can act defensively to protect nationals is not new to international law. As will be outlined, although the exercise of a right to intervene to protect nationals has been controversial, criticized, and often downplayed in terms of importance, there has remained a level of tolerance for this form of self-help particularly in situations where there is no effective territorial government. The multiplicity of legal approaches and bases used to justify the rescue of nationals will be explored to highlight the unsettled nature of this area of law.

As will be discussed in Chapter 10, a unique aspect of these defensive actions is that they can transcend the full conflict spectrum from international armed conflict to law enforcement. Two iconic counterterrorism missions—the 1976 Israeli rescue of hostages at Entebbe and the 2000 British operation to free soldiers captured by the West Side Boys, an organized armed group in Sierra Leone—will be looked at as case studies. Hostage rescue is also addressed in the context of law enforcement using examples from Columbia, Nigeria, and Somalia. As will be discussed, many of the capabilities that make Special Forces units so effective for international hostage rescue are also applicable in missions to capture insurgents. This indicates there is a convergence of tactics and facilitates a shift toward a law enforcement approach. Ultimately, the chapter suggests it is the effectiveness of the law enforcement option, rather than the categorization of the violence as an armed conflict, that should become the limiting factor in its employment in a counterterrorism role.

Chapter 11, "Law Enforcement and 'Self-Defense,'" explores the relationship between human rights–based law enforcement and the two other areas of law that have dominated the discussion concerning international operations: the State exercise of self-defense and humanitarian law. This analysis will establish that although the law governing national self-defense and law enforcement share common terminology (i.e., necessity, proportionality, and imminence), the principles operate differently within each body of law. The chapter compares the law enforcement and conduct of hostilities normative frameworks by exploring their shared just war roots in terms of the use of force in self-defense. This State authority to use force will be seen to have developed differently, depending upon whether force is used internally or externally. It will be established that the law enforcement paradigm relies not only on a relatively narrow authority to act in self-defense or defense of others but also authorizes the use of force for mission accomplishment (i.e., enforce the law). A particular limiting factor regarding the human rights–based law enforcement approach will be seen to be its principle of imminence. The dominant influence of the national self-defense principles will be considered in the context of general rules of engagement doctrine, "robust" UN peacekeeping, and the U.S. Standing Rules of Engagement. This dominance will be seen to sometimes exert too much influence on the operation of humanitarian law and human rights law in respect of controlling the use of force. Among the examples considered is the November 2004 U.S.-led assault on Fallujah. Finally, the application of law enforcement rules in the international security environment is discussed to highlight some of the pressures being exerted on a body of law that was primarily designed to regulate internal State order during times of relative peace.

Chapter 12, "The Narrow Operational and Normative Gap," explores the narrow normative gap that can occur when engaged in conduct of hostilities and law enforcement roles during a number of contemporary operations. As is discussed, the employment of military forces in a direct or support law enforcement role occurs during traditional counterinsurgency and counterterrorism operations, including their role as mentors, training and advising local police forces. The analysis highlights the degree to which hostage rescue missions and countering the threat of explosive devises such as IEDs and suicide bombers can cause military forces to adopt tactics more consistent with law enforcement norms, even during armed conflict. Conversely, these unique threats can also result in law enforcement authorities applying force in a manner that approximates the use of lethal force under conduct of hostilities rules. An outline of the 1997 rescue of hostages held at the Japanese Embassy in Lima, Peru, is provided to highlight the risks, elevated levels of violence, and controversy that can arise in respect of hostage rescues. Further, the legal test for using lethal force under the human rights law enforcement framework will be considered. What becomes evident is that the dividing line between conduct of hostilities– and policing–based uses of force are not always as a clear-cut as some might suggest.

With the increased recognition that law enforcement makes up a significant part of international operations, Chapter 13, "Limits of Law Enforcement and Human Rights Law," looks at the effectiveness of human rights law in controlling the use of force during armed conflict. The limits of that body of law are explored from three perspectives: practical limitations, overreach in application, and limits caused by interpretation. Practical limitations are assessed in the context of the long-running insurgency in Northern Ireland, which the UK government chose to address as a domestic law enforcement matter. The scope of the violence during the "Troubles," the government response, and unique factors impacting on the success of that approach are looked at to determine the degree to which that conflict can underpin an argument that terrorism is primarily amenable to a law enforcement response, or serve as a template for counterinsurgency and counterterrorism operations in the world's ungoverned spaces. The analysis will also explore the "shoot to kill" controversy that arose during the "Troubles" to consider the point at which a law enforcement approach may no longer be effective and whether this reality helps determine when a conduct of hostilities response is required. Next, the impact of developing militarized police forces to extend the application of a law enforcement role during higher end security operations will be discussed. The problem of overreach in the context of the European Court of Human Rights application of human rights law principles to restrict aerial bombing and other warfare-related violence is assessed. Finally, interpretive limits on the use of force will be looked at in the context of the different post-9/11 U.S. and Canadian, German, and Australian responses to the threat posed by hijacked aircraft.

Finally, Part V, "The Way Ahead," discusses the approaches that can be applied to address the evolving, diverse, and unique security threats facing the international community. Chapter 14. "A Holistic Solution," outlines the application of a holistic, operational law approach that is designed to address the evolving, diverse, and unique security threats facing the international community. Divided into four parts, the analysis first places this legal challenge in context by establishing that both the military and legal communities must situate the solution for countering contemporary security threats in an analytical framework that encompasses conventional conflict, irregular warfare, and criminal activity regardless of how an armed conflict is categorized legally. The second part addresses two key issues regarding operational law–based decision making. There is consideration of how the law governing the recourse to warfare, humanitarian law and human rights law, interacts in responding to transnational attacks and threats to nationals. This includes an assessment of the boundaries between these areas of law, consideration of when armed conflict commences in the context of armed attacks, and the identification of criteria to be applied in separating an armed conflict from situations requiring law enforcement. The analysis also focuses on applying a human rights based response as a matter of law, as State policy, or as a discretionary decision by the chain of command. The manner in which law enforcement is privileged as a response to attacks at home and abroad are explored. The third part provides an outline of a decision-making process that will assist in reconciling the application of the relevant bodies of international and domestic law. This process is summarized in a "decision tree" analysis set out in Appendix I. Particular attention will be paid to the role that responding States should play in acting as a proper authority in the functional absence of a territorial State when responding to threats in ungoverned spaces. The final area of analysis highlights the link between policing and "normalcy" in reaching the conclusion that the ability to apply such a human rights–based approach provides an important indicator of success in respect of conflict with non-State actors.

The fifteenth and final chapter, "Preparing for Twenty-First-Century Conflict," assesses the challenges facing the international legal community in preparing legal advisors to provide operational law advice in this complex security environment. It focuses on the type of education, training, and other preparation that should be considered in order to produce State legal advisors who can assess legal issues in a holistic fashion and provide effective operational law advice. Included in this discussion is the need to ensure that advice keeps within the bounds of legal theory but also that the work of theorists is better aligned with the practical realities of contemporary conflict. A key message is that in order for contemporary security threats to be defeated there will be a requirement for State legal advisors to be as prepared to apply various bodies of law, including human rights law (international and domestic), as they have traditionally been to give humanitarian law advice. The result is that a holistic application of relevant international and domestic laws is needed to ensure that legal boundaries do not become barriers to operational success or to the protection of civilians regardless of where they live.

II The Interaction Between Normative Frameworks

War is nothing but a duel on a larger scale.[1]

CARL VON CLAUSEWITZ

2 Controlling State Involvement in Conflict

2.1. Introduction

To understand how international "conflict" law came to be divided into distinct subject areas, as well as the impact of that division on contemporary twenty-first-century warfare, it is important to consider its development from the mid-nineteenth century onward. That history has been one of an obsession with the State and its tremendous destructive power. The result has been the development of two separate normative frameworks specifically designed for the regulation of inter-State conflict. The first body of law governs the recourse to war. It is discussed most often in a contemporary context as the exercise of a State or national right to act in self-defense. The other legal framework applies to the conduct of warfare (i.e., humanitarian law). Influenced by the efforts of the antiwar movement, the development of these normative frameworks has overall been a story of success, particularly in terms of limiting the recourse to war. This effort to limit warfare directly resulted in the creation of the UN Charter and, over the past half-century, a significant diminution of inter-State violence.

However, what remains to be considered is how this twentieth-century advance in the law relates to contemporary conflict, which predominantly involves States fighting non-State actors. The changing nature of warfare is forcing a reassessment of many commonly held beliefs. Among the questions that arise is the degree to which two foundational bodies of international law, the law of national self-defense and humanitarian law, can or should retain their traditional emphasis on inter-State conflict. States and many international

[1] CARL VON CLAUSEWITZ, ON WAR 75 (Michael Howard & Peter Paret eds. & trans., 1989).

lawyers have long preferred to view conflict with non-State actors as a uniquely domestic matter. This has occurred even though the historic recognition of some internal conflicts as taking on the attributes of an international conflict (i.e., "recognition of belligerency") and the action taken following World War II to enact treaty provisions applying humanitarian law to such conflicts (i.e., Common Article 3 to the 1949 Geneva Conventions and their two 1977 Additional Protocols) highlight that hostilities with non-State actors have always presented a more complex situation.

As will be discussed, the international rules governing the conduct of hostilities have, particularly since the 1990s, been interpreted as having greater application to conflict with non-State actors.[2] Adding to the analytical challenge is the ease with which some armed conflicts are viewed as changing from international to noninternational ones (i.e., Iraq and Afghanistan) and the ability of non-State actors to carry out attacks that not only transcend borders but in some situations whole continents. Importantly, contrary to long-held views, the law governing the recourse to war is now increasingly seen to apply to the response to attacks from these transnational non-State actors and not just to inter-State conflict. In effect, there is a recognition that the actors in the international "duel" that is war are no longer exclusively States.[3]

2.2. A Focus on the State

Perhaps it is because of the horrific death and destruction resulting from the two world wars of the twentieth century that there is a tendency, even in contemporary thinking, to consider "war" particularly in terms of States fighting States.[4] This focus on inter-State conflict has been prevalent in scholarly research[5] and in the study of warfare by military professionals.[6] Such work often emphasizes the nineteenth-century theories of Carl Von Clausewitz and Alfred Thayer Mahan, which concentrate on conventional warfare between States.[7] The pull of conventional warfare doctrine has been so strong that it has been attributed to skewing the American approach during the Vietnam Conflict. A "Clausewitzian" approach was adopted toward fighting a guerrilla war that emphasized conventional forces, physical destruction, and "the use of force directly against the presumed external sponsoring power to persuade it both to refrain it from conventional invasion and put its guerrillas

---

[2] Prosecutor v. Tadić, Case No. IT-94-1-A, Decision On The Defence Motion For Interlocutory Appeal On Jurisdiction, ¶¶ 96-127 (Int'l Crim. Trib. for the Former Yugoslavia, 1995), http://www.icty.org/x/cases/tadic/acdec/en/51002.htm.

[3] Hugo Grotius, 1 The Rights of War and Peace 135 (2005) (where war is described as the "State or Situation of those . . . who dispute by Force of Arms" with the root of the word *bellum* (war) being derived from the "old word *duellem* (a duel).").

[4] John Keegan, A History of Warfare 50 (1993) (where it is estimated that 20,000,000 people died in the First World War and 50,000,000 in the Second World War).

[5] Jack S. Levy & William R. Thompson, Causes of War 2 (2010).

[6] Robert M. Cassidy, Counterinsurgency and the Global War on Terror 117 (2008).

[7] Clausewitz, *supra* note 1, and Alfred Thayer Mahan, The Influence of Sea Power Upon History 1660–1783 (1889) (2009). *See also* Levy & Thompson, *supra* note 5, at 7 ("Most books on the history of war in the modern area (which historians date from about 1500 on) focus on interstate wars, with particular attention to interstate wars between the great powers, the most powerful states in the system.").

on a leash."[8] This State-centric gravitational pull is not unique to the twentieth century. There remains a very real danger that military doctrine developed by States will overemphasize conventional warfare threats as readjustments are made to its security policy in the aftermath of the withdrawal of the international community from fighting in the two major insurgencies of the post-9/11 period (i.e., Iraq and Afghanistan).

This emphasis on the State is also reflected in the treatment of "war" under international law. Attempts to regulate State-related conflict have resulted in three formal legal categories: international armed conflict, noninternational armed conflict, and internal "situations of internal disturbances and tensions."[9] The definition of "international armed conflict" is uniquely restricted to hostilities between States.[10] The historical emphasis by the legal community on inter-State war is clearly demonstrated by the fact that of the two types of armed conflict it is the former which is as a matter of treaty law subject to the most comprehensively established international humanitarian law regime.

The attention paid to inter-State war can, at times, mask the reality it is not, and never has been, the predominate form of warfare.[11] Inter-State warfare is in significant decline,[12] although given the number of twenty-first-century conflicts, such as in Afghanistan, Iraq, and the Congo, it would be premature to suggest State versus State conflict is a relic of the past.[13] The risk of inter-State armed conflict is evident in the rising tension between Israel, the United States, and other Western States with Iran, which has been described in its present form as a "shadow war"[14] or "twilight" war.[15] It can also be seen in the international discord caused by the 2014 Russian annexation of the Crimea.[16] Further, the potential exists for inter-State warfare to erupt in the Pacific Rim over excessive maritime claims.[17] The Turkish shootdown of a Russian warplane along the Turkey/Syria border in November 2015 highlights that inter-State tensions will continue to arise.[18] However, the

---

[8] Larry E. Cable, Conflict of Myths: The Development of American Counterinsurgency Doctrine and the Vietnam War 6 (1986).

[9] AP II, art. 1(2).

[10] Legal Consequences Of The Construction Of A Wall In The Occupied Palestinian Territory, 2004 I.C.J. ¶ 139 at 194, *available at* http://www.icj-cij.org/docket/files/131/1671.pdf. *See also* Prosecutor v. Tadić, No. IT–94–1–A, Appeals Judgment, ¶84, at 34 (Int'l Crim. Trib. for the Former Yugoslavia, July 15, 1999), http://www.icty.org/x/cases/tadic/acjug/en/tad-aj990715e.pdf.

[11] David Kilcullen, Counterinsurgency ix–x (2010). *See also* The National Military Strategy of the United States of America 2015, at 4 (June 2015) [hereinafter The National Military Strategy], http://www.jcs.mil/Portals/36/Documents/Publications/2015_National_Military_Strategy.pdf ("VEOs [violent extremist organizations] . . . pose an immediate threat to transregional security by coupling readily available technologies with extremist ideologies.").

[12] Levy & Thompson, *supra* note 5, at 12.

[13] Colin S. Gray, Another Bloody Century: Future War 170 (2005) ("it is far too early to write the obituary for . . . [interstate war]".).

[14] Yaakov Katz & Yoaz Hendel, Israel vs. Iran: The Shadow War 208 (2012).

[15] David Crist, The Twilight War: The Secret History of America's Thirty-Year Conflict with Iran 5 (2012).

[16] Shaun Walker & Ian Traynor, *Putin Confirms Crimea Annexation as Ukraine Soldier becomes First Casualty*, The Guardian, Mar. 19, 2014, http://www.theguardian.com/world/2014/mar/18/putin-confirms-annexation-crimea-ukrainian-soldier-casualty.

[17] James Kraska, Maritime Power and the Law of the Sea 312–29 (2011).

[18] *Turkey's Downing of Russian Warplane - What We Know*, BBC News (Nov. 30, 2015), http://www.bbc.com/news/world-middle-east-34912581.

assessment remains in *The National Military Strategy of the United States of America 2015* that "the probability of U.S. involvement in interstate war with a major power is assessed to be low but growing."[19] This is still a period during which many Western military forces are withdrawing from over a decade of engagement in counterinsurgency and counterterrorism operations in Afghanistan and Iraq. Naturally, States are recalibrating where their doctrinal priorities should lie: conventional conflict or counterinsurgency/counterterrorism.

Notwithstanding the recalibration of doctrinal priorities by military forces, the majority of world conflict is and is likely to continue to be in the form of "small wars." Such wars were historically defined as "campaigns undertaken to suppress rebellions and guerrilla warfare in all parts of the world where organized armies are struggling against opponents who will not meet them in the open field."[20] It was estimated in 2012 that "[g]lobally, there are well over one hundred ongoing small, irregular, asymmetric, and revolutionary wars, in which violent nonstate actors are helping their own organizations or political patrons bring about radical change or acquire power."[21] These small wars have also been referred to by various other titles such as low-intensity conflicts or military operations other than war.[22] Although historically conceived of in the context of colonial wars and interventions in other States, "small wars" is a term that has come to describe any warfare between a State and non-State actor.[23] This inevitably places conventional State military forces in combat against groups that apply more "primitive" tactics in the form of guerrilla warfare, which have proven over time to be exceedingly effective. To be successful, the conventional force must in a number of ways adopt the tactics of this more indirect form of warfare.[24]

Unfortunately, it remains that warfare is "a constantly recurring and universal characteristic of human existence,"[25] which can take place between the armed forces of two or more States, or more likely involve a non-State actor either resisting or challenging a sovereign's authority. Given the essential part that the State plays both in international armed conflict and in its more prevalent noninternational counterpart, understanding the role of that entity remains key to assessing how law impacts on warfare. This is because it has mainly been States and their military establishments that have set the rules governing armed conflict.[26] The State serves as the primary actor in determining what norms are applied within those conflicts. Further, it is the unique role performed by States as the "proper authority" in controlling internal violence that raises the issue of the application of human rights law (see Chapter 4, section 4.3, *Impact of the Proper Authority Principle*). Humanitarian law and human rights law are in a number of important respects different, particularly as the former body of law provides both State rights and individual human rights.[27]

---

19 The National Military Strategy, *supra* note 11, at 4.

20 C. E. Callwell, Small Wars: Their Principles and Practice 21 (3rd ed. 1996) (1906).

21 Max G. Manwaring, The Complexity of Modern Asymmetric Warfare 3 (2012).

22 Max Boot, The Savage Wars of Peace: Small Wars and the Rise of American Power xv (2002).

23 Small Wars Manual, United States Marine Corps ¶¶ 1-1. *a.*, at 1 (2009) (1940).

24 Lawrence H. Keeley, War Before Civilization: The Myth of the Peaceful Savage 81 (1996).

25 Anthony Stevens, The Roots of War & Terror 7 (2004). *See also* Azar Gat, War in Human Civilization 662 (2006).

26 Kilcullen, *supra* note 11, at x.

27 Yoram Dinstein, The Conduct of Hostilities Under the Law of International Armed Conflict ¶ 45, at 20 (2nd ed. 2010).

René Provost emphasizes their difference, indicating that the former body of law has its basis in hostility, while the human rights law is based on a harmonious relationship between the State and those under its jurisdiction.[28] However, those bodies of law share a common focus on the State and many of the same normative principles (see Chapter 5, section 5.3, *A More Intimate History*). As will be discussed, States themselves have jealously guarded their unique status, often resisting attempts to introduce international law to regulate internal conflicts or legitimize the participation of non-State actors in armed conflict.

So what brought about the near obsession exhibited in international law with regulating inter-State armed conflict instead of its more prevalent internal counterpart? What impact does this obsession have on contemporary efforts to ensure the normative framework governs the full scale of armed conflict, both internal and external? That State focus appears to be the result of three factors. First, by the nineteenth century there was a growing and well-placed concern over the destructive power of the modern State. Second, there was a strongly held view that only "civilized" States and those who fought in their conventionally organized armies could participate in hostilities. Finally, there was a reluctance to interfere with the role of the State as a "civilizing" influence for limiting internal strife and insecurity. Inherent in each of these factors was the function the State performed as the "proper" or "right" authority for governing and ultimately waging war. The 2015 *Department of Defense Law of War Manual* highlights the status of the State when it refers to it as the "competent authority" for resorting to military force; when distinguishing whether an armed conflict is international (i.e., inter-State) or noninternational (i.e., intra-State) in character; and for determining who is entitled to lawful belligerent status.[29] It has been noted elsewhere that "*all warfare is precisely, and ultimately, about governance.*"[30] Importantly, in performing this role the State has also been the subject of efforts to limit its destructive power. This led directly to rules, which limited the application of force by means of a normative framework that by the early twentieth century was at its heart "antiwar."

2.3. The Destructive Power of the State

2.3.1. REINTRODUCTION OF JUST WAR THEORY

It seems to be no coincidence that the international effort to control armed conflict developed in the mid to late nineteenth century just as States were beginning to demonstrate their tremendous destructive capability. That era saw what has been termed the first modern conflict, the exceedingly destructive U.S. Civil War.[31] That conflict included savage guerrilla warfare "that made the war a far bloodier affair than armies alone could have

---

[28] René Provost, International Human Rights and Humanitarian Law 8 (2002).

[29] Department of Defense Law of War Manual, Office of the General Counsel ¶ 3.5.3, at 88 (June 2015), http://www.dod.mil/dodgc/images/law_war_manual15.pdf.

[30] Brian Orend, The Morality of War 3 (2006).

[31] Leo Braudy, From Chivalry to Terrorism: War and the Changing Nature of Masculinity 268 (2003). *See also* Orend, *supra* note 30, at 20 ("The US Civil War was a major spark in this regard.").

done."[32] This civil war was a harbinger of a period where in less than one hundred years a technological and scientific-based revolution would lead to the dropping of an atomic bomb on Hiroshima and Nagasaki.[33] At the same time the U.S. Civil War was to spawn the first comprehensive attempt to codify the rules governing the conduct of hostilities. This codification came in the form of the 1863 *Instructions for the Government of Armies of the United States in the Field*, written by Francis Lieber.[34] It is not without some irony that the *Lieber Code* was written in the context of a civil war rather than a conflict between States. That it remains the only comprehensive code prepared on behalf of a government for the governance of internal conflict highlights the degree to which subsequent codification efforts were to focus on inter-State warfare and in many respects avoid dealing with internal conflicts.

Significantly, from the middle of the nineteenth century there was an increasing effort to limit war. This was reflected in the adoption of the 1864 Geneva Conventions sponsored by the Red Cross,[35] as well as in "several meetings of the Universal Peace Congress, the creation of the Nobel Peace Prize (1897, first awarded 1901), and in 1899 the first Hague conference."[36] The period marked the beginning of efforts to codify secular just war theory, which had fallen into disfavor with the rise of the State in the post-Westphalian period.[37] This modern just war theory replaced a "balance of power" doctrine, which rested "on the political *status quo* which was assumed to represent a 'Balance of Power' or *principe d'equilibre* between the various states or group of states."[38] Instead, the focus of international lawyers would become one of constraining the right to go to war.

The reintroduction of just war principles from the late 1800s onward represented, at least initially, a holistic approach toward regulating state involvement in war. Notably, that effort addressed both State recourse to war and conduct of hostilities issues,[39] although there have also been suggestions this marked the point where the two bodies of law became separate.[40] In pragmatic terms it has been observed that the combining of the two ideas:

> enables lawyers and humanitarians to regulate and mitigate the evils of war twice over; once through restrictions on the recourse to war, and again (with the knowledge that the first set of restrictions will often be ineffective) through restrictions on the conduct of war.[41]

---

[32] DANIEL E. SUTHERLAND, A SAVAGE CONFLICT: THE DECISIVE ROLE OF GUERRILLAS IN THE AMERICAN CIVIL WAR 277 (2009).

[33] BRAUDY, *supra* note 31, at 267–75.

[34] *U.S. War Department, Instructions for the Government of Armies of the United States in the Field, General Orders No. 100*, Arts. (April 24, 1863), *in* RICHARD SHELLY HARTIGAN, LIEBER'S CODE AND THE LAW OF WAR 45, 45–71 (1983) [hereinafter the *Lieber Code*].

[35] Adam Roberts, *Against War*, *in* THE OXFORD HISTORY OF MODERN WAR 317, 320 (Charles Townshend ed. 2005).

[36] BRAUDY, *supra* note 31, at 417. *See also* Lindsay Moir, *The Historical Development of the Application of Humanitarian Law in Non-International Armed Conflicts to 1949*, 47 INT'L & COMP L.Q. 337, 351–52 (1998).

[37] OREND, *supra* note 30, at 10.

[38] IAN BROWNLIE, INTERNATIONAL LAW AND THE USE OF FORCE BY STATES 14 (1963).

[39] BRAUDY, *supra* note 31, at 417.

[40] JUDITH GARDAM, NECESSITY, PROPORTIONALITY AND THE USE OF FORCE BY STATES 49 (2004).

[41] DAVID RODIN, WAR & SELF-DEFENSE 167 (2002).

This effort was significantly influenced by the growth of pacifism[42] and pacificism,[43] which were part of a broader movement that fostered "widespread questioning of the belief that war was an elemental part of political, economic, and social life."[44] Further, it was affected by powerful ideas of religiously based humanism, ultimately "making possible the Geneva Conventions, the League of Nations and the United Nations, and so on."[45]

While recourse to war issues were initially addressed contemporaneously with rules governing the conduct of hostilities, their codification was anything but simultaneous. The development of conventions governing the conduct of hostilities is perhaps best represented in the 1907 Hague Land Warfare Regulations,[46] which predated the codification of restrictions on the resort to war. However, as Thomas Franck has noted, "[n]either the *jus ad bellum* [law governing the recourse to war] nor the *jus in bello* [humanitarian law] regarded recourse to violence as a wrong per se."[47] The effort to prohibit war, subject to the right to act in self-defense, did not itself see codification until after World War I with the enactment of the 1928 Kellogg-Briand Pact.[48] That effort was directly linked to the post–World War I efforts to "outlaw" inter-State war.[49] It is this connection with an antiwar motivation, as well as its exclusive application to States in the twentieth century, that has helped set the law governing the recourse to war apart from the body of law regulating the actual conduct of hostilities. These bodies of law were to be increasingly considered separately.[50] At the heart of this separation was the development of the "equal application principle" where humanitarian law applies to all participants in hostilities regardless of the rightness or the wrongness of the steps that resulted in their becoming embroiled in the conflict. In this respect, "[t]he fundamental postulate of the *jus in bello* [humanitarian law] is the equal application of its legal norms to all Belligerent Parties, regardless of their relative standing in the eyes of the *jus ad bellum* [the justness of the cause]."[51] One historical basis for that separation has been suggested to be that humanitarian law is the product of a medieval chivalric code, while the law governing the recourse to war was an invention of churchmen and lawyers challenging the assumption that warfare is an acceptable activity.[52]

---

42 Braudy, *supra* note 31, at 417 (where based on an interpretation by Martin Ceadel, elaborating on a suggestion from the historian, A. J. P. Taylor, "pacifism" is "reserved for the absolute belief that wars are always wrong.").

43 *Id.* ("pacificism" is identified with the idea that "some wars may be necessary....").

44 *Id.*

45 Stevens, *supra* note 25, at 7–8.

46 *See* Roberts, *supra* note 35, at 320.

47 Thomas M. Franck, Recourse to Force: State Action Against Threats and Armed Attacks 10 (2002).

48 Albrecht Randelzhofer, *Article 2(4) in* I The Charter of the United Nations 112, ¶ 10, at 16 (Bruno Simma et al. eds., 2nd ed. 2002) [hereinafter referred to as The UN Charter Commentary].

49 Yoram Dinstein, War Aggression and Self-Defence ¶ 228, at 85 (5th ed. 2011).

50 Robert Kolb, *Origin of the Twin Terms Jus Ad Bellum/Jus in Bello*, 320 Int'l Rev. Red Cross (1997), https://www.icrc.org/eng/resources/documents/misc/57jnuu.htm (where it is indicted it was "extremely rare to find terms *jus ad bellum* and *jus in bello* used before 1930."). *But see* Manual of Military Law 234 (His Majesty's Stationary Office 1916) (1914) (one example of an earlier use is seen in the 1914 reference that "[t]he present laws of war are a result of a slow and gradual growth. Isolated milder war practices became... so-called *usus in bello*, manner of warfare, and these usages were developed into legal rules by custom and treaties.").

51 Dinstein, *supra* note 27, ¶7, at 3. *See also* Eyal Benvenisti, *Rethinking the Divide Between Jus ad Bellum and Jus in Bello in Warfare Against Nonstate Actors*, 34 Yale J. Int'l L. 541, 545 (2009).

52 Orend, *supra* note 30, at 31.

2.3.2. INTER-STATE CONFLICT: LIMITING THE RECOURSE TO WAR

The early antiwar movement did not succeed in preventing the two World Wars. However, its efforts ultimately led to the development of a prohibition on the use of force by states in the 1949 UN Charter, which "could be more appropriately referred to as *ius contra bellum* [law on the prevention of war]."[53] It was disgust over the First World War that had "inspired the efforts of some nations and individuals in the 1920s and 1930s to limit war."[54] The focus of the legal effort on preventing further conflict during those years, which was particularly active in the United States,[55] included the development of "jural [legal] instruments" designed to control war.[56] One of the first significant instruments was the 1919 Covenant of the League of Nations.[57] Although a flawed and ultimately unsuccessful document, one of its most significant effects was to "make any war between states a matter of international concern."[58] In this respect, "[w]ar was no longer to have the aspect of a private duel, but of a breach of the peace which affected the whole community."[59]

Tragically, the next world war was to be one where "forces of hitherto unknown magnitude had been unleashed upon the world."[60] It would result in 17.7 million combatant deaths and over 39 million civilians being killed.[61] Unregulated, States proved what destruction they were capable of. The revulsion against this death and destruction ultimately resulted in the 1949 UN Charter, which introduced a legal framework that significantly limits the recourse to war by States. That document was the realization of a goal for a generation of international lawyers.[62] The Charter prohibited "the threat or use of force against the territorial integrity or political independence of any state"[63] and only permitted hostilities between states under the auspices of a Chapter VII determination of the UN Security Council[64] or in the exercise of self-defense.[65] The UN Charter did not prohibit war altogether, although there was opinion expressed in the early years of its operation that such conflict "in a legal sense has been in large measure outlawed."[66] Reflective of that trend has been the near disappearance of the word "war" from the legal lexicon, although reference to it can still be found in foundational international documents such as the 1949 Geneva Conventions and the

---

[53] Marco Sassoli, *Ius ad Bellum and Jus in Bello—The Separation between the Legality of the Use of Force and Humanitarian Rules to Be Respected in Warfare: Crucial or Outdated?*, *in* INTERNATIONAL LAW AND ARMED CONFLICT: EXPLORING THE FAULTLINES 241, 242 (Michael Schmitt & Jelena Pejic eds., 2007).

[54] BRAUDY, *supra* note 31, at 418.

[55] D. W. BOWETT, SELF-DEFENSE IN INTERNATIONAL LAW 125 (The Lawbook Exchange Ltd., 2009) (1958).

[56] Quincy Wright, *Changes in the Conception of War*, 18 AM. J. INT'L L. 755, 767 (1924).

[57] *Covenant of League of Nations, Adopted by the Peace Conference at Plenary Session, April 28 1919*, No. 2, 13 AM. J. INT'L L. 128 (Apr. 1919).

[58] BROWNLIE, *supra* note 38, at 57.

[59] *Id.*

[60] RUPERT SMITH, THE UTILITY OF FORCE: THE ART OF WAR IN THE MODERN WORLD 141 (2005).

[61] *Id.*

[62] The link between this effort and "just war" theory during the interwar period is reflected in Joachim von Elbe, *The Evolution of the Concept of the Just War in International Law*, 33 AM. J. INT'L 1 (1939), and Joseph Kunz, *Bellum Justum and Bellum Legale*, 45 AM. J. INT'L L. 528 (1951), and Wright, *supra* note 56.

[63] UN Charter, art. 2(4).

[64] *Id.* art. 42.

[65] *Id.* art. 51.

[66] Quincy Wright, *The Outlawry of War and the Law of War*, 47 AM. J. INT'L L. 365, 365 (1953).

UN Charter.[67] This tension between limiting conflict and outlawing war appears to reflect the different approaches inherent in pacifism and pacificism. The former belief sees war as always wrong, while the latter approach accepts "some wars may be necessary, even while they stress the need for peaceful arbitration before that point is reached."[68] However, viewed together these approaches are often collectively framed as being "antiwar."

The influence of just war doctrine on the UN Charter has been the subject of debate, particularly in the context of whether the purpose of the Charter is to seek peace, justice, or both. One view expressed was that the maintenance of international peace and security was prioritized over justice,[69] while another took a nuanced approach seeing both peace and justice realized.[70] Perhaps the best indication that such a link to just war theory exists is found in the widespread use of the just war concept, *jus ad bellum*, to describe the Charter-based framework for controlling the recourse to war. One commentator described the continuing relationship as follows:

> Though it substitutes the notion *bellum legale* (lawful war) for that of *bellum justum* (just war), international law is recognizably a continuation of the just war tradition, providing a treatment of the *jus ad bellum* [law governing the recourse to war], contiguous with earlier discussions.[71]

However, it has also been suggested the body of law governing the recourse to war had changed under the Charter system, with "the substitution of a new yardstick—self-defense, the *jus ad bellum* of the Charter—against which the means and overall 'evil' of the use of force are measured."[72] This narrower version of the law governing the recourse to war, in the form of a State right to act in self-defense in response to an armed attack, was no doubt facilitated by the view it represents a body of law separate from humanitarian law.

Ultimately, the efforts of the twentieth-century antiwar movement had a profound effect on inter-State conflict. Yet, the impact of the post–World War II law governing the recourse to war had its limits, specifically because of its focus on such conflict. For example, the prohibition on the threat or use of force against the territorial integrity or political independence of a State found in Article 2(4) of the UN Charter has no application to conflicts within the territory of a State. Further, intervention in the internal affairs of a State was specifically not authorized in Article 2(7) of the Charter, except for "the application of enforcement measures under Chapter VII." While such intervention has occurred, it has not been a regular occurrence and has been the subject of considerable debate.[73] Further, the success gained by the antiwar movement, a general prohibition on the use of force by States, has had little impact on armed conflict internal to States. This is an area where the preeminent

---

67 UN Charter, *Preamble*.

68 BRAUDY, *supra* note 31, at 417.

69 Kunz, *supra* note 62, at 533.

70 FRANCK, *supra* note 47, at 138–39.

71 RODIN, *supra* note 41, at 104.

72 Judith Gardam, *Proportionality and Force in International Law*, 87 AM. J. INT'L L. 391, 403 (1993).

73 Georg Nolte, *Article 2(7)* in THE UN CHARTER COMMENTARY, *supra* note 47, 148 ¶51(aa), at 164–165 (where he discusses the debate in the late 1990s over whether the United Nations had, under the application of Art. 2(7), a role in the settlement of "domestic political problems" regarding the "Kosovo crisis".).

status ascribed to States as the proper authority to use force internally has blunted efforts to introduce international regulation.

Perhaps the best indication of the significant effect that the UN security system has had on constraining conflict is the reality that "[t]he starting point for any examination of the law is the prohibition of the use of force."[74] The restrictive impact of this State-centric recourse to war framework is also seen in the ongoing debate regarding the exact scope of the right of self-defense. This debate has included disagreement regarding the circumstances, under which the right to self-defense by States can be claimed,[75] whether there is a residual customary law right to self-defense,[76] as well as suggestions that the use of force can occur within the concept of countermeasures.[77] More recently, it has been stated that only a minority of experts accept that force can be used as part of such countermeasures.[78] Yoram Dinstein has introduced the concept of "defensive armed reprisals" to described State responses to armed attack that are delayed but not punitive.[79] This discussion has extended to questions about whether the prohibition on the use of force is absolute,[80] permitting, perhaps, unilateral humanitarian intervention.[81] There has also been ongoing debate regarding the existence of an authority to act to protect nationals threatened in foreign lands as part of the exercise of the State right of self-defense (see Chapter 10, *Self-Defense and the Protection of Nationals*). These debates among international lawyers regarding extraordinary State actions demonstrate that notwithstanding the certainty with which the theoretical boundaries surrounding the law governing the recourse to war are sometimes drawn, those frontiers are more uncertain or at least more flexible than commonly discussed. In this regard, many contemporary security operations such as hostage rescue, noncombatant evacuation operations, maritime interdiction operations, and counterpiracy, which often occur at the lower end of the conflict scale, raise challenging questions concerning the scope of the exercise of State self-defense and the defense of nationals.

Questions regarding the exercise of State self-defense will also be seen to be at the heart of the discussion concerning the response to transnational terrorism, and in particular targeted killing through the use of drones (see Chapter 8, section 8.5, *Legal Frameworks for Responding to Transnational Terrorism*) and the conduct of hostage rescues (see Chapter 10, section 10.3, *Protecting Nationals and Hostage Rescue*). The conduct of these contemporary operations often demand the simultaneous consideration of various other bodies of law including humanitarian law, and international and domestic human rights law. Notwithstanding the theoretical separation of the law governing the recourse to war and humanitarian law, practitioners must frequently consider their interaction.

---

[74] Christine Gray, International Law and the Use of Force 6 (3rd ed. 2008).

[75] *See, e.g.*, Dinstein, *supra* note 49, ¶555, at 210, and Gray, *supra* note 74, at 181 (for differing views regarding the threshold for an "armed attack").

[76] Gray, *supra* note 74, at 117–19.

[77] Antonio Cassese, International Law 371–72 (2nd ed. 2005), and Franck, *supra* note 47, at 109–12 (where the right to use forceful measures in response to attacks below the threshold of "armed attack" is recognized.). *But see* Lindsay Moir, Reappraising the Resort to Force: International Law, *Jus ad Bellum* and the War on Terror 29 (2010) (where he notes that other commentators have taken the view "any such activities were violations of the *jus ad bellum*.").

[78] Tallin Manual on the International Law Applicable to Cyberwarfare ¶5, at 38 (Michael Schmitt ed., 2013).

[79] Dinstein, *supra* note 49, at ¶¶ 646–75, at 244–55.

[80] Gray, *supra* note 74, at 30–33.

[81] *Id.* at 51.

2.4. Noninternational Armed Conflict

2.4.1. DOMESTIC AUTHORITY, INTERNATIONAL LAW, AND NON-STATE ACTORS

A particular contemporary challenge for the international legal community has been dealing with how the law governing the recourse to war relates to armed conflict against non-State actors. This issue has gained prominence as such conflict has become increasingly cross-border in nature. Traditionally, the international legal regime governing the recourse to war was considered to apply solely to conflicts between States. As has been noted, Article 2(4) of the UN Charter "prohibits the threat or use of force in international relations, not in domestic situations."[82] The concept of the exercise of State self-defense was only relevant when tied to "offensive operations by the forces of a state."[83] For example, as Ian Brownlie suggested in 1963, it was conceivable that a coordinated and general campaign by non-State actors "*with obvious or easily proven complicity of the government of a state* from which they operate would constitute an 'armed attack,'" however, this would not extend to "[s]poradic operations by armed bands."[84] Establishing a connection between non-State actors and States was to become an enduring requirement for traditionalist interpretations of international law.

Marco Sassoli suggests that "[t]echnically, no international *ius ad bellum* [law governing the recourse to war] exists concerning non-international armed conflicts, since such conflicts are neither justified nor prohibited by international law."[85] Rather, the authority to engage in such conflict is seen as having a domestic source since "we may assume that the national legislation of all States prohibits anyone acting under their jurisdiction to wage armed conflict...."[86] This approach reflects a classic notion of internal armed conflict. When a State battles insurgents within its borders, it is national law that regulates the violence. This is an interpretation that favors a decidedly one-sided approach toward "legitimacy," with only the State being properly engaged in using force against internal enemies. It is also a result that is consistent with how international law historically developed, as States gained a more prominent role in the post-Westphalian period. International law regulated inter-State conflicts, while States were recognized as the "proper authority" for maintaining order and controlling internal conflicts (see Chapter 4, section 4.4.1, *Public, Private, and Mixed War*).

However, armed conflict with non-State actors has always been more complex than this traditional view of international law suggests. This will be explored in four contexts: the historical recognition that some conflicts with non-State actors can be international in character; the impact of conflicts transforming from international to noninternational; the increasing recognition that the law governing the recourse to

[82] MALCOLM N. SHAW, INTERNATIONAL LAW 1149 (6th ed. 2008).

[83] BROWNLIE, *supra* note 38, at 278

[84] *Id.* at 278–79 (emphasis added).

[85] Sassoli, *supra* note 53, at 254. *See also* SHAW, *supra* note 82, at 1149 ("There is no rule against rebellion in international law.").

[86] Sassoli, *supra* note 53, at 255, and SHAW, *supra* note 82, at 1149 ("the matter is purely within the hands of the authorities of the country concerned....").

war can apply to govern defensive responses to non-State actor attacks; and the struggle international lawyers are currently experiencing in describing how these bodies of international law interact.

2.4.2. INTERNAL CONFLICTS WITH AN "INTERNATIONAL CHARACTER"

There is actually nothing new in stating that some State versus non-State actor conflict can take on the attributes of an international armed conflict, although States have consistently downplayed these situations. It is evident in the historical concept of "recognition of belligerency," where an internal conflict could develop to the point that rebels are accorded recognition as long as certain conditions are fulfilled.[87] This most obviously occurred when a conflict developed to the point of being a civil war.[88] Among the necessary conditions for recognition was that hostilities had to be conducted "by organized armed groups operating under a *responsible* authority."[89] At the point the recognition of a belligerency occurred, the rebels "become subjects of international law and responsible in international law for all their acts."[90]

While recognition of belligerency fell into disuse prior to World War II,[91] the possibility of the application of some aspects of international law to regulate non-State actor conflict did not disappear. This most obviously occurred in the immediate postwar period with the development of the Common Article 3 provisions of the four 1949 Geneva Conventions. Significantly, an extension of international law to conflicts involving non-State actors was also seen in the two Additional Protocols developed in 1977 to update the Geneva Conventions. Neither Common Article 3 nor Additional Protocol II applicable to noninternational armed conflicts extended any form of legitimacy to non-State actors.[92] Further, it appears that the Additional Protocol II reference in Article 3 to nothing in that document affects "the sovereignty of a State or the responsibility of the government, by all legitimate means, to maintain or re-establish law and order..." actually provides a form of international treaty recognition of the State as a "proper authority" to deal with internal conflict.[93]

This extension of international humanitarian law to conflicts with non-State actors effectively meant that broader just war–based concepts—the corpus of the laws and customs of war and recognition that the State was the "proper authority"—were acknowledged to be applicable to internal armed conflict. The question remains whether such change was limited to only these aspects of just war theory. In this regard, Additional Protocol I presented a different proposition than either Common Article 3 or Additional Protocol II. Reflecting the challenges presented by post–World War II incidents of guerrilla-based revolutionary warfare, Additional Protocol I more extensively applied humanitarian law

[87] SANDESH SIVAKUMARAN, THE LAW OF NON-INTERNATIONAL ARMED CONFLICT 9–20 (2012). *See also* Moir, *supra* note 36, at 339–50.

[88] SIVAKUMARAN, *supra* note 87, at 11.

[89] SHAW, *supra* note 82, at 1149 (emphasis added).

[90] *Id.* at 1150.

[91] Moir, *supra* note 36, at 352–53.

[92] *See, e.g.*, GC IV, art. 3, and AP II, art. 3.

[93] AP II, art. 3.

to what have been called "wars of national liberation" (i.e., "peoples fighting against colonial domination and alien occupation and against racist regimes *in the exercise of their right of self-determination*") involving States and non-State actors.[94] This recognition of wars of national liberation provides an interesting historical example where conflict with some non-State actors was viewed as being significantly different than ordinary non-State actor conflict. This is because this Protocol not only raised issues regarding the application of humanitarian law but also the status of non-State participants in conflict related to self-determination.

If these non-State groups were considered to be participants in an international armed conflict such that the 1949 Geneva Conventions were fully applicable, and the non-State actors were to be treated in a manner similar to States in terms of the application of humanitarian law,[95] the question remains whether, in theory at least, the law governing the recourse to war might also have some application. One analysis of self-determination seems to have indirectly raised this possibility in terms of an authority for non-State actors to have recourse to hostilities. A key element of the recognition of such armed groups is that "peoples" were fighting for "self-determination."[96] The Additional Protocol I provisions were part of a broader international recognition of the principle of self-determination "as an *anti-colonist standard*, as a *ban on foreign military occupation*, and as a requirement that all *racial groups be given access to government*."[97] Antonio Cassese suggested that the principle of self-determination had a twofold impact regarding the use of force. First, it extended "the general ban on force (previously existing for States in relations with other States)" to prohibit its use for denying self-determination.[98] Second, it "resulted in granting to liberation movements *a legal licence to use force* for the purpose of reacting to a forcible denial of self-determination... [that] amounts to a derogation from the customary ban on the use of force."[99]

As it turns out, the idea that national liberation movements might be authorized to use force did not have to be considered. The potential recognition of wars of national liberation prompted significant State criticism. This reaction was consistent with the traditional State rejection of any form of legitimacy being extended to non-State actors. It was particularly concerns over possibly providing legitimacy to non-State actors that prompted the United States to refuse to ratify Additional Protocol I.[100] At this stage, this provision relating to national liberation movements appears, in many respects, to be a historic relic of the 1960s and 1970s. It is now generally considered highly unlikely, given the narrow interpretation normally applied to Article 1(4) of Additional Protocol I, that it will ever be invoked.[101] This is what undoubtedly has led the majority of world nations to ratify the Protocol.

---

[94] AP I, art. 1(4) (emphasis added).

[95] Ronald Reagan, *Letter of Transmittal*, 81 AM. J. INT'L L. 910, 911 (1987) ("One of its provisions, for example, would automatically treat as an international conflict any so-called 'war of national liberation.'").

[96] AP I, art. 1(4).

[97] CASSESE, *supra* note 77, at 61.

[98] *Id.* at 63.

[99] *Id.* (emphasis added).

[100] Reagan, *supra* note 95, at 912 ("The repudiation of Protocol I is one additional step, at the ideological level so important to terrorist organizations, to deny these groups legitimacy as international actors.").

[101] Theodor Meron, *The Time Has Come for the United States to Ratify Geneva Protocol I*, 88 AM. J. INT'L L. 678, 683 (1994).

However, what this treaty recognition of wars of liberation does indicate is that the application of just war principles may not necessarily be uniquely limited to inter-State conflict. As is reflected in Additional Protocol I, as well as in the historical application of the recognition of belligerency, there are forms of conflict with non-State actors that can begin to take on an "international character." As a result, conflict with non-State actors can be far more complex than general statements of international law might suggest. Even in practical terms such conflicts may take on the attributes of inter-State hostilities and even involve conventional warfare. Simply put, not all conflicts with non-State actors are the same, and the law at times tries to account for those differences. It has been suggested that "the legality of wars of national liberation in certain conditions and, as a consequence, the legality of external assistance to such conflicts" [102] was one of the areas of controversy in what was otherwise a "rather tidy legal regime"[103] that was put in place with the enactment of the UN Charter. In the twenty-first century, conflict with non-State actors has proven to be anything but "tidy."

2.4.3. CONFLICT EVOLUTION

Conflicts of an international character are not the only ones involving non-State actors that introduce a potential interface between the law governing the recourse to war and humanitarian law. In a classic sense noninternational armed conflicts occur as an insurgency originating from within the borders of a State. However, these internally generated conflicts only represent one way in which noninternational armed conflicts can develop. As can be seen in the cases of Iraq and Afghanistan, the situation may occur where an international armed conflict transforms into a noninternational one. It can be argued that in the case of Iraq the 2003 invasion actually created the subsequent insurgency. The reality is that the international roots of these conflicts and the reasons why they occurred in the first place cannot be easily discarded once a transformation to a noninternational armed conflict is suddenly declared to have taken place. Reliance on a domestic legal source as the basis for operations in such conflict also seems all the more challenging in an era where multinational coalitions may be fighting non-State actors.[104] Not only does this internationalize the conflict, consideration often has to be given to differing views concerning the status of these conflicts. Regarding the post-9/11 Afghan conflict, it was suggested as late as 2008 that it was an inter-State war.[105] Notwithstanding that perspective, the International Committee of the Red Cross (ICRC) took the view that with the establishment of the Afghan transitional government on June 19, 2002, the conflict had changed from an international armed conflict to one that was noninternational in character.[106] By 2015, viewing that conflict as a noninternational one had become the prevailing opinion.

---

102 Ian Brownlie, Principles of Public International Law 741 (7th ed. 2008).

103 *Id.* at 739.

104 *ISAF Troop Contributing Nations, available at* http://www.nato.int/isaf/structure/nations/ (last visited Apr. 18, 2013) (As of April 2013 there were twenty-eight NATO and fourteen non-NATO countries participating in the ISAF mission in Afghanistan.).

105 Yoram Dinstein, *Terrorism and Afghanistan*, 85 Int'l L. Stud. 43, 51 (2008).

106 Françoise J. Hampson, *Afghanistan 2001–2010, in* International Law and the Classification of Conflicts 242, 251 (Elizabeth Wilmshurst ed. 2012).

The Afghan situation raises the question of whether noninternational armed conflicts must find their legitimacy uniquely in the authority of the territorial State. It has been argued that the United States claim to be acting in self-defense "ended in 2002 when the *loya jurga* established new leadership for Afghanistan in place of Mullah Omar and his Taliban loyalists."[107] It is then suggested such participation had to be based on the request for assistance by the Afghan government; although it is also argued such an invitation at most only offered "a colorable base for intervention."[108] Presumably, it could also have been based on the UN Security Council mandate that established the International Security Assistance Forces (ISAF) in December 2001, at least in respect of "the maintenance of security in Kabul and its surrounding areas."[109] Notwithstanding the emphasis this argument places on the 2002 creation of an Afghan government, the complexity of the origins of the war and the multifaceted nature of the conflict (i.e., three conflicts involving ISAF, the U.S.-led Operation Enduring Freedom (OEF), and between Afghanistan and the Taliban[110]) make it difficult to argue that foreign State participation in the Afghan conflict had to be based specifically on that government's consent, rather than, say, the exercise of State self-defense.

The complexity of the Afghan conflict, particularly during its early years, is evident when the expansion of the UN mandate is considered. As that mandate was "operationalized," ISAF activities were initially limited to providing security in Kabul. That organization did not extend its area of responsibility to all of Afghanistan until October 2006.[111] Until that occurred, operations in the remainder of the country largely depended upon legal bases other than UN authority. A Canadian court acknowledged in 2008 there were at that time three concurrent legal bases for that country's participation in the conflict.[112] These included the exercise of the individual and collective right of self-defense in response to the attacks of 9/11, the consent of the Afghan government, and the authority provided by the UN Security Council.[113] This was in a context where, for the purposes of the case, the armed conflict was viewed as a noninternational one.[114] This continuing application of the law governing the recourse to war to what had increasingly come to be accepted as a noninternational armed conflict highlights the challenge of operating in these complex contemporary security situations. In addition, the multiple legal bases reflected the different reasons for and the variety of military operations conducted within Afghanistan. Such

---

[107] Mary Ellen O'Connell, *Drones Under International Law*, International Debate Series 4 (Oct. 8, 2010), *available at* http://law.wustl.edu/harris/documents/OConnellFullRemarksNov23.pdf.

[108] *Id.*

[109] S.C. Res. 1386 ¶ 1 (Dec. 20, 2001).

[110] Hampson, *supra* note 106, at 256.

[111] Gray, *supra* note 74, at 203–07.

[112] The use of multiple legal bases is not unique to Afghanistan. *See PM Statement Responding to FAC report on Military Operations in Syria*, Oral Statement to Parliament (Nov. 26, 2015), https://www.gov.uk/government/speeches/pm-statement-responding-to-fac-report-on-military-operations-in-syria (where in seeking to justify air strikes against the Islamic State in Syria in November 2015 reliance was placed on collective and individual self-defense under Article 51 of the UN Charter, the "unwilling or unable" test, and support gained from the unanimous adoption of S.C. Res. 2249, UN Doc. S/RES/2249 (Nov. 20, 2015))[hereinafter *PM Statement*].

[113] Amnesty International Canada v. Canada (Chief of the Defence Staff) (F.C.), 2008 FC 336, [2008] 4 F.C.R. 546, ¶¶ 20–21, *available at* http://reports.fja-cmf.gc.ca/eng/2008/2008fc336.html.

[114] *Id.* at ¶ 17.

operations did not have a homogeneous application with "the bulk of ISAF being committed to counter-insurgency whereas large elements of the American presence had been committed to counter-terror operations."[115]

Assessing whether a self-defense basis for acting has continuing application to what is considered to be a noninternational armed conflict is made all the more complex in situations where there is no obvious consensus as to whether a conflict is international or noninternational in character. Further, in certain situations both international and noninternational armed conflicts may be viewed as occurring simultaneously within the same territory. States themselves are often reluctant to categorize conflicts, and "the evidence may indicate a result or conclusion without establishing whether that was the party's view of what was legally required or simply what they decided to do."[116] Many practitioners simply look for the indicia of an armed conflict without focusing on further categorization. For some States, reliance on internationally based sources of legitimacy such as the exercise of State self-defense or a UN Security Council authorization are more likely than a domestically sourced authority even for what are increasingly viewed as noninternational armed conflicts. In this respect, operational reality is often more challenging than how the situation is often viewed through a theoretical lens that focuses on the domestic authority being provided by the territorial State. These international bases for acting raise significant questions regarding the continuing application of principles associated with State self-defense (see Chapter 3, section 3.6.3, *Applying Self-Defense Principles Across the Armed Conflict Spectrum*).

2.4.4. CROSS-BORDER ATTACKS BY NON-STATE ACTORS

Notwithstanding a traditional view that domestic law acts as the authority for engaging in internal conflict, there remains a general question regarding the application of the inherent right of self-defense against non-State actors that operate transnationally. It is this issue that has caught the attention of the international community in the context of States using drones as part of counterterrorist operations (see Chapter 8, section 8.5, *Legal Frameworks for Responding to Transnational Terrorism*). Considered more broadly, the inadequacy of relying on a domestic law to control all forms of contemporary noninternational armed conflict is particularly evident in the example of the long-running U.S. conflict with Al Qaeda. Assuming that the U.S. Supreme Court decision in *Hamdan v. Rumsfeld* reflects an accurate statement of international law, a plurality of the justices ruled in 2006 that the conflict with Al Qaeda is one that is not of an international character.[117] Under a strict interpretation of the "domestic source" theory, the almost unthinkable conclusion from an international law perspective would be that these transnational hostilities only need to be authorized by the 2001 U.S. congressional Authorization to Use Military Force and the U.S. Constitution, rather than under the UN Charter. Such a conclusion would be inconsistent with both the genesis of that conflict in the exercise of self-defense pursuant to Article 51 of the UN Charter, as well as its overall transnational character.

---

115 Richard Connaughton, Modern Warfare: The True Story of Conflict from the Falklands to Afghanistan 353 (2008).

116 Hampson, *supra* note 106, at 249.

117 Hamdan v. Rumsfeld, 45 I.L.M. 1130, 1154 (2006) (U.S. Supreme Court 2006).

The events of 9/11 have brought the question of the exercise of self-defense against non-State actors to the forefront of legal discussion.[118] The first line of defense for those seeking to restrict State recourse to war is to deny the authority to act in response to transnational attacks made solely made by non-State actors. In respect of the exercise of national self-defense, the traditional and very conservative view has been that in order for an "armed attack" to be considered to have occurred, such acts of terrorism must be attributable to a State.[119] This attribution may be seen to occur in a couple of ways. First, the non-State actor carrying out the attack may be viewed as being controlled by another State (see Chapter 9, section 9.3.1, *Non-State Actors Controlled by a State*).[120] Second, a host State, by refusing to meet its obligations under international law, may be seen to not allow its territory to be used for attacks by the non-State actor against another State. It is argued that the host State would then become responsible for the actions of the terrorist group (see Chapter 9, section 9.3.2, *Harboring and Active Support*).[121] This was the argument that was put forward in 2001 as the basis for the claim that Afghanistan was responsible for the Al Qaeda attacks of 9/11, leading to an invasion of that country and the removal of the Taliban as the governing authority.[122] It should be noted though that there has also been a certain resistance by international lawyers toward accepting this second form of attribution beyond the unique situation of Afghanistan.[123]

The requirement for State attribution reflects a traditional twentieth-century State-centric approach to the law governing the recourse to war that is so strongly held by some analysts it has been stressed to the point of stating, "there is no need for lengthy discourse on the matter."[124] This conservative interpretation finds support in the 2005 decision of the International Court of Justice in *Armed Activities on the Territory of the Congo (Democratic Republic of the Congo v. Uganda)*, which noted a lack of attribution to a State in determining action taken by Uganda was not justified as self-defense.[125] However, it seems problematic there has been a reluctance to accept a right of State self-defense against attacks by non-State actors since the test for assessing the international law right to act in self-defense has historically been based on the *Caroline* Case, an incident involving just such a group.[126] That said, the State-centered approach is favored by a number of analysts, notwithstanding the fact that Article 51 of the UN Charter does not restrict the exercise of the right

---

[118] GRAY, *supra* note 74, at 198–207.

[119] Albrecht Randelzhofer, *Article 51*, *in* THE UN CHARTER COMMENTARY, *supra* note 47, 788, ¶ 34 at 802, and OLIVIER CORTEN, THE LAW AGAINST WAR: THE PROHIBITION ON THE USE OF FORCE IN CONTEMPORARY INTERNATIONAL LAW 193 (2012).

[120] DINSTEIN, *supra* note 49, ¶¶ 586–92, at 221–24 (for a discussion of the "control" criterion).

[121] *Id.* ¶ 693, at 261.

[122] GRAY, *supra* note 74, at 200.

[123] *Id.* at 201.

[124] Maurice Kamto, *L'agression en droit international*, A. Pedrone, Paris, 145–46 (2010), trans. provided in Zakaria Dabone, *International Law; Armed Groups in a State-centric System*, 93 INT'L REV. RED CROSS 395, 403 (2011).

[125] Case Concerning Armed Activities on the Territory of the Congo (DRC v. Uganda), [Dec. 19, 2005] 45 ILM 271, ¶¶146–47, at 306 (2006). *See also* Theresa Reinhold, *State Weakness, Irregular Warfare, and the Right to Self-Defense Post 9/11*, 105 AM. J. INT'L L. 244, 259–60 (2011), and Mary Ellen O'Connell, *Dangerous Departures*, 107 AM. J. INT'L L. 380, 383 (2013).

[126] GEORGE F. G. STANLEY, CANADA'S SOLDIERS: THE MILITARY HISTORY OF AN UNMILITARY PEOPLE 203–04 (3rd ed. 1974) (for an outline of the *Caroline* incident).

of self-defense solely to an armed attack from a State. As with the *Lieber Code*, normative principles developed in a context other than inter-State wars appear to have been overshadowed by the passage of time and the overarching influence that an almost unitary focus on inter-State conflict has had on international law.

The reluctance to apply the law governing the recourse to war to action taken against non-State actors is maintained regardless of the ability of a territorial State to discharge its sovereign obligations and stop transnational non-State actors using its territory. It is suggested this is because a policy choice has been made "to accept blatant discrepancies between de jure [legal] and de facto [factual] sovereignty"[127] in order to "prevent interventionist powers from exploiting the vulnerability of weak states, thereby jeopardizing the supreme value of international peace."[128] Given the success since World War II of the UN Charter provisions in restricting inter-State conflict, it is difficult to criticize such a policy choice in that context. However, that approach does not seem to adequately address the more pressing security challenge of the twenty-first-century, transnational attacks by non-State actors in situations where the territorial State is unwilling or unable to act. To the extent the State-centric interpretation of the right to self-defense is based on a policy choice, the winds of change have been forcing a different assessment.

The traditional approach requiring a State connection to the non-State actors in order for an armed attack to be considered to occur has come under considerable pressure in the aftermath of the attacks of 9/11. As was stated by Daniel Bethlehem in addressing the strategic and operational reality that States faced in 2012, "[i]t is by now reasonably clear and accepted that states have a right of self-defense against attacks by nonstate actors—as reflected, for example, in UN Security Council Resolutions 1368 and 1373 of 2001, adopted following the 9/11 attacks in the United States."[129] This statement is made without any reference to a requirement for attribution to a State in respect of the non-State actor's action. Instead, he addresses host State involvement in terms of consent; whether it is a colluding or harboring State or is a reluctant host.[130] It is recognized "[t]here is little intersection between the academic debate and the operational realities,"[131] with the differences being described as a doctrinal divide "between those who favor a restrictive approach to the law on self-defense and those who take the view that the credibility of the law depends ultimately upon its ability to address effectively the realities of contemporary threats."[132] The disconnect between traditional theorists and practitioners of international law continues to be problematic.[133]

For a variety of reasons these "Bethlehem principles" have not been particularly well received by the international legal community.[134] However, one area where the traditional,

---

[127] Reinhold, *supra* note 125, at 250.

[128] *Id.*

[129] Daniel Bethlehem, *Principles Relevant to the Scope of a State's Right as Self-Defense Against an Imminent or Actual Armed Attack by Nonstate Actors*, 106 AM. J. INT'L L. 770, 774 (2012).

[130] *Id.* Rules 10–12, at 776.

[131] *Id.* at 773.

[132] *Id.*

[133] TOM RUYS, ARMED ATTACK AND ARTICLE 51 OF THE UN CHARTER: EVOLUTIONS IN CUSTOMARY LAW AND PRACTICE 179–80 (2010).

[134] Dire Tladi *The Nonconsenting Innocent State: The Problem with Bethlehem's Principle 12*, 107 AM. J. INT'L L. 570 (2013), Dapo Akande & Thomas Liefländer, *Clarifying Necessity, Imminence, and Proportionality in the Law of Self-Defense*,

conservative, and at times formalistic State-centric approach toward the exercise of self-defense has changed is in respect of the extension of the Article 51 right to self-defense against attacks by non-State actors. It has been acknowledged that the reaction by States to terrorist attacks "may be seen as raising the question whether there has been a significant change in the law."[135] Further, "such responses did not seem to be ruled out completely."[136] A detailed analysis of the international reaction to non-State actor threats predating 9/11, as well as its aftermath, first published in 2010, reached the conclusion "that today non-State actors can (exceptionally) commit 'armed attacks' in the international meaning of that phrase."[137] Elsewhere, it has been stated more strongly that proponents of the traditionalist approach are "compelled to concede that the response of the international community to 9/11 has left its mark on customary international law."[138] The right to take defensive action against non-State actors has been recognized in the 2005 Chatham House *Principles of International Law on the Use of Force by States in Self-Defence*[139] and the 2010 *Leiden Policy Recommendations on Counter-Terrorism and International Law.*[140] It seems that, as David Kretzmer indicates, the requirement that an attack must be attributable to a State in order for a victim State to use extraterritorial force against a non-State actor "is no longer the majority view."[141] Reinforcement for this position is evident in State reliance on self-defense as providing a legal basis for conducting air strikes against the Islamic State in Syria in 2015.[142]

Given the change toward responding to attacks by terrorist groups, perhaps the manner by which the focus on the State will be maintained in the future is to emphasize that action taken in self-defense against non-State actors remains exceptional in nature.[143] Viewing the transborder threat posed by terrorist groups as an exceptional event is not new. Even where it has been suggested that terrorist acts must be attributed to a State in order for an armed attack to be said to have occurred, it was suggested, "[a] special situation arises, if a State is not reluctant but incapable of impeding acts of terrorism committed by

---

107 AM. J. INT'L L. 563 (2013), Mahmoud Hmoud, *Are New Principles Really Needed? The Potential of the Established Distinction Between Responsibility for Attacks by Nonstate Actors and the Law of Self Defense*, 107 AM. J. INT'L L. 576 (2013).

135 GRAY, *supra* note 74, at 198.

136 MOIR, *supra* note 77, at 30.

137 RUYS, *supra* note 133, at 493.

138 DINSTEIN, *supra* note 49, ¶602, at 227–28. *See also* MOIR, *supra* note 77, at 54, and Michael N. Schmitt, *Drone Attacks Under the Jus Ad Bellum and Jus In Bello: Clearing the "Fog of Law,"* 13 Y.B. INT'L L. 311, 317 (2010).

139 ELIZABETH WILMSHURST, PRINCIPLES OF INTERNATIONAL LAW ON THE USE OF FORCE BY STATES IN SELF-DEFENCE Rule 6, at 11 (Oct. 2005), http://www.chathamhouse.org/publications/papers/view/108106 [hereinafter CHATHAM HOUSE STUDY].

140 *Leiden Policy Recommendations on Counter-Terrorism and International Law, in* COUNTER-TERRORISM STRATEGIES IN A FRAGMENTED INTERNATIONAL LEGAL ORDER: MEETING THE CHALLENGES, Annex, ¶ 38, at 715–16 (Larissa Van Den Herik & Nico Schrijver eds., 2013) [hereinafter *Leiden Recommendations*].

141 David Kretzmer, *The Inherent Right to Self-Defence and Proportionality in Jus Ad Bellum*, 24 EUR. J. INT'L L. 235, 273 (2013).

142 *See e.g., France Launches New Air Strike in Syria Against IS: minister*, REUTERS (Oct 9, 2015, 8:11 AM EDT), http://www.reuters.com/article/2015/10/09/us-france-syria-airstrike-idUSKCN0S30HJ20151009 (the first strike was on Sept. 27, 2015), Matthew Doran, *Australian Fighter Jet Carries Out First Air Strike Inside Syria*, ABC (Sep. 15, 2015, 8:29 PM), http://www.abc.net.au/news/2015-09-16/australian-fighter-jets-first-air-strikes-syria-us-military/6779104., and *PM Statement, supra* note 112 (outlining the United Kingdom position).

143 Hmoud, *supra* note 134, at 577.

making use of its territory."[144] In this respect, "a so-called failed State would turn out to be a safe haven for terrorists, certainly not what Arts. 2(4) and 51 of the Charter are aiming at."[145] However, given the scope and scale of emerging jihadist threats, such as the Islamic State and Al Qaeda movement (see Chapter 6, section 6.4.3, *Terrorism and Insurgency: Al Qaeda and an Islamic State*), it has become exceedingly difficult to maintain that defensive State action against non-State actors should be assessed as an "exceptional" occurrence.

2.4.5. AN INTERPRETIVE STRUGGLE

The complexity of the debate concerning contemporary conflict is reflected in the terminology that has been adopted when dealing with non-State actors. For example, Yoram Dinstein uses the term "extra-territorial law enforcement" to "describe the phenomenon of recourse in self-defense to cross-border counter-force against terrorists and armed bands" that are not organs of the territorial State.[146] While applied in the context of enforcing international law extraterritorially,[147] that terminology evokes a perception of parallelism with the role performed by the State as the "proper authority" for countering criminal activity within its borders. Although the responding State is doing what the territorial state should have done (i.e., maintain law and order), this act of self-defense may be carried out using aircraft, missiles, or the dispatch of military forces to destroy the infrastructure of the armed group.[148] These are not ordinary law enforcement means. Whether considered as an exercise of traditional State self-defense or as a special form of international "law enforcement," it seems clear that the role performed by the responding State as the proper authority impacts significantly on this dialogue.

Overall, the international legal community is still in the process of coming to grips with the use of force against non-State actors generally and, in particular, in the context of States exercising their inherent right of self-defense. In taking action against non-State actors it turns out that the source of the authority to go to "war" can be either domestic or international. Transnational threats rising to the level of an armed attack conducted by non-State actors can result in a noninternational armed conflict occurring in two ways. First, a demand by the targeted State for a host State to maintain law and order within its territory could lead to an internal armed conflict as the host State meets its legal obligations and seeks to stop the activities of a non-State actor armed group located on its territory. Depending upon the security and governance situation within the State, the threat may also be amenable to being dealt with as a matter of law enforcement (see Chapter 13, section 13.2.1, *The "Troubles"*). This is where a domestic law–based approach toward noninternational armed conflict fits most comfortably.

Second, when the transnational threat results in a cross-border response by the targeted State, it appears that an application of the international law governing the recourse to war is more appropriate. This occurs even when that response is viewed as occurring in the context of

---

144 Randelzhofer, *supra* note 119, ¶ 36, at 802.
145 *Id.*
146 DINSTEIN, *supra* note 49, ¶720, at 272.
147 *Id.*
148 *Id.* ¶715, at 270.

a noninternational armed conflict. The result is that contemporary conflict is forcing a change in the application of twentieth-century-based notions of the right of State self-defense. These two approaches for dealing with the non-State actor continue to be decidedly "State-centric" with the right authority being either the territorial or threatened State. It is very likely that States embracing this approach do so because, like the use of force domestically, it authorizes only one actor, the State, to act. For opponents, the problem is that it can appear to be a significant expansion of the authority of the State to use cross-border force when part of the ideological underpinning for the traditional approach toward controlling international conflict has historically been decidedly "antiwar" in nature. Almost any use of force by States, regardless of whether it is against other States or non-State actors, is viewed as being unacceptable.

This application of the law governing the recourse to war to attacks by non-State actors is presenting other significant challenges to the international community. Unlike inter-State conflict, where it is relatively easy to identify the opposing armed forces, non-State actor violence can involve ordinary criminals, members of terrorist groups, civilian participants in hostilities, and members of organized armed groups. This raises the question of at what point the initiating non-State actor violence is raised to the threshold of an armed attack. One assessment of the right to act in self-defense has suggested that "a series of suicide bombers" would reach that threshold, although barely.[149] Some analysis refers to "large-scale" attacks as the appropriate threshold.[150] Another looks for "one or more truly large-scale attacks or is engaged in a protracted campaign (of possibly less substantial) attacks."[151] Unfortunately, this is an area that has largely been unexplored in respect of "one-off" attacks by non-State actors (see Chapter 14, section 14.3. *Responding to Transnational Attacks and Threats to Nationals*).

To the extent a response in self-defense may be applied, the question is also raised whether the persons involved are engaged in criminal activity, are civilians taking a direct part in hostilities, or both. Similarly, if they are members of a group, there is the issue of whether it is a criminal enterprise such as a drug cartel or an organized armed group that is involved in an armed conflict. The difficulty of addressing this issue is evident in Daniel Bethlehem's treatment of the application of the law governing the recourse to war to non-State actors. There is reliance placed on the term "taking a direct part in hostilities" to describe the persons engaged in or providing material support to the armed attack in the context of the exercise of State self-defense.[152] The indication that "[t]he concept of direct participation [in a self-defense context] draws on, but is distinct from, the *jus in bello* [humanitarian law] concept of direct participation in hostilities" places an exceptionally thin veil between these two bodies of law.[153] It presents some potential for confusion and an overly broad cross-pollination of legal terms, particularly if a connection is also made between the term "material support" and "direct participation in hostilities" in a targeting context (see Chapter 8, section 8.5.2, *The Permissive "Conduct of Hostilities" Approach*).

---

149 Akande & Liefländer, *supra* note 134, at 567.

150 CHATHAM HOUSE STUDY, *supra* note 139, Rule 6, at 11, and the *Leiden Recommendations, supra* note 140, Annex, ¶ 39, at 716.

151 RUYS, *supra* note 133, at 531.

152 Bethlehem, *supra* note 129, at 775 (see Principle 6).

153 *Id.* n. c).

What the action taken in response to threats from non-State actors does draw attention to is that not all such uses of force by States necessarily occurs in the context of an armed conflict. Much depends upon how the scope of the right to act in self-defense is interpreted. In reality States use force internationally in defense of nationals in a wide variety of circumstances, although this too is an area of controversy for many traditional interpreters of international law (see Chapter 10, section 10.2.1, *A State Responsibility*).[154] Where the threat originates from purely criminal organizations, the force used must be governed by law enforcement norms (see Chapter 10, section 10.3.5, *Rescues and Law Enforcement*). Even where force is applied in the context of an armed conflict, it may occur in what is increasingly viewed as an insurgency with global, regional, or at least transnational reach. That in turn raises the question of whether the State response should be governed by the "police primacy" principle commonly applied in traditional counterinsurgency doctrine used in internal conflict (see Chapter 7, section 7.3, *The Primacy of Law Enforcement*). In other words, a policy choice to favor a human rights–based law enforcement response.

In summary, it is not that relying on a domestic source for the authority to engage in noninternational armed conflict is not a valid concept. Rather, it is simply too narrow an approach to be applied to all such forms of such conflict. As has been noted, conflicts with non-State actors can evolve from an existing international armed conflict or involve the exercise of the right of national self-defense. The law governing the recourse to war under international law is relevant to these conflicts. The result is that in analyzing contemporary conflict, practitioners cannot permit themselves to be too narrowly constrained by overly conservative interpretations of the law. The boundaries associated with those interpretations were established to meet a particular security challenge in the twentieth century, which was decidedly inter-State-focused. Instead, analysts must be prepared to more broadly assess the nature of the noninternational conflict in order to determine what law properly governs the recourse to the use of force as well as the conduct of operations. The contemporary security situation is simply not as "tidy" as the binary notion of international and noninternational armed conflicts suggest. International law is in the process of adjusting to that reality. One area where that adjustment has to occur is resolving how the law governing the recourse to war interacts with that governing armed conflict across a spectrum of conflict that includes international, transnational, and internal armed conflict. It is to that issue that the analysis now turns.

---

[154] SHAW, *supra* note 82, at 1143.

> It is true that states must be accorded the right to decide on defensive necessity in the first instance but in making this ex parte decision they should be inhibited by a rule which is related to facts which have objective characteristics and not to mere estimates of intention.[1]

3 Applying the Self-Defense Principles During Armed Conflict

3.1. Introduction

A fundamental question is whether it can be claimed that the law governing the recourse to war and international humanitarian law operate completely independently of one another during either inter-State, or State versus non-State actor conflict. While the traditional approach by international lawyers has been to emphasize their separation, these bodies of law share a common theoretical underpinning in just war theory. This suggests that the existence of a more complex and ultimately nuanced interrelationship than is commonly discussed.

What has been missing from much of the legal analysis is consideration of the degree to which the two frameworks actually interface and interact with one another across the full spectrum of conflict. It is this interaction that helps form one of the bases for a more holistic application of international law in the twenty-first century. In this analysis particular emphasis is placed on the interface between the self-defense legal principles (i.e., necessity, proportionality, imminence, and immediacy) and humanitarian law, which primarily controls the use of force during operations in armed conflict. This chapter will establish that the boundary between the law governing the exercise of State self-defense and that controlling operations is not as firmly drawn as traditional theories appear to suggest. The former body of law will be seen to have a strategic influence on the use of force by States, rather than exerting control over the actual conduct of operations at the tactical level.

[1] Ian Brownlie, International Law and the Use of Force by States 260 (1963).

3.2. The Continuing Application of Self-Defense Principles

In addition to whether the law governing State self-defense applies to conflicts with non-State actors, there is the question of how that body of law, which shares a common theoretical underpinning in just war theory with humanitarian law, interacts with the normative rules governing the conduct of hostilities. This issue takes the form of assessing the degree to which the law governing the recourse to war continues to apply once hostilities commence. In assessing the exercise of State self-defense with a holistic approach it is essential to first consider what principles govern its application. As will be seen, those principles figure prominently in any consideration of the interface between State self-defense and humanitarian law.

It was established in the 1996 International Court of Justice (ICJ) case, *Advisory Opinion on the Legality of the Threat or Use of Nuclear Weapons*, that the exercise of the right of self-defense is subject to the conditions of necessity and proportionality, both as a matter of customary law and "under Article 51 of the [UN] Charter, whatever the means of force employed."[2] Additional principles are the "imminence" of an armed attack and the "immediacy" with which a State must respond. It has been indicated that imminence might be viewed as spanning the "deep chasm between interceptive and anticipatory self-defence," although it has also been suggested the usefulness of the term "imminent" is doubtful.[3] This term has been widely used within the academic community[4] and is a concept that has dominated much of the post-9/11 discussion concerning the defensive use of force by States against non-State actors.[5]

The discussion about imminence normally focuses on whether defensive action can be taken before an armed attack actually materializes. Imminence is a concept with which the international legal community has struggled, particularly because the immediate post-9/11 period witnessed an attempt to expand anticipatory self-defense to include the notion of "preemptive self-defense." Preemptive self-defense, which underpinned the 2002 U.S. National Security Strategy—or as it came to be known, the Bush Doctrine—has been widely and rightly criticized.[6]

Notwithstanding the concern over preemptive self-defense, there is a wide although not universal level of acceptance within the international community that States can exercise

---

[2] Advisory Opinion on the Legality of the Threat or Use of Nuclear Weapons, 1996 I.C.J. 226, ¶ 41, at 245.

[3] Yoram Dinstein, War Aggression and Self-Defence ¶ 543, at 205 (5th ed. 2011).

[4] *See, e.g.*, Christine Gray, International Law and the Use of Force 160–65 (3rd ed. 2008), and Lindsay Moir, Reappraising the Resort to Force: International Law, *Jus ad Bellum* and the War on Terror 57–58, 75–82 (2010).

[5] Elizabeth Wilmshurst, Principles of International Law on the Use of Force by States in Self-Defence 8–9 (Oct. 2005), http://www.chathamhouse.org/publications/papers/view/108106 [hereinafter Chatham House Study], *Leiden Policy Recommendations on Counter-Terrorism and International Law*, *in* Counter-Terrorism Strategies in a Fragmented International Legal Order: Meeting the Challenges, Annex, ¶¶ 41, 45–48, at 716–18 (Larissa Van Den Herik & Nico Schrijver eds., 2013) [hereinafter *Leiden Recommendations*], and the Tallin Manual on the International Law Applicable to Cyberwarfare ¶¶ 2–7, at 63–66 (Michael Schmitt ed., 2013) [hereinafter Tallin Manual].

[6] Dinstein, *supra* note 3, ¶¶ 514–16, at 194–95, Gray, *supra* note 4, at 213, Moir, *supra* note 4, at 75–82, A More Secure World: Our Shared Responsibility, Report of the High-level Panel on Threats, Challenges and Change ¶¶ 188–92, at 63 (2004), https://www.un.org/en/peacebuilding/pdf/historical/hlp_more_secure_world.pdf, and *Leiden Recommendations*, *supra* note 5, at Annex, ¶ 45, at 717.

anticipatory self-defense, provided strict requirements of necessity and proportionality are applied.[7] At times that acceptance has even extended to acquiescence, or at least not publically challenging, the use of quite anticipatory force to remove threats posed by certain States (i.e., a 2007 Israeli attack on a Syrian nuclear reactor).[8] One challenge in assessing the application of the self-defense principles of necessity, proportionality, and imminence is the degree to which they are viewed as being interrelated. Necessity is linked to proportionality,[9] and the use of force in self-defense is justified when it "is necessary to bring an attack to an end, or to avert an imminent attack. . . ."[10] Finally, there is the requirement of immediacy, which "signifies that action taken in self-defence must be taken in a reasonable timeframe in relation to the occurrence of the attack."[11]

The proportionality principle is often discussed in terms of its impact on defensive responses and the armed conflict that ensues. In this context that principle has a different meaning than is ascribed to what is commonly, although misleadingly, referred to as a targeting proportionality test under humanitarian law. Under this latter body of law the proportionality requirement of the targeting precautions prohibits collateral civilian casualties and damage resulting from an attack that are "excessive,"[12] or even "clearly excessive."[13] In contrast, "the proportionality requirement in self-defence is more concerned with the preservation of international order and the minimization of the use of force."[14]

An issue that the international legal community has struggled with in controlling armed conflict is the degree to which these self-defense principles have a continuing impact on hostilities once they commence. Adherence to the view that there is a complete separation between the law governing the recourse to war and humanitarian law suggests there is no room for one body of law to affect the other. Certainly, the rules governing the exercise of State self-defense precede the application of humanitarian law and therefore are, at least initially, sequential in application. However, it is not clear that the dividing line between these two bodies of law is as exact as is often claimed. What is interesting is that there has been relatively little attention paid to the interaction between these two bodies of law. Where

---

[7] CHATHAM HOUSE STUDY, *supra* note 5, at 4, and NOAM LUBELL, EXTRATERRITORIAL USE OF FORCE AGAINST NON-STATE ACTORS 59 (2010).

[8] DAN RAVIV & YOSSI MELMAN, SPIES AGAINST ARMAGEDDON 321-22 (2012) and MICHAEL BAR-ZOHAR & NISSIM MISHAL, MOSSAD: THE GREATEST MISSIONS OF THE ISRAELI SECRET SERVICE 279-87 (2014) (describing the Israeli discovery of and subsequent 2007 attack on a nuclear reactor being constructed in Syria. Both countries were muted in their public comments). In contrast, *see* MICHAEL BAR-ZOHAR & NISSIM MISHAL, NO MISSION IS IMPOSSIBLE: THE DEATH-DEFYING MISSIONS OF THE ISRAELI SPECIAL FORCES 231 (2015) (referring to the international condemnation following the 1981 Israeli destruction of the Iraqi Osirak nuclear reactor).

[9] CHATHAM HOUSE STUDY, *supra* note 5, at 7.

[10] *Leiden Recommendations*, *supra* note 5, ¶ 41, at 716. *See also* CHATHAM HOUSE STUDY, *supra* note 5, at 7.

[11] Terry Gill, *Legal Basis of the Right of Self-Defence under the UN Charter and under Customary International Law*, *in* THE HANDBOOK OF THE INTERNATIONAL LAW OF MILITARY OPERATIONS 187, ¶ 6, at 197 (Terry Gill & Dieter Fleck eds., 2010). *See also* TALLIN MANUAL, *supra* note 5, ¶ 8, at 66. Confusingly, "immediacy" is also sometimes used in the context of establishing an attack is about to happen or is imminent. *See Leiden Recommendations*, *supra* note 5, ¶ 46, at 717. However, in this book "immediacy" is restricted to the timing of the defensive response.

[12] AP I, art. 57(2)(iii).

[13] The 1998 Rome Statute, art. 8(2)(b)(iv). *See also* YORAM DINSTEIN, THE CONDUCT OF HOSTILITIES UNDER THE LAW OF INTERNATIONAL ARMED CONFLICT ¶ ¶ 322, at 131 (2nd ed. 2010).

[14] Christopher Greenwood, *Self-Defense and the Conduct of International Armed Conflict*, in INTERNATIONAL LAW AT A TIME OF PERPLEXITY: ESSAYS IN HONOUR OF SHABTAI ROSENNE 278 n.22 (Yoram Dinstein ed., 1989).

it has been analyzed, there appears to be two schools of thought, both involving at least to some degree a continuing application of the self-defense principles.

One approach applies the self-defense principles in an "overarching" framework that exerts a continuing influence on any armed conflict resulting from the exercise of State self-defense.[15] Alternatively, a more "limited" approach seeks to remain more faithful to the separation of the law regarding the recourse to war from the operation of humanitarian law. This approach does not accept that the self-defense principles, such as proportionality, continue to apply once significant hostilities commence that reach the scale of a "traditional" inter-State war. Instead, humanitarian law exclusively governs the conduct of those hostilities, particularly as they take on the scale of a traditional inter-State "war." However, even with this more restricted approach, it appears that self-defense principles have application to a more isolated and lower-scale defensive response by a State to an armed attack. To the extent the hostilities resulting from such an attack and defensive response constitute an armed conflict, however limited, there would remain a requirement to reconcile the application of self-defense norms with the law governing the conduct of hostilities. In addition to these "overarching" and "limited" theoretical frameworks, questions concerning the application of the self-defense principles can also arise in the context of reporting such State action to the UN Security Council, or when a case is brought before the ICJ. It is to the application of self-defense principles under the "overarching" and "restricted" theories, as well as the additional international restraining mechanisms, that the analysis now turns.

3.3. An "Overarching" Application

In 1989 Christopher Greenwood suggested that the use of force during an international armed conflict initiated as an exercise of the right of self-defense by States is constrained by the principles of proportionality and necessity throughout the existence of the conflict.[16] Under this theory, those self-defense principles are viewed as affecting the temporal and geographic scope of any resulting conflict, the choice of legitimate military targets, the choice of weapons, belligerent reprisals, the conduct of an occupation, and dealings with neutrals.[17] The result is an "overarching" application of these principles, which impacts on virtually all aspects of how, where, and for how long hostilities are conducted. This approach does not mean that humanitarian law itself is altered. For example, the determination of what constitutes a military objective would remain based on the criteria found in Article 52 of Additional Protocol I. However, the self-defense principles of proportionality and necessity would dictate whether a specific military objective could be targeted. It could also impact on decisions by commanders as to what weapons systems are used. The influence of these principles is variable. For example, "[t]he more extensive the violence becomes, the less will be the influence that the principles of self-defence exercise in the choice of weaponry."[18]

---

[15] Greenwood, *supra* note 14, at 274, and JUDITH GARDAM, NECESSITY, PROPORTIONALITY AND THE USE OF FORCE BY STATES 156 (2004).

[16] Greenwood, *supra* note 14, at 274–75.

[17] *Id.* at 275.

[18] *Id.* at 280.

This theory also affects where warfare is conducted. Traditionally, humanitarian law has prescribed the areas of war. Under that normative framework, "[w]ar can be waged over large portions of the planet and beyond."[19] While hostilities by the belligerents may in fact be restricted to "a fairly narrow theatre of operations," the potential always exists for combat to occur over a wider area.[20] Hostilities may occur on the territories of the belligerent parties (e.g., land areas, internal waters, archipelagic waters, the territorial sea, the related subsoil and submarine areas, and the superjacent airspace), the high seas and exclusive economic zone (including those of neutral countries), and outer space.[21] Humanitarian law limitations on the location of conflict are based principally on the concept of neutrality and neutralized zones.[22] However, an approach adopting the continuing application of self-defense principles contemplates a more restricted geographical scope. Another proponent of an "overarching" approach, Judith Gardam, has suggested that "[u]nder the Charter system, it is the accepted view that, generally speaking, proportionality requires that forceful actions in self-defence must be confined to the area of attack that they are designed to repel."[23] The previous freedom to operate across a wider geographical area is limited. This occurs to the point "[a]s some writers describe it, armed conflict became limited as opposed to total."[24] However, it is acknowledged that each case will differ, and the "repulsion of an attack, particularly in the sense of repelling an invader, will sometimes warrant the invasion of the territory of the aggressor state."[25]

The "overarching" approach to the application of self-defense principles does not appear to be as restrictive as the theory might at first glance suggest. The potential exists for a broadly defined area of war, including in the territories of the parties of the conflict based on national boundaries, as well as on the high seas and in exclusive economic zones. [26] This could occur where the conflict between the States has reached the level of a total war, with the result that "the proportionality requirement would no longer dictate that defensive measures should be confined to a restricted geographical area."[27] However, before that level of conflict is reached, any justification of an expansion beyond where operations are actually taking place by a belligerent, the "area of operations" or "theatre of war," would "depend upon whether it is necessary for him to do so in order to exercise his right of self defence."[28] Thus a degree of tension is established between this approach limiting State action in armed conflict through a continuing application of self-defense principles and a more traditional view that relies exclusively on humanitarian law to prescribe geographic zones of conflict.

A specific example of how such an overarching theory might be applied is found in Christopher Greenwood's reference to the sinking of the Argentine warship, the *General*

---

[19] Dinstein, *supra* note 3, ¶ 49, at 19.
[20] *Id.*
[21] *Id.* ¶¶ 50–63, at 19–24.
[22] *Id.* ¶¶ 52–58, at 20-22, and ¶¶ 64–78, at 25–29.
[23] Gardam, *supra* note 15, at 163.
[24] *Id.*
[25] *Id.* at 164.
[26] Christopher Greenwood, *Scope of Application of Humanitarian Law*, *in* The Handbook of International Humanitarian Law 59–62 (Dieter Fleck ed., 2nd ed. 2008).
[27] Greenwood, *supra* note 14, at 278.
[28] *Id.* at 62.

Belgrano, during the 1980 Falklands/Malvinas conflict. That ship was attacked in the South Atlantic outside a British-established temporary exclusion zone. From the perspective of the Argentinian captain, the warship was not threatening the British units as it was "pointing straight at the Argentine coast on a bearing which she had been following for hours. She was not 'closing on' the British Task Force but sailing just the other way."[29] However, from a British perspective, the attack was justified on the basis that their position was made clear defensive measures could be taken outside the exclusion zone; the *Belgrano* and its escorts were a threat, if not an immediate one; and the result of the sinking was that an opposing surface fleet stayed in port for the rest of the conflict.[30] Further, the Argentinians did admit the warship was providing "aircraft direction for their air force,"[31] and it has been suggested to be incontrovertible that "the British strategic purpose was to defeat the enemy's air and sea forces before the amphibious landing force was committed."[32] The Greenwood analysis of the *Belgrano* incident states that "[o]n balance, it is thought that the sinking was lawful," although it is also noted that "the concept of self defense may impose serious restrictions upon the right of a state to attack, what in terms of the *jus in bello* [humanitarian law], is a legitimate military target."[33] In contrast, it is suggested that had a British warship encountered an Argentine warship in the Pacific at the time of the invasion of the Falklands/Malvinas, it "would not have been a lawful measure of self-defence for one ship to engage the other unless there was clear evidence the other ship was about to launch an attack."[34]

A particular challenge for States in applying this suggested approach is that it is not clear what criteria are to be applied in limiting the use of force. The example of two ships encountering one another in the Pacific indicates a requirement for some form of evidence the enemy warship intends to take aggressive action in order for a use of force to be justified. This appears to introduce the controversial prewar concept of "anticipatory" self-defense into the midst of an armed conflict. However, this is a situation that did not arise during the conflict. In any event it is fairly easy to conclude the overall British action during the Falklands/Malvinas conflict involved a proportionate response to the threat posed by Argentina. In this respect, the land battle was focused on the islands away from the Argentinian mainland, and the sea battle was largely restricted to the exclusion zone.[35] What is less clear is whether the decision to geographically limit the hostilities was because the United Kingdom viewed itself subject to legal restrictions or because it was founded on policy.[36]

The imposition of geographic limitations based on self-defense principles also suggests that strategic uses of force in situations other than "total war" might be at risk of being considered an unlawful exercise of self-defense. Examples of force that can involve the extension of warfare to areas remote from the primary area of hostilities include a strategic aerial bombing campaign carried out on the territory of an opponent or a blockade conducted off

---

29 Desmond Rice & Arthur Gavshon, The Sinking of the Belgrano 112 (1984).

30 Greenwood, *supra* note 14, at 279.

31 Max Hastings & Simon Jenkins, The Battle for the Falklands 148 (1983).

32 *Id.*

33 Greenwood, *supra* note 14, at 279.

34 *Id.* at 277.

35 Gardam, *supra* note 15, at 164.

36 *Id.* at 163.

the coast of that State. The negative impact that these strategic uses of force have historically had on civilians has meant these operations are not necessarily viewed positively from a humanitarian perspective. Indeed, there was considerable debate between the two world wars of the twentieth century as to whether aerial bombing should be confined to the zone of military operations, although it was noted even at that time that "one may be permitted to doubt whether the practical way to give effect to it is to prohibit completely all air bombardment outside zones of operations."[37] What this does highlight is the close interaction that can arise between the law governing the recourse to war, with its antiwar bias seeking to limit or prohibit conflict, and humanitarian law that has alleviation of suffering as a goal.

However, the reality of warfare between States is such that strategic uses of force cannot be restricted solely on the basis of geographic factors. Factors to be considered would also be the nature of opponent (e.g., another State with advanced armed forces) and what action is required strategically, operationally, and tactically to defeat the enemy (see section 3.6.2, *Targeting: A Matter of Strategy and Tactics*, below). Of course, any operations would also be subject to the humanitarian law–based restrictions on the conduct of hostilities. Inter-State warfare in particular has become "industrialized,"[38] and the technological advancement of twentieth-century warfare including the development of airpower has enabled States to conduct attacks on targets located in an opponent's "home front" that provide both direct and indirect support to their involvement in the conflict.[39] This has broadened the geographic area over which war is fought and expanded the range of possible targets.[40] In practical terms, even limited armed conflict may necessitate striking at strategic targets. For example, even if a conflict is occurring on an offshore island, there can be significant military advantage gained from attacking port facilities, logistics warehouses, and railroad lines on the mainland that are facilitating an opponent's participation in the hostilities.

Technologically driven "distance" warfare has had a direct impact on humanitarian law and the law governing the recourse to war. The conduct of aerial warfare in the twentieth century was connected to the development of the humanitarian law concept of proportionality, although its theoretical roots are historically based in the moral doctrine of "double effect." That doctrine, which was incorporated into just war theory, permitted an act with "evil consequences" by weighing those consequences against the desired "good effect."[41] It has been suggested "the main developments in the doctrine were a direct consequence of the advent of these new methods of warfare, whose use resulted in widespread damage to civilian populations."[42] Similarly, the conduct of blockades has affected perceptions of how

---

[37] J. M. Spaight, Air Power and War Rights 250 (3rd ed. 1947).

[38] Marcus Jones, *Fighting "This Nation of Liars to the Very End," The German Army in the Franco-Prussian War, 1870–1871*, *in* Hybrid Warfare: Fighting Complex Opponents from the Ancient World to the Present 171, 173–178 (Williamson Murray & Peter R. Mansoor eds., 2012), and Emily Crawford, The Treatment of Combatants and Insurgents under the Law of Armed Conflict 7–9 (2010).

[39] Kenneth Watkin, *Targeting in Air Warfare*, 44 Isr. Y.B. on Hum. Rts. 1 (2014) (for a discussion of the impact of airpower theory on air targeting).

[40] Leo Braudy, From Chivalry to Terrorism: War and the Changing Nature of Masculinity 459 (2003).

[41] Michael Walzer, Just and Unjust Wars 151–54 (1977). *See also* James Turner Johnson, Morality and Contemporary Warfare 131–32 (1999), and Brian Orend, The Morality of War 115–18 (2006).

[42] Judith Gardam, *Proportionality and Force in International Law*, 87 Am. J. Int'l L. 391, 400 (1993).

broadly warfare can touch civilian populations. By 1995 a handbook written by experts in the law of naval warfare, the *San Remo Manual Applicable to International Conflicts at Sea*, inserted a proportionality requirement in its Rules prohibiting blockades, if "the damage to civilian population is, or may be expected to be, excessive in relation to the concrete and direct military advantage anticipated from the blockade."[43] This rule of proportionality more readily evokes consideration of the strategic impact of a blockade than is often associated with targeting rules. It is the strategic impact of large-scale civilian casualties and damage that appears to influence what might constitute a disproportionate exercise of the right to self-defense by a State.

The idea of States acting proportionately offers another area of analysis where the two legal regimes can interact. Concern over the overall collateral impact of military attacks provides a link between the State self-defense and humanitarian law concepts of proportionality. While proportionality serves a different purpose under humanitarian law than as a self-defense-based principle, it has been suggested "[t]he two tests are nevertheless closely related."[44] As a general legal principle, proportionality developed from the end of the nineteenth century onward centered on "the need for and the will to protect human rights from the powers of the state."[45] This goal is evident under both humanitarian law and that governing the right of State self-defense. During conflict, integral to decisions regarding the choice of methods and means of warfare, there "must be a consideration of the *anticipated overall scale of civilian casualties*, the level of destruction of the enemy forces, and finally damage to territory, the infrastructure of the target State and the environment generally."[46] It makes sense that "States must thus consider the humanitarian implications in order to determine the security standard they may pursue using military action."[47] An example where allegations of excessive collateral damage under humanitarian law were used as the basis to claim a disproportionate response in self-defense was in respect of Israel's 2006 operations in Lebanon.[48] Factors suggested to support that claim included "the scale of its action, which considerably exceeded what was deemed necessary for repelling the attack; the fact that the response involved the destruction of military and civilian infrastructures located hundreds of miles from the area attacked, which were therefore unrelated to the defensive objective of the action; and the threat to and harm sustained by civilians."[49]

This reference to the 2006 Israel/Lebanon conflict, which involved hostilities between a State and a non-State actor, raises the question about whether the "overarching" theory extends the application of self-defense principles even further to the full range of international security operations such as counterinsurgency, counterterrorism, protection of

---

[43] San Remo Manual on International Law Applicable to Armed Conflicts at Sea, Rule 102(b), at 179 (Louise Doswald-Beck ed., 1995) [hereinafter San Remo Manual].

[44] Greenwood, *supra* note 14, at 278 n.22.

[45] Aharon Barak, Proportionality: Constitutional Rights and Their Limitations 177 (2012).

[46] Gardam, *supra* note 15, at 168 (emphasis added).

[47] Enzo Cannizzaro, *Contextualizing Proportionality: Jus ad Bellum and Jus in Bello in the Lebanese War*, 88 Int'l Rev. Red Cross 779, 792 (2006). *See also* Thomas M. Franck, *On Proportionality of Countermeasures in International Law*, 102 Am. J. Int'l L. 715, 732–34 (2008) (for an overview of the claims of disproportionate actions in respect of the 2006 Israel-Lebanon War).

[48] Cannizzaro, *supra* note 47, at 784.

[49] *Id.*

nationals, noncombatant evacuation, international hostage rescue, and peace support operations. This issue arises because the UN Charter–based right to act in self-defense is limited to responding to an "armed attack."[50] In his analysis, Christopher Greenwood noted that it was a matter of controversy whether operations such as the protection of nationals abroad, humanitarian intervention, and the protection of vital national or international interests "are recognized by international law except in so far as they can be brought within the scope of self defence."[51] However, he then went on to state, "even if international law does permit the use of force on any of these grounds, it does so subject to the same limitations that exist in the case of self-defence."[52] This position was taken on the basis that "the right to use force is said to derive from the need to achieve a specific objective and is therefore limited to what is needed to achieve that objective."[53] Such an interpretation highlights the powerful influence of recourse to war principles, which are designed to limit the use of force by States. The ultimate goal of restraining State action remains an overriding consideration when taking action against non-State actors.

For those international lawyers who accept that the law governing the recourse to war applies to operations undertaken in response to attacks by non-State actors, the application of the self-defense principles can occur in both an indirect and direct fashion. An example of indirect application is if an "overarching" application of the self-defense principles is applied to the post-9/11 Afghanistan conflict. Assuming that the character of that conflict changed instantaneously in June 2002, from an international to a noninternational armed conflict, it would be difficult to argue that the limiting influences of the self-defense principles should automatically have ceased at that time. In this respect, while the Afghan security forces were to become progressively more effective as a result international training, equipment, and mentorship, neither the security situation on the ground nor the heavy combat involvement by international forces in Afghanistan fundamentally changed. Although the conflict has evolved, that does not mean that self-defense principles such as necessity and proportionality would be any less relevant or necessary in restraining the violence (see Chapter 2, section 2.4.3, *Conflict Evolution*). A more direct application of the self-defense principles in operations against non-State actors occurs in the context of responding directly to transnational terrorist attacks. It is logical that an "overarching" application of self-defense principles would apply to the use of force in response to such attacks.

3.4. A More "Limited" Theory

The second theory, which adopts a more limited application of self-defense principles, makes a distinction between traditional armed conflict between States and isolated defensive responses and border skirmishes. This theory relies on a bifurcated approach. Yoram Dinstein, who is a leading proponent of this approach, agrees with the application of the

---

[50] UN Charter, art. 51.

[51] Greenwood, *supra* note 14, at 274.

[52] *Id.*

[53] *Id.*

self-defense principles[54] to certain more limited defensive reactions by States (i.e., incidents "short of war"[55]), but not in the case of a "war" launched in self-defense.[56] State action in "short of war" situations is constrained by an interpretation of the self-defense principle of proportionality, which is applied to the conflict in its totality and is one that focuses on a comparison between "the quantum of force used and counter-force used, as well as the casualties and damages sustained."[57] These "closed" incidents, which do not reach the higher threshold of a "war" between States, can involve border patrols exchanging fire, vessels being torpedoed, and planes being shot down.[58]

In contrast to these less serious skirmishes, "war" is suggested to involve a comprehensive use of force assessed against four factors, applied individually or more commonly in combination. Those factors are that force is employed "across sizeable tracts of land or far-flung corners of the ocean"; "over a protracted period of time"; "entailing massive military operations"; and "inflicting extensive human casualties and destruction to property."[59] Under the Dinstein model there is a form of proportionality assessment applied to defensive war, but not one based on the self-defense principle.[60] Further, "the comparative evaluation of force and counter-force has to take place not at the termination of the exercise of self-defense but at its inception."[61] Applying General Rupert Smith's analysis of the post–World War II period to this concept of "war," it most closely equates to traditional "industrial war" occurring between States. In contrast, "short of war" situations would appear to relate to the "crisscrossing between confrontation and conflict" that defined much of Cold War.[62] In these latter situations, as well as for much of contemporary conflict, the overall military objective is "expressed as a condition: 'maintain law and order', ensure a safe, secure environment', maintain a no-fly zone'."[63] The self-defense principles are applied to these more limited inter-State conflicts with the goal of avoiding escalation into broader conflict and ultimately "war" on an industrial scale.

It is acknowledged that one challenge in applying this theory is determining when a "war," as opposed to a situation "short of war," is in existence. The relative difference is said to be that a war setting involves "more force, employed over a longer period of time, within a larger theatre of operations."[64] Even applying this guidance, it appears that "short of war"

---

54 Dinstein, *supra* note 3, ¶ 676, at 255

55 *Id.* ¶¶ 640–86, at 242–59 (Incidents "short of war" consist of "on-the-spot" reactions, defensive armed reprisals, and the protection of nationals abroad).

56 *Id.* ¶ 697, at 262.

57 *Id.*

58 *Id.* ¶ 24, at 11.

59 *Id.* ¶ 28, at 12.

60 *Id.* ¶ 697, at 262. *See also* Michael N. Schmitt, *Responding to Transnational Terrorism under the Jus ad Bellum: A Normative Framework, in* International Law and Armed Conflict: Exploring the Faultlines 157, 175 (Michael Schmitt & Jelena Pejic eds., 2007).(where in a slightly different context necessity and proportionality are viewed as being of relevance to defensive action take in respect of continuing attacks by a terrorist group, but not in respect of an all-out "war" of self defense by another State as "concerns about proportionality recede as the State's survival is placed at risk.").

61 Dinstein, *supra* note 3, ¶ 699, at 263 ("The decision has to be predicated on the gravity of the isolated armed attack and the degree to which it affects the victim state.").

62 Rupert Smith, The Utility of Force: The Art of War in the Modern World 181 (2005).

63 *Id.* at 182.

64 Dinstein, *supra* note 3, ¶ 30, at 12–13.

situations, to which there would be a continuing application of self-defense principles, can encompass a broad range of conflict. When applying these criteria to the warfare that has occurred since 1945, it seems there may be a relatively limited number of armed conflicts that would meet the definition of "war." Much appears to depend on what criteria for defining war are applied. A narrow definition of war would result in a greater number of inter-State conflicts to which the self-defense principles would continue to apply even after hostilities commence. The "limited" theory permits an exemption from the constraining influences of the self-defense principles once a defensive war between States occurs. This exemption is claimed in absolute terms with armed conflict at this elevated scope and intensity being governed exclusively by humanitarian law. It is an approach that reflects a general reluctance by many States to apply what are viewed as just war principles during the actual conduct of hostilities. It is one that is more consistent with the traditional "equal application theory," where humanitarian law is interpreted as applying equally to both parties without consideration of the justness of their cause.

Looked at from one perspective, the relative freedom of action provided to States when an armed with another State reaches the level of a defensive war could be seen by critics to be a return to the pre-Charter era. In this regard D. W. Bowett's explanation of the difference between the historic right to self-defense based upon protecting certain rights during peacetime and State engagement in war has a certain resonance:

> In contrast war, or rather the state's freedom to go to war, was not dependent on any prior breach of a duty by the victim state. Once undertaken, the aim of complete subjugation of the victim was apparently permissible, and the reaction of the victim was legally indistinguishable from the attack of the "aggressor". Once war *de jure* [at law] existed the two states engaged in action of a like kind. In war the duty of non-intervention disappeared, the substantive rights upon which self-defence depended were non-existent, and the "right" of self-defence became meaningless.[65]

However, there is a significant difference between this view of nineteenth-century warfare and the post-Charter approach. The commencement of any contemporary conflict between States is limited by the UN Charter, which broadly characterizes the use or threat of force "as either a delict, or a sanction, or self-defence."[66] Further, as has been noted even in the case of a defensive war, the "limited" theory requires some form of proportionality assessment to be made at the commencement of the hostilities.

The "limited" approach differs from its "overarching" counterpart in that the issue of a State response to cross-border attacks made by organized armed groups or terrorists is specifically addressed. As has been noted, Yoram Dinstein refers to such action as "extra-territorial law enforcement." This defensive reaction constitutes an extraordinary case of the exercise of self-defense pursuant to Article 51 of the UN Charter, which requires the application of the necessity, proportionality, and immediacy principles.[67] It is the exercise

---

65 D. W. Bowett, Self-Defense in International Law 118 (The Lawbook Exchange Ltd., 2009) (1958).

66 *Id.* at 155.

67 Dinstein, *supra* note 3, ¶ 728, at 275.

of this right under the Charter that is the basis for many contemporary counterterrorist operations.[68] In justifying a response against non-State actors, reliance is placed on the *Caroline* incident, the classic description of the principles applying to self-defense, as a historical example of this form of operation.[69] It has been indicted that the application of these principles places significant limits in terms of "time, space, and purpose" on cross-border intrusions by States when acting in self-defense against terrorist attacks.[70] Michael Schmitt suggests in respect of cross-border action against terrorists "the operation must be limited geographically to the minimum territorial infringement consistent with mission success."[71] A 2008 strike by Columbian forces killing a rebel leader located in a Revolutionary Armed Forces of Colombia (FARC) guerrilla camp in Ecuador appears to provide an example of such a limited strike against an organized armed group, which involved an isolated incursion into another State.[72]

While the classic conceptualization of the exercise of State self-defense against non-State actors involves isolated, temporally restricted, and geographically restrained responses, that is not always the case. For example, the 1982 incursion by Israel into Lebanon is identified as "a singular illustration of extra-territorial law enforcement."[73] While there was no exchange of fire with Lebanese armed forces, Israel sent a sizable expeditionary force into that country in an effort "to destroy a vast complex of Palestinian bases from which multiple armed attacks across the international border had originated."[74] Adding to the complexity of the security situation, Israel also clashed with Syrian armed forces in Lebanon in the continuation of an international armed conflict between those two countries.[75] This intervention ultimately led to an occupation of Southern Lebanon[76] that lasted seventeen years.[77] After the initial invasion, which devastated the Palestinian Liberation Organization, another non-State actor emerged, Hezbollah, which pushed Israel back, "first to a security zone along the border in 1985, then completely out of Lebanon in 2000."[78]

Thomas Franck identified the 1982 occupation of Lebanon, with its prolonged invasion of sovereign territory and high civilians casualties, as one that is likely to be viewed as "an unacceptable response."[79] Those hostilities were followed by another significant military engagement with Hezbollah in 2006, when Israel asserted its right to act in self-defense in response to attacks by that terrorist organization.[80] The nature of that conflict with Hezbollah was such that it has been suggested it went "beyond the strict confines of an extra-territorial law enforcement" because, while Hezbollah did not control the government, it

---

[68] Schmitt, *supra* note 60, at 176–82.

[69] DINSTEIN, *supra* note 3, ¶ 726, at 274, and Schmitt, *supra* note 60, at 177.

[70] Schmitt, *supra* note 60, at 182.

[71] *Id.*

[72] Dana Priest, *Covert Action in Columbia*, WASH. POST, Dec. 21, 2013, http://www.washingtonpost.com/sf/investigative/2013/12/21/covert-action-in-colombia/.

[73] DINSTEIN, *supra* note 3, ¶ 723, at 273.

[74] *Id.*

[75] *Id.* and GIL MEROM, WHY DEMOCRACIES LOSE SMALL WARS 156 (2003).

[76] EYAL BENVENISTI, THE INTERNATIONAL LAW OF OCCUPATION 181–82 (1993).

[77] RAVIV & MELMAN, *supra* note 8, at 187, 203.

[78] DANIEL BYMAN, A HIGH PRICE: THE TRIUMPHS & FAILURES OF ISRAELI COUNTERTERRORISM 9 (2011).

[79] THOMAS M. FRANCK, RECOURSE TO FORCE: STATE ACTION AGAINST THREATS AND ARMED ATTACKS 65 (2002).

[80] Franck, *supra* note 47, at 732.

was represented in the Lebanese cabinet.[81] Called the "Second Lebanon War," it was said to "have at least certain dimensions of an inter-State armed conflict."[82] However, fundamentally, it remained an armed conflict with a non-State actor. It was a defensive response that prompted a mixed reaction from uninvolved States, but one that ultimately attracted considerable criticism as being a disproportionate response.[83]

Even with respect to contemporary counterterrorist operations, the post-9/11 defensive response undertaken by the United States against Al Qaeda, which commenced as an exercise of self-defense pursuant to Article 51 of the UN Charter, has continued for well over a decade. From an American perspective it is treated as an ongoing armed conflict that transcends borders and involves operations not only in Afghanistan but also countries such as Pakistan, Yemen, and Somalia.[84] With extraterritorial law enforcement being discussed in terms of a "spillover" conflict,[85] it would not encompass the U.S. Supreme Court classification of the conflict with Al Qaeda as a noninternational armed conflict.[86] However, transnational operations against organized armed groups and terrorist organizations, whether called extraterritorial law enforcement or noninternational conflict, which rise to the level of an armed conflict are governed by humanitarian law, as well as by the application of self-defense principles.[87] These situations are similar to those occurring under the "overarching" theory, at least to the extent that the simultaneous operation of the law governing the recourse to war and humanitarian law must be reconciled.

In respect of conflicts between States and transnational non-State actors, a number of issues still remain. Where hostilities with groups such as Al Qaeda, Hezbollah, or the Taliban extend in time and/or scope, it becomes increasingly challenging to apply self-defense principles to regulate a State response in the same way as is suggested for more isolated uses of force. These defensive acts put in play a basic tension between States and authors who utilize a restrictive approach toward the scope of the right of self-defense and others who prefer a more liberal, permissive interpretation of that right.[88] It is also a discussion where those applying a permissive interpretation rely on customary law, while advocates of the restrictive approach utilize a more literal interpretation of Article 51 of the UN Charter.[89] At the restrictive end of the spectrum is the view that each attack by a terrorist group has to be treated as an isolated incident, with any response taking place as the attack

---

[81] Dinstein, *supra* note 3, ¶ 723, at 273.

[82] *Id.*

[83] Gray, *supra* note 4, at 241–44.

[84] *Obama's Speech on Drone Policy*, N.Y. Times, May 23, 2013, http://www.nytimes.com/2013/05/24/us/politics/transcript-of-obamas-speech-on-drone-policy.html, and *Department of Justice White Paper*, 4 (undated), http://msnbcmedia.msn.com/i/msnbc/sections/news/020413_DOJ_White_Paper.pdf (last visited May 27, 2015) [hereinafter *Department of Justice White Paper*] ("The United States is in an armed conflict with al-Qa'ida and its associated forces . . . .").

[85] Yoram Dinstein, Non-International Armed Conflicts in International Law ¶ 82, at 26–27 and ¶ 170, at 53–54 (2014).

[86] Hamdan v. Rumsfeld, 45 I.L.M. 1130, 1154 (2006) (U.S. Supreme Court 2006). *But see* Dinstein, *supra* note 3, ¶ 130, at 56 (where the *Hamdan v. Rumsfeld* approach applicable to a worldwide cross-border "war on terrorism" is rejected as a fiction).

[87] Dinstein, *supra* note 3, ¶ 728, at 275.

[88] Gill, *supra* note 11, at 189 n.7.

[89] Gray, *supra* note 4, at 117–18.

is unfolding. No follow-on strike can occur until another armed attack by that group is actually taking place. That is because in any effort to deter and prevent future terrorists attacks "it is difficult, if not impossible, to employ these central criteria of self-defense [i.e., necessity and proportionality] in the absence of detailed evidence about a specific threatened attack."[90] It is an approach that appears to look at each individual attack in isolation and does not embrace the theory of anticipatory self-defense.

Countering this view is a perspective that sees terrorism, with its potential for the use of weapons of mass destruction and massive civilian casualties, combined with the uncertainty of when an attack will take place, as more of an ongoing threat. It makes sense to require that a defensive act be "interceptive" to the extent it allows military action "after the other side has committed itself to an armed attack in an ostensibly irrevocable way."[91] The challenge for the State in applying the narrower interpretation of the right of self-defense[92] is determining when to strike with a goal of "nipping an armed attack in the bud."[93] Terrorist members of armed groups are seen to be "continuously planning attacks."[94] This is an approach that leads to interpretations of imminence that permit anticipatory defensive action "even if there is no specific evidence of where such an attack will take place or the precise nature of the attack,"[95] or to the view that strikes can occur "when the attacker is clearly committed to launching an armed attack and the victim State will lose its opportunity to effectively defend itself unless it acts."[96] This "limited window of opportunity approach" is reflected in the 2011 U.S. *Department of Justice White Paper* on the legality of drone strikes.[97]

An example of an altered application of self-defense principles can be seen in Michael Schmitt's analysis of counterterrorism operations. He suggests in respect of multiple terrorist strikes by the same organization, "it is more appropriate to characterize them as a continuous attack, much as individual and distinct tactical engagements coalesce into a military campaign."[98] The result is that "the immediacy criterion applies only to the first in an anticipated series of attacks," with the rest of the terrorist campaign "entitling the State to an extended period of self-defence." [99] Although it is indicated that the self-defense

---

[90] *Id.* at 203.

[91] DINSTEIN, *supra* note 3, ¶ 542, at 204–05.

[92] LUBELL, *supra* note 7, at 58–59 (where it is noted the concept of "interceptive" self-defense "is part of 'classic' self-defense, rather than anticipatory.").

[93] DINSTEIN, *supra* note 3, ¶ 546, at 206.

[94] *Memorandum for the Attorney General, Re: Applicability of Federal Criminal Laws and the Constitution to Contemplated Lethal Operations Against Shaykh Anwar al-Aulaqi*, Office of the Assistant Attorney General, July 16, 2010, at 27, https://www.aclu.org/sites/default/files/assets/2014-06-23_barron-memorandum.pdf.

[95] *Testimony of Attorney-General Lord Goldsmith*, 660 *Hansard H.L.* (Apr. 21, 2004) 370 (U.K.), *available at* http://hansard.millbanksystems.com/lords/2004/apr/21/international-self-defence.

[96] TALLIN MANUAL, *supra* note 5, ¶ 4, at 64, and *Department of Justice White Paper, supra* note 84, at 7.

[97] *Department of Justice White Paper, supra* note 84, at 7. *See also* Louise Arimatsu & Michael N. Schmitt, *Attacking "Islamic State" and the Khorasan Group: Surveying the International Law Landscape*, COLUM. J. TRANSNAT'L L. BULLETIN 1, 17 n.65 (2014), http://jtl.columbia.edu/attacking-islamic-state-and-the-khorasan-group-surveying-the-international-law-landscape// (outlining the genesis of the concept of "last window of opportunity" in Michael Schmitt's scholarship).

[98] Schmitt, *supra* note 60, at 175.

[99] *Id.* at 175.

principles of proportionality and necessity would continue to be relevant, their application is ultimately closely linked to the availability and efficacy of a law enforcement response.[100] This interpretation, combined with a reference to the "mere existence of the [terrorist] group means the dispute between it and the State(s) will remain violent," suggests that barring disbandment of the terrorist group or its alteration to a political organization, the response to the threat is likely to include the repeated use of military force.[101] The question of whether defensive action against Al Qaida groups located in countries other than Afghanistan can also be viewed under this broader concept of "continuous attack" is at the heart of the debate regarding the use of drones by the United States (see Chapter 8, section 8.5.2, *The Permissive "Conduct of Hostilities" Approach*).

Questions relating to the scope of non-State actor conflict also have to address the reality that the scale of conflict with some non-State actors can begin to approximate inter-State hostilities, if not technically at law, then in a factual sense. This can be seen in the 2006 Israel/Lebanon conflict and the hostilities against the Islamic State commencing in 2014. Regarding operations in Afghanistan, what started out as an aerial bombardment supported by limited numbers of Special Forces and CIA personnel changed to a broader ground operation involving not only the new government of Afghanistan but also a diverse coalition of foreign military forces. It is the transition from a traditional view of self-defense that is focused on isolated attacks to its application in an ongoing armed conflict, which presents unique challenges to the international community. Under these circumstances, the application of self-defense principles begins to be linked to questions regarding the broader strategic impact of the hostilities, rather than being focused on the attack and response associated with more isolated acts of violence.

On its face, the "limited" theory for the application of self-defense principles appears to present a significantly different model than the "overarching" one. While there are differences between the two approaches, they are not necessarily as absolute or as sharply drawn as it might first appear. For example, since the "overarching" theory appears to accept that the expansion of a conflict operates to lessen its influence,[102] at some point the two theories attain a degree of congruence. This would particularly be the case if the armed conflict were to take the form of a total war.[103] Another similarity is that defensive operations falling below the threshold of defensive war will be, in a manner similar to the "overarching" theory, subject to a simultaneous application of the self-defense principles and humanitarian law.[104] The likelihood that "short of war" situations between States will continue to take place in the twenty-first century; the lack of certainty as to when "war" actually commences; and the more frequent engagement in "extra-territorial enforcement" or defensive responses against terrorist groups points to a need to resolve how the recourse to war and the humanitarian law normative regimes interact.

---

[100] *Id.* at 174.

[101] *Id.* at 175.

[102] Greenwood, *supra* note 14, at 278 (for reference to a lessened impact on the areas where warfare is conducted).

[103] Dinstein, *supra* note 3, ¶ 29, at 12.

[104] *Id.* ¶ 41, at 16 (As Yoram Dinstein notes "the actual fighting will be governed by the *jus in bello*" when States "resort to a limited use of force 'short of war'.").

3.5. Additional Restraining Mechanisms

The restraining effect of the self-defense-based "proportionality" principle does not arise solely in the context of the "limited" and "overarching" theories. It can also be realized through operation of the political organs of the United Nations, the Security Council and the General Assembly, or because of cases being brought before its judicial organ, the International Court of Justice (ICJ).[105] Even if an act of self-defense is categorized as a defensive war under the "limited" theory, and therefore not subject to its governing principles, that conflict remains subject to another potential constraint found within Article 51 of the UN Charter. That article provides for the immediate reporting to the Security Council of the exercise of the right of self-defense by States. The inherent right to act in self-defense can be exercised only "until the Security Council has taken the measures necessary to maintain international peace and security,"[106] and that "Council can contain the hostilities."[107] In this regard, the Security Council may pass binding resolutions that Member States must follow.[108]

Unfortunately, the Security Council has not offered a robust mechanism for restraining conflict. One challenge has been that "since the SC [Security Council] has, for a long time, been far from performing its intended function, self-defense has become the regular course of action."[109] There have been relatively few Security Council resolutions that expressly refer to Article 51. The exercise of self-defense and declarations regarding the legality of specific uses of force that do exist are not viewed as being able "to resolve the doctrinal controversies as to the *scope* of the right of self-defence."[110] Further, the exercise of this power does not necessarily mean that the right to self-defense will be terminated. As was evidenced in response to the 1990 Iraq invasion of Kuwait and in the aftermath of 9/11, the Security Council "recognized the right of the state to defend itself with the help of its allies and specifically reaffirmed the right after the Council began to order the taking of collective measures against the attackers."[111]

However, there is the question of whether the decisions of the Security Council are themselves governed by the self-defense principles of necessity, proportionality, imminence, and immediacy. It has been suggested that not only the use of force in self-defense but also collective security action under Chapter VII of the UN Charter "is limited by the customary law requirement that it be proportionate to the unlawful aggression that gave rise to the right."[112] This is an issue that is open to considerable debate, although given the history of the development of the Charter and its link to just war thinking, it is difficult to completely divorce Security Council powers from the context in which they were developed. It has been noted that the issue of proportionality was part of the discourse during the

---

[105] FRANCK, *supra* note 79, at 67 (where this bodies are referred to as quasi-juries).

[106] UN Charter, art. 51.

[107] DINSTEIN, *supra* note 3, at 265.

[108] ANTONIO CASSESE, INTERNATIONAL LAW 186 (2nd ed. 2005).

[109] Albrecht Randelzhofer, *Article 51*, *in* I THE CHARTER OF THE UNITED NATIONS 788, ¶ 41, at 804 (Bruno Simma et al. eds., 2nd ed. 2002).

[110] GRAY, *supra* note 4, at 121.

[111] FRANCK, *supra* note 79, at 49.

[112] Gardam, *supra* note 42, at 391.

debates regarding the 1990 authorization for States to use "all necessary means" to remove Iraq from Kuwait. So was the question whether the Coalition that formed in the aftermath of Iraq's invasion were "using disproportionate military means in excess of those authorized by the Council became the subject of repeated debate."[113] The result is that notwithstanding the scope and scale of the defensive response, States must be prepared to justify their actions according to commonly accepted self-defense principles. This conclusion is consistent with the use of just war principles in the 2001 report, *The Responsibility to Protect*, to guide Security Council action regarding humanitarian intervention.[114]

The restraining nature of the reporting process is also reflected in the comment that "[s]ince the judgment in the *Nicaragua Case* it is noticeable that states on the whole do comply with the Article 51 requirement" to report, and "it is clear that states have taken seriously the Court's message that failure to do this will weaken any claim to be acting in self-defence."[115] This reference to the ICJ also highlights the potential role that court can perform in prescribing the circumstances under which force may be used in self-defense. The fact the ICJ may ultimately be asked to rule on the legality of the recourse to self-defense provides another element of restraint. Decisions of that court, such as the *Military and Paramilitary Activities in and against Nicaragua Merits*, "suggests that the principle of proportionality imposes an inescapable constraint on the level of permitted force, even when exercised in reliance on the legitimate right to self-defence."[116] In terms of proportionality, Thomas Franck observed when discussing the impact of Israeli targeting decisions during the 2006 Israel-Hezbollah war on the international perception of the legitimacy of the conflict:

> [w]hatever the technical legal position, proportionality has become an undeniable part of the public assessment of state conduct. A government contemplating a military action, as a matter of calibrating its strategy in cost/benefit terms, needs to keep this element—how that action is likely to be perceived—in clear focus, whether or not it is ultimately determinative.[117]

It seems likely that if the Security Council decided to intervene in a particular conflict, it too would have to undertake a similar strategic cost-benefit analysis in determining how to act.

However, the potential remains for a broad use of force by a State in almost any conflict, and in particular an inter-State one. This can raise significant concerns for those who seek to limit armed conflict. It has been argued that exercising the right of self-defense means "[u]nless the Security Council adopts a binding cease-fire resolution, a war of self-defence—once lawfully started—can be fought to the finish. . . ."[118] Such an interpretation is not new.

---

[113] Franck, *supra* note 47, at 732.

[114] The Responsibility to Protect: Report of the International Commission on Intervention and State Sovereignty, ICSS, XII, Dec. 2001, http://responsibilitytoprotect.org/ICISS%20Report.pdf (the principles to be applied were just cause, right intention, last resort, proportionate means, reasonable prospects of success, and acting under the direction of the right, or proper authority).

[115] Gray, *supra* note 4, at 121.

[116] Franck, *supra* note 47, at 720.

[117] *Id.* at 734.

[118] Dinstein, *supra* note 3, ¶ 703, at 265.

Josef Kunz's 1947 description of the then newly introduced Article 51of the UN Charter is very similar. As he noted, the exercise of self-defense provides the authority to

> resort to a justified war, to carry this war to victory, to impose a peace treaty upon the vanquished aggressor, always presupposing that the Security Council has failed and continues to fail of taking the measures necessary to maintain international peace and security. The right of self-defense is, in such cases, a right to resort to war.[119]

It must be noted though that this view was expressed with the belief "the conditions of necessity, reasonableness, and a certain proportionality, which the municipal law prescribes for the exercise of the right of self-defense, are lacking in Art. 51."[120] Given the widespread acknowledgment that the principles of necessity, proportionality, imminence, and immediacy do apply to the right of State self-defense, this assessment, articulated in the immediate post–World War II era, appears dated.

Over time the perspective of a broad "right to resort to war" in self-defense, fettered solely by humanitarian law, has shifted to a complete rejection of that approach (i.e., the "overarching" theory), or to one that claims that freedom in a narrower set of armed conflicts (i.e., the "limited" theory and war of self-defense). Even when not considered in terms of those theoretical constructs, the issue of proportionality looms large in the context of a possible assessment of State conduct by the Security Council during the course of a conflict, or because of the potential for a case to be brought to the ICJ when hostilities cease. The question remains how a general self-defense principle like proportionality affects the conduct of hostilities, which are simultaneously governed by the separate humanitarian law framework. In other words, how those two bodies of law interact in what appears to be a situation of overlapping application. Given the continuing application of the self-defense principles, as well as what appears to be a relatively small number of conflicts for which the "overarching" and "limited" theories actually diverge, this issue is of particular importance for the legal practitioner. In this regard, the scene is set to consider the degree to which, in respect of contemporary conflict, the principle of proportionality acts to constrain the use of force when acting in self-defense.

3.6. The Self-Defense Principles: A Strategic or Tactical Issue?

3.6.1. SELF-DEFENSE AND THE LEVELS OF WAR

The question of how the principles associated with the law governing the exercise of State self-defense interact with those applicable under humanitarian law does not appear to have been the subject of concerted study, although it has generated comment in an academic

---

119 Josef L. Kunz, *Individual and Collective Self-Defense in Article 51 of the Charter of the United Nations*, 41 AM. J. INT'L L. 872, 876–77 (1947).

120 *Id.* at 877.

context.[121] Where the ICJ has been seen to enter into this area, it has been heavily criticized. For example, in the *Advisory Opinion on the Legality of the Threat or Use of Nuclear Weapons*, the ruling of a badly divided court has been interpreted to indicate the use of nuclear weapons would be lawful where the threatened State is acting "in the extreme circumstance of self-defence, in which the very survival of the a State would be at stake."[122] This has also been suggested to mean that "the adversary of that Party—the aggressor—cannot employ nuclear weapons no matter what: not even if its survival is at stake."[123] The incorporation by the court of the criterion for acting in self-defense in extreme circumstances as a basis for determining the lawfulness of the employment of nuclear weapons has been criticized as "a dangerous departure from the concept that the *jus in bello* [humanitarian law] applies equally to all Belligerent Parties, irrespective of their status pursuant to the *jus ad bellum* [law governing the recourse to war]."[124]

Criticism of the ICJ is not limited to the *Advisory Opinion on the Legality of the Threat or Use of Nuclear Weapons*. It has also been suggested the court incorporated an element of the law governing the recourse to war into humanitarian law in the *Case Concerning Oil Platforms*.[125] The court held the "United States must also show that its actions were necessary and proportional to the armed attack made on it, and that the [Iranian] platforms were a legitimate military target open to attack in the exercise of self-defence."[126] The judgment indicated it was not sufficiently established there was a sufficient military presence and activity on one platform; no such evidence was presented regarding the military aspects of other platforms that were attacked; and that "the United States forces attacked the R-4 platform as a 'target of opportunity', not one previously identified as an appropriate military target."[127]

One assessment of this case considers that requiring a target must be of a military nature before it can be lawfully attacked was an improper incorporation of elements of humanitarian law into the norms of the law governing the recourse to war.[128] However, this view does not appear to be universally held. Lindsay Moir has suggested there is an alternate interpretation that "the Court was simply incorporating this requirement [legitimate military objective] into the concept of necessity."[129] This could be seen in the assertion that "[o]ne aspect of these criteria [self-defense necessity and proportionality] is the nature of the target of the force used avowedly in self-defence."[130] Regardless of which interpretation is correct, it is evident from both the *Advisory Opinion on the Legality of the Threat or Use of Nuclear Weapons* and the *Case Concerning Oil Platforms* decisions that there is a considerable lack of clarity surrounding the court's handling of the interface between the two bodies of law.

---

[121] *See, e.g.*, Greenwood, *supra* note 14, at 278–79, GARDAM, *supra* note 15, at 168–73, Cannizzaro, *supra* note 47, at 785–89, and Franck, *supra* note 47, at 728–34.

[122] Advisory Opinion on the Legality of the Threat or Use of Nuclear Weapons, 1996 I.C.J. 226, ¶ 105(2) E., at 266.

[123] DINSTEIN, *supra* note 3, ¶ 462, at 172–73.

[124] *Id.* ¶ 463, at 173.

[125] Case Concerning Oil Platforms, 2003 I.C.J. 158.

[126] *Id.* ¶ 51, at 187.

[127] *Id.* ¶ 76, at 198.

[128] DINSTEIN, *supra* note 3, ¶ 464, at 174.

[129] MOIR, *supra* note 4, at 125.

[130] *Id.* and Case Concerning Oil Platforms, 2003 I.C.J. 161, ¶ 74, at 196.

The manner in which self-defense and humanitarian law principles interact was addressed during the development of the 1995 *San Remo Manual*. It was reportedly the subject of controversy with a disagreement among legal experts about whether "the principles of necessity and proportionality are applicable in a strategic sense only, or also on a tactical level."[131] For proponents of having the self-defense principles control the conduct of hostilities at all levels of war this would mean:

> they would restrict the choice of targets, the use of methods and means of warfare and measures taken against neutral shipping taken not only in accordance with the rules of international humanitarian law, but also whether they would be necessary and proportionate to the needs of the parties.[132]

In contrast, the perspective of those advocating a solely strategic impact was that "the principles of necessity and proportionality cannot affect tactical behaviour once an action in self-defence has commenced and that these principles rather affect only strategic decisions on how to achieve the needs of self-defence."[133]

The *San Remo Manual* presents two different visions of how self-defense principles interact with traditional humanitarian law rules. It is one thing to indicate the law governing the exercise of State self-defense exerts a strategic influence on how operations are conducted, and entirely another to view that body of law as exerting a more controlling function over those activities at the tactical level. It is likely a concern that self-defense principles would be applied to tactical decision making, which is the genesis of the discord between advocates of broadly applying self-defense principles to contemporary hostilities and those who favor an unfettered application of humanitarian law once an armed conflict has commenced. In a sense this tension reflects the point of divergence between the "overarching" and "limited" theories, although the difference between those two approaches appears to be relevant only in the context of a defensive war. Two countervailing pressures influence this debate. First, there is the issue of the effectiveness of humanitarian law in controlling conflict. Second, the question arises as to whether such an application of self-defense principles represents an overreach by advocates for a body of law that traditionally has been focused on prohibiting any use of force by States.

Humanitarian law regulates the conduct of hostilities. However, not all scholars view humanitarian law, on its own, as offering adequate restraint on the conduct of warfare. This appears evident in Judith Gardam's comments regarding targeting proportionality. She states that "[i]t is becoming apparent that to rely solely on IHL [international humanitarian law] to regulate the consequences of armed conflict fails to acknowledge the greater potential of *ius ad bellum* [law governing the recourse to war] to achieve that goal."[134] Cynicism regarding the effectiveness of humanitarian law is not a new phenomenon. A critical view

---

[131] San Remo Manual, *supra* 43, ¶ 4.5, at 77.

[132] *Id. See also* Gardam, *supra* note 15, at 156 ("any ensuring forceful action will need to be monitored closely to ensure that the strategic objectives and *the methods chosen to achieve them* remain proportionate to the aim of the response.") (emphasis added).

[133] San Remo Manual, *supra* 43, ¶ 4.5, at 77.

[134] Gardam, *supra* note 15, at 209.

of humanitarian law was also evident at the end of World War I. It has been noted, "much of the terrible slaughter of the trenches in the war had been technically in accord with the Hague regulations . . . governments were not interested in further refining the *jus in bello* [humanitarian law], but sought rather to prevent war altogether through the mechanisms of the League of Nations, including disarmament and collective security . . . ."[135] It was this perspective that ultimately led to the reintroduction of the just war–based provisions that were to become such an important part of the UN Charter security framework.

A concern that an overemphasis on self-defense principles during an armed conflict may lead to their controlling the use of force at the tactical level of operations may be well founded. That concern is based on the self-defense concept of proportionality being given a disproportionate weight, such that it interferes with the conduct of operations necessary to provide security. In this respect, linked to the self-defense principles are the historically pacifist underpinnings of the law governing the recourse to war. Michael Walzer discussed this possibility in the context of twenty-first-century efforts to expand noncombatant immunity associated with protecting uninvolved civilians during attacks:

> The easiest way to resist is to make noncombatant immunity into a stronger and stronger rule, until it is something like an absolute rule: all killing of civilians is (something close to) murder; therefore any war that leads to the killing of civilians is unjust; therefore every war is unjust. So pacifism reemerges from the very heart of the theory that was originally meant to replace it. This is the strategy adopted most recently by many opponents of the Afghanistan war.[136]

With this approach, the law governing the exercise of State self-defense would fundamentally alter traditional interpretations of proportionality under humanitarian law, since the latter body of law only prohibits excessive collateral civilian casualties and damage.[137] In effect, the law governing the recourse to war would alter or even displace humanitarian law. It is an approach that reignites the "pacifist" antiwar debate regarding the "outlawing" of war that underpinned much of the lead-up to the adoption of the UN Charter.

Notwithstanding these different perspectives, what is striking in the *San Remo Manual* is the acknowledgment of the "levels of war" and their possible relevance to how principles related to the State exercise of self-defense might impact on the conduct of hostilities. It is within this framework that the two different approaches to the application of self-defense principles can be best analyzed and resolved. In looking at this issue it is important to first consider the conceptual framework of the various levels of war in terms of how warfare is conducted.[138] As originally envisaged by Clausewitz, there are two levels of war: strategy and tactics. Tactics involve "the use of armed forces in the engagement," while strategy is described as "the use of engagements for the object of the war."[139] This earlier notion of

---

[135] Adam Roberts, *Against War*, *in* The Oxford History of Modern War 320–21 (Charles Townshend ed., 2005).

[136] Michael Walzer, Arguing About War 13 (2004).

[137] AP I, art. 57(2)(a)(iii), and the 1998 Rome Statute, art. 8(2)(b)(iv).

[138] Kenneth Watkin, *Military Advantage: A Matter of "Value", Strategy and Tactics*, 17 Int'l Y.B. Hum. L. 319-23 (2014)(forthcoming)(for a discussion of the "levels of war") (forthcoming).

[139] Carl Von Clausewitz, On War 128 (Michael Howard & Peter Paret eds. & trans., 1989).

strategy was focused on the art of the commander. As such, it was more closely related to what would now be considered the operational level of war, which is situated between modern strategy and tactics.[140]

As a concept, strategy came to reflect "the complexity of situations in which states and organizations found themselves."[141] Reflecting that complexity, the strategic level has been further divided into higher (grand strategy) and lower (strategy) planes.[142] Grand strategy is "policy in execution" with its role being "to co-ordinate and direct all the resources of a nation, or band of nations, towards the attainment of the political object of the war—the goal defined by fundamental policy."[143] In this respect, "fighting power is but one of the instruments of grand strategy . . . ."[144] In contrast, lower-level strategy has been referred to as the "art of the general,"[145] concerned with "distributing and applying military means to fulfill the ends of policy."[146] The distinction between these two levels of strategy is that, "while the horizon of strategy is bounded by the war, grand strategy looks beyond the war to subsequent peace."[147] Grand strategy is visionary, while "military strategy look[s] to the near term and work[s] with specific situations."[148] Within this construct, policy objectives, which at the higher level look toward the attainment of peace, are translated into strategic military objectives, which in turn are executed at the tactical level to meet the policy goals. It is also the level at which decisions may be made to favor law enforcement when responding to terrorist acts that are threatened, or occurring even in what could be determined to be an armed conflict (see Chapter 13, section 13.2.1, *The "Troubles"*). Taken together, the concept of modern strategy had evolved to locate strategy "not in the art of the commander but in the links that tie war to policy."[149]

It has been suggested that in the post–World War II period the impact of the threat of nuclear war changed the strategic dialogue to one "divorced from actual fighting: deterrence theory, strategic thought's principle output of the 1960s, was largely shaped by civilian strategists."[150] This was to lead to the development of an intermediate "level of war," which more closely reflected Clausewitz's original notion of strategy.[151] It is at this operational level of war where the translation of general strategic objectives to more concrete military direction takes place. The operational level is where campaigns and major operations are planned and carried out to fulfill strategic objectives. An example of an operational level

---

140 Hew Strachan, The Direction of War: Contemporary Strategy in Historical Perspective 57 (2013).

141 Lawrence Freedman, Strategy: A History 72 (2013).

142 B. H. Liddell Hart, Strategy 321–22 (2nd ed. 1991). *See also* Strachan, *supra* note 140, at 31–33.

143 Liddell Hart, *supra* note 142, at 322, and Peter Paret, *Introduction*, *in* Makers of Modern Strategy: From Machiavelli to the Nuclear Age 3 (Peter Paret ed., 1986) (where only the term "strategy" is used; however, it is acknowledged two exist in two ways: "the use of armed force to achieve the military objectives and, by extension, the political purpose of the war" and in a broader inclusive sense that includes "the development, intellectual mastery, and the utilization of all the state's resources for the purpose of implementing its policy in war.").

144 Liddell Hart, *supra* note 142, at 322.

145 *Id.*

146 *Id.* at 321.

147 *Id.*

148 Strachan, *supra* note 140, at 235.

149 *Id.* at 57.

150 *Id.* at 212.

151 *Id.*

of war is the International Security Assistance Forces headquarters in Kabul, which had a principle role of setting the conditions for success at the tactical level by linking "strategic direction with traditional tactical activity."[152]

Unfortunately, for the overall conduct of warfare, the operational level of war is often a preferred area of concentration for military commanders, who see it is as "a policy free zone, in which military expertise was unfettered and where armies reasserted their authority over war's conduct."[153] The proper conduct of warfare requires clear strategic direction. Strategy is located "along the fault-line between policy and the operational level of war."[154] The successful conduct of warfare is not solely about military decisions focused on applying a maneuver versus attritional philosophy.[155] It requires an iterative and ongoing interaction between political direction and military execution throughout the course of the conflict. Strategy has been described as "a dialogue where ends also reflect means, and where the result—also called strategy—is a compromise between the ends of policy and the military means available to implement it."[156]

As is reflected in the position of some experts reported in the *San Remo Manual* treatment of this subject, self-defense principles appear particularly well suited to operate at the strategic level.[157] Focused on the actions of the State, where policy interfaces with law, the just war doctrine upon which the right to exercise self-defense is based "was designed to ensure states would not resort to unprincipled and brutal violence under cover of redressing an alleged wrong."[158] At its heart, that doctrine reflected "a need to balance the overall utility of the war with the damage it may inflict."[159] In this respect, the self-defense principles look beyond the war toward peace. In contrast, under humanitarian law, targeting proportionality is focused on the conduct of hostilities, although consideration must be given to the strategic impact of tactical-level targeting decisions (see section 3.6.2, *Targeting: A Matter of Strategy and Tactics*, below).

The interaction between law and policy, which occurs at the strategic level of war, is reflected in the comments of two of the proponents of the "overarching" theory for applying the self-defense principles. In the Greenwood analysis of the decision by the UK government not to extend the war to Argentina during the 1982 Falklands/Malvinas conflict, it was noted:

> To what extent the decision was based on a belief that an extension of the conflict to the mainland would be disproportionate and therefore illegal and how far it merely reflected political and military considerations is unclear. Nevertheless the decision not to extend the conflict beyond the Islands and the surrounding waters

---

[152] RICHARD CONNAUGHTON, MODERN WARFARE: THE TRUE STORY OF CONFLICT FROM THE FALKLANDS TO AFGHANISTAN 349 (2008).

[153] STRACHAN, *supra* note 140, at 213.

[154] *Id.* at 66.

[155] FREEDMAN, *supra* note 141, at 242.

[156] STRACHAN, *supra* note 140, at 45. *See also* FREEDMAN, *supra* note 141, at 242 ("The idea of a military strategy separate from a political strategy was not only misleading but also dangerous.").

[157] SAN REMO MANUAL, *supra* 43, ¶ 4.5, at 77.

[158] Franck, *supra* note 47, at 719.

[159] BARAK, *supra* note 45, at 176.

undoubtedly assisted the United Kingdom in portraying its actions as confined within the boundaries of self-defence.[160]

This law and policy interplay is an inherent part of higher strategy. Law and policy inform one another, often making it difficult to ascertain which factor is actually being reflected in the decision that is reached. The political interplay with the law governing the recourse to war is also reflected in Judith Gardam's acknowledgment that "[i]n a highly politicized and sensitive area as the resort to and the scale of force, it is to be expected that there would be considerable differences of opinion on how to measure proportionality."[161]

These references to political and legal interaction provide strong support for the proposition that, at a minimum, State self-defense principles operate at the strategic level of war. Christopher Greenwood reinforces this point when he suggests that "the limits inherent in the concept of self-defence add a requirement of *strategic* proportionality; the attack itself must be a necessary and proportionate measure of self-defence."[162] Interestingly, he also distinguishes between the concept of strategic proportionality and what is described as "*tactical* proportionality." The latter form of proportionality is based in humanitarian law.[163] In this respect, the attack may meet the "tactical proportionality" requirements because a substantial military advantage is provided, however, it will "fail to meet the strategic requirement, because the attack represents an escalation of the conflict which cannot be justified as a necessary measure of self-defence."[164] In this example the term "attack" is discussed in the singular, leaving unanswered how the interaction between the law governing State self-defense and humanitarian law operates across the spectrum of conflict where in some circumstances literally thousands of attacks may take place during the defensive operations carried out in response to an armed attack.[165]

This analysis views self-defense proportionality and its humanitarian law counterpart operating at opposite ends of the levels of war hierarchy. This raises a number of fundamental issues about how warfare is conducted and the degree to which targeting proportionality is restricted to the tactical level. It also raises questions as to how these bodies of law interact during defensive operations that can take place across the spectrum of conflict. For example, is there any difference in how that interaction occurs during the larger scale 1982 or 2006 Israeli operations in Lebanon, in comparison to the more limited late 1980s U.S. attack on Iranian oil platforms? These issues are the subjects of analysis in the next section.

3.6.2. TARGETING: A MATTER OF STRATEGY AND TACTICS

A key question to be addressed in resolving the interface between self-defense and humanitarian law principles is whether targeting proportionality must be assessed solely at the

---

[160] Greenwood, *supra* note 14, at 277.

[161] GARDAM, *supra* note 15, at 155.

[162] Greenwood, *supra* note 14, at 279.

[163] *Id.* at 278–79.

[164] *Id.* at 279.

[165] *See, e.g.*, ANTHONY H. CORDESMAN, GEORGE SULLIVAN, & WILLIAM D. SULLIVAN, LESSONS OF THE 2006 ISRAELI-HEZBOLLAH WAR 121 (2007) (where it is noted that in thirty-four days of the 2006 Israel/Lebanon conflict 15,500 sorties were flown [of which 10,000 were fighter sorties] and 7,000 targets were struck.).

tactical level. If the assessment of proportionality under humanitarian law were not limited to that level of war, the application of the self-defense principle of proportionality would have to be more directly reconciled with the broader effects of a targeting campaign. At the heart of this issue is the question of whether the military advantage to be gained from an attack, a key criterion in the proportionality assessment in that it is weighed against the likely collateral civilian casualties or damage, needs to be assessed strategically.[166] Unfortunately, this is yet another humanitarian law issue that has not been resolved within the international legal community. There is a split in opinion on whether targeting proportionality should be considered on a broader strategic basis.

Restricting the assessment of proportionality during targeting to the tactical level of war is a theme reflected in one strongly held view of humanitarian law. For example, in assessing the targeting provisions found in Article 57 of Additional Protocol I, the ICRC *Additional Protocol Commentary* indicates "military advantage"[167] should to be limited to "ground gained and in annihilating or weakening the enemy armed forces."[168] Further, it is noted targeting is not "concerned with strategic objectives but with the means to be used in a specific tactical operation."[169] This is an approach that has been strongly advocated as a method of limiting the collateral impact of targeting. The targeting practices of the Coalition during the 1991 Gulf War were criticized because attacks on targets such as water treatment and power plants were calculated on their cumulative effects "in terms of their contribution to the weakening of the overall military forces of Iraq, rather than by taking into account the long- and short-term civilian casualties and balancing this factor against a direct immediate military advantage."[170]

It has been suggested that the very nature of the "equal application principle" means "proportionality cannot logically be measured by reference to the ultimate goals of the military mission, but instead to the more immediate aims of each single military action."[171] This is said to be because the law governing the recourse to war confers upon the State a "superior power to take defensive action, and the proportionality requirement serves only to determine the degree to which other values can be sacrificed to that higher value."[172] In contrast, under humanitarian law "there is by definition no higher value, as the offensive or defensive character of the military action does not count as such for assessment of the proportionality thereof."[173]

A major problem with such a tactically focused approach is that it does not fully reflect how or why wars are fought. This approach downplays the strategic importance of the crucial components of a State's capability to wage war: its industry, oil and electrical power systems,

---

166 AP I, arts. 52(2) and art. 57(2)(a)(iii) (assessing "military advantage" is a key element of the determination of whether a military object is a valid military objective and the proportionality of an attack).

167 Commentary on the Additional Protocols of 8 June 1977 to the Geneva Conventions of 12 August 1949, Protocol I (Yves Sandoz et al., 1987), *available at* http://www.cicr.org/ihl.nsf/COM/470-750073?OpenDocument.

168 *Id.* art. 57, ¶ 2218, at 685.

169 *Id.* art. 57, ¶ 2207, at 683.

170 Gardam, *supra* note 42, at 409.

171 Cannizzaro, *supra* note 47, at 786.

172 *Id.* at 787.

173 *Id.*

transportation and logistics systems, and so forth.[174] These objects have been indispensable parts of a State's military capability since the nineteenth-century technological revolution led to the industrialized form of warfare associated with inter-State conflict. An essential aspect of degrading the ability of an opponent to wage war, particularly in an inter-State conflict, can involve targeting key components of its strategic resources. The conduct of strategic warfare is particularly linked to the technological advances of airpower with its ability to strike at war supporting resources located on the home front.[175] As has been noted, it is this strategic aspect of warfare that makes imposing geographic restrictions on conflict so challenging (see section 3.3, *An "Overarching" Application*, above).

The use of airpower over the years has been a matter of considerable controversy from the perspective of its humanitarian impact. Unfortunately, the history of the strategic use of airpower has been replete with claims of effectiveness that have overreached the ability of military forces to attain its theoretical goals.[176] In addition, strategic targeting and the use of aerial bombing generally have considerably less relevance to fighting small wars, insurgencies, or terrorism.[177] However, the strategic use of airpower has also undergone significant change since the early twentieth century, shifting away "from war-supporting industry to war-supporting command—from economic warfare to control warfare." [178] As a method of warfare, strategic targeting remains a crucial aspect, although only one, of the use of airpower in inter-State conflict.[179]

The importance of the strategic level of warfare is further reflected in the enduring theme that all wars are conducted for political reasons.[180] Forcing an opponent to change its thinking has been and remains the ultimate goal of warfare.[181] A strategic targeting campaign against military objectives can have the additional goal of undermining confidence among the supporters of the target State regarding its ability to provide them protection. This is an aspect of modern warfare that is particularly controversial because of the potential for civilians to be put at risk. Under humanitarian law it is not permitted to either target the civilian population or conduct attacks solely for demoralizing the population. Article 51(2) of Additional Protocol I states: "[a]cts or threats of violence *the primary purpose of which is to spread terror* among the civilian population are prohibited."[182] However, the foregoing provision does leave space for attacks against military objectives designed to have an

---

[174] Michael W. Lewis, *The Law of Aerial Bombardment in the 1991 Gulf War*, 97 AM. J. INT'L L. 481, 48–96 (2003) (where it is indicated these categories of targets were attacked during the 1991 Gulf War).

[175] BRAUDY, *supra* note 40, at 459.

[176] I. B. Holley, *Reflections on the Search for Airpower Theory*, *in* THE PATHS OF HEAVEN: THE EVOLUTION OF AIRPOWER THEORY 579, 597 (Phillip S. Meilinger ed., 1997) [hereinafter THE PATHS OF HEAVEN].

[177] JOHN A. WARDEN, THE AIR CAMPAIGN: PLANNING FOR COMBAT, 147 (1988), *available at* http://www.au.af.mil/au/awc/awcgate/warden/ward-toc.htm ("Air is of marginal value in a fight against self-sustaining guerrillas who merge with the population. In this case no significant target exists for air attack.").

[178] David S. Fadok, *John Boyd and John Warden: Airpower's Quest for Strategic Paralysis*, *in* THE PATHS OF HEAVEN, *supra* note 176, at 357, 385.

[179] Rebecca Grant, *The Redefinition of Airpower*, 86 AIR FORCE: J. OF THE AIR FORCE ASSOC. 32, 36 (2003).

[180] CLAUSEWITZ, *supra* note 139, at 86–87.

[181] Kenneth Watkin, *Assessing Proportionality: Moral Complexity and Legal Rules*, 8 Y.B. OF INT'L HUM. L. 17–23 (2005) (for a discussion of strategic goals and targeting in the warfare.).

[182] AP I, art. 51(2) (emphasis added).

indirect strategic impact on the decision makers or the population of the state by causing them to believe they cannot win militarily.

Therefore, it remains permissible to seek to influence that population by striking at lawful military objectives, although a particular challenge for those seeking to carry out a strategic air campaign on this basis is that they will be hard pressed to present examples of where such an approach has actually worked.[183] It should be remembered that a focus on the strategic impact of operations is not limited to the use of airpower. It has long been an aspect of maritime warfare as well, particularly in the context of conducting a blockade and economic warfare.[184] A focus on the strategic level is also represented in the concept of "effects based targeting,"[185] which forms the basis of contemporary targeting doctrine applicable to all operational environments: land, sea, and air.[186] The result is that, as an operational concept, strategic targeting remains an essential part of the conduct of armed conflict, particularly in the context of inter-State warfare.

Mirroring the level of war framework, national objectives for the conflict are identified at the level of grand strategy, and then campaign objectives are broadly defined at the military strategic level. The subsequent planning process involves a sophisticated targeting process that looks at the relationship between potential targets and what is to be achieved in any attack. The result is a link between the tasked targets and the overall strategic effects being sought in conducting the operations. Strategic objectives ultimately influence what targets are of military value, with the subsequent attacks conducted at the tactical level. Successful completion of those attacks contributes to the overall military and ultimately political success of the military campaign. The idea that tactical-level actions can have strategic effect is not limited to targeting. It was evident in the post–Cold War conceptualization of the "strategic corporal," where the ultimate outcome of a conflict was viewed as hinging "on decisions made by small unit leaders, and by actions taken at the *lowest* level."[187]

Importantly, resistance to a uniquely tactical-level assessment of targeting proportionality is reflected in a number of States Party to Additional Protocol I having indicated that the expression "military advantage" in Article 57 "refers to the advantage anticipated from the military attack *considered as a whole* and not only from isolated or particular parts of that attack."[188] Similarly, this broader view of targeting is reflected in Article 8(2)(b)(iv) of the Rome Statute, which refers to "the concrete and *overall* military advantage anticipated."[189] Although it has been suggested in the ICRC *Customary Humanitarian Law Study* that the inclusion of the word "overall" in this instance did not change the existing law, it is

---

[183] Colin S. Gray, Another Bloody Century: Future War 101 (2005) (noting that the 1999 Kosovo air campaign did not on its own bring about success). *See also* William H. Boothby, The Law of Targeting 501 (2012).

[184] Freedman, *supra* note 141, at 119–20.

[185] Michael N. Schmitt, *Effects-Based Operations and the Law of Aerial Warfare*, 5 Wash. Glob. Stud. L. Rev. 274–76 (2006).

[186] Joint Targeting, United States Joint Publication 3-60, viii (Jan. 31, 2013) ("Targeting links the desired effects to actions and tasks.").

[187] Charles C. Krulak, *The Strategic Corporal: Leadership in the Three Block War*, Marines Mag. (Jan. 1999), *available at* http://www.au.af.mil/au/awc/awcgate/usmc/strategic_corporal.htm.

[188] Customary International Humanitarian Law Study 49 (Jean-Marie Henckaerts & Louise Doswald-Beck eds., 2005) (emphasis added).

[189] *Id.* at 49–50 (emphasis added).

difficult to find support for that conclusion on a plain reading of the statute.[190] Since a targeting campaign may have strategic, operational, and tactical-level effects, it can be expected that when an armed conflict is total in the sense of seeking complete subjugation of an enemy State or by a commitment of the full resources of a State to the enterprise, the result will be warfare of greater scope and intensity. Collectively, this will likely also result in a greater legitimate collateral civilian damage or injury than in comparison to a limited border engagement. Such a conclusion is not inconsistent with the "overarching" approach for assessing the application of the self-defense principles, which accepts the proportionality principle will no longer geographically restrict where hostilities take place as a conflict expands toward total war.[191] Inherent in such an expansion is greater potential for the conduct of strategic warfare, and with it elevated collateral civilian impact.

Another factor to consider is that during large-scale conflict, elevated levels of collateral civilian casualties and damage do not necessarily have to result from a culmination of numerous smaller attacks. This can be seen in the *Advisory Opinion on the Legality of the Threat or Use of Nuclear Weapons*, where the ICJ did not reject the idea that nuclear weapons could be resorted to in exercising self-defense when the very survival of a State would be at stake.[192] However, it is the overall cumulative effect of the targeting decisions that normally raises self-defense proportionality issues, rather than individual targeting choices at the tactical level. Each attack considered separately may not result in significant collateral civilian casualties. Even if they do, those casualties must be weighed in the context of the overall strategic goals of the campaign. However, low levels of collateral casualties arising from individual attacks may cumulatively come to be viewed as disproportionate when assessed against the goals of the conflict.[193]

While there is a strongly held view that targeting proportionality must only be assessed at the tactical level, this approach does not reflect how warfare is conducted. Targeting is carried out not only for the purpose of gaining a military advantage at the tactical level but also for the advantage that can be realized strategically from the destruction of that target or a series of interrelated targets. In strategic terms, it is often the cumulative effect of the targeting process where the military advantage is realized, and the very reason for which the attacks are carried out. The impact of the strategic approach to warfare means that there will not always be a direct comparison of the effects of each attack at the tactical level (i.e., "tactical proportionality") against strategic-level self-defense-based proportionality (i.e., "strategic proportionality"). Rather, the manner in which the tactical-level effects of an attack come to be assessed at the strategic level for self-defense purposes depends upon the scope and scale of the defensive response. As will be seen, isolated attacks lend themselves to a more direct comparison of tactical affects against the self-defense principle of proportionality. A broader defensive response, involving large-scale attacks and strategic targeting, is

---

[190] *Id.*

[191] Greenwood, *supra* note 14, at 278.

[192] Advisory Opinion on the Legality of the Threat or Use of Nuclear Weapons, 1996 I.C.J. 226, ¶ 96, at 263.

[193] Cannizzaro, *supra* note 47, at 792 (This appears to have occurred during the 2006 Israel/Lebanon conflict when "[i]n a number of reactions by third states and international organizations the existence of collateral damage, in particular the high toll of victims among civilians, was invoked as proof of the disproportionality of Israel's self-defence reaction.").

more likely to result in the effects of those operations being assessed on a cumulative basis. Given the wide range of armed conflict that can result from an exercise of State self-defense, the analysis of this issue will benefit from consideration of how the principle of proportionality is assessed throughout hostilities across the armed conflict spectrum.

3.6.3. APPLYING SELF-DEFENSE PRINCIPLES ACROSS THE ARMED CONFLICT SPECTRUM

One aspect of applying State self-defense principles is determining when a proportionality assessment has to be made. As was seen in respect of the "limited" theory, a characteristic that distinguishes defensive wars from less serious inter-State incidents is the manner in which proportionality is assessed. For these latter situations there is a retroactive consideration of self-defense proportionality,[194] while it is suggested a different proportionality analysis is carried out at the onset of a war of self-defense.[195] However, regardless of whether a "limited" approach or the broader "overarching" theory is applied, a retroactive assessment of self-defense proportionality is a reality in respect of many armed conflicts. This is particularly evident when a conflict is placed under scrutiny through a postconflict inquiry or investigatory process. At a minimum, those planning and conducting operations must keep in mind that it is likely there will be some form of postconflict review of whether the scope and scale of the defensive response was justifiable as being proportionate. This generally will result in the consideration of the proportionality of a defensive response throughout the conflict. That assessment has to be carried out in a complex and oftentimes constrained security environment. Organizations focused on State accountability, such as human rights groups, will likely make a critical appraisal of the State response, as may the Security Council. The use of force could be subjected to investigation by a UN-appointed body or a domestic inquiry or commission. Further, State advisors must remain mindful that the question of the proportionality of defensive action may ultimately be the subject of litigation, either in the ICJ, or the International Criminal Court.

This raises the question of how the strategic application of self-defense principles impacts on the conduct of hostilities. While not involving the exercise of State self-defense, the manner in which limits imposed at the level of high strategy affect the conduct of operations can be seen in how decisions of the UN Security Council are "operationalized."[196] In this respect, UN Security Council Resolution 1973 relating to the 2011 NATO campaign in Libya provides insight into how strategic direction operates on the higher plane.[197] That Resolution placed limits on the scope of State action in authorizing the use of necessary means to protect civilians and civilian populated areas, "while excluding a foreign occupation force of any form on any part of Libyan territory."[198] What is notable is the general nature of this limitation, which would need to be incorporated into the military strategic level direction developed as a result of that Security Council authorization. That strategic

---

[194] DINSTEIN, *supra* note 3, ¶ 697, at 262.

[195] *Id.* ¶ 699, at 263.

[196] GRAY, *supra* note 4, at 120–21.

[197] UNSCR Res. 1973 (2011) (17 March 2011).

[198] *Id.* ¶¶ 4–12.

direction in turn would impact operational planning, and then the tactical execution by the military forces involved. In carrying out the Libyan bombing campaign, it would have been humanitarian law that governed the identification of military objectives, the selection of targets, and actual conduct of hostilities necessary to meet those objectives. However, in terms of accountability, it would be open to an inquiry or other investigatory body to assess whether the manner in which the subsequent campaign was conducted, including targets struck and resulting collateral effects, were consistent with the broad strategic direction provided by the Security Council.

Applying "strategic proportionality" in a self-defense context would have a similar constraining influence on targeting decisions, although the impact would normally be felt in even broader terms than the limitations set out in the Libya example. In part, this is because as a legal principle proportionality "has mostly eluded definition in any but the most general terms."[199] It also occurs because of the nature of the principles intended to restrict the exercise of State self-defense. The post–World War I effort to "outlaw" war, which resulted in the enactment of the self-defense provisions in the UN Charter, was designed to limit the use of force at the inter-State or transnational level. It is inherently strategic or general in nature. As a body of law, it has been successful in influencing the actions of States. However, the self-defense principles were not developed with specific conflicts or tactical-level engagements in mind. Further, while both the law governing the recourse to war and humanitarian law are part of the same just war tradition, the humanitarian law targeting rules were developed specifically to control hostilities. Unlike the self-defense normative framework this latter body of law includes specific direction on precautions to be applied in an attack in order to lessen its humanitarian impact.[200] Further, these unique rules are designed to be applied below the strategic level and even govern the actions taken by individual soldiers and units.

Self-defense proportionality can affect the conduct of hostilities, without usurping the role of humanitarian law, by influencing the boundaries of the acceptable overall level of collateral civilian casualties and damages. Concerns over the effect of operations in terms of collateral civilian injury and damage are a relevant consideration in the recourse to war context where the self-defense principles are seen to have a continuing influence once hostilities commence. As has been noted, it is in this context where targeting and self-defense proportionality are closely related.[201] It is at the strategic level where self-defense limitations can be reconciled with the assessment of the overall effect of military operations and targeting conducted under the humanitarian law normative framework. It is also at that level of war where the political and legal interface so integral to recourse to war decision making is found. The State self-defense-based principles influence the development of targeting direction through its interface with strategic-level military decision making. As a result, the nature and scope of the justifiable defensive response will determine the range of military action taken by the State. Once those goals are set, humanitarian law governs targeting planning and decision making at the strategic, operational, and tactical levels. Finally, the effects of those attacks are assessed against the strategic limitations imposed by the self-defense principles.

---

[199] Franck, *supra* note 47, at 716.

[200] AP I, art. 57.

[201] Greenwood, *supra* note 14, at 278 n.22.

The self-defense proportionality principles can have an important overarching effect on humanitarian law–based decisions.[202] In this regard, applying those principles may ultimately serve to restrict what valid military objectives are struck, the number of attacks carried out, and their location. However, issues such as what constitutes a lawful military objective, the framework for assessing collateral civilian casualties and damage, as well as the rules governing the use of weapons must remain firmly grounded in humanitarian law. If the strategic goals of a campaign are properly identified, then the choice of targets, and the resulting attacks, both individually and collectively, will fall within any self-defense established limitations on the use of force. This means that in some circumstances certain lawful military objectives may not be attacked in order to meet the strategic self-defense-based restrictions. Further, operations that result in less collateral death and damage may be privileged over other lawful attacks for which it is anticipated there could be greater impact on the civilian population. As a result, targets and attacks generated throughout the strategic, operational, and tactical levels of war will not only meet the humanitarian law–based standards, but they will also ultimately comply with any self-defense-based limitations on the use of force.

However, the impact of the self-defense principles is very context based. In situations where there is an isolated armed attack, and the defensive response is limited to an individual or unit firing on the attacker, consideration of whether the self-defense norms have been complied with will be more directly linked to the specific hostilities. This is because situations involving such isolated and low-level uses of force may not even directly engage the operational and strategic levels of command in terms of planning or decision making. In those circumstances, the issues involving higher levels of command may be limited to what preexisting standard operating procedures and rules of engagement were in place to guide the tactical-level commander regarding the use of force in *extremis*. It is this type of limited defensive engagement at the tactical level that is most likely to result in tactical proportionality being assessed directly against the strategic self-defense principle. Another way a direct assessment may arise is with the use of a nuclear weapon or other weapon of mass destruction. Any use of those weapons is more likely to directly raise strategic questions concerning the law governing the recourse to war, even though humanitarian law is applied to govern any attack. Even when such weapons are described as "tactical nukes," they have a significant strategic impact that is more likely to engage a concurrent assessment of humanitarian law and the self-defense principles.[203] It is the unique nature of such weapons, with their elevated potential for an adverse impact on the civilian population, that could have led the ICJ in the *Advisory Opinion on the Legality of the Threat or Use of Nuclear Weapons* to simultaneously address the law governing the recourse to war and humanitarian law when assessing the legality of the use of such weapons.[204]

However, when defensive responses begin to engage more complex military operations involving a larger number of units, extended operational planning at higher levels

---

[202] Eyal Benvenisti, *Rethinking the Divide Between Jus ad Bellum and Jus in Bello in Warfare Against Nonstate Actors*, 34 YALE J. INT'L L. 541, 544-44 (2009)(for an approach that suggests a party pursuing unrelated goals, or prolonging a conflict would be limited in its ability to justify collateral casualties in pursuing military objectives.).

[203] *See, e.g.*, Advisory Opinion on the Legality of the Threat or Use of Nuclear Weapons, 1996 I.C.J. 226, ¶ 43, at 245.

[204] *Id.*

of command, and the execution of a targeting campaign with multiple and diverse target sets, the impact of the State self-defense principles will have less direct influence on tactical-level targeting decisions. The results of individual attacks are also less likely to raise self-defense proportionality concerns. This is particularly evident in circumstances where those principles are said to apply throughout the course of a conflict. A case in point is the post-9/11 invasion of Afghanistan. That operation was conducted pursuant to the exercise of the State right of self-defense claimed under Article 51 of the UN Charter. However, the self-defense principles did not exert direct control over the individual targeting decisions that were made in the conduct of that campaign. In this regard it is reported that during the first 76 days of Operation Enduring Freedom alone some 6,500 aerial strike sorties were flown, dropping approximately 17,500 munitions.[205] It was humanitarian law that directly governed those attacks, and each attack was not individually assessed against State self-defense principles.

Another example of a larger-scale defensive response can be found in looking once again at Israel's operations against Hezbollah following that group's attack on a small patrol on July 12, 2006. That attack had resulted in the death of eight soldiers, the abduction of two more, and the destruction of a Merkava tank.[206] Israel viewed the incident as the latest in a series of attacks that justified an exercise of its right to act in self-defense.[207] It is reported in the opening stages of the conflict that the Israeli government approved attacks on Hezbollah rockets and a limited bombing of infrastructure, including the "Beirut airport tarmac and the Beirut-Damascus highway—putatively in order to block the kidnapper's escape routes beyond the country's borders."[208] The initial response by Israel involved destroying fifty-nine Hezbollah stationary rocket launchers concealed in homes with a reported twenty Lebanese civilian casualties.[209] Those strikes were followed by attacks against longer-range Hezbollah missiles located in the Beirut area,[210] as well as on Hezbollah's headquarters and the television station in the Dahia district located in southern Beirut.[211]

At first there appears to have been widespread, if somewhat cautious, support for Israel's defensive action.[212] This included support from Sunni Arab States, which were concerned about rising Iran-backed Shi'ite power and its impact on internal opposition forces, "especially as Hezbollah gained enthusiastic support even among the vast Sunni population of the Arab world."[213] However, the scope and scale of the response came under increasing scrutiny as attacks were undertaken against housing on the outskirts of Beirut in an attempt to hit missile-launching sites (e.g., Qana),[214] as well as on lines of communications such as the

---

205 Benjamin S. Lambeth, Airpower Against Terror: America's Conduct of Operation Enduring Freedom 248 (2005).

206 Amos Harel & Avi Issacharoff, 34 Days: Israel, Hezbollah and the War in Lebanon 1–15 (2008).

207 *Id.* at 87.

208 *Id.* at 86.

209 *Id.* at 91.

210 *Id.* at 92.

211 *Id.* at 99–100.

212 Michael N. Schmitt, *"Change Direction" 2006: Israeli Operations in Lebanon and the International Law of Self-Defense*, 84 Int'l L. Stud. 265, 273 (2008).

213 Augustus Richard Norton, Hezbollah: A Short History 137 (2007).

214 Harel & Issacharoff, *supra* note 206, at 161–62, and Schmitt, *supra* note 212, at 273.

Beirut International Airport and Lebanese bridges and roads.[215] Israel also established air and naval blockades.[216] As a result of the Israeli operations, "southern Lebanon was largely emptied of its civilian population—900,000 people were evacuated, or nearly one-quarter of Lebanon's population."[217] Further, Israel expanded its ground operations late in the conflict. From initial fighting near the Lebanese border on August 11, 2006, "the decision was made to fully commit its active and reserve forces in an attempt to go far beyond the border area and to seize Southern Lebanon up to the Litani River line."[218] With the international community pushing for an end to hostilities, a ceasefire was agreed to on August 13.[219]

The 2006 Israel/Lebanon conflict prompts a number of observations. First, there appears to have been ongoing discussions within the Israeli government concerning the scope and scale of the response, including the expansion of Israeli operations away from the Lebanese border and the impact of the collateral civilian casualties.[220] Second, both the support for Israeli action and the growing condemnation of the hostilities being a disproportionate use of force as they continued was expressed internationally at the political level.[221] This highlights the political/legal interface at the strategic levels of war. Finally, a distinction remains regarding the application of humanitarian law and that governing the exercise of national self-defense. Questions can be raised regarding the scope and scale of the State response, such as whether airfields or bridges should be bombed that are geographically distant from the location of the enemy military forces. However, whether those targets are valid military objectives remains a question of humanitarian law.[222] The result is that issues related to geography may be assessed twice. First, under humanitarian law to determine if it is a valid military objective in that the bridge makes a valid contribution to military action, and whether its destruction offers a definite military advantage.[223] Then there is the question of whether, from a self-defense perspective, the overall destruction, including the types of targets struck and where attacks took place, represents a disproportionate expansion of the conflict. However, the self-defense concept of proportionality does not govern the conduct of individual attacks but rather is assessed on a strategic basis.

Between these larger-scale post-9/11 Afghanistan and 2006 Israel/Lebanon conflict examples on the one hand, and isolated responses by individuals and small units on the other, there exists a broad range of potential State defensive action that engages the levels of war above the tactical level. The attacks that formed the factual basis for the *Case Concerning Oil Platforms* decision provide an example of a State response that would also require consideration of the higher-level strategic assessment of targeting decisions and their

---

[215] Schmitt, *supra* note 212, at 283.

[216] *Id.*

[217] Norton, *supra* note 213, at 142 (on the Israeli side of the border approximately 500,000 were displaced because of Hezbollah action).

[218] Cordesman, Sullivan, & Sullivan, *supra* note 165, at 5.

[219] Harel & Issacharoff, *supra* note 206, at 100.

[220] *Id.*

[221] Gray, *supra* note 4, at 241–43 (for an outline of the increasingly negative State reaction to the Israeli use of force, although the United States and United Kingdom are reported to have refused to condemn Israel's actions as being disproportionate).

[222] Dinstein, *supra* note 14, ¶ 249, at 102 (assessing bridges as military objectives).

[223] AP I, art. 52(2).

broader effects before assessing the impact in terms of self-defense proportionality. These included the October 19, 1987, attacks on the Iranian oil platforms in response to a missile attack on a U.S. flagged Kuwaiti tanker three days earlier, and the April 14, 1988, use of naval and air assets to attack two oil complexes four days after an American warship struck a mine would have required considerable planning and command involvement.[224] The ICJ noted in respect of the second incident, "the attacks on the Salman and Nasr platforms were not an isolated operation, aimed simply at the oil installations, as had been the case with the attacks of 19 October 1987; they formed part of a much more extensive military action, designated 'Operation Praying Mantis'. . . ."[225] This defensive response also resulted in the destruction of two Iranian frigates, as well as other naval vessels and aircraft.[226]

This approach toward assessing the self-defense principle of proportionality at the strategic, rather than the tactical level, as is reflected in the foregoing Afghanistan, Lebanon, and Iran examples, finds support in a major post-9/11 academic study. After reviewing the application of the principle of proportionality, the 2005 Chatham House study, *Principles of International Law on the Use of Force by States in Self-Defence*, indicates that "[t]he force used must take into account the self-defence operation 'as a whole'. It does not relate to specific incidents of targeting (which is a matter for international humanitarian law)."[227] What is left unsaid is that the interaction between the law governing the recourse to war and humanitarian law must often be resolved at the strategic level. However, the distinction between the overall (i.e., strategic) impact of the self-defense principle in comparison to "specific incidents of targeting" offers a clear starting point for resolving the interaction between these two bodies of law.

3.7. Less Certain Boundaries

As has been established in this chapter, since the late nineteenth century the focus of the international legal community has been on controlling the State recourse to war. This led to the separation of just war–based concepts into two separate areas: the law governing the recourse to war and humanitarian law governing the conduct of hostilities. While the trend in the post–World War II period was to emphasize the separation between these foundational bodies of law, the changing nature of the contemporary threats facing States has forced a reconsideration of this approach. A growing recognition that the prevalent security threat is posed by non-State actors requires a reassessment of the inter-State focus of these bodies of law. A significant change is that not all conflict with non-State actors is now seen as an internal State matter. States are being forced to react to transnational terrorist threats. They have also become involved in conflicts that quickly evolve from technically being

224 Case Concerning Oil Platforms, 2003 I.C.J. 161, ¶ 25, at 175–176; ¶¶ 63-68, at 191–194.

225 *Id.* at 68.

226 *Id.*

227 CHATHAM HOUSE STUDY, *supra* note 5, at 10. *See also* Case Concerning Oil Platforms, 2003 I.C.J. 161, ¶ 77, at 198 ("the Court cannot assess in isolation the proportionality of that action to the attack to which it was said to be a response; it cannot close its eyes to the scale of the whole operation, which involved, inter alia, the destruction of two Iranian frigates and a number of other naval vessels and aircraft.").

international ones to noninternational in character. However, little else changes beyond the formal legal categorization. Neither the participants nor the practical security challenges are altered by this transition. This means that contemporary non-State actor threats are increasingly being viewed through an international law lens. While the State remains the proper authority for countering transborder threats by organized armed groups and terrorist organizations, any resulting armed conflict between these groups is increasingly involving international law concepts once applied solely to hostilities between States. This is particularly evident with the greater recognition that States may exercise a right of self-defense against non-State actors whose actions are not attributable to a State.

The changing nature of warfare has also made it increasingly difficult to argue that the laws governing State involvement in conflict and humanitarian law can operate completely separately from one another. While conceptually helpful and operationally effective for controlling inter-State conflict in the twentieth century (i.e., the Cold War), the boundary between these bodies of law is less starkly drawn regarding contemporary conflict. Although States have largely been successful in avoiding non-State actors being viewed as legitimate participants in conflict, the law governing the State right to act in self-defense is being applied as conflict with non-State actors unfolds. The additional restraining influence of self-defense principles is particularly felt where States intervene on the territory of another State in order to respond to a terrorist threat. These types of conflicts are also forcing consideration of how the law governing State self-defense and that controlling the conduct of hostilities interact.[228] This is required since both are seen to continue to operate simultaneously when responding to contemporary threats, whether viewed under a "limited" or "overarching" analytical approach.

A key issue to be resolved is how the law governing the exercise of State self-defense and humanitarian law interact with one another. The law governing the exercise of national self-defense is neither superior to, nor does it usurp, humanitarian law. Put in more practical terms, the law governing the recourse to war does not act in such a way as to provide tactical direction to military commanders. Instead, it influences targeting at the strategic level with the interaction between these two bodies of law being highly context-based. It is the overall effects of humanitarian law–based decision making that are ultimately assessed against the operation of self-defense principles. Where there are isolated or limited responses to armed attacks, the assessment of tactical targeting in terms of the self-defense principle of proportionality will inevitably appear more direct than in the case for a more robustly planned and executed defensive response. The result is that the requirement to operate within self-defense-imposed limitations strategically influences operational decision making, however, it does not exert "control" at the tactical level.

---

[228] Benvenisti, *supra* note 202, at 544–45 (where it is noted observers linked State self-defense and humanitarian law concepts in assessing Israeli actions in Gaza in 2008–2009).

A state that uses a monopoly on force to protect its citizens from one another may be the most consistent violence reducer that we have encountered . . . .[1]

4 States, "Proper Authority," and Conflict

4.1. Introduction

The interaction between the body of international law governing the State recourse to war and humanitarian law is not limited to the use of force in self-defense. This occurs because of the unique status of the State under international law. As just war theory, and with it international law, developed, particular importance was placed on the State and its role as the proper authority. While the authority to wage war was initially based in religion, its demise has been "traced to the medieval redefinition of the church's authority over war to that of a judge, not executive. By contrast, the secular ruler could be his own judge in matters affecting his state as well as executive over his own armed forces."[2] In respect of international conflict this involved warfare between two governing powers.[3] However, that civil power (in its modern form, the "State") was, and remains, responsible for maintaining internal order within its territory. The result is that engaging in conflict, whether internal or external, was to become an inherent part of the State governance role.[4] This makes the State a common focal point for an analysis of the authority to engage in either international or noninternational armed conflict. It also provides a point of reference for considering not only how the State recourse to war and humanitarian law intersect, and interacts, but also more broadly their interface with human rights law.

---

[1] Steven Pinker, The Better Angels of Our Nature: Why Violence Has Declined 680 (2011).

[2] James Turner Johnson, The Holy War Idea in Western and Islamic Traditions 88 (2001).

[3] Hugo Grotius, 1 The Rights of War and Peace 240 (2005) ("a publik War, which is made on each Side by the Authority of the Civil Power.").

[4] *Id.* at 162 ("The Civil Power is that which governs the *State*.").

This chapter explores the impact of the special status provided to States as a proper authority regarding its effect on the application of international humanitarian and human rights law. The criminalization of non-State actor participation in both international and internal conflict permeates this normative discussion highlighting the role of human rights in contemporary conflict. In doing so, it will be established that State control over the use of force introduces and mandates consideration of human rights norms during noninternational armed conflict. The linkage between State governance and human rights will be seen to impact directly on the historical reluctance by States to have humanitarian law regulate internal hostilities. While international law has made greater inroads in regulating internal conflict since World War II, the State preference largely remains to apply domestic law in dealing with terrorist and insurgent challengers. This in turn has meant that human rights law is usually the normative framework of choice for countering internal non-State actor challenges to the State governance role. This is followed by a discussion of State efforts to fill a perceived void, or at least uncertainty, regarding the application of humanitarian law to noninternational armed conflicts. This is done through means of a policy approach that applies the principles of that body of law on all international operations, which include conflict with non-State actors. Finally, the issue of "ungoverned spaces" will be explored since it is from those areas that non-State actors often present the greatest threat to recognized States.

4.2. Humanitarian Law and "Proper Authority"

The State, as the proper authority, enjoys a privileged position under humanitarian law. The simplicity of that statement masks the profound and far-reaching effect this special status has had on the law governing the conduct of hostilities. For example, the just war principle of "proper authority" directly impacts one of the foundational humanitarian law principles: the distinction between those who fight and civilians who do not. In this regard, "[t]he separation of armies and peaceful inhabitants into two distinct classes is perhaps the greatest triumph of International Law."[5]

However, only those who fight on behalf of States enjoy combat immunity, have the right to take a direct part in hostilities, and are granted prisoner of war (POW) status. These legitimate combatants are distinguished from civilians who must be protected at all times unless and for such time as they take a direct part in hostilities.[6] Simply stated, the privileging of State participants over their non-State protagonists creates an imbalance that permeates humanitarian law. Some participants, no matter the degree to which they comply with the requirements of humanitarian law, will not be eligible for POW status solely because of their membership in a group not affiliated with or belonging to a State. This approach undermines to a certain degree the equal application principle upon which the operation of humanitarian law is based. In this respect, Emily Crawford has noted the uniform application of the law "cannot be achieved if certain fighters have specific rights and

[5] James Molony Spaight, War Rights on Land (1911) 37 (1911). *See also* AP I, arts. 48, 50, 51, 52(2), and 57.

[6] AP I, arts. 50 and 51.

responsibilities, while others do not."[7] In reality, the law only applies equally if you are a member of the long-standing privileged class of warriors that fights for States.[8] Historically, those who failed to practice open and public warfare were viewed as murderers and brigands.[9] The continuing operation of the just war proper authority principle highlights that claims of a completely formal separation between the law governing the recourse to war and humanitarian law are in some fundamental respects difficult to sustain.

It is not only classic terrorist and isolated insurgent groups that are affected. The development of humanitarian law, beginning in its initial stages in the late nineteenth century, witnessed a struggle among States over which persons fighting on their behalf may legitimately participate in conflict and which may not.[10] At the turn of the twentieth century there was a significant disagreement with dominant military powers pitted against those concerned with potentially being invaded. During the deliberations leading up to the adoption of the 1907 Hague Land Warfare Regulations, the dominant military State of the day, Prussia, sought to restrict less powerful "patriotic" States in their championing of the traditional levée en masse and authority for individual citizens to repel invading forces.[11] The result was a limited compromise in which the 1907 Hague Land Warfare Regulations provided belligerent status to armies and militia that met specific criteria. Less dominant military powers gained protection for inhabitants of a territory that is not occupied, "who on the approach of the enemy, spontaneously take up arms to resist the invading troops without having time to organize themselves."[12] Those who were excluded could be dealt with as illegitimate actors. Ultimately, this required the consideration of other law, such as domestic law, in addition to humanitarian law, when dealing with these nonprivileged participants.

At least up to the mid twentieth century those who fought without appropriate authority, thereby not meeting the requirements of belligerency, continued to be viewed as robbers, bandits, and war criminals.[13] This ultimately meant being dealt with some form of detention and trial process. World War II did see a subtle change in approach toward unprivileged participation in hostilities largely because the Allied Powers had relied so heavily

---

[7] Emily Crawford, The Treatment of Combatants and Insurgents under the Law of Armed Conflict 48 (2010) ("Uniform application of the law requires uniform treatment of all participants in armed conflict.").

[8] G. I. A. D. Draper, *Combatant Status: A Historical Perspective*, 11 Mil. L. & L. War Rev. 135, 140 (1972). *See also* Department of Defense Law of War Manual, Office of the General Counsel ¶ 5.5.5, at 88 (June 2015) [hereinafter DOD Law of War Manual], http://www.dod.mil/dodgc/images/law_war_manual15.pdf.

[9] G. I. A. D. Draper, *The Status of Combatants and the Question of Guerrilla Warfare*, 24 Brit. Y.B. Int'l L. 172, 173 (1971) (for a discussion of the development of a privileged class of warriors finding its roots in the Codes of Chivalry of the Middle Ages [the jus militaire]), Francis Lieber, *Guerrilla Parties Considered with Reference to the Law and Usages of War, in* Richard Shelly Hartigan, Lieber's Code and the Law of War 34–37 (1983), and Emer de Vattel, The Law of Nations § 68, at 508 (Knud Haakonssen ed., 2008) (1797).

[10] Kenneth Watkin, *Warriors Without Rights? Combatants, Unprivileged Belligerents, and the Struggle over Legitimacy*, 2 Occasional Paper Series (Program on Humanitarian Policy and Conflict Research Harvard University, Winter 2005), http://www.hpcrresearch.org/sites/default/files/publications/OccasionalPaper2.pdf.

[11] Spaight, *supra* note 5, at 48–54.

[12] 1907 Hague Land Warfare Regulations, art. 2. *See also* Spaight, *supra* note 5, at 54–55.

[13] Manual of Military Law ¶ 19, at 238 (His Majesty's Stationary Office 1916) (1914) [hereinafter Manual of Military Law 1914] ("If [peaceful inhabitants]... make an attempt to commit hostile acts, they are not entitled to the rights of armed forces, and are liable to execution as war criminals."), and Lester Nurick & Roger W. Barrett, *Legality of Guerrilla Forces Under the Laws of War*, 40 Am. J. Int'l L. 563 (1946).

on resistance movements during the war.[14] In 1942, in *Ex Parte Quirin*, the U.S. Supreme Court equated unlawful combatants to spies who are not illegal under international law, although it also indicated that such combatants were "offenders against the law of war."[15] By 1950, in *The Hostage Case*, the Nuremburg Tribunal noted a person "may act lawfully for his country and at the same time be a war criminal to the enemy, so guerrillas may render great service to their country and, in the event of success, become heroes even, still they remain war criminals in the eyes of the enemy and may be treated as such."[16] However, of note, a criminal aura was still attached to such participation at least in terms of what action the opposing power could take against those participants.

Concern over the activities of non-State actors was reflected in the limits on how far States were willing to go at the end of World War II in "legitimizing" the involvement of organized resistance movements. Part of that concern may have been based on the fact that a number of the resistance movements during that conflict were not necessarily fighting to re-establish the displaced regime, but rather install a new government.[17] In any event, the conditions placed on organized resistance movements in the Third Geneva Convention for attaining combatant and therefore POW status have long been viewed as being unrealistic.[18] Faced with such limitations, Richard Baxter, who advocated for enhanced protection, equated participation in guerrilla warfare to spying, and suggested "it is highly unreal to regard them as international criminals."[19] The result, it was suggested, was that "'[u]nlawful belligerency' is actually 'unprivileged belligerency'."[20] However, while international law does not condemn these participants in warfare, neither does it offer them protection from prosecution under the domestic laws of the capturing power.[21] This has led to a generally accepted view that such captured belligerents have not committed an international crime, however, they may, like spies, be prosecuted under the domestic laws of the capturing State.[22] Notwithstanding this viewpoint, the issue of whether unprivileged

---

14 Richard R. Baxter, *So-called "Unprivileged Belligerency": Spies, Guerrillas, and Saboteurs*, *in* HUMANIZING THE LAWS OF WAR: SELECTED WRITINGS OF RICHARD BAXTER 37, 48 (Detlev F. Vagts et al. eds., 2013) (originally published in 1951 BRIT. Y.B. INT'L L. 323) ("Only a rigid formalism could lead to the characterization of the resistance against Germany, Italy, and Japan as a violation of international law.").

15 Ex Parte Quirin 317 U.S. 1, 31 (1942).

16 The Hostage Case, Trials of War Criminals, Washington: Government Printing Office 759, 1245 (1950).

17 John Shy & Thomas W. Collier, *Revolutionary War*, *in* MAKERS OF MODERN STRATEGY: FROM MACHIAVELLI TO THE NUCLEAR AGE 815, 833 (Peter Paret ed., 1986) (e.g., Tito's partisans in Yugoslavia).

18 GC III. *See also* GERHARD VON GLAHN, THE OCCUPATION OF ENEMY TERRITORY 51–52 (1957), Colonel G. I. A. D. Draper, *The Legal Classification of Belligerent Individuals*, *in* REFLECTIONS ON THE LAW AND ARMED CONFLICT 196, 201 (Michael A. Meyer & Hilaire McCoubrey eds., 1998) (where it was noted "[i]f memory be short, so is gratitude."), and HOWARD LEVIE, PRISONERS OF WAR IN INTERNATIONAL ARMED CONFLICT, 59 INT'L L. STUD. 39–40 (1977).

19 Baxter, *supra* note 14, at 50.

20 *Id.* at 55.

21 *Id.* at 51 ("What formulation of law is necessary to permit his punishment is essentially a matter of domestic law or practice.").

22 YORAM DINSTEIN, THE CONDUCT OF HOSTILITIES UNDER THE LAW OF INTERNATIONAL ARMED CONFLICT ¶¶ 83–93, at 35–39 (2nd ed. 2010), ALLAN ROSAS, THE LEGAL STATUS OF PRISONERS OF WAR: A STUDY IN INTERNATIONAL HUMANITARIAN LAW APPLICABLE IN ARMED CONFLICT 305 (1976), and LESLIE GREEN, THE CONTEMPORARY LAW OF ARMED CONFLICT 176 (3rd ed. 2008). *See also* DOD LAW OF WAR MANUAL, *supra* note 8, at ¶ 3.4.1.2., at 103 ("members of a non-State group may be subject to prosecution under a State's domestic law for their participation in hostilities against it."), and *id.* at ¶ 17.4.1.1, at 1025.

belligerents are to be treated as "war criminals" has been resurrected by the United States in its post-9/11 establishment of military commissions to prosecute "enemy unprivileged belligerents" for such crimes.[23] This is a clear contemporary example of a continuing dominant State resistance to providing any sense of legitimacy to those who do not fight on behalf of a State.

A softer approach toward unlawful belligerency was reflected in the establishment of a security detention regime, including internment, in the Fourth Geneva Convention, which mirrors the framework applicable to prisoners of war. This was an approach that had been applied to the detention of civilians in Britain during World War I.[24] The Fourth Geneva Convention detention regime was not linked to "legitimacy" in the sense that detained persons threatening the security of the occupying power were provided conditions and standards of treatment that were effectively the same as POWs' without their having to be a member of an armed force of a State. Further, although it was clear that saboteurs "should cease to be entitled to treatment provided for law abiding citizens,"[25] and certain restrictions were applied to spies, saboteurs, or persons under suspicion of activity hostile to the security of an occupying power, they nevertheless had to be treated with humanity and provided the rights of a fair and regular trial.[26]

Although the security detention regime does not have universal application, with some captured participants in conflict still falling outside the jurisdiction of the Fourth Geneva Convention,[27] it continues to provide a useful policy template for the handling of any security detainees. Somewhat as an aside in a book about the use of force, it needs to be stated that this is one area where a straightforward and, importantly, humane policy decision to apply the internment and treatment provisions of the Fourth Geneva Convention could have averted much of the post-9/11 controversy regarding detainees.[28] That the clear proscriptions against torture and the abuse of detainees found in both international humanitarian law and human rights law were not observed by some elements of the U.S. government is, in a word, unconscionable.

The continuing impact of this proper authority–based struggle over legitimacy can be seen in the resistance exhibited by States toward legitimizing non-State participation in international armed conflict, notwithstanding the 1977 expansion of the categories of

---

[23] Military Commission Act of 2009, Pub. L. No. 111-84, § 948(a)(7), 123 Stat. 2190 (codified at 10 U.S.C 47A (2006)) (for the definition of "unprivileged enemy belligerent"). *See also* DOD LAW OF WAR MANUAL, *supra* note 8, at ¶ 4.3., at 103.

[24] *See* the quote from Judge Low in The King v. Superintendent of Vine Street Station, 1 K.B. 277-278 (1916), *in* WILLIAM E. S. FLORY, PRISONERS OF WAR 25–26 (1942).

[25] 11 FINAL RECORD OF THE DIPLOMATIC CONFERENCE OF GENEVA OF 1949 § A, at 621 (1949), *available at* http://www.loc.gov/rr/frd/Military_Law/pdf/Dipl-Conf-1949-Final_Vol-2-A.pdf.

[26] GC IV, art. 5. *See also* GC IV, art. 76 (In occupied territory persons who were "accused of offences" must, if possible, be separated from other detainees and be kept under specified conditions "at least equal to those obtaining to prisons in the occupied territory.").

[27] Baxter, *supra* note 14, at 41. *See also* THE MANUAL OF THE LAW OF ARMED CONFLICT, United Kingdom Ministry of Defence ¶ 11.2, at 279 (2004).

[28] JEREMY SCAHILL, DIRTY WARS: THE WORLD IS A BATTLEFIELD 26–30 (2013) (outlining the development of the rendition and "black site" detention program applied by the administration of President George W. Bush). *See also* ALFRED W. MCCOY, A QUESTION OF TORTURE: CIA INTERROGATION FROM THE COLD WAR TO THE WAR ON TERROR 219–25 (2006) (for reference to the struggle between the neoconservative political appointees and U.S. military lawyers who resisted the imposition of treatment other than those required under the Geneva Conventions.).

legitimate actors[29] to include members of national liberation movements.[30] The idea that only States may legitimately take part in international armed conflict was directly challenged by the development of Additional Protocol I, which encompassed those movements. However, even though 89 percent of States have formally adopted its provisions, there remains a strong resistance exhibited toward recognizing the legitimacy of non-State actor participation in armed conflict.[31] This resistance is most obviously reflected in decision of a number of States, such as the United States, to not become parties to that Protocol.[32] On its face this suggests the most powerful military States may be isolated in the stance they have taken against such recognition. However, it is also subtly reflected in the emphasis placed by others on interpretations of the law that non-State actors are not likely to be able to meet the criteria for a national liberation movement.[33] For State Parties, such as Canada, who have adopted Additional Protocol I, it is relatively easy to agree to its provisions if its scope is ultimately seen to be limited to a historically unique and discrete group of non-State actors.

Further, notwithstanding its codification in Additional Protocol I, there has been a tendency to omit any reference to non-State actors being potential Parties to international armed conflict in general statements of the law. International armed conflict is widely referred to as simply being a conflict between States.[34] The impact of Additional Protocol I has also been somewhat ambiguously explained as creating "situations in which LOIAC [Law of International Armed Conflict] will apply in intra-State armed conflicts *as if they were international in character*. . . ."[35] On one level this interpretation appears to situate Additional Protocol I–based conflicts between States and non-State actors within the category of noninternational armed conflicts, even if they have an international character. However, as will be discussed in Chapter 9, the connection between conflicts of an "international character" and noninternational armed conflict is reflective of a broader struggle international law has experienced in categorizing non-State actor conflict (see Chapter 9, section 9.3, *Categorizing Conflict with Non-State Actors*). The other situation in which internal armed conflict is identified as being "treated as if it were an international armed conflict" is where there is recognition of belligerency from the central government.[36]

---

[29] AP I, arts. 44.

[30] AP I, art. 1(4).

[31] *State Parties to the Following International Humanitarian Law and Other Related Treaties as of 6 Jan 2014*, International Committee of the Red Cross, http://www.icrc.org/applic/ihl/ihl.nsf/vwTreaties1949.xsp?redirect=0 [hereinafter API States Parties] (where it is indicated there are 174 States Party to AP I and 168 States Party to AP II), and *Growth in United Nations Membership, 1945–present*, United Nations, http://www.un.org/en/members/growth.shtml (last visited June 5, 2015) (where it is identified there are 193 member States in the United Nations).

[32] Ronald Reagan, *Letter of Transmittal*, 81 Am. J. Int'l L. 910–12 (1987). Other States that are not Parties to AP I include India, Pakistan, Iran, South Korea, North Korea, and Israel. *See* AP I States Parties, *supra* note 31.

[33] George H. Aldrich, *Prospects for United States Ratification of Additional Protocol I to the 1949 Geneva Conventions*, 85 Am. J. Int'l L. 1, 4–6 (1991); Hans-Peter Glasser, *An Appeal for Ratification by the United States*, 81 Am. J. Int'l L. 912, 916–17 (1987); and Theodor Meron, *The Time Has Come for the United States to Ratify Geneva Protocol I*, 88 Am. J. Int'l L. 678, 683 (1994). *See also* Christopher Greenwood, *Terrorism and Humanitarian Law—The Debate over Additional Protocol I*, 19 Isr. Y.B. Hum. Rts. 187, 194–95 (1989).

[34] Yoram Dinstein, War Aggression and Self-Defence ¶ 8, at 6 (5th ed. 2011).

[35] Dinstein, *supra* note 22, ¶ 68(i), at 28 (emphasis added).

[36] *Id.* ¶ 68(ii), at 28. *See* Sandesh Sivakumaran, The Law of Non-International Armed Conflict 9–20 (2012) (for a discussion of the recognition of belligerency principle). *See also* Yoram Dinstein, Non-International Armed Conflicts in International Law ¶¶ 338–54, at 108–14 (2014).

The problem with an interpretation of Additional Protocol I suggesting qualified non-State actors are still participating in a noninternational armed conflict is that Article 1(4) specifically indicates that the situations referred to in Common Article 2 to the 1949 Geneva Conventions (i.e., international armed conflict) include those involving what are commonly referred to as national liberation movements. However, States continue to jealously guard their unique status. It is a status that finds clear recognition in Article 3 of Additional Protocol II dealing with civil wars, which specifically refers to it being the responsibility of governments to maintain or re-establish law and order in the State or defend national unity and the territorial integrity of the State (see Chapter 2, section 2.4.2, *Internal Conflicts with an "International Character"*).[37] This provision was added to the Protocol to "quieten the anxieties of certain States which were hesitant about adopting rules for conflicts of a non-international character at all. . . ."[38] It is evident that the debate carried out by the international legal community remains far removed from Richard Baxter's 1951 vision that "it is possible to envisage a day when the law will be so tailored as to place all belligerents, however garbed, in a protected status."[39]

4.3. Impact of the Proper Authority Principle

As has been noted, the preeminent status for the State as "the" legitimate participant in armed conflict has a strong historical basis. Significantly, this traditional State (i.e., proper authority) bias against unprivileged participation in hostilities virtually guarantees the application of human rights norms for the treatment of these "criminals." In occupied territory such prosecution may occur before "non-political military courts" of the occupier sitting in the occupied territory as contemplated under Articles 64 and 66 of the Fourth Geneva Convention. Again, notwithstanding the international legal basis for such activity, and the legal guarantees set out in that Convention, the national laws governing the courts will also be relevant. Such involvement of domestic laws and courts is not unique as it has been noted that, in respect of war crimes, "[i]international crimes are primarily intended to be prosecuted at the domestic level."[40] This reference to domestic courts further reflects the cross-boundary application of international criminal law.

With participation without State authority being criminalized in international armed conflict, it should be no surprise that such action is also unlawful in respect of conflicts of a noninternational character. The invocation of domestic law has particular relevance to those conflicts since there is no combatant immunity at all for non-State actors, or

---

[37] AP II, art. 3. *See also* Commentary on the Additional Protocols of 8 June 1977 to the Geneva Conventions of 12 August 1949, art. 3, ¶ 4500, at 1362 (Yves Sandoz et al., 1987), *available at* http://www.cicr.org/ihl.nsf/COM/470-750073?OpenDocument ("[t]his is the responsibility of governments and is expressly recognized here.").

[38] Michael Bothe, Karl Josef Partsch, & Waldemar A. Solf, New Rules for Victims of Armed Conflicts: Commentary on the Two 1977 Protocols Additional to the Geneva Conventions of 1949, at 633 (1982).

[39] Baxter, *supra* note 14, at 56.

[40] Robert Cryer et al., An Introduction to International Criminal Law and Procedure 64 (2nd ed. 2010).

POW status for any belligerent.[41] In contrast, security forces who fight on behalf of a State remain privileged participants even in noninternational armed conflict, although they would be subject to prosecution for breaches of their own national law or humanitarian law.[42] Those who fight for non-State actors against a State are normally considered to be criminals subject to domestic prosecution.[43] Much like unprivileged belligerents in international armed conflict, "[w]hile the law does not expressly prohibit or criminalize participation in non-international armed conflict, international law does not immunize such participation from the operation of domestic law."[44]

This aspect of noninternational armed conflict is an area where humanitarian law has had limited reach. States seeking to reinforce the criminal nature of non-State actors; human rights advocates uniquely championing that body of law; and the technical textual limitation of the 1949 Geneva Conventions to international armed conflict have stilted the expansion of the protection for detainees through the means of humanitarian law in internal conflict. As a result the discussion in the aftermath of 9/11 has, not surprisingly, often centered on whether human rights law applied to their treatment.[45] Further, law enforcement methods are often used to counter these "criminals." For example, in declaring an emergency in May 2013 in respect of the insurgent and terrorist threat posed by Boko Haram, the Nigerian president indicated that security forces would take necessary action, including exercising the authority to arrest and detain suspects, conduct searches, and apprehend persons in illegal possession of weapons.[46] A categorization as criminals throws the door wide open for consideration of human rights norms, both international and domestic, regarding the treatment and prosecution of persons who do not fight for a proper authority.

International lawyers tend to view the treatment and trial of these captured non-State actors in terms of the interface between international human rights and humanitarian law.[47] However, the prosecution of detained Al Qaeda and Taliban personnel under U.S. national law, which is subject to American constitutional oversight, highlights the domestic legal aspect of such legal proceedings.[48] The result is that consideration of the law

---

[41] LIESBETH ZEGVELD, THE ACCOUNTABILITY OF ARMED OPPOSITION GROUPS IN INTERNATIONAL LAW 36 (2002), and CRAWFORD, *supra* note 7, at 78.

[42] Watkin, *supra* note 10, at 65. *See* Ian Henderson, *Civilian Intelligence Agencies and the Use of Armed Drones*, 13 Y.B. INT'L HUM. L. 133. 149–50 (2010), and Kenneth Anderson, *Readings: Civilian Intelligence Agencies and the Use of Armed Drones by Ian Henderson*, LAWFARE (June 27, 2014, 3:00 PM), http://www.lawfareblog.com/2014/06/readings-civilian-intelligence-agencies-and-the-use-of-armed-drones-by-ian-henderson/ (where the same position is adopted). *See also Memorandum for the Attorney General, Re: Applicability of Federal Criminal Laws and the Constitution to Contemplated Lethal Operations Against Shaykh Anwar al-Aulaqi*, Office of the Assistant Attorney General, July 16, 2010, at 33–34, https://www.aclu.org/sites/default/files/assets/2014-06-23_barron-memorandum.pdf.

[43] LINDSAY MOIR, THE LAW OF INTERNAL ARMED CONFLICT 60 (2002).

[44] CRAWFORD, *supra* note 7, at 78. *See also* DOD LAW OF WAR MANUAL, *supra* note 8, *id.* at ¶ 17.4.1, at 1025 ("a State may use not only its war powers to combat non-State armed groups, but it may also use its domestic law, including its ordinary criminal law, to combat non-State armed groups.").

[45] John B. Bellinger & Vijay M. Padmanabhan, *Detention Operations in Contemporary Conflicts: Four Challenges for the Geneva Conventions and Other Existing Law*, 105 AM. J. INT'L L. 201 (2011) (for a discussion of the debate regarding the treatment of detainees in the aftermath of 9/11.).

[46] *Nigeria: Read Full Text of Jonathan's Declaration of Emergency Rule*, VANGUARD, May 14, 2013, ¶ 14, http://www.vanguardngr.com/2013/05/read-full-text-of-jonathans-declaration-of-emergency-rule/.

[47] *See, e.g.*, CRAWFORD, *supra* note 7, at 118–51 (discussing the role of human rights law).

[48] *See, e.g.*, Hamdan v. Rumsfeld, 45 I.L.M. 1130 (2006) (U.S. Supreme Court 2006).

applicable to the treatment and trial of unprivileged belligerents, in both international and noninternational armed conflict, inevitably involves an assessment not only of international humanitarian and human rights law, but domestic law as well. For example, in *Hamdan v. Rumsfeld*, the U.S. Supreme Court considered both international humanitarian law and U.S. military law in determining that the U.S. Commissions as originally constituted were unconstitutional.[49]

Requiring that only those operating under the proper authority (i.e., the State) can legitimately participate in hostilities is also an integral part of the discussion as to what law generally applies to conflicts not of an international character and other situations where the State is put in a situation of governance. The analysis now turns to how the linkage between the recourse to war principle of "proper authority" and governance sets the scene for a discussion regarding the application of human rights law and, ultimately, its interface with international humanitarian law.

4.4. The State and Governance

4.4.1. PUBLIC, PRIVATE, AND MIXED WAR

As society developed and became organized, the authority to use violence was concentrated first in the hands of the sovereign. Linked directly to being the appropriate authority to use violence, this was the basis for Grotius's distinction between public, private, and mixed war.[50] The former referred to war between sovereigns, private war involved fighting between private individuals, and its mixed version concerned conflict "made on one Side by public Authority, and on the other by mere private Persons."[51] These categories resonate with the contemporary terminology of international armed conflict, domestic law enforcement, and "armed conflict not of an international character." As has been noted, this authority finds modern recognition in Article 3 of Additional Protocol II, where the State is identified as having responsibility for maintaining, or re-establishing internal order, and defending national unity and territorial integrity (see section 4.2, *Humanitarian Law and "Proper Authority,"* above). Those specific State responsibilities were referred to because it was felt the concept of sovereignty was open to various interpretations, and as a result "more concrete examples of the exercise of sovereignty were given . . ."[52]

The exclusive role allotted to the governing authority to suppress all types of war was an integral part of the development of the post-Westphalian State.[53] However, a key distinction between inter-State and private conflict was the prohibition placed on private individuals using violence in a domestic context other than in the exercise of self-defense.[54] In effect, it

---

[49] *Id.*

[50] GROTIUS, *supra* note 3, at 240. *See also* DE VATTEL, *supra* note 9, § 2, at 469 (for reference to public and private war).

[51] GROTIUS, *supra* note 3, at 240.

[52] BOTHE, PARTSCH & SOLF, *supra* note 38, at 633 (this was to avoid infringement on the sovereignty of the State, not only by other States "but also international or private organizations.").

[53] IAN BROWNLIE, INTERNATIONAL LAW AND THE USE OF FORCE BY STATES 3–13 (1963). *See also* MARTIN VAN CREVELD, THE RISE AND DECLINE OF THE STATE 156–64 (1999).

[54] GROTIUS, *supra* note 3, at 240–42.

was the "outlawing" of internal violence. With the centralization of the power in the hands of the State came the development of separate police forces for internal security and prisons to house those who breached the law.[55] The power exercised by the sovereign resonates with the idea that the law governing internal armed conflict is centered on the State. There is a clear connection between the creation of the State, with its role of internal governance, and the development of international law regulating the relationship among States. In this regard, "[i]t was the evolution of the concept of an international community of separate, sovereign, if competing states, that marks the beginning of what is understood by international law."[56]

Unlike its inter-State counterpart, internal conflict is a form of war for which no "balance of power" doctrine developed. The State and the status of those being governed remained unequal. Private individuals had recourse to the laws and the courts established by the sovereign to resolve disputes.[57] It was through the development of courts, police, prisons, State-run education systems, and social legislation governing work and health that State authorities maintained control over populations under their jurisdiction.[58] It is the success of States in imposing internal order within "civilized" society that has been identified as a major reason for the decline in overall mortality rates, although that obviously has not meant an end to all internal violence.[59] The framework within which force is used internally to maintain public order is most directly associated with law enforcement. It is domestic law, including human rights law, that has played the most direct and pervasive role in maintaining order and regulating conflict within a State.

The concentration of this power in the hands of State authorities raised other concerns. Danger remained from the "imposition of absolute control by coercive government."[60] Human rights developed to protect individuals from the potential abuse of power by the State. The unique and integral relationship between the State and human rights is reflected in the fact that the State is "both the guardian of basic rights and . . . the behemoth against which one's rights need to be defended."[61] In this respect, human rights are concerned with the exercise of State power in relation to the individual. These human rights "found their natural expression in domestic constitutional law."[62] As will be discussed, the application of these norms, albeit not necessarily universally applied in all states, predated modern treaty-based human rights law (see Chapter 5, section 5.3.1, *Prior to the Mid-Twentieth Century*). They continue to constitute part of the default legal framework governing internal conflicts. Given the importance of the role played by human rights and the need for a State to govern even during armed conflict, it is also evident that domestic human rights law cannot be simply discarded in favor of humanitarian law when violence increases within a State.

---

55 Van Creveld, *supra* note 53, at 165–70.

56 Malcolm N. Shaw, International Law 21 (6th ed. 2008).

57 Grotius, *supra* note 3, at 240–41.

58 Van Creveld, *supra* note 53, at 205–24.

59 Azar Gat, War in Human Civilization 408 (2006).

60 Pinker, *supra* note 1, at 58.

61 Micheline R. Ishay, The History of Human Rights: From Ancient Times to the Globalization Era 8 (2004)

62 Robert Kolb, *The Relationship Between International Humanitarian Law and Human Rights Law: A Brief History of the 1948 Universal Declaration of Human Rights and the 1949 Geneva Conventions*, 38 Int'l Rev. Red Cross 409, 410 (1998).

4.4.2. GOVERNANCE

The link between the State as the right authority to use force and the exercise of governance raises the question of what it means to "govern." Martin Van Creveld has indicated that government and the State are not the same. He suggests:

> The former [government] is a person or group which makes peace, wages war, exacts laws, exercises justice, raises revenue, determines currency, and looks to the internal security on behalf of society as a whole, all the while attempting to find a focus for people's loyalty and, perhaps, a modicum of welfare as well. The latter [the State] is merely one of the forms which, historically speaking, the organization of government has assumed, and which, accordingly, need not be considered eternal and self-evident any more than were previous ones.[63]

However, as James Crawford notes, there is a close relationship "between the concepts of government and statehood."[64] There is also "a good case for regarding government as the most important single criterion of statehood since all others depend upon it."[65] It is clear the State exists to "govern." An "effective government, with central administrative and legislative organs is the best indication of a stable political community," and such a community "supporting a legal order, in a certain area" provides "[t]he shortest definition of a state."[66] Importantly, "international law defines 'territory' not by adopting private law analogies but by reference to the extent of governmental power exercised, or capable of being exercised, with respect to some territory and population."[67]

At its simplest, governance means to "control, influence, or regulate (a person, action, or course of events)."[68] It has been suggested that "governance (as opposed to 'good' governance) can be defined as the rule of the rulers, typically within a given set of rules."[69] Good governance, on the other hand "is defined in terms of the *mechanisms* thought to be needed to promote it," which in various places "has been associated with democracy and good civil rights, with transparency, with the rule of law, and with efficient public services."[70] When considered in its broadest sense, governance is associated with "the different ways that organizations, institutions, businesses, and governments manage their affairs."[71] To

---

[63] VAN CREVELD, *supra* note 53, at 415.

[64] JAMES CRAWFORD, THE CREATION OF STATES IN INTERNATIONAL LAW 34 (2nd ed. 2006).

[65] *Id.* at 56.

[66] IAN BROWNLIE, PRINCIPLES OF PUBLIC INTERNATIONAL LAW 71 (7th ed. 2008).

[67] CRAWFORD, *supra* note 64, at 56.

[68] OXFORD DICTIONARY, http://www.oxforddictionaries.com/definition/english/govern?searchDictCode=all (last visited Dec. 18, 2014).

[69] *What Is Governance?*, The World Bank, *available at* http://web.worldbank.org/WBSITE/EXTERNAL/COUNTRIES/MENAEXT/EXTMNAREGTOPGOVERNANCE/0,,contentMDK:20513159~pagePK:34004173~piPK:34003707~theSitePK:497024,00.html (last visited May 1, 2013).

[70] *Id.* (Good governance has also been associated with specific outcomes such that people have a voice, and they receive just and equitable treatment).

[71] *Global Governance*, World Health Organization, http://www.who.int/trade/glossary/story038/en/ (last visited Dec. 20, 2014).

govern ultimately involves the exercise of control.[72] It is in the hands of the State, either directly through the exercise of the sovereign right of wielding authority over all individuals living in its territory,[73] or indirectly through membership in the United Nations, where that power is concentrated.[74] Since there is no international government, "global governance typically involves a range of actors including states, as well as regional and international organizations."[75] In an international law context, despite periodic predictions of its demise and the proliferation of other actors on the international scene, the State remains the central and most effective instrument of governance.

The most obvious exercise of State governance is in respect of its own sovereign territory.[76] However, it is not limited to circumstances where it is the lawful power. A responsibility for governance can occur when operating in a territory where there is no functioning government. For example, with respect to belligerent occupation that responsibility is reflected in numerous provisions of the Fourth Geneva Convention. It is found in the 1907 Hague Land Warfare Regulations Article 43 obligation of the Occupier to "take all the measures in his power to restore, and ensure, as far as possible, public order and safety, while respecting, unless absolutely prevented, the laws in force in the country," and in Article 64 of the Fourth Geneva Convention, which states "[t]he Occupying Power may, however, subject the population of the occupied territory to provisions which are essential to enable the Occupying Power to fulfill its obligations . . . to maintain the orderly government of the Territory . . . ."

Responsibilities for governance are further reflected in provisions of the Fourth Geneva Convention, which refer to the maintenance in force of penal laws,[77] the effective administration of justice,[78] the maintenance of the status of public officials or judges,[79] ensuring and maintaining medical, public health, and hygiene,[80] and the provision of food and medical supplies.[81] The effect of these international law provisions is that there is a privileging of the practical exercise of the governance by the occupying State in the stead of the displaced lawful government. This responsibility of governance is placed in the hands of the occupier even though "[o]ccupation does not create any change in the status of the territory . . . . The former sovereign remains sovereign and there is no change in the nationality of the inhabitants." [82]

---

[72] OXFORD DICTIONARY, *supra* note 68.

[73] ANTONIO CASSESE, INTERNATIONAL LAW 49 (2nd ed. 2005).

[74] BROWNLIE, *supra* note 66, at 79 (where it is noted that Article 4 of the UN Charter indicates membership in that organization is open to States and "[a]dmission to membership is to be by decision of the General Assembly upon the recommendation of the Security Council.").

[75] *Global Governance*, *supra* note 71.

[76] BENJAMIN WITTES & GABRIELLA BLUM, THE FUTURE OF VIOLENCE: ROBOTS AND GERMS, HACKERS AND DRONES 95 (2015) ("The quintessential modern state exists when a government controls a bounded territory with nearly exclusive security and policing powers.").

[77] GC IV, art. 64.

[78] *Id.*

[79] GC IV, art. 54.

[80] GC IV, art. 56.

[81] GC IV, art. 55.

[82] GREEN, *supra* note 22, at 285. *See also* RULES OF LAND WARFARE 1914 ¶ 287, at 105 (photo, reprint 2009) (1917).

In situations of occupation the legal obligation to "govern" is established once the factual requirements of Article 42 of the Hague Land Warfare Regulations are met:

> Territory is considered occupied when it is actually placed under the authority of the hostile army. The occupation extends only to the territory where such authority has been established and can be exercised.

As Leslie Green indicated, "[t]erritory is occupied only when it is actually under the control and administration of an occupant and extends only to those areas in which he is actually able to exercise such control."[83]

The International Committee of the Red Cross (ICRC) has traditionally favored a broad test for establishing the existence of an occupation, extending the provisions of the Fourth Geneva Convention to the invasion phase even before a stable regime of occupation is established.[84] This would include situations when a patrol "penetrates into enemy territory without any intention of staying there."[85] However, it appears this historical approach may be undergoing some change to the extent there has been an acknowledgment that "Article 42 of the *1907 Hague Regulations* can be regarded as the only legal basis on which the determination of the existence of a state of occupation can be made."[86] A key element for the determining the establishment of an occupation is whether the State is exercising "effective control."[87] Once again there is a clear link between a State exercising control and the notion of governance.

Not only is effective control important in terms of establishing the existence of an occupation, that term has also been relied on by the European Court of Human Rights as a test for the determining the extraterritorial application of treaty-based human rights provisions.[88] It is no coincidence that the same form of test, effective control, would be applied in the context of a State exercising its human rights responsibilities extraterritorially. As is reflected in the Fourth Geneva Convention, even in the situation of international armed conflict, the exercise of effective control is intimately linked to human rights and law enforcement. Since "IHL [international humanitarian law] regulates some conflicts with non-state actors, the complementary nature of different bodies of international

---

[83] *Id. See* YORAM DINSTEIN, THE INTERNATIONAL LAW OF BELLIGERENT OCCUPATION ¶86–95, at 38–42 (2009).

[84] THE GENEVA CONVENTIONS OF 12 AUGUST 1949: COMMENTARY, (IV) GENEVA CONVENTION RELATIVE TO THE PROTECTION OF CIVILIAN PERSONS IN TIME OF WAR, Art. 6, at 60 (Jean S. Pictet ed., 1958).

[85] *Id.*

[86] Tristan Ferraro, *Determining the Beginning and End of an Occupation under International Humanitarian Law*, 94 INT'L REV. RED CROSS 133, 139 (2012) and INTERNATIONAL HUMANITARIAN LAW AND THE CHALLENGES OF CONTEMPORARY ARMED CONFLICT, Report 32nd International Conference of the Red Cross and Red Crescent EN 32IC/15/11, at 11-12 (2015), *available at* https://www.justsecurity.org/wp-content/uploads/2015/11/2015-ICRC-Report-IHL-and-Challenges-of-Armed-Conflicts.pdf. *See also* Sargsyan v. Azerbaijan, Eur. Ct. H. R., Application No. 40167/06, ¶ 94, at 28 (June 16, 2015), http://hudoc.echr.coe.int/eng?i=001-155662.

[87] *Id.* at 143.

[88] *See, e.g.*, Al-Skeini v. The United Kingdom, IV Eur. Ct. H. R. Rep. 99, ¶¶ 139–40, at 169–70 (2011), Hassan v. The United Kingdom, Eur. Ct. H. R., Application No. 29750/09, ¶ 74, at 39–40 (Sept. 16, 2014), http://hudoc.echr.coe.int/sites/eng/pages/search.aspx?i=001-146501, Jaloud v. The Netherlands, Eur. Ct. H. R., Application No. 47708/08, ¶ 139, at 61–62 (2014), http://hudoc.echr.coe.int/sites/eng/pages/search.aspx?i=001-148367.

law suggests that human rights law also contributes to regulation of these conflicts."[89] The issue of control is not limited to "effective control," considered in a broader jurisdictional sense. As will be discussed, the ability to exercise physical control over an area where an operation is being contemplated is a key factor in determining whether a law enforcement–based approach seeking to arrest rather than kill a member of a non-State actor organized armed group can be undertaken during armed conflict (see Chapter 7, section 7.4.2, *The "Targeted Killing" Case and Control*). In that situation what is required is a more tactical-level assessment of a State's ability to actually exercise a level of physical control.

Nor is the link between States and governance restricted to occupation and responsibilities exercised on a State's own territory. It has also arisen in the context of States controlling territory under a UN mandate. This has been referred to as humanitarian occupation, or "the assumption of governing authority over a state or a portion thereof, by an international actor for the express purpose of creating a liberal, democratic order."[90] It should come as no surprise that analogous arguments might be made regarding how a State should act toward the local population when it is operating as the only functioning "authority" present in a territory or as a member of a Coalition of States exercising authority, inside another failed state. The question thus remains as to what role a State, which is threatened by transnational non-State actors, should play when it is operating extraterritorially in a territory where the lawful State is unable or unwilling to act. In practical terms, the issue is whether the nonterritorial State is only bound by humanitarian law, or whether it should apply human rights–based governance-related provisions, either as a matter of law or policy, when dealing with the terrorists who may be hiding among the foreign civilian population.

4.4.3. A RELUCTANT APPLICATION OF HUMANITARIAN LAW

As has been noted, in acting as the proper authority to maintain order within their territory, States have resisted the application of humanitarian law to internal conflict. This limitation is reflected in the observation by Leslie Green that, historically, international law was viewed as only being concerned with the relations between States.[91] As a result "non-international conflicts were considered to be subject to the domestic jurisdiction of the sovereign concerned."[92] International law was not concerned with internal conflicts, or "with a conflict between an imperial power and a colonial territory."[93] Exceptions to the sanctity of this form of absolute sovereign authority occurred when belligerents themselves behaved as if they were involved in an international conflict, or third States treated a conflict as if it was international in character (i.e., recognition of belligerency).[94]

Therefore, it was not that the rules governing armed conflict never applied in what were fundamentally internal conflicts. It has been noted that by the early nineteenth century the doctrine of "recognition of belligerency" could be applied in situations where "the conflict

---

[89] Bellinger & Padmanabhan, *supra* note 45, at 209–10.

[90] Gregory H. Fox, Humanitarian Occupation 4 (2008).

[91] Green, *supra* note 22, at 66.

[92] *Id.*

[93] *Id.*

[94] *Id.*

had attained such a sustained level that both sides were entitled to be treated in the same way as belligerents in an international armed conflict."[95] Recognition by third States meant the customary international law of neutrality was applicable[96] and "the parent State brought into effect the *jus in bello* [international humanitarian law] in its entirety between it and the rebels."[97] This doctrine was linked to significant civil wars and is most readily identified with the U.S. Civil War with its participants being viewed as having "two widely divergent outlooks: almost two distinct civilizations."[98]

A reflection of the reluctance by States to accept international involvement in internal conflict is that the recognition of belligerency doctrine had fallen into disuse by World War II.[99] That said, the doctrine had "nevertheless represented a shift in state practice, eroding the impermeability of state sovereignty in international law."[100] In any event, given the high thresholds for recognition of belligerency, other internal conflict such as rebellion and insurgency remained outside the scope of the application of humanitarian law. Rebellion and insurgency still had the potential to be violent in their own right. Rebellion was viewed as "a modest, sporadic challenge by a section of the population intent on attaining control," while an insurgency has been referred to as "a more substantial attack against the legitimate order of the State, the rebelling faction being sufficiently organised to mount a credible threat to the government."[101] However, it was the domestic laws of the State that fundamentally provided the normative framework for controlling these internal conflicts.

The rules governing war were also not readily applied in other situations of conflict against "non-State actors." In an era of colonial conquest and colonization, the British *Manual of Military Law 1914* viewed "wars with uncivilized States and tribes" to be outside the scope of the application of "rules of International Law."[102] These situations involved the entry of Western powers into what would then have been viewed as the "ungoverned" spaces of the period. The rules of international law were only seen as applying to "warfare between civilized nations, where both parties understand them and are prepared to carry them out."[103] Instead, it was the discretion of the commander and general notions of

---

95 Lindsay Moir, *The Historical Development of the Application of Humanitarian Law in Non-International Armed Conflicts to 1949*, 47 INT'L & COMP L.Q. 337, 339 (1998). *See also* SIVAKUMARAN, *supra* note 36, at 9–20.

96 Moir, *supra* note 95, at 341.

97 *Id.* at 343.

98 WILLIAM MCELWEE, THE ART OF WAR: WATERLOO TO MONS 148 (1974).

99 SIVAKUMARAN, *supra* note *36, at 17–20,* Moir, *supra* note 95, at 352, *and* DINSTEIN, *supra* note 36, ¶ 347, at 111. But see Iain Scobbie, *Gaza*, *in* INTERNATIONAL LAW AND THE CLASSIFICATION OF CONFLICT 280, 301-02 (Elizabeth Wilmshurst ed., 2012) (where it is suggested Israel's imposition of a blockade in January 2009 "consitutted a recognition of the belligerency of Hamas".).

100 ANTHONY CULLEN, THE CONCEPT OF NON-INTERNATIONAL ARMED CONFLICT IN INTERNATIONAL HUMANITARIAN LAW 23 (2010).

101 Moir, *supra* note 95, at 338.

102 MANUAL OF MILITARY LAW 1914, *supra* note 13, ¶ 7, at 235. *See also* HEW STRACHAN, THE DIRECTION OF WAR: CONTEMPORARY STRATEGY IN HISTORICAL PERSPECTIVE 201(2013) (for reference to the British 1909 *Field Service Regulations* and the instruction in a chapter on "warfare against the uncivilized enemy" that the principles of regular warfare would have to modified "[i]n campaigns against savages . . . ."), GREEN, *supra* note 22, at 66 ("[the law of armed conflict] was not in any way concerned . . . with a conflict between an imperial power and a colonial territory."), and MAX BOOT, INVISIBLE ARMIES: AN EPIC HISTORY OF GUERRILLA WARFARE FROM ANCIENT TIMES TO THE PRESENT 175 (2013) ("On the Northwest Frontier, as on most other imperial battlefields, local fighters had no knowledge of the 'laws of war' invented in the West, and Westerners had no intention of applying those laws to 'savages.'").

103 MANUAL OF MILITARY LAW 1914, *supra* note 13, ¶ 7, at 235.

justice and humanity that were to guide military action.[104] The question remains as to what norms of conduct restrained State action outside its borders when it was placed in these situations of governance. However, by the turn of the twentieth century public opinion was forming to suggest that at a minimum the laws and customs of war may apply.[105]

There was also pressure to apply the principle of minimum force when dealing with local populations. This is a principle that is now readily associated with human rights–based law enforcement. The British approach initially was to apply the principle of minimum force only in situations up to and including riots but not insurgencies. However, as a result of military forces massacring Indian civilians at Amritsar, India, in 1919, there was a change in approach. In that incident, "[i]n April 1919, Brigadier General Dyer ordered his men to open fire on a crowded assembly in Jallianwala Bagh in Amritsar. Within ten minutes, the soldiers expended 1650 rounds, killing 379 and wounding over 1200."[106] From that point onward, at least in respect of the British Army, "this principle [minimum force] was applied to most forms of internal unrest except those types of counterguerrilla operations that approximated conventional combat."[107] By 1934 a leading British text made reference to a use of the minimum amount of military force that the situation demanded in colonial situations, as it "should always be borne in mind that the hostile forces are fellow citizens of the Empire . . . ."[108] This included the army being used in a policing role against revolutionary movements.[109]

The association of the State with governance impacted significantly on the normative regime applied when using force. It also reflected a distinction between using force against "outsiders" and those fully within the control of the State. While there could be a requirement to apply "all the ordinary methods of warfare" and operate against guerrilla forces,[110] it is the use of "minimum force" that is most readily equated to a domestic law enforcement regime. To the extent this principle was adhered to, it reinforces the conclusion that the maintenance of order in situations of governance often requires a nuanced interface between conduct of hostilities norms and those now associated with human rights–based law enforcement. It also highlights a form of extraterritorial application of domestic standards, and with it the normative framework associated with human rights law.

Notwithstanding the decline of the recognition of belligerency doctrine, the general movement to codify the law governing armed conflict witnessed efforts in the early nineteenth century to extend humanitarian protection to the victims of internal conflicts. Led by the ICRC, it had, by 1949, attained at least a partial success in the form of Common Article 3 of the 1949 Geneva Conventions. However, during the deliberations concerning those Conventions, and in subsequent efforts to codify the law governing internal conflicts, States remained particularly sensitive toward challenges to their status as the proper authority. The rejection of any potential legitimization of non-State actors even extended

---

104 *Id.*

105 *Id.* at 235 (e).

106 Robert M. Cassidy, Counterinsurgency and the Global War on Terror 93 (2008). *See also* Charles W. Gwynn, Imperial Policing 34–64 (1934).

107 Cassidy, *supra* note 106, at 93.

108 Gwynn, *supra* note 106, at 14.

109 *Id.* at 10–11.

110 *Id.* at 11.

to discussions regarding organized resistance movements in situations of occupation, where the prisoner of war status being provided to those armed groups hinged on that provision not being seen as "a 'jus insurrectionis' [right of insurrection] for the inhabitants of occupied territory."[111] One assessment in the late 1950s noted "[g]overnments by tradition and inclination, regard rebels and traitors as worse offenders than ordinary criminals."[112] A half century later the U.S. administration "downplayed the extent of the uprisings in Iraq, saying they were being driven by fallen regime 'dead-enders,' 'criminals,' 'looters,' 'terrorists,' 'anti-Iraqi forces' and 'those influenced by Iran.'"[113] States can also "resent any attempt by outside bodies, including the International Committee of the Red Cross, to intercede on behalf of such rebels."[114]

Concerns over a broad application of the Geneva Conventions to internal conflicts met with objections during its negotiation that any "such proposals giving insurgents a legal status, and consequently support, would hamper the Government in its measures of legitimate repression."[115] As has been noted by Rob McLaughlin, the objection included both the proposed application of inter-State humanitarian law to internal conflict and to any definitive establishment of a threshold between law enforcement and situations governed by the law governing armed conflict.[116] The scope of Common Article 3 was originally envisioned to have "a field of application similar to civil war [such that] the majority of contemporary non-international armed conflicts would be excluded from the remit of international humanitarian law."[117] Further, it has been noted the absence of a formula for determining the applicability of Common Article 3 "also had the effect of weakening its implementation."[118] This is seen in "the tendency of states engaged in non-international armed conflict not to recognise the applicability of this provision."[119]

The application of Common Article 3 is now often linked to the criteria set out by the ICTY Appeal Chamber case, *Prosecutor v. Tadić*, of "protracted armed violence between governmental authorities and organized armed groups or between such groups within a State." [120] However, there has been a general failure of the international legal community in the post-9/11 period to reach consensus on when a noninternational armed conflict is in existence. Questions remain regarding the application of Common Article 3 particularly because of the increase of noninternational armed conflicts with an extraterritorial

---

111 The Geneva Convention Relative to the Treatment of Prisoners of War: (III) Commentary, art. 4, at 57–58 (Jean Pictet ed., 1960) [hereinafter ICRC Commentary, GC III].

112 G. I. A. D. Draper, The Red Cross Conventions 14 (1958).

113 Scahill, *supra* note 28, at 112. *See also* Noah Barkin & Sabine Siebold, *Germany to Support Military Campaign Against IS After French Appeal*, Reuters, Nov. 26, 2015, 8:11 PM GMT, http://uk.reuters.com/article/uk-mideast-crisis-germany-france-idUKKBN0TF0ZU20151126 (where the German Defense Minister referred to the Ismalic State as a "murderous gang").

114 Draper, *supra* note 112, at 14.

115 ICRC Commentary, GC III art. 3, *supra* note 111, at 32.

116 Rob McLaughlin, *Legal Policy Considerations and Conflict Characterizations at the Threshold between Law Enforcement and Non-International Armed Conflict*, 13 Melb. J. Int'l L. 11 (2012).

117 Cullen, *supra* note 100, at 60.

118 *Id.* at 187.

119 *Id.*

120 Prosecutor v. Tadić, Case No. IT-94-1-A, Decision On The Defence Motion For Interlocutory Appeal On Jurisdiction, ¶ 70 (Int'l Crim. Trib. for the Former Yugoslavia, 1995), http://www.icty.org/x/cases/tadic/acdec/en/51002.htm.

element.[121] Further non-State actors are capable of carrying out "one-off" armed attacks that mirror the violence associated with State armed forces. It is also difficult to see how a criteria too closely associated with the violence having to be protracted adequately addresses the levels of violence and the nature of the threat posed by many terrorist and other organized armed groups (see Chapter 9, section 9.3, *Categorizing Conflict with Non-State Actors*, and section 9.4. *Determining the Existence of a Noninternational Armed Conflict*). To the extent the identified organizational criteria for such groups too closely mirrors that associated with conventional state armed forces, such as requiring "the existence of a headquarters,"[122] many less hierarchically organized groups participating in significant armed conflict may be left to be uniquely dealt with under the human rights–based legal regime.[123] The problem is there are limits to what levels and type of force human rights law can justify, or what it should be asked to do (see Chapter 13, section 13.6, *Limits and the Need for Clarity*).

For example, in April 2012, in respect of armed violence in Syria, a UN commission reported it was unable to state there was an armed conflict in existence due to an inability to verify the degree of organization of the opposition armed forces. This resulted in Syria's obligations being assessed at that point according to international human rights law.[124] It may be the analysis concentrated on looking for a traditional form of military organization to establish the existence of a conflict. However, that type of organization does not always reflect the networked form of warfare practiced by non-State actors (see Chapter 9, section 9.4.2, *Group Organization*).[125] By July 2012 it was determined that an armed conflict was occurring in Syria, which raised the issue of the operation of the protective provisions of humanitarian law and the obligations placed on participants regarding the conduct of hostilities.[126] This conclusion was reached even though the military forces of the opposition continued to not be organized along traditional lines. It was reported in July that although there were attempts to establish a cohesive command structure, at that point at least eighty armed groups were involved in the fighting as part of small personal networks in one part

---

[121] INTERNATIONAL HUMANITARIAN LAW AND THE CHALLENGES OF CONTEMPORARY ARMED CONFLICTS, REPORT 31ST INTERNATIONAL CONFERENCE OF THE RED CROSS AND RED CRESCENT, EN 31IC/11/5.1.2 9 (28 November–1 December 2011), http://www.icrc.org/eng/assets/files/red-cross-crescent-movement/31st-international-conference/31-int-conference-ihl-challenges-report-11-5-1-2-en.pdf.

[122] *Id.* ¶ i, at 8.

[123] DINSTEIN, *supra* note 36, ¶ 113, at 37 ("Below-the-threshold violence [all forms of disorganized—individual or mob—violence] lies outside the scope of application of LONIAC [humanitarian law] and is controlled by the law enforcement paradigm . . . .")

[124] Human Rights Council, *Report of the independent international commission of inquiry on the Syrian Arab Republic*, UN Doc. A/HRC/19/69, ¶13, at 6 (Feb. 22, 2012), http://daccess-dds-ny.un.org/doc/UNDOC/GEN/G12/106/13/PDF/G1210613.pdf?OpenElement. *But see* Laurie R. Blank & Geoffrey S. *Corn, Syria Must be Held to the Law of War*, CNN (Apr. 4, 2012), http://www.cnn.com/2012/04/04/opinion/blank-corn-syria-war/index.html (for criticism of the conclusion the situation in Syria was not at that time an armed conflict).

[125] Stanley A. McChrystal, *It Takes a Network: The New Frontline of Modern Warfare*, FOREIGN POL'Y (Mar./Apr. 2011), http://foreignpolicy.com/2011/02/21/it-takes-a-network/. *See also* PETER L. BERGEN, MANHUNT: THE TEN-YEAR SEARCH FOR BIN LADEN FROM 9/11 TO ABBOTTABAD 153–57 (2012) (outlining how U.S. Joint Special Operations Command was transformed in Iraq to make it "more like al-Qaeda").

[126] Louis Charbonneau, *Syria Conflict Now a Civil War, UN Peacekeeping Chief Says*, REUTERS (June 12, 2012), http://www.reuters.com/article/2012/06/12/us-syria-crisis-un-idUSBRE85B1BI20120612, and *Syria in Civil War, Red Cross Says*, BBC NEWS MIDDLE EAST (July 15, 2012), http://www.bbc.co.uk/news/world-middle-east-18849362.

of the country alone.[127] In the case of Syria, the levels of violence appear to have become a more significant contributor to the determination than the organization of the groups involved. It also highlights a political element of the decision as to when an armed conflict is stated to be in existence, and perhaps a preference to default toward the application of a human rights regime. Setting the bar high in respect of the organization of the group opposing the State in order for claim an armed conflict exists is entirely consistent with the bias in international law favoring States and the role they perform as the proper authority.

States, both dominant military powers and less powerful ones, have remained reticent to extend the application of humanitarian law to conflicts in their territory.[128] It is reported that in the 1970s even the newly formed states, while supportive of broadening the application of Additional Protocol I to certain conflicts, were concerned the controversial wording of that Protocol be carefully drafted so secessionist movements within their own territories would not be able to claim to be national liberation movements.[129] It is also indicated that in later sessions those States "turned their attention to dismantling Protocol II."[130] It has been suggested that the fact that Additional Protocol II governing conflicts not of an international character included a high threshold for application "serves to strengthen the discretionary power of states to deny the Protocol's applicability."[131] Denial of the application of that Protocol limits the introduction of a more expanded international treaty law regime into internal armed conflicts. It is also entirely consistent with the just war–based notion of proper authority and the unique status the States claim under international law as being responsible for their internal security. However, it once again raises the profile of human rights law, both domestic and international, as the default regime governing the regulation of such violence.

4.4.4. A POLICY SOLUTION FOR LEGAL UNCERTAINTY

Efforts have been made to fill the void in legal guidance that resulted from a reluctance to address the application of humanitarian law in small wars. Guidance regarding what humanitarian law rules apply during armed conflict can be seen in the Appeal Court of the ICTY decision in *Prosecutor v. Tadić*,[132] the 2005 ICRC *Customary International Humanitarian Law Study*,[133] and by States through the development of military manuals on humanitarian law.[134] However, there remains considerable uncertainty regarding what

[127] C. J. Chivers, *Fresh From Syria, Rebel Commanders Unite in Frustration*, N.Y. Times, July 13, 2012, http://www.nytimes.com/2012/07/14/world/middleeast/commanders-of-syrian-rebels-unite-in-frustration.html?pagewanted=all.

[128] Mary Ellen O'Connell, *Combatants and Combat Zones*, 43 U. Rich. L. Rev. 845, 860 (2009).

[129] Keith Suter, An International Law of Guerrilla Warfare 147 (1984).

[130] *Id.* at 148.

[131] Cullen, *supra* note 100, at 113.

[132] Prosecutor v. Tadić, Case No. IT-94-1-A, Decision On The Defence Motion For Interlocutory Appeal On Jurisdiction, ¶¶ 96–127 (Int'l Crim. Trib. for the Former Yugoslavia, Appeal Chamber, 2 October 1995), http://www.icty.org/x/cases/tadic/acdec/en/51002.htm.

[133] Customary International Humanitarian Law Study xvi (Jean-Marie Henckaerts & Louise Doswald-Beck eds., 2005) (where the need to address noninternational armed conflict is explained in the foreword by Dr. Yves Sandoz.).

[134] *Report of the International Commission of Inquiry on Darfur*, Pursuant to Security Council Resolution 1564 of 18 September 2004, Report to the Secretary General, Geneva, ¶159, at 45 (Jan. 25, 2005), http://www.un.org/news/dh/

rules of international humanitarian law apply to conflicts that do not qualify as international armed conflicts. It has been noted that there is a "growing recognition among government and international organization officials and among IHL experts that the black-letter rules of the venerable Geneva Conventions often do not provide clear guidance for states engaged in conflicts with non-state actors."[135] Importantly, there is also a significant lack of consensus regarding the threshold for establishing a noninternational armed conflict is in existence (see Chapter 9, section 9.4, *Determining the Existence of a Noninternational Armed Conflict*).

The result has been that some States have adopted policy direction to broadly apply international humanitarian law to contemporary military operations. For example, U.S. government direction indicates its forces comply with the law of war during all armed conflicts, however such conflicts are characterized, and in all other military operations.[136] As is reflected in a 2001 manual, Canadian military direction to its deployed personnel was that "as a minimum, apply the spirit and principles of the LOAC [law of armed conflict] during all operations other than domestic operations."[137] This provision was developed in the aftermath of the 1993 torture and killing of a Somali youth in the custody of the Canadian Forces.[138] Not specifically focused on the use of force, the provision was inserted with the goal of providing persons detained on any international operation with the highest standard of care and protection. The United Nations also has extended compliance with humanitarian law by its forces to "enforcement actions, or in peacekeeping operations when the use of force is permitted in self-defence."[139]

The positive result of these approaches is the extension of humanitarian protections of the law governing warfare to all types of conflict. It also helps ensure there is clear direction to military personal regarding standards of conduct during what can be complex security situations. Such direction can be as simple as "[t]reat all detained persons humanely in accordance with the standard set by the Third Geneva Convention. Any form of abuse, including torture, is prohibited."[140] Adopting this approach can be particularly important when operating in failed and failing states where there is no functioning justice system or domestic security apparatus. However, it can be exceeding problematic if such an approach does not, at a minimum, also account for the application of human rights–based law enforcement norms, either through an extraterritorial application of treaty law or as an integral part of

---

sudan/com_inq_darfur.pdf (where reference is made to the German and UK manuals on humanitarian law in reaching the conclusion that "a body of customary rules regulating internal armed conflicts has thus evolved in the international community...."). *See also* CRAWFORD, *supra* note 7, at 30–31.

135 Bellinger & Padmanabhan, *supra* note 45, at 202.

136 DoD Law of War Program, Department of Defense, Directive Number 2311.01E, May 9, 2006 (Incorporating Change 1, November 15, 2010, Certified Current as of February 22, 2011) ¶ 4.1, at 2, http://www.dtic.mil/whs/directives/corres/pdf/231101e.pdf. *See also* DOD LAW OF WAR MANUAL, *supra* note 8, ¶ 3.1.1.2, at 71–72.

137 LAW OF ARMED CONFLICT AT THE OPERATIONAL AND TACTICAL LEVELS, B-GJ-005-104/FP-021 ¶ 1702. 2., at 17-1 (2001-08-13).

138 I DISHONOURED LEGACY: LESSONS OF THE SOMALIA AFFAIR, Report of the Commission of Inquiry into the Deployment of Canadian Forces to Somalia 318–24 (1997).

139 *Secretary-General's Bulletin: Observance by United Nations Forces of International Humanitarian Law*, *in* BRUCE OSWALD, HELEN DURHAM & ADRIAN BATES, DOCUMENTS ON THE LAW OF UN PEACE OPERATIONS 201, ¶ 1.2, at 202 (2010) [hereinafter *Secretary-General's Bulletin*].

140 *CF Code of Conduct CF Personnel*, Office of the Judge Advocate General B-GG-005-027/AF-023 2-9 (undated).

humanitarian law (see Chapter 5, section 5.3.2.3, *Divergence and Growing Reconciliation*). Further, not all international security operations are carried out in the context of armed conflict. For example, hostage rescues may be carried both on the high seas and in foreign territory to rescue nations seized by criminal gangs (see Chapter 10, section, 10.3.5.3, *States Acting Unilaterally to Rescue Their Own Nationals: Somalia*). Since in situations where there is no armed conflict humanitarian law does not apply as a matter of law, it is difficult to see how it can be justified as a matter of policy to control the use of force (see Chapter 8, section 8.5.3, *The "Self-Defense" Option*).

A particular risk is that having the default being a policy application of humanitarian law means that States will sometimes apply conduct of hostilities rules in security situations overseas where they might be reticent to do so within their own territory. While foreign troops operating in a territory may be inclined to broadly apply international humanitarian law to address insurgent activity, the host nation might look at the same security threat through a more restrictive domestic lens. The territorial State may prefer a less violent law enforcement–based approach be applied in respect of its own citizens. In particular, a human rights approach ordinarily limits the collateral effects that are permitted when using force. This can create a legal, cultural, and doctrinal tension between local authorities and foreign military forces. Such tension could be seen in a long-standing Afghan government concern over levels of collateral civilian casualties and the conduct of night raids by Coalition forces.[141] The latter raids have been suggested to "more than any other single issue, turned the population against them."[142]

However, concerns expressed by the territorial government have not always led to a change in tactics. It is indicated that notwithstanding a public campaign in 2009 by the American Coalition commander suggesting reduced use of air strikes and the development of new rules for house raids, by December of that year the raids carried out by Special Operations forces had increased fourfold.[143] Further, after a reported 2013 ban on such raids imposed by President Hamid Karzai, the new Afghan administration that came to power in late 2014 is said to have quietly reinstituted them in order to better counter Taliban night activities. However, it was also acknowledged "their resumption is likely to be controversial among Afghans, for whom any intrusion into private homes is considered offensive."[144] These raids were said to include American Special Forces personnel in an advisory role and "combat enabler" support (i.e., air support, transportation, communications) provided to the Afghan forces involved.[145]

Negative reactions toward house searches have not been limited to Afghanistan. They can even arise when a law enforcement approach is the favored security response in countering insurgent threats. In Northern Ireland during the mid-1970s a conscious effort was made by policing authorities to make better use of information in order reduce the

---

[141] Matthew Rosenburg & Helene Cooper, *Karzai Lashes Out at NATO over Civilian Deaths*, N.Y. Times, Mar. 16, 2012, http://www.nytimes.com/2012/03/17/world/asia/karzai-lashes-out-at-united-states-over-inquiry-on-massacre.html.

[142] Jean MacKenzie, *Afghanistan—Part II, Counterinsurgency in Afghanistan: Myth or Reality?*, *in* Hearts and Minds: A People's History of Counterinsurgency 200, 220 (Hannah Gurman ed., 2013).

[143] Scahill, *supra* note 28, at 330–35.

[144] Rod Nordland & Taimoor Shah, *Afghanistan Quietly Lifts Ban on Nighttime Raids*, N.Y. Times, Nov. 23, 2014, http://www.nytimes.com/2014/11/24/world/asia/afghanistan-quietly-lifts-ban-on-night-raids.html.

[145] *Id.*

number of searches of private homes, as they "understood that better intelligence could help the security forces in their relations with the nationalist community, which bitterly resents house searches."[146] This indicates that resistance to intervention by security forces in the private lives of the inhabitants of a territory may be as much a universal privacy, and therefore a human rights issue, as one based on religion or culture. However, the negative reaction is heightened when, as one local Afghan leader is reported to have said, "there will be losses again of women and children when Taliban shoot from people's houses, and in reaction the foreigners will bomb or kill them . . . ."[147] The challenges of fighting at the legal boundaries between the conduct of hostilities and law enforcement paradigms are evident in this scenario. Importantly, an approach favoring the application of humanitarian law rules can potentially lead to an actual or a perceived overuse of force that can adversely affect the overall counterterrorism effort (see Chapter 7, section 7.6.2.2, *Proportionality and Counterinsurgency*).

4.4.5. SMALL WARS AND "UNGOVERNED SPACES"

As Robin Geiβ has noted, "[i]n a primarily state-centered international system that after two world wars was primarily designed to curb powerful states, the conceptualization and integration of overly weak states has proved quite problematic."[148] The fact that conflicts between State armed forces and organized groups fighting on behalf of non-State actors often occur among the civilian population and increases the likelihood that the interface between international humanitarian law and human rights law will have to be addressed. The way that States have fought small wars, whether in times of conquest, to establish or maintain colonial power, or in response to internal threats, has not always met the humane standards of contemporary society. Policies such as the use of airpower by Britain for colonial policing during the interwar years[149] and the executions and application of torture by French forces to gain intelligence during the 1950s in the Algerian conflict highlight this fact.[150] However, there was also increasing recognition in the last half of the twentieth century that protecting the population and the primacy of policing over the use of military power are key aspects of defeating an insurgency or confronting terrorist groups.[151] That concept of "police primacy" is intimately linked to governance.

---

146 MARK URBAN, BIG BOYS' RULES: THE SECRET STRUGGLE AGAINST THE IRA 108 (1992).

147 Nordland & Shah, *supra* note 144.

148 Robin Geiβ, *Armed Violence in Fragile States*, 91 INT'L REV. RED CROSS 127, 132 (2009).

149 *See, e.g.*, John Pilmott, *The British Experience*, *in* THE ROOTS OF COUNTER-INSURGENCY: ARMIES AND GUERRILLA WARFARE, 1900–1945, 17, 23–27 (Ian F. W. Beckett ed., 1988). *See also* PHILIP ANTHONY TOWLE, PILOTS AND REBELS: THE USE OF AIRCRAFT IN UNCONVENTIONAL WARFARE 1918–1988, at 9–55 (1989), and GWYNN, *supra* note 106, at 9 and 222–23.

150 PAUL AUSSARESSES, THE BATTLE OF THE CASBAH: COUNTERTERRORISM AND TORTURE IN ALGERIA (Robert L. Miller trans., 2006). *See also* ALISTAIR HORNE, A SAVAGE WAR OF PEACE: ALGERIA 1954–1962, at 195–207 (2006) (1977).

151 TONY GERAGHTY, THE IRISH WAR: THE MILITARY HISTORY OF A DOMESTIC CONFLICT 41 (1998) (where it as noted in respect of Northern Ireland "[t]he new Irish strategy , slow and painful in its evolution, finally combined painstaking, forensic police investigation . . . with massive surveillance of suspects as a substitute for forces resettlement.").

This focus on governance is evident in the writings of a number of influential theorists. As Sir Robert Thompson noted in 1966, "[a]n insurgent movement is a war for the people. It stands to reason that government measures must be directed to restoring government authority and law and order throughout the country so that control over the population can be regained and its support won."[152] Frank Kitson stated in respect of the government, "in order to eliminate the subversive party and its unarmed and armed supporters, it must gain control of the population."[153] In the words of Bernard Fall, the goal is "to establish a competitive system of control."[154] An insurgency problem "is militarily only in a secondary sense, and politically, ideologically, and administratively in a primary sense."[155] He also sagely suggested, "*[w]hen a country is being subverted it isn't being out-fought; it's being out-governed.*"[156]

One contemporary analyst, David Kilcullen, has expanded on Bernard Fall's terminology to explain the pattern "in which nonstate armed groups create predictability and order as a way of generating popular support, and build incentive systems to corral target populations" as the theory of "competitive control."[157] In this context, several actors, including governments, compete to control the population in a violent and contested environment through the use of a range of incentives and disincentives.[158] The populations in disputed areas respond to a predictable, ordered normative system. Non-State actor versions of "control systems" can be seen "in operation at the hands of criminal groups in the "System" of Jamaica garrison communities, in the actions of clan militias in Somalia, and in the operations of Taliban courts in Afghanistan."[159] Fundamentally, winning an insurgency is about governance.

In terms of the State exercise of control, the peaceful maintenance of order through policing, an operational prison system, and the other enforcement structures of modern "Rule of Law"–based societies can be used to measure effective governance. If a State cannot enforce its laws or maintain order, then its role as the proper authority is not only being challenged, the government is well on the path toward losing to its non-State challenger. Indicia of problems in being able to exert authority can include a lack of government presence, police regularly requiring military support to simply perform basic functions like patrolling, or security forces needing to be prepared to fight their way into an area to attempt to make an arrest or conduct a search. The challenges to State authority can come from a wide range of non-State actors, including insurgents, terrorist organizations, and criminal gangs. Throughout this book the term "ungoverned spaces" is used to describe areas where the State, as the legitimate governing authority, either does not exercise or does not have

---

152 ROBERT THOMPSON, DEFEATING COMMUNIST INSURGENCY 51 (1966). *See also* DAVID GALULA, COUNTERINSURGENCY WARFARE: THEORY AND PRACTICE 89–90 (1964). For an overview of the history of counterinsurgency theory, *see Introduction, in* COUNTERINSURGENCY IN MODERN WARFARE 13–19 (Daniel Marston & Carter Malkasian eds., 2008).

153 FRANK KITSON, LOW INTENSITY OPERATIONS: SUBVERSION INSURGENCY & PEACEKEEPING 49 (1971)

154 Bernard Fall, *The Theory and Practice of Insurgency and Counterinsurgency*, 17 NAV. COL. REV. 21, 22 (1965).

155 *Id.*

156 *Id.* at 34.

157 DAVID KILCULLEN, OUT OF THE MOUNTAINS: THE COMING OF AGE OF THE URBAN GUERRILLA 125 (2013).

158 *Id.* at 131.

159 *Id.* at 126–27.

sufficient control to carry out the required policing and enforcement duties for which it is responsible. In other words, where that State cannot or will not act as the proper authority.

It has been suggested that "policy makers are wrong to be concerned about the existence of *ungoverned spaces*; their concern should be about *improperly governed spaces*."[160] This is because a complete lack of functioning State institutions is rare, and actually may be quite problematic for the non-State actor.[161] That non-State actor may also be forced to deal with "local bandits, the cost of corruption in neighboring states, and the ability of Western forces to act in ungoverned spaces."[162] While this observation has merit, there are such a broad range of circumstances in which legitimate governing authorities (i.e., States) do not exercise sufficient control over a territory that the generic term "ungoverned spaces" will be used to include areas where there may be some but inadequate governance exercised by a legitimate territorial State. [163] The key element is that the non-State actor is located in a place where it can operate with relative impunity because the territorial State ultimately does not exercise sufficient control or demonstrate the will to police its territory. Non-State actors such as the Islamic State or Al Qaeda can take advantage of a "safe haven" from which to carryout an insurgency, or terrorist action.[164]

Historically, this lack of State control, or an inability to govern effectively, can have a physical and even social geography component. In terms of physical geography, insurgency and guerrilla warfare has been associated with the areas of woods, forests, jungles, mountains, deserts, marshes, or swamps,[165] or even sparsely populated and underdeveloped areas.[166] However, there are limits as the resulting separation from civilian supporters can also be problematic for insurgent forces particularly in terms of being provided support.[167] In some cases the physical control exercised by the central government over its territory may be limited but capable of being exercised at least periodically. It was noted in 2011 in Yemen that security forces were able to "project themselves throughout the country on an ad hoc basis, but to operate effectively in many places they require the support or acquiescence of the tribes."[168] This lack of persistent physical control is compounded by the social integration of jihadist forces. While Yemeni tribes do not have an affinity with Al Qaeda, many Yemeni youth have engaged in jihad. The connection to Al Qaeda expands within

---

[160] Jacob N. Shapiro, The Terrorist's Dilemma: Managing Violent Covert Organizations 266 (2013) (emphasis added).

[161] Geiß, *supra* note 148, at 129 (where it is indicated to be quite rare to have an inability to perform any basic State functions or a loss of complete control over territory). *See also* Crawford, *supra* note 64, at 719–23 (for a critical discussion of the use of the term "failed State").

[162] Shapiro, *supra* note 160, at 267. *See also* Geiß, *supra* note 148, at 130 (for reference to rudimentary political orders emerging to fill the governance void).

[163] Witte & Blum, *supra* note 76, at 255 (where reference is made to failed or fragile states as "de facto ungoverned spaces" in the context of being "natural havens for perpetrators of crime and violence. . . .?").

[164] S.C. Res. 2249, UN Doc. S/RES/2249 ¶ 5 (Nov. 20, 2015)(where where it is noted that the Islamic State operates from safe haven in significant parts of Iraq and Syria).

[165] John Ellis, From the Barrel of a Gun: A History of Guerrilla, Revolutionary and Counter-Insurgency Warfare, from the Romans to the Present 237 (1995), and C. E. Callwell, Small Wars: Their Principles and Practice 127 (3rd ed. 1996) (1906) ("Hilly and broken ground or districts clothed with jungle growth and thickets especially lend themselves to these very desultory operations.").

[166] Carl Von Clausewitz, On War 480 (Michael Howard & Peter Paret eds. & trans., 1989).

[167] Ellis, *supra* note 165, at 238.

[168] Edmund J. Hull, High-Value Target: Countering Al Qaeda in Yemen xxvi (2011).

families.[169] As a result, "[t]his long-standing social geography makes Yemen fertile ground for al Qaeda recruitment."[170] This leaves Al Qaeda free to operate within Yemen, which has been identified by a jihadist strategist as one of the "regions of savagery."[171]

Abu Bakr Naji's *The Management of Savagery: The Most Critical Stage Through Which the Umma Will Pass* has been described as the most accurate description of Al Qaeda's global strategy[172] and "could clearly be seen in both its [Al Qaeda in Iraq] military and media strategies.[173] It is a jihadist tract that has "heavily influenced ISIS's strategy across multiple fronts."[174] That work indicates that the jihadist area of "Yemen" is not restricted by UN-assigned boundaries, but rather incorporates areas identified as "Yemen, the Hijaz, and Oman."[175] Abu Bakr Naji indicates such regions of savagery are the priority areas from which an Islamic State will develop.[176] Among the factors to be considered in identifying a "region of savagery" are the "weakness of the ruling regime and the weakness of the centralization of its power in the peripheries of the borders of its state and sometimes in internal regions, particularly those that are overcrowded," topography and geographic depth within a State, the nature of the people, and the prevalence of weapons.[177]

Pakistan is also an area where non-State actors have been able to take advantage of a safe haven. The retreat by Al Qaeda to Pakistan's tribal region following the 2001 invasion of Afghanistan was said to be with the goal of finding "a natural fortress from which it could incite the whole region to evolve into a theater of war."[178] It was "[n]ot long after their retreat from Afghanistan, the Taliban and Al Qaeda announced the establishment of the Islamic State of North Waziristan, *with a Taliban police to combat vice and crime.*"[179] It was from that time that "the tribal areas underwent a rapid 'Talibanization' as the Pakistan Army, under pressure from the US, attempted to flush out the Islamist influence."[180] This saw the growth of the Tehrek-e-Taliban Pakistan (TTP), with a power base in the Pashtun dominated tribal areas (i.e., the NWFP and FATA) that are subject to limited oversight and control by the Pakistan government.[181] The exercise of Taliban power included the establishment of "courts, a police system, and a tax collection arrangement in South Waziristan,

---

[169] *Id.* at xxvi–xxviii.

[170] *Id.* at xxix.

[171] *Id. See also* ABU BAKR NAJI, THE MANAGEMENT OF SAVAGERY: THE MOST CRITICAL STAGE THROUGH WHICH THE UMMA WILL PASS, at [16] (William McCants trans., 2006) [hereinafter MANAGEMENT OF SAVAGERY], https://azelin.files.wordpress.com/2010/08/abu-bakr-naji-the-management-of-savagery-the-most-critical-stage-through-which-the-umma-will-pass.pdf.

[172] MICHAEL W. S. RYAN, DECODING AL-QAEDA'S STRATEGY: THE DEEP BATTLE AGAINST AMERICA 143 (2013) (where it is also indicated it was Al-Muqrin's military doctrine, *A Practical Course for Guerrilla Warfare* [reproduced in NORMAN CIGAR, AL-QA'DA'S DOCTRINE FOR INSURGENCY 82–175 (2009)] that led directly to the development of Naji's MANAGEMENT OF SAVAGERY, *supra* note 171).

[173] JESSICA STERN & J. M. BERGER, ISIS: THE STATE OF TERROR 24 (2015).

[174] *Id.* at 115.

[175] MANAGEMENT OF SAVAGERY, *supra* note 171, at [16].

[176] *Id.* (this is a reference to a generic Islamic State).

[177] *Id.*

[178] SYED SALEEM SHAHZAD, INSIDE AL-QAEDA AND THE TALIBAN: BEYOND BIN LADEN AND 9/11, XIX (2011).

[179] *Id.* at 20.

[180] ABDEL BARI ATWAN, AFTER BIN LADEN: AL QAEDA, THE NEXT GENERATION 153 (2012). *See also* SHAHZAD, *supra* note 178, at 25–34.

[181] ATWAN, *supra* note 180, at 152–54.

while Pakistan's security forces remained passive in their military headquarters."[182] A degree of government presence remained, however, the area was, in reality, an "ungoverned space" in terms of State involvement. One weakness in the State governance apparatus has been the relative ineffectiveness of Pakistan's police forces. This is suggested to be due to a low civilian police to population ratio, inadequate material and armaments, poor training, corruption, and a culture of being partial to elite authority arising from its colonial history, which contributes to public mistrust.[183]

Afghanistan provides another example of a partially "ungoverned space" despite the existence of a central government. The international boundaries of Afghanistan are often viewed as being arbitrary, and lines dividing "communities that continue to see themselves as one."[184] Rather than depend upon a centralized government, its four distinct regions (i.e., Herat, Kandahar, Mazar-i-shaif, and Kabul) survive and re-emerge "no matter the changes in political organization, arrivals of new populations or religions, or attempts to impose larger and more uniform identities on them."[185] Confronted with an inefficient, ineffective, and corrupt central government, local populations are attracted to systems of governance that can provide the essentials of order and justice. It was just such a local effort to provide a secure environment in two Kandahar provincial districts in 1994 that led to the creation of the Taliban.[186] The Afghan Taliban has continued to demonstrate a level of effective governance in large areas of Afghanistan, notwithstanding the existence of a recognized national government that has received the full support of the international community.[187] The existence of this shadow Taliban governing structure indicates there are significant "ungoverned spaces" within parts of Afghanistan where the State government is failing to meet its obligations.[188]

The lack or loss of State control clearly does not need to be throughout the whole territory. It can be localized to one region, or even a city. The loss of control over a city was evident in Fallujah in 2004 when Sunni-based groups displaced the Coalition forces responsible for maintaining public order. This necessitated a large-scale and violent armed confrontation to re-establish control.[189] Such localized loss of control by a State is consistent with the jihadist theory that identifies "regions of savagery" to include "a city, or a village, or two cities, or a district, or part of a large city."[190] With the increasing urbanization of

---

182 SHAHZAD, *supra* note 178, at 29.

183 *See, e.g.*, Adnan Naseemullah, *Police Capacity and Insurgency in Pakistan*, *in* POLICING INSURGENCIES: COPS AS COUNTERINSURGENTS 177, 186–89 (C. Christine Fair & Sumit Ganguly eds., 2014).

184 THOMAS BARFIELD, AFGHANISTAN: A CULTURAL AND POLITICAL HISTORY 47 (2010).

185 *Id.* at 48–49.

186 JAMES FERGUSSON, TALIBAN: THE TRUE STORY OF THE WORLD'S MOST FEARED GUERRILLA FIGHTERS 14–20 (2010), and HASSAN ABBAS, THE TALIBAN REVIVAL: VIOLENCE AND EXTREMISM ON THE PAKISTAN-AFGHANISTAN FRONTIER 62–63 (2014).

187 Joseph Goldstein & Taimoor Shah, *Taliban Present Gentler Face but Wield Iron Fist in Afghan District*, N.Y. TIMES, Aug. 14, 2015, http://www.nytimes.com/2015/08/15/world/asia/taliban-baghran-district-afghanistan.html?smprod=nytcore-iphone&smid=nytcore-iphone-share&_r=0.

188 ANTONIO GIUSTOZZI, KORAN, KALASHNIKOV AND LAPTOP: THE NEO-TALIBAN INSURGENCY IN AFGHANISTAN 110–18 (2008), and KILCULLEN, *supra* note 157, at 149–50.

189 BING WEST, NO TRUE GLORY: A FRONTLINE ACCOUNT OF THE BATTLE FOR FALLUJAH (2005) (for a detailed account of the battle for Fallujah), and DICK CAMP, OPERATION PHANTOM FURY: THE ASSAULT AND CAPTURE OF FALLUJAH, IRAQ (2009).

190 MANAGEMENT OF SAVAGERY, *supra* note 171, at [16].

society, the question of the governance of urban areas and the degree to which mega-cities will become safe havens for crime and insurgency will become an even greater security issue. It has been noted that virtually all this urbanization will occur in the poorest areas of the world.[191] Since many of these places are majority Muslim this raises the concern "that *takfiri* extremists—successors and imitators of Osama bin Laden—will undoubtedly keep threatening their own societies and the world at large."[192]

In exceptional circumstances territory may be subject to a relatively well-developed social, political, legal, and commercial infrastructure overseen by a non-State actor. The level of control exercised by a non-State actor can even be carried to an extreme, such as with Hezbollah in Lebanon. On the eve of the 2006 Israel/Hezbollah conflict, that non-State actor exercised elected political power within the Lebanese government[193] but also had its own military organization controlling the south of Lebanon.[194] To the extent that it was Hezbollah that controlled southern Lebanon, and not the Lebanese government, the territory was a space insufficiently governed by a recognized State. Like Hezbollah, the Hamas, which seized control over all of Gaza in 2007, rules over a unique space that is technically not governed by a State. The territory is not controlled by a State that has been recognized definitively at law. While there is considerable political pressure to recognize the State of Palestine,[195] Gaza itself remains under the control of a group that has been widely viewed as a terrorist organization.[196]. The lack of legitimate control by Hamas is particularly evident if the view is accepted that Gaza has remained occupied, even though Israel physically withdrew in 2005.[197] These areas are "ungoverned" in a formal technical legal sense and functionally provide safe haven to groups dedicated to carrying out terrorist activity.

The key issue, given the importance of the State generally under international law, and more particularly as a proper authority, is that a recognized State and not a non-State actor must govern territory. Some non-State organizations such as Hezbollah and Hamas are often capable of policing their territory because of the level of physical control they exercise. They may also have a relatively sophisticated governing framework. What they cannot be counted on is to act consistent with their international legal obligations to not use that territory for acts of aggression against another State. They do not "police" their territory in the unbiased fashion or act pursuant to recognized human rights norms as would normally be expected of recognized States within the international community. Further, their hostility toward Israel makes it extremely difficult for that State to even consider applying a uniquely

---

[191] KILCULLEN, *supra* note 157, at 28–29.

[192] *Id.* at 29–30.

[193] AMOS HAREL & AVI ISSACHAROFF, 34 DAYS: ISRAEL, HEZBOLLAH AND THE WAR IN LEBANON 51–55 (2008).

[194] *Id.* at 48 ("[t]he southern region of Lebanon, contained the 'operational core' (the offensive and defensive zone) of Hezbollah organization.").

[195] *See e.g.*, *State of Palestine Flag to Fly at United Nations Headquarters, Offices as General Assembly Adopts Resolution on Non-Member Observer States*, SIXTY-NINTH GENERAL ASSEMBLY MEETINGS COVERAGE AND PRESS RELEASES (Sep. 10, 2015), http://www.un.org/press/en/2015/ga11676.doc.htm, and Henry Austin, *Vatican Treaty Recognizes a Palestinian State*, NBC NEWS (Jun. 26, 2015, 9:16 AM), http://www.nbcnews.com/news/mideast/vatican-treaty-recognizes-state-palestine-n382341.

[196] *See e.g.*, MATTHEW LEVITT, HAMAS: POLITICS, CHARITY, AND TERRORISM IN THE SERVICE OF JIHAD 50 (2006), and *Hamas*, COUNCIL ON FOREIGN RELATIONS BACKGROUNDERS (Aug. 1, 2014), http://www.cfr.org/israel/hamas/p8968.

[197] DINSTEIN, *supra* note 83, ¶¶ 28–33, at 12–15.

law enforcement approach when responding to external threats emanating from the territory these non-State actors control. In this respect, the situations involving Hezbollah, Hamas, and the Islamic State represent a degree of attainment of the ultimate goal of insurgent forces: to exclusively control territory.

The existence of "ungoverned spaces" is not limited to situations of occupation or other armed conflict. This has been evidenced in the garrison districts of Kingston, Jamaica, where political leaders and their armed partisans have controlled whole neighborhoods. In these districts "the formal institutions of the Jamaican State were almost entirely absent, but nonstate armed groups (initially licensed by the dominate political party, but increasingly independent over time) exercised informal governance responsibilities, including law enforcement."[198] In 2010 an operation by the Jamaican Defence Force to seize and extradite the overlord of a gang to the United States on drugs and weapons trafficking charges involved heavy fighting, including the use of mortars and bulldozers to break down barricades.[199] The violence spread throughout the capital and operation went on for weeks with at least twenty-three civilians and six members of the security forces being killed.[200]

This phenomenon of gangs exerting control over territory can also been seen in other States such as Mexico, Guatemala,[201] and Brazil.[202] In 2011 Brazilian Special Forces and police occupied the La Rocina *favela* in Rio de Janeiro, a crowded slum of 350,000 people, in an effort to control crime and drug trafficking.[203] The result has been described as "an operational pattern that looks a lot like a police-led version of urban counterinsurgency, Baghdad style."[204] The battle to control populations has been suggested by David Kilcullen to apply "to insurgents, terrorists, drug cartels, street gangs, organized crime syndicates, pirates, and warlords, and it provides useful explanations and insights for law enforcement, civil war, and diffuse social conflict—not just for insurgency."[205] The challenge is that unless the violence has risen to the level of an armed conflict, then it is a human rights–based law enforcement approach that must be applied.

Overall there can be an extremely broad range of spaces that are ungoverned by a proper authority. These include loss of control over isolated cities, parts of territories within traditional State borders, and cross-border regions controlled by insurgent forces. A contemporary example of this latter situation is the 2014 claim of the creation of the "Islamic State," which encompasses large segments of Iraq and Syria. However, this example is far from unique, with Boko Haram operating in the border regions, controlling territory, or conducting operations in Nigeria, Chad, Cameroon, and Niger.[206] Affected States have reacted by

---

[198] KILCULLEN, *supra* note 157, at 94–95.

[199] *Id.* at 89 (information provided in a U.S. Department of State cable).

[200] *Id.* at 92.

[201] MAX G. MANWARING, THE COMPLEXITY OF MODERN ASYMMETRIC WARFARE 103 (2012).

[202] *Id.* at 69–73.

[203] KILCULLEN, *supra* note 157, at 235–36. *See also* John Sullivan & Adam Elkus, *State of Siege: Mexico's Criminal Insurgency*, *in* JOHN P. SULLIVAN & ROBERT J. BUNKER, MEXICO'S CRIMINAL INSURGENCY 7, 13 (2012) (This has been a long-standing challenge in Brazil as can be seen from this reference to cartel attacks on Brazilian public security officials in São Paulo in 2006.).

[204] KILCULLEN, *supra* note 157, at 236.

[205] *Id.* at 115.

[206] *Nigerian Islamic Extremists Pose Regional Threat*, INT'L N.Y. TIMES, Dec. 21, 2014, http://www.nytimes.com/aponline/2014/12/21/world/africa/ap-af-nigeria-violence.html?_r=0.

conducting multinational operations that transcend their borders.[207] State armed forces are frequently confronted with a "mosaic war" involving varying levels of control being exercised by non-State actors in situations of armed conflict.[208] The resulting small wars or insurgencies present complex security challenges. In some areas State armed forces may be able to exercise sufficient control to consider utilizing a law enforcement response, while in other territory non-State actors can deny access to their opponents, except when significant military operations are conducted (see Chapter 7, section 7.4.2, *The "Targeted Killing" Case and Control*).

The conduct of a counterinsurgency campaign and other operations aimed at maintaining order in a civilian population present a unique security challenge, since a premium is placed on the employment of police and intelligence services (see Chapter 7, section 7.3, *The Primacy of Law Enforcement*). Even in terms of gathering intelligence it is the police that have been identified as the best organization responsible for "internal security intelligence."[209] Limited police resources or an increased insurgent threat means military forces may also be tasked with performing policing tasks. The essential link between counterinsurgency operations, governance, and the maintenance of law and order is clearly reflected in contemporary military doctrine.[210] For example, it has been suggested that in order to avoid failure in Afghanistan there is a need "to focus on governance, rule of law, anti-corruption doctrine, and protecting the people at the local level."[211]

It is of particular importance for State security officials to realize that successful governance and maintenance of the rule of law are intimately linked with the application human rights norms. A "desired outcome has been expressed in one way or another as a stable state, governed democratically, in which the rule of law functions within international norms, society is developed with due reference to human rights and the economy is run in such a way that the fiscal and monetary arrangements are reliable."[212] The link between governance, human rights, and law and order holds true whether fighting insurgents in a noninternational armed conflict, as an occupying power during international conflict, or seeking to maintain control of areas infested with transnational drug gangs, even when the situation has not reached the level of an armed conflict.

4.5. The State, Governance, and the Intersection of Laws

As with international armed conflict, the State remains the focal point of analysis in considering how order is maintained in conflicts with non-State actors. In part this is because the nature of violence with non-State actors can transcend the assigned categories of conflict.

---

[207] *Anti-Boko Haram Force "to Cross Borders to Fight,"* BBC NEWS (July 29, 2015), http://www.bbc.com/news/world-africa-33702738.

[208] THE U.S. ARMY, MARINE CORPS, COUNTERINSURGENCY FIELD MANUAL, ¶ 1-37, at 14 (2007) [hereinafter COUNTERINSURGENCY MANUAL].

[209] THOMPSON, *supra* note 152, at 85, 104–06. *See also* COUNTERINSURGENCY MANUAL, *supra* note 208, ¶ 6-91, at 229.

[210] COUNTERINSURGENCY MANUAL, *supra* note 208, at ¶¶ 1-23 to1-24, at 8–9. *See also* DOD LAW OF WAR MANUAL, *supra* note 8, at ¶ 17.2.2.4, at 1021.

[211] DAVID KILCULLEN, COUNTERINSURGENCY 161 (2010).

[212] RUPERT SMITH, THE UTILITY OF FORCE: THE ART OF WAR IN THE MODERN WORLD 377 (2005).

For example, the similarity between aspects of international and noninternational armed conflict, and in particular the connection between the State, governance, and insurgency, is evident in the 2006 U.S. *Counterinsurgency Manual*. That doctrine defines "insurgency" as "an organized, protracted politico-military struggle designed to weaken the control and legitimacy of an *established government, occupying power, or other political authority* while increasing insurgent control."[213] The operation of security forces, both police and military, in such complex security environments inevitably requires the application of both humanitarian and human rights norms.

The performance of a governance role by a State means that national laws may also impact on international operations. For example, this could occur as a result of the extraterritorial exercise of disciplinary jurisdiction over military forces,[214] the application of domestic war crimes legislation,[215] or petitioners seeking redress before domestic and regional courts.[216] Conversely, the attacks of 9/11, which were conceived in a foreign jurisdiction but carried out within the territory of the United States, highlights that even internally within a State the exercise of domestic legal authorities relating to national security, investigation, and trial in response to an attack may be of particular importance. This occurs notwithstanding the claim under international law to be acting in self-defense.[217]

The powers and responsibility that come with the legitimate governance of territory is another reflection of the State-centric bias in international law. The special role and responsibility the State has in maintaining internal order also inevitably means that human rights norms will be relevant to situations of armed conflict. Combined with the traditional focus of international law on States in limiting the recourse of war, and the State-dominant approach taken toward legitimate participation in hostilities, these State-centric concepts create considerable space for human rights to be applied in contemporary armed conflict. In order to understand how international law, in its fullest sense, applies to such conflict it is necessary to consider the application not only of international humanitarian and human rights law but also in certain circumstances domestic law as well. An outstanding question is how the humanitarian law and human rights (either domestic or international) based legal frameworks interact in terms of providing a normative framework for the governance of such conflict.

---

[213] COUNTERINSURGENCY MANUAL, *supra* note 208, ¶ 1-2, at 2 (emphasis added).

[214] *Secretary-General's Bulletin, supra* note 139, § 2, at 202 (where it is noted the provisions of the *Bulletin* "do not prejudice the application thereof, nor do they replace the national laws by which military personnel remain bound throughout the operation.").

[215] *E.g.*, in respect of Canada, the *Crimes Against Humanity and War Crimes Act* (S.C. 2000, c. 24); the United Kingdom, the *International Criminal Court Act, 2001*, c. 17; for Australia, the *War Crimes Act, 1945*, Act no. 48 of 1945 as amended; and the United States, the *War Crimes Act of 1996* (P.L. 104-192.).

[216] *See, e.g.*, Al-Skeini v. The United Kingdom, IV Eur. Ct. H. R. Rep. 99 (2011).

[217] THE 9/11 COMMISSION REPORT: FINAL REPORT OF THE NATIONAL COMMISSION ON TERRORIST ATTACKS UPON THE UNITED STATES 326–28 (where reference is made to police investigations, immigration detentions, and collection of intelligence in the immediate aftermath of the 9/11 attack).

International humanitarian law is human rights law, and it is the protection of individuals from death and suffering that must be the central concern of the law.[1]

5 The Humanitarian Law and Human Rights Law Interface

5.1. Introduction

What appears to be a forced separation between the law governing the recourse to war and that controlling the conduct of hostilities is not a unique phenomenon in international law. A similar exclusionary approach arises in contemporary dialogue regarding the application of international humanitarian law and human rights law. The nature of the boundary between these two bodies of law has been an issue that has garnered considerable attention in the post-9/11 period. It raises questions that have particular relevance to whether operations are conducted under a hostilities or a law enforcement paradigm. One of the challenges in discussing the interface between these two bodies of law is that the advocates of each one sometimes appear to want to avoid any consideration of the other legal framework. It is the existence of this phenomenon, the underlying history of the relationship between these two bodies of law, and increasing evidence of an effort to span this divide that is the subject of this chapter.

The following analysis highlights what has been all too often an exclusionary interpretation of the two foundational bodies of law regulating the use of force in modern society. In this regard, much of the contemporary debate has focused on whether it is humanitarian law or human rights law that is the appropriate body of law to govern the use of force during armed conflict. This chapter explores the history of the relationship between these

[1] Richard Baxter, *Humanitarian Law or Humanitarian Politics?*, *in* HUMANIZING THE LAWS OF WAR ¶ 3, at 307 (Detlev F. Vagts et al. eds., 2013) (the text first appeared in 16 HARV. IN'TL L.J. 1 (1976)).

two bodies of law, their common roots, as well as the impact that the periods of neglect, forced integration, divergence, and now growing reconciliation have on how they interface with one another. The analysis highlights that masked behind the exclusionary arguments is the reality that human rights norms have always been an integral part of humanitarian law and military forces have long had to apply an enforcement approach during armed conflict. Increasingly, as States have been required to address threats posed by organized armed groups operating among the people, often in the ungoverned spaces of the world, they have been required not only to consider the overlap of these two bodies of law but also their convergence in application.

5.2. Exclusionary Arguments and Operational Reality

5.2.1. THE "SPECIAL LAW" DEBATE

The resistance to applying human rights law by advocates of humanitarian law is reflected in the preference often voiced by States, and in particular some military legal advisors, for using the terms "law of armed conflict" or "law of war" rather than "international humanitarian law."[2] It has been suggested the use of international humanitarian law[3] can lead to the false impression that the laws governing war have a predominantly humanitarian purpose or may be confused with human rights law.[4] It has even been argued that the conflation of terminology demonstrates a preference for a "pacific" approach.[5] Efforts to exclude consideration of human rights can sometimes be detected in broad claims that the "law of armed conflict" acts exclusively in wartime, thereby replacing human rights law altogether. This discussion occurs on the basis that the law of armed conflict acts as the special body of law (i.e., *lex specialis*) governing such conflict.[6]

---

[2] John B. Bellinger III & William J. Haynes II, *A US Government Response to the International Committee of the Red Cross Study Customary International Humanitarian Law*, 89 INT'L REV. RED CROSS 443, 443 (2007) (where it is noted the United States prefers to use the "law of war" or the "laws and customs of war").

[3] Christopher Greenwood, *The Definition of the Term "Humanitarian Law," in* THE HANDBOOK OF INTERNATIONAL HUMANITARIAN LAW ¶ 1, at 11 (Dieter Fleck ed., 2nd ed. 2008) [hereinafter HUMANITARIAN LAW HANDBOOK] (where international humanitarian law is expansively defined to comprise "all those rules of international law which are designed to regulate the treatment of the individual—civilian or military, wounded or active—in international armed conflicts"). The increasing acceptance of the term is reflected in the *Advisory Opinion on the Legality of the Threat or Use of Nuclear Weapons*, 1996 I.C.J. 226, ¶ 75, at 256. *See also* DEPARTMENT OF DEFENSE LAW OF WAR MANUAL, Office of the General Counsel ¶ 1.3.1.2, at 8 (June 2015) [hereinafter DOD LAW OF WAR MANUAL], http://www.dod.mil/dodgc/images/law_war_manual15.pdf (where it is indicated that international humanitarian law can have the same substantive meaning as the law of war, however, "[i]n other cases, international humanitarian law is understood more narrowly than the law of war (*e.g.*, by understanding international humanitarian law not to include the law of neutrality).").

[4] YORAM DINSTEIN, THE CONDUCT OF HOSTILITIES UNDER THE LAW OF INTERNATIONAL ARMED CONFLICT ¶ (ii), at 19 (2nd ed. 2010). *See also* DOCUMENTS ON THE LAWS OF WAR 2 (Adam Roberts & Richard Guelff eds., 3rd ed. 2003) (where it is noted "it could be seen as implying that the laws of war have an exclusively humanitarian purpose, when their evolution has in fact reflected various practical concerns of states and their armed forces on grounds other than those which may be considered humanitarian.").

[5] GARY D. SOLIS, THE LAW OF ARMED CONFLICT: INTERNATIONAL HUMANITARIAN LAW IN WAR 23 (2010).

[6] Andrea Gioia, *The Role of the European Court of Human Rights in Monitoring Compliance with Humanitarian Law in Armed Conflict, in* INTERNATIONAL HUMANITARIAN LAW AND INTERNATIONAL HUMAN RIGHTS LAW 201, 213 (Orna Ben-Naftali ed., 2011). *See also* Michael J. Dennis, *Application of Human Rights Treaties Extraterritorially in Times of Armed Conflict and Military Occupation*, 99 AM. J. INT'L L. 119, 141 (2005).

There is always a chance that terminology will unduly influence consideration of the scope and complexity of the underlying law. However, it is not clear why the use of the term "humanitarian" itself should be viewed as problematic.[7] In that regard the iconic Martens clause, found in the Preamble to a seminal document on the conduct of hostilities, the 1907 Hague Land Warfare Regulations, provides for "the protection and the rule of the principles of the law of nations, as they result from the usages established among civilized peoples, from *the laws of humanity* and the dictates of public conscience." The term "laws of humanity" is recognized in a foundational "law of war" document. It is also an accurate descriptor of the nature of the movement commenced in the nineteenth century to limit not only the horror of war but also recourse to such conflict (see Chapter 2, section 2.3.1., *Reintroduction of Just War Theory*). Significantly, as will be discussed, by downplaying such a link through the use of the term "law of armed conflict," proponents of that legal framework actually may be offering increased space within which human rights law can be said to operate.

There is some indication of the emergence of a less defensive approach toward the use of the term "humanitarian law." In the 2010 *Commentary on the HPCR Manual on International Law Applicable to Air and Missile Warfare*, terms such as "international humanitarian law," "*jus in bello*," "law of war," and the "law of armed conflict" were used synonymously, as their use was seen as a "a semantic rather than a substantive issue."[8] Notwithstanding this approach there still appears to be reluctance to address the full range of international law applicable to contemporary conflict. For example, the *Law of Air Warfare Commentary* avoided considering human rights law on the basis that "most members of the Group of Experts believe it has only minimal bearing on air and missile warfare in international armed conflicts because the law of armed conflict is *lex specialis* [the special law]."[9]

This raises the issue of the impact of the debate about specialized bodies of law. The rationale of the *lex specialis* principle was described in the 2006 study, *Fragmentation Of International Law: Difficulties Arising From The Diversification And Expansion Of International Law*, as follows:

> That special law has priority over general law is justified by the fact that such special law, being more concrete, often takes better account of the particular features of the context in which it is to be applied than any applicable general law. Its application may also often create a more equitable result and it may often better reflect the intent of the legal subjects.[10]

---

[7] René Provost, International Human Rights and Humanitarian Law 55 (2002) ("The language and philosophy of human rights has permeated international law as whole, and humanitarian law is no exception.").

[8] Commentary on the HPCR Manual on International Law Applicable to Air and Missile Warfare, Program on Humanitarian Policy and Conflict Research at Harvard University, ¶ 3, at 41 (2010) [hereinafter Air and Missile Warfare Manual Commentary], http://ihlresearch.org/amw/Commentary%20on%20the%20HPCR%20Manual.pdf.

[9] *Id. Introduction* D(d), at 6.

[10] *Fragmentation Of International Law: Difficulties Arising From The Diversification And Expansion Of International Law*, Rep. of the Int'l L. Comm'n, 58th session, May 1–June 9, July 3–Aug. 11, 2006, UN Doc. A/CN.4/L.702, ¶ 14(7), at 9 (July 18, 2006).

However, a particular challenge for advocates of an exclusionary approach toward interpreting this principle is that the study also recognized the "application of the special law does not normally extinguish the relevant general law" and that body of law continues "to give direction for the interpretation and application of the relevant special law and will become fully applicable in situations not provided for by the latter."[11] This suggests that the general law, human rights law, as well as the normative principles associated with that body of law used to interpret humanitarian law could continue to have relevance, even in respect of the conduct of air warfare.

It is resistance to any residual operation of human rights law that is at the heart of the exclusionary approach. This is an approach that views "the specialty of IHL [international humanitarian law] operating at a very general level, so that IHL would replace IHRL [international human rights law] altogether in times of armed conflict."[12] The approach has generally been attributed to states such as the United States and Israel, [13] although in 2015 a U.S. Army doctrine manual, the *Operational Law Handbook*, indicated the country no longer embraced this "displacement" theory.[14]

Concerns about the potential overreach of the human rights–based framework may not be entirely unfounded. Interestingly, those advocating for an exclusive human rights approach also focus on the concept of there being a special body of law governing conflict. As a result, the approach suggested by the International Court of Justice in the 1996 *Advisory Opinion on the Legality of the Threat or Use of Nuclear Weapons* to assess the arbitrary deprivation of life during conflict has increasingly become a focal point for the dispute over, and ultimately control of, which paradigm should apply during armed conflict.[15] While not all commentators adopt an exclusive human rights approach,[16] the special law principle is often interpreted in a manner that downplays the scope and impact of humanitarian law. This occurs particularly in respect of noninternational armed conflict.

The view that a significant part of the human rights law community favors an exclusionary legal interpretation was undoubtedly enhanced by the traditional approach adopted by the European Court of Human Rights of uniquely applying human rights law when assessing the use of force in armed conflict.[17] This approach is in the process of changing. In *Hassan v. the United Kingdom*[18] the court recognized that international humanitarian law could be used to interpret rights found in the European Convention on Human

---

[11] *Id.* at ¶ (9), at 9–10.

[12] Gioa, *supra* note 6, at 213.

[13] *Id.* (where this approach has been called "radical").

[14] Operational Law Handbook, §V. A. ¶ 1., at 54 (Lieutenant-Commander David Lee, ed., 2015), http://www.loc.gov/rr/frd/Military_Law/pdf/operational-law-handbook_2015.pdf ("The United States embraced this view until very recently."). *But see* DOD Law of War Manual, *supra* note 3, ¶ 1.6.3.1, at 22 ("apparent conflicts may be resolved by the principle that the law of war is the *lex specialis* during situations of armed conflict, and, as such, is the controlling body of law with regard to the conduct of hostilities and the protection of war victims.").

[15] Advisory Opinion on the Legality of the Threat or Use of Nuclear Weapons 1996 I.C.J. 226, ¶ 25, at 240.

[16] *E.g.*, Gioia, *supra* note 6, at 214 (where it is suggested what can be used in most cases is the *lex specialis completat legi generali* principle, where both branches of the law are applied simultaneously, such that human rights norms "have to be interpreted in light of IHL norms.").

[17] *Id.* at 247 (where it is stated "[i]t is, therefore, regrettable that the Court has so far largely ignored IHL when deciding cases where the interested state was clearly involved in an armed conflict.").

[18] Hassan v. The United Kingdom, Eur. Ct. H. R., Application No. 29750/09, ¶ 103, at 53 (Sept. 16, 2014), http://hudoc.echr.coe.int/sites/eng/pages/search.aspx?i=001-146501.

Rights. This decision aligned the position of that court more closely with the International Court of Justice (ICJ), which in addition to the *Advisory Opinion on the Legality of the Threat or Use of Nuclear Weapons* had recognized the application of both humanitarian and human rights law in the *Legal Consequences Of The Construction Of A Wall In The Occupied Palestinian Territory*,[19] and the *Case Concerning Armed Activities on the Territory of the Congo (DRC v. Uganda)*.[20]

However, a number of questions remain. Because the *Hassan v. The United Kingdom*[21] case dealt with a period of occupation during an international armed conflict, it is not clear whether the European court will make a similar finding with respect to noninternational armed conflict. It is also not evident how it will reconcile this approach with earlier decisions that have applied human rights law principles in cases involving the use of aerial bombing and other weapons of war in internal conflict.[22] Of course, the use of airpower is a quintessential combat activity that historically is regulated by international humanitarian law. The result is that the European Court of Human Rights may continue to favor human rights law, rather than humanitarian law during internal or other noninternational armed conflicts. Since these are the dominant forms of armed conflict, the impact of the European Court decision may not be as significant as it might first appear. The result may be a significant overreach on the part of human rights courts and advocates when attempts are made to regulate the levels of violence associated with combat (see Chapter 13, section 13.4, *Human Rights Overreach*). Further, decisions by the European Court of Human Rights are not necessary determinative of the special law issue, and courts in other jurisdictions may choose to handle this issue differently by continuing to broadly apply human rights law. Human rights advocates may also continue to argue for an exclusionary application of human rights law, particularly in respect of noninternational armed conflict.

In terms of favoring a human rights approach, one commentator has even suggested that body of law can effectively exclude humanitarian law in respect of noninternational armed conflicts since "[t]he rationale that makes resort to humanitarian law as *lex specialis* [specialized law] appealing—that the rules have greater specificity—is missing in internal armed conflicts . . . the humanitarian law of internal armed conflict is quite spare and seldom specific."[23] It has even been referred to as the "human rights law of internal armed conflict"

---

[19] Legal Consequences Of The Construction Of A Wall In The Occupied Palestinian Territory, 2004 I.C.J. ¶¶ 105–06, at 177–78.

[20] Case Concerning Armed Activities on the Territory of the Congo (DRC v. Uganda), [December 19, 2005] 45 ILM 271, ¶¶ 215–20, at 317–19 (2006).

[21] Hassan v. The United Kingdom, Eur. Ct. H. R., Application No. 29750/09 ¶ 104, at 53 (Sept. 16, 2014), http://hudoc.echr.coe.int/sites/eng/pages/search.aspx?i=001-146501. *But see* Case of Cruz Sánchez et al. v. Peru, Series C no. 292, Preliminary Objections, Merits, Reparations and Costs, Inter-Am. Ct. Hum. (2015), http://www.corteidh.or.cr/docs/casos/articulos/seriec_292_esp.pdf (where the Inter-American Court of Human Rights applied humanitarian law to interprete human rights law in respect of an internal armed conflict).

[22] *See, e.g.*, Isayeva v. Russia, Application Nos. 57947/00, 57948/00, 57949/00, Eur. Ct. H. R. (2005), http://hudoc.echr.coe.int/sites/eng/pages/search.aspx?i=001-68379, and Isayeva v. Russia, Application No. 57959/00, Eur. Ct. H. R. (2005 http://hudoc.echr.coe.int/sites/eng/pages/search.aspx?i=001-68381, and Kerimova v. Russia, Application Nos. 17170/04, 20792/04, 22448/04, 23360/04, 5681/05 and 5684/05 Eur. Ct. H. R. (May 3, 2011), http://hudoc.echr.coe.int/sites/eng/pages/search.aspx?i=001-104662.

[23] William Abresch, *A Human Rights Law of Internal Armed Conflict: The European Court of Human Rights in Chechnya*, 16 Eur. J. Int'l L. 741, 747 (2005).

and a "construct of scholars" that can be broken down into two schools of thought.[24] The first is the fully exclusionary one mentioned above, while the second school applies human rights law to low-intensity conflict and humanitarian law to a high-intensity one, with Additional Protocol II acting as the threshold between the two.[25]

A particular challenge in applying either of the human rights-based exclusionary approaches is that armed conflict can occur between States and non-State actors (e.g., applying the degree of organization and intensity of violence criteria for noninternational armed conflict) in situations well below the Additional Protocol II threshold, including when acting in self-defense (see Chapter 14, section 14.3. *Responding to Transnational Attacks and Threats to Nationals*). Along similar lines, in respect of procedural guarantees for arrested persons, it is suggested that human rights law must prevail since it is considered such rules do not exist under humanitarian law. Human rights rules are seen as "more precise and more restrictive."[26] However, in most cases that conclusion would undoubtedly require consideration of human rights case law and policy documents, rather than simply the text of the treaties themselves. It is also an area where the laws and policies of domestic jurisdictions could have particular relevance.

The debate also has to address the existence of customary international humanitarian law applicable to noninternational armed conflicts, which would act as the special law. However, one approach has been to prefer rights articulated in human rights "texts" generated in the post–World War II period over reliance on interpretations of customary law. Questioning the scope and content of customary international humanitarian law makes greater room for the application of what is viewed as more specific textual references in human rights law.[27] This extends to describing a customary rule prohibiting arbitrary detention identified in the ICRC *Customary Humanitarian Law Study* somewhat tentatively as an "alleged" humanitarian law rule.[28] Ironically, the resistance sometimes shown by States toward articulating customary international humanitarian law standards, particularly in respect of internal conflicts, helps create the perception of a gap in the application of humanitarian law within which such a debate can take place.[29]

The traditional State-centric resistance to introducing humanitarian law to internal conflicts increases the likelihood that human rights law, either international or domestic, will be seen to govern military operations. It is an approach that also appears to run counter to the broad policy-based application of humanitarian law that guides a number of States in respect of their international operations, including involvement in noninternational armed

---

[24] Sandesh Sivakumaran, *Re-evisaging the International Law of Internal Armed Conflict*, 22 EUR. J. INT'L L. 219, 235 (2011).

[25] *Id.* at 235–36.

[26] Marco Sassoli, *The Role of Human Rights and International Humanitarian Law in New Types of Armed Conflicts*, *in* INTERNATIONAL HUMANITARIAN LAW AND INTERNATIONAL HUMAN RIGHTS LAW, *supra* note 6, at 92.

[27] Abresch, *supra* note 23, at 750 ("When . . . states are reluctant to acknowledge the applicability of treaty-based humanitarian law, it is difficult to credit that they consider themselves bound under customary humanitarian law.").

[28] Sassoli, *supra* note 26, at 92.

[29] *See, e.g.*, Bellinger & Haynes, *supra* note 2, at 448 (in this article two senior U.S. government lawyers express concern regarding the conclusions of the CUSTOMARY INTERNATIONAL HUMANITARIAN LAW STUDY (Jean-Marie Henckaerts & Louise Doswald-Beck eds., 2005), which were seen to generally seek to expand AP I and make certain rules binding in internal armed conflict).

conflicts (see Chapter 4, section 4.4.4, *A Policy Solution for Legal Uncertainty*).[30] If a State takes the position there is no extraterritorial application of human rights treaty law, this policy-based approach appears to be designed to fill a perceived void in the application of international law. However, such policies cannot be applied as part of the specialized law debate. That debate is about the law, not policy. A policy application of humanitarian law cannot exclude or trump human rights law.

In what appears, on its face, to be a reversal of the special law and general law relationship for these two bodies of law, as articulated by the ICJ, it has been suggested that human rights law acts as its own form of special law.[31] In this approach the special law concept is not applied as a principle of interpretation where a general rule is interpreted in light of a more specific one, but rather "as a rule governing conflicting norms."[32] In effect, it is viewed almost as a contest between two forms of specialized law: international humanitarian law and human rights law, where "the lex specialis [special law] principle . . . gives precedence to the rule that is most adapted and tailored to the specific situation."[33] Given these variety of approaches, it is fairly said that the special law principle framed in its "legalese Latin" is "descriptively misleading, vague in meaning, and of little practical use in application,"[34] and that "[s]ome way needs to be found to develop a coherent approach to the problem."[35]

There are a number of problems with a conflict of norms approach. An obvious challenge is the *Advisory Opinion on the Legality of the Threat or Use of Nuclear Weapons* identification of international humanitarian law as the special law in relation to the general human rights rule. Reference was also made to this principle in the subsequent *Legal Consequences Of The Construction Of A Wall In The Occupied Palestinian Territory* decision.[36] The International Court of Justice seems to have favored the first principle of interpretation rather than the suggested "conflicting norms" approach. This is also the approach that appears to have been adopted in *Hassan v. the United Kingdom*.[37] Specific reference in that case is made to the UN *Analytical Report of the Fragmentation of International Law Study* where it is noted in respect of the interaction between humanitarian law (as the special law) and human rights law:

> [The special law] did hardly more than indicate that though it might have been desirable to apply only human rights, such a solution would have been too idealistic,

---

[30] For example, *see* the DoD Law of War Program, Department of Defense, Directive Number 2311.01E, May 9, 2006 (Incorporating Change 1, November 15, 2010, Certified Current as of February 22, 2011) ¶ 4.1, at 2, http://www.dtic.mil/whs/directives/corres/pdf/231101e.pdf, and the Canadian LAW OF ARMED CONFLICT AT THE OPERATIONAL AND TACTICAL LEVELS, B-GJ-005-104/FP-021 ¶ 1702. 2., at 17–1 (2001-08-13).

[31] Cordula Droege, *Elective Affinities? Human Rights and Humanitarian Law*, 90 INT'L REV. RED CROSS 501, 536 (2008), and Sassoli, *supra* note 26, at 86 (where the author suggests international human rights law applies as *the lex specialis* where government forces control the place where an attack occurs when acting on their own territory.).

[32] Droege, *supra* note 31, at 524. (This interpretation is based on Martti Koskenniemi, *Study on the Function and Scope of the Lex Specialis Rule and the Question of "Self Contained Regimes,"* 4 UN Doc. ILC(LVI)/SG/FIL/CRD.1 and Add. (2004)).

[33] *Id. See also* Abresch, *supra* note 23, at 749.

[34] Marko Milanović, *Norm Conflicts, International Humanitarian Law and Human Rights Law*, *in* INTERNATIONAL HUMANITARIAN LAW AND INTERNATIONAL HUMAN RIGHTS LAW, *supra* note 6, at 124.

[35] Françoise J. Hampson, *The Relationship between International Humanitarian Law and Human Rights from the Perspective of a Human Rights Treaty Body*, *in* 90 INT'L REV. OF THE RED CROSS 549, 562 (2008).

[36] Legal Consequences Of The Construction Of A Wall In The Occupied Palestinian Territory, 2004 I.C.J. Reports 136.

[37] Hassan v. The United Kingdom, Eur. Ct. H. R., Application No. 29750/09, ¶¶ 102–104, at 53–53 (Sept. 16, 2014), http://hudoc.echr.coe.int/sites/eng/pages/search.aspx?i=001-146501.

> bearing in mind the speciality and persistence of armed conflict. So the Court created a systemic view of the law in which the two sets of rules related to each other as today's reality and tomorrow's promise, with a view to the overriding need to ensure "the survival of a State".[38]

Further, "[t]he important point to retain here is that when *lex specialis* [the special law] is invoked as an exception to the general law then what is being suggested is that the special nature of the facts justifies a deviation from what otherwise would be the 'normal' course of action."[39] Human rights best represents normality (i.e., the general law), and not a special body of law (see Chapter 14, section 14.6, *Law Enforcement and the Search for "Normalcy"*). With an emphasis being placed on pragmatism, this interpretation appears most consistent with the ICJ jurisprudence.[40]

One argument that has tried to limit the impact of the *Advisory Opinion on the Legality of the Threat or Use of Nuclear Weapons* case is to suggest that since the court made no reference to the special law principle in *Case Concerning Armed Activities on the Territory of the Congo (DRC v. Uganda)*,[41] "it is not clear whether the omission was deliberate and shows a change in the approach of the Court."[42] However, it is not immediately evident how not referring to the previous cases constitutes a reversal of those decisions. It has been rightly suggested, "this lapse does not prove much."[43] Further, the reference to the three International Court of Justice decisions in *Hassan v. United Kingdom* suggests this line of argument has not gained traction.[44]

Second, it is not clear human rights law always provides a level of detail that, from either a conventional or customary law perspective, means it should act to exclude humanitarian law as a form of special law. As has been noted, "the Universal Declaration's regulation of the right to life in Articles 3 and 29 remains too general to provide practical guidance for the lawful use of lethal force under the law enforcement paradigm."[45] The detail is left to soft law instruments such as the *Basic Principles on the Use of Force and Firearms by Law Enforcement Officials*,[46] or decisions rendered largely by regional or domestic tribunals or courts. One of the challenges with "soft" law is that it is not law in the sense of being legally binding.[47]

---

[38] *Id.* ¶ 38, at 20-21. *See also Analytical Report, Fragmentation Of International Law: Difficulties Arising From The Diversification And Expansion of International Law*, Report of the Study Group of the International Law Commission Fifty-eighth session Geneva, 1 May–9 June and 3 July–11 August 2006, UN Doc. A/CN.4/L.682 13 April 2006, ¶ 104, at 57 (hereinafter *Analytical Report*).

[39] *Analytical Report, supra* note 38, ¶ 105, at 57.

[40] *Id.* at 57–58.

[41] Case Concerning Armed Activities on the Territory of the Congo (DRC v. Uganda), [December 19, 2005] 45 ILM 271 (2006).

[42] Droege, *supra* note 31, at 522. *See also* Nancie Prud'homme, *Lex Specialis: Oversimplifying a More Complex and Multifaceted Relationship?*, 40 ISR. L. REV. 356, 385 (2007) (where it is suggested the *Congo Case* is a "jurisprudential retreat from the principle of *lex specialis*.").

[43] Dinstein, *supra* note 4, ¶ 55, at 23.

[44] Hassan v. The United Kingdom, Eur. Ct. H. R., Application No. 29750/09 ¶¶ 35–37, at 19–20 (Sept. 16, 2014), http://hudoc.echr.coe.int/sites/eng/pages/search.aspx?i=001-146501.

[45] Nils MELZER, TARGETED KILLING IN INTERNATIONAL LAW 196 (2008).

[46] *UN Basic Principles on Use of Force.*

[47] *Frequently Asked Questions on International Humanitarian, Human Rights and Refugee Law in the context of armed conflict*, Inter-Agency Standing Committee Task Force on Humanitarian Action and Human Rights (2004), http://

Therefore, like the State approach of applying policy to fill normative voids, human rights advocates relying on soft law appear to be embarked on a similar exercise. Their claim for special law status based on a "soft" law underpinning appears similarly tenuous. Finally, the conflict of norms approach seems fundamentally exclusionary in nature. One body of law displaces the other. Unfortunately, it does not appear to account for human rights norms being part of humanitarian law (see Chapter 5, section 5.3.2.3, *Divergence and Growing Reconciliation*).

The complexity of the special law debate is reflected in the numerous interpretive theories that have been developed. One analysis suggests three possible models exist:

> First is the Displacement Model. The Displacement Model has the virtue of simplicity: During an armed conflict, humanitarian law displaces human rights law. When no armed conflict exists, human rights law displaces humanitarian law. Second is the Complementarity Model. Complementarity is relatively simple in theory, though substantially more complicated in practice. In the Complementarity Model, as in all the models, when there is no armed conflict, only human rights law applies. When there is an armed conflict, however, human rights law and humanitarian law are applied and interpreted harmoniously. The two bodies of law thus have what this Article terms a "relationship of interpretation." Third is the Conflict Resolution Model. In the Conflict Resolution Model, when an armed conflict is present, the decision maker must evaluate the relationship between human rights law and humanitarian law. If they are, in fact, complementary, then both are applied. If they conflict, however, the model offers three possible decision rules—event-specific displacement, reverse event-specific displacement, and specificity—for deciding the appropriate body of law to be applied.[48]

In favoring the conflict resolution model it was argued that "humanitarian law is often—but not always—the body of law most specifically tailored to armed conflict situations. Merely because a given situation occurs within an armed conflict zone does not necessarily preclude the application of human rights law."[49]

It seems much of the reason for this complexity has been the focus on how the two bodies of law differ combined with significant efforts by advocates of each to exclude the operation of the other. It also has been suggested that the characteristics of the special law model, "and the environments it operates in, leads quite easily to a clear conclusion as to its inability to articulate an agreeable and legally sound theoretical model for the parallel application of international humanitarian law and international human rights law."[50] Having questioned the utility of this approach, the suggestion is made to utilize other criteria such as the type

---

www.refworld.org/docid/4a54bc03d.html ("'Soft' international law is an important body of non-treaty standards usually adopted within the framework of the United Nations system (declarations, bodies of principles, standard minimum rules, etc.). Although not legally binding, soft law serves to interpret and elaborate treaty provisions and to develop new standards in emerging areas of international law.").

48 Oona A. Hathaway et al., *The Relationship Between International Humanitarian Law and Human Rights Law in Armed Conflict*, 96 MINN. L. REV. 1883, 1886 (2012).

49 *Id.* at 1943.

50 Prud'homme, *supra* note 42, at 384.

of conflict, protected person, and right involved.[51] However, this theory, in effect, ends up with an exclusionary model where "the closer one is to non-international armed conflict and peace-time situations, the more arguments will exist in favor of applying international human rights law, or the most protective rule available."[52] A problem with this approach is that it must be reconciled with the view that "rather than stretching human rights law to armed opposition groups, efforts should be directed towards promoting the application of Common Article 3 and Protocol II in all situations that arguably constitute internal armed conflicts within the meaning of these rules."[53]

This does not mean that States cannot simply choose as a matter of policy to apply a human rights–based law enforcement approach to deal with contemporary insurgencies or terrorism (see Chapter 14, section 14.3.3.2, *A Policy Choice by States*). In the case of Northern Ireland, the UK government chose to apply a human rights (in this case domestic law) based law enforcement approach to address an insurgency being conducted by the Irish Republican Army.[54] However, making such a choice must be done with a full understanding of the limitations and challenges of applying a human rights–based approach in terms of the ability of that normative paradigm to counter the types of groups and levels of violence that are common in contemporary conflict. For example, Northern Ireland is all too often held up as an example where a law enforcement approach was successfully applied to counter an insurgency/terrorism without a deeper discussion of the unique circumstances that resulted in success in that situation (see Chapter 13, section 13.2.1, *The "Troubles"*).

Ultimately, this debate about special law and exclusion appears to depend upon the two bodies of law, humanitarian law and human rights law, being viewed as fundamentally different and needing to be applied completely separately from one another. The international legal community will likely continue to struggle in coming to terms with what it means to have a special body of law (i.e., humanitarian law) apply in time of armed conflict. It does appear that the approach of treating the special law principle as an interpretive tool can provide the needed coherent legal approach for resolving the technical legal interface between humanitarian law and human rights law.[55] What it is not evident is that focusing on these differences or seeking a technical legal solution will fundamentally address a key operational challenge presented in contemporary conflicts. The reality is that military commanders must be prepared not only to fight but also to maintain order and security. As will be addressed in the next section, when the similarities between humanitarian and human rights law are considered, the debate over what normative framework to apply can become far less polarized.

---

[51] *Id.* at 391.

[52] *Id.*

[53] Liesbeth Zegveld, The Accountability of Armed Opposition Groups in International Law 53 (2002).

[54] Steven Haines, *Northern Ireland 1968–1998*, *in* International Law and the Classification of Conflicts 117, 130 (Elizabeth Wilmshurst ed., 2012).

[55] Marko Milanović, Extraterritorial Application of Human Rights Treaties: Law, Principles, and Policy 251 (2011).

5.2.2. THE MISSING DISCUSSION: SIMILARITIES AND OVERLAP

Overall, what appears to be missing from the dialogue about the interface between humanitarian and human rights law is a focus, not on the differences between the two normative regimes, but rather their similarities and considerable overlap. It is not obvious to what extent certain "norms" are actually in conflict, particularly in respect of the performance of law enforcement duties. An area of interest has been the use of force against non-State actors.[56] To be sure, the differences between conduct of hostilities–based norms governing the use of force and the human rights norms applicable to law enforcement are well established. Further, each body of law plays its own essential role in terms of States being able to maintain order. The normative framework associated with human rights law is intimately linked to governance and policing where a State exercises effective control over a society. This is a specific area of concentration for that body of law.

The law enforcement framework seeks to limit the use of force to situations of absolute necessity;[57] to only permit the use force when it is strictly unavoidable;[58] to apply the principles of necessity and proportionality to any use of force; and to only allow the use of firearms "in self-defense or the defense of others against the imminent threat of death or serious injury; to prevent a particularly serious crime involving grave threat to life; to arrest a person presenting such a danger and resisting their authority; or to prevent his or her escape."[59] Importantly, under the law enforcement paradigm the preference is to capture rather than kill a suspect,[60] and there is a well-established accountability regime seeking investigations of unlawful killings.[61]

Of particular note, a "shoot to kill" policy is to be strictly avoided. This term was used in the context of allegations of excessive use of force by security forces in Northern Ireland,[62] as well as to describe a means of counterterrorism that is contrasted with a police focused investigatory, forensic, and surveillance effort.[63] When looked at in the context of post-9/11 counterterrorism operations, it has been suggested "[t]he use of shoot-to-kill tactics

---

[56] Sassoli, *supra* note 26, at 78 (where he notes that the areas where the two branches of law appear to offer divergent solutions "are when a fighter may be attacked, and when and under what procedures a fighter may be detained."), and Droege, *supra* note 31, at 527–39.

[57] McCann v. United Kingdom, Application No. 18984/91, Eur. Ct H. R. ¶ 148 (1995), http://hudoc.echr.coe.int/sites/eng/pages/search.aspx?i=001-57943, and *UN Basic Principles on the Use of Force*, at ¶ 9.

[58] *UN Basic Principles on Use of Force*, at ¶ 9.

[59] 1 PUBLIC COMMISSION TO EXAMINE THE MARITIME INCIDENT OF 31 MAY 2010 at 232 (2011) [hereinafter FIRST TURKEL REPORT], and the *UN Basic Principles on Use of Force*, at ¶ 9.

[60] Public Committee Against Torture in Israel v. Israel, Israel Supreme Court [16 December 2006], 46 ILM 375, ¶40, at 393–94 (2007) (discussing the application of human rights based principles regarding law enforcement and the preference for arrest, investigation, and trial.).

[61] *United Nations Principles on the Effective Prevention and Investigation of Extra-legal, Arbitrary and Summary Executions*, E.S.C. res. 1989/65, annex, 1989 U.N. ESCOR Supp. (No. 1) at 52, UN Doc. E/1989/89 ¶ 9 (1989).

[62] McKerr v. United Kingdom, Application No. 28883/95, ECtHR, Judgment ¶ 100 (2001), http://hudoc.echr.coe.int/sites/eng/pages/search.aspx?i=001-59451. *See also* PETER HARCLERODE, SECRET SOLDIERS: SPECIAL FORCES IN THE WAR AGAINST TERRORISM 147, 535, 542–43 (2000) (for reference to concern a "shoot to kill" policy was being applied by United Kingdom security forces against the Irish Republican Army in the 1980s), and MELZER, *supra* note 45, at 23–24.

[63] TONY GERAGHTY, THE IRISH WAR: THE MILITARY HISTORY OF A DOMESTIC CONFLICT 75 (1998) (where reference is made to a senior officer indicating a "shoot to kill" policy was favored by a majority of officers as the principle means of defeating the IRA, rather than a post-1975 attrition strategy designed to "play it long.").

also imports, either consciously or otherwise, the language of international humanitarian law into situations which are essentially matters of law enforcement that international law requires be dealt with within the framework of human rights."[64] However, this appears to be a mischaracterization to the extent the phrase "shoot to kill" is not found in either humanitarian or human rights law. Further, there is a difference between a broad-based authority to shoot an enemy due to their membership in an organized armed group or their direct participation in hostilities, and the question of whether the authority to use deadly force under a human rights–based law enforcement regime is being too liberally exercised (see Chapter 13, section 13.2.2, *Shooting to Kill*).[65]

Unfortunately, an exclusionary approach that situates the law enforcement paradigm uniquely within human rights law does not adequately recognize the application of human rights norms within the framework of humanitarian law (e.g., while maintaining order during periods of occupation) (see Chapter 7, section 7.3, *The Primacy of Law Enforcement*).[66] What reference to the principles governing law enforcement does do is help identify a point of interface between the application of conduct of hostilities norms and their human rights–based law enforcement counterpart. However, this interface occurs not only in terms of the two bodies of law, human rights law and humanitarian law, but also internally within humanitarian law. For example, the conduct of hostilities rules governing operations against insurgents in occupied territory must ultimately be reconciled with the humanitarian law obligations of an occupying power to maintain law and order.[67] Other more pragmatic terminology that has been used to address the point of division is the application of deadly force "as a measure of first resort."[68] This is intended to represent force used outside a law enforcement context. It offers a useful contrast to the law enforcement approach that seeks to apply force as "a last resort." However, its use must be approached with caution as it might create the impression force can be used precipitously. This is because of the need to also recognize the obligation to apply the precautionary measures under humanitarian law before launching an attack in the "first resort" (see Chapter 12, section 12.2.1, *The Military Task: Applying Force of a First and Last Resort*).

In contrast to the law enforcement paradigm, the use of force to kill an opponent during the conduct of hostilities is an inherent part of military operations and is governed by humanitarian law. That law relies on the principle of distinction and a different concept of proportionality that accepts a level (i.e., less than excessive) collateral civilian death, injury, or destruction.[69] These targeting provisions are only applicable against combatants,

---

64 Extrajudicial, summary or arbitrary executions, *Civil And Political Rights, Including The Questions Of Disappearances And Summary Executions*, Comm'n. on Human Rights, UN Doc. E/CN.4/2006/53 ¶ 46, at 18 (Mar. 8, 2006) (by Philip Alston), http://www.unhcr.org/refworld/pdfid/45377b100.pdf.

65 James Adams et al., Ambush: The War Between the SAS and the IRA 89 (1988) (outlining that there was not a deliberate shoot to kill policy).

66 Kenneth Watkin, *Use of Force during Occupation: Law Enforcement and Conduct of Hostilities*, 94 Int'l Rev. Red Cross 267, 310–14 (2012). *See also* Melzer, *supra* note 45, at 157–67 (discussing the obligations of belligerent occupation).

67 Hague Land Warfare Regulations, art. 43.

68 Geoffrey Corn, *Mixing Apples and Hand Grenades: The Logical Limit of Applying Human Rights Norms to Armed Conflict*, in 1 J. Int'l Hum. Leg. Stud. 52, 74–75 (2010).

69 AP I, arts. 51(5)(b) and 57.

members of organized armed groups, civilians taking a direct part in hostilities, and other "military objectives."[70] As a general statement of law, the consideration of capture is not required before killing when the targeted person is a lawful military objective.[71] Consistent with a strategic legal dialogue that is often exclusionary in nature, humanitarian law advocates frequently, but quite wrongly, seem to begin and end their discussion on the use of force by stating this legal basis for a broader right to kill.

Like human rights law, accountability is also an important aspect of humanitarian law. As Article 87 of Additional Protocol I notes, military commanders have an obligation "to prevent and, where necessary, to suppress and to report to competent authorities breaches of the Conventions and of this Protocol." The international criminal law framework, dealing with alleged breaches of international law as war crimes and crimes against humanity, is unique, as it derives its origins from humanitarian law, human rights law, and national criminal law.[72] That accountability is maintained primarily by nationally based humanitarian law investigations conducted under military and civilian justice systems, although increasingly international tribunals are making an impact. [73] Therefore, in terms of accountability there are various bodies of law, including the domestic human rights principles applicable to national military and civilian justice systems, which are relevant to the conduct of contemporary operations in armed conflict.

The application of human rights norms found in humanitarian law can be seen in the obligations regarding treatment of uninvolved civilians during armed conflict. In the conduct of military operations the rules governing the use of force against civilians not taking a direct part in hostilities, such as during the maintenance of "public order and safety" in occupied territory, are, by default, law enforcement related even though the obligation is a humanitarian law one.[74] In this respect international humanitarian law incorporates substantive rights, such as the right to life, which are the basis for the law enforcement model.[75] Such norms are also incorporated into customary international humanitarian law based on the "the general principles of law recognized by civilized nations."[76] This is reflected in the ICRC commentary to the Third Geneva Convention, wherein the rights contained in Common Article 3 are described as "certain rules, which were already recognized as essential in all civilized countries, and embodied in the national legislation of the States in question, long before the Convention was signed."[77]

---

[70] *See, e.g.*, AP I, art. 52(2) (identifying military objects).

[71] W. Hays Parks, *Memorandum of Law: Executive Order 12333 and Assassination*, THE ARMY LAWYER, 7 n.6 (1989), http://www.loc.gov/rr/frd/Military_Law/pdf/12-1989.pdf (they canot be targeted if they have surrendered or are "hors de combat").

[72] ANTONIO CASSESE, INTERNATIONAL CRIMINAL LAW 6 (2008).

[73] ROBERT CRYER ET AL., AN INTRODUCTION TO INTERNATIONAL CRIMINAL LAW AND PROCEDURE 64 (2nd ed. 2010) (where it is noted that such crimes "have been regulated in domestic law the longest and have been prosecuted most often.").

[74] 1907 Hague Land Warfare Regulations, art. 43. *See* Watkin, *supra* note 66.

[75] *See, e.g.*, 1907 Hague Land Warfare Regulations, art. 46 (this provision includes the requirement to expect family honor and rights, the lives of persons, religious convictions, and private property). GC IV, art. 27, AP I, art. 75, AP II, art. 4.

[76] MALCOLM N. SHAW, INTERNATIONAL LAW 98 (6th ed. 2008) (discussing this general principle of law.).

[77] THE GENEVA CONVENTIONS OF 12 AUGUST 1949: COMMENTARY, (III) GENEVA CONVENTION RELATIVE TO THE TREATMENT OF PRISONERS OF WAR, 36 (Jean Pictet ed., 1958). *See also* Richard B. Jackson, *Perfidy in Non-International Armed Conflict*, 88 INT'L L. STUD. 237, 250–51 (2012) (where he notes "the national security risks entailed

Human rights law remains the general law with human rights norms relevant to law enforcement also being reflected in the "special law" of international humanitarian law. However, the incorporation of human rights norms into humanitarian law means there are areas of considerable overlap with human rights law. This situation of overlap does not appear to support a conclusion there is necessarily a "conflict of norms" in respect of rights recognized under both bodies of law. This remains the case even where human rights law is used to interpret a humanitarian law–based norm because of a paucity of precedent.[78] The norm itself is part of humanitarian law even if it is imported from "the general law of international human rights."[79] If a conflict is to arise, it would seem most likely to occur where there is an attempt to apply conduct of hostilities norms to a situation of law enforcement, or human rights–based law enforcement norms during hostilities occurring at a level of intensity far beyond that ordinarily found in a domestic context.

In many respects it seems that battle lines have been drawn with a strategic legal conflict for "control" being waged by proponents of the two governing legal frameworks. In this struggle the *lex specialis* (i.e., special law principle) has been identified as the high ground upon which success or failure will rest. Unfortunately, as has been noted the discussion among international lawyers on this issue often tends to be exclusionary in nature with considerable disagreement in the theory and application of the substantive law. In this atmosphere of disagreement it is easy to be misled into thinking the interaction of human rights and humanitarian law is a modern phenomenon and that each body of law is far removed from the other. They are not. In order to more fully assess the relationship between these two normative frameworks the analysis will turn toward its history of neglect, integration, and divergence.

5.3. A More Intimate History

5.3.1. PRIOR TO THE MID-TWENTIETH CENTURY

The horrific human rights abuses of World War II, and the almost concurrent drafting of the 1948 Universal Declaration of Human Rights and the 1949 Geneva Conventions, often result in the relationship between human rights and humanitarian law being assessed in the context of events occurring during the last half of the twentieth century.[80] Unfortunately,

---

in a non-international armed conflict do not require abandonment of societal norms intended to provide minimal protections to the populace.").

[78] Prosecutor v. Kunarac, No. IT-96-23-T & IT-96-23/1-T, Judgement, ¶ 467, at 160 (Int'l Crim. Trib. for the Former Yugoslavia Feb. 22, 2001), http://www.icty.org/x/cases/kunarac/tjug/en/kun-tj010222e.pdf (where it is noted "[b]ecause of the paucity of precedent in the field of international humanitarian law, the Tribunal has, on many occasions, had recourse to instruments and practices developed in the field of human rights law."). *See also* Theodor Meron, *The Humanization of Humanitarian Law*, 94 AM. J. INT'L L. 239, 266 (2000) (where it is indicated that judicial guarantees in common Article 3 to the 1949 Geneva Conventions and essential guarantees of independence and impartiality found in GC III, art. 84 "should inevitably be interpreted and applied drawing on human rights.").

[79] AHARON BARAK, PROPORTIONALITY: CONSTITUTIONAL RIGHTS AND THEIR LIMITATIONS 204–05 (2012) (discussing the importation of norms from international human rights law). *But see* Sassoli, *supra* note 26, at 74 (where it is suggested that when determining what the judicial guarantees are that are referred to in Common Article 3 to the Geneva Conventions "IHRL specifies, as *lex specialis*, what those guarantees are.").

[80] PROVOST, *supra* note 7, at 2 ("Links between the two bodies of law had been discussed from the end of the Second World War, following successive adoptions of the 1948 Universal Declaration of Human Rights and the 1949 Geneva Conventions."). *See also* Prud'homme, *supra* note 42, at 356–57.

this sometimes creates an impression that human rights are a relatively recent concept and are almost exclusively based on the written international human rights texts of the last half of the twentieth century. However, the interface between these two normative frameworks is a more nuanced and intimate one than interpretations of the post–World War II treaty law suggest. As Leslie Green has noted, while the intention behind the creation of these two documents was to distinguish between Geneva law as applicable in armed conflict and the law of human rights, which concerned peacetime, "[i]t may be questioned, however, whether this is a truly substantive distinction or merely semantic in nature, especially as history shows that there were regulations seeking to show some respect for human rights in the widest sense of the term in both peace and war long before the creation of either the United Nations or the International Committee of the Red Cross."[81]

That is not to say that human rights law and humanitarian law are the same. René Provost has noted there is a "real and meaningful difference between the normative frameworks of human right law and humanitarian law,"[82] with the difference resting on human rights law being "centred, indeed built, on granting rights to the individual."[83] In contrast, "humanitarian law is constructed around the direct imposition of obligations on individuals who wield some power over others."[84] Further, it is suggested human rights attach to individuals as human beings, while "[h]umanitarian law, on the contrary, is a patchwork of variable protections granted to individuals in some specific situations, if they meet criteria linked to conduct and nationality or group membership."[85] However, when the dialogue is about "norms" rather than bodies of law, these differences are reduced.

Like humanitarian law, human rights law shares a common basis in religious humanism.[86] International humanitarian law has its roots firmly entrenched in world religion and accompanying moral prescriptions.[87] In this respect, "despite differing historical origins, and differing temporal applicability, what both IHL and IHRL have in common is a shared philosophical underpinning."[88] Common ground can also be found in the struggle to deal with armed conflict, governance, and justice. As Micheline Ishay notes in her history of human rights, the promotion of justice was carried out in different ways where "[s]ome traditions stress preaching or educating, others invoke the sword, and others insist on passive resistance."[89] Included were the concepts based in various religions that wars, when fought, be waged justly. These early "visions" were encapsulated in the 1864 Geneva Convention, "an international convention that required, for the first time, the protection of prisoners

---

[81] Leslie C Green, *The Relations between Human Rights Law and International Humanitarian Law: A Historical Overview*, *in* TESTING THE BOUNDARIES OF INTERNATIONAL HUMANITARIAN LAW 49, 50 (Susan C. Breau & Agnieszka Jachec-Neale eds., 2006).

[82] PROVOST, *supra* note 7, at 13.

[83] *Id.*

[84] *Id.* at 116.

[85] *Id.* at 41–42.

[86] MICHELINE R. ISHAY, THE HISTORY OF HUMAN RIGHTS: FROM ANCIENT TIMES TO THE GLOBALIZATION ERA 19 (2004).

[87] LESLIE GREEN, THE CONTEMPORARY LAW OF ARMED CONFLICT 26–29 (3rd ed. 2008).

[88] EMILY CRAWFORD, THE TREATMENT OF COMBATANTS AND INSURGENTS UNDER THE LAW OF ARMED CONFLICT 122 (2010).

[89] ISHAY, *supra* note 86, at 40.

of war, *recognition of their rights*, and assistance for the wounded regardless of the side for which they fought."[90]

In suggesting international human rights and humanitarian law are discrete bodies of law it is not uncommon to see reference to human rights as a domestic internal matter prior to the development of the 1948 Universal Declaration.[91] In effect, human rights are not viewed as being "international" in nature until after World War II.[92] This approach appears unduly limiting since foundational rights documents such as the 1776 U.S. Declaration of Independence and the 1789 French Declaration of the Rights of Man and Citizen were based on transnational theories of rights.[93] As has been stated, "[d]eclarations—in 1776, 1789, and 1948—provided a touchstone for those rights of humanity. . . ."[94] Seeing pre–World War II rights as domestic in nature also downplays the evolution of human rights through history as witnessed by the abolition of torture as part of the European judicial process by the end of the eighteenth century.[95]

The exclusionary approach is also one that seems to de-emphasize the natural law basis of human rights law, preferring instead the positivist view that the articulation of rights emanates from the State such as in the form of a declaration or treaty. However, as Michael Walzer has noted, "[i]t is enough to say that they are somehow entailed by our sense of what it is to be a human being . . . natural or invented, they are a palpable feature of our moral world."[96] It is difficult to see how considering human rights to be universal and extraterritorial in application can be any better phrased than, those rights "are held by individuals simply because they are part of the human species."[97] As such they are, according to the U.S. Declaration of Independence, "truths to be held self-evident."[98] Those rights are transnational in scope regardless of whether they find their voice under domestic or international law. Like human rights law, humanitarian law also had domestic roots with the *Leiber Code* being perhaps one of the most obvious examples.[99] It is undeniable that humanitarian law was expressed in international treaties before its human rights counterpart; however, this fact does not make the common norms upon which they are based any less universal in nature.[100] It seems exceedingly formalistic to erect strict boundaries between domestic and

---

90 *Id.* (emphasis added).

91 Droege, *supra* note 31, at 503.

92 Robert Kolb, *The Relationship Between International Humanitarian Law and Human Rights Law: A Brief History of the 1948 Universal Declaration of Human Rights and the 1949 Geneva Conventions*, INT'L REV. RED CROSS (1998), *available at* https://www.icrc.org/eng/resources/documents/misc/57jpg2.htm. *See also* G. I. A. D. DRAPER, THE RED CROSS CONVENTIONS 14 (1958).

93 ISHAY, *supra* note 86, at 96–97.

94 LYNN HUNT, INVENTING HUMAN RIGHTS 214 (2007).

95 NIGEL S. RODLEY & MATT POLLARD, THE TREATMENT OF PRISONERS UNDER INTERNATIONAL LAW 9 (3rd ed. 2009). *See also* HUNT, *supra* note 94, at 81–82.

96 MICHAEL WALZER, JUST AND UNJUST WARS 54 (3rd ed. 1977).

97 ISHAY, *supra* note 86, at 3. *See also* Brian J. Brill, *Human Rights: Time for Greater Judge Advocate Understanding*, THE ARMY LAWYER 54, 54 (2010).

98 THE DECLARATION OF INDEPENDENCE (U.S. 1776).

99 Green, *supra* note 81, at 62–67 (for a discussion of the various domestic Articles of War and other ordinances governing the activities of armed forces), and RICHARD SHELLY HARTIGAN, LIEBER'S CODE AND THE LAW OF WAR 1–21 (1983) (for an outline of the history of the development of the *Lieber Code*.).

100 ALLAN ROSAS, THE LEGAL STATUS OF PRISONERS OF WAR: A STUDY IN INTERNATIONAL HUMANITARIAN LAW APPLICABLE IN ARMED CONFLICT 57 (1976) ("The philosophical premises of the new law and practice [relating to prisoners of war] were formulated by such philosophers of the Enlightenment as *Montesquieu* and *Rousseau*.").

international human rights law, and then between that law and its humanitarian counterpart, regarding their historical development and scope of impact. Their interface and interaction is too intimate and overlapping for that to be the case.

Even today the domestic and international interface of human rights and humanitarian law cannot always easily be delineated. This is evident in respect of the prosecution of war crimes, which relies predominately on domestic military and civilian courts.[101] Further, one of the sources of international law is "the general principles of law recognized by civilized nations" such as those appearing in municipal law.[102] While there is considerable debate regarding the scope of this term, there is also an acknowledgment "[t]he most fertile fields, however, for the implementation of municipal law analogies have been those of procedure, evidence and the machinery of the judicial process,"[103] the classic purview of international human rights law. In this respect the use of institutional mechanisms, such as a judicial system, associated with situations of political stability is a founding premise of human rights law.[104] Further, it is not clear what is to be gained by focusing primarily on the post–World War II human rights treaties and differences in treaty law when assessing the interface between humanitarian and human rights law. The two bodies of law are different, however, that "has not prevented the principle of humanity from becoming the common denominator of both systems."[105]

The interrelationship between humanitarian law and human rights law cannot be fully assessed by restricting the analysis to "international" human rights treaties. It has to be considered in a context where "a bright-line separation between foreign and domestic affairs has proved problematic. . . . In an era of globalization, the interdependence between nations causes virtually every issue of domestic affairs to bear on the external affairs of the nations and on their national security policies."[106] For practitioners of operational law there is a need to consider not only the international law of armed conflict and human rights but also frequently the domestic human rights provisions of their own State. Consideration also has to be given to the human rights laws of the country in which they are operating. This is particularly evident when a State is an occupying power, or in situations where international forces assist indigenous security forces enforce their own domestic laws as they jointly conduct counterinsurgency operations in a noninternational armed conflict. The requirement to make these bodies of law work together is the essence of the "operational law" discipline.

Further, it cannot be denied that international humanitarian treaty law makes substantive reference to human rights norms. For example, those norms are reflected in prohibitions against torture,[107] the requirement for the humane treatment of detainees,[108]

---

[101] 2 PUBLIC COMMISSION TO EXAMINE THE MARITIME INCIDENT OF 31 MAY 2010 Chap. B, at 152–264 (2013) (for an overview of the military and civilian justice systems applied for the investigation and prosecution of war crimes by the United States, Canada, Australia, the United Kingdom, Germany, and the Netherlands.).

[102] SHAW, *supra* note 76, at 98.

[103] *Id.* at 100.

[104] PROVOST, *supra* note 7, at 344.

[105] Meron, *supra* note 78, at 245.

[106] OREN GROSS & FIONNUALA NÍ AOLÁIN, LAW IN TIMES OF CRISIS: EMERGENCY POWERS IN THEORY AND PRACTICE 208 (2006).

[107] The 1949 Geneva Conventions, Common Art. 3(1)(a); GC I, art. 12, GC II, art. 12, GC III, arts. 17 and 87, GC IV, art. 32, AP I, art. 75(2)(ii), and AP II, art. 4(2)(a).

[108] Common art. 3(1) to the four GCs; GC I, art. 12; GC II, art. 12; GC III, art. 13; GC IV, arts. 27 and 37; AP I, art. 75(1), and AP II, arts. 4(1), 5(3) and 7(2).

protection from violence to life or health of persons,[109] discrimination on the basis of race, sex, language, or religion,[110] and due process.[111] Importantly, the Fourth Convention includes an international treaty requirement for a belligerent to maintain in force the domestic penal laws of the occupied territory, except where "they constitute a threat to its security or an obstacle to the application of the present Convention."[112] As a result the occupying power is directly responsible for the application of human rights norms through the domestic laws of the occupied territory. This serves as further evidence of the requirement to apply various bodies of law, international and domestic, in the context of an armed conflict. Significantly, the Fourth Geneva Convention has been described as providing a bill of rights for an occupied population,[113] and Common Article 3 to the four Conventions is recognized as ensuring fundamental human rights.[114] In the aftermath of their development it was even considered in 1958 that "[i]n respect of both legal definition and enforcement they have gone considerably further than the European Convention of Human Rights of 1950."[115] This view of these Conventions has not withstood the test of time, however, that does not make them any less incorporative of human rights.

Although there is considerable debate as to whether there is a hierarchy of rights, it is clear that the "right to life" is a fundamental norm for both international humanitarian and human rights law.[116] What is often different is how and under what circumstances the right to life is limited by these bodies of law. However, this does not mean that the normative prescription of the "right to life" is not as important to humanitarian law as it is in the human rights framework. To suggest otherwise has the potential to significantly undermine a fundamental principle of humanitarian law: the principle of distinction. The focus placed by the Geneva Conventions on humanity is perhaps best reflected in the commentary to Common Article 2, which notes that the Conventions were regarded less as agreements of reciprocal actions between States and more as solemn affirmations of principles done so "out of respect for the human person as such."[117] Similarly, the use of force in individual self-defense, often associated with human rights principles of necessity and proportionality, is also provided for under the law of armed conflict.[118]

---

[109] AP I, art. 75 (2)(a).

[110] Meron, *supra* note 78, at 266.

[111] *See, e.g.*, GC III, arts. 99–108; GC IV, arts. 5 and 71–75; AP I, arts. 75(3), (4) and (7); and AP II, art. 6(2).

[112] GC IV, art. 64.

[113] Eyal Benvenisti, The International Law of Occupation 105 (1993). *See also id.* at 9 (where Article 43 of the 1907 Hague Land Warfare Regulations is described as "a miniconstitution for the occupation administration"). *See* Meron, *supra* note 78, at 246 ("The Fourth Geneva Convention establishes a new balance between the rights of the occupant and the rights of the population of the occupied country.").

[114] Meron, *supra* note 78, at 246, and Baxter, *supra* note 1, at 292.

[115] Draper, *supra* note 92, at 24.

[116] Meron, *supra* note 78, at 266–67.

[117] The Geneva Conventions of 12 August 1949, Commentary: (I) Geneva Convention For The Amelioration Of The Condition Of The Wounded And Sick In Armed Forces In The Field, art. 2, 28–29 (Jean S. Pictet ed., 1952).

[118] Prosecutor v. Kordić, No. IT-95-14/2-T, Judgment, ¶ 451 (Int'l Crim. Trib. For the Former Yugoslavia Feb 26, 2001), http://www.icty.org/x/cases/kordic_cerkez/tjug/en/kor-tj010226e.pdf (noting that the principle of self-defense enshrined in the ICC Statute, at art. 31(1)(c) "reflects provisions found in most national criminal codes and may be regarded as constituting a rule of customary international law."). *See also* GC I, art. 22(1) and (2) and AP I, arts. 65(1) and 67 regarding the arming of military and civilian personnel for self-defense purposes.

This does not mean that all the norms found in human rights law are fully addressed in humanitarian law.[119] However, it has been suggested there are only a few examples of conflict between international humanitarian law and human rights law, although "there is no denying the incompatible approach of LOIAC [law of international armed conflict] and human rights law to some vital issues."[120] Perhaps the most obvious one is the collateral killing of civilians.[121] A common, if not universally held, view is that human rights law does not accept any collateral killing of civilians.[122] The emphasis that has been placed on viewing human rights and humanitarian law as separate bodies of law reflects the powerful influences of the exclusionary theories and might be described as the war and peace camps. The practical application of these bodies of law requires a more careful assessment of their interface and overlap. In normative terms the interaction between humanitarian and human rights law did not start with the termination of World War II. However, the past sixty years have witnessed an often changing relationship between the two bodies of law that sometimes demonstrated neglect, other times convergence and integration, and more recently conflict. It is to that relationship that this analysis will now turn.

5.3.2. IN THE AFTERMATH OF WORLD WAR II

5.3.2.1. Neglect

Human rights law and, in particular, the 1948 Universal Declaration of Human Rights were taken into account during the drafting of the 1949 Geneva Conventions, although "references to human rights were few and far between."[123] However, as has been discussed, it is also evident the immediate postwar period saw a paralysis in the development of international humanitarian law. Perhaps the most significant harbinger of change was the adoption of the quite limited Common Article 3 to the Geneva Conventions. As Geoffrey Best has noted, the doctrine of the sovereignty of States had competition and "[a]longside it, and preconfigured in the Charter itself, there ran a parallel legislative stream of humanitarian and human rights rules and standards which States undertook at least to take note of and which, if words mean anything, they should in some last resort be required to observe."[124]

---

119 Adam Roberts, *Transformative Military Occupation: Applying the Laws of War and Human Rights*, 100 AM. J. INT'L L. 580, 594 (2006) ("international human rights standards may not merely fill in gaps in the laws of war, but also provide procedures for assisting in the implementation of key provisions of those laws.").

120 Dinstein, *supra* note 4, ¶ 57, at 24.

121 *Id. See also* Meron, *supra* note 78, at 240.

122 INTERNATIONAL HUMANITARIAN LAW AND THE CHALLENGES OF CONTEMPORARY ARMED CONFLICT, Report 32nd International Conference of the Red Cross and Red Crescent EN 32IC/15/11, at 16 (2015), *available at* https://www.justsecurity.org/wp-content/uploads/2015/11/2015-ICRC-Report-IHL-and-Challenges-of-Armed-Conflicts.pdf (where it is noted the use of IHL norms would allow for collateral damage, "which would not be the case if the rules on law enforcement are relied on."). *But see* Gabrella Blum & Philip Heymann, *Law and Policy of Targeted Killing*, 1 HARV. NAT. SEC. J. 145, 148 (2010), Andronicou v. Cyprus, Application Nos. 86/1996/705/897, Eur Ct. H. R. ¶ 194 (Oct. 9, 1997), http://hudoc.echr.coe.int/sites/eng/pages/search.aspx?i=001-58102, Kerimova v. Russia, Application Nos. 17170/04, 20792/04, 22448/04, 23360/04, 5681/05 and 5684/05 Eur. Ct. H. R. ¶ 246 (May 3, 2011), http://hudoc.echr.coe.int/sites/eng/pages/search.aspx?i=001-104662 (where some level of potential collateral casualties are contemplated under a human rights based analysis), and Finogenov v. Russia, VI Eur. Ct. H. R. ¶ 236, at 409 (2011)(extracts).

123 Kolb, *supra* note 92.

124 GEOFFREY BEST, WAR AND LAW SINCE 1945, 79 (2002) (1994).

However, the State obsession with not "legitimizing" its non-State opponents effectively forestalled the development of treaty law governing the conduct of internal conflicts for much of the 1950s and 1960s. The State was to remain the sole "proper" authority for waging war and the consideration of what laws applied in war, to the extent there were any, focused on conventional warfare in an inter-State context.

It was not that deficiencies in the law were not recognized, and attempts were made by the ICRC to introduce draft rules governing armed conflict in 1957.[125] However, they failed to gain the support of States.[126] The result was "this failure made the ICRC reluctant to try to update the law in the 1960s."[127] A large part of reason for the lack of effort on amending the laws governing armed conflict following World War II was the impact of the development of the UN system for controlling conflict. The International Law Commission created in 1947 to work on the progressive development and codification of international law noted it was opposed to studying the law of war since "public opinion might interpret its action as showing a lack of confidence in the efficiency of the means at the disposal of the United Nations for maintaining peace."[128] The "clear association of international humanitarian law with the International Movement [of the Red Cross and Red Crescent], and of international human rights law with the United Nations, provided a clear basis for the compartmentalization of the two fields . . . ."[129]

Further, member States themselves were not interested in developing the law of war. By the late 1940s it was assessed "the emotional drive and political commitment were almost completely spent by the time Universal Declaration was adopted in 1948."[130] Instead the international legal community, and in particular the United Nations, focused on the development of human rights law, which by the mid-1960s comprised an impressive list of treaties. With the United Nations concentrating on human rights, and the ICRC having been unsuccessful in its efforts during the 1950s to address international humanitarian law, consideration of the relationship between the two bodies of law fell into a period of "neglect." However, this situation was to change as the international community was forced to deal with the "small wars" that increased both in number and ferocity in the postwar period. Of particular note, in respect of the interface between human rights and humanitarian law, was that the pressure for change came from the human rights community rather than institutional actors such as the United Nations or the ICRC. It was members of what would today be called "civil society," such as human rights activist Sean McBride and the International Commission of Jurists, who forced the updating of international humanitarian law.[131] This ultimately led to a greater integration of human rights norms into that law.

---

125 Josef L. Kunz, *The Chaotic Status of the Laws of War and the Urgent Necessity for Their Revision*, 45 AM. J. INT'L L. 37 (1951), and H. Lauterpacht, *The Problem of the Revision of the Law of War*, 29 BRIT. Y.B. INT'L L. 360, 365–66 (1952).

126 Baxter, *supra* note 1, at 289.

127 KEITH SUTER, AN INTERNATIONAL LAW OF GUERRILLA WARFARE 43 (1984).

128 *Id.* at 42–43 (quoting the YEARBOOK OF THE INTERNATIONAL LAW COMMISSION: 1949 52 (1956)).

129 Prud'homme, *supra* note 42, at 360.

130 SUTER, *supra* note 127, at 43 (relying on MOSES MOSKOWITZ, THE POLITICS AND DYNAMICS OF HUMAN RIGHTS 85 (1968)).

131 *Id.* at 24–35.

5.3.2.2. Forced Integration

The significant number of guerrilla wars in the post–World War II period did not allow the neglect of humanitarian law to continue for long. Just as the small wars of the twenty-first century are now forcing consideration of the interface between humanitarian and human rights law, such was the case over forty years ago. A milestone often referenced as the starting point for updating the laws governing armed conflict was the 1968 International Conference on Human Rights held in Tehran as part of the International Year for Human Rights. As Richard Baxter noted, the UN Human Rights Commission and the Third Committee had completed work on the international Covenants dealing with human rights in time of peace "and it was natural that those persons in the United Nations systems who were interested in human rights should turn to the protection of such rights in war time."[132] However, Geoffrey Best interpreted the situation somewhat differently, indicating that "[f]rustration at the apparent impotence of *both* systems, humanitarian and human rights law alike, contributed to bring about a new initiative in 1968."[133]

The impetus to look toward the regulation of internal conflict has also been associated with Article 2(7) of the UN Charter, which "paved the way for the evolution of norms applicable to internal conflicts and eventually for the resolutions by which the Security Council has authorized forcible interventions response to internal atrocities."[134] However, action in that regard was largely to await the end of the Cold War.[135] It has been suggested at the point when the 1968 Conference should have provided ideas for the next stage of human rights development, it is "generally thought to have achieved very little."[136] In part, the reason for the lack of progress was the "three armed conflicts of varying intensities which were then ensuing or had just been waged: the 1967 Middle East conflict, and those of southern Africa and Vietnam."[137] It was the impact of those conflicts that took up much of the discussion at the Conference. This did lead to "a climate of opinion which boosted the success of the passage through the conference of the resolution on human rights in armed conflicts."[138] However, unlike the situation in the twenty-first century, the efforts of human rights advocates in the 1960s and 1970s to seek more human rights protection during armed conflict was to be directly reflected in the development of humanitarian law.[139]

It was this push by civil society that ultimately caused the United Nations, and then the ICRC, to commence the process which resulted in the two 1977 Additional Protocols.[140] That process was a difficult one. However, significantly, it had the support of the United

---

132 Baxter, *supra* note 1, at 290. *See also* PROVOST, *supra* note 7, at 3, and Meron, *supra* note 78, at 247.

133 BEST, *supra* note 124, at 73

134 Meron, *supra* note 78, at 247.

135 Jochen Abr. Frowein & Nico Krisch, *Article 42*, *in* THE CHARTER OF THE UNITED NATIONS: A COMMENTARY 749, 751–52 (Bruno Simma et al. ed., 2002) (for an outline of the post–Cold War invocation of Article 42.).

136 SUTER, *supra* note 127, at 23.

137 *Id.*

138 *Id.* at 24.

139 *Id.* at 23–33.

140 *Id.* at 29 (where a letter from Sean McBride is quoted indicating that the ICRC although invited were not represented at the Tehran conference. This may be an indication that humanitarian law was not expected to be the focus of discussion at the conference.).

States, which attached importance to "the preparation of new law on internal conflicts."[141] What is clear is that those Protocols saw an unprecedented incorporation of the human rights norms reflected in the newly developed human rights treaties directly into the framework of international humanitarian law, although it has also been suggested " '[h]uman rights' as such were less in evidence at the end of the process than they had been at the beginning."[142] This post–World War II influence has been referred to as a "parallelism and growing convergence" that enriches both bodies of law.[143] While it must be stated the two remain separate bodies of law—each with its own goals, strengths, and weaknesses—that incorporation also represents an *integration* of specific human rights norms into humanitarian law.

The extent of that integration is perhaps best reflected in Article 75 of Additional Protocol I and Article 4 of Additional Protocol II, which were clearly influenced by human rights documents such as the 1948 Universal Declaration and 1966 International Covenant on Civil and Political Rights.[144] At the time, it was suggested by one delegate that the provision that was to become Article 75 of Additional Protocol I was in effect a "mini-convention and protection of the human person, in so far as that element had been omitted from the scope of the Geneva Conventions of 1949, or, at any rate, inadequately covered in the fourth Convention."[145] However, another delegate was quick to argue "the purpose of the Conference was not to draft a Convention on Human Rights."[146] Whatever its status, it has been noted that the post–World War II relationship between international humanitarian law and human rights law "is expressed in the adoption of major human rights principles in Article 75 AP I."[147]

The impact of human rights law is reflected in the comment that "the universality of the protections granted in Article 75 stands in stark contrast to the patchwork approach adopted in the 1949 Geneva Conventions."[148] Significantly, the *Commentary on the Additional Protocols of 8 June 1977 to the Geneva Conventions of 12 August 1949* in dealing with Article 75 of Additional Protocol I, indicate "[m]ost of the guarantees listed in sub-paragraphs (a)–(j) [related to due process] are contained in the Conventions and the Covenant on Human rights," although a difference is that, unlike the Covenant, the Additional Protocol provision is not subject to derogation.[149] The list of incorporated rights is lengthy and includes

---

[141] Baxter, *supra* note 1, at 292.

[142] Best, *supra* note 124, at 73.

[143] Meron, *supra* note 78, at 266.

[144] *See, e.g.*, Rosas, *supra* note 100, at 290 (where it is also noted that the draft provisions of AP II regarding detained persons "lays down certain principles similar to those contained in the 1966 Covenant on Civil and Political Rights." These provisions, relating to application of the death penalty and granting of amnesties, were ultimately codified in AP II, art. 6. Similarly, AP II, art. 4 mirrors the AP I, art. 75 protections.).

[145] XV Official Records of the Diplomatic Conference on the Reaffirmation and Development of International Humanitarian Law Applicable in Armed Conflicts, Geneva (1974–1977) ¶ 30, at 31, CDDH/III/SR. 43 (1978) (the Belgian delegate).

[146] *Id.* ¶ 21, at 49, CDDH/III/SR. 44 (where the concern of the Canadian delegate was that the Protocol provision might impact on the operation of Canadian domestic law).

[147] Christopher Greenwood, *Scope of Application of Humanitarian Law*, *in* Humanitarian Law Handbook, *supra* note 3, Rule 254 at 74.

[148] Provost, *supra* note 7, at 55.

[149] Commentary on the Additional Protocols of 8 June 1977 to the Geneva Conventions of 12 August 1949 art. 75, ¶3092 (Yves Sandoz et al., 1987). *See also id.* ¶3101, at 881 ("This provision reproduces almost word-for-word paragraph 1 of Article 17 of the Covenant on Civil and Political Rights"), and *id.* ¶ 3116, 884 ("Once again the drafters of the Article have tried to stay as close as possible to the Covenant (Article 14, paragraph 7).").

prohibition of discrimination; the prohibition against violence to life, health, or physical or mental well-being, torture, humiliating and degrading treatment; and threats to carry out such acts.[150] Given the common roots of these rights, their scope, and the intended effect, it is difficult to see how there is not a significant overlap in the protection offered by these humanitarian law provisions and those set out in international human rights law.

For military forces of the majority of the world's States that are Parties to the Additional Protocols (89 percent of States for AP I and 84 percent for AP II)[151] or accept the provisions of Article 75 of Additional Protocol I or Article 4 of Additional Protocol II as reflecting customary law, the existing debate regarding the applicability of human rights norms during armed conflict should engender surprise. Those norms apply as a matter of treaty or customary humanitarian law. Similarly, since numerous human rights–based norms are also a fundamental part of the Geneva Conventions, it should also surprise countries that are not parties to the Protocols. They are required to apply "human rights" in the conduct of military activity such as in their treatment of prisoners of war and detained civilians. This should cause pause for proponents of the exclusionary approach focusing on the "law of armed conflict," which seeks to even avoid using the term "humanitarian." The law governing armed conflict has always included protection of individual "rights" regardless of whether they are actually called "human" rights or are framed as being part of "human rights law."

5.3.2.3. Divergence and Growing Reconciliation

At least from the perspective of treaty law the adoption of the 1977 Additional Protocols may be viewed as a high-water mark for the integration of human rights norms into humanitarian law. However, since that time the two exclusionist approaches have developed in a way that has significantly limited the realization of the goal of those Protocols in further humanizing the law governing war. In terms of the human rights exclusionary approach this has included seeking to significantly narrow the applicability of humanitarian law even in respect of the conduct of hostilities. The reasons for the divergence in approach appear to be diverse, but the result is one that, at times, seems to manifest itself in a virtual paralysis in the dialogue that should be occurring between the human rights and humanitarian law communities.

Perhaps the most obvious indication of that paralysis has been a lack of effort to update the Additional Protocols, or other humanitarian law treaties, in order to address the new security challenges of the twenty-first century. As Yoram Dinstein has noted, this is "in sharp contrast with the experience of the twentieth century (when the Geneva Conventions were reviewed on no less than four occasions: 1906, 1929, 1949 and 1977)."[152] Following the

---

[150] AP I, art. 75(1) and (2).

[151] *State Parties to the Following International Humanitarian Law and Other Related Treaties as of 6 Jan 2014*, International Committee of the Red Cross, http://www.icrc.org/applic/ihl/ihl.nsf/vwTreaties1949.xsp?redirect=0 (where it is indicated there are 174 States Party to AP I and 168 States Party to AP II), and *Growth in United Nations Membership, 1945–present*, United Nations, http://www.un.org/en/members/growth.shtml (last visited June 5, 2015) (where it is identified there are 193 member States in the United Nations).

[152] DINSTEIN, *supra* note 4, ¶ 743, at 295–96. *See also* Richard Baxter, *Human Rights in War*, *in* HUMANIZING THE LAWS OF WAR, *supra* note 1, at 244 (where it was noted in 1977 there "seems to be a twenty-year cycle in the revision of the humanitarian law of war.").

negotiation of the 1977 Additional Protocols it might have been expected a significant review of the Geneva Conventions would have been completed by the turn of the century. The attack of 9/11 no doubt could have impacted on the timing of any review as States and interested parties struggled with the aftermath of that act of transnational terrorism. However, significantly, there seems to have been no overt pressure prior to that point from the institutional actors (e.g., the ICRC and the United Nations), States, or human rights advocates to comprehensively review the major documents of the Geneva law. The work that was done, while significant in its humanitarian effect, was primarily focused on banning or limiting certain weapons[153] and enhancing accountability for the commission of crimes committed during armed conflict.[154]

It has been suggested that in the mid 1990s and early 2000s international criminal law replaced international human rights law as the reinvigorating force for international humanitarian law.[155] However, in terms of codification it has a relatively limited scope and while a very important step in enforcing accountability is one that carries with it the inherent limitations that criminal law has in impacting more broadly on normative behavior.[156] This suggestion serves to highlight the degree to which human rights law and its advocates have taken a different path than that followed in previous decades. Unfortunately, what has been missing is the pressure from the nongovernmental organization (NGO) community to change humanitarian law that was crucial in the lead up to the Additional Protocols. There has been no twenty-first-century Sean McBride.

Neither the institutional actors nor States were the primary instigators of the major review of humanitarian law in the 1960s and 1970s. The question remains as to why the previous agents for change, human rights advocates, did not force further change. There are a number of reasons why the human rights community may have become disinterested in amending existing treaty law. First, there is the issue of how much that community wants to engage in applying human rights within humanitarian law. It has been suggested that having addressed the similarities, by the 1980s "careful people found it desirable to recall the differences as well as the similarities, lest the specificities of each be lost sight of in a cloud of universalized goodwill."[157] This comments resonates with the past thirty years where the analysis has focused to a large extent on the differences between human rights law and its international humanitarian law counterpart.[158] It remains to be been seen if this approach can continue. The effort to harmonize the two bodies of law, brought about in the late 1960s by the reality of noninternational armed conflicts, now needs to be considered once again as a result of the challenges of contemporary conflict. Attempts to maintain a

---

[153] *Id.* (where he notes "[t]angible advances in this field only serve to spotlight the lethargy in others.").

[154] SHAW, *supra* note 76, at 399–430 (Post–World War II saw the creation of the Nuremburg, Tokyo, and other war crimes tribunals. The 1990s saw the development of ad hoc tribunals: the International Criminal Tribunal for the Former Yugoslavia and the International Criminal Tribunal for Rwanda, as well as the development of the ICC. Mixed tribunals to prosecute domestic/international crimes have been set up in Kosovo, Bosnia Herzegovina, East Timor, Sierra Leone, Cambodia, and Lebanon.).

[155] Sivakumaran, *supra* note 24, at 233.

[156] *Id.* at 238–40 (for a discussion of the limitations of the scope of international criminal law in relation to humanitarian law).

[157] BEST, *supra* note 124, at 69.

[158] PROVOST, *supra* note 7, at 349–50.

bright line separation between human rights and humanitarian law are not, in practical terms, sustainable.

Second, the Additional Protocols, while representing a significant advancement in the direct incorporation of human rights norms, did not address the vast majority of conflicts in which the human rights community was interested. The expansion of Additional Protocol I to wars of national liberation can now rightly be viewed as covering a relatively small number of conflicts, which are viewed in the twenty-first century as a historical anomaly. Further, the threshold for the application of Additional Protocol II was set very high "as virtually to be confined to a civil war situation in which both the government and the rebel forces are in control of part of the national territory."[159] States, acting in a fashion entirely consistent with their status as the proper authority, have demonstrated a continued preference for not developing treaty law governing internal conflict. As a result, in terms of treaty law the vast majority of conflicts remained primarily "under cover (such as it was) of Common Article 3."[160] Importantly, customary international humanitarian law would also apply, although significant debates remain concerning its content and scope of application.

In respect of the guerrilla or small wars that many human rights NGOs and humanitarian organizations were interested in, it is not clear their considerable efforts of the 1970s would necessarily be viewed as having been successful at the end of the negotiation process for the Additional Protocols.[161] As a result, "[t]he later 1970s and 1980s therefore witnessed the growth of interest in the direct application of human rights law to every sort of conflict that was not simply, solely and unarguably 'international'."[162] In this regard, by fighting the adoption of Additional Protocol I States seem to have been the authors of the very situation that proponents of the exclusive application of the law governing armed conflict often chafe against: greater recognition of human rights law. Importantly, the reliance by human rights advocates on an approach favoring that body of law for regulating internal conflict is entirely consistent with an international governance system that views states as the "proper authority." In that respect it has been human rights law, in its domestic form, which has had the longer historical claim to regulating state action in such armed conflict (see Chapter 4, section 4.4.3, *A Reluctant Application of Humanitarian Law*). Notwithstanding this fact, that body of law is not necessarily well suited to deal with many aspects of an internal armed conflict, particularly because of its unitary focus on the state and the challenges of applying it to non-State actors.[163] The challenge is that, in respect of an armed conflict, neither body of law can be applied in isolation.

Third, the two Additional Protocols have not enjoyed universal acceptance, particularly from states that were actually involved in many of the armed conflicts since 1977. The exclusionary approach by some militarily dominant States has likely affected the willingness

---

159 GREEN, *supra* note 87, at 63.

160 BEST, *supra* note 124, at 74.

161 *Id.* ("By the time they finished, they were wondering whether the victims of civil wars would be better off than they had been before they began.").

162 *Id.*

163 INTERNATIONAL HUMANITARIAN LAW AND THE CHALLENGES OF CONTEMPORARY ARMED CONFLICTS, Report 31st International Conference Of The Red Cross And Red Crescent EN 31IC/11/5.1.2 18 (2011), http://www.icrc.org/eng/assets/files/red-cross-crescent-movement/31st-international-conference/31-int-conference-ihl-challenges-report-11-5-1-2-en.pdf [hereinafter 2011 CHALLENGES OF CONTEMPORARY ARMED CONFLICT REPORT].

of human rights advocates to work within the humanitarian law community. While objecting States have not necessarily been forthcoming regarding what provisions of the Protocols are considered to be acceptable, it is clear that there has been significant opposition to many of the treaty provisions. The U.S. objection that Additional Protocol I would legitimize "terrorist" groups stands out, not only as a consistent dominant military state approach of criminalizing non-State participation in warfare but also as one that inevitably shapes the discussion regarding the governing legal regime. The result, somewhat ironically given the exclusionary approach favoring humanitarian law adopted by some States, is the requirement to consider human rights–based notions of authority to detain, standards of treatment, fair trial, and punishment. In other words, what does it mean to treat someone humanely if they do not fall within the scope of the Third and Fourth Geneva Conventions and under what body of law are those standards found? This inevitably leads to a discussion of human rights norms.

The perception created by objecting States has been one of deadlock with the result being uncertainty as to whether the Additional Protocols will ever be universally accepted. Given the lack of progress on these previous attempts to regulate internal armed conflicts it is unclear how motivated human rights groups would be in embarking on another attempt to further change law that had already not been accepted by a number of militarily powerful States. Rather than advancing the law it could reasonably be viewed that any opening up of the existing provisions to further discussion might result in a rolling back of gains already made a quarter century earlier. In any event, a common position has been that what is needed is for existing law to be put into practice. Unfortunately, it sometimes appears as a mantra that conveniently masks the need to look at necessary changes to humanitarian law.[164] That said, there has been some indication that at least the law governing detention should be further developed.[165]

This does not mean there has been no attempt to advance the law governing armed conflict during the interim. One significant effort was the 2007 ICRC *Customary International Humanitarian Law Study*, which included a chapter on human rights–based "Fundamental Guarantees."[166] However, this Study has been controversial. It was generally criticized by the United States as an attempt to expand the application of the Additional Protocols through customary law rules to States that declined to become parties to those treaties.[167] Further, the Study was interpreted as seeking to extend the Geneva Conventions and Additional Protocols to internal armed conflict.[168] Those concerns may have some foundation. It is indicated in the introduction to that Study that "the gaps in the regulation of the conduct of hostilities in Additional Protocol II have largely been filled through State practice, which has led to the creation of rules parallel to those in Additional Protocol I, but applicable as customary law to non-international armed conflicts."[169] Of particular note is the criticism

---

[164] DINSTEIN, *supra* note 4, ¶ 741, at 295 (where it is stated "[i]t is nevertheless a gross mistake to believe (as some observers do) that better implementation is the sole major item on the present agenda of LOIAC.").

[165] 2011 CHALLENGES OF CONTEMPORARY ARMED CONFLICT REPORT, *supra* note 163, at 18.

[166] CUSTOMARY INTERNATIONAL HUMANITARIAN LAW STUDY, *supra* note 29, at 299–383.

[167] Bellinger & Haynes, *supra* note 2, at 448.

[168] *Id.*

[169] CUSTOMARY INTERNATIONAL HUMANITARIAN LAW STUDY, *supra* note 29, at xxix.

that nonbinding recommendations from human rights instruments were used "to fill perceived gaps in the customary law and to help them interpret terms in the law of war."[170]

Considering the past positions adopted by the United States, a response criticizing the *Customary Law Study* could not have been unexpected. It is the response of a powerful State not unlike that of the nineteenth-century "dominant military powers" that sought to restrict the extension of legitimate belligerent status to persons who might take up arms without being organized into regular armed forces or militias (see Chapter 4, section 4.3, *Impact of the Proper Authority Principle*). The desire to privilege the State as the proper authority responsible for maintaining internal order remains a strong one. It is a reality that has significant impact on international humanitarian law generally, and the law specifically applied to noninternational armed conflict.

The challenge for States and advocates of humanitarian law is that such generalized criticism can support a broader discounting of customary international law rules by advocates seeking to exclusively apply human rights law in internal conflicts. Therefore, what appears to be a continuation of the historical resistance expressed by States toward the application of humanitarian law in conflicts not of an international character can have unintended strategic consequences. A lack of clearly articulated customary international law rules means it is more likely that human rights law will be determined, by default, to be applicable. Given that effect, it may very well be time for States to closely consider this legal reality if they believe human rights law itself does not adequately govern significant aspects of internal conflict. The focus of international attention is increasingly being directed specifically at what law does apply to such operations. In this respect, statements that military forces apply the laws governing warfare simply as a matter of policy without addressing what law actually applies is not likely to appear adequate when exposed to the enhanced scrutiny of the specialized law debate.

This does not mean there have been no indications of change in the approach taken by States. For example, in 2011 Hillary Clinton as the United States Secretary of State stated that country would, out of a sense of "legal obligation," apply the norms of Article 75 of Additional Protocol I in international armed conflict in the treatment of detainees.[171] There was no indication as to what parts of Article 75 constituted the norms that were to be applied, or whether that article was to be viewed as customary international law. However, the statement by such a senior administration official will undoubtedly be viewed as expressing an indication that the State is bound by those norms as a matter of law (*opinio juris*). Harold Koh, the Department of State legal advisor, acknowledged that likelihood.[172]

---

[170] Bellinger & Haynes, *supra* note 2, at 447.

[171] Hillary Rodham Clinton, *Reaffirming America's Commitment to Humane Treatment of Detainees* Press Statement (Mar. 7 2011), http://m.state.gov/md157827.htm. *But see* John B. Bellinger & Vijay M. Padmanabhan, *Detention Operations in Contemporary Conflicts: Four Challenges for the Geneva Conventions and Other Existing Law*, 105 AM. J. INT'L L. 201, 207 (2011) (The authors note "the administration neither stated that Article 75 is customary international law nor agreed to apply Article 75 in noninternational armed conflicts, such as the conflict with Al Qaeda.").

[172] Harold Koh, Legal Adviser, Department of State, *Responses to Questions Submitted by Senator Richard G. Lugar, Libya and War Powers: Hearing Before the Committee on Foreign Relations, U.S. Senate*, 112th Congress, First Session, 53, 57 (June 28, 2011) ("The U.S. statement, coupled with a sufficient density of State practice and opinio juris, would contribute to creation of the principles reflected in Article 75 as rules of customary international law, which all States would be obligated to apply in international armed conflict. (The 168 States that are party to Protocol I are of course already required to comply with Article 75 as a matter of treaty law.")). *See also* DOD LAW OF WAR MANUAL, *supra*

In 2014, the U.S. Department of Defense provided perhaps the strongest indication of the general reliance by that country on the human rights–based provisions of humanitarian law when, in its directive on the *DoD Detainee Program* reference was made, albeit as matter of policy, to Common Article 3 of the Geneva Conventions, Articles 4 to 6 of Additional Protocol II, and Article 75 of Additional Protocol I, forming the minimum standards of treatment for detainees.[173] This was followed by the release of a joint manual on *Detainee Operations*, which follows the wording of the Clinton press release in specifically referring to applying Article 75 to international armed conflict "out of a sense of legal obligation."[174]

That Department of Defense detainee directive, the *Detainee Operations* manual, the Clinton statement, the *Department of Defense Law of War Manual*, combined with the judgment in *Hamdan v. Rumsfeld* (where a plurality of the U.S. Supreme Court ruled Article 75 of Additional Protocol I applied as a matter of customary international law to the noninternational armed conflict with Al Qaeda[175]) provide a strong indication that the United States has substantively moved toward the application of human rights norms as part of humanitarian law regardless of how a conflict is characterized. Those continuing to claim these human rights obligations should be viewed simply as a policy choice will face increasing pressure to reconsider the reasons why they take that position. One reason could be U.S. sensitivities toward acknowledging the applicability of the Additional Protocols as a matter of law. If that is the case, the above noted 2014 references to the Additional Protocols in official government publications appear to suggest this is a misplaced concern. However, it may be that the position is based on a bias that favors an exclusionary application of conduct of hostilities rules completely separate from any reference to human rights. If so, this is an approach that appears to inadequately address the nature of contemporary conflict, and one that has been overtaken by operational requirements.

In any event, the recognition of these provisions, as a matter of law or policy, represents a significant practical injection of human rights norms into humanitarian law as was envisaged when Additional Protocol I and Additional Protocol II were drafted in 1977. From an American perspective it has been suggested "the contentious issues should not blind us to the reality that the great bulk of human rights law is not contentious and that the United States remains an important and effective advocate for the promotion of human rights worldwide."[176] As early as 1996, operational necessity had led to the development of training for U.S. Army military lawyers on "Civilian Protection Law" (CPL), where "international human rights law provides the cornerstone of CPL, serving as the starting point for almost any CPL discussion."[177] With that law being "made up of a wide array of both customary

---

note 3, ¶ 8.1.4.2, at 490 ("This statement was intended to contribute to the crystallization of the principles contained in Article 75 as rules of customary international law applicable in international armed conflict.").

173 *DoD Detainee Program*, Department of Defense Directive, Number 2310.01E, ¶3 a., at 2 (Aug. 19, 2014), http://www.dtic.mil/whs/directives/corres/pdf/231001e.pdf.

174 DETAINEE OPERATIONS, JP3-63, 1-2 (Nov. 14, 2014), http://www.dtic.mil/doctrine/new_pubs/jp3_63.pdf.

175 Hamdan v. Rumsfeld, 45 I.L.M. 1130, 1155 (2006) (U.S. Supreme Court 2006).

176 Brill, *supra* note 97, at 64.

177 Richard M. Whitaker, *Civilian Protection Law in Military Operations: An Essay*, DEPARTMENT OF THE ARMY PAMPHLET 27-50-288, 3, at 10 (Nov. 1996), http://www.loc.gov/rr/frd/Military_Law/pdf/11-1996.pdf.

and conventional legal regimes (treaties and international agreements) and domestic law and policy," as well as host nation law, it stands out as an early form of operational law.[178]

In effect, fully considering the import of Article 75 of Additional Protocol I changes the nature of the strategic dialogue between the humanitarian and human rights law communities. It certainly will not end the debate since the scope of Article 75 is narrower than human rights treaties, and the two bodies of law are in a number of respects viewed as being fundamentally different.[179] However, the issue will not only be how the two normative frameworks relate to one another but also how human rights norms interact with conduct of hostilities norms within the context of humanitarian law itself. Recognition of the applicability of the norms incorporated into Article 75 of Additional Protocol I and Article 4 of Additional Protocol II has the effect of expanding the scope of customary international humanitarian law in respect of both international and noninternational armed conflict.

This renewed interest in the application of human rights norms has arisen in lockstep with the involvement of states conducting counterinsurgency and counterterrorism operations in fighting "small wars." Such conflicts have historically increased civil society engagement when questions are raised, particularly in democratic societies, concerning the legal norms applied by State armed forces.[180] The heightened interest in applying human rights law during such conflicts is perhaps best reflected in the issue of the extraterritorial application of human rights treaties and domestic constitutional law. This issue has been raised, although not always successfully, in a number of forums including European Court of Human Rights,[181] national courts,[182] and the Human Rights Committee.[183] Canadian courts specifically rejected the extraterritorial application of domestic human rights protection to persons being transferred by the Canadian Forces to Afghan authorities in Afghanistan and held, "[t]here is no legal vacuum, considering that the applicable law is international humanitarian law."[184]

The United States and Israel have traditionally taken the position there is no extraterritorial application of human rights treaty law. However, in 2014 the United States announced a change, which accepted that the Convention Against Torture and Other Cruel, Inhuman or Degrading Treatment or Punishment[185] had limited extraterritorial application. The acting Legal Advisor for the U.S. Department of State indicated that Convention "extended to

---

[178] *Id.*

[179] PROVOST, *supra* note 7, at 7.

[180] GIL MEROM, HOW DEMOCRACIES LOSE SMALL WARS 136–51 (2003) (for a discussion of the political impact within France stemming from the brutal counterinsurgency conducted in Algeria).

[181] *See, e.g.*, Al-Skeini v. The United Kingdom, IV Eur. Ct. H. R. Rep. 99 (2011), *available at* http://www.echr.coe.int/Documents/Reports_Recueil_2011-IV.pdf.

[182] *See, e.g.*, Amnesty International Canada v. Canada (Chief of the Defence Staff), 2008 FCA 401, [2009] 4 F.C.R. 149, *available at* http://reports.fja-cmf.gc.ca/eng/2009/2008fca401.html, and (F.C.), Amnesty International Canada v. Canada (Chief of the Defence Staff), 2008 FC 336, [2008] 4 F.C.R. 546, *available at* http://reports.fja-cmf.gc.ca/eng/2008/2008fc336.html.

[183] *General Comment No. 31, The Nature of the General Legal Obligation Imposed on States Parties to the Covenant*, ¶¶10 and 11 CCPR/C/21/Rev.1/Add.13 26 May 2004, *available at* http://www.unhchr.ch/tbs/doc.nsf/(Symbol)/CCPR.C.21.Rev.1.Add.13.En?Opendocument (last visited Aug. 11, 2012).

[184] Amnesty International Canada v. Canada (Chief of the Defence Staff), 2008 FCA 401, [2009] 4 F.C.R. 149, ¶36, *available at* http://reports.fja-cmf.gc.ca/eng/2008/2008fc336.html.

[185] Convention Against Torture and Other Cruel, Inhuman or Degrading Treatment or Punishment, Dec. 10, 1984, 1465 U.N.T.S. 85, 113.

certain areas beyond the sovereign territory of the State Party, and more specifically to 'all places that the State Party controls as a governmental authority.' We have determined that the United States currently exercises such control at the U.S. Naval Station at Guantanamo Bay, Cuba, and with respect to U.S. registered ships and aircraft."[186] Further, while there was reference to the law of armed conflict (i.e., humanitarian law) being the controlling body of law during armed conflict:

> a time of war does not suspend operation of the Convention Against Torture, which continues to apply even when a State is engaged in armed conflict. The obligations to prevent torture and cruel, inhuman, and degrading treatment and punishment in the Convention remain applicable in times of armed conflict and are reinforced by complementary prohibitions in the law of armed conflict.[187]

While this statement did not indicate there was an unlimited extraterritorial application of this human rights–based treaty, it nevertheless constituted a significant shift in the U.S. position. The reference to these human rights law provisions being reinforced by the complementary prohibitions under humanitarian law, combined with an acknowledgment elsewhere that "[w]here the LOAC [law of armed conflict] is silent or its guidance inadequate, specific provisions of applicable human rights law may supplement the LOAC,"[188] suggests an acceptance of greater interaction between the two normative frameworks.

There has also been a substantial acceptance by States,[189] courts (both international and domestic),[190] and academics[191] that human rights law can apply as a matter of customary law. Even as the United States struggles with accepting that human rights law treaties having broad extraterritorial application, the 2015 *Operational Law Handbook* specifically refers to fundamental human rights found in customary international law binding "a State's forces during all operations, both inside and outside of the State's territory."[192] The importance

---

186 *Opening Statement Mary E. McLeod Acting Legal Adviser U.S. Department of State Committee Against Torture*, Mission of the United States, Geneva, Nov. 12–13, 2014 [hereinafter *Acting Legal Advisor Opening Statement*], https://geneva.usmission.gov/2014/11/12/acting-legal-adviser-mcleod-u-s-affirms-torture-is-prohibited-at-all-times-in-all-places/.

187 *Id.*

188 Operational Law Handbook, *supra* note 14, § Part V. A. ¶ 3, at 54.

189 2 Restatement of the Law: The Foreign Relations Law of the United States 161–75 (1987) (for an outline of a United States perspective on the scope of customary human rights law). *See also* Operational Law Handbook, *supra* note 14, § Part IV A., at 53, and Amnesty International Canada and British Columbia Civil Liberties Union v. Chief of Defence Staff for the Canadian Forces et al., Case File No. T-324-07, Respondent's Factum ¶83, at 26 (Jan. 18, 2008), http://web.ncf.ca/fk624/data/Factum%20-%20Crown%20(Charter,%20Jan%2008).PDF (where the Canadian Government argued common Article 3 of the Geneva Conventions and Article 75 of Additional Protocol I apply as a matter of treaty or customary law to guarantee fundamental rights of a detained person charged with a criminal offence that are virtually identical to norms found in international human rights law.).

190 Crawford, *supra* note 88, at 123 ("Domestic courts have also looked to the international human rights regime in times of armed conflict for guidance as to how to balance the needs of the State in conducting an armed conflict against individuals who may be affected by that State in the execution of its military objectives.").

191 Yoram Dinstein, The International Law of Belligerent Occupation ¶ 166, at 71 (2009), Noam Lubell, Extraterritorial Use of Force Against Non-State Actors 235 (2010), and Crawford, *supra* note 88, at 120–22.

192 Operational Law Handbook, *supra* note 14, § IV. B. 1, at 53.

of customary international human rights law cannot be overstated. It has been noted that "as a practical matter, not all States are parties to international human rights instruments" and that "the Asia-Pacific region has no unique regional human rights instrument nor any regional human rights court or commission . . . ."[193] Further, "a considerable number of States in the Asia-Pacific region are not party to any of the major international human rights documents."[194] This does not mean inhabitants of these countries do not enjoy human rights protection under international law. Given the paucity of treaty protection, those rights must be found under customary international law.

There remains the issue of whether customary international human rights law has extraterritorial application for a State that denies a treaty-based obligation or is not a party to such as treaty. In this regard it has been suggested "it does not seem at all possible to disentangle the territorial limitations on human rights as prescribed in treaties from any customary substantive rules of human rights law."[195] Further, "it is quite unlikely that states have assumed more extensive obligations under customary human rights law that they have under treaty law."[196] However, considering that a foundational aspect of such law is that it is of a universal in nature and the *erga omnes* [rights owed to everyone] status of many human rights obligations, it is difficult to support this approach.[197] It is also not clear the degree to which this approach is influenced by negative views of customary international law. In this respect it has been suggested it is difficult to determine the content of customary rules from the practice of states, and that customary human rights law "rarely has a forum for enforcing such obligations directly."[198] These factors would have to be weighed against the significant advantages of inculcating compliance with customary human rights obligations within State security forces and the extension of those protections to civilians in conflict.

The argument that customary human rights law does not apply should prompt a critical moral discussion as to why all persons should not enjoy such basic rights; and a State, as a proper authority, under international law would not be required to recognize and protect those rights wherever it is operating. In any event, even if it were argued there is no extraterritorial application of customary human rights law, human rights norms would have to be considered to the extent they are part of humanitarian law. Because of the reality of contemporary military operations, with their inevitable interface with local populations, the issue is not whether human rights norms apply, but rather how and to what degree.

There are also considerable advantages from an accountability perspective in recognizing the overlapping application of human rights and humanitarian law. For those States that do not recognize an extraterritorial application of human rights law there remains the obligation to comply with humanitarian law. The resulting reliance on humanitarian law mechanisms of accountability does not preclude the application of regional and domestic human rights law bodies where they have jurisdiction.[199] However, it does help avoid the potential

---

193 CRAWFORD, *supra* note 88, at 120.

194 *Id.*

195 MILANOVIĆ, *supra* note 55, at 3.

196 *Id.*

197 LUBELL, *supra* note 191, at 233–35 (for arguments supporting the application of customary human rights law).

198 MILANOVIĆ, *supra* note 55, at 3.

199 ANTONIO CASSESE, INTERNATIONAL LAW 389 (2nd ed. 2005) (who notes that regional supervisory mechanisms include judicial bodies such as the European Court of Human Rights, the Inter-American Commission, and the

gaps in the application of human rights norms that arises from human rights tribunal reliance on "exclusive control" and "state agent control" criteria as the bases for extraterritorial applicability of human rights treaties.[200] Further, an illegal use of force is amenable to prosecution as a war crime as a breach of humanitarian law.[201]

Importantly, this more holistic approach also allows for easier integration and socialization within the doctrine and training of military forces. This can help avoid the potential confusion of that can result from trying to simultaneously apply two bodies of law. Human rights training is a crucial part of preparing security forces for contemporary operations. Stressing that human rights obligations are an integral part of humanitarian law or the law of armed conflict (if that is the term used) helps reinforce that the warrior ethos extends to providing such protection. Finally, unlike human rights law, the norms contained in humanitarian law are more clearly applicable to the non-State opponent.[202] To the extent a non-State opponent seeks to gain recognition or a form of legitimacy through adherence to humanitarian law, this must also include respect for its fundamental human rights provisions.

5.4. Overlap and Convergence

One of the most challenging aspects of resolving the humanitarian law and human rights law interface is the significant gap that can exist between the legal debate at the strategic level and the reality facing military commanders on the ground. At the theoretical and national strategic level, the discussion is often about the exclusion of one body of law or the other, the impact of special laws, and the degree to which human rights law has extraterritorial application. In contrast, at the operational and tactical levels, military commanders and other State security personnel face daily dilemmas regarding the use of force potentially involving both of these normative paradigms. It is the practical challenges that appear to have led security forces to more readily embrace the notion there can be a concurrent application of human rights and humanitarian law obligations. The most obvious contemporary challenges arose in respect of counterinsurgency operations in Iraq and Afghanistan.

Inter-American Court of Human Rights. There is also the African Court on Human and People's Rights, as well as the monitoring body, the African Commission on Human Rights and the Rights of Peoples. However, not all countries are subject to these human rights bodies.).

200 Al-Skeini v. The United Kingdom, IV Eur. Ct. H. R. Rep. ¶¶ 131–40, at 166–70 (2011), and Hassan v. The United Kingdom, Application No. 29750/09, Eur. Ct. H. R. ¶¶ 75–80 (2014), http://hudoc.echr.coe.int/sites/eng/pages/search.aspx?i=001-146501.

201 Meron, *supra* note 78, at 266 (where it is noted "the offences included in the ICC Statute under crimes against humanity and common Article 3 are virtually indistinguishable from major human rights violations. They overlap with violations of some fundamental human rights law, which thus become criminalized under an instrument of international humanitarian law.").

202 Sassoli, *supra* note 26, at 87 ("Both parties must be equal as far as the applicable IHL is concerned, but they are not equal as far as IHRL is concerned."), and Louise Doswald-Beck, *The Right to Life in Armed Conflict: Does International Humanitarian Law Provide All the Answers?*, 88 Int'l Rev. Red Cross 881, 890 (2006) ("On the other hand, it is normal that governments have more responsibilities under international law than their citizens, in this case the restrictions of human rights law."). *But see also* Crawford, *supra* note 88 at 127 (where it is suggested that non-State insurgent groups can bear some human rights obligations.).

However, questions regarding what use of force options are available to commanders can rise in a variety of other situations including hostage rescue, the use of drones (see Chapter 8, *Counterterrorism and the "Away Game")*, and even "robust peacekeeping" (see Chapter 11, section 11.3.2, *United Nations Operations and "Robust Peacekeeping"*).

The acknowledgment in U.S. military doctrine that customary human rights law applies to foreign operations and that human rights can be used to interpret humanitarian law obligations seems at odds with what was, as late as April 2014, a State position denying the extraterritorial application of human rights treaty law obligations,[203] and one embracing the traditional special status of humanitarian law.[204] The fact that the United States has altered its position regarding the concurrent application of the Convention Against Torture and Other Cruel, Inhuman or Degrading Treatment or Punishment and the law governing warfare provides an indication that this gap is closing. However, it is likely premature to state the distance between high-level rhetoric and the practical reality of contemporary operations has completely closed. The remaining distance can be seen in the State Department reference to the U.S. reliance on presidential executive orders, and manuals and policies to extend protection against torture and in humane treatment to all circumstances where a person is detained.[205] It is reported that "[t]he State Department, which had proposed changing the Bush-era position on the scope of the cruelty ban, faced resistance from military and intelligence lawyers, who raised unspecified operational concerns."[206]

Notwithstanding the resistance frequently demonstrated by both the humanitarian and human rights communities toward a concurrent application of these bodies of law, reference to an overlap of normative regimes is definitively set out in *Legal Consequences Of The Construction Of A Wall In The Occupied Palestinian Territory*, although this largely remains an underexplored area of the law. In that case, the court noted there are thus three possible situations concerning the relationship between these two bodies of law, "some rights may be exclusively matters of international humanitarian law; others may be exclusively matters of human rights law; *yet others may be matters of both these branches of international law.*"[207] As has been suggested in this chapter, a particular area of overlap occurs between the norms found in human rights law and those same norms contained in both treaty and customary humanitarian law. During armed conflict both bodies of law continue to apply, and the norms found therein logically overlap. This is particularly evident when the operational tasking is to maintain law and order, which is fundamentally a human rights–based law enforcement task.

---

[203] H.R. Committee, Concluding observations on the fourth periodic report of the United States of America, Human Rights Committee, UN Doc CCPR/C/USA/CO/4, ¶ 4. (Apr. 23, 2014). *See also* The Use of Force in Armed Conflicts: Interplay Between the Conduct of Hostilities and Law Enforcement Paradigms, ICRC at 5–6 n.21 (Nov. 2013), http://www.icrc.org/eng/assets/files/publications/icrc-002-4171.pdf, and Dennis, *supra* note 6, at 141.

[204] Operational Law Handbook, *supra* note 14, Part V. A. 1., at 52 (setting out the traditional "displacement view" that is indicated to have been previously "embraced" by the United States).

[205] *Acting Legal Advisor Opening Statement, supra* note 185.

[206] Charlie Savage, *U.S. Tells U.N. Panel of Steps to Revise Interrogation Policy*, N.Y. Times, Nov. 12, 2014, http://www.nytimes.com/2014/11/13/us/us-vows-to-stop-using-torture-against-terrorism-suspects.html?_r=0.

[207] Legal Consequences Of The Construction Of A Wall In The Occupied Palestinian Territory, 2004 I.C.J. Reports 136, ¶106, at 178 (emphasis added).

However, the interface between the normative regimes is not simply a question of overlap and concurrent application. The nature of contemporary insurgencies and terrorism also creates a situation where the application of these frameworks results in a considerable degree of convergence. This takes place because so much of the violence occurs at the boundary between hostilities and law enforcement. The intensity of violence is one of the key threshold criteria for establishing the existence of an armed conflict, and therefore a potential dividing line between the two legal frameworks. This can be seen in Article 1(2) of Additional Protocol II with that Protocol being indicated as not applying "to situations of internal disturbances and tensions, such as riots, isolated and sporadic acts of violence and other acts of a similar nature, as not being armed conflicts." The threat on the ground is enhanced by the increasing use of explosives, often in the form of improvised explosive devices (IEDs) and suicide bombing (see Chapter 12, section 12.3.1, *A Ubiquitous Weapon*).

Police and military forces both frequently have to confront threats occurring at the intersection of law enforcement and hostilities. These scenarios may arise when military forces are tasked with carrying out policing functions as they attempt to bring order to a territory (see Chapter 12, section 12.2.1, *The Military Task: Applying Force of First and Last Resort*). Conversely, police forces can be required to address threats from organized groups engaged in armed conflict. This latter situation can occur because of a conscious decision by a State to apply a law enforcement response to an insurgency threat, or because the police are simply the first responders. In terms of training, equipment, and culture, those law enforcement authorities are likely to apply a human rights–based approach regardless of whether the threat is an insurgent or a terrorist engaged in armed conflict.

Adding to the complexity of the security challenge is the requirement to counter transnational criminal actors that are organized and equipped for warfare and criminals who provide support to non-State actors engaged in insurgency and terrorism. Further, there are operations, such as hostage rescue, which can occur across the conflict spectrum. These rescues are often positioned at the fault line between armed conflict and criminal activity. They may be conducted by either police or military forces that can effectively perform the same function, receive similar training, and which often use the same tactics (see Chapter 10, section 10.4, *Converging Tactics: Special Forces, Law Enforcement, and Hostilities*). The pressure that is being placed on human rights–based rules governing the use of force is perhaps most graphically illustrated in the 2011 European Court of Human Rights decision, *Finogenov v. Russia*.[208] In that case the court accepted that more than 120 collateral civilian deaths,[209] which occurred during a counterterrorist operation against Chechen rebels in a Moscow theater, were not disproportionate.[210] Traditionally, a fundamental distinguishing feature between humanitarian law and human rights law has been the specific provision in the former body of law that collateral civilian deaths can be justified (see Chapter 11, section 11.2.3, *Law Enforcement and Its Unique Role*).[211]

---

208 Finogenov v. Russia, VI Eur. Ct. H. R. 365 (2011) (extracts).

209 *Id.* ¶ 34, at 381 (of the 129 victims it appears 125 did not die from bullet wounds.).

210 *Id.* ¶¶ 227-36, at 407–09 (where the court concluded for a variety of reasons, including that the gas used to incapacitate the terrorists was not intended to kill over 120 hostages, meant that its use was not disproportionate.).

211 AP I, art. 57(2)(a)(iii).

Perhaps the most frustrating aspect of the exclusivity debate about the application of human rights and humanitarian law is that military forces have long had to apply human rights–based concepts, regardless of whether a conflict is categorized as international or noninternational in character. For example, that requirement can arise in the context of a blockade, or when conducting visit and search on the high seas during an international armed conflict where the ships being searched have civilian crews and passengers. As was noted in the Israeli *Public Commission to Examine the Maritime Incident of 31 May 2010* investigating the Israeli action to maintain the Gaza blockade, to the extent the crew and passengers on merchant vessels are not taking a direct part in hostilities any force used against them would be proscribed by human rights–based law enforcement–like norms.[212] This requirement to limit the use of force is not restricted to the high seas. This was evident during the extensive "brown water warfare" operations in South Vietnam where inspections of civilian watercraft, averaging 16,000 boardings per month between June 1966 and June 1968, were conducted to interdict enemy personnel and cargo transiting internal, territorial, and adjacent waters.[213]

Even the most limited of military operations carried out during an international armed conflict can require armed forces to interface with and potentially use force against civilians. This is particularly true of contemporary counterinsurgency and counterterrorism operations focused on the threats posed by organized armed groups fighting "amongst the people."[214] Such fighting reflects:

> the reality in which the people in the streets and houses and fields—all the people anywhere—are the battlefield. Military engagements can take place anywhere: in the presence of civilians, against civilians, in defence of civilians. Civilians are the targets, objectives to be won, as much as an opposing force.[215]

The force used by security forces is often directly linked to the type of threat being posed. The simple requirement to man checkpoints, whether for vehicle control, vital point security, conducting searches, or separating civilians from areas of military operations, can present challenging scenarios.

Security personnel, both military and police, employed on checkpoint duties must be trained in human rights–based standards and be ready to use force associated with law enforcement including a graduated and proportionate response. Force is used in the last resort and primarily in self-defense. With a default position to apply law enforcement norms, the escalation of force can act as a form of threat discriminator. At the same time where the threat is posed by a member of an organized armed group the response at law may be one grounded in the conduct of hostilities rules found in humanitarian law, regardless of whether the force is being applied by police or military forces.[216] The change between normative regimes may have to be made on the basis of a split-second decision. Given the nature

---

212 First Turkel Report, *supra* note 59, at 234.

213 Thomas J. Cutler, Brown Water, Black Berets 81, 133–34 and 166 (1988).

214 Rupert Smith, The Utility of Force: The Art of War in the Modern World 3 (2005).

215 *Id.* at 3–4.

216 Watkin, *supra* note 66, at 311–12 (for a more detailed discussion of this issue).

of contemporary operations, it is clear that State security forces do not necessarily have the luxury of operating in one "world" or the other.

It has been noted there is an increasing awareness among military forces that the authority for dealing with nonbelligerents "is operationally similar to the authority of police officers interacting with the public during times of peace."[217] The challenge for States and international law is how to maintain law and order in its broadest sense during conflicts in which the violence occurs among the very people these political entities can have a responsibility for governance (either in fact or law). It cannot be forgotten that "IHL is meant to protect human rights in situations of armed conflict."[218] What is certain is that the operating environment facing military commanders and their legal advisors is too complex to allow the luxury of an approach that attempts to keep the governing law separated into different "silos." It is from just such complex operating environments that the discipline of operational law developed.

However, this is only part of the story in the contemporary operating environment. The use of force issue is not just about killing. Increasingly, the debate is about when a suspect has to be or should be captured. In respect of both counterinsurgency and counterterrorism, operations resulting in capture can significantly exceed those where killing is the intended outcome.[219] This may occur, in part, because civilians not taking a direct part in hostilities are only subject to what is essentially a law enforcement response, even if they are providing indirect support to an organized armed group. Whether the security forces involved rely on human rights law or human rights norms found in humanitarian law to arrest these supporters, the obligations remain the same.[220]

Further, operations directed at killing insurgents may be preceded by the arrest or capture of persons who cannot be targeted. Operations may also concurrently involve such activity. A decision may also be made to arrest rather than kill a suspected terrorist. Finally, even under humanitarian law, a targeting assessment can in particular circumstances reach the conclusion an attack is not warranted leaving open the option of seeking to capture an individual (see Chapter 7, section 7.6.1, *Insurgents and Target "Value"*). From a practical perspective it matters little whether the overlap of the two normative frameworks is because of law or policy if the direction provided is to privilege the capture of a terrorist. Contemporary commanders are being required to apply both the conduct of hostilities and human rights normative frameworks. This makes the adoption of exclusionary approaches toward the application of human rights and humanitarian law appear badly out of touch with existing security challenges and fundamentally inadequate to deal with existing operational challenges. Indeed, if legal advisors are not as intimately knowledgeable about human rights law as they are about the law governing the conduct of hostilities, it is difficult to see how they can properly advise their client commanders.

---

[217] Corn, *supra* note 68, at 61.

[218] BARAK, *supra* note 79, at 205.

[219] MARK URBAN, TASK FORCE BLACK 257 (2010) (Where it is noted that A Squadron, SAS, operating in Iraq from May to November 2007 "mounted raids almost nightly, during which it arrested 335 people and killed 88. The latter figure in particular marks a stark contrast from the squadron's deployment of late 2005, when it took just one life.").

[220] *See, e.g.*, AP I, art. 75, or AP II, art. 4.

III The Threat, the State Response, and Legal Uncertainty

Surrounded by a vast organization, which prepares his [the terrorist's] task and assists him in its execution, which assures his withdrawal and his protection, he runs practically no risks—neither that of retaliation by his victims nor that of having to appear before a court of justice.[1]

6 Contemporary Threats

INSURGENCY AND TERRORISM

6.1. Introduction

While the potential continues to exist for inter-State warfare the most widespread and persistent security challenges of the twenty-first century involve insurgencies and terrorism. Even where the conventional type of threats exist, such as with the Hezbollah use of anti-tank missiles and rockets, the conflict can also be viewed as being mixed with aspects of guerrilla warfare (e.g., the use of tunnels) and terror (e.g., kidnapping).[2] The first decade of this century has focused attention on non-State conflict and, in particular, the two operational doctrines developed to fight such small wars: counterinsurgency and counterterrorism. The two types of operations are closely linked but are also shaped by the different nature of the threats they are respectively intended to address. Consistent with the general approach under international law for categorizing conflicts, both insurgency and terrorism are provided definitional boundaries within which the threat is identified and State responses assessed. However, attempts at categorization are challenged to provide a well-defined construct for the assessment of these security threats. This chapter will explore the threat of insurgency and contrast it with how terrorism and terrorists are described. The analysis will also look at the emerging threats posed by criminal insurgencies and the links between terrorism, insurgency, and crime. As a result, this chapter is largely descriptive in

[1] Roger Trinquier, Modern Warfare: A French View of Counterinsurgency 16 (2006) (1964).

[2] Yaakov Katz & Yoaz Hendel, Israel vs. Iran: The Shadow War 30 (2012).

nature, focusing primarily on the nature of the insurgent and terrorist threat rather on the legal issues surrounding their activities. Importantly, it sets the scene for the subsequent analysis of the law governing counterinsurgency and counterterrorism. In doing so the analysis will place particular emphasis on Al Qaeda, the Islamic State, and other Salafi jihadists as the dominant transnational "terrorist" threat of this century.[3]

6.2. Insurgency

6.2.1. CRITERIA FOR AN INSURGENCY

The categorization challenge regarding insurgency arises because of the difficulty in establishing clear thresholds, including where the line is drawn between ordinary crime and more threatening insurgency, as well as the overlapping nature of insurgency and terrorism. In respect of the overlap between terrorism and insurgency it has been noted that "[t]errorism is a component in almost all insurgencies, and insurgent objectives (i.e., a desire to change the status quo through subversion and violence) lie behind almost all nonstate terrorism."[4] To understand the interaction between the two threats, as well as the doctrines designed to defeat them, it is important to first consider what makes each form of violence unique. For insurgency this involves looking at it in its classic form and as an element of criminal activity, such as in transnational criminal organizations and even social banditry.

Much like the thresholds for noninternational armed conflict there are two factors historically linked to an insurgency: levels of violence and the degree of organization of the insurgent forces. Insurgency is associated with guerrilla warfare, implying "a certain level of organized violence which is above and beyond isolated acts of sniping, kidnapping and robbery."[5] It is closely associated with guerrilla or "small" wars, which often involve "the use of hit-and-run tactics by an armed group directed primarily against a government and its security forces for political or religious reasons."[6] A key attribute of an insurgency has been the existence of a military organization that can fight and hold territory, but also one that can eventually be converted into "a more conventional military force capable of taking on the authorities on their own terms."[7]

As Lindsay Moir has identified, under classic international law, there were three stages of violence identified as being used against the established authorities "depending on the scale and intensity of the conflict: rebellion, insurgency and belligerency."[8] Rebellion represented

---

[3] Jessica S. Stern & J. M. Berger, ISIS: The State of Terror 268 (2015) (where Jihadi Salafism is explained).

[4] David Kilcullen, Counterinsurgency 184 (2010). *See* Audrey Kurth Cronin, *ISIS Is Not a Terrorist Group: Why Counterrorism Won't Stop the Latest Jihadist Threat*, For. Aff. 87, 88 (Mar./Apr. 2015)(with respect to ISIS it is argued, "although it uses terrorism as a tactic, it is not really a terrorist organization at all." Rather ISIS "is a pseudo-state led by a conventional army.").

[5] John Ellis, From the Barrel of a Gun: A History of Guerrilla, Revolutionary and Counter-Insurgency Warfare, from the Romans to the Present 14 (1995).

[6] Max Boot, Invisible Armies: An Epic History of Guerrilla Warfare from Ancient Times to the Present xxii (2013).

[7] Ian F. W. Beckett, Modern Insurgencies and Counter-insurgencies: Guerrillas and Their Opponents since 1750, at vii (2001). *See also* Ellis, *supra* note 5, at 14.

[8] Lindsay Moir, *The Historical Development of the Application of Humanitarian Law in Non-International Armed Conflicts to 1949*, 47 Int'l & Comp L.Q. 337, 338 (1998).

a sporadic and more modest challenge to the control over the population, "[p]rovided the uprising could be dealt with swiftly and effectively in the normal course of internal security, the conflict remained fully domestic."[9] As such the rebels had no recognition under international law and remained punishable under municipal law.[10] In contrast, an insurgency represented a more substantial challenge against the order of the State with "the rebelling faction being sufficiently organised to mount a credible threat to the government."[11] This forced a form of acknowledgment of the existence of an insurgency, although it did not accord any belligerent rights.[12] In contrast, the recognition of belligerency by the parent government or a third State "amounted to a declaration by the recognising party that the conflict had attained such a sustained level that both sides were entitled to be treated in the same way as belligerents in an international conflict."[13]

A key factor in determining which type of conflict directed against the established authority (i.e., the "proper authority") was in existence has been the level of violence. The criterion of "*protracted armed violence* between governmental authorities and organized armed groups or between such groups within a State" was particularly relied on in early contemporary legal assessments as to when a noninternational armed conflict is in existence.[14] However, it has subsequently been noted that the requirement of protracted violence "refers more to the intensity of the armed conflict rather than its duration."[15] Distinctions based on levels of violence can also be seen in the threshold increasingly adopted in treaty law to determine whether a matter is uniquely criminal, as opposed to an insurgency or other armed conflict. One distinction between criminal activity and armed conflict is articulated as a threshold of "internal disturbances and tensions, such as riots, isolated and sporadic acts of violence and other acts of a similar nature, as not being armed conflicts."[16] This is a standard that has been identified as having been "based on or inspired by Common Article 3" (see Chapter 9, section 9.4.3, *The "Totality of the Circumstances" and Other Criteria*).[17]

---

[9] *Id.*

[10] *Id.*

[11] *Id.*

[12] *Id.* at 339.

[13] *Id.*

[14] Prosecutor v. Tadić, Case No. IT-94-1-A, Decision On The Defence Motion For Interlocutory Appeal On Jurisdiction, ¶ 70 (Int'l Crim. Trib. for the Former Yugoslavia, Appeal Chamber, 2 October 1995), http://www.icty.org/x/cases/tadic/acdec/en/51002.htm.

[15] *Report of the Special Rapporteur on extrajudicial, summary or arbitrary executions*, UN Doc. AA/68/382, (Sept. 13, 2013) (by Christof Heyns) ¶ 57, at 12/24. *See also* Prosecutor v. Haradinaj, Case No. IT-04-84-T, ¶ 49, at 27-28 (Int'l Crim. Trib. for the Former Yugoslavia, Judgment, 2008), http://www.icty.org/x/cases/haradinaj/tjug/en/080403.pdf, and Prosecutor v. Limaj, Case No. IT-03-66-T, ¶ 90 (Int'l Crim. Trib. for the Former Yugoslavia 2005), http://www.icty.org/x/cases/limaj/tjug/en/lim-tj051130-e.pdf.

[16] AP I, art. 1(2), 1996 Protocol on Prohibitions or Restrictions on the Use of Mines, Booby-Traps and Other Devices (Protocol II), entered into force Dec. 2, 1983, 1342 U.N.T.S. 168, art. 1(2), the 1998 Rome Statute, art. 8(2)(d) and (f), the 1999 Second Protocol To The Hague Convention For The Protection Of Cultural Property In The Event Of Armed Conflict, 249 U.N.T.S. 240, art. 22, entry into force Aug. 7, 1956, and the 2001 Amendment To The Geneva Convention On Prohibitions Or Restrictions On The Use Of Certain Conventional Weapons Which May Be Deemed To Be Excessively Injurious Or To Have Indiscriminate Effects, 2260 U.N.T.S. 82 (as amended Dec. 21, 2001), art. 2.

[17] Liesbeth Zegveld, The Accountability of Armed Opposition Groups in International Law 147 (2002).

6.2.2. INSURGENT ORGANIZATION

A particularly challenging aspect of insurgency is the wide range of organized groups that engage in fighting against State authorities. Armed groups can range from small organizations operating individually, or as part of a broader coalition of groups, to armed forces that are organized along conventional lines.[18] An example of a small insurgent organization is the ten-person independent nationalist Sunni group in Iraq armed with hand grenades, AK-47 machine guns, RPG-7s, and a 62 mm mortar that was reported to be fighting in the aftermath of the 2003 U.S.-led invasion. By June of that year the group had growing ties with Islamic and ex-Ba'athist groups, as well as members of the Fedayeen Saddam.[19] This was just one form of a wide variety of armed groups fighting the Coalition and Iraqi armed forces. The insurgent groups included combat cells ranging from large and well-developed organizations described as squadrons, battalions, or brigades, to part-time participants.[20] These groups were far from homogeneous. They included Arab nationalists, disgruntled Muslims, foreign fighters, and Sunni extremists.[21] Further, while there were dominant Sunni and Shiite groups, "there were many other 'resistance' organizations as well—by one count, fifty-six in all."[22]

The small Sunni resistance groups in Iraq can be contrasted with the military forces of the South West Africa People's Organization (SWAPO) fighting South African forces in Nambia in the 1970s. By late 1976, SWAPO was well established in southern Angola with "most of their guerrilla forces were being reorganized and trained on a conventional basis."[23] This included battalion bases, incorporating logistics, that "were established safely out of the way of South African attacks, some 200-km or more north of the SWA/Namibian border."[24] Nor has such a conventional organization been uncommon, particularly as an attempt is made to transition from guerrilla warfare to hostilities that permit the holding of territory. The Sudanese People's Liberation Army (SPLA) was organized like a conventional military force including platoons, companies, battalions, and brigades. There were disciplined artillery, infantry, signals, engineers, and logistics divisions led by professional officers "who had formalized training throughout their career."[25] By late 1989 the SPLA not only had effective control of large areas of southern Sudan, the regional battalions "also constituted the civil-military administration for these areas."[26] Similarly, the 1980s Salvadoran FMLN guerrilla organization included "mobile forces, organized into columns, battalions,

---

[18] Kenneth Watkin, *Opportunity Lost: Organized Armed Groups and the ICRC "Direct Participation in Hostilities" Interpretive Guidance*, 42 N.Y.U. J. Int'l L. & Pol. 641 (2010).

[19] Zaki Chehab, Inside the Resistance: The Iraqi Insurgency and the Future of the Middle East 6–8 (2005) (for a description of a meeting with five members of the organized armed group in June 2003.).

[20] Ahmed S. Hashim, Insurgency and Counter-insurgency in Iraq 158–60 (2006).

[21] *Id.* at 138–39.

[22] Boot, *supra* note 6, at 545 (relying on Mohammed M. Hafez, Suicide Bombers in Iraq: The Strategy and Ideology of Martyrdom 243–49 (2007)).

[23] Peter Stiff, The Silent War: South African Recce Operations 1969–1994, at 192 (2009).

[24] *Id. See* A. J. Venter, Gunship Ace: The Wars of Neall Ellis, Helicopter Pilot and Mercenary 72 (2011).

[25] Anthony Vinci, Armed Groups and the Balance of Power: The International Relations of Terrorists, Warlords and Insurgents 18 (2009)

[26] *Id.* at 19.

and brigades consisting of as many as 800 experienced, full-time troops capable of conducting major combat operations throughout the country."[27]

The adoption of conventional military organizations by non-State actors is not limited to this time period. After Hamas seized power in Gaza in 2007, its military organization changed from serving as an underground terrorist group to a small army complete with specialized sniper, antitank, combat intelligence, infantry, antiaircraft, and artillery professions, as well as naval commandos. Gaza was divided into five sections, each corresponding to a brigade that consisted of battalions and specialized units.[28]

Insurgent groups can also take on a hybrid organization. For example, the Afghan guerrilla leader, Ahmad Shah Masoud, who was killed by Al Qaeda just two days prior to 9/11, was a mujahideen commander who took up arms to resist the Soviet invasion of 1979. He very effectively maintained control of the Panjshir Valley despite repeated Soviet assaults.[29] In addition to administering a "'liberated zone' with its own schools, courts, mosques, prisons, a French-operated hospital, a military training center," he also "divided his forces into mobile groups of full-time fighters (*moutarik*) and a local militia of part-time helpers who would defend their villages (*sabet*)."[30] The *moutarik* were organized into 120-man companies with an assortment of weapons including DshK 12.7-milimeter machine guns and ZPU-2 antiaircraft guns.[31]

In Lebanon, Hezbollah also applied a unique "hybrid" organizational model during its 2006 conflict with Israel. It employed the "so-called elite or regular fighters, who number about 1,000 men and who were often given advanced weapons training; and village fighters, whose numbers are difficult to estimate because the often include local men only loosely affiliated with Hezbollah."[32] One account has estimated the number of village guards to be 5,000 to 10,000 fighters in 2006.[33] Israeli sources are reported to have indicated the numbers have increased to 30,000 guerrilla fighters.[34] During the 2006 conflict Hezbollah offered "neither an ideal model of 'guerrilla' warfare nor one of 'conventional' warfighting, but its approach and proficiency nonetheless place it well within a band that has characterized many past state militaries in interstate conflicts."[35] Needless to say, the forces engaged in insurgencies can be extremely diverse.

The diverse nature of insurgency is also reflected in the type of groups and operations undertaken by forces fighting a State (i.e., the "proper authority"). One study has identified four types of insurgent groups in terms of how they "combine central and local

---

[27] Todd Greentree, Crossroads of Intervention: Insurgency and Counterinsurgency Lessons from Central America 94 (2008).

[28] Katz & Hendel, *supra* note 2, at 125.

[29] Boot, *supra* note 6, at 490-91.

[30] *Id.* at 489.

[31] *Id.*

[32] Anthony H. Cordesman, George Sullivan, & William D. Sullivan, Lessons of the 2006 Israeli-Hezbollah War 80 (2007).

[33] Matthew Levitt, Hezbollah: The Global Footprint of Lebanon's Part of God 13 (2013).

[34] Katz & Hendel, *supra* note 2, at 27.

[35] Stephen Biddle & Jeffrey A. Friedman, The 2006 Lebanon Campaign and the Future of Warfare: Implications for Army and Defense Policy xv (2008), http://www.strategicstudiesinstitute.army.mil/pdffiles/pub882.pdf.

control: integrated, vanguard, parochial and fragmented. . . ."[36] Although the underlying structure of a group may not be easily changed, "[l]eaders may decide to combine forces, enter into an umbrella agreement, or merge their structures."[37] The scope and intensity of operations can vary from isolated guerilla warfare attacks by small groups of fighters to, at the civil war end of the conflict scale, military forces holding territory and performing a governance role. Successful guerrilla leaders realize the need to develop the "organizational machinery capable of handling the training, disciplining, equipping, and strategic and tactical direction of such an army."[38] During the course of a conflict the fortunes of the insurgent force may wax and wane. After the U.S.-led invasion of Afghanistan in 2001, the Taliban were quickly defeated. However, "[t]he Taliban leadership fled to Pakistan and revived the movement, this time as a force for guerrilla insurgency rather than the quasi-conventional combat of the 1990s."[39] While it has waged an effective war since that time, the conflict has not been without its setbacks.[40] For example, it is suggested the Taliban may have attempted a transition too quickly to a Maoist-type third phase of a conventional style of operation in Kandahar in 2006,[41] which resulted in a major defeat at the hands of Canadian and other Coalition forces.[42]

The success or failure of the insurgent group can have a profound impact on the organization and tactics that are adopted. While insurgents can be successful without a safe haven, the ability to control "a portion of contested territory where an organization's rivals cannot intervene with enough force to disturb its operations" has been identified as a key determinant of an insurgent organization's ability to win.[43] Safe havens exist in "ungoverned spaces." It has been suggested that the preference is for the safe haven to be within the contested territory as "[h]avens across the border are rarely safe for long because finicky sponsors may interrupt operations at will or distract the organization from its original goals by making it a tool to project influence in the neighbouring country."[44] Where organizations do not have a safe haven they "must adopt a non-centralized structure to increase their odds of outlasting their rivals."[45] The structure of the insurgent organization and the tactics adopted will often reflect the degree of security that groups have in the areas where they conduct operations. The result may be that groups adopt a cellular organization in some areas, often in urban settings, while conducting more large-scale warfare in other locations that approaches that of conventional operations.[46]

---

[36] Paul Staniland, Networks of Rebellion: Explaining Insurgent Cohesion and Collapse 5 (2014).

[37] *Id.* at 25.

[38] Ellis, *supra* note 5, at 242.

[39] Staniland, *supra* note 36, at 137.

[40] *Id.*

[41] Antonio Giustozzi, Koran, Kalashnikov and Laptop: The Neo-Taliban Insurgency in Afghanistan 128–29 (2008).

[42] Karl Forsberg, The Taliban's Campaign for Kandahar, Afghanistan Report 3, 24–27 (2009), *available at* http://www.understandingwar.org/sites/default/files/The_Talibans_Campaign_For_Kandahar.pdf.

[43] Abdulkader H. Sinno, Organizations at War in Afghanistan and Beyond 13 (2008).

[44] *Id.*

[45] *Id.* at 17.

[46] Carlos Marighella, Minimanual of the Urban Guerrilla 32–33 (2008) (1972) (where the basic organization of the "urban guerrilla" is described as a team of four or five called a "firing group" with a decentralized command structure."), and Walter Laqueur, Guerrilla Warfare: A Historical & Critical Study 348 (2006) (1976).

The iconic example of a cellular guerrilla organization was that used in the Autonomous Zone of Algiers, which operated as part of the National Liberation Front (FLN) fighting the French in Algeria during 1956–57. Both the political (FLN) and military organizations [National Liberation Army (ALN)] were subdivided into small groups. In each district the ALN, consisting of thirty-five armed men, included a district commander and deputy "at the head of three armed groups, each headed by a leader and deputy leader of three cells of three men each."[47] It included a distinct "bomb throwing network."[48] As will be discussed it is common for such groups to be referred to as terrorists, both because of their small organization and the types of indiscriminate attacks they carry out. However, groups with a cellular organization can also represent a fundamental part of an insurgency challenging the authority of the State.

Insurgent groups may also have to alter their tactics over the course of a conflict. The Al Shabaab demonstrated this in 2011 when a combination of losing control of the Bakarra Market with its tax source in Mogadishu, a crippling drought for tax-paying farmers in the south, and finally a monsoon that limited sea trade, meant it "could literally no longer afford the war it had fought for years."[49] Al Shabaab fighters conducted an organized retreat to their strongholds in the south of the country. By October 2012 they lost control of the last major city they had held, Kismayu.[50] However, it was noted, "the group's fighters are still highly active in much of the countryside in southern and central Somalia and have carried out suicide bombings and other attacks in cities they no longer control."[51] The insurgency has survived and adapted to address the tactical setback. In July 2013 Al Shabaab was reported to be a military force of some 5,000 fighters, which "remains arguably intact in terms of operational readiness, chain of command, discipline and communication capabilities."[52] It continued to control most of southern and central Somalia.[53] Despite targeted killings of leaders such as Ahmed Abdi Gadone,[54] the group demonstrated in 2015 a continuing ability carry out high-profile terrorist attacks in Somalia[55] and Kenya.[56]

---

47 Trinquier, *supra* note 1, at 10.

48 *Id.* at 11.

49 Scahill, *supra* note 29, at 491.

50 *Somali Troops Take Key Port of Kismayo after Al Shabab Rebels Retreat*, L.A. Times, Oct. 1, 2012, http://latimesblogs.latimes.com/world_now/2012/10/somalia-troops-occupy-kismayo-al-shabab-retreat.html.

51 *Somalia Islamists Abandon Kismayo Amid AU Attack*, BBC News (Sept. 29, 2012), http://www.bbc.co.uk/news/world-africa-19769058.

52 Report of the Monitoring Group on Somalia and Eritrea pursuant to Security Council resolution 2060 (2012): Somalia, *Letter to the Security Council*, UN Doc S/2013/413, 7 (July 12, 2013), http://www.securitycouncilreport.org/atf/cf/%7B65BFCF9B-6D27-4E9C-8CD3-CF6E4FF96FF9%7D/s_2013_413.pdf.

53 *Id.* ¶ 12, at 12.

54 David Smith, Abdalle Ahmed, & Tom McCarthy, *Al-Shabaab Leader Ahmed Abdi Godane Killed by US Air Strike in Somalia*, The Guardian, Sept. 5, 2014, http://www.theguardian.com/world/2014/sep/05/al-shabaab-leader-godane-killed-us-airstrike-somalia.

55 *Somalia Hotel Rocked by Bomb and Gunfight in Mogadishu*, BBC News (Mar. 27, 2015), http://www.bbc.com/news/world-africa-32091804.

56 Jeffrey Gettleman, *Shabab Militants Learning to Kill on a Shoestring*, N.Y. Times, Apr. 6, 2015, http://www.nytimes.com/2015/04/07/world/africa/setbacks-press-shabab-fighters-to-kill-inexpensively.html?smprod=nytcore-iphone&smid=nytcore-iphone-share&_r=0.

6.2.3. PURPOSE OF THE CONFLICT

What international law does not address is another essential element of "war."[57] That is fighting is for a political purpose. As Clausewitz famously noted, "war is not merely an act of policy but a true political instrument, a continuation of political intercourse, carried on by other means."[58] The importance of a "purpose" is not limited to inter-State conflict. As David Galula, one of the leading counterinsurgency theorists, noted in the 1960s:

> Paraphrasing Clausewitz, we might say that "Insurgency is the pursuit of the policy of a party, inside a country, by every means." It is not like an ordinary war—"continuation of the policy by other means"—because an insurgency can start long before the insurgent resorts to the use of force."[59]

This political dimension is also reflected in the U.S. military *Counterinsurgency Manual*, where insurgency is defined as "an organized, protracted politico-military struggle designed to weaken the control and legitimacy of an established government, occupying power, or other political authority while increasing insurgent control."[60] Stated more directly, "[a]chieving victory still depends on a group's ability to mobilize support for its political interests (often religiously or ethnically based) and to generate enough violence to achieve political consequences."[61]

A fundamental aspect of insurgency is the violent challenge to the existing governing authority. The definition in the *Counterinsurgency Manual* highlights that insurgency is not limited to noninternational armed conflict. It can take place in the context of a belligerent occupation, which is part of international armed conflict. The similarity between insurgency during occupation and that occurring in noninternational armed conflict was evident in Iraq during 2004 when the occupation had legally ended, but the threat facing Coalition and Iraqi forces did not factually change.[62] Neither the insurgent forces, nor their tactics, were altered. As occurred in Afghanistan in 2002, when that the conflict began to be viewed as having changed from an international to noninternational armed one, a modification in conflict categorization does not mean the hostilities or threat posed by enemy forces are necessarily altered. This is an essential point for theorists to remember as they seek to fit armed conflict within established international law boundaries. The categorization of

---

[57] YORAM DINSTEIN, NON-INTERNATIONAL ARMED CONFLICTS IN INTERNATIONAL LAW ¶ 55, at 17 (2014) ("In a legal analysis, the nuts and bolts of a NIAC are not the subjective motives of an insurgency but its objective manifestations.").

[58] CARL VON CLAUSEWITZ, ON WAR 87 (Michael Howard & Peter Paret eds. & trans., 1989).

[59] DAVID GALULA, COUNTERINSURGENCY WARFARE: THEORY AND PRACTICE 3–4 (1964). *See also* MAO TSE-TUNG, ON GUERRILLA WARFARE 71–72 (Samuel B. Griffith trans., 2000) (1961) ("Without a political goal guerrilla, warfare must fail. . . ."), and CHARLES TOWNSHEND, BRITAIN'S CIVIL WARS: COUNTERINSURGENCY IN THE TWENTIETH CENTURY 14 (1986) ("Like open war they are an extension of politics by 'other means'.").

[60] THE U.S. ARMY, MARINE CORPS, COUNTERINSURGENCY FIELD MANUAL ¶ 1-2, at 2 (2007) (hereinafter THE COUNTERINSURGENCY MANUAL).

[61] *Id.* ¶ 1-1, at 1.

[62] Adam Roberts, *Transformative Military Occupation: Applying the Laws of War and Human Rights*, 100 AM. J. INT'L L. 594, 617 (2006) ("even if the occupation was theoretically over, the likelihood remained that uses of force, perhaps even exercises of administrative authority, that closely resembled a situation of occupation would occur".).

conflict as "noninternational" does not mean the threat and accompanying levels of violence are altered. Further, regardless of the nature of the conflict within which insurgency occurs, the "battle" over control of the people is essentially about governance.

The political dimension of insurgency is not reflected in the criteria applied under international humanitarian law for the determination of when a noninternational armed conflict is in existence. Instead, reliance is placed on the level of organization of an armed group and the intensity of the violence. A governance element might be implied from Article 1(1) of Additional Protocol II. That provision contains additional threshold criteria for what is viewed as a "civil war" standard, including organized armed groups "which, under responsible command, exercise such control over territory as to enable them to carry out sustained and concerted military operations and to implement this Protocol."[63] However, this standard remains fixed on military operations rather than governance over the controlled territory.

The legal criteria of violence and organization are obviously essential, however, the political aspect of insurgency is also very relevant. Its absence from the humanitarian law test has likely occurred because of adherence to the "equal application theory" for humanitarian law. The purpose of the conflict, often historically expressed as its "justness" (i.e., just cause), is viewed not only as irrelevant but also a problematic invocation of just war principles. This is reflected in the International Tribunal for the Former Yugoslavia decision, *Prosecutor v. Limaj*, where the court stated, "the purpose of the armed forces to engage in acts of violence or also achieve some further objective is, therefore, irrelevant."[64] The irrelevance of the just war principle is because "[g]overnments have been known to deny the applicability of international humanitarian law on the grounds of not recognizing the cause of an organized armed group."[65]

However, if a political purpose test was to use objective criteria, it is not clear why the existence of such a purpose could not be used as it is with identifying terrorism (see Chapter 6, section 6.4.1, *Terrorism*). The key would be simply establishing that such a purpose exists, rather than considering its nature or justness. In any event, as will be explored in respect of criminal insurgency, the issue of identifying a political goal has attained new prominence. In this context the question of whether there is a true political purpose exhibited by a "gang" has become a key element in determining the degree to which the activity of such a group is criminal in nature, rather than constitute a broader challenge to established authority entering the realm of armed conflict. This is because traditionally in terms of assessing participation in "small wars," "[b]andits in search of nothing more than lucre are excluded; they are usually not interested in shaking up the established order, just in profiting from it."[66] In this regard, it is "the belonging to a cause which is national or even international in scope

---

[63] Lindsay Moir, The Law of Internal Armed Conflict 106 (2002) (for reference to criticism that the threshold would only be applicable to "civil war in the classic sense."). *See also* Zegveld, *supra* note 17, at 143–45.

[64] Prosecutor v. Limaj, Case No. IT-03-66-T, ¶ 170 (Int'l Crim. Trib. for the Former Yugoslavia 2005), http://www.icty.org/x/cases/limaj/tjug/en/lim-tj051130-e.pdf.

[65] Anthony Cullen, The Concept of Non-International Armed Conflict in International Humanitarian Law 130 (2010).

[66] Boot, *supra* note 6, at xxii, and Randall Wilson, Blue Fish in a Dark Sea: Police Intelligence in a Counterinsurgency 14 (2013) ("their basic motivation is profit through illicit activity rather than gaining political power for its own sake.").

which distinguishes an insurgent cell from a criminal gang."[67] Once again the boundaries established in the immediate World War II period are increasingly being questioned as the international community struggles to deal with twenty-first-century conflict.

6.3. Insurgency and Crime

6.3.1. THE NATURE OF THE THREAT

A unique feature of twenty-first-century insurgency is its connection to criminal activity, either directly through its own activities or by association with transnational criminal gangs. In part, this is because the proceeds of crime are of considerable importance to the conduct of any military activity. Insurgencies, or terrorism for that matter, are just as reliant on logistics and funding as conventional military operations. As the Al Qaeda theorist, Abu Mus'Ab al-Suri, identified, an important factor for operational activity is "high financial capabilities," which enables cells to operate within the Arab and Islamic world, as well as in American and allied countries.[68] There is a particular synergy between insurgency and criminal activity. Not only do criminal gangs thrive in the ungoverned spaces associated with insurgencies, but insurgent groups often turn to criminal activity in order to fund their activities. The dividing line between terrorism, insurgency, and criminal activity can become a rather amorphous one.[69] In this respect, "terrorist groups around the world have been known to use all forms of criminal activity to raise money, including drug trafficking, extortion, kidnapping, human trafficking, all forms of fraudulent schemes and counterfeiting."[70]

As has been noted, the connection between finances and insurgent success was evident in Al Shabaab having to withdraw from Mogadishu in 2011 after it lost sources of funding.[71] The group Al Qaeda in Iraq (AQI) relied extensively on funding from foreign supporters and through fundraising from local Iraqis.[72] In Iraq, "[b]y its peak in 2005–2006 AQI was raising nearly $4.5 million a year, primarily from criminal rackets such as gasoline smuggling, car theft, and extortion."[73] In 2014 the newly minted "Islamic State" was identified as possibly making as much as $3 million a day in illicit oil sales.[74] It has also "been levying taxes on everybody from vegetable sellers in the market to mobile phone and construction companies."[75] In Afghanistan the drug trade has been a recurring issue for those fighting the insurgency, where the Taliban reaps profits from taxation of producers,

---

67 WILSON, *supra* note 66, at 14.

68 BRYNJAR LIA, ARCHITECT OF GLOBAL JIHAD: THE LIFE OF AL-QAIDA STRATEGIST ABU MUS'AD AL-SURI 430–31 (2008).

69 STEVEN METZ, RETHINKING INSURGENCY, Strategic Studies Institute, U.S. Army War College 30 (2007) ("Iraq is a classic case of preexisting organized crime initially developing a partnership with insurgents followed by a melding where the insurgents themselves became criminal organizations or, at least, barely distinguishable from organized crime.").

70 JUAN C. ZARATE, TREASURY'S WAR: UNLEASHING OF A NEW ERA OF FINANCIAL WARFARE 23 (2013).

71 SCAHILL, *supra* note 29, at 491.

72 SETH G. JONES, HUNTING IN THE SHADOWS: THE PURSUIT OF AL QA'IDA SINCE 9/11, 154 (2012).

73 BOOT, *supra* note 6, at 532.

74 Keith Johnson, *The Islamic State Is the New Petrostate*, FOREIGN POL'Y (July 28, 2014), http://www.foreignpolicy.com/articles/2014/07/28/baghdadis_hillbillies_isis_iraq_syria_oil_terrorism_islamic_state.

75 PATRICK COCKBURN, THE JIHADIS RETURN: ISIS AND THE NEW SUNNI UPRISING 33 (2014).

conducting protection rackets and running heroin labs.[76] It would appear that "[t]he seemingly unbreakable allure of poppy profits—for producers and traffickers, government officials and Taliban commanders alike—has kept fighting opium at the heart of efforts to improve security."[77] Similarly, it is reported that "Al Qaeda in the Islamic Maghreb [AQIM] took home dozens of millions of dollars a year from ransom kidnappings; over a decade, the group possibly netted as much as $200 million."[78] Indeed, kidnapping has proven to be so lucrative that "the second in command of Al Qaeda's central leadership recently [in 2014] described [money from kidnapping] accounting as for as much as half of his operating revenue."[79] Further, Al Qaeda Central is said to oversee negotiations with AQIM, AQAP, and Al Shabaab, reportedly "coordinating their efforts and abiding by a common kidnapping protocol."[80]

Hezbollah and other radical Islamic groups have been linked to the diamond trade to fund their terrorist activities.[81] In this respect, Hezbollah stands out as a group engaged in criminal activity in North America, Africa, and South America in order to support and fund its military and terrorist campaigns.[82] For example, in the tri-border area of Brazil, Argentina, and Paraguay, Hezbollah supporters were reported to be "actively involved in drug smuggling, arms trafficking, money laundering, fraud and intellectual property piracy."[83] However, the involvement in criminal activity does not need to be on a grand scale, as is reflected in reports of Israeli criminals working through Israeli Arabs in selling arms to Palestinian terrorist groups;[84] Hezbollah cells engaged in illegal cigarette sales in the United States,[85] as well as efforts to acquire sophisticated electronic equipment and weapons in Canada; [86] or Al Shabaab pursuing "illicit trade in elephant ivory, the smuggling of charcoal, and the expropriation of cash intended for respectable Islamic charities".[87] It is also reported in respect of the Iraq insurgency following the 2003 invasion that "[v]irtually all the foreign fighters entered Iraq through Syria, with the aid of smuggling and criminal networks rather than al Qa'ida's own personnel."[88] One impact of this mix of insurgent

---

76 Gretchen Peters, How Opium Profits the Taliban 17–21 (2009), *available at* http://www.usip.org/files/resources/taliban_opium_1.pdf, and Abdel Bari Atwan, After Bin Laden: Al Qaeda, The Next Generation 147 (2012). *See also* Zarate, *supra* note 70, at 23, and Johnson, *supra* note 74 (were it is indicated the Taliban "raked in between $100 million and $200 million annually from the drug trade and smuggling timber and minerals.").

77 Allisa J. Rubin & Matthew Rosenburg, *U.S. Efforts Fail to Curtail Trade in Afghan Opium*, N.Y. Times, May 26, 2012, http://www.nytimes.com/2012/05/27/world/asia/drug-traffic-remains-as-us-nears-afghanistan-exit.html?pagewanted=all&_r=0.

78 Johnson, *supra* note 74. *See* Robert R. Fowler, A Season in Hell: My 130 Days in the Sahara with Al Qaeda 309–11 (2011) (for a critical discussion of national policies concerning the payment of ransoms).

79 Rukmini Callimachi, *Paying Ransoms, Europe Bankrolls Qaeda Terror*, N.Y. Times, July 29, 2014, http://www.nytimes.com/2014/07/30/world/africa/ransoming-citizens-europe-becomes-al-qaedas-patron.html?_r=0.

80 *Id.*

81 Venter, *supra* note 24, at 186. *But see* Zarate, *supra* note 70, at 36 (where it is suggested Al Qaeda investment in blood diamonds from West Africa is said to be unproven.).

82 *See, e.g.*, Levitt, *supra* note 33, at 246–76 (for a discussion of Hezbollah's logistics and finance operations in Africa).

83 Katz & Hendel, *supra* note 2, at 90.

84 Daniel Byman, A High Price: The Triumphs & Failures of Israeli Counterterrorism 139 (2011).

85 Levitt, *supra* note 33, at 156–56 (213), and Byman, *supra* note 84, at 228.

86 Levitt, *supra* note 33, at 163–68.

87 David Smith, *Al-Shabaab Rebuilds Forces in Somalia as African Union Campaign Stalls*, The Guardian, Oct. 28, 2013, http://www.theguardian.com/world/2013/oct/28/al-shabaab-somalia-african-union.

88 Jones, *supra* note 72, at 154.

engagement in both hostilities and criminal activity is that States must consider the application of human rights–based law enforcement norms when countering this threat. This is because such logistic and financial support does not fall generally within the narrower category of support activities associated with "direct participation in hostilities."

It is of interest that in May 2015 the United States conducted a "capture or kill" operation in an attempt to seize an Islamic State financier: "Abu Sayyaf is a midlevel leader in the organization—one terrorism analyst compared him to Al Capone's accountant."[89] He was also identified as a participant in military operations, which would make him a direct participant in hostilities and a lawful target. The fact that the mission was focused on capturing him for the purposes of interrogation, as well as the collection of computer and other records, highlights the importance of financing to the Islamic State effort.[90] Even if the operation were conducted as a form of law enforcement, the attempt to fight the special operations forces likely created a situation where they could use deadly force in self-defense.[91]

The crime-based security challenge facing States is even more evident regarding the emerging threat posed by transnational criminal organizations. They have exhibited a degree of organization that is similar to many traditional insurgent groups. Groups in Central and South America provide clear examples of this phenomenon. For instance, the Zetas, a group formed from former Special Forces soldiers, are described as "Mexico's most blood-thirsty gang."[92] This group was initially aligned with the Gulf Cartel, however, "they have morphed into a cartel in their own right."[93] The Zetas have been one of the seven major dominate organizations that are viewed as having fragmented to between nine and as many as twenty major organizations.[94]As an organization, the Zetas represent a new trend in security threats.

The Zetas not only operate across Mexico's northern and southern frontiers, they take advantage of Guatemala's weak law enforcement and corruption to reach into Guatemala's northeastern and eastern provinces.[95] The group is basically organized into a pyramidal structure with a small command structure providing strategic- and operational-level guidance; a second leadership level overseeing "areas of intelligence, operational planning, financial support, and recruitment and training"; a third level of cell members; and finally, a fourth level of aspirants; as well as specialists (including logistics and communications).[96] However, the group also has attributes of a business organization "that can quickly, flexibly,

---

[89] Helen Cooper & Eric Schmitt, *ISIS Official Killed in U.S. Raid in Syria, Pentagon Says*, N.Y. Times, May 16, 2015, http://www.nytimes.com/2015/05/17/world/middleeast/abu-sayyaf-isis-commander-killed-by-us-forces-pentagon-says.html?hp&action=click&pgtype=Homepage&module=first-column-region&region=top-news&WT.nav=top-news&_r=0.

[90] *Id.* (outlining the intelligence goals of the mission to capture Abu Sayyaf).

[91] *Id.* ("Abu Sayyaf 'tried to engage' the commandos, the Defense Department official said, and was shot and killed.").

[92] Ioan Grillo, El Narco: Inside Mexico's Criminal Insurgency 94 (2011).

[93] John P. Sullivan, *Criminal Insurgencies in the Americas*, *in* Mexico's Criminal Insurgency 71, 73 (John P. Sullivan & Robert J. Bunker eds., 2013).

[94] S. Beittel, *Mexico's Drug Trafficking Organizations: Source and Scope of the Rising Violence*, Congressional Research Service 9–10 (Apr. 15, 2013), http://fpc.state.gov/documents/organization/155587.pdf (the seven dominant organizations were Sinaloa, Los Zetas, Tijuana/AFO, Juárez/CFO, Beltrán Leyva, Gulf, and La Familia Michoacana. Of these, "the Sinaloa and Los Zetas— appear to be preeminent.").

[95] Sullivan, *supra* note 93, at 73–74.

[96] Max G. Manwaring, Gangs Pseudo-Militaries and Other Modern Mercenaries 135–36 (2010).

and effectively respond to virtually any opportunity, challenge, or changing situation."[97] The Zetas actively recruit soldiers and ex-soldiers. The regional Mexican "cells" have included local men and are considered "akin to franchises."[98] This has included members of the Guatemalan "Kaibil commandos that tore through rebel villages in that nation's civil war."[99] The Zetas at first modeled their chain of command on the Mexican military, however, "the war evolved their structure to become closer to Latin America's guerrilla armies or right wing paramilitaries, who use autonomous cells to coordinate thousands of men at arms."[100]

Cartels like the Zetas, La Familia, the Sinola cartel, and the Beltrán-Leyva organization "and their enforcer gangs—which amount to virtual private armies—threaten the stability of the state."[101] For example, in response to the Zetas, the Sinola cartel created its own "cells of combatants with heavy weaponry and combat fatigues," and one Beltrán Leyva death squad was armed with automatic rifles, pistols, grenade launchers and flak jackets.[102] The organizational structure of La Familia is derived from the Zetas, and "Plaza heads run cells that are semiautonomous."[103] Their activity is not limited to drugs, as "they have branched into other profitable crimes such as kidnapping, assassination for hire, auto theft, controlling prostitution, extortion, money-laundering, software piracy, resource theft, and human smuggling."[104]

Nor are such powerful criminal organizations limited to Mexico. As Max Manwaring has identified, Colombia has faced some of the worst political violence in the western hemisphere, confronting three different protagonists, the "narcos" (*naroctraficantes ilegales trasnacionales*); the insurgents consisting of the Revolutionary Armed Forces of Colombia (FARC) and the smaller National Liberation Army (ELN); and the United Self-Defence Groups of Colombia (AUC groups), or paramilitaries.[105] With the weakening of FARC due to Colombian military successes there has been both disarray and fragmentation, such that "disaffected units and individuals are breaking away from the insurgent organization and merging with AUC groups or drug cartels for self-enrichment."[106]

The AUC presents a unique challenge as a vigilante-based organization, where since 2005 some former members upon demobilization "are operating in an outsourcing mode as 'subsidiaries', 'pseudo-paramilitaries,' or 'gangs' (banda criminales)—renting their services to the highest bidder."[107] The Columbian government is reported to view the "BACRIM" (the Spanish acronym for *banda criminales*) as its most serious security threat. Many of these groups are involved primarily in drug trafficking, while others such as "*Los Rastrojos*

---

97 *Id.* at 135.

98 GRILLO, *supra* note 92, at 104–05.

99 *Id.* at 105.

100 *Id.* at 211.

101 Sullivan, *supra* note 93, at 73.

102 GRILLO, *supra* note 92, at 106.

103 *Id.* at 211.

104 Beittel, *supra* note 94, at 20.

105 MANWARING, *supra* note 96, at 58–63.

106 *Id.* at 62.

107 *Id.* at 63. *See also* JAN RÖMER, KILLING IN A GRAY AREA BETWEEN HUMANITARIAN LAW AND HUMAN RIGHTS 20 (2010) (where it is noted the gangs are also called "emerging gangs" by the Columbian government ("*bandas emergentes*")).

have political interests, and stands accused of financing candidates in congressional elections in 2010."[108] Further, "[i]t operates an armed wing, and consists of 1500 members."[109] To a certain extent the involvement in the drug trade by AUC-related gangs is not a new phenomenon.[110] However, these groups have branched out and changed their focus. In some cases they join with leftist guerrillas, and in other areas there have been clashes.[111] These new FARC and AUC gangs "seem to be more than bandas criminales: they are re-shaping the narco-terrorist-insurgent-criminal world in Columbia and are exacerbating threats that are already eroding Colombian democracy and the Colombian state."[112]

The threat of transnational gangs is being felt across the Southern Hemisphere as "Guatemala and Honduras, Panama and Costa Rica, indeed all Central America, are currently at risk of being caught in the 'cross-fire' of the regions' drug wars."[113] They are joined by "third generation" gangs "like Mara Salvatrucha (MS-13) and Brazil's urban drug gangs that have transcended operating on localized turf with a simple market focus to challenge political structures—are both partners and foot soldiers for the dominate cartels."[114] The generational categorization of gangs reflects an increasing sophistication and scale of operation. The most numerous first generation are street gangs, while the second-generation gangs are more complex, usually involved in illegal activity like drug trafficking and operate across multiple cities or even internationally. The second-generation gang leadership is more centralized, has links to transnational criminal organizations, and "[t]heir violence is systemic and meant to protect or expand market share."[115] While the third-generation gangs "participate in many of the same activities as first- and second-generation gangs, they operate on a grander scale."[116]

The third-generation gangs are the transnational criminal organizations that "stand astride the line separating crime and insurgency" and operate in a way to protect their illicit economic activity by using "intense calculated violence to carve out geographic zones where they can dominate the population and operate completely free of state control."[117] Such gangs are not limited to Mexico, or the drug cartels, and it is suggested, "a small number in the United States, Canada, Central and South America as well as South Africa have

---

[108] Felicity Szesnat & Annie R. Bird, *Columbia*, *in* INTERNATIONAL LAW AND THE CLASSIFICATION OF CONFLICTS 203, 213 (Elizabeth Wilmhurst ed., 2012) ("Many groups are engaged primarily in drug trafficking."). *But see also Profiles: Colombia's Armed Groups*, BBC NEWS (Feb. 8, 2013), http://www.bbc.co.uk/news/world-latin-america-11400950 [hereinafter *Profiles, Colombia's Armed Groups*] (where it is indicated these groups "combine control of cocaine production and smuggling with extreme violence, but do not have any apparent political agenda.").

[109] Szesnat & Bird, *supra* note 108, at 213.

[110] ANGEL RABASA & PETER CHALK, COLOMBIAN LABYRINTH 58–59 (2001) (where there is acknowledgment of it occurring in 2000).

[111] *Profiles, Colombia's Armed Groups*, *supra* note 87.

[112] MANWARING, *supra* note 96, at 71.

[113] Sullivan, *supra* note 93, at 73.

[114] *Id.*

[115] Hal Brands, *Third-Generation Gangs and Criminal Insurgency in Latin America*, *in* MEXICO'S CRIMINAL INSURGENCY, *supra* note 93, 43, at 44. *See also* MANWARING, *supra* note 96, at 129–30 (outlining the involvement of the Central American Mara Salvatruch 13 and Mara Salvatrucha 18 (referred to as the "Maras") in gang activity in Mexico.).

[116] Brands, *supra* note 115, at 44–45.

[117] *Id.* at 45.

acquired third generation characteristics."[118] These gangs are said to operate as drug-oriented enterprises in up to thirty-five States, "controlling entire housing projects, schools and blocks, that conduct overt political activity while actively seeking to infiltrate and co-opt local police and contract security forces."[119] The internationalization of these gangs is reflected in U.S. groups linking up with Mexican cartels in order to perform enforcement slayings invoking comparisons to "netwarriors."[120] In addition, the flow of drugs is directed toward Europe as well as the United States.[121] The Mexican Sinola cartel is indicated to have a substantial international presence throughout South America, Europe, West Africa and Southeast Asia.[122] While cocoa production has declined in Columbia, "there has been an increasing flow of Colombian cocaine to other regions such as West Africa and Europe."[123]

It is reported that "[t]hanks to its support networks on both sides of the Atlantic, Hezbollah has a natural advantage vis-à-vis Latin American drug kingpins looking to transport their product to or through Africa."[124] In addition to hostage taking, Islamic rebels in Africa familiar with the Sahara, Al Qaeda in the Islamic Maghreb (AQIM), "have struck deals under which they provide armed security escorts for drug traffickers in return for a slice of their profits" as they smuggle drugs into Europe through Africa.[125] In April 2013 it was reported that the FARC has engaged in a drugs for weapons exchange with AQIM.[126] Islamic cadres linked to Al Qaeda were also tied to rebel diamond operations in Sierra Leone and the arming of the Revolutionary United Front (RUF).[127] The impact of gang criminal activity is being felt in other locations as well. It has been assessed that by 2007 Iraq had "essentially been overrun by 3rd generation gangs and their criminal-soldier equivalents."[128] Not only do transnational criminal gangs challenge the political authority in their own right, they also form linkages with insurgent groups seeking arms and funding.

6.3.2. CHALLENGING THE STATE

While the threat is by no means limited to drug cartels, Mexico provides a clear example of the security challenge that criminal groups pose to States. It is a violent challenge, although

---

[118] John P. Sullivan, *Future Conflict: Criminal Insurgencies, Gangs and Intelligence*, *in* MEXICO'S CRIMINAL INSURGENCY, *supra* note 93, 29, at 36.

[119] *Id.* at 37.

[120] *Id.*

[121] Andrew O'Reilly, *Mexican Drug Cartels Join Forces with Italian Mafia to Supply Cocaine to Europe*, FOX NEWS LATINO (June 21, 2012), http://latino.foxnews.com/latino/news/2012/06/21/mexico-drug-cartels-supply-italian-mafia-with-cocaine-for-europe/.

[122] Beittel, *supra* note 94, at 12.

[123] *Id.* at 9.

[124] LEVITT, *supra* note 33, at 275.

[125] Robin Yapp, *South American Drug Gangs Funding al-Qaeda terrorists*, THE TELEGRAPH, Dec. 29, 2010, http://www.telegraph.co.uk/news/worldnews/southamerica/colombia/8230134/South-American-drug-gangs-funding-al-Qaeda-terrorists.html.

[126] James Bargent, *Arrests Point to FARC Ties to Al Qaeda in North Africa*, INSIGHTCRIME: ORGANIZED CRIME IN THE AMERICAS (Apr. 8, 2013), http://www.insightcrime.org/news-briefs/arrests-point-to-farc-ties-to-al-qaeda-in-north-africa.

[127] VENTER, *supra* note 24, at 186.

[128] John P. Sullivan, *Iraq & The Americas: 3 Gen Gangs Lessons and Prospects*, *in* MEXICO'S CRIMINAL INSURGENCY, *supra* note 93, 1, at 2.

the reported statistics vary considerably. The Calderón, Mexico, government "reported that between December 2006 and September 2011 more than 47,500 killings were organized crime-related homicides."[129] A report from the Trans-Border Institute (TBI) at the University of San Diego in California[130] outlines a steady increase in annual deaths from 2007 to 2010, with the statistics for 2011 and 2012 varying greatly, indicating that the number of deaths either leveled off or declined significantly.[131] A range of between 9,744 and 12,390 organized-crimes-style killings were reported by media sources in 2012.[132] It was also reported that media sources, "*Reforma* and *Milenio* estimate the number of organized-crime-style 'executions' (*ejecucciones*) from 2007 to 2012 at 47,509 and 54,047 people, respectively."[133] That represented approximately half of all homicides in Mexico during that period.[134] It has also been indicated that "[i]n the first three months of 2013, reportedly homicides related to drug trafficking and organized crime remained stubbornly high at slightly under 1,000 killings per month, indicating little change from the recent monthly tallies under the prior [Calderón] administration."[135]

The violence has been both intra- and inter-cartel, however, it has also been directed against the State.[136] In the first four years of the Calderón administration (lasting from 2006–12) "cartel gunmen slayed more than twenty-five hundred public servants, including twenty-two hundred policemen, two hundred soldiers, judges, mayors, a leading gubernatorial candidate, the leader of a state legislature, and dozens of federal officials."[137] A 2013 report indicates "it appears that 2010 was the peak of violence targeting these government personnel, as was also the case for mayors," although it is suggested this "may also reflect the withdrawal of federal forces from the provision of domestic security in key areas over the last two years, such as the state of Chihuahua." [138] An example of the degree to which governance is impacted at the local level can be seen in the 2014 example of the small town of Nueva Italia. It is reported that one hundred vigilantes consisting of local residents entered the town, exchanged gunfire with the Knights Templar gang controlling the area, and disarmed the local police. However, when the vigilantes entered another town, local residents set fire to their vehicles in a failed attempt to expel them. The vigilante groups have been accused of siding with a rival drug cartel,[139] although that allegation was fiercely denied by

---

[129] Beittel, *supra* note 94, at 23.

[130] Cory Molzahn, Octavio Rodriguez Ferreira, & David A. Shirk, Drug Violence in Mexico: Data and Analysis Through 2012, Trans-border Institute, Feb. 15, 2013, http://justiceinmexico.files.wordpress.com/2013/02/130206-dvm-2013-final.pdf.

[131] *Id.* at 18 ("Thus, depending on the source, both overall homicide and organized-crime-style homicides grew dramatically from 2007 to 2010, but leveled off or declined substantially in 2011 and 2012.").

[132] *Id.* at 14.

[133] *Id.*

[134] *Id.* at 16 (as the author of the study indicates estimates from various sources range from 40 to 60 percent.). *See also id.* at 13 ("Using either measure, the authors estimate that the total number of homicides during the Calderón administration was likely around 120,000 to 125,000 people killed . . . .").

[135] Beittel, *supra* note 94, at 1.

[136] John P. Sullivan & Adam Elkus, *Cartel v. Cartel: Mexico's Criminal Insurgency*, *in* Mexico's Criminal Insurgency, *supra* note 93, 59, at 60. *See also* Beittel, *supra* note 94, at 22–23 (providing an overview of the violence).

[137] Grillo, *supra* note 92, at 11.

[138] Molzahn et al., *supra* note 130, at 32.

[139] *Mexico Vigilantes Clash with Knights Templar Cartel in Michoacan*, BBC News (Jan. 13, 2014), http://www.bbc.co.uk/news/world-latin-america-25708297.

the vigilantes.[140] The police and federal troops remained in a town that was the headquarters of the New Templar gang, while the Mexican government announced the deployment of federal troops to quell the violence.[141]

What the Mexican gangs seek to do is "intimidate the state and the citizenry into submission and win a free hand in pursuing their lucrative business dealings."[142] The tactics are often designed to terrorize, including through the use of violent acts involving decapitations and grenade attacks.[143] Sometimes victims are publicly displayed, including being hung from bridges. Other times they are dissolved in acid or disposed of in anonymous mass graves.[144] The attempt to arrest members of cartels can also be very violent. In 2009, when seeking to capture a "narco" leader, Arturo Beltrán Leyva, Mexican Marines had to storm the apartment where after a two-hour firefight the trapped criminals who fought back with gunfire and grenades were ultimately killed.[145] The almost "war-like" violence can also be seen in the 2015 downing of a police helicopter with a rocket-propelled grenade, and a firefight with members of the "Jalisco New Generation cartel, one of the most powerful and fastest-growing organized crime groups to surface in recent years."[146] The firefight resulted in forty-two suspected criminals and one federal agent killed.[147] These are levels of violence that resemble and could be said in some respect to exceed the violence of the Northern Ireland conflict (see Chapter 13, section 13.2.1, *The "Troubles"*).

In terms of conflict thresholds, many of the transnational criminal paramilitary groups appear to have the degree of organization necessary to meet the *Prosecutor v. Tadić* criteria requiring an elevated threshold for a noninternational armed conflict.[148] In addition, the levels of violence caused by those groups appear to attain the threshold associated with that decision and exceed "situations of internal disturbances and tensions, such as riots, isolated and sporadic acts of violence and other acts of a similar nature" (see Chapter 9, section 9.4.3, *The "Totality of the Circumstances" and Other Criteria*).[149] However, the general approach is to consider these conflicts as being internal law enforcement matters. There appear to be three factors that come into play in the determination that an armed conflict is not occurring.

---

[140] *Mexican Vigilantes Clash with Soldiers in Michoacan State*, BBC News (Jan. 14, 2014), http://www.bbc.co.uk/news/world-latin-america-25739937.

[141] *Id.*

[142] Brands, *supra* note 115, at 48. *See also* Grillo, *supra* note 92, at 218–19 ("The enemies they see first and foremost are rival cartels. When they hit police or civilians, it is often to hurt these rivals by breaking their system of protection.").

[143] Grillo, *supra* note 92, at 219 (for a description of a grenade attack that killed eight civilians on Independence Day "shortly after the state governor had rung the bell for independence."). *See* John P. Sullivan, *Explosive Escalation? Reflections on the Car Bombing in Ciudad Juarez, in* Mexico's Criminal Insurgency, *supra* note 93, 81, at 83 (outlining the use of grenades by Mexican Cartels.).

[144] Beittel, *supra* note 94, at 3.

[145] Grillo, *supra* note 92, at 126.

[146] *Investigators Seek to Sort Out Mexico's Deadly Shootout that Killed 43*, CTV News (May 23, 2015, 6:24 PM), http://www.ctvnews.ca/world/investigators-seek-to-sort-out-mexico-s-deadly-shootout-that-killed-43-1.2387634.

[147] *Id.*

[148] Prosecutor v. Tadić, Case No. IT-94-1-A, Decision On The Defence Motion For Interlocutory Appeal On Jurisdiction, ¶ 70 (Int'l Crim. Trib. for the Former Yugoslavia, Appeal Chamber, 2 October 1995), http://www.icty.org/x/cases/tadic/acdec/en/51002.htm. Römer, *supra* note 107, at 21 (where it is indicated a "few" of the criminal gangs carry out military operations mostly against the FARC, "it is unclear how far their degree of organization goes in providing the concerned group with their ability to fulfill the obligations under IHL [international humanitarian law].").

[149] AP II, art. 1(2).

These are the reluctance of States generally to admit such a conflict exists; the bias under the international tribunal system toward applying human rights norms in respect of internal conflict, which makes the violence appear as a law enforcement issue; and the view that transnational criminal organizations do not fight for a political purpose. This latter point is consistent with the view that "[t]he nature of criminal organizations is capitalistic and largely apolitical, while that of insurgent organizations is political by definition."[150]

First, consistent with their role as the "proper authority," States have generally been very reluctant to admit there is an armed conflict in existence within their borders, and therefore admit "that few, if any, situations of internal crisis fall to be regulated by Common Article 3 or Protocol II."[151] Two prime examples are Colombia and Mexico. It was only in May 2011 that President Santos of Colombia altered the characterization of the conflict with the FARC by referring specifically to an armed conflict rather than a fight against terrorism, which had been the approach of his two predecessors.[152] While the change was reportedly made to ensure Colombia armed forces were to be assessed under humanitarian law, it is reported that the Santos' predecessor, ex-President Uribe, was of the view "that the law should not recognize drug dealing terrorists as belligerents, and pointed out that his government demobilized more than 50,000 members of terrorist groups without the need to declare an armed conflict."[153] However, the use of drone strikes against insurgent leaders patterned after the American success against the Iraqi Al Qaeda leader, Abu Musab al-Zarqawi, suggests a far more complex situation where hostilities were being conducted, but politically the government sought to avoid any acknowledgment that was the case.[154] In part, the concern over recognizing an armed conflict was in existence was linked to the traditional State concern of bestowing legitimacy on the rebels.[155] Portraying the FARC as "narco-terrorists" was a hallmark of the post-9/11 approach by U.S. and Colombian officials. It was an approach aided by the mainstream media, which focused on FARC human rights violations and in particular kidnapping in portraying the group as a terrorist organization.[156]

Even then, the recognition of the existence of an armed conflict in Colombia did not extend to privileging the BACRIMs, which continue to be viewed as a criminal organization subject to law enforcement. The use of that separate term was "precisely to contrast them with illegal armed groups."[157] In Mexico, the Calderón administration also resisted

---

[150] WILSON, *supra* note 66, at 15.

[151] OREN GROSS & FIONNUALA NÍ AOLÁIN, LAW IN TIME OF CRISIS: EMERGENCY POWERS IN THEORY AND PRACTICE 360 (2006).

[152] Szesnat & Bird, *supra* note 108, at 214–15, and Toni Peters, *Santos Upholds Redefinition of Colombia's "Armed Conflict,"* COLOMBIA REPORTS (May 10, 2011), http://colombiareports.com/santos-upholds-redefinition-of-colombias-armed-conflict/.

[153] Vanessa Paz Lecompte, *Santos, Uribe Clash Over Colombia Conflict*, INSIGHTCRIME (May 5, 2011), http://www.insightcrime.org/news-briefs/santos-uribe-clash-over-colombia-conflict.

[154] Dana Priest, *Covert Action in Columbia*, WASH. POST, Dec. 21, 2013, http://www.washingtonpost.com/sf/investigative/2013/12/21/covert-action-in-colombia/. *See also* Szesnat & Bird, *supra* note 108, at 232–39.

[155] GARRY LEECH, THE FARC: THE LONGEST INSURGENCY 102–03 (2011).

[156] *Id.* at 103.

[157] DISMANTLING COLOMBIA'S NEW ILLEGAL ARMED GROUPS: LESSONS FROM A SURRENDER, International Crisis Group Latin America Report N°41, at 3 (June 8, 2012) [hereafter DISMANTLING COLOMBIA'S ARMED GROUPS], http://www.crisisgroup.org/~/media/Files/latin-america/colombia/41-dismantling-colombias-new-illegal-armed-groups-lessons-from-a-surrender.pdf.

attempts to have the violence in that country referred to as an "insurgency." The cartels were "said to represent the forces of organized crime and nothing more, even though some hundreds of 'zones of impunity' are recognized to exist...."[158] This led to some tension between Mexican and American government authorities when Secretary of State Hillary Clinton referred in 2010 to the conflict morphing into an insurgency. This was rejected by the Mexican government.[159]

Second, this preference for treating internal conflict generally, and transnational criminal organizations more specifically, as a criminal matter is reinforced by an international human rights tribunal framework that privileges that normative regime even in the context of violence associated with armed conflict. This is particularly the case with the European Court of Human Rights, which has favored a human rights law approach applied in assessing internal State conduct (see Chapter 5, section 5.2, *Exclusionary Arguments and Operational Reality*). The Inter-American human rights system has recognized that humanitarian law can be used in the interpretation of human rights law.[160] However, the normative influence exhibited by those tribunals in viewing armed conflict action, either exclusively or primarily through a human rights lens, as well as seeking "to order and supervise national prosecutions when states have been unwilling or unable to act," has had a powerful influence on the States subject to their jurisdiction.[161] Adopting a human rights approach, whether by the State or a supervising court, normally means criminal gangs are treated as a threat subject to law enforcement, although in Columbia "their combat power, their organisational consolidation and the growing negative humanitarian impact of their operations have fuelled discussions about whether they should be recognised as part of the conflict."[162]

Finally, there is issue of the political impact that criminal insurgency has on State governance. This has become a factor in the rejection of an armed conflict status even though technically that criterion is not part of the textually or judicially articulated humanitarian law threshold for internal conflicts. Indeed, a key question is what the impact is when cartels "tax the local population, viewing themselves as benefactors offering basic services, protection and spiritual guidance."[163] In rejecting a traditional insurgency label, the distinction is made that the motive for the violence is one of profit rather than the attainment of political control. Unlike the criminal insurgent, who is viewed as being "apolitical," the goal of "both classical and modern political insurgents—is the removal of foreign forces, the satisfaction of discrete political demands, or regime change."[164]

---

[158] Robert J. Bunker, *The Mexican Cartel Debate: As Viewed Through Five Divergent Fields of Security Studies*, *in* MEXICO'S CRIMINAL INSURGENCY, *supra* note 93, 107, at 114.

[159] Beittel, *supra* note 94, at 5, and GRILLO, *supra* note 92, at 203–05.

[160] Lindsay Moir, *Law and the Inter-American Human Rights System*, 25 HUM. RTS. Q. 182, 212 (2003). *See also* RÖMER, *supra* note 107, at 35.

[161] Alexandra Huneeus, *International Criminal Law by Other Means: The Quasi-Criminal Jurisdiction of the Human Rights Courts*, 107 AM. J. INT'L L. 1, 1 (2013).

[162] DISMANTLING COLOMBIA'S ARMED GROUPS, *supra* note 157, at 2.

[163] David Dilegge, *Foreword: Mexico's Criminal Insurgency and SWJ El Centro*, *in* MEXICO'S CRIMINAL INSURGENCY, *supra* note 93, at xi.

[164] John P. Sullivan & Adam Elkus, *State of Siege: Mexico's Criminal Insurgency*, *in* MEXICO'S CRIMINAL INSURGENCY, *supra* note 93, 7, at 12.

There is considerable merit to making a distinction between criminal and ordinary insurgent groups on the basis of "political purpose." Some groups simply do not want to "take over the government or force it to accede to ideological demands. They want a piece of the state that they can use to develop parallel structures for profit."[165] The groups are viewed as sitting astride the dividing line between crime and insurgency, although "[t]hese organizations go to such lengths to protect their highly lucrative economic activities that they end up undermining the authority and legitimacy of the state."[166] They are seen as profit-driven brutal criminal networks seeking to "create 'ungoverned spaces' where the rule of law is weak or non-existent; they open the door for anarchy and murder and strike direct at the foundations of civil order, an environment in which terrorism thrives."[167] In respect of the Mexico cartels, it has also been noted that an explicit commercial motive is also a political one. In this case, though, the goal is not necessarily to destroy the Mexican State, but rather seek to incrementally capture it.[168]

While the distinction between classic and even contemporary insurgency on the one hand, and criminal insurgency on the other, may be a subtle one, it is relied on by States to identify the normative legal framework applicable to its eradication. In Colombia the motives of the criminal organizations are also seen as being economic rather than political.[169] The International Crisis Group has indicated that recognizing these criminal organizations are part of armed conflict would pave the way for a more active military role in combating them.[170] It is also suggested that a determination of the existence of an armed conflict would provide enhanced protection for the civilian population.[171] However, that is not necessarily the approach that is being followed. Instead, the members of these groups face criminal investigations and trial before ordinary courts with no special legal considerations.[172] For the military, the complex operating environment involving both insurgent forces engaged in crime and paramilitary criminal organizations has created a challenging situation within which to assess whether the use of force is governed by humanitarian law or human rights–based law enforcement.[173] As Jan Römer has noted in respect of the law applying to Colombian National Police Force operations, "the requirements of each branch of the law converge more easily if it can be expected that the attacked individual or group will show aggression."[174]

Faced with challenges of police corruption and the need for justice reform, the strategy followed by the Mexican government has been to employ military forces to counter

---

165 *Id.* at 20.

166 Brands, *supra* note 115, at 45.

167 Robert Killebrew, *A Volatile Brew*, *in* MEXICO'S CRIMINAL INSURGENCY, *supra* note 93, 53, at 56. *See also* John P. Sullivan, *Future Conflict: Criminal Insurgencies, Gangs and Intelligence*, *in* MEXICO'S CRIMINAL INSURGENCY, *supra* note 93, at 29, 30 (where it is suggested the goal of the criminal gangs is to "hollow out" the State.).

168 MANWARING, *supra* note 96, at 138.

169 Juan Carlos Gomez, *Twenty-First-Century Challenges: The Use of Military Forces to Combat Criminal Threats*, 88 INT'L L. STUD. 279, 285 (2012).

170 DISMANTLING COLOMBIA'S ARMED GROUPS, *supra* note 157, at 2.

171 Szesnat & Bird, *supra* note 108, at 226.

172 DISMANTLING COLOMBIA'S ARMED GROUPS, *supra* note 157, at 4.

173 Gomez, *supra* note 169, at 285–86 (outlining a two-card system for Rules of Engagement: blue for law enforcement and red for operations against military objectives).

174 RÖMER, *supra* note 107, at 166.

the drug cartels. In pursuing a policy of "going after the high-value targets: the leadership of the major DTOs [drug trafficking organizations],"[175] the strategy maintained that the only viable option applied was to "confront well-armed DTOs equipped with military style weapons was to use the well-armed military."[176] It is also viewed as having increased the fragmentation of the drug cartels as well as the overall level of violence.[177] However, it has been suggested that the power to deal with the kinds of "criminal-military threats to traditional military values does not lie in hard military firepower or even in police power but instead requires a whole-of-government approach."[178] Effectively, the solution lies in good governance. Such an approach may invoke images of counterinsurgency doctrine highlighting the close interface been the criminal threats posed by transnational criminal organizations and traditional insurgency. However, it can just as easily suggest the approach adopted by the United Kingdom in Northern Ireland of exclusively applying a law enforcement framework, albeit altered with significant support provided by military forces, and the invocation of special powers in order to address the unique violence and threats that State was facing (see Chapter 13, section 13.2.1, *The "Troubles"*).

6.3.3. OTHER CRIMINAL ACTIVITY

Insurgent groups in crime and transnational criminal organizations do not represent the only groups operating in the grey area between armed conflict and law enforcement. Other criminal activities relate to warlords and the effects of tribalism that occurs in some ungoverned or weakly governed spaces of the world.[179] For example, during conflict, warlords may team up with "war profiteers and 'rogue capitalists' in order to develop their own shadow logistics and financial services."[180] In the aftermath of a conflict "warlords either turn into 'mafiosi', or are forced to gradually drop their dangerous liaisons and clean up their act if they want to reinvent themselves as politicians or statesmen."[181] Another phenomenon is "social banditry," where individuals or groups seek to redress injustice in society. This violence has broader goals than pecuniary gain, causing it to resemble violence for a political purpose.[182] Such banditry includes a subset of actors identified as the "avengers." These avengers "feeling terribly wronged, strike back without restraint against their tormentors."[183] Avenger violence is often unrestrained by traditional ethical norms. Historically, one example where there has been such "explosions of cruelty" is when militants are forced back into a life of

---

[175] Beittel, *supra* note 94, at 33.

[176] *Id.* at 35. *See also* GRILLO, *supra* note 92, at 285.

[177] Beittel, *supra* note 94, at 18–23.

[178] MANWARING, *supra* note 96, at 146. *See also* Sullivan & Elkus, *supra* note 136, at 64 ("The dynamics on intraconflict policing demand more than COIN [counterinsurgency].").

[179] MARK MAZZETTI, THE WAY OF THE KNIFE: THE CIA, A SECRET ARMY, AND THE WAR AT THE ENDS OF THE EARTH 35 (2013) (where it is reported that an Afghan tribal leader, Jalaluddin Haqqani, who had once been a CIA asset during the war in Afghanistan had "pledged loyalty to Al Qaeda and had built up a sprawling criminal empire from his base in Miranshah, North Waziristan.").

[180] ANTONIO GIUSTOZZI, EMPIRES OF MUD: WARS AND WARLORDS IN AFGHANISTAN 19 (2009).

[181] *Id.* at 20 (the example provided is that of Charles Taylor of Liberia as "his failure to clean up led to his quick downfall.").

[182] JOHN ARQUILLA, INSURGENTS, RAIDERS, AND BANDITS: HOW MASTERS OF IRREGULAR WARFARE HAVE SHAPED OUR WORLD 242 (2011).

[183] *Id.*

"outlaws and robbers" under circumstances where a peasant insurrection or rebellion has failed to turn into a social revolution. Another is "that second generation of 'children of violence' who graduate from the ashes of their homes, the corpses of their fathers and the raped bodies of their mothers and sisters to a life of outlawry. . . ."[184]

One modern example of such an avenger, who was a master of the art of the irregular warfare, is said to be the Indian "Bandit Queen," Phoolan Devi. She was a poor young woman who was robbed, raped, and otherwise violently abused, becoming an avenging *dacoit*,[185] a folk hero, and eventually even being elected to the Indian Parliament in 1996.[186] She was murdered in 2001, reportedly in retribution for revenge killings she had carried out on twenty-two villagers.[187] A connection has been made between armed groups in Afghanistan available to fight for hire and the Indian concept of *dacoity* Such organized banditry "is so common in parts of South Asia that it is considered as kind of profession—and not necessarily an unrespectable one."[188] John Arquilla links the exceptional violence associated with "avengers" to modern terrorism, including that exhibited by Al Qaeda and its affiliates.[189] He notes, "[w]hat we call the age of terror is one replete with examples of insurgents and terrorists."[190] In dealing with these insurgents, terrorists, and criminals what is particularly noteworthy is that States often rely on an application of human rights–based law enforcement as being a more effective and relevant option than engaging in warfare.

6.4. Terrorism and Terrorists

6.4.1. TERRORISM

If an insurgency is a protracted politico-military struggle focused on weakening the control and legitimacy of government,[191] "terrorism" and "terrorists" have traditionally presented a different challenge, not only for security forces but also for international law. Looked at simplistically, terrorism is a tactic, while those who carry out such acts are terrorists. However, terrorism covers a wide range of activity that can be carried out by individuals, criminal gangs, and insurgents, as well as States. It has proven exceeding difficult to define, and in legal terms, while "some regional treaties have adopted general definitions, the variation in these definitions militates against the emergence of any shared international conception of terrorism."[192] Consensus on a definition cannot be reached by looking at UN Security Council practice, judicial decisions, or national definitions of terrorism.[193]

---

[184] Eric Hobsbawm, Bandits 73 (reprint 2012) (2011).

[185] Arquilla, *supra* note 182, at 242-43 ("Given the sharp class and caste differences that evolved in Indian society, bandits tended to follow the 'noble robber' pattern, these *dacoits* targeting upper-caste Brahims in particular. . . .").

[186] *Id.* at 249–50.

[187] *Id.* at 248 and 251.

[188] James Fergusson, Taliban: The True Story of the World's Most Feared Guerrilla Fighters 137 (2010).

[189] Arquilla, *supra* note 182, at 242.

[190] *Id.* at 252.

[191] Counterinsurgency Manual, *supra* note 60, ¶1-2 at 2.

[192] Ben Saul, Defining Terrorism in International Law 190 (2006).

[193] *Id.* at 270.

As Richard Baxter once noted, "[w]e have cause to regret that a legal concept of 'terrorism' was ever inflicted upon us."[194]

One proposed legal definition of "terrorism" for international criminal law purposes refers to serious, violent criminal acts (including against property), committed outside of armed conflict; conducted for a political, ideological, religious, or ethnic purpose; and intended to cause extreme fear in order to seriously intimidate a population or government.[195] The exclusion of acts committed in armed conflict is on the basis that there are specialized provisions on terrorism under humanitarian law.[196] However, this approach also recognizes terrorist acts during hostilities must be distinguished "from other terrorist acts committed in wartime, but unrelated to the conflict, or outside the theatre of hostilities."[197] This latter distinction appears too limiting given the global reach of transnational terrorist groups. Importantly, in the post-9/11 security, environment terrorism, as a concept, has increasingly captured the attention of States as the prevailing threat. In the contemporary dialogue, terrorism has migrated away from its previous connection to be being primarily a criminal matter and has instead become deeply intertwined with the concept of "war."[198]

From a U.S. perspective this connection to war is evident in President Barak Obama's May 2013 speech on the use of drones. He purposely noted the United States was at war with organizations (i.e., Al Qaeda, the Taliban, and their associated forces), rather than with terrorism itself. This comment can be juxtaposed with his predecessor, George Bush, whose administration focused on a Global War on Terrorism (GWOT). While wars do end, it is carefully noted in the *Drone Speech* that no president "can promise the total defeat of terrorism." [199] This is a dialogue that appears to be tempered by a view that law enforcement is the appropriate means for dealing with such threats. This leaves open the possible outcome that reducing the Al Qaeda threat to a level where it is amenable to law enforcement can be a desirable or at least attainable goal for the United States (see Chapter 14, section 14.6 *Law Enforcement and the Search for "Normalcy"*). As such, it highlights the ubiquitous nature of terrorism as well as provides a perspective on how that threat is being viewed a decade after the attacks of 9/11. In terms of the use of force it means that attaining that goal could be directly dependent upon the ability of law enforcement to contain such terrorism. The capacity of a law enforcement approach to fulfill that role is in turn dependent in part upon the levels of force permitted under this human rights–based framework.

The challenge of dealing with terrorism in the twenty-first century has not been helped by the United States' early use of the term GWOT. Other countries avoided using this language. For example, Canada referred to its operations as part of a broader campaign against terrorism (the CAT). As a more inclusive and less bellicose term, the CAT attracted

---

194 Richard Baxter, *A Skeptical Look at the Concept of Terrorism*, *in* HUMANIZING THE LAWS OF WAR 211 (Detlev F. Vagts et al. eds., 2013) (the text first appeared in 7 AK. L. REV. 380 (1974)).

195 SAUL, *supra* note 192, at 65–66.

196 *Id.* at 63.

197 *Id.* at 312.

198 PHILIP B. HEYMANN, TERRORISM AND AMERICA 4 (1998) (where it is indicated the academic definition of terrorism often has a far more moral or criminal flavor than state definitions, which suggest that terrorists are hostile forces acting for political purposes).

199 *Obama's Speech on Drone Policy*, N.Y. TIMES, May 23, 2013, http://www.nytimes.com/2013/05/24/us/politics/transcript-of-obamas-speech-on-drone-policy.html.

less attention. However, it has also been suggested, given the broad spectrum of the efforts to counter the terrorist threat (including the citizenry being involved to foil attacks), that words such as "war" and "campaign" are "inapt".[200] The GWOT language not only covered a wide range of both international (i.e., Iraq) and noninternational conflicts (i.e., Afghanistan), but has also been critiqued as a being misnomer, since it was a tactic rather than a cause.[201] The idea of a global war involving terrorism attracted considerable attention and criticism, including a concern it suggests military means alone might prevail in neutralizing the threat.[202] At the heart of the debate about State reaction to contemporary terrorism is whether it is an activity most properly dealt with by law enforcement or is a form of armed conflict. Like most of the debate concerning the law governing contemporary conflict, it tends to be both polarized and exclusionary in nature. However, practical reality demands a more holistic approach. The need for such an approach will be highlighted by looking first at the phenomenon of modern terrorism and then at those who carry it out.

A unique challenge in grappling with terrorism is coming to terms with the fact there is no widely accepted definition. This problem is not limited to the legal sphere. It has been defined "as the deliberate creation and exploitation of fear through violence or the threat of violence in the pursuit of political change."[203] Another definition is "the use of violence by nonstate actors directed primarily against noncombatants (mostly civilians but also including government officials, policemen, and off-duty soldiers) in order to intimidate or coerce them and change their government's policies or compositions."[204] Yet another approach specifically excludes attacks on soldiers in a combat zone with the result that "Hizbollah's shooting at Israeli military forces in Lebanon is not terrorism."[205] The definitions of terrorism are many, and as has been noted an actual consensus on what it means has eluded the international community. This outcome is perhaps best described as occurring because "[t]errorism is intended to be a matter of perception and is thus seen differently by different observers and at different points in history."[206] Unlike armed conflict there is no attempt to "objectify" the criteria for terrorism. The relationship between violence and a political purpose is a foundational one, perhaps reflecting a grounding not only in a non-State actor versus State context but also what is fundamentally viewed as criminal activity.

---

[200] Benjamin Wittes & Gabriella Blum, The Future of Violence: Robots and Germs, Hackers and Drones 205 (2015)(the United States governing engaged "in what was really a broad-based effort to force, urge, incentivize, and nudge a great many different actors into postures that, cumulatively, would create a global environment more secure from terrorists.").

[201] Bernd Horn, *Defining Terrorism*, *in* The Difficult War: Perspectives on Insurgency and Special Operations Forces 111, 113 (Emily Spencer ed., 2009).

[202] Paul Reynolds, *Declining Use of the "War on Terror,"* BBC News (Apr. 17, 2007), http://news.bbc.co.uk/2/hi/uk_news/politics/6562709.stm.

[203] Bruce Hoffman, Inside Terrorism 40 (2006). *See also* Bard E. O'Neill, Insurgency & Terrorism: From Revolution to Apocalypse 33 (2005) (where a similar definition of terrorism is applied. It is "defined as the threat or use of physical coercion, primarily against noncombatants, especially civilians, to create fear in order to achieve various political objectives.").

[204] Boot, *supra* note 6, at xxii.

[205] Byman, *supra* note 84, at 7.

[206] Audry Kurth Cronin, How Terrorism Ends: Understanding the Decline and Demise of Terrorist Campaigns 7 (2009).

Although the genesis of terrorism is found in the French Revolution, where *terreur* was associated with State action,[207] the term is now most often associated with non-State actors.[208] Bruce Hoffman has identified that over the years terrorism has been linked to the actions of State and non-State actors and associated with a wide range of causes, such as revolts by "indigenous nationalist/anti-colonial groups,"[209] "nationalist and ethnic separatist groups,"[210] "radical, entirely ideologically motivated organizations,"[211] and "a type of covert or surrogate warfare whereby weaker states could confront larger, more powerful rivals without the risk of retribution."[212] In the late 1960s a concept of the "urban terrorist" was increasingly recognized. The development of this category was influenced particularly by an attempted shift in Latin America from rural to urban guerrilla operations.[213] Terrorism is also linked to "gray area phenomenon," exemplified by the Columbian (and now the Mexican) drug cartels, where criminal organizations are associated with terrorist and guerrilla organizations.[214] Religiously motivated groups like Al Qaeda have clearly been identified as terrorist organizations.[215]

By the end of the twentieth century, terrorism was seen to have changed. In 2002 Paul de B. Taillon indicated, "we are experiencing what has been described as hyperterrorism, superterrorism and megaterrorism."[216] This typology relied on the interest and capability of the terrorist "to orchestrate devastating attacks" and involved individuals who are prepared "go beyond conventional terrorism and, unlike most terrorist groups, as well as nations, are apparently willing to employ weapons of mass destruction."[217] This focus on the level of violence and destruction was a harbinger of the post-9/11 debate that has gripped the international legal community regarding the point at which attacks by non-State actors cross the line to become armed attacks under Article 51 of the UN Charter and armed conflicts under humanitarian law. A similar approach for delineating an insurgency from terrorism has been suggested by a police intelligence officer experienced in counterinsurgency. As Randall Wilson notes, "only when the level of threat posed presented by armed opponents of the state reaches levels sufficient enough to threaten the fabric of or continued existence of that state does one graduate from terrorism to insurgency."[218] This perspective has considerable resonance with the traditional categories of rebellion, insurgency and belligerency, with "[t]he principle distinction between the three related to the intensity of the violence."[219]

---

[207] PAUL J. SMITH, THE TERRORISM AHEAD: CONFRONTING TRANSNATIONAL VIOLENCE IN THE TWENTY-FIRST CENTURY 8–9 (2008) (where it is suggested the difference between State and non-State terror is enforcement terror versus agitational terror.).

[208] *Id.* at 8. *See also* BOOT, *supra* note 6, at 205 (where terrorism is defined in such a way that it is restricted to non-State groups).

[209] HOFFMAN, *supra* note 203, at 16.

[210] *Id.*

[211] *Id.*

[212] *Id.* at 17.

[213] LAQUEUR, *supra* note 46, at 343–52.

[214] HOFFMAN, *supra* note 203, at 18.

[215] *Id.* at 82.

[216] J. PAUL DE B. TAILLON, HIJACKING AND HOSTAGES: GOVERNMENT RESPONSES TO TERRORISM 61 (2002)

[217] *Id.*

[218] WILSON, *supra* note 66, at 137.

[219] SANDESH SIVAKUMARAN, THE LAW OF NON-INTERNATIONAL ARMED CONFLICT 9 (2012).

In light of the diverse definitions, one suggested approach is to deal with terrorism in terms of its characteristics. As Audrey Kurth Cronin indicates:

> "[A]t a minimum, the concept has the following characteristics: a fundamentally political nature, the symbolic use of violence, purposeful targeting of noncombatants, carried out by nonstate actors."[220]

The first characteristic highlights a common element in definitions of terrorism, the killing for a political purpose. In this regard it appears that "[p]olitical murder comes with the emergence of political authority. . . ."[221] It is this characteristic that reveals the potentially broad application of "terrorism." An array of people and organizations may kill for a political purpose. In this regard the fact that insurgency and terrorism both have a goal of political change complicates the comparison between the two activities, as well as their respective relationship to armed conflict.

Terrorist activity is said to be different from that of ordinary criminals and "lunatic assassins" because of the different purposes for which violence is used.[222] Criminals use violence as a means to an end,[223] while the "lunatic assassin" does so for egocentric goals.[224] The terrorist is an altruist in that "he believes that he is serving a 'good' cause designed to achieve a greater good for a wider constituency—whether real or imagined—that the terrorist and his organization purport to represent."[225] Terrorism can occur as a criminal matter in peacetime or be a part of the conduct of hostilities. This depends upon whether an organized armed group is involved and the context in which it occurs. Further confusing how terrorism is viewed is the reality that insurgency and terrorism share characteristics in that they "employ the same tactics (assassinations, kidnapping, hit-and-run attack, bombings of public gathering places, hostage-taking, etc.) for the same purposes (to intimidate or coerce, thereby affecting behavior through arousal of fear) as terrorists."[226] Insurgents and terrorists also operate out of uniform, hiding among the people.[227]

Causing terror is not a new phenomenon in history. It has been suggested that the Ismaili Assassins, part of a Shiite sect operating in Syria in the eleventh and twelfth centuries, "may well be the first terrorists."[228] However, in terms of impact, "terrorism is strikingly modern," being "made possible by the spread of four phenomena: destructive and portable weaponry, the mass media, literacy and secular ideologies."[229] While post–World War II anticolonial terrorist campaigns had an international orientation, it was the 1968 hijacking

---

[220] CRONIN, *supra* note 206, at 7. *See also* Horn, *supra* note 201, at 114–15 (where the core components are identified as it being unlawful, politically motivated (i.e., not for criminal gain), premeditated, directed against innocents, meant to cause fear and terror, violence directed to impact others than the victims, and involve actions outside the accepted limits of the use of force in warfare (i.e., targeting noncombatants)).

[221] BERNARD LEWIS, THE ASSASSINS: A RADICAL SECT IN ISLAM 125 (1967).

[222] HOFFMAN, *supra* note 203, at 36–37.

[223] *Id.* at 36.

[224] *Id.* at 37.

[225] *Id.* at 36–37.

[226] *Id.* at 35.

[227] *Id.*

[228] LEWIS, *supra* note 221, at 129–30.

[229] BOOT, *supra* note 6, at 209.

by the Popular Front for the Liberation of Palestine (PFLP) of an Israeli El Al flight in 1968 that has been identified as the beginning of modern terrorism.[230] Perhaps one of the most dramatic incidents was the September 1972 Munich Olympic tragedy, which resulted in the deaths of nine Israeli athletes.[231] Facilitated by technological advances (e.g., international commercial air travel, the speed and quality of television coverage), this form of terrorism involved travel to other countries to conduct attacks, which at times targeted civilians who were not connected to the terrorist cause.[232] These technical advances made the threat a global one.[233] It has been noted "[w]hat the world was seeing was the evolution of a virtually professional terrorist."[234]

"Modern" terrorism defied clear categorization, even during its heyday from the 1960s to the1980s. Countering terrorism included the adoption of rigorous security measures, screening of luggage, and the introduction of "air marshals," which continues to apply to international air travel.[235] For much of the world the view was that "[t]he response to terrorism should be primarily a police responsibility."[236] This focus on law enforcement as the appropriate means to counter terrorism can be linked to the focus on countering Western, "home-grown" terrorist groups. Examples of 1970s terrorist groups include "Action Directe (France), the Baader-Meinhof Gang, the Japanese Red Army, the Provisional IRA, the Basque ETA, the Greek Revolutionary Organization 17 November, the Quebec Liberation Front, the Black Panthers, the Weathermen, the Symbionese Liberation Army . . . ."[237] Such groups cooperated and carried out international attacks as was evident in the "Entebbe" hijacking of an Air France flight in 1976 by members of a German terrorist group acting in conjunction with the PFLP (see Chapter 10, section 10.3.2, *"Iconic" Counterterrorism: A Thunderbolt Strikes Entebbe*).[238] Terrorism was primarily viewed and dealt with as a law enforcement matter,[239] although the response to counter terrorist hijacking included the development of both elite military and police counterterrorism units (see Chapter 10, section 10.4, *Converging Tactics: Special Forces, Law Enforcement, and Hostilities*).[240]

The degree to which terrorism was seen as a domestic matter is reflected in the following 1990 assessment of Canadian, U.S., and UK responses to terrorism:

> Although Canada and the United States share the benefits of relative isolation from the centres of terrorist violence in Europe and the Middle East, they have overseas interests which offer lucrative targets. Both North American nations also have a

---

[230] HOFFMAN, *supra* note 203, at 63–64, and DE B. TAILLON, *supra* note 216, at 17.

[231] HOFFMAN, *supra* note 203, at 67–68.

[232] *Id.* at 64.

[233] DE B. TAILLON, *supra* note 216, at 50.

[234] *Id.* at 25.

[235] SIMON DUNSTAN, ISRAEL'S LIGHTNING STRIKE: THE RAID ON ENTEBBE 1976, at 5 (2009) (for an outline of security measures adopted by Israel following the advent of terrorist hijacking in the 1970s). *See also* G. DAVIDSON SMITH, COMBATING TERRORISM 21–26 (1990) (for changes made to government and business operations due to the advent of international terrorism).

[236] GRANT WARDLAW, POLITICAL TERRORISM 100 (2nd ed. 1990).

[237] BOOT, *supra* note 6, at 456.

[238] SAUL DAVID, OPERATION THUNDERBOLT: FLIGHT 139 AND THE RAID ON ENTEBBE AIRPORT, THE MOST AUDACIOUS HOSTAGE RESCUE MISSION IN HISTORY 8–56 (2015).

[239] DE B. TAILLON, *supra* note 216, at 78.

[240] *Id.* at 76–79.

> number of large ethnic communities that are potential sources of terrorist populations, either in terms of support infrastructure or as centres of unrest. As well, the "successes" of some recognized terrorist groups in achieving limited gains have tended to engender a certain admiration among issue groups in all three nations, and could promote an increase in acts of domestic terrorism.[241]

It was noted that "[e]ach nation [Canada, the United States, and the United Kingdom] has wisely chosen to treat terrorism as criminal behaviour, and to approach and oppose such activity under the concept of the rule of law."[242] Such an approach highlights that the States were acting as the proper authority within their territory with the result that internal security was viewed primarily through a law enforcement lens.

However, since the 1990s there has been a shift in thinking, as well as in the threat. The link between terrorism and domestic jurisdiction can be overstated as "most terrorist incidents on European soil since 9/11 have *not* been purely home-grown but have drawn on sponsorship, support or guidance from AQ [Al Qaeda]."[243] The threats have not been the 9/11 "expeditionary" model (clandestine insertion of a team to conduct an attack), but rather "a 'guerrilla' model in which local clandestine cells are recruited and trained in the target country."[244] In this context "[s]ome cells are directly linked to AQ [Al Qaeda]; others receive training from AQ affiliates, or are inspired by AQ propaganda."[245] While many commentators only see a limited link between international terrorism and some attacks, such as the 2004 Madrid bombing,[246] it is clear that twenty-first-century terrorism has a global reach. Any approach favoring law enforcement must realistically address the nature of the international threat posed by a number of contemporary terrorist groups.

Even in the 1960s through to the 1980s not all terrorism was uniquely dealt with in a law enforcement matter. The Israeli response to international terrorism included retaliatory attacks that appear to have been neither grounded in regular notions of law enforcement nor traditional interpretations of the right to self-defense.[247] Such action prompted significant debates in the international community regarding the legality under international law of responses to attacks carried out not only by Israel but also by other affected States like the United States. It was a discussion that was primarily framed in terms of whether those responses were punitive rather than defensive.[248] Questions were also raised as to whether

---

241 DAVIDSON SMITH, *supra* note 235, at 233.

242 *Id.*

243 DAVID KILCULLEN, THE ACCIDENTAL GUERRILLA: FIGHTING SMALL WARS IN THE MIDST OF A BIG ONE 246 (2009).

244 *Id.* (examples provided include the 7/7 London bombing and the Madrid bombing).

245 *Id.*

246 *See, e.g.*, PETER L. BERGEN, THE LONGEST WAR: THE ENDURING CONFLICT BETWEEN AMERICA AND AL-QAEDA 204 (2011) ("the Madrid attacks were largely the work of leaderless jihadis who financed the bombings with the proceeds of substantial drug deals, and only some of whom were affiliated with a known jihadist organization, the Moroccan Islamic Combatant Group.").

247 DUNSTAN, *supra* note 235, at 5–6 (for an example of an Israel Defence Forces operation at the Beirut International Airport on December 28, 1968, which destroyed thirteen planes belonging to Middle Eastern airlines (and mistakenly one Ghanaian airlines plane) in retaliation for an attack on an El Al airliner at Athens airport in Greece.).

248 CHRISTINE GRAY, INTERNATIONAL LAW AND THE USE OF FORCE 195–97 (3rd ed. 2008) (where she indicates that claims by both Israel and the United States to be acting in self-defense against specific terrorist attacks on nationals were not "necessary" since the attacks had already taken place. Instead the attacks were aimed at retaliation and

such techniques were as "successful a deterrent as many seem to believe when measured in terms of changes in the level of terrorist activity."[249]

One aspect of post–World War II political violence that has complicated the assessment of whether a group is engaged in insurgency or terrorism, and as a result how it might be confronted, is the migration of conflict toward operations in an urban setting.[250] The manner in which urban violence has been viewed provides a good example of how the focus on terrorism as law enforcement during the 1960s and 1970s has undergone change as the nature of such violence has altered. This issue appears to have first arisen in the late 1960s when insurgency was attempted in an exclusively urban setting.[251] During those years urban violence was more readily associated with terrorism than guerrilla warfare or insurgency. The association between such violence and terrorism, rather than insurgency, has been linked to theorists such as Carlos Marighella and to the urban *focoist* [a small band of revolutionaries providing the focus for revolution[252]] approach adopted in Latin America.[253] The concept of "urban guerrilla warfare" was even referred to as "terrorism in a new dress."[254]

Carlos Marighella's work, the *Minimanual of the Urban Guerrilla*, although written in the context of a campaign intended to wage an urban guerrilla campaign in Brazil, was ultimately to inspire left-wing radicals in European universities in the 1960s.[255] It is from this environment that what were then viewed as "modern" terrorist groups, such as the Italian Red Brigades, were to develop.[256] However, this connection to Marighella and *focoism* is not unique to the terrorist movements of the late twentieth century. It has also been associated post-9/11 with insurgent and terrorist groups aligned with Al Qaeda.[257] This raises questions as to whether urban violence carried out by such contemporary groups should continue to be uniquely viewed as part of terrorism or is more properly situated within the context of an "insurgency."

Much of the criticism of Marighella's approach, and that of the urban *focoist* theory, has centered on the size of the organizations involved and the reliance of violence to motivate

---

deterrence). *But see* YORAM DINSTEIN, WAR, AGGRESSION AND SELF-DEFENCE ¶¶ 646–75, at 244–55 (5th ed. 2011) (suggesting that "defensive armed reprisals" where an attacked State responds at a time and place different from the original attack fall within the self-defense provisions of Article 51 of the UN Charter).

[249] WARDLAW, *supra* note 236, at 203.

[250] *Id.* at 160–62.

[251] BECKETT, supra note 7, at 174 (where it is noted Latin American revolutionaries shifted their emphasis from rural to urban settings because of the increasing urbanization of the population).

[252] *Id.* at 170.

[253] LEQUEUR, *supra* note 40, at 343–52.

[254] *Id.* at xix (urban guerrilla warfare is also referred to as a misnomer). *See also* ELLIS, *supra* note 5, at 14, and ANTHONY JAMES JOES, URBAN GUERRILLA WARFARE 7 (2007) (where he notes for students of successful past insurgencies, or Clausewitz and/or Mao, "might well have been tempted to conclude the phrase '"urban guerrilla war' was close to being an oxymoron.").

[255] MICHAEL BURLEIGH, BLOOD & RAGE: A CULTURAL HISTORY OF TERRORISM 193 (2009).

[256] *Id.* at 189–206 (for a history of the birth of the Italian "Red Brigades") and specifically at *id.* at 233 (e.g., in the formative stages of the creation of Baader-Meinhof gang (the Red Army Faction) Andreas Baader "talked incessantly . . . about the Russian nihilist terrorist Nechaev, Lenin and the Brazilian urban guerrilla Carlos Marighella.").

[257] KILCULLEN, *supra* note 4, at 177 (As David Kilcullen has observed, insurgent and terrorist groups aligned with Al Qaeda not only have Islamic influences, they are also impacted by other theories such as communist revolutionary theory in acting as a vanguard party, and "Che Guevara's concept of 'focoist insurgency', while Carlos Marighela's *Mini-manual of the Urban Guerrilla* has also been influential."). *See also* KILCULLEN, *supra* note 243, at 86 (where a similar point is made in respect of the Taliban).

the citizenry. In terms of organization, Marighella was of the view urban guerrillas were to be organized in small groups of four to five members.[258] The connection between urban warfare and terrorism is consistent with the general observation that "the fundamental structure of most terrorist groups is based on a cell system: that is, individual aggregations of from three to seven persons known only to each other or the cell leader by whom they were originally recruited."[259] The result was that while insurgency was becoming an urban phenomenon, it was viewed as "shading into urban terrorism."[260] Further, in that setting violence tended to be applied indiscriminately.[261] The violence was used "without the same intention of taking over the state apparatus," thereby distinguished them from insurgent groups.[262] As a tactic it was decidedly unsuccessful, as have been most efforts to conduct urban guerrilla warfare.[263] It was also a tactic for which law enforcement was seen as the primary means of dealing with the threat.

Rather than being viewed as an insurgency this urban violence was assessed as falling within a different category. It was "an alternative to or in conjunction with guerrilla warfare."[264] One historical assessment of urban guerrilla warfare suggests it is difficult to separate conflicts in Algiers, Montevideo, Sao Paulo, Saigon, and Belfast from terrorism as:

> Terrorists cut themselves off from the masses through their secrecy and anonymity; they substitute activism for analysis, the individuals for the mass; through their bombings of public places and attacks on urban infrastructure they inconvenience, horrify, or kill members of the proletariat, and so on.[265]

Following this analytical approach, it is relatively easy to see how urban-based attacks on government security forces and interests were viewed as "terrorism."

However, in respect of contemporary conflict the increasing urbanization of society has altered the view that an insurgency cannot occur within an urban environment. It has been suggested that by 2000 the percentage of the world population living in cities "had reached at least 45 percent."[266] As of April 2008, "the world had passed the 50 percent urbanization mark . . . ."[267] Many of these cities are sprawling urban areas consisting of shantytowns found

---

258 MARIGHELLA, *supra* note 46, at 32.

259 DAVIDSON SMITH, *supra* note 235, at 16.

260 BECKETT, *supra* note 7, at vii.

261 *Id.* at 176 ("The principle techniques would be the use of urban terror such as letter and parcel bombs and assassination, politically motivated kidnapping, manipulation of riots and strikes, and so on.). *See also* MARIGHELLA, *supra* note 46, at 48–49 (for an outline of the types and nature of missions for the urban guerrilla.).

262 BECKETT, *supra* note 7, at viii.

263 *Id.* at 180 (after reviewing urban insurgencies in Latin American notes "[u]rban insurgency has not proved very successful where governments were prepared to abandon restraint. For liberal democracies . . . violence could often be reduced to a sufficiently acceptable level as to amount to a kind of normalcy."). *See also* JOES, *supra* note 254, at 7 (in discussing the success rate of this form of warfare, notes: "it will surprise no reader of this book that all of the urban-based insurgencies examined in it [Warsaw 1944, Budapest 1956, Algiers 1957, Montevideo and São Paulo in the 1960s, Saigon 1968 (the Tet Offensive), Northern Ireland (mainly Belfast) 1970–1998, and Grozny 1994–1996] suffered utter defeat.").

264 LEQUEUR, *supra* note 40, at xix. *See also* HOFFMAN, *supra* note 203, at 53–58 (where he refers to urban terrorism as part of the 1950s guerrilla warfare campaigns in Cyprus and Algeria).

265 JOES, *supra* note 254, at 6.

266 *Id.* at 1–2.

267 DAVID KILCULLEN, OUT OF THE MOUNTAINS: THE COMING OF AGE OF THE URBAN GUERRILLA 29 (2013).

in less-developed countries that are growing larger and more ungovernable.[268] Further, this growth is not evenly spread, with cities tending to be "concentrated in coastal (littoral) areas, within a few dozen miles of the sea."[269] In a war increasingly occurring among the people, conflict inevitably is now taking place in urban environments and will increasingly do so in the future.

It has been noted that while the mujahideen returning from Afghanistan in the 1990s trained primarily in rural guerrilla warfare, a "new generation of jihadists will have acquired in Iraq invaluable firsthand experience in urban warfare."[270] Similarly, while the Hezbollah were able to claim considerable success against the Israeli military during the 2006 Lebanon War by fighting from tunnels located in what were called "nature reserves," it is anticipated that future conflict will not occur in open fields but commence in urban settings where its missiles are hidden and then move through tunnels to the forests. The fighting will involve "a combination of urban and guerrilla warfare."[271] This trend toward urban warfare has prompted the suggestion that a particular challenge for soldiers, diplomats, and aid workers who have "had their heads in the Afghan mountains" will be to prepare for security challenges in the future that will be "urban, networked, and coastal."[272]

As was demonstrated in Chechnya, Somalia, Israel, and Iraq, contemporary insurgents have become increasingly adept at operating in an urban environment.[273] It has been noted that "[t]he Chechens simply applied their mastery at the art of forest warfare, so evident in the eighteenth and nineteenth centuries, to the urban forests of Grozny and other cities."[274] In Somalia, Al Shabaab fighters used a honeycomb of underground tunnels and trenches to move from building to building while fighting in the capital Mogadishu and "[b]y some accounts those tunnels stretched continually for miles."[275] One commentator indicated that the battlefield was reminiscent of mine warfare in Flanders in World War I and that Mogadishu was "an African version of Stalingrad."[276] In fighting against Palestinian guerrillas in Nablus in 2002, the Israel Defence Forces units were forced to avoid moving in streets, roads, alleys, or courtyards, but rather transit through holes blasted into walls, ceilings, and floors.[277] Similarly, in 2004 in Fallujah, Iraq, jihadists provided stiff resistance against efforts by United States and other forces to seize the city.[278] The conduct of such warfare, albeit in cities, makes it less likely those insurgent operations will be able to be assessed uniquely as "terrorist" in nature solely because they are urban based.

---

[268] JOES, *supra* note 254, at 2.

[269] KILCULLEN, *supra* note 267, at 30.

[270] HOFFMAN, *supra* note 203, at 293.

[271] KATZ & HENDEL, *supra* note 2, at 28.

[272] KILCULLEN, *supra* note 267, at 22.

[273] RICHARD H. SHULTZ & ANDREA J. DEW, INSURGENTS, TERRORISTS AND MILITIAS: WARRIORS OF CONTEMPORARY COMBAT 265–66 (2006). *See also id.* at 265 ("In the 1990s for example, Chechen warriors were extremely adroit at innovating and extending the fight to the cities—a new operational environment for them. They adapted their small-team approach—ambushes, hit-and-run tactics, surprise attacks, and quick withdrawals—to the capital city of Grozny with lethal efficiency.").

[274] ROBERT M. CASSIDY, COUNTERINSURGENCY AND THE GLOBAL WAR ON TERROR 66 (2008).

[275] SCAHILL, *supra* note 29, at 475. *See also* JAMES FERGUSSON, THE WORLD'S MOST DANGEROUS PLACE: INSIDE THE OUTLAW STATE OF SOMALIA 18 (2013).

[276] FERGUSSON, supra note 275, at 18.

[277] KILCULLEN, *supra* note 267, at 110 (relying on Eyal Weizman, *The Art of War*, 99 FRIEZE (May 2006)).

[278] SHULTZ & DEW, *supra* note 273, at 254.

Terrorism is a concept that exists in its own right. However, it is also a form of conflict that is undergoing significant change. Terrorism is not simply domestic or "homegrown" attacks on political authority. Neither is all urban violence necessarily uniquely terrorism. While the treatment of terrorism as a law enforcement matter has continuing application in the twenty-first century,[279] the attacks of 9/11 represent an even more sophisticated and destructive form of what has come to be called transnational terrorism.[280] This means there "are members of organizations currently conducting insurgent activities in other parts of the world but whose members and actions are considered terroristic when found in identical situations of intent and motivation in the West."[281] As a result of effective governance that some States can exhibit (i.e., policing), the difference is seen as "one of scale caused by the more stable society, more efficient security services and a much more shallow pool within which to hide.[282] Transnational terrorism is being conducted as part of a globalized form of insurgency, which has impacted on a wider number of States and dominated much of the international security dialogue. The reality is that terrorism can exist on its own caused by groups uniquely focused on that mode of conflict, or can be conducted by insurgents during the course of an insurgency. Significantly, terrorism is intimately associated with groups such as the Al Qaeda and Islamic State jihadist movements that exhibit attributes of both "terrorists" and "insurgents."

6.4.2. THE "TERRORISTS"

A common approach in assessing terrorism is to consider the differences between terrorist and insurgent groups. At first glance this might seem odd given the traditional use of language suggesting criminality (i.e., bandits, rebels, marauders) to describe all non-State actors challenging the authority of the State. Indeed, one question is the degree to which "terrorists" form a unique group of participants in conflict at all. In this respect it has been noted that attempts to look at terrorist groups by compiling a list of the specific acts they carry out (i.e., assassination, hijacking, kidnapping) "soon peters out: too many terrorist actions duplicate either military or criminal acts."[283]

However, it appears that even within the dialogue about the rebels and others with which they are at "war," States have imposed a hierarchy within which participants in conflict against States are assessed. This is particularly evident in the post-9/11 international efforts to deal with terrorism, such in the realm of international intelligence sharing, tracking of terrorist financing, enhanced border security, and law enforcement cooperation.[284] To be branded a "terrorist" or a "terrorist" organization has meaning and ultimately consequences. However, given its political connotations, being labeled a "terrorist" is also a concept that is open to considerable manipulation and ultimately abuse. Terrorists have been distinguished from insurgents by Bruce Hoffman on the basis that they do not function

---

279 KILCULLEN, *supra* note 4, at 187.

280 DUNSTAN, *supra* note 235, at 61.

281 WILSON, *supra* note 66, at 140.

282 *Id.*

283 CHARLES TOWNSHEND, TERRORISM: A VERY SHORT INTRODUCTION 5 (2002).

284 *See, e.g.*, Kristen Archick, *U.S.-EU Cooperation Against Terrorism, Congressional Research Service* (2013), http://www.fas.org/sgp/crs/row/RS22030.pdf.

as armed units operating in the open, do not seize or hold territory, avoid combat, "are constrained numerically and logistically from undertaking concerted mass political mobilization efforts, and exercise no direct control or governance over a populace at either the local or the national level."[285] It is also suggested that insurgencies primarily target military or government forces rather than civilians "because they realize the backlash and implosion in targeting illegitimates."[286]

Max Boot suggests a similar analysis when he compares terrorism to guerrilla warfare, which is an integral way by which insurgency is normally conducted. The differences are said to be:

1. Guerrillas often try to hold territory and terrorists do not;
2. Guerrilla armies can number in the tens of thousands, while most terrorist organizations have a few hundred adherents;
3. Terrorists focus attacks on the "home front" where no formal state of war exists and guerrillas usually limit operations to a recognized "war zone"; and
4. Guerrillas seek a physical defeat of an enemy and terrorists hope for spectacular attacks to trigger a revolution.[287]

He also suggests there is a continuum of armed conflict with "terrorism at the bottom, next are guerrillas, then conventional forces, and finally nuclear weapons and other weapons of mass destruction."[288] This reference to conventional forces and weapons of mass destruction appears to relate to traditional inter-State warfare. A continuum approach does help situate this mode of violence in relation to more traditional forms of hostilities. However, such an approach should not be restricted to the relationship between terrorism and other forms of conflict. Within terrorism itself the degree of the threat may be quantified. This is reflected in the *Drone Speech* of 2013 where attacks on U.S. interests in the 1980s and 1990s were indicated to be less of a threat than the attack of 9/11.[289]

It has also been suggested that unencumbered by the obligations and responsibilities of governance, purely terrorist organizations do not have political or humanitarian wings. Further, in response to the effective action by State security forces, terrorist groups are increasingly adopting a leaderless network strategy "to thwart law enforcement and intelligence agency efforts to penetrate them."[290] While not all networked organizations are necessarily terrorists, terrorist organizations are increasingly seen as network based. Unlike during an insurgency, the struggle against terrorism seems less about governance in the sense of winning control over territory or the institutions of the State. This is particularly true where the terrorist threat emanates from outside the State.

---

[285] HOFFMAN, *supra* note 203, at 35. *See also* LAQUEUR, *supra* note 46, at xix ("there have been guerrilla units of ten thousand men and women but an urban terrorist unit seldom, if ever, comprises more than a few people, and urban terrorist 'movements' rarely consist of more than a few hundred members.").

[286] CRONIN, *supra* note 206, at 165

[287] BOOT, *supra* note 6, at xxiii.

[288] *Id.*

[289] *Drone Speech*, supra note 199.

[290] HOFFMAN, *supra* note 203, at 38.

However, as much as there are differences, there can also be similarities and considerable overlap that complicates any assessment. For example, the adoption of a cellular network is not unique to terrorism. It was three-man Russian partisan groups of World War II that were said to have inspired the use of cells by Algerian terrorists fighting the French during the battle of Algiers.[291] As discussed, insurgent groups have commonly adopted such networks when operating in cities and outside of safe havens. Groups assessed as being terrorist organizations, such as Hamas and Hezbollah, may have political or humanitarian wings, or at least engage in humanitarian activities.[292] For example, the Pakistani militant group Lashkar-e-Tabia (LeT), which dispatched a "fidayeen squad"[293] to carry out the 2008 Mumbai attacks, was under the organizational control of the Markaz al-Dawa wal-Irshad (MDI), later to be called the Jamaat-ul-Dawa or JuD.[294] That organization ran extensive educational, medical, and missionary services in Pakistan.[295] The difficulty in separating humanitarian action from terrorism is reflected in Hezbollah's operations in Iraq supporting Shi'a military groups being reported to involve the establishment of charitable organizations to assist in recruiting.[296] Similarly, the Islamic State expanded its reach into communities and "recruited followers under the pretense of opening a Dawah office, an Islamic missionary center."[297] The ability to establish and run a large-scale humanitarian operation is dependent upon controlling territory, or at least having a sanctuary from which to operate. Even then, as is evident with the Israeli view of Hezbollah and Hamas, this does not mean an opponent will accept such an internal division of responsibility actually changes the ultimate terrorist designation or represents an actual functional separation within the group.[298]

Another issue is that while some groups may be engaged full time in terrorism and can clearly be considered "terrorist groups," there can also be insurgent groups that use terrorism as part of a broader strategy challenging the State. Insurgent groups can also have members and subunits who individually carry out terrorist acts. It was suggested in 2009 that the Liberation Tigers of Tamil Eelam (LTTE or Tamil Tigers) "is as much a large insurgency against the Sri Lankan military as it is a classic terrorist campaign targeting civilians." [299]

---

[291] Paul Aussaresses, The Battle of the Casbah: Counterterrorism and Torture xx (2006).

[292] Zaki Chehab, Inside Hamas: The Untold Story of the Militant Islamic Movement 30–33 (2007), Matthew Levitt, Hamas: Politics, Charity, and Terrorism in the Service of Jihad 9 (2006), and Levitt, *supra* note 33.

[293] Stephen Tankel, Storming the World Stage: The Story of Lashkar-e-Taiba 207 (2013) (these attacks involve small groups of fighters intending to inflict as much damage as possible before the fighters are "martyred"). *See also* Cathy Scott-Clark & Adrian Levy, The Siege: 68 Hours Inside the Taj Hotel 106–07 (2014).

[294] Tankel, *supra* note 293, at 115–17. (The replacement of MDI by JuD has been described as a cosmetic change designed post-9/11 to remove Pakistan's legal liability and requirement to go after the organization assets.).

[295] *Id.* at 67–102 (for an outline of the organization of MDI and the education, medical, and social services offered within Pakistan). *See also* Mazzetti, *supra* note 179, at 260.

[296] Levitt, *supra* note 33, at 297–98 (as the author notes, this was based on reporting by controversial group, Mujahedin-e-Kahlq, which itself was designated by the United States as a terrorist group until 2012).

[297] Christoph Reuter, *The Terror Strategist: Secret Files Reveal the Structure of Islamic State*, Spiegel On Line Int'l (Apr. 18, 2015), http://www.spiegel.de/international/world/islamic-state-files-show-structure-of-islamist-terror-group-a-1029274.html.

[298] James Kantor & Judi Rudoren, *European Union Adds Military Wing of Hezbollah to List of Terrorist Organizations*, N.Y. Times, July 22, 2013, http://www.nytimes.com/2013/07/23/world/middleeast/european-union-adds-hezbollah-wing-to-terror-list.html?_r=0. *See also* Byman, *supra* note 84, at 317.

[299] Cronin, *supra* note 206, at 57.

The question becomes what scale or frequency of terrorist acts turns an insurgent group into a terrorist one. This challenge is compounded by assessments that Al Qaeda and the Taliban have become "inextricably involved organizationally, operationally and ideologically."[300]

The Taliban itself has emerged on both sides of the Afghanistan/Pakistan border. The Tehrek-e-Taliban Pakistan (TTP) is reported to be an umbrella organization of twenty-eight like-mind jihadi organizations. It is separate from its Afghanistan counterpart, having its genesis in the "lawlessness, tribalism, and Pakistan's perennial neglect of the Federally Administered Tribal Areas."[301] It is said their goal is not "to take over Islamabad and govern there: their preferred path is to make Waziristan the capital of their cherished Islamic State."[302] There is also another Taliban movement in the Punjab incorporating major jihadi groups.[303] A particularly noteworthy Punjab-based group is LeT, which has been recognized to operate as a Pakistani proxy focused primarily on liberating Kashmir from Indian administration but also engaging post-9/11 in the global jihad against the West.[304] It is LeT that carried out the 2008 Mumbai attack.[305]

The broader Taliban organization has also been described as the traditional Afghan Taliban, which lives mostly in southwest Afghanistan and southwest Pakistan, and the Neo-Taliban owing first allegiance to the Taliban, but ultimately linked to Al Qaeda that has been centered on the Federally Administered Trial Areas of Pakistan.[306] By late October 2015 the "weight" of the Al Qaeda leadership was reported to be migrating to Afghanistan, Syria and Yemen.[307] While both Taliban groups fight in Afghanistan, "for the traditional Taliban the war begins and ends in Afghanistan, the Neo-Taliban's war starts from Central and South Asia and ends with the emergence of the Global Caliphate."[308] While there has been a tendency to treat the Mumbai attack as a separate incident, it has also been suggested that the attack on India reflects a broader effort to coordinate Muslim resistance movements around the world and "extend its war against the West all the way to the Middle East for the final 'End of Time' battles to liberate Palestine."[309] While it has also been noted that LeT would reject the appellation as an Al Qaeda affiliate[310] that group has also expanded its activities from focusing on Kashmir to fighting in Afghanistan and "contributing to the

---

[300] ATWAN, *supra* note 76, at 127.

[301] HASSAN ABBAS, THE TALIBAN REVIVAL: VIOLENCE AND EXTREMISM ON THE PAKISTAN-AFGHANISTAN FRONTIER 221 (2014).

[302] *Id.*

[303] ATWAN, *supra* note 76, at 150.

[304] TANKEL, *supra* note 293, at 1–8.

[305] *Id.* at 207–21, and KILCULLEN, *supra* note 267, at 52–60 (for an outline of the LeT attack in Mumbai).

[306] SYED SALEEM SHAHZAD, INSIDE AL-QAEDA AND THE TALIBAN: BEYOND BIN LADEN AND 9/11, at 221–22 (2011). *See also* ASSAF MOGHADAM, THE GLOBALIZATION OF MARTYRDOM, 140 (2008) (where reference is made to U.S. officials suggesting Al Qaeda "had coordinated closely—indeed, established a virtual merger—with the Taliban and established a haven along the Afghani-Pakistani border . . . .").

[307] Bill Roggio & Thomas Joscelyn, *Al Qaeda operates in southern Helmand province*, THE LONG WAR JOURNAL (Oct 24, 2015), http://www.longwarjournal.org/archives/2015/10/al-qaeda-operates-in-southern-helmand-province.php.

[308] SHAHZAD, *supra* note 306, at 221–22. *See also* GIUSTOZZI, *supra* note 41, at 13,138–39 (for reference to the Neo-Taliban fighting in Afghanistan being more integrated into the international jihadist movement).

[309] SHAHZAD, *supra* note 306, at xx.

[310] TANKEL, *supra* note 293, at 1.

fight against America and its allies."[311] While the jihad has been expanded to include the West, for the LeT leadership "the road to re-establishing the Caliphate ran through South Asia" and "India would remain the primary enemy."[312]

In September 2014 there was an announcement by Al Qaeda that it was forming a new group, Al Qaeda in the Indian Subcontinent (AQIS), which encompassed the territory in Burma, Bangladesh, Assam, Gujarat, Ahmedabad, and Kashmir[313] and led by a member of the Pakistan Taliban (TPP).[314] While doubts have been expressed on the feasibility of such a group, the threat is being taken seriously by Indian intelligence services.[315] By October 2015 it was reported that "[f]oreign jihadists, including members of al Qaeda in the Indian Subcontinent (AQIS), are training at facilities in southern Helmand province in Afghanistan."[316] What this initiative does reflect is the global aspirations of the overall jihadist movement, the close relationship between its various groups, and the different paths considered for the establishment of a pan-Islamic caliphate.

The Afghan Taliban provides a clear example of a militant group that employs terror tactics including beheadings, the public display of bodies, and indiscriminate suicide attacks. However, it is a group that, in addition to its military activities, also established a "no-frills" administration based on the judiciary and eventually a form of shadow government seeking to establish popular support for the movement in the areas over which it gains a degree of de facto control.[317] The question remains as to whether the Taliban should be considered more as a terrorist group, as opposed to an insurgent one. It is reported that Canada became the first NATO country to designate the Afghan Taliban as a terrorist organization in May 2013.[318] The justification for that designation shows a bifurcated reasoning. The Taliban's main objectives were "the removal of all foreign forces from Afghanistan, and the restoration of the Islamic Emirate of Afghanistan through the overthrow of the current government."[319] However, it is also a group that uses terrorist tactics, such as attacking girl's schools and relying on improvised explosive devices (IEDs) that cause a large number of civilian casualties.[320]

The status of the Taliban is more complicated internationally than this designation suggests. The United States does not list the Afghan Taliban as a Foreign Terrorist Organization, although the TTP was listed as such a group in September 2010.[321] Related

---

311 *Id.* at 5.

312 *Id.* at 41.

313 Rupam Jain Nair, *Indian States on Alert after Al Qaeda Announces Local Wing*, REUTERS (Sept. 4, 2014), http://www.reuters.com/article/2014/09/04/us-india-qaeda-idUSKBN0GY2EM20140904.

314 Ishaan Thahoor, *Why Al-Qaeda Is Opening a New Wing in South Asia*, WASH. POST, Sept. 4, 2014, http://www.washingtonpost.com/blogs/worldviews/wp/2014/09/03/why-al-qaeda-is-opening-a-new-wing-in-south-asia/.

315 Jayaraj Sivan, *Al-Qaida Won't Find India Easy: Intelligence Agencies*, THE TIMES OF INDIA, Sept. 6, 2014, http://timesofindia.indiatimes.com/india/Al-Qaida-wont-find-India-easy-Intelligence-agencies/articleshow/41819288.cms?intenttarget=no.

316 Roggio & Joscelyn, *supra* note 307.

317 GIUSTOZZI, *supra* note 41, at 107–18, and ATWAN, *supra* note 76, at 147.

318 Stewart Bell, *Canadian Government Adds Taliban to Its List of Terrorist Organizations*, NATIONAL POST, May 15, 2013, http://news.nationalpost.com/2013/05/15/canadian-government-adds-taliban-to-its-list-of-terrorist-organizations/.

319 *Currently Listed Entities*, Public Safety Canada, *available at* http://www.publicsafety.gc.ca/cnt/ntnl-scrt/cntr-trrrsm/lstd-ntts/crrnt-lstd-ntts-eng.aspx#2044 (last visited June 10, 2015).

320 *Id.*

321 *Foreign Terrorist Organizations*, U.S. Department of State, http://www.state.gov/j/ct/rls/other/des/123085.htm (last visited June 10, 2015).

to efforts to seek a negotiated settlement of the conflict, there have also been suggestions that at least some members of the Afghan Taliban might be delisted from the UN list at the request of Afghanistan and the United States. Part of the rationale was reported to be that "while al Qaeda has remained focused on its worldwide jihad against the West and establishment of a religious state in the Muslim world, the Afghan Taliban has shown little interest in attacking targets outside of Afghanistan and has concentrated on fighting occupying forces within its own country."[322] In other words, their activities are primarily of an insurgent nature. However, such efforts have been resisted by Russia, China, and India.[323]

Controversy surrounding designations are not limited to the Taliban. The European Union maintains "two lists of terrorist individuals and organizations, one prepared under the authority the Sanctions Committee of the Security Council and endorsed by the EU, the UN/EU list; and the other, the EU list, prepared unanimously by the Council of the EU."[324] The European Union was reluctant to add all, or part (i.e., the military wing) of Hezbollah to its common terrorist list. It was only in July 2013 that it was agreed to add the military wing of Hezbollah to the EU list of terrorist organizations.[325] A similar distinction had been made by the European Union with respect to Hamas in 2001, however, by 2003 there had emerged a European consensus that such distinctions was not tenable "given available evidence of the interdependence between such groups."[326] The result was that "all Hamas wings, including charitable and political wings" were banned.[327] However, this did not end the issue. In December 2014 the European Union General Court upheld a Hamas challenge to the designation on procedural grounds.[328] The European Union appealed that decision in January 2015, and as a result Hamas remained "on the EU's terrorism list and its assets will remain frozen pending a judgment by the Court of Justice, the highest legal authority."[329]

Another example of the complex relationship between insurgent and terrorist designations occurred during the Vietnam conflict. While the South Vietnamese government believed the Viet Cong were traitors and criminals, civil resources could not handle the civilian criminal incarceration and trial of the large number of detainees. The United States adopted the policy that all combatants captured during military operations regardless of the type of unit to which they belonged were to be accorded prisoner of war status. It was the "[t]errorists, spies and saboteurs" (fundamentally the traditional unprivileged belligerents) who were not accorded that status (see Chapter 4, section 4.3, *Impact of the Proper*

---

322 *UN Proposes Removing Taliban from Terror List to Promote Dialogue*, Reuters (June 17, 2011), http://www.dw.de/un-proposes-removing-taliban-from-terror-list-to-promote-dialogue/a-6550683.

323 *Id.*

324 *European Union: Lists of Terrorist Persons or Organizations*, Library of Congress, June 2007, http://www.loc.gov/law/help/eu_lists_of_terrorist.php.

325 *EU Ministers Agree to Blacklist Hezbollah's Armed Wing*, BBC News (July 22, 2013), http://www.bbc.co.uk/news/world-middle-east-23397003.

326 Levitt, *supra* note 292, at 50.

327 *Id.*

328 *EU Court Takes Hamas Off Terrorist Organisations List*, BBC News (Dec. 17, 2014), http://www.bbc.com/news/world-middle-east-30511569.

329 Adrian Croft, *EU to Appeal Ruling that Hamas Should Be off Terror List*, Reuters (Jan. 19, 2015, 9:29 AM EST), http://www.reuters.com/article/2015/01/19/us-eu-hamas-courts-idUSKBN0KS0N820150119.

Authority Principle).[330] The result was that only "[s]uspected Viet Cong captured under circumstances not warranting their treatment as prisoners of war were handled as civilian defendants."[331] The Viet Cong were not treated exclusively as a terrorist organization.

While there are clear examples of groups uniquely committed to terrorism it also appears groups may be both terrorist and insurgent organizations. Further, ultimately the designation as a "terrorist" group may depend more upon perception, politics, and practical reality.[332] While the old adage "one person's terrorist is another's freedom fighter" may seem somewhat tired in the twenty-first century, a designation as a terrorist group must be viewed in the light that "States have not been slow to brand violent opponents with this title, with its clear implications of inhumanity, criminality, and—perhaps most crucially—lack of political support."[333]

The changing security environment of the twenty-first century, with insurgency becoming more global and a battle for ungoverned spaces, as well as its growing synergy with criminal activity, is providing further challenges in the effort to categorize State opponents. In respect of transnational criminal organizations, the issue of whether they are terrorist organizations appears somewhat more straightforward. The very reason they are not considered participants in an armed conflict, their lack of political purpose, also appears to stand in the way of the consideration of such organizations as "terrorists." The lack of a political purpose would disqualify them from having that status. It has been noted that "[t]he DTOs, however, appear to lack a discernible political goal or ideology, which is one element of a widely recognized definition of terrorism."[334] However, this conclusion must be considered in light of two other factors. First, the "political purpose" basis for not considering transnational criminal organizations to be terrorist groups can provide a subtle and somewhat formalistic basis for viewing them as "merely" criminals.[335] Second, there are groups such as the FARC, ELN, and the AUC that have been considered to be terrorist organizations, but which have also engaged heavily in narcotic and other criminal activity.[336]

Separating the successor BACRIMs from the activities and methods of the AUC appears to provide a particular challenge. Complicating the assessment is the reality that groups like FARC are also primarily insurgent organizations.[337] Further impacting on the analysis is the subtle difference between traditional insurgency and transnational criminal

---

330 GEORGE S. PRUGH, LAW AT WAR 1964–1973, at 66 (1975).

331 *Id.*

332 ELLIS, *supra* note 5, at 260 (Some groups like the Palestine Liberation Organization were from their origins easily assessed to be a terrorist organization.).

333 TOWNSHEND, *supra* note 283, at 3.

334 Beittel, *supra* note 94, at 40.

335 JOHN ROLLINS & LIANA SUN WYLER, TERRORISM AND TRANSNATIONAL CRIME: FOREIGN POLICY ISSUES FOR CONGRESS, Congressional Research Service 8 (June 11, 2013), https://www.fas.org/sgp/crs/terror/R41004.pdf ("Most observers recognize that few instances of Mexican DTO violence appear to be motivated by ideology or a desire to overthrow the Mexican government.").

336 *See, e.g., id.* at 25 (where in quoting § 305 of P.L. 107-206 it is noted, "[i]n 2002, for example, Congress first authorized DOD to support a 'unified campaign against narcotics trafficking . . . [and] activities by organizations designated as terrorist organizations such as the Revolutionary Armed Forces of Colombia (FARC), the National Liberation Army (ELN), and the United Self-Defense Forces of Colombia (AUC)'.").

337 Beittel, *supra* note 94, at 25 ("While the FARC and other insurgents turned to drug trafficking to help finance their cause, their goal was to overthrow the sovereign state.").

organizations. As the use of the term "insurgency" to describe these criminal gangs suggests, it is challenging to separate the BACRIM from their insurgent counterparts. Even with the FARC, the primary, if not the exclusive, legal regime applied to counter this insurgent terrorist group is human rights.[338] Given the focus on the criminal nature of these various groups, the growing threat of criminal gangs means that the application of law enforcement norms will gain even greater prominence in the years to come.

6.4.3. TERRORISM AND INSURGENCY: AL QAEDA AND AN ISLAMIC STATE

Al Qaeda and the other jihadist groups have provided a particularly interesting categorization challenge. Given 9/11 and numerous terrorist attacks perpetrated by or in the name of Al Qaeda, it is not difficult to reach the general conclusion that Al Qaeda is a terrorist organization. However, in terms of organizational structure, that group, considered holistically, does not resemble the stereotype of a traditional small cellular terrorist organization. It has proven very challenging to reach a consensus about the structure of Al Qaeda. In part, this is because, in much the same way that reference to the United States often means the "West," "Al Qaeda" has to a certain extent become a figurehead or shorthand for the larger jihadist movement.[339] In reality that organization is only one of a number of Salafi jihadist–based groups[340] that have the goal of restoring the caliphate as a form of global Islamic state.[341] This was most graphically demonstrated by the "Islamic State" announcement in the summer of 2014 of the establishment of a caliphate that occupies land in Syria, Iraq, and Libya.[342]

Al Qaeda's organization is somewhat unique in that it has changed over time from a group dependent upon centralized leadership to an ideological (i.e., Islamist) and political structure that is horizontal in nature.[343] A common way of categorizing Al Qaeda has been to divide it conceptually into three components: a core group operating out of the FATA area, associated groups that are regionally based, and networked local groups that are mainly ideologically linked.[344] While the core element and the networked local groups, or adherents (i.e., some "homegrown" cells), individually reflect the attributes of a small and committed terrorist organization, many of the associated groups often appear to more closely resemble insurgent organizations than purely terrorist groups.[345] Groups such as: Al

---

[338] Gomez, *supra* note 169, at 283 (where it is noted after discussing the use of both human rights and humanitarian law to counter illegal and criminal groups, "in many instances the more appropriate response is law enforcement actions designed to arrest and prosecute individual members.").

[339] Horn, *supra* note 201, at 117 ("In fact, AQ [Al Qaeda] is now the face of over a dozen jihadist organizations that act alike.").

[340] Mark E. Stout et al., Terrorist Perspectives Project: Strategic and Operational Views of Al Qaida and Associated Movements 5–7 (2008).

[341] *Id.* at 19.

[342] Stern & Berger, *supra* note 3, at 278–79.

[343] Atwan, *supra* note 76, at 14.

[344] National Strategy for Counterterrorism, The White House 2 (June 2011), http://www.whitehouse.gov/sites/default/files/counterterrorism_strategy.pdf (which refers to "al-Qa'ida and its affiliates and adherents"), Cronin, *supra* note 206, at 170, Jason Burke, Al Qaeda: The True Story of Radical Islam 8 (2004), and Jones, *supra* note 72, at 228–29.

[345] Stern & Berger, *supra* note 3, at 59 ("The affiliate structure immediately began to shift al Qaeda's focus away from global terrorism towards local insurgencies").

Qaeda in the Arabian Peninsula (AQAP) in Yemen, Al Shabaab in Somalia, and AQIM in the African Maghreb and Sahel have all engaged in insurgent activities in their own right or as part of a broader insurgent challenge to State authority. The AQIM is itself an umbrella group formed in 2007 "dominated by members of the Algerian Salafist Group for Preaching and Combat (GSPC) and including elements from the Libyan Islamic Fighting Group (LIFG) as well as the Moroccan Islamic Combatant Group (MICG) and Tunisian *jihadis* . . . ."[346]

Al Qaeda has also been actively engaged in fighting in Libya and Syria. This has included fighters from the Pakistani Taliban who have deployed from Afghanistan and Pakistan to the Syrian conflict.[347] Al Qaeda–associated jihadists were participants in the fighting that saw the end of the Gadhafi regime and continue to operate in Libya postconflict.[348] As will be discussed in respect of AQI (also known as the Al Qaeda in the Land of Two Rivers, or Al Qaeda in Mesopotamia[349]), its affiliation with Al Qaeda did not necessarily mean subservience or a permanent relationship. The ultimate transformation of AQI into a separate organization, the "Islamic State," highlights the variations in personality-driven allegiances, goals, and tactics that can exist within the jihadist community.[350]

The organization of Al Qaeda and its relationship to other jihadist groups is far more complex than a three-level structure suggests. There are broader affiliations between the traditional Al Qaeda organization and other groups that have also resulted in reference to an Al Qaeda "network." In 2006 Bruce Hoffman identified four dimensions to Al Qaeda: Al Qaeda Central, Al Qaeda Affiliates and Associates, Al Qaeda Locals, and the Al Qaeda Network. [351] The network consisted of homegrown Islamic radicals who have no direct connection with Al Qaeda or other identifiable terrorist group, but "who none the less are prepared to carry out attacks in solidarity with or support of al Qaeda's radical jihadist agenda."[352] Elsewhere the entire structure beyond Al Qaeda "central" has been viewed as a complex network of franchises, allies, affiliated groups, sleeper cells of homegrown terrorists and so-called lone-wolf attackers.[353]although the latter participants have also been viewed as not being part of the larger Al Qaeda organization.[354] As can be seen with "allies,"

---

[346] ATWAN, *supra* note 76, at 170.

[347] Thomas Joscelyn, *Pakistani Taliban Leader Discusses "Global Jihad," Syria in Al Qaeda Video*, THE LONG WAR JOURNAL (July 24, 2013), http://www.longwarjournal.org/archives/2013/07/_umar_is_the_author.php ("In addition to the Pakistani Taliban, other Central Asian and Pakistani jihadist groups have reportedly joined the fight and swelled al Qaeda's ranks.")

[348] ATWAN, *supra* note 76, at 194–98, and Nic Robertson & Paul Cruickshank, *Al Qaeda Leader Sends Veteran Jihadists to Establish Presence in Libya*, CNN.COM (Dec. 29, 2011), http://www.cnn.com/2011/12/29/world/meast/libya-jihadists/index.html?iphoneemail.

[349] ATWAN, *supra* note 76, at 214.

[350] Ben Hubbard, *ISIS Threatens Al Qaeda as Flagship Movement of Extremists*, N.Y. TIMES, June 30, 2014, http://www.nytimes.com/2014/07/01/world/middleeast/isis-threatens-al-qaeda-as-flagship-movement-of-extremists.html?_r=0.

[351] HOFFMAN, *supra* note 203, at 285–89.

[352] *Id.* at 287.

[353] ATWAN, *supra* note 76, at 14.

[354] *Jeh Johnson Speech at the Oxford Union*, LAWFARE (Nov. 30, 2012, 12:01 PM), https://www.lawfareblog.com/jeh-johnson-speech-oxford-union (where it is indicated that self-radicalized lone wolves inspired by Al Qaeda, but who never become members are not part of the enemy force and therefore subject to a civilian law enforcement response. Unfortunately, this analysis does not address the issue of whether they are direct participants in hostilities).

such as the Neo-Taliban, the relationship with Al Qaeda can be a closely integrated one.[355] The network also includes Southeast Asia's Jamaa al-Islamiya, the Islamic Movement of Uzbekistan, the North Caucasus mujahideen, Palestinian groups, and a number of smaller groups, such as the Chinese Uyghurs.[356] Boko Haram, the Nigerian Islamist group, initially had ties with Al Qaeda and affiliates such as Al Shabaab and AQIM,[357] although in 2015 it swore allegiance to the Islamic State.[358] It has been estimated there are forty jihadi groups that have some connection to Al Qaeda.[359]

As a concept, the "network" is an even larger jihadist organization than Al Qaeda. Its genesis is linked, in part, to the association between terrorism and law enforcement, as well as the adoption of the network as theoretical model to describe nontraditional and particularly non-State belligerents in the context of emerging and future modes of warfare.[360] However, in jihadist thinking an almost defining characteristic is suggested to be "the creation and growth of a movement with a revolutionary vanguard marching at its head."[361] In this context, networks and cells "are only structural aspects of the vanguard, which itself is the most ideologically evolved element of the overall movement."[362] The expansive scope of these ideologically linked jihadist groups has led to the term Al Qaeda and Associated Movements (AQAM) being used to capture the breadth of significant change in the nature and organization of the group since 9/11.[363] Given the rise of the Islamic State and the prominence it has attained in the public discourse, the term "Jihadist Movements" might be more appropriate.[364]

The relationship between AQI and its successor organizations with Al Qaeda Central, and ultimately the withdrawal of that Iraqi-generated organization as an al Qaeda–based affiliate, demonstrates the loose nature of associations that can exist within the jihadist community. The history of the Al Qaeda presence in Iraq also reflects the robust and transitional nature of these organizations. AQI was a product of the U.S. invasion of Iraq in 2003. It grew out of a jihadist cell led by a Jordanian, Abu Musab al-Zarqawi, which perhaps most

---

355 SHAHZAD, *supra* note 306, at 221–22.

356 ATWAN, *supra* note 76, at 236-37 (outlining Uyghur involvement in the broader jihadist movment). *But see* MARTIN I. WAYNE, CHINA'S WAR ON TERRORISM: COUNTER-INSURGENCY, POLITICS AND INTERNAL SECURITY 31, 53 (2008) (where it is noted the motivation for the Uyghur insurgency is complex, involving "ethnic nationalism, the desire for greater autonomy or an independent state, and the goal of establishing an Islamic capliphate stretching across Central Asia." It has remained primarily an indigenous effort).

357 Robin Simcox, *Boko Haram and Defining the "Al-Qaeda Network"*, AL JAZEERA (June 6, 2014), http://www.aljazeera.com/indepth/opinion/2014/06/boko-haram-al-qaeda-201463115816142554.html.

358 *Islamic State "Accepts" Boko Haram's Allegiance Pledge*, BBC NEWS (Mar. 13, 2015), http://www.bbc.com/news/world-africa-31862992.

359 ATWAN, *supra* note 76, at 212.

360 STOUT ET AL., *supra* note 340, at 33 ("terrorists violate American law when they conduct their operations in the United States," and "[b]roadly speaking, criminal groups can be described as networks because their membership is well defined, whereas few if any criminal groups have pretensions to the status of a social movement."). *See also* JOHN ARQUILLA & DAVID RONFELDT, THE ADVENT OF NETWAR (1996) (where the reference to networks and contemporary armed conflict is viewed as having been first introduced).

361 STOUT ET AL., *supra* note 340, at 34.

362 *Id.*

363 ATWAN, *supra* note 76, at 15.

364 STERN & BERGER, *supra* note 3, at 271 ("In the context of jihadi Salafism, *jihad* most frequently refers to physical warfare or armed struggle.").

famously was responsible for the August 2003 bombing of UN Headquarters in Baghdad.[365] In October 2004 al-Zarqawi agreed to come under the command of Al Qaeda, and AQI was born.[366] However, a disregard for collateral civilian casualties, violence directed at Sunni Sheikhs, the fueling of a sectarian conflict by attacking Shi'ite Muslims, and attacks in neighboring countries (e.g., a 2005 triple suicide bombing in Amman, Jordan) undermined Al Qaida support.[367]Al-Zarqawi was criticized by senior Al Qaeda leaders and was ultimately demoted by Osama bin Laden in April 2006.[368] By June of that year he had been located and killed by U.S. forces.[369]

After 2006 AQI was rebranded as the Islamic State of Iraq (ISOI) "and placed under an indigenous leadership."[370] In claiming to be a "State" "al Qaida also sought to gain legitimacy."[371] That leadership found its genesis in an umbrella group of insurgent groups called the Mujahideen Shura Council.[372] From 2006 until the withdrawal of U.S. forces from Iraq in 2011 the fortunes of ISOI waxed and waned.[373] The Al Qaeda group was, in hindsight, optimistically viewed as being decimated by Coalition Special Forces activity by 2009, although this is not the only time that the demise of a jihadist group was prematurely or at least inaccurately predicted.[374] Overall, 2006 to 2013 saw "the Al Qaeda-led insurgent umbrella, the Islamic State of Iraq (ISOI), step up attacks on government and domestic security targets as well as escalating sectarian tensions in the country."[375] By 2011, ISOI was also beginning to flow across the border into Syria, which once again graphically demonstrated the transnational goals of the jihadist movement. In effect, "the infrastructure which had been established to facilitate the flow of *jihadis* into Iraq via Syria was now in service the other way."[376] ISOI appears to have created a mujahideen group called the al Nusra Front, which commenced operations against the Syrian government.[377] This expansion into Syria prompted ISOI to change its name to ISIS (Islamic State in Iraq and al Sham).[378] However, al Nusra, which is viewed as a more moderate jihadist organization, resisted attempts by ISIS to take over that organization.

---

365 Jones, *supra* note 72, at 141–44.

366 *Id.* at 153. *See also id.* at 155 (It is reported the AQI had "fifteen brigades operating under its banner, including two 'martyr' brigades dedicated to suicide operations.").

367 *Id.* at 250–51.

368 Atwan, *supra* note 76, at 214. Jones, *supra* note 72, at 248–50.

369 Jones, *supra* note 72, at 256.

370 Atwan, *supra* note 76, at 262. *But see id.* at 215 (where it is indicated "the leadership was mostly foreign and at least 595 foreign *jihadis* came to Iraq to join the fight to oust invading forces.").

371 Jones, *supra* note 72, at 256. *See also* Atwan, *supra* note 76, at 214 ("the change heralded a new military campaign and the aim of transforming Iraq into an Islamic emirate.").

372 Atwan, *supra* note 76, at 214.

373 *Id.* at 215–17.

374 Mark Urban, Task Force Black 272–73 (2010) (discussing the effect of the strategy against AQI), and Jones, *supra* note 72, at 256–60.

375 Atwan, *supra* note 76, at 18.

376 *Id.* at 51. Jones, *supra* note 72, at 147 (outlining the basis of al Zarqawi support moving through Syria.).

377 *Al Nusra Front*, Mapping Military Organizations, Stanford University, *available at* http://web.stanford.edu/group/mappingmilitants/cgi-bin/groups/view/493 (last visited June 10, 2015) ("Al-Nusra Front (also the Nusra Front or Jabhat al-Nusra) was formed in late 2011, when Al Qaeda in Iraq (AQI) emir Abu Bakr al-Baghdadi sent operative Abu Muhammad al-Julani to Syria to organize jihadist cells in the region.").

378 *ISIS*, Mapping Military Organizations, Stanford University, *available at* http://web.stanford.edu/group/mappingmilitants/cgi-bin/groups/view/1 (last visited June 10, 2015).

This dispute ultimately led to al Nusra pledging allegiance to Al Qaeda in April 2013[379] and direction from Al Qaeda to ISIS for the latter group to limit its activities to Iraq, which that group refused to do.[380] With al Nusra's pledge there were at that time two Al Qaeda–linked groups fighting as part of the Syria opposition forces.[381] However, the dispute over al Nusra ultimately led Al Qaeda to officially break ties to ISIS.[382] Prior to its split with Al Qaeda, ISIS expanded its operations in Syria and remained active in Iraq.[383] Part of its resurgence in Iraq included systemic attacks on prisons using mortar attacks and suicide bombers, "culminating in a successful assault on Abu Graib and Taji prisons in the summer of 2013 when at least 500 captives, many of them experienced fighters escaped."[384] The effectiveness of the group was further evidenced by the ability of ISIS to wrestle control of Fallujah from the Iraqi government early in 2014.[385] In a move that was clearly unanticipated by Iraq and many Western States, ISIS launched an offensive in June 2014, seizing Tikrit and Mosul and gaining control of significant portions of western and central Iraq in addition to areas it already captured.[386]

While the operational advances likely could not have been made without support from the broader Sunni community in Iraq, it is clear ISIS was seeking to place itself in a governance role.[387] ISIS, adopting the simpler and symbolically more significant name, the "Islamic State," provided evidence of this goal. It has also adopted a governance structure of "four high councils, each of which oversees a different area: shura, or policy; Shariah, or law; internal security; and military."[388] Further, "ISIS's organizational structure in Iraq and Syria was based on the *wilayat*, essentially provincial subdivisions each with its own governor."[389] The concept of *wilayat* has also been extended to areas further afield (e.g., Afghanistan, Algeria, Egypt, Libya, Nigeria, Pakistan, Saudi Arabia, and Yemen).[390] This

---

379 *Syria Crisis: Al-Nusra Pledges Allegiance to Al-Qaeda*, BBC News (Apr. 10, 2013), http://www.bbc.co.uk/news/world-middle-east-22095099.

380 Ben Hubbard, *Al Qaeda Breaks with Jihadist Group in Syria Involved in Rebel Infighting*, N.Y. Times, Feb. 3, 2014, http://www.nytimes.com/2014/02/04/world/middleeast/syria.html.

381 Liz Sly, *Al-Qaeda Expands in Syria via Islamic State*, Wash. Post, Aug. 12, 2013, http://www.washingtonpost.com/world/al-qaeda-expands-in-syria-via-islamic-state/2013/08/12/3ef71a26-036a-11e3-9259-e2aafe5a5f84_story.html.

382 Hubbard, *supra* note 380.

383 Henry Austion, *Global Jihadis or Al Qaeda Wannabes: Who Are the Islamic State of Iraq and the Levant?*, NBC News (Jan. 11, 2014), http://worldnews.nbcnews.com/_news/2014/01/11/22243203-global-jihadis-or-al-qaeda-wannabes-who-are-the-islamic-state-of-iraq-and-the-levant?lite.

384 Cockburn, *supra* note 75, at 61, and *Iraq Jailbreaks: Hundreds Escape in Taji and Abu Ghraib*, BBC News (July 22, 2013), http://www.bbc.co.uk/news/world-middle-east-23403564.

385 *Iraqi PM Says He Won't Order Assault on Fallujah*, CBS News (Jan. 12, 2014), http://www.cbsnews.com/news/iraqi-pm-says-he-wont-order-assault-on-fallujah/.

386 *Iraq Crisis: Militants "Seize Tikrit" after Taking Mosul*, BBC News (June 11, 2014), http://www.bbc.com/news/world-middle-east-27800319.

387 Cockburn, *supra* note 75, at 60–63 (explaining the alienation of the Sunni population by the Iraqi government.).

388 Jess Bravin, *Probe into War Crimes Targets Islamic State Militants*, Wall St. J. (Sept. 4, 2014), http://online.wsj.com/news/article_email/probe-into-war-crimes-targets-islamic-state-militants-1409864371-lMyQjAxMTA0MDAwNDEwNDQyWj?tesla=y. *See also* Ben Hubbard & Eric Schmitt, *Military Skill and Terrorist Technique Fuel Success of ISIS*, N.Y. Times, Aug. 27, 2014, http://www.nytimes.com/2014/08/28/world/middleeast/army-know-how-seen-as-factor-in-isis-successes.html ("Mr. Baghdadi's deputies include 12 walis, or local rulers; a three-man war cabinet; and eight others who manage portfolios like finance, prisoners and recruitment.").

389 Stern & Berger, *supra* note 3, at 184.

390 Karen Youish, Derek Watkins, & Tom Giratikanon, *Where ISIS Has Directed and Inspired Attacks Around the World*, N.Y. Times, June 17, 2015, http://www.nytimes.com/interactive/2015/06/17/world/middleeast/map-isis-attacks-around-the-world.html?src=mv, and Stern & Berger, *supra* note 3, at 185.

development "appeared to be an expansion of the proto state beyond its borders."[391] The governance function being developed within the Islamic State is also reflected in the call by the self-styled caliph, Abu Bakr al-Baghdadi, for judges, doctors, engineers, and administrative specialists to help run this new caliphate.[392] This convoluted evolution of the Islamic State from a small terrorist group operating as part of the post 2003 Iraq invasion conflict, to being of the most violent al Qaeda affiliates, and finally its emergence as a separate organization calling itself a caliphate spanning the territory of two countries, highlights the diverse and interlinked nature of the jihadist movement.

In terms of the broader jihadist movement, the exact nature of the linkage within Al Qaeda and externally with other jihadist groups is not always immediately evident. The fact that al Nusra did not change its name is said to be part an emerging trend of avoiding overt identification with Al Qaeda. This trend is seen in Al Shabaab, an AQAP offshoot, Ansar al Shari'a, and "certain groups in Gaza."[393] The lack of an overidentification with Al Qaeda is also suggested not to mean the groups are no longer part of AQAM. As is noted, one needs "look instead for a common purpose and those deep-rooted links of ideology, intelligence, ambition, allegiance, weapons supplies, finance and modus operandi that connect up an increasingly widespread collection of Salafi-*jihadi* groups."[394] In this respect ISIS has correctly been identified as an "Al-Qa'ida-type" group.[395] It is clear that the Islamic State and Al Qaeda are at odds with one another. This has extended to open fighting in Syria between al Nusra and Islamic State forces,[396] and "the two groups are now locked in a battle for supremacy and for the loyalties of unaffiliated groups and the members of existing organizations."[397]

However, the United States recognizes the common cause of these groups. As was indicated by Stephen W. Preston, the U.S. Department of Defense General Counsel, "[a] power struggle may have broken out within bin Laden's jihadist movement, but this same enemy of the United States continues to plot and carry out violent attacks against us to this day."[398] Further, as Michael Morrell, a retired Deputy Director of the Central Intelligence Agency has observed:

> [a]lthough there is a rift between the leadership of ISIS, it is important to note that ISIS is effectively al Qa'ida. ISIS shares Bin Ladin's long-term goal of establishing a global caliphate, it sees both the West and its allies in the Middle East as its primary enemies, and it sees violence as the most effective means of achieving its goals.[399]

---

391 Stern & Berger, *supra* note 3, at 186.

392 Ed Adamcyzk, *ISIS Leader to World's Muslims: "Build Islamic State,"* UPI (June 1, 2014), http://www.upi.com/Top_News/World-News/2014/07/01/ISIS-leader-to-worlds-Muslims-Build-Islamic-State/8971404235322/.

393 Atwan, *supra* note 76, at 262.

394 *Id.* at 262–63.

395 Cockburn, *supra* note 75, at 27.

396 Stern & Berger, *supra* note 3, at 43.

397 *Id.* at 178. *See also* Shiv Malik et al., *How Isis Crippled Al-Qaida*, The Guardian (June 10, 2015, 22:30 BST), http://www.theguardian.com/world/2015/jun/10/how-isis-crippled-al-qaida.

398 Stephen W. Preston, *The Legal Framework for the United States' Use of Military Force Since 9/11*, Remarks by the General Counsel of the Department of National Defense on the Legal Framework for the United States' Use of Force Since 9/11, at 6 (Apr. 10, 2015), http://www.defense.gov/Speeches/Speech.aspx?SpeechID=1931.

399 Michael Morell & Bill Harlow, The Great War of Our Time: The CIA's Fight Against Terrorism from Al Qa'ida to ISIS 307 (2015).

While giving credit to Osama bin Laden for having "ownership" of the jihadist movement seems displaced, it is clear that the various participants in that movement ultimately work toward similar ends. [400] There is also a significant danger that the focus on any one group will mask the overall danger presented by the jihadist movement. Given the falling out with Al Qaeda, it may be that the Islamic State will be viewed as operating outside even the loose affiliations of AQAM. This should not end the analysis of that group as posing a threat as part of a broader grouping of common enemies. It is, and remains, part of the broader jihadist movement.

As can been seen in historical examples from World War II, the 1980s Afghanistan conflict, and during the 2003 occupation of Iraq, an insurgency can comprise a diverse number of groups that not only are seeking different goals but also may periodically fight with one another. During World War II, "guerrilla operations world-wide varied considerably in size, composition, motivation, mission, and effectiveness."[401] In France there were five major distinct resistance groups with the Communist-supported Franc-Tireurs et Partisans "furnishing nearly a third of the maquis."[402] That group did not give up control of its units, but did agree to join other organizations in establishing a secret army.[403] It was "a cooperation prompted more by need for recognition, arms and money than by patriotism."[404] The two main resistance movements in Yugoslavia, nationalist Chetniks and Tito's Communist Partisans (which were also fighting to take over power from the exiled regime[405]), started out the resistance together, although irretrievably broke apart.[406] Eventually the Partisans fought with the Chetniks,[407] including when the latter group began collaborating with Italian occupiers.[408] In any event, "[t]he political allegiances of the resistance movement had little influence on the military operations conducted by the occupying powers,"[409] which took the view "all in arms against them were enemies . . . ."[410]

---

[400] S.C. Res. 2249, UN Doc. S/RES/2249 ¶ 5 (Nov. 20, 2015)(where even though the Resolution was prompted by attacks in Tunisia, France, Turkey, the Sinai and Beirut by the Islamic State it was also directed at the "ANF [Al Nusra Front], and all other individuals, groups, undertakings, and entities associated with Al-Qaida, and other terrorist groups, as designated by the United Nations Security Council, and as may further be agreed by the International Syria Support Group (ISSG) and endorsed by the UN Security Council . . . .").

[401] 1 Robert B. Asprey, War in the Shadows: The Guerrilla in History 409 (1975).

[402] *Id.* at 434–35.

[403] *Id.* at 435. *See also* Ian Ousby, Occupation: The Ordeal of France 1940–1944, at 240 (1997) (where it is indicated that the inaugural meeting of the Conseil National de la Résistance saw sixteen organizations represented including political parties, the communists, as well as the leading resistance movements), and John F. Sweets, The Politics of Resistance in France, 1940–1944, at 33–69 (1976) (for a description of the major resistance organizations in the south of France).

[404] *Id.* at 435.

[405] John Shy & Thomas W. Collier, *Revolutionary War*, *in* Makers of Modern Strategy from Machiavelli to the Nuclear Age 833 (Peter Paret ed., 1986).

[406] Asprey, *supra* note 394, at 472.

[407] *Id.* at 472–73, Robert M. Kennedy, Hold the Balkans: German Antiguerrilla Operations in the Balkans, 1941–1944, at 20 (2000), and Matthew Bennett, *The German Experience*, *in* The Roots of Counterinsurgency: Armies and Guerrilla Warfare 1900–1945, 60 at 74 (Ian F. W. Beckett ed., 1988) [hereinafter The Roots of Counterinsurgency].

[408] Kennedy, *supra* note 407, at 21–22.

[409] *Id.* at 20.

[410] *Id.*

A decentralized structure for groups fighting in a similar cause can be seen in more modern examples. The Haraakat-I Inquilab Islam-I was a loose affiliation of local units formed around madrasa educated clerics based in the Pashtun areas of east and south Afghanistan. This parochial organization was a major player in the resistance to the Russian presence in Afghanistan in the 1980s, although with a weak central command "Soviet counterinsurgency and rivalry amongst militants led to the fragmentation and ultimate collapse of the Harakat."[411] Similarly, the Iraq occupation and insurgency that followed the 2003 invasion, and which spawned AQI, involved a wide variety of groups fighting against the Coalition forces and ultimately a nascent Iraqi government.[412] Although the resistance was primarily Sunni-based, the Coalition forces fought with Shi'ite militias,[413] Arab nationalists, disgruntled Muslims, foreign fighters,[414] detained members of the Iranian Quds force, and Hezbollah supporting the Shi'ite militia,[415] and even exchanged cross-border fire with Iranian forces.[416]

It has been suggested groups having global reach such as Al Qaeda, the Islamic State and their affiliated groups do not have 'a "unitary" organization required by a non-State party opposing one or more States', and therefore cannot form the basis of "an armed conflict of global dimensions".[417] However, this view does not reflect either the historical precedent of diverse groups acting in common cause, or the nature of the threat the jihadist movements pose. Like States, non-State actors may enter into coalitions to fight what is perceived to be their common enemy. As a result, international lawyers need to remain sensitive to the reality that groups fighting State armed forces may not have a unitary, ordered structure where their components are linked together under a common command structure. In this regard Jason Burke notes the Al Qaeda leadership had always tried to tread a careful path between a "local strategy entirely focused on establishing individual 'fronts'" and "the global strategy focused on provoking a broad uprising of Muslims. . . ."[418] While the 9/11 strategy had moved Al Qaeda toward a global strategy, both bin Laden and Al-Zawahiri recognized the need for a base from which to operate.[419] In the context of the overall jihadist threat it may matter little if the Islamic State is part of AQAM or not, in terms of the threat those organizations collectively pose to local or regional States that are viewed as "apostate" (i.e., the

---

[411] STANILAND, *supra* note 36, at 121.

[412] Matt Sherman & Josh Paul, *The Role of Police in Counterinsurgency Operations in Iraq, 2003–2006*, *in* POLICING INSURGENCIES: COPS AS COUNTERINSURGENTS 227, 230 (C. Christine Fair & Sumit Ganguly eds., 2014). *See also* STANLEY MCCHRYSTAL, MY SHARE OF THE TASK: A MEMOIRE 175 (2013) ("In truth, more than fifty named insurgent groups fought at one time or another. . . . By the numbers, AQI never constituted the majority of the insurgency.").

[413] THOMAS E. RICKS, FIASCO: THE AMERICAN MILITARY ADVENTURE IN IRAQ 338 (2007). *See also* HASHIM, *supra* note 20, at 256–64, and PATRICK COCKBURN, MUQTADA AL-SADR AND THE SHIA INSURGENCY IN IRAQ 172–86 (2008).

[414] HASHIM, *supra* note 20, at 138–39.

[415] DAVID CRIST, THE TWILIGHT WAR: THE SECRET HISTORY OF AMERICA'S THIRTY-YEAR CONFLICT WITH IRAN 511–37 (2012), and LEVITT, *supra* note 33, at 285–310.

[416] CRIST, *supra* note 415, at 525–26.

[417] INTERNATIONAL HUMANITARIAN LAW AND THE CHALLENGES OF CONTEMPORARY ARMED CONFLICT, Report 32nd International Conference of the Red Cross and Red Crescent EN 32IC/15/11, at 18 (2015), *available at* https://www.justsecurity.org/wp-content/uploads/2015/11/2015-ICRC-Report-IHL-and-Challenges-of-Armed-Conflicts.pdf.

[418] JASON BURKE, THE 9/11WARS 388 (2011).

[419] *Id.*

"near" enemy[420]), or in the longer term to the United States and its Western allies (i.e., the "far" enemy[421]).

In comparison to traditional terrorist groups, "AQAM is enormous, diverse, and widely distributed, and therefore, practically speaking, impossible to map fully."[422] As such, it and the broader jihadist movement do not represent a "traditional" terrorist organization. AQAM has been called a terrorist organization with clout, being endowed with many of the resources of a State actor, including armies, weapons, the apparatus of government, and its own judicial system, funding, and fiscal policies.[423] While it has yet to control an entire country, it "has emirates in many locations . . . including the two-thirds of Afghanistan under Taliban control, much of southern and central Yemen and most of the Sahel."[424] To this list of AQAM-controlled territory can be added the territory administered by the Islamic State and the previous claims by Boko Haram that it had established an Islamic State in the territory it controls.[425] These groups, in one form or another, all share the radical jihadist agenda.

In assessing AQAM and other jihadist groups, such as the Islamic State, the concept of what constitutes a traditional terrorist group itself must be critically analyzed. Even during the 1970s and 1980s, when much of the world saw international terrorism as an almost exclusively law enforcement matter, the Palestine Liberation Organization (PLO) presented a complex organization with groups emerging and disappearing. The "Palestinians regularly fought one another, and Arab states exploited these rivalries for their own purposes."[426] The PLO also developed a large military wing, which at the time of the 1982 invasion of Lebanon "had fifteen thousand full-time military personnel . . . . They also had a token tank force and several hundred pieces of artillery."[427] This combination of more hierarchically organized military forces and smaller cells or structures focused on terror appears to be a not uncommon occurrence. Therefore, in terms of structure, groups such as Al Qaeda may not be entirely unique, although the geographic scope of the movement is unprecedented. Further, "[j]ihadist groups have a long history of splintering and separation."[428]

The doctrinal underpinnings of the jihadist movement are consistent with its participants being considered part of a global insurgency. For example, a jihadist theorist linked to that group, Abu Mus'ab al Suri, contemplated a broad range of action in his 2005 work, *The Military Theory of the Global Islamic Resistance Call*, involving three schools: "secret military organizations (regional—secret—hierarchical)," "open fronts and overt confrontations," and

---

420 Fawaz Gerges, The Far Enemy: Why Jihad Went Global 1 (2005) (where reference is made to "jihadis who have used violence against . . . their own governments (the near enemy) . . . ."). *See also* Stout et al., *supra* note 340, at 234.

421 Gerges, *supra* note 420, at 1 ("*The Far Enemy*, or al-Adou al-Baced, is a term used by jihadis to refer to the United States and its Western Allies."), and Stout et al., *supra* note 336, at 234.

422 Stout et al., *supra* note 340, at 32.

423 Atwan, *supra* note 76, at 267.

424 *Id.*

425 *Nigeria's Boko Haram "Seize" Bama Town in Borno*, BBC News Africa (Sept. 2, 2014), http://www.bbc.com/news/world-africa-29021037.

426 Byman, *supra* note 84, at 65.

427 *Id.* at 67.

428 Stern & Berger, *supra* note 3, at 177.

"individual jihad and small cell terrorism."[429] In his view it was the success of the international security efforts post-1990 and 9/11 (including America's stunning technical superiority, strategic air strikes, and "complete control over space and the electronic world"[430]) that led to the defeat of the first two organizational structures. This forced the jihadist movement to adopt individual and small cell jihad as a primary means to conduct operations.[431] Al Suri's doctrine envisaged a wide range of military options with the one associated with terrorism being forced upon the jihadist movement. It also included an apocalyptic discourse.[432] His philosophy "reverberates throughout the training centers run by the Pakistani Taliban."[433]

Nor was Al Suri's the only jihadist doctrine that reflected an insurgent theme. 'Abd Al-'Aziz Al-Muqrin, a longtime jihadi militant and a founding member of AQAP,[434] produced a doctrine manual in 2003, *A Practical Course for Guerrilla Warfare*,[435] which provided an outline for the conduct of an unconventional guerrilla war. That war involved a final phase where the guerrilla army "will be capable (God willing) of confronting conventional forces using the same tactics as the latter."[436] Other Al Qaeda groups similarly applied this revolutionary warfare model.[437] AQAP was also not the only Al Qaeda–linked group espousing traditional goals of an insurgency. In contrast to Al-Suri's distributed form of warfare, Abu Musab al-Zarqawi, the leader of Al Qaeda in Iraq (AQI), favored a more traditional approach of guerrilla operations and "saw the establishment of physical enclaves as the key goal of any militant movement."[438] The actions of al-Zarqawi were not necessarily divorced from Al-Suri's vision of jihad. Zarqawi is viewed as having "set about fulfilling al Suri's [apocalyptic] prophesies", and his successor Al Baghdadi "is taking the fulfillment of apocalyptic portents even more seriously than his predecessor."[439]

In Iraq, the establishment of AQI enclaves manifested itself early on, perhaps most graphically in the 2004 battles for Fallujah. When the U.S. forces captured the city in November of that year the insurgent militants numbered an estimated one thousand "hard core" and two thousand "part-timers."[440] They were defeated in an urban environment as a result of combat involving the wide-scale use of conventional munitions[441] and involving combat causing of significant casualties.[442] This desire by Iraqi jihadists to control territory was subsequently reflected in the actions of ISOI, ISIS, and their successor organization, the Islamic State. It has been suggested that the Islamic State has progressed beyond an

---

[429] LIA, *supra* note 68, at 352–68 (these "schools" of jihadi organizations are set out in excerpts from Abu Mus'ab Al-Suri, *The Military Theory of the Global Islamic Resistance Call* reproduced in Chapter 10).

[430] *Id.* at 362.

[431] *Id.* at 363–68.

[432] STERN & BERGER, *supra* note 3, at 223.

[433] ABBAS, *supra* note 301, at 221.

[434] NORMAN CIGAR, AL-QA'DA'S DOCTRINE FOR INSURGENCY 6 (2009).

[435] *Id.* at 82–175.

[436] *Id.* at 94.

[437] MICHAEL S. RYAN, DECODING AL-QAEDA'S STRATEGY: THE DEEP BATTLE AGAINST AMERICA 230 (2013).

[438] BURKE, *supra* note 417, at 167.

[439] STERN & BERGER, *supra* note 3, at 224.

[440] BING WEST, NO TRUE GLORY: A FRONTLINE ACCOUNT OF THE BATTLE OF FALLUJAH 256 (2005).

[441] *Id.* at 315–16 (During the November 2004 assault there were 500 air strikes and 14,000 artillery shells, and 2,500 tank main gun rounds expended. Of Fallujah's 39,000 buildings, some 18,000 were damaged or destroyed.).

[442] *Id.* at 316 (where it is estimated the twenty-month struggle for Fallujah resulted in 151 Americans being killed and 1,000 wounded.).

insurgency to "a conventional civil war between a breakaway territory and a weak central state."[443] The writings and actions of these Islamic jihadists do not always reflect the traditional view of terrorists, at least in this context. That said, actions like beheadings and the use of indiscriminate violence against other Sunnis clearly identify al- Zarqawi and those serving him as terrorists.[444] The involvement of the Islamic State in executions, rape, and child abduction have demonstrated a similar propensity to engage in terrorism.[445]

Given the similarities and overlap between terrorism and insurgency in respect of Al Qaeda, the Islamic State and other groups the question remains as to how those seeking to address these threats can tell the difference. After discussing the U.S. State Department approach toward classification, Bruce Hoffman indicates that the definitional rule of thumb in the aftermath of the 2003 Iraq invasion was that secular Ba'athist Party loyalists and other groups associated with the regime that conducted guerrilla-like attacks or carried out IED attacks were insurgents. The foreign jihadists and domestic Islamic extremists, such as AQI, who were largely responsible for suicide attacks and beheading hostages, were labeled terrorists.[446] Here the distinction appears to have been made on the basis of jihadist motivation combined with the type of violence that was used. As is often the case, the motivation and nature of the opposing forces presented a complex and interrelated situation. [447] Ahmed Hashim has observed that the "characterization of the insurgents as regime dead-enders was simply not accurate."[448] The insurgency was primarily, if not exclusively, Sunni based, which was the group that had held power under Saddam Hussein. The origins and motives of the insurgency in Iraq were diverse, ranging from a desire to protect Sunni Arab identity to a traditional tribal reluctance to submit to strong central authority.[449] While there was Ba'athist motivation, the fight against Coalition occupation was also based on alienation of former military personnel; fear of alienation; nationalism, honor, revenge, and pride; as well as religion.[450] There was a deepening religious radicalization of the Sunni Arabs "as many in the community sought to find answers to their marginalization."[451] This included turning toward Salafi Islam.[452]

The wide variety of groups engaged in fighting the Coalition and Iraqi government forces had organizational structures that were horizontal, centralized, or a combination of both to varying degrees.[453] To the extent small groups with a cellular structure are a hallmark

---

[443] Cronin, *supra* note 4, at 95.

[444] Boot, *supra* note 6, at 531.

[445] Dominic Evans & Maggie Fick, *U.N. Accuses Islamic State of Executions, Rape, Forced Child Recruitment in Iraq*, Reuters (July 18, 2014), http://www.reuters.com/article/2014/07/18/us-iraq-security-idUSKBN0FN27N20140718.

[446] Hoffman, *supra* note 203, at 36.

[447] *See e.g.*, Wayne, *supra* note 356, at 53 (where it is noted in respect of the Uyghar insurgency in China "[e]ven as links to the global jihad, material and ideational, have been formed by Xinjiang's insurgency, this insurgency remained at its core an indigenous phenomenon.").

[448] Hashim, *supra* note 20, at 60.

[449] *Id.* at 60–120. *See, e.g.*, Chehab, *supra* note 19, at 18 (where it is noted a fight between American and Iraqi insurgents was "more of a tribal war than a resistance force," caused by what was perceived to be poor treatment of the tribes.).

[450] Hashim, *supra* note 20, at 60–120.

[451] *Id.* at 116.

[452] *Id.* at 108–20.

[453] *Id.* at 152–55. *See also id.* at 155 (where it is noted the Fedayeen Saddam exhibited both horizontal and hierarchical characteristics).

of terrorist organizations, those attributes were also evident in the Fayadeen Saddam[454] and the specialized "technical" IED and suicide bomber cells linked generally to the insurgency.[455] A prime goal of the insurgency was to "eject the foreign presence in Iraq," however, there was also a desire by some Sunni Islamist insurgents to establish an Islamic government.[456] In this respect, although the Islamists were associated with "terrorism," they had a goal that is a key indicator of an insurgency.

Adding to the complexity of a jihadist-based insurgency is the reality that it can consist of a mix of more permanent and local guerrillas. David Kilcullen has highlighted the hybrid nature of the Al Qaeda insurgent organization in what he describes as the "accidental guerrilla" syndrome. It is a situation where local forces fight alongside the extremist forces. However, those local forces do so not because they support the extremist ideology, but "because they oppose outside interference in their affairs, because they are rallied to support local tribal or community interests, or because they are alienated by heavy-handed actions of the intervening force."[457] Needless to say, the forces engaged in insurgencies can be extremely diverse in structure and motivation.

It is evident from the Iraq war and other examples that the relationship between insurgency as a form of warfare and terrorism is an exceedingly complex one. It has been estimated that "nearly a third of the thirty-seven groups on the [2003] U.S. State Department's 'Designated Foreign Terrorist Organizations' list could just as easily be categorized as guerillas."[458] As a result, "[t]errorism is often confused or equated with, or treated as synonymous with, guerrilla warfare and insurgency."[459] To the extent groups such as "Al Qaeda Central" remain small in size, follow a *focoist* or vanguard approach, and appear to claim no specific ambitions to take on a governance role, they fit into the terrorist model. Such groups may be particularly susceptible to military action seeking to kill or detain their members. However, it is also evident that not all associated Al Qaeda or other jihadist groups necessarily fit that model. These latter groups also demonstrate attributes of an insurgency, in terms of their size, organization, and political goals. At the same time they can commit horrendous terrorist acts. The issue is how State security forces counter such threats, particularly regarding the use of force.

6.5. Counterinsurgency and Counterterrorism: An Integrated Response

Given the changing nature of twenty-first-century insurgencies and terrorism, updated doctrinal approaches have been developed in order to counter the unique, but interrelated, threats they pose. The doctrines are respectively called "counterinsurgency" and "counterterrorism." Counterinsurgency has a relatively long lineage with the genesis of its modern

---

454 *Id.* at 155 (where it is indicated the Fayadeen groups had specialized cells for combat, procurement, construction of homemade bombs, and operational security, which were "kept separate and know a little as possible about each other. . . .").

455 *Id.* at 160–62.

456 HASHIM, *supra* note 20, at 121–24.

457 KILCULLEN, *supra* note 243, at 38.

458 HOFFMAN, *supra* note 203, at 35–36.

459 *Id.* at 35.

form stretching back to the late nineteenth century.[460] It has been described as "the range of military, political and socio-economic measures adopted by security forces in response to the outbreak of revolutionary guerrilla warfare or insurgency...."[461] At its heart counterinsurgency is population-centric to the point of "putting the well-being of noncombatant civilians ahead of other consideration, even—in fact especially—ahead of killing the enemy."[462] It is not that there will be no killing, as "counterinsurgency is war."[463] However, security forces "kill only those active, irreconcilable, combatants who must be killed or captured, and where possible they avoid making more insurgents in the process."[464]

The learning of the lessons from counterinsurgency experiences in the post–World War II period and their incorporation into doctrine applied in twenty-first-century operations through such means as the development of the *Counterinsurgency Manual*, has had a significant impact on contemporary operations in Iraq and Afghanistan.[465] The *Counterinsurgency Manual* has been identified, along with the Sunni tribes turning against AQI, with gaining the relative military success attained in Iraq in 2009.[466] It was an approach that required a significant deployment of military forces; the need for the military to take on a full range of tasks, including nonmilitary ones; an acceptance of the complexity of counterinsurgency; and a focus on strategic objectives including the requirement that less force be used.[467] It is the emphasis on minimum force that results in a privileging of the application of law enforcement–type norms over those traditionally associated with military operations. For the U.S. military in particular, which has historically suppressed the study of counterinsurgency operations in favor of conventional inter-State warfare, this proved to be a difficult but necessary process.[468]

What has also been developed in conjunction with a revised counterinsurgency doctrine is the concept of counterterrorism. Given the historical connection between terrorism and law enforcement, a counterterrorism strategy would be expected to include policing as part of proposed State action. For example, the 2011 U.S. *National Strategy for Counterterrorism* contemplates a wider range of initiatives than military strikes to counter terrorism, including focused foreign and development assistance, blocking the flow of financial resources, use of the domestic legal system, increasing Homeland defenses, denial of safe havens, and building the security capacity of other countries.[469]

The 2014 U.S. *Counterterrorism* manual acknowledges that "[e]ffective CT [counterterrorism] requires the sustained global CT effort of all relevant USG departments

---

460 Ian F. W. Beckett, *Introduction*, *in* The Roots of Counterinsurgency, *supra* note 407, at 6–15.

461 *Id.* at 6.

462 Kilcullen, *supra* note 4, at 4.

463 *Id.*

464 *Id.* at 5.

465 The Counterinsurgency Manual, *supra* note 60.

466 Manwaring, *supra* note 96, at ix, and Boot, *supra* note 6, at 541–45 (outlining the implementation of the counterinsurgency doctrine.).

467 David H. Ucko, The New Counterinsurgency Era 110–12 (2009).

468 Cassidy, *supra* note 274, at 99–126, and Boot, *supra* note 6, at 399 (who notes while not unique to the U.S. armed forces, "[t]he problem was more acute"). *See also* John A. Nagl, Learning to Eat Soup with a Knife: Counterinsurgency Lessons from Malaya and Vietnam (2005) (for one of the early groundbreaking assessments of the need to change military culture and training to address the counterinsurgency challenge).

469 National Strategy for Counterterrorism , *supra* note 344, at 8-9.

and agencies and PNs, each with unique capabilities, perspectives, and authorities."[470] However, emphasis is also placed on military action. Counterterrorism has been defined as "[a]ctions taken directly against terrorist networks and indirectly to influence and render global and regional environments inhospitable to terrorist networks . . . ."[471] Further, "[t]he ways of CT operations are to capture, kill, or otherwise neutralize terrorist leadership and key subordinates, isolate terrorists from their supporting administrative and logistic infrastructure, and dismantle their capabilities and bases."[472] Such actions, as has been demonstrated with the use of drones, have been considered to be "an almost exclusively strike-oriented endeavor."[473] Further, "counterinsurgency tended to be overshadowed by counterterrorism."[474] Counterterrorism focuses on the enemy organization, both in respect of individual terrorists, and its networked activity. This "enemy-centric" approach "seeks to destroy this network, proceeding from the assumption that removing the network removes the problem."[475] It has also been described as a top-down rather than a bottom-up approach.[476] As such, it fits well within the preferred course of action for traditionally focused military forces, which tend not to embrace counterinsurgency or a law enforcement approach.

While conceptually separate doctrines, the interrelated nature of contemporary threats means counterinsurgency and counterterrorism action will often be taken concurrently. It has been noted that these "two techniques merge and blur at the tactical level, even though the approaches are still seen as competing paradigms by the public, news organizations, and many policy-makers."[477] In a sense this phenomenon reflects the tendancy in legal theory to separate the bodies of law applicable to armed conflict while at a practical level there is a requirement to apply them more holistically. The relationship between terrorist groups and insurgents makes the combined threat they pose both more significant and challenging. It has been observed that the "linkage between [terrorist] groups does represent a danger, but terrorism linked to the wings of a formalized insurgent movement constitutes a greater peril."[478] Certainly, the State response to insurgent and terrorist groups must be focused on the full spectrum of threats they pose. In some circumstances this means primarily a counterterrorism approach. In other situations a counterinsurgency approach appears to offer the best means to respond to the threat.

Given the diversity of groups and the relationships between them, the response by a State will likely involve a mixture of the two types of operations. For example, counterinsurgency operations, such as those conducted in Afghanistan and Iraq, have applied elements of

---

470 COUNTERTERRORISM, JP 3-26, ¶ 1 a., at V-1 (Oct. 24, 2014).

471 DEPARTMENT OF DEFENSE DICTIONARY OF MILITARY AND ASSOCIATED TERMS, Nov. 8, 2012 (as amended through Nov. 15, 2014), http://www.dtic.mil/doctrine/dod_dictionary/index.html.

472 COUNTERTERRORISM, *supra* note 470, ¶ 1 c., at V-1.

473 UCKO, *supra* note 467, at 88.

474 *Id.*

475 KILCULLEN, *supra* note 243, at xv. *See also* COUNTERTERRORISM, *supra* note 470, at vii ("CT activities and operations are taken to neutralize terrorists, their organizations, and networks in order to render them incapable of using violence to instill fear and coerce governments or societies to achieve their goals.").

476 WAYNE, *supra* note 356, at 10.

477 KILCULLEN, *supra* note 4, at 187 n *.

478 DAVIDSON SMITH, *supra* note 235, at 17–18.

counterterrorism. In 2009, when the United States was considering whether to surge troops for a broadening of a counterinsurgency campaign, a suggested alternative approach was a "counterterrorism plus" mission that involved only 20,000 instead of 40,000 personnel.[479] Instead, "Obama had ended up splitting the difference and authorizing thirty thousand soldiers for the surge . . . ."[480] In Iraq, following the 2003 invasion, the conflict had elements of both an insurgent and a terrorist threat. As a result, "[t]he American military was fighting two wars really, in Iraq: one to tamp down a roiling Sunni-Shia rivalry that risked civil war, the other a terrorist conflict inspired by Al Qaeda."[481]

At the beginning of 2013 it was reported that options for a continued presence of American troops in Afghanistan following a projected 2014 pullout would see forces ranging from 6,000 to 20,000. A smaller force was viewed as being engaged in counterterrorism and would primarily involve "Special Operations commandos who would hunt down insurgents. There would be limited logistical support and training for Afghan security forces."[482] A larger force would have included conventional units engaged in limited patrolling potentially engaged in supporting a counterinsurgency effort.[483] In this context, counterterrorism is directly linked to a strike-oriented doctrine. The switch to such a doctrine may be reflected in the 2013 UN Assistance Mission in Afghanistan (UNAMA) report that the number of drone strikes by Coalition forces rose sharply in that "506 weapons were released by drones in 2012, compared with 294 the previous year."[484] In March 2015 it was announced that there would be a delay in the anticipated withdrawal of U.S. forces until the end of the year "in order to bolster American counterterrorism efforts in that country, including the Central Intelligence Agency's ability to conduct secret drone strikes and other paramilitary operations from United States military bases."[485] The U.S. forces remained focused on participating in a training mission and "continue[] to engage in counterterrorism activity in Afghanistan to target the remnants of al-Qa'ida and prevent an al-Qa'ida resurgence or external plotting against the homeland or U.S. targets abroad."[486] By October 2015 the United States indicated it would be maintaining a training and counterterrorism presence (i.e. drones and Special Forces personnel) in Afghanistan.[487]

---

479 Peter L. Bergen, Man Hunt: The Ten Year Search for Bin Laden from 9/11 to Abbottabad 173 (2012).

480 *Id.*

481 Eric Schmitt & Thom Shanker, Counterstrike: The Untold Story of America's Secret Campaign Against Al Qaeda 162 (2011).

482 Elizabeth Bumiller & Eric Schmitt, *Afghan War Commander Gives Options for after '14*, N.Y. Times, Jan. 2, 2013, http://www.nytimes.com/2013/01/03/world/middleeast/afghan-war-commander-gives-options-for-after-2014.html?ref=global-home.

483 *Id.*

484 *US Drone Strikes in Afghanistan Rose Sharply Last Year, UN Reports*, The Guardian, Feb. 19, 2013, http://www.guardian.co.uk/world/2013/feb/19/us-drone-strikes-afghanistan-un-report ("The US air force central command also recorded an increase, giving the numbers of weapons released by drones as 243 in 2009, 277 in 2010, 294 in 2011 and 494 in 2012.").

485 Michael D. Shear & Mark Mazetti, *U.S. to Delay Pullout of Troops from Afghanistan to Aid Strikes*, N.Y. Times, Mar. 24, 2015, http://www.nytimes.com/2015/03/25/world/asia/ashraf-ghani-of-afghanistan-wants-us-troops-to-stay-longer.html?_r=0.

486 Preston, *supra* note 398, at 6.

487 Matthew Rosenberg & Michael D. Shear, *Obama Announces Halt of U.S. Troop Withdrawal in Afghanistan*, N.Y. Times, Oct. 15, 2015, http://www.nytimes.com/2015/10/16/world/asia/obama-troop-withdrawal-afghanistan.html?smprod=nytcore-iphone&smid=nytcore-iphone-share&_r=0.

In situations where the threat is primarily a terrorist one, such as that presented by groups acting transnationally, factors such as the conduct of operations in the "ungoverned spaces" far from the populations of the targeted State will more likely favour a counterterrorist response. However, it can also be difficult to separate counterterrorism from counterinsurgency. It has been noted that "[o]ne problem Israel has in common with other democracies is that it focuses narrowly on its foes' use of terrorism and ignores their wider strategies."[488] While most groups Israel faces, such as Hezbollah and Hamas, have carried out terrorist acts against civilians, they are also broader social and governing organizations. As a result, it is suggested Israel needs to take lessons from counterinsurgency, "which addresses not only the military (or "kinetic" in American soldier parlance) dimensions but also the political, economic, and social ones as well."[489]

It is reported that one of the issues the United States struggled with in deciding to accelerate their drone strikes in Yemen in 2011 was that many of the intended senior targets came from a wing of AQAP "that was not 'externally focused', not interested in attacking the United States, and therefore according to some not targetable."[490] The AQAP was a terrorist group, but it was also an insurgency with tribal elements occurring in another country.[491] Further, engaging in such strikes represented "a 'slippery slope' to counterinsurgency" in Yemen, an activity in which the United States did not want to become engaged.[492] More globally, Al Qaeda is only one of a number of Salafi jihadist groups, and because of its ready association with "terrorism" this masks the nature of the overall threat. Given the ultimate goal of such groups, there will remain a requirement to consider counterinsurgency principles from a strategic perspective, even in the conduct of a counterterrorist campaign against transnational terrorists. Counterterrorist responses to this transnational terrorist organization must be conducted with a continuing sensitivity toward the role that group plays as part of an overarching global insurgency.

Counterinsurgencies by their nature require a more personnel and resource intensive effort on the part of the State or States fighting them. Gil Merom has identified that a challenge for democracies when fighting small wars is finding "a winning balance . . . between acceptable levels of casualties and acceptable levels brutality."[493] Disenchantment with the human and material costs of the counterinsurgencies in Iraq and Afghanistan has undoubtedly fueled a move by the United States toward a counterterrorism policy largely focused on Al Qaeda, the Islamic State and Salafi jihadist groups with transnational reach.[494] Even though counterinsurgency and counterterrorism operations are different, the similarity of the legal issues they raise is noteworthy. These issues include the interface between law enforcement and the conduct of hostilities, whether it is appropriate to capture or arrest rather than kill suspected terrorists, the potential for collateral killing or injury of civilians,

---

488 BYMAN, *supra* note 84, at 376.

489 *Id.*

490 DANIEL KLAIDMAN, KILL OR CAPTURE 254 (2012).

491 *Id.* at 254.

492 *Id.* at 255.

493 GIL MEROM, WHY DEMOCRACIES LOSE SMALL WARS 24 (2003).

494 *But see* NATIONAL SECURITY STRATEGY FOR COUNTERTERRORISM, *supra* note 344, at 18 (where Hezbollah, Hamas, and the Revolutionary Armed Forces of Colombia (FARC) were also identified as a significant threat to U.S. strategic interests).

and the weighing of the elevated levels of risk to which security personnel may have to be exposed to lessen collateral civilian casualties. Neither of these two operational doctrines by themselves fully addresses the twenty-first-century security challenges since an inherent part of both insurgency and terrorism is the challenge of gangs and criminal activity.

Further, perhaps one of the greatest challenges for State military forces in the contemporary operating environment, especially during counterinsurgency and counterterrorism, has been to participate in law enforcement–based activities. Such operations are likely to become even more prevalent given the rise of transnational criminal gangs. The State approach toward the criminal gang phenomenon has been and likely will continue to be firmly rooted in law enforcement. This ultimately means that seeking to cut off insurgents and terrorists from sources of criminal sources of funding either carried out by them or through their relationship with transnational criminal organizations must be addressed as a "policing" issue. In any event there are "strong analogies between police work, counterinsurgency and counterterrorism."[495] Police gang suppression approaches are "analogous to counterinsurgency, especially in urban environments," while "some terrorist networks share structural and operational similarities with organized crime networks . . . ."[496] With the law enforcement approach, "the key objective is to apprehend the perpetrators of terrorist attacks."[497] In the context of counterterrorism this has ignited a debate regarding the capture or killing of members of the enemy forces. In these cases, however, the elevated potential for violence when dealing with contemporary terrorists can present significant challenges for a human rights law enforcement framework based on the concept of applying minimum force.

Addressing these security challenges requires the application of a range of legal frameworks that is entirely consistent with a holistic approach reflected in the Koh speech, which raised the issue of the application of a post-9/11 legal framework.[498] In many instances, security forces will apply a primarily counterinsurgency approach and in others, particularly against transnational actors, the focus will be on counterterrorism. However, many security situations, such as those presented in Afghanistan and Iraq, will require the application of both doctrines simultaneously. No one body of law is likely to prevail when States are confronted with such diverse threats. The threats posed by non-State actors and in particular the jihadist movement are not only diverse, but also transnational and even global in nature.[499] The very nature of counterinsurgency and counterterrorist operations often necessitates consideration of the use of force in national self-defense, the law governing the conduct of hostilities, and the application of law enforcement standards most commonly associated with human right, both international and domestic. Added to this is the requirement to apply various accountability mechanisms including international and national criminal law.

---

495 KILCULLEN, *supra* note 4, at 183 n *, Author's Note.

496 *Id.*

497 *Id.* at 187.

498 Harold Hongju Koh, *The Obama Administration and International Law* (Mar. 25, 2010), http://www.state.gov/s/l/releases/remarks/139119.htm.

499 S.C. Res. 2249, UN Doc. S/RES/2249 *Preamble* (Nov. 20, 2015)(where reference is made to the global threat posed by the Islamic State).

The civilized societies in which we best like to live are governed by law, which means they are policed, and policing is a form of coercion.[1]

7 Counterinsurgency and Converging Norms

7.1. Introduction

This chapter explores a number of unique challenges that can arise regarding counterinsurgency operations. It starts by addressing the continuing relevance of counterinsurgency notwithstanding the large-scale withdrawal of Western military forces from such activity with the winding down of the post-9/11 Iraq and Afghanistan deployments. Impacting significantly on this discussion is the traditional reluctance of State militaries to engage in such operations. Counterinsurgency will be considered in terms of the application of "police primacy" principle and the resulting requirement to resolve the overlapping application of humanitarian and human rights law. The analysis also looks at the key role that the ability to maintain sufficient physical control over an area of operations plays when determining if a law enforcement approach is feasible. The focus then turns to counteracting the threat posed by improvised explosive device (IED) and suicide bombing cells. The group nature of this threat and factors affecting the responses available to State security forces will be considered. This includes an assessment of the nature of the participation of persons who support or facilitate the use of these explosive means. This is done in order to establish who must be captured and those who may be targeted as direct participants in hostilities.

The analysis then addresses the effect on targeting of the counterinsurgency goal of winning and maintaining the support of the population. A particular emphasis is placed on assessing how the "value" of a target can affect targeting decisions during counterinsurgency operations. As will be explored, causing collateral civilian casualties and damage can

[1] John Keegan, A History of Warfare 386 (1993).

particularly impact strategic assessments of the military advantage to be gained from an attack. Further, concerns over the collateral effects of operations influences the application of the humanitarian law precaution of proportionality. The strategic goal of limiting such collateral effects during counterinsurgency generally results in a narrower acceptable "zone of proportionality." Finally, it is evident that States default to a law enforcement model domestically even when confronted with significant security threats. This option is explored in terms of considering that same approach when conducting extraterritorial operations.

7.2. Counterinsurgency in the Twenty-First Century

With insurgencies forming a major threat in the twenty-first century, counterinsurgency has, along with counterterrorism, dominated much of the security dialogue since 9/11. That discussion arose primarily because in the Afghanistan and Iraq conflicts where U.S.-led international coalitions became heavily engaged, often reluctantly, in counterinsurgency operations. In both cases the conflicts began with swift military victories,[2] only to be followed by increasing engagement in difficult and drawn-out battles with insurgents.[3] This was a role for which the State military forces were largely not organized, equipped, or trained to carry out. Consistent with the traditional preference of military forces to engage in conventional conflict, the armed forces from largely Western States have generally been hesitant to involve themselves in counterinsurgency operations.[4] As one American general officer noted, the military strengths of the U.S. forces were focused on "short, decisive conventional conflicts waged for limited ends. . . ."[5] Elsewhere it has been stated, "[t]he result of poor planning and training . . . was a military that was spinning its wheels, and this deeply impacted on its ability to carry out the responsibilities of an occupier under international law."[6] Perhaps the most obvious example of the lack of preparedness and an actual unwillingness to act was the failure of United States and other Coalition military forces to take over responsibility for the maintenance of order in the aftermath of the toppling of the Iraq regime in 2003.[7]

---

[2] In respect of the 2003 Iraq invasion, *see*, JOHN KEEGAN, THE IRAQ WAR 1 (2004) ("The [2003 Iraq] war was not only successful but peremptorily short, lasting only twenty-one days, from 20 March to 9 April."), and regarding Afghanistan, *see* JASON BURKE, THE 9/11 WARS 51–71 (2011) (where it is indicated bombing the Taliban commenced on October 7, 2001, and by December 22, Hamid Karzai was sworn in as the chairman of the Interim Government of Afghanistan).

[3] *See, e.g.*, THOMAS E. RICKS, FIASCO: THE AMERICAN MILITARY ADVENTURE IN IRAQ 392–93 (2007) (where it is noted that by August 2004 the American Commander in Iraq had issued a counterinsurgency-based campaign plan).

[4] ROBERT M. CASSIDY, COUNTERINSURGENCY AND THE GLOBAL WAR ON TERROR: MILITARY CULTURE AND IRREGULAR WAR 100 (2008) ("the U.S. Army has consistently refused to seriously consider any type of war except a European-style conventional war.") and MARK GALEOTTI, *SPETSNAZ*: RUSSIA'S SPECIAL FORCES 30 (2015) (the reluctance to engage in unconventional warfare can be seen in willingness with which the Russian military hierarchy sought to abandon the operational innovations pioneered during the Afghanistan war).

[5] DANIEL P. BOLGER, WHY WE LOST: A GENERAL'S INSIDE ACCOUNT OF THE IRAQ AND AFGHANISTAN WARS 429 (2014).

[6] Matthew R. Hover, *The Occupation of Iraq: A Military Perspective on Lessons Learned*, 94 INT'L REV. RED CROSS 339, 344 (2012).

[7] DAVID H. BAYLEY & ROBERT M. PERITO, THE POLICE IN WAR: FIGHTING INSURGENCY, TERRORISM, AND VIOLENT CRIME 6 (2010) ("[o]nce it became clear that US soldiers were not going to intervene, public exuberance, joy at liberation, and economic opportunism quickly darkened into a systemic effort to strip the capital's stores and public institutions of everything of value"). *See also* KEEGAN, *supra* note 2, at 206–07.

It has been noted that in Iraq the "U.S. occupation forces were expected to fill the vacuum; but military forces typically focus on the maintenance of order at the macro level rather than the enforcement of law at the micro level."[8] There was also little support in the U.S. government following the invasion to provide the international police augmentation considered necessary to assist the Iraqi police.[9] The only other available security forces, the military, were not ready or trained to fill the void. This reality was particularly disconcerting since policing a civilian population is an activity that is an inherent part of international armed conflict. Policing is clearly the operational and legal responsibility of an occupying power.[10] In this regard, the descent into chaos in Iraq was ultimately a product of failed strategic thinking.[11] The failure to prepare for the day after "victory" in the conventional war was the beginning of a spiral into disorder and lawfulness. The security vacuum not only permitted organized crime to thrive, it also helped fan the flames of insurgency.

The experiences in both the Afghanistan and Iraq conflicts forced considerable change. Confronted with growing insurgencies in those two countries, the U.S. military ultimately demonstrated considerable flexibility in adjusting to the post-9/11 security challenge. The broader U.S.-led counterinsurgency response included enhancing its Special Operations Forces counterterrorism forces, employing drones to carry out targeted killings, and in the publication of doctrinal guidance in the form of the *United States Army/Marine Corps Counterinsurgency Manual*.[12] Reflecting an effort that unfortunately appears to have been more driven from the bottom than strategically directed, in 2004 the U.S. military in Iraq instituted lessons learned from earlier operation, and developed a refined "clear and hold" counterinsurgency approach. There was even a counterinsurgency school established "in country" that all incoming regimental and battalion commanders had to attend.[13]

In Afghanistan the application of counterinsurgency theory was evident in the 2009 direction provided by U.S. International Security Assistance Force (ISAF) commanders regarding restrictions on the use of force, which were designed to limit the collateral impact on the civilian population.[14] This did not necessarily mean that the tactical directive, which focused on limiting "[t]he use of air-to ground munitions and indirect fire [artillery and mortars] against residential compounds,"[15] was well received by commanders and soldiers tasked with countering the insurgent threat.[16] However, whatever the problems with

---

[8] Paul Williams, Criminals, Militia, and Insurgents: Organized Crime in Iraq 37 (2009).

[9] Bayley & Perito, *supra* note 7, at 10–11 (outlining the efforts to provide international policing assistance immediately following the 2003 invasion of Iraq).

[10] The 1907 Land Warfare Regulations, art 43, and GC IV, art. 64.

[11] Hover, *supra* note 6, at 340–42.

[12] The U.S. Army, Marine Corps, Counterinsurgency Field Manual (2007) [hereinafter The Counterinsurgency Manual].

[13] Carter Malkasian, *Counterinsurgency in Iraq: May 2003–January 2010*, *in* Counterinsurgency in Modern Warfare 287, 296–301 (Daniel Marston & Carter Malkasian eds., 2008). *But see id.* at 300 (However, operational challenges continued with some commanders "still focused on mechanized sweeps or air assaults," and efforts to obtain a consistent counterinsurgency approach appear to have been, somewhat counterintuitively, hampered by "the decentralized command and control structure developed for conventional war.").

[14] ISAF Commander's Tactical Directive (July 6, 2009), ISAF Headquarters, http://www.nato.int/isaf/docu/official_texts/Tactical_Directive_090706.pdf [hereinafter ISAF Commander's Tactical Directive].

[15] Bolger, *supra* note 5, at 305 (quoting the tactical directive issued by General Stanley McChrystal). *See also* ISAF Commander's Tactical Directive, *supra* note 14.

[16] *Id.* at 336–39 (outlining some of the criticism of the directive and how it was implemented at the tactical level).

implementation, this higher-level direction reflected the goal of limiting civilian casualties. This has been and remains one of the fundamental tenets of counterinsurgency theory.

The conduct of counterinsurgency operations and the limited degree of interest within State military forces to engage in that activity, continue to be the subject of considerable analysis. With the winding down of post-9/11 Coalition operations in Iraq and Afghanistan, it is not surprising that military forces are reconsidering the emphasis placed on counterinsurgency since the beginning of the twenty-first century. For the U.S. military that review has looked at confronting more traditional foes (i.e., nuclear armed China, Russia, Korea, and Iran)[17] and redeveloping traditional war fighting skills such as "air-defense suppression," "anti-submarine and anti-mine missions," "amphibious landings," "major combined-arms armored tactics and the ability to mass the fires of dozens of tubes of artillery on a single target."[18] State armed forces must be prepared to conduct operations across the broad range of threats, from guerrilla war to conventional inter-State conflict.

The post-Iraq/Afghanistan period has witnessed a negative discussion about the efficacy of counterinsurgency doctrine and how it was applied since 9/11. This has included an analysis that criticized the role played by Special Forces and downplayed the unique challenges of counterinsurgencies and small wars. Instead it championed the use of conventional military forces.[19] Another review has emphasized the value of applying a counterterrorism approach that targets Al Qaeda by using much smaller forces, rather than engaging in "the folly of trying to build Afghanistan into a modern state overnight."[20] The concept of counterinsurgency has been suggested to be a complete failure, and it was indicated that the military should not "be transformed into an all-purpose constabulary force tailored for national building at the expense of an army that can conduct sustained combat at the higher end of the conflict spectrum."[21] It indicated a particular lesson learned is that "American strategy has failed in Afghanistan (and Iraq) because it was founded on an illusion—that American style counterinsurgency could win Muslim hearts and minds at gunpoint and create viable nation-states on the Western model virtually from scratch overnight."[22] For the most part these critiques largely favor the status quo of maintaining conventional forces, and represent to various degrees a mainstream aversion to small wars and irregular operations. A common theme has been a negative or at least skeptical assessment of the counterinsurgency doctrine found in the *United States Army/Marine Corps Counterinsurgency Manual*, particularly in terms of its intellectual focus rather than practical applicability.[23]

---

17 The National Military Strategy of the United States of America 2015, at 2 (June 2015) [hereinafter The National Military Strategy], http://www.jcs.mil/Portals/36/Documents/Publications/National_Military_Strategy_2015.pdf.

18 Bolger, *supra* note 5, at 433.

19 Douglas Porch, Counterinsurgency: Exposing the Myths of the New Way of War (2013) (This book, written in the aftermath of the Iraq drawdown, presents a scathing critique of this form of warfare, the role performed by Special Forces, and the unique doctrine that has developed from the reality of contemporary conflict).

20 Gian Gentile, Wrong Turn: America's Deadly Embrace of Counter-Insurgency 135 (2013).

21 *Id.* at 8.

22 *Id.* at 135.

23 Bolger, *supra* note 5, at 238 ("Compared to the standard, terse military jargon and acronyms, this stuff sang a pleasing intellectual tune. It is hard to say what the hell it meant to a platoon sergeant or a lieutenant trading bullets west of Samarra or ducking EFPs on the fringes of Sadr City."). *See also* Gentile, *supra* note 20, at 25 (who questions the validity of the historical experiences upon which it is based (i.e., Malaya, Vietnam and Algeria), as well as its scholarly rigor).

In a British context another postconflict assessment has leveled criticism more at the military itself than counterinsurgency doctrine. It is suggested that the armed services failed to produce an officer corps, particularly at the general rank, which was sufficiently prepared for the conflicts of the post-9/11 period.[24] The result was that "[t]he British were at sea in both places [Iraq and Afghanistan], devoid of viable doctrine, without awareness of their environment, lacking adequate forces and minus any coherent strategy to pursue."[25] A key problem was treating the Iraq and Afghanistan conflicts as though the military was "in the midst of insurgencies," rather than "civil wars (into which 'we' crashed as participants)."[26] Also missing from that effort was the realization that "there is one principle that must be paramount—the overriding principle of all humanitarian work: 'do no harm'."[27] Fealty to that principle requires restraint on the use of force,[28] and one positive result of the "new" counterinsurgency doctrine "is that emphasis has started to shift away from 'kinetic' operations."[29] Yet another assessment has criticized the Coalition effort in Afghanistan, in part, because of an overreliance on a counterterrorism approach "with its night raids and body-count metrics, and calling it counterinsurgency."[30]

Whatever the merits of these criticisms, and there are a number, what does not change is the continuing requirement for military forces to not only engage in counterinsurgency and counterterrorism but also be prepared to adjust to a constantly changing operational environment. The potential for "small wars" with non-State actors is vast, ranging from "one-off" 9/11 style attacks, the more likely violence associated with fidayeen assaults (e.g., the 2014 Peshawar School Massacre[31]), to civil wars and situations where such groups control territory transcending national borders. A key component of the State response to such threats will continue to be the conduct of counterinsurgency and associated counterterrorism operations. A fundamental aspect of counterinsurgency, and one that must be emphasized in any assessment of contemporary conflict is that the concept must also be looked at from a broad perspective. As has been noted, "[c]ounterinsurgency is not a strategy. Insurgency-counterinsurgency, like conventional war, is a type of conflict."[32] It is a form of conflict that is not going to disappear for much of the world's armed forces no matter the preference of dominant military powers.

---

[24] Frank Ledwidge, Losing Small Wars: British Military Failure in Iraq and Afghanistan 262–64 (2011).

[25] *Id.* at 261.

[26] *Id.* at 255.

[27] *Id.*

[28] *Id.* at 171–91.

[29] *Id.* at 191.

[30] Jean MacKenzie, *Afghanistan Part II—Counterinsurgency in Afghanistan: Myth or Reality?*, *in* Hearts and Minds 200, 219 (2013).

[31] Jon Boone & Ewen MacAskill, *More than 100 Children Killed in Taliban Attack on Pakistan School*, The Guardian, Dec. 16, 2014, http://www.theguardian.com/world/2014/dec/16/taliban-attack-army-public-school-pakistan-peshawar (where it is indicated that six Taliban members, with some wearing suicide vests, carried out such an attack in December 2014 at military-run school in Peshawar, Pakistan), and Carmen Fishwick, *Peshawar School Massacre: "This Is Pakistan's 9/11—Now Is the Time to Act,"* The Guardian, Dec. 19, 2014, http://www.theguardian.com/world/2014/dec/19/peshawar-school-massacre-pakistans-911 (reporting that 141 people were killed, including 132 children).

[32] Christopher D. Kolenda, The Counterinsurgency Challenge: A Parable of Leadership and Decision Making in Modern Conflict xiv (2012).

As a form of conflict, insurgency/counterinsurgency is a much broader overarching construct, which ultimately involves the engagement of different aspects of State power. If insurgents are to be beaten, the effort requires a wide range of political, military, and socioeconomic measures that extend far beyond the issues dealing with direction regarding the use of military force, or which type of unit (i.e., Special Forces or conventional units) is best suited to the task. From a security perspective the State response must also focus on the performance of a policing function, and ultimately the creation of an effective police force in the territory where insurgents operate.[33] One problem is that the conduct of counterinsurgency operations has not necessarily been addressed from a strategic perspective. For Hew Strachan, much of the "strategy" for the Afghanistan conflict articulated in 2009 by the Commander of ISAF "was shaped from the bottom up; without clear articulation by NATO or the United States of their political objectives and hence their strategies, it could not be anything else."[34] It was "essentially operational, not strategic, even if they posed as strategic."[35]

As its own unique category of conflict, an insurgency/counterinsurgency presents a varied and diverse set of security challenges for States, ranging from hostilities that are essentially conventional in nature, to the lawless crime-based insecurity common in the breakdown of social order associated with such violence. What sets insurgency-based conflict apart from its more conventional inter-State counterpart has been the greater requirement to deal with governance-related issues (see Chapter 14, section 14.2, *From "Conventional" to "Hybrid" to "Holistic"*). This places a premium on a policing response and the development of a human rights–based justice system. Further, insurgents also carry out terrorist acts. Therefore, it should come as no surprise to see a State response to insurgency that includes counterterrorist operations and tactics. Engaging in noninternational armed conflict is a highly complex endeavor.

Military counterinsurgency doctrine can and should be critiqued. The manner in which a particular counterinsurgency is conducted by military forces may be called into question on the basis of its being too limiting in terms of authority to use force, for not relying enough on conventional forces, for failing to recognize the civil war nature of a conflict, or for relying too much on a counterterrorism approach. However, as has been perceptively noted by Frank Ledwidge, "the proponents and opponents of 'population-centered' counterinsurgency have been talking past each other."[36] In certain environments population-centric counterinsurgency can work, while in others such as Iraq and Afghanistan, which might be viewed more "as the nature of occupations and complicated by significant civil wars," such an approach is more challenging.[37] He suggests in terms of counterinsurgency there "was little attempt to 'protect the people' in either theatre."[38] However, what does

---

[33] David P. Fidler, *Police in Counterinsurgency: The Challenge of Comprehensive Reformation*, *in* POLICING INSURGENCIES: COPS AS COUNTERINSURGENTS 317, 324 (C. Christine Fair & Sumit Ganguly eds., 2014) [hereinafter POLICING INSURGENCIES].

[34] HEW STRACHAN, THE DIRECTION OF WAR: CONTEMPORARY STRATEGY IN HISTORICAL PERSPECTIVE 226 (2013).

[35] *Id.* at 227.

[36] LEDWIDGE, *supra* note 24, at 169.

[37] *Id.*

[38] *Id.*

not change is the fundamental difference in the threat that insurgents present in terms of their challenge to State governance when compared to a conventional conflict between two States. When dealing with a non-State actor threat, the response by State armed forces must therefore uniquely address the challenge of fighting "amongst the people."[39]

States, such as the United States, may choose to have their military forces concentrate on traditional conventional operations, or even strike-based counterterrorism. Their preference may also be to have other nations carry out what military traditionalists sometimes refer to pejoratively, or at least skeptically, as "nation building."[40] By 2015 *The National Military Strategy of the United States* placed counterinsurgency toward the bottom of its Joint Force Prioritized Missions.[41] What this means is that the military forces of States, such as the United States, that are pulling back from the full spectrum of counterinsurgency operations may only be able to perform a limited role in defeating an insurgent threat. However, this change in operational focus by some military forces does not alter the fundamental nature of an insurgency/counterinsurgency conflict or what action is required by security forces to defeat an insurgency. In this regard any desire to degrade and destroy the Islamic State by means of a counterterrorism strategy is inconsistent with its being "far more significant as an insurgency than as a terrorist group."[42] The military forces of some State will still have to engage in counterinsurgency operations that address the full range of potential threats from that enemy. This includes conducting operations in a manner consistent with proper governance. A failure to do so will adversely affect the ability of that State to control the territory in dispute. In Iraq the manner in which the predominately Shi'a governing authorities and security establishment address sectarian-based violence and the treatment of minority Sunnis will be key factors in any success that State has in battling the Islamic State.[43] Unless the minority population feels it is living in a secure environment, it is difficult to see how the insurgency will be defeated. The same can be said for Afghanistan. While as of late 2015 the United States planned to only maintain training and counterterrorism forces in Afghanistan it is clear that host nation will need to conduct effective counterinsurgency operations in order to successfully defeat the insurgent forces (i.e. the Taliban, Al Qaeda and the Islamic State) arrayed against it.[44]

The territorial State is the most obvious and normally most effective candidate to carry out the counterinsurgency role.[45] However, even where that State can field a counterinsurgency

---

[39] Rupert Smith, The Utility of Force: The Art of War in the Modern World 3 (2005).

[40] Bolger, *supra* note 5, at 44 ("In [General] Frank's view [the Commander of U.S. CENTCOM overseeing the 2001 invasion of Afghanistan], and that of many members of the U.S. military, Americans kicked down doors; let others build nations."). *See also* Hover, *supra* note 6, at 341 ("the military did not want to be involved in stability operations, and it was attempting to avoid nation-building at all costs.").

[41] The National Military Strategy, *supra* note 17, at 11.

[42] Jessica S. Stern & J. M. Berger, ISIS: The State of Terror 49 (2015).

[43] Missy Ryan, *U.S. Sees Bigger Test for Iraq after Tikrit Battle*, Wash. Post, Mar. 7, 2015, http://www.washingtonpost.com/world/us-sees-even-bigger-test-for-iraq-and-iran-in-the-aftermath-of-tikrit-battle/2015/03/07/9b82a0f0-c487-11e4-ad5c-3b8ce89f1b89_story.html.

[44] Matthew Rosenberg & Michael D. Shear, *Obama Announces Halt of U.S. Troop Withdrawal in Afghanistan*, N.Y. Times, Oct. 15, 2015, http://www.nytimes.com/2015/10/16/world/asia/obama-troop-withdrawal-afghanistan.html?smprod=nytcore-iphone&smid=nytcore-iphone-share&_r=0.

[45] John A. Nagl, Learning to Eat Soup with a Knife: Counterinsurgency Lessons from Malaya and Vietnam xiv (2005) (for reference to the advantages that local forces have in fighting an insurgency).

response, other countries may still add capabilities to assist in that endeavor. The type of external support that can be provided, which reflects the adoption of an approach favoring conventional operations and counterterrorism, can be seen in the foreign State response to the rise of the Islamic State in Iraq in 2014. The United States and its Coalition partners have primarily restricted their military operations to aerial bombing and the provision of limited primarily Special Forces support in training Kurdish and Iraqi government military forces. Those countries have also provided arms, ammunition, and humanitarian support.[46]

Conventional military support such as aerial bombing is an important addition to the counterinsurgency effort, however, it is unlikely, on its own, to defeat the military forces of the Islamic State or result in control being established over disputed territory. In the final analysis it is the Iraqi government that will need to provide "boots on the ground" and apply appropriate counterinsurgency methods in order for the collective State response to be successful. Notwithstanding any move away from traditional counterinsurgency operations, the supporting States will still have to situate their operations within the overall territorial State response. This means that not only must foreign military advisors be fully familiar with counterinsurgency theory in order to give effective advice, but any kinetic support provided, such as bombing, will also have to remain sensitive to the unique requirements of such conflicts. This is particularly in terms of the impact on the civilian population.[47] A doctrinal turn away from engagement in a broader counterinsurgency effort does not mean the State military forces involved can ignore the existence and nature of such conflict. The requirement to engage in counterinsurgency operations is not going to disappear because it is not incorporated into military doctrine and training or is no longer the study area of choice at military professional institutions.

Further, it is not likely that State armed forces preferring to operate in a narrow conventional or uniquely counterterrorist role will always be able to. For example, as happened to the United States and its Coalition partners in Iraq, even during inter-State conflicts an occupying power may be required to conduct counterinsurgency operations. Regardless of the doctrinal approach favored by the military forces of a State, a political decision may also be made in the future to involve those armed forces in such operations. These political or operational factors could cause the pendulum to swing once again away from the "preferred" conventional operations. For example, if in acting against the Islamic State non-territorial States put "boots on the ground" in Iraq or Syria (e.g. U.S., U.K., France, Russia) they are likely to be confronted with an insurgency.

Finally, looking more globally, the Iraq and Afghanistan operations of the post-9/11 period must be put in context. In this regard, they only constitute a fraction of an estimated over one hundred global conflicts involving insurgencies or potential insurgencies.[48] Whether the threat comes from the Boko Haram in Nigeria, the Al Shabaab in Somalia, the Taliban in Pakistan, the Revolutionary Armed Forces of Colombia (FARC) in Columbia,

---

[46] Justine Drennan, *Who Has Contributed What in the Coalition Against the Islamic State*, Foreign Pol'y (Nov. 12, 2014, 2:50 PM), http://foreignpolicy.com/2014/11/12/who-has-contributed-what-in-the-coalition-against-the-islamic-state/.

[47] Murray Brewster, *Strikes on ISIS in Iraq Sparing Civilians: Canadian Commander*, CBC News (Jan. 8, 2015, 2:54 PM), http://www.cbc.ca/news/politics/strikes-on-isis-in-iraq-sparing-civilians-canadian-commander-1.2894014.

[48] Max G. Manwaring, The Complexity of Modern Asymmetric Warfare 3 (2012).

or Maoist extremists in India,[49] there are numerous States faced with combating these internal threats. Added to these groups is the growing threat posed by "criminal insurgencies" in States such as Mexico, Columbia, Jamaica, and Guatemala that increasingly appear to straddle the line between criminal activity and traditional notions of insurgency (see Chapter 6, section 6.3.1, *The Nature of the Threat*). Combating these threats will continue to require military forces capable of applying a policing-based approach.

States may struggle with the question of whether or how much their military forces should be prepared and equipped to conduct counterinsurgency operations. However, given the nature of contemporary conflict and the preponderance of small wars in the twenty-first century it is highly likely that they will be placed in a situation of having to do just that. This makes knowledge of, and a commitment to, a counterinsurgency doctrine and the legal issues that arise from such conflict an essential requirement of contemporary military study, training, planning, and operational activity. The result is that unique aspects of counterinsurgency operations, such as placing a premium on law enforcement; acknowledging the link between applying a capture over killing approach and establishing control over an area of operations; and taking additional care to limit civilian casualties during targeting, will have to be understood and incorporated into military operations. It is to those issues that the analysis now turns.

7.3. The Primacy of Law Enforcement

As Western States in particular begin to readjust their priorities as a result of the withdrawal from Iraq and Afghanistan, it appears that, as adaptable and flexible as many State regular armies may be, counterinsurgency and counterterrorism offer operational challenges far different from those presented on the conventional battlefield. Perhaps one of the most significant challenges for conventionally trained military forces in the twenty-first century has been their involvement in security operations that place a premium on the application of human rights–based law enforcement norms. At the same time those forces have to remain prepared to engage in hostilities involving violence associated with traditional combat. The requirement to apply force across a broader spectrum of operations than during conventional armed conflict has not only made military forces alter doctrine and training, it has forced a more nuanced discussion regarding the two governing normative regimes: human rights and humanitarian law.

Insurgencies in their traditional form, and increasingly in the context of global challenges to State authority (either jihadist- or crime-based), present a significant security challenge. Counterinsurgency operations require the application of both human rights and humanitarian law under circumstances where they not only interface but also overlap. As is evident in the debate about preferring "capture" rather than "killing," confronting an insurgency operating among the civilian population places unique pressures on interpretations of humanitarian law. The result has been the recognition of a convergence of norms where humanitarian law–based outcomes have increasingly approximated to those associated

---

[49] Arvind Verma, *The Police and India's Maoist Insurgency*, *in* POLICING INSURGENCIES, *supra* note 33, at 291, 291–318.

with law enforcement. The successful prosecution of a counterinsurgency strategy requires the involvement of both military and police forces, as well as the use of "soft power" tools such as "governance, religion and culture, economic development, and education."[50]

There must be no misunderstanding that military operations will not remain an essential part of that strategy. A 2013 study of seventy-one insurgencies, from World War II until 2010, identified that "in no case did the COIN [counterinsurgency] force win unless it overmatched the insurgents and could force them to fight as guerrillas by the decisive phase of the conflict."[51] However, a key element also appears to have been the quality of the counterinsurgent forces including both police and paramilitary forces, although it has been suggested these latter forces have been too often inadequately trained, or equipped.[52] The result is that "in the 23 cases in which police or paramilitaries were *not* ineffective, COIN forces won 69 percent of the time."[53]

The emphasis placed on law enforcement during counterinsurgency operations is directly linked to doctrine based on a "police primacy principle."[54] It has been suggested to be puzzling that most of the attention, and much of the literature, doctrine, and studies have concentrated on the role of armed forces[55] when "the task of restoring public order invariably involves careful and sustained police work."[56] The concentration on military forces is often translated into the development of national military forces at the expense of a police capacity. This can have a significant negative impact on operational capabilities. In Afghanistan this meant no competent indigenous police force was available that "could develop personal relationships with locals, who would then report terrorists in their neighborhoods. The Afghan army, on the other hand, whizzing through cities in their closed military vehicles and using massive firepower to kill rather than attempting to obtain information, could never develop a comparable network of informants."[57]

The maintenance of security through policing is a key aspect of winning support of the people. In Afghanistan, Iraq, and elsewhere where insurgencies exist, the allegiance of the population is often based on which group (i.e., the government or non-State actors) can maintain a semblance of law and order. Those allegiances can shift, or be tempered, by a belief that one or the other group (i.e., the government or the insurgents) can best provide

---

[50] Martin I. Wayne, China's War on Terrorism: Counter-insurgency, politics and internal security 145 (2008)(where it is noted that providing security is a broader concept than simply killing terrorists/insurgents. In this regard "[c]ounterinsurgents must use the tools of soft power to convince society that meaning full security, that a better future locally understood, resides with the state.").

[51] Christopher Paul et al. et al., Paths to Victory, Rand xxviii (2013).

[52] *Id.*

[53] *Id. See also* Wayne, *supra* note 50, at 87-88 (where it is noted that even though China's counterinsurgency efforts against the Uyghurs in the Xinjiang region have been harsh and draconian it has been the shift away from a military response to one involving paramilitary, police and civilian engagement that has limited the insurgent threat).

[54] Counterinsurgency Manual, *supra* note 12, ¶ 6-90, at 229 (2007) ("The primary frontline COIN force is often the police—not the military."). *See also* Bayley & Perito, *supra* note 7, at 68–69.

[55] C. Christine Fair & Sumit Ganguly, *Introduction: The Police in Counterinsurgency Operations*, *in* Policing Insurgencies, *supra* note 33, at 1, and Randall Wilson, Blue Fish in a Dark Sea: Police Intelligence in a Counterinsurgency 2–3 (2013).

[56] Fair & Ganguly, *Introduction*, *supra* note 55, at 1.

[57] Terry Gould, Worth Dying For: Canada's Mission to Train Police in the World's Failing States 44 (2014) (referring to arguments made by an RCMP CivPol leader, Superintendent Joe McAllister, to Canadian generals at Kandahar airfield).

security at its most basic level.[58] This often means providing an adequate policing response by State authorities. Further, since many insurgent groups engage in criminal activity in order to finance their activities there is a natural requirement to involve law enforcement as part of the security solution.

Countering such criminal activity by ordinary policing can be an integral and essential part of a counterinsurgency campaign. However, performance of the policing role during an insurgency can be anything but "ordinary." The implementation of a counterinsurgency approach can be a complex endeavor, which often results in a blending of responses by security forces. The nature of the threat, combined with the often limited availability of trained law enforcement personnel, means the police are not the only forces that will be engaged in law enforcement (see Chapter 12, section 12.2.1, *The Military Task: Applying Force of First and Last Resort*). Military forces will still be employed in the conduct of hostilities; however, they will also likely fill gaps in policing capacity and provide support to police operations.[59] At the same time police forces may be developed with special capabilities, sometimes being trained and deployed as what has been termed, "little soldiers."[60] One of the challenges of counterinsurgency is that "[w]hen the insurgency manifests itself as something beyond a mere law-and-order problem, governments tend to militarize their responses . . ."[61] (see Chapter 13, section 13.3, *The Militarization of the Police*). Included in the conceptual struggle of how to respond to an insurgency is the degree to which the constabulary element of the State security apparatus should be a "police service" or a "police force."[62] As occurred in Iraq there can be the creation of specialist units, such as the Police Commandos, which "were identified as a police unit but were not trained or legally empowered to play a police role."[63] The result is that the activities of military and police forces will often overlap.

The overlapping employment of security forces reflects the continuum of threat posed by insurgents and the resulting broad range of security operations required to counter those groups during these small wars. Within this continuum it can become difficult to categorize the appropriate response as being purely law enforcement or conduct of hostilities related. Counterinsurgency operations are by their nature "population-centric," with tactics involving flooding an area with military forces "to deter enemy interference with the population."[64] This often places military forces in direct contact with the civilian population and the security challenges faced by them. Such operations are "less like conventional warfare and more like police work: cops patrolling the beat to prevent violent crime."[65] The blending of hostilities and law enforcement for military forces is particularly evident when suicide bombers seek to pass by checkpoints[66] and can occur in situations where insurgents

---

[58] David Kilcullen, Counterinsurgency 152 (2010).

[59] Bayley & Perito, *supra* note 7, at 69.

[60] *Id.* at 76.

[61] Fidler, *supra* note 33, at 323.

[62] *Id.*

[63] Matt Sherman & Josh Paul, *The Role of Police in Counterinsurgency Operations in Iraq, 2003–2006*, *in* Policing Insurgencies, *supra* note 33, 227, at 238.

[64] Kilcullen, The Accidental Guerrilla 130 (2009).

[65] *Id.*

[66] Kenneth Watkin, *Use of Force during Occupation: Law Enforcement and Conduct of Hostilities*, 94 Int'l Rev. Red Cross 267, 310–14 (2012) (for a discussion of the practical issues that can arise in resolving the interface between humanitarian law and human rights law during an insurgency, including the manning of checkpoints).

fire from a crowd of demonstrators (see Chapter 12, section 12.3.3, *Confronting the Explosive Threat*).[67] The challenge for security forces is responding to such threats with an appropriate level of force.

Not all operations demand a law enforcement focus. Where members of insurgent armed forces are identified they can be targeted under humanitarian law as legitimate military objectives. Further, depending upon the relative success of the insurgency, full-scale combat can occur between State armed forces and conventionally structured armed groups. Successful insurgencies ultimately involve the seizure of territory and can include the employment of more traditionally organized military forces. Countering such military threats inevitably requires the application of force under humanitarian law conduct of hostilities rules. However, even when conducting hostilities, security forces are still often required to take on more subtle and difficult challenges uniquely related to insurgencies and small wars taking place among a civilian population. This can include identifying members of the insurgent organized armed groups as such, and then engaging them in circumstances that minimize collateral civilian damage.

The involvement of insurgent forces in criminal activity does not make the counterinsurgency task necessarily an easy or straightforward one. For example, the Taliban has been closely associated with the drug trade in Afghanistan.[68] In 2008 this prompted a significant debate within NATO as to whether Taliban participants in the illegal narcotics trade in Afghanistan were also direct participants in hostilities such that they could be targeted under a conflict of hostilities paradigm.[69] It quickly became evident that a more nuanced assessment of what constitutes direct participation in hostilities, the identification of military objects, and when involvement in the ordinary law enforcement was required.[70] Similar issues regarding the status of workers arose with the commencement of the 2014 bombing campaign against the oil-production facilities controlled by the Islamic State in Syria.[71]

The requirement that soldiers operate in an environment that spans a broad threat spectrum means it is not uncommon for States to issue a blended rules of engagement (ROE) card, although separate cards may also be used.[72] "Blended" ROE cards often include both conduct of hostilities–based and human rights law enforcement–focused rules (see

---

[67] *See, e.g.*, VIOLENT RESPONSE: THE U.S. ARMY IN AL-FALLUJA, 15 HUMAN RIGHTS WATCH, Iraq, No. 7 (E), June 2003, http://www.hrw.org/reports/2003/iraqfalluja/.

[68] ABDEL BARI ATWAN, AFTER BIN LADEN: AL QAEDA, THE NEXT GENERATION 147 (2012).

[69] Michael N. Schmitt, *Targeting Narcoinsurgents in Afghanistan: The Limits of International Humanitarian Law*, 12 Y.B. INT'L HUM. L. 1, 1–4 (2009). *See also* TONY GERAGHTY, BLACK OPS: THE RISE OF SPECIAL FORCES IN THE C.I.A., THE S.A.S., AND MOSSAD 143–44 and 347–48 (2010).

[70] Schmitt, *supra* note 69, at 1–4.

[71] Kenneth Watkin, *Targeting "Islamic State" Oil Facilities*, 90 INT'L L. STUD. 499, 509–10 (2014).

[72] Kenneth Watkin, *Maintaining Law and Order During Occupation: Breaking the Normative Chains*, 41 ISR. L. REV. 178 (2008) (for a discussion of the "blended" versus "situation based" approach to applying force in occupied territory.). *But see* Juan Carlos Gomez, *Twenty-First-Century Challenges: The Use of Military Forces to Combat Criminal Threats*, 88 INT'L L. STUD. 279, 285–86 (2012) (for an outline of the approach adopted by the Columbian armed forces using separate blue card outlining human rights based rules for the use of force for law enforcement missions, and a red card based on humanitarian law for combat missions. This approach can be effective as long as it is remembered that even when a red card is applied there remains the obligation to apply blue card, human rights based, rules to uninvolved civilians.).

Chapter 11, section 11.3.1, *Rules of Engagement Doctrine*). The rules particular to each normative regime are applied depending upon the nature of the operation and the threat presented to the security forces. However, consistent with the "police primacy" principle, the default remains the application of minimum force and law enforcement–type rules, unless there is the positive identification of an insurgent member of an armed group or a civilian taking a direct part in hostilities. From both a legal and practical perspective, support for the law enforcement dominant approach is strengthened by the increasing recognition that human rights–based norms, reflected in Article 75 of Additional Protocol I and Article 4 of Additional Protocol II, are part of humanitarian law. If those provisions do not apply as a matter of treaty law, then there is a need to consider them as customary international law (see Chapter 5, section 5.3.2.3, *Divergence and Growing Reconciliation*). This approach not only addresses an operational requirement permitting military forces to better interface with the civilian population. It also reflects the reality there is a substantial parity in the protection provided in human rights law–based provisions and those found in humanitarian law documents.[73]

There are multiple advantages to applying human rights based–law enforcement norms as part of humanitarian law. For example, it helps avoid the difficult debate regarding the extraterritorial application of human rights law due to the universal applicability of humanitarian law during an armed conflict. This approach also limits the risks inherent in an exclusionary approach when applying either body of law to operational situations for which they were not designed. The sole application of conduct of hostilities norms carries with it the danger of an overapplication of force in situations more amenable to law enforcement. At the same time, a traditional policing approach can be woefully inadequate in the face of an organized and well-armed insurgent opponent. The integration of the two normative regimes during counterinsurgency operations is enhanced when human rights law is used to interpret humanitarian law based–human rights norms. The same "soft law" instruments outlining the principles governing law enforcement applied under the human rights accountability regime can be used.[74] Where gaps exist they can be addressed by applying human rights law, as a matter of either treaty or customary law.[75] In addition, the decisions of domestic courts and regional human rights bodies assessing law enforcement activities under human rights law may also be considered (see Chapter 5, section 5.4, *Overlap and Convergence*).

However, recognition that there is a simultaneous application and often overlap of the two normative regimes is only the beginning of the discussion. Questions remain regarding the circumstances under which those regimes apply and how fighting an insurgency influences their application to resolve security challenges in a manner consistent with the rule of law. Importantly, in deciding to employ police or military forces there is the necessity to consider the limits of each normative regime. For example, it is one thing to adopt a policy

---

[73] Emily Crawford, The Treatment of Combatants and Insurgents under the Law of Armed Conflict 125 (2010).

[74] *See, e.g., UN Basic Principles on the Use of Force* (which outlines human rights based principles regarding the use of force).

[75] Cordula Droege, *Elective Affinities? Human Rights and Humanitarian Law*, 90 Int'l Rev. Red Cross 501, 521–22 (2008) (for a discussion of the meaning of "complementarity").

of preferring capture rather than killing an opponent, which is a fundamental aspect of policing, and quite another to articulate when such a tactic can and should be used. As will be discussed next, the ability to apply a capture policy is directly impacted by the capacity to physically control the area where an operation is taking place.

7.4. Control and Operational Choices

7.4.1. THE NATURE OF CONTROL

Not coincidentally, the goal of establishing of control over territory is a fundamental aspect of counterinsurgency operations. It is an issue intimately linked to geography. Geography and war are raised in a number of other contexts (see Chapter 3, section 3.3, *An "Overarching" Application*, and Chapter 8, section 8.5, *Legal Frameworks for Responding to Transnational Terrorism*). Unique terms, such as the "geography of war"[76] and "hot battlefields" have been introduced in recent years.[77] Much of the contemporary discussion has centered on attempts to have geography set limits, or "boundaries," often based on international borders, for when humanitarian law could be applied to carry out drone strikes. However, in a counterinsurgency context the ability to exercise control over territory ultimately helps determine whether a law enforcement or conduct of hostilities approach can be applied. David Kilcullen has noted, "whoever does better at establishing a resilient system of control, that gives people order and a sense of security where they sleep, is likely to gain their support and ultimately win the competition for government."[78] This theory of "competitive control" over a population focuses on its functional basis (i.e., incentives and disincentives), rather than a structural (i.e., networks, or movements) or spatial one (i.e., contested areas) (see Chapter 4, section 4.4.5, *Small Wars and Ungoverned Spaces*).[79] However, the battle for the physical control of territory inevitably becomes one of the key factors in winning that competition.

In contrast to purely criminal gangs or even traditional terrorist groups, the goal of insurgents is to physically control territory and begin to establish the structures of governance. This is evident with Salafi jihadist organizations, such as Al Qaeda, which seek to create a caliphate and the Islamic State that claims to have done so. In effect, "the traditional supranational Islamic form of government," in order to rule over vast territories, which are presently under the juridical and physical control of the Westphalian State structure.[80] The exclusive control of territory exercised by either States or non-State actors can have a direct effect on how operations are conducted by a State. Some non-State actors exercise sufficient control over territory that they can block or at least seriously obstruct incursions by opposing State armed forces.

---

[76] Michael N. Schmitt, *Charting the Legal Geography of Non-International Armed Conflict*, 90 INT'L L. STUD. 1 (2014).

[77] Ashley S. Deeks, *Pakistan's Sovereignty and the Killing of Osama Bin Laden*, 15 AM. SOC. INT'L L. INSIGHTS (May 5, 2011), http://www.asil.org/insights/volume/15/issue/11/pakistans-sovereignty-and-killing-osama-bin-laden.

[78] KILCULLEN, *supra* note 58, at 152.

[79] *Id.* at 131, 152–53.

[80] ASSAF MOGHADAM, THE GLOBALIZATION OF MARTYRDOM 68 (2008).

Where non-State actors exercise exclusive control, the conflict between the State and non-State actor can take on semblance of inter-State warfare involving significantly elevated levels of violence. The result is that the ability by State armed forces to apply a law enforcement response in such non-State actor controlled territory is likely to be exceedingly limited. This has been perhaps most graphically demonstrated in the case of the Hezbollah in Lebanon and the Hamas in Gaza. Both of these non-State actors exercise control and even significant governance functions over relatively well-defined territories. They have maintained military forces that are sufficiently organized to effectively counter, to varying degrees, Israel's ability to enter and maintain a presence on the land those groups control. Their armed forces have a more conventional or at least "hybrid" organization. For example, efforts by Israel to enter either Lebanon or Gaza over the past decade have required military operations involving significant ground forces (i.e., armored formations, artillery), naval operations, and the use of airpower. The number of aerial attacks alone during these conflicts suggests levels of violence and the potential for collateral damage that is not well suited to regulation under a human rights–based paradigm.[81]

Rather than point toward a humanitarian law response, when a State governs and exercises adequate control the opposite frequently occurs. States' ability to broadly enforce an elevated level of control, particularly in a domestic context, most often leads directly to the law enforcement option being applied. States generally prefer to use a policing response even when dealing with alleged participants in armed conflict who are operating within their own borders (see Chapter 4, section 4.4.3, *A Reluctant Application of Humanitarian Law*). Attacks involving war-like violence may still occur,[82] but what makes these isolated eruptions of violence likely to be viewed primarily as policing matters is the ability of the security authorities to exert sufficient geographic control over the area thereby enabling them to contain the impact of the incident, apply an effective policing response, and react quickly and effectively to these threats.

Between those situations where States and non-State actors exercise what is effectively an exclusive control function, there can be a complex and violent space within which these protagonists fight for control over territory and the populations that live there. That territory can be urban or rural. The borders of a State may identify the territory in question; it could transcend international borders to "regions of savagery";[83] or may simply be limited

---

[81] Benjamin S. Lambeth, Air Operations in Israel's War Against Hezbollah: Learning from Lebanon and Getting It Right in Gaza, Rand Project Air Force 251 (2011), http://www.rand.org/content/dam/rand/pubs/monographs/2011/RAND_MG835.pdf (there were 18,900 aerial sorties during the thirty-four-day 2006 Israeli-Hezbollah conflict that struck 7,000 targets. During the twenty-three-day 2008 conflict with Hamas 5,650 sorties were carried out with 3,430 targets being hit.). *See also* Ben Hartman, *50 Days of Israel's Gaza Operation, Protective Edge—By the Numbers*, The Jerusalem Post (Aug. 28, 2014, 0425 AM), http://www.jpost.com/Operation-Protective-Edge/50-days-of-Israels-Gaza-operation-Protective-Edge-by-the-numbers-372574. (In 2014, in over fifty days of conflict, Hamas fired 4,564 rockets at Israel, and that country responded with strikes at "5,226 targets in Gaza, including 1,814 targets linked to rocket firing, 109 weapons depots, and 85 weapon-making facilities.").

[82] *See, e.g.*, Isabel Teotonio, *Toronto 18*, The Star, http://www3.thestar.com/static/toronto18/ (last visited June 26, 2015) (One example was the "Toronto 18," a homegrown cell of jihadists in Canada whose plan in 2006 was "to storm Parliament Hill, behead politicians and force the withdrawal of troops from Afghanistan. Also on his hit list were Canadian military bases in Edmonton and Trenton, the Pickering power plant, and the CBC building and CSIS offices in Toronto.").

[83] Abu Bakr Naji, The Management of Savagery: The Most Critical Stage Through Which the Umma Will Pass, at [16] (William McCants trans., 2006), https://azelin.files.wordpress.com/2010/08/abu-bakr-naji-the-management-of-savagery-the-most-critical-stage-through-which-the-umma-will-pass.pdf.

to the geographic limits of a city. A term found in the *United States Army, Marine Corps Counterinsurgency Field Manual*, "mosaic war," offers the best description of this intervening space where "[i]nsurgents may use guerrilla tactics in one province and an urban approach in another. There may be differences in political activities between villages in the same province."[84] This struggle can occur during occupation, as an element of a civil war, or otherwise as part of an internal insurgent battle within States. While the term "mosaic war" appears to have been created in the context of conflict within a State, it can easily be extended to transnational "regions of savagery" where control over territory is being contested.

The mosaic nature of this struggle relates not only to the threat but also to the State response. In countering the insurgent threat, and in exercising governance generally, States will inevitably have to use both law enforcement and combat-related approaches. As will be discussed, the practical choice made by the State of whether to employ law enforcement or conduct of hostilities methods can hinge on the question of the degree of physical control over the immediate area of operations. Control is a concept inherently linked to law and order. In legal terms, the rule of law "is the ultimate normative system of control."[85] The existence of "effective control" is often discussed in jurisdictional terms regarding the applicability of a human rights law regime,[86] or to establish when territory is occupied.[87] However, in looking at the concept of control, care must be taken to avoid extreme interpretations of the law that suggest human rights jurisdiction extends to the outermost range of a weapon system (i.e., cruise missile, sniper bullet). That form of "control" would be exercised literally at the end of a gun.[88] The problem is that these approaches appear divorced from the reality of attempting to operationalize the rights involved.

Theories about jurisdictional reach should not be confused with what can be practically and normatively carried out. When assessing the practicality of applying a law enforcement approach, "control" is more likely to be considered in terms of physical control, rather than as a jurisdictional legal sense. The maintenance of physical control over territory has developed as an important criterion for resolving the question of whether human rights, or conversely conduct of hostilities norms can be applied. This has occurred particularly in the context of assessing if an attempt can be made to carry out the capture of an insurgent, or whether a targeted killing will be ordered. This was highlighted in an Israeli case, *Public Committee Against Torture in Israel v. Israel* (also commonly referred to as the *Targeted Killing Case*).[89] Put in practical terms, the physical control necessary to successfully, and with a reasonable degree of safety for the security forces involved and civilians in the area, affect an arrest or detention.

---

[84] THE COUNTERINSURGENCY MANUAL, *supra* note 12, ¶ 1-37, at 14.

[85] *Id.* at 153.

[86] *See, e.g.*, Al-Skeini v. The United Kingdom, IV Eur. Ct. H. R. Rep. 99 (2011).

[87] 1097 Hague Land Warfare Regulations, art. 42 (such as "where such authority has been established and can be exercised.").

[88] NOAM LUBELL, EXTRATERRITORIAL USE OF FORCE AGAINST NON-STATE ACTORS 223–24 (2010). *But see* LEDWIDGE, *supra* note 24, at 2 (where reference is made to British forces controlling, in a practical sense, "only the area a few hundred metres from the barrels of the machine guns guarding our beleaguered bases.").

[89] Public Committee Against Torture in Israel v. Israel, Israel Supreme Court [16 December 2006], 46 ILM 375.

7.4.2. THE "*TARGETED KILLING*" *CASE* AND CONTROL

The 2006 *Public Committee Against Torture in Israel v. Government of Israel*[90] case arose in the context of targeted killings conducted by Israeli government security forces in occupied territory. This included a 2002 strike that killed Saleh Shehada, a Hamas founder and leader. He was killed when an Israeli F-16 struck his apartment building and other adjacent structures with a 2,000-pound bomb. The attack "killed fourteen civilians, including Shehada's daughter and nine other children as well as Shehada's wife and aide."[91] The strike occurred after previous attempts by Israeli authorities to arrest Shehada had failed.[92]

In a unique extension of human rights–based principles to an analysis of humanitarian law–based targeted killing, the Israeli Supreme Court identified "control" as one of the factors that impacted on whether law enforcement norms could be applied.

The court found that the use of arrest, prosecution, and trial might be particularly practical and "realizable possibilities" during belligerent occupation in circumstances "in which *the army controls the area* where the operation is taking place . . . ."[93] The risks to soldiers and civilians were the other relevant factors.[94] Logic suggests that these factors are closely interrelated. The *Targeted Killing Case* also introduced the human rights law notion that the right to life of the criminal suspect had to be factored into the use of force assessment.[95] Under a traditional humanitarian law–based conduct of hostilities assessment, the lawful target can be killed unless they have surrendered or are otherwise *hors de combat*.

The *Targeted Killing Case* suggestion that "control" should one of the factors used to determine if a capture rather than kill approach is to be applied to terrorists has not been universally accepted. Ryan Goodman takes the view this approach is too limiting, and there is "no substantial precedent for this [control-based] constraint."[96] Instead, it is argued that a least-restrictive-means (LRM) approach should be used in armed conflict against "enemy combatants."[97] The preferred LRM approach indicates if combatants can be captured, they should not be injured. If they can be wounded, then combatants should not be killed. Grave injuries should be avoided when light injuries can be used to put the combatant out of action.[98]

The use of "control" as a factor to determine levels of force to be used during military operations is identified in the LRM analysis as having been explicitly introduced "by the [2009 ICRC] *Interpretive Guidance*—essentially as a sliding scale in which the 'practical

---

[90] *Id. See also* KENT ROACH, THE 9/11 EFFECT: COMPARATIVE COUNTER-TERRORISM 124–28 (2011) (for a discussion of the *Targeted Killing Case*).

[91] DANIEL BYMAN, A HIGH PRICE: THE TRIUMPHS & FAILURES OF ISRAELI COUNTERTERRORISM 309 (2011).

[92] *Id.*

[93] Public Committee Against Torture in Israel v. Israel, Israel Supreme Court [16 December 2006], 46 ILM 375, ¶40, at 393–94 (2007) (emphasis added).

[94] *Id.* ¶40, at 394. *See also* Aerial Drone Deployment on 4 October 2010 in Mir Ali/Pakistan, 157 I.L.R. 722, 742 (2013) (where it is noted in respect of a drone strike in Pakistan "it would have been impossible to perform a military capture and arrest operation without a heightened risk for the soldiers involved or for the civilian population.").

[95] Public Committee Against Torture in Israel v. Israel, Israel Supreme Court [16 December 2006], 46 ILM 375, ¶40, at 393–94 (2007).

[96] Ryan Goodman, *The Power to Kill or Capture Enemy Combatants*, 24 EUR. J. INT'L L. 819, 827 (2013).

[97] Goodman, *supra* note 96, at 819.

[98] *Id.* at 819–20.

importance' of RUF [restraints on the use of force] 'increase[s] with the ability of a party to the conflict to control the circumstances and area in which military operations are conducted'."[99] That portion of the *Interpretive Guidance on the Notion of Direct Participation in Hostilities Under International Humanitarian Law* had itself been directly impacted by the decision in the *Targeted Killing Case.*[100] As is rightly pointed out, the effect of applying a control criterion could be to "limit the application of the rule to situations that more closely approximate to law enforcement operations rather than active battlefields in standard international armed conflicts."[101] However, neither the Israeli decision, nor the *Interpretive Guidance* specifically limits the application of the control criterion to law enforcement operations, although both realistically recognize that a human rights–based approach that favors capture rather than killing is more likely to be instituted in a situation of governance.[102]

Rather than find its basis in the theory underpinning the *Interpretive Guidance*, the LRM analysis relies on two principles reflected in Additional Protocol I: the scope of the *hors de combat* principle, which indicates once a combatant has that status they "cannot be subject to attack but can be apprehended and detained";[103] and the prohibition against employing methods of warfare that cause superfluous injury or unnecessary suffering.[104] This approach has garnered considerable criticism.[105] For example, it has been argued, "limitations on specific means and methods do not equate to a general principle requiring that the least restrictive means must be used in every (or any) combat engagement."[106] Significantly, in rejecting the *Targeted Killing Case* approach the LRM theory only provides protection to combatants in what is admitted to be "well specified and narrow circumstances."[107] This has been critiqued as constituting a "rare" occurrence."[108] Problematically, and perhaps most important, is that the situation posited by the LRM theory is not the most significant challenge facing contemporary commanders or represents the greatest risk to civilians. Rather, it is the one addressed by the *Targeted Killing Case* and the *Interpretive Guidance.* That is how to deal with insurgents hiding among the people.

In any event it is not clear that the issue of control can be easily dismissed. While the LRM theory outwardly rejects the control-based test for restraining the use of force, the

---

99 *Id.* at 827. *See* the INTERPRETIVE GUIDANCE ON THE NOTION OF DIRECT PARTICIPATION IN HOSTILITIES UNDER INTERNATIONAL HUMANITARIAN LAW, International Committee of the Red Cross (2009) [hereinafter INTERPRETIVE GUIDANCE].

100 ICRC INTERPRETIVE GUIDANCE, *supra* note 99, at 81 n.220. *See also* Goodman, *supra* note 96, at 827 n.32 (for reference to *Public Committee Against Torture in Israel v. Israel*, Israel Supreme Court [16 December 2006], 46 ILM 375, ¶40, at 394 (2007), and the issue of "control").

101 Goodman, *supra* note 96, at 827.

102 *See, e.g.*, ICRC INTERPRETIVE GUIDANCE, *supra* note 99, at 80–81, 81 n.220.

103 Goodman, *supra* note 96, at 830. *See* AP I, art. 41(2)(a).

104 Goodman, *supra* note 96, at 848–52. *See* AP I, art. 35(2).

105 *See, e.g.*, Geoffrey S. Corn, Laurie R. Blank, Chris Jenks, & Eric Talbot Jensen, *Belligerent Targeting and the Invalidity of a Least Harmful Means Rule*, 89 INT'L L. STUD. 536 (2013), and Michael N. Schmitt, *Wound, Capture, or Kill: A Reply to Ryan Goodman's "The Power to Kill or Capture Enemy Combatants,"* 24 EUR. J. INT'L L. 855, 857 (2013). *See also* Ryan Goodman, *The Power to Kill or Capture Enemy Combatants: A Rejoinder to Michael N. Schmitt*, 24 EUR. J. INT'L L. 863 (2014).

106 Corn et al., *supra* note 105, at 592–96.

107 Goodman, *supra* note 96, at 819.

108 Schmitt, *supra* note 105, at 857.

exercise of an elevated level of control appears to be a fundamental requirement for the operation of the theory. It is found in the suggested obligation "to not use lethal force when a soldier could clearly be rendered *hors de combat* just as easily—and *without endangering the attacking party*—by injury or capture rather than death." [109] The attacking party would not be "endangered" only where there is an almost absolute level of control being exercised. As the *Targeted Killing Case* identifies, control and risk (to both soldiers and civilians) are intimately intertwined.[110] The issue of immediate physical control appears to have been a key aspect of Michael Schmitt's critique of the LRM theory when he accepts that an enemy is in the power of the attacking force "if *the house and the area* in which it was located were under the *full control* of friendly forces, and apprehension could be easily effect *without risk* to the team or civilians. . . ."[111] Of note, this scenario requires complete control and no risk being posed to the soldiers involved in order for the obligation to arise. In this respect, it sets a standard that is likely impossible to achieve in all but the most exceptional of circumstances.

The importance of the *Targeted Killing Case* is that it provides an analytical framework that assists commanders in reconciling a more pressing and widespread operational issue: the application of humanitarian law rules while fighting among the people. The need to reconcile the risk posed to both soldiers and civilians most accurately reflects the decisions that have to be made during contemporary counterinsurgency operations. At the heart of the practical application of a capture rather than kill approach is the reality that such activity (i.e., what is conceptually a law enforcement approach) is dependent upon security forces being able to exert significant levels of physical control over the immediate area where the incident occurs. However, such control does not have to be absolute. That is why risks to soldiers and civilians are also factors to be assessed. Ultimately, an inability to establish adequate physical control could result in a decision to use deadly force against a lawful target. Such distinctions are not foreign to law enforcement. For example, during a hostage situation, the inability to arrest or detain the hostage taker may result in the decision to use deadly force, often by use of sniper fire.[112] This can mean a use of force that in many circumstances does not involve an attempt to wound, but rather is intended to kill (see Chapter 12, section 12.3.4, *Strictly Necessary, Proportionate, and Overwhelming*). However, many capture operations will be feasible in situations where the ability to control the area is temporally limited and the risks may be relatively elevated but the goal of the operation makes attempting it worthwhile.

The unsettled security situation present in many small wars, such as insurgencies, presents additional challenges. It will often be difficult to determine the degree of control exercised by an insurgent group in comparison to the State armed forces. The control exercised can fluctuate considerably over the course of a conflict. However, it has been noted "[w]hen the Government's armed forces have been thoroughly evicted, the insurgents' control can

---

[109] Goodman, *supra* note 96, at 822.

[110] Public Committee Against Torture in Israel v. Israel, Israel Supreme Court [16 December 2006], 46 ILM 375, ¶40, at 393–94 (2007).

[111] Schmitt, *supra* note 105, at 861 (emphasis added).

[112] *See, e.g., Toronto Hostage-taker Had History of Domestic Violence: Reports*, CBC (Aug. 26, 2004), http://www.cbc.ca/news/canada/story/2004/08/25/hostage_union040825.html.

scarcely be denied even if their presence on the ground is light."[113] While security forces applying a law enforcement paradigm have confronted terrorist groups using significant levels of violence it is not common for the police to have to fight their way to or from an objective when making an arrest. Security forces also may have to be concerned about being ambushed while conducting the operation. Significant security threats confronted in a counterinsurgency context have included the use of improvised explosive devices (IEDs), mortars, and other weapons of war.[114] Rocket-propelled grenades (RPGs)[115] and antiaircraft missiles have been used to shoot down supporting aircraft.[116] These threats impact directly on the feasibility of an arrest, or detention.

Regardless of whether a policing approach embarked upon as a matter of policy or is required at law, it remains to be determined if it can be practically carried out in the existing security environment. The ability to conduct law enforcement in areas where there is competition for physical control will be highly context dependent. For example, Israel's experience in seeking to maintain order in the territories it occupies points to a number of the challenges that can arise when conducting arrest operations. The Israeli security response has varied from carrying out a policing function, to the conduct of hostilities. Israeli security forces maintained order in the Occupied Territories during the first intifada (1987–93) by at times using clubs, CN and CS gas, rubber and plastic bullets, and increasingly restrictive rules of engagement.[117] The use of the Israel Defence Forces (IDF) in this role was controversial, with criticism of what was viewed as a "heavy-handed approach"[118] and conversely concern being expressed that the military was becoming a police force.[119]

While the levels of violence occurring within the Occupied Territories at various times were very elevated, this did not necessarily preclude the conduct of arrest operations, although not always without the danger of violence.[120] Israeli methods have at times been very controversial. They included the use of local Palestinian residents, who are often neighbors or family members, to provide an "early warning" to the occupants of a house believed to be containing a suspect.[121] In 2005, the Israeli Supreme Court ultimately ruled in *Adalah v. GOC Central Command* that the procedure violated international law in requiring civilians of occupied territory to aid the security forces, failed to separate the civilian from the zone of hostilities, real consent on the part of the person could not be guaranteed, and the

---

113 YORAM DINSTEIN, NON-INTERNATIONAL ARMED CONFLICTS IN INTERNATIONAL LAW ¶144, at 45 (2014).

114 Al-Skeini v. The United Kingdom, IV Eur. Ct. H. R. Rep. ¶ 23 at 121 (2011).

115 Ray Rivera, Alissa J. Rubin, & Thom Shanker, *Copter Downed by Taliban Fire; Elite U.S. Unit Among Dead*, N.Y. TIMES, Aug. 6, 2011, http://www.nytimes.com/2011/08/07/world/asia/07afghanistan.html?pagewanted=all.

116 MARK URBAN, TASK FORCE BLACK 155 (2010).

117 SERGIO CATIGNANI, ISRAELI COUNTER-INSURGENCY AND THE INTIFADAS 82–83 (2008).

118 *Id.* at 82.

119 *Id.* at 92 (quoting Martin Van Creveld *in* RUTH LINN, CONSCIENCE AT WAR: THE ISRAELI SOLDIER AS A MORAL CRITIC 72 (1996)).

120 AARON COHEN, BROTHERHOOD OF WARRIORS 199 (2008) ("The whole reason the [Duvdevan undercover] unit was created during the First Intifada—the entire methodology behind our training—was to take targets alive.").

121 Roland Otto, *Neighbours as Human Shields? The Israel Defense Forces' "Early Warning Procedure" and International Humanitarian Law*, 86 INT'L REV. RED CROSS 771 (2004), and Adalah v. GOC Central Command, IDF, HCJ 3799/02 ¶¶ 3–4 (judgement of Vice President M. Chislin) (June 23, 2005), http://elyon1.court.gov.il/Files_ENG/02/990/037/A32/02037990.a32.pdf (for reference to the pressures that can be brought to bear on family members to conduct the warning). *See also* BYMAN, *supra* note 91, at 150–51.

danger to the person could not be accurately assessed in advance.[122] If the procedure had been reviewed under a human rights lens, it is also highly likely it would have been found to be contrary to that body of law.

The elevated levels of violence and actions by organized armed groups can result in a very inhospitable environment within which to operate. While arrests may be able to be carried out, they sometimes can result in a loss of life.[123] During the Al-Aqsa intifada (2000–05) members of the resistance used roadside bombs, mortar attacks, large-scale ambushes, and gunmen among crowds of civilians, and snipers to attack Israeli security personnel in what was effectively urban guerrilla warfare.[124] Israel reacted with increasing firepower, viewing itself as being in a state of war.[125] The challenge in trying to enter areas being defended by guerrilla forces was graphically demonstrated by a 2002 Israeli operation in Jenin. As part of Operation Defensive Shield the Israeli Defense Force sought to seize weapons, arrest terrorists and their support network, destroy weapons factories and suicide bomb warehouses, as well as kill suicide bombers.[126]

The resistance in Jenin, carried out by perhaps 200 to 300 fighters,[127] included the extensive use of booby-traps, ambushes, and snipers.[128] The fighting involved intense urban warfare resulting in extensive property damage. There was an apparent misuse of civilians by both sides.[129] Twenty-three Israeli soldiers were killed.[130] A 2002 Human Rights Watch report indicates 52 Palestinians died, of which 22 were civilians.[131] There was also considerable international criticism of Israeli tactics.[132] However, as one commentator has noted, "[a]ssessments of the appropriateness of the devastation in Jenin depend upon whether one views such operations as police or military actions."[133] As a police operation, the casualties are problematic. In hostilities the deaths occurring in this urban setting are lower than other situations that have occurred. What this operation does demonstrate is that attempts to enter urban areas to conduct arrests and deal with weapons facilities can, depending upon the circumstances, be extremely dangerous for the security services involved, as well as civilians in the area.

This cost-benefit analysis when deciding how to proceed against enemy groups is not unique to the Israeli/Palestinian conflict. A similar approach appears to have been adopted in counterterrorism operations conducted during the Iraq insurgency. In order to defeat

---

122 Adalah v. GOC Central Command, IDF, HCJ 3799/02 ¶ 24 (June 23, 2005), http://elyon1.court.gov.il/Files_ENG/02/990/037/A32/02037990.a32.pdf.

123 Yaakov Lapin, *IDF Kills Palestinian during Overnight Arrest Raid in Ramallah*, Jerusalem Post, Sept. 10, 2014, http://www.jpost.com/Arab-Israeli-Conflict/IDF-kills-Palestinian-during-overnight-arrest-raid-in-Ramallah-374921.

124 Catignani, *supra* note 117, at 104–06.

125 *Id.* at 111.

126 *Id.*

127 Byman, *supra* note 91, at 144 (estimates 200 Palestinian fighters). *But see* Catignani, *supra* note 117, at 112 (puts the number at around 300 guerrillas).

128 Byman, *supra* note 91, at 146.

129 *Id.* at 149–51.

130 *Id.* at 154.

131 Israel, The Occupied West Bank and Gaza Strip, and the Palestinian Authority Territories, Jenin: IDF Military Operations, Human Rights Watch, Vol. 14, No. 3 (E) 12 (May 2002), http://www.hrw.org/reports/2002/israel3/israel0502.pdf.

132 Byman, *supra* note 91, at 153–54, and Catignani, *supra* note 117, at 114.

133 Byman, *supra* note 91, at 151.

al-Zarqawi and Al Qaeda in Iraq (AQI) a requirement was identified to "carry out nightly raids into gritty neighborhoods to systemically dismantle his network and capture of insurgents who hardly appeared to be high-value targets."[134] It has been noted that "[t]o many in the elite units, and to some critics outside the command, these less glorious tasks were better left to police or conventional forces."[135] This demonstrates once again that the line between what constitutes the conduct of hostilities and policing activity can be a fine one in a counterinsurgency context. The targeting of terrorists did occur, as was demonstrated in the June 7, 2006, killing of al-Zarqawi. However, even in this instance there was a strong desire to capture and interrogate the terrorist leader. Ultimately, a strike was carried out, because "a ground raid would be difficult, with a high probability of failure."[136] The two bombs dropped by an F-16 aircraft destroyed the house where he was located. Significantly, this attack was followed up with a helicopter-borne team tasked to "secure the site and *arrest* anyone who escaped the blast. . . ."[137] The goal of arresting terrorists again reflects the hybrid nature of these contemporary counterinsurgency/counterterrorism operations. There was targeting with conduct of hostilities rules but also an application of human rights law–based concepts in seeking to carry out "arrests."

The resistance presented to the security forces also does not have to involve large-scale violence. The attempt to detain Saddam Hussein's two sons, Uday and Quasay, in July 2003 provides a graphic example of the challenge that can be presented when operating in a counterinsurgency environment. Despite knowing the location of the targets in Mosul, Iraq, and having the ability to control the immediate area of operations, simply waiting the suspects out who were holed up in a house was not viewed as an option because "[a] siege situation made the US forces vulnerable to sniper fire from the surrounding area."[138] Further, an initial capture attempt resulted in three soldiers being wounded. The result was a significant amount of force being used, two collateral civilian deaths, and ultimately the deaths of the two brothers, Quasay's son, and a bodyguard.[139] However, simply because the security situation is unsettled does not mean a law enforcement approach can never be applied. Operations in Northern Ireland presented significant challenges to British forces seeking to act within a law enforcement normative framework, but a law enforcement approach was capable of being applied in preference to one relying on conduct of hostilities rules (see Chapter 13, section 13.2.1, *The "Troubles"*).

To date, the *Targeted Killing Case* requirement that capture must be considered, as a matter of law, appears not to have been widely applied. Further, the applicability of the *Targeted Killing Case* criteria across the broader conflict spectrum is unclear, although the German Federal Prosecutor did just that in an assessment of a 2010 American drone strike in Pakistan that killed a German citizen.[140] The *Targeted Killing Case* criteria, which

---

134 Stanley McChrystal, My Share of the Task: A Memoire 1749 (2013).

135 *Id.*

136 *Id.* at 229.

137 *Id.* (emphasis added).

138 David C. Isby, Leave No Man Behind: Liberation and Capture Missions 346 (2004).

139 *Id.* at 347 (among the weapons eventually used were TOW antitank missiles, helicopters firing 2.75-inch rockets, .50 Cal machine guns, and MK-19 40 mm grenade launchers). *See also* Williamson Murray & Robert H. Scales, The Iraq War: A Military History 235 (2003) ("The air force offered the finality of a few JDAMs, but [Brigadier-General] Helmick refused, preferring to capture the brothers alive if possible.").

140 Aerial Drone Deployment on 4 October 2010 in Mir Ali/Pakistan, 157 I.L.R. 722, 749 (2013).

places a premium on control and risk, appear to be particularly relevant to the situations of occupation that underpin that decision. Such situations are ones where an occupying power, much like a governing power in an internal armed conflict, can often exercise a high level of physical control.[141] In an Israeli context, the *Targeted Killing Case* obligations are more likely to lead to capture over killing in the Occupied Territories of the West Bank where physical control is more easily exercised than in Gaza, which is presently controlled by Hamas. The situation in Gaza is less about a State operating in a governance role where human rights norms have a practical ability to be enforced.

While the approach adopted in the *Targeted Killing Case* is not a statement of broadly accepted international law (i.e., the *lex lata*), it does reflect the unique challenges facing courts and international lawyers when they attempt to articulate the rules applicable to security situations occurring at the interface, or "grey zone" between human rights–based law enforcement and the conduct of hostilities under humanitarian law. It is an approach that has garnered significant criticism, particularly from exclusionists who prefer to keep the law of armed conflict separate from human rights considerations. However, the announced policy of the Obama administration to require consideration of the capture prior to killing of "all potential terrorist targets, regardless of whether or not they are American citizens," indicates the *Targeted Killing Case* was the harbinger of a more integrated normative approach toward the conduct of contemporary operations.[142]

7.5. The IED and Suicide Bombing Threat: Arrest or Kill?

The group aspect of insurgency has a unique impact on the levels of violence threatened, and what responses are available to a State. This is because some of the most violent attacks by insurgents, such as the frequent use of IEDs and suicide bombing, are fundamentally group rather than individual efforts (see Chapter 12, section 12.3, *Explosives: The Game Changer*). The impact of such group violence extends to the question of whether the restrictive interpretation on the imminence principle under human rights law affects the utility of applying a law enforcement approach when dealing with these contemporary threats (see Chapter 11, section 11.2.4, *Imminence and the Nature of the Threat*). The ability to deal with IED and suicide bombing cells is also affected by the operating environment. The environment within which a law enforcement approach, which privileges capture over killing an opponent, must be assessed is often an exceedingly complex one.

Among the issues to be addressed is whether a law enforcement response is required at law or is simply a policy choice. The analysis requires consideration of who may be kinetically targeted under humanitarian law and who may not. Those who may not be targeted

---

[141] Legal Consequences Of The Construction Of A Wall In The Occupied Palestinian Territory, 2004 I.C.J. ¶ 139, at 194 (where the court distinguished the Palestinian terrorist threat from that posed by Al Qaeda during the 9/11 attacks on the basis that Israel exercises "control" over the Occupied Territories).

[142] *Obama's Speech on Drone Policy*, N.Y. Times, May 23, 2013 [hereinafter the *Drone Policy*], http://www.nytimes.com/2013/05/24/us/politics/transcript-of-obamas-speech-on-drone-policy.html. *See also*, *Letter Attorney General Eric Holder to The Honorable Patrick Leahy*, May 22, 2013, at 3 [hereinafter *US Attorney General Letter*], http://www.justice.gov/slideshow/AG-letter-5-22-13.pdf.

must be arrested. This issue requires consideration of the structure of an organized armed group and its supporting infrastructure. What will become evident is that sometimes as a matter of law, and in other situations as a matter of policy, a law enforcement approach will very likely have to be applied in contemporary counterinsurgency and counterterrorism operations to counter the threat posed by explosive devices.

The group nature of IED attacks can be seen in the description of specialized technical cells in Iraq. Such specialized cells could include "technical and bomb-making, suicide-bomber support or facilitator cells, reconnaissance and operational security."[143] The IED and suicide bomber controlled "technical" cells have been identified as the two most important ones providing "critical and direct support to *combat* units."[144] They are also identified as "essentially combat units themselves."[145] The IED cells can consist of six to eight people, a planner/financier, the bomb maker, the "emplacer," the triggerman, and often a cameraman to video the results for propaganda purposes.[146] Suicide bombers also require an organization providing logistical infrastructure, often including "the provision of a safe house and food, and the explosive-laden vehicle or suicide belt."[147] The use of suicide vests is a cost-effective weapon "costing as little as $50 to $150."[148]

A fundamental humanitarian law requirement is that a person being targeted is taking a "direct part in hostilities."[149] One question to be addressed is whether the role being performed by the person in facilitating IED or suicide bombing attacks meets the legal criteria of direct participation. Finding the answer to that question has been made easier as a result of the International Committee of the Red Cross (ICRC) efforts to clarify the meaning of that term in the *Interpretive Guidance*.[150] There is now a general consensus, although not a unanimous one,[151] among international lawyers that persons may be targeted on the basis of their membership in an organized armed group.[152] However, disagreement remains regarding how such membership is determined and the degree to which it extends to persons fulfilling certain support functions. One controversial issue has been the very restrictive interpretive criteria developed by the ICRC to assess direct participation. In an effort to limit the size of the group against which deadly force may be used, the ICRC introduced a number of new concepts in its *Guidance*, such as persons having to perform a "continuous combat function,"[153] a "revolving door" of protection for periodic civilian

---

143 Ahmed S. Hashim, Insurgency and Counter-Insurgency in Iraq 160 (2006).

144 *Id.* at 161 (emphasis added).

145 *Id.* at 160.

146 Ricks, *supra* note 3, at 218. *See also* Hashim, *supra* note 143, at 161.

147 Hashim, *supra* note 143, at 162. *See also* McChrystal, *supra* note 134, at 170–73 (for a description of the suicide bomber "pipeline" from Syria to Iraq and the point of detonation), and Jessica Stern, Terror in the Name of God: Why Religious Militants Kill 51 (2003) ("The most important factor is organization: almost nobody does this as an individual; candidates are almost always trained.").

148 Moghadam, *supra* note 80, at 32.

149 AP I, art. 51.

150 Interpretive Guidance, *supra* note 99.

151 *Report of the Special Rapporteur on extrajudicial, summary or arbitrary executions, Philip Alston*, UN Doc. A/HRC/14/24/Add.6, (May 28, 2010) ¶¶ 57–69, at 19-21, http://www2.ohchr.org/english/bodies/hrcouncil/docs/14session/A.HRC.14.24.Add6.pdf.

152 Interpretive Guidance, *supra* note 99, at 34. *See also* Aerial Drone Deployment on 4 October 2010 in Mir Ali/Pakistan, 157 I.L.R. 722, 747–48 (2013).

153 Interpretive Guidance, *supra* note 99, at 27–35.

participants,[154] and a "one causal step" limitation for determining when an act results in harm.[155] What is unclear is their basis in law and, equally important, their suitability in practice.

All these new ICRC suggested criteria raise concerns. The one causal step approach, if followed, would impact directly on the targeting of IED or suicide bombing cells. Under the *Interpretive Guidance* it is not sufficient there is a causal connection between the act being carried out and harm produced. Referring specifically to IEDs, "it is noted the assembly and storing of an improvised explosive device (IED) in a workshop, or the purchase or smuggling of its components, may be connected with the resulting harm through an uninterrupted causal chain of events, but, unlike the planting and detonation of that device, do not cause that harm directly."[156] Those engaged in a continuous function of preparing, executing, or commanding such acts would be targetable under the ICRC "continuous combat function" criteria,[157] as would a person acting as a lookout.[158] However, persons who build, store, transport, recruit, or train would not.[159]

This approach is consistent with the statement in the *Interpretive Guidance* that "[i]ndividuals who continuously accompany or support an organized armed group, but whose function does not involve direct participation in hostilities, are not members of that group within the meaning of IHL [international humanitarian law]."[160] In effect, the logistics and training roles performed for a non-State actor are equated in status to private contractors and civilians accompanying State armed forces.[161] The problem is while civilians may be employed in that capacity, that is not the only way in which logistics support is provided to organized armed groups. Such groups usually have their own integral logistics component even within cellular organizations. This can be seen in Al Qaeda–related doctrine, which contemplates the need for a logistics element, or team of two to four individuals within small independent units operating within urban areas "to supply everything that the other units need in terms of weapons, tools, equipment, documents, safe houses, vehicles, etc."[162] Under the ICRC theory, those providing this support to the organized armed group would only be amenable to a law enforcement approach, which would focus on the capture, detention, and possible trial of those persons.

---

[154] *Id.* at 70–71. *See* William Boothby, The Law of Targeting 160–62 (2012), and Department of Defense Law of War Manual, Office of the General Counsel ¶ 5.9.4.2, at 231–32 (June 2015) [hereinafter DOD Law of War Manual], http://www.dod.mil/dodgc/images/law_war_manual15.pdf (for a rejection of the "revolving door" theory). *See also* Kenneth Watkin, *Opportunity Lost: Organized Armed Groups and the ICRC "Direct Participation in Hostilities" Interpretive Guidance*, 42 N.Y.U. J. Int'l L. & Pol. 641, 686–90 (2010).

[155] Interpretive Guidance, *supra* note 99, at 53.

[156] *Id.* at 54.

[157] *Id.* at 34. *See* Aerial Drone Deployment on 4 October 2010 in Mir Ali/Pakistan, 157 I.L.R. 722, 748 (2013) (while the continuous combat function test was adopted, "members of such a group will not be entitled to re-claim the protected status of a civilian until such time as they cease to assume their continuous combat role *in a lasting and conclusive manner.*") (emphasis added).

[158] Interpretive Guidance, *supra* note 99, at 53 ("although the recruitment and training of personnel is crucial to the military capacity of a party to the conflict, the causal link with the harm inflicted on the adversary will generally remain indirect.").

[159] *Id.*

[160] *Id.* at 34.

[161] *Id.*

[162] Norman Cigar, Al-Qa'da's Doctrine for Insurgency 123 (2009).

From a security perspective, the ICRC approach creates a number of problems. In an IED cell, the bomb builder is not only the most technically skilled person but "also doubles as training instructor."[163] While eliminating that person may not completely cease operations of the cell, since it can be ultimately reconstituted, his removal has far more impact on the ability to thwart the insurgent efforts than trying to strike at individual bombers at the point of attack. The latter approach, which is reflected in the *Interpretive Guidance* focus on the tactical level (i.e., "one causal step"), places the security forces in a decidedly reactive mode. The removal of a terrorist threat at its source can be a particularly attractive option. It is operationally more sound to attack those who build, train, and distribute IEDs as part of an organized armed group's activities, rather than restrict a response to persons at the point they lay or detonate the explosives.[164] The removal of the bomb maker focuses on the "brains" of the IED threat. Striking at the leadership and technical experts in an IED cell is particularly effective since the bomber "is often the cheapest and least important part of this entire network."[165] It also means that instead of trying to intercept multiple operatives once they are dispatched on a mission, a single or relatively limited number of attacks can be used to remove the threat at is source.

Attacking a threat at its source logically helps to end, reduce, or at least disrupt to some degree future attacks, given the importance of a skilled bomb maker to the organization's operations. However, not everyone agrees that the best military advantage can be gained from such an approach. It has been suggested the "attack-the-leaders" strategy is unproductive since frequently more aggressive, smarter, and younger candidates will replace the leaders. Further, it means the risk to those actually carrying out the attacks was "close to zero!"[166] Instead, it is suggested to be far better to concentrate lower down on the bomb planters and triggermen, "who were susceptible to increased risk."[167] However, this theory does not appear to adequately take into account the proliferation of IEDs in locations such as Afghanistan, or the difficulties in locating and targeting the bomb layers and triggermen.[168] What may be most problematic from a humanitarian perspective is that more civilians can be placed at risk if the decision is made to wait until an attack is attempted, or focus is maintained on the point of attack. The threatened State is left trying to remove multiple threats at or near the time that an attack is likely to occur, often in the midst of a heavily populated

[163] HASHIM, *supra* note 143, at 161.

[164] INTERPRETIVE GUIDANCE, *supra* note 99, at 54. *See also* Watkin, *supra* note 154, at 680–81 (for a critique of this approach based, in part, on the integrated nature of IED cells), and HASHIM, *supra* note 143, at 160–62 (for reference to the bomb builder of an IED cell often providing the training and the essential role performed by the logistic network.).

[165] BYMAN, *supra* note 91, at 101. *See also* Seán D. Naylor, *Killing Abu Ghadiya*, FOREIGN POL'Y (Aug. 31, 2015), https://foreignpolicy.com/2015/08/31/deltaforce-seals-syria-iraq-terrorism/?utm_source=Sailthru&utm_medium=email&utm_campaign=New%20Campaign&utm_term=%2AEditors%20Picks (where it is reported a senior Al Qaeda militant was killed in a 2008 raid on a safe house in Syria. Abu Ghadiya ran a foreign fighter network that included twenty to thirty suicide bombers entering Iraq each month).

[166] ANDREW COCKBURN, KILL CHAIN: THE RISE OF THE HIGH-TECH ASSASSINS 157–59 (2015).

[167] *Id.* at 158.

[168] TOBY HARNDEN, DEADMEN RISEN: THE WELSH GUARDS AND THE REAL STORY OF BRITAIN'S WAR IN AFGHANISTAN 212–13 (2011) (where it is noted that following a tactic of focusing on the devices was not working. The number of IEDs and the difficulty in discovering them being laid would appear to undermine the "attack the emplacers/triggerman" theory).

civilian environment and under extremely compressed time lines. Unfortunately, the locations of attacks can frequently be found in marketplaces, business areas, public transit, tourist attractions, and other gathering points for uninvolved civilians (see Chapter 12, section 12.3.2, *Suicide Missions*).

The narrow view of support functions suggested by the ICRC is fundamentally at odds with how armed groups organize themselves and conduct hostilities. In an apparent contrast to the *Interpretive Guidance*, it has been noted a number of States appear to have a "shared sense" that criteria such as "combat support and combat service support functions if performed for a regularly constituted armed force and carrying arms openly, exercising command over the group or one of its units, or conducting planning related to the conduct of hostilities" can be used to determine membership in an organized armed group.[169] While, the author of the *Interpretive Guidance* does not share this view, it is one that most accurately reflects how military forces, including those belonging to non-State actors (i.e. organized armed groups), conduct operations.[170] Further, the *Guidance* contemplates an ability to transfer from a continuous combat function to a function where participation would be periodic. As William Boothby has rightly noted, "[o]ne frankly doubts the realism of this, and indeed other aspects, of the ICRC's stated position and it is highly unlikely that this approach will meet the general approval of States."[171]

However, this does not mean that States will not have to clearly justify who is targeted. There will be persons indirectly involved in the insurgent IED and suicide bombing effort that may not meet the criteria of direct participation. For example, the question can be asked whether the Pashtun tribesman known as "al-Qaeda's tailor," who made his living making suicide vests, is actually part of the jihadist organization or simply a civilian indirectly supporting hostilities in much the same way as weapons production occurs in a State.[172] The integral nature of his activities to the operations of that terrorist group points to his effectively being a member of that organized group. However, additional information such as his acting under the direction of Al Qaeda leadership would further solidify this conclusion. In this respect, being able to equate the function performed by a bomb maker to the combat support or combat service support functions performed in State armed forces; assisting in the planning of an attack; or preparing the bomber for the operation are additional criteria that would assist in making a membership assessment. However, if it cannot be established the bomb maker is a member of the organized armed group, or not otherwise taking a direct part in hostilities, then security forces could only apply a law enforcement model to deal with that person.

---

169 Stephen Pomper, *Toward a Limited Consensus on the Loss of Civilian Immunity in Non-International Armed Conflict: Making Progress through Practice*, 88 INT'L LEG. STUD. 181, 189 (2012).

170 Nils Melzer, *Keeping the Balance Between Military Necessity and Humanity: A Response to Four Critiques of the ICRC's Interpretive Guidance on the Notion of Direct Participation in Hostilities*, 42 N.Y.U. J. OF INT'L L. AND POL. 831, 848–49 (2010) (where it is suggested in response to criticism of the narrow approach to membership in organized groups in the ICRC *Interpretive Guidance* that combat support activities like logistics "would almost invariably constitute an integral part of combat operations, because they generally involve direct support to combat units" it would not extend to combat service support. This approach appears to view the provision of combat service support as an exclusively civilian activity, however, that simply is not the case.).

171 BOOTHBY, *supra* note 154, at 162.

172 JOBY WARRICK, THE TRIPLE AGENT 150–51 (2011) (describing how the suicide vests are made by "al-Qaeda's tailor").

Another limitation on the conduct of hostilities approach, which ultimately would leave the security forces considering a law enforcement option, is the potential for excessive collateral damages should a strike be made on an IED or suicide bombing cell. This unique aspect of fighting insurgences conducted "amongst the people" could lead security forces to apply a law enforcement approach where feasible (see section 7.6.2.2, *Proportionality and Counterinsurgency*, below). Further, the *Interpretive Guidance* was correct about the logistics issue to the extent it recognized that organized armed groups may employ local civilians to provide support to their operations in a manner similar to State armed forces.[173] Some support activities may not constitute direct participation in hostilities, while others might.

For example, Al Qaeda made extensive use of criminal smuggling organizations to bring suicide bombers to Iraq.[174] The "supply lines of material, money recruiters, handlers, and, most importantly, volunteers, stretched to Riyadh and Aleppo, Tunis and Hamburg."[175] Not all such activity would constitute direct participation by the civilians involved, any more than geographically distant provision of contractor logistics support would in an inter-State context.[176] The need for "[e]nhanced cooperation with international law enforcement agencies"[177] was identified in order to address criminal activity. However, as happened in Liberia, armed groups may employ or even force women and children to carry ammunition, food, and other supplies to the front lines.[178] This is an example where a legal conclusion may be reached that such participation is sufficiently direct to permit targeting, however, morality can dictate a solution other than an attack.

The provision of logistical support is not the only circumstance where involvement in a conflict is periodic or fueled by a desire for financial reward. In Afghanistan the planting of bombs, which constitutes direct participation in hostilities, has been done on a contractual basis. It is reported that the Taliban have paid "civilians $10 a time to dig a hole."[179] Similarly, in Iraq the participation in the planting and igniting of IEDs in some situations became a commercial enterprise where the "reward for an IED team was $50 for a lookout, $100 to dig a hole and $200 to trigger a device."[180] While persons carrying out these actions could be legally targeted, and with repeated involvement may come to be viewed as members of the organized armed group, there can still be a reluctance to engage local participants in this form of hostilities. The calculation of "counterinsurgency math," referring to such deaths motivating insurgent recruiting, appears to work as much when local insurgents are

---

173 Interpretive Guidance, *supra* note 99, at 34.

174 Seth G. Jones, Hunting in the Shadows: The Pursuit of Al Qa'ida Since 9/11, at 154 (2012). *But see also* Jamie Smith, Gray Work: Confessions of an American Paramilitary Spy 101–02 (2015) (where it is noted rebels fighting against Muammar Gaddafi in 2011 used sea-smuggling routes from Malta "to illegally bring in weapons and fighters, while bringing in medical supplies and reporters on the same boats. The reality of things is that espionage requires dealing with liars, smugglers, traitors and criminals of all sorts . . . .").

175 McChrystal, *supra* note 134, at 168, and Williams, *supra* note 8, at 165 (for a discussion of the border security in Iraq following the disbandment of the Iraq Border Army after the 2003 invasion).

176 Williams, *supra* note 8, at 231–32 (referring to AQI oil smuggling and an arrest of a key Iraqi smuggler and AQI financier).

177 *Id.* 265.

178 James Brabazon, My Friend the Mercenary 59–60 (2010).

179 Andy Bloxham, *Soldiers Told Not to Shoot Taliban Bomb Layers*, The Telegraph, July, 8, 2011, http://www.telegraph.co.uk/news/newstopics/onthefrontline/8626344/Soldiers-told-not-to-shoot-Taliban-bomb-layers.html.

180 Richard Connaughton, Modern Warfare: The True Story of Conflict from the Falklands to Afghanistan 303 (2008).

killed as it does with respect to uninvolved civilians.[181] This makes sense given that the surviving family members are unlikely to appreciate the fine legal distinction between their loved one being killed as a direct participant in hostilities or as a collateral casualty. As a result, a policy decision may be made, based on strategic counterinsurgency goals, that the importance of not shooting local civilians outweighs the advantage gained in ceasing the hole-digging activity by kinetic means.[182] Once again, capture according to a human rights–based "arrest" approach would become the means applied to stop such participation (see section 7.6.2.2, *Proportionality and Counterinsurgency*, below).

What these limitations requiring direct participation in hostilities mean is that during armed conflict there will be some facilitators or personnel providing indirect support to armed groups, who are not directly participating in hostilities and therefore cannot be targeted. Those civilians are only subject to arrest or detention and processing under a law enforcement or security detention paradigm. This requirement exists regardless of whether or not those persons are operating in areas under the control of the State. Further, as has been noted, restrictions on targeting resulting from an application of the targeting precautions may similarly require military commanders to consider a law enforcement–based capture rather than kill approach. In these situations, and where a policy choice is made to apply a law enforcement approach, the ability to carry out arrests or detentions will hinge to a great extent on the factors identified in the *Targeted Killing Case*: the risk to security personnel and civilians, as well as the ability to "[control] the area in which the operation takes place . . . "[183] As Gabriella Blum and Philip Heymann have noted when analyzing the targeted killing paradigm, "[t]he limits set by the Israeli Supreme Court—ironically, within the paradigm of wartime operations—are a good place to start."[184]

In summary, in situations where relatively peaceful access to a location is sufficient to enable an arrest operation is not guaranteed, commanders are placed in the position having to weigh risk to soldiers and civilians against the possible reward of arresting members of a bombing cell or their supporters. Even during armed conflict there can be definite advantages to capturing rather than killing a member of a cell, which by definition is taking a direct part in hostilities. For example, it is suggested that in the Israeli experience the use of targeted killings "had galvanized rather than deterred the terrorists," and that using intelligence-driven arrests "has been a much more effective tool at foiling Palestinian terror."[185] However, in these situations, "the decision to safeguard one's own soldiers and security officers from harm has always been the determining factor of whether or not to carry out an arrest or targeted killing."[186] Importantly, the authority to attack insurgents or terrorists does not mean people providing support can be targeted if they are not direct participants in hostilities. They must be captured. However, those civilians would remain at risk under a collateral damage analysis if a weapons facility or group compound were subjected to a strike.

---

[181] KOLENDA, supra note 32, at 19.

[182] HARNDEN, *supra* note 168, at 398–99 (for reference to a situation where holding off on a requested air strike avoided killing or injuring a civilian woman and child).

[183] Public Committee Against Torture in Israel v. Israel, Israel Supreme Court [16 December 2006], 46 ILM 375, ¶40, at 393–94 (2007).

[184] Gabriella Blum & Philip Heymann, *Law and Policy of Targeted Killing*, 1 HARV. NAT. SEC. J. 145, 169 (2010).

[185] CATIGNANI, *supra* note 117, at 168.

[186] *Id.*

7.6. Military Advantage and Counterinsurgency

7.6.1. INSURGENTS AND TARGET "VALUE"

Another unique aspect of counterinsurgency is the impact this form of conflict has on how the rules governing targeting are applied. The law relating to targeting must be applied in a contextual manner. The application of targeting precautions during the conduct of a strategic air campaign that is designed to degrade the war-making capacity of a State occurs in a context having far different facts, military goals, and attendant challenges than an operation focused on an individual fighter located in an isolated civilian compound. This latter situation is more apt to occur during an insurgency. Conflict with non-State actors frequently can involve situations where violence is used against individuals or relatively small groups of fighters operating among the people. However, considerable combat can also occur in isolated areas of a territory. It is not that conflicts with non-State actors do not periodically take on the attributes of inter-State conflict. As has been noted, this can occur when the non-State actors control territory, adopt a governance structure, and organize their armed forces along conventional lines, or use a hybrid military organizational structure.[187] However, as was evident during both the Afghanistan and Iraq insurgencies of the post-9/11 period, these quasi-conventional-type conflicts are not the norm.

Most conflicts with non-State actors involve indirect methods of traditional guerrilla warfare, where those challenging the State largely operate among the civilian population. With that population increasingly living in an urban environment, conflict has also migrated to cities and inhabited areas.[188] The prosecution of these small wars frequently requires a different operational approach. For example, "[b]ombing civilians, or targeting insurgents and terrorists in urban areas with resulting civilian casualties, generally works to the propaganda advantage of the rebels."[189] The goal of denying such an advantage to the enemy can alter the assessment of what is to be gained militarily from the use of force.

One of the most interesting outcomes during counterinsurgency is that the traditional gap between the hostilities and law enforcement paradigms is often significantly narrowed. This occurs because the use of violence often becomes, as has been noted, more individualized. The capture rather than killing of an opponent, most often associated with human rights–based law enforcement, has also become a conduct of hostilities issue. The result is an application of the targeting precautions set out in humanitarian law does not involve a "one size fits all" approach to the use of force.[190]

In this section, "military advantage" will be looked at in terms of the relative "value" to be gained when targeting members of organized armed groups during a counterinsurgency.[191]

---

[187] *See, e.g.*, YAAKOV KATZ & YOAZ HENDEL, ISRAEL VS. IRAN: THE SHADOW WAR 27 (2012) (for a discussion of Hezbollah and its approximation of the structure of State militaries).

[188] ANTHONY JAMES JOES, URBAN GUERRILLA WARFARE 2 (2007) ("It seems clear that, at least for the intermediate term, the future belongs to urban guerrilla warfare.").

[189] JAMES S. CORUM & WRAY R. JOHNSON, AIRPOWER IN SMALL WARS: FIGHTING INSURGENTS AND TERRORISTS 429 (2003).

[190] AP I, art. 57(2) (setting out what are commonly referred to as "targeting precautions").

[191] For a more detailed discussion regarding "military value" and the assessment of proportionality in the context of counterinsurgency *see* Kenneth Watkin, *Military Advantage: A Matter of "Value", Strategy and Tactics*, 17 INT'L Y.B. HUM. L. 277 (2014).

The next section looks more broadly at military advantage and proportionality in the context of counterinsurgency operations. The term "military advantage" is used in a variety of contexts under humanitarian law (i.e., the proportionality test, a test for indiscriminate attack, and the definition of military objective).[192] In respect of the public dialogue, "military advantage" has a particular relevance to the weighing of the anticipated military gains to be attained from carrying out an attack against the collateral effects on the civilian population and property (i.e., death, injury, or damage). This is often described as the proportionality test, although that term is not actually found in humanitarian treaty law.[193] Military advantage is linked to military necessity[194] and has been interpreted by some legal experts to mean "any consequence of an attack which directly enhances friendly military operations or hinders those of the enemy."[195]

Importantly, from both an operational and accountability perspective, it is the military commander who identifies the "military advantage" to be gained from an attack as a crucial preliminary step in the application of targeting precautions. In terms of accountability, the unique role performed by military commanders is reflected in the *Targeted Killing Case*, where the Israeli court ruled "the decision, made on the basis of the knowledge of the military profession, to perform a preventative act which causes the deaths of terrorists in the area . . . is the responsibility of the executive branch."[196] The court in turn "will ask itself if a reasonable military commander could have made the decision which was made."[197] In this regard, "special weight is to be granted to the military opinion of the official who bears the responsibility for security. . . ."[198] It is that military advantage, identified in the legal test, which is then assessed against the collateral effects of an attack. This means that while a military commander has an essential role to play in articulating what the military advantage is to be gained from an attack, the advantage is ultimately assessed as part of the legal test applied to determine whether such an attack could be justified.[199]

As is set out in Article 57(2)(a)(iii) of Additional Protocol I, the collateral civilian effects of an attack are weighed against "the concrete and direct military advantage anticipated." The ICRC *Commentary* to Additional Protocol I suggests "that the words 'concrete and direct' impose stricter conditions on the attacker than those implied by the criteria defining

[192] Kenneth Watkin, *Assessing Proportionality: Moral Complexity and Legal Rules*, 8 Y.B. of Int'l Hum. L. 17–23 (2005) (for a discussion of the term "military advantage").

[193] AP I, art. 57(2)(a)(iii) (where the collateral effects of an attack are described in terms of their being "excessive" in relation to the "concrete and direct military advantage anticipated."). *See* the 1998 Rome Statute, art. 8(2)(b)(iv) (for reference to the collateral effects being "clearly excessive").

[194] *See also* Neta Crawford, Accountability for Killing 167 (2013) ("military necessity has come to mean having a military objective, doing something that is useful to win the war.").

[195] Commentary on the HPCR Manual on International Law Applicable to Air and Missile Warfare, Program on Humanitarian Policy and Conflict Research at Harvard University ¶ 3, at 45 (2010), http://ihlresearch.org/amw/Commentary%20on%20the%20HPCR%20Manual.pdf.

[196] Public Committee Against Torture in Israel v. Israel, Israel Supreme Court [16 December 2006], 46 ILM 375, ¶ 57, at 399.

[197] *Id.*

[198] *Id.* ¶ 57, at 400.

[199] Yishai Beer, *Humanity Considerations Cannot Reduce War's Hazards Alone: Revitalizing the Concept of Military Necessity*, 26 EUR. J. INT'L L. 801 (2015) (linking the military necessity criterion (assessing military advantage) to an application of military professional standards that constrain the use of force and enhance humanitarian protection).

military objective in Article 52."[200] That article refers to a "definite" military advantage. However, the interpretative complexity of having a different standard for military advantage when determining what constitutes military objective than when assessing proportionality during a specific attack in highly problematic. It is a standard that has to be applied by individual soldiers, as well as planners working on a strategic air campaign involving thousands of targets. It would be virtually unworkable to have to apply two different standards of "military advantage" in the process of conducting an attack. This is particularly so in the often stressful time constrained environment of targeting. A potential complication in carrying out targeting decisions is that, unlike the definition of military objects,[201] there is no reference to military advantage in identifying when people are "targets."[202] However, this potential gap is closed with military advantage being part of the targeting precautions of Article 57(2)(a)(iii) of that Protocol, which applies to equally to people and things.

The military advantage to be gained from a targeting decision has to be weighed in respect of the tactical level effect of a strike, as well as its impact strategically. This is because all targeting is ultimately carried out for a strategic purpose (see Chapter 3, section 3.6.2, *Targeting: A Matter of Strategy and Tactics*). No commander, or individual soldier for that matter, gets to operate solely within a tactical environment. Any tactical-level decision can have strategic consequences. In assessing military advantage a common aspect of the contemporary targeting is the "value" that is placed on a person or object that is a potential target. In this context value is considered in terms of the importance, usefulness, or worth to be gained from attacking a target, rather than a principle, a standard of behavior, or a numerical amount.[203] The morality of targeting is particularly evident in the decisions that must be made to assess the military value of a target against the lives of uninvolved civilians. The moral and legal complexity of decision making inherent in many contemporary targeting analyses is reflected in the use of terms such as "high value target" to emphasize a target's "worth." [204] This moral challenge is not fundamentally altered by contemporary efforts to apply an objective computer modeling approach for assessing collateral damage,[205]

---

200 COMMENTARY ON THE ADDITIONAL PROTOCOLS OF 8 JUNE 1977 TO THE GENEVA CONVENTIONS OF 12 AUGUST 1949, Protocol I, art. 57, ¶ 2218, at 685 (Yves Sandoz et al., 1987) [hereinafter ICRC ADDITIONAL PROTOCOL COMMENTARY], *available at* http://www.cicr.org/ihl.nsf/COM/470-750073?OpenDocument.

201 AP I, art. 52 (2) (the reference to "military advantage" in this article only applies to military objectives that are objects).

202 AP I, arts. 43 and 51(3) (make no reference to military advantage).

203 OXFORD DICTIONARIES, http://www.oxforddictionaries.com/definition/english/value (last visited Apr. 4, 2015) (the meaning of "value" includes: "[t]he regard that something is held to deserve; the importance, worth, or usefulness of something"; "[p]rinciples or standards of behaviour; one's judgement of what is important in life"; or "[t]he numerical amount denoted by an algebraic term; a magnitude, quantity, or number").

204 DEPARTMENT OF DEFENSE DICTIONARY OF MILITARY AND ASSOCIATED TERMS, Nov. 8, 2012 (as amended through Mar. 15, 2015), http://www.dtic.mil/doctrine/dod_dictionary/index.html (where "high-value target" is defined as:

> A target the enemy commander requires for the successful completion of the mission. Also called HVT. See also high-payoff target; target. (JP 3-60).

Interestingly, there appears to be no definition for "medium-value" or "low-value" targets.).

205 George S. McNeal, *Targeted Killing and Accountability*, 102 GEO. L.J. 681, 740–45 (2014), *available at* http://papers.ssrn.com/sol3/papers.cfm?abstract_id=1819583 (where he outlines a multistep process followed by the U.S. military that he calls "Collateral Damage Methodology" (CDM). He indicates *id.* at 741, "[t]he CDM and the weapons-effect data contained in FAST-CD are based on empirical data gathered in field tests, probability, historical observations

or by setting numerical limits on the collateral casualties that can be approved at various levels of command.[206]

Therefore, the relative importance of a target, such as an insurgent, must be weighed against a different "value"-based assessment, the anticipated collateral death or injury of uninvolved civilians. This values-based approach can be seen in the *Targeted Killing Case* description of the proportionality test. It focused on "a balancing between conflicting values and interests," where "the benefit stemming from the attainment of the proper military objective is proportionate to the damage caused to innocent civilians harmed by it."[207] The difficult nature of this decision is an integral part of the application as a legal rule of what was historically a moral doctrine (e.g., the doctrine of "double effect").[208] Yet there is an aspect of this values-based test that is often overlooked. If a certain "value" can be assigned to a person or object in order to indicate its relatively greater importance from a targeting perspective, then it would also appear a target having a "lower" value could lead to the conclusion more limited or even no collateral casualties or damage may be justified.

For example, in a situation where the target is a single relatively low-level insurgent hiding in a building in a crowded urban area, the proportionality assessment can involve consideration of the appropriateness of the use of violence in the form of significant kinetic force (i.e., aerial or artillery bombardment). Operations in built-up areas will generally mean an increased potential for collateral civilians casualties. In considering an attack against an insurgent, the risk to civilians (including possibly members of that insurgent's family) has to be assessed, as well as the danger posed to the security forces. Ultimately, the question that has to be asked is at what point the life of a civilian trumps that of a "low-value" target. Logically, the answer cannot be "never." The assessment of the risk to a civilian could even extend beyond the use of high-explosive weapons, such as aerial delivered ordinance, to high-powered firearms and the dangers associated with firefights in an urban environment.[209]

This does not mean that aerial delivered or other explosive armaments (i.e., artillery or mortar fire) might never be used in an urban setting. Given the importance of Osama bin Laden, the risk to members of his family and nonbelligerent associates caused by the assault on the Abbottabad compound was justified. The "high value" of that target could properly have figured prominently in the justification of an attack, regardless of whether it was

---

from weapons employed on the battlefield, and physics-based computerized models for collateral damage estimates."). *See also* CRAWFORD, *supra* note 194, at 350–53 (outlining the development of a computerized approach toward collateral damage assessment).

[206] Douglas Jehl & Eric Schmitt, *Errors Are Seen in Early Attacks on Iraqi Leaders*, N.Y. TIMES, June 13, 2004 ("General Moseley, the top Air Force commander during the war who is now the Air Force vice chief of staff, said in the interview last summer that commanders were required to obtain advance approval from Mr. Rumsfeld if any planned airstrike was likely to result in the deaths of 30 more civilians. 'More than 50 such raids were proposed, and all were approved,' General Moseley said."). *See also* McNeal, *supra* note 205, at 750 (In an American context it has been noted, "depending on a host of variables, the approval authority for targeted killings may be a General Officer, the Director of the CIA, the Secretary of Defense, or even the President.").

[207] Public Committee Against Torture in Israel v. Israel, Israel Supreme Court [16 December 2006], 46 ILM 375, ¶ 45, at 395–96.

[208] MICHAEL WALZER, JUST AND UNJUST WARS 151–53 (3rd ed., 2000) (for a discussion of the doctrine of double effect). *See also* Watkin *supra* note 192, at 26–30.

[209] Watkin, *supra* note 191 at 296–98 and 305–06 (for a discussion of targeting low level soldiers).

carried out through bombing or in the form of an assault using firearms.[210] Interestingly, however, it was because of the elevated importance of bin Laden and with it the desire to confirm he had been killed that weighed heavily in the decision to enter the compound with "boots on the ground" in the form of a raid by special operations forces.[211] Indeed, this operation highlights that attacking a "high-value" target does not automatically mean its elevated military advantage will result in higher civilian casualties. In that case, the nature of the advantage being sought resulted in an approach that did have a complementary effect of limiting the risk to civilians present at the compound. The limitation of collateral civilian casualties was a factor considered during mission planning, but it was not the only factor impacting on operational decision making.[212]

In any event, regardless of the value of the target and the anticipated military advantage, attacking forces have to consider "all feasible precautions in the choice of means and methods of attack with a view to avoiding, and in any event minimizing incidental loss or civilian life." [213] The scenario may arise where the use of aerial delivered ordinance, such as a missile fired from a drone into a residential compound, might be viewed as creating excessive collateral damage when the relative importance of the target is assessed against the risk to civilians. In this case, consideration could turn to an assault, surreptitious entry to capture the individual, or even an operation to surround a building or compound to seek that person's surrender. Where feasible, the latter option can provide the advantage of limiting the risk to both surrounding civilians and members of the assaulting force, although that is far from guaranteed. However, the fact that a particular action (such as capture) is potentially possible does not mean it will be carried out since factors such as the degree of control, risk and even the time available on the objective must still be considered. [214]

Seeking the surrender of a suspected insurgent located in a residential compound is an approach that has been adopted during contemporary counterinsurgency operations. In order to maintain the support of the population in Afghanistan, the mandating of less-aggressive capture options became so pervasive that the ISAF Commander imbedded the unique terminology of "soft knock" and "hard knock" in his December 2011 *Night Operations Tactical Directive*.[215] A "soft knock" was used to describe a search and potential

---

210 Peter L. Bergen, Manhunt: The Ten-Year Search for Bin Laden from 9/11 to Abbottabad 174–86 (2012) (for an outline of the various courses of action considered for the bin Laden raid, which included the use of B-52s, drone strikes, and a Special Forces raid.).

211 David E. Sanger, Confront and Conceal: Obama's Secret Wars and Surprising Use of American Power 86 (2012). *See also* Charlie Savage, Power Wars: Inside Obama's Post 9/11 Presidency 262 (2015).

212 Bergen, *supra* note 210, at 178, Eric Schmitt & Thom Shanker, Counterstrike: The Untold Story of America's Secret Campaign Against Al Qaeda 261 (2011), and Daniel Klaidman, Kill or Capture 242 (2012) (e.g., concern over risk to U.S. troops). *See also* Watkin, *supra* note 191 at 298–99.

213 AP I, art. 57(2)(ii).

214 International humanitarian law and the challenges of contemporary armed conflict, Report 32nd International Conference of the Red Cross and Red Crescent EN 32IC/15/11, at 35 (2015), *available at* https://www.justsecurity.org/wp-content/uploads/2015/11/2015-ICRC-Report-IHL-and-Challenges-of-Armed-Conflicts.pdf (where it is noted 'the degree of control over a specific area or circumstance, and the intensity of the hostilities at the time and place of a particular operation, constitute relevant factors, among others, to assess what is "feasible" in terms of the application of IHL rules on precautions in attack . . . .').

215 COMISAF Night Operations Tactical Directive, Headquarters International Security Assistance Force 2 (Dec. 2011), *available at* http://www.isaf.nato.int/images/docs/20111105%20nuc%20night%20operations%20tactical%20directive%20(releaseable%20version)%20r.pdf [hereinafter Night Operations Tactical Directive]. *See*

capture option, which involved an Afghan security force–led "call-out" of the suspect. It was an approach that "should include clear instructions, and the occupants given sufficient time to comprehend, acknowledge, and comply."[216] Where the "soft knock" was unsuccessful, or the tactical situation precluded its use, then a "hard knock" such as aggressively breaching the door to the building could be applied.[217] The result was that the military advantage of capturing a "target" made the operation begin to resemble law enforcement activity.

The adoption of a "soft," or even "hard" knock approach, which seeks to capture an opponent, should not come as a surprise given the importance of "police primacy" approach to counterinsurgency success (see section 7.3, *The Primacy of Law Enforcement*, above). One of the particular criticisms leveled at the Coalition effort was the use of culturally insensitive night raids as "part of the kill/capture program in which Special Operations forces parachute into a village, knock down doors, search homes, and detain or sometimes kill the inhabitants . . . ."[218] Concerns were also expressed regarding the ability of the counterinsurgency forces to identify insurgents. One report estimated 80 percent of those captured were subsequently released.[219] Another factor that impacts on the decision to implement a capture policy during counterinsurgency is that not all capture operations involve insurgents. There may be a requirement to deal with transnational criminal gangs, other criminal elements, and persons providing support to the insurgency who are not taking a direct part in hostilities. Further, the military forces involved in the counterinsurgency mission may simply be supporting local police in their effort to generally maintain law and order. Where the mission is law enforcement related, or does not involve persons who may be targeted as lawful military objectives, then detention or arrest must be applied as a matter of law (see Chapter 14, section 14.4.1, *As a Matter of Law*).

The adoption of a less-aggressive counterinsurgency policy in Afghanistan, which limited the use of kinetic means, was not necessarily popular with all military personnel.[220] There was greater higher command scrutiny of tactical-level decision making. It has been noted that "[i]f Americans entered an Afghan house, they could expect a lot of scrutiny and second-guessing from senior officers. Raiding by conventional battalions got very, very difficult."[221] Such an approach was also sometimes viewed as an improper integration of

---

also LEDWIDGE, *supra* note 24, at 200 (where it is noted "hard knock" was a term used in Northern Ireland for a "forced entry").

216 NIGHT OPERATIONS TACTICAL DIRECTIVE, *supra* note 200, at 2. *See also* Elliot D. Woods, *The Soft Knock*, VA. Q. REV. (Oct. 9, 2009), http://www.vqronline.org/news/soft-knock.

217 Brian Hutchinson, *Assignment Kandahar: Soft Knocking the Bazaar*, NAT. POST, Aug. 8, 2011, http://news.nationalpost.com/2011/04/21/assignment-kandahar-soft-knocking-the-bazaar/. *See also* COHEN, *supra* note 120, at 199 (where in respect of Israeli undercover military arrests it was noted "one of two things would happen: we would knock on the door, announcing ourselves to be the Israeli army and ordering the occupants to come outside, or else a demolition expert would put a small charge on the door and blow it off its hinges.").

218 MacKenzie, *supra* note 30, at 219. *See also* SEAN NAYLOR, RELENTLESS STRIKE: THE SECRET HISTORY OF JOINT SPECIAL OPERATIONS COMMAND 362, 366–67 (2015) (referring to the overuse of house searches, and the tension even within the Special Operations community regarding the frequency and effect of night raids).

219 Gareth Porter, *Ninety Percent of Petraeus's Captured "Taliban" Were Civilians*, IPS (June 12, 2011), http://www.ipsnews.net/2011/06/ninety-percent-of-petraeuss-captured-taliban-were-civilians/ (*referred to in* MacKenzie, *supra* note 30, at 220). *See also* NAYLOR, *supra* note 218, at 367 (referring to concerns about detaining the wrong people).

220 BOLGER, *supra* note 5, at 305–06. *See also* NAYLOR, *supra* note 218, at 430 (where it is indicated that the requirement to "call out" before assaulting a target at night was very unpopular with "strike forces".)

221 BOLGER, *supra* note 5, at 305–06.

human rights law into operations governed by conduct of hostilities norms. However, this approach was not necessarily adopted because of a human rights–based privileging of capture over killing. Even in respect of insurgents, the application of targeting precautions under humanitarian law can result in a decision to seek the capture of an insurgent.[222] The capture option may present the least risk to civilians in the area, while still permitting the insurgent to be removed as a threat.

However, as with a law enforcement–based operation, an attempt to capture is not a guarantee that deadly force may not ultimately have to be used. Even with respect to a surreptitious entry or other effort to capture an insurgent, targeting precautions would have to be applied when considering the tactics to be followed and weapons to be used (i.e., methods and means of attack). The potential to use deadly force may remain very high.[223] In addition, a crucial factor in the ability to carry out a capture would be the existence of a relatively permissive environment in the immediate surrounding area of the operation. Barring an ability to attempt a capture operation without unacceptable risk to either the civilian population, or the military forces involved, the option left to the military planners may ultimately be to suspend any attack and wait until the conditions are right to carry out an attack. This could include seeking to locate the target in a place where the likely collateral damage does not preclude a targeted killing.

As has been noted, the conduct of operations must be considered across the tactical, operational, and strategic levels of warfare (see Chapter 3, section 3.6.1, *Self-Defence and the Levels of War*). The strategic goals of a campaign can have a direct impact on the choice of targets and the way an attack is carried out at the tactical level. This issue can be particularly relevant in a counterinsurgency and in associated counterterrorist-based attempts to decapitate the leadership of an enemy force. It is this element of contemporary counterinsurgency campaigns, particularly through the use of drones, that has attracted attention and controversy (see Chapter 8, section 8.4.2, *Drones: Special Terms and High-Profile Attacks*). There are scenarios where attacking the leadership of an insurgent or terrorist organization can effectively interrupt its operations. For example, the 1998 killing by Philippine police of the charismatic Islamist leader of the Abu Sayyaf group has been identified as just such an attack. The result was "a shift in the group's mission, as it split into smaller factions and subsumed its political agenda to criminal greed."[224] The military advantage to be gained from killing an enemy leader can raise the question of whether elevated collateral civilian casualties or damage can be justified during an attack. However, damaging the operational effectiveness of an insurgent group by killing a leader is not guaranteed. The decision to conduct such attacks requires an analysis of "the second- and third-order effects of the removal of terrorist leaders."[225]

In seeking the strategic military goal of ending a conflict, it has been recognized that killing the leaders of a non-State actor may also be counterproductive. That action may remove

---

[222] AP I, art. 57(2)(a)(ii).

[223] COHEN, *supra* note 120, at 199.

[224] AUDRY KURTH CRONIN, HOW TERRORISM ENDS: UNDERSTANDING THE DECLINE AND DEMISE OF TERRORIST CAMPAIGNS 27 (2009).

[225] *Id.* at 23. *See also* NAYLOR, *supra* note 218, at 288 (where it is noted that the 2006 killing of the Al Qaeda in Iraq leader, al Zarqawi, did not lessen the civilian death toll and instead it actually increased into 2007).

"useful negotiating partners in the future."[226] Another challenge in a counterinsurgency context is that the size and organizational strength of insurgent groups, as opposed to small terrorist organizations, may operate to insulate those groups from the negative impact of a targeted killing campaign. It has been noted that killing senior leaders may not damage an insurgency, especially if they are replaced "by younger, hungrier, more radical and more operationally experienced leaders from the next generation."[227] There is some evidence that "the deaths of leaders of groups such as Hamas and Palestinian Islamic Jihad apparently inspired more young Palestinians to join the movement."[228] There is also no guarantee that the leadership replacement will be an improvement or that the fight for succession will not ignite further attacks.[229]

The suggestion that armed groups will be unable to replace its leadership may be overrated. It is reported that by mid-2013 the U.S. drone campaign has killed over fifty senior leaders of Al Qaeda and the Taliban, out of some 3,300 operatives who have died.[230] However, neither of these two groups resembles the types of hierarchical, small, and relatively newly established terrorist groups that are historically vulnerable to decapitation strikes.[231] The Pakistani journalist Syed Saleem Shahzad indicated that in 2005–06, the Afghan Taliban, with an indigenous strength of 20,000 and with new recruits from Pakistan's tribal areas, increased in numbers to the same size as the 80,000 NATO forces operating in Afghanistan at the time.[232] Further, while in 2005 eighty-six drone strikes and other operations undertaken in Pakistan had killed eighteen top militant commanders, it was noted, "the military operations were unable to dent their long-term purpose."[233] In a similar vein, when the Pakistani Taliban leader, Hakimullah Meshud, was killed in November 2013, it is reported there were four candidates thought to be in the running to replace him.[234]

As with many other parts of the jihadist movement, the Afghanistan and Pakistan Taliban are large and resilient organized groups. So, it would appear, is Al Qaeda, and the same could also be said for the Islamic State. This does not mean that members of organized armed groups should not be targeted. However, the varying outcomes suggest that a

---

[226] Steven R. David, *Fatal Choices: Israel's Policy of Targeted Killing*, *in* DEMOCRACIES AND SMALL WARS 138, 145 (Efraim Inbar ed., 2003). *See also* BYMAN, *supra* note 91, at 56 (discussing the possibly negative impact that the Israeli killing of Abu Jihad in Tunis in 1988 may had given that Yasser Arafat was left alone in charge of the PLO), KILCULLEN, *supra* note 58, at 45 (where he notes "particularly late in the campaign, you may need to negotiate with the enemy"), and CRONIN, *supra* note 224, at 26.

[227] KILCULLEN, *supra* note 58, at 75.

[228] CRONIN, *supra* note 224, at 26.

[229] *Id.*

[230] Daniel Byman, *Why Drones Work: The Case for Washington's Weapon of Choice*, 92 FOREIGN AFF. 33 (July/August 2013) (relying on data from the New America Foundation). For more detail, *see Pakistan Drones*, NEW AMERICA FOUNDATION, http://securitydata.newamerica.net/drones/pakistan-analysis.html.

[231] Audrey Kurth Cronin, *Why Drones Fail: When Tactics Drive Strategy*, 92 FOREIGN AFF. 44, 45 (May/June 2013) ("Al Qaeda, by contrast, is a resilient, 25-year-old organization with a broad network of outposts. The group was never singularly dependent on Osama bin Laden's leadership, and it has proved adept at replacing dead operatives.").

[232] SYED SALEEM SHAHZAD, INSIDE AL-QAEDA AND THE TALIBAN: BEYOND BIN LADEN AND 9/11, at 154–55 (2011).

[233] *Id.* at 192.

[234] Declan Walsh, Ismail Khan, & Salman Masood, *Pakistani Taliban Gather to Select Successor to Leader Killed in U.S. Drone Strike*, N.Y. TIMES, Nov. 2, 2013, http://www.nytimes.com/2013/11/03/world/asia/pakistani-taliban-gather-to-select-successor-to-leader-killed-in-us-drone-strike.html.

belief that the dismantling of an organized group can occur by killing leadership targets can offer an oversimplification of the effect of such action. Further, identifying a target as being "high value," based simply on their place on an enemy organizational chart, may be problematic. The "value" assigned to a target requires a more sophisticated analysis. It has been suggested that enhanced drone strikes conducted from 2009 onward has hurt the Al Qaeda organization.[235] However, other analysis acknowledges the impact but concludes "[t]argeted killings have not thwarted the group's ability to replace dead leaders with new ones."[236] The killing of even high-level members of these organizations will not automatically justify elevated collateral civilian casualties or damage.

It was also recognized very early on in the post-9/11 operations in Yemen that "eliminated terrorists could quickly be replaced unless the general population was turned against Al Qaeda."[237] This once again highlights the importance of gaining the support of the civilian population during counterinsurgency operations, which can include counterterrorism-related strikes. The ability of relatively large insurgent/terrorist organizations to weather a targeted killing campaign should not come as a surprise when those organized armed groups are compared to State armed forces. There is significant annual turnover of senior leaders in conventional military forces through retirement and postings. This loss of leadership does not adversely affect their operational effectiveness. Similarly, it is not immediately evident that leadership replacement on the scale being conducted through targeted killing in large insurgent or terrorist forces like the Taliban and Al Qaeda's affiliated groups will necessarily present insurmountable long-term obstacles to those organizations.

It has been noted, "decisions about whether and how to target leaders reflect philosophies in a state's counterterrorism approach, and . . . tactical choices have significant strategic implications for the outcomes of campaigns."[238] The United States is said to prefer strike oriented endeavors (see Chapter 8, section 8.3 *Counterterrorism and "Direct Action"*).[239] However, States must be careful to avoid embracing targeted killing without fully assessing the potential military advantage, or disadvantage, of such attacks. An alternative to conducting such targeted killings is to seek the capture of leaders of an organized armed group. It has been suggested that "[o]n the basis of the historical record, arresting a leader is more effective in damaging a group than by killing him."[240] Sergio Catignani indicates that in the Israeli-Palestinian context "[t]he greater use of intelligence-driven arrests, instead of targeted killings, which had galvanized rather than deterred potential terrorists, has been a much more effective tool at foiling Palestinian terror."[241] The military advantage of conducting an arrest where possible, which also includes a potential for minimizing collateral casualties, has been viewed as outweighing the advantage to be gained from carrying out a targeted killing. However, consideration of the risk to security personnel remains a factor in determining whether arrests or targeted killing is to be carried out.[242]

---

[235] Byman, *supra* note 230, at 33.

[236] Cronin, *supra* note 231, at 46. *See* Atwan, *supra* note 68, at 259.

[237] Edmund J. Hull, High-Value Target: Countering Al Qaeda in Yemen 15 (2011).

[238] Cronin, *supra* note 224, at 16.

[239] David H. Ucko, The New Counterinsurgency Era 88 (2009).

[240] Cronin, *supra* note 224, at 32.

[241] Catignani, supra note 117, at 168. *See also* Cronin, *supra* note 224, at 21.

[242] Catignani, supra note 117, at 168.

In terms of military advantage it may also, at times, be more effective to target the middle-level staff, such as the planners, technical specialists, trainers, and so forth.[243] The approach adopted by the Coalition Special Forces in Iraq in 2005 was to direct operations against the middle ranks of the Al Qaeda network in an effort to cause the network to collapse.[244] This focus may increase their "value" as targets. Of note, such operations have included capture rather than kill missions. This broader targeting has been credited as one of the factors that brought about the decline of Al Qaeda in Iraq. This will not always be the case. It has been suggested that "constraints, from the political to operational ones posed by more limited intelligence, prevent the application of the JSOC [Joint Special Operations Command] Iraq model in many parts of the world."[245] When this broader targeting approach was attempted during the 2009 surge in Afghanistan, it appears not to have had as obvious a positive impact.[246] Similarly, expanded targeting in Pakistan in 2011 incorporating military-age males as part of the target set, which appears to have led U.S. officials to claim drone strikes had not killed civilians, was to result in a significant negative reaction from the Pakistan government and its population, particularly because of collateral civilian casualties.[247]

Contemporary conflict can require consideration of factors reflecting a more complex and nuanced assessment of what constitutes a military advantage than might be required if the targets were simply enemy soldiers belonging to State armed forces engaged in conventional warfare. The relative "value" assigned to individual members of insurgent organizations as targets may depend upon an in-depth consideration of the function they perform, their position in an organization, and the consequences their removal may have in terms of its impact on the ability of the organization to conduct the insurgency. However, as will be discussed in the next section, the military advantage to be gained from an attack can also require an assessment of broader strategic issues. Among those issues are the second- and third-order effects of such targeting, particularly in terms of potentially losing the support of the civilian population. Because of the critical strategic requirement to establish and maintain that support there is often a military advantage gained if State security forces restrict how operations against insurgents are conducted in order to lessen the impact on that population. The result can be that fewer targets will be attacked, and restrictions will be placed on the types of weapons and armaments used. This issue is also directly tied to the assessment of the military advantage anticipated from an attack and the collateral impact on civilians that is at the heart of the proportionality debate.

---

[243] KILCULLEN, *supra* note 58, at 75–76 ("killing or capturing the insurgent 'middle management' tier can do significant damage to the organization, while leaving senior leaders intact and perhaps even convincing them over time that their campaign is futile, and without killing large numbers of lower-tier fighters."). *See also* Byman, *supra* note 230, at 33, and McNeal, *supra* note 205, at 712–21 (for a discussion of the targeting of low level terrorists). *See also* COCKBURN, *supra* note 166, at 57–59 (where it is suggested it is better to target lower ranking personnel as they were viewed as more risk adverse than the leaders who being driven by ideology were more willing to die for the cause).

[244] URBAN, supra note 116, at 80. *See also* NAYLOR, *supra* note 218, at 270–71 (discussing a switch in effort to one "eviscerating Al Qaeda in Iraq's middle level leaders.").

[245] *Id.* at 273.

[246] JEREMY SCAHILL, DIRTY WARS: THE WORLD IS A BATTLEFIELD 331 (2013) (where it is suggested this period "would see a notable rise in support for the Taliban and a record number of US soldiers killed.").

[247] MARK MAZETTI, THE WAY OF THE KNIFE 290–91 (2013). *See also* Byman, *supra* note 230, at 39.

7.6.2. STRATEGIC ADVANTAGE, PROPORTIONALITY, AND COUNTERINSURGENCY

7.6.2.1. Strategic Military Advantage and Civilian Casualties

A unique aspect of an insurgency-based conflict is the effect that a key strategic operational goal, the avoidance of collateral civilian casualties and damage, has on how operations are conducted by counterinsurgents. [248] This impact is felt in two ways. First, in achieving a military advantage the manner by which counterinsurgency operations are conducted is heavily influenced by this sensitivity toward civilian casualties and damage. Second, there is the question of how this counterinsurgency unique determination of military advantage affects the targeting proportionality test. Ultimately, the strategic goal of limiting collateral effects on civilians can affect the tactical application of the proportionality test under humanitarian law. It can create a situation where there is a narrower acceptable range of collateral effects that can be justified.

Counterinsurgency operations are unique. During an insurgency/counterinsurgency, military action "is constantly directed toward a political goal."[249] Further, as Hew Strachan has noted, "[i]n counterinsurgency the distinctions between levels of war, tactical, operational, strategic and political are much less clear than in major war."[250] The result is that counterinsurgencies "can establish a direct link between the low-level, comparatively minor actions and significant political outcomes which is very rare in major war."[251] This ultimately means that "[p]olitical effects are therefore part of the immediate framework of military action."[252] The effects of an attack are often tied to collateral civilian casualties and damage.

United States joint doctrine acknowledges that counterinsurgency "differs in many respects from a traditional military force-on-force operation and requires a special mindset. It is the population-centric nature of COIN that distinguishes it from most traditional military force-on-force operations."[253] This recognition of insurgency/counterinsurgency as a type of warfare that is different from total or even more limited conventional conflict is reflected in other military doctrine.[254] In addition, maintaining the support of the people is also a key aspect of counterterrorism, which often forms a significant element of a counterinsurgency response. As a former U.S. Ambassador to Yemen indicated, counterterrorism "[i]n many ways resembled counterinsurgency, success of which depended primarily on winning the hearts and minds of the population."[255]

The broader strategic political and military impact of causing collateral civilian casualties and damage cannot be overstated. It is the strategic effect of a decision to use force that

---

248 Watkin, *supra* note 191, at 350–56. (where the impact of insurgency/counterinsurgency and counterterrorism operations on targeting proportionality is further discussed).

249 David Galula, Counterinsurgency Warfare: Theory and Practice 89 (1964).

250 Strachan, supra note 34, at 219.

251 *Id.*

252 *Id.*

253 Counterinsurgency, Joint Publication 3-24, ¶ 2, at I-2 (Nov. 22, 2013), http://www.dtic.mil/doctrine/new_pubs/jp3_24.pdf.

254 The Counterinsurgency Manual, *supra* note 12. *See also* Matthew Beran, *How Counterinsurgency Changes "Military Advantage"*, The Army Lawyer 4–5 (2010), *available at* http://www.loc.gov/rr/frd/Military_Law/pdf/08-2010.pdf.

255 Hull, *supra* note 237, at 15.

is particularly relevant in a counterinsurgency context. The collateral effect of targeting, in terms of its impact on the war effort, cannot be measured solely in respect of the numbers of civilians killed and wounded, or the extent of collateral property damage. It can extend to the negative effect that targeting decisions can have on support for government military forces. The military advantage gained from an attack is frequently viewed from an "effects-based" perspective. A key consideration is how commanders assess the effects of a decision to use force. Those effects can be direct or indirect, with the indirect collateral "effects" impacting both on the military result, as well as the traditional assessment of civilian casualties.[256] As a result, the military advantage need not solely be assessed by its direct effect on military action. The indirect effects of an attack can impact significantly on the military effort. Consistent with how wars are fought those effects can be felt at the tactical, operational, or strategic levels (i.e., high-strategy and military strategic levels of war) (see Chapter 3, section 3.6.1, *Self-Defense and the Levels of War*).

As the 2014 U.S. *Counterterrorism* manual notes, "[w]hile effects may be created by the conduct of missions at the tactical level, their contribution to achieving an objective is measured at the operational and higher levels."[257] For example, it has been suggested that as a result of the U.S. bombing campaign during the Vietnam War, "[c]ivilian casualties in Cambodia drove an enraged populace into the arms of an insurgency that had enjoyed relatively little support until the bombing began, setting in motion the expansion of the Vietnam War deeper into Cambodia, a coup d'état in 1970, the rapid rise of the Khmer Rouge, and ultimately the Cambodian genocide."[258] Similarly, one of the criticisms associated with the counterterrorist drone campaign in Yemen has been the negative impact it has had on the local population.[259] These adverse effects are not just felt at the strategic level. Another effect that can occur when targeting locally based insurgents is the alienation of the immediate population and the creation of "blood feuds that perpetuate the conflict."[260]

Significant collateral civilian casualties can have a negative impact on political-strategic outcomes, even if an attack is viewed as tactically and operationally effective.[261] In Pakistan the impact of collateral civilian casualties has been suggested to aid opposition recruiting.[262] Even Coalition military forces engaged in counterinsurgency in Afghanistan raised questions regarding the compatibility of targeted killings with such operations.[263] Being

---

[256] Michael N. Schmitt, *Effects-Based Operations and the Law of Aerial Warfare*, 5 WASH. U. GLOBAL STUD. L. REV. 265, 275–76 (2006).

[257] COUNTERTERRORISM, JP 3-26, ¶ 5 e., at IV-8 (Oct. 24, 2014), http://www.dtic.mil/doctrine/new_pubs/jp3_26.pdf.

[258] Taylor Owen & Ben Kiernan, *Bombs Over Cambodia*, THE WALRUS (Nov. 2006).

[259] Charlie Savage, *Drone Strikes Turn Allies into Enemies, Yemeni Says*, N.Y. TIMES, Apr. 23 2013, http://www.nytimes.com/2013/04/24/world/middleeast/judiciary-panel-hears-testimony-on-use-of-drones.html?ref=middleeast&_r=0 (outlining how a death from a drone strike turned Yemeni villagers against the United States).

[260] KILCULLEN, *supra* note 58, at 74.

[261] CATIGNANI, *supra* note 117, at 166–67.

[262] *Pakistan Official Slams Drones ahead of CIA Talks, Says Attacks Recruit New Militants*, ASSOC. PRESS (July 27, 2012), http://www.foxnews.com/us/2012/07/27/pakistan-official-slams-drones-ahead-cia-talks/ ("In a frank debate Friday with White House war adviser Douglas Lute, Ambassador Sherry Rehman said the drone attacks have already succeeded in damaging al-Qaida but are now only serving to recruit new militants.").

[263] CHRISTOPHER J. LAMB & EVAN MUNSING, SECRET WEAPON: HIGH-VALUE TARGET TEAMS AS AN ORGANIZATIONAL INNOVATION, Strategic Perspectives 4, Institute for National Strategic Studies 14 (Mar. 2011) ("The collateral political and sometimes physical damage complicated counterinsurgency efforts and made conventional force commanders less willing to collaborate with the high-value target teams.").

concerned about the impact of operations on a local population is in the direct security interests of the military force involved. Avoiding collateral civilian deaths and damage through the application of a more restrained use of force can limit the risk to one's own soldiers. The use of what is viewed among the population as excessive force or a mistreatment of civilians can be counterproductive.[264] What is perceived as excessive force can increase animosity toward the counterinsurgency forces and aggravate the security and political situation.[265] These tactical-level decisions do not exist in isolation. From a military perspective the purpose of an attack has to be nestled within the overall operational and strategic goals of a campaign. When targeting decisions are made there is a clear military advantage in avoiding such strategic alienation. Therefore, a negative strategic impact resulting from such targeting decisions can subsequently be felt at the tactical level as attacks increase against counterinsurgent forces.

It is concern over the broader negative effect of certain tactical-level decisions regarding the use of force that results in decisions by senior commanders to issue directives such as the 2011 Afghanistan *Night Operations Tactical Directive*.[266] It also can create a tension concerning what have been described as political generals, who have been suggested, "do better in counter-insurgency than 'gung-ho' warriors."[267] The adoption of a more restrictive approach toward the use of force by military commanders during counterinsurgency operations raises the question of whether that occurs simply as a matter of the exercise of command discretion (a form of policy choice), or results from an application of law. It has been suggested that winning the "hearts and minds," a classic goal of counterinsurgency, "does not qualify as advantage vis-à-vis the definition of military objective. Rather, military advantage is purely military in nature; there must be some direct contribution to military operations. Political, economic or social advantage does not suffice."[268] Further, "[w]hile true that collateral damage motivates civilian sympathy for the enemy, such general effects are too attenuated."[269]

A similar assessment takes the view that military advantage is limited to tactical gains. Therefore, "the proportionality-calculus in counterinsurgency is strictly policy-based and powered by strategic imperatives, and results in restrictions on causing collateral damage not required by the LOAC-based principle of proportionality."[270] Such a focus on tactical gains is consistent with the ICRC approach toward targeting.[271] However, this argument does not appear to reflect the broader position adopted by many States[272] or Article 8(2)(b)(iv) of the 1998 Rome Statute, which makes reference to the "overall" military advantage

---

264 CORUM & JOHNSON, *supra* note 189, at 426.

265 CATIGNANI, *supra* note 117, at 166–67 (outlining the Israeli experience with the impact that operations causing significant collateral damage have had on political-strategic outcomes).

266 NIGHT OPERATIONS TACTICAL DIRECTIVE, *supra* note 200.

267 STRACHAN, *supra* note 34, at 219.

268 Michael Schmitt, *Targeting and International Humanitarian Law in Afghanistan*, 85 INT'L L. STUD. 307, 323 (2009).

269 *Id.*

270 Eric Pouw, International Human Rights Law and the Law of Armed Conflict in the Context of Counterinsurgency 297 (Nov. 14, 2013) (unpublished Phd. Dissertation, University of Amsterdam) (on file with the author).

271 ICRC ADDITIONAL PROTOCOL COMMENTARY, *supra* note 200, art. 57, ¶ 2218.

272 CUSTOMARY INTERNATIONAL HUMANITARIAN LAW STUDY 49 (Jean-Marie Henckaerts & Louise Doswald-Beck eds., 2005).

that is anticipated.[273] Following this approach, concrete and direct military advantage of an attack should not be assessed only at the tactical level (e.g., by looking at numbers of insurgents killed). Warfare is simply not conducted solely at that level (see Chapter 3, section 3.6.2, *Targeting: A Matter of Strategy and Tactics*). Acknowledging the strategic level of war brings into play a broader perspective on the effects of an attack and its military consequences.

With the function of strategy being to apply "military means to fulfill the ends of policy,"[274] there is inevitably an interface between military action and intended political goals. In terms of counterinsurgency, in particular, the link between collateral civilian casualties and damage, and an adverse impact on the conduct of military operations, is not too remote. While the impact of those casualties can be political, it also has direct and tangible military consequences. The result is that it is to the military advantage of a commander to alter the manner in which military operations are conducted and degree of force used. A failure to do so can carry with it the danger of significant adverse military consequences for a counterinsurgency operation. This does not mean that certain types of force, such as aerial bombing or the use of artillery, should be banned. They can be extremely effective in countering an insurgent force, particularly where it has begun to adopt a conventional organization and approach toward the hostilities. However, the military advantage to be gained from their use must be weighed against the strategic consequences that can result. This means "[c]ounterinsurgents must therefore be strategic in applying force and sensitive to its second-order political and military effects."[275] In this context it is difficult to see how avoiding collateral civilian effects is not linked, on both utilitarian and legal grounds, to military advantage.

In the final analysis counterinsurgency operations are conducted differently than conventional warfare precisely because of the different nature of insurgency/counterinsurgency warfare. An analysis that is generally critical about an application of the *Targeted Killing Case* to warfare acknowledges that its "infusion of human rights principles [in the context of a decades long occupation in the Occupied Territories] . . . is not overly controversial."[276] However, the principles found in that case of control and the assessment of risk, which underpin the application of a capture approach, potentially have a wider, if context specific, impact since conflict during an occupation is fundamentally insurgency/counterinsurgency warfare. The different nature of such conflict changes the calculus that has to be performed in seeking military success. Consistent with a general approach in targeting of seeking to leverage the language of science there is frequently reference to the increased support for an insurgency being the result of "counterinsurgency math."[277] It has been suggested that

---

[273] 1998 Rome Statute, art. 8(2)(b)(iv).

[274] B. H. Liddell Hart, Strategy 321 (2nd ed. 1991).

[275] Sarah Sewall, *Introduction to the University of Chicago Press Edition*, *in* The Counterinsurgency Manual, *supra* note 12, at xxv.

[276] Corn et al., *supra* note 105, at 606.

[277] Kolenda, *supra* note 32, at 19 (referring to COIN or counterinsurgency math). *See also* the ISAF Commander's Counterinsurgency Guidance ISAF Headquarters 2 (2009), http://www.nato.int/isaf/docu/official_texts/counter-insurgency_guidance.pdf [hereinafter the Commander's Counterinsurgency Guidance] (where reference is made to "attritional math" being turned on its head because of the intricate familial, tribal, and clan connections of Afghan society), and Ledwidge, *supra* note 24, at 189–90.

even the killing of two insurgents during a counterinsurgency can create twenty or more new recruits, and "[i]f civilians casualties occurred, that number will be much higher."[278] Whatever the precision of such "calculations," the acknowledgment that a military advantage can be gained or lost by how the local population or an opponent is treated and operations are conducted also places that decision within the legal framework governing the use of force.

The result is that humanitarian concerns impact twice on targeting decisions. First, strategically, there is the manner in which operations are planned and executed to generally avoid an adverse impact on the civilian population. This may include strategic or operational level direction on how those operations are to be carried out, including what types of weapons or armaments are to be used and under what circumstances. Depending upon the operational situation, the capture of an insurgent may be required or at least the preferred course of action. Second, having decided to conduct operations in a certain way, any individual attack that is conducted must still be assessed in respect of its compliance with the proportionality precautions provided for under humanitarian law. This can require the weighing of the anticipated military advantage of using force in a way that limits civilian casualties.[279] In this context the strategic, operational, and tactical goals of conducting operations in a way that limits collateral civilian effects will be seen to alter what outcomes are justified at law, and often narrow the range of acceptable collateral effects.

7.6.2.2. Proportionality and Counterinsurgency

As has been noted, targeting decisions made at the tactical level do not take place in an operational vacuum. The assessment of the military advantage to be gained from an attack under Article 57(2)(a)(iii) of Additional Protocol I ordinarily must take into account the impact a use of force will have on the conduct of military operations at the operational and strategic levels of war. With an overall goal of limiting collateral civilian casualties and damage, it is inevitable that increased value will be assigned to military courses of action that do just that. The overall effect of adopting that operational approach, which is designed to bring about a military advantage specific to counterinsurgency, is often a more restricted "zone of proportionality" than occurs in the context of a conventional strategic targeting campaign.[280] This occurs because the type of military advantage being sought affects the method and means of attack that is chosen. That decision constitutes a preliminary step in the proportionality assessment, and one that establishes the scale of potential collateral effects. At that point, "[o]nce that minimum collateral damage method has been identified it is then assessed to establish its proportionality."[281] Where the collateral effects are intended to be very limited as a result of the method chosen to conduct the operation, the resulting proportionality assessment will undoubtedly be similarly affected.

---

[278] Commander's Counterinsurgency Guidance, *supra* note 270, at 2.

[279] AP I, art. 57(2)(a)(iii).

[280] Public Committee Against Torture in Israel v. Israel, Israel Supreme Court [16 December 2006], 46 ILM 375, ¶58, at 400 (2007). *See also* Watkin, *supra* note 191, at 310–19 (for a discussion of the "zone of proportionality").

[281] BOOTHBY, *supra* note 153, at 125.

Seeking to restrict the use of force, and even favoring the capture of insurgents, does not mean the outcome of a conduct of hostilities targeting assessment will necessarily mirror the use of force restrictions inherent in a law enforcement normative framework. There will still remain a humanitarian law–based zone of proportionality within which commanders "must then consider whether the potential harm to civilians creates an unacceptable policy risk even if it is lawful."[282] However, the more restricted range of acceptable collateral damage that occurs in a counterinsurgency context, whether considered from a practical, moral, or legal perspective, arises from the link to governance and the role performed by the State as a "proper authority." In this sense it does remain an outcome that is heavily influenced by norms associated with law enforcement and the expectations of how a properly functioning State will maintain internal order.

Notwithstanding situations where the application of the two normative frameworks, law enforcement and conduct of hostilities, may still have the same result the capture option, or an even more limited use of force may not be viable during counterinsurgency operations. For example, an operation may involve clearing an urban area held by a large group of insurgents, rather than attempting to neutralize one or two isolated insurgents in a house or compound. This situation occurred in Fallujah, Iraq, in November 2004, where significant violence was required to take back control of the city.[283] That conflict was assessed as approximating a "conventional confrontation,"[284] and as such the application of humanitarian law conduct of hostilities norms in this context would have a result more closely aligned with combat between State armed forces.

This does not mean that the military advantage to be gained from lessening the collateral effects on the civilian population during the Fallujah operation was eliminated. Although a large portion of the civilian population had left the city,[285] there was a concerted effort to avoid collateral casualties through the issuance of restrictive rules of engagement.[286] This urban operation was still conducted in the context of an insurgency/counterinsurgency. However, the nature of the military task in Fallujah also broadened the acceptable zone of proportionality by introducing other operational goals to be considered, along with the collateral effect on the civilian population when determining the military advantage of an attack (see Chapter 11, section 11.3.3, *United States Rules of Engagement and "Self-Defense"*).

An effort to apprehend isolated suspects does not mean the operation will be free from violence. Much depends upon the mission and the context within which it is performed. Arguments favoring the application of law enforcement norms over the humanitarian law

---

[282] Geoffrey S. Corn & Gary P. Corn, *The Law of Operational Targeting: Viewing the LOAC Through an Operational Lens*, 47 TEX. INT'L L.J. 337, 369 (2012) (it is also noted that "a commander may not want to alienate the civilian population by causing casualties that, while lawful, would still be perceived as unjustified"). *See also* DoD LAW OF WAR MANUAL, *supra* note 154, ¶ 17.5.1.3, at 1030 (where it is noted "[t]he sympathy and support of the civilian population are frequently important objectives in non-international armed conflict. In order to ensure such support, commanders and their forces may operate under rules of engagement that are more restrictive than what the law of war requires.").

[283] BING WEST, NO TRUE GLORY: A FRONTLINE ACCOUNT OF THE BATTLE OF FALLUJAH 315–16 (2005) (where it is reported that during the November 2004 attack there were 540 air strikes, 14,000 artillery and mortar shells fired, as well as 2,500 tank main gun rounds. Some 18,000 of Fallujah's 39,000 buildings were damaged or destroyed).

[284] BURKE, *supra* note 2, at 173.

[285] RICKS, *supra* note 3, at 399.

[286] DICK CAMP, OPERATION PHANTOM FURY: THE ASSAULT AND CAPTURE OF FALLUJAH, IRAQ 151–52 (2009) (see 1st Marine Division Rules of Engagement).

targeting provisions usually present somewhat artificial scenarios of unarmed or nonresisting insurgents sitting in restaurants or visiting family members in the territory in conflict, in occupied territory, or in a third State removed from the area of combat.[287] This is an issue that has been the subject of considerable debate about the morality of being able to kill potentially unarmed opponents.[288] However, it is one about which states are unlikely to readily cede their operational flexibility during an armed conflict by accepting that only a human rights–based law enforcement regime can be applied. Part of that unwillingness undoubtedly stems from a concern of overreaching in terms of the utility and effectiveness of applying a law enforcement approach.

It is not that a human rights law enforcement approach does not accept that the use of deadly force may be required. For example, in respect of hostage situations a domestic law enforcement paradigm does not prohibit the killing of hostage takers during an operation. This has been particularly evident in situations where security forces have been required to conduct an assault on a building in order to free hostages. The threat posed by the hostage takers is often such that most if not all of them are killed. For example, during the 1980 Libyan embassy siege in London all but one of the hostage takers was killed,[289] while a 2012 rescue of an American citizen and a Danish citizen in Somalia resulted in the death of nine abductors and the wounding of five more.[290] Neither, apparently, would a law enforcement approach necessarily preclude a certain level of collateral injury and damage in order to end the criminal activity.[291] This was evident in the 2002 Moscow Theatre incident, where the European Court of Human Rights accepted that 129 hostages dying during the hostage rescue operation was not a disproportionate measure.[292]

However, there are fundamental differences in the scope and scale of violence normally associated with and generally seen as being manageable by the two normative regimes. Nowhere is this more evident than in respect of a decision to shoot down a plane hijacked by terrorists. The necessity of shooting down an aircraft posing a 9/11-type threat will very likely result in significant potential collateral deaths involving the crew, passengers, and even persons on the ground, as well as property damage, on a scale that significantly exceeds most

---

287 INTERPRETIVE GUIDANCE, *supra* note 99, at 81 (the examples given are an unarmed civilian sitting in a restaurant providing tactical targeting intelligence, an insurgent military commander visiting his family, and unarmed civilians blocking passage across a bridge), Droege, *supra* note 75, at 529 (where reference is made to a FARC member shopping in Bogota, a suspected PKK member taking part in a demonstration, and a Chechen rebel at home), and ANTONIO CASSESE, INTERNATIONAL LAW 423 (2nd ed. 2005) (referring to not firing at a terrorist in a taxi or in his home).

288 WALZER, *supra* note 208, at 138–43 (perhaps the most famous discussion is that presented by Michael Walzer regarding the morality of the decision whether to shoot a naked soldier taking a bath). *See also* Gabriella Blum, *The Dispensable Lives of Soldiers*, 2 J. LEGAL ANALYSIS 69 (2010).

289 *See, e.g.*, PETER HARCLERODE, SECRET SOLDIERS: SPECIAL FORCES IN THE WAR AGAINST TERRORISM 383–408 (2000) (for an outline of the May 1980 UK Special Forces hostage rescue operation at the Iranian Embassy in London in which a lone terrorist was arrested after hiding among the hostages).

290 JESSICA BUCHANAN, ERIK LANDEMALM & ANTHONY FLACCO, IMPOSSIBLE ODDS: THE KIDNAPPING OF JESSICA BUCHANAN AND HER DRAMATIC RESCUE BY SEAL TEAM SIX 250 (2013).

291 Andrea Gioia, *The Role of the European Court of Human Rights in Monitoring Compliance with Humanitarian Law in Armed Conflict*, *in* INTERNATIONAL HUMANITARIAN LAW AND INTERNATIONAL HUMAN RIGHTS LAW 201, 231 (Orna Ben-Naftali ed., 2011) (where it is suggested "since strict proportionality does not imply that 'incidental damages' are not acceptable, this perceived difference [between international humanitarian and human rights law] appears to be rather relative.").

292 Finogenov v. Russia, VI Eur. Ct. H. R. ¶ 236, at 409 (2011) (extracts).

interpretations of what is traditionally acceptable under a law enforcement paradigm.[293] In 2005 the German Constitutional Court struggled with the issue of such a shoot-down and finally determined the law authorizing such activity was unconstitutional.[294] What is not clear is where this leaves State authorities confronted with such a decision regarding a threat that has already proven to be a realistic one (see Chapter 13, section 13.5, *The Shoot-Down of Hijacked Aircraft*).

At some point it seems clear that a human rights law enforcement paradigm will have to undergo significant alterations regarding its approach toward collateral injury and damage if that framework is to be generally applied in regulating armed conflict. Humanitarian law appears better situated to address the broad range of collateral effects that can result from military operations during armed conflict. Even in its noninternational form, the hostilities can involve significant levels of violence. This reality also highlights that a "zone of proportionality" approach applied under a humanitarian law framework must be contextual. For example, whereas the *Targeted Killing Case* indicated an attack could be disproportionate "if the building is bombed from the air *and scores of its residents and passersby* are harmed,"[295] the shoot-down of an aircraft could result in hundreds of passengers alone being killed. Depending upon the importance of the target being prosecuted and the military advantage to be gained, the number of collateral civilian casualties that are justified can be quite elevated. Ultimately, assessing the limits on the boundaries of a zone of proportionality depends upon the context within which force has to be used.

7.7. Overlapping Normative Frameworks

There also remains the question of how to resolve the interface between, and at times the overlap, of the human rights and humanitarian legal frameworks during counterinsurgency operations. That overlap may arise simply from the fact an insurgent can be the subject of an attack during the conduct of hostilities but also remains amenable to human rights–based law enforcement as a "criminal." As has been noted, this overlap is a direct result of the State-centric effort to "de-legitimize" certain participation in armed conflict (see Chapter 4, section 4.3, *Impact of the Proper Authority Principle*). However, it is more likely to occur because military forces engaged in counterinsurgency often find themselves performing policing duties. These forces operate among the civilian population. Given the tendency of criminal organizations to thrive in situations of insecurity, as well as the involvement of

---

[293] Sarah Lyall, *A New Amenity May Show Up on Some London Roofs for the Olympics: Missiles*, N.Y. TIMES, May 12, 2012, http://www.nytimes.com/2012/05/13/world/europe/london-wary-of-rooftop-missiles-plan-during-olympics.html?pagewanted=all (where it is noted one of the issues regarding the use of ground-based antiaircraft missiles for the defense of the 2012 London Olympic Games was the collateral damage caused by a shoot-down of an aircraft being used as a weapon).

[294] Dr. H. et al. v. s.14.3 of the Aviation Security Act of 11 January 2005 (2006), File no: 1 BvR 357/05. (German Federal Constitutional Court) ¶¶ 121–153, *available at* http://www.bundesverfassungsgericht.de/entscheidungen/rs20060215_1bvr035705en.html.

[295] Public Committee Against Torture in Israel v. Israel, Israel Supreme Court [16 December 2006], 46 ILM 375, ¶46, at 396 (2007) (emphasis added).

insurgents in criminal activity in order to fund their operations, security forces may find a significant portion of their activities directed toward law enforcement.

The fact that there is concurrent and even overlapping application of the two legal regimes is not remarkable with that outcome being reflected in the International Court of Justice case, *Legal Consequences Of The Construction Of A Wall In The Occupied Palestinian Territory*.[296] While exclusionists may seek to uniquely apply either international human rights law or humanitarian law as exclusive specialized bodies of law, a more realistic approach is one that recognizes there may, in certain circumstances, be a choice in the application of the appropriate body of law (see Chapter 5, section 5.2.1, *The "Special Law" Debate*).[297] However, in adopting this approach, the decision regarding which body of law to apply must be the result of a principled application of the appropriate normative framework. Of primary importance is the fact that the conduct of hostilities norms found in the law of armed conflict can only be applied as a matter of law when such a conflict is determined to exist.[298] If there is no armed conflict, then it is human rights law that applies.[299]

Further, where there is an overlap of legal regimes it may possible for a State to apply the more restrictive law enforcement paradigm as a matter of policy, even during an armed conflict (see Chapter 14, section 14.4.2 *A Policy Choice by States*). This is true whether operating within that State or when assisting another State fighting an insurgency. In this context, the choice of legal regime may directly result from a "threat"-based analysis of the type and levels of force needed to neutralize the opposing force. In situations where there is effective governance and sufficient control is established so that law enforcement can be effectively carried out, then the default position might be the application of the human rights–based legal regime (see Chapter 13, section 13.2.1, *The "Troubles"*). In any event, such an approach is consistent with the police primacy principle, which is an integral part of counterinsurgency doctrine.

Depending upon the situation, the conduct of law enforcement activities could be controlled by domestic human rights law, by its international counterpart, or by the application of human rights norms reflected in humanitarian law provisions such as Article 75 of Additional Protocol I (see Chapter 5, section 5.3.2.3, *Divergence and Growing Reconciliation*). With a declared commitment to a law enforcement approach, any departure, such as by applying conduct of hostilities norms, would from a public (both domestic and international) accountability perspective likely have to be explained and justified. Reasons for such a departure could include those outlined in the *Targeted Killing Case*, such as a lack

[296] Legal Consequences Of The Construction Of A Wall In The Occupied Palestinian Territory, 2004 I.C.J. ¶106, at 178.

[297] Marko Milanović, *Norm Conflicts, International Humanitarian Law and Human Rights Law*, *in* INTERNATIONAL HUMANITARIAN LAW AND INTERNATIONAL HUMAN RIGHTS LAW, *supra* note 291, 95, at 124 (where it is suggested that while humanitarian law and human rights law treaties can be reconciled and harmoniously interpreted, "there will be instances—there *are* instances—where this quest for harmony will fail, when the two bodies of law cannot be reconciled, when all legitimate methods of norm avoidance and resolution will be exhausted, and when ultimately a political choice will have to be made to describe the relationship between IHL and IHRL [international human rights law] as a whole.").

[298] Prosecutor v. Tadić, Case No. IT-94-1-A, Decision On The Defence Motion For Interlocutory Appeal On Jurisdiction, ¶ 67 (Int'l Crim. Trib. for the Former Yugoslavia, 1995), http://www.icty.org/x/cases/tadic/acdec/en/51002.htm. *See also* DINSTEIN, *supra* note 113, at ¶ 113, at 37.

[299] DINSTEIN, *supra* note 113, at ¶ 113, at 37.

of physical control at the site of the attack or the potential risk to security personnel and civilians if capture was to be attempted.[300] A policy application of human rights norms may also be used to restrict the use of force under a humanitarian law framework. That seems to be the approach adopted by the Obama administration in respect of targeting terrorists operating "outside the United States and areas of active hostilities."[301] What is less clear is why such a policy choice would not extend to inside "areas of active hostilities" as well where the tactical situation permits it.

There may be instances where insurgents or members of opposing armed forces are unarmed and located in crowded public spaces or even in private residences. In these situations physical capture could normally be attempted with low risk to the soldiers involved or to civilians located in the area. In such cases a human rights–based law enforcement regime could be applied as the default legal paradigm, notwithstanding the existence of an armed conflict. Applying such a policy-based approach has the advantage of not requiring a fundamental change to existing humanitarian law. It also retains flexibility in the face of a complex security situation that does not necessarily lend itself to the exclusive application of either of these bodies of law. Further, the linkage of human rights to situations of governance helps avoid overreach in the application of law enforcement norms. The result is a "continuum" of rights protection linked to the nature of the threat and the operating environment within which armed forces have to maintain security. Increasing recognition of such a continuum has led to a suggestion there is a fluidity and exchange between human rights and humanitarian law that "acknowledges the malleable aspect of emergency situations as a practical matter and their capacity to move upwards and downwards in intensity and effects."[302]

It has been argued that a domain approach (i.e., law enforcement, emergency, armed conflict for civilians, and armed conflict for combatants) with its separate rules for law enforcement and hostilities targeting should be abandoned in favor of applying one set of "functional" principles to be applied across the spectrum of security situations.[303] This approach offers a useful analytical tool that reflects the often uncertain interface between law enforcement and conflict norms, particularly in respect of lower intensity conflict. However, it also presents a number of challenges from both an implementation and a broader public policy perspective. There is considerable value in maintaining a separation between law enforcement and conduct of hostilities frameworks due to the exceptional levels of violence that can occur even in internal conflict. To a significant extent, aspects of armed conflict are unlikely to be amenable to a human rights law approach. It is also unlikely that the targeting provisions of international customary and treaty law could be readily changed to adopt a more homogenous framework. In this regard the construct is helpful from a theoretical level, but practical reality will continue to demand reliance on existing frameworks.

---

300 Public Committee Against Torture in Israel v. Israel, Israel Supreme Court [16 December 2006], 46 ILM 375, ¶40, at 394 (2007).

301 *US Attorney General Letter, supra* note 142, at 3.

302 Oren Gross & Fionnuala Ní Aoláin, Law in Time of Crisis: Emergency Powers in Theory and Practice 364 (2006).

303 Monica Hakimi, *A Functional Approach to Targeting and Detention*, 110 Mich. L. Rev. 1365 (2012) (where the functional principles are suggested to be liberty–security, mitigation, and mistake).

While States can be criticized for resisting recognition of the existence of an internal armed conflict, it remains a reality that the law enforcement paradigm has overall helped ensure life within States is more secure and less violent (see Chapter 4, section 4.4, *The State and Governance*). Too early a recognition of the existence of an armed conflict can result in greater potential for the use of violence. Further, it has been suggested "[h]uman rights bodies need to be cautious in labeling a situation an armed conflict, because it could be taken by the state concerned as a political justification for its policies violating human rights."[304] In short, while there can be controversy and disagreement as to when an armed conflict is in existence, there is value in the debate in terms of defining and ultimately limiting the scope of that conflict. Similarly, there is value in having a regulatory framework governing the use of force that matches the nature of the security situation being addressed. This results in the application of separate, albeit overlapping normative frameworks.

However, this does not mean that the two normative frameworks can, or should, operate in isolation from one another. The continuum approach, applied even using existing legal frameworks, avoids gaps in accountability and helps "to map out a set of legal structures that accommodate the varying experiences of state practice."[305] There appears to be no reason why such an approach cannot accommodate an overlap of existing legal regimes, particularly during lower intensity conflict. Situations involving governance and lower levels of threat during counterinsurgency and periods of occupation could result in the invocation of a default position of a preference for, and expectation of, the adoption of the human rights–based law enforcement regime to govern the use of force. As the threat increases due to the activities of organized armed groups, and the ability to "govern" is affected by an absence or loss of control, it is conduct of hostilities norms that provides the rights-based framework necessary to regulate the activities of states, as well as their opponents.

A lesson can be learned from the position most often adopted by States to deal with internal order, even when faced with potentially significant security threats. While not a counterinsurgency situation, the default application of a law enforcement approach is consistent with the 2002 arrest of Jose Padilla, an Al Qaeda–linked enemy belligerent, on U.S. territory. Federal Bureau of Investigation (FBI) agents arrested him on a material witness warrant when he got off a plane in Chicago.[306] It was only later that the Department of Defense, on orders from the White House, "took Padilla in the middle of the night from a federal lockup in New York and put him in a military brig, designating him an enemy combatant."[307] Decades earlier, capture by civilian police authorities was also how Richard Quirin and seven other German saboteurs were dealt with during World War II. In 1942 they were arrested in the United States for operating "in civilian dress within zones of military operations and elsewhere," having passed "through the military and naval lines and defenses of the United States."[308] In each of these situations there was little doubt that a law enforcement approach was applicable and up to the task.

---

[304] René Provost, International Human Rights and Humanitarian Law 336 (2002).

[305] Gross & Ní Aoláin, *supra* note 302, at 364.

[306] Terry McDermott & Josh Meyer, The Hunt for KSM 219 (2012), and Jones, *supra* note 174, at 154.

[307] McDermott & Meyer, *supra* note 306, at 219.

[308] Louis Fisher, Nazi Saboteurs on Trial 51–52 (2nd ed. 2005). *See also*, *id.* at 66–68 (indicating that all eight were convicted and six were executed).

However, a decision to apply a law enforcement approach often depends upon the situation that is presented to the security authorities. Had an enemy "belligerent," such as Jose Padilla, been involved in an attack, it is possible that the application of conduct of hostilities norms might have been considered to counter the violence. An attack on a State might resemble a Mumbai-style attack involving trained members of an organized armed group. In the 2008, ten operatives of the Pakistan terrorist group, Lashkar-e-Taiba (LeT), landed in a sophisticated seaborne infiltration of Mumbai and attacked hotels and other targets with firearms, grenades, and improvised explosive devices.[309] There were 172 people killed in the attack "including 16 police, 27 hotel staff, 2 commandos, and 9 out of the 10 LeT terrorists, while another 304 were injured."[310] Such an attack can quickly overwhelm traditional law enforcement. While a preference may be to deal with such threats exclusively through law enforcement means, the nature and intensity of the violence may preclude such action. Similarly, it is reported that the French Minister of Defense described a January 2013 hostage taking by an estimated forty Al Qaeda–linked terrorists at an Algerian gas plant as an " 'act of war' because of the large number of people involved."[311] Increasingly these low-level "one-off" attacks are raising questions about whether they occur in the context of an armed conflict (see Chapter 14, section 14.3.1, *Armed Conflict or Law Enforcement?*).

The key point is that State authorities often use a law enforcement paradigm to deal with "belligerents" (e.g., the arrest of Padilla and Quirin) even in the context of what may be viewed as armed conflict. Give the level of control exercised domestically in the United States examples; the capabilities of policing authorities; and the low risk presented by Jose Padilla, or sixty years earlier by the German saboteurs, it would have been an exceptional situation had a decision been made to engage those "enemy belligerents" under conflict of hostilities norms. The question remains why such an approach should not be also applied extraterritorially in situations where the security forces (international or domestic) maintain an adequate level of control, even if the control exercised were limited to a particular locale. To an increasing extent that is the situation, as is reflected in the police primacy approach associated with counterinsurgency operations. In addition, the principles articulated in the *Targeted Killing Case*,[312] as well as the views expressed by President Obama regarding consideration of the capture before killing a U.S. citizen or any terrorist overseas, provide additional evidence of such a changing normative landscape.[313]

---

309 David Kilcullen, Out of the Mountains: The Coming of Age of the Urban Guerrilla 52–66 (2013), and Burke, *supra* note 2, at 395–400 (for background to the LeT and the Mumbai attack).

310 Kilcullen, *supra* note 309, at 60.

311 Amal Belalloufi, *Five Kidnappers Taken Alive in Algeria: TV*, A.F.P. (Jan. 21, 2013), https://sg.news.yahoo.com/seven-hostages-dead-army-storms-algeria-complex-134857787.html.

312 Public Committee Against Torture in Israel v. Israel, Israel Supreme Court [16 December 2006], 46 ILM 375, ¶40, at 394 (2007).

313 *Drone Policy Speech*, *supra* note 142.

> "They (Frontier Wars) are but the surf that marks the edge and advance of the wave of civilization."
>
> Lord Salisbury, Guildhall, 1892[1]

8 Counterterrorism and the "Away Game"

8.1. Introduction

In many respects counterterrorism operations have dominated the security dialogue in the post-9/11 period. With terrorism having been widely seen as a law enforcement matter prior to the twenty-first century, the attacks by Al Qaeda fundamentally changed the operational perspective toward one that favored a militarization of the counterterrorist response. Indeed, the term "counterterrorism" has become synonymous with covert Special Forces raids and the use of drones firing missiles at Islamic jihadists in foreign lands. However, like counterinsurgency, counterterrorism requires a multidisciplinary effort much broader than the intelligence-driven, military-conducted operations that have gained such widespread publicity.[2] Even within Western States the action taken to address the contemporary terrorist threat has been suggested to take on elements of counterinsurgency doctrine. For example, one analysis of the UK response for dealing with an "internal" jihadist terrorist threat linked to external insurgencies "has been to adopt an approach which, however, inadvertently, greatly resembles the classic counterinsurgency strategy of community involvement, social responsiveness, and treatment of the insurgency as merely criminal instead of the more lofty perch to which its adherents aspire."[3]

---

1 Winston Churchill, The Story of the Malakind Field Force: An Episode of Frontier War (Octopus Publishing Group 1989) (1898) (found on title page).

2 Edmund J. Hull, High-Value Target: Countering Al Qaeda in Yemen 15 (2011) (regarding counterterrorism efforts in Yemen, "it resembled counterinsurgency, success of which depended primarily on winning the hearts and minds of the population.").

3 Randall Wilson, Blue Fish in a Dark Sea: Police Intelligence in a Counterinsurgency 144–45 (2013).

The employment of military forces to deal with terrorists is not a new phenomenon. This has been evident in the historical response to terrorist attacks by the United States and Israel.[4] However, the ability of non-State actors to strike North America as exhibited at the beginning of this century has stirred a particularly strong military response. The result has been that the conflict with terrorists has been increasingly viewed as a transnational "war," rather than internal violence solely involving criminal activity. This change toward seeing terrorism as part of an armed conflict has been aided in large part by the increasing levels of violence brought about by the terrorist use of explosives and suicide bombing tactics (see Chapter 12, section 12.3.1, *A Ubiquitous Weapon*). Many of these terrorist attacks are not wholly international in their execution. Perhaps what has made viewing terrorism as a transnational threat particularly remarkable is that while the planning originated overseas, the 9/11 attacks physically originated on planes that took off from U.S. domestic airports. It "was foreign—but from foreigners who had infiltrated into the United States."[5] This was an attack where the first line of defense might ordinarily have been expected to be the domestic security forces of the United States, the police.

The twenty-first-century transnational terrorist threat has provided unique challenges for State security structures that previously were compartmentalized to deal with either foreign or domestic threats. It soon became evident that if the "foreign" transnational terrorist threat was to be dealt with at its source, largely in the ungoverned spaces of the world, it was necessary to invoke a military response. In other words, conducting counterterrorism operations in places where the governing authority is not policing its own territory. This is because the threat for Western States, as the "far enemy," has remained largely focused on jihadist groups like Al Qaeda or the Islamic State headquartered in distant lands. These groups often use territory created by internal conflict in other States (i.e., the battle against the near enemy) as safe havens from which to plan, launch, and command transnational attacks. Even though the threat may come from cells that are established within a threatened State, the primary source of the problem ultimately remains an external one.[6]

This chapter will look at two methods of conducting counterterrorism operations: Special Forces raids and the use of drones. Questions regarding their use are often interrelated in terms of assessing the capabilities they provide. Under certain circumstances Special Forces units provide a capability to attempt to capture a terrorist suspect, while drones can significantly extend the reach of the security forces of a State to kinetically remove the threat posed by non-State actors, literally in the safe havens where they hide. Due to the legal controversy that has arisen around the use of drones, it is the latter weapon system that will be a significant focus of this chapter. This analysis will consider the extent to which drones actually represent a unique form of warfare. The use of drones also provides a particularly helpful framework within which to consider the interrelation

---

[4] W. Michael Reisman & James E. Baker, Regulating Covert Action 108–111 (1992) (for reference to a 1986 U.S. attack on Libya in response to a Berlin disco bombing), and *id.* at 86 (referring to a 1985 raid on PLO Headquarters in Tunis).

[5] The 9/11 Commission Report: Final Report of the National Commission on Terrorist Attacks Upon the United States 263 (2004).

[6] Greg Miller, *Islamic State Working to Establish Cells outside Middle East, U.S. Says*, Wash. Post, Aug. 14, 2014, http://www.washingtonpost.com/world/national-security/islamic-state-working-to-establish-cells-outside-iraq-and-syria-us-says/2014/08/14/639c32b0-23f5-11e4-8593-da634b334390_story.html.

of the various bodies of law governing counterterrorism. For practitioners and theorists, this issue, perhaps more than any other, is forcing consideration of how State self-defense, humanitarian law, human rights law, and the domestic law interrelate in controlling armed conflict. In doing so, limitations regarding the scope of traditional interpretations of international law are being exposed.

As with the general human rights and humanitarian law debate, the interface between the areas of law relevant to counterterrorism will be seen to be largely unsettled. As this chapter explores, there appear to be three main theories presented by legal commentators as to how the law governs the application of force through the use of drones. The first theory, which will be referred to as the restrictive "law enforcement" approach, relies on a traditional State-centric interpretation of international law that seeks to limit the State reaction to transnational terrorism to a human rights–based policing paradigm. The second theory is the permissive "conduct of hostilities" approach. Although this theory also has traditionalist roots, it is an approach that favors a military response. Finally, there is a third, more narrowly constructed "self-defense" paradigm. This last theory is uniquely framed in terms of the application of State self-defense principles; however, in the final analysis it largely resembles the "conduct of hostilities" approach in its effect.

8.2. A Lack of Consensus

The discussion about drones and counterterrorism, and to a lesser extent the use of Special Forces, is one where there appears to be a fundamental lack of consensus among international lawyers on even basic legal issues that are essential to determining how law regulates "targeted killing." At the heart of this debate is the question of whether these uses of force will be governed almost exclusively by conduct of hostilities norms, or restricted in their entirety because of a requirement to apply a law enforcement response that itself is limited by a need to respect State sovereignty. The degree to which there is a lack of consensus regarding what body of law to apply is perhaps best demonstrated in two major 2012 human rights studies of drone strikes where a definitive statement regarding what law applies appears to have been avoided.

One study, *Living Under Drones: Death, Injury, and Trauma to Civilians From US Drone Practices in Pakistan*, adopted the approach of analyzing the use of drones under both humanitarian law and human rights law, with the applicability of former body of law to strikes in Pakistan being viewed as dependent upon meeting the criteria for the existence of an armed conflict.[7] It notes that "US officials have been quick to apply IHL [international humanitarian law] without establishing that the requisite threshold for its application has been met," and "numerous experts have raised questions about whether the US is, in fact, in an armed conflict with all of the groups whose members the US has targeted."[8] However, the analysis then proceeds to consider the applicability of humanitarian law with a reference

---

[7] Living Under Drones: Death, Injury, and Trauma to Civilians from US Drone Practices in Pakistan, Stanford Law School/NYU School of Law 110–11 (Sept. 2012) [hereinafter Living Under Drones], http://www.livingunderdrones.org/report/.

[8] *Id.* at 111.

to "*[i]f* there is an armed conflict...."[9] The result is that the use of drones is analyzed under a number of potential governing legal paradigms: humanitarian law, human rights,[10] as well as assessing the impact of U.S. domestic law.[11]

The other study, *The Civilian Impact of Drones: Unexamined Costs, Unanswered Questions*, acknowledges the application of humanitarian and human rights law but also seems to avoid taking a definitive position as to which body of law governs the use of force. The study indicates, "the complex legal issues raised by covert drone strikes cannot be resolved solely by reference to the laws of war. Other bodies of law place significant limits on targeting operations, and there are important debates about, for instance, the applicability of international human rights law."[12] The analysis ultimately uses what could be described as a general normative approach, which seeks to "identify concerns at the juncture of ethics and the law."[13] While this study relies on law of war principles and its norms, the analysis also discusses "alternative" approaches such as capture that are clearly based on human rights principles. The issue of capture inevitably raises questions regarding how that is to be done, and what forces are to be used (usually Special Forces). The human rights influence is evident when it is stated the force used "must not be greater than absolutely necessary."[14] As with the *Living Under Drones*, study this legal debate is not resolved. So it appears that a full decade after the international legal community first turned its attention to drone strikes, an agreement on the legal framework governing targeted killing by drones and on issues such as who may be targeted remains elusive.[15]

However, as often happens in discussions about contemporary international security issues where consensus has appeared particularly difficult to reach, the proponents of the various legal theories often speak with a level of certainty that is difficult to reconcile either with the complexity of the issues or the fluid nature of much of this area of law. A particular challenge for advocates of the law enforcement and the more permissive humanitarian law–based theories (i.e., the "conduct of hostilities" and "self-defense" based approaches) is that, perhaps due to their exclusionary roots, it will be seen that these theories do not appear to fully address the complexity or overlapping nature of the security threats posed by transnational terrorism. The lack of consensus seems to be a direct result of the disagreement and endless debates within the international legal community. This also serves as an indication that none of these legal theories individually appear to supply an adequate answer to this pressing security challenge. It is the assessment of these theories and the unresolved issues

---

9 *Id.* at 112 (emphasis added).

10 *Id.* at 100 and 117–18

11 *Id.* at 118–22.

12 The Civilian Impact Of Drones: Unexamined Costs, Unanswered Questions, Center for Civilians in Conflict/Columbia Law School 67–68 (2012) [hereinafter Civilian Impact of Drones], http://web.law.columbia.edu/human-rights-institute/counterterrorism/drone-strikes/civilian-impact-drone-strikes-unexamined-costs-unanswered-questions.

13 *Id.* at 67.

14 *Id.* at 72.

15 Kenneth Watkin, *Opportunity Lost: Organized Armed Groups and the ICRC "Direct Participation in Hostilities" Interpretive Guidance*, 42 N.Y.U. J. Int'l L. & Pol. 641, 641–42 (2010) (for reference to the background of the development of the Interpretive Guidance on the Notion of Direct Participation in Hostilities Under International Humanitarian Law, International Committee of the Red Cross (2009) [hereinafter the Interpretive Guidance]).

that they raise that sets the scene for consideration what needs to be done in order to provide a more coherent and integrated operational law approach for addressing twenty-first-century security threats. There does appear to be a growing, if grudging, acceptance that action can be taken by a threatened State in ungoverned spaces when it is established the territorial State is unwilling or unable to stop armed groups from plan and launch attacks. However, as this approach gains traction there needs to be a more nuanced discussion of how the law governing State self-defense, human rights law (both international and domestic), and humanitarian law all impact on decisions to use force.

8.3. Counterterrorism and "Direct Action"

The use of State security forces at long distances from the borders of Western States, which have felt particularly threatened by transnational terrorism, has presented a range of new challenges for international law. The response has often been controversial, including the capture and rendition of suspects under circumstances that telegraphed the manipulation of and even a disregard for international borders;[16] off-shore detention in Guantanamo Bay, Cuba, dubbed a "black hole";[17] the inter- and intra-State transfer of detainees between national authorities;[18] the invocation of what has now been acknowledged to be torture in order to gain intelligence;[19] covert Special Forces conducting nighttime raids in far-off lands;[20] the leveraging of technology to maintain around-the-clock surveillance on possible targets;[21] as well as the capability of killing enemy personnel, even when they are asleep in bed with their families.[22]

---

[16] *See, e.g.*, Stephen Grey, Ghost Plane: The True Story of the CIA Torture Program 190–213 (2006) (for an outline of the seizure of Osama Nasr ("Abu Omar") in Milan Italy on February 17, 2004).

[17] The phrase "black hole" was used in the UK Court judgment of *Abassi and Anor. v. Secretary of State for Foreign and Commonwealth Affairs and Secretary of State for the Home Department* ¶ 64 (2002), *available at* http://www.bailii.org/cgi-bin/markup.cgi?doc=/ew/cases/EWCA/Civ/2002/1598.html ("Mr Abbasi is at present arbitrarily detained in a 'legal black-hole'."). *See also* Johan Steyn, *Guantánamo Bay: The Legal Black Hole* (the 27th FA Mann Lecture, Nov. 25, 2003), *reprinted in* 53 Int'l & Comp. L. Q. 1, 1 (2004).

[18] *UK "Ready to Hand over Afghan Detainees,"* BBC (May 29, 2013), http://www.bbc.co.uk/news/uk-22704484. *See also* Munaf v. Geren, 47 I.L.M. 708 (2008) (for a U.S. Supreme Court decision dealing with the transfer of detainees to Iraqi authorities) and Amnesty International Canada v. Canada (Chief of the Defence Staff) (F.C.A.), 2008 FC 336, [2008] 4 F.C.R. 546, ¶ 36, *available at* http://reports.fja-cmf.gc.ca/eng/2009/2008fca401.html (for a Canadian court ruling on the law governing the transfer of detainees in Afghanistan).

[19] *Obama's Speech on Drone Policy*, N.Y. Times, May 23, 2013 [hereinafter *Drone Policy*], http://www.nytimes.com/2013/05/24/us/politics/transcript-of-obamas-speech-on-drone-policy.html. *See also* Alfred W. McCoy, A Question of Torture: CIA Interrogation from the Cold War to the War on Terror 124–25 (2006) (for reference to 14,000 Iraqi security detainees, high-value prisoners (1,100) held at Guantanamo Bay and Bagram, and other detainees who were subjected to harsh interrogation and torture by U.S. military and CIA personnel).

[20] Jean MacKenzie, *Afghanistan, Part II—Counterinsurgency in Afghanistan: Myth or Reality*, *in* Hearts and Minds 200, 219–21 (Hannah Gurman ed., 2013) (for a critical analysis of nighttime raids in Afghanistan).

[21] Matt J. Martin & Charles W. Sasser, Predator Remote-Control Air War Over Iraq and Afghanistan: A Pilot's Story 104 (2010) (for a description of the operational capabilities of drones).

[22] Adam Entous, *Special Report: How the White House Learned to Love the Drone*, Reuters (May 18, 2010), http://www.reuters.com/article/2010/05/18/us-pakistan-drones-idUSTRE64H5SL20100518 (for reference to family members who live and travel with CIA targets being killed). *See also* Daniel Byman, A High Price: The Triumphs & Failures of Israeli Counterterrorism 308 (2011) (where at 2002 drone strike killed a Hamas leader [Salah Shehada], as well as his wife, daughter, and other persons in the vicinity).

While some of these issues clearly raise "red line" questions of legality (e.g., the torture and abuse of detainees in particular), not all of these responses, such as the use of raids and drones immediately do. However, in certain circumstances these issues can be interrelated. What could be thought of as a less controversial activity, such as the conduct of Special Forces raids to "arrest" or kill terrorism suspects, might be linked to questions of detention, torture, or abuse to obtain intelligence; cross-border activity to kill or capture a terrorist; and the use of drones or other "kinetic" options (i.e., bombing by conventional aircraft). For example, such factors arose in one form or another in respect of the 2011 raid in Pakistan that killed Osama bin Laden.[23] For State legal advisors this also means it is no longer practically possible in the post-9/11 period to rely on traditional exclusionary views of the law or to provide relatively simple answers based on the "black letter" law of the 1949 Geneva Conventions governing inter-State conflict. The "counterterrorism" operating environment is simply too complex to permit such a simple, even static response.

While considerable media interest has been directed toward the use of drones, Special Forces personnel have become a key resource in the conduct of counterterrorist activity. Counterterrorist operations have often been linked to "direct action" missions conducted by Special Forces units.[24] Doctrinally, a direct action mission is defined as "[s]hort-duration strikes and other small-scale offensive actions conducted as a special operation in hostile, denied, or diplomatically sensitive environments and which employ specialized military capabilities to seize, destroy, capture, exploit, recover, or damage designated targets."[25] The emphasis placed in this definition on seizure, capture, exploitation, and recovery is noteworthy. Counterterrorist missions include raids designed to kill, capture, and, increasingly, to arrest opponents. They have occurred in context of occupation, counterinsurgencies, in the territory of other States, and even on the high seas.

Primary counterterrorist means, such as drone strikes and Special Forces raids, provide unique operational choices for commanders. While drone or other aerial strikes offer the opportunity to neutralize a target at long distances, Special Forces units not only present a deadly threat, they allow for consideration of the capture of the person being targeted. Such raids were not only a key part of the efforts to degrade and destroy Al Qaeda in Iraq during that insurgency,[26] they have also been used to deal with high-value targets in other

---

23 Peter L. Bergen, Manhunt: The Ten Year Search for Bin Laden From 9/11 to Abbottabad 176, 186, & 199 (2012) (for reference to consideration of the use of a drone to conduct a missile strike, the development of contingency plans for his surrender and interrogation, and concerns that the raid might result in a permanent rupture of American/Pakistani relations). *See also* Michael Morell & Bill Harlow, The Great War of Our Time: The CIA's Fight Against Terrorism from Al Qa'ida to ISIS 272 (2015) (where it is stated information that led to the courier tied to Bin Laden was obtained through the use of "enhanced interrogation techniques"). *But see* The Report of the Constitution Project's Task Force on Detainee Treatment, The Constitution Project 11 (2013), *available at* http://detaineetaskforce.org/pdf/Full-Report.pdf (where it is concluded "there was no noteworthy connection between information gained from such interrogations and the finding of Osama bin Laden.").

24 David H. Ucko, The New Counterinsurgency Era 92–93 (2009) (for a discussion of the U.S. military Special Operations Force prioritization on "direct action and strike missions.").

25 Department of Defense Dictionary of Military and Associated Terms, Nov. 8, 2012 (as amended through Mar. 15, 2015), http://www.dtic.mil/doctrine/dod_dictionary/index.html.

26 Mark Urban, Task Force Black 80 (2010). *See also* Sean Naylor, Relentless Strike: The Secret History of Joint Special Operations Command 255-66 (2015) (discussing the highly effective use of raids by special operations forces in confronting the insurgency, and in particular Al Qaeda in Iraq).

countries. For example, among the options considered by the United States in September 2009 to deal with Saleh Ali Saleh Nabhan, a Kenyan member of Al Qaeda's East Africa cell operating in Somalia, was in order of "risk": the firing of missiles, Navy SEALs flying by helicopter to kill the target and collect DNA evidence, or capturing him for interrogation.[27] As events unfolded, the Special Forces units deployed, but helicopters strafing the terrorists' vehicles killed Nabhan, along with three Al Shabaab operatives.[28] In that case, as with any decision to deploy personnel in or over hostile territory, there was a necessary balancing of risk, both to military personnel and the civilians located in the area. Whether it was a malfunctioning missile launcher on a plane designated for a kinetic strike,[29] or the cloud cover obscuring the target, the resulting available option was to use Special Forces personnel and helicopters.[30]

A hallmark of twenty-first-century operations has been the development of Special Forces personnel who are specially trained and equipped to embark on operations designed to kill or capture designated targets. Those personnel are chosen for their maturity and trained to a high level of tactical proficiency. Their special training and tactics also reflect the requirement for State military forces to often adopt the same unconventional or irregular warfare tactics as their non-State enemy.[31] In terms of tactics, it is not uncommon to find Special Forces units trained in covert and overt entry of operational areas including compounds and buildings to affect the capture of a target in hostile environments, as well as conduct hostage rescue operations. It has been noted that the personnel involved in carrying out these latter missions must "possess exceptional shooting skills," which enable them to "burst into a darkened room, rapidly identify hostile targets and immediately neutralize them with head shots from a sub-machine gun or pistol."[32] These skills are particularly relevant to contemporary counterterrorism operations, regardless of whether they are used in the context of an armed conflict or other international situations requiring a robust law enforcement solution (see Chapter 12, section 12.2.2.1, *Unique Challenges and Responses*).

A key point is that notwithstanding the prevalent use of the term "counterterrorism" in relation to direct action taken against terrorist groups, the threat posed by non-State actors spans the conflict spectrum beyond counterinsurgency and counterterrorism to defending nationals in circumstances that solely involve a law enforcement response. Special Forces units are often the force of choice for many operations exactly because they can conduct operations focused on capturing rather than killing a target. Significantly, a frequent goal of these operations is to collect actionable intelligence on the targets they are hunting. The operational requirement to consider capture can occur when conducting operations applying humanitarian law–based precautions in the attack,[33] or while carrying out law enforcement operations in or outside the context of armed conflict. Operations include the rescue

---

[27] Mark Mazetti, The Way of the Knife 246 (2013)

[28] Naylor, *supra* note 26, at 408–11.

[29] Mazetti, *supra* note 27, at 246–47 (where it is reported the missile launcher malfunctioned on the plane necessitating the use of the naval commandos.).

[30] Naylor, *supra* note 26, at 409–10.

[31] J. Paul de B. Taillon, Hijacking and Hostages: Government Responses to Terrorism 78 (2002).

[32] Leroy Thompson, Ragged War: The Story of Unconventional and Counter-Revolutionary Warfare 143 (1994).

[33] AP I, art. 57.

of hostages abducted by criminal groups with no or only tangential connection to insurgents (see Chapter 10, section 10.3.5, *Rescues and Law Enforcement*).[34] Often, due to the nature of counterterrorism operations, there is a requirement to resolve the interface and, frequently, overlap between human rights and humanitarian law. In an operating environment such as an occupation where there is an obligation to maintain law and order, or when dealing with transnational criminal organizations and other criminal groups, that capability is not only necessary but also highly valued as representing a pinnacle of professional ability.

Of note, some counterterrorism activity may be carried out as an integral part of counterinsurgency operations. While not all military forces are equipped, trained, or regularly tasked to engage in what is viewed as a specialized function, operational pressure can force change in the perceived roles and missions for military units. For example, in early 2007 in Iraq the "white" or overt Special Forces Group (viewed as Tier 2 special operations forces in comparison to Tier 1 or "black" units, such as U.S. Army Delta and Navy SEAL Team 6) became increasingly involved supporting the hunt for Al Qaeda.[35] Further reflecting a broadening of roles, the mentoring of indigenous security forces, which has often been thought of as a Special Forces mission, was widely supported by personnel from regular military units in Afghanistan.[36] Similarly, military police[37] and conventional military units assisted Afghan forces in carrying out the search for intelligence on insurgents in the ordinary course of their operations (see Chapter 12, section 12.2.1, *The Military Task: Applying Force of First and Last Resort*).[38] Some of this activity can extend into what might be classified as counterterrorism.

As with the need for States to adopt counterinsurgency methods, counterterrorism has also demanded a change in the focus and the expectations within many Western military forces. In this regard it has been suggested that a new paradigm is required to fight the global insurgency that "will center on dispersive and adaptive teams operating on the 'fringes.'"[39] Those fringes include operating at the boundaries of armed conflict and law enforcement. Employing military units in these roles demands additional training, including in cultural sensitivity;[40] the development of supporting infrastructure such as proper detention facilities; and careful monitoring to avoid situations of insufficient screening of who should be detained.[41] As is evident from a number of contemporary transnational operations, the capacity exists within a number of nations to extraterritorially deploy security forces in a law enforcement context both on the high seas and in the territory of other States

---

34 *SAS Rescue Girl Was "Hours from Death": Prime Minister Gave Mission the Go-Ahead after Taliban Threat*, Mail Online (Aug. 30, 2012), http://www.dailymail.co.uk/news/article-2154224/Helen-Johnston-SAS-rescue-girl-hours-death.html (In May 2012 the kidnapping of four aid workers was carried out by "bandits, a criminal gang in contact with Taliban fighters, demanded a £7million ransom and the release from jail of a colleague held for murder and drugs.").

35 Urban, *supra* note 26, at 37.

36 *Afghan National Security Forces (ANSF): Training and Development*, Media Backgrounder NATO (Feb. 12, 2012), http://www.nato.int/nato_static/assets/pdf/pdf_topics/20120202_120202-Backgrounder_ANSF.pdf.

37 Gordon Cucullu & Chris Fontana, Warrior Police 250 (2011).

38 Elliot D. Woods, *The Soft Knock*, Va. Q. Rev. (Oct. 9, 2009), http://www.vqronline.org/news/soft-knock.

39 Robert M. Cassidy, Counterinsurgency and the Global War on Terror 163 (2008).

40 Emily Spencer, *The People Puzzle*, *in* The Difficult War: Perspectives on Insurgency and Special Operations Forces 79, 105 (Emily Spencer ed., 2009).

41 MacKenzie, *supra* note 20, at 220.

(see Chapter 10, section 10.2.2, *Taking Action Across the Conflict Spectrum*). Such operations frequently occur in situations where there is no effective governing authority. This includes capturing terrorists and rescuing hostages.

In some cases transnational counterterrorism operations are carried out at the invitation of the territorial State. In others they have been conducted unilaterally. Unilateral action usually occurs when the territorial State is considered to be unwilling or unable to take the necessary action to remove the security threat. As with internal conflict and occupation, there appears to be no reason why a law enforcement approach favoring capture could not be applied, assuming there is the legal authority to enter the sovereign territory of the other State (e.g., acting in defense of nationals); the significant logistical and other operational challenges can be overcome; and the security environment is a sufficiently permissive one. In other words, when the non-State actors do not exert sufficient control to challenge, however fleeting, the intervening State freedom action in the immediate vicinity where the operation is conducted. In any event, even during armed conflict capture rather than kill operations have to be considered as a matter of law for those civilians located in another State or territory who are not taking a direct part in hostilities. It would likely be considered in some circumstances where drones, other aerial strikes, naval gunfire, or even artillery could not be used without causing an unacceptable level of collateral civilian casualties. The capture option might also be applied as a matter of policy because of the military advantage that may be realized for reasons such as intelligence gathering or maintaining public support both at home and abroad.

The United States has demonstrated a willingness to adopt a modified approach favoring the capture of terrorists in its 2013 drone policy.[42] However, as was seen in the 2009 operation against Saleh Ali Saleh Nabhan, capture was an option considered for dealing with terrorists prior to the release of that formal policy.[43] The capture option was applied with the 2011 seizure on the high seas by a U.S. Navy SEAL team of the Somali, Abdulka-dir Warsame, the principle liaison between Al Shabaab and AQAP.[44] After considering capture or kill missions in Somalia, and drone strikes in Yemen against Warsame,[45] his arrest was undertaken with a particular goal of extracting intelligence to foil future plots and "take even more dangerous terrorists off the battlefield."[46] The "hybrid military-criminal approach" of initial intelligence based interrogation and subsequent prosecution before a civilian court applied in this case became the model for the handling of a number of high profile terrorism suspects captured away from a traditional war zone.[47] It has been referred to as "a pillar of its [the U.S.] national security legal policy."[48] Warsame pled "guilty in a closed court proceeding in Manhattan in December 2011, about five months after he was

---

[42] *Drone Policy, supra* note 19.

[43] Daniel Klaidman, Kill or Capture 122–27 (2012). *See id.* at 126 (there was also concern "[t]he inability to detain terror suspects was creating perverse incentives that favoured killing or releasing suspected terrorists over capturing them."). *See also* Jeremy Scahill, Dirty Wars: The World Is a Battlefield 294–97 (2013).

[44] Klaidman, *supra* note 43, at 237–39. *See also* Naylor, *supra* note 26, at 414–15.

[45] Klaidman, *supra* note 43, at 237–38.

[46] *Id.* at 238.

[47] Charlie Savage, Power Wars: Inside Obama's Post 9/11 Presidency 340-47 (2015).

[48] *Id.* at 344.

brought to New York."[49] That advantage of that capture was evident in that he "met weekly with the government for hours at a time, disclosing intelligence information about his Shabab and Qaeda co-conspirators...."[50]

Operations to capture suspected terrorists and bring them to trial in the United States have continued to be carried out, or at least attempted. In October 2013 Abu Anas al-Libi was seized in a joint special operations forces Delta/SEAL Team 6 operation for his alleged involvement in the 1998 Al Qaeda bombings of two U.S. embassies in East Africa.[51] In June 2014, Ahmed Abu Khattala, a suspected ringleader of the 2012 attack on the U.S. embassy in Benghazi, was captured by another joint military/police team in that city.[52] There will be debate about whether the attacks on the embassies prompting these arrests occurred in the context of an armed conflict. As this book argues, when a "totality of the circumstances" analytical approach is applied the intensity of the violence, the participation of an organized armed group, weapons used by the "terrorists" and the military tactics used by the attackers suggest the Benghazi attack was such a conflict (see Chapter 9, section 9.4.3, *The "Totality of the Circumstances" and Other Criteria*).[53] However, the use of specialized police, as well as military teams, to conduct these "arrest" operations, the transfer of the detained persons to a naval ship for onward transportation to the United States, and ultimately bringing the terrorists before a civilian court also indicates the subsequent State action may be comfortably situated from a policy perspective within the context of a law enforcement operation.

At the same time as the Libyan operation to arrest Abu Anas al- Libi, U.S. special operations forces were unsuccessfully attempting to capture an Al Shabaab leader, Abdikadir Mohamed Abdikadir (known as Ikrimah). That operation was terminated out of concern for the presence of women and children after a gunfight ensued.[54] In September 2014 the overall al Shabaab leader, Ahmed Abdi Godane, was killed in a drone strike in Somalia.[55] These cases provide another example of the hybrid nature of many contemporary counterterrorism operations, which may contemplate an "arrest" or a kill approach. In this complex security environment, factors weighing on the option chosen can include questions regarding the effectiveness of a drone/missile strike in ensuring the target will be killed.[56] Further, there is also the issue of whether such an operation is politically feasible in the eyes of the territorial government that is not an adversary, but which might not be capable of

---

[49] Benjamin Weiser, *Terrorist Has Cooperated with U.S. Since Secret Guilty Plea in 2011, Papers Show*, N.Y. TIMES, Mar., 25, 2013, http://www.nytimes.com/2013/03/26/nyregion/since-2011-guilty-plea-somali-terrorist-has-cooperated-with-authorities.html.

[50] *Id. See also* SAVAGE, *supra* note 47, at 347.

[51] NAYLOR, *supra* note 26, at 430.

[52] *Id.* at 431-32 and Karen DeYoung, Adam Goldman & Julie Tate, *U.S. Captures Benghazi Suspect in Secret Raid*, The WASH. POST, June 17, 2014, http://www.washingtonpost.com/world/national-security/us-captured-benghazi-suspect-in-secret-raid/2014/06/17/7ef8746e-f5cf-11e3-a3a5-42be35962a52_story.html.

[53] MITCHELL ZUCKOFF, 13 HOURS: THE INSIDE ACCOUNT OF WHAT REALLY HAPPENED IN BENGHAZI 211–20, at 254–80 (2014).

[54] NAYLOR, *supra* note 26, at 430–31.

[55] Richard Engel, Charlene Gubash, & Erin McClam, *Terror Leader Ahmed Abdi Godane Killed in U.S. Strike: Source*, NBC NEWS (Sept. 5, 2014), http://www.nbcnews.com/news/world/terror-leader-ahmed-abdi-godane-killed-u-s-strike-source-n196601.

[56] KLAIDMAN, *supra* note 43, at 123–24.

dealing with the situation.[57] In another operation in Syria in May 2015, U.S. special operations forces attempted to seize Abu Sayyaf with a plan to "interrogate him in Iraq and then send him to the United States for criminal prosecution if authorities believed they could build a strong case against him."[58] He died resisting that capture. The United States did not coordinate the attack with Syrian authorities or inform them about it, but had the consent of the Iraq government.[59]

However, it needs to be clearly stated. The capture approach by Special Forces will not be available in all circumstances. An important factor that undoubtedly weighs heavily in determining if capture is feasible is the tremendous operational capability required to complete these types of arrest operations. The reality is that only a few countries possess the personnel, equipment, or the global reach to even consider such action, although States might get support from various countries that makes it a possible option. Ultimately, the Abu Godane operation highlights that, like internal insurgency situations where there is competition for the physical control over territory, capturing transnational terrorists will not always be feasible in the "ungoverned spaces" where they are situated. The potential resistance at the site of the operation can be a determinative factor as to whether a capture operation can be carried out. As a result, when there is an armed conflict, the targeting option may have to be considered, although only where the person being attacked is a direct participant in hostilities.

8.4. New Technology/New Terminology: Placing Drones in Context

8.4.1. A UNIQUE WEAPON?

Drones have come to be associated particularly with post-9/11 counterterrorist operations. This has primarily occurred because they entered the public consciousness during operations conducted in the immediate aftermath of those transnational attacks. As such they have also been intimately linked to cross-border non-State actor conflict. The attention paid to drones as a long-endurance stand-off weapon has also resulted in their often being looked upon as a unique means of conducting warfare. It is a weapons system frequently viewed as being focused on targeting the leadership of enemy groups. There is no doubt this air asset offers special capabilities. In a low-threat environment drones can provide a persistent presence; a very capable intelligence-gathering platform; and with the addition of armament an accurate, although not perfect integral strike capability.

As will be discussed, the use of drones has even spawned its own terminology, as well as calls for a weapon-specific analytical approach. However, this aura of uniqueness must ultimately be reconciled with their use across a wide spectrum of contemporary operations,

---

[57] *Id.* at 237 (where it was noted the capture of Warsame was not feasible in Yemen "because of the political backlash it would cause.").

[58] Helen Cooper & Eric Schmitt, *ISIS Official Killed in U.S. Raid in Syria, Pentagon Says*, N.Y. TIMES, May 16, 2015, http://www.nytimes.com/2015/05/17/world/middleeast/abu-sayyaf-isis-commander-killed-by-us-forces-pentagon-says.html?hp&action=click&pgtype=Homepage&module=first-column-region&region=top-news&WT.nav=top-news&_r=0.

[59] Carl de Vries, *Carter: ISIS Raid a "Significant Blow" to Terror Group*, CNN (May 16, 2015, 4:17 PM), http://www.cnn.com/2015/05/16/politics/abu-sayyaf-isis-white-house-raid-syria/index.html.

ranging from assisting in the conduct of air strikes during international armed conflict (i.e., Libya) to international law enforcement (i.e., surveillance platform). The generality of their use, and the roles they perform, which are often matched by other means of warfare (i.e., other aircraft) argues against a weapons-specific approach toward regulation.[60] The challenge for State legal advisors and theorists alike is to separate myth from reality and to determine whether this weapon system truly presents a special means of conducting warfare such that it warrants special attention or its own set of rules governing its use.

The following analysis will look at drones from three perspectives. First, there are factors that reflect their being viewed as a unique weapon. These are the application of special terminology regarding their use and the involvement of senior levels of leadership in targeting decisions. The second area of inquiry will look at the more general use of drones on contemporary operations in order to put the higher profile transnational counterterrorism missions in context. Finally, the normative framework that has developed regarding drone strikes will be examined. As will be established, while specific rules and policies have been developed regarding some drone use, their impact is likely to be limited to particular operational contexts. Importantly, these same issues can affect other methods and means of warfare including the use of Special Forces raids. Therefore attempts to apply drone-specific rules should be subjected to a critical analysis as to whether the issue is truly unique to that weapon system.

8.4.2. DRONES: SPECIAL TERMS AND HIGH-PROFILE ATTACKS

In certain circumstances drones can provide a significant capability for identifying targets and seemingly personalize the attacks. One indication of how drones are often viewed as being unique is found in the "new" terminology that such attacks have spawned. Attempts by commentators to reconcile attacks where the identity of the target is specifically known, with those where there may not be the same level of intelligence available, has led to the use of terms such as "personality" strikes and "signature" strikes.[61] The former term refers to "targeting named, allegedly high-value leaders of armed, non-state groups."[62] It is noteworthy that personality strikes are normally focused on "high-value leaders." In contrast, signature strikes involve "the targeting of groups of men who bear certain signatures, or defining characteristics associated with terrorist activity, but whose identities aren't necessarily known."[63]

It is personality strikes and the use of drones to attack the senior leadership of terrorist and insurgent organizations that initially attracted the attention of legal analysts and human rights advocates. As a weapon system that integrates impressive communications

---

[60] John P. Abizaid & Rosa Brooks, Recommendations And Report Of The Task Force On Us Drone Policy, The Stimson Center 10 (2014) [hereinafter Drone Task Force Report], http://www.stimson.org/images/uploads/task_force_report_final_web_062414.pdf (outlining the Task Force conclusions that drones are not a "super-weapon," and the United States does not have monopoly on their use.).

[61] Living Under Drones, *supra* note 7, at 12, Civilian Impact of Drones, *supra* note 12, at 8 and Savage, *supra* note 47, at 254–57.

[62] Living Under Drones, *supra* note 7, at 12. *See also* Klaidman, *supra* note 43, at 41 (where it is suggested for "personality" or "high value" strikes, "a terrorist leader was positively identified before the missile was launched.").

[63] Klaidman, *supra* note 43, at 41.

technology, drones have provided an operational capability that has led to the involvement of high-level (i.e., strategic) political and military leaders in the United States in operational decision making. It has been reported the U.S. government has prepared and regularly updates a list of identified terrorist leaders who may be targeted.[64] In early 2010 there were "at least three entities within the US government that were maintaining kill lists: the National Security Council . . . the CIA; and the US military."[65]

The adoption of special terms such as "signature strikes", "personality strikes", or even "*terrorist attack disruption strikes*, or TADs"[66] to describe attacks linked primarily to the use of drones is a not uncommon phenomenon when grappling with conflicts outside the realm of traditional inter-State armed conflict. In this regard, the term "targeted killing" appears to have a similar genesis. The dressing up of this issue in new terminology is also not a trend that is unique to non-government observers. It can also been seen in suggestions that targeting is justified by government authorities when it is directed at "military age" males.[67] The problem is that inventing new terms for the public dialogue is not necessarily helpful in providing a comfort level that the law has been fully applied to targeting decisions. This is because, while the technology and capability may be new, the legal rules governing targeting are reasonably well established.

It is clear drone use against the leadership of transnational terrorist groups has led to new and thought-provoking analysis. For example, one approach sees attacks using drones as being governed by a discrete body of law that focuses on "individualized" targeting.[68] The use of this technology has also led to suggestions that the ability to focus on high-level targets provides a morally superior means of warfare, since one "can discriminate between low-level and high-level combatants and minimize the loss of life to foot soldiers on the other side by concentrating fire on the leaders."[69] This targeting process further lends itself to suggestions there can be greater oversight of this individualized targeting decision making.[70] It has also been observed "the fact that all targeted killing operations in combatting terrorism are directed against particular individuals makes the tactic more reminiscent of a law enforcement paradigm, where power is employed on the basis of individual guilt rather than status (civilian/combatant)."[71] This extends to the preference for capture over killing, which is a "more subtle, outcropping of the emerging norm of individualization."[72]

---

[64] Scott Shane, *Election Spurred a Move to Codify US Drone Policy*, N.Y. TIMES, Nov., 24, 2012, http://www.nytimes.com/2012/11/25/world/white-house-presses-for-drone-rule-book.html?pagewanted=all.

[65] SCAHILL, *supra* note 43, at 351. *See also* Gregory S. McNeal, *Targeted Killing and Accountability*, 102 GEO. L.J. 681, 701–29 (2014).

[66] SAVAGE, *supra* note 47, at 257 (where it is indicated the term "terrorist attack disruption strikes" referencing some evidence of a threat was developed for use instead of "signature" strikes).

[67] Jo Becker & Scott Shane, *Secret "Kill List" Proves a Test of Obama's Principles and Will*, N.Y. TIMES, May 29, 2012, http://www.nytimes.com/2012/05/29/world/obamas-leadership-in-war-on-al-qaeda.html?pagewanted=all, and CIVILIAN IMPACT OF DRONES, *supra* note 12, at 32.

[68] Gabriella Blum & Philip Heymann, *Law and Policy of Targeted Killing*, 1 HARV. NAT. SEC. J. 145, 148 (2010).

[69] Samuel Issacharoff & Richard H. Pildes, *Targeted Warfare: Individuating Enemy Responsibility*, 88 N.Y.U. L. REV. 1521, 1573 (2013).

[70] *Id.* at 1596–97. *See also* DRONE TASK FORCE REPORT, *supra* note 60, at 43–45 (calling for the creation of an oversight Commission "outside of hot battlefields.").

[71] Blum & Heymann, *supra* note 68, at 148.

[72] Issacharoff & Pildes, *supra* note 69, at 1583.

However, what remains to be seen is whether this analysis, which is linked to the emphasis that has been placed on high-profile individualized strikes, is warranted. The challenge for proponents of an individualized approach toward targeted killing appears to be twofold. First, as the use of the two terms, personality and signature strikes, suggest, not all counterterrorism action viewed as targeted killings rely on such individualization. While personality strikes point to an individual approach, the use of signature strikes itself suggests drones are often used on a more impersonal basis to prosecute a wider range of targets. The second related point is that targeted killings frequently take place in the context of a broader counterterrorism effort, as well as counterinsurgency. The majority of drone attacks are not discretely limited to the more isolated and the higher profile activity that have involved senior civilian and military leaders or have caught the attention of many human rights advocates and other commentators. The publicity this senior-level engagement has attracted can leave the impression that most drone strikes have been against high-level terrorist or insurgent leaders. However, despite the focus on personality strikes most persons killed by drone strikes are said to remain relatively low-level fighters.[73] Importantly, the same fundamental legal questions regarding the use deadly force have to be resolved if a sniper is used to kill an insurgent or terrorist in an urban environment during an insurgency.

8.4.3. THE "ORDINARINESS" OF DRONES

From both a legal and policy perspective, the significant intelligence resources made available for high-profile personality strikes, the development of preapproved target lists, and the practice of often retaining authorization for such strikes at the strategic level makes personality-based targeting appear different. However, as will be discussed, the vast majority of targeting with drones takes place outside the context of this strategically determined process (i.e., at the operational and tactical level). In this context for both military commanders and their advisors neither signature strikes nor personality strikes present a particularly unique form of warfare or a different targeting challenge. The fact that a particular weapon system, such as a drone, is used does not make "signature" attacks exceptional. Such strikes are neither unique to what has been termed "covert" drone warfare nor to locations like Pakistan, Yemen, or Somalia.[74]

The "ordinariness" of drones is seen in three ways: the degree to which drone-specific terminology really reflects targeting criteria already integrated into operations; the use of drones across a broad spectrum of military activity; and the control over drones that is exercised at lower levels of command. Regarding terminology, drones have been used in many counterinsurgency situations (e.g., Afghanistan and Iraq) where emphasis has to be placed on identifying a target based on a number of preestablished criteria, rather than

---

[73] Civilian Impact Of Drones, *supra* note 12, at 9–10. *See also Obama's Lists: A Dubious History of Targeted Killings in Afghanistan*, Spiegel OnLine (Dec. 28, 2014), http://www.spiegel.de/international/world/secret-docs-reveal-dubious-details-of-targeted-killings-in-afghanistan-a-1010358.html (for reference to the targeting of a Taliban mid-level leader in Afghanistan who was 3,673rd on the kill list and who had been provided "a priority level of three on a scale of one to four.").

[74] Living Under Drones, *supra* note 7, at 7–18 (for a discussion of the U.S. covert drone campaign, including in Pakistan, Yemen, and Somalia).

exact knowledge of who the enemy might be. Therefore what has relatively recently been described in drone-specific terms as signature strikes already has a technical language of its own integrated into military jargon and targeting doctrine. Such terms include "pattern of life" and "positive identification" or "PID"[75] The connection to signature strike terminology can be seen in the definition of positive identification which "is defined as the 'reasonable certainty that a functionally and geospatially defined object of attack is a legitimate military target in accordance with the Law of War and applicable Rules of Engagement.'"[76] For example, in Iraq "[s]trict ROE mandated that a Predator could not fire unless there was positive identification [PID] of a weapon, accompanied by hostile intent or action, and with the additional stipulation that collateral damage be restricted."[77]

Importantly, regardless of how an attack is described (i.e., "signature" or "personality"), or whether other terminology is applied (i.e., "PID" or "pattern of life"), the humanitarian law rules governing targeting remain the same. Indeed, the precautions identified in Article 57 of Additional Protocol I apply to the individual use of force by a soldier, as much as to the conduct of a strategic air campaign. The difference between a sniper killing a known enemy and the use of drones is not an issue of different law being applied, but rather one of the means used to carry out an attack.[78] The true challenge in the contemporary debate about drones is not one of terminology, but rather determining whether the appropriate legal framework is actually being applied to govern drone use. This has often been put in doubt on a most basic level with claims such as those made by a senior U.S. official in June 2011 "that there had not been a single collateral death in the previous year. . . ."[79] The reaction has rightly been that such claims "strained credibility—and badly undermined U.S. credibility."[80]

The second factor to be considered when assessing drones and targeting is the narrow scope of the covert drone effort in comparison to other operations. This is particularly evident when such use is considered in respect of contemporary air campaigns, such as those carried out in Iraq in 2003 or Libya in 2011. For example, between March 19/20 and April 25, 2003, the Coalition overall flew nearly 50,000 fixed wing missions in Iraq of which 36 percent were strike sorties that dropped 28,820 munitions.[81] In the 2011 Libya conflict

---

75 L. S. Turner et al., *Optimizing Deadly Persistence in Kandahar: Armed UAV Integration in the Joint Tactical Fight*, 13 Can. Army J. 117, 122–23 (2010).

76 Civilian Impact of Drones, *supra* note 12, at 73 (Relying on the *Joint Targeting Cycle and Collateral Damage Estimation Methodology (CDIM)*, General Counsel to the Chairman of the Joint Chiefs of Staff 26 (Nov. 10, 2009)).

77 Martin & Sasser, *supra* note 21, at 238. *See also* James Fergusson, A Million Bullets: The Real Story of the British Army in Afghanistan 120 (2008) (where the same strict PID requirement was applied to helicopter air support).

78 In terms of the use of snipers, *see* Byman, *supra* note 22, at 312 ("Israel regularly uses snipers, though they receive less publicity."), Michael Bar-Zohar & Nissim Mishal, Mossad: The Greatest Missions of the Israeli Secret Service 287–90 (2014) and Dan Raviv & Yossi Melman, Spies Against Armageddon 322 (2012) (for reference to an Israeli sniper on a naval vessel killing a Syrian general in 2008 while he was eating dinner at a coastline villa. There were no collateral injuries among his dining companions). *See also* Mark Galeotti, Spetnaz: Russia's Special Forces 36 E(3) (2015)(where it is noted "[s]nipers have become increasingly central to Russian counterinsurgency tactics . . . .").

79 Audrey Kurth Cronin, *Why Drones Fail: When Tactics Drive Strategy*, 92 Foreign Aff. 44, 47 (July/Aug. 2013).

80 *Id.*

81 Walter J. Boyne, The Influence of Airpower Upon History 379 (Pen & Sword Aviation, 2005) (2003). *See also* Williamson Murray & Robert H. Scales, Jr., The Iraq War 170–171 (2003) (where it is noted over a

it is reported NATO flew 17,939 armed sorties and dropped 7642 air-to-surface weapons.[82] Unmanned aerial vehicle missions were reported to comprise 509 of the 25,944 air sorties made during the Libya operation. Of the armed sorties, some 250 missions involved drones conducting intelligence, surveillance, and reconnaissance operations missions.[83] It is a drone strike that is reported to have struck Omar Qaddafi's convoy in October 2011 leading to his capture and eventual death.[84] Comparatively, there were significantly fewer "covert" drone strikes carried out in Pakistan, with 399 taking place over ten years (2004 to June 2015),[85] and 125 attacks over twelve plus years in Yemen (2002 to June 2015).[86] Drones are also not the only means of warfare used to conduct either "personality" or "signature" strikes. For example, in terms of personality strikes, C-130 gunships were used by the United States to target an Al Shabaab leader in Somalia in 2007.[87] In 2008 American aircraft (a "Chain Shot" Navy P-3 Orion sub-hunter) provided video surveillance prior to an attack on an Al Shabaab leader with Tomahawk missiles.[88] As has been noted, in 2002 Israel employed F-16 aircraft to kill the Hamas military leader, Saleh Shehada, in Gaza.[89]

Further, drones provide a capability that can be used in support of both conventional and Special Forces operations. It may be that the status attributed to drones as a strike weapon is aided by the use of terms such as "unmanned combat aerial vehicles" and "unmanned combat aerial systems."[90] These terms highlight a kinetic role. However, it is in surveillance missions where the drone has proven to be a particularly useful asset. As a sensor platform, drones can be used to facilitate the carrying out of attacks by other weapons systems such as aircraft,[91] or very accurate long-range artillery firing GPS-guided munitions.[92] When used

two-day period 1,500 to 2,000 sorties were launched per day. Of these attacks only 500 targets were preplanned in each twenty-four-hour period.).

[82] *Letter Peter Olson to Judge Kirsch*, OLA (2012) 006, at 4–5, Jan. 23, 2012 [hereinafter *NATO Response Letter*], http://www.nato.int/nato_static/assets/pdf/pdf_2012_05/20120514_120514-NATO_1st_ICIL_response.pdf.

[83] *Id.* at 4.

[84] David E. Sanger, Confront and Conceal: Obama's Secret Wars and Surprising Use of American Power 354 (2012).

[85] Drone Wars Pakistan: Analysis, International Security Program, New America Foundation, http://securitydata.newamerica.net/drones/pakistan-analysis.html (last visited June 14, 2015).

[86] Drone Wars Yemen: Analysis, International Security Program, New America Foundation, http://securitydata.newamerica.net/drones/yemen-analysis.html (last visited June 14, 2015) (there were an additional fifteen air strikes by other than drones also reported).

[87] Mazetti, *supra* note 27, at 150.

[88] Naylor, *supra* note 26, at 339.

[89] Byman, *supra* note 22, at 307–10.

[90] HPCR Manual on International Law Applicable to Air and Missile Warfare, Program on Humanitarian Policy and Conflict Research at Harvard University, Rule 1(ee), at 55 (2009), http://ihlresearch.org/amw/Commentary%20on%20the%20HPCR%20Manual.pdf (for the definition of unarmed combat aerial vehicle), and Michael N. Schmitt, *Unmanned Combat Aircraft Systems and International Humanitarian Law: Simplifying the Oft Benighted Debate*, 30 B.U. Int'l L.J. 595, 595 (2012) (for reference to unarmed combat aerial system).

[91] Fergusson, *supra* note 77, at 59–62 (setting out how the laser guidance system on a Predator drone was used to improve the accuracy of A-10 Warthog ground attack aircraft during the siege of a British held compound in Now Zad, Helmand Province, Afghanistan). *See also NATO Response Letter*, *supra* note 82, at 7 (where in respect of a May 1, 2011, attack on a command building in Tripoli it is noted "[f]ull motion video acquired by manned aircraft and UAVs at the time of the strike indicated that no civilians were in the target area.").

[92] Dan Lamothe, *Long Shot: Artillery Battery Sets Lethal Record*, Marine Corps Times, June 30, 2013, http://www.marinecorpstimes.com/news/2012/06/marine-record-setting-long-shot-artillery-strike-m777-063012w/.

as a surveillance platform, it also can play an important role when dealing with transnational criminal organizations and other engaged in criminal activity.

Even in the context of targeting high-profile individuals, drones may only be used to provide situational awareness to other military forces that actually conduct raids to capture or kill the target.[93] Nor is the way they are use for targeting necessarily unique. The use of drones to strike at targets throughout an enemy organization is consistent with the approach of using the Special Forces teams to degrade non-State actor armed groups. For example, supported by intelligence analysis, such teams conducted dozens of raids each night in Afghanistan. These teams targeted "Taliban shadow governors, midlevel insurgent commanders, and individuals who handled finances and logistics for the Taliban."[94] The leadership of an organized armed group and those providing direct support within an organization such as logistics personnel, are lawful military objectives (see Chapter 7, section 7.5, *The IED and Suicide Bombing Threat: Arrest or Kill?*). Similarly, "drones also hurt terrorist organizations when they eliminate operatives who are lower down on the food chain but who boast special skills: passport forgers, bomb makers, recruiters, and fundraisers."[95] This of course requires they meet the criteria of being members of an organized armed group or are otherwise direct participants in hostilities.

The use of drones is also far more ubiquitous than the counterterrorist focus on strikes in Pakistan, Yemen, or Somalia suggests, and for a broader range of operations. They are not only deployed by the United States. It was reported in 2013 that nearly ninety other countries have surveillance drones.[96] Like the United States, the United Kingdom has developed the capability to pilot drones from that country for the conduct of operations, including missile strikes in Afghanistan.[97] In late 2014 British drones were being operated in Iraq against the Islamic State.[98] This was extended to a strike against two British jihadists in Syria in September 2015 based on the exercise of State self-defense.[99] Drones have also

---

93 Martin & Sasser, *supra* note 21, at 265 (outlining the use of a Predator drone to assist in a Special Forces raid on a house in Husabayah on the Iraq/Syrian border). *See* Phillip Alston, *The CIA and Targeted Killings Beyond Borders*, 2 Harv. Nat'l. Sec. J. 283, 297 (2011) (where it is noted that targeted killing has been used to describe a wide range of situations and "[t]he means and methods of killing vary, and include shooting at close range, sniper fire, firing missiles from helicopters or gunships, firing from UAVs, the use of car bombs, and poison."), and Kimberley Dozier, *Who Will Drones Target? Who in the US Will Decide?*, Assoc. Press (May 21, 2012), http://bigstory.ap.org/content/who-will-drones-target-who-us-will-decide (outlining that the U.S. government may use a Special Operations team or even the FBI to attempt to capture the "target." In this context "targeting" will not necessarily mean the use of kinetic force in the sense of carrying out an "attack" as defined under AP I, art. 49).

94 Eric Schmitt & Thom Shanker, Counterstrike: The Untold Story of America's Secret Campaign Against Al Qaeda 259 (2011).

95 Daniel Byman, *Why Drones Work: The Case for Washington's Weapon of Choice*, 92 Foreign Aff. 32, 33 (July/Aug. 2013).

96 *Id.* at 41. *See also* Simon Jones, *Drones by Country: Who Has All the UAVs?*, Guardian Datablog (Aug. 3, 2012, 17:00 BST), http://www.guardian.co.uk/news/datablog/2012/aug/03/drone-stocks-by-country#data.

97 Nick Hopkins, *UK Starts Controlling Drones in Afghanistan from British Soil*, BBC News (Apr. 25, 2013), http://www.guardian.co.uk/world/2013/apr/25/uk-controlling-drones-afghanistan-britain.

98 Andrew Chuter, *UK Reapers Up the Intensity Against IS in Iraq*, Defense News, Nov. 18, 2014, http://www.defensenews.com/article/20141118/DEFREG01/311180035/UK-Reapers-Up-Intensity-Against-Iraq.

99 Nicholas Watt, Patrick Wintour and Vikram Dodd, *David Cameron Faces Scrutiny Over Drone Strikes Against Britons in Syria*, The Guardian (Sep. 8, 2015, 11:08 BST), http://www.theguardian.com/world/2015/sep/07/david-cameron-justifies-drone-strikes-in-syria-against-britons-fighting-for-isis.

been used in support of UN operations.[100] Some non-State actors, such as Hezbollah and Hamas, have developed a drone capability.[101] This non-State actor use graphically reflects the degree to which the technology is proliferating and drones are becoming a normal part of contemporary operations.

It seems at times that the ability to strike at leadership targets from long distances has created the perception drones are the counterterrorism weapons of choice. However, drones are not a panacea weapons system and have definite operational limitations. For example, their use was ultimately rejected for the attack on the bin Laden compound because of concerns that his death could not be guaranteed with certainty due to the small warhead size.[102] It also could not necessarily be confirmed that he had in fact been killed. [103] As a result, it was a Special Forces raid that ultimately killed that terrorist leader. That "boots on the ground" option enabled the necessary verification that the mission was successful,[104] while at the same time limiting the potential for collateral civilian casualties and obtaining valuable intelligence.[105] Drones also do not provide the operational flexibility of Special Forces assets, which can attempt to capture an enemy. Further, unmanned aircraft do not have an obvious or ready kinetic role in dealing with transnational criminal organizations or other criminal gangs. However, this may be changing as an aerial strike by aircraft does appear to have been used by EU forces in order to destroy pirate infrastructure in Somalia.[106] It is obvious that drones could also be used to perform such a mission.

Another potential myth about drones is that it is usually the highest levels of the U.S. government, including the president, which directly control their use. Such a perception could be created as a result of strikes in Pakistan, Yemen, and even Somalia, where the targeting authority does appear to be centered at the national level in the United States. This has even led to the suggestions that domestic judicial or other authorities should control drone strikes.[107] However, just as drone strikes constitute a relatively small number of attacks overall, the involvement of the strategic-level leadership in targeting decisions is even more limited. In contrast to the public perception of political figures (or even national-level military commanders) having their finger on the trigger, the use of drones in contemporary operations and the need for timely decisions means drone strikes are, other than

---

[100] Performance Peacekeeping: Final Report of the Expert Panel on Technology and Innovation in UN Peacekeeping, Under-Secretary Department of Peacekeeping Operations & Under-Secretary Field Operations 134 (Dec. 22, 2014), and Samuel Oakford, *Drones, Drones, Everywhere: UN Ramping Up Peacekeeper Surveillance Flights*, Aljazeera America (Aug. 27, 2014), http://america.aljazeera.com/articles/2014/8/27/united-nations-drones.html.

[101] *Hezbollah Admits Launching Drone over Israel*, BBC News (Oct. 11, 2012), http://www.bbc.co.uk/news/world-middle-east-19914441, *Hezbollah Drone Flies over Israel*, BBC News (Nov. 7, 2004), http://news.bbc.co.uk/2/hi/middle_east/3990773.stm (where it was reported as early as 2004 that the Hezbollah was flying drones over Israel), and Martin & Sasser, *supra* note 21, at 218. *See also*, *Israel Says It Knocked Out Hamas Drone Program*, CBS News (Nov. 16, 2012), http://www.cbsnews.com/8301-202_162-57551216/israel-says-it-knocked-out-hamas-drone-program/.

[102] Bergen, *supra* note 23, at 200, and Schmitt & Shanker, *supra* note 94, at 260.

[103] Savage, *supra* note 47, at 261–62.

[104] Sanger, *supra* note 84, at 78, and Morell & Harlow, *supra* note 23, at 157.

[105] Morell & Harlow, *supra* note 23, at 165, Bergen, *supra* note 23, at 178, and Schmitt & Shanker, *supra* note 94, at 261.

[106] *See also* Byman, *supra* note 95, at 41 (where it is indicated Beijing had announced "(although it later denied) that it had considered sending a drone to Myanmar . . . to kill a wanted drug trafficker hiding there.").

[107] *Drone Speech*, *supra* note 23.

exceptional circumstances, much more likely to be authorized lower within the military chain of command.

For example, in Afghanistan the authority to strike was "exercised either by designated senior officers in operations centers (e.g., a NATO or Joint Special Operations Task Force Operations Center) or, in certain defensive situations, ground commanders engaged with enemy forces."[108] The operational tempo in Iraq after the 2003 invasion and during the counterinsurgency in Afghanistan inevitably limited the involvement of senior leaders in targeting decisions, even if that had been a desired course of action.[109] For example, Special Forces operations in Iraq "went from half a dozen operations a month in Iraq in the spring of 2004 to three hundred a month by the summer of 2006."[110] By the end of that year they had killed 2,000 members of the Sunni jihadist groups with an even higher count in 2007.[111] Not all of these operations would have involved drones.[112] However, that is the point. Drones offer a platform that may be used by military commanders at various levels across a broad range of military operations. Much depends upon the operational goals and the situation on the ground.

8.4.4. CONTROLLING DRONE USE

Drone strikes have raised important humanitarian law questions regarding exactly who is being targeted. Those questions center on direct participation, membership in organized armed groups, and what types of support make a person a lawful target (see Chapter 7, section 7.5, *The IED and Suicide Bombing Threat: Arrest or Kill?*). They have also attracted the attention of various human rights groups and UN investigatory bodies.[113] Yet at times drones appear to be treated by much of the international community outside the military as new "bright shiny objects." As such, they appear to have captivated governments and their critics alike. Drone capabilities, and their use against high-value targets have to a considerable degree skewed the discussion regarding the law governing their use. There is a real danger that too much emphasis being placed on their use against transnational terrorists runs the risk of setting legal standards and expectations that are neither realistic nor effective in controlling the use of force in broader conflict. Just as they do not offer an all-encompassing solution for dealing with contemporary security threats, these limited drone strikes should not alter the legal standards for targeting.

---

108 Schmitt, *supra* note 90, at 600.

109 Stanley A. McChrystal, *It Takes a Network*, Foreign Pol'y (Mar./Apr. 2011), *available at* http://foreignpolicy.com/2011/02/21/it-takes-a-network/ ("finding just one initial target could lead to remarkable results: The network sometimes completed this cycle three times in a single night in locations hundreds of miles apart—all from the results of the first operation.").

110 Bergen, *supra* note 23, at 156.

111 Jason Burke, The 9/11 Wars 278 (2011).

112 *Id.* at 154–55 (where it is noted initially the Joint Special Operations Command only had one Predator in Iraq, but it "acquired its own dedicated drones so that it could maintain an 'unblinking eye' 24/7 over targets.").

113 Human Rights Organizations' and United Nations' Recommendations for Targeted Killings, Stimson: Managing Across Borders Initiative 2–5 (May 2014), http://www.stimson.org/images/uploads/human_rights_groups_and_un_special_rapporteur_recommendations.pdf (for a summary of Human Rights and UN reports regarding the use of drones and targeted killings).

This does not mean that the bodies of law governing contemporary operations will not have to be specifically applied regarding their use. The same has to occur with any new or emerging military capability. Evidence of new approaches being developed to govern drone strikes are reflected in the 2013 U.S. drone policy.[114] That policy indicates capture would be considered prior to targeting,[115] and further states that attacks would not take place unless there was a "near certainty" no civilians will be killed or injured. [116] This policy restriction does not reflect traditional humanitarian law rules. In contrast, under international criminal law the liability for a war crime arises when collateral casualties are "clearly excessive in relation to the concrete and direct overall military advantage anticipated," rather than no civilian injury or death occurring at all. Further, the drone policy "near-certainty" test is more stringent than the established legal standard of the decision maker having a "reasonable belief" a certain outcome will occur.[117] It is this latter test that is applied as a matter of law even in a human rights–based law enforcement context.[118] The standards set for drone use in what the United States views as being part of an existing armed conflict are therefore exceptionally restrictive. It is an approach that reflects, and in some respects exceeds, the requirements of human rights law.

However, the U.S. policy framework for controlling drones is not entirely unique. The 2006 *Israeli Targeted Killing Case*[119] established a legal requirement for the security forces of that State to consider capturing rather than killing a terrorist suspect (see Chapter 7, section 7.4.2, *The "Targeted Killing" Case and Control*).[120] That decision also set out a requirement for an investigation after each strike against a "civilian" (including a terrorist).[121] Both these requirements reflect an integration of human rights–based principles to govern targeted killing. However, it is worth noting the 2002 incident that seems to have prompted this litigation involved the use of F-16 aircraft to bomb Hamas military-wing leader, Salah Shehada, rather than drones.[122] This suggests that the legal restrictions are more closely aligned with the operational situation than the specific type of weapon system used.

These legal (Israel) and policy (United States) restrictions should be considered in the context in which they arose. Israel's application of *Targeted Killing Case* requirements appear to have been restricted to the unique situation of occupied territory or to other areas where control can be exercised by security forces to a level sufficient to conduct an arrest.[123]

---

[114] *Drone Speech*, *supra* note 23.

[115] *Id.* ("America does not take strikes when we have the ability to capture individual terrorists; our preference is always to detain, interrogate, and prosecute.").

[116] *Id.* (President Obama noted "before any strike is taken, there must be a near-certainty that no civilians will be killed or injured—the highest standard we can set.").

[117] Prosecutor v. Galić, Case No. IT-98-29-T, Judgment and Opinion, ¶¶ 50, 51 & 55 (Int'l Crim. Trib. for the Former Yugoslavia, Trial Chamber, December 2003), http://www.un.org/icty/galic/trialc/judgement/gal-tj031205e.pdf (for reference to a reasonable belief being required).

[118] McCann v. United Kingdom, 21 Eur. H.R. Rep. 97, ¶ 200, at 173, ¶ 5, at 179 (1995), http://hudoc.echr.coe.int/sites/eng/pages/search.aspx?i=001-57943.

[119] Public Committee Against Torture in Israel v. Israel, Israel Supreme Court [16 December 2006], 46 ILM 375 (2007).

[120] *Id.* ¶40, at 393–94.

[121] *Id.* ¶40, at 394.

[122] Byman, *supra* note 22, at 308.

[123] Public Committee Against Torture in Israel v. Israel, Israel Supreme Court [16 December 2006], 46 ILM 375, ¶40, at 394 (2007).

These restrictions do not seem to have been applied to broader armed conflict (i.e., the 2008 Gaza, or subsequent conflicts).[124] Further, there appears to be only one reported investigation of the type contemplated in that decision.[125] Similarly, the U.S. drone policy has been developed in the unique context of dealing with a strike against a U.S. citizen "beyond the Afghan theatre" in Pakistan, Yemen, and Somalia.[126] The United States was quick to point out that the 2013 drone policy was not applicable to the more traditional air campaign commenced against Islamic State targets in Syria and Iraq in 2014.[127] The reality is that the deliberate nature of the attacks contemplated by the Israeli decision and the United States policy; the amount of intelligence that can realistically be collected, or made available; and the time available to make a targeting decision do not apply in the vast majority of circumstances within which targeting assessments have to be made during armed conflict.

This does not mean that the issues raised by the Israeli domestic jurisprudence and the U.S. national drone policy and guidance they provide do not have a broader application. The capture or arrest before killing a person taking a direct part in hostilities; assessing the element of control exercised over territory before capture is considered; and a sensitivity toward the risk of death or injury to both soldiers and uninvolved civilians are key factors to consider in seeking to apply a holistic legal approach more broadly to counterinsurgency and counterterrorism operations. The same can be said regarding suggestions made by the UN Special Rapporteur on Extrajudicial, Summary or Arbitrary Executions when he commented on the requirement to holistically apply the laws governing the State use of force, humanitarian law, and human rights law.[128] However, drones are not weapons requiring specialized legal terminology or unique international legal rules by which to be regulated.

8.5. Legal Frameworks for Responding to Transnational Terrorism

8.5.1. THE RESTRICTED "LAW ENFORCEMENT" THEORY

The use of drones to respond to transnational terrorist attacks has stirred considerable legal controversy. Consideration of their use in this context requires an assessment of additional bodies of law: State self-defense and human rights law. In effect, a "holistic" legal lens has been directed at the analysis of drone strikes. This response to transnational terrorism will be explored in terms of three separate theories governing the use of force. They are the restricted "law enforcement" theory, a permissive "conduct of hostilities" approach, and the

---

[124] *See, e.g.*, BENJAMIN S. LAMBETH, AIR OPERATIONS IN ISRAEL'S WAR AGAINST HEZBOLLAH: LEARNING FROM LEBANON AND GETTING IT RIGHT IN GAZA 251 (2011), *available at* http://www.rand.org/content/dam/rand/pubs/monographs/2011/RAND_MG835.pdf (for reference to 5,650 aerial sorties flown during the 23-day 2008 conflict with Hamas in Gaza with 3,430 targets being hit.),

[125] SALAH SHEHADEH-SPECIAL INVESTIGATORY COMMISSION, Israel Ministry of Foreign Affairs, ¶1 (Feb. 21, 2011), http://mfa.gov.il/MFA/AboutIsrael/State/Law/Pages/Salah_Shehadeh-Special_Investigatory_Commission_27-Feb-2011.aspx.

[126] *Drone Speech, supra* note 23.

[127] Ken Dilanian, *U.S.-led Airstrikes Launched Amid Significant Intelligence Gaps*, CTV NEWS (Oct. 1, 2014), http://www.ctvnews.ca/world/u-s-led-airstrikes-launched-amid-significant-intelligence-gaps-1.2033338.

[128] *See also Report of the Special Rapporteur on extrajudicial, summary or arbitrary executions*, UN Doc. AA/68/382, ¶ 24, at 6 (Sept. 13, 2013) (by Christof Heyns).

"self-defense" option. An outline will be provided of these theoretical approaches, followed by an assessment of a number of issues that each option raises.

The first theory advocates for a human rights–based law enforcement approach. Before embarking on an outline of this highly restrictive approach, it is important to indicate this theory does not ban all uses of "weaponized" drones. A leading proponent, Mary Ellen O'Connell, clearly acknowledges that drones are lawful battlefield weapons. Examples provided of their lawful use include during conflicts in Iraq and Afghanistan.[129] The position taken, though, is that "[d]rones are not lawful for use outside combat zones,"[130] with strikes in Pakistan, Somalia, and Yemen being questioned as to their legality.[131] Outside of these conflict zones, "police are the proper law enforcement agents and police are generally required to warn before using lethal force."[132] This use of force is governed by human rights–based law enforcement standards. As is noted, in law enforcement situations "[w]hat drones cannot do is comply with police rules for the use of lethal force away from the battlefield."[133] The result is that the use of drones is theoretically restricted to what has increasingly although artificially been called "hot" battlefields. It should be noted that the term "hot battlefield" is itself problematic, as it has no practical or legal meaning. Even within geographic boundaries there can be some areas where armed conflict is taking place and others that are peaceful.[134] In effect, this law enforcement–based theory seeks to restrict noninternational armed conflict within the geographic boundaries of a territory in which it is most obviously occurring, although not necessarily where a threat may additionally exist.

The law enforcement–based theory also has variations. For example, not all interpretations limit the use of drone strikes conducted in the post-9/11 Afghanistan conflict to the geographic borders of that country. A 2011 analysis by the Oxford Research Group accepts there is also a "spillover" to the adjacent country of Pakistan on the basis that extremists in the northwest area of that territory were fighting NATO forces in Afghanistan.[135] In that respect, "[a] Non-International Armed Conflict exists in Pakistan which is part and parcel of the Non-International Armed Conflict in Afghanistan."[136] However, this aspect of the

---

[129] Mary Ellen O'Connell, *Lawful Use of Drones*, Congress of the United States, Hearing: Rise of the Drones II: Examining the Legality of Unmanned Targeting 1 (Apr. 28, 2010), https://www.fas.org/irp/congress/2010_hr/042810oconnell.pdf.

[130] *Id.*

[131] *Id.* at 2–3.

[132] *Id.* at 1.

[133] *Id.* at 2. ("In law enforcement it must be possible to warn before using force, in war-fighting this is not necessary, making the use of bombs and missiles lawful.").

[134] The U.S. Army, Marine Corps, Counterinsurgency Field Manual ¶ 1-37, at 14 (2007) (for reference to the term "mosaic wars").

[135] Susan Breau, Marie Aronsson, & Rachel Joyce, Discussion Paper 2: Drone Attacks, International Law, and the Recording of Civilian Casualties or Armed Conflict, Oxford Research Group 12 (June 2011), http://www.oxfordresearchgroup.org.uk/sites/default/files/ORG%20Drone%20Attacks%20and%20International%20Law%20Report.pdf [hereinafter Oxford Research Paper]. *See* Yoram Dinstein, Non-International Armed Conflicts in International Law ¶¶ 77–82, at 25–27 (2014) (where it is indicated extraterritorial spillover can extend to the high seas, cyberspace, and cross-border hostilities on land), and Michael N. Schmitt, *Charting the Legal Geography of Non-International Armed Conflict*, 90 Int'l L. Stud. 1, 16–17 (2014). *See also* Aerial Drone Deployment on 4 October 2010 in Mir Ali/Pakistan, 157 I.L.R. 722, 742 (2013) (where the German Federal Prosecutor General took the view the Afghanistan conflict spilled over into Pakistan).

[136] Oxford Research Paper, *supra* note 135, at 1.

Afghan conflict was considered to be limited to the FATA, and the NWFP territories in Pakistan. Attacks outside those areas are governed by the law enforcement model and "raise questions on state sovereignty as well as unlawful killings."[137] This analysis did not support the use of drone strikes in Yemen, as they were not part of any noninternational armed conflict in that country.[138]

Interestingly, the Oxford Research Group approach also did not preclude the use of drones under a human rights–based paradigm, including "as an exceptional use of force in self-defence alongside peacetime law enforcement."[139] However, the analysis suggests "the possibility of using targeted killings as a counterterrorism measure . . . is more limited than under the law of armed conflict."[140] In order for targeted killings to be justified, the United States would have to show an operation "is lawful under the domestic law of homicide as well as human rights law and that they are carried out with respect for the sovereignty of other States."[141] This analysis is interesting in two aspects. First, it acknowledges a form of international law enforcement that at least potentially involves the use of drones. Second, this approach considers various legal regimes (i.e., both international and domestic) when attempting to reconcile the use of drones in contemporary operations. An approach more commonly applied by the international legal community has been for human rights–based analyses to focus exclusively on international law.

It appears the use of drone strikes can occur under the restricted "law enforcement" theory in two scenarios. First, assuming compliance with humanitarian law targeting precautions, the attacks can take place in respect of an established battlefield that is territorially limited. That restricted territory might encompass a "spillover" noninternational armed conflict in an adjacent State, but perhaps only to part of that State. Second, the use of drones by a State can be justified as a lawful exercise of the right of self-defense in respect of the use of force between States, with one interpretation acknowledging a potential ability to use drones under a law enforcement paradigm.[142] However, the basis for the use of drone-launched missiles in practical terms seems almost exclusively grounded in the existence of an armed conflict. Absent such a conflict, the applicable normative regime is a law enforcement one based on a human rights law framework.

The basis for this restrictive "law enforcement" theory can be found in very traditional and long-standing twentieth-century interpretations of international law. There is also an underlying theme influencing the legal interpretation, which ultimately defines this restricted approach. The theory seeks to remove or at least significantly limit any State use of military force outside of what are considered to be established battlefields, regardless of whether such use is considered in a State self-defense or humanitarian law context. The result is this interpretation is particularly focused on geographically limiting the use of drones. To better

---

137 *Id.* at 12.

138 *Id.* at 9 (this was on the basis that drone strikes in Yemen were based on "an argument of a 'war on terror' which is not included in any definition of conventional armed conflict.").

139 *Id.* at 11.

140 *Id.* at 9. *See also* Noam Lubell, Extraterritorial Use of Force Against Non-State Actors 176–77 (2010) (suggesting in very limited circumstances a drone strike might be carried out under human rights law).

141 Oxford Research Paper, *supra* note 135, at 9.

142 *Id.* at 11.

understand the restricted "law enforcement" approach, it is helpful to look at this theory both in terms of its reliance on the traditional view of State self-defense and in respect of the emphasis placed on legal thresholds for defining when armed conflict is occurring.

What stands out regarding the restricted approach is its very traditional and conservative interpretations of the law governing State self-defense. The theory depends not only on a limited authority to act in self-defense (see Chapter 2, section 2.3.2, *Inter-State Conflict: Limiting the Recourse to War*), but it also places strong reliance on the governing State as the proper authority for controlling violence within its territory (see Chapter 4, *States, "Proper Authority," and Conflict*). This aspect of the "State-centric" approach focuses on limiting the transnational use of military force by States. The argument relies heavily on the position that the rules governing self-defense only "applies to inter-state uses of force."[143] As Mary Ellen O'Connell states, the use of force against non-State actors is not justified in situations where there is no "significant armed attack ordered by the state or by a state-controlled group that carried it out."[144]

A particular challenge with the restricted theory is that it does not reflect what has increasingly become the accepted view within the broader international legal community since the attacks of 9/11. That States may act in self-defense against non-State actors acting on their own (see Chapter 2, section 2.4.4, *Cross-Border Attacks by Non-State Actors*). However, it is an interpretation that is entirely consistent with what had traditionally been, at least until 9/11, a widely held view of international lawyers. It is also an interpretation that has had a positive effect on the maintenance of international security in the post–World War II era, at least in terms of inter-State conflict.

A resort to drones in order to conduct targeted killing outside of established conflict zones is viewed as being without authority under international law.[145] Perhaps because the action by terrorist groups is seen as occurring outside the context of inter-State conflict, the issue of whether terrorist attacks could rise to the level of an "armed attack" under Article 51 of the UN Charter is not addressed.[146] There appears to be a presumption that such armed groups are small in size, and the violence, however destructive, would not meet the required level of intensity for an armed conflict even if a State were responsible for the attack. It is noted such "low-level attacks or border incidents do not give rise to the right to use force in self-defense on the territory of the responsible state."[147] The law governing the exercise of State self-defense "is not designed for responding to the violent criminal actions of individuals or small groups."[148] The conclusion is that "[p]eacetime criminal law, not the law of armed conflict, is the right choice against sporadic acts of terrorist violence."[149] This indicates the level of violence would also not rise to the level of an armed conflict.

143 Mary Ellen O'Connell, *The Choice of Law Against Terrorism*, 4 NAT'L SEC. L. & POL'Y 343, 358 (2010).

144 *Id.* at 359.

145 *Id.* at 362 ("the United States has targeted and killed persons under the combatant's privilege to kill [in Pakistan, Yemen, and Somalia] where it did not possess that privilege.").

146 *But see*, Prosecutor v. Boškoski, Case No. IT-04-82-T 90, Judgement, ¶ 190 (Int'l Crim. Trib. for the Former Yugoslavia, Trial Chamber, 10 July 2008) (where it is noted "national courts and UN bodies have not discounted acts of a terrorist nature in their consideration of acts amounting to armed conflict.").

147 O'Connell, *supra* note 143, at 359.

148 *Id.*

149 *Id.* at 368.

The restrictive "law enforcement" approach also places emphasis on identifying territorial limitations to state action.[150] It is argued that the U.S. claim to be acting in self-defense offers "no justification for using force or exercising wartime privileges beyond Afghanistan."[151] As has been noted, a particular feature of this law enforcement theory is its reliance on the territorial State as the proper authority for dealing with transnational terrorism. This approach focuses on the responsibility of that State either through the concept of sovereignty or by default through a lack of authority for the threatened State to act transnationally in another territory against non-State actors. The territorial State is left to address the threat. In either case there is a strong territorial aspect to these interpretations, which extends to the point of privileging State sovereignty over the authority of a threatened State to act in self-defense. As has previously been mentioned, it is argued that the right to act in self-defense by the United States and members of the Coalition in Afghanistan ended when the Karzai government came to power in 2002 (see Chapter 2, section 2.4.3, *Conflict Evolution*). This leaves the authority to attack or detain Al Qaeda dependent upon the invitation of Afghanistan to participate in the counterinsurgency in that country.[152] There are further limits attached to this theory. It is also indicated that the territorial State cannot grant permission to conduct drone strikes within territorial boundaries where there is no armed conflict.[153]

The reliance on territorial limits in the restrictive theory may be a response to the U.S. position, "that conflict follows the members of . . . [a non-State armed group] wherever they go as long as the group's members continue to engage in hostilities against that state (either on the "hot battlefield" or from their new location)."[154] In contrast, the restricted "law enforcement" approach suggests hostilities must occur "within limited zones, referred to as combat zones, theaters of operations, or similar terms."[155] As such, it is an argument that is not solely based on State sovereignty.

The argument that the recourse to war concept of self-defense must be limited to combat zones is stated to be based on a parallel principle found in the humanitarian law.[156]Like the suggestion there are geographic limitations arising in the application of self-defense proportionality and necessity principles (see Chapter 3, section 3.3, *An "Overarching" Application*), it is argued that under humanitarian law "[c]ombatants may not kill the enemy wherever they find him, but only when reasonably necessary."[157] This reflects a cross-boundary linkage between these two bodies of law. This once again highlights the pressure contemporary

[150] *Id.* at 357. (For instance, it is suggested that since the United States did not argue other States might be responsible for the 9/11 attacks, there is "no right under jus ad bellum to use force against other states, besides Afghanistan.").

[151] *Id.* at 358.

[152] *Id.* at 358.

[153] O'Connell, *supra* note 129, at 2 ("States cannot, however, give consent to a right they do not have."), and Mary Ellen O'Connell, *Unlawful Killing with Combat Drones: A Case Study of Pakistan, 2004–2009*, at 16 (Notre Dame Law School Legal Studies Research Paper No. 09-43, Final Draft, July 2010), https://www.law.upenn.edu/institutes/cerl/conferences/targetedkilling/papers/OConnellDrones.pdf. *See also* Sanger, *supra* note 84, at 257–59 (outlining the reliance placed by the United States on Pakistani, Yemeni permission to conduct drone strikes).

[154] Ashley S. Deeks, *Pakistan's Sovereignty and the Killing of Osama Bin Laden*, 15 Am. Soc. Int'l L. Insights (May 5, 2011), http://www.asil.org/insights/volume/15/issue/11/pakistans-sovereignty-and-killing-osama-bin-laden.

[155] O'Connell, *supra* note 143, at 355.

[156] *Id.* at 355 n.64 ("A parallel principle is found in the *jus in bello.*").

[157] *Id.*

conflict is placing on interpretations of international law that traditionally have preferred to see these bodies of law as completely conceptually isolated. In seeking to limit combat to specific zones, the restrictive "law enforcement" approach mirrors the self-defense-based arguments (i.e., the *Belgrano* example occurring in the Falklands/Malvinas conflict) applied to geographically limit inter-State armed conflict. However, it is used here in respect of conflict between States and non-State actors. As such, it represents a broadening of the traditional interpretations of the law governing the recourse to war so that they impact on other than international armed conflict, albeit under the guise of a humanitarian law argument. In doing so, the contemporary dialogue has introduced new terminology: "the geography of war."

Another key aspect of the restrictive approach is its reliance on humanitarian law–based thresholds for the establishment of an armed conflict in order to further limit when force can be used under a conduct of hostilities paradigm. This occurs through what are identified as minimum criteria for a noninternational armed conflict: the presence of organized armed groups and intensive intergroup fighting as the basis for arguing that the isolated terrorist attacks did not constitute an armed conflict.[158] It leads to an interpretation that "the isolated terrorist attack, regardless of how serious its consequences, is not an armed conflict."[159] Instead, the "terrorists" are dealt with as criminals subject to law enforcement, although it is acknowledged, "in some circumstances it [terrorism] may be carried out so continuously as to be the equivalent of an armed conflict."[160] This approach places particular emphasis on a high threshold being established for an armed conflict when considering the criteria established in the *Prosecutor v. Tadić* decision.[161]

Reliance is placed on an International Law Association report concerning the meaning of armed conflict. It is noted the *Prosecutor v. Tadić* case "is widely relied on as authoritative for the meaning of armed conflict in both international and non-international armed conflicts."[162] That case certainly provides an appropriate starting place for the discussion of what constitutes the criteria for the existence of a noninternational armed conflict. However, there is a danger in exclusively relying on a high threshold interpretation of the *Tadić* case and other International Criminal Tribunal for the Former Yugoslavia (ICTY) jurisprudence. There is certainly no unanimity among international lawyers on this issue regarding the existence of a definitive standard regarding the categorization of conflict, particularly given the complex nature of contemporary conflict (see Chapter 9, section 9.4, *Determining the Existence of a Noninternational Conflict*).[163] In this respect the *Oxford Research Paper* rightly acknowledges "[t]here is no agreed legal definition of a Non-International Armed Conflict within International Humanitarian Law."[164]

---

158 Prosecutor v. Tadić, Case No. IT-94-1-A, Decision On The Defence Motion For Interlocutory Appeal On Jurisdiction, ¶ 70 (Int'l Crim. Trib. for the Former Yugoslavia, Appeal Chamber, 2 October 1995).

159 O'Connell, *supra* note 143, at 355.

160 *Id.* at 356.

161 Prosecutor v. Tadić, Case No. IT-94-1-A, Decision On The Defence Motion For Interlocutory Appeal On Jurisdiction, ¶ 70 (Int'l Crim. Trib. for the Former Yugoslavia, Appeal Chamber, 2 October 1995).

162 International Law Association, Final Report of the Use of Force Committee: The Meaning of Armed Conflict 14 (Aug. 2010), http://www.ila-hq.org/en/committees/index.cfm/cid/1022 [hereinafter ILA Use of Force Report]. *See also* O'Connell, *supra* note 143, at 355.

163 Theodor Meron, *The Humanization of Humanitarian Law*, 94 Am. J. Int'l L. 239, 261 (2000).

164 Oxford Research Paper, *supra* note 135, at 4.

This theory does not adequately address the issue of short duration, but very intense "one-off" style attacks by organized armed groups. This would include attacks linked to an armed conflict, but which some analysts view as occurring too geographically or temporarily remote from those hostilities. Examples of such attacks are the 2008 Mumbai attack,[165] the September 11, 2012, attack on the U.S. diplomatic facilities in Benghazi,[166] and the September 21–24, 2013, Westgate Mall attack.[167] Law enforcement authorities do confront terrorists in these types of incidents (often in conjunction with military forces), however, that does not mean the violence does not occur within the factual and, as will be argued, legal context of an armed conflict (see Chapter 14, section 14.4.4, *A Law Enforcement Response to Attacks at Home and Abroad*). As will be discussed, the *Prosecutor v. Tadić* criterion of protracted armed violence, which has been equated subsequently to the intensity of violence,[168] should not preclude significant violence lasting even a relatively a short period of time from crossing the threshold into armed conflict (see Chapter 9, section 9.4, *Determining the Existence of a NonInternational Armed Conflict*).[169] The challenge for security forces is having to deal with armed conflict–related violence as it unfolds, rather wait weeks, months, or even years for a formal assessment by a court or the attainment of a degree of international consensus as to whether the attack constituted part of an armed conflict at law.

Significantly, the restrictive "law enforcement" theory does not seem to adequately consider the *Abella v. Argentina* precedent,[170] or the International Law Association, *Use of Force Report* indication that "[t]he two concepts, intensity and protraction, are linked and a lesser level of duration may satisfy the criterion if the intensity level is high."[171] To suggest that extremely violent short duration attacks by organized armed groups, which may involve the use of mortars, crew-served weapons, and the like, can only be dealt with by means of a human rights–based law enforcement response looks uncomfortably like adherence to a strict formalist interpretation of law focused on avoiding inter-State conflict, rather than an approach grounded in the factual realities of the tactical situation. This is a point where legal theory can become disconnected from the nature of the violence occurring between organized armed groups and State armed forces.

---

[165] David Kilcullen, Out of the Mountains: The Coming of Age of the Urban Guerrilla 52–66 (2013).

[166] Zuckoff, *supra* note 53, at 254–80 (outlining the most violent part of the attack). *See also* Perm. S. Com. on Intelligence, 113th Cong., Investigative Rep. on the Terrorist Attacks on U.S. Facilities in Benghazi, Libya, September 11–12, 2012, at 3–12 (Nov. 21, 2014), https://intelligence.house.gov/sites/intelligence.house.gov/files/documents/Benghazi%20Report.pdf.

[167] Daniel Howden, *Terror in Westgate Mall: The Full Story of the Attacks that Devastated Kenya*, The Guardian, Oct. 4, 2013, http://www.theguardian.com/world/interactive/2013/oct/04/westgate-mall-attacks-kenya-terror#undefined.

[168] Prosecutor v. Haradinaj, Case No. IT-04-84-T Judgement ¶ 40 at 22 (Int'l Crim. Trib. For the Former Yugoslavia-Trial Chamber I, Apr. 2008), http://www.icty.org/x/cases/haradinaj/tjug/en/080403.pdf (The Trial Chamber consequently interpreted the term "protracted armed violence" to refer to the "intensity of the conflict").

[169] Sandesh Sivakumaran, The Law of Non-International Armed Conflict 168 (2012). *But see* Dinstein, *supra* note 135, ¶ 101, at 33 (where it is accepted there is no minimum period for assessing what constitutes "protracted violence," however, it is stated "[w]hatever the shortest admissible space of time may be it cannot plummet down to just a few hours or even a few days.").

[170] Case 11.137, Abella v. Argentina, Inter-Am. Y.B. on H.R. ¶ 1 (1997), http://www1.umn.edu/humanrts/cases/1997/argentina55-97a.html (Commission report). *But see* Dinstein, *supra* note 136, ¶ 102, at 33–34 (where the author indicates he does not agree with this decision).

[171] ILA Use of Force Report, *supra* note 162, at 30.

Another essential element of the restrictive "law enforcement" approach is equating terrorism with crime carried out by small terrorist groups. The acts perpetrated by Al Qaeda and other terrorist organizations are seen fundamentally as "isolated acts of terrorism" carried out by local groups active in a variety of countries.[172] This aspect of the restrictive approach appears to equate terrorism uniquely with terrorist threats of the 1970s and 1980s. Whenever terrorist groups do present such a threat they must be dealt with by means of a law enforcement response. However, the "isolated acts of terrorism" theory is problematic for three reasons. First, the organization of groups associated with twenty-first-century jihadist terrorism alone is far more complex than this theory seems to suggest. Second, reliance on the number of terrorist attacks either attempted, or successful, in order to establish the existence of an armed conflict does not necessarily reflect the overall intensity of the jihadist conflict. Finally, a particular challenge for proponents of the law enforcement approach, who prefer to compartmentalize the threat by country, is the large number of States within which these terrorist groups are operating. Their link to a common cause (i.e., the jihadist movement) suggests the threat is neither organizationally nor operationally "localized" (see Chapter 6, section 6.4.3. *Terrorism and Insurgency: Al Qaeda and an Islamic State*).

Like other aspects of the law enforcement approach there is a theoretical basis for this separation of Al Qaeda into small distinct groups. The idea that Al Qaeda consists of small-disconnected groups is reminiscent of a theory that was popular as the conflict with this group first developed. It was based on a view that the organization was engaged in a "leaderless jihad." Under this theory, Al Qaeda's central core was seen as having been degraded to the point of being contained;[173] the affiliated "little Al Qaedas" operated with minimal central direction;[174] and where the resulting threat was viewed as "basically anti-Western political violence, which is the lowest common denominator of the various ideologies of global Islamist terrorism."[175] Under the "leaderless jihad" theory the threat from networked terrorists was seen as being manageable through increased homeland security; by applying greater restraint in military action to avoid inciting young Muslims; by using military force only to deny terrorists sanctuary; and by reducing terrorists to the status of common criminals.[176]

When first introduced, the "leaderless jihad" theory was the subject of vigorous debate. However, it is a perspective that has largely receded into the background,[177] even though it continues to be referred to in the context of the inspired network of "like-mindeds" that have acted without direct involvement of Al Qaeda Central or its affiliates.[178] The attacks

---

172 O'Connell, *supra* note 143, at 356 ("Canada, France, Germany, Indonesia, Morocco, Saudi Arabia, Spain, the United Kingdom, Yemen, Kenya, Uganda, and elsewhere.").

173 Marc Sageman, Leaderless Jihad: Terror Networks in the Twenty-First Century 128–29 (2008).

174 *Id.* at 129.

175 *Id.* at 144. *See also* Fawaz A. Gerges, The Rise and Fall of Al-Qaeda 190 (2011) (this author also viewed Al Qaeda as leaderless and in disarray).

176 Sageman, *supra* note 173, at 147–54. *But see* Michael W. S. Ryan, Decoding Al-Qaeda's Strategy: The Deep Battle Against America 238–39 (2013) (where it is noted that "Abu Mus'ab al-Suri's concept of individual and small unit jihad is not the same as Sageman's 'leaderless jihad' or the recent popular characterization of 'home-grown terrorism'. . . .").

177 Bruce Hoffman, *The Myth of Grass-Roots Terrorism Subtitle: Why Osama bin Laden Still Matters*, Foreign Aff. 133 (May 2008–June 2008) (The "leaderless jihad" argument was forcefully countered by Bruce Hoffman). For a reply and counter-reply *see* Marc Sageman & Bruce Hoffman, *Does Osama Still Call the Shots? Subtitle: Debating the Containment of al Qaeda's Leadership*, 87 Foreign Aff. 163 (July/Aug. 2008).

178 Seth G. Jones, Hunting in the Shadows: The Pursuit of Al Qa'ida Since 9/11, 287–310 (2012).

in Western countries (i.e., the far enemy) are consistent with Abu Mus'Ab al-Suri's concept of individual jihad and small cell terrorism, which is part of a larger conflict.[179] Further, notwithstanding claims of Al Qaeda's early demise, the United States and other governments have continued to be confronted by a degraded, but apparently resilient Al Qaeda Central. It was indicated that "[b]y 2007 Al-Qaeda's ability to wreak havoc nearly rivaled the group's pre-2002 peak," and that "[i]n some ways the threat was even worse: Al-Qaeda had effectively merged with some of Pakistan's extremist groups, while spawning new chapters in North Africa, Iraq and the Arabian Peninsula."[180] There have also been numerous domestic terrorist threats in the West that have a connection to that jihadist cause, ranging from the receipt of Al Qaeda training and resources to the simple provision of motivation and "how to" bomb making instructions on the Internet.[181] By 2014, concern over the jihadist threat had expanded to include the threat posed by the Islamic State.

Although wrong on the facts, the narrative of Al Qaeda as a small terrorist group has served a number of interests. That of the United States, which wanted to project success, particularly with regard to an intensified drone campaign against Al Qaeda. However, focusing on this "story" appears to have contributed to the dramatic strategic surprise that occurred when the Islamic State emerged in 2014 to claim control of large portions of Syria and Iraq. Not unrelated, it had also been suggested that the portrayal of Al Qaeda as a small and centralized organization is also favored by that organization, as that group has long derived benefits from masking its strengths and intentions.[182] Finally, this "terrorist" label underpins the argument of those who wish to restrict State military action, such as those advocating the restricted "law enforcement" theory. The challenge is Al Qaeda, the Islamic State, and other groups are much more diverse organizations, which are part of a broader jihadist movement.

Here, the restrictive "law enforcement" approach seems to run afoul of Sun Tzu's principle of knowing your enemy.[183] For example, with respect to the non-State actor threat, Al Qaeda and the Islamic State do not comprise a small number of autonomous local groups constrained to various countries. These groups may not agree with one another, but they remain entities of what is essentially a common cause focused on removing existing governments in order to develop a broader caliphate. The Islamic State, an offshoot of this jihadist movement has demonstrated its willingness to take just such as step (see Chapter 6, section 6.4.3, *Terrorism and Insurgency: Al Qaeda and an Islamic State*). Crucially, the restricted law enforcement approach does not address the distinction between international terrorism previously carried out by such now defunct groups as the Baader-Meinhof Group, the Italian Brigades, the Japanese Red Army, and contemporary terrorist groups acting as a

---

179 *Id.* at 289. *See also* Brynjar Lia, Architect of Global Jihad: The Life of Al-Qaeda Strategist Abu Mus'ab al-Suri 363–67 (2008).

180 Warrick, The Triple Agent 11 (2011).

181 Andre Mayer, *Does al-Qaeda Have a Network in Canada?*, CBC News (Apr. 24, 2013), http://www.cbc.ca/news/canada/story/2013/04/23/f-al-qaeda-canada.html.

182 Syed Saleem Shahzad, Inside Al-Qaeda and the Taliban: Beyond Bin Laden and 9/11, at 225 (2011) (outlining Al Qaeda motivation to keep its activities hidden from the West), and David Gartenstein-Ross & Bridget Moreng, *Al Qaeda Is Beating the Islamic State: Get Ready for the Clash of Caliphates*, Politico Mag. (Apr. 14, 2015), http://www.politico.com/magazine/story/2015/04/al-qaeda-is-beating-the-islamic-state-116954.html#.VTDiRlwnhPp.

183 Sun Tzu, The Art of War 84 (Samuel B. Griffith trans., 1971) (1963) ("Know the enemy and know yourself; in a hundred battles you will never be in peril.").

component of broader insurgencies. Distinguishing between criminals, terrorists, and insurgents in the contemporary context is a far more challenging task than an argument equating terrorism to crime suggests.

The restricted approach also places particular reliance on jihadist violence being both limited and restricted by national borders. The violence in this conflict, particularly in the minds of its perpetrators, is neither limited by international borders nor bounded by time in the manner suggested in the phrase "single, isolated act of terrorism."[184] Consideration of transnational terrorism, as opposed to truly domestically motivated threats, requires an assessment of the totality of the violence. This means viewing the acts of violence from a multinational, rather than a merely national perspective. These attacks also need to be assessed as to whether they are individual acts or represent an "accumulation of events" such that a series of "pinprick" attacks can be considered to constitute an armed attack under Article 51 of the UN Charter.[185]

The threat from radical jihadists writ large has been a continuous one for nearly two decades. For example, there have been numerous Al Qaeda–linked attacks in multiple countries involving significant levels of violence. One accounting between 1996 and 2005 identified 130 Al Qaeda "victories" that "yielded dead and gutted tourists, demolished embassies, and ongoing—indeed strengthening—Islamic insurgencies around the world."[186] Another study suggests that between 2004 and 2008 there were 3,010 casualties resulting from 313 attacks by Al Qaeda or Al Qaeda–affiliated groups.[187] This latter statistic raises the issue of whether the intensity of violence is assessed by the number of attacks or by the number of casualties. The jihadist threat is not simply confined within territorial borders. The Islamic State has also embarked on a "program of inspiring, and sometimes helping, ISIS sympathizers to conduct attacks in the West," and worldwide have carried out major attacks in twenty-one countries and territories.[188] In 2015 there were high profile attacks by that group in France, Tunisia, Kuwait,[189] and Egypt.[190]

---

[184] O'Connell, *supra* note 143, at 356.

[185] Christine Gray, International Law and the Use of Force 155–56 (3rd ed. 2008), and Yoram Dinstein, War Aggression and Self-Defence ¶ 584, at 220–21 (5th ed. 2011). *See also* Lubell, *supra* note at 140 (while doubting Al Qaeda is sufficiently organized he notes that "[i]f numerous incidents round the world classified as terrorism could be attributed to the same entity then one could argue that the threshold for conflict has been crossed.").

[186] Michael Scheuer, Through Our Enemies' Eyes: Obama bin Laden, Radical Islam, and the Future of America 209 (2006)

[187] Scott Helfstein, Nassir Abdullah, & Muhammad al-Obaidi, Deadly Vanguards: A Study of al-Qa'ida's Violence Against Muslims, Combating Terrorism Center at West Point, Occasional Paper Series 6 (Dec. 2009), http://www.ctc.usma.edu/wp-content/uploads/2010/10/deadly-vanguards_complete_l.pdf.

[188] Karen Youish, Derek Watkins, & Tom Giratikanon, *Where ISIS Has Directed and Inspired Attacks Around the World*, N.Y. Times, June 17, 2015, http://www.nytimes.com/interactive/2015/06/17/world/middleeast/map-isis-attacks-around-the-world.html?src=mv (the listed countries and territories are: Australia, Algeria, Canada, the United States, the West Bank, Saudi Arabia, France, Libya, Morocco, Belgium, Germany, Israel, Bangladesh, Lebanon, Spain, Egypt, Denmark, Tunisia, Yemen, Afghanistan, and Malaysia.).

[189] Greg Miller, *Three Attacks Deepen Fears about Islamic State's Global Reach*, Wash. Post., June 27, 2015, http://www.washingtonpost.com/world/national-security/three-attacks-deepen-fears-about-islamic-states-reach/2015/06/27/04b82dd2-1c1d-11e5-ab92-c75ae6ab94b5_story.html, and *Three Hours of Terror in Paris, Moment by Moment*, N.Y. Times, Nov. 15, 2015, ¶ http://www.nytimes.com/interactive/2015/11/13/world/europe/100000004036728.app.html?nytapp=ipad&_r=0.

[190] *Russia Plane Crash: 'Terror Act' Downed A321 over Egypt's Sinai*, BBC News (Nov. 17, 2015), http://www.bbc.com/news/world-europe-34840943.

There have also been numerous planned and executed attacks specifically against American interests globally and in the United States. A number of those attacks predated the attacks of 9/11 and have continued since that time.[191] For example, in 2007 the FBI thwarted a planned 2007 attack on a U.S. military installation at Fort Dix, New Jersey, by a group inspired by Al Qaeda[192] and a plot to blow up jet fuel supply tanks and pipelines at John F. Kennedy airport in New York.[193] It is reported that in 2009 there were "eight homegrown plots and two successful attacks in the United States."[194] The successful jihadist-linked attacks involved a mass shooting at Fort Hood, Texas, and a shooting at an Arkansas recruiting center.[195] There was also an attempted destruction of an aircraft nearing Detroit with a bomb hidden in the Nigerian terrorist's underwear.[196] In 2010 there was an attempted car bombing in Times Square,[197] as well as the placement of explosives on cargo aircraft bound to the United States.[198] Perhaps one of the most widely reported attacks was the 2013 Boston Marathon bombing by jihadist-inspired individuals.[199]

The number of attacks successfully completed or even attempted by jihadists do not represent the full extent of the threat. Most drone strikes are carried out against the leaders and planners of organized armed groups in order to disrupt and deter attacks. It is difficult, if not impossible, to quantify how many attacks may have been thwarted by such preemptive killing.[200] The overall assessment of the violence must also include the strikes carried out by the United States. As Ken Anderson has noted, conflicts cannot be assessed as starting or stopping based on whether one side or the other wants to fight at a particular point.[201] Allowing "such a rule would give either party the ability to call 'game on' or 'game off' as strategic considerations dictated: lawfare par excellence."[202] The nature of the violence must be considered in its entirety. In this regard, the threat posed by the jihadist movement as represented by Al Qaeda, the Islamic State, and numerous other groups is real, persistent, exceptionally violent, and not yet easily placed in the category of isolated acts of terrorism.

The use of drone strikes under the restrictive approach is dependent upon an armed conflict being categorized as noninternational in character and then geographically prescribed by international boundaries, or at least by operational zones. Outside that context

---

191 John K. Cooley, Unholy Wars: Afghanistan, America and International Terrorism 218–47 (2000) (for a discussion of attacks prior to 9/11).

192 Jones, *supra* note 178, at 287–303 (for an outline of the planned 2007 terrorist attack on Fort Dix).

193 *Id.* at 303–09.

194 Catherine Herridge, The Next Wave: On the Hunt for Al Qaeda's American Recruits 63 (2011). *See also* Abdel Bari Atwan, After Bin Laden: Al Qaeda, The Next Generation 18–19 (2012), and Schmitt & Shanker, *supra* note 94, at 210–12.

195 Herridge, *supra* note 194, at 64.

196 *Id.* at 30–32.

197 Schmitt & Shanker, *supra* note 94, at 212.

198 Herridge, *supra* note 194, at 218–20.

199 Jos Levs & Monte Plott, *Boy, 8, One of 3 Killed in Bombings at Boston Marathon; Scores Wounded*, CNN (Apr. 18, 2013), http://www.cnn.com/2013/04/15/us/boston-marathon-explosions/.

200 Herridge, *supra* note 194, at 121. *But see* Morell & Harlow, *supra* note 23, at 137 ("There is no doubt in my mind that these strikes have prevented another attack on the scale of 9/11.").

201 Kenneth Anderson, *Targeted Killing and Drone Warfare: How We Came to Debate Whether There Is a "Legal Geography of War,"* Am. U. Wash. Col. L. 5 (College of Law Research Paper No. 2011-16, 2011), https://www.law.upenn.edu/live/files/3484-anderson-k-targeted-killing-and-drone-warfare.

202 *Id.*

drone use is prohibited, or at least constrained to exceptional circumstances. However, the localized nature of the conflict can be challenged in terms of an emphasis being placed on the international boundary separating Afghan from Pakistani territory as constituting the limit for the conflict occurring in Afghanistan. In this instance, the scope of this conflict is constrained by strict interpretations of international law determined by the Durand line, which in practical terms is often treated by the locals simply as a line on the map. Like many international boundaries, that line is an artifact of the colonial era. In this respect, "[t]he world may see the Durand line as a border between two nations, but the Pashtuns sure don't—particularly the Taliban."[203] The Durand line represents not only a notoriously porous border but also not even one that successive Afghan governments have accepted as the demarcation line between that country and Pakistan.[204] It is on such formal but frequently shaky foundations that the restrictive "law enforcement" approach is based.

In reality, jihadist based violence can transcend multiple geographic borders in a number of areas. The jihadist threat is increasingly being seen as one of more global proportions. Whole regions (e.g., "regions of savagery"[205]) are facing the reality of significant conflict (see Chapter 4, section 4.4.5, *Small Wars and "Ungoverned Spaces"*). For example, the area of AQIM's southern influence "now extends throughout most of the Sahel, from the Algerian Sahara to Mali, Niger, Nigeria, Libya, Mauritania and Chad."[206] However, the threat is a broader one. The global nature of the threat posed by the Islamic State was recognized in a November 20, 2015 United Nations Security Council Resolution 2249.[207] In this respect the Islamic State mirrors the threat posed by Al Qaida.[208]

There is no doubt that law enforcement measures provide an effective and preferred method for dealing with violence internal to a country (see Chapter 4, section 4.4.1, *Public, Private, and Mixed War*). It is also evident that under the right circumstances an extraterritorial application of a law enforcement approach by capturing suspects can offer an effective means by which to deal with a transnational terrorist threat. However, the challenges in applying the restrictive approach are numerous. First, the threatened State must have the capability to do so. As President Obama noted in the 2013 *Drone Speech*, it is "not possible for America to simply deploy a team of Special Forces to capture every terrorist."[209] Second, even in this situation the human rights–based paradigm governing a law enforcement approach will likely have to be interpreted in a fashion that permits elevated levels of

---

[203] Sanger, *supra* note 84, at 9. *See also* Hassan Abbas, The Taliban Revival: Violence and Extremism on the Pakistan-Afghanistan Frontier 35–40, 49–50 (2014), and James Fergusson, Taliban: The True Story of the World's Most Feared Guerrilla Fighters 25 (2010).

[204] Abbas, *supra* note 203, at 49

[205] Abu Bakr Naji, The Management of Savagery at [16] (William McCants trans., 2006), https://azelin.files.wordpress.com/2010/08/abu-bakr-naji-the-management-of-savagery-the-most-critical-stage-through-which-the-umma-will-pass.pdf.

[206] Atwan, *supra* note 194, at 187.

[207] S.C. Res. 2249, UN Doc. S/RES/2249 *Preamble* (Nov. 20, 2015)(which notes "its recruitment and training of foreign terrorist fighters whose threat affects all regions and Member States, even those far from conflict zones, the Islamic State in Iraq and the Levant (ISIL, also known as Da'esh), constitutes a global and unprecedented threat to international peace and security".).

[208] *Id.* (where the threat posed by "the Al-Nusrah Front (ANF) and all other individuals, groups, undertakings and entities associated with Al-Qaida" was also referred to).

[209] *Drone Speech*, *supra* note 23.

violence to be used against terrorists due to the armaments often used by such groups (see Chapter 12, section 12.3.4, *Strictly Necessary, Proportionate, and Overwhelming*).

Third the security forces of the threatened State must gain access to the State where the non-State actors are located. The restrictive "law enforcement" approach relies on obtaining the consent of the territorial State in order to conduct an arrest, although it has also been suggested that the "[r]escue and apprehension of international criminal suspects where the state carrying out the operation lacks a territorial state's consent may still be lawful under the law of countermeasures."[210] In doing so, "[t]he apprehension must follow law enforcement rules, which restrict the use of lethal force, ensuring no breach of Article 2(4) [of the UN Charter]."[211] Without such consent, or in circumstances where the law enforcement approach cannot be physically carried out because of a lack of State capacity to do so, or because the risks to the security personnel and civilians are too elevated, the threatened State would not be able to respond to imminent threats. By placing emphasis on sovereignty and erecting a barrier to State action in the form of international borders, the handling of such threats seems fundamentally to be left to the territorial State. The fourth and probably the most significant hurdle for those advocating the restrictive theory is making a credible argument the law enforcement approach will always provide a viable and effective model by which the threatened State can respond to violent acts emanating from another country.

The Oxford Research Group approach, which accepts the conflict can be extended to at least a portion of another State, only provides a partial solution if the terrorist group can still take advantage of sanctuary elsewhere. If the territorial State does not take action itself or consent to cross-border law enforcement is not provided, there does still remain the option of the harmed State taking nonforceful measures, or perhaps even bringing the matter before the United Nations.[212] However, given the consequences of a transnational terrorist act and legitimate concerns regarding the short-term effectiveness of such alternate means, it appears that applying the law enforcement approach realistically means no strike will be carried out. This has the potential to leave terrorist groups considerable freedom to act with relative impunity within the host State. The threatened State is left to wait to respond to an attack on its own territory, or perhaps where possible on the high seas. That means action to thwart such an attack may only practically be carried out as it unfolds thereby increasing the potential risk to civilians in the targeted State.

Operations against members of organized groups, which are often located in remote areas, can offer a significantly better outcome in terms of its overall humanitarian impact than waiting for innocent civilians to be put at risk at the point terrorist attacks are unleashed. Imposing the specter of geographic "no go" areas on a State faced with an unaddressed serious threat appears formalistic and ultimately inadequate. States that have the ability to act should be able to do so when another State that is responsible to address the threat cannot

---

210 Mary Ellen O'Connell, *The True Meaning of Force*, AJIL UNBOUND (Aug. 4, 2014, 9:56 AM), http://www.asil.org/blogs/true-meaning-force.

211 *Id.*

212 MALCOLM N. SHAW, INTERNATIONAL LAW 1128–29 (6th ed. 2008) (outlining retorsion as a possible means of self-help. It can include the severance of diplomatic relations, or various travel and economic restrictions).

or is unwilling do so. The type of reaction will be dependent upon the nature of the threat, its imminence, and consideration of the consequences of not acting. In this regard, a reaction beyond territorial borders against non-State actors would not be dissimilar to traditional State responses to threats on the high seas. In this context D. W. Bowett noted:

> the State cannot await the arrival of a danger to its security within its own territorial jurisdiction, but must take measures to prevent that danger from materializing while still outside its territorial jurisdiction.[213]

The principle of the freedom of the high seas cannot "be of any greater sanctity than that of territorial integrity which . . . is subject to qualification in the interests of security of a neighbouring state."[214] Nor should the sanctity of an inadequately policed territory trump the interests of a threatened State. It is unlikely that powerful States will allow themselves to be constrained by the formalism of borders that are not backed by effectively governed territory.[215] Again, the specter of ungoverned spaces looms large in this assessment. What ultimately has to be avoided is a security "black hole."

8.5.2. THE PERMISSIVE "CONDUCT OF HOSTILITIES" APPROACH

The restrictive "law enforcement" approach can be contrasted with a more permissive one based on a humanitarian law response. This "conduct of hostilities"–based approach broadly authorizes states to take action when a territorial state is unwilling or unable to deal with transnational threats originating within its borders.[216] This theory is linked to the view that groups like Al Qaeda and the Islamic State pose a global, regional, or at least a transnational threat in the context of an ongoing armed conflict. As such it represents an interpretation of international law that proponents of the restrictive theory specifically seek to counter. Each approach is addressing the same threat, but they appear to be on a collision course, having started at different ends of the conflict spectrum.

Like the restricted theory, with its acknowledgment that drones can lawfully be used in Afghanistan, the conduct of hostilities approach also has more depth than simply arguing for the use of military force in the context of a broader conflict against terrorism. The two theories share some common ground. There is an acceptance that the more permissive theory has a geographical component, as it would be "unlawful to deploy forces into locations void of terrorists or insurgents."[217] It is also noted that should a situation not rise to the level

---

[213] D.W. BOWETT, SELF-DEFENSE IN INTERNATIONAL LAW 66 (The Lawbook Exchange Ltd., 2009) (1958).

[214] *Id.*

[215] BENJAMIN WITTES & GABRIELLA BLUM, THE FUTURE OF VIOLENCE: ROBOTS AND GERMS, HACKERS AND DRONES 257 (2015).

[216] Ashley S. Deeks, *"Unwilling or Unable": Toward a Normative Framework for Extra-Territorial Self-Defense*, VA. J. OF INT'L L. 483, 486 (2012), and *Memorandum for the Attorney General, Re: Applicability of Federal Criminal Laws and the Constitution to Contemplated Lethal Operations Against Shaykh Anwar al-Aulaqi*, Office of the Assistant Attorney General, July 16, 2010, at 25 [hereinafter *Department of Justice Drone Opinion*], https://www.aclu.org/sites/default/files/assets/2014-06-23_barron-memorandum.pdf.

[217] Michael N. Schmitt, *Drone Attacks Under the Jus ad Bellum and Jus in Bello: Clearing the "Fog of Law,"* 13 Y.B. INT'L HUM. L. 310, 317 (2010).

of an armed conflict, operations "will be governed by applicable domestic and human rights norms."[218] Further, it is acknowledged that the ability of one State to act in self-defense against terrorists or insurgents taking sanctuary in another State is not unfettered. There must be a balancing of that right with the "territorial state's undeniable right to control access to, and activities on, its territory."[219]

Therefore, the conduct of hostilities approach is also "State-centric" in its initial reliance on the territorial State to deal with security matters within its borders. Even if the threat occurs in the context of an armed conflict, the territorial State retains the responsibility to act first. However, this also marks the point of departure from the restrictive "law enforcement" theory. If there is a failure or inability on the part of the territorial State to take measures against the terrorists engaged in an armed conflict with the victim State, then the latter State can "act militarily in self-defense, including through the use of drones, to put an end to the unlawful activities."[220] At that point there appears to be a cessation of the contemplation of a law enforcement role. Notably, it is not suggested there is an obligation on the part of the threatened State to apply law enforcement norms, such as the consideration of capture assuming that is a feasible option. If drone strikes are carried out during an armed conflict, "cross border military operations, regardless of platforms or forces employed, must comport with international humanitarian law."[221] The response at this point is firmly rooted in conduct of hostilities norms.

This permissive approach differs from its more restrictive counterpart in a number of other material respects. There is broad reliance on the view that acts of self-defense by States can be directed against non-State actors. This interpretation relies on the changed security and legal landscape after the attacks of 9/11 in order to provide legitimacy for the use of military force to counter transnational terrorism (i.e., non-State entities) (see Chapter 2, section 2.4.4, *Cross-Border Attacks by Non-State Actors*). As Thomas Franck has noted, more generally, notwithstanding a "narrowly dogmatic insistence on a traditional armed attack by a national army as the sole justification for an armed response in self-defense," there has been recognition a victim State may invoke Article 51 of the UN Charter to take armed countermeasures "against any territory harboring, supporting or tolerating activities that culminate in, or are likely to give rise to, insurgent infiltration or terrorist attack."[222]

It is noteworthy the degree to which the more permissive approach relies not only on an emerging contemporary interpretation of the self-defense but also on law of armed conflict principles to justify such intervention in another State. This linkage can be seen in Michael Schmitt's outline of the application of State self-defense and humanitarian law principles to drone strikes. He indicates that when an international armed conflict is in existence the law of neutrality provides a basis for cross-border strikes, and "it is reasonable to extend this general approach into non-international armed conflicts that spill-over into neighboring

---

[218] *Id.* at 319.
[219] *Id.* at 316.
[220] *Id.*
[221] *Id.* at 319.
[222] Thomas Franck, Recourse to Force: State Action Against Threats and Armed Attacks 67 (2002).

states. The legal basis for doing so is the law of self-defence."[223] The reliance on neutrality[224] invokes principles from an area of law that is identified as a long-standing customary law norm reflected in humanitarian law treaties.[225] This is supported by the observation that "[s]oon after the 'unwilling or unable' test took root in the context of international armed conflict, it migrated into rules governing a state's use of force extraterritorially against non-state actors."[226] Importantly, this analogous application of neutrality law to a situation involving noninternational armed conflict is specifically relied on in the U.S. *Department of Justice Drone Opinion* justifying a strike against one of its citizens, Anwar Aulaqi, located in Yemen.[227]

This incorporation of principles from the law of neutrality to justify an act in self-defense is not the only way in which humanitarian law concepts work their way into the legal analysis of drone strikes. State acts based in self-defense are restricted by the principles of necessity, proportionality, imminence, and immediacy. A traditional conservative interpretation of international law, as is reflected in the *Caroline* incident, requires a State to "show a necessity of self-defence, instant and overwhelming, leaving no choice of means, and no moment for deliberation."[228] The *Department of Justice Drone Opinion* highlights that the targeted terrorist poses a "'continued and imminent threat' of violence to United States persons and interests . . . and continues to plot attacks intended to kill Americans from his base of operations in Yemen."[229] In a Department of Justice white paper, the issue of imminence was addressed by indicating "the threat posed by al-Qa'ida and its associated forces demands a broader concept of imminence in judging when a person *continually planning terror attacks* presents an imminent threat . . . ."[230]

---

[223] Schmitt, *supra* note 217, at 318. *But see* Laurie R. Blank, *Defining the Battlefield in Contemporary Conflict and Counterterrorism: Understanding the Parameters of the Zone of Combat*, GEO. J. INT'L & COMP. L. 1, 25 (2010) (where it is suggested applying neutrality law could lead to perverse results providing protection to terrorist groups operating from safe havens in nonbelligerent States) and INTERNATIONAL HUMANITARIAN LAW AND THE CHALLENGES OF CONTEMPORARY ARMED CONFLICT, Report 32nd International Conference of the Red Cross and Red Crescent EN 32IC/15/11, at 16 (2015), *available at* https://www.justsecurity.org/wp-content/uploads/2015/11/2015-ICRC-Report-IHL-and-Challenges-of-Armed-Conflicts.pdf (where the application of neutrality law to target persons in third States is dismissed on the basis that "[t]he law of neutrality does not apply to the NIAC scenario posited above.").

[224] Deeks, *supra* note 216, at 497–506.

[225] Schmitt, *supra* note 217, at 318. The law of neutrality referred to in leading humanitarian law texts. *See* LESLIE GREEN, THE CONTEMPORARY LAW OF ARMED CONFLICT 297–305 (2nd ed. 2000), and Michael Bothe, *The Law of Neutrality*, *in* THE HANDBOOK OF INTERNATIONAL HUMANITARIAN LAW, Rule 11 571–604 (Dieter Fleck ed., 2nd ed. 2008). *See also* YORAM DINSTEIN, THE CONDUCT OF HOSTILITIES UNDER THE LAW OF INTERNATIONAL ARMED CONFLICT ¶22, at 9–10 (2nd ed. 2010), DOCUMENTS ON THE LAWS OF WAR 2 (Adam Roberts & Richard Guelff eds., 3rd ed. 2003), and *Advisory Opinion on the Legality of the Threat or Use of Nuclear Weapons*, 1996 I.C.J. 226, ¶¶ 88–89, at 260–261 (where neutrality is referred to as one of the fundamental principles of humanitarian law).

[226] Deeks, *supra* note 216, at 501.

[227] *Department of Justice Drone Opinion, supra* note 216, at 25 (where an analogy is made to States taking action in international armed conflict when a neutral State cannot prevent a violation of its neutrality).

[228] DINSTEIN, *supra* note 185, ¶ 726, at 274 (quoting American Secretary of State Daniel Webster). *See also* Schmitt, *supra* note 135, at 17 ("concerns as to the unconstrained spread of hostilities should be assuaged in part by the effect of legal norms lying outside IHL [law governing the recourse to war].").

[229] *Department of Justice Drone Opinion, supra* note 216, at 21.

[230] *Department of Justice White Paper*, 7 (undated) (emphasis added), http://msnbcmedia.msn.com/i/msnbc/sections/news/020413_DOJ_White_Paper.pdf (last visited June 15, 2015).

The invocation of a continuous notion of imminence is consistent with status-based targeting found in humanitarian law. The status of a targeted individual is an issue that has been the subject of considerable discussion during the post-9/11 period. Even the ICRC *Interpretive Guidance*, which adopts a narrower interpretive approach toward direct participation than many States, indicates that "individuals whose *continuous function* involves the preparation, execution, or command of acts or operations amounting to direct participation in hostilities are assuming a continuous combat function."[231] In this respect, the conduct of hostilities-based drone theory appears to rely heavily, albeit indirectly, on concepts analogous to continuously performing a combat function. This should not be surprising for a normative framework governing the use of drones that sees the armed conflict tied personally to the individual member of an opposing force wherever they may go. What it does demonstrate again is the subtle merging of recourse to war and humanitarian law concepts by the international community as attempts are made to resolve challenging contemporary operational issues.

The conduct of hostilities approach also raises questions regarding who may be targeted, and under what circumstances. The basis for strikes, at least in respect of a U.S. citizen, has been discussed in the context of targeting a senior member of Al Qaeda in the Arabian Peninsula (AQAP). However, as has been noted previously, reports suggest personnel much lower in the organizational hierarchy have been targeted.[232] Given this broader use of drones it is not clear, even from the perspective of the arguments presented in the *Department of Justice Drone Opinion*, that the legal analysis regarding the conduct of hostilities is limited to senior Al Qaeda leaders. However, focusing on the leadership of the group in the legal and public dialogue arguably makes it easier to justify collateral civilian casualties. This is because killing senior leaders is often viewed as offering a greater military advantage, although in reality such targeted killings may not actually provide the desired military effect (see Chapter 7, section 7.6.1, *Insurgents and Target "Value"*).[233]

The conduct of hostilities approach involves a number of unresolved issues, specifically because of its reliance on neutrality principles and the concept of functional participation in the armed conflict. While the application of legal principles is presented as a simple test as to whether the territorial State is "unable or unwilling to act," neutrality law seems far more nuanced and complex than the permissive drone theory appears to suggest. Among the issues that need to be explored is the content of neutrality law in the post–UN Charter era and what activity by "neutrals" is actually prohibited. Further, under that body of law it matters what the person under threat of attack in neutral territory is actually doing, and on whose behalf it is being done.

There appears to be a consensus that, since the enactment of the UN Charter, neutrality law has continued relevance to modern armed conflict.[234] However, that law is viewed

---

231 Interpretive Guidance, *supra* note 15, at 34 (emphasis added).

232 Peter Bergen & Jennifer Rowland, *CIA Drone War in Pakistan in Sharp Decline*, CNN (Mar. 28, 2012), http://www.cnn.com/2012/03/27/opinion/bergen-drone-decline ("the CIA was primarily killing low-level militants in its drone strikes.").

233 Naylor, *supra* note 26, at 288 (where it is noted the killing of al Zarqawi in Iraq in 2006 did not bring about a hoped for downturn in violence).

234 Bothe, *supra* note 225, at 574–75 (where it is noted that in the post-Charter era, States continue to rely on the law of neutrality, with both international and national court decisions having recognized its continued operation). *See also* Advisory Opinion on the Legality of the Threat or Use of Nuclear Weapons, 1996 I.C.J. 226, ¶¶ 88–89, at 259–261.

as having undergone change, since "[t]he traditional law of neutrality with its duty of impartiality, i.e. the prohibition of discrimination between the parties to the conflict, seems to be incompatible with this development which outlaws the aggressor."[235] It is recognized in an inter-State context that should the UN Security Council fail to act or not become fully seized of a matter, an exercise of self-defense can raise the question of the application of the law of neutrality.[236] However, the Security Council performs the role of determining when there has been aggression, after which a member State, in theory, should no longer claim impartiality.[237] The result is a narrower application of neutrality law than previously had been the case; although when the Security Council has acted, some States have continued to maintain support for, or at least act impartially toward the transgressor State.[238] However, whatever its imperfections in application, the key point is that neutrality principles are generally more limited in scope than the very general reliance by the conduct of hostilities approach on an "unable or unwilling" standard suggests.

In the aftermath of the 9/11 attacks, UN Security Council Resolutions have provided an often-cited basis for arguing there is international acceptance that the right of self-defense can be exercised against non-State actors.[239] It was that exercise of self-defense that marked the initiation of the armed conflict upon which the permissive approach to targeting is based. However, this does not end the analysis. As is reflected in UN Security Council Resolution 1973 (2001), which mandated international reactions to terrorism, those responses included far more than the exercise of self-defense. In referring to taking all necessary steps to suppress terrorist activities, including suppressing terrorist financing; refraining from providing support; and denying safe haven, the Security Council highlighted that methods such as criminal law enforcement are, at a minimum, to be exercised concurrently with any act of self-defense.[240] This would seem to undermine any argument made by proponents of the permissive approach to the use of drones that the targeting of terrorists under conduct of hostilities norms operates to the exclusion of or completely independent of human rights–based law enforcement. It also acts as a subtle reminder that in the post-Charter era the neutrality-based argument is affected by factors other than reliance on historic notions of neutrality under humanitarian law.

Under the permissive "conduct of hostilities" approach an important factor leading to a determination of whether self-defense can be exercised is consideration of what the person being targeted may be doing. The participation of someone in planning or performing other functions reflecting membership in an armed group forms the basis of the lawful use of drone strikes. However, some interpretations of neutrality law have adopted the view a

---

[235] Bothe, *supra* note 225, at 574.

[236] *Id.* ¶¶ 1–2, at 574–75.

[237] Wolff Heintschel von Heinegg, *Current Legal Issues in Maritime Operations: Maritime Interception Operations in the Global War on Terrorism, Exclusion Zones, Hospital Ships and Maritime Neutrality*, 80 INT'L L. STUD. 207, 224 (2006). *See also* Blank, *supra* note 223, at 25.

[238] GREEN, *supra* note 225, at 298.

[239] Louise Arimatsu & Michael N. Schmitt, *Attacking "Islamic State" and the Khorasan Group: Surveying the International Law Landscape*, COLUM. J. TRANSNAT'L L. BULLETIN 1, 19–20 (2014), http://jtl.columbia.edu/attacking-islamic-state-and-the-khorasan-group-surveying-the-international-law-landscape//.

[240] S. C. Res. 1373, ¶¶ 3–4, UN Doc. S/RES/1373 (Sept. 28, 2001).

belligerent's acts must cause "material prejudice" to the offended belligerent.[241] While the determination can be challenging, there would remain an obligation to establish actual prejudice. If that is the rule for international armed conflict, it is not clear why it would not also apply to noninternational armed conflict as well. It is likely resistance to drone strikes will be more muted where it is established the person targeted is a major figure actively engaged in terrorist action while enjoying safe haven. A clear example of a more reserved international reaction was the 2011 operation that killed Osama bin Laden. In contrast, the reaction to the strike against Anwar Al Awlaqi was more vocal, although that case also included the controversial aspect of his being an American citizen.[242] Criticism mounts when the threat posed by a person being targeted cannot be concretely identified, or the strikes are expanded in scope and over time to include lower ranking personnel causing the incidents to look like an expansion of an existing armed conflict.

The requirement for establishing a more concrete threat before justifying intervention in a neutrality law context is reflected in Paragraph 22 of the *San Remo Manual on International Law Applicable to Armed Conflicts at Sea*.[243] That manual indicates a belligerent can take action if a violation of neutral waters "constitutes a *serious and immediate threat* to the security of the opposing belligerent."[244] Thus it appears not every entry by a belligerent into neutral "space" would justify action by an opponent. Even when action is justified there is a further limitation. Where such a violation occurs a state "may, in the *absence of any feasible and timely alternative*, use such force as is *strictly necessary* to respond to the threat posed by the violation."[245] This wording introduces a condition of "strict" necessity in respect of neutral territory that would not ordinarily apply in other areas of operations during international armed conflict. Yoram Dinstein suggests a similar restrictive approach regarding the exercise of extraterritorial law enforcement against non-State actors, including consideration of alternative measures.[246]

The reference to feasible and timely alternatives raises the question of whether capture must be considered. The *Department of Justice Drone Opinion* notes that capture was considered, but was not a feasible option in the case of Awlaqi.[247] The Drone Policy certainly addresses this issue and notes the preference for capture and respect for State sovereignty.[248] The challenge is that entering another country to carry out an arrest may actually be more invasive of territorial sovereignty than a drone strike. As was demonstrated in the attempt to capture a senior Al Shabaab leader in Somalia in 2013, such an operation may become quite violent.[249] However, overall, "arrest" operations are traditionally viewed as being less

---

241 Deeks, *supra* note 216, at 500 (where it is noted some scholars have argued action may only be taken when the belligerent's acts have caused "material prejudice").

242 SCAHILL, *supra* note 43, at 32–33.

243 SAN REMO MANUAL ON INTERNATIONAL LAW APPLICABLE TO ARMED CONFLICTS AT SEA, Rule 22, 101 (1995) [hereinafter SAN REMO MANUAL].

244 *Id.* (emphasis added).

245 *Id.* (emphasis added).

246 *See also* DINSTEIN, *supra* note 185, ¶ 729(v), at 275.

247 *Department of Justice Drone Opinion, supra* note 216, at 40.

248 *Drone Policy, supra* note 19.

249 Peter Beaumont, *US Forces Target Leading al-Shabaab Militant in Somalia Raid*, THE GUARDIAN, Oct. 6, 2013, http://www.theguardian.com/world/2013/oct/05/somalia-kenya-al-shabaab-westgate mall ("This time the Seal team members encountered fiercer resistance than expected, and after a 15–20 minute firefight, the unit leader decided to

violent, and normally attract less controversy due to a generally more limited potential for collateral civilian casualties.

It is evident from the *San Remo Manual* analysis that the conduct of hostilities in neutral waters appears to have more conditions attached to it than in non-neutral operational areas. This reflects an approach where the existing armed conflict is seen to not enter unfettered into neutral territory. A minimalist reaction is required when responding to an established threat. These restrictions were built into the *San Remo Manual* notwithstanding the view expressed by one group of experts that "the use of force is governed by the law of armed conflict, the issue of justification of self-defence having been resolved at an earlier stage."[250] The result is that if truly applied by analogy to noninternational armed conflict or counterterrorism operations, neutrality law may not as readily empower the conduct of hostilities approach toward drone strikes as some interpretations of the permissive "conduct of hostilities" theory seem to suggest.

It also appears in the context of international armed conflict not everything that is done in a neutral State in support of a belligerent justifies intervention. Of particular relevance for the application of the "unwilling and unable" test is that historically the provision of support to a belligerent party from neutral territory was not absolutely prohibited. This aspect of neutrality law has a very nuanced application. For example, as a general proposition during international armed conflict, the government of a neutral state cannot furnish supplies of any type. However, this statement has to be placed in historical context. The considerable support provided by the United States to other nations at war with Germany prior to its declaration of war "did not render the United States a party to the conflict. . . ."[251] This has led to the conclusion, "[o]nly where a hitherto neutral state *participates to a significant extent* in hostilities is there a change in status."[252]

Neutrality law is no less nuanced regarding private individuals operating within that neutral State. During an international armed conflict the neutral state is "not obligated to prevent private individuals from selling and exporting arms, ammunition and war materials to Belligerent Parties."[253] There appears to be no requirement that the private individuals be citizens of the neutral State. Such activity can be banned by the State, or any barrier to private trade may be erased, but any prohibition must be applied to both Parties.[254] However, where such sales and exports do take place, the neutral State "must be on the alert not to become a base of military operations against one of them."[255] This highlights that carrying out a support role does not always involve direct participation in military operations.

abort the mission and they swam away, the official said."), and Simon Tisdale, *Al-Shabaab Target May Explain US Secrecy over Failed Somali Raid*, THE GUARDIAN, Oct. 6, 2013, http://www.theguardian.com/world/2013/oct/06/al-shabbab-target-secrecy-somali-raid ("Obama will have personally given the go-ahead for both raids. His orders were reportedly to capture, if possible, rather than kill.").

250 SAN REMO MANUAL, *supra* note 243, ¶ 22.3, at 102.

251 Bothe, *supra* note 225, ¶ 3, at 580–81. *See also* DINSTEIN, *supra* note 185, ¶¶ 74–76, at 28–29 and ¶¶ 477–482 at 178–80.

252 DINSTEIN, *supra* note 185, ¶¶ 74–76, at 28–29 (emphasis added).

253 *Id* ¶ 73, at 28. *See also* Article 7 of both The Hague Convention (V) Respecting the Rights and Duties of Neutral Powers and Persons in Case of War on Land, and the 1907 Hague Convention (XIII) Concerning the Rights and Duties of Neutral Powers in Naval War.

254 DINSTEIN, *supra* note 185, ¶ 74, at 28.

255 *Id.* ¶ 77, at 29.

Applying this law by analogy to a noninternational armed conflict could underpin an argument that simply providing private support to a non-State actor would not automatically indicate a failure of territorial State's legal obligations such that drone strikes would be justified. This does not mean there is a carte blanche for support to be provided to Al Qaeda. In the case of international terrorism and crime, and in particular Al Qaeda, there are international agreements to address financing and other support provided to groups engaged in such activity.[256] The requirement for international cooperation is set out in Security Council direction.[257] However, the key point is that such obligations do not necessarily arise in the context of an application of humanitarian law–based neutrality principles.

The result is that to the extent drone strikes are justified under an "unable and unwilling" standard, based on an analogy to neutrality law, the authority to use force may be considerably less permissive than has been argued to date. It can also place the State using drones at odds with interpretations of neutrality law, which would not prohibit all forms of support provided to a "belligerent." This is particularly evident if the threatened State takes a broad view of what constitutes planning, support, or other activity-enhancing threats it is facing. It is an issue that takes on even more challenging aspects when the use of drones is expanded to include lower ranking members of a terrorist organization. This can extend to questions of whether the simple provision of financing and other means of indirect support makes a person a valid target, thereby justifying transnational drone strikes. This is a crucial area where advocates of the "permissive" self-defense theory may find themselves particularly criticized because of the lack of transparency and clarity about the criteria being applied by U.S. views to target "terrorists.".

As the ICRC *Interpretive Guidance* process established, the issue of who is a valid target under humanitarian law is an area of considerable controversy in the international law community. Perhaps the most positive outcome from that document has been a general consensus that members of an organized armed group and individual civilians who take a direct part in hostilities may be targeted.[258] What constitutes support for non-State actor armed groups is a complex topic (see Chapter 7, section 7.5, *The IED and Suicide Bombing Threat: Arrest or Kill?*). It can cover a broad range of activity, including financing and the provision of armaments. Financing itself has been identified as a crucial asset for the conduct of terrorist activities.[259] However, funding for insurgences and terrorism can be derived from a variety of sources, including criminal activity,[260] through the Internet,[261] Islamic-based

---

[256] *Fighting International Terrorism and Crime*, Permanent Mission of Canada to the United Nations, http://www.canadainternational.gc.ca/prmny-mponu/canada_un-canada_onu/positions-orientations/peace-paix/terrorism-terrorisme.aspx?lang=eng (last visited June 15, 2015) (for an outline of the numerous international conventions on terrorism and crime to which Canada is party).

[257] *See, e.g.*, S. C. Res. 1373, ¶¶ 3–4, UN Doc. S/RES/1373 (Sept. 28, 2001).

[258] INTERPRETIVE GUIDANCE, *supra* note 15, at 34.

[259] Holder v. Humanitarian Law Project 130 S. Ct. 2705, 2725–2727 (2010). *See also* NORMAN CIGAR, AL-QA'DA'S DOCTRINE FOR INSURGENCY 108 (2009) (It has been noted that "[m]oney is the nervous system of *jihad*.").

[260] GARRY LEECH, THE FARC: THE LONGEST INSURGENCY 56–74 (2011), and VICTORIA BRUCE, KARIN HAYES, & JORGE ENRIQUE BOTERO, HOSTAGE NATION: COLUMBIA'S GUERRILLA ARMY AND THE FAILED WAR ON DRUGS 50–52 (2010) (regarding the link between the Columbian insurgent group FARC and drug production).

[261] BRUCE HOFFMAN, INSIDE TERRORISM 211 (2006).

"charitable giving,"[262] participation in commercial and real estate investments,[263] front companies, expatriate remittances, and Mafia-style shakedowns.[264] Participation in such activity does not make someone automatically a lawful military objective.

A key targeting issue is whether involvement in financing constitutes direct participation in hostilities. Participation in financial activity when integrated directly in the command structure of the organization is different from simply providing money to support a rebel cause. The difference between the provision of civilian financial support and integration within a guerrilla or terrorist movement will not always be straightforward.[265] It has been suggested, "the links between members of the Hamas political wing and terrorist activities are symbiotic and have been intertwined since the founding of Hamas in 1987."[266] Much depends upon the degree of organization and, ultimately, the political maturity of the insurgent group. Groups such as Al Qaeda, which adopt a vanguard or *focoist* approach, are more likely not to have truly separate organizational "wings" (e.g., political, military, or humanitarian), as compared to an insurgent organization engaged in governance.[267] This is likely to occur if the groups are engaged in an insurgency that controls territory. For example, the Taliban has exhibited a capacity to set up governing structures in the areas of Afghanistan it operates, although it has been suggested this should be considered "more in terms of influence than direct political control . . . ."[268] This has included operating sharia courts, collecting taxes, and even developing an ombudsman system to handle complaints regarding the actions of local commanders (see Chapter 6, section 6.4.2, *The "Terrorists"*).[269] The transition toward governance and a more diverse organization can be seen in the 2014 Islamic State call for doctors, judges, engineers, and administrative specialists to help run the new caliphate,[270] the operation of banks,[271] and efforts made to establish its own currency.[272]

To determine what constitutes direct participation in hostilities reliance should be placed on criteria such as operating within a command structure; performing combat, combat

---

[262] Juan C. Zarate, Treasury's War: The Unleashing of a New Era of Financial Warfare 70–72 (2013) (referring to the reliance by Al Qaeda on the Islamic obligation of *zakhat* (charitable giving), and the Hezbollah and Hamas reliance on charitable operations, and Dawa (fundraising) committees).

[263] Hoffman, *supra* note 261, at 259 (outlining the Abu Nibal terrorist group financing operations in the 1980s).

[264] Matthew Levitt, Hezbollah: The Global Footprint of Lebanon's Party of God 254–61 (2013).

[265] Dinstein, *supra* note 135, ¶ 129, at 42 ("Some scholars attach great significance to this profiling [separate political and military wings], but in many instances telling the two wings apart may be a chimera.").

[266] Matthew Levitt, Hamas: Politics, Charity, and Terrorism in the Service of Jihad 33 (2006). *See also* Nicholas Blanford, Warriors of God: Inside Hezbollah's Thirty-year Struggle Against Israel 79–83 (2011) (for a discussion of the development of Hezbollah's social welfare network.).

[267] Regis Debray, Revolution in the Revolution? 106 (Bobye Ortiz trans., 1967) (for a discussion about *focoist* groups acting as nucleus and their relationship with a governing army and party), and Walter Laqueur, Guerrilla Warfare: A Historical & Critical Study 331 (2006) (1976).

[268] Abbas, *supra* note 203, at 186.

[269] David Kilcullen, Counterinsurgency 157–58 (2010), Antonio Giustozzi, Koran, Kalashnikov and Laptop: The Neo-Taliban Insurgency in Afghanistan 110–18 (2008), and Atwan, *supra* note 194, at 147.

[270] Ed Adamcyzk, *ISIS Leader to World's Muslims: "Build Islamic State,"* UPI (June 1, 2014), http://www.upi.com/Top_News/World-News/2014/07/01/ISIS-leader-to-worlds-Muslims-Build-Islamic-State/8971404235322/.

[271] Jamila Trimble, *No Blacklist Yet for Islamic State Banks*, Foreign Pol. (Nov. 12, 2014), http://www.foreignpolicy.com/articles/2014/11/12/iraq_syria_blacklist_sanction_islamic_state_banks_ISIS_IS.

[272] *Isis to Introduce Its Own Currency, the Islamic Dinar*, The Guardian, Nov. 14, 2014, http://www.theguardian.com/world/2014/nov/14/isis-own-currency-islamic-dinar.

support, or combat service support functions; carrying arms openly; maintaining communications within the group; intelligence gathering; being subject to discipline; or carrying out logistical staff duties, such as managing internal group financing or supply, which are designed to facilitate the conduct of hostilities.[273] Therefore, a person could be seen as working within a non-State armed group if they were acting as the director of its finances.[274] In other words, handling finances in a roughly analogous manner to State military forces. This would include members of organized armed groups carrying out a "finance" logistics function, such as internally managing those resources for the armed group. It is alleged that was the role performed by the Islamic State financier, Abu Sayyaf, who was killed by U.S. Special Forces in May 2015 in Syria during an attempted capture operation.[275] However, simply making a financial contribution to Al Qaeda, the Islamic State, or other jihadist group; facilitating the transfer of funds; or even operating a bank are activities for which a person should only be arrested or detained. In this regard it would be difficult to argue that everyone engaged in financial transactions relating to governing the territory controlled by the Islamic State are members of its organized armed group or otherwise taking a "direct" part in hostilities.

A challenge in understanding the U.S. approach to this issue is that both national legislation and case law appear to suggest overly broad criteria for membership in an organized armed group. For example, the empowering legislation for military commissions provides a definition of "unprivileged belligerent" that extends to persons who have "purposefully *and materially supported* hostilities against the United States or its coalition partners."[276] As was noted in *Holder v. Humanitarian Law Project*, "material support" is very broadly worded under U.S. domestic legislation to include any property, tangible or intangible, or service.[277] United States domestic detainee litigation has also applied an expansive interpretation of what constitutes both membership in and support to those groups: substantial, material, direct, or otherwise.[278] This breadth of activity must be put in context. Historically, being assessed to be an "unprivileged belligerent" or "unlawful combatant" appears to have focused on the person performing hostile acts against the enemy,[279] and not all forms of

---

[273] *See, e.g.*, Watkin, *supra* note 15, at 691. *See also* DEPARTMENT OF DEFENSE LAW OF WAR MANUAL, Office of the General Counsel ¶ 5.9.3 at 224–26 (June 2015), http://www.dod.mil/dodgc/images/law_war_manual15.pdf (for suggested U.S. criteria for direct participation in hostilities).

[274] H. VON DACH, TOTAL RESISTANCE: SWISS ARMY GUIDE TO GUERRILLA WARFARE AND UNDERGROUND OPERATIONS 3, 90, & 140–51 (1965) (outlining a theory of guerrilla warfare where the "civilian resistance movements" that carry out a variety of activities including sabotage, raids, and eventually general uprisings in occupied territory also have a finance section tasked with procuring funds.).

[275] Cooper & Schmitt, *supra* note 58. *See also* Eric Schmitt, *Leader of Qaeda Cell Killed in Airstrike in Syria, Pentagon Says*, N.Y. TIMES, Oct. 18, 2015, http://www.nytimes.com/2015/10/19/world/middleeast/leader-of-qaeda-cell-killed-in-airstrike-in-syria-pentagon-says.html?smprod=nytcore-iphone&smid=nytcore-iphone-share&_r=0 (Sanafi al-Nasr of the Khorasan Group appears to be the type of financier and recruiter who would be targetable as a member of an Al Qaeda organized armed group).

[276] Military Commission Act of 2009, Title XVIII of the National Defense Authorization Act for Fiscal Year 2010 (Pub. L. 111-84, H.R. 2647, 123 Stat. 2190 enacted October 28, 2009) §948a Definitions (7) (emphasis added).

[277] Holder v. Humanitarian Law Project 130 S. Ct. 2705, 2708 (2010).

[278] *See, e.g.*, Al-Bihani v. Obama, 590 F.3d 866, 872 (2010), en banc rehearing denied, 619 F.3d 1 (D.C. Cir. 2010) (where the court made reference not only to "substantial support" and "direct support" but also the Military Commission Act based criteria of "purposefully and materially supported hostilities").

[279] Richard R. Baxter, *So-called "Unprivileged Belligerency": Spies, Guerrillas, and Saboteurs, in* HUMANIZING THE LAWS OF WAR: SELECTED WRITINGS OF RICHARD BAXTER 37, 40 (Detlev F. Vagts et al. eds., 2013) (originally published

support that would constitute "material support."[280] Importantly, in respect of the scope of "material support" in an armed conflict context, a 2012 appellate decision, *Hamdan v. The United States*, indicated such activity carried out for terrorism cannot be considered a war crime.[281]

However, there remain questions as to the degree to which domestic law notions of support such as substantial or material support reflect established international law standards or are actually applied to targeting decisions.[282] In dealing with detention, the U.S. courts have applied a broad zone of deference to the decision making of military commanders.[283] It has been suggested that "[i]n its current form, it is somewhat difficult to draw from the [detention] caselaw more than broad guidance about the boundaries of that zone, and there is a great deal that is left unsaid about the specific factors that a specific decision maker in a specific set of circumstances should weigh in taking a targeting decision."[284] However, at least one author has considered the phrase "purposely and materially supported hostilities" as a basis for targeting criteria.[285] Similarly, the *Department of Justice Drone Opinion* relied heavily on domestic cases dealing with detainees when indicating "[t]he DoD operation, therefore, would be carried out against someone whom Congress has authorized the use of necessary and appropriate force."[286]

What the unresolved issues associated with the "conduct of hostilities" theory highlight is that justification of a permissive targeting approach requires a far deeper analysis of the applicable law. A threatened State should not simply be able to apply the "unwilling or unable" analogy from neutrality law, adopt broad interpretations of what constitutes support to identify a target, and then launch a drone strike. Relying on analogies to the law governing operations in neutral territory should require a more constrained use of force than in other operational contexts. If the law of neutrality applicable in international armed conflict is to be applied by analogy in noninternational conflict, then it is best done to the full extent possible. Neutrality law generally attributes a unique status to the uninvolved State. To rely simply on a threshold that the territorial State is unable and unwilling to act, without addressing the underlying nuances of that body of law, makes the conduct of hostilities framework begin to look like a very selective application of the legal rules.

---

in 1951 BRIT. Y.B. INT'L L. 323) ("persons who in the face of the enemy or behind his lines, have committed hostile acts without meeting the qualifications prescribed for lawful belligerents."). *See also* THE MANUAL OF THE LAW OF ARMED CONFLICT, United Kingdom Ministry of Defence ¶ 11.14, at 279–80 (2004).

[280] J. M. SPAIGHT, AIR POWER AND WAR RIGHTS 47 (3rd ed. 1947) (Although attempting to justify attacks on factory workers as quasi-combatants, a concept that has been rejected under international law, James Spaight identified limits on the concept of "material support" that resonant with more modern notions of direct participation.).

[281] Hamdan v. United States of America, 696 F. 3d 1250 (2012).

[282] Al-Bihani v. Obama, 590 F.3d 866, en banc rehearing denied, 619 F.3d 1 (D.C. Cir. 2010) (where "material support" is applied in the context of detention).

[283] Stephen Pomper, *Toward a Limited Consensus on the Loss of Civilian Immunity in Non-International Armed Conflict: Making Progress through Practice*, 88 INT'L L. STUD. 181, 185 (2012).

[284] *Id.*

[285] Eric T. Jensen, *Targeting of Persons and Property*, *in* MICHAEL W. LEWIS, THE WAR ON TERROR AND THE LAWS OF WAR 37, 57 (2009) (suggesting that standard could be used to overcome the challenge of targeting the person who constructs IEDs, or recruits and pays suicide bombers). *See also* SAVAGE, *supra* note 47, at 117–20 (which appears to suggest that decisions regarding detention without trial and targeted killing were assessed under the same criteria known as the "March 13 Standard." This standard included persons who provided *substantial support*).

[286] *Department of Justice Drone Opinion*, *supra* note 216, at 21, and 21–22 n.28.

The legal framework should result in the adoption of well-articulated criteria for the determination of when a person is "targetable" pursuant to humanitarian law standards, an outline of the material prejudice that person's actions is causing, and an application of the principles reflected in Paragraph 22 of the *San Remo Manual* necessitating that only in the absence of feasible and timely alternatives would such force as is strictly necessary be used. These more nuanced rules reflect a convergence of norms similar to how the law is applied during counterinsurgency operations (see Chapter 7, section 7.7, *Overlapping Normative Frameworks*). To the extent targeting is based on a status test, that person must be performing a combat function closely aligned with the activity carried out by members of State armed forces or otherwise taking a direct part in hostilities in a context where such activity is more narrowly prescribed than might be the case when considering the detention of a suspect. Finally, in respect of Al Qaeda, the Islamic State or other groups engaged in terrorism, it is difficult to see how the permissive approach can operate entirely to the exclusion of law enforcement operations or norms. Just as counterinsurgency operations occur under a "police primacy" construct, so must counterterrorism activities, whether the contemplated activity involves raids by Special Forces or the conduct of drone strikes (see Chapter 6, section 6.5, *Counterinsurgency and Counterterrorism: An Integrated Response*).

8.5.3. THE "SELF-DEFENSE" OPTION

The idea there exists a third legal paradigm governing drone strikes has been read into the March 2010 statement by the US. State Department legal advisor, Harold Koh, where he indicated "a state that is engaged in an armed conflict *or* in legitimate self-defense is not required to provide targets with legal process before the state may use lethal force."[287] It has been suggested this means a resort to the use of force "in ways in which the means and levels of force used are not part of an armed conflict, as a matter of the technical law of war."[288] While the reference to acting in legitimate self-defense might actually have been in respect of a "one-off" exercise of force involving an isolated armed conflict, rather than the ongoing one against Al Qaeda, the idea of a separate self-defense paradigm has gained some support. This separate paradigm raises fundamental questions regarding the scope of self-defense and the rules governing the use of force.

Ken Anderson called the authority to use force that would not necessarily be part of an armed conflict, "naked" self-defense.[289] Framing this discussion as a use of force outside the

---

[287] Harold Hongju Koh, *The Obama Administration and International Law*, Mar. 25, 2010, http://www.state.gov/s/l/releases/remarks/139119.htm (emphasis added). *See* Ian Henderson, *Civilian Intelligence Agencies and the Use of Drones*, 13 INT'L Y.B. HUM. L. 133, 159 (2010).

[288] Anderson, *supra* note 201, at 8 ("The US government addressed the issue of self-defense straightforwardly in the 1980s, notably in an important 1989 speech (later published in the *Military Law Review*) by then State Department legal advisor Abraham Sofaer at the US Army JAG School."). *See* Geoffrey S. Corn, *Self-defense Targeting: Blurring the Line between the Jus ad Bellum and the Jus in Bello*, 88 INT'L L. STUD. 57, 76–77 (2012) (for a very helpful analysis of the Harold Koh statement), and Jordan J. Paust, *Self-Defense Targetings of Non-state Actors and Permissibility of U.S. Use of Drones in Pakistan*, 19 J. TRANS. L. & POL. 237, 260 n.54. (2009–10). *See also* Abraham D. Sofaer, *The Sixth Annual Waldemar A. Solf Lecture in International Law: Terrorism, the Law, and the National Defense*, 126 MIL. L. REV. 89 (1989).

[289] Anderson, *supra* note 201, at 8.

context of armed conflict results in a search to ascertain what law applies to govern that use of force. As has been noted, traditionally the view has been there are two legal frameworks governing such force: international human rights law and international humanitarian law (the law of armed conflict), with the latter body of law only applying when an armed conflict is in existence. Notwithstanding this well-established approach, advocates of a separate "self-defense" theory appear to have adopted the position that humanitarian law principles govern the drone strikes even if no armed conflict technically exists. Interestingly, two of the main commentators, Ken Anderson and Jordan Paust, who have referred to the existence of a unique self-defense option, approach this problem from quite different perspectives.

Ken Anderson acknowledges the use of force "binary": law enforcement or armed conflict. Regarding the law enforcement model, he suggests that only in the most limited cases, such as hostage rescue, would shooting rather than capturing the criminal be condoned. Further, Professor Anderson notes that if the critics of the U.S. armed conflict approach "have bite," and "[i]f one accepts the binary, presumably the legal situation reverts to proceeding under the standards of law enforcement."[290] Consistent with the restrictive "law enforcement" approach, the conclusion is reached that the targeted killing program "using drone warfare, or any targeted killing under US national security law, in practical terms could not survive as a lawful practice."[291]

In contrast, his "naked" self-defence paradigm is "required to meet the standards of necessity, distinction, and proportionality... *even if not formally part of an armed conflict*."[292] The "naked" self-defence model contemplates circumstances where a resort to force is made "against nonstate actors, such as individual terrorists, which do not (yet) rise to the NIAC threshold."[293] These standards apply not "in the technical law of armed conflict," but rather as a matter of customary law.[294] The result is that this theory fills a perceived void, expanding the application of humanitarian law norms to uses of force outside the context of established armed conflicts. What is not clear is whether the reference to the "technical law" is made in context of it being treaty law, or because an actual armed conflict "technically" needs to exist for it to apply. Further, the interaction between the State self-defense principles (i.e., necessity, imminence, and proportionality) and those of humanitarian law (i.e., necessity, distinction, and proportionality) is not specifically discussed. This theory has garnered some support with the suggestion "in this third legal paradigm the legal concepts of necessity and proportionality as those concepts are understood in the law relating to national self-defence provide both the *jus ad bellum* [law governing the recourse to war] and also the *lex specialis* [special law] against which to judge whether the use of lethal force was lawful under human rights standards."[295]

To put this approach in context, it appears there were a number of unique factors impacting on its development: the Bush administration's perspective of a Global War on Terrorism and its intelligence-driven counterterrorism program, as well as the countervailing

---

[290] *Id.* at 4.

[291] *Id.*

[292] *Id.* at 8 (emphasis added). *See also* Henderson, *supra* note 287, at 160–64 (for analysis of Ken Anderson's theory).

[293] Anderson, *supra* note 201, at 8

[294] *Id.*

[295] Henderson, *supra* note 287, at 164

development of the theory of a "geography of war." The spatial theory, which is at the heart of the restrictive "law enforcement" approach to drone strikes, is seen as a reaction to a perception the United States "intended the laws of war as a rubric for something quite different from the conduct of hostilities."[296] It was also one where the armed conflict is seen to follow the individual being targeted and "[t]he world was our battlespace."[297] The naked self-defense framework appears to have been developed specifically to address these issues, and was developed uniquely in that context. In effect, it appears as an attempt to find a middle ground between the restrictive "law enforcement" approach and a State leadership that did not appear to want to be bound by the traditional laws governing armed conflict.

The naked self-defense theory underwent alteration with the end of the Bush presidency. It is indicated that "the Obama administration's rejection of the global war on terror permits the traditional conduct of hostilities standard to reemerge," and "a formal legal geography of war recede along with the global war on terror. . . ."[298] As a result, "this moves the legal argument back to the traditional standard of the conduct of hostilities, which explicitly abandons the Bush administration's position."[299] The changes put in place by the Obama administration led the author of "naked" self-defense to adopt the view that targeting occurring "in places such as Yemen are part of the preexisting and ongoing NIAC under the traditional standard of 'conduct of hostilities,' and thus it is not necessary at this point to reach to naked self-defense."[300]

This shift, brought about by a changed U.S. approach toward the jihadist threat and the election of a government viewed as being more committed to meeting traditional international law standards, appears then to lean toward acceptance of the permissive "conduct of hostilities" theory. As a result, naked self-defense is said to have a more narrow and prospective application, being preserved for use by "future presidents, in circumstances we cannot not forsee. . . ."[301] It exists therefore as a form of safeguard in the context of a State propensity to undertake intelligence-driven uses of force short of war. If that were to happen, "[i]t is better to regulate the intelligence agencies' conduct under customary principles than ignore it—and better to regulate this as its own legal and operational category of the use of force in both *jus ad bellum* [State self-defense] and *jus in bello* [humanitarian law]."[302] In a sense the principle of naked self-defense appears to cut across the boundaries of these traditional bodies of law. The ultimate effect seems to be that humanitarian law principles are seen to apply as a matter to custom, even in circumstances where force is used against non-State actors where the armed conflict threshold has not been met.[303]

Jordan Paust provides another interpretation of the self-defense paradigm. Unlike the naked self-defense theory, he does not treat the paradigm as an exceptional framework to be applied when governments seek to avoid legal oversight. He suggests drone attacks against

---

[296] Anderson, *supra* note 201, at 13.

[297] *Id.* at 16.

[298] *Id.* at 14.

[299] *Id.* at 16.

[300] *Id.* at 14.

[301] *Id.* at 9.

[302] *Id.* at 16.

[303] *Id.* at 8.

members of Al Qaeda operating in Pakistan creates neither a state of war within that territorial State nor a noninternational armed conflict with the non-State actor. An armed conflict against a non-State actor is not necessarily considered to exist, "since the minimal level of war or armed conflict under international law involves an armed conflict not of an international character or an insurgency and some non-state actors, such as al Qaeda, do not meet the test for insurgent status."[304] In adopting this approach, a high threshold for both insurgency and insurgent status is relied on by using criteria often associated with civil wars.[305] Of interest, this threshold appears to be higher than the one applied by proponents of the restrictive "law enforcement" theory.

The reliance on this high threshold for the establishment of an armed conflict means that the application of humanitarian law, as a matter of either treaty or customary law, is significantly restricted. Logically, the resulting void would then have to be filled. It appears there are three options: by human rights law; by the adoption of an approach where humanitarian law norms are applied a matter of policy (see Chapter 4, section 4.4.4, *A Policy Solution for Legal Uncertainty*); or by means of a new interpretation of the self-defense paradigm. It appears the latter avenue is followed in creating what might be called the "robust" paradigm. The application of human rights law is excluded on the basis that the persons being targeted are not within the territorial jurisdiction of the responding state or "within the actual 'power and control' of the responding state."[306] In effect, there is no extraterritorial application of international human rights treaty law, and no reference is made to customary human rights law or even peremptory (i.e., *jus cogens*) legal principles. It is stated to be evident that "human rights protections do not pertain and that a human rights paradigm is not directly relevant."[307] Further, to the extent it is assumed human rights law does apply to a particular person, the loss of life would not be arbitrary "if it occurs as part of a permissible self-defense response, such a targeting will be reasonably necessary, rational and not arbitrary."[308]

The perceived gap between the law applicable to an armed conflict and the restricted view of the application of human rights law might have been attempted to be filled by a policy approach. Instead, the robust version of the "self-defense" paradigm claims a legal "home" in the law governing State self-defense. Any void is filled by reliance on "the test with respect to permissibility of particular measures of self-defense . . . ."[309] The question, then, is what standard of necessity and proportionality was to be applied. Here, there seems to be a broad importation of the humanitarian principles of necessity and proportionality in place of the standard self-defense ones. It is indicated that it is the "[g]eneral principles of reasonable necessity and proportionality":

> integrated into several provisions of Geneva law applicable to an international armed conflict that will condition the permissibility of actual measures taken in

---

[304] Paust, *supra* note 288, at 259–60.

[305] *Id.* at 260 n.52 and n.53 (particular reliance is placed on criteria associated with AP II, art. 1).

[306] *Id.* at 265.

[307] *Id.*

[308] *Id.* at 266–67.

[309] *Id.* at 269.

self-defence during an international armed conflict and *they provide useful guidance with respect to methods and means of self-defense in other contexts because all measures of self-defense must comply with the same general principles.*[310]

The interface between the humanitarian law principles and the State self-defense principles of necessity or proportionality seems unclear.

Of particular interest is the suggestion that the principles of the humanitarian law govern, as a matter of guidance, even in situations where there is no armed conflict. In contrast, the naked self-defense theory relies on customary law, albeit when conflict has not reached the threshold of a noninternational armed conflict. The more robust paradigm has the same practical effect as a policy-based approach (see Chapter 4, section 4.4.4, *A Policy Solution for Legal Uncertainty*). It is humanitarian law–based principles rather than human rights–based law enforcement that applies to the use of force by a State against a non-State actor even though no armed conflict is considered to be in existence.

Like the restricted "law enforcement" and permissive conduct of hostilities paradigms, the "naked" and "robust" self-defense approaches also raise a number of unresolved issues. Geoffrey Corn is critical of the concept of a separate framework of self-defense targeting. From his perspective, the "self-defense" paradigm represents an improper injection of recourse to war principles for the governance of activity that clearly falls within the law governing the conduct of hostilities. This is because such attacks are conducted as part of an armed conflict. In his analysis, the State self-defense principles are equated to self-defense in criminal law[311] (i.e., police powers[312]) and that those principles have "never before been viewed as principles to regulate operational and tactical execution."[313] This criticism is made notwithstanding the apparent reliance in the "robust" version of the theory on the principle of distinction, a humanitarian law concept, and the reference to the principles of necessity and proportionality (as articulated in the "Geneva Law") being applied as a matter of guidance. However, the lack of clarity in just how the established State self-defense and humanitarian law principles interact with one another under the self-defense-based theories makes this an understandable concern. It appears that the Corn critique is focused on identifying clear bodies of law rather than relying on customary norms or the application of principles in unique circumstances in order to outline the rules governing the use of drones.

That analysis also highlights a particular challenge presented by the two self-defense-related theories. Any approach that seeks to apply humanitarian law outside the context of an armed conflict enters into the territory of overreach, leaving it open to considerable criticism. It is problematic to argue that the "law of armed conflict" should apply outside of armed conflict.[314] At the heart of the issue of whether a separate self-defense paradigm is necessary is

---

310 *Id.* at 270 (emphasis added).

311 Corn, *supra* note 288, at 67.

312 *Id.* at 74 ("For example, are there situations where a State when asserting the right of national self-defense is obligated by the *jus ad bellum* [State self-defense] proportionality requirement to rely on police powers instead of combat power?").

313 *Id.* at 59. *See also id.* at 67 ("as operational parameters, they impose a peacetime self-defense model onto wartime employment of combat power.").

314 Prosecutor v. Tadić, Case No. IT-94-1-A, Decision On The Defence Motion For Interlocutory Appeal On Jurisdiction, ¶ 67 (Int'l Crim. Trib. for the Former Yugoslavia, Appeal Chamber, 2 October 1995), http://www.icty.org/x/cases/

the question of when an armed conflict is actually in existence. As has been noted, this issue is open to considerable debate among international lawyers. The naked self-defense theory appears to have been subsequently altered and ultimately relegated to the status of an exception by the acceptance that an armed conflict exists in territory such as Yemen. The robustness of the second version of the "self-defense" paradigm is itself directly tied to the setting of a high bar for the existence of an armed conflict. If the threshold for armed conflict were established at law as being significantly lower than a civil war, this theory would also have to be considered as having more of an exceptional status. Ultimately, the two "self-defense" theories appear to be affected by the long-standing inability of the international community to reach consensus regarding when and where noninternational armed conflict is occurring.

Interestingly, Geoffrey Corn's analysis places significant emphasis on what constitutes an armed conflict. In doing so, he introduces the perspective of a practitioner. A key element is the suggestion that an armed conflict is governed by the decision of a State to use combat power against "non-State operatives".[315] It could be called a "means and authority" test. This involves a broader assessment of what constitutes an armed conflict than either the naked self-defense or robust approach. It is presented as a pragmatic approach since "the nature of these operations makes them inconsistent with peacetime law enforcement legal principles."[316] Therefore, when a State "employs combat power in a manner that indicates it has implicitly invoked LOAC principles (by employing deadly force as a measure of first resort), it is engaged in an armed conflict."[317] It is indicated that while not every State employment of military units qualifies as an armed conflict, the use of those forces combined with "[LOAC-based]engagement authority indicates they are using the 'tools' of war, and must respect, at a minimum, *the core principles of the 'rules' of war*."[318]

The employment of the "combat power" test for establishing the existence of an armed conflict is itself open to criticism.[319] Accepting that, in practical terms, the use of military forces can be a useful indicator of an armed conflict, this approach appears to leave that determination uniquely in the hands of the acting State. It does not address the problem of an overreaction by that State, or consider the influence that State self-defense principles are seen to have by other analysts even in "short of war" situations (see Chapter 3, section 3.4, *A More "Limited" Theory*). It also does not overtly apply the established *Prosecutor v. Tadić* criteria for establishing the existence of an armed conflict or the existence of an organized armed group, and intensity of armed violence.[320] However, when

---

tadic/acdec/en/51002.htm ("International humanitarian law governs the conduct of both internal and international armed conflicts. Appellant correctly points out that for there to be a violation of this body of law, there must be an armed conflict."). *But see* ILA Use of Force Report, *supra* note 162, at 31 (where it is suggested in a self-defense context "[e]ven in the absence of armed conflict, a member of the armed forces may invoke IHL to justify the use of lethal force.").

315 Corn, *supra* note 288, at 64.

316 *Id.* at 63.

317 *Id.* at 74.

318 *Id.* (emphasis added).

319 Dinstein, *supra* note 135, ¶ 109, at 36 (where it is noted that the deployment of military forces is not necessarily indicative that an armed conflict is in existence).

320 Prosecutor v. Tadić, Case No. IT-94-1-A, Decision On The Defence Motion For Interlocutory Appeal On Jurisdiction, ¶ 70 (Int'l Crim. Trib. for the Former Yugoslavia, Appeal Chamber, 2 October 1995), http://www.icty.org/x/cases/tadic/acdec/en/51002.htm.

the involvement of military forces is used as one of the factors to be considered, along with the nature of the enemy force, it is a criterion that is very helpful in establishing the existence of an armed conflict. The suggestion that this criterion be applied as one of the factors when considering if an armed conflict is occurring helps bridge the theoretical and practical approaches and reintroduces a factor that historically has been considered part of the "armed conflict" test (see Chapter 9, section 9.4.3, *The "Totality of the Circumstances" and Other Criteria*).[321]

As has been noted, the robust "self-defense" option is particularly affected by its being based on the view there is a very high threshold for the existence of an armed conflict. In his analysis of the third self-defense paradigm regulating the use of force, Ian Henderson applies a similarly high threshold. However, he also notes "there is no legal reason why counter-terrorist operations in response to ongoing terrorist-style attacks cannot amount to an armed conflict."[322] The key issue is what constitutes "protracted armed violence."[323] As was discussed in respect of the restricted "law enforcement" theory, this issue requires more attention. Even "one-off" terrorist attacks by organized armed groups can be of sufficient intensity to warrant their being considered to have occurred in the context of an armed conflict (see Chapter 9, section 9.4, *Determining the Existence of a Noninternational Armed Conflict*). The adoption of a lower threshold for armed conflict not only reflects the reality on the ground, it also undercuts the requirement for a separate self-defense option.

Even though there are differences between naked self-defense, the robust paradigm, and the Corn critique, there is also a commonality to these analyses. They adopt humanitarian law as the appropriate legal framework to govern drone strikes, as opposed to human rights–based law enforcement norms. There is an acknowledgment that a self-defense-based "use of armed force might be subject to law enforcement–type use of force authority," however, it is viewed as not providing a fully adequate framework to address the threat of transnational terrorism.[324] To a large extent the self-defense-based approaches are focused on "gap filling" brought on by a view that human rights–based law enforcement is not up to the task of protecting the interests of the threatened State. It could be argued such an interpretation appears to downplay the potential effectiveness of a law enforcement approach when it is applied with legal interpretations that permit it to address the upper ends of violence while at the same time not distorting its fundamentally restrictive character (see Chapter 12, section 12.3.4, *Strictly Necessary, Proportionate, and Overwhelming*).

Further, the "self-defense" option does not provide a complete answer for dealing with conflicts where the violence occurs at the boundaries of or in the overlap between the law enforcement and conduct of hostilities paradigms. This discussion surrounding the

---

321 Commentary, Convention (IV) relative to the Protection of Civilian Persons in Time of War. Geneva, 12 August 1949 Common Article 3, ¶ 1. A. 2., at 35 (Jean S. Pictet ed., 1960). *See also* Prosecutor v. Boškoski, Case No. IT-04-82-T 90, Judgement, ¶ 190 (Int'l Crim. Trib. for the Former Yugoslavia, Trial Chamber Decision, 2008), http://www.icty.org/x/cases/boskoski_tarculovski/tjug/en/080710.pdf (emphasis added) (for reference to the engagement of the armed forces with terrorists as a factor to be considered in determining there is an armed conflict).

322 Henderson, *supra* note 287, at 163.

323 *Id.* at 164.

324 Corn, *supra* note 288, at 74.

"self-defense" theory appears to be exclusionary in nature. In any event, the U.S. policy decision to consider capture before killing terrorists when they are located in areas away from an area of active hostilities points to a different solution for reconciling the perceived gaps between the application of human rights–based law enforcement and conduct of hostilities norms than is presented by these analyses.[325] It is an approach that more readily relies on domestic law and a policy solution for bridging any perceived gap.

8.6. Seeking a Solution

The certainty with which the three drone-related theories—the "law enforcement," "conduct of hostilities," and "self-defense" approaches—are described reflect the existence of fundamental and long-standing disagreements concerning the interpretation of international law in the realm of State security. These theories appear to be affected by the same factors that influenced the twentieth-century recourse to war debate. For example, the restricted approach seems to have an "antiwar" influence, which was fundamental in the creation of the legal framework governing the recourse to war found in the UN Charter (see Chapter 2, section 2.3.2, *Inter-State Conflict: Limiting the Recourse to War*). Significantly, it also places reliance on the proper authority principle, which privileges States dealing with threats emanating from within their borders (see Chapter 4, section 4.2, *Humanitarian Law and "Proper Authority,"* and section 4.3, *Impact of the Proper Authority Principle*). While the restrictive "law enforcement" approach has a solid basis in twentieth-century legal interpretations, it also carries with it some inherent limitations. Being built upon such a restrictive foundation, questions remain regarding the ability of this theoretical construct to permit States to adequately deal with twenty-first-century security threats originating from the "ungoverned spaces" of the world. Restrictive recourse to war thinking enjoyed considerable success in the last century when dealing with inter-State security threats. However, contemporary threats are presenting new challenges at the lower end of the conflict scale, which ultimately involve taking action to defend against nontraditional but exceedingly dangerous non-State actor threats.

The more permissive "conduct of hostilities" and "self-defense" theories also appear to be firmly rooted in their own traditional space. In this respect, they focus on a countervailing argument that States are permitted under the UN Charter to act in self-defense. However, these more permissive theories ultimately favor the exercise of the right of self-defense and the application of humanitarian law virtually to the exclusion of its human rights–based law enforcement counterpart. Further, these theories appear to be grounded in the view that either the principles governing the exercise of self-defense have limited or no application to armed conflict once it begins, or they rely on a broad interpretation of the freedom of action that arises from a claimed policy or customary law perspective once the Article 51 UN Charter right to act in self-defense is invoked. On a strategic level these two theories raise questions about the degree to which, after World War II, there has remained a desire by States to have the freedom to engage in armed conflict when they consider it necessary,

---

[325] *Drone Speech*, *supra* note 23.

albeit bounded by humanitarian law. This restrictive versus permissive debate regarding State authority to use force is as old as the UN Charter itself.[326]

Although consistent with the view often favored by powerful States, the permissive "conduct of hostilities" and "self-defense" drone theories are in some respects at odds with the overarching legal regime established in the UN Charter. That regime was fundamentally devised to end transnational State violence. While most widely held interpretations of the right to self-defense seek to limit the State use of military force, these drone-based theories endeavor to expand the envelope of authority for action, notwithstanding that overarching pressure. It is seen in some quarters as an overexpansive use of military force, and one that has attracted considerable criticism and controversy. These are theories that would benefit from greater consideration of the potential for a concurrent application of both the law governing self-defense and humanitarian law when action is taken against non-State actors (i.e., much like what is required in respect of inter-State uses of force; see Chapter 3, section 3.6.3, *Applying Self-Defense Principles Across the Armed Conflict Spectrum*).

The use of drones in the territory of other States causes a "perfect storm" of issues to coalesce: collateral civilian casualties, the use of airpower, and the geographic scope of conflict. Such issues have attracted the attention of the human rights and humanitarian legal communities since the advent of the modern technologically driven warfare at the beginning of the twentieth century. The result has been claims by human rights groups, by the media, and sometimes by the territorial States themselves that there have been unacceptable levels of collateral casualties. The concern over collateral casualties and the suggested "brutality" of the option chosen (i.e., kinetic strikes rather than capture) for dealing with transnational terrorism raises issues that historically have caused significant problems for democracies engaged in small wars.[327] This in turn has led to political pressure to restrain such military violence.

While the "law enforcement" approach may be viewed as being too restrictive, both the "conduct of hostilities" and "self-defense" paradigms carry with them the perception of not only being overly destructive but also too handy a tool that expands the ongoing conflict. There has even been criticism that, although drones can be more discerning in terms of targeting, the perceived lack of disincentives in using these weapons systems will make it "too easy" to go to war.[328] The issue of collateral casualties alone raises questions concerning the degree to which the permissive approaches are compatible with the more restrained uses of violence that have become associated with contemporary counterinsurgency and counterterrorism (see Chapter 7, section 7.6.2.2, *Proportionality and Counterinsurgency*).

The question remains about how to reconcile the inherent limitations of the restrictive "law enforcement" theory with the perceived overreach of the permissive "conduct of hostilities" and unique "self-defense" theories. Perhaps the answer lies in considering whether

---

[326] Josef L. Kunz, *Individual and Collective Self-Defense in Article 51 of the Charter of the United Nations*, 41 AM. J. INT'L L. 872, 876–77 (1947).

[327] GIL MEROM, WHY DEMOCRACIES LOSE SMALL WARS 24 (2003).

[328] Kenneth Anderson, *Efficiency in Bello and ad Bellum: Targeted Killing Through Drone Warfare* 7 (2011), *available at* http://papers.ssrn.com/sol3/papers.cfm?abstract_id=1812124 (last viewed June 16, 2015) (for a vigorous rebuttal of this line of argument.). *See also* Aerial Drone Deployment on 4 October 2010 in Mir Ali/Pakistan, 157 I.L.R. 722, 756–57 (2013) (where the argument that a remote operation of drones will lead to a hasty decision to attack is rejected).

stark lines can be drawn between the approaches dealing with transnational terrorism. For example, it is evident in Geoffrey Corn's critique of the naked self-defense theory that the relationship between terrorism and law enforcement in a transnational context is a complicated one. It is noted by relying on the *Targeted Killing Case*[329] "there are plausible arguments that the nature of some self-defense missions might justify a more restrictive operational framework based on a hybrid of LOAC and law enforcement principles."[330] The application of such a hybrid approach depends on "the nature of the threat/response continuum" rather than an application of self-defense principles.[331] This comment highlights that the boundaries between suggested frameworks governing drone strikes may not be as absolute as they appear at first glance. Further, the practical response may result from an application of an amalgam of legal principles, whether as a matter of law or policy. The response available to and taken by States will often depend upon where the threat is situated on the conflict spectrum. Some terrorist threats, even those occurring in the context of an armed conflict, are fully amenable to a law enforcement response.

A key factor influencing the counterterrorism discussion has been the responsibility of States that arises from the performance of a governance or proper authority role. Its most obvious manifestation in the context of drones is whether a territorial State can or is willing to carry out its responsibilities for governance and therefore ultimately "police" its own territory. The question remains as to whether the requirement to apply governance principles begins and ends with the territorial boundaries of a State. It appears not necessarily, since the United States itself has articulated a requirement to consider capture before killing a U.S. citizen who is a terrorist because of domestic legal obligations that are viewed as being international in reach. That responsibility toward its own citizens arises even if they are operating from ungoverned spaces of the world. As President Obama noted, he "did not believe it would be constitutional for the government to target and kill any U.S. citizen—with a drone, or a shotgun—without due process."[332]

However, the United States has gone further and extended the protection of capture being considered before conducting a drone strike to any individual terrorist involved with Al Qaeda and its associated forces. This extension does not appear to be based on an acknowledgment of constitutional obligation,[333] but rather because it "respects the inherent dignity of every human life."[334] The result is very similar to that set out in the Israeli *Targeted Killing Case*.[335] It is not clear the unique protection provided U.S. citizens by their country would be broadly accepted as being based in international law or would necessarily be supported by all domestic jurisdictions.[336] As presented, the reference to respecting the

---

329 Public Committee Against Torture in Israel v. Israel, Israel Supreme Court [16 December 2006], 46 ILM 375 (2007).

330 Corn, *supra* note 288, at 75.

331 *Id.*

332 *Drone Speech*, *supra* note 23.

333 *Id.* (see where it is stated "[o]f course the targeting of any Americans raises constitutional issues that are not present in other strikes. . . .").

334 *Id.*

335 Public Committee Against Torture in Israel v. Israel, Israel Supreme Court [16 December 2006], 46 ILM 375 (2007)

336 Issacharoff & Pildes, *supra* note 69, at 1587 (where it is noted "as a matter of international law and the domestic law of some countries, providing greater protections to one's own citizens in the terrorism context can be a reason to condemn, not praise. . . .").

dignity of all human life provides a powerful normative basis for that policy decision. The concept of the dignity of human life is one that applies in both human rights and humanitarian law. In this context, a requirement to consider capture before killing of a terrorist has a clear human rights connotation. The position adopted by the United States does not appear to be a statement that human rights law governs drones strikes. However, it does reflect the advancement of a general trend, first seen in the *Targeted Killing Case*,[337] but then also reflected in Part IX of the *ICRC Interpretive Guidance* of incorporating human rights–based law enforcement norms into interpretations of humanitarian law under defined operational situations (see Chapter 7, section 7.4.2, *The "Targeted Killing" Case and Control*).[338] Such an incorporation of human rights–based law enforcement norms might not have been necessary in the case of drone strikes if the "conduct of hostilities" theory was more clearly bound to the full nuance of neutrality law by incorporating its more restrictive aspects regarding the use of force.

The U.S. position on drone strikes closely resembles the human rights law–based enforcement paradigm. This is reflected in President Obama's statement that American citizenship "should no more serve as a shield than a sniper shooting down on an innocent crowd should be protected from a swat team."[339] A more restrictive approach can also be seen in his indication no strikes would occur unless there was a "near certainty" there would be no collateral civilian casualties.[340] This outcome also has a resonance with the framework suggested by the *Oxford Research Paper*, which foresaw a use of drones in a law enforcement context.[341] What is not clear is whether the stated goal of being able to restrict collateral casualties to this extent is a realistic or attainable one. It is an approach that is relatively easy to adopt in a context where it is believed the threat has been significantly downgraded, such as in Pakistan. However, within three months of it being publicly announced there were already indications that drones would continue to be used in Yemen.[342] To the extent the targeting restrictions are viewed as being policy based, if the "U.S. government wants to up the pressure and return to the 2009–2010 heyday of the decade-long drone war, there is nothing stopping it."[343] As a narrative produced to answer criticism of the drone campaign this revised counterterrorism approach does not appear to have survived first contact with a misunderstood and regenerated enemy. As has been noted, it was disavowed in respect of the 2014 air campaign against the Islamic State.

The "virtual" adoption of a human rights–based approach, albeit within a conduct of hostilities paradigm, raises a fundamental question as to why greater reliance could not be placed on human rights law in a counterterrorism context. Certainly, the limits on the use of deadly force under that normative paradigm could be a deciding factor. It may be that the approach toward the use of drones remains anchored in the armed conflict paradigm

---

337 Public Committee Against Torture in Israel v. Israel, Israel Supreme Court [16 December 2006], 46 ILM 375 (2007).

338 Interpretive Guidance, *supra* note 15, at 77–82.

339 *Drone Speech, supra* note 23.

340 *Id.*

341 Oxford Research Paper, *supra* note 135, at 9.

342 Elias Groll, *The Sudden and Unexpected Return of the Drone War*, Foreign Pol. Passport (Aug. 8, 2013), http://blog.foreignpolicy.com/posts/2013/08/08/the_sudden_and_unexpected_return_of_the_drone_war_yemen.

343 *Id.*

because of a view that law enforcement is seen as being more tied to and limited by sovereignty to the territorial State. As was noted in the *Drone Speech*, the United States could not place "sole reliance on law enforcement in territories that have no functioning police or security services—and indeed no functioning law."[344] The effectiveness of law enforcement within the borders of a State has an effect on choices made by States.

Looking at law enforcement in too limited a fashion is problematic. While the contemporary discussion has been largely framed by very traditional views of the State self-defense and humanitarian law, there has been a long history of the extraterritorial use of force in defense of the State and its people. Many of these operations occur at the lower end of the conflict spectrum. In this regard, "[c]ontemporary international practice abounds with incidents in which one country uses force within the territory of another, in order to protect or rescue nationals, while invoking self-defence."[345] Such activity has included the extraterritorial exercise of a policing function (see Chapter 10, section 10.2.2, *Taking Action Across the Conflict Spectrum*).

The analysis also needs to consider the full impact of the human rights–based law enforcement paradigm. Looking closely at the ability of the law enforcement framework to regulate violence at the lower end of the conflict scale highlights there is also a significant overlap between human rights and humanitarian law. At this point the options available to a threatened State are no longer exclusionary in nature, but rather allow conscious decisions to be made as to which body of law to apply. To the extent human rights law can effectively regulate terrorist activity not only at home but also in the "ungoverned spaces" of the world, a greater range of less violent options is available to States seeking to counter these threats (see Chapter 14, section 14.5, *Confronting Transnational Violence: A Holistic Approach*). A better understanding of the robustness of law enforcement also increases the likelihood that States will be able to adopt a police primacy approach, not only when dealing with insurgences but also in countering acts of cross-border violence by terrorist organizations.

8.7. An Emerging Broader "Unwilling or Unable" Justification

With a continuing transnational terrorist threat and the increasing availability of drones, what remains to be determined is which of these three legal bases for drone strikes will be accepted in the long run by the international community. This choice is important, not only in respect of drones but also because it forms the basis for other counterterrorism action such as Special Forces raids and a broader use of airpower. The "self-defense" option appears to have been developed because of uncertainty over the application of the two other theories. This approach is consistent with the common State approach of applying a policy solution using humanitarian law principles to fill perceived voids in international law regulation (see Chapter 3, section 3.3.4, *A Policy Solution for Legal Uncertainty*). However, the "self-defense" option is likely to lose considerable traction, assuming it gains a sufficient following, if greater consensus forms around the other two theories, which rely directly on more

---

[344] *Drone Speech*, *supra* note 23.

[345] DINSTEIN, *supra* note 185, ¶ 677, at 256.

widely accepted law to regulate the use of force. The greatest challenge for the restrictive "law enforcement" theory is the security void that it appears to leave in place if the formal criteria for its invocation are not met. This leaves the "unwilling or unable" justification.

One indication that the unwilling or unable criterion is gaining broader application, although far from complete acceptance, is the reliance placed on that theory as the basis for conducting an air campaign against the Islamic State in Syria. In this instance it is relied on for a much broader military effort, rather than be limited to counterterrorism. As such it also appears to move the issue away from reliance on neutrality-based legal analogies for isolated applications of force toward more general participation in an armed conflict on Syrian territory. There has been significant debate regarding what should be the legal base for Coalition air strikes. The options discussed include a Security Council Resolution, consent from the Syrian government, self-defense (including anticipatory and collective self-defense), or acceptance the threat posed by the Islamic State as part of an ongoing armed conflict against Al Qaeda and "affiliated" groups.[346] Each of these legal bases presents their own challenges. Political reasons have precluded consent being sought or provided by Syria for military action other than by Russia.

The option of a UN Security Council authorization was initially viewed as highly unlikely given the "toxic" relationship between the United States and Russia,[347] although the 2015 Islamic State terrorist attacks in Tunisia, Egypt, Turkey, Lebanon and France resulted in a November 2015 UN Security Council Resolution 2249 authorizing the use of "all means necessary" to eradicate the safe haven created in Syria and Iraq.[348] Debate erupted over what action that Resolution actually authorized with the view being expressed it was constructively ambiguous allowing States to rely on previously stated legal positions for acting.[349] This debate highlights not only the political complexity of the situation, but also the degree to which many international lawyers apply a "formalist" traditional approach to interpreting use of force authorizations. It is one that places a premium on restricting legal authority for the State use of force, and as such reflects the historical underpinnings of the recourse to war provisions of the UN Charter. Therefore, although Resolution 2249 called upon States with sufficient capacity "to take all necessary measures", the lack of reference to "acting under Chapter VII"[350] and "decide" has been suggested to mean there was no specific authorization for the use of force.[351]

What is clear is that taking no action against the Islamic State, Al Qaeda and other terrorist groups in Syria is also highly problematic and States have been increasingly ready to act under a number of legal bases. Permitting the Islamic State to conduct operations in

---

[346] Ashley Deeks, *U.S. Airstrikes Against ISIS in Syria? Possible International Legal Theories*, LAWFARE (Aug. 23, 2014, 3:04 PM), http://www.lawfareblog.com/us-airstrikes-against-isis-syria-possible-international-legal-theories. *See also* Arimatsu & Schmitt, *supra* note 239.

[347] Deeks, *supra* note 346.

[348] S.C. Res. 2249, UN Doc. S/RES/2249 (Nov. 20, 2015).

[349] Dapo Akande & Marco Milanović, *The Constructive Ambiguity of the Security Council's ISIS Resolution*, EJIL: *TALK!* (Nov. 21, 2015,), http://www.ejiltalk.org/the-constructive-ambiguity-of-the-security-councils-isis-resolution/.

[350] *Id.*

[351] Ashley Deeks, *Threading the Needle in Security Council Resolution 2249*, LAWFARE (Nov. 23, 2015, 3:23 PM), https://www.lawfareblog.com/threading-needle-security-council-resolution-2249 (where reference is also made to the word "decides" not being used "as the active verb in the paragraph that authorizes force.").

Iraq while maintaining a safe haven in Syria makes it extremely difficult to counter that organized armed group.[352] The problem is that traditional interpretations of international law limiting State action can create a situation much like the restricted "law enforcement" drone theory where a non-State actor is left free to prepare for and conduct attacks from a territorial safe haven.[353] The application of the "unwilling or unable" criterion to justify strikes against the Islamic State and Al Qaeda–linked groups more closely reflects the practical challenge facing States. It is the unwillingness or inability of a territorial State to act that creates the conditions for which defensive action by other States is considered necessary.

That is the basis that the United States presented to the United Nations for the military action it has taken in Syria since 2014.[354] It has been suggested the "unwilling and unable" approach is unlikely to find broad or strong acceptance among many of the United States' allies.[355] A number of participants in the Coalition against the Islamic State initially restricted their military activities to Iraq, where operations were being conducted with the consent of that government. However, what stood out was the limited international condemnation of the U.S. approach.[356] The fact that the bombing campaign in Syria initially included other Coalition partners, such as Bahrain, Jordan, Saudi Arabia, and the United Arab Emirates, provided an indication that there is a growing acceptance of the legal authority to do so.[357] A March 2015 decision by the Canadian government to expand its aerial bombing operations into Syria,[358] followed by Australia[359] and France,[360] provided yet further evidence. In addition to acting in self-defense Turkey, Canada and Australia also made reference to "the controversial 'unwilling or unable' test in their letters to the

---

352 Abdulkader H. Sinno, Organizations at War in Afghanistan and Beyond 13 (2008) (discussing the importance of safe havens).

353 Mary Ellen O'Connell, *Obama's Illegal War*, Politico Mag. (Sept. 11, 2014), http://www.politico.com/magazine/story/2014/09/obamas-illegal-war-110863.html#.VHXWn4uBBRc ("The only lawful way around obtaining consent from the head of state is if the United Nations Security Council authorizes strikes against ISIL in Syria.").

354 *Letter to United Nations Secretary General from United States Ambassador Samantha J. Power, dated 23 September, 2014, reproduced in* Marty Lederman, *The War Powers Resolution and Article 51 Letters Concerning Use of Force in Syria Against ISIL and the Khorasan Group [UPDATED to add statement of the U.N. Secretary-General]*, Just Security (Sept. 23, 2014, 3:21 PM), http://justsecurity.org/15436/war-powers-resolution-article-51-letters-force-syria-isil-khorasan-group/.

355 Colum Lynch, *Obama Hints at Legal Rationale for Air Strikes in Syria*, Foreign Pol'y (Aug. 28, 2014, 05:11 PM), http://thecable.foreignpolicy.com/posts/2014/08/28/us_airstrikes_in_syria_would_raise_legal_questions.

356 Ryan Goodman, *Australia, France, Netherlands Express Legal Reservations about Airstrikes in Syria [Updated]*, Just Security (Sept. 25, 2014, 8:01 AM), http://justsecurity.org/15545/australia-france-netherlands-express-legal-reservations-airstrikes-syria/.

357 *Nov. 24: Military Airstrikes Continue Against ISIL in Syria and Iraq*, U.S. Central Command News Release (Nov. 24, 2014), http://www.centcom.mil/en/news/articles/nov.-24-military-airstrikes-continue-against-isil-in-syria-and-iraq.

358 *MPs Vote to Expand Canada's Military Mission Against ISIS into Syria*, CTV News (Mar. 30, 2015 10:22 PM), http://www.ctvnews.ca/politics/mps-vote-to-expand-canada-s-military-mission-against-isis-into-syria-1.2303656.

359 Matthew Doran, *Australian Fighter Jet Carries Out First Air Strike Inside Syria*, ABC (Sep. 15, 2015, 8:29 PM), http://www.abc.net.au/news/2015-09-16/australian-fighter-jets-first-air-strikes-syria-us-military/6779104. *See also* Shalailah Mededhora, *Australian Jets Diverted from Syria as Russia's Entry Complicates Mission*, The Guardian (Oct. 7, 2015, 3:42 BST), http://www.theguardian.com/australia-news/2015/oct/07/australian-airstrikes-in-syria-on-hold-as-russias-entry-complicates-mission (where it is reported Australia paused operations in part because they were not needed).

360 *France Launches New Air Strike in Syria Against IS: Minister*, Reuters (Oct 9, 2015, 8:11 AM EDT), http://www.reuters.com/article/2015/10/09/us-france-syria-airstrike-idUSKCN0S30HJ20151009 (the first strike was on Sept. 27, 2015).

UN."[361] In November 2015 the British Prime Minister relied on that test in explaining its rationale for acting in collective and individual self-defense under Article 51 as the bases for the carrying out proposed air strikes in Syria. This action was "further underscored by the unanimous adoption of UN Security Council Resolution 2249."[362]

Controversy remained. A new Canadian government elected in November 2015 promised during the election campaign to withdraw from participation in air strikes in both Iraq and Syria.[363] However, the decision to withdraw from air operations in both countries suggests more of an ideological problem with air strikes than simply a disagreement regarding the legal basis for attacks in Syria. Such a decision will not necessarily avoid Canadian ground participation in combat (even if acting in self-defence), or make the Islamic State view Canada as anything other than the "far enemy".[364] A similar reticience to be involved in "actual" combat occurred when Germany voted in December 2015 to provide military support including reconnaissance and refueling aircraft to assist air strikes, but not actually drop bombs.[365] However, it is difficult to see how refueling strike aircraft is not "participating in the air strikes".[366] In contrast, the United Kingdom commenced air strikes that same month.[367]

The use of airpower to strike at Islamic State targets highlights once again that drones are not necessarily that unique a weapons system. Neither is the "unable or unwilling" test specifically a drone-related issue. This was evident when the U.S. Special Forces units entered Syria in August 2014 in an unsuccessful attempt to rescue one of its citizens being held hostage by the Islamic State[368] and the May 2015 operation seeking to capture Abu Sayyaf.[369]

---

361 Jonathon Horowitz, *A Legal Map of Air Strikes in Syria (Part 1)*, JUST SECURITY (Dec. 7, 2015, 1:20 PM, https://www.justsecurity.org/28167/legal-map-airstrikes-syria-part-1/. *See also* Letter dated 24 July 2015 from the Chargé d'affaires a.i. of the Permanent Mission of Turkey to the United Nations addressed to the President of the Security Council, U.N. Doc. S/2015/583 (July 24, 2015)("the regime in Syria is neither capable of nor willing to prevent these threats emanating from its territory"), Letter dated 31 March 2015 from the Chargé d'affaires a.i. of the Permanent Mission of Canada to the United Nations addressed to the President of the Security Council, U.N. Doc. S/2015/221 (Mar. 31, 2015) ("States must be able to act in self-defence when the Government of the State where a threat is located is unwilling or unable to prevent attacks emanating from its territory.") and Letter dated 9 September 2015 from the Permanent Representative of Australia to the United Nations addressed to the President of the Security Council, U.N. Doc. S/2015/693 (Sept. 9, 2015)("Syria, demonstrated that it is unwilling or unable to prevent those attacks [in Iraq]".).

362 *PM Statement Responding to FAC Report on Military Operations in Syria*, Oral Statement to Parliament (Nov. 26, 2015), https://www.gov.uk/government/speeches/pm-statement-responding-to-fac-report-on-military-operations-in-syria.

363 Catherine Cullen, *Justin Trudeau Says Canada Will 'Do Its Part' in Anti-ISIS Coalition*, CBC NEWS (Nov 16, 2015 8:02 PM), http://www.cbc.ca/news/politics/paris-justin-trudeau-isis-mission-canada-g20-1.3320526.

364 *Canadian forces strike back against ISIS offensive in northern Iraq*, CBC NEWS (Dec. 16, 2015, 1:16 AM ET), http://www.cbc.ca/news/politics/canada-iraq-isis-offensive-1.3370820.

365 Alison Smale, *German Parliament Votes to Send Military Assistance to Fight ISIS*, INT'L N.Y. TIMES, Dec. 4, 2015, http://www.nytimes.com/2015/12/05/world/europe/german-parliament-military-isis-syria.html?_r=0.

366 *German Bundeswehr takes part in first Syrian missions*, DW (Dec. 16, 2015), http://www.dw.com/en/german-bundeswehr-takes-part-in-first-syrian-missions/a-18920073.

367 Patrick Wintour, *Britain Carries Out First Syria Airstrikes After MPs Approve Action Against ISIS*, THE GUARDIAN (Dec. 3, 2015, 8:53 GMT), http://www.theguardian.com/world/2015/dec/02/syria-airstrikes-mps-approve-uk-action-against-isis-after-marathon-debate?CMP=Share_iOSApp_Other.

368 Michael D. Shear & Eric Schmitt, *In Raid to Save Foley and Other Hostages, U.S. Found None*, N.Y. TIMES, Aug. 20, 2014, http://www.nytimes.com/2014/08/21/world/middleeast/us-commandos-tried-to-rescue-foley-and-other-hostages.html.

369 Cooper & Schmitt, *supra* note 58. *See also* NAYLOR, *supra* note 26, at 313–24 (this is not the only time that United States special operations forces are said to have entered Syria. For example, on October 28, 2008 a raid into Syria killed Abu Ghadiya, who was running a network funnelling foreign fighters, including suicide bombers, into Iraq).

These operations highlight once again that counterterrorism can involve a much broader range of military assets than unmanned aircraft. Further, the expansion of military action to strategic air strikes against Islamic-controlled oil production facilities provides a further indication that insurgency and terrorism can be closely related.[370] In this case, the Islamic State not only carries out terrorist acts but also fields a significant insurgent military force. In the final analysis, like the recognition that States can act in self-defense against non-State armed groups, this conflict provides an indication the "unwilling or unable" justification may very well gain broader, if grudging, acceptance within the international community. Traditionalists will be fighting a rear guard action in seeking to deny its applicability.[371] What needs to be reinforced is that such action is not unfettered. The overarching application of State self-defense principles; consideration of feasible alternatives; applying a human rights–based policing approach where required (i.e. persons not taking a direct part in hostilities) or feasible and a sensitivity toward the narrower acceptance of collateral casualties when targeting occurs in a counterinsurgency or counterterrorist context—all need to be brought to bear in determining how these operations are carried out and how States are held accountable.

---

370 Kenneth Watkin, *Targeting "Islamic State" Oil Production Facilities*, 90 INT'L L. STUD. 499 (2014).

371 Jonathon Horowitz, *A Legal Map of Airstrikes in Syria (Part 2)*, (Dec. 8, 2015, 1:55 PM), https://www.justsecurity.org/28199/legal-map-airstrikes-syria-part-2/ (where in respect of Australia, Turkey, Canada and the US relying on the "unable or unwilling" test to justify airstrikes in Syria it is stated "[w]hether this is indicative of a broader trend has yet to be fully seen, but if it is, then a considerable amount of work must be done to define its parameters."

Despite the continued existence of large conventional forces, most post-1945 conflicts have been fought along very different lines from those which characterized World War II.[1]

9 Non-State Actors and Armed Conflict

9.1. Introduction

Sorting out the different legal bases for responding to transnational terrorism (see Chapter 8, section 8.5, *Legal Frameworks for Responding to Transnational Terrorism*) does not present the only area of legal uncertainty regarding conflict with non-State actors. International lawyers have found themselves struggling in the post-9/11 period to categorize contemporary conflict between States and their non-State actors protagonists. Despite a tendency to think of violence occurring between States and non-State actors uniquely under the rubric of an internal noninternational armed conflict, it actually spans the spectrum from international armed conflict to criminal acts subject to law enforcement. Increasingly, attention is being paid to non-State actors carrying out cross-border attacks against States, as well as the corresponding defensive action taken by States against terrorists and other groups operating outside its territory.

Transnational conflict has taken on a particularly high profile. In respect of such conflict with non-State actors it is important to consider that the operations associated with the trans-border violence may unfold inside a targeted State (i.e., the 9/11 attacks involving aircraft hijacked in the United States), or from outside the borders of a State (i.e., attacks on Coalition forces in Afghanistan by Taliban operating out of Pakistan). External to the targeted State, attacks can also occur in the territorial/host State (e.g. Embassies) where the non-State actors are located, in third States, or even on the high seas (see Chapter 14, section

[1] KEITH SUTER, AN INTERNATIONAL LAW OF GUERRILLA WARFARE 1 (1984).

14.5, *Confronting Transnational Violence: A Holistic Approach*). The transnational aspect of this violence is reflected in a State having to react to threats beyond its borders. Therefore, the defensive response by the threatened State may take place within its own borders or outside its territorial limits on land, at sea, or in the air. Cross-border defensive action can include activity such as counterterrorism raids by Special Forces units, drone strikes, non-combatant evacuation operations, and hostage rescue operations.

The relevance of a law enforcement–based response in the context of non-State actor violence, whether applied as a matter of law or policy, has also attracted considerable attention in the post-9/11 period. In large part this is because those operations have increasingly focused on counterinsurgency and counterterrorism. Frequently, both conduct of hostilities and law enforcement responses must be simultaneously considered and reconciled with one another. This can occur in a number of ways. For example, even when the violence is associated with an armed conflict, the State response may be limited to law enforcement action as a matter of policy. This could even include preferring an exclusively human rights law–based policing response during a counterinsurgency when dealing with attacks within the threatened State (see Chapter 13, section 13.2.1, *The "Troubles"*). A law enforcement response is also a legal requirement even during armed conflict where the actions of the terrorist group do not meet the threshold of armed conflict, or the persons being targeted are neither members of organized armed groups nor direct participants in hostilities.

The range of international operations that have historically been conducted against terrorists and other non-State actors reflect their diverse nature. In respect of hostage rescues alone, such operations have extended from the 1976 Entebbe raid to free Israeli hostages held by Palestinian terrorists who were being assisted, indeed protected, by the government of Uganda[2] (see Chapter 10, section 10.3.2, *"Iconic" Counterterrorism: A Thunderbolt Strikes Entebbe*) to the use of deadly force in the twenty-first century by U.S. Special Operations Forces to rescue citizens seized by Somali pirates engaged in criminal acts on the high seas.[3] What is unique about many counterterrorism operations is the use of military forces to carry out operations that unfold at the interface between armed conflict and law enforcement (see Chapter 10, section 10.2.2, *Taking Action Across the Conflict Spectrum*). Questions about whether an armed conflict is in existence or if law enforcement principles must be applied as a matter of law in controlling the use of force have become key operational law issues.

As will be discussed in this chapter, the challenges facing international lawyers when dealing with this State versus non-State actor violence are twofold. First, there is a significant lack of consensus on how such violence is to be categorized. Second, there is the requirement to determine when the thresholds for the existence of an armed conflict are met. Regarding conflict categorization, it will be demonstrated existing interpretive methodologies are varied. A key issue is whether the armed conflict is an international or non-international one. The categories of non-State actor conflict may include violence being viewed as part of an international armed conflict, where the actions of a terrorist group are attributable to the host State; a similar categorization simply because a foreign State

[2] PETER HARCLERODE, SECRET SOLDIERS: SPECIAL FORCES IN THE WAR AGAINST TERRORISM 345–66 (2000).

[3] Robert D. McFadden & Scott Shane, *In Rescue of Captain, Navy Kills 3 Pirates*, N.Y. TIMES, Apr. 13, 2009, http://www.nytimes.com/2009/04/13/world/africa/13pirates.html?pagewanted=all.

enters the territorial State without the latter's consent to attack the non-State actor; conflict between a State and non-State actor being seen as having an "international character"; as a unique framework of extraterritorial law enforcement; a "noninternational armed conflict" being considered to be in existence, even though it crosses international borders; and a possible assessment the violence falls solely within the law enforcement paradigm.

As will be demonstrated, the complexity of non-State actor conflict has resulted from an array of legal interpretations. The result is that correspondingly different normative rules may have to be applied with some violence with non-State actors being qualified as international in character, while other situations are not. These various legal interpretations are provided in circumstances where, for the practitioner, the threat posed by the organized armed groups often remains substantively the same regardless of the way the conflict is categorized. An overall and in many ways disconcerting lack of agreement among theoreticians on the approach to be adopted when assessing non-State actor conflict points to the very real possibility that the traditional interpretations of threshold criteria developed pre-9/11 to categorize conflict are ill-suited to assess this form of contemporary violence.

What will become evident is that there are common themes in State versus non-State actor conflict, which impact directly on how the violence is viewed regardless of how the conflict is ultimately categorized (i.e., international or noninternational in character). This commonality is centered on that different power relationship between a State on the one hand and a terrorist group or other a non-State actor on the other. This is a relationship that for States is focused on legitimacy and in respect of the non-State actor is often situated within an overarching atmosphere of "criminality." It is the association between non-State actors and criminal activity that has influenced how those groups have been dealt with under international law since the end of the nineteenth century (see Chapter 4, section 4.3, *Impact of the Proper Authority Principle*). In its contemporary form that relationship is also increasingly reflected in the increased profile that law enforcement has taken on in regulating non-State actor violence. Given the diverse nature of the transnational terrorist and insurgent threat, as well as the various theories developed to categorize that threat, the challenge for practitioners is to find an approach that permits States to maintain security across the broad spectrum of potential violence taking place in this complex operating environment.

The second issue dealt with in this chapter is determining when a noninternational armed conflict is in existence. In order for humanitarian law–based conflict of hostilities norms to be applied an armed conflict must, as a matter of law, be taking place.[4] If not, then the use of force by a State will be governed by human rights–based law enforcement norms.[5] This makes the assessment of whether there is an armed conflict occurring an essential part of any operational law analysis. As will be discussed, identifying the threshold at which a noninternational armed conflict exists has proven challenging, and at times controversial. Ultimately, it will be established that the criteria used to determine when such an armed conflict is occurring: intensity of violence (including protracted armed violence) and group organization should be applied with a "totality of the circumstances approach."

---

[4] Prosecutor v. Tadić, Case No. IT-94-1-A, Decision On The Defence Motion For Interlocutory Appeal On Jurisdiction, ¶ 67 (Int'l Crim. Trib. for the Former Yugoslavia, 1995), http://www.icty.org/x/cases/tadic/acdec/en/51002.htm.

[5] Yoram Dinstein, Non-International Armed Conflicts in International Law ¶¶ 98, at 32–33 (2014).

That approach includes matching the threat to the State response necessary to defeat it. Applying this approach results in a threshold for such conflict that is lower than is sometimes suggested to be the case. This has the effect of increasing the situations of overlap between humanitarian and human rights law. The impact of the present lack of theoretical consensus on how conflicts are categorized will be explored in terms of practitioners focusing on whether an armed conflict is in existence, rather than engaging in highly technical, confusing, and frequently divisive debates concerning conflict categorization as being non-international in character.

9.2. Inter-State Conflict

There is a well-developed legal framework governing inter-State conflict. That framework provides a useful means by which to look at the interaction of various bodies of law (e.g., the recourse to war, humanitarian, and human rights law) in such conflict in order to provide a baseline for comparison with the regulatory structure governing conflict involving transnational terrorists and insurgents. A key issue to be resolved is the circumstances under which an armed conflict between States is considered to be in existence. What immediately becomes evident is the exceptionally wide scope given to the concept of inter-State armed conflict. Common Article 2, the four 1949 Geneva Conventions introduced the term "armed conflict" but did not provide a definition.[6] However, the *ICRC Commentary* to Common Article 2 indicates that provision encompasses "*[a]ny difference* arising between two States and leading to the intervention of members of the armed forces."[7] It does not matter how long a conflict lasts between States, how many are killed, or even if there is no fighting: the Geneva Conventions apply.[8] An international armed conflict will be in existence.

This approach reflects the degree to which, post–World War II, the international community was focused on taming inter-State warfare. This had been a traditional recourse to war goal. Enacted in the same era as the reintroduction of the just war provisions of the UN Charter (see Chapter 2, section 2.3.2, *Inter-State Conflict: Limiting the Recourse to War*), this broad understanding of what constitutes armed conflict was designed to help ensure any use of force between States fell within the regulatory framework of humanitarian law. The application of humanitarian law was even extended to "all cases of partial or total occupation of the territory of a High Contracting Party, even if the said occupation meets with no armed resistance."[9] This provision, combined with the development of a separate convention on the protection of civilians, had the effect of significantly increasing the circumstances under which humanitarian law overlaps with its human rights counterpart. In part, that is because at the heart of the Fourth Geneva Convention is the governance role

---

[6] 1949 Geneva Conventions, Common art. 2.

[7] International Committee of the Red Cross, Geneva Convention Relative to the Treatment of Prisoners of War: Commentary, art. 2, at 23 (Jean Pictet ed., 1960) [hereinafter Geneva Convention III Commentary] (emphasis added).

[8] *Id.*

[9] GC IV, art. 2.

performed by an occupying power.[10] The application of human rights–based norms associated with law enforcement and the maintenance of an effective justice system therefore has a particular relevance to conflict with non-State actors (i.e., insurgents, terrorists, or criminals) that can occur in situations of occupation.

Notably, unlike noninternational armed conflict, there is no requirement to assess criteria such as the degree of organization of an armed group or the intensity of violence when considering the threshold for establishing the existence of an international armed conflict.[11] There has been some suggestion that such criteria might be used to determine when an armed conflict occurs between non-State actors who do not belong to a "Party to a Conflict" and an occupying power.[12] Those criteria would be applied even though an occupation is part of an international armed conflict. However, such a threshold would have little relevance to large-scale violent occupations as occurred in Iraq or during World War II. In that situation, the criteria applicable to noninternational armed conflict might best be viewed as providing a practical means, rather than a legal standard, for determining if hostilities have reignited in a part of occupied territory where there is sporadic or renewed hostilities.[13] The involvement of two or more States is sufficient for establishing an international armed conflict is occurring, including during an occupation phase.

The broad interpretation of Common Article 2 of the Geneva Conventions virtually guarantees that an international armed conflict will be in existence when a State takes action in self-defense against another State. It has been noted "the correct legal position appears to be that whenever force is employed in international relations, States are obligated to carry out the norms of the *jus in bello* [humanitarian law] that befit that situation."[14] The wide scope of Common Article 2 is also evident when it is considered in comparison to how Article 51 of the UN Charter has been interpreted by the International Court of Justice. As is set out in *Military and Paramilitary Activities in and against Nicaragua Merits*, the right to self-defense can only be invoked in response to an attack from a State involving "the most grave forms of the use of force."[15] Unfortunately, there is no clear consensus on exactly what uses of force would fall below that threshold. The court did make reference in *Military and Paramilitary Activities in and against Nicaragua Merits* to mere frontier incidents not meeting the required threshold.[16] However, in the *Case Concerning Oil Platforms (Islamic Republic Of Iran V. United States Of America)*, it was ruled that the single act of mining a warship might constitute an armed attack (see Chapter 10, section

---

[10] GC IV, art. 64 (where it is noted the population may be subjected to provisions "which are essential to enable the Occupying Power to fulfil its obligations under the present Convention to *maintain the orderly government of the territory*. . . .") (emphasis added).

[11] Prosecutor v. Tadić, Case No. IT-94-1-A, Decision On The Defence Motion For Interlocutory Appeal On Jurisdiction, ¶ 70 (Int'l Crim. Trib. for the Former Yugoslavia, 1995), http://www.icty.org/x/cases/tadic/acdec/en/51002.htm.

[12] *Third Meeting Of Experts: The Use Of Force In Occupied Territory 29–30 October 2009*, Geneva, *in* OCCUPATION AND OTHER FORMS OF ADMINISTRATION OF FOREIGN TERRITORY, ICRC 109, 122–24, 132–36 (Tristan Ferraro ed., 2012) (for a discussion of hostilities in occupied territory).

[13] Kenneth Watkin, *Use of Force during Occupation: Law Enforcement and Conduct of Hostilities*, 94 INT'L REV. RED CROSS 267, 291 (2012).

[14] YORAM DINSTEIN, WAR AGGRESSION AND SELF-DEFENCE ¶42 at 17 (5th ed. 2011).

[15] Military and Paramilitary Activities in and against Nicaragua Merits, 1986 I.C.J. ¶ 191, at 101 ("it will be necessary to distinguish the most grave forms of the use of force (those constituting an armed attack) from other less grave forms.").

[16] *Id.* ¶195, at 103–04.

10.2.1 *A State Responsibility*).[17] In any event there is ongoing debate within the academic community regarding what circumstances constitute such an armed attack.[18] However, the fact there is a threshold for the establishment of a legitimate exercise of the right of self-defense does not affect the issue of whether humanitarian law, acting as the specialized body of law applicable during armed conflict, governs the inter-State use of force falling below that threshold.

This categorization is so broad that a single use of force involving soldiers from one unit firing in response to a cross-border incident or the shooting down of a military pilot could be considered an armed conflict,[19] regardless of whether the acts were considered to meet the threshold of an "armed attack." Similarly, while there is an accepted view that no basis exists for a State confronted with less grave uses of force to respond in self-defense,[20] should a State do so, humanitarian law would still govern its actions since an armed conflict would be in existence between two States. Further, to the extent there is some support for the view that forceful countermeasures can be employed by one State against another the result would be an international armed conflict.[21]

There is not even a requirement for a forceful response by the State being attacked for an armed conflict to be considered in existence, since an actual use of force is not a prerequisite to the operation of Common Article 2. An armed conflict would exist under humanitarian law where members or units of the opposing armed forces of two States use no violence at all as one country occupies the territory of another.[22] It has also been suggested that even where there is a declaration of war but no engagement by the armed forces of the State in hostilities, then obligations under humanitarian law would arise.[23] Regardless of whether the hostilities between States are legal or illegal (in a recourse to war context), the law governing armed conflict still applies.[24] In can be argued such a broad application of humanitarian law is consistent with the traditional position that the law governing the recourse to war and humanitarian law operate independently of one another. What is noteworthy though is that consistent with their shared history both bodies of international law have a similar goal of regulating State involvement in armed conflict (see Chapter 2, section 2.3 *The Destructive Power of the State*).

There is the issue as to whether limited levels of force associated with minor border incidents should be thought of as "hostilities" at all. In State self-defense terms it has been suggested "if a rifle shot is fired by an Arcadian soldier cross the border of Utopia and the bullet hits a tree or cow, no armed attack has been perpetrated since the action is below

---

[17] Case Concerning Oil Platforms (Islamic Republic Of Iran V. United States Of America), 2003 I.C.J. ¶ 72, at 195.

[18] CHRISTINE GRAY, INTERNATIONAL LAW AND THE USE OF FORCE 189–92 (3rd ed. 2008).

[19] Christopher Greenwood, *The Scope of Application of Humanitarian Law*, *in* THE HANDBOOK OF INTERNATIONAL HUMANITARIAN LAW Rule 45, 47, ¶ 3 at 48 (Dieter Fleck ed., 2nd ed. 2008).

[20] GRAY, *supra* note 18, at 147–48.

[21] DINSTEIN, *supra* note 14, at ¶¶ 550–54, at 208–10 (for a discussion of forceful countermeasures and responses to less grave uses of force).

[22] GENEVA CONVENTION III COMMENTARY, *supra* note 7, art. 2, at 23 ("Even if there has been no fighting, the fact that persons covered by the Convention are detained is sufficient for its application.").

[23] Greenwood, *supra* note 19, ¶ 1 at 47.

[24] LESLIE GREEN, THE CONTEMPORARY LAW OF ARMED CONFLICT 91 (3rd ed. 2008).

the *de minimis* [the law does not concern itself with trifles[25]] threshold. . . ."[26] What is less clear is whether there could be a similar limitation in the terms of establishing the existence of an international armed conflict. Such a limitation would not be inconceivable. It has been noted that countries have not always treated isolated events, such as border clashes and naval incidents, as armed conflicts,[27] although that would not necessarily preclude them from having that status at law. It may be difficult to separate State associated uses of force from individual unauthorized acts, which in substance verge more on the criminal. However, it is possible that where the facts establish an individual criminal rather than a State attributed act, the response would have to be consistent with human rights–based legal principles governing individual acts of self-defense. Although these individual self-defense norms rely on the principles of necessity, proportionality, and imminence, they are not substantively the same as the recourse to war principles, which unfortunately have the same appellation (see Chapter 11, section 11.2.1, *The Terminology Challenge*).[28] In any event, such acts of individual violence would remain isolated.

One outcome from the broad categorization of inter-State action as an armed conflict is that humanitarian law has become a body of law regularly applied to regulate international military operations. It is also the normative regime most often associated with defensive action taken by States in response to an armed attack. This defaulting to an armed conflict–type of response is reflected in the approach adopted by States that relies on humanitarian law as a matter of policy when it is unclear what body of law applies in respect of external deployments of military forces (see Chapter 4, section 4.4.4, *A Policy Solution for Legal Uncertainty*). As has been suggested, this does not mean that human rights–based law enforcement norms will never be applied during inter-State violence. However, particularly during international armed conflict, humanitarian law has often been, as a matter of law and policy, prioritized over human rights law enforcement for the purposes regulating violence. Frequently, in the context of an armed conflict the application of human rights law must be assessed in light of the special legal status of humanitarian law (see Chapter 5, *The Humanitarian Law and Human Rights Law Interface*).

That said, the priority normally assigned to humanitarian law–based responses is altered when States are confronted with a different challenge: fighting non-State actors operating among the people. Again, the security aspects of being an occupying power must be addressed. Although humanitarian law is the specialized body of law applicable during armed conflict, the battle for civilian support, such as when fighting insurgencies, normally results in the privileging of a policing response (see Chapter 7, section 7.3, *The Primacy of Law Enforcement*). It is the question of when human rights–based law enforcement rules must or should be applied to govern the use of force that has particularly captured the attention of the international law community as conflict with terrorists and insurgents has gained a greater profile in this century. Further, a challenging

---

[25] Bryan A. Garner, *Garner's Dictionary of Legal Usage*, Oxford Dictionaries, http://www.oxforddictionaries.com/secondary/garner_dict_legal_usage/de-minimis?q=de+minimis (last visited Mar. 5, 2015) (taken from the Latin maxim *de minimis non curat lex*).

[26] Dinstein, *supra* note 14, ¶ 555, at 210.

[27] Greenwood, *supra* note 19, ¶ 3, at 48.

[28] *See, e.g., UN Basic Principles on the Use of Force.*

issue in determining the appropriate security response is the extent to which terrorist and insurgent groups are themselves engaged in or benefit from criminal activity even in the midst of an international armed conflict. Criminal activity is associated with law enforcement–based responses.

For example, due to the general unrest and lack of security associated with armed conflict, there can be an increase in attacks by purely criminal gangs on citizens of an embattled State, as well as on foreign State nationals present in that State. This not only complicates the assessment of the threat posed by insurgent groups and terrorists, it creates challenges in ensuring the appropriate use of force framework is selected by the State involved. A distinction must be made by State security forces between insurgents and simple criminals when considering the application of conduct of hostilities rules.[29] Ultimately, the responsible State (e.g. an occupying power) must apply domestic law enforcement against the ordinary criminals and may only use conduct of hostilities targeting rules when acting against organized armed groups or civilians taking a direct part in hostilities. At the same time, depending upon the security situation and the ability and willingness of that State to act, a foreign State may want to intervene to protect its own nationals in the midst of the conflict. This could be particularly true where the control of the territory in question is in a state of flux. Where such an intervention occurs, either with the consent of the State controlling the territory or otherwise, it must be governed by human rights–based law enforcement norms, unless the violence against those foreign nationals rises to the level of an armed conflict (see Chapter 10, section 10.3.5.3, *States Acting Unilaterally to Rescue Their Own Nationals: Somalia*).

Finally, the bias that privileges State participants in armed conflict will largely relegate non-State participants, at best, to the status of unprivileged belligerents. This result is one which States have been careful to ensure by incorporating a high threshold for prisoner of war status into the Third Geneva Convention. The non-State belligerents will not likely be eligible to be afforded prisoner of war status and therefore remain liable to domestic prosecution by a capturing State for their participation in the conflict (see Chapter 4, section 4.3, *Impact of the Proper Authority Principle*). The involvement of non-State actors in an international armed conflict inevitably requires greater consideration of law enforcement norms than would be the case if the violence were only occurring between States. An effective response to these security threats can frequently require consideration of both the conduct of hostilities and law enforcement normative regimes.

The result is that in inter-State conflict there is a close relationship between the main bodies of international law leading to and applied during armed conflict: the law governing the recourse to war, humanitarian law, and human rights law. Further, the primacy assigned to either humanitarian or human rights law is very much context dependent. A broad application of humanitarian law cannot be divorced from the overall desire by the international community at the end of World War II to limit conflict between States. However, a factor that significantly alters the traditional emphasis placed on humanitarian law during international armed conflict is hostilities with non-State actors. In those situations the application of human rights law, and a law enforcement approach can increasingly take priority.

---

29 AP I, art. 48.

It therefore should come as no surprise that conflicts not of an international character, which involve States fighting non-State actors, also require particular emphasis be placed on human rights norms and law enforcement.

9.3. Categorizing Conflict with Non-State Actors

9.3.1. NON-STATE ACTORS CONTROLLED BY A STATE

In assessing conflict with non-State actors it is important to remember that States battle non-State actors in a wide array of conflicts. While, factually, the nature of much of the violence with non-State actors in any armed conflict remains the same, attempts by the international community to categorize such violence have created a patchwork of conflict descriptions. This variety of categorizations highlights the degree to which there is lack of consensus among theoreticians. This sets the scene for a potential disconnect between practitioners and those seeking to create order on a higher theoretical plane. In order to understand the degree to which the international community is struggling with non-State actor conflict it is useful to assess the various "legal" theories relating to such violence. This assessment starts with hostilities with non-State actors that qualify as an international armed conflict, and then shifts to different versions of noninternational armed conflict.

Given the high profile of contemporary transnational terrorist attacks carried out by "stateless" groups such as Al Qaeda, there is a tendency to think of an armed conflict with non-State actors as being exclusively noninternational in character. However, it is important when considering the scope of State versus non-State actor conflict to remember that an armed attack against a State by a non-State actor sent by, or operating under the control of another State will be attributable to that controlling State.[30] In this situation, the non-State actor would be viewed as an organ of or acting on behalf of the State.[31] This results in the violence being classified as an international armed conflict to which humanitarian law will apply. Significantly, not all transnational non-State actor attacks need to be viewed as being cross-border in the style of a 9/11 or Mumbai attack. For example, if a State "controlled" the type of attacks attributed to non-State actors that have occurred at sea;[32]

---

[30] Military and Paramilitary Activities in and against Nicaragua Merits, 1986 I.C.J. ¶115, at 64–65 (where the ICJ established "effective control" rather than simply general control would have to be exercised by a State to have the actions of a non-State actor attributed to it), and Prosecutor v. Tadić, No. IT-94-1-A, Appeals Judgment, ¶145, at 62 (Int'l Crim. Trib. for the Former Yugoslavia, July 15, 1999), http://www.icty.org/x/cases/tadic/acjug/en/tad-aj990715e.pdf (the ICTY requires a different standard of "overall control," which basically did not require the issuing of specific orders or instructions regarding military action). *See also* DINSTEIN, *supra* note 14, ¶¶ 586–92, at 221–24 (for a discussion of the "control" criterion), JUDITH GARDAM, NECESSITY, PROPORTIONALITY AND THE USE OF FORCE BY STATES 144 (2004), and GRAY, *supra* note 18, at 200 (setting out the U.S. and U.K. claims).

[31] Military and Paramilitary Activities in and against Nicaragua Merits, 1986 I.C.J. ¶ 109, at 62, and Prosecutor v. Tadić, No. IT–94–1–A, Appeals Judgment, ¶84, at 34 (Int'l Crim. Trib. for the Former Yugoslavia, July 15, 1999), http://www.icty.org/x/cases/tadic/acjug/en/tad-aj990715e.pdf.

[32] Jon Henley & Heather Steward, *Al-Qaida Suspected in Tanker Explosion*, THE GUARDIAN, Oct. 7, 2002, http://www.guardian.co.uk/world/2002/oct/07/alqaida.france, Duncan Gardham, *Al-Qaeda Attack on Supertanker*, THE TELEGRAPH, Aug. 6, 2010, http://www.telegraph.co.uk/news/worldnews/middleeast/unitedarabemirates/7930642/Al-Qaeda-attack-on-supertanker.html, and *Algeria "Foils al-Qaida Attack on Ships,"* UPI (Jan. 31, 2012), http://www.upi.com/Top_News/Special/2012/01/31/Algeria-foils-al-Qaida-attack-on-ships/UPI-45111328039154/.

at diplomatic facilities in the aggressor or in third States;[33] or against nationals abroad,[34] then it could be concluded that an international armed conflict is in existence between the controlling and threatened States.

This sets up a unique scenario regarding what constitutes an international armed conflict. It is possible, in theory at least, for all of the violence to occur between the threatened State and the non-State actor. The attack itself would still be treated as if it were one carried out on behalf of the controlling State. Attribution to a State results in the violence between the threatened State and the transnational terrorist or insurgent group being governed by the full framework of legal rules applicable to international armed conflict. A transnational attack may be of sufficient gravity to constitute an armed attack. It has been suggested that for an attack by a non-State actor to constitute such an attack it must be of a "large-scale."[35] However, theoretically an armed conflict could be in existence even in respect of limited uses of force by a transnational terrorist or insurgent group, or possibly no use of force at all (i.e., threats, or simply the trans-border movement of an organized armed groups). This is because of the operation of Common Article 2 of the Geneva Conventions. The actual hostilities between State and non-State actor would fall within the broad scope of an international armed conflict between States.

Further, since the threshold for international armed conflict is so low, these attacks might, in theory at least, occur with an intensity of violence below that normally associated with a noninternational armed conflict. In other words, involving violence that might ordinarily be amenable to a law enforcement resolution when taking place in an internal State context. Since an international armed conflict would technically be in existence, humanitarian law might be used to govern the response rather than human rights–based law enforcement. Such is the theoretical impact of a determination that an inter-State conflict is occurring even though a non-State actor is the other significant participant. Mitigating a conduct of hostilities response when the violence occurs within the threatened State is the reality that many of the first responders to such attacks will be security forces trained and equipped for law enforcement (see Chapter 14, section 14.4.4, *A Law Enforcement Response to Attacks at Home and Abroad*). However, as was evidenced with the 9/11 defensive acts that included the scrambling of military aircraft to intercept the hijacked aircraft, the response to transnational attacks by terrorists and insurgents can easily become more militarized.[36] The violence associated with the 9/11 attacks comfortably reached the level of an "armed attack" and can be viewed as the initiating incident for the subsequent international armed conflict.[37]

It is likely that the "controlling" State will be in some fashion drawn into the conflict. However, this type of conflict characterization introduces a level of complexity that is not

---

33 DINSTEIN, *supra* note 14, ¶¶ 566–70, at 214–16.

34 GRAY, *supra* note 18, at 156–60 & 202 (although it is argued that taking action to protect nationals abroad does not have wide support).

35 ELIZABETH WILMSHURST, PRINCIPLES OF INTERNATIONAL LAW ON THE USE OF FORCE BY STATES IN SELF-DEFENCE Rule 6, at 13 (Oct. 2005), http://www.chathamhouse.org/publications/papers/view/108106.

36 THE 9/11 COMMISSION REPORT: FINAL REPORT OF THE NATIONAL COMMISSION ON TERRORIST ATTACKS UPON THE UNITED STATES 37–38 (2004) [hereinafter 9/11 COMMISSION REPORT].

37 Whether the attacks of 9/11 constituted the initiating incident of the international armed conflict with Afghanistan and Al Qaeda or involved earlier actions such as the 2000 attack on the USS Cole in Yemen is open to considerable debate.

immediately evident when two States engage in hostilities. There remains a danger that one State might be overzealous in claiming attribution to another State when it wants to use military force to respond to terrorism. Given the significant recourse to war pressures to limit inter-State conflict, such claims are likely to be met with a certain amount of skepticism. The debate within the international community regarding any attack will focus on whether the requisite "control" over a non-State actor was being exercised by the territorial/host State. This is a scenario that is factually, legally, and politically complex and one that will not always lead to a ready consensus. Until it is possible to clearly establish the necessary control by another State, any attack by a non-State actor would be treated as if a transnational terrorist group were acting on its own. In this respect the determination that a conflict is international in character may not be immediately obvious or verifiable.

What being engaged in an international armed conflict will not likely do is result in the transnational terrorists or insurgents being viewed as lawful participants. It is here that the difference between State participation in armed conflict and transnational violence by non-State actors will be evident. As occurs in internal conflict, non-State participants in this form of international armed conflict are very likely to be viewed and treated as criminals subject to prosecution. This effectively means that, notwithstanding of the application of a conduct of hostilities normative structure governing the use of force, those participants are likely be dealt with according to human rights–based law enforcement norms for potential breaches of international law or under the domestic law of the opposing State (see Chapter 4, section 4.3, *Impact of the Proper Authority Principle*).

9.3.2. HARBORING AND ACTIVE SUPPORT

The situation could also unfold where the territorial State does not control the non-State actor, but an attack is carried out from its territory in circumstances where that State either actively supports it or fails to meet its international legal obligations to end such activity. Discussed in the post-9/11 era as "harboring" terrorists, this is an area of law where it has been suggested there is considerable uncertainty as to what degree of State involvement is necessary to implicate it in an armed attack by non-State actors. One critique of this approach has suggested that its supporters cannot adduce State practice, other than from the Operation Enduring Freedom response to the 9/11 attacks. As a result, they appear to be relying on policy considerations to permit action against "non-state actors operating from a state which has tolerated their activities or is unable to control them . . . ."[38] However, another analysis has suggested "the reaction of the international community to these events [the Taliban regime harboring of Al Qaeda in Afghanistan] was generally positive, which provides evidence of a shift that appears likely to continue in the governing legal principles in relation to State-sponsored terrorist activities."[39]

A State, by failing to take action to meet its obligations under international law to not allow its territory to be used for actions against another State, can become responsible for

---

[38] GRAY, *supra* note 18, at 201.

[39] GARDAM, *supra* note 30, at 145.

the actions of the terrorist group.[40] This approach is reflected in the 1980 International Court of Justice decision, *Case Concerning United States Diplomatic and Consular Staff in Tehran* (United States v. Iran).[41] In that decision the court ruled that the initial seizure of U.S. diplomatic premises and detention of diplomatic staff hostages by militants was not attributable to Iran, however, the subsequent approval by Iranian officials of the actions "and the decision to perpetuate them translated continuing occupation of the Embassy and detention of the hostages into acts of that State."[42] The challenge is deciding at what point the line is crossed to one of support for the non-State actor, particularly in States that are the subject of a weak or failing governance structure.

The taking of action against a harboring State was the rationale for Coalition forces to respond in self-defense against the Taliban-led government of Afghanistan.[43] This is suggested to occur because although the Taliban had not participated in the 9/11 attacks "before and during the act, they became accessories after the fact."[44] This conclusion was reached on the basis that the Taliban refused to act against Al Qaeda and continued to offer them shelter.[45] With Al Qaeda having carried out an armed attack against the United States, this created a situation of an international armed conflict between the Afghanistan and the Coalition forces invoking the right of self-defense.[46] Interestingly the response was not limited to the Taliban but also extended to Al Qaeda, with it being named as a separate entity in State explanations for the taking of self-defense action.[47] The result was that international humanitarian law governed the resulting conflict with the Taliban and Al Qaeda (as the "special law" applicable in armed conflict). However, human rights–based law enforcement was also relied on. For example, within North America the defensive response by Canada and the United States has remained sensitive to having to apply law enforcement criteria in order to respond to inattentive pilots or uniquely criminal acts even though reliance continues to be placed on a military defense treaty (i.e., NORAD) to react to airborne threats from hijacked aircraft (see Chapter 13, section 13.5 *The Shoot-Down of Hijacked Aircraft*).

This was not the only introduction of human rights norms or law. To an even greater degree than occurs with situations where States exert control over non-State groups, the less direct connection between the non-State actor and the harboring State makes it unlikely that prisoner of war status or protection from prosecution for participation in hostilities (i.e.,

---

40 DINSTEIN, *supra* note 14, ¶ 693, at 261.

41 Case Concerning United States Diplomatic and Consular Staff in Tehran (United States v. Iran), 1980 I.C.J. 3.

42 *Id.* ¶ 74 at 35. *See also* DINSTEIN, *supra* note 14, ¶ 600, at 226–27.

43 GRAY, *supra* note 18, at 200.

44 DINSTEIN, *supra* note 14, ¶ 693, at 261. *See also* Stephen W. Preston, *The Legal Framework for the United States' Use of Military Force Since 9/11*, Remarks by the General Counsel of the Department of National Defense on the Legal Framework for the United States' Use of Force Since 9/11, at 3 (Apr. 10, 2015), http://www.defense.gov/Speeches/Speech.aspx?SpeechID=1931 (where it is noted "it was not Afghanistan that had launched the attack. It was bin Laden and his terrorist organization.").

45 DINSTEIN, *supra* note 14, ¶ 693, at 261.

46 *Id.* ("The military operations in Afghanistan were a classical State versus State exercise of self-defence.").

47 Kirby Abbott, *"Terrorists: Criminals, Combatants or . . . .": The Question of Combatancy*, *in* THE MEASURES OF INTERNATIONAL LAW: EFFECTIVENESS, FAIRNESS AND VALIDITY 366, 372 (2004) (where it is noted in respect of Canada its actions were "directed against Osama bin Laden's Al-Qaeda terrorist organization and the Taliban regime that is supporting it").

combatant immunity) will be available to those not fighting directly for that State. This is also less likely where the acts are viewed as terrorist in nature. For example, prisoner of war status was not extended to Al Qaeda members fighting in 2001 in Afghanistan. This included those who participated in the 055 Brigade, a brigade of Afghan Arabs integrated into the Army of the Islamic Emirate of Afghanistan from 1997 to 2001 to fight the Northern Alliance.[48] Once again, the human rights–based law enforcement standards take on even greater relevance for the treatment and potential prosecution of those non-State participants.

Further, even though this form of the conflict involving a "harbored" non-State actor is international in character the involvement of a group like Al Qaeda created a different atmosphere. That atmosphere was one where concern over criminality could be seen to have an important influence on the normative framework that security forces applied during the course of operations. The participation of members of such a non-State group in the armed conflict was obviously viewed as "illegal." The operational goal was not simply to kill or detain members of Al Qaeda but also arrest, interrogate, and bring them to trial for their heinous terrorist acts. In contrast, during traditional international armed conflict the penal consequences for members of State armed forces are largely reserved for war crimes committed during the conflict.

9.3.3. ENTERING THE TERRITORY OF ANOTHER STATE: INTERNATIONAL ARMED CONFLICT

There is a third way in which, theoretically at least, States can become involved in an international armed conflict in situations where the fighting occurs against a non-State actor. This theory suggests any time a foreign State acts against a non-State actor in another State without the consent of the latter State the result is that an international armed conflict is in existence. This conclusion is based on the argument that such a use of force is "*against* the territorial State."[49] Such a situation is said to arise "even if the use of force is not directed against the governmental structures of the territorial State, or the purpose of the use of force is not to coerce the territorial State in any particular way."[50] In making this argument, reliance is placed on viewing the State as encompassing more than the government. It extends to the people and the territory controlled by the State (see Chapter 4, section 4.4.2, *Governance*).[51] As a result, an intervention on the territory of another State without that State's permission is viewed as automatically creating an international armed conflict between the two States.[52]

This argument emphasizes that the foreign State breaches its obligations regarding the sanctity of the borders of other States.[53] The idea that the borders of States are inviolable

---

[48] DILIP HIRO, WAR WITHOUT END: THE RISE OF ISLAMIST TERRORISM AND GLOBAL RESPONSE 330 (2002), and SETH G. JONES, HUNTING IN THE SHADOWS: THE PURSUIT OF AL QA'IDA SINCE 9/11, at 67 (2012).

[49] Dapo Akande, *Classification of Armed Conflicts: Relevant Legal Concepts*, *in* INTERNATIONAL LAW AND THE CLASSIFICATION OF CONFLICTS 32, 73 (Elizabeth Wilmshurst ed., 2012).

[50] *Id.* at 73–74.

[51] *Id.* at 75.

[52] *Id.* at 74.

[53] *Id.*

is a theme that runs throughout much of the traditional approach toward inter-State relations applied by many international lawyers in the post–UN Charter context. It is also a theory that appears to underpin criticism of the right to defend nationals (see Chapter 10, section 10.2.1, *A State Responsibility*)[54] and the use of drones in cross-border operations in States where non-State actors are located (see Chapter 8, section 8.5.1, *The Restricted "Law Enforcement" Theory*).[55] It is an argument that is entirely consistent with the territorial State being viewed as the proper authority for dealing with violence within its borders. However, the problem that ultimately arises is when that territorial State is unwilling or unable to deal with the non-State actor operating from within its borders. Inviolable borders cannot become a source of impunity for non-State actors. It is such impunity that forces States to consider cross-border operations.

On the positive side, this theory provides a clear normative framework for the conduct of hostilities. However, since it reflects the view that intervention in a territorial State by a foreign State is a "wrong" under international law,[56] it is unlikely to be embraced by either the intervening State or other States that feel they may have to act this way in the future. This is the likely result even though a conflict with a non-State actor being viewed as an international one might, on one level, be seen by that State as an attractive outcome to the extent that reliance could more readily be placed on conduct of hostilities rules for the purposes of targeting. Significantly, the theory represents a potentially broad expansion of hostilities that can be classified as international armed conflicts. This is particularly the case since it does not depend upon a clash between the armed forces of the two States. Actual violence between States is technically not a prerequisite for the existence of an armed conflict.[57]

This theory does appear highly formalistic because of its simple reliance on what are basically geographic criteria. The nonpermissive entry of foreign State forces within the territorial jurisdiction of another State to act against a non-State actor creates the international armed conflict. What is conceptually difficult is that what may be framed as a breach of international law involving no violence between the two States is actually premised on considerable hostilities possibly occurring between intervening State armed forces and a non-State actor. That result is likely to be seen as particularly problematic to the extent this theoretical approach extends the normative framework governing inter-State conflict to hostilities with non-State actors acting on their own. As seen with the Additional Protocol I, controversy this is an outcome that has not been previously embraced by the international community (see Chapter 2, section 2.4.1, *Domestic Authority, International Law, and Non-State Actors*).

---

[54] Tom Ruys, "Armed Attack" and Article 51 of the UN Charter: Evolutions in Customary Law and Practice 249 (2010) (where it is suggested a rescue operation by a State to rescue its own nationals is only permissible where the territorial State consents), and Olivier Corten, The Law Against War: The Prohibition on the Use of Force in Contemporary International Law 89 (Christopher Sutcliffe trans., 2010) (where emphasis is placed on intervention without consent, such as hostage rescue, being prohibited under international law in most circumstances. Intervention is permitted only under very strict terms).

[55] Mary Ellen O'Connell, *The Choice of Law Against Terrorism*, 4 Nat'l Sec. L. & Pol'y 343, 357 (2010) (where particular emphasis is placed on limiting State intervention in the post-9/11 period in respect of States other than Afghanistan).

[56] Akande, *supra* note 49, at 74.

[57] GC IV, art. 2.

Further, in determining that a cross-border conflict with the non-State actor is an international one, the historical reluctance by States to apply humanitarian law within their own borders may also become an issue (see Chapter 4, section 4.4.3, *A Reluctant Application of Humanitarian Law*). It is not clear that a territorial State will embrace the classification of the violence with a non-State actor within its borders as an international armed conflict regardless of whom it is against. It has been acknowledged that the classification of these cross-border attacks against non-State actors as an armed conflict is not "the majority view in the existing literature."[58] Given the much broader contemporary acceptance that such conflicts are noninternational in character, it is unlikely that the international armed conflict approach will gain significant traction (see section 9.3.4.3, *A Contrast: The American Approach*, below). However, what this discussion does highlight is how unsettled and complex this area of law has become in the post-9/11 period.

9.3.4. THE ISRAELI SUPREME COURT AND CROSS-BORDER TERRORISM

9.3.4.1. Targeted Killing, Occupation, and "Conflicts of an International Character"

Perhaps the clearest example of the evolving nature of transnational terrorism and its impact on conflict characterization can be seen in Israeli High Court of Justice (Supreme Court) jurisprudence. That court has been challenged in identifying the normative framework to be applied in the struggle against the varied and powerful terrorist groups aligned on Israel's borders. The problem for the court is that the nature of the conflict in the region has changed from one, which in 1948 was primarily focused on existential threats to Israel from neighboring States, to a situation where non-State actors, such as Hezbollah and Hamas, are seen as posing a significant threat.[59] As a result the Israeli Supreme Court has been forced to assess how to characterize warfare with non-State actors that exercise governance over territory adjacent to that country's borders. These groups can field increasingly technologically advanced military forces, as well as carry out acts of terrorism; have fostered low-level conflict, or *intifadas* ("from the Arabic, meaning 'to shake off' or 'to wake up.' "[60]) in territory occupied by Israel following the 1967 war; and have regularly conducted attacks across Israel's border, prompting defensive action by that State. These are not the traditional protagonists in an international armed conflict.

In the 2006 *Public Committee Against Torture in Israel v. Israel* decision (referred to as the *Targeted Killing Case*), the court dealt specifically with violence between Israel and various terrorist groups in the "area" of Judea and Samaria (the Israeli term for the West Bank) and the Gaza Strip. This territory was occupied by Israel during the 1967 Six Day War. Sorting out an agreed status for the Occupied Territories has been complicated by the Israeli government position that the Geneva Conventions do not apply to the those territories, although the Supreme Court noted, "Israel honors the humanitarian provisions

---

[58] Akande, *supra* note 49, at 75.

[59] SERGIO CATIGNANI, ISRAELI COUNTER-INSURGENCY AND THE INTIFADAS: THE DILEMMAS OF A CONVENTIONAL ARMY 24–25 (2008).

[60] DANIEL BYMAN, A HIGH PRICE: THE TRIUMPHS & FAILURES OF ISRAELI COUNTERTERRORISM 72 (2011).

of that convention."[61] However, it seems clear that the Fourth Geneva Convention should apply as a matter of law. In any event, the issues before the court were not those ordinarily associated with a World War II–type of belligerent occupation. The court had to deal with the reality of cross-border attacks into Israel proper by terrorist organizations, which presented a significant threat to the Israeli population. [62] The issue was not only one of the usual relationships between an occupier and insurgents but also transnational terrorism.

The novelty of the legal challenge was evident in the decision made by the Israeli government to alter its approach toward classification of these conflicts in the course of legal argument. In this case the government had initially argued that the conflict between Israel and the terrorists was part of an international armed conflict. However, that position was subsequently amended to indicate that the classification of the conflict as an international or noninternational one or an even a new form conflict between a State and a terrorist organization was not important since the laws of war permitted the targeting of a "combatant" in any armed conflict.[63]

There are two possible interpretations of the *Targeted Killing Case* regarding the approach taken by the High Court of Justice toward conflict classification. A traditional, more formalist interpretation views the violence in the "area" as being governed by international humanitarian law because it occurs between organized armed groups and the occupying power. Following this approach the conflict is viewed as an international one. It does not matter that the terrorist groups are not aligned with the State involved in the war that resulted in the territory first being considered occupied, or that those non-State actors are carrying out attacks in the territory of the Occupier State. A second approach concentrates more specifically on the trans-border aspect of the violence when looking at the "character" of the conflict. This interpretation also concludes the conflict is "international in character," but points toward a new approach in suggesting such a categorization applies to non-State actor conflict.

In addressing the issue of trans-border violence, the *Targeted Killing Case* focused on one of the factors that go to the heart of the changing nature of the contemporary terrorist threat. After indicating that the violence occurring within occupied territory fell within the framework of an international armed conflict, the Israeli Supreme Court added: "[t]his law applies in any case of an armed conflict of international character—in other words, one that *crosses the borders of the state—whether or not the place in which the armed conflict occurs is subject to belligerent occupation*."[64] This wording highlights that, unlike traditional assessments of what constitutes an international armed conflict, the *Targeted Killing Case* emphasized the trans-border nature of the conflict, rather than the requirement that it occur between two or more States.[65] This decision opened the door to an application of the

[61] Public Committee Against Torture in Israel v. Israel, Israel Supreme Court [16 December 2006], 46 ILM 375, ¶ 20, at 383 (2007).

[62] *Id.* at ¶ 16, at 381.

[63] *Id.* at ¶ 11, at 378–79.

[64] *Id.* at ¶ 18, at 382 (emphasis added).

[65] Marko Milanović, *Lessons for Human Rights and Humanitarian Law in the War on Terror: Comparing Hamdan and the Israeli Targeted Killings Case*, 89 Int'l Rev. Red Cross 373, 384 (2007) (where it is indicated the decision of the Supreme Court is remarkable since "the single defining characteristic of international armed conflicts has not been their cross-border, but their interstate, nature.").

second interpretative approach, which views cross-border non-State actor violence as something different from a noninternational armed conflict. The adoption of this approach by the Israeli Court was referred to by the Israeli *Public Commission to Examine the Maritime Incident of 31 May 2010* in its 2011 analysis of the Israeli blockade established off the coast of Gaza.[66]

9.3.4.2. A Unique Security Threat

Support for the interpretation that the conflict is one of an international character is found the security context within which this non-State actor violence was taking place during the period the case was argued and the judgment was delivered. This decision clearly was not delivered in a vacuum. Israel had witnessed terrorist violence associated with the first and second intifadas (1987–93 and 2000–05)[67] in the adjacent occupied territories, as well as a significant suicide bombing campaign within Israel[68] and rocket attacks against Israel originating from outside that country.[69] Following a fierce debate, Israel withdrew from Gaza in September 2005, more than a year before the release of the *Targeted Killing Case* judgment in December 13, 2006.[70] From that point the factual situation in Gaza was not like the other territories referred to collectively in the *Targeted Killing Case* as the "area."[71]

In a subsequent 2008 decision, *Al-Bassiouni v. Prime Minister*, the Supreme Court interpreted the Israeli withdrawal from Gaza to mean that territory was no longer occupied as of 2005, although Israel continued to have humanitarian obligations toward the Gaza population.[72] This view has not been universally shared. There are UN organizations as well as humanitarian and human rights groups that have argued Gaza remains occupied territory.[73] That position is supported by the argument that Gaza remains occupied by Israel due in part to the control that country exercises over the land border, as well as air and sea access to that territory.[74] What is clear is that the complexity of contemporary conflict with these non-State actors has made reaching consensus on the categorization of this conflict very difficult.

Given the passage of time since 1967 when Israel actually occupied Gaza and its subsequent peace treaty between with Egypt, it does appear to have become increasingly difficult to suggest that the stronger legal framework is one based on an international armed

---

66 1 PUBLIC COMMISSION TO EXAMINE THE MARITIME INCIDENT OF 31 MAY 2010, ¶ 41 at 47 (2011) [hereinafter FIRST TURKEL REPORT].

67 CATIGNANI, *supra* note 59, at 1.

68 BYMAN, *supra* note 60, at 141–142, and Public Committee Against Torture in Israel v. Israel, Israel Supreme Court [16 December 2006], 46 ILM 375, ¶ 16, at 381 (2007). *See also* CATIGNANI, *supra* note 59, at 101–07.

69 BYMAN, *supra* note 60, at 182.

70 *Id.* at 174–77.

71 Public Committee Against Torture in Israel v. Israel, Israel Supreme Court [16 December 2006], 46 ILM 375, ¶ 16, at 381 (2007).

72 Al-Bassiouni v. Prime Minister, HCJ 9132/07 ¶ 12 (unpublished, Jan. 30, 2008), and FIRST TURKEL REPORT, *supra* note 66, ¶ 41, at 47.

73 FIRST TURKEL REPORT, *supra* note 66, ¶ 41, at 48.

74 *See, e.g.*, YORAM DINSTEIN, THE INTERNATIONAL LAW OF BELLIGERENT OCCUPATION ¶¶ 664–73, at 276–80 (2009).

conflict between two or more States.[75] Contemporary attacks related to Gaza do occur in a cross-border context, rather than as violence taking place uniquely within occupied territory. In terms of State control, the Israeli withdrawal would seem to place Gaza in a unique category, since at that point the territory did not then fall under the responsibility of either Israel or Egypt. By January 2006 Hamas had beaten Fatah in the Palestinian Legislative Council elections, and a subsequent internal Palestinian struggle led to Hamas seizing power in Gaza in June 2007. This put that territory firmly under the governance of a terrorist group committed to attacking Israel.[76] In a number of respects the argument that the territory remains occupied often appears more closely connected to a desire to ensure the humanitarian protections of the Fourth Geneva Convention continue to apply. It is an approach that seems less intended to provide an accurate reflection of the type of hostilities taking place on the ground (i.e., cross-border).

The *Targeted Killing Case* did not just focus on cross-border terrorism. It specifically addressed the issue of targeted killing from the perspective of a territory being subject to belligerent occupation.[77] That clearly remained the situation in the West Bank. The decision noted in respect of the requirement to consider the capture of a person being targeted that "[i]t might actually be *particularly practical* under the conditions of belligerent occupation, in which the army controls the area in which the operation takes place, and in which arrest, investigation, and trial are at times realizable possibilities."[78] While this wording refers specifically to occupation, it can also be read to suggest the decision was not limited to that situation since it implies the capture of a terrorist might be practical under other circumstances as well. This could be the case in Gaza where Israel has continued to make military incursions. However, as a 2014 raid on a rocket site in Gaza during Operation Protective Edge, which resulted in Israeli naval commandos being wounded, demonstrated that State's lack of physical control over Gaza could limit the feasibility of any capture operation, thereby opening the door to a strike decision.[79]

As is evident with Hezbollah in Lebanon, Gaza is not the only situation where a terrorist group on the border of Israel controls enclaves outside the authority of the territorial State.[80] In 2006 Israel was engaged in a short month-long but intensive conflict with

---

[75] *Id.* ¶¶ 28–33, at 12–15 (outlining the history of the control and occupation of Gaza. Note that Professor Dinstein remains firmly of the view that Israel continued as an occupying power). *But see* Adam Roberts, *Prolonged Military Occupation: The Israeli-Occupied Territories Since 1967*, 84 AM. J. INT'L L. 75–79 (1990) (setting out the complexity of the status of the Occupied Territories, with specific reference to Egypt and Jordan and the possibility of Palestinian self-determination).

[76] BYMAN, *supra* note 60, at 177–82, and ZAKI CHEHAB, INSIDE HAMAS: THE UNTOLD STORY OF THE MILITANT ISLAMIC MOVEMENT 1–14 (2007).

[77] *Aerial Strikes Against Terrorists: Some Legal Aspects*, Israel Defence Forces, International Law Department (Mar. 2, 2014), http://www.mag.idf.il/163-6582-en/patzar.aspx (this analysis found on the Israeli Military Advocate General website appears to narrowly take the position the *Targeted Killing Case* dealt with strikes on civilians taking a direct part in hostilities "in areas where Israel applied the laws of belligerent occupation").

[78] Public Committee Against Torture in Israel v. Israel, Israel Supreme Court [16 December 2006], 46 ILM 375, ¶ 40, at 394 (2007) (emphasis added).

[79] Or Heller, *Open Mind & Out-of-the-Box Thinking*, ISRAEL DEFENSE (Aug. 19, 2015), http://www.israeldefense.co.il/en/content/open-mind-out-box-thinking ("This operation demonstrated the difficulties facing special unit operations in a place like Gaza in the course of a war, when the other side is alert and ready for IDF [Israel Defence Forces] attacks to begin with.").

[80] DINSTEIN, *supra* note 14, ¶ 579, at 219 (referring to the Hezbollah/Lebanon government relationship in 2006).

Hezbollah, which had resulted from an attack by that terrorist group on Israeli forces near the border with Lebanon.[81] Even though there were Hezbollah members in the Lebanese Cabinet and that group exercised government functions in the south, it has been generally assessed that Lebanon, as the territorial State, was not responsible for the attack. This situates the conflict within the context of warfare between a State and a non-State actor.[82] This armed conflict predated the Israeli judgment on targeted killing. However, the threats posed by both Hamas and Hezbollah were clearly known to the Israeli Supreme Court in December 2006. It seems that this reality could very well have influenced the degree to which emphasis in the *Targeted Killing Case* was placed on cross-border threats posed by terrorist groups, rather than make reference solely to occupied territory.[83]

In the final analysis *Targeted Killing Case* represents a significant and laudatory attempt to deal with what has been recognized in the post-9/11 environment to be as a very difficult challenge: conflict categorization. As was noted by Antonio Cassese when discussing Israeli drone strikes against organized terrorist groups, "admittedly there may be arguments to support the view that a new type of armed conflict is *de facto* [in fact, as opposed to at law] developing as a historical phenomenon."[84] While in his subsequent analysis he relied on the law of occupation when categorizing the conflict "in the territories" as an international one, this acknowledgment reflects the unique pressure that has been placed on traditional interpretations of the law as a result of the changing contemporary security environment.[85] In the post-9/11 era, trans-border violence by and against terrorist groups has monopolized the discussion about conflict characterization. The result is that in the twenty-first century, States, courts, and international lawyers are now being forced to pay greater attention to these situations in an attempt to find solutions to international legal issues that have previously escaped detailed scrutiny.

9.3.4.3. A Contrast: The American Approach

It is not just the Israeli Supreme Court that has been required to look at conflict classification. In contrast to the Israeli jurisprudence, the U.S. Supreme Court has adopted a different approach toward categorizing non-State actor armed conflict. In the 2006 *Hamdan v. Rumsfeld* case that Court viewed the transnational violence with Al Qaeda as a noninternational armed conflict. Notably, it was a conflict seemingly unbounded by State borders. As the court indicated, "[t]he term 'conflict not of an international character' is used here in contradistinction to a conflict between nations."[86] The approach in *Hamdan v. Rumsfeld* is fundamentally different from more traditional interpretations of humanitarian law, which

[81] Anthony H. Cordesman, George Sullivan, & William D. Sullivan, Lessons of the 2006 Israeli-Hezbollah War (2007) and Amos Harel & Avi Issacharoff, 34 Days: Israel, Hezbollah and the War in Lebanon (2008)(for analyses of this conflict).

[82] Michael N. Schmitt, *"Change Direction" 2006: Israeli Operations in Lebanon and the International Law of Self-Defense*, 84 Int'l L. Stud. 265, 274–79 (2008).

[83] Public Committee Against Torture in Israel v. Israel, Israel Supreme Court [16 December 2006], 46 ILM 375, ¶ 21, at 383 (2007) (for reference to such unaligned terrorist groups).

[84] Antonio Cassese, International Law 420 (2nd ed. 2005).

[85] *Id.* at 420–21.

[86] Hamdan v. Rumsfeld, 45 I.L.M. 1130, 1154 (2006) (U.S. Supreme Court 2006).

have viewed noninternational armed conflict as being solely intra-State[87] and "between a central government and a group of insurgents belonging to the same State (or between two or more insurrectional groups belonging to that State). . . ."[88] In this case the focus of the U.S. Supreme Court was on the status of the Parties to the conflict.

The *Hamdan v. Rumsfeld* case has been criticized regarding its determination that a noninternational armed conflict can have a reach beyond the territorial boundaries of a State.[89] However, it is also an approach that has gained considerable traction in contemporary legal analysis.[90] The legal challenge is how to characterize conflict that is not between one or more States or one that is internal to a State. Notably, if the struggle between the State and the non-State actor does not qualify on a theoretical level for either international or noninternational armed conflict status, then it must be dealt with under a human rights–based law enforcement paradigm. However, this latter paradigm clearly may not adequately address the extent and nature of security threat. The result is that the threatened State is placed at a distinct disadvantage. By removing reference to territorial limitations and still calling the violence a noninternational armed conflict, there is a closer affinity to the transnational nature of contemporary non-State actor conflict, even if it is a significant departure from prior interpretations of the law.

The question can also be raised regarding what option was available to categorize armed conflicts with non-State actors having the capabilities of a Hamas, Hezbollah or the Islamic State organized armed group and for which there was no obvious connection to another State. For example, if the *Hamdan v. Rumsfeld* approach is not applied to the situation of terrorist attacks emanating from an unoccupied Gaza, thereby assessing the conflict as a noninternational one, it appears the only remaining option is to characterize the armed conflict as having an international character. On this basis there is considerable logic to an interpretation of the *Targeted Killing Case* that does just that in the context of having viewed the previous occupation by Israel as having been terminated. It is helpful to recall that a narrow form of international armed conflict between States and non-State actors operating with territorial boundaries has already received international treaty recognition in Additional Protocol I. Although this approach has failed to gain universal recognition, in part, because of concerns over potentially legitimizing the non-State actor (see Chapter 2, section 2.4.2, *Internal Conflicts with an "International Character"*), that is not an issue with these cross-border situations. For example, given the clear association of Hamas with terrorism, it is obvious that the Israeli Supreme Court was not raising the issue of legitimacy in its discussion of such armed conflict being one of an international character.

Therefore, in making a determination that the attacks by terrorist groups constituted an armed conflict and by placing a particular emphasis on the cross-border aspect of attacks by terrorist groups, it is possible to interpret the Israeli Supreme Court as having simply reached a conclusion that the conflict with Hamas and like terrorist organizations has an

---

87 DINSTEIN, *supra* note 14, ¶ 8, at 5–6.

88 CASSESE, *supra* note 84, at 420.

89 YORAM DINSTEIN, THE CONDUCT OF HOSTILITIES UNDER THE LAW OF INTERNATIONAL ARMED CONFLICT ¶ 130, at 56 (2nd ed. 2010), and DINSTEIN, *supra* note 5, ¶ 83, at 27. *See also* O'Connell, *supra* note 55, at 356–57.

90 *Department of Justice White Paper*, 3 (undated) (emphasis added), http://msnbcmedia.msn.com/i/msnbc/sections/news/020413_DOJ_White_Paper.pdf (last visited July 23, 2013).

international or cross-border character.[91] It is a result that stands in contrast to the approach applied by its U.S. counterpart. While no doubt traditionalists will be unhappy with either option, the *Hamdan v. Rumsfeld* decision or the *Targeted Killing Case*, these approaches provide examples of courts having to grapple with the theoretical limits of established international law in the contemporary security environment. These courts are being asked to decide legal issues at the boundaries of existing legal interpretation. They often decide difficult legal questions that have escaped resolution either through treaty negotiation or what appears at times to be endless academic debate. As a source of international law such judicial decisions are providing the impetus to advance twentieth-century interpretations of the international law in attempt to address the reality of contemporary warfare.[92] As such, the debate about the categorization of transnational terrorism is likely far from being over.

9.3.4.4. The Practical Impact

It is not just courts that are struggling with the issue of conflict categorization. State practitioners have also been challenged in the attempt to categorize the complex operational situations that arise because of cross-border terrorist action. As the Israeli jurisprudence indicates, these situations simply do not fit neatly into traditional interpretations of international law. As is reflected in the two possible interpretations of the *Targeted Killing Case*, as well as the contrasting approach applied in *Hamdan v. Rumsfeld* by the U.S. Supreme Court, there is considerable room for debate as to what category of conflict is in existence. Continuing the approach adopted in the *Targeted Killing Case*, the Israeli government position appears to have focused less on the category of conflict and more on the normative rules to be applied. That approach can lead to a decision to effectively apply "blended" humanitarian law standards rather than focus on different rules that might apply to international, or noninternational armed conflicts.

The application of more generic humanitarian law standards can be seen in respect of Operation Cast Lead, Israel's 2008–09 incursion into Gaza to stop Hamas attacks on its territory. In the aftermath of the *Al-Basiouni* decision, Israel indicated the "Gaza Strip is neither a State nor a territory occupied or controlled by Israel."[93] It was argued by the government that Israel "as a matter of policy applies to its military operations in Gaza the rules of armed conflict governing both international and non-international armed conflicts."[94] Therefore, in practical terms, the "classification of the armed conflict between Hamas and Israel as international or non-international in the current context is largely of theoretical concern, as many similar norms and principles govern both types of conflicts."[95] An

---

[91] Public Committee Against Torture in Israel v. Israel, Israel Supreme Court [16 December 2006], 46 ILM 375, ¶ 16, at 381 (2007) (for an outline of the violence and casualties suffered in the conflict).

[92] Statute of the International Court of Justice, Apr. 18, 1946 art. 38(1)(d) (where reference is made to the application of "judicial decisions" in deciding disputes under accordance with international law). *See also* MALCOLM N. SHAW, INTERNATIONAL LAW 109–11 (6th ed. 2008).

[93] THE OPERATION IN GAZA 21 DECEMBER, 2008–18 JANUARY 2009, LEGAL AND FACTUAL ASPECTS, The State of Israel ¶ 30, at 11 (2009) [hereinafter GAZA REPORT].

[94] *Id. See also* FIRST TURKEL REPORT, *supra* note 66, ¶ 40, at 47 (2010) (referring to the testimony of the Israeli Military Advocate General, Major-General Avichai Mandelblit), and Public Committee Against Torture in Israel v. Israel, Israel Supreme Court [16 December 2006], 46 ILM 375, ¶ 11, at 378–79 (2007).

[95] GAZA REPORT, *supra* note 93, ¶ 30, at 14.

approach recognizing the convergence of the law applicable to noninternational and international armed conflict has some support in the International Criminal Tribunal for the Former Yugoslavia case law,[96] at least "as far as certain war crimes are concerned."[97] Further, the same humanitarian targeting rules reflected in Additional Protocol I (which applies to international armed conflict) are widely considered to apply to noninternational armed conflicts. In this regard, the effect of the Israeli policy approach is largely the same whether the conflict was considered to be of an international or a noninternational character.

Yoram Dinstein has rightly warned that it is far from immaterial whether an armed conflict is a noninternational or international one.[98] One example provided is the danger of importing noninternational armed conflict threshold criteria into the determination of when an international armed conflict is in existence.[99] Legal advisors must remain cognizant of areas of humanitarian law that may not have ready application to one form of conflict or the other. One obvious difference between the two bodies of law is the status of conflict participants, with the prisoner of war provisions applicable in international armed conflict not being relevant from a legal perspective to its noninternational counterpart. However, even then the humanitarian standards found in the 1949 Geneva Conventions for the treatment of prisoners of war and civilians are amenable to being applied as a matter of policy in noninternational armed conflict. Further, the conduct of hostilities rules, particularly in respect of targeting, can be applied across the armed conflict spectrum regardless of how the conflict is categorized.[100]

9.3.4.5. Unlawful Combatants

It must also be noted that in respect of counterterrorism issues, the *Targeted Killing Case* analysis was not limited to conflict categorization. As with other instances where a State is confronted with threats from non-State actors, the question of the status of members of those groups was also dealt with. It should come as no surprise given the strict standards for obtaining prisoner of war status under humanitarian law that the Israeli Supreme Court determined "the terrorists and the organizations which send them to carry out attacks are unlawful combatants. They do not enjoy the status of prisoners of war. They can be tried for their participation in hostilities, judged, and punished."[101] However, the court did find

---

[96] Prosecutor v. Tadić, Case No. IT-94-1-A, Decision On The Defence Motion For Interlocutory Appeal On Jurisdiction, ¶ 94 (Int'l Crim. Trib. for the Former Yugoslavia, 2 October 1995), http://www.icty.org/x/cases/tadic/acdec/en/51002.htm, and Prosecutor v. Halilović, Case No. IT-01-48-T, Judgement, ¶ 25 (Int'l Crim. Trib. for the Former Yugoslavia, 16 November 2005), http://www.icty.org/x/cases/halilovic/tjug/en/tcj051116e.pdf (where it was noted that given the applicability of the provisions of Common Article 3 in both types of conflicts, "it is immaterial whether the armed conflict was international or non-international in nature," and "there is no need for the Trial Chamber to define the nature of the conflict in the present case."). *See also* DINSTEIN, *supra* note 5, ¶ 162, at 50–51.

[97] DINSTEIN, *supra* note 5, ¶ 162, at 51.

[98] *Id. But see* James Stewart, *Towards a Single Definition of Armed Conflict in International Humanitarian Law: A Critique of Internationalized Armed Conflict*, 85 INT'L REV. RED CROSS 315, 344–49 (2003) (for a discussion of the adoption of a single body of international humanitarian law applicable to all armed conflict), and Lindsay Moir, *Towards the Unification of International Humanitarian Law*, *in* INTERNATIONAL CONFLICT AND SECURITY LAW 108 (Richard Burchill, Nigel D. White, & Justin Morris eds., 2005).

[99] *Id.* ¶ 163, at 51.

[100] This was the approach applied by the present author, and it was this author's experience that many State military legal advisors have adopted this approach in the post-9/11 period.

[101] *Id.* ¶ 25, at 386.

there was no separate formal legal category of "unlawful combatant" under humanitarian law and their technical legal status was that of civilian. This reasoning continued the long-standing trend under international law that views non-State actors as having an illegitimate or at least an unprivileged status under international law.

After determining the "terrorists" were civilians, the Israeli court went one step further. As has been noted, the court considered the human rights of the targeted person in concluding that capture should be contemplated prior to carrying out a targeted killing .[102] Therefore, notwithstanding a determination that the conflict between terrorist groups and Israel was a conflict that is international in character, the issue of criminality and with it law enforcement appears to have heavily influenced the approach adopted by the Israeli Supreme Court. This case represents a significant shift in focus away from the traditional unitary application of conduct of hostilities norms toward a more law enforcement–based approach when dealing with "terrorists." With this approach, consideration of law enforcement norms arises in the context of both the targeting and treatment of unlawful civilian participants in hostilities. However, as has been noted, viewing non-State actor participants in hostilities as criminals is a concept that is not foreign to international armed conflict. In this respect the Israeli court appears to have simply extended the application of human rights–based norms, rather than introduced them as an entirely new concept.

9.3.5. EXTRATERRITORIAL LAW ENFORCEMENT

Yoram Dinstein has suggested another theory with which to assess cross-border defensive action by States against "terrorists and armed bands." He refers to these operations as "extra-territorial law enforcement."[103] They are uniquely directed at non-State actors and limited to situations where the actions of a terrorist group are not attributable to the host State. In such cases, none or limited responsibility for the terrorist attacks arises on the part the "host" State if it does not know its territory is being used by terrorist groups to launch an attack; it may know, but "is incapable of putting an end to the their activity"; or such attacks are not condoned, but the State is "too weak (militarily, politically or otherwise) to prevent these operations."[104] Such cross-border defensive action is viewed as "an extraordinary case demanding, and getting an extraordinary solution in international law."[105]

Whether viewed as being extraordinary or not, such operations against transnational threats have become a prevailing part of contemporary conflict. Yoram Dinstein's analysis is particularly helpful as he highlights the exceptional diversity of conflict with non-State actors. This is a reality that is often glossed over or not even referred to by advocates of an exclusively human rights–based law enforcement approach when dealing with noninternational armed conflict and terrorism. The examples of extraterritorial law enforcement provided include incursions by Turkish troops into Northern Iraq in response to attacks by the PKK, the 2008 Columbian military raid on a Revolutionary Armed Forces of Colombia (FARC) camp in Ecuador, and the 2011 U.S. operation in Pakistan to kill Osama bin

---

[102] *Id.* ¶ 40, at 394.

[103] DINSTEIN, *supra* note 14, ¶ 720, at 272.

[104] *Id.* ¶¶ 712–14, at 269–70.

[105] *Id.* ¶ 715, at 270.

Laden.[106] Reactions by States to cross-border attacks from terrorist groups regularly involve the use of force by military units. That response can include strikes with aircraft and missiles (including drone strikes), as well as more orthodox action such as the dispatch of military units "with a view to destroying the infrastructure of the armed group or terrorists."[107] These are uses of force that have traditionally been associated with armed conflict.

The complex situation that can arise during intervention in another State is evident when it is noted "[a] singular illustration of extra-territorial law enforcement was provided by the Israeli incursion into Lebanon, in 1982, designed to destroy a vast complex of Palestinian bases from which multiple armed attacks across the international frontiers had originated."[108] As Gil Merom has described the situation, "Israel conducted three wars in Lebanon, in two phases. In the first phase, it fought two wars of territorial acquisition: one against the PLO and one against the Syrians. In the second phase, Israel was involved in a protracted guerrilla war against various groups in Lebanon."[109] The operation did not amount to a war with Lebanon, as the territorial State, although there was a continuation of the ongoing conflict with Syria.[110] In this context extraterritorial law enforcement took place contemporaneously with international armed conflict. What is clear is that as a concept extraterritorial law enforcement is not limited to isolated drone strikes or raids. It can extend to operations of significant duration involving elevated levels of violence. These operations also can transcend counterterrorism into the realm of counterinsurgency.

Extraterritorial law enforcement is viewed as an act of self-defense that is governed by the State self-defense principles of necessity, proportionality, and immediacy.[111] Since these extraterritorial operations are not directed at another State, they constitute neither inter-State war nor incidents "short of war" between States. The question then is what normative regime governs this use of force during the actual conduct of hostilities. Consistent with the position that a noninternational armed conflict occurs where "the fighting is taking place within the territory of a single country (although the armed conflict may have spill-over effects in a neighbouring country and perhaps there is a parallel non-international armed conflict),"[112] an extraterritorial law enforcement action may be part of an "internal" conflict. In meeting that threshold the conflict has to transcend internal situations of disturbance and sporadic violence.[113] In this situation it appears the humanitarian law is applicable to that noninternational armed conflict.

As noted above, extraterritorial law enforcement is discussed in terms of a spillover of a noninternational armed conflict to an adjacent State.[114] Yoram Dinstein specifically rejects

---

[106] *Id.* ¶ 721, at 272–73.

[107] *Id.* ¶ 715, at 270.

[108] *Id.* ¶ 723, at 273.

[109] Gil Merom, How Democracies Lose Small Wars 156 (2003) (where it is also noted the 1982 invasion of Lebanon involved some 80,000 soldiers organized in six to seven divisions supported by the Israeli Air Force), and Sally V. Mallison & W. Thomas Mallison, Armed Conflict in Lebanon, 1982: Humanitarian Law in a Real World Setting 11 (1985) (for an outline, based on newspaper accounts, of wide scale destruction and casualties).

[110] Dinstein, *supra* note 14, ¶ 723, at 273.

[111] *Id.* ¶ 728, at 275.

[112] *Id.* ¶ 8, at 5–6.

[113] *Id.* ¶ 8, at 5.

[114] Dinstein, *supra* note 5, ¶ 82, at 26–27 and ¶ 170, at 53–54.

the argument in *Hamdan v. Rumsfeld* that "the cross-border worldwide 'war on terrorism' is a non-international armed conflict."[115] A question that arises is what type of conflict would be in existence if a technologically capable non-State actor strikes from a non-adjacent State and the threatened State responds with military force against the group in that geographically separated host country. With non-State actors fielding increasing sophisticated weapons, such as "suicide drones," it is conceivable they could strike across multiple borders.[116] If attacks from a nonadjacent State are considered to be a form of "spillover" noninternational armed conflict, then the extraterritorial law enforcement operation could still be conducted according to the appropriate humanitarian law rules. On its face this seems to create a situation that begins to resemble the *Hamdan v. Rumsfeld*[117] approach. However, it may be that responding to a specific attack or threat of an attack from a nonadjacent State under an extraterritorial law enforcement model makes the operation appear less global in concept.

A particular challenge that arises when considering the normative framework applicable to extraterritorial law enforcement is the degree of autonomy that some terrorists groups may enjoy. For example, Hezbollah's integration into the Lebanese government as well as its role in governance has prompted the comment that during the 2006 conflict with Israel the hostilities had "at least certain dimensions of an inter-State conflict"[118] In this regard, the Second Lebanese War was a conflict that the international legal community has struggled to categorize as either international or noninternational in character.[119] Further, adding to the complexity of cross-border counterterrorist operations is the situation where the threatened State does intervene against such an armed group and ends up also engaging the armed forces of that State. Consistent with the broad application of humanitarian law to inter-State uses of force, it is noted that in "any military confrontation between the foreign expeditionary force and local military units [of the territorial State], the *jus in bello* [humanitarian law] will obviously apply." [120] In this situation there would be an international armed conflict between the two States.

Another challenging aspect of the categorization of cross-border self-defense action against terrorists and armed bands as extraterritorial "law enforcement" is the potential for terminological confusion. Under this conceptual framework it is not ordinary domestic-style crimes that are being addressed in the reference to law enforcement, but rather international laws that are being enforced.[121] Further, in the context of this "law enforcement" framework, force will primarily be governed by humanitarian law–based conduct of hostilities rules. Extraterritorial law enforcement does not appear to extend to counterterrorism operations occurring outside the context of an "armed conflict." For example, if a "one-off" attack were to be viewed as not meeting the threshold for a noninternational armed

---

[115] DINSTEIN, *supra* note 89, ¶ 130, at 56.

[116] Rowan Scarborough, *Iran Creating "Suicide" Drones that Threaten Israel, U.S. Navy: Pentagon*, THE WASH. TIMES, Apr. 8, 2015, http://www.washingtontimes.com/news/2015/apr/8/iran-creating-suicide-drones-us-army-report-warns/?page=all.

[117] Hamdan v. Rumsfeld, 45 I.L.M. 1130 (2006) (U.S. Supreme Court 2006).

[118] DINSTEIN, *supra* note 14, ¶ 724, at 273.

[119] Andreas Paulus & Mindia Vashakmadze, *Asymmetrical War and the Notion of Armed Conflict—A Tentative Conceptualization*, 91 INT'L REV. RED CROSS 95, 111–12 (2009).

[120] DINSTEIN, *supra* note 14, ¶ 733, at 277.

[121] *Id.* ¶ 720, at 272.

conflict (i.e., "protracted armed violence"),[122] then it would have to be dealt with under a law enforcement paradigm.[123]

In any event, State military forces responding by means of extraterritorial law enforcement must still be prepared to apply both conduct of hostilities and law enforcement norms during operations. For example, civilians, including supporters of the terrorists groups that did not qualify as members of an organized armed group, or direct participants in hostilities would have to be treated according to human rights–based law enforcement norms. As with the other than international armed conflict categorizations, the likelihood of having to apply law enforcement norms is elevated because there is often an integral connection between insurgents, terrorists, transnational criminal organizations, and even ordinary criminals (see Chapter 6, section 6.3, *Insurgency and Crime*). Further, the participation of non-State actors in activity that prompts defensive action by a State under the extraterritorial law enforcement theory would place them firmly in the category of illegitimate participants in hostilities. This would ultimately mean the application of human rights–based law enforcement standards and treatment regarding any detention, investigation, or prosecution for their participation in hostilities. As a result, even though extraterritorial law enforcement might be considered primarily in terms of applying humanitarian law in the context of an armed conflict, international and domestic human rights–based law enforcement rules may also have to be applied.

9.3.6. TRANSNATIONAL ARMED CONFLICT

Another analytical framework for categorizing State versus non-State actor conflict is "transnational armed conflict." Developed in the shadow of the 9/11 attacks, one advocate, Geoffrey Corn, was particularly concerned about limitations associated with Common Articles 2 and 3 of the Geneva Conventions regarding "triggers" for determining the existence of an armed conflict. With Common Article 2 limited to inter-State armed conflicts[124] and Common Article 3 historically restricted to internal conflicts (i.e., within geographic borders),[125] a regulatory gap was seen to exist. In particular, conflict with terrorist organizations such as U.S. hostilities with al Qaeda[126] or hostilities between Israel and Hezbollah in Lebanon were viewed as not falling under either legal framework.[127] While not necessarily warmly embraced by the international legal community,[128] this proposal for a new way of looking at non-State actor conflict once again reflects the unique pressures that are forcing change in how armed conflict is viewed in the twenty-first century.

---

[122] DINSTEIN, *supra* note 5, ¶ 97 (ii), at 32 ("A NIAC cannot burst in one fell swoop").

[123] *Id.* ¶ 98, at 33.

[124] Geoffrey Corn, *Hamdan, Lebanon, and the Regulation of Hostilities: The Need to Recognize a Hybrid Category of Armed Conflict*, 40 VAND. J. TRANSNAT'L L. 295, 301–03 (2007).

[125] *Id.* at 303–09.

[126] *Id.* at 312–14.

[127] *Id.* at 305–06 and 320.

[128] *See, e.g.*, DINSTEIN, *supra* note 5, ¶ 73, at 24 ("The doctrinal impulse to break with tradition may be potent, but it is not backed by the general practice of States. Whenever States have an opportunity to pore over the matter, they reaffirm the existing dual taxonomy of IACs and NIACs.").

The transnational armed conflict theory addresses a key challenge arising from contemporary conflicts with non-State actors where the threat is not limited by geographic boundaries. Since the theory was proposed it has been acknowledged that the *Hamdan v. Rumsfeld*[129] case addresses this issue from a U.S. perspective. That decision expanded the concept of noninternational armed conflict beyond the borders of a State and with it the application of Common Article 3 to hostilities with al Qaeda. [130] However, concerns remained. It has been argued this Geneva Convention provision only specifically dealt with humanitarian protection, but did not refer to foundational principles governing the conduct of hostilities.[131] As a result, the principles of necessity, distinction, proportionality, and the prohibition of unnecessary suffering, which were "clearly essential to regulate the application of combat power," were left unaddressed.[132]

The suggested solution was to legally acknowledge an approach that some States and, in particular, the United States have long followed in applying humanitarian law principles whenever military combat power is deployed in operations against a transnational non-State actor.[133] This form of conflict was called transnational armed conflict to distinguish it from traditional international and noninternational conflict. The criteria relied on to determine whether an armed conflict exists against a terrorist group were twofold. First, there was the nature of the means deployed by the State to respond to the threat. The second criterion was the type of engagement authority provided by the State to the armed forces (i.e., status-based rules of engagement).[134] This criterion will be referred to as a "means and authority" test.

Importantly, the transnational armed conflict model acknowledges the possibility of the application of domestic law enforcement means by the State during transnational operations. However, it was viewed as "less likely that the state will be able to resort to its own domestic law enforcement capabilities in such a context."[135] The association of military forces with armed conflict and this separate reference to domestic law enforcement capacity has particular relevance to U.S. practice. While not having an aversion to applying law enforcement means internationally, there has been a historical reluctance by that State to employ military forces on such "policing" duties. This reluctance is reflective of the general trend displayed by many military forces favoring conventional warfare over counterinsurgency operations (see Chapter 7, section 7.2, *Counterinsurgency in the Twenty-First Century*). Of course, counterinsurgency doctrine privileges policing activity, and with it the application of human rights-based law enforcement norms.[136]

---

[129] Hamdan v. Rumsfeld, 45 I.L.M. 1130 (2006) (U.S. Supreme Court 2006).

[130] Corn, *supra* note 124, at 325–26.

[131] *Id.* at 314 and 326.

[132] *Id.* at 326–27. *See also* Geoffrey Corn & Eric Talbot Jensen, *Transnational Armed Conflict: A "Principled" Approach to the Regulation of Counter-Terror Combat Operations*, 42 Isr. L. Rev. 1 (2009).

[133] Corn, *supra* note 124, at 327.

[134] *Id.* at 341–46. In particular *see id.* at 346 ("the employment of combat forces by the state in response to a threat and the authorization for these forces to engage an enemy not exclusively in response to hostile act or intent, but based on status identification—provides an effective means of determining the existence of any armed conflict.").

[135] *Id.* at 343.

[136] Robert M. Cassidy, Counterinsurgency and the Global War on Terror 99–126 (2008), David H. Ucko, The New Counterinsurgency Era 126 (2009), and Max Boot, Invisible Armies: An Epic History of Guerrilla Warfare from Ancient Times to the Present 399 (2013).

The transnational armed conflict theory appears to be closely related to noninternational armed conflict. Its author relies on the ICRC Common Article 3 *Commentary* as an interpretive aid in highlighting a reference in that explanatory note to the use of "regular military forces" by the legal government as being indicative of an armed conflict.[137] This is viewed as a critical factor under the transnational armed conflict model for establishing when an armed conflict, rather than an internal disturbance, is in existence.[138] Notwithstanding the similarities, this theory can be distinguished from noninternational armed conflict. Specifically, in terms of the conflict thresholds it does not rely on the noninternational armed conflict–based *Prosecutor v. Tadić* criteria of the intensity of violence (protracted armed violence[139]) occurring between a State and organized armed groups, or the lower *Abella v. Argentina* criteria (see section 9.4, *Determining the Existence of a Noninternational Armed Conflict*, below).[140] The transnational armed conflict theory appears to be weighted toward assessing the response required to counter the threat that is presented to the State.

As a result, this conflict model concentrates on a key but largely overlooked issue that is crucial when assessing whether an armed conflict exists in relation to non-State actors, that is considering the "means" necessary to remove or at least neutralize the threat. As was discussed in the context of drone strikes, the threshold for armed conflict is too often assessed on the basis of the violence inflicted by the non-State actor. It is the overall context that matters, including the levels of force necessary to resolve the situation. In applying Common Article 3 criteria the intensity of violence must be assessed in its entirety, including what action is necessary by the State to successfully counter the threat. Put in practical terms, it may be that the delict resulting in an international operation against non-State actors involves little or no violence, for example, when foreign nationals are seized as hostages in territory that is not adequately governed by the territorial State. In responding to a threat to its nationals the State must consider whether law enforcement– or conduct of hostilities–based norms can be applied to rescue those hostages. As a matter of law, an armed conflict must be in existence for humanitarian-based norms to be applied.

However, the factors to be assessed include not only the violence associated with the kidnapping but also the nature of the organization that seized the hostages and their capacity for violence in resisting any operation by the State to whom the nationals belong. Hostage rescue is an increasingly common type of operation conducted in a counterterrorism and counterinsurgency context. However, defense of nationals is an issue that has been largely sidelined at the theoretical level because of the twentieth-century obsession with inter-State conflict and conflicts internal to States (see Chapter 10, section 10.2, *Self Defense and the Extraterritorial Protection of Nationals*). In a rescue operation, force far beyond that ordinarily considered to fall within the human rights–based law enforcement

---

[137] Corn, *supra* note 124, at 342, and THE GENEVA CONVENTIONS OF 12 AUGUST 1949: COMMENTARY, GENEVA CONVENTION (I) FOR THE AMELIORATION OF THE CONDITION OF THE WOUNDED AND SICK IN ARMED FORCES IN THE FIELD Art. 3, ¶ 1A(2) at 49 (Jean S. Pictet (ed.), 1958) [hereinafter GENEVA CONVENTION I COMMENTARY].

[138] Corn, *supra* note 124, at 342.

[139] Prosecutor v. Tadić, Case No. IT-94-1-A, Decision On The Defence Motion For Interlocutory Appeal On Jurisdiction, ¶ 70 (Int'l Crim. Trib. for the Former Yugoslavia, 1995), http://www.icty.org/x/cases/tadic/acdec/en/51002.htm.

[140] Case 11.137, Abella v. Argentina, Inter-Am. Y.B. on H.R. ¶ 1 (1997) (Commission report), http://www1.umn.edu/humanrts/cases/1997/argentina55-97a.html.

paradigm may be required to effect a release of hostages (see Chapter 10, section 10.3.3, *Modern Counterterrorism: Sierra Leone*). Resolving this issue places the spotlight directly on the question of what the limits are regarding the effectiveness of human rights–based law enforcement and at what point the confrontation between a State and non-State actor may only be amenable to an armed conflict–based response.

More problematic is the second part of the transnational armed conflict test that places a particular premium on the State making the decision that the use of armed conflict means is required. This deference toward State decision making was also noted in the analysis of Geoffrey Corn's critique of the "naked self-defense" theory applicable to drone strikes and his championing of the "means and authority" criteria (see Chapter 8, section 8.5.3, *The "Self-Defense" Option*). The "authority" aspect of the transnational armed conflict theory does not on its face appear to address the possibility that a State may overreact in its application of combat power. The danger of there being such an overreaction is enhanced where States have a long-standing policy of applying humanitarian law to regulate all international military operations (see Chapter 4, section 4.4.4, *A Policy Solution for Legal Uncertainty*). This means the default position is one of applying humanitarian law–based conduct of hostilities principles at a time when greater recognition is being given to the application human rights–based law enforcement norms during extraterritorial deployments.

The transnational armed conflict theory would be enhanced by a clearer indication of what level of threat or circumstances justifies a State response with combat forces applying a conduct of hostilities response. Some terrorist threats are sufficiently developed to rise to the level of an armed conflict. Not all threats to States and their citizens, including those posed by terrorists, occur in that context. This raises the question of whether the "means" test is a trigger for the existence of an armed conflict or is better viewed more narrowly as a response to a triggering event. In other words, it is one albeit an important criterion to consider. The application of the traditional Common Article 3 criteria of intensity of violence and group organization would go some way in addressing this issue. So would reference to the *Abella v. Argentina* decision with its reliance on more temporally limited violence rising to the level of an armed conflict.[141] However, it must also be remembered it is the adequacy of those criteria that has precipitated considerable debate, discussion, and disagreement regarding whether the United States is in an armed conflict with Al Qaeda. As Geoffrey Corn has rightly noted, "characterizing the contention between al Qaeda and the United States as an armed conflict defies indicators traditionally applied to identify the existence of non-international armed conflicts."[142] However, the challenge is that to the extent the criteria may be seen by some as under inclusive, the "means and authority" test has the potential, if improperly applied, to produce the opposite result.

The transnational armed conflict paradigm provides an important addition to the categorization of conflict debate. It helps avoid the linguist challenge introduced in *Hamdan v. Rumsfeld*[143] of having the term "non-international armed conflict" apply to conflicts that cross borders. As has been noted, terminological issues in this area can also arise from the use

---

141 *Id.*

142 Geoffrey S. Corn, *Self-defense Targeting: Blurring the Line between the Jus ad Bellum and the Jus in Bello*, 88 Int'l L. Stud. 57, 63 (2012).

143 Hamdan v. Rumsfeld, 45 I.L.M. 1130 (2006) (U.S. Supreme Court 2006).

of the term "extra-territorial law enforcement." Further, like the latter theory it highlights the limitations of traditional analytical frameworks used to categorize responses to cross-border terrorist attacks. Perhaps most important, the transnational armed conflict theory introduces the crucial factor of what means are required to resolve the security situation. In doing so it places a spotlight directly on the inadequacies of the predominately law enforcement response favored by some commentators (see Chapter 8, section 8.5.1, *The Restricted "Law Enforcement" Theory*). Finally, to the extent it mirrors the *Hamdan v. Rumsfeld*[144] decision in result, this theory further reflects the need for an approach that does not rely exclusively on twentieth-century interpretations of international law. However, given the broad reliance on and apparent preference for just two categories of armed conflict, the transnational armed conflict theory is unlikely to gain sufficient traction to displace the more traditional approaches. That said, it will remain influential in forcing change in the way contemporary conflict is viewed.

9.3.7. NONINTERNATIONAL ARMED CONFLICT AND THE PROPER AUTHORITY

9.3.7.1. Not of an International Character: *Hamdan v. Rumsfeld*

Notwithstanding questions as to whether operations against terrorists constitute "conflicts of an international character," "extraterritorial law enforcement," or "transnational armed conflicts," increasingly such hostilities are being viewed as noninternational conflicts. While noninternational armed conflict has long been considered to be restricted to violence that is internal to States, the advent of transnational terrorism has broadened the acceptance of this category of conflict being applied to incidents involving cross-border hostilities with non-State actors. Under this theory, which is most prominently reflected in the *Hamdan v. Rumsfeld* decision, an attack not attributable in some fashion to another State (i.e., a non-State actor attack controlled, harbored, or assisted by a State) is considered to be a conflict "not of an international character."[145] This approach is both traditional and radical at the same time. In a traditional sense it limits international armed conflict to inter-State warfare. In terms of a new interpretive approach it appears to vastly expand the concept of what constitutes a noninternational armed conflict.

Perhaps the most confusing aspect of viewing transnational terrorism in this way is that the actual "character" of the conflict has little to do with it being "noninternational" in a geographic sense. It is inescapable that factually the attacks on the United States on 9/11 were transnational and ultimately "international" from that perspective. However, since being "international" in a legal sense has been long associated with conflicts between nations (excepting, of course, "wars of national liberation" in Additional Protocol I), a non-State actor attack not attributable to a State is relegated to the status of being a "conflict not of an international character."[146] In contrast, the character of inter-State conflict does

[144] *Id.*

[145] *Id.* at 1154.

[146] *See, e.g., id. See also* Geoffrey S. Corn & Eric Talbot Jensen, *Untying the Gordian Knot: A Proposal for Determining Applicability of the Laws of War to the War on Terror*, 81 Temple L. Rev. 787, 801–02 (2008).

not appear to hinge on whether it involves a trans-border use of force between the attacking and defending States. For example, an attack by one State on an Embassy or the armed forces of another State located within the borders of the first State, similar targets situated in a third State, or against military forces on the High Seas can precipitate an international armed conflict.[147] Although inter-State violence occurs most often in a cross-border context, the conflict is fundamentally international in character because it takes place between two "proper authorities" (i.e., States). While this interpretation is well established in international law, the approach precipitates its own linguistic complexities in order to make cross-border attacks carried out solely by non-State actors "not" of an international in character. This complexity is evident in the use of terms such as "extra-territorial law enforcement,"[148] "transnational armed conflict,"[149] and even "multi-national NIAC"[150] when referring to cross-border non-State actor conflict. Analysts appear to go to considerable lengths to avoid using the word "international." Categorizing the two types of conflicts as "State armed conflict" (i.e. international armed conflict) and "non-State armed conflict" (i.e. non-international armed conflict) respectively would more accurately reflect their differences, rather than trying to suggest they are "international" or not. However, international law is far too tradition bound to see a change in terminology any time soon.

The emerging theory that a State versus non-State actor armed conflict is noninternational in character regardless of where it occurs has the advantage of standardizing how conflict is characterized. Like international armed conflict, the focus is on the status of the Parties and not their physical location or that of the attack. There is a basis for such an approach found in one interpretation of Common Article 3 of the Geneva Conventions. It has been suggested there is no territorial limitation to the operation of Common Article 3 "providing that it applies to armed conflicts 'not of an international character occurring in the territory of *one* of the High Contracting Parties'."[151] In this interpretation "one" is equated to "a," thereby allowing for conflicts to spill over into neighboring States while still be considered noninternational in character. The result is that greater emphasis is placed on the "parties involved rather than by the territorial scope of the conflict."[152] The *Hamdan v. Rumsfeld*[153] decision might be interpreted as being consistent with this approach to the extent the original attack on 9/11 occurred in the United States and therefore in the territory of "one" of the High Contracting Parties. However, since that time the violence has often taken place in States not connected to the original attack (i.e., Somalia, Yemen,

---

[147] DINSTEIN, *supra* note 14, ¶¶ 566-71, at 214-16 (these would be "armed attacks" which by definition would create a situation of an international armed conflict between two States).

[148] *Id.* ¶¶ 711–32, at 268–77.

[149] Corn, *supra* note 124, at 299 ("a term used to represent the extraterritorial application of military combat power by the regular armed forces of a state against a transnational non-state armed enemy.").

[150] SUSAN BREAU, MARIE ARONSSON, & RACHEL JOYCE, DISCUSSION PAPER 2: DRONE ATTACKS, INTERNATIONAL LAW, AND THE RECORDING OF CIVILIAN CASUALTIES OR ARMED CONFLICT, Oxford Research Group 12 (June 2011), http://www.oxfordresearchgroup.org.uk/sites/default/files/ORG%20Drone%20Attacks%20and%20International%20Law%20Report.pdf.

[151] LIESBETH ZEGVELD, THE ACCOUNTABILITY OF ARMED OPPOSITION GROUPS IN INTERNATIONAL LAW 136 (2002) (emphasis added).

[152] *Id.*

[153] Hamdan v. Rumsfeld, 45 I.L.M. 1130 (2006) (U.S. Supreme Court 2006).

Pakistan). This would appear to extend the concept of "spillover" beyond its normal application in an adjacent State, although the 9/11 attacks would still seem to qualify under this theory, at least as a triggering event.

9.3.7.2 A Concentration of Power and Responsibility

Under the model suggested by the U.S. Supreme Court in *Hamdan v. Rumsfeld*,[154] the conflict between a State and a non-State actor can been viewed as being more about the exercise of power and control by the State than geographical boundaries. It is such a power relationship, based on legitimacy, that underpins the just war principle of "proper authority." Even with respect to internal order, much of the ability to control banditry and the misuse of territory for purposes outside the law has depended upon the concentration of power within authority of the territorial State.[155] It is as a result of that power being placed in the hands of the State that the human rights legal framework, judicial systems, and law enforcement capabilities were developed to control the relationship between those who govern and those who are governed (see Chapter 4, section 4.4.1, *Public, Private, and Mixed War*).

Further, as was evidenced with the acknowledgment by the highest U.S. officials that domestic constitutional protection extends to its own citizens operating against its interests from the "ungoverned spaces," this power relationship does not necessarily change based on where that actor is located or because borders must be crossed.[156] Significantly, an unequal legal relationship exists between State and non-State protagonists, even when the latter group is operating outside the territory of the threatened State and within the borders of another State. Except in exceptional circumstances, such as is provided for in Article 1(4) of Additional Protocol I, or the very limited circumstances under the Third Geneva Convention where members of organized armed groups might gain prisoner of war status during international armed conflict, non-State actors do not enjoy a privileged status under international humanitarian law analogous to that granted to States.[157] Therefore, at law, the position of non-State participants engaged in transnational terrorism is relatively easy to discern. As with internal conflicts, there is no prisoner of war status, and non-State actor participants are liable to be treated as criminals. This immediately evokes questions of law enforcement and efforts to bring the perpetrators of an attack to "justice." [158] Bringing the terrorists to justice has become a commonly used phrase in the contemporary lexicon governing activity, ranging from arrest to targeted killing. Similar to other efforts to categorize non-State actor conflict, the issue of criminality significantly influences the normative discussion within the transnational noninternational armed conflict context.

---

[154] *Id.*

[155] Eric Hobsbawm, Bandits 11 (reprint 2012) (2011).

[156] *Obama's Speech on Drone Policy*, N.Y. Times, May 23, 2013 [hereinafter *Drone Policy*], http://www.nytimes.com/2013/05/24/us/politics/transcript-of-obamas-speech-on-drone-policy.html.

[157] GC III, art 4. A. (2).

[158] *US Forces Kill Osama bin Laden in Pakistan*, NBC News (Aug. 14, 2011), http://www.nbcnews.com/id/42852700/ns/world_news-death_of_bin_laden/t/us-forces-kill-osama-bin-laden-pakistan/#.UgyqJhbxaao (where in the aftermath of the operation that killed Osama bin Laden justice being done was a theme of President Obama and two of his immediate predecessors, Presidents Clinton and Bush).

9.3.7.3. Borders or Barriers?

Using power and status to define the parameters of noninternational armed conflict rather than focusing on territory is problematic when some analysts seek to impose geographic limits on the ability of a State to react to a non-State actor threat. Attempts to establish territorial limits have prompted considerable debate within the legal community (i.e., the geography of war).[159] A cynical view is that a rigid territorial approach results is an asymmetrical situation where the threat posed by non-State actors seems to know no boundaries, while international borders would seemingly limit a State response. Importantly, this situation occurs even if the host territorial State does not exercise effectively control over the area where the non-State actor is operating (see Chapter 8, section 8.5.1, *The Restricted "Law Enforcement" Theory*). However, States themselves may rely on borders to specifically differentiate between operational approaches. This has been the situation regarding the different types of operations being conducted in Afghanistan (i.e., counterinsurgency) and Pakistan (i.e., counterterrorism).[160]

A geography-based test that limits State action would be uncontroversial if territorial States did what they are supposed to do by exercising their role as "proper authorities." States have the responsibility to control and ultimately police their own territory. Unfortunately, for a variety of reasons that does not always happen including because a territorial State may actually govern in name only. As has been evidenced in some of the "ungoverned spaces" of the world, such as in parts of Pakistan, Yemen, Somalia, as well as the Maghreb and Sahel Regions of Africa, having the responsibility to govern does not necessarily equate to a capability or willingness to do so (see Chapter 4, section 4.4.5, *Small Wars and "Ungoverned Spaces"*). Long associated with the notion of State governance is the idea of law enforcement. To the same extent that borders should not offer an absolute bar to a State acting in self-defense by applying hostilities rules, neither should that boundary set similar limits for when a threatened State responds through use of law enforcement means. In effect, performing a form of governance in the "void" of the territorial State meeting its responsibilities.

Reliance on the noninternational armed conflict framework, even in the context of transnational terrorism, could mean attention will be increasingly placed on the role law enforcement can play when responding to cross-border acts of violence by non-State actors. To date State responses appear to have often relied on counterterrorist drones strikes or Special Forces raids applying humanitarian law. A change in approach would be based on the connection between governance and the traditional role human rights–based law enforcement has played during such conflict. There are analogies that can readily be made to other conflict between State and non-State actors. When a territorial State acts against internal threats posed by a non-State actor, it typically constrains its options through the application of domestic law, which focuses on exercising policing powers. There is also a

---

[159] Kenneth Anderson, *Targeted Killing and Drone Warfare: How We Came to Debate Whether There Is A "Legal Geography of War,"* AM. U. WASH. COL. L. (College of Law Research Paper No. 2011-16, 2011).

[160] George S. McNeal, *Targeted Killing and Accountability*, 102 GEO. L.J. 681, 756 (2014), *available at* http://papers.ssrn.com/sol3/papers.cfm?abstract_id=1819583.

heavy reliance on a police primacy approach during counterinsurgency operations (see Chapter 7, section 7.3, *The Primacy of Law Enforcement*).

Further, often referred to as "terrorists," non-State actors engaged in cross-border attacks against States can range in organization from sophisticated insurgent forces approximating regular armed forces to more traditional small-cell-based terrorist groups. For many groups associated with the jihadist movement the threat can be both insurgent and terrorist in nature (see Chapter 6, section 6.4.3, *Terrorism and Insurgency: Al Qaeda and an Islamic State*). The reality on the ground is that, like insurgents, terrorists also operate among the people either in the State they are attacking or when located in a sanctuary in another State. In this regard, the Afghan Taliban is no less an insurgent group when it operates in Pakistan than when it conducts operations in Afghanistan. It is not automatically a terrorist group on one side of the border and an insurgent group on the other. As is evidenced by the concern over collateral damage that has arisen as a result of drone strikes carried out in countries such as Pakistan, the conflict that unfolds in many "ungoverned spaces" occurs no less among the inhabitants of that territory. Strikes against "terrorist groups" located in cross-border sanctuaries often involve fighting a "war amongst the people." Factually, it is simply another form of "small war."

A critical question then becomes why a frequently broad claim for the application of humanitarian law during a transnational noninternational armed conflict can or should be made to the exclusion of human rights–based law enforcement. While it has been rightly noted that a State is generally less likely to be able to apply a law enforcement response during transnational operations, it does not mean that will always be the case.[161] For example, the 1982 Israeli invasion of Lebanon ultimately created a situation where some of the fighting might be more accurately described as taking place in an insurgency/counterinsurgency context, rather than simply striking back at terrorists.[162] Counterinsurgency doctrine relies heavily on policing in addition to the conduct of military operations. Even in respect of more limited operations the same tactical situation that permits a law enforcement response as a matter of policy on one side of an international border may exist just 10 kilometers inside an adjacent State.

In situations where the borders between two countries are relied on by States to delineate different operational approaches, the challenge may become one of justifying the difference. Afghanistan has been fought as a counterinsurgency with a particular emphasis placed on avoiding civilian casualties, while Pakistan is normally seen as a strike-oriented counterterrorism effort. In respect of the targeting carried out by the United States in Pakistan, it has been suggested, "winning hearts and minds is not a central component of the overall strategy."[163] The result has been that the targeting proportionality assessment is applied differently in each country.[164] However, a strong argument can be made that it is the operational situation and not simply an international boundary that should govern the security response. For example, the border separating Afghanistan and Pakistan does not necessarily change the organization, goals, or threat that the Taliban poses to the stability of the

---

161 Corn, *supra* note 124, at 343.

162 Merom, *supra* note 109, at 156 ("Israel was involved in a protracted guerrilla war against various groups in Lebanon.").

163 McNeal, *supra* note 160, at 756.

164 *Id.*

former country. Nor does it mean that the Pakistani civilians are any less entitled to protection from the collateral effects of targeting than their Afghan counterparts. If "soft" or "hard" knock approaches associated with counterinsurgency, which are designed to capture a suspected terrorist while limiting collateral effects, are feasible on the Afghanistan side of the Durand Line, they might also be capable of being applied in a similarly situated scenario in Pakistan.[165] Whether they would be or not involves legal (i.e., lawful intervention), political (i.e., "boots on the ground" concerns), and operational considerations (i.e., element of risk and exercise of control in area of operations).

Doctrinally, the conduct of counterinsurgency operations may require humanitarian and human rights law to be applied under circumstances where they both interface and overlap. Certainly not all circumstances will be amenable to law enforcement. In such situations insurgents or terrorists who are engaged in armed conflict can be targeted under humanitarian law rules. However, reflecting the same pressures that led to the development of "police primacy" doctrine for counterinsurgency, counterterrorism might also be considered in specific situations as a law enforcement dominant activity. As has been noted regarding the counterterrorism campaign against Al Qaeda, "[a] light-footprint approach that relies on clandestine law enforcement, intelligence, and special operations forces to support foreign governments and conduct precision targeting has been more effective."[166]

Looked at it from another perspective, given the focus on the power relationship between the State and non-State actors it is not immediately clear why, other than because of a formal attachment to international boundaries, that the threatened State should not be expected to act in the place of a nonfunctioning "proper authority" when the circumstances reasonably allow that to occur. An international border should not be a barrier to requiring the application of the principles of good governance and the maintenance of law and order. For example, such an approach could be applied where sufficient control is exercised over the immediate territory where the non-State actor is operating and the risk to both security personnel and uninvolved civilians is be mitigated sufficiently to allow an operation to be carried out according to human rights–based law enforcement standards. Of note, the criteria of risk and control are the same ones identified in the *Targeted Killing Case* to be applied in determining when law enforcement concepts such as arrest, investigation, and trial should be incorporated into decision making regarding targeted killing[167] (see Chapter 7, section 7.4.2, *The "Targeted Killing" Case and Control*).

9.3.7.4. Acting in Place of the Territorial State

Adopting a "police primacy" approach for counterterrorism carried out in the context of a noninternational armed conflict better reflects the different power relationship that exists at law between the State and the non-State actors. Relying on a policing approach is

---

165 COMISAF Night Operations Tactical Directive, Headquarters International Security Assistance Force (Dec. 2011), *available at* http://www.isaf.nato.int/images/docs/20111105%20nuc%20night%20operations%20tactical%20directive%20(releaseable%20version)%20r.pdf (for a discussion of "soft" and "hard" knock approaches for searching Afghan residences).

166 Jones, *supra* note 48, at 438.

167 Public Committee Against Torture in Israel v. Israel, Israel Supreme Court [16 December 2006], 46 ILM 375, ¶ 40, at 393–94 (2007).

also consistent with the well-established condition precedent for conducting cross-border counterterrorist activity. The territorial State must be "unable or unwilling" to police its own territory. If the terrorist threat is such that it is amenable to law enforcement action by the territorial State, then it makes sense that a similar approach might be asked of the threatened State acting in a cross-border capacity when it is feasible to do so in the circumstances. A counterargument may be that unlike with respect to its own citizens the threatened State has no governance responsibility toward civilians located in another State. Further, it may be difficult to have the citizens of one State identify with those of another State in the context of an armed conflict. This could occur particularly when some of those foreign civilians are carrying out terrorist acts or killing security personnel of the first State. These concerns have to be balanced against the humanitarian imperatives of both international and domestic law (see Chapter 14, section 14.5, *Confronting Transnational Violence: A Holistic Approach*).

Various theories have been identified for protecting civilians.[168] It has been suggested that if an enemy cannot identify with an opponent's civilians, they "should at least tolerate them."[169] Such toleration is enough in war, although it "does not necessarily have to involve recognition of enemy civilians as being 'like us' . . . "[170] In any event, it "must involve respect for people's basic rights."[171] Professor Eyal Benvenisti has also identified obligations that arise with respect to what he calls symmetric warfare (e.g., inter-State armed conflict) and transnational asymmetric warfare (e.g., against a non-State actor). He makes a distinction between respecting and ensuring rights with the latter only arising in a context where control is established (e.g., occupation). Ensuring rights is identified as a positive obligation to protect enemy citizens, while respecting rights entails a duty to avoid unnecessary harm.[172] For example, an attacking army has an obligation to "respect" the citizens of the enemy state until such time as effective control is exercised, and then there is a requirement to "*ensure* the rights of enemy civilians."[173] However, he also suggests "[t]he vertical power relations that exist in transnational asymmetric conflicts, particularly against non-state actors, seem to call for recognizing positive duties towards those civilians, like in an occupation," even if such an obligation would be quite demanding.[174] This is a reference that resonates with the concept of the proper authority.

Professor Benvenisti anticipates that what is called the LOAC "camp" will resist imposing a duty to ensure the rights of civilians in context of transnational asymmetric conflict on the basis "that there is no moral or legal basis for the obligation to consider other-regarding

---

168 Hugo Slim, Killing Civilians: Method, Madness, and Morality in War 259–93 (2008) (these include respect for the preciousness of life, mercy, fairness, their innocence, emotional identification with enemy civilians, toleration, and a range of self-interested factors such as self-interest, law and punishment, consideration of rights, the rewards of peace, and the global movement to protect civilians.).

169 *Id.* at 278.

170 *Id.*

171 *Id.*

172 Eyal Benvenisti, *The Legal Battle to Define the Law on Transnational Asymmetric Warfare*, 20 Duke J. Comp. & Int'l L. 339, 351 (2010).

173 *Id.*

174 *Id.* ("Such a duty will reflect the nature and scope of the power that the 'attacking' army (during an on-going, indefinite 'attack') has over the attacked population.").

considerations in the absence of reciprocity and mutuality of obligations, when there is no assurance that others are equally committed to act selflessly."[175] However, it might also be argued that prominent among human rights, whether it is "respected" or "ensured," is the right to life. That right applies equally to civilians regardless of where they reside and must be protected under both humanitarian and human rights law. For example, no distinction is made on the basis of nationality or the geographic location of the civilians regarding the requirement to protect the civilian population and individual civilians "against dangers arising from military operations."[176] This remains a legal obligation even for the most ardent advocates of what has been described as the "LOAC camp." In both humanitarian and human rights law terms the lives of uninvolved civilians located in another country where military operations are being carried out must rate as highly as those of the threatened State.

It is in this context that President Obama's reference in his 2013 *Drone Speech* to the extension of the consideration of capture before killing any terrorists as one that "respects the inherent dignity of every human life" has a particular resonance.[177] Ultimately, it is this resonance with enemy civilians on a human level that will provide the most consistent humanitarian protection. The question that needs to be asked when undertaking targeting in the context of a counterinsurgency or counterterrorist operation is: Would any resulting collateral casualties be acceptable if an attack was being conducted among the civilians located in the territory of the State undertaking a strike? This concept can simply be referred to as the "cold stark mirror of reciprocity." It is an approach that can lead to a sobering analysis and a more realistic view of the strategic impact of targeting in contemporary conflict.

9.4. Determining the Existence of a Noninternational Armed Conflict

9.4.1. CONFLICT CRITERIA AND THE INTENSITY OF THE VIOLENCE

Unlike the inter-State warfare with its generally expansive and ultimately low threshold for determining the existence of a conflict, the assessment of whether violence with a non-State group is a transnational noninternational armed conflict depends upon different factors. It is in terms of thresholds that some of most significant differences between international and noninternational armed conflict may be felt. This section will explore the threshold for noninternational conflict in terms of the two foundational criteria set out in the *Prosecutor v. Tadić* decision: intensity of violence (protracted armed violence) and group organization.[178] However, before doing so the problem of applying a State policy favoring the use of humanitarian law to govern all international operations will be discussed.

---

[175] *Id.* at 356.

[176] AP I, art. 51(1). *But see also* Benvenisti, *supra* note 172, at 356 (where it is noted "it is not entirely clear whether the law governing conventional international armed conflict expects the 'reasonable military commander' to be 'nationality-blind' in order to avoid criminal or state responsibility.").

[177] *Drone Policy, supra* note 156.

[178] Prosecutor v. Tadić, Case No. IT-94-1-A, Decision On The Defence Motion For Interlocutory Appeal On Jurisdiction, ¶ 70 (Int'l Crim. Trib. for the Former Yugoslavia, 1995), http://www.icty.org/x/cases/tadic/acdec/en/51002.htm.

In operational terms the governing body of law sets the parameters for the use of force by State security forces. If the situation does not constitute an armed conflict, then only human rights law applies. In this scenario human rights law cannot be displaced by a policy decision to apply humanitarian law rules. This means that, at law, even on international operations States may be limited to an application of more restrictive law enforcement principles when considering the use of force. A particularly challenging question is what this threshold means in terms of transnational non-State actor violence. The adoption of a legal threshold for such armed conflict appears to run counter to the policy adopted by some States of applying humanitarian law principles whenever they conduct international operations regardless of whether they are categorized as an armed conflict (see Chapter 4, section 4.4.4, *A Policy Solution for Legal Uncertainty*).

Perhaps one of the most interesting aspects of this policy is that its reliance on geography in determining the choice of law (i.e., operations external to the State) now seems inconsistent with the *Hamdan v. Rumsfeld* decision, which adopted a status-based approach for categorizing conflict (i.e., State versus non-State actor).[179] With the noninternational armed conflict designation comes not only the application of a law enforcement paradigm for operations below the conflict threshold but also potentially greater reliance on that human rights approach even during the conduct of hostilities. To the extent States apply a policy favoring the application of humanitarian law on international operations, that option needs to be carefully reconsidered, at least in respect of the use of force.

Certainly there are advantages to applying a humanitarian law paradigm for international operations in terms of clarity, training, and even humanitarian protection. Defaulting to humanitarian law as a matter of policy is generally protective of civilians as it can help ensure military forces apply the highest standards for detainee treatment. For example, a policy that all detainees will be treated to the standards applicable to prisoners of war, regardless of the nature of the operation, is a positive step in ensuring persons arrested or taken into custody receive humane treatment. It also eases the training burden by setting up only one very high humanitarian standard to be met regardless of the type of operation. However, particular care needs to be taken to ensure that conduct of hostility rules regarding the use of force, which permit targeting on the basis of status and have a greater acceptance of collateral civilian casualties, are not used in non-armed conflict situations.

Importantly, international law has progressed since the policy of generally applying humanitarian law to international operations was first introduced in the 1990s. This occurred during a period when there was even less agreement regarding the threshold criteria for what constituted an armed conflict with non-State actors. It is no coincidence that this approach was developed at a time when the State military forces were struggling with doctrinal concepts such as military operations other than war (MOOTW) in an effort to separate post–Cold War irregular warfare from traditional conventional inter-State conflict (see Chapter 14, section 14.2, *From "Conventional" to "Hybrid" to "Holistic"*).[180] The uncertainty during this period about when a conflict was in existence was also reflected

---

[179] Hamdan v. Rumsfeld, 45 I.L.M. 1130, 1154 (2006) (U.S. Supreme Court 2006).

[180] Joint Doctrine for Military Operations Other Than War, Joint Pub 3-07, ¶ 2, at I-1, (16 June 1995), http://www.bits.de/NRANEU/others/jp-doctrine/jp3_07.pdf (for an explanation of MOOTW).

in a UN context with that organization ultimately adopting the 1999 *Secretary-General's Bulletin: Observance by United Nations Forces of International Humanitarian Law* (see Chapter 11, section 11.3.2, *United Nations Operations and "Robust Peacekeeping"*) in order to provide guidance to deployed forces on when humanitarian law rules applied.[181]

Since that time there has been a greater acceptance by States that human rights law also applies during armed conflict. These conflicts include those that are transnational in nature. In effect, there should be no legal void that has to be filled by means of a policy application of humanitarian law particularly regarding the use of force. The increased profile of human rights law during international operations is a direct result of the efforts by the human rights community to extend the operation of that body of law extraterritorially. However, importantly, from a State perspective it is also a product of the reality faced by military forces when placed in a governance role whether as an occupying power or when assisting other States in their internal conflicts. Having been thrust into a policing role, State military forces had to adjust to applying law enforcement rules governing the use of force (see Chapter 12, section 12.2.1, *The Military Task: Applying Force of First and Last Resort*). The post-9/11 period has witnessed a considerable shift in the debate about what law applies to international operations. It is a discussion that increasingly will focus on the threshold for transnational noninternational armed conflict.

However, very problematically, disagreements continue regarding what should constitute the threshold for noninternational armed conflict, whether internal or transnational in character. Setting a high threshold for the existence of an armed conflict favors the application of human rights–based law enforcement. However, such a threshold also potentially excludes the operation of humanitarian law in situations where it is warranted due to the nature of the threat. Conversely, applying too low a threshold for armed conflict can potentially lead to an overuse of force and damage counterinsurgency and counterterrorism efforts. It is in establishing the noninternational armed conflict threshold that theoretical approaches and practice have the potential to diverge. However, divergence between theory and practice is best avoided by clarifying the threshold criteria for transnational armed conflict, rather than by masking this dispute or avoiding the debate through reliance on a policy approach that favors the application of humanitarian law.

The question that has to be addressed is how well a chosen threshold reflects the contemporary non-State actor threat. An essential task for the international community and State security forces is to establish a threshold for conflict that matches the reality of the violence being faced on the ground. As the following analysis establishes, that threshold for a noninternational armed conflict is context dependent. Further, it will be argued the threshold is much lower than is often claimed particularly by those advocating for a broad and exclusionary application of human rights law.

As has been noted, the discussion regarding thresholds centers on the intensity of the violence and the degree of organization of the non-State armed groups. Turning first to the intensity of violence criterion, the 1995 *Prosecutor v. Tadić* Appeal Chamber decision

---

[181] Secretary-General's Bulletin: Observance by United Nations Forces of International Humanitarian Law, *in* Bruce Oswald, Helen Durham, & Adrian Bates, Documents on the Law of UN Peace Operations 201–05 (2010) [hereinafter Secretary-General's Bulletin].

established that a noninternational armed conflict is in existence when there is "*protracted armed violence* between governmental authorities and organized armed groups or between such groups within a State."[182] A similar violence-related criterion is also reflected in Article 8(2)(f) of the Rome Statute of the International Criminal Court, which sets out offences that apply to "armed conflicts that take place in the territory of a State when there is *protracted armed conflict* between governmental authorities and organized armed groups or between such groups."[183] There has been some debate whether the reference in the former definition to protracted armed violence, rather an armed conflict means the Rome Statute is intended to create a different standard "with an additional qualifying element of duration . . . ."[184] However, there are strong arguments against further complicating the categorization of conflict issue in this way, and the better approach is to consider there is no material distinction in their meaning.[185]

Interestingly, the violence criterion was subsequently referred to as "intensity of violence" in a 1997 *Prosecutor v. Tadić* Trial Chamber decision. Notably, this term was introduced in that decision under the heading of "Protracted armed violence between governmental forces and organized armed groups."[186] This has raised questions regarding the interaction between the references to "intensity of violence" and "protracted armed violence" in the case law. As a result, "[d]espite the *Tadić* definition, uncertainty continues to surround the identification of non-international armed conflict."[187] Much of that debate centers on the concept of "duration." One approach adopts the position that "[t]he precondition of intensity . . . comes on top of the precondition of protracted hostilities . . . . It is wrong to look at the intensity of the violence (as some commentators do) as an alternative to protected hostilities."[188] Therefore, "[v]iolence that is protracted but not intense, or intense but not protracted, does not amount to a NIAC."[189] This seems to suggest that a "one-off" attack carried out independently by an organized armed group, no matter how violent, would fail to qualify as a noninternational armed conflict. Applying the duration criterion means, "a short period of fighting, for example some 30 hours in the case of *La Tablada* [the *Abella v. Argentina* decision], would not amount to a non-international armed conflict."[190]

However, what is clear is the legal criterion applied to conflict categorization are not limited to consideration of the initial *Prosecutor v. Tadić* criteria of organization and duration. Another view is that duration alone cannot be determinative and other criteria such as

---

182 Prosecutor v. Tadić, Case No. IT-94-1-A, Decision On The Defence Motion For Interlocutory Appeal On Jurisdiction, ¶ 70 (Int'l Crim. Trib. for the Former Yugoslavia, 1995), http://www.icty.org/x/cases/tadic/acdec/en/51002.htm (emphasis added).

183 The 1998 Rome Statute, art. 8(2)(f) (emphasis added).

184 Sandesh Sivakumaran, The Law of Non-International Armed Conflict 192 (2012). *See also* Anthony Cullen, The Concept of Non-International Armed Conflict in International Law 174 (2010).

185 Cullen, *supra* note 184, at 177–79.

186 Prosecutor v. Tadić, Case No. IT-94-1-T, Opinion and Judgement, ¶ 562 (Int'l Crim. Trib. for the Former Yugoslavia, May 7, 1997), http://www.icty.org/x/cases/tadic/tjug/en/tad-tsj70507JT2-e.pdf. *See also* Dinstein, *supra* note 5, ¶ 103, at 34.

187 Sivakumaran, *supra* note 184, at 156.

188 Dinstein, *supra* note 5, ¶ 106, at 34–35.

189 *Id.* ¶ 106, at 35.

190 *Id.* at 167. *See also* Case 11.137, Abella v. Argentina, Inter-Am. Y.B. on H.R. ¶ 1 (1997) (Commission report), http://www1.umn.edu/humanrts/cases/1997/argentina55-97a.html.

the number of deaths, injuries, damage to property, "the number of people involved in the fighting and the military hardware used" can also be considered.[191] Therefore, the intensity criterion could include an assessment of the duration, as well as other factors such as the magnitude or scale of the violence. Importantly, these factors have come to be looked upon as being linked. As a result "violence of a relatively brief duration may still amount to a non-international armed conflict provided that other indicia suggesting intensity are present to a significant degree."[192] Reliance on a broader set of indicia in addition to duration is actually supported by the 1997 *Prosecutor v. Tadić* decision where reference is made to the factors set out in the Pictet *Geneva Convention I Commentary*,[193] which included the use of regular military forces.[194] This approach resonates with the "means" suggested in the context of transnational armed conflicts (see section 9.3.6, *Transnational Armed Conflict*, above).

Complicating the analysis of this issue is the fact that both sides of the debate can rely on case law to support their position.[195] It is not immediately clear how these two approaches toward what is meant by the "intensity of violence" will ultimately be resolved. However, it is essential that the approach adopted take into account the practical realities of modern conflict, and the nature of the existing threat. Particularly, if the position adopted means that the law enforcement paradigm is to take on a greater role in controlling the use of force, the supporting analysis must include an assessment of whether a policing approach will realistically enable the threatened State to counter the violence posed by the non-State actor. This requires consideration of the limitations of the law enforcement option in respect of the use of force (e.g., see Chapter 13, *Limits of Law Enforcement and Human Rights Law*). Of particular concern for those advising States is that it is difficult to see how the violence associated with the *Abella v. Argentina* case (i.e., 30 hours), [196] the 2000 Sierra Leone hostage rescue (4 hours),[197] the hijacked aircraft September 11, 2001, attacks (21 to 45 minutes),[198] the November 26–28, 2008 Mumbai

---

[191] Sivakumaran, *supra* note 184, at 167.

[192] *Id.* at 168.

[193] Prosecutor v. Tadić, Case No. IT-94-1-T, Opinion and Judgement, ¶ 562 (Int'l Crim. Trib. for the Former Yugoslavia, May 7, 2007), http://www.icty.org/x/cases/tadic/tjug/en/tad-tsj70507JT2-e.pdf.

[194] Geneva Convention I Commentary, *supra* note 137, ¶ 1A(2) at 49.

[195] *See, e.g.*, Prosecutor v. Haradinaj, Case No. IT-04-84-T, Judgement, ¶ 40 (Int'l Crim. Trib. for the Former Yugoslavia, Apr. 3, 2008), http://www.icty.org/x/cases/haradinaj/tjug/en/080403.pdf ("The [Prosecutor v. Tadić] Trial Chamber consequently interpreted the term 'protracted armed violence' to refer to the 'intensity of the conflict'"), conversely Prosecutor v. Gombo, Case No. ICC-01/05-01/08, Decision Pursuant to Article 61(7)(a) and (b) of the Rome Statute ¶¶ 231–35 (International Criminal Court, 15 June 2009), http://www.icc-cpi.int/iccdocs/doc/doc699541.pdf (where reference is made to both intensity of violence and protracted armed conflict). *See also* Dinstein, *supra* note 5, ¶ 106, at 35 (where Prosecutor v. Haradinaj case is distinguished, and the Prosecutor v. Gombo case is relied on).

[196] Case 11.137, Abella v. Argentina, Inter-Am. Y.B. on H.R. ¶ 1 (1997) (Commission report), http://www1.umn.edu/humanrts/cases/1997/argentina55-97a.html, and International Law Association, Final Report of the Use of Force Committee: The Meaning of Armed Conflict 30 (August 2010), http://www.ila-hq.org/en/committees/index.cfm/cid/1022 (where reliance is placed on the *Abella* case in reaching the conclusion "[t]he two concepts, intensity and protraction, are linked and a lesser level of duration may satisfy the criterion if the intensity level is high.").

[197] Richard Connaughton, Modern Warfare: The True Story of Conflict from the Falklands to Afghanistan 283 (2008).

[198] 9/11 Commission Report, *supra* note 36, at 32–33 (where the timings of the hijackings and subsequent crashes of the four civilian airliners are established. American Airlines Flight 77 appears to have been in the air the longest with the hijacking estimated to have occurred at 8:51–8:54 a.m., and it crashing into the Pentagon at 9:37:46 a.m.).

attack (60 hours),[199] the September 11, 2012, attack on the U.S. diplomatic facilities in Benghazi (i.e., 13 hours),[200] the September 21–24, 2013, Kenyan Westgate Mall attack (i.e., 80 hours),[201] or the 2015 Paris assaults (3 hours)[202] does not factually take on the attributes of an armed conflict and can only be dealt with by means of human rights–based law enforcement.

Each of these situations meets the indicia of intense but shorter periods of armed conflict. It might also be argued that an attack like the Nairobi Westgate Mall attack is distinguishable because it was an extension (i.e., spillover) of the conflict between Kenyan forces operating in Somalia and the terrorist group, Al Shabaab.[203] Similarly, the Paris attacks were claimed, in a statement attributed to the Islamic State, to be linked to France's operations against that organized armed group in Syria, as well as within its own borders.[204] The President of France called the attacks an "act of war".[205] Given French involvement in the Coalition fighting the Islamic State in Iraq and Syria it can be argued the violence was an extension of that conflict into France.[206] However, there are still those who view such geographically or temporarily separated attacks as occurring outside of a "hot battlefield".

---

199 DAVID KILCULLEN, OUT OF THE MOUNTAINS: THE COMING OF AGE OF THE URBAN GUERRILLA 52–66 (2013). *See also* CATHY SCOTT-CLARK & ADRIAN LEVY, THE SIEGE: 68 HOURS INSIDE THE TAJ HOTEL 278 (2013) (where reference is made to "the Pradham Commission, a two-man panel formed in Mumba . . . to explore the 'war-like' attacks on the city.").

200 MITCHELL ZUCKOFF, 13 HOURS: THE INSIDE ACCOUNT OF WHAT REALLY HAPPENED IN BENGHAZI 254–80 (2014) (describing the third and final attack, which included the use of mortars by an organized armed group on the CIA Annex in 2012 in Benghazi), and PERM. S. COM. ON INTELLIGENCE, 113th CONG., INVESTIGATIVE REP. ON THE TERRORIST ATTACKS ON U.S. FACILITIES IN BENGHAZI, LIBYA, SEPTEMBER 11–12, 2012, 3–12 (Nov. 21, 2014), https://intelligence.house.gov/sites/intelligence.house.gov/files/documents/Benghazi%20Report.pdf. *See also* MICHAEL MORELL & BILL HARLOW, THE GREAT WAR OF OUR TIME: THE CIA'S FIGHT AGAINST TERRORISM FROM AL QA'IDA TO ISIS 205–10 (2015). *But see* DINSTEIN, *supra* note 5, ¶ 160 at 50 (where the position appears to be taken that this violence occurred at a "sub-NIAC level of intensity (below-the-threshold)".).

201 Daniel Howden, *Terror in Westgate Mall: The Full Story of the Attacks that Devastated Kenya*, THE GUARDIAN, Oct. 4, 2013, http://www.theguardian.com/world/interactive/2013/oct/04/westgate-mall-attacks-kenya-terror#undefined.

202 *Three Hours of Terror in Paris, Moment by Moment*, N.Y. TIMES, Nov. 15, 2015, http://www.nytimes.com/interactive/2015/11/13/world/europe/100000004036728.app.html?nytapp=ipad&_r=0.

203 Afua Hirsch, *Westgate Mall Attack Has Roots in Kenya's Intervention in Somalia*, THE GUARDIAN, Sept. 23, 2013, http://www.theguardian.com/world/2013/sep/23/kenya-westgate-mall-attack-somalia.

204 Ben Doherty, Angelique Chrisafis & Peter Walker, *Paris terror attacks: Islamic State says France is 'top target' for actions in Syria*, THE GUARDIAN (Nov. 14, 2015, 12:01 GMT), http://www.theguardian.com/world/2015/nov/13/paris-attacks-shootings-explosions-hostages (where ISIS is reported to have stated "France and those who follow her voice must know that they remain the main target of Islamic State and that they will continue to smell the odour of death for having led the crusade, for having dared to insult our prophet, *for having boasted of fighting Islam in France and striking Muslims in the caliphate with their planes*".) (emphasis added).

205 Aurlien Breeden, *French Leader Calls Attacks Act of War and Says ISIS is to Blame*, INT'L N.Y. TIMES (Nov. 14, 2015, 5:00 AM ET), http://www.nytimes.com/live/paris-attacks-live-updates/french-leader-calls-attacks-an-act-of-war-and-says-isis-to-blame/#.

206 INTERNATIONAL HUMANITARIAN LAW AND THE CHALLENGES OF CONTEMPORARY ARMED CONFLICT, Report 32nd International Conference of the Red Cross and Red Crescent EN 32IC/15/11, at 14 (2015), *available at* https://www.justsecurity.org/wp-content/uploads/2015/11/2015-ICRC-Report-IHL-and-Challenges-of-Armed-Conflicts.pdf (the ICRC views assisting a host nation in an armed conflict against non-state actors means IHL applies to the territory of the assisting State. This is because "assisting States involved in an extraterritorial NIAC should not be able to shield themselves from the operation of the principle of equality of belligerents". However, under this approach the November 2015 Islamic State attacks on civilian targets in Paris would remain war crimes, as well as crimes under domestic law).

As a result making a separate armed conflict determination remains helpful.[207] Whether a "one off" short term conflict, or the extension of an existing one from outside the borders of a State the nature of violence is the same. Most of these intense attacks involved non-State actor use of weapons and levels of violence ordinarily associated with warfare (see Chapter 14, section 14.3, *Responding to Transnational Attacks and Threats to Nationals*). This suggests that the second broader interpretive approach, which relies on a multiplicity of factors for determining the intensity of violence related to a noninternational armed conflict, and accepts the existence of shorter duration armed conflicts, more accurately reflects the practical challenges facing State security forces.

Another key issue is that most existing theoretical analysis, and certainly court decisions, look retrospectively on the violence. In this context the duration criterion may be more readily applied. However, this is not a luxury that State security personnel and their legal advisors normally have. Decisions on the level of force that can be used to confront the violence most often have to be made under severe time constraints at the time of, or just prior to an attack. As Noam Lubell has noted:

> [t]he criterion of "protracted" may not have been filled in such as short period of time, but the demand of a lengthy time period is problematic, since IHL must be applied from the very start of a conflict situation, and at the moment of the outbreak of hostilities it is not always possible to know how long they will last.[208]

In this context it is unrealistic to place disproportionate weight on the violence being protracted or of long duration. The fact that the exact nature of the conflict will likely be determined after the operational decision is made to use force undoubtedly serves to cause security forces to incorporate more restrictive norms associated with law enforcement to temper the amount and type of force used even as the violence escalates.

The idea that an armed conflict with non-State actors can occur in the context of a single attack will undoubtedly be viewed as controversial by advocates of a higher threshold for the existence of a noninternational armed conflict. It is a view that will be compounded by traditionalists who seek to restrict the expansion of a State right to use force connected to armed conflict under any circumstances. However, the security forces should not be put in a position of having to hope that a court is willing to apply an expanded notion of human rights law to govern a use of force in response to an attack that is clearly related to the conduct of hostilities. The security forces have to react at the time of the attack, and the force used must permit the successful resolution of the security situation. Theory must match reality.

Recognition of this lower conflict threshold does represent change for advocates of the protracted violence threshold and for many human rights advocates. It will also run counter to traditional interpretations of international law, which seek to limit the cross-border

---

[207] Noah Barkin & Sabine Siebold, *Germany to Support Military Campaign Against IS After French Appeal*, Reuters, Nov. 26, 2015, 8:11 PM GMT, http://uk.reuters.com/article/uk-mideast-crisis-germany-france-idUKKBN0TF0ZU20151126 (where the German Defense Minister indicated "Germany was not at war because it is not fighting a sovereign country but a 'murderous gang,' referring to Islamic State.").

[208] Noam Lubell, Extraterritorial Use of Force Against Non-State Actors 106–07 (2010).

use of force by States in any context. However, contemporary violence is occurring in a complex environment within which traditional views of the law, such as States not being permitted to exercise self-defense against non-State actors not attributable to a State, have already been forced to change. Of course, where an armed conflict is determined to exist that does not preclude the use of a law enforcement–style response (in whole or in part) depending upon its effectiveness in dealing with the security threat. In this respect, what has to occur is the setting up a normative framework that does not artificially bar the existence of an armed conflict, but rather expects a law enforcement approach to be considered first, and only not applied when it is not feasible to do so. In other words, requiring a more holistic application of the available bodies of law governed by a "police primacy" approach. This approach allows for a more effective matching of the response to the threat and provides for an escalation of means as a situation deteriorates.

9.4.2. GROUP ORGANIZATION

9.4.2.1. Varied Organizational Types

The second major criterion applied in assessing the threshold for noninternational armed conflict is whether the armed groups involved have a sufficient level of organization. This raises questions regarding what factors are to be considered in assessing what constitutes adequate organization in this legal test. For example, it has been suggested that groups like Al Qaeda causing what is viewed as sporadic violence are criminal terrorist organizations.[209] While group organization and the intensity of violence are ultimately linked, each should first be assessed in its own right and then considered holistically. The assessment of how a group is organized can be overly influenced by reliance on certain *Prosecutor v. Tadić* and other ICTY criteria, such as looking for an official command structure, headquarters, uniforms, discrete roles and responsibilities for different entities, modes of communications, and military training that would exclude smaller or less hierarchically organized groups from consideration. Other criteria relied on by the ICTY jurisprudence typically includes internal regulations, disciplinary procedures, the control of territory, undertaking negotiations with third parties, the distribution of arms, and recruiting new members.[210] Similarly, requiring that the group on being sufficiently organized to "enable it to implement international humanitarian law" also appears out of step with the terrorist nature of many armed groups. [211] These are useful and relevant criteria for many armed groups. However, they do

[209] O'Connell, *supra* note 55, at 355–57.

[210] SIVAKUMARAN, *supra* note 184, at 170–71 (for a detailed listing of the criteria identified in a number of ICTY cases). *But see* the U.N. Secretary General, Report of the International Commission of Inquiry on Darfur, Geneva ¶ 75, at 26–27 (25 January 2005), http://www.un.org/news/dh/sudan/com_inq_darfur.pdf (where reliance was made on rebels controlling territory as part of the rationale for determining an armed conflict was in existence), and CULLEN, *supra* note 184, at 138 n.114 (where it is suggested "[t]his appears to set a much higher threshold for the application of international humanitarian law than that which would ordinarily be required for the application of common Article 3.").

[211] Jelena Pejic, *The Protective Scope of Common Article 3: More than Meets the Eye*, 93 INT'L REV. RED CROSS 189, 192 (2011) (emphasis is also placed on the group on being sufficiently organized to "enable it to implement international humanitarian law."). *See also* Aerial Drone Deployment on 4 October 2010 in Mir Ali/Pakistan, 157 I.L.R. 722, 742 (2013) ("Features recognised as possible indicators for this include the existence of a headquarters or the ability to recruit, train, and arm one's own fighters.").

not necessarily apply to all of the organized armed groups that are integral participants in contemporary non-State actor warfare.

The structure of organized groups that participate in armed conflict is far more varied than this association with regular armed force structure suggests (see Chapter 6, section 6.2.2, *Insurgent Organization*). Organized armed groups include vanguard, parochial, integrated, and fragmented structures, and combine both centralized and local control.[212] The organization of the armed groups may range from a traditional hierarchical approach resembling conventional armed forces to a horizontal cellular structure frequently associated with terrorism and urban guerrillas. The potential for an overly narrow application of the armed conflict threshold increases when reliance is placed on a listing of criteria such as relying on the existence of a headquarters.

When assessing the nature of organized armed groups, a degree of caution needs to be applied. Simply because a group is organized along military lines does not mean it is a participant in an armed conflict. Armed groups may organize themselves in various ways ranging from hierarchical military organizations to small cells of mission-specific personnel. In this respect, Mexican drug cartels are often organized along military lines and have engaged to a limited degree in the use of improvised explosives. However, without engagement in challenging the political authority of the State, such criminal groups remain solely amenable to a law enforcement response (see Chapter 6, section 6.3.1, *The Nature of the Threat*). Further, as was seen in the case of the IRA and the Loyalist groups operating in Northern Ireland, the adoption of a military-style structure will not necessarily mean a conduct of hostilities approach will need to be applied (albeit as a matter of policy) unless the operational situation reaches a critical tipping point.

Further, the challenge in ascertaining whether a group is an organized one involved in an armed conflict is complicated by the frequent adoption by insurgents, terrorists, and transnational criminals of a similar cellular structure (see Chapter 6, section 6.2.2, *Insurgent Organization*, and section 6.4.1, *Terrorism*). Sandesh Sivakumaran has rightly noted, "decentralization does not mean that groups do not meet the international humanitarian law requirement of organization."[213] However, the adoption of a cellular organization may make it more difficult to conclude the group is engaged in such conflict. The use of this common organizational structure also points to why participants in armed conflict, like members of criminal groups, may often be dealt with through a law enforcement response.[214] The organizational similarities between criminal groups and some insurgent forces can be seen in contemporary descriptions of the enemy. Regarding the Sunni-based insurgents, it has been suggested "[t]he insurgent groups were like local gangs, while AQI—richer, crueler, and better linked across the country—was the mafia."[215]

What is clear is that some of the most destructive activity, such as the use of improvised explosive devices (IEDs) and the conduct of suicide attacks, are fundamentally group activities, and ones commonly associated with or connected to armed conflict. There can be individuals who engage in "lone wolf"-style terrorist attacks, or small terrorist groups

---

[212] Paul Staniland, Networks of Rebellion: Explaining Insurgent Cohesion and Collapse 5 (2014).

[213] Sivakumaran, *supra* note 184, at 173.

[214] Randall Wilson, Blue Fish in a Dark Sea: Police Intelligence in a Counterinsurgency 149 (2013).

[215] Stanley McChrystal, My Share of the Task: A Memoire 175 (2013).

unconnected to an armed conflict, that remain subject solely to a policing response.[216] It is these unique acts of violence and limited organizational structure that help make those threats a law enforcement matter. In contrast, the intensity of the violence inflicted by groups for a political purpose, even in terms of "one-off" attacks such as the 2008 Mumbai attack, can quickly take on the attributes of an armed conflict.[217] That threshold is easily crossed when the levels of violence reached that experienced in Iraq. In 2006–07 a cavalry squadron commander reported his unit was subjected in one year to "over 350 improvised explosive devices (IED) attacks, averaging close to one a day. Some of them proved to be quite deadly."[218] Attacks were also carried out by means of "sniper-fire, mortar fire, rocket-propelled grenades attacks and suicide car bombings."[219] Even decentralized and cellular groups can have sufficient structure to enable them to "carry out hostilities, enforce orders, mete out internal discipline, and the like."[220]

Further, during the course of a conflict the group may fluctuate between these two structures, often depending upon how well they are doing. As was seen in Iraq, the size and organization of insurgent groups may vary considerably, ranging from combat cells to large military-style organizations.[221] The armed resistance may involve numerous and diverse small insurgent organizations, rather than one group that takes on the attributes of a traditional armed force. The success or failure of armed groups can have a significant effect on whether they adopt a more conventional or cellular organization (see Chapter 6, section 6.2.2, *Insurgent Organization*). The attainment of a more conventional structure can occur and is reflected in traditional insurgency doctrine. However, it represents only one stage of armed conflict and, accordingly, just one type of organizational structure armed groups may adopt when engaged in armed conflict (see Chapter 14, section 14.2, *From "Conventional" to "Hybrid" to "Holistic,"* Figure 14.1). Ultimately, the intensity and scope of the violence and the context in which it occurs (e.g., a political purpose) may provide more influential criteria than group organization.

9.4.2.2. The Kosovo Liberation Army

The Kosovo Liberation Army (KLA) provides an interesting example of where the organization of an armed group may fluctuate over time and actually be less straightforward than conventional wisdom might indication. It is this organized group that was the subject of analysis by the ICTY.[222] As will be established, organization is a much more diverse

---

[216] Jessica Stern, Terror in the Name of God: Why Religious Militants Kill 10–31 (2003) (for a description of the Christian-based terrorist movement whose adherents included Timothy McVeigh. He killed 168 people in the 1995 bombing of the Oklahoma City Federal Building).

[217] Kilcullen, *supra* note 198, at 52–66 (for a description of the 2008 Mumbai attack).

[218] Gian Gentile, Wrong Turn: America's Deadly Embrace of Counter-insurgency xiv (2013).

[219] *Id.*

[220] Sivakumaran, *supra* note 184, at 173.

[221] Ahmed S. Hashim, Insurgency and Counter-insurgency in Iraq 158–60 (2006).

[222] *See, e.g.*, Prosecutor v. Milŏsević, Case No. IT-02-54-T, Decision on Motion for Judgment of Acquittal (Int'l Crim. Trib. for the Former Yugoslavia, June 16, 2004), http://www.icty.org/x/cases/slobodan_milosevic/tdec/en/040616.pdf, Prosecutor v. Haradinaj, Case No. IT-04-84-T, Judgement (Int'l Crim. Trib. for the Former Yugoslavia, Apr. 3, 2008), http://www.icty.org/x/cases/haradinaj/tjug/en/080403.pdf, and Prosecutor v. Limaj, Case No. IT-03-66-T 37, Judgement (Int'l Crim. Trib. for the Former Yugoslavia, Nov. 30, 2005), http://www.icty.org/x/cases/limaj/tjug/en/lim-tj051130-e.pdf.

concept than is suggested in the case law. In assessing that jurisprudence it is important to highlight that at the time the matter got before the ICTY the court was dealing with a relatively established conflict in Kosovo. It was not assessing the exact point at which in an emerging conflict or when the exercise of self-defense that the legal threshold of "armed conflict" might actually be crossed.[223] Importantly, it does not address the point at which State armed forces can begin to use conduct of hostilities rules to combat the insurgent or terrorist threat. In any event, the focus of a court is likely to be more narrowly directed toward the time of the alleged offences brought before it.

As has been noted, "the indicia are taken from instances of high-intensity non-international armed conflicts and may not be features of low-level conflicts."[224] In this respect an analysis of the 2004 *Prosecutor v. Milŏsević*[225] decision provides useful guidance on the application of the *Prosecutor v. Tadić*–based criteria in that the former decision found there was no requirement for an armed group control territory or be organized under a civil authority. A closer look at the KLA, the organized armed group in question, also establishes that its organizational structure actually changed over time. In a number of ways the KLA, in terms of organizational structure, was unremarkable in comparison to other groups engaged in insurgency and terrorism around the world.

The trial court in *Prosecutor v. Milŏsević* determined that there was sufficient evidence, particularly relating to the 1998 to 1999 period, to conclude the KLA was an "organised military force, with an official joint command structure, headquarters, designated zones of operation, and the ability to procure, transport, and distribute arms."[226] However, it is important to put this ruling in context. That may have been the case as the KLA matured as an organized armed group, however, as is noted by James Pettifer, in early 1998 the KLA consisted of only about 500–600 organized fighters.[227] It did not fundamentally change its organizational structure from being horizontal to one more hierarchical in structure until after a major defeat at Rahavec in July of that year.[228] In the intervening period the conflict had witnessed long-distance gun battles between villages "across the ethnic divide," where the lack of KLA uniforms caused the media to mistake recruits to be "an informal village defence force."[229]

The KLA had also lost battles because it had tried to fight from formal positions.[230] This highlights that the fortunes of the KLA waxed and waned over the course of the conflict. For a significant period its organization could not be readily equated to a "regular" armed

[223] Sivakumaran, *supra* note 184, at 167 ("Both the ICTY and the International Criminal Tribunal for Rwanda (ICTR) were dealing with large-scale non-international armed conflicts and, as such, many of the same indicia may not be present in smaller and shorter conflicts.").

[224] *Id.* at 168.

[225] Prosecutor v. Milŏsević, Case No. IT-02-54-T, Decision on Motion for Judgment of Acquittal (Int'l Crim. Trib. for the Former Yugoslavia, June 16, 2004), http://www.icty.org/x/cases/slobodan_milosevic/tdec/en/040616.pdf. *See also* Cullen, *supra* note 184, at 128–29.

[226] Prosecutor v. Milŏsević, Case No. IT-02-54-T, Decision on Motion for Judgment of Acquittal ¶ 23 (Int'l Crim. Trib. for the Former Yugoslavia, June 16, 2004), http://www.icty.org/x/cases/slobodan_milosevic/tdec/en/040616.pdf.

[227] James Pettifer, The Kosova Liberation Army: Underground War to Balkan Insurgency, 1948–2001, at 121 (2013).

[228] *Id.* at 146.

[229] *Id.* at 130.

[230] *Id.* at 132.

force. Nor could it engage in open warfare. This is evident in the *Prosecutor v. Limaj*, where at the time relevant to the indictment in that case, May to July 1998, the KLA "General Staff" (also referred to as the General Headquarters) functioned as an underground organization were in constant risk of capture and communicated primarily by telephone and fax.[231] By the standards of regular armed forces it appears to have been a "headquarters" in name only.

The *Prosecutor v. Milŏsević* decision also relies on violence committed by the KLA in the 1996 to 1997 time period as part of its overall assessment of the "[l]ength or protracted nature of the conflict and seriousness and increase in armed clashes."[232] Therefore, the organization of the group during that period is also relevant. While it was suggested by a witness that the KLA at that time was "still a disorganized and somewhat mysterious organization,"[233] the situation seems to have more accurately reflected an insurgent group that was in transition, emerging from the shadows into the public domain.[234] The emergence of that group from those shadows included a propaganda effort that "placed the organisation firmly within the Balkan insurgent tradition of mysterious clandestine revolutionary organizations, and the dress style and behaviour . . . borrowed the messages of the Special Forces units of top armies like the British SAS."[235] Another account notes "between 1994 and 1997, when according to Serbian statistics 136 attacks were mounted by the KLA against security forces and installations, the organisation of the KLA was horizontal, where every soldier could be a leader in his own locality."[236] The number of attacks during that period suggests levels of violence beyond those ordinarily associated with ordinary crime and law enforcement.

Looked at as a whole, it appears that an armed conflict existed in Kosovo prior to the KLA adoption of formal headquarters, command structures, or uniforms that are often overly relied on as distinguishing characteristics of an organized armed group. Further, the KLA appeared to have attributes of a "hybrid" organization with both central and more independent local control being exercised. In this respect, the criteria to establish the existence of an organized group set out in the ICTY jurisprudence must be placed in context.

---

231 Prosecutor v. Limaj, Case No. IT-03-66-T 37, Judgement ¶ 104 (Int'l Crim. Trib. for the Former Yugoslavia, Nov. 30, 2005), http://www.icty.org/x/cases/limaj/tjug/en/lim-tj051130-e.pdf, and *id.* at ¶ 132 ("The members of the General Staff did not meet regularly because of the security situation and identified themselves not by their names but by numbers for the same reason."). *See also* Prosecutor v. Haradinaj, Case No. IT-04-84-T, Judgement, ¶ 68 (Int'l Crim. Trib. for the Former Yugoslavia, Apr. 3, 2008), http://www.icty.org/x/cases/haradinaj/tjug/en/080403.pdf ("until August 1998, the KLA had no rigid hierarchical structure, but a horizontal command structure, and communication between KLA groups was limited.").

232 Prosecutor v. Milŏsević, Case No. IT-02-54-T, Decision on Motion for Judgment of Acquittal ¶ 28 (Int'l Crim. Trib. for the Former Yugoslavia, June 16, 2004), http://www.icty.org/x/cases/slobodan_milosevic/tdec/en/040616.pdf (the heading of this section of the judgment). *See also* Prosecutor v. Kordić, Case No. IT-95-14/2-A 90, Judgement ¶ 341 (Int'l Crim. Trib. for the Former Yugoslavia, Dec. 17, 2004), http://www.icty.org/x/cases/kordic_cerkez/acjug/en/cer-aj041217e.pdf (where it was established in other circumstances that armed conflict can be occurring even before the period ruled relevant by a court. It was noted that violence occurring prior to the period in which the trial court has assessed as the commencement of an armed conflict in April 1993 was not a mere civil disturbance or single acts of terrorism. As a result "in the time following October 1992 there was serious fighting for an extended period of time.").

233 Prosecutor v. Milŏsević, Case No. IT-02-54-T, Decision on Motion for Judgment of Acquittal ¶ 28 (Int'l Crim. Trib. for the Former Yugoslavia, June 16, 2004), http://www.icty.org/x/cases/slobodan_milosevic/tdec/en/040616.pdf.

234 Pettifer, *supra* note 227, at 107.

235 *Id.*

236 *Id.*

In reality, a much looser concept of organization can meet the ICTY criterion. As the court in *Prosecutor v. Limaj* indicated, "*some degree of organisation* by the parties will suffice to establish the existence of an armed conflict."[237] It is also noteworthy that decision relied on a 1999 ICRC study, which indicated "the term 'armed conflict' presupposes the existence of hostilities between armed forces organised to *a greater or lesser extent*; there must be the opposition of armed forces and *a certain intensity* of the fighting."[238] The ICTY cases do not preclude a horizontally organized group, a "hybrid" structure, or even an organization based on a cellular framework from meeting the legal standard that forms part of the threshold for the existence of an armed conflict.

In assessing whether a noninternational armed conflict is in existence, care must also be taken to avoid completely separating the assessment of the intensity of violence from the organized armed group indicia. Just as factors such as the intensity and duration of the violence are linked, so is the "armed" aspect of the organized group.[239] In this respect the use of a cellular or hierarchical structure may be less important in assessing if an armed conflict is occurring than how those groups are armed. The use of weapons of war and, perhaps even more importantly, explosives, by armed groups is more likely to result in violence that is commonly associated with armed conflict (see Chapter 12, section 12.3.1, *A Ubiquitous Weapon*). This approach can be seen in the *Prosecutor v. Haradinaj* decision where the court referred specifically to the type of weapons and other military equipment used, as well as the number and caliber of munitions fired as criteria that are indicative of an armed conflict.[240]

9.4.3. THE "TOTALITY OF THE CIRCUMSTANCES" AND OTHER CRITERIA

It is not only how the organized group is armed that is important. A fundamental threshold question is what security forces and associated armaments the State must use to successfully counter the non-State actor threat. While frequently associated with the intensity of

[237] Prosecutor v. Limaj, Case No. IT-03-66-T 37, Judgement ¶ 89 (Int'l Crim. Trib. for the Former Yugoslavia, Nov. 30, 2005), http://www.icty.org/x/cases/limaj/tjug/en/lim-tj051130-e.pdf. *See also* Robin Geiβ, *Armed Violence in Fragile States*, 91 INT'L REV. RED CROSS 127, 135–36 (2009) (where reference is made to this quotation in the *Prosecutor v. Limaj*, as well as the *Abella Case* [e.g., "the famous *Tablada* decision"] as supporting the situation in Somalia being viewed a noninternational armed conflict).

[238] Prosecutor v. Limaj, Case No. IT-03-66-T 37, Judgement ¶ 89 (Int'l Crim. Trib. for the Former Yugoslavia, Nov. 30, 2005), http://www.icty.org/x/cases/limaj/tjug/en/lim-tj051130-e.pdf (quoting an ICRC, Working Paper, 29 June 1999). *See also* Prosecutor v. Akayesu, Case No. ICTR-96-4-T, Judgement ¶ 620 (Int'l Crim. Trib. for Rwanda, Sept. 2, 1998), http://www.unictr.org/sites/unictr.org/files/case-documents/ictr-96-4/trial-judgements/en/980902.pdf ("The term, 'armed conflict' in itself suggests the existence of hostilities between armed forces organized to a greater or lesser extent.").

[239] SIVAKUMARAN, *supra* note 184, at 168. *See, e.g.*, the Prosecutor v. Limaj, Case No. IT-03-66-T 37, Judgement ¶¶ 135–67 (Int'l Crim. Trib. for the Former Yugoslavia, Nov. 30, 2005), http://www.icty.org/x/cases/limaj/tjug/en/lim-tj051130-e.pdf (outlining the weaponry used by both sides of the conflict in terms of levels of violence and their effects. The weapons included artillery, tanks, grenades, rockets, machine guns, etc.), and Prosecutor v. Haradinaj, Case No. IT-04-84-T, Judgement, ¶¶ 90–99 (Int'l Crim. Trib. for the Former Yugoslavia, Apr. 3, 2008), http://www.icty.org/x/cases/haradinaj/tjug/en/080403.pdf (the weapons used included firearms, hand grenades, and other explosive devices; armored vehicles and heavy weapons; antiaircraft guns and howitzers; and helicopters and jet aircraft. However, it appears the violence was not viewed as meeting the threshold for armed conflict until the heavier indirect support weapons were used).

[240] Prosecutor v. Haradinaj, Case No. IT-04-84-T, Judgement, ¶ 49 (Int'l Crim. Trib. for the Former Yugoslavia, Apr. 3, 2008), http://www.icty.org/x/cases/haradinaj/tjug/en/080403.pdf.

violence criterion, the type of State response can also be connected to armed group organization. There is a danger that by focusing too much on the actions and organization of a non-State actor armed group, another important determinative factor will be missed. That is the use of military forces by States to counter those organized armed groups (e.g., the Corn "means" test[241]). It is evident from the use of State military forces to conduct domestic and international law enforcement is that their deployment does not, on its own, mean an armed conflict is occurring.[242] Such an approach is open to abuse by States, which may overreact to perceived security threats. However, the connection between the nature of the armed group and the military response is evident in *Prosecutor v. Boškoski*, where the court indicated "while isolated acts of terrorism may not reach the threshold of armed conflict, when there is protracted violence of this type, *especially where they require the engagement of the armed forces in hostilities*, such acts are relevant to assessing the level of intensity with regard to the existence of an armed conflict."[243] This approach not only links the nature of the two forces engaged in combat, it also reflects the application of a "totality of the circumstances" approach in assessing the evidence.[244] Taking into account the totality of the circumstances provides distinct advantages in terms of accurately reflecting the nature of the violence. There must be a rational connection between the nature of the perceived threat and the scale and type of response permitted by the State.

In terms of accountability, the State should have to explain why the use of conduct of hostilities means were reasonably necessary. In doing so consideration should be made to the professional judgment of the security personnel (both police and military forces) as to whether that was the case.[245] However, if the organized armed group is armed with mortars, heavy caliber machine guns (i.e., crew-served weapons), and uses explosives; carries out planned, coordinated attacks; organizes for defensive action, with armed guards and prepared positions; and applies military tactics, that connection can readily be made.[246] As was seen in Northern Ireland, the use by a terrorist group of mortars,

---

[241] Corn, *supra* note 124, at 342. *See also* GENEVA CONVENTION I COMMENTARY, *supra* note 137, ¶ 1A(2) at 49.

[242] SIVAKUMARAN, *supra* note 184, at 169, DINSTEIN, *supra* note 5, ¶ 109, at 36, and Kenneth Watkin, *Controlling the Use of Force: A Role for Human Rights Norms in Contemporary Armed Conflict*, 98 AM. J. INT'L L. 1, 26 (2004).

[243] Prosecutor v. Boškoski, Case No. IT-04-82-T 90, Judgement ¶190 (Int'l Crim. Trib. for the Former Yugoslavia, July 10, 2008), http://www.icty.org/x/cases/boskoski_tarculovski/tjug/en/080710.pdf (emphasis added). *See also* Laurie R. Blank, *Defining the Battlefield in Contemporary Conflict and Counterterrorism: Understanding the Parameters of the Zone of Combat*, GEO. J. INT'L & COMP. L. 1, 34–35 (2010) (referring to the government response as a factor to be applied).

[244] Prosecutor v. Boškoski, Case No. IT-04-82-T 90, Judgement ¶ 257 (Int'l Crim. Trib. for the Former Yugoslavia, July 10, 2008), http://www.icty.org/x/cases/boskoski_tarculovski/tjug/en/080710.pdf (where the term "totality of the circumstances" is use to describe an assessment of weakness of a witnesses description of the KLA).

[245] Yishai Beer, *Humanity Considerations Cannot Reduce War's Hazards Alone: Revitalizing the Concept of Military Necessity*, 26 EUR. J. INT'L L. 801 (2015)(where is argued that professional military standards should be applied to narrow the military objectives that can be targeted). *See also* Kenneth Watkin, *Military Advantage: A Matter of "Value", Strategy and Tactics*, 17 Int'l Y.B. Hum. L. 277, 313 (2014) [where the "zone of proportionality" standard for targeting set out in Public Committee Against Torture in Israel v. Israel, Israel Supreme Court [16 December 2006], 46 ILM 375, ¶58, at 400 (2007) is assessed in terms of the reasonable military commander].

[246] Case 11.137, Abella v. Argentina, Inter-Am. Y.B. on H.R. ¶ 155 (1997), http://www1.umn.edu/humanrts/cases/1997/argentina55-97a.html ("More particularly, the attackers involved carefully planned, coordinated and executed an armed attack, i.e., a military operation, against a quintessential military objective—a military base."). *But see* Aerial Drone Deployment on 4 October 2010 in Mir Ali/Pakistan, 157 I.L.R. 722, 741 (2013) (where it is wrongly suggested that "[a]s a general rule, the methods and means which the actors use to make war are no more relevant than their motives

explosives, and similar weapons associated with an armed conflict does not absolutely preclude the adoption of a law enforcement approach (see Chapter 13, section 13.2.1, *The "Troubles"*). However, it provides a strong indication that the violence has risen to the level of an armed conflict for which the State may choose to apply a conduct of hostilities paradigm.

Importantly, the Additional Protocol II criteria of "internal disturbances and tensions, such as riots, isolated and sporadic acts of violence and other acts of a similar nature" establishes a clear treaty-based line below which armed conflict is well understood does not exist.[247] In addition, reliance on conditions found in the ICRC *Commentary* to Common Article 3 of the Geneva Conventions, such as the one that refers to the type of State action (i.e., recourse to regular military forces), is particularly helpful as that article has historically been viewed as setting a relatively low threshold for the existence of an armed conflict. The combination of this treaty wording, the historical approach reflected in the ICRC *Commentary*, and instances where the violence points to situations of hostilities rather than traditional law enforcement (i.e., *Abella v. Argentina*[248]) provides considerable support for the argument that the threshold for armed conflict is significantly lower than some interpretations of the ICTY jurisprudence suggest.

Depending upon the circumstances, this lower armed conflict threshold can create a significant overlap of the operation of human rights and humanitarian law. However, in doing so it also acknowledges a situation that more accurately reflects the intertwined history of humanitarian and human rights law, as well as the practical circumstances experienced on the ground (see Chapter 5, section 5.3, *A More Intimate History*). When confronted with an overlap of humanitarian and human rights law, States have to make policy choices as to which body of law should be applied in countering the insurgent or terrorist threat. Therefore, when asked to justify the use of force, it is not unreasonable to require a State to explain why force related to armed conflict was used if a law enforcement approach appeared feasible and effective. It seems highly formalistic and exclusionary for a State to respond humanitarian law was applied simply because it was permitted "at law." Such a response might be sufficient for some lawyers, however, is unlikely to be persuasive in the public sphere, nor before courts, as they have been increasingly confronted with resolving long-standing international law disagreements.[249] Historically, the choice made by States, when acting internally, has been to favor a human rights–based approach by treating the threat as a domestic criminal one (see Chapter 4, section 4.4.3, *A Reluctant Application*

---

and objectives when it comes to classifying an armed conflict."). *See also* Rob McLaughlin, *Legal Policy Considerations and Conflict Characterizations at the Threshold between Law Enforcement and Non-International Armed Conflict*, 13 MELB. J. INT'L L. 1, 5 (2012) (where it is indicated States can influence the characterization of armed conflict through a "labelling choice" as a coordinated attack, or military-style offensive, rather than an isolated terrorist incident).

[247] AP II, art. 1(2). *See also* Prosecutor v. Akayesu, Case No. ICTR-96-4-T, Judgement ¶ 620 (Int'l Crim. Trib. for Rwanda, Sept. 2, 1998), http://www.unictr.org/sites/unictr.org/files/case-documents/ictr-96-4/trial-judgements/en/980902.pdf (where it is noted that hostilities between armed forces "consequently rules out situations of internal disturbances and tensions.").

[248] Case 11.137, Abella v. Argentina, Inter-Am. Y.B. on H.R. ¶ 1 (1997), http://www1.umn.edu/humanrts/cases/1997/argentina55-97a.html.

[249] *See, e.g.*, Public Committee Against Torture in Israel v. Israel, Israel Supreme Court [16 December 2006], 46 ILM 375 (2007).

of Humanitarian Law). The question to be asked is why that approach should not also be favored when it is feasible to do so as military forces deploy to deal with internal threats in other states or operate in "ungoverned spaces." Counterinsurgency operations rely on a police primacy approach for sound operational reasons.

That said, setting the threshold for armed conflict requires a realistic assessment of the intensity of violence (including any potential violence in resolving the incident), as well as the organizational structure and capabilities of its participants. It should not be based on a goal of favoring a law enforcement response that is applied at the expense of an objective assessment of the threat being faced by State security forces. Nor should a comfort with humanitarian law as a default position for international operations overly influence the analysis. An advantage in recognizing the potential for a lower threshold for noninternational armed conflict is it would more closely mirror international armed conflict. That will reduce the challenge of combating non-State actors when there are concurrent international and noninternational armed conflicts taking place in the same territory, or ones that evolve from one form of conflict to the other (see Chapter 2, section 2.4.3, *Conflict Evolution*). Finally, the resulting lower threshold lessens the likelihood for confusion when State military forces respond to transnational attacks by non-State actors, which may or may not be controlled or harbored by other States. Whether the threat posed by non-State actors occurs in situations of international or noninternational armed conflict, the nature of the threat is fundamentally the same (see Chapter 14, section 14.3.1, *Armed Conflict or Law Enforcement?*). It is one that largely falls within the historical framework of indirect "small wars." The State response often requires consideration of the application of human rights and humanitarian law, as well as State self-defense where appropriate.

9.5. Addressing the Challenge

What is evident, more than a decade after the attacks of 9/11, is that the international legal community is far from reaching a consensus on the approach to be applied when dealing with the threat of transnational terrorism and cross-border insurgency. Numerous tests exist for the categorization of conflicts in which States fight non-State actors. These include: considering the degree of control exerted by States over non-State actors; determining whether active support is provided to such actors or there is the harboring of terrorist groups; looking simply at one State entering the territory of another State to combat a non-State actor; assessing terrorist attacks that cross borders as a conflict that is international in character; applying unique "extraterritorial law enforcement" or "transnational armed conflict" frameworks to categorize defensive action taken in response to cross-border attacks by terrorists and armed bands; and looking at such violence as constituting a noninternational armed conflict, albeit one not restricted by State borders.

Combined with these approaches for conflict determination are the more historical characterization questions arising from the recognition in Additional Protocol I that non-State actor conflicts can have an international character, as well as the now largely dated concept of the recognition of belligerency. To complicate matters, conflict can change from having an international to noninternational character (see Chapter 2, section 2.4, *Noninternational Armed Conflict*), can transition from a noninternational to an international one,[250] or simultaneous international and noninternational conflicts can exist at the same time.[251] Nowhere, is the complexity of contemporary conflict categorization more evident than when considering the type of conflict that results when States intervene on the side of either a State or its non-State adversary in the context of an internal armed conflict.[252] As has been noted, "[i]n practice, the dividing line between inter-State and intra-State armed conflicts cannot always be delineated in a few easy strokes."[253] Adding to this complexity are the additional theories relating to drone strikes, including the restrictive "law enforcement" and more permissive "conduct of hostilities" and "self-defence" approaches (see Chapter 8, section 8.5, *Legal Frameworks for Responding to Transnational Terrorism*).

Reflected in these many theories is a significant and disconcerting lack of consensus among international lawyers. Some situations in which States and non-State actors combat one another are viewed as occurring in the context of an international armed conflict. Others are seen as "noninternational" in character, and some involve the conclusion there is no armed conflict in existence at all. For practitioners, the many and often subtle differences arising in the various legal theories on the characterization of conflict can seem to add an unnecessary level of complexity. From a security perspective, the terrorist or insurgent group basically pose the same threat regardless of how the conflict is classified. As a result, the point can be readily reached where legal theory begins to diverge from the practical reality being faced on the ground.

A very real problem arises when the methods and forms of categorization become so formalistic and dogmatic that they become an obstacle to addressing what can be an exceptionally serious security threat. It is unlikely that military commanders confronted with very real and complex security threats will be overly interested in the subtle differences and legal interpretations found in various categories of conflict, but ones for which same legal rules often appear to apply. It is the significant degree of commonality in these armed conflicts, and the requirement for operational consistency that has likely driven practitioners to favor an approach that sees many aspects international and noninternational armed conflict through the same lens. This can even be seen in the UN *Secretary-General's Bulletin*, where reference is made generically to "situations of armed conflict" without distinguishing whether they are international or noninternational in character.[254] At this point in the twenty-first century, the international legal community has not been able to provide clear guidance for practitioners. This is perhaps its greatest failing and the most important challenge facing that community in terms of regulating transnational violence.

---

[250] DINSTEIN, *supra* note 14, ¶¶ 13–14, at 7–8.

[251] *Id.* ¶ 11, at 6–7.

[252] Hans-Peter Gasser, *Internationalized Non-International Armed Conflicts: Case Studies of Afghanistan, Kampuchea, and Lebanon*, 33 AM. U. L. REV. 145, 147 (1983).

[253] DINSTEIN, *supra* note 14, ¶ 12, at 7.

[254] SECRETARY-GENERAL'S BULLETIN, *supra* note 181, § 1, ¶ 1.1, at 202. *See also* DINSTEIN, *supra* note 5, ¶ 290, at 94 (where it is noted the phrase "situations of armed conflicts" covers both international and noninternational armed conflict).

Part of the challenge is that many of the interpretations of the law appear to remain grounded in theories developed uniquely in the context of either high-end armed conflict or relatively low-level law enforcement scenarios. These are at the extremes of the conflict spectrum. They represent older twentieth-century views of international law that focus on inter-State and internal State conflict. To further complicate matters, these theories are often exclusionary in nature, not fully appreciating that both the conduct of hostilities and law enforcement norms must be applied to meet the threats from across the conflict spectrum. The twentieth-century focus on the State as the greatest threat to international security has been overshadowed by other events. The very effective legal approach developed in that context in many ways appears to not be fully adequate to a contemporary security challenge centered on non-State actors. The method of addressing the non-State actor threat in the twenty-first century is increasingly holistic rather than exclusionary, incorporating all bodies of the law into the operational discussion.

IV Applying Force Across the Conflict Spectrum

Self-Defense is nature's oldest law.[1]

10 Self-Defense and the Protection of Nationals

10.1. Introduction

Although counterterrorism and, in particular, the use of drones have dominated much of the contemporary security dialogue, the challenge presented to the international community of acting in another country to protect its interests is actually not a new one. States have long taken action to respond to threats against their nationals that spans the armed conflict and law enforcement paradigms. This has occurred regardless of whether those threats come from insurgents, terrorist organizations, or criminal gangs. A study of these operations highlights that reacting to threats, particularly in the ungoverned spaces of the world, is an activity the international community has consistently accepted or at least tolerated. Such tolerance points to a more ready acceptance of States exercising force against non-State actors when the levels of violence are kept to a minimum. Categorizing these lower-level threats posed by non-State actors under international law and identifying an appropriate State response has presented a particular challenge for the international legal community. It is an area where that community continues to struggle.[2] Many of these contemporary

---

[1] ROBERT DEBS HEINL, JR., DICTIONARY OF MILITARY AND NAVAL QUOTATIONS 82 (1966) (quote by JOHN DRYDEN, ABSALOM AND ACHITOPHEL (1681).

[2] *See, e.g.*, Tom Ruys, *The Meaning of "Force" and the Boundaries of the Jus Ad Bellum: Are "Minimal" Uses of Force Excluded from UN Charter Article 2(4)?*, AM. J. INT'L L. 159 (2014)[hereinafter *The Meaning of "Force" and the Boundaries of the Jus Ad Bellum*], Mary Ellen O'Connell, *The True Meaning of Force*, AJIL UNBOUND (Aug. 4, 2014, 9:56 AM), http://www.asil.org/blogs/true-meaning-force [hereinafter *The True Meaning of Force*], Tom Ruys, *The True Meaning of Force—A Reply to Mary Ellen O'Connell*, AJIL UNBOUND (Sept. 3, 2014, 9:30 AM), http://www.asil.org/blogs/true-meaning-force-reply-mary-ellen-o'connell, Mary Ellen O'Connell, *The True Meaning of Force: A Further*

threats do not always readily reflect the traditional interpretations of State versus State recourse to war paradigm focused on "armed attack" under Article 51, or what might constitute a use of force for the purposes of Article 2(4) of the UN Charter.

A defining characteristic of the contemporary security environment is that many of the threats occur at a point on the conflict spectrum where the armed conflict and law enforcement paradigms intersect and often overlap. Increasingly, technology is enabling the cross border acts by a wide range of individuals further blurring the traditional idea that policing is primarily domestic in nature.[3] The type and extent of violence, as well as the organization of the groups involved, often make it difficult to separate the insurgents and terrorists engaged in armed conflict from criminal gangs seeking economic gain. The difficulty in separating the threats is compounded by the categorization of insurgents and terrorists as "criminals" and their frequent involvement in criminal activity to finance their activities. For example, one study of kidnapping in Iraq indicates that hostage taking fell into one of two categories: economic or political.[4] It was not simply political action associated with terrorism. The groups involved included former regime elements, former convicts, unemployed youths and young men, opportunistic amateurs, insurgents and jihadist groups, and militias and militia factions.[5] Delineating the participants was challenging as "[s]ome groups are almost certainly hybrids, and some kidnappers probably move from one group to another in what may be a constantly shifting kaleidoscope of allegiance, membership, and motives."[6]

Perhaps one of the most interesting areas of inquiry is the impact the nature of these operations has on the tactics applied even during armed conflict. Ultimately, States must be prepared to respond in defense of their nationals across the full conflict spectrum. One example is operations conducted extraterritorially to rescue hostages. Due to the need to protect the lives of hostages these operations can force the adoption of more restrictive tactics than otherwise might occur during hostilities. This will be seen in the analysis of two iconic counterterrorism missions discussed in this chapter: the 1976 Israeli raid on Entebbe and the 2000 British rescue of hostages in Sierra Leone. It will be argued these hostage rescue operations were conducted in the context of armed conflicts that were limited in time, scale, and the amount of force used. There is also an increasing recognition that many of the threats to nationals are uniquely posed by criminal gangs operating in the vacuum left by insecurity caused by armed conflict or disorder internal to States. Operations against these gangs require a law enforcement response. However, questions can be raised regarding the impact of these extraterritorial operations on the human rights–based normative regime as it is applied internationally in situations involving considerable violence.

A unique feature of hostage rescue and other contemporary counterterrorism operations is that, overall, a commonality of approach in terms of tactics, weapons, and levels of force

---

Response to Tom Ruys in the Interest of Peace, AJIL Unbound (Sept. 4, 2014, 10:04 AM), http://www.asil.org/blogs/true-meaning-force-further-response-tom-ruys-interest-peace.

[3] Benjamin Wittes & Gabriella Blum, The Future of Violence: Robots and Germs, Hackers and Drones 168 (2015) (where it is noted that technology is empowering individuals to carryout attacks on an international scale that blur the divisions between "foreign versus domestic, public versus private, or criminal versus belligerent").

[4] Paul Williams, Criminals, Militia, and Insurgents: Organized Crime in Iraq 109–11 (2009).

[5] *Id.* at 113–15.

[6] *Id.* at 115.

has developed for countering many of these threats. As will be discussed, this has involved the development of both police and military units that can perform substantively the same role. There is often a reliance on the same specially trained security forces to carry out these operations across the conflict spectrum regardless of whether the operation is armed conflict or law enforcement based. That skill set is transferable to other operations. An example that will be explored is the 2011 raid on the bin Laden compound in Pakistan, which highlights the ability of specialized military forces to conduct operations in such a manner that capture rather than killing a terrorist is a viable option.

The involvement of military forces in international law enforcement, and the development of skills and tactics associated with these types of operations have provided States with a broader range of force options for counterterrorism missions. Ultimately, the development of this more sophisticated approach enhances the operational choices available to States when confronting transnational terrorism. It also sets the scene to consider not just whether an armed conflict may be in existence but also the degree to which applying a more restrictive law enforcement normative framework can in the right circumstances better meet operational goals.

10.2. Self-Defense and the Extraterritorial Protection of Nationals

10.2.1. A STATE RESPONSIBILITY

The protection of nationals is a core State responsibility arising directly from the intimate relationship between a government and the governed. The determination of statehood itself depends upon qualifying factors such as a permanent population, a defined territory, a government, and the capacity to enter into relations with other States.[7] These criteria are interrelated, as the population criterion "is intended to be used in association with that of territory, and connotes a stable community."[8] In this respect it has been said, "the requirement of community is essential, and without nationals, without community, the state ceases to exist."[9] Internally, the protection of citizens is carried out in a variety of ways including by the maintenance of law and order. The establishment of a human rights regime, with its focus on the protection of minorities, provides another example of how a population is protected within a governance structure. The creation of clear international boundaries and the defense of sovereign territorial integrity has also been a hallmark of the stability that the post–World War II recourse to war legal regime has brought to international security.[10] Therefore, "[i]t cannot be doubted that the state may defend its nationals or any person within its territorial jurisdiction from hostile attack."[11] The defense of nationals outside of that sovereign territory has tended to be more controversial.

---

[7] Ian Brownlie, Principles of Public International Law 70 (7th ed. 2008) (based on Article I of the Montevideo Convention on the Rights and Duties of States, Signed 26 Dec. 1993; Hudson, Int. Legis., vi. 620).

[8] *Id.* at 70–71.

[9] D. W. Bowett, Self-Defense in International Law 91 (The Lawbook Exchange Ltd., 2009) (1958).

[10] Joshua S. Goldstein, Winning the War on War 278 (2011).

[11] Bowett, *supra* note 9, at 87.

One area where the general effort to restrict transnational use of force by States has had a particular but contentious impact is on the question of whether a State can act extraterritorially to defend its nationals. It has been acknowledged that in the nineteenth century it was lawful for a State to take such action.[12] Regarding nationality of the person, "it is commonly supposed that the state not only has a right but a duty to extend its protection to that subject."[13] This is a proposition, which with the adoption of the UN Charter "has become rather more controversial since of necessity the 'territorial integrity and political independence' of the target state is infringed."[14] It is also an area where the near obsession with restricting inter-State use of force has created a theoretical State-centric framework that appears, at times, to have called into question the right of States to protect their nationals.

There is significant disagreement among legal scholars as to whether the protection of nationals abroad can be looked upon as protecting the State.[15] That debate started almost immediately after the adoption of the recourse to war provisions of the UN Charter. It is a discussion that has not produced universal agreement on the legal basis for or even the lawfulness of such action. Some jurists regard it "as an autonomous right of intervention, others as a form of legitimate defence."[16] Conversely, it has been suggested there are "few states accepting a legal right to protect nationals abroad."[17] It has been said, "it is impossible to assert that there exists *de lege lata* [the law as it is][18] a customary right of forcible protection of nationals."[19] For others it is described as "extremely tenuous" if considered as a right of intervention.[20] Further, "instruments expressly prohibiting intervention both before and after 1945 may be considered to have rendered intervention for the projection of nationals illegal."[21]

However, as Antonio Cassese suggested, "it would seem that the objections of many States have not led to the obliteration of general rule on the matter,"[22] although when considered as part of the right to act in self-defense such action is subjected to the debate over how restrictively its criteria are to be applied. There has been a particular sensitivity toward States using claims to be acting in "defence of nationals" as a pretext for intervening in another State. This concern is not without some basis. Many interventions have involved the use of force by States where self-defense was invoked only as a partial justification for

---

[12] Ian Brownlie, International Law and the Use of Force by States 289 (1963), Malcolm N. Shaw, International Law 1143 (6th ed. 2008), and Thomas M. Franck, Recourse to Force: State Action Against Threats and Armed Attacks 76 (2002).

[13] Bowett, *supra* note 9, at 95.

[14] Shaw, *supra* note 12, at 1143. *See also* Terry D. Gill & Paul A. L. Ducheine, *Rescue of Nationals*, *in* The Handbook of The International Law of Military Operations 217, ¶ 12.02 1., at 217 (Terry Gill & Dieter Fleck eds., 2010).

[15] Yoram Dinstein, War, Aggression and Self-Defence ¶ 575, at 218 (5th ed. 2011).

[16] Brownlie, *supra* note 12, at 292.

[17] Christine Gray, International Law and the Use of Force 157 (3rd ed. 2008). *See also* Antonio Cassese, International Law 367–68 (2nd ed. 2005).

[18] Oxford Reference, http://www.oxfordreference.com/view/10.1093/oi/authority.20110803100103408 (last visited July 23, 2015) ("[Latin: the law that has been borne] What the law is.").

[19] Tom Ruys, "Armed Attack" and Article 51 of the UN Charter: Evolutions in Customary Law and Practice 243 (2010).

[20] Brownlie, *supra* note 12, at 298.

[21] *Id.*

[22] Cassese, *supra* note 17, at 368.

action.[23] Emblematic of the struggle the international legal community has had in dealing with this issue is the fundamental disagreement as to whether, or the degree to which, the protection of nationals abroad is authorized under Article 51 of the UN Charter. The basis for such operations has been claimed under both Article 51 and customary international law.

It has been indicated "those States that have actually carried out such rescue operations have always refrained from accusing the territorial State of an 'armed attack'. Instead they have usually referred for a justification to customary law, existing unaffected by Art. 51."[24] As such, these operations would, at one level, fly below the legal radar designed to track inter-State conflict. However, it has also been noted that a general rule permitting intervention, "which might be subsumed under the general notion of self-defence pursuant to Article 51 of the UN Charter—may only be resorted to *under very strict conditions*, dictated by the UN Charter system for the maintenance of peace and security."[25] Acceptance of this interpretation can be made somewhat easier by viewing it as exceptional and "limited to those cases in which no other means of protection are available."[26]

The controversy surrounding interventions to protect nationals is undoubtedly influenced by the fact most operations are carried out by Western Powers on the territory of a developing power.[27] It remains mostly Western States that have expressed the view those operations are justified under either Article 51 of the UN Charter or a customary international law right to self-defense.[28] The United Kingdom Manual on *The Law of Armed Conflict* states in its opening chapter that "[s]elf- defence may include the rescue of nationals where the territorial state is unable or unwilling to do so."[29] The United States similarly claims that right based on the "unwilling or unable" test.[30] Military intervention to remove citizens from a hostile environment can only practically be carried out by a limited number of States, particularly when acting on their own. The logistical requirements alone of an air- or sea-based evacuation or rescue operation are considerable even in a relatively permissive environment. A hostile environment immediately raises the specter of levels of violence associated with armed conflict. The practical limitations on the ability of a State to act in a foreign land must to be kept in mind when a law enforcement approach is suggested to govern a use of force. As has been discussed, policing-type operations require a significant level of control to be exercised on the ground (see Chapter 7, section 7.4.2, *The "Targeted*

---

[23] Gray, *supra* note 17, at 88–92.

[24] Albrecht Randelzhofer, *Article 51*, *in* I The Charter of the United Nations 788, ¶ 27, at 799 (Bruno Simma et al. eds., 2nd ed. 2002). *See also* Gill & Ducheine, *supra* note 14, ¶ 12.02 2., at 218 (where it is stated, "the most logical and coherent legal basis for this justification is that it forms part of the right of self-defence under customary international law.").

[25] Cassese, *supra* note 17, at 368.

[26] Bowett, *supra* note 9, at 88.

[27] Cassese, *supra* note 17, at 367.

[28] *Id.*

[29] The Manual of The Law of Armed Conflict, United Kingdom Ministry of Defence ¶ 1.5 at 2 (2004).

[30] Department of Defense Law of War Manual, Office of the General Counsel ¶ 1.11.5.3., at 48 (June 2015), http://www.dod.mil/dodgc/images/law_war_manual15.pdf ("A State's right to use force in self-defense may be understood to include the right to use force to protect its nationals abroad. The United States has taken action to protect U.S. nationals abroad when the government of the territory in which they are located was unwilling or unable to protect them. A State need not await actual violence against its nationals before taking such action if an attack against them is imminent.").

Killing" Case and Control). Establishing that required degree of control on extraterritorial operations may be beyond the capability of the most militarily capable States, let alone the broader international community.

One approach taken by legal scholars appears to avoid discussing the issue by downplaying the frequency of such rescues, relying on States not having complained about such activity within their borders, or note the matter has not been raised at the United Nations.[31] The issue has even been suggested to be one of limited historical interest, suggesting the need no longer arises.[32] It has also been argued that many cases occur where there is no effective government and third States make no comment regarding the evacuation of nationals. A concern is only raised "with regard to those rescue missions where the territorial state objects to the intervention or where the protection of the nations was just a pretext for an invasion with wider objections."[33] In effect, there is a level of toleration for this form of self-help particularly in situations where there is no effective territorial government.

There is also the issue of consent. Where provided, the intervention is lawful and is likely less controversial.[34] When territorial States consent to operations such as hostage rescues, it has been noted "they cannot be counted as genuine precedents for the application of the inter-State law of self-defence."[35] However, it is not always clear such consent will not be more than a fig leaf if the political leadership of the territorial State is dependent upon the intervening State, is itself acting under questionable authority, or does not really control the country.[36] This can make the consent issue at times appear as a rather artificial discriminator on whether action is taken in self-defense. This may be particularly the case where consensual interventions involve foreign military forces conducting the operations substantively unassisted by the territorial State.

The nationality of a person attacked or abducted is the factor relied on to justify intervention. If the nationals are diplomatic envoys or visiting dignitaries, it is suggested the "answer is certainly affirmative" to the question of whether forcible measures can be exercised in self-defense.[37] There appears to be less of a consensus that an attack on a citizen simply because of nationality is an armed attack on the State.[38] However, "[t]he allegation that an attack against nationals abroad can never be regarded as an attack against the State itself swings away from reality; it carries the legal fiction of the State to extreme and illogical lengths."[39] The question can also arise as to whether, when a territorial State acts against the national of another State, the matter is simply one of domestic law enforcement. Consistent with the State being the right authority for law enforcement action taken on its own territory, in jurisdictional areas such as the coastal Exclusive Economic Zone or in respect of the continental shelf ,it has been argued there is a presumption such action does not affect

---

[31] CASSESE, *supra* note 17, at 159–60.

[32] GRAY, *supra* note 17, at 160.

[33] *Id. See also* FRANCK, *supra* note 12, at 77.

[34] Charles C. Joyner, *Reflections on the Lawfulness of Invasion*, 78 AM. J. INT'L L. 133, 138 (1984).

[35] DINSTEIN, *supra* note 15, ¶ 677, at 256.

[36] *Id.* (questioning the authority of Grenada's Governor-General to authorize the 1983 U.S. invasion of Grenada).

[37] *Id.* ¶ 575, at 218.

[38] *Id.*

[39] *Id.*

international relations.[40] However, it is also suggested that criteria such as the political context, the gravity or intensity of the acts, whether they are isolated or recurring, the actions are carried out by military rather than police units, and at what level they are ordered could be important factors in deciding if forcible action against nationals arises in the context of a dispute between States rather than a matter of law enforcement.[41]

As has been noted, not everyone agrees that States may act when nationals are attacked. It has also been said that linking nationality to an armed attack on the State is "rightly rejected by the majority of commentators on the grounds that the security or existence of a State is not under threat if its nationals are assaulted in another country."[42] However, a strong argument can be made that is exactly the case when a group like Al Qaeda attacks the citizens of Western States in a foreign land. These attacks, conducted as part of a transnational insurgency, are clearly carried out for a political purpose. They are not the random, often senseless acts of criminal violence associated with ordinary lawfulness and disorder. In this respect, the proliferation of non-State actors operating from ungoverned spaces is rapidly changing the operational reality for States.

A particularly strong sentiment expressed by portions of the legal community is that breaching the sovereignty of States by entering their territory without consent cannot be tolerated. As a result the view has been adopted that the sovereignty principle should not be violated,[43] or perhaps only done so in very limited circumstances.[44] The problem with these interpretations is that they do not necessarily address the reality of contemporary transnational terrorist threats or situations where a host State is unwilling or unable to deal with the crisis threatening the nationals of another State. Such a restrictive interpretation of the law must be considered in a broader context. Particularly in the post-9/11 period there has been a debate regarding the scope of the prohibition on the use of force under Articles 2(4), and the meaning of 51 of the UN Charter. The focus of that discussion has been whether those provisions stand in the way of dealing with threats to nationals by terrorist groups and other non-State actors. This is a debate that has been about "thresholds" for the application of terms such as "armed attack" in Article 51, and as a result the "right" to act in self-defense or if the use of certain types and levels of force are prohibited under Article 2(4).

Most famously, the threshold issue arose in 1986 in an International Court of Justice (ICJ) case, *Military and Paramilitary Activities in and against Nicaragua Merits*.[45] The court ruled that there could be uses of force that might breach the prohibition on the use of force but were of "lesser gravity" than the threshold of an "armed attack" set out in Article 51. In that circumstance the exercise of self-defense would not be justified.[46] However, the identification of a "gravity" standard has not resulted in unanimity among the international

---

[40] Ruys, *The Meaning of "Force" and the Boundaries of the Jus Ad Bellum*, *supra* note 2, at 206.

[41] *Id.* at 207.

[42] Randelzhofer, *supra* note 24, ¶ 27, at 798.

[43] Ruys, *supra* note 19, at 249.

[44] Olivier Corten, The Law Against War: The Prohibition on the Use of Force in Contemporary International Law 89 (Christopher Sutcliffe trans., 2010) ("[a] priori, any military operation in the territory of another State without its consent violates its sovereignty, unless the intervening State can validly invoke circumstances precluding wrongfulness (distress, state of necessity, countermeasures, etc.) on the very stringent terms provided for by international law.").

[45] Military and Paramilitary Activities in and against Nicaragua Merits, 1986 I.C.J. 13 (1986).

[46] *Id.* ¶ 247, at 127.

community regarding exactly what that threshold might be, or even whether States will accept it exists. For example, the United States, the world's strongest military power, has consistently taken the view it does not accept the restrictive interpretation of the right of self defense reflected in the ICJ jurisprudence.[47] Another approach is to set a very low *de minimis* standard (the law does not concern itself with trifles[48]) below which the right of self-defense could not be engaged. In other words, there is a broad right to act in self-defense, except in benign situations such as when a soldier fires across a border "and the bullet hits a tree or cow . . . ."[49] In effect, the author of this approach, Yoram Dinstein, applies a relatively broad interpretation of the authority of a State to defend nationals in response to an armed attack.[50]

Even subsequent judicial pronouncements leave much room for debate. While the *Military and Paramilitary Activities in and against Nicaragua Merits* case made reference to mere frontier incidents not meeting the required threshold,[51] the International Court of Justice also subsequently indicated in the *Case Concerning Oil Platforms (Islamic Republic Of Iran V. United States Of America)* that the single act of mining a warship might constitute an armed attack.[52] The International Court of Justice cases have led to considerable debate within the academic community regarding what circumstances constitute such an armed attack.[53] Complicating this debate is that it does not actually appear an act of violence has to occur in order to have such an attack. The establishment of a blockade by one State against another, albeit backed up with the threat of force, has been interpreted to constitute an armed attack potentially justifying a response in self-defense.[54]

The question of thresholds has also arisen in the context of the operation of Article 2(4) of the UN Charter. For example, it has been suggested there is a *de minimis* exception to that article. In this exception the legal regime of countermeasures is viewed as providing adequate authority for the "minor uses of force that states engage in all of the time," such as dealing with "navies pursuing fishing vessels, soldiers wandering across borders, accidental air incursions, and unapproved cross-border law enforcement operations."[55] Significantly, it

---

47 Harold Koh, Legal Advisor, U.S. Department of State, Remarks at USCYBER-COM Inter-Agency Legal Conference (Sept. 18, 2012), http://www.state.gov/s/l/releases/remarks/197924.htm:

> the United States has for a long time taken the position that the inherent right of self-defense potentially applies against any illegal use of force. In our view, there is no threshold for a use of deadly force to qualify as an "armed attack" that may warrant a forcible response. But that is not to say that any illegal use of force triggers the right to use any and all force in response—such responses must still be necessary and of course proportionate. We recognize, on the other hand, that some other countries and commentators have drawn a distinction between the "use of force" and an "armed attack," and view "armed attack"—triggering the right to self-defense—as a subset of uses of force, which passes a higher threshold of gravity. My point here is not to rehash old debates, but to illustrate that States have long had to sort through complicated jus ad bellum questions.

See also Sean D. Murphy, *U.S. Reaction to ICJ Judgment in Iranian Oil Platforms Case*, 98 AM. J. INT'L L. 597 (2004).

48 Bryan A. Garner, *Garner's Dictionary of Legal Usage*, OXFORD DICTIONARIES, http://www.oxforddictionaries.com/secondary/garner_dict_legal_usage/de-minimis?q=de+minimis (last visited Mar. 5, 2015).

49 DINSTEIN, *supra* note 15, at ¶ 555, at 210.

50 *Id.* ¶¶ 575–77, at 217–19.

51 Military and Paramilitary Activities in and against Nicaragua Merits, 1986 I.C.J. ¶195, at 103–04.

52 Case Concerning Oil Platforms (Islamic Republic Of Iran V. United States Of America), 2003 I.C.J. ¶ 72, at 195.

53 GRAY, *supra* note 17, at 189–192.

54 Randelzhofer, *supra* note 24, ¶ 23, at 797.

55 O'Connell, *The True Meaning of Force*, *supra* note 2.

is suggested "it is the regime of countermeasures that permits rescue attempts and attempts to apprehend criminal suspects such as Osama bin Laden or [2012 U.S. Benghazi diplomatic compound attack suspect] Abu Khattala."[56] This approach reflects a school of thought that prefers to see the categorization of force as an exercise of law enforcement or police powers as the appropriate means for States to react to low-level threats, although a wider use has also been suggested elsewhere.[57] Some uses of force (i.e., fisheries law enforcement) would readily fall within the law enforcement paradigm. However, that is not the case when State warships or aircraft attack one another. This approach has been rightly criticized on that basis.[58]

A particular challenge for anyone seeking to claim the use of force between State security forces as a form of law enforcement (in the sense of countering criminal activity) is the low threshold for the existence of an international armed conflict. There does not have to be a use of force for an armed conflict between two States to be in existence (see Chapter 9, section 9.2, *Inter-State Conflict*).[59] Technically, the specialized body of law that applies is humanitarian law, although the State could always choose to apply a more restrictive law enforcement approach. It may be that, in this context, the term "law enforcement" is actually being applied as a substitute for the principles governing State self-defense (e.g., necessity, proportionality, imminence, immediacy) and not a human rights law–based use of force. Overall, this discussion lacks a holistic perspective that would require reconciling the application of the law governing the recourse to war, humanitarian law and human rights law.

The challenge of attempting to apply a human rights–based law enforcement framework also arises in respect of use of force against non-State actors. One approach is to treat terrorist groups as a private group subject to police action. Olivier Corten suggests that the State does "not act against any other State and so there is no need for it to rely on Article 51 of the Charter."[60] Similarly, for Mary Ellen O'Connell, Abu Khattala, who is implicated in the 2012 attack on U.S. diplomatic facilities in Benghazi,[61] is an international criminal suspect subject to "police-type law enforcement rules" applied by military forces that are justified as countermeasures.[62] However, what needs to be done is more closely assess whether the violence associated with the 2012 Benghazi attack constituted an armed attack by a non-State actor. A review of the facts indicates that it met the legal threshold for both an armed attack (i.e., attack on diplomatic personnel and property), and when a "totality of the circumstances" test is applied, as outlined in this book, the criteria for an armed conflict (see Chapter 9, section 9.4.3, *The "Totality of the Circumstances" and Other Criteria*).[63] The

---

[56] *Id.*

[57] Ruys, *The Meaning of "Force" and the Boundaries of the Jus Ad Bellum, supra* note 2, at 179 (referring to opinions that attacks of "lesser intensity" by military forces are to be countered with police measures). *See also* CORTEN, *supra* note 44, at 52–66.

[58] Ruys, *The Meaning of "Force" and the Boundaries of the Jus Ad Bellum, supra* note 2, at 180–82.

[59] DINSTEIN, *supra* note 15, ¶ 22 at 10.

[60] CORTEN, supra note 44, at 170.

[61] *US Seizes Benghazi Raid "Ringleader" Ahmed Abu Khattala*, BBC NEWS (June 18, 2014), http://www.bbc.com/news/world-us-canada-27893831.

[62] O'Connell, *The True Meaning of Force, supra* note 2.

[63] MITCHELL ZUCKOFF, 13 HOURS: THE INSIDE ACCOUNT OF WHAT REALLY HAPPENED IN BENGHAZI 254–80 (2014), and PERM. S. COM. ON INTELLIGENCE, 113th CONG., INVESTIGATIVE REP. ON THE TERRORIST ATTACKS ON U.S. FACILITIES IN BENGHAZI, LIBYA, SEPTEMBER 11–12, 2012, 3–12 (Nov. 21, 2014), https://intelligence.house.

violence certainly met the factual intensity of an armed conflict rather than "situations of internal disturbances and tensions, such as riots, isolated and sporadic acts of violence and other acts of a similar nature, as not being armed conflicts."[64] This helps explain how the United States could take the position that the action to apprehend Abu Khattala involved an Article 51 exercise of self-defense,[65] particularly given that nation's reluctance to accept a gravity threshold under that provision.[66] That the State security forces involved in the Abu Khattala situation chose to apply more restrictive "law enforcement–type" approach involving capture does not alter the legal categorization of the attack.

As has been noted, what is problematic is when the restrictive principles of human rights–based law enforcement are loosely applied as form of proxy for a "recourse to war"–type restriction on the use of State force. Unfortunately, this is done without any analysis of the differences between human rights–based law enforcement principles (i.e., necessity, proportionality, imminence) and those traditionally applicable to defensive uses of force by States (i.e., necessity, proportionality, imminence, and immediacy) (see Chapter 11, section 11.2, *Distinguishing Law Enforcement*). What becomes evident is that the twentieth-century recourse to war legal debate, which from a theoretical level has been grounded in restricting any use of force, continues to heavily influence the analysis of many international lawyers. In this regard it has been suggested "the possible extension of the scope of application of 2(4) and 51 of the Charter does risk giving arguments to States wishing to conduct military actions in the territory of other States in circumstances that are incompatible with the regime as it is laid down in the UN Charter."[67] The impact of the traditional highly restrictive recourse to war approach is particularly obvious in the efforts to find alternative bases for States to act such as law enforcement or forceful countermeasures. However, these alternative approaches are not without their own challenges or controversy.

With respect to forceful countermeasures, Judge Bruno Simma did suggest the possibility of defensive action by force in response to a "smaller-scale use of force" in his separate judgment in *Case Concerning Oil Platforms (Islamic Republic Of Iran V. United States Of America)*.[68] While this approach has garnered some support,[69] there is also a broadly held view that countermeasures cannot include armed force.[70] Reliance on forceful

gov/sites/intelligence.house.gov/files/documents/Benghazi%20Report.pdf (for an outline of the significant levels of force used by an organized armed group, including the use of mortars, over a thirteen-hour period).

[64] AP II, art. 1(2).

[65] Letter from the United States Ambassador to the United Nations to The President of the United Nations Security Council (June 17, 2014), https://www.documentcloud.org/documents/1201215-power-letter-to-un-about-khattala.html.

[66] *But see also* O'Connell, *The True Meaning of Force*, *supra* note 2 (where the basis for the ambassador's letter is described as a "mystery").

[67] CORTEN, *supra* note 44, at 172 ("The pacifying object of the rule pleads therefore in favour of a restrictive interpretation of its scope of application.").

[68] Case Concerning Oil Platforms (Islamic Republic Of Iran V. United States Of America), 2003 I.C.J. ¶ 12, at 332.

[69] CASSESE, *supra* note 17, at 371–72, and FRANCK, *supra* note 12, at 109–12 (recognizing the use of forceful measures in response to attacks below the threshold of "armed attack." This includes a reference to State practice claiming the exercise of "self-help".).

[70] JAMES CRAWFORD, STATE RESPONSIBILITY 690–91 (2013) ("The Charter era provides ample authority to support an exclusion of forcible countermeasures."), DINSTEIN, *supra* note 15, ¶ 552 at 209, and ¶ 673 at 254 (where the approach is taken that the use of force falls within the self-defense framework). *See* TALLIN MANUAL ON THE INTERNATIONAL LAW APPLICABLE TO CYBER WARFARE RULE 9, ¶ 5, at 38 (Michael N. Schmitt ed., 2013) (where it is noted the majority of the experts involved in writing the manual were of the view "cyber countermeasures may not involve the

countermeasures appears to be championed by some international lawyers in this complex contemporary debate as a means of insulating traditional State versus State uses of force rules from being watered down by these "lesser" threats. However, forceful countermeasures as envisioned by Judge Simma still involve "*proportionate defensive measures* on the part of the victim, equally short of the quality and quantity of action in self-defence expressly reserved in the United Nations Charter."[71] Any substantive distinction between "proportionate defensive measures" and self-defense is difficult to discern. Further, explaining to a military commander how these two terms might differ and any practical effect that might have on the conduct of operations could prove challenging.

Another approach for dealing with the challenge of the protection of nationals sees a particular military doctrine for the rescue of nationals aboard, noncombatant evacuation operations (NEOs), as falling outside the scope of Article 2(4) of the UN Charter. This is "because they are essentially harmless and can therefore be deemed not to reflect a hostile intent."[72] Noncombatant evacuation operations are not designed to target the territorial State, with the doctrine only permitting the protection of the military personnel "and the individuals being evacuated."[73] The development of NEO doctrine by countries such as the United States, the United Kingdom, Canada, and Australia provides an indication that a number of countries view intervention to protect nationals as justifiable under international law.[74] These interventions have been argued to represent something new. They are also suggested to be the law as it ought to be (i.e., *lex ferenda*), rather than being reflective of existing law.[75] However, what is being offered with the NEO approach is not necessarily new or radically different from that proposed by the law enforcement, forceful countermeasures, or self-defense theories. This approach is reflected in the long-standing principles governing intervention created by Sir Humphrey Waldock in the 1950s. Those principles themselves reflect the State self-defense principles of necessity, proportionality, and immediacy.[76]

The multiplicity of approaches and interpretations reflects the unsettled nature of this area of law. However, there is clear recognition across a wide spectrum of legal viewpoints that in dealing with non-State actors some ability to use force to intervene in another State has to be justifiable. That authority extends to States being able to exercise a law enforcement power. For example, the rescue of nationals from threats posed by criminal gangs or

threat or use of force."), and Kenneth Watkin, *The Cyber Road Ahead: Merging Lanes and Legal Challenges*, 89 INT'L L. STUD. 472, 500–03 (2013). *See also* Draft Articles on Responsibility of States for Internationally Wrongful Acts (2001), art. 51(a), http://legal.un.org/ilc/texts/instruments/english/commentaries/9_6_2001.pdf.

[71] Case Concerning Oil Platforms (Islamic Republic Of Iran V. United States Of America), 2003 I.C.J. ¶ 12, at 332 (emphasis added).

[72] Ruys, *The Meaning of "Force" and the Boundaries of the Jus Ad Bellum*, *supra* note 2, at 198.

[73] *Id.*

[74] RUYS, *supra* note 19, at 244–48. *See also* Gill & Ducheine, *supra* note 14, ¶ 12.01, at 217 ("Rescue of nationals is generally referred to as 'non-combatant evacuation' in military doctrine.").

[75] *Id.* (where it is argued the right to protect nationals is reflective of the law as it ought be (i.e. *de lege ferenda*), rather what it actually is (i.e. *lex lata*)). *See also* DINSTEIN, *supra* note 15, at ¶ 577, at 218–19 (where the view is taken there is no new rule that is becoming part of customary law regarding the defense of nationals. Instead forcible measures taken in a foreign territory to rescue nations "must be based on self-defence in response to an armed attack." This position is facilitated by a very broad view of what constitutes an armed attack in the context of exercising the right of self-defense).

[76] DINSTEIN, *supra* note 15, ¶ 676, at 255–56.

the enforcement of laws on the high seas. There are strong legal and policy arguments supporting the right to defend nationals abroad. However, such action needs to be undertaken within an overall restrictive framework. The use of a simplified "unwilling or unable" argument regarding a host State's failure to act, in order to justify large-scale or repeated interventions is difficult to support at law particularly if it is suggested humanitarian law alone regulates the use of force (see Chapter 8, section 8.5.2, *The Permissive "Conduct of Hostilities" Approach*). The use of force during these cross-border interventions must also remain significantly restricted by self-defense principles whatever their legal genesis.[77]

The challenge from a theoretical perspective is to determine exactly how such action will be categorized. For example, will it be viewed as law enforcement, forceful countermeasures, self-defense in response to an armed attack, proportionate defensive measures, NEO operations, or simply the defense of nationals? As with many international law issues, the existing lack of agreement among theorists is not particularly helpful. Even though there is powerful normative pressure brought to bear by proponents of the narrow strict interpretation of the recourse to war regime, interventions to protect nationals are a reality of the international security environment. Given the controversy surrounding the defense of nationals issue, it should come as no surprise that the acceptability of intervention within another State, politically and legally, often appears to be case specific.[78] One approach for dealing with this clash between that reality and narrow interpretations of the law appears to see State action "condoned as legitimate in specific mitigating circumstances, even though the recourse is still recognized as technically illegal."[79] This reflects the powerful pull of the traditionalist view of the law governing the recourse to war. However, international law must be able to provide better authority for such action than suggesting action to defend nationals is legitimate but ultimately illegal. Using such terminology is not only confusing, it would not provide appropriate advice for military commanders who are required to follow the law.

An interpretation of the law accepting intervention as legitimate, even if illegal, is no doubt frustrating for recourse to war traditionalists focused on maintain international order.[80] However, it also highlights the challenge that international law faces when the boundaries of a particular area of law do not fully encompass the security situation facing States. In this respect there is a "divergence of pragmatic practice from pure theory."[81] The lack of agreement among theoreticians often leaves practitioners focusing on the substantive issues rather than on what may appear to be more esoteric and often intractable arguments

---

77 *See, e.g.*, San Remo Manual on International Law Applicable to Armed Conflicts at Sea Rule 22, at 101 (1995) (for a description of the limits on entry into neutral waters based in humanitarian law), and Dinstein, *supra* note 15, ¶ 729(v), at 275 (where in respect of "extra-territorial law enforcement" that it must be clearly demonstrated "attacks by organized armed groups or terrorists cannot be defeated through recourse to alternative measures that are less intrusive in their effects on the territorial sovereignty of [the Host State].").

78 Lindsay Moir, Reappraising the Resort to Force: Inter-National Law, *Jus ad Bellum* and the War on Terror 30–31 (2010).

79 Franck, *supra* note 12, at 96.

80 Wittes & Blum, *supra* note 3, at 257 (where it is stated "the state of affairs in which vast parts of the world enjoy sovereignty and immunity from external intervention, even while local authorities cannot prevent local actors from using their territories to export harm to others, is unsustainable.").

81 Franck, *supra* note 12, at 77.

(see Chapter 9, section 9.5, *Addressing the Challenge*). For those practitioners, there will remain the requirement to act.[82] It may be that the categorization "defense of nationals" as an umbrella remains the most helpful, leaving the theoreticians to argue exactly what legal basis (e.g., Article 51, customary law, countermeasures, NEO) provides the technical justification for what is a permissible and understandable defensive act carried out by a State.

10.2.2. TAKING ACTION ACROSS THE CONFLICT SPECTRUM

The continuing debate and degree of equivocation exhibited by many legal scholars toward a right to intervene itself provides an indication of the principle's continued applicability in the aftermath of the creation of the UN Charter. Framing in terminology that resonates with the basis for contemporary drone strikes, D. W. Bowett indicated in 1958 that the threats to nationals can come "from the acts of individuals which the territorial state is *unable, or unwilling,* to prevent."[83] A particular concern has been that States must be able to act when a territorial State cannot protect foreign nationals within its borders. In this regard, there is a link between the territorial State offering justice for those who fall within its jurisdiction and the power of a national's State to intervene. As has been stated, "the price of inviolability of any territory is the maintenance of justice therein. Accordingly, when that price is not paid in relation to foreign life and property, the landing of forces for their protection is to be anticipated."[84]

There are a wide range of State practices that have been claimed on the basis of the defense of nationals. Examples include aerial bombing;[85] missile strikes;[86] hostage rescue operations;[87] the diversion of aircraft carrying terrorists;[88] and larger-scale interventions, some of which involved significant combat with the armed forces of the territorial State.[89] Many of these operations are similar to those conducted against non-State actors in the context of what is now called "counterterrorism." This is not to say that all of the operations taken in defense of nationals were seen as being justified by the international community or were based solely on self-defense. For example, the 1965 U.S. intervention in the Dominican Republic has been referred to as being "questionable" and the 1983 action in Grenada as "dubious."[90]

---

82 WITTES & BLUM, *supra* note 3, at 236 ("Without the ability to rely consistently on other sovereigns to police their territories and citizens effectively, some states will find it necessary to pursue their own security interests in defiance of other states' sovereignty, boundaries, or political independence.").

83 BOWETT, *supra* note 9, at 88.

84 CHARLES CHENEY HYDE, 1 INTERNATIONAL LAW: CHIEFLY AS INTERPRETED AND APPLIED BY THE UNITED STATES 649 (1945).

85 FRANCK, *supra* note 12, at 89–91.

86 *Id.* at 94–96 (by the United States in Iraq (1993), Afghanistan and the Sudan (1998)).

87 *Id.* at 82–86 (Israeli rescue operation in Uganda (1976)), DINSTEIN, *supra* note 15, ¶¶ 681–86, at 257–59, and CASSESE, *supra* note 17, at 367 n.16 (referring to a failed 1978 Egyptian commando rescue attempt in Cyrus. *See also* FRANCK, *supra* note 12, at 85–86, and PETER HARCLERODE, SECRET SOLDIERS: SPECIAL FORCES IN THE WAR AGAINST TERRORISM 298 (2000).

88 FRANCK, *supra* note 12, at 88–89.

89 *Id.* at 78–82, 86–88 and 91–94 (in Belgium in the Congo (1960, 1964), Turkey in Cyrus (1964); as well as the United States in the Dominican Republic (1965), Grenada (1983), and U.S.-Panama (1989)). *See also* DINSTEIN, *supra* note 15, ¶ 679(ii), at 257 (where reference is also made to claims for acting in protection of nationals by Russia in Georgia (2008)), and GRAY, *supra* note 17, at 159 (for reference to a series of interventions by France in Africa).

90 DINSTEIN, *supra* note 15, ¶678, at 256.

However, limited State action to protect its nationals has been an enduring part of international relations. Such action raises the issue of the degree to which the contemporary debate about the legality of defensive drone strikes carried out against non-State actors is in some respects old news. It may be that conducting an isolated or limited drone strike focused on responding to a specific threat emanating from an "ungoverned space" would attract a similar level of acceptance or at least tolerance within the international community as intervention to protect nationals. However, as has been seen with the U.S. drone campaign, such a defensive response is more likely to be challenged when the strikes reflects a pattern of activity suggesting involvement in a broader conflict. These drone strikes are not a limited defensive response flying below the threshold of the political "radar" applied by the international community for tracking State involvement in armed conflict.

Acceptance of a right to intervene does not mean there is a broad authority to breach the sovereignty of other nations. The exercise of that right "is conditioned by the same requirements which international law imposes upon any exercise by the state of its right of self-defence."[91] It has been suggested, "it would seem preferable to accept the validity of the rule in carefully constructed situations consistent with the conditions laid down in the *Caroline* incident."[92] The oft-cited conditions restricting such action were outlined by Sir Humphrey Waldock as: "[t]here must be (1) an imminent threat of injury to nationals, (2) a failure or inability on the part of the territorial sovereign to protect them and (3) measures of protection strictly confined to the object of protecting them against injury."[93] The historical basis for such conditions has been questioned with States in the nineteenth century viewed as having acted generally to protect citizens and their interests "with no reference to the *Caroline* incident or the conditions suggested."[94] However, States have relied on similar restrictions on the use of force. For example, they were referred to by the United Kingdom in 1956[95] and 1993.[96] A leading academic and jurist, the late Antonio Cassese, also relied on these types of restrictions to justify the use of force to protect nationals.[97] Not only do these conditions reflect the principles of necessity, proportionality, and imminence associated with the *Caroline* incident, there is a resonance with the "unwilling or unable" criteria applied in contemporary counterterrorist operations.[98] As there has been a "remarkable increase in episodes of the taking of hostages and incidents of transnational terrorism,"[99] it is likely the need for States to act to defend their nationals will continue and even expand, rather than fade away as this century progresses.

What is unique about States acting to protect nationals, whatever the legal basis, is that it transcends the full conflict spectrum. The motivation to protect one's citizens is a powerful one, even in the context of an international armed conflict. Nationals of one State may be

---

[91] BOWETT, *supra* note 9, at 96. *See also* Gill & Ducheine, *supra* note 14, ¶ 12.02, at 219.

[92] SHAW, *supra* note 12, at 1145.

[93] DINSTEIN, *supra* note 15, ¶ 676, at 255–56.

[94] BROWNLIE, *supra* note 12, at 299.

[95] GRAY, *supra* note 17, at 158.

[96] SHAW, *supra* note 12, at 1145, and CASSESE, *supra* note 17, at 367 n.17.

[97] CASSESE, *supra* note 17, at 368.

[98] DINSTEIN, *supra* note 15, ¶ 677, at 256.

[99] *Id.* ¶ 576, at 218.

seized by another State with which it is at war. This can include civilians who are interned or military personnel that are made prisoners of war or otherwise detained. While these latter seizures fall within the scope of an international armed conflict, this does not mean that rescue missions will not be attempted, particularly if those nationals are being abused. In some cases threats may come directly from States. The 1975 seizure of the merchant vessel *SS Mayaguez* by the Cambodian government is one example of such State action.[100] Regarding the actions of non-State actors, there are immediate parallels between traditional action taken to protect nationals and State reaction to contemporary transnational terrorism threats. Complicating the security situation is that non-State actors do not always act alone. States may be involved in varying degrees. Such State involvement occurred in the case of the seizure of the U.S. embassy in Tehran in November 1979 by Muslim fundamentalist revolutionaries (see Chapter 9, section 9.3.2, *Harboring and Active Support*),[101] and in the context of the Ugandan government providing direct support to terrorist hijackers in 1976 in Entebbe.[102]

The broad range of extraterritorial threats highlights the limitations of the post–World War II recourse to war construct, which has traditionally been focused solely on international armed conflict. The void between the State-centric interpretations of the right to act in self-defense and the much broader scope of threats to nationals is even more apparent in situations where those threats emanate from criminal elements. For example, a State assisting a territorial government in fighting a counterinsurgency may also be confronted with threats to its own citizens who could be providing humanitarian assistance, working as journalists, or otherwise acting independently of the State. [103] A State may also decide to act to protect its nationals in situations where the territorial State is simply failing to police its territory or take action to address lawlessness on the high seas.[104] For example, a hostage rescue operation was conducted by the United States against Somali pirates who abducted the American Captain of the *SS Maersk Alabama* in the Indian Ocean in 2009[105] and against a group that seized hostages within Somalia itself in 2012.[106] This highlights

---

[100] THE SEIZURE OF THE MAYAGUEZ: A CASE STUDY OF CRISIS MANAGEMENT, U.S. Government Accountability Office (May 11, 1976), http://www.gao.gov/assets/240/233910.pdf [hereinafter MAYAGUEZ REPORT], and CLAYTON K. S. CHUN, THE LAST BOARDING PARTY: THE USMC AND THE SS MAYAGUEZ 1975 (2011).

[101] DEREK LEEBAERT, TO DARE & CONQUER: SPECIAL OPERATIONS AND THE DESTINY OF NATIONS, FROM ACHILLES TO AL QAEDA 555–59 (2006). *See also* CHARLES A. BECKWITH & DONALD KNOX, DELTA FORCE: THE U.S. COUNTER-TERRORISM UNIT AND THE IRANIAN HOSTAGE RESCUE MISSION 186–284 (1983).

[102] HARCLERODE, *supra* note 87, at 345–66. *See also* SIMON DUNSTAN, ISRAEL'S LIGHTNING STRIKE: THE RAID ON ENTEBBE 1976, 14–16 (2009) (for an outline of Idi Amin's support for the Palestinian cause).

[103] Sarah Hampson, *How Mellissa survived 28 days of captivity in Afghanistan*, THE GLOBE AND MAIL, May 11, 2011, http://www.theglobeandmail.com/life/relationships/how-mellissa-fung-survived-28-days-of-captivity-in-afghanistan/article624659/.

[104] Ruys, *The Meaning of "Force" and the Boundaries of the Jus Ad Bellum*, *supra* note 2, at 208 ("the unauthorized boarding of foreign vessels in situations other than those listed in the convention [the UN Convention on the Law of the Sea] has, on occasion, been framed as an application of self-defense.").

[105] MARTIN N. MURPHY, SOMALIA: THE NEW BARBARY?: PIRACY AND ISLAM IN THE HORN OF AFRICA 104–05 (2011), and Robert D. McFadden & Scott Shane, *In Rescue of Captain, Navy Kills 3 Pirates*, N.Y. TIMES, Apr. 13, 2009, http://www.nytimes.com/2009/04/13/world/africa/13pirates.html?pagewanted=all.

[106] Jeffrey Gettleman, Eric Schmitt, & Thom Shanker, *U.S. Swoops in to Free 2 From Pirates in Somali Raid*, N.Y. TIMES, Jan. 25, 2012, http://www.nytimes.com/2012/01/26/world/africa/us-raid-frees-2-hostages-from-somali-pirates.html?pagewanted=all.

that acting in defense of nationals can extend beyond armed conflict to ordinary lawlessness. This in turn indicates that some State reactions are more appropriately framed as law enforcement, rather than participation in armed conflict. Even for legal analysts who adopt a very narrow interpretation of State self-defense there must remain a residual authority to act to protect nationals, however, that authority is categorized (e.g., forcible countermeasures, proportionate defensive measures).

It is also not uncommon for States to act in a coalition when carrying out operations to defend nationals, or for an agreement to be reached that more militarily capable States will rescue the citizens of another nation. For example, it is reported that on March 3, 2011, a British C-130 aircraft landed at a desert airstrip in Libya to evacuate 150 expatriate oil workers (of whom only 20 were British) who had been threatened by gunmen and bandits as that country became engulfed in a civil war.[107] Technically, the rescue of other nationals is not normally included within the concept of self-defense. In this respect a State may afford protection to the nationals of other States, however, "[w]hat is denied is that, in the absence of the nexus of nationality, the use or threat of force in the protection of persons is part of the concept of self-defence."[108] The legal basis suggested for such action may be found in treaty, as a result of a "request from the state whose nationals those persons are, or even under the principle of humanitarian intervention."[109] What all of this indicates is that as powerful as the pressures have been to restrictively apply the right of self-defense, there has also been a strong movement to continue to find a way to meet these real-world security challenges. Foreign intervention for the protection of nationals appears to still be lawful, with a particular emphasis being placed on the restrictive interpretation of principles associated with State self-defense being applied.[110]

There can be little question regarding the positive impact the "antiwar"-based UN Charter–based normative framework has had on world security. However, on the issue of the defense of nationals one problem is that framework did not fundamentally address the full spectrum of security challenges facing States even at that time, let alone how the international security environment has developed since then. The predominant form of warfare has always been between State and non-State actors. Further, with the control over territory exercised by some States either having not been fully established or increasingly being challenged, the resulting security void may have to be filled by other States when necessary. Nearly all interventions have occurred where "there was a real breakdown in the territorial system of public order, or inability on the part of local government to prevent the perpetration of unlawful acts against foreigners."[111] In effect, the need has predominately arisen in the "ungoverned spaces" of the world.

---

[107] Mark Urban, *Inside Story of the UK's Secret Mission to Beat Gaddafi*, BBC News (Jan. 12, 2012), http://www.bbc.co.uk/news/magazine-16573516.

[108] Bowett, *supra* note 9, at 95. *See also* Dinstein, *supra* note 15, ¶ 686, at 259.

[109] Bowett, *supra* note 9, at 95. *See also* Gill & Ducheine, *supra* note 14, ¶ 12.02, at 217 (where reference is made to it being permissible to evacuate the nationals of third States "in so far as they wish to be evacuated.").

[110] Cassese, *supra* note 17, at 368.

[111] *Id.* at 367.

10.3. Protecting Nationals and Hostage Rescue

10.3.1. RESCUES DURING ARMED CONFLICT

There is a broad range of circumstances under which a State may feel compelled to act in defense of its nationals. Within this range of operations, "hostage rescue" provides a useful model for assessing the legal frameworks governing such action in situations ranging from armed conflict, to counterterrorism, and finally to law enforcement. Circumstances where rescue operations have been attempted are as varied as the nature of the violence itself. The decision to conduct rescue operations can arise during armed conflict because of concerns over the abuse and murder of prisoners and internees by their captors.[112] In many cases the captured military and civilian personnel are used, and sometimes abused, for "propaganda" purposes, in order to influence the conduct of their nations. Contemporary examples of the abuse of detainees has been the beheading of captured personnel by Al Qaeda in Iraq for the purposes of intimidating Iraqi government and Coalition authorities[113] and the public executions including death by immolation by the Islamic State.[114]

Rescue operations are not limited to conflict with non-State actors. Persons may be detained by the enemy during an international armed conflict under a variety of circumstances including as prisoners of war and civilian internees. Conducted against a State opponent during armed conflict these operations do not raise the same issues regarding legal authority to enter territory controlled by an enemy. Complicating a historical analysis of the rescue issue is the fact that seizing civilians as hostages was not necessarily contrary to international law prior to the end of World War II,[115] although horrendous excesses during that conflict led to significant war crimes trials.[116] Hostage taking was a practice that ultimately was to be outlawed under humanitarian law.[117] States will on occasion take action to rescue prisoners and civilians in the hands of the enemy. During World War II there were some raids to rescue military prisoners conducted in both the Pacific[118] and European

---

[112] WILLIAM H. MCRAVEN, SPEC OPS: CASE STUDIES IN SPECIAL OPERATIONS WARFARE: THEORY AND PRACTICE 248 (1995) (where it is indicated in respect of the Cabanatuan POW camp by January 1945, over 3,000 men had died and "[s]aving the remaining 512 would require swift action on the part of the Rangers. . . ."), GORDON L. ROTTMAN, THE LOS BAÑOS PRISON CAMP RAID: THE PHILIPPINES 1945, 15–16 (2010) (the internees were being starved and concerns were raised that mass graves were being dug.), and DAVID C. ISBY, LEAVE NO MAN BEHIND: LIBERATION AND CAPTURE MISSIONS 22 (2004) (where it is reported "140 prisoners had been massacred on the island of Palawan. . . .").

[113] JASON BURKE, THE 9/11 WARS 165–66 (2011), and BING WEST, THE STRONGEST TRIBE: WAR, POLITICS AND THE ENDGAME IN IRAQ 72 (2009).

[114] Suleiman Al-Khalidi, *Islamic State Shows Burning of Hostage, Jordan Vows "Earth-shaking" Response*, REUTERS (Feb. 3, 2015), http://www.reuters.com/article/2015/02/03/us-mideast-crisis-killing-idUSKBN0L71XE20150203, and *Islamic State Militants "Burn to Death 45 in Iraq"*, BBC NEWS (Feb. 17, 2015), http://www.bbc.com/news/world-middle-east-31502863.

[115] Ellen Hammer & Marina Salvin, *The Taking of Hostages in Theory and Practice*, 38 AM. J. INT'L L. 20 (1944). *See also* Lord Wright, *The Killing of Hostages as a War Crime*, 25 BRIT. Y.B. INT'L L. 297 (1948) (for a critical review of the practice).

[116] *See, e.g.*, The Hostage Case, Trials of War Criminals, Washington: Government Printing Office 759 (1950).

[117] *See, e.g.*, The 1949 Geneva Conventions, Common Art. 3, GC IV, arts. 34 and 147, AP I, art. 75(2)(c), AP II, art. 4(2)(c), and the 1998 Rome Statute, arts. 8(2)(a)(viii) and (c)(iii). *See also* CUSTOMARY INTERNATIONAL HUMANITARIAN LAW STUDY Rule 96, 334–36 (Jean-Marie Henckaerts & Louise Doswald-Beck eds., 2005).

[118] MCRAVEN, *supra* note 112, at 245–86 (for an outline of a Japanese-run POW camp in January 1945 at Catanatuan in the Philippines during World War II), and ISBY, *supra* note 112, at 39 (for reference to Office of Strategic Studies (OSS) using airborne raiders to secure Allied prisoners in Manchuria and northern China).

theaters.[119] These types of operations were not limited to rescuing military personnel. For example, in October 1944 U.S. military units carried out a successful raid to rescue Dutch and Javanese civilians held in New Guinea.[120] At the beginning of February 1945, Army units rescued 3,700 civilian internees at Santos Tomás University in Manila,[121] and later that month liberated 2,100 internees at an internment camp at Los Banõs 40 miles southeast of the capital city.[122]

In Korea, an airborne operation had the rescue of POWs as one of its mission objectives.[123] During the Vietnam conflict, which the United States considered to be an international armed conflict,[124] "[o]ver ninety-one POW rescue attempts had been mounted between 1966 and 1970."[125] It has been reported that in Vietnam "[o]f the 91 rescue operations, 20 succeeded—in rescuing 318 South Vietnamese soldiers and 60 civilians."[126] The justification for the November 1970 Special Forces raid to rescue American POWs believed to be at Son Tay Prison near Hanoi was based, in part, on the receipt of information about the deaths of prisoners in the custody of the North Vietnamese.[127] During that war there was widespread torture used by the North Vietnamese against captured American POWs in order to obtain false confessions of criminal activity. Those confessions were intended to demonstrate the illegality of the U.S. intervention and bombing campaign.[128] The abuse of prisoners increased the pressure to attempt rescues.

At the close of the Vietnam War, the United States became engaged in armed conflict with Cambodian military forces as a result of their seizing the *SS Mayaguez*, although it has been suggested the United States erred in exercising a right of self-defense in respect of this merchant vessel.[129] The efforts to free the crew of the merchant vessel included an assault by U.S. Marines on Cambodian military positions on Koh Song island (where it was wrongly believed the crew might have been taken). The overall operations were aided by close air support and naval gunfire,[130] and involved the retaking of the Mayaguez by naval boarding,[131] as well as the bombing of the Cambodian mainland.[132] The Mayaguez crew was actually being held elsewhere and was released concurrently with the conduct of the military operations.[133]

---

119 Benjamin F. Schemmer, The Raid 59 (1976) (for reference to a raid ordered by General Patton in Germany to liberate a camp in which his son-in-law was being held).

120 McRaven, *supra* note 112, at 254.

121 Rottman, supra note 112, at 11.

122 *Id.* at 4. *See also* Isby, *supra* note 112, at 21–41(for a description of the Los Banõs raid).

123 Isby, *supra* note 112, at 42–62.

124 George S. Prugh, Law at War 1964–1973, at 63 (1975).

125 McRaven, *supra* note 112, at 319 n *.

126 Schemmer, *supra* note 119, at 237 ("But of the 45 raids mounted to rescue American prisoners only one succeeded.").

127 *Id.* at 227, and Isby, *supra* note 112, at 63–91 (for an outline of the Son Tay raid).

128 John G. Hubbell, P.O.W.: A Definitive History of the American Prisoner-of-War Experience in Vietnam, 1964–1973, at 475–78 (1976) (outlining an incident in which then Commander James Stockdale was tortured to obtain a false confession wherein he acknowledged criminal activity).

129 Dinstein, *supra* note 15, ¶ 574, at 217.

130 Chun, *supra* note 105, at 48–70, and Isby, *supra* note 112, at 92–130.

131 Chun, *supra* note 105, at 42–46.

132 *Id.* at 46–48.

133 *Id.* at 45–46. *See also* Mayaguez Report, *supra* note 105, at 101 (where it is indicated the Koh Song assault and mainland bombing did not influence the release of the crew, however, other action such as the sinking of Cambodian gunboats and the U.S. air activity probably did).

The 1989 U.S. intervention in Panama witnessed a number of different types of "rescue" operations. One involved a special operations force assault on the Modelo Prison to release a CIA operative who had been arrested by Manuel Noriega and who had been threatened with execution if there was an invasion.[134] This operation was planned as one of the overall "Operation Just Cause" mission objectives.[135] It involved significant fighting when entering the prison and during withdrawal with the freed hostage.[136] A second assault was made at the Renacer Prison to release a number of prisoners, including two Americans.[137] Another incident involved two American women who were taken by Panamanian Defence Force (PDF) personnel. The hostage situation unfolded as a U.S. Ranger unit attempted to secure the passenger terminal of the Torrijos-Tocumen Airport.[138] The PDF personnel eventually surrendered and the hostages were safely released when the American military personnel gave the soldiers an ultimatum to surrender or be killed.[139] There was also a significant operation carried out by 82nd Airborne Division personnel to rescue American, foreign, and Panamanian hostages being held at a hotel in Panama City.[140] During the 2003 invasion of Iraq United States special operations forces were tasked with rescuing a captured soldier, Private Jessica Lynch, being held at a hospital in Nasiriyah.[141]

A point to note is that across this range of armed conflict–related operations the tactics and force used was affected by the unique circumstances of the mission and by the need to protect the hostages during any assault.[142] A requirement to gain surreptitious access to the location of the hostages may itself limit the number and types of forces involved, as well as avoid activity that could alert the hostage takers. For example, in 1945, after a brief skirmish during the Santos Tomás raid, the American forces negotiated with guards giving the Japanese safe passage away from the camp.[143] Subsequently during the Los Banõs raid the soldiers deploying to the compound were told not to return fire if guards fired at their noise and to take any casualties with them to avoid being discovered. Of primary concern was that the Japanese guards would start a massacre.[144]

During the 1976 Mayaguez operation concern over the safety of the hostages and a desire to maintain surprise led to a decision not to use close air or naval gunfire to support the initial Marine helicopter assault.[145] There was also a use of riot control agents in an attempt to stop Cambodian vessels transporting the crew[146] and to retake the Mayaguez that is

---

134 TOM CLANCY, CARL STINER, & TONY KOLTZ, SHADOW WARRIORS: INSIDE THE SPECIAL FORCES 314 (2002).

135 *Id.* at 322.

136 *Id.* at 341–43.

137 ISBY, *supra* note 112, at 222 (the prisoners included "Diego Jiminex of Miami and another American, five anti-Noriega Panamanians and a larger number of criminals.").

138 CLANCY ET AL., *supra* note 134, at 355–56.

139 *Id.* at 356, and ISBY, *supra* note 112, at 236.

140 ISBY, *supra* note 112, at 247–50.

141 SEAN NAYLOR, RELENTLESS STRIKE: THE SECRET HISTORY OF JOINT SPECIAL OPERATIONS COMMAND 211-12 (2015).

142 *See, e.g.*, MCRAVEN, *supra* note 112, at 265 (during the 1945 raid on the POW camp at Cabanatuan in the Philippines an airplane was used to circle the camp in order to divert attention away from the approaching raiding force).

143 ROTTMAN, *supra* note 112, at 11–12.

144 *Id.* at 45. *See also* ISBY, *supra* note 112, at 30.

145 CHUN, *supra* note 105, at 31.

146 *Id.* at 38–39 (warning shots from 20 mm guns on-board U.S. aircraft were also used, however, they struck the ship, which caught fire and sank). *See also* ISBY, *supra* note 112, at 102.

reflective of law enforcement practices.[147] It is evident that while still involving combat against enemy units, the military personnel engaged in the rescue can receive direction to limit their use of force and even conduct negotiations with those forces. These operations can be far more restrained than ordinary combat. While they might not entirely resemble a law enforcement rescue, there can be similarities, at least to the extent there is a requirement to take actions to avoid causing death or injury to the hostages.[148] However, what these restrictions do not change is that the applicable specialized law applicable during armed conflict governing these operations is humanitarian law.

The defense of nationals in occupied territory is another situation that can provide unique challenges for security forces. The occupying power acts in place of the territorial State with a specific legal obligation to maintain public order and security.[149] Nationals seized by a variety of resistance movements, organized armed groups, and criminal gangs may include not only inhabitants of the occupied territory but also those of third States and the occupying power itself. As a result, there may be operations conducted by the occupying power, possibly assisted by indigenous security personnel, to rescue those detainees. Where more than one State is involved as an occupying power, the rescue operation may involve the military personnel of more than one nation. In these situations the issue of breach of sovereignty does not arise, since the occupying power acts in place of territorial State. The fact that the rescues are attempted in occupied territory will not necessarily make the operational environment any more permissive since resistance forces may contest control over the territory. The fulfillment of the occupying power's responsibility can result in rescue operations being conducted across the spectrum of conflict. What this means is that even though the operations take place in the context of an international armed conflict, the hostage rescues will not necessarily take place as part of the hostilities. They could involve the rescue of nationals and others seized by uniquely criminal elements. This means occupation forces could have to apply human rights–based law enforcement norms when attempting a rescue.

The aftermath of the 2003 invasion of Iraq provides an example of how kidnapping can be used both "as a highly profitable activity and a form of asymmetric warfare for the weak against the strong."[150] Kidnappings were identified as having followed two major streams, commencing with abductions for criminal purposes, to which after 2004 was added the seizure of foreigners.[151] It was an activity that transcended the change in the categorization of the conflict in June 2004 from an international one to an insurgency as part of an internal armed conflict.[152] The seizure of Iraqis constituted the greatest number of kidnappings, with the taking of foreigners being "both clustered and relatively short-lived."[153] The death of al-Zarqawi, whose group Al Qaeda in Iraq was particularly active in seizing foreigners;

[147] CHUN, *supra* note 105, at 44, and ISBY, *supra* note 112, at 112.

[148] NAYLOR, supra note 141, at 374 (where it is suggested there is no hostage rescue scenario during which grenades should be used). *But see also* DUNSTAN, *supra* note 102, at 46-47 (where grenades were reported to have been used by IDF forces during the 1976 Entebbee operation).

[149] 1907 Hague Land Warfare Regulations, art. 43, and GC IV art. 64.

[150] WILLIAMS, *supra* note 4, at 105.

[151] *Id.* at 116–34.

[152] Al-Skeini v. The United Kingdom, IV Eur. Ct. H. R. Rep. ¶ 144, at 170 (2011), *available at* http://www.echr.coe.int/Documents/Reports_Recueil_2011-IV.pdf (July 7, 2011).

[153] WILLIAMS, *supra* note 4, at 134.

the exodus of foreigners who were part of the nongovernmental organization community; and the "improvements in the security situation, especially in the latter half of 2007" all appear to have contributed to a reduced number of foreigners being kidnapped.[154] However, it has also been noted that although "[t]he kidnapping of Iraqis has been more sustained and systematic than the kidnapping of foreigners,"[155] by March 2006 "around two hundred foreigners had been kidnapped in Iraq."[156] Disrupting kidnapping gangs "was the exception rather than the rule, and in most cases little help was given by either the police or coalition forces."[157] That said, rescues were attempted against criminals, including the August 2003 arrest of a nine-member kidnapping gang.[158]

In a noninternational armed conflict, kidnapping and the taking of hostages can occur in a wide variety of circumstances. Periodically such threats can lead to military intervention by a State to rescue the hostages. For example, in November 1964 Belgian Para-Commandos transported by U.S. military aircraft carried out Operation Dragon Rouge, an airborne assault on Stanleyville in the Congo to rescue foreign hostages held by Simba rebels. They rescued 1,600 hostages and refugees with 33 hostages being killed. Two Belgian soldiers were killed and three wounded during the operation.[159] The rebel casualties were assessed to be "minimal,"[160] although there was significant military engagement, particularly when intervention forces were protecting an evacuation airfield.[161] Kidnappings can also occur as part of internal conflicts. A raid into Dagestan, Russia, in 1996 by Chechen guerrillas involved the seizure of approximately 3,000 hostages.[162] In Columbia, "1998 was notable for the number of kidnappings that took place throughout its duration: a total of 2,000, of which FARC [Revolutionary Armed Forces of Columbia] was reported as being responsible for 60 per cent."[163]

Hostage taking has taken a new profile in post-9/11 hostilities. This can be seen in the January 2015 hostage takings by jihadists in Paris,[164] and by an Al Qaeda affiliated group in Bamako, Mali in November 2015.[165] It is an integral part of terrorist tactics as is evidenced in the jihadist ideological work, the *Management of Savagery: The Most Critical Stage Through Which the Umma Will Pass*, which refers to the taking of hostages to be exchanged in return for captured "mujahids."[166] Further, "[t]he policy of violence must also be followed such that if the demands are not met, the hostages should be liquidated in a terrifying manner, which

---

154 *Id.*

155 *Id.* at 146.

156 Mark Urban, Task Force Black 126 (2010).

157 Williams, *supra* note 4, at 120.

158 *Id.*

159 Thomas P. Odom, Dragon Operations: Hostage Rescues in the Congo, 1964–1965, at 159 (1988).

160 *Id.*

161 *Id.* at 111–12, and 118–20 (this included the use of mortars by the rebel forces).

162 Robert M. Cassidy, Counterinsurgency and the Global War on Terror 67 (2008).

163 Harclerode, *supra* note 87, at 250.

164 *The Special Forces Behind France's Rescue Operations,* CBS News (Jan. 9, 2015, 8:09 PM)[hereinafter *France's Rescue Operations*], http://www.cbsnews.com/news/the-special-forces-behind-frances-rescue-operations/.

165 *Mali Hotel Attack: 'No more hostages' after special forces raid,* BBC News (Nov. 20, 2015), http://www.bbc.com/news/world-africa-34879955.

166 Abu Bakr Naji, The Management of Savagery: The Most Critical Stage Through Which the Umma Will Pass, at [33] (William McCants trans., 2006), https://azelin.files.wordpress.com/2010/08/abu-bakr-naji-the-management-of-savagery-the-most-critical-stage-through-which-the-umma-will-pass.pdf.

will send fear into the hearts of the enemy and his supporters."[167] As noted, such action has become an integral part of Islamic State messaging through the release of beheading videos.[168] The Islamic State has "also raised millions by ransoming Western hostages."[169] As a result, taking action to rescue hostages has become a regular part of contemporary military operations that often garners significant international attention. For example, on July 4, 2014, U.S. military forces conducted an unsuccessful raid near Raqqa, Syria, in an attempt to rescue hostages held by the Islamic State.[170] Nor are the situations prompting rescue operations limited to the seizure of civilians. In March 2013 Afghan and NATO forces are reported to have killed twenty-four insurgents as part of an operation to rescue two captured Afghan soldiers in Logar Province, Afghanistan.[171] An October 2015 raid on an Islamic State prison by United States Delta Force, special operations helicopters, and Kurdish Special Forces are reported to have rescued over 70 ISIS prisoners, including 20 Iraqi security personnel; killed at least 10 insurgents; and detained other ISIS personnel.[172]

While the requirement to protect nationals may arise because of criminal activity, the rescue operations may still be considered part of State involvement in the overall counterinsurgency effort. This is because counterinsurgency is ultimately a governance function. Which State conducts the rescue could depend upon the scope and nature of the insurgency; the basis for intervention or the extent of consent provided by the territorial State; and the ability of the territorial State to take action on their own. A foreign State providing logistical or training assistance to a counterinsurgency effort may not have the military capability within the country to carry out hostage rescue. In that situation the State might then have to rely on indigenous forces or separately deploy mission-capable troops to intervene in the country. The host State may want to deal with the incident on its own. However, the State whose nationals are affected might also consider acting unilaterally, knowing that the territorial State can provide little or no assistance. Further, even though the intervening State may be supporting the territorial one, there can be security concerns about sharing intelligence or the tactics used by rescue forces. As a result, the intervening State may not want any involvement of the host nation beyond it granting consent or staying clear of the operation. Depending upon their relationship, the taking of action to rescue nationals could put the intervening State at odds with the territorial State and raise significant international law concerns regarding the breach of territorial sovereignty.

---

167 *Id.*

168 Jessica Stern & J. M. Berger, ISIS: The State of Terror 120–23 (2015).

169 *Id.* at 46.

170 Oliver Holmes & Jason Szep, *U.S. Hostage Rescuers Dropped from Night Sky: Syria Activist*, Reuters (Aug. 22, 2014), http://www.reuters.com/article/2014/08/23/us-syria-usa-raid-idUSKBN0GM26620140823 and Nicholas Schmidle, *Inside the Failed Raid to Save Foley and Sotloff*, The New Yorker, Sep. 5, 2014, http://www.newyorker.com/news/news-desk/inside-failed-raid-free-foley-sotloff.

171 Rahim Faiez, *Afghanistan: Joint NATO-Afghan Operation Kills More Than 20 Insurgents*, Huffington Post (Mar. 27, 2013), http://www.huffingtonpost.com/2013/03/27/afghan-nato-operation_n_2961054.html.

172 *One US Soldier Killed in Iraq During Rescue of More Than 70 ISIS Hostages*, Fox News (Oct. 22, 2015), http://www.foxnews.com/world/2015/10/22/us-special-ops-rescue-hostages-in-iraq-defense-official-says/, and Missy Ryan, Mustafa Salem & Thomas Gibbons-Neff, *First American Soldier is Killed in Combat in Iraq Since 2011 troop exit*, Wash. Post, Oct. 22, 2015, https://www.washingtonpost.com/world/reports-us-iraqi-commandos-raid-islamic-state-outpost-hostages-freed/2015/10/22/f0586b60-78bd-11e5-bc80-9091021aeb69_story.html.

The status of the insurgent group as a participant in an armed conflict can determine the legal framework governing the use of force. Humanitarian law–based conduct of hostilities rules may apply even though an insurgent armed group abducts a foreign national for the purpose of seeking a ransom.[173] Those kidnappers remain targetable at all times under humanitarian law as long as they are members of an organized armed group. If the kidnappers were not members of the armed group but rather civilians who periodically take a direct part in hostilities, then they would have to be dealt with under a law enforcement paradigm for any period they are not engaged in hostilities.[174]

As with international armed conflict, the force used may have to be adjusted to match the operational situation. The need to keep the hostage safe and a requirement to often conduct operations in civilian areas can have a significant impact on the levels of force employed. There have also been hostage seizures in conflicts where a State resists categorizing the violence as an armed conflict. For example, it was only in 2011 that Colombia publicly recognized it was in an armed conflict with FARC rather than simply fighting terrorism.[175] Such a situation could lead to greater expectations that a law enforcement framework will be used when conducting rescue operations, although the level of force employed technically could be tied to the threat posed by the non-State actor (i.e., a direct participant in an armed conflict). The State may also make a policy choice that human rights–based law enforcement will be applied even if technically a conduct of hostilities approach could be used (see Chapter 14, section 14.4.2, *A Policy Choice by States*).

10.3.2. "ICONIC" COUNTERTERRORISM: A THUNDERBOLT STRIKES ENTEBBE

While the attacks of 9/11 have focused attention on transnational terrorism, it is not a new phenomenon. Further, countering such terrorism has not, nor does it now, solely involve "attacks" such as drone strikes. Operations are frequently undertaken in defense of nationals who are taken hostage by terrorist groups. *Operation Thunderbolt (Yonatan),* [176] the Israeli rescue of hostages at Entebbe in 1976, is still viewed in many circles as the iconic transnational counterterrorist rescue operation.[177] It provides one of the best examples of the potential political and legal complexity of these types of operations. The Entebbe operation also raises fundamental questions regarding the legal framework governing the inter-State use of force against transnational terrorist groups. In this regard, the operation appears to almost be transitional in character in its straddling the armed conflict and law enforcement paradigms.

---

[173] URBAN, *supra* note 156, at 126 ("the greatest return could come from selling them to extremists).

[174] AP I, art. 51(3) (where it is noted civilians enjoy protection "unless and for such time as they take a direct part in hostilities.").

[175] Felicity Szesnat & Annie R. Bird, *Columbia, in* INTERNATIONAL LAW AND THE CLASSIFICATION OF CONFLICTS 203, 214–15 (Elizabeth Wilmhurst ed., 2012).

[176] DUNSTAN, *supra* note 102, at 23 (called *Operation Thunderbolt* at the time of the mission, it was renamed *Operation Yonatan* in honor of its commander, Lieutenant-Colonel Yonatan Netanyahu, who was killed during the assault).

[177] SAUL DAVID, OPERATION THUNDERBOLT: FLIGHT 139 AND THE RAID ON ENTEBBE AIRPORT, THE MOST AUDACIOUS HOSTAGE RESCUE MISSION IN HISTORY 369-373 (2015)(discussing the impact of the Entebbe raid on counterterrorism operations).

In June 1976, hijackers acting on behalf of the Popular Front for the Liberation of Palestine (PFLP), but including members of a small German radical left wing terrorist organization[178] , seized an Air France commercial airliner that had departed Athens, Greece.[179] The plane eventually landed at Entebbe, Uganda,[180] where the four hijackers met up with other accomplices, ultimately detaining the hostages in an old airport passenger terminal.[181] The execution of this hijacking was international in scope. It had required the cooperation of Palestinian sympathizers permitting the terrorists to board the plane; the provision of refueling facilities in Libya; the agreement of Ugandan President Idi Amin to land at Entebbe; the provision of support from that country for the detention of the hostages; and permission for the overall terrorist leader to set up headquarters in Somalia.[182] Further, reflecting the international connection, the terrorists released a communiqué indicating that hostages would be executed if demands to release fifty-three terrorists held in five countries (i.e., Israel, West Germany, Kenya, Switzerland, and France), as well as provide $5 million U.S. dollars were not met.[183]

The living conditions for the hostages in the terminal quickly deteriorated.[184] Eventually, the hostages were separated into two groups with approximately eighty Israelis and other Jews being placed in one hall.[185] It is reported that the Ugandan military personnel assisted in this separation.[186] Then, President Amin, as a "goodwill" gesture, flew two groups, numbering fort-seven and one hundred and one, mainly French non-Jewish hostages to France.[187] This left over 100 hostages including the civilian aircrew at the terminal.[188] One hostage, Dora Bloch had been previously taken to a hospital.[189] This was not the sole Ugandan government involvement. It was assessed that, in addition to the terrorists, the terminal building was guarded by between sixty and one hundred Ugandan soldiers, who when "not on duty were generally stationed on the second floor of the old terminal."[190] By the time of the attack "only half, or maybe one-third, of the guards were believed to be on duty at any one time."[191] Further, near the airport was a military base with approximately two battalions of Ugandan troops.[192] Attempts at negotiation with the hostage takers were drawn out, with a decision nearly being reached to accede to the hijacker's requests. However, at that point the decision was made to carry out the only alternative, which was the conduct of a rescue operation.[193]

---

178 DAVID, *supra* note 177, at 46-52 (identified as members of the "Revolutionary Cells").

179 *Id.* at 8-56.

180 HARCLERODE, *supra* note 87, at 346 (for the identities and affiliations of the hijackers).

181 DUNSTAN, *supra* note 102, at 14.

182 HARCLERODE, *supra* note 87, at 351.

183 *Id.* at 352.

184 DUNSTAN, supra note 102, at 22.

185 *Id.* at 19–20 and DAVID, *supra* note 177, at 107–09.

186 FRANCK, *supra* note 12, at 82.

187 DUNSTAN, *supra* note 102, at 22 and 24. *See also* DAVID, *supra* note 177, at 131–33, 154–57, 197–99, 201–202 and 217–221.

188 DUNSTAN, *supra* note 102, at 24 and IDDO NETANYAHU, YONI'S LAST BATTLE: THE RESCUE AT ENTEBBE, 1976, at 48 (2002).

189 DAVID, *supra* note 177, at 236–37.

190 MCRAVEN, *supra* note 112, at 346. *See also* DAVID, *supra* note 177, at 234.

191 *Id.* at 354.

192 *Id.* at 347.

193 HARCLERODE, *supra* note 87, at 356–60, and DUNSTAN, *supra* note 102, at 34.

On July 4, 1976, an audacious plan was executed where Israel Defence Forces (IDF) special operations units, supported by air assets and intelligence personnel, landed under the cover of darkness at the Entebbe airport in a raid to free the hostages.[194] The hostilities began when the arriving Israeli commandos, some of who were dressed in Ugandan military uniforms, or at least uniforms resembling that nation's camouflage pattern, attempted to drive past Ugandan soldiers. Those soldiers were standing guard on the taxiway outside the old terminal where the hostages were being held.[195] The lead Israeli personnel were driving a Mercedes car disguised to resemble a Ugandan staff car.[196] The Israeli soldiers fired on two Ugandan sentries who attempted to intervene.[197] Subsequently, numerous Ugandan soldiers were killed in and around the terminal building. In addition, significant firefights developed between the Israeli soldiers and Ugandans military personnel located on the control tower.[198]

The terrorists were killed in the terminal. They had exchanged fire with Israeli soldiers, and two who feigned surrender were subsequently killed as one of them pulled out a grenade.[199] Three hostages, as well as one Israeli soldier, the raid commander, Lieutenant Yoni Netanyahu, were killed during the operation.[200] Two soldiers were wounded.[201] Seven hijackers and numerous Ugandan soldiers were killed.[202] Of the hostages, two were killed during the operation,[203] while the other subsequently died in Nairobi from wounds suffered during the assault.[204] The planes left Entebbe with all the hostages on-board except Dora Boch[205] who was killed by "Ugandan thugs" after the operation.[206] The Israelis also destroyed a number of Ugandan Air Force jet fighters on the ground.[207] It is now reported this was done at the request of Kenyan authorities to avoid a revenge attack by Uganda because of the assistance provided to Israel (landing and refueling at Nairobi airport).[208]

---

194 DAVID, *supra* note 177, at 290-316 (for a detailed outline of the rescue at the Entebbe airport). *See also* DUNSTAN, *supra* note 102, at 38–54.

195 DUNSTAN, *supra* note 102, at 33 (for reference to the uniforms worn), MCRAVEN, *supra* note 112, at 339, 376, and MUKI BETSER, SECRET SOLDIER 318 (1996). *See also* NETANYAHU, *supra* note 188, at 48 (2002) (where it is noted "[t]he Unit's men would wear camouflage fatigues resembling Ugandan army uniforms to further confuse the Ugandan soldiers and the terrorists.").

196 BETSER, *supra* note 195, at 314, 326–28, and NETANYAHU, *supra* note 188, at 47–48.

197 NETANYAHU, *supra* note 188, at 180–83, BETSER, *supra* note 195, at 326–28, and DUNSTAN, *supra* note 102, at 42.

198 HARCLERODE, *supra* note 87, at 361–64, and MCRAVEN, *supra* note 112, at 354, 358, 362–65. *See also* DUNSTAN, *supra* note 102, at 42–55, and BETSER, *supra* note 195, at 334.

199 BETSER, *supra* note 195, at 336, MCRAVEN, *supra* note 112, at 362, and DUNSTAN, *supra* note 102, at 47.

200 DUNSTAN, *supra* note 102, at 53 and 55 (noting that three hostages died) and DAVID, *supra* note 177, at 294 and 306 (outlining Lieutenant-Colonel Netanyahu's death).

201 *Id.* at 303 and 305 (One soldier was wounded by an Israeli grenade and an Israeli paratrooper who was seriously wounded clearing the New Terminal building).

202 *Id.* at 299–303 (detailing the killing of the 7 terrorists and several Ugandan soldiers) and MICHAEL BAR-ZOHAR & NISSIM MISHAL, NO MISSION IS IMPOSSIBLE: THE DEATH-DEFYING MISSIONS OF THE ISRAELI SPECIAL FORCES 18 (2015)(where it is indicated that 12 Ugandan soldiers were killed).

203 DAVID, *supra* note 177, at 299-300 and 310 (in one case the hostage was mistaken to be a terrorist).

204 *Id.* at 325.

205 DUNSTAN, *supra* note 102, at 53.

206 HARCLERODE, *supra* note 87, at 365. *See also* DAVID, *supra* note 177, at 359–61.

207 MCRAVEN, *supra* note 112, at 364–65, and NETANYAHU, *supra* note 188, at 203.

208 DAVID, *supra* note 102, at 246-48. *See also* TONY WILLIAMSON, COUNTER STRIKE ENTEBBE 47 (1976) and DUNSTAN, *supra* note 102, at 53 (where it is suggested this was done to prevent pursuit of the withdrawing Israeli planes).

Until the Entebbe raid, international hijackings were largely dealt with through negotiations, exchanges of prisoners, the payment of ransom, or by means of special military or paramilitary teams assaulting the aircraft with the permission of the territorial State.[209] Terrorism was largely viewed as a law enforcement matter to be dealt with as a police responsibility.[210] In contrast, the Entebbe rescue operation involved Israeli military forces fighting not only the terrorists but also Ugandan forces. In assessing the legal context within which the operation was conducted, it has already been noted that the activities of a State can be a key issue in determining if acts of non-State actors are attributed to that State. The activities of Ugandan authorities point toward such an attribution. One factor to be considered is that Uganda had increasingly aligned itself with the Palestinian terrorist movement. Idi Amin had terminated a previous close relationship with the Israeli government and subsequently developed relations with the Palestine Liberation Organization. This included permitting a training camp to be established in Uganda. Further, Libya supplied, "Amin with a Palestinian 300-man bodyguard, MiG fighter aircraft and Soviet-manufactured armoured vehicles."[211]

Allowing a hijacked plane to land, providing supplies to aid the hostages, and even refueling the aircraft is not an uncommon State practice. It has clear humanitarian goals. Had Uganda limited its activities to these actions, it would likely not have been implicated in the hijacking. Indeed, in the early stages of the planning for the rescue operation it was not clear the extent to which Uganda was supporting the hijackers. Much of the initial planning, which included consideration of both covert and large-scale parachute drops by Israeli forces, was "based on the assumption that the Ugandan authorities would be amenable to such rescue attempts and allow everyone to leave."[212] Consideration was even given to having the commando force surrender to the Ugandan government once the hostages were freed.[213] However, as the incident progressed it became evident the Ugandan president was personally involved, and his soldiers were providing more than logistical support to the terrorists.[214]

If Uganda had not been so involved in the hijacking, consideration would have had to be given to whether consent should be sought from that State, simple notification would be provided, or the operation would be kept as covert as possible. The obtaining of such consent is not an absolute requirement since the right to exercise self-defense, where justified, can override some of the sovereign rights of the territorial State. An intervention itself is normally based on a determination the territorial State is unwilling or unable to act to protect the hostages.[215] Factors to be considered in seeking consent could include

---

[209] DUNSTAN, *supra* note 102, at 4–11(for an outline of the history of aerial hijacking leading up to the Entebbe operation in 1976). *See also* MICHAEL BURLEIGH, BLOOD & RAGE: A CULTURAL HISTORY OF TERRORISM 167 (2009) (for reference to the creation of specialized antiterrorist units to rescue hostages).

[210] GRANT WARDLAW, POLITICAL TERRORISM 100 (2nd ed. 1990).

[211] HARCLERODE, *supra* note 87, at 350.

[212] DUNSTAN, *supra* note 102, at 19. *See also* HARCLERODE, *supra* note 87, at 356–57.

[213] DUNSTAN, *supra* note 102, at 19, and MCRAVEN, *supra* note 112, at 335.

[214] WILLIAMSON, *supra* note 208, at 17, and J. PAUL DE B. TAILLON, HIJACKING AND HOSTAGES: GOVERNMENT RESPONSES TO TERRORISM 110 (2002). *See also* NETANYAHU, *supra* note 188, at 39 (where it is noted the degree of the Ugandan army collaboration with the terrorists became evident when the released hostages were debriefed in Paris that the Ugandan army was in control of the building where they had been held and were helping guard the prisoners).

[215] DINSTEIN, *supra* note 15, ¶ 677, at 256.

the degree to which the territorial State government sympathizes with the hostage takers and whether that State can keep the operation confidential or might obstruct its execution. Such an assessment can be complicated by the fact that not all parts of the territorial State governing apparatus may be trusted.

In conducting a rescue operation, the intervening State cannot take action against the local population and ordinarily must avoid engaging the territorial State armed forces. Conversely, the territorial State must not interfere with a legitimate exercise of self-defense by the intervening State.[216] A primary concern in either seeking consent or providing notice would be to avoid unintended hostilities between the territorial and intervening States. However, with the complexity of such operations problems can also arise with entry into or flight over third States. Such a concern arose in respect of the Entebbe operation regarding what role Kenya might play.[217] Operational security matters ultimately weigh heavily in the determination as to whether the territorial State or third States are informed of the operation prior to its commencement.

The result of Ugandan government action supporting the terrorists was that the Israeli planning had to recognize that "any military operation would have to consider the Ugandan forces as part of the problem"[218] and "[a]ny action against the hijackers must, therefore, also be an action against Uganda."[219] Once it became evident that the Ugandans were working with the hijackers,[220] "[t]he only plan that counted involved landing at the airport, freeing the hostages, and flying out.[221] In this regard, if the reasoning of the subsequent 1980 *Case Concerning United States Diplomatic And Consular Staff In Tehran (United States Of America V. Iran)*,[222] or the approach adopted post-9/11 by the international community toward the Taliban harboring of Al Qaeda were applied to the Uganda situation, the acts of the terrorists could readily be identified with those of the Ugandan State (see Chapter 9, section 9.3.2, *Harboring and Active Support*). On the basis of the collaboration between Ugandan authorities and the hijackers, it can readily be concluded that actions of the Israel were justifiable as self-defense against the host State and the terrorists.[223]

The hostilities at Entebbe occurred in the context of an international armed conflict between States,[224] although it only lasted one hour and thirty-nine minutes.[225] Other legal factors influencing the operation were that it had to be carried out within the limitations of

---

216 *Id.* ¶ 732, at 277.

217 DUNSTAN, *supra* note 102, at 30.

218 MCRAVEN, *supra* note 112, at 338.

219 WILLIAMSON, *supra* note 208, at 19 and DAVID, *supra* note 177, at 229 (where it is recorded that the Israeli Chief of Staff was advised if the IDF soldiers encountered resistance from Ugandan forces they should "Shoot to Kill").

220 DE B. TAILLON, *supra* note 214, at 121 (where it is reported that a briefing provided by Israeli Major-General Gazit indicated the "IDF knew that guards were a mix of terrorists (ten or eleven) plus Ugandan soldiers.").

221 BETSER, supra note 195, at 306, and NETANYAHU, *supra* note 188, at 39 (As an Israeli Colonel is reported to have stated "the only choice was an all-out military operation . . . it couldn't be a small group of men arriving, overpowering the terrorists, and then handing themselves over to the Ugandans.").

222 Case Concerning United States Diplomatic And Consular Staff In Tehran (United States Of America V. Iran), 1980 I.C.J. 3, ¶ 74 at 35.

223 DINSTEIN, *supra* note 15, ¶ 681, at 258.

224 CORTEN, supra note 44, at 87 (while discussed in terms of the use of force pursuant to Article 2(4) of the UN Charter reference is made to combat occurring between the armed forces of two States during the Entebbe operation).

225 DUNSTAN, *supra* note 102, at 54.

the self-defense principles of necessity, immediacy, and proportionality. This is an example where, even under the limited application of the State self-defense theory (see Chapter 3, section 3.4, *A More "Limited" Theory*), the conduct of an operation "short of war" (i.e., hostilities between Israel and Uganda) remained subject to these principles, notwithstanding the application of a humanitarian law framework governing the use of force. Consistent with that limitation the Israeli forces did not strike at all available targets, but rather focused on neutralizing immediate threats such as a column of Ugandan troops approaching the airfield.[226]

There was controversy and debate following the Entebbe operation in the United Nations and elsewhere.[227] One view has been that since a majority of States denounced the operation as a violation of international law, the operation could not be justified.[228] However, to the extent this denunciation of the Israeli raid was based on the perspective that Uganda had played a positive role in negotiating with the hijackers, it is based on a factor that has been discredited over time.[229] The incident provides an example where greater consideration should been given to Waldock's prerequisite regarding "the inability or unwillingness of the territorial State to protect foreign nationals."[230] However, the "unwillingness" on the part of Uganda went far beyond disinterest, reluctance, or providing a level of support to the terrorists. That State took an active part in the delict.

At the heart of the international legal debate about Entebbe was the issue of whether the prohibition on the use of force found in Article 2(4) of the UN Charter could be narrowly interpreted when the collective security system provided for in the Charter is ineffective.[231] Thomas Franck took the view that "there was no agreement as to whether a state might use force to protect its citizens' lives overseas, in situations where neither their host-state nor the international system had been able to offer effective protection."[232] Undoubtedly influencing the subsequent discussion regarding the legality of the Israeli action was the unpopularity of some of Israel's policies, such the maintenance of Occupied Territories. This often results in criticism of that country being grounded in a broader geopolitical context, which can unfortunately impact the analysis of specific legal issues. Other factors relevant to any analysis should include the lack of regional support for Idi Admin regime;[233] the urgency of the hostage situation; the care taken by Israel to avoid collateral damage; and the fact "there was little question of Israel's pursuing any national interest other than rescue."[234] Interestingly, the operation was provided with assistance by "several 'friendly' governments" spanning

---

226 *Id.* at 53 and DAVID, *supra* note 177, at 315.

227 FRANCK, *supra* note 12, at 82–86, SHAW, *supra* note 12, at 1144, and GRAY, *supra* note 17, at 32–33. *See also* HUGH M. KINDRED ET AL. ET AL., INTERNATIONAL LAW: CHIEFLY AS INTERPRETED AND APPLIED IN CANADA 1134–37 (6th ed. 2000) (for excerpts of the statements made by representatives of Israel, the United Republic of Cameroon, and the United States).

228 RUYS, *supra* note 19, at 228, and CORTEN, *supra* note 44, at 87.

229 RUYS, *supra* note 19, at 228 ("Many were convinced that Uganda had in fact played a positive role in negotiating with the hijackers . . . .).

230 *Id. See also* DINSTEIN, *supra* note 15, ¶ 676, at 255–56.

231 FRANCK, *supra* note 12, at 83–84 (relying on D. P. O'CONNELL, INTERNATIONAL LAW 304 (2nd ed. 1970)).

232 *Id.* at 84.

233 NICHOLAS STÜRCHLER, THE THREAT OF FORCE IN INTERNATIONAL LAW 142–46 (2007) (for an outline of the tensions between Kenya and Uganda, and the lack of support for the Amin government).

234 FRANCK, *supra* note 12, at 85.

"the spectrum from international moral support to having a foreign military paramilitary representative providing intelligence for the Israeli rescue force."[235] This included Kenyan "assurances that an Israeli strike force would be allowed to land at Nairobi to refuel and care for the wounded on its homeward trip from Uganda."[236]

The Entebbe operation has been contrasted with the U.S. operation in Grenada in 1983, which attracted considerable international criticism.[237] Such a comparison also goes to the heart of the State self-defense proportionality question. It cannot be said the Grenada "island-wide operation could be justifiably related to the protection of nationals for whom an Entebbe-style rescue would appear to have been adequate."[238] It has also been argued the attack on the Israeli hostages was clearly an attack on their government, and the hostages were taken to Uganda in violation of international law.[239] Meeting the requirements for self-defense under international law also appears to have been a primary concern of Israeli legal advisors.[240] In the final analysis, with the passage of time, the operation has been more widely recognized as a lawful exercise of the right of self-defense.[241]

Finally, the Israeli hostage rescue operation did occur at a time the terrorist threat was primarily viewed by the world as a criminal matter. However, the nature of the operation and the levels of force used by the Israeli forces, including against the terrorists, fit comfortably within a conduct of hostilities paradigm. It was clearly anticipated there would be combat with Ugandan armed forces protecting the airport terminal and significant exchanges of gunfire resulted between military forces. This was particularly evident in the use of .50 caliber machine guns and rocket-propelled grenades (RPGs) against Ugandan troops in the control tower.[242] At the same time, the need to keep the hostages safe, while also protecting them from terrorist efforts to kill them once the operation was discovered, affected the tactics used. As one participant noted, what makes a hostage rescue so unique is the first concern must be eliminating the terrorists. The Israelis were not "going all the way to fight Ugandans."[243] The protection of the hostages required restricted applications of force within the terminal. The hostage takers were killed. However, that is not an uncommon occurrence even during domestic law enforcement–based rescue operations. Concern over hostages possibly being killed during the raid resulted in megaphones being provided to Israeli soldiers to warn the captives to keep down.[244] Direction was also given to keep safety catches engaged on weapons to avoid civilians being killed in a second terminal where it was

---

[235] DE B. TAILLON, *supra* note 214, at 108. *See also id.* at 116–17 (where it is noted vital intelligence was provided by the United Kingdom, United States, France, West Germany, and Canada.).

[236] *Id.* at 116.

[237] DINSTEIN, *supra* note 15, ¶ 684, at 258.

[238] RICHARD CONNAUGHTON, MODERN WARFARE: THE TRUE STORY OF CONFLICT FROM THE FALKLANDS TO AFGHANISTAN 121 (2008).

[239] DINSTEIN, *supra* note 15, ¶¶ 682–683, at 258.

[240] WILLIAMSON, *supra* note 208, at 96–97.

[241] CASSESE, *supra* note 17, at 368 n.19, and DINSTEIN, *supra* note 15, ¶ 685, at 258.

[242] DUNSTAN, *supra* note 102, at 51 and DAVID, *supra* note 177, at 307.

[243] BETSER, supra note 195, at 312. *See also* BAR-ZOHAR & MISHAL, *supra* note 202, at 19 (where it is noted the Israeli commandoes allow some Ugandan soldiers to escape and in one instance locked two soldiers in the terminal office).

[244] DUNSTAN, *supra* note 102, at 33, and HARCLERODE, *supra* note 87, at 362. *See also* DAVID, *supra* note 177, at 300 (where it was noted that during the assault there was a delay in the use of the megaphone warning hostages to lie down).

not expected there were Ugandan soldiers.[245] Further, the "break-in" crews were signaled "to set their weapons to single shot for selective shooting."[246]

Not all the force used by the Israeli soldiers resembled that normally applied during law enforcement. For example, a hand grenade was thrown into the terrorist's living quarters.[247] Nor, at law, did law enforcement norms have to be applied since both the Ugandan forces and the terrorists were participants in an armed conflict. In a similar setting the U.S. special operations force Delta Force plan to rescue hostages in Tehran in 1980 included a proactive use of force where "a select group of operators would drive up to the embassy in the Datsun pickup and with .22-caliber suppressed (with silencers) handguns take down the two guard posts and the walking guards . . . ." [248] However, the overall need to apply surgical force to remove the terrorist threat within the terminal meant that aspects of the Entebbe operation took on elements of a domestic hostage rescue operation. It was not simply an assault on military objective.

10.3.3. MODERN COUNTERTERRORISM: SIERRA LEONE

The dawn of the twenty-first century witnessed another daring hostage rescue operation. In 2000, British forces assaulted two encampments in Sierra Leone in order to rescue some of their own military personnel and a Sierra Leonean soldier who had been taken hostage by a 600-strong, armed group known as the West Side Boys. The leader of that organized armed group was twenty-four-year-old, self-styled "Brigadier" Foday Kallay.[249] The British operation, *Operation Barras*, is noteworthy in that it highlights the complex security situation within which rescue missions may have to be performed, as well as the significant violence that sometimes is required to confront the non-State actors involved.

Sierra Leone and its organized armed groups were not entirely unknown to British forces. The United Kingdom was a former colonial power.[250] Despite the presence of the United Nations Mission in Sierra Leone (UNMASIL), Britain had intervened previously in May 2000 to conduct a noncombatant evacuation operation. *Operation Palliser* had evacuated 353 "entitled persons," although 453 so registered decided to stay.[251] That intervention ultimately had the effect of stabilizing the security situation; at least to the extent it prevented the capture of Freetown by the main rebel group, the Revolutionary United Front (RUF).[252] British military forces were subsequently withdrawn, except for a training team left behind to assist the Sierra Leone army.

---

245 NETANYAHU, *supra* note 188, at 204, and DUNSTAN, *supra* note 102, at 50.

246 BETSER, *supra* note 195, at 326, DAVID, *supra* note 177, at 292 and NETANYAHU, *supra* note 188, at 70 (indicating that only single rounds could be used to avoid hitting the hostages).

247 DUNSTAN, *supra* note 102, at 46–47 and DAVID, *supra* note 177, at 302.

248 BECKWITH, *supra* note 101, at 254.

249 CONNAUGHTON, *supra* note 238, at 266.

250 WILLIAM FOWLER, OPERATION BARRAS THE SAS RESCUE MISSION: SIERRA LEONE 2000, 19–30 (2005) (for an outline of British involvement in Sierra Leone).

251 *Id.* at 76–105 (the entitled persons were British, EU, and Commonwealth passport holders.).

252 CONNAUGHTON, *supra* note 238, at 263–65, and WILL FOWLER, CERTAIN DEATH IN SIERRA LEONE: THE SAS AND OPERATION *BARRAS* 2000, at 7–9 (2010) (for an outline of the UK intervention in 2009.).

By 2000, both UNMASIL and British forces were operating in Sierra Leone, a country with a weak government and one that had been wracked by a diamond-fueled civil war. In July 2000 the 1st Battalion of the Royal Irish Regiment staffed the British training mission.[253] The Royal Irish "would also mount vehicle and foot patrols to liaise with UNMASIL forces and give reassurance to the population."[254] It was one of those patrols, consisting of three Land Rover vehicles manned by eleven British and one Sierra Leonean soldier, that was stopped on August 25, 2000, by what has been described as "a notoriously fickle and dangerous band of irregulars who terrorized the west side of Sierra Leone."[255] The West Side Boys was a well-armed group fueled by heroin, cocaine, and alcohol, whose activities included murder, rape, torture, and mutilation.[256] They had supported the government,[257] and "the British forces had enjoyed a reasonable working relationship with the loosely structured group."[258] However, on this occasion the military hostages were detained and taken to the West Side Boys' headquarters at Gberi Bana.[259]

From the beginning the British government tried to negotiate the release of the hostages. The regular negotiations involved the Commanding Officer of the Royal Irish and other personnel assisted by two officers from London Metropolitan Police Hostage Crisis Negotiation Unit.[260] Two SAS members with negotiation skills eventually joined this group as preparations were being made for a rescue operation.[261] Selected hostages were brought to the negotiation sessions as "proof of life."[262] Initially, the negotiations proved fruitful as five of the eleven British soldiers were released on August 31, 2000.[263] However, the hostages were being subjected to abuse and, it is reported, mock executions.[264] The Sierra Leonean soldier was particularly badly abused.[265] In the meantime a hostage rescue force was inserted into Sierra Leone consisting of a combined SAS/SBS group of Special Forces soldiers, based on D Squadron SAS[266] and A Company, 1st Battalion, The Parachute Regiment (1 Para). The latter force included a patrol platoon, heavy machine gun elements, and a Mortar Section with 81 mm mortars.[267] Air support consisted of "three [British] Chinook, two Army Lynx and a Mi-24 gunship of the Sierra Leone Air Force operated by its former Executive Outcomes [private military contractor] crew." [268] Finally naval support consisted of the frigate *HMS Argyll* and two Royal Fleet Auxiliaries to provide medical and other support offshore.[269]

---

[253] CONNAUGHTON, *supra* note 238, at 265.

[254] FOWLER, *supra* note 252, at 9.

[255] CONNAUGHTON, *supra* note 238, at 266.

[256] *Id.*

[257] *Id.*

[258] FOWLER, *supra* note 252, at 11.

[259] *Id.* at 12.

[260] *Id.* at 13, and CONNAUGHTON, *supra* note 238, at 273.

[261] CONNAUGHTON, *supra* note 238, at 273, and FOWLER, *supra* note 252, at 18.

[262] FOWLER, *supra* note 252, at 14.

[263] *Id.*

[264] *Id.* at 15, and CONNAUGHTON, *supra* note 238, at 271.

[265] CONNAUGHTON, supra note 238, at 271–72.

[266] *Id.* at 268.

[267] FOWLER, *supra* note 252, at 23.

[268] CONNAUGHTON, supra note 238, at 281.

[269] FOWLER, *supra* note 252, at 35.

Concerns quickly grew about the abuse of the hostages, that the West Side Boys were not negotiating in good faith, and it was feared the soldiers might be moved or split up.[270] Therefore a decision was made to mount a rescue operation. The resulting heliborne assault was one of three options that had been considered. The other two, over land and inserting a rescue force by river, were rejected as unfeasible.[271] The Special Forces group was to fast-rope from a hovering Chinook into the village of Gberi Bana, which held approximately 200 West Side Boys members, their commander, and their families. Preparation for the assault was assisted by an SAS observation patrol that had been inserted covertly to gather intelligence.[272] At the same time, the 1 Para group was to carry out a helicopter assault on the village of Magbeni, which had also been under observation by another SAS patrol.[273] That village had a population of 400 and was located across the 250-yard-wide Rokel Creek. The Magbeni assault was planned because of concern that fire directed from that location by the West Side Boys could imperil the SAS/SBS rescue operation at Gberi Bana on the other side of the river.[274]

The mission has been described as one "to defeat, not to destroy the enemy. This latitude allowed commanders to fight the battle in such a way as to permit the enemy to slip away if that were deemed necessary. Barras was not about settling scores but saving lives."[275] However, the plan also involved an aggressive use of force including the use of attack helicopters operating at Magbeni to "rake the village and sites identified as being locations of heavy weapons capable of putting supportive fire down on Gberi Bana."[276] There could be no doubt it was a high-risk operation conducted against a very well-armed and more numerous armed group. In addition to assault rifles, heavy machine guns, and RPGs, the West Side Boys had 2 inch, 60mm, and 81 mm mortars.[277] The armaments at Magbeni included "a captured Bedford 4-ton vehicle with twin 4.5 mm machine guns, at least three technicals and pick-up trucks with heavy machine guns mounted on the rear."[278] The assault plan included the British use of helicopter gun support, mortars, and 66 mm antitank weapons.[279] In effect, the plan anticipated the conduct of hostilities. The Sierra Leone government was kept informed of the planned operation,[280] and "[i]n Freetown President Kabbah had already given the JTFHQ carte blanche to conduct an operation against the West Side Boys at a time of their choosing."[281] Written authorization for the operation was obtained from the British head of the Sierra Leone Police Authority, and "the action against the West Side Boys became a military rather than a police responsibility."[282]

---

[270] *Id.* at 32. *See also* CONNAUGHTON, *supra* note 238, at 274.

[271] CONNAUGHTON, *supra* note 238, at 270.

[272] *Id.* at 282. *But see* FOWLER, *supra* note 252, at 36 (where it suggested reports estimated there were fifty to one hundred West Side Boys in each village).

[273] FOWLER, *supra* note 252, at 26.

[274] CONNAUGHTON, *supra* note 238, at 270.

[275] *Id.* at 281. *See also* FOWLER, *supra* note 252, at 27.

[276] CONNAUGHTON, *supra* note 238, at 276.

[277] *Id.* at 284.

[278] *Id.*

[279] *Id.*

[280] FOWLER, *supra* note 252, at 32, and CONNAUGHTON, *supra* note 238, at 272.

[281] FOWLER, *supra* note 252, at 32.

[282] *Id. See also* FOWLER, *supra* note 250, at 136 (where it is indicated a British Army Legal Corps Major had flown out with a document seeking authorization from the Sierra Leone police).

In the assault that followed on September 10, 2000, all of the hostages were rescued. The Special Forces operation at Gberi Bana saw RPG-7s fired at one of hovering Chinook helicopters and considerable armed resistance from West Side Boy fighters that was quickly overcome.[283] As the West Side Boys attacked they were "struck down by the accurate fire of the SAS/SBS and the machine guns of the Lynx attack helicopters."[284] The assault on Magbeni witnessed a sustained firefight that involved the use of air support from attack helicopters, including the Sierra Leone Air Force M-24 Hind gunship.[285] The command element of the Para Company was hit by mortar fire that wounded seven men.[286] The firefight lasted four hours, "at the end of which 10 per cent of the attackers had been wounded."[287] The UK Ministry of Defence press release indicated that one British soldier was killed and another twelve were wounded. It was also reported that "[d]etails of West Side Boys casualties remain unclear, however, we can confirm 25 dead including 3 females."[288] Other reports have suggested the death toll among the hostage takers was much higher.[289] For example, the pilot of the Mi-24 estimated he had killed fifty to sixty rebels, and the British gunships and main attack force another forty rebels.[290] A number of prisoners were taken who were eventually passed to the Sierra Leonean police.[291] *Operation Barras* was a very daring operation, which when objectively viewed can be seen to have occurred in the context of an "armed conflict".

10.3.4. SHORT-TERM CONFLICTS

It might be argued that the operation was not an exercise of national self-defense because authority to act was sought from Sierra Leonean authorities. Further, it might be suggested that obtaining that consent meant it was conducted as a law enforcement matter. Sierra Leone did consent to the operation, although given that government's reliance on the United Kingdom it is doubtful the consent would have been withheld. Sierra Leone also did not have the capability to mount such a rescue. This demonstrates the artificiality (or, at least, the "formalist" quality) of the analysis that can occur under international law where an operation by a State to defend its nationals in another State might be viewed as being outside the realm of self-defense because of such consent. The UK Ministry of Defence press release following the operation made reference to the assistance provided by the Sierra Leone army and the UN mission in Sierra Leone, "particularly the Jordanian battalion,

---

[283] CONNAUGHTON, *supra* note 238, at 282. *See also* FOWLER, *supra* note 250, at 139 ("As the Chinook carrying the SAS hostage-snatch and fire teams swept in and hovered above the palms in the half-light of dawn, the West Side Boys at Gberi Bana opened fire.").

[284] CONNAUGHTON, *supra* note 238, at 282.

[285] AL J. VENTER, GUNSHIP ACE: THE WARS OF NEALL ELLIS, HELICOPTER PILOT AND MERCENARY 261–62 (2011) (for the Hind this included firing volleys of four to eight rockets and using the "Gatling").

[286] FOWLER, *supra* note 252, at 47.

[287] CONNAUGHTON, *supra* note 238, at 283.

[288] FOWLER, *supra* note 252, at 58 (setting out a Ministry of Defence press release).

[289] *See also* Tim Butcher, *SAS Vengeance on the West Side Boys*, MAIL ONLINE (Aug. 29, 2010), http://www.dailymail.co.uk/news/article-1307151/SAS-vengeance-West-Side-Boys.html.

[290] VENTER, *supra* note 285, at 262. *See also* FOWLER, *supra* note 250, at 149 (where it was noted one source indicated if blood trails were followed into the jungle, the official West Side Boys death toll of twenty-five could be quadrupled.).

[291] FOWLER, *supra* note 252, at 51 (this was done through the UNAMSIL Jordanian Battalion).

which provided security along the Maslaka highway."[292] Notwithstanding this reference, the operation was conducted almost exclusively as a British one. Active involvement of the Sierra Leone military, other than the air support provided by the MI-24 with its experienced ex-private security company pilot, was rejected during the planning stage.[293]

A suggestion that this rescue operation could constitute a form of law enforcement and thereby should be governed solely by human rights law is factually a difficult argument to make. That the raid was carried out with the permission of State territorial authorities does not necessarily make the operation a policing one. It is a hallmark of both counterinsurgency and counterterrorism that security forces must operate within the law even during armed conflict. For example, during the Malaysia counterinsurgency "every action taken by Government officials, soldiers and policemen was within the law; stringent laws could be as robust as the situation required provided they were properly observed."[294] However, this did not change the reality that the patrolling, ambushes, and even aerial bombing against the Communist guerrillas during that "Emergency" took place in the context of an armed conflict.

Another factor separating the Sierra Leone hostage taking from ordinary criminal law enforcement was the political motive expressed by the hostage takers. During the course of the incident the hostage takers became emboldened, indicating "the hostages were taken in order to get Britain to pressure President Kabbah to recognize the West Side Boys as a legitimate group, to free their arrested leaders from prison, and to form a new government with seats for their leaders.[295] The West Side Boys leader also reportedly had a Plan B in which the hostages would be killed, thereby hopefully creating a "Somalia effect" causing a British withdrawal and the collapse of UNMSIL.[296] From both a factual and legal perspective *Operation Barras* was not an ordinary criminal law enforcement matter as that type of operation might commonly be understood.

It might be suggested that the overall operation reflected a "law enforcement" paradigm since the force used was defensive in nature and limited to the protection of the hostages and the military personnel conducting the operation. Looked at broadly, the force was applied in the defense of nationals and therefore had to engage the principles associated with State self-defense law (i.e., proportionality, necessity, imminence, and immediacy). However, an argument that force was narrowly used in "individual" self-defense in response to threats from the West Side Boys is difficult to substantiate. Arguments that military units engage in individual self-defense in respect of attacks from other hostile forces are hard to sustain.[297] In this regard State self-defense does share common roots with individual self-defense, but serves different goals (see Chapter 11, section 11.2.2, *Self-Defense: Common Roots and Different Goals*). It is not even clear which side fired first. Hostilities commenced as the Chinook helicopters arrived at Gberi Bana to being their hover in order to allow the rescue team to deploy. One account indicates that members of the SAS observation team at

---

292 *Id.* at 58.

293 CONNAUGHTON, *supra* note 238, at 279.

294 E. D. SMITH, COUNTER-INSURGENCY OPERATIONS: 1, MALAYA AND BORNEO 44 (1985).

295 FOWLER, *supra* note 252, at 16.

296 CONNAUGHTON, *supra* note 238, at 271.

297 Tom Ruys, *The True Meaning of Force—A Reply to Mary Ellen O'Connell*, *supra* note 2.

Gberi Bana "opened fire on men near the White House preventing them from killing the hostages" before the West Side Boys started to fire at the helicopters.[298] What is clear is that the West Side Boys fired two RPG-7 rockets at the helicopters at approximately that same time.[299] Across the river at Magbeni it is reported the West Side Boys fired first, aiming at one of the helicopters. As has been noted, this action prompted a vigorous response by the supporting helicopters.[300] This was not traditional "law enforcement."

The question of which side fired first should not be determinative of whether there is an armed conflict or the matter is one of law enforcement. The British government and military commanders anticipated high levels of violence being used against an organized armed group. They planned for the use of and actually applied levels of force that were of armed conflict proportions. As a reference point, it can be argued the limited armed conflict during *Operation Barras* in September 2000 began with the first use of violence during the assault. However, it might also be argued that the hostilities began with the hostage taking. It is not the government forces that precipitated the armed conflict, but rather the actions of the organized armed group (e.g., the attack on the foreign State nationals) (see Chapter 14, section 14.3.2, *Commencement of Armed Conflict*). If that violence had continued between the British forces and the West Side Boys beyond the four hours of that operation, it would have become an armed conflict of even greater magnitude. As an armed conflict humanitarian law, and not human rights law, primarily governs the use of force. However, operational planners must keep in the mind the potential uncertainty regarding any after action categorization of the conflict; the presence of civilians (i.e., family members), as well as the requirements of self-defense proportionality. This in turn would ordinarily lead to a much more restrained use of force.

What made *Operation Barras* an "armed conflict" was the existence of intense hostilities during the rescue operation with a well-armed organized non-State group, the West Side Boys. The action taken to protect the nationals involved reflects the reality that an armed conflict can be in existence even during limited self-defense operations. As the International Criminal Tribunal for the former Yugoslavia indicated in the *Prosecutor v. Haradinaj* case, "[t]he criterion of protracted armed violence has therefore been interpreted in practice, including by the *Tadić* Trial Chamber itself, as referring more to the intensity of the armed violence than to its duration."[301] That decision also noted:

> indicative factors include the number, duration and intensity of individual confrontations; the type of weapons and other military equipment used; the number

---

[298] FOWLER, *supra* note 252, at 39. *See also* FOWLER, *supra* note 250, at 139 (in this earlier work the author had indicated the SAS patrol broke cover before the helicopter was fired on and indicated they went to the huts "and 'discouraged and interference' until the main body of soldiers arrived." There was a firefight where "only two people escaped alive from the West Side Boys' huts . . . .").

[299] CONNAUGHTON, *supra* note 238, at 282.

[300] FOWLER, *supra* note 250, at 144 (where it is indicated the Lynx helicopters raked with 12.7 mm machine gun fire "the cluster of WSB [West Side Boys] around the heavy machine gun that had opened fire at the 1 Para Chinook . . . .").

[301] ICTY, Prosecutor v. Haradinaj, Case No. IT-04-84-T, Judgement, ¶ 49 (Int'l Crim. Trib. for the Former Yugoslavia, Apr. 3, 2008), http://www.icty.org/x/cases/haradinaj/tjug/en/080403.pdf. *See also* ICTY, Prosecutor v. Limaj, Case No. IT-03-66-T, Judgement, ¶90 (Int'l Crim. Trib. for the Former Yugoslavia, Nov. 30, 2005), http://www.icty.org/x/cases/limaj/tjug/en/lim-tj051130-e.pdf.

> and calibre of munitions fired; the number of persons and type of forces partaking in the fighting; the number of casualties; the extent of material destruction; and the number of civilians fleeing combat zones. The involvement of the UN Security Council may also be a reflection of the intensity of a conflict.[302]

Many of these factors were clearly evident during the *Operation Barras* rescue (e.g., the intensity of the confrontation; the weapons and munitions used; the casualties; and the extent of material destruction).

During the planning stage for any operation, the political context within which a mission is to be conducted; the degree of violence likely to result and therefore planned for; the level of organization and the size of the armed group that abducted the hostages; its armaments and participation in warfare are all relevant in assessing whether "hostilities" will take place. In respect of this mission those factors all pointed toward the requirement to conduct hostilities in order to release the hostages. Added to this list is the fact that the State armed forces did not exercise a level of control over the scene of the rescue attempt that was consistent with a law enforcement approach (see Chapter 7, section 7.4.2, *The "Targeted Killing" Case and Control*). Not only did the rescue forces have to assault the location where the hostages were located, the operation included a separate deployment of troops to neutralize the threat posed by significant elements of the organized armed group located across the river. The military forces had to fight their way onto both objectives. Viewing the limited conflict in Sierra Leone as an armed conflict is consistent with the "totality of the circumstances" approach that has been suggested for considering when such hostilities are in existence (see Chapter 9, section 9.4.3, *The "Totality of the Circumstances" and Other Criteria*). An armed conflict can exist without lasting weeks or months. *Operation Barras* was certainly not a hostage rescue operation like the one conducted against a small criminal gang, which had seized a couple of aid workers such as occurred in Somalia in January 2012.[303]

Factually, the operation was also transnational in character for the British units. It lasted two and a half hours longer than the conflict at Entebbe. While the latter conflict was inter-State in character, *Operation Barras*, with its State and non-State protagonists, fits comfortably within a noninternational armed conflict framework that is based on the status of the participants (see Chapter 9, section 9.3.7.1, *Not of an International Character: Hamdan v. Rumsfeld*). At some point the formalism of categorization (i.e., armed conflict or law enforcement) has to match the factual reality being experienced by those engaged in the violence. To have human rights law govern the force used would, at a minimum, have necessitated the incorporation of "conduct of hostilities like" norms within that body of law (i.e., collateral damage). This would, in turn, fundamentally transform the basis of a claim the operation was governed by human rights law as it is normally applies in a law enforcement context. The successful execution of *Operation Barras* involved a use of force far beyond

---

302 ICTY, Prosecutor v. Haradinaj, Case No. IT-04-84-T, Judgement, ¶ 49 (Int'l Crim. Trib. for the Former Yugoslavia, Apr. 3, 2008), http://www.icty.org/x/cases/haradinaj/tjug/en/080403.pdf.

303 *Details of Somalia Hostage Rescue Emerge*, L.A. TIMES, Jan. 25, 2012, http://latimesblogs.latimes.com/world_now/2012/01/somalia-rescue-hostages-pirates.html [hereinafter *Details Emerge*].

that applied by EU forces in a 2012 helicopter raid on Somali pirate infrastructure,[304] or in "individualized" aerial bombing incidents related to Chechnya dealt with by the European Court of Human Rights.[305] A human rights tribunal very likely would look at such conflict exclusively through a human rights lens. However, the choice made by a court or tribunal to interpret its mandate literally and narrowly as being solely amenable to human rights law jurisdiction does not make incidents, such as those that occurred on *Operation Barras*, any less an armed conflict at law. This is particularly relevant to those States whose legal approach is not as restricted by the decisions of such tribunals.

That an armed conflict is considered to be in existence does not mean there is no concurrent application of human rights law or even of human rights norms through humanitarian law itself. As will be discussed regarding the British approach toward the IRA during the Northern Ireland "Troubles," a State can also choose to apply human rights–based law enforcement, even in the face of significant violence inflicted by an organized group using "weapons of war" (see Chapter 13, section 13.2.1, *The "Troubles"*). That conflicts can involve different levels of violence is reflected in the comments of an officer who served in Northern Ireland and the Falklands/Malvinas war. He noted that "Northern Ireland had its very violent periods and some prolonged operations, but none with the full orchestration of war . . . ."[306] *Operation Barras* appears to have more closely resembled an armed conflict requiring such an orchestration. However, in this instance, the intensity of the violence occurred over a relatively short period of four hours.

There will be international lawyers who will not embrace the conclusion that the Sierra Leone hostage rescue operation involved an armed conflict. However, the analysis points to the utility of the "means" test, which is part of the transnational armed conflict theory (see Chapter 9, section 9.3.6, *Transnational Armed Conflict*). There was a broader engagement of two armed groups in fighting that took on the attributes of warfare. In this case it occurred in the context of a State exercising the right to protect its nationals seized by an organized armed group having significant military capacity. The resulting hostilities are reminiscent of Clausewitz's reference to war being "nothing but a duel on a larger scale."[307] The scale and ferocity of the *Operation Barras* "duel" moved it from the realm of law enforcement into that of a "small war."

Such a limited, but intense engagement, as occurred during Sierra Leone operation, highlights the potential for underinclusiveness if the *Prosecutor v. Tadić* case criteria of protracted violence with an organized armed group sets too high a threshold for the existence of an armed conflict.[308] That threshold was developed in a different context, and was not considered in the context of a "one-off" self-defense operation against a well-armed and

---

[304] *Somali Piracy: EU Forces in First Mainland Raid*, BBC NEWS AFRICA (May 15, 2012), http://www.bbc.co.uk/news/world-africa-18069685.

[305] *See, e.g.*, Isayeva v. Russia, Application Nos. 57947/00, 57948/00, 57949/00, Eur. Ct. H. R. (2005), http://hudoc.echr.coe.int/sites/eng/pages/search.aspx?i=001-68379, and Isayeva Application No. 57959/00, Eur. Ct. H. R. (2005), http://hudoc.echr.coe.int/sites/eng/pages/search.aspx?i=001-68381.

[306] CONNAUGHTON, *supra* note 238, at 38 (quoting John Crosland, Officer Commanding B Coy (OC B Coy), 2 Para.).

[307] CARL VON CLAUSEWITZ, ON WAR 75 (Michael Howard & Peter Paret eds. & trans., 1989).

[308] ICTY, Prosecutor v. Tadić, Case No. IT-94-1-A, Decision On The Defence Motion For Interlocutory Appeal On Jurisdiction, ¶ 70 (Int'l Crim. Trib. for the Former Yugoslavia, 1995), http://www.icty.org/x/cases/tadic/acdec/en/51002.htm.

determined non-State organized armed group capable of inflicting "State like" violence. Every action must be considered on its own merits. This potential for short and sharp incidents to factually constitute an armed conflict, even in the situation of hostage rescue, provides a possible indication why some States have leaned toward a policy of applying humanitarian law on international operations. However, the challenge for States adopting such approach is that not all operations against non-State actors will rise to the level of an armed conflict (see section 10.3.5, *Rescues and Law Enforcement*, below).

Like *Operation Thunderbolt* at Entebbe, humanitarian law logically governed the hostilities that occurred during *Operation Barras*. However, both of these rescue operations were also constrained by principles associated with the right of self-defense, and by extension the defense of nationals. While the principles governing State self-defense are not the same as those applicable to law enforcement, their overarching application to a defensive act such as these operations have a considerable restraining influence. The assaults only occurred when it was felt negotiations were going to lead to an unacceptable result and it looked like the hostages were at significant risk. The operations focused on rescuing the hostages and defeating, rather than destroying, the armed groups or military forces holding them. In this sense these assaults offered an option of last resort, and ones that addressed the requirements of proportionality, at least in a defense of nationals context. Cleary, *Operation Thunderbolt* and *Operation Barras* as planned contemplated the use of lethal force. However, this is not an uncommon requirement for operations designed to rescue hostages held by determined groups, whether they are conducted under an armed conflict or law enforcement paradigm.

Finally, a general word about accountability regarding rescue operations in a foreign country that is not directed at the specifics of these two iconic operations. There is a tendency for these operations to occur "below the international radar." A successful operation undoubtedly brings about a tremendous sense of relief. However, applying broader conduct of hostilities norms to hostage rescues conducted during armed conflict does not remove the requirement to account for what force is used. For example, ensuring the principle of discrimination is applied and the degree of collateral damage is justified in what is intended to be a limited operation remains extremely relevant. It has been suggested in respect of *Operation Barras* that its conduct outside of the United Kingdom made it easier to put into action. Unlike an operation in London where they would have faced forensic questions, "no one was ever going to be brought to book for what they had done out there on the ground."[309] However, accountability during an operation conducted outside a State, even during armed conflict, is every bit as important as in a domestic law enforcement context.

What may be different are how the investigation is conducted, the information available, and even the ability to access the site of the incident.[310] In this respect, "[o]ne veteran

---

309 Damien Lewis, Operation Certain Death 380 (2004) (it was noted in particular that for an operation located in the center of London "they would have faced all sorts of forensic questions afterward."). *See also* Urban, *supra* note 156, at xvi (where it is noted "Iraq presented a completely different arena from Northern Ireland where considerations such as the legality of the use of force ranked much higher."), and James Adams, Robin Morgan, & Anthony Bambridge, Ambush: The War Between the SAS and the IRA 50 (1988) (where in referring to greater operational freedom for SAS operations in Oman or Malaya it is noted "few people cared about the circumstances in which unnamed jungle revolutionaries died . . . .")

310 Kenneth Watkin, Controlling the Use of Force: A Role for Human Rights Norms in Contemporary Armed Conflict 1, 32–34 (2004).

SAS officer characterized the Sierra Leone operation as 'not a clinical, black balaclava, Princess Gate type operation. It was a very grubby, green operation with lots of potential for things to go wrong.'"[311] This statement captures a fundamental difference between a law enforcement–based operation in a controlled environment in a London embassy and a complex rescue operation against a large well-armed organized armed group located in two locations (i.e., its headquarters and a separate garrison village) along the Rokel Creek in Sierra Leone.[312] However, this does not mean that postincident accountability for such operations is not required. A key aspect of maintaining discipline with the military forces engaged in such activity is ensuring there is a culture of accountability regardless of where and under what circumstances operations are conducted.[313]

10.3.5. RESCUES AND LAW ENFORCEMENT

10.3.5.1. State-Conducted Hostage Rescue: Columbia

While the 1976 raid on Entebbe and the 2000 assault in Sierra Leone can be rightly viewed as iconic hostage rescue missions, they provide only a narrow slice of the types of threats posed to the foreigners located in other States. Given the intimate relationship between insurgency, terrorism, and crime, it is clear that threats to nationals in foreign countries span the conflict spectrum (see Chapter 6, section 6.3, *Insurgency and Crime*). A hostage may be taken by a criminal gang taking advantage of the lawlessness that can be pervasive in situations of conflict, or by an insurgent or terrorist organization. The wide spectrum of threats against nationals, ranging from other States to transnational criminal organizations and even small gangs, raises fundamental questions regarding the limits on the right to defend nationals. Can States act in self-defense against any organized group and in respect of any threat to its nationals in a foreign land? In a perfect world the answer appears to be a straightforward one. Such action would not be necessary. The territorial State is responsible for criminal acts occurring in and emanating from its borders. However, the reality of failed and failing States is that there are isolated geographic areas in which law and order is not maintained adequately. For example, the proliferation of piratical acts on the high seas off the coast of Somalia suggests some territorial States simply are incapable of maintaining order within their jurisdiction.

While nationals outside their home country can be the victims of the type of "ordinary" crime found in any society, that risk is elevated in areas where the territorial State demonstrates a diminished capacity or a complete lack of the ability to govern. Ineffective governance is also a factor closely connected to internal armed conflict. Disorder and lawlessness can be both a precursor of such conflict and one of its byproducts. The threats to nationals can include those emanating from relatively small gangs operating on their own, or from areas controlled by criminal organizations that have effectively displaced the State

---

311 FOWLER, *supra* note 252, at 59.

312 *Id.* at 12, and CONNAUGHTON, *supra* note 238, at 270 (outlining the geographic dispersion of the camps).

313 Kenneth Watkin, Warriors, *Obedience and the Rule of Law*, 3 THE ARMY DOCTRINE AND TRAINING BULLETIN 24 (Winter 2000/Spring 2001).

governing apparatus. As has been noted, non-State actors may also seize foreign nationals in order to finance participation in an armed conflict through the payment of ransoms.

One example of the type of crime that can thrive in situations of unrest, insecurity, and armed conflict is the drug trade. Another activity that frequently occurs is the taking of hostages for ransom. In Mexico "[m]any thugs linked to drug cartels are directly involved in kidnapping."[314] It has been noted that kidnapping of both Mexicans and other nationals is occurring "on an industrial scale."[315] In Columbia, both armed groups and drug traffickers perpetrate such acts. The Columbian Defence Ministry has indicated, "of nearly 2,800 cases recorded in 2007, more than 1,100 people had been freed, 211 had died, and 614 cases were unclear."[316] This criminal activity has occurred within a wider context of violence associated with "three simultaneous ongoing and interrelated wars involving three major violent, internal nonstate groups: FARC (Revolutionary Armed Forces of Columbia), the paramilitary/vigilante AUC (United Self-Defense Groups of Columbia), and the illegal transnational drug industry."[317] Complicating this situation is the development of *bandas criminals* (criminal gangs), which devolved from the demobilization and reintegration program instituted for the AUC (see Chapter 6, section 6.3.1, *The Nature of the Threat*). There has also been the possible federation of some of those gangs with the other non-State actors, leading to "a significant terrorist-criminal-insurgent non-state actor in the Colombian malaise."[318]

Colombia offers one example where hostage rescue has remained within the control of that State and its security forces, although other States have provided resources, training, and other assistance.[319] It is important to note that not every hostage incident ends with an assault to free the hostages. Negotiations or the payment of a ransom can result in their release.[320] On other occasions hostages have escaped.[321] In 2008, the Columbian government obtained the release of fifteen hostages by subterfuge, tricking the guerrillas into handing them to armed forces personnel.[322] However, direct action has also been undertaken to release hostages. Such assaults can be high-risk activity, with four hostages taken by FARC being killed by the hostage takers in 2011 during an assault by Columbian security forces.[323] Successful rescues by specially trained police units have also been credited in helping to significantly reduce incidents of hostage taking. It is reported that in 2007 "Colombian forces rescued 136 hostages . . . and only one hostage was killed during an operation conducted by a GAULA [Unified Action Groups for Personal Liberty] unit

---

[314] Ioan Grillo, El Narco: Inside Mexico's Criminal Insurgency 261 (2011).

[315] *Id.* at 266.

[316] *Q&A: Colombia Hostage Situation*, BBC (Feb. 14, 2011) [hereinafter *Q&A Columbia Hostage Situation*], http://www.bbc.co.uk/news/10308177.

[317] Max G. Manwaring, Gangs Pseudo-Militaries and Other Modern Mercenaries 56 (2010).

[318] *Id.* at 57.

[319] Linda Robinson, *The Future of Special Operations: Beyond Kill and Capture*, Foreign Aff. 114–15 (2012).

[320] *Q&A Columbia Hostage Situation, supra* note 316.

[321] Geoffrey Ramsey, *Escape of ELN Hostage Highlights Obstacle to Colombia Peace Talks*, InSight Crime, (Sept. 11, 2012), http://www.insightcrime.org/news-analysis/eln-kidnapping-obstacle-peace.

[322] Garry Leech, The FARC: The Longest Insurgency 137–38 (2011.

[323] *Rescued Colombia Farc Hostage Reunited with Family*, BBC News (Nov. 27, 2011), http://www.bbc.co.uk/news/world-latin-america-15914232.

that had not yet undergone training."[324] By aggressively addressing the kidnapping problem Columbia went from being the country with the worst kidnapping rate to being in ninth place.[325] The result has been that most kidnappings in Columbia began to take place "in war-wracked corners of countryside. Kidnappings for ransom in the capital, Bogota, have been reduced to close to zero."[326]

Human rights law does impact significantly on the whole range of security operations in Columbia. The legal principles applied by the Colombian armed forces depend upon the nature of the mission and the operational context within which it takes place. The rules governing the use of force are similarly dependent upon the type of mission assigned to the security forces.[327] In areas where the State exercises control and the military does not need to act in a traditional military capacity, "criminal activity, including terrorist acts, can now be dealt with as a law enforcement matter."[328] In addition, consistent with counterinsurgency doctrine generally, the Columbian approach has developed to one where even when humanitarian law can be applied, "the use of force must be more restrictively applied as a last resort and should only be used when necessary in legitimate defense of the military force and when there is no other alternative to accomplish the assigned mission."[329] A restrictive use of force approach puts a premium on applying law enforcement norms, as "[i]t can be opportune to do so, in particular, in situation of a gray area, in order to make sure the requirements of human rights law are respected."[330] This shift toward a law enforcement approach, to the extent that force used during hostilities is tempered by that normative structure, highlights the nuanced nature of contemporary counterterrorism operations. It also indicates that counterterrorist operations, such as hostage rescues, are often amenable to law enforcement action. This shift also demonstrates a State preference for using human rights law to regulate internal security, a choice that is not unique to Columbia (see Chapter 14, section 14.4.2, *A Policy Choice by States*).

10.3.5.2. Hostage Rescue and Foreign Involvement

Not all States conduct hostage rescue operations unaided within their own territory. This was perhaps most famously demonstrated in the 1977 operation by Germany's Border Protection Group 9 (GSG-9) to free hijacked passengers held on a Lufthansa commercial airliner in Mogadishu, Somalia.[331] During the course of that operation, as the hijacked plane flew to various locations across the Middle East and Africa, considerable cooperation occurred with the territorial States where the plane landed. This extended to the GSG-9 and SAS operatives training members of the Dubai Defence Force and Palace Guard in

---

[324] Kenneth Whitelaw, *Inside Colombia's War on Kidnapping*, U.S. NEWS (Feb. 27, 2008), http://www.usnews.com/news/world/articles/2008/02/27/inside-colombias-war-on-kidnapping.

[325] GRILLO, *supra* note 314, at 287 (Mexico tops the list followed by Iraq and India.).

[326] *Id.*

[327] Juan Carlos Gomez, *Twenty-First-Century Challenges: The Use of Military Forces to Combat Criminal Threats*, 88 INT'L L. STUD. 279, 285–86 (2012).

[328] *Id.* at 284.

[329] *Id.* at 285.

[330] JAN RÖMER, KILLING IN A GRAY AREA BETWEEN HUMANITARIAN LAW AND HUMAN RIGHTS 163 (2010).

[331] CHRIS MCNABB, STORMING FLIGHT 181: GSG 9 AND THE MOGADISHU HIJACK 1977 (2011), and DE B. TAILLON, *supra* note 214, at 125–38.

assault technics so they could assist in a possible raid.[332] In Somalia the assessment was made that Somali troops were not up to the task of assaulting the aircraft.[333] However, the Somali military ultimately provided assistance by maintaining perimeter security to seal off the airport;[334] starting a large bonfire to distract the attention of the terrorists on-board the aircraft;[335] and by agreeing to take over responsibility once the assault was complete, thereby freeing up GSG-9 to concentrate on the assault.[336] International cooperation has been a key aspect of many successful hostage rescue missions. It has been suggested that if there is no capable indigenous force available, "and a capable foreign counterterrorism team is available, especially if the hostages are citizens of a foreign country, it is politically and morally responsible for the host nation to set aside sovereignty issues and allow the best counterterrorist unit to deal with the crisis at hand."[337]

A contemporary example of a State permitting a hostage rescue attempt is found in a March 2012 incident in Nigeria. British Special Boat Squadron personnel carried out an operation in Nigeria in an attempt to rescue two engineers (British and Italian) alleged to have been kidnapped by Boko Haram, a fundamentalist Nigerian Islamist group. Like *Operation Barras* this rescue operation was carried out with the consent of the territorial State. As a result, that operation falls into a category where some legal analysts may not view it as a true "defense of nationals" exercise of the right of self-defense. However, factually that is exactly what it was. The Nigerian operation provides insight into the dividing line between rescues that occur in the context of law enforcement and those that constitute an armed conflict. It is reported:

> British commandos stormed the kidnappers' hide-out while Nigerian forces formed a cordon farther back, and that the kidnappers resisted, leading to a brief gunfight in which one of the kidnappers was killed and two others captured. But when the commandos entered the building, the officials said, they found the two hostages dead, apparently shot by their captors as the troops moved in.[338]

Adding to the sensitivity of the operation was the fact the British government had not informed Italy of the operation involving one of their citizens. This reportedly was because of the latter State's preference to pay ransoms in order to gain the release of hostages.[339] However, it has also been suggested that a ransom raised by the kidnapped men's families had been paid[340] and some communications occurred between the State forces and

---

332 McNabb, *supra* note 331, at 36.

333 *Id.* at 47.

334 *Id.* at 49.

335 *Id.* at 61.

336 *Id.* at 74.

337 de B. Taillon, *supra* note 214, at 136 (reference is made to *Special Operations*, University of New Brunswick, Centre for Conflict Studies 119).

338 Elisabetta Povoledo & Johan F. Burns, *Failed Raid to Rescue Hostages in Nigeria Stirs Italy's Anger*, N.Y. Times, Mar. 9, 2012, http://www.nytimes.com/2012/03/10/world/africa/britain-italy-in-spat-over-failed-rescue-try-in-nigeria.html.

339 *Id.*

340 Andrew Walker, What Is Boko Haram?, Special Report, United States Institute for Peace 10 (2012), *available at* http://www.usip.org/files/resources/SR308.pdf.

kidnappers. The "[t]roops reportedly asked the men to release the hostages and only entered after the kidnappers refused."[341]

Questions can be asked regarding the legal regime governing the use of force during the operation. The Nigerian government has been engaged in fighting with Boko Haram since 2009.[342] From a Nigeria perspective, the hostage taking might be seen as part of an ongoing conflict with that "terrorist" group. However, it should also be noted the Nigerian government has been very reluctant to call this internal violence an armed conflict.[343] In any event, from an external State perspective a number of factors point toward the operation falling more comfortably within a law enforcement response. There appears to have been no indication that the United Kingdom was engaging itself more broadly in the internal Nigerian violence. The rescue operation itself was narrowly focused on rescuing British and Italian nationals working as contractors in Nigeria for an Italian company. Unlike the British hostage rescue operation in Sierra Leone in 2000, the nature of the threat posed by the hostage takers in the Nigeria situation, the intensity of violence involved, and the weapons and tactics employed by the security forces did not raise the specter of participation in an armed conflict. The tactical situation seems to have resembled the type of hostage incidents handled under a domestic law enforcement paradigm.

A "totality of the circumstances" assessment, including an application of the various factors identified in the *Prosecutor v. Haradinaj* decision, points toward a law enforcement situation rather an than armed conflict.[344] In addition, the fact the United Kingdom is subject to the legal oversight of the European Court on Human Rights, a court that has consistently applied human rights even during armed conflict, means the default position regarding the use of force in legal grey areas would likely mean that human rights law regulatory framework will be applied. As a result, the rules governing the use of force by British forces during the operation probably reflected the domestic and international human rights legal obligations of the United Kingdom. That human rights law is capable of regulating such hostage rescue operations provides an indication that body of law can on occasion accommodate a robust use of force when necessary (see Chapter 12, section 12.2.2, *Force and Hostage Rescue*).

In contemporary security situations the dividing line between criminal and insurgent can be an uncertain one. It has been suggested the group that carried out the hostage seizure in Nigeria, like in a number of other incidents in Mali and Niger, may have been a criminal gang conducting kidnappings for the purpose of ransom or selling them to Al Qaeda in the Islamic Maghreb.[345] There have been other situations, such as in Afghanistan, Somalia, and

---

341 *Id.*

342 *Boko Haram*, NGEX, http://www.ngex.com/nigeria/bokoharam.htm (last visited June 2, 2015) (for a timeline of Boko Haram attacks and activities). *See also* WALKER, *supra* note 340, at 3–6 (for a history of Boko Haram).

343 *Nigeria: Read Full Text of Jonathan's Declaration of Emergency Rule*, VANGUARD, 14 May 2013, http://www.vanguardngr.com/2013/05/read-full-text-of-jonathans-declaration-of-emergency-rule/, and Onyebuchi Ezigbo et al., *FG: We Are Fighting Terrorism, Not Civil War*, THIS DAY LIVE (Dec. 4, 2013), http://www.thisdaylive.com/articles/fg-we-are-fighting-terrorism-not-civil-war/165863/.

344 Prosecutor v. Haradinaj, Case No. IT-04-84-T, Judgement, ¶ 49 (Int'l Crim. Trib. for the Former Yugoslavia, Apr. 3, 2008), http://www.icty.org/x/cases/haradinaj/tjug/en/080403.pdf.

345 WALKER, *supra* note 340, at 10–11.

Columbia, where the hostage taking is carried out by gangs not directly connected to an ongoing insurgency or armed conflict, but willing to sell hostages for profit. For example, it is reported that a Columbian criminal gang sold hostages to the ELN guerrilla group for up to $125,000 for each kidnap victim.[346] Similarly, in 2008 a gang seeking ransom seized a freelance Canadian journalist, Amanda Lindhout, in Somalia. She was warned by her captors that Al Shabaab would be happy to "sweep us up if we caught their attention,"[347] and she was threatened to be sold to that group as negotiations dragged on.[348] In 2011 another Canadian journalist, Melissa Fung, was kidnapped in Afghanistan by "a petty band who ran a hostage business of sorts."[349]

In some cases the identity of the hostage takers may not be immediately clear. In November 2013 it was reported that persons of uncertain affiliation, who may have been members of Al Qaeda or a Tuareg splinter group, kidnapped two French journalists in Mali. Those journalists were subsequently murdered as French and government military forces gave chase.[350] Hostages have also been taken by militant groups, as was evidenced by the 2013 abduction of seven foreign construction workers by Ansaru (an offshoot of Boko Haram), which was linked to Al Qaeda in northwest Africa. Tragically, in that case it is reported all the hostages were murdered when British military planes were seen at a Nigerian airport prompting the hostage takers to fear a rescue attempt.[351] The frequent lack of clarity regarding the identity, status, and motivation of the hostage takers highlights the challenge of attempting to draw too fine a distinction between criminal acts and participation in hostilities in many operational situations. However, it cannot simply be assumed that hostage takers are participants in armed conflict. Even in situations where an armed conflict is taking place, there remains an obligation to positively distinguish between "the civilian population and combatants."[352] Unless it is established otherwise, the default position is that the operation involves civilians rather than members of organized armed groups or individuals taking a direct part in hostilities. This means the operation would be governed by human rights–based law enforcement rules unless it is established the hostage takers are enemy fighters.

10.3.5.3. States Acting Unilaterally to Rescue Their Own Nationals: Somalia

Somalia has been associated with number of high-profile kidnappings in a context where that State has been unable to police its territory or offshore waters. It should be noted the

---

[346] *Colombian Gang "Sold Kidnap Victims to ELN Guerrilla,"* BBC NEWS (Feb. 23, 2015), http://www.bbc.com/news/world-latin-america-31588781?utm_source=Sailthru&utm_medium=email&utm_term=%2AMorning%20Brief&utm_campaign=2014_MorningBrief%202.23.15.

[347] AMANDA LINDHOUT & SARA CORBETT, A HOUSE IN THE SKY 190 (2013) ("our captors were apparently a renegade group, unaffiliated with the dominant extremists.").

[348] *Id.* at 320.

[349] Hampson, *supra* note 103. *See also* WILLIAMS, *supra* note 4, at 129–34 (outlining the complex relationship between insurgent groups and criminal gangs in Iraq).

[350] Adam Nossiter, *Two French Journalists Are Kidnapped and Killed by Gunmen in Mali*, N.Y. TIMES, Nov. 2, 2013, http://www.nytimes.com/2013/11/03/world/africa/two-french-journalists-are-kidnapped-and-killed-by-gunmen-in-mali.html?_r=0.

[351] *British Hostage "Killed" in Nigeria in "Act of Cold-blooded Murder,"* BBC NEWS (Mar. 10, 2013), http://www.bbc.co.uk/news/uk-21734118.

[352] AP I, art. 48.

vast majority of pirate kidnappings are not resolved with a hostage rescue operation. The victims may not be of a high enough profile, are nationals of a country that does not have the capacity to mount a rescue, have no family money for ransom, or are employed by firms that treat them as expendable.[353] The result is that while marine insurance companies regularly paid out large ransoms, which also helped fuel piracy,[354] "trawler crews from the Far East could languish for years in pirate captivity."[355] However, States have acted in defense of their nationals seized by these criminal elements. The Somalia-related examples are unique in that the rescues have happened both at sea and on land.

Piracy emanating from Somalia is directly linked to the weak governance situation in that State caused by the armed conflicts occurring in one form or another over the past two decades. Between 1991 and 2004 alone there were fifteen failed attempts to form a national government.[356] Included as a participant in that conflict has been the Al Qaeda–linked militant group, Al Shabaab.[357] However, piracy is largely a commercial criminal endeavor that has been carried out separately from those conflicts.[358] It has been suggested a deal was struck in 2011 in southern Somalia where Al Shabaab permitted the pirate gangs to continue to operate for 20 to 25 percent of every ransom paid. However, this has not changed the broader criminal nature of the activity.[359] Notably, piracy is "not a criminal fraternity hiding in the midst of an otherwise law-abiding society. It constitutes a significant part of that society." [360] In this respect, "Somali piracy is a rational response that satisfies an economic need by exploiting a security weakness."[361] This keeps it well within the realm of criminal activity, although it is possible the violence could reach levels associated with an armed conflict. In this regard UN Security Council Resolution 1851 contemplated that any measures taken by States shall be consistent with applicable international humanitarian and human rights law. [362] This suggests that armed conflict between States and pirate groups might be a possibility, although that does not appear to have happened to date.

One of the most daring examples of a rescue operation occurred in 2009 when Somali pirates seized the U.S. flagged commercial container ship *Maersk Alabama* 400 nautical miles east of Mogadishu and took the ship's captain, a U.S. citizen, as a hostage.[363] Given that the incident happened on the high seas, the container ship was American flagged, and the nationality of the hostage, the jurisdictional basis for the United States to act appears straightforward.[364] With the *Maersk Alabama* flying a U.S. flag, the operation also included

---

[353] James Fergusson, The World's Most Dangerous Place: Inside the Outlaw State of Somalia 183 (2013) (These hostages were "impoverished migrant workers of a sort often found below decks in the Far-East's under-regulated and poorly paid fisheries industry.")

[354] *Id.* at 187.

[355] *Id.* at 183.

[356] Murphy, *supra* note 105, at 164.

[357] *Id.* at 139.

[358] *Id.*

[359] Abdel Bari Atwan, After Bin Laden: Al Qaeda, The Next Generation 124–25 (2012).

[360] Murphy, *supra* note 105, at 178.

[361] *Id.*

[362] S.C. Res. 1851, art. 6. UN Doc. S/RES/1851 (Dec. 16, 2008).

[363] Murphy, *supra* note 105, at 104.

[364] Peter T. Leach, *Maersk Alabama Crew Sues Owner, Operator*, J. of Comm. Online News Story (May 29, 2012), *available at* http://shippingblog.com/current-news/.

defending the property of nationals, which has been viewed historically as an accepted interpretation of scope of self-defense.[365] The claim of acting in defense of nationals has particular strength in this instance given the incident happened on the high seas where the role of a territorial State is not in issue. There is no doubt "that the armed forces of the flag state may use reasonable force to defend vessels from attack whether by pirates or forces acting with or without the authority of any state."[366] If the incident had happened in Somali territorial waters, the issue of consent from the Somalia government would have arisen. Even then, since that State did not have the ability to conduct the operation obtaining such consent likely would have largely been a formality. After parachuting a naval Special Forces team into the water near a U.S. warship and taking up positions observing a lifeboat where the hostage was being held, snipers ended the hostage incident by killing three of the Somali pirates with three exceptional simultaneous shots.[367]

Notwithstanding the policies of States, such as the United States, to broadly apply humanitarian law as a matter of policy during international operations, the piracy threat clearly indicates that not all action taken internationally to protect their citizens falls within the framework of an armed conflict. Therefore it cannot be resolved with conduct of hostilities rules governing the use of force. The incident involving the *Maersk Alabama* fell within the law enforcement paradigm. The pirates were clearly involved in criminal activity. They were not part of an organized armed group engaged in an armed conflict for a political purpose. The law enforcement nature of the situation is reflected in the direction provided to the SEALs engaged in the operation, which was reported to have been "only act if Phillips's life was in imminent danger."[368] It was an exercise of extraterritorial criminal law enforcement. This conclusion is aided by the fact that in addition to any municipal laws and UN Security Council resolutions dealing with the subject,[369] pirates "as *hostes humanai generis* [enemies of all mankind], come under a universal jurisdiction and are punishable under customary international law."[370]

Not all operations undertaken to counter piracy on the high seas have been successful. In February 2011 an attempt to rescue U.S. citizens on the yacht, *SV Quest*, which had been seized by pirates 275 miles off the coast of Oman, ended in failure. There was a breakdown in negotiations, and the pirates began shooting. This included an RPG 7 being fired at a Navy ship and firing within the pirate vessel. After the assault by Navy SEALs, the hostages were found dead and two of the pirates had been killed.[371] This use of force, with its tragic ending, does not remove the incident from the realm of high seas law enforcement.

State action to protect nationals and others from criminal activity has also not been limited to offshore operations. It is reported that in January 2011 "US soldiers conducted

---

[365] Bowett, *supra* note 9, at 104. *See also* Shaw, *supra* note 12, at 1143 (for reference to the historical protection of property).

[366] Brownlie, *supra* note 12, at 305.

[367] Mark Owen, No Easy Day: The Autobiography of a Navy Seal 89–98 (2012).

[368] Owen, *supra* note 367, at 96. *See also* Jeremy Scahill, Dirty Wars: The World Is a Battlefield 275 (2013) (outlining the decision-making process involving the president of the United States before the shooting was carried out, which included questions about risks to U.S. forces and collateral casualties).

[369] *See, e.g.*, S.C. Res. 1851, UN Doc. S/RES/1851 (Dec. 16, 2008).

[370] Bowett, *supra* note 9, at 74.

[371] Scahill, *supra* note 368, at 482–83. *See also* Naylor, *supra* note 141, at 412–13.

a counter piracy incursion inland, snatching three young Somali men and bringing them aboard a ship for questioning."[372] A year later, in January 2012 U.S. Special Forces freed two hostages, American and Danish aid workers, who had been captured on land by a criminal gang and were being held near the city of Adado, Somalia.[373] It is reported that Somali elders described the adductors as members of a pirate gang, while "American officials argued that the kidnappers were criminals with no direct links to any of the pirate bands that have attacked shipping lanes off Somalia . . . ."[374] It appears that the abductors were not from Al Shabaab since one of the threats made during the course of ransom negotiations was that the hostages would be sold to that terrorist organization.[375] The raid occurred after ransom talks had dragged on and concerns arose regarding the health of the female hostage.[376] The assaulting force parachuted into Somalia, trekked to the camp where the hostages were held, and engaged the hostage takers in a firefight. [377] The kidnappers were fired on after one of them began to raise the alarm.[378] It has been reported that nine of the abductors were killed and five were wounded in the course of a mission that successfully freed the hostages.[379]

It is not known if during the 2012 rescue of the two hostages there was time to seek the consent of the governing Somali Transitional Federal Government, or if any attempt was made to do so. That government is "an entity that controlled barely any territory, had few forces of their own and was kept in power by foreign troops operating under an AU mandate." [380] Given the limitations on Somali State governance, if it had been approached such consent would likely have represented a legal band-aid to cover the operation. A lack of consent also very likely would not have impacted on whether an operation was conducted. As President Obama has indicated, the United States will "not tolerate the abduction of our people, and will spare no effort to secure the safety of our citizens and to bring their captors to justice."[381] However, it should also be noted that Somali forces have not been completely inactive. They cooperated with Kenyan forces in carrying out rescue of four aid workers who had been seized in Kenya by gunmen in 2012 and then taken across the border into Somalia.[382]

American forces are not the only ones to have conducted hostage rescue operations in and near Somalia. France has demonstrated considerable resolve in their pursuit of pirates and by the conduct of hostage rescue operations both at sea and on Somali territory. French military forces have acted on a number of occasions in respect of hostage incidents. In 2008,

---

[372] SCAHILL, *supra* note 368, at 482.

[373] JESSICA BUCHANAN, ERIK LANDEMALM, & ANTHONY FLACCO, IMPOSSIBLE ODDS: THE KIDNAPPING OF JESSICA BUCHANAN AND HER DRAMATIC RESCUE BY SEAL TEAM SIX 215–16 (2013).

[374] Gettleman, Schmitt, & Shanker, *supra* note 106. *See also* BUCHANAN ET AL., *supra* note 373, at 231–32 ("Criminals belonging to three of the Habr Gidir subclans, Sa'ad, Sulieman, and Ayre, were believed to have cooperated in arming and manning the operation.").

[375] BUCHANAN ET AL., *supra* note 373, at 209.

[376] *Id.* at 233–234.

[377] Gettleman, Schmitt, & Shanker, supra note 106. *See also* NAYLOR, *supra* note 141, at 425–27.

[378] BUCHANAN ET AL., *supra* note 373, at 250.

[379] *Details Emerge, supra* note 303.

[380] MURPHY, *supra* note 105, at 172.

[381] *Details Emerge, supra* note 303 (reported as a statement made by U.S. President Barak Obama.).

[382] *Four Foreign Aid Workers Rescued in Somalia*, REUTERS (July 2, 2012), http://www.reuters.com/article/2012/07/02/us-somalia-hostages-idUSBRE86109920120702.

the crew of a seized ship, the *Ponant*, was taken to a Somali port and then released following the payment of a ransom. French commandos then intercepted the pirates in Somali, capturing six of the escaping criminals.[383] During the same time period that the *Maersk Alabama* incident unfolded, French Special Forces boarded a yacht, the *Tanit*, containing two couples and a child that had been seized by pirates. During that 2009 operation a French hostage and two pirates were killed and another three criminals were captured.[384] Similar to the U.S. approach, the French Defence Minister indicated France would not give in to blackmail, would prosecute the criminal acts, and "liberate the hostages every time that a ship under a French flag is captured."[385] Further, preceding a January 2013 French operation in Mali to assist that government in battling Islamist opponents, French Special Forces were sent into Somalia in a failed attempt to rescue an intelligence operative being held by Al Shabaab.[386] The Somali government is reported to have condemned this raid.[387]

Other nations have also participated in rescue operations. In February 2010, Danish Special Forces operating from the Danish warship *Absalon* rescued the crew of a merchant vessel that had locked themselves in a safe room on that ship. The pirates had left before the Danish boarding party arrived.[388] On May 6, 2010, Russian commandos were reported to have assaulted a Liberian-flagged, Russian-owned vessel, the *Moscow University*, which had been seized by pirates. The twenty-three Russian crewmen, who had similarly hidden in a safe room after disabling the ship, were all rescued, with one pirate being killed and ten others captured.[389] Controversially, those pirates were subsequently reported in the Russian media to have died after having been released in their vessel at sea.[390] Indian Navy commandos have also been deployed on counterpiracy operations in the Gulf of Aden presumably ready to conduct a hostage rescue if required.[391] These operations against pirates carried out or contemplated by numerous States fall within the realm of criminal law enforcement (i.e., outside the context of an armed conflict). Further, like the United Kingdom, Russia, and France, "all European nations participating in naval operations off the shores of Somalia are bound not only by international law but also by the European Convention on Human Rights. . . ."[392] This makes it highly likely those States would be held accountable according to human rights standards. Overall, the result is that the use of force against pirates is most

---

383 *France Raid Ship after Crew Freed*, BBC News (Apr. 12, 2008, 16:24 UK), http://news.bbc.co.uk/2/hi/africa/7342292.stm.

384 *Frenchman Dies in Somali Rescue*, BBC News (Apr. 11, 2009), http://news.bbc.co.uk/2/hi/in_depth/7994201.stm.

385 *Id.*

386 Peter Beaumont, *French Secret Service Hostage and Soldiers Killed in Somali Rescue Mission*, The Guardian, Jan. 12, 2013, http://www.guardian.co.uk/world/2013/jan/12/french-hostage-soldiers-killed-somalia.

387 *Federal Govt Condemns France Military Operation in Somalia*, AllAfrica (Jan. 13, 2013), http://allafrica.com/stories/201301130152.html.

388 *NATO Troops Free Ship off Somalia after Pirate Attack*, BBC (Feb. 5, 2010), http://news.bbc.co.uk/2/hi/africa/8500611.stm

389 *Russian Special Forces in Dawn Raid on Oil Tanker Pirates*, The Telegraph, May 6, 2010, http://www.telegraph.co.uk/news/worldnews/europe/russia/7684434/Russian-special-forces-in-dawn-raid-on-oil-tanker-pirates.html.

390 *Freed Somali Pirates "Probably Died"—Russian Source*, BBC (May 11, 2010), http://news.bbc.co.uk/2/hi/8675978.stm.

391 Rajat Pandit, *Commandos Answer SOS from Saudi, Indian Vessels, Scare Off Pirates*, Times of India, Nov. 11, 2008, http://articles.timesofindia.indiatimes.com/2008-11-11/india/27942549_1_merchant-vessel-indian-frigate-indian-warship.

392 Stefano Piedimonte Bodini, *Fighting Maritime Piracy under the European Convention on Human Rights*, 22 Eur. J. Int'l L. 829, 830 (2011).

likely to be viewed as being regulated by a human rights–based law enforcement paradigm, rather than a conduct of hostilities framework.

As can be seen in the examples of Columbia, Nigeria, and Somalia, hostage rescue is frequently conducted under a law enforcement framework. While that might be expected for States dealing with such incidents within their own borders, this approach is also increasingly being applied in rescue operations in foreign States and on the high seas. In many situations, such as in Columbia or Mexico, it is not likely that another State will intervene to rescue its nationals from criminal elements, even if the foreign State had the ability to do so. In those situations the capacity of the territorial States' security forces, their willingness to address the challenge, and the prevailing political situation will operate to avoid foreign intervention.

However, as is evident in failed States (e.g., Somalia) and in other States where there is a local limited capacity to deal with such challenges (e.g., Nigeria, Afghanistan, and Mali), hostage rescue operations may still be conducted under a law enforcement framework. In these latter situations, operations have been carried out jointly between the territorial State and the States whose nationals are at risk. In this regard, the foreign State may intervene to assist the territorial State in fulfilling its governance role, thereby protecting its citizens. When necessary the foreign State may conduct an operation unilaterally in order to fulfill out its own responsibility toward its nationals. These hostage rescues may be conducted in some cases without territorial State assistance, consent, or sometimes its knowledge. The fulfillment of these national obligations of governance is regular feature of the twenty-first-century international security environment.

10.3.5.4. Law Enforcement and the Protection of Nationals

Hostage rescue operations present both a challenge and an opportunity for international law. The challenge is in coming to grips with how these operations fit within a narrowly prescribed largely inter-State self-defense framework favored by many international lawyers. The opportunity comes from the realization that these situations offer a chance to change the dialogue concerning international security operations. That change shifts the perspective from one that appears overly armed conflict based to a more nuanced discussion that matches the legal framework to the nature of the security threat. That nuance arises from the adoption of the law enforcement framework in the governance hostage rescue operations, both as a matter of law and policy.

As has been noted, some incidents fall uniquely within the law enforcement paradigm. From a self-defense perspective, an argument might be made that a threat to any national is an attack on that person's State. However, on its face that connection to State interest may appear hard to maintain when the seizure is by a criminal gang acting for "commercial" reasons rather than by a non-State actor in conflict with the national's State. At some point the act fails to have the "political" connection that underpins all armed conflict. In these situations the attack on the citizen of the State generally would not be because of his or her nationality, although where a hostage is taken by insurgent groups the nationality of that person may be a factor that is difficult to separate from the desired pecuniary gain.[393] The

[393] DINSTEIN, *supra* note 15, ¶ 575, at 218 (discussing that attacks on foreigners because of their nationality could give rise to a lawful exercise of self-defense).

result is that States engage in human rights law–based international criminal law enforcement when seeking to rescue their nationals.

Care must still be used when considering what activity falls as a matter of law within the law enforcement paradigm. A problem that may arise from a theoretical level is when overreaching occurs and the "law enforcement" title is applied to even inter-State uses of force (see section 10.2.1, *A State Responsibility*, above).[394] Further, while it is not evident that a single kidnapping or even multiple seizures of hostages for ransom necessarily rises to the level of an "armed attack," as was demonstrated in 2000 in Sierra Leone, an armed conflict can result from even one act of kidnapping (see section 10.3.3, *Modern Counterterrorism: Sierra Leone*, above). Much depends on the context. However, what is interesting is the lack of public outcry concerning operations designed to rescue nationals seized by what are identified as "criminal gangs." Perhaps, as Christine Gray has noted, the discussion of action to defend nationals in States where there is no effective government is just avoided, with controversy only being aroused when there is an objection, or the action is seen as pretext for intervening for some other purpose.[395]

Notwithstanding this observation, the legal basis for acting is not dependent upon a lack of objection. It must exist independent of such concerns being raised by other States. Its most likely basis is found in customary international law, if it cannot be interpreted to exist within the UN Charter–based State self-defense framework. D. W. Bowett relied on Elihu Root's 1910 analysis of the State right to protect citizens to suggest that notions of State responsibility and protection were dependent upon "the failure of the territorial state to conform to minimum standards based upon general international law, to a 'standard of justice, very simple, very fundamental, and of such general acceptance of all civilized nations. . . .' "[396] Such an intervention cannot be justified for just any incident where a national might be put at risk. Crime and legal inequities occurs within any State. However, the statement that "every citizen is entitled to have police protection accorded to him commensurate with the exigency under which he may be placed" is as true in the twenty-first century as it was at the turn of the previous one.[397]

The right of a State to intervene in order to protect its nationals clearly must extend to situations where there is little or no ability on the part of the territorial State to provide the necessary protection. In this regard, there appears to be little difference between the threats posed by many non-State actors (including criminal gangs) in "ungoverned spaces" on land and those presented by pirates operating on the high seas. After any legitimate concerns about alternative motives by an intervening State are addressed, it is left to that State to carry out its responsibilities toward its citizens. In effect, there is no one else. Appealing to the UN Security Council is not likely to help because of the relative seriousness of incident (i.e., criminality) and the relative weakness of collective enforcement.[398]

---

394 Ruys, *The Meaning of "Force" and the Boundaries of the Jus Ad Bellum*, *supra* note 2, at 180–82.

395 Gray, *supra* note 17, at 159–60.

396 Bowett, *supra* note 9, at 89. *See also* Elihu Root, *The Basis of Protection to Citizens Residing Abroad*, 4 Proc. of the Am. Soc. of Int'l L. at Its Annual Meeting 16, 21 (1910).

397 Root, *supra* note 396, at 22–23.

398 Cassese, *supra* note 17, at 368.

Factors that keep action to defend of nationals from attracting the criticism of even some of the most stalwart protectors of a narrowly prescribed State self-defense legal framework are twofold. First, these operations are frequently conducted against what are widely accepted as criminal organizations or gangs operating beyond the reach of territorial State authorities. Although there is increasing recognition that transnational criminal organizations can have an "insurgency" effect, there remains a reluctance to cross over into the realm of armed conflict. Second, hostage rescue is often carried out using policing rather than conduct of hostilities norms even during what might be claimed to be armed conflict. The preference to use human rights–based law restrictions on the use of force can occur because of the need to protect the hostages. Further, since hostage rescues often take place "amongst the people," the goal of many States will be to significantly limit collateral casualties or damage arising from any use of force. This is a context wherein human rights–based law enforcement norms can offer a less violent framework within which to operate. The human rights legal framework has been applied successfully within States and internationally to deal with criminals threatening considerable violence. The result is that, as a matter of policy, law enforcement–type rules are often applied even in armed conflict situations (see Chapter 14, section 14.4.2, *A Policy Choice by States*).

It is the general acceptance of law enforcement–based operations to rescue hostages that may provide the greatest lesson for advocates of cross-border transnational terrorist operations. The concern of many legal analysts and particularly those schooled in traditional self-defense theory is limiting States going to war. Enforcing the law extraterritorially in the limited situations where governance is weak or nonexistent attracts far less attention or criticism. When carrying out these operations, the security forces of the intervening State must also be protective of the human rights of the citizens of the territorial State and even those of the criminals themselves. A key point is that the operations against criminal elements are, as a matter of law, governed by policing norms. This occurs regardless of whether the requirement to apply human rights law to the extraterritorial operations of the intervening State is viewed as arising as a matter of treaty or customary law.

However, given the potential for significant levels of violence emanating from well-organized criminal gangs, the question that remains as to how effective the law enforcement paradigm is in meeting the challenge of intervention in foreign territory or on the high seas. When the law enforcement framework is assessed as not providing an effective means of maintaining security, questions are immediately raised as to whether the intensity of the violence and the nature of the organized threat points to the possible existence of an armed conflict and an accompanying requirement to conduct hostilities (see Chapter 13, section 13.6, *Limits and the Need for Clarity*). However, even at that point States may alter the manner in which force is used during armed conflict to more closely reflect a law enforcement model.

10.4. Converging Tactics: Special Forces, Law Enforcement, and Hostilities

A unique aspect of operations to rescue kidnapped nationals is the concentration of skills required for these missions in the hands of both specialized military and police personnel. For military Special Forces units, an ability to conduct such rescue operations in both armed conflict and law enforcement situations has become a hallmark of their professionalism. In this regard, the Entebbe rescue operation has been called "one of the greatest Special Forces missions ever."[399] Similarly, the British forces raid in 2000 to rescue its soldiers in Sierra Leone is referred to "one of the most highly successful rescue missions in recent military history."[400] Special Forces operations have also included parachuting into the waters of Somalia to end a hostage crisis[401] and being inserted by helicopter into remote areas of Afghanistan to rescue hostages seized by a criminal gang.[402] Notably, such operations can uniquely involve law enforcement missions carried out by military forces. These operations involved highly trained Special Forces and support personnel employing stealth, overwhelming force, and, when necessary, restraint on the use of force in the conduct of counterterrorism operations. In the post-9/11 environment this skill set has been relied to an exceptional degree with the U.S. special operations budget alone being reported to have increase from $2.3 billion in 2001 to $10.5 billion in 2012.[403]

However, one of the most interesting aspects of modern hostage rescue operations is that similarly trained military, paramilitary units, or police may be employed under either paradigm. In the 1970s as part of the international response to terrorism, and in particular the 1972 Munich Olympic tragedy, Western governments started the development of specialized counterterrorist units.[404] That high-risk hostage rescue capability can exist within both military[405] and police units.[406] The law enforcement nature of many hostage rescues is highlighted by the development of the specialized police or paramilitary units to conduct such missions. These units include Germany's Border Protection Group 9 (GSG-9);[407] France's National Gendarmerie Intervention Group (GIGN),[408] Research, Assistance, Intervention,

---

399 DUNSTAN, *supra* note 102, at 62. *See also* TONY GERAGHTY, BLACK OPS: THE RISE OF SPECIAL FORCES IN THE C.I.A., THE S.A.S., AND MOSSAD 195 (2010), and DE B. TAILLON, *supra* note 214, at 108.

400 CONNAUGHTON, *supra* note 238, at 250.

401 OWEN, *supra* note 367, at 89–98.

402 Sean Rayment, *How the British Hostages Were Rescued in Afghanistan*, THE TELEGRAPH, June 3, 2012, http://www.telegraph.co.uk/news/worldnews/asia/afghanistan/9307833/How-the-British-hostages-were-rescued-in-Afghanistan.html.

403 Robinson, *supra* note 319, at 110.

404 BRUCE HOFFMAN, INSIDE TERRORISM 68 (2006).

405 HARCLERODE, *supra* note 87, at 266–72 (outlining the development of the counterterrorism role for the United Kingdom Special Air Service and Special Boat Service), *id.* at 288–97 (the counterterrorism role and early missions of the Israeli Sayeret Matkal and Mistaf'arvim units), and *id.* at 409–25 (outlining the early history of counterterrorist capability with the development of the U.S. Army Delta Force and the Navy SEAL Team 6 maritime counter measures units). *See also* DUNSTAN, *supra* note 102, at 12, BETSER, *supra* note 195, at 277–78 (for an outline of the development of the IDF's Sayeret Matkal as a hostage rescue force); and *Joint Task Force (2)*, National Defence, http://www.forces.gc.ca/en/operations-special-forces/jtf2.page (last visited Aug. 11, 2015) (the JTF 2 counterterrorism mandate was inherited from the Royal Canadian Mounted Police, Special Emergency Response Team).

406 HARCLERODE, *supra* note 87, at 283.

407 MCNABB, *supra* note 331, at 11–12 ("Although part of the police, GSG-9 was essentially a fully militarized special forces unit."). *But see also* LEROY THOMPSON, THE RESCUERS: THE WORLD'S TOP ANTI-TERRORIST UNITS 66 (1986) (where it is noted that "GSG-9's philosophy is to attempt to take prisoners if possible.").

408 MARTEN C. AROSTEGUI, TWILIGHT WARRIORS: INSIDE THE WORLD'S SPECIAL FORCES 82–90 (1996) (for an outline of the February 1976 hostage rescue where members of the GIGN, supported by French Foreign Legion soldiers, rescued French schoolchildren held on a bus in Djibouti by terrorists from the Front de Libération de la Côte de Somalie (FLCS)), and YAROSLAV TROFIMOV, THE SIEGE OF MECCA: THE FORGOTTEN UPRISING IN ISLAM'S HOLIEST SHRINE AND THE BIRTH OF AL QAEDA 188–219 (2007) (outlining GIGN involvement in the operation to end the 1979 siege of Mecca). *See also* Pierre Bienamié, *These Are the Elite French Police Units that Have Been Deployed*

Deterrence (RAID), National Police Intervention Groups (GIPN), and the Research and Intervention Brigade (BRI);[409] the U.S. Federal Bureau of Investigation Hostage Rescue Team (HRT);[410] the Israeli police "Yamam" tactical hostage rescue team,[411] Columbia's GAULA,[412] the Philippines National Police Special Action Force (SAF),[413] and the London Metropolitan Police, Specialist Firearms Command (SCO) 19.[414] India has developed a hybrid military/police unit, the National Security Guard (NSG), also known as the "Black Cats."[415] Russia's premier counterterrorism unit is the Federal Security Service (FSB) Alpha group,[416] which is "recruited from the military, MVD [Ministry of Internal Affairs], FSB, and other militarized services or military academies. . . ."[417]

The decision on whether to use military or police forces domestically and in some situations internationally in a hostage rescue/law enforcement role can depend upon a number of factors including operational capability, geographic location, nature of the threat, an ability to be deployed in a timely fashion, and even domestic constitutional restrictions. Some of these factors, reflecting on what units are deployed, were evident in a May 2009 piracy incident involving a German merchant vessel. It is reported that after pirates seized a freighter off the coast of Somalia, the preference of the German government was to deploy its elite counterterrorism police unit, the GSG-9. That unit had more experience with kidnappings, and the view was that the job of law enforcement was to hunt down criminals. Further, German law would have required parliamentary approval for the use of military units.[418] However, the rescue plan was thwarted because Germany did not have the logistics or the

---

to Fight Terrorists in the Country, BUSINESS INSIDER (Jan. 16, 2015, 5:10), http://www.businessinsider.com/these-are-frances-elite-police-units-2015-1 (where in addition to the GIGN it is noted there are two additional police hostage rescue units: RAID (Research Assistance Intervention Dissuasion) and BRI (Research and Intervention Brigade)).

409 Tyler Rogoway, *Meet France's Elite Revolver-Toting Counter-Terrorism Units*, FOX TROT ALPHA (Jan. 12, 2015, 3:45 PM), http://foxtrotalpha.jalopnik.com/meet-frances-elite-revolver-toting-counter-terrorism-un-1678793691.

410 CHRISTOPHER WHITCOMB, COLD ZERO: INSIDE THE HOSTAGE RESCUE TEAM 124–61 (2001) (for an outline of FBI HRT training, including sniper training carried out at the U.S. Marine Corps Scout/Sniper School) and *Agents of the Hostage Rescue Team*, https://www.fbijobs.gov/explore-careers/sa-hostagerescue.asp (last visited Sept. 1, 2015) (where military combat arms, piloting, or Special Forces tactical operations experience is indicated as being sought as recruits).

411 Or Heller, *Open Mind & Out-of-the-Box Thinking*, ISRAEL DEFENSE (Aug. 19, 2015), http://www.israeldefense.co.il/en/content/open-mind-out-box-thinking.

412 Whitelaw, *supra* note 324.

413 *Philippine National Police Special Action Force*, http://saf.pnp.gov.ph/index.php/about-us (last visited Aug. 11, 2015).

414 *Armed Response Units & Specialist Firearms Officers*, Metropolitan Police, http://content.met.police.uk/Article/Armed-Response-Units--Specialist-Firearms-Officers/1400013622269/1400013622269 (last visited Jan. 12, 2015). *See also* Vikram Dodd, *Scotland Yard Creates SAS-style Unit to Counter Threat of Terrorist Gun Attack*, THE GUARDIAN (June 29, 2015, 19:23 BST), http://www.theguardian.com/uk-news/2015/jun/29/scotland-yard-creates-sas-style-unit-to-counter-threat-of-terrorist-gun-attack (where reference is made to using minimum force and not having orders to shoot to kill).

415 *National Security Guard, available* at http://nsg.gov.in/organisation_history.php (last visited July 18, 2015) ("The NSG was modeled on the pattern of the SAS of the UK and GSG-9 of Germany. It is a task-oriented Force and has two complementary elements in the form of the Special Action Group (SAG) comprising Army personnel and the Special Ranger Groups (SRG), comprising personnel drawn from the Central Para Military Forces/State Police Force. All the personnel are on deputation."). *See also* CATHY SCOTT-CLARK & ADRIAN LEVY, THE SIEGE: 68 HOURS INSIDE THE TAJ HOTEL 248 (2013).

416 MARK GALEOTTI, RUSSIAN SECURITY FORCES AND PARAMILITARY FORCES SINCE 1991, at 35–40 (2013).

417 *Id.* at 39.

418 *Mission Impossible: German Elite Troop Abandons Plan to Free Pirate Hostages*, SPIEGEL ONLINE (May 4, 2009) [hereinafter *Mission Impossible*], http://www.spiegel.de/international/germany/mission-impossible-german-elite-troop-abandons-plan-to-free-pirate-hostages-a-622766.html.

available aircraft and ships to allow GSG-9 to deploy swiftly enough.[419] What is important to emphasize is that regardless of whether this counterpiracy hostage rescue mission had been conducted by a military or police unit, the operation would have fallen under a human rights law enforcement framework.

Some missions may involve the deployment of both military and police counterterrorist units. For example, following attacks by jihadists on journalists and police in Paris, France in January 2015, both the military gendarmerie-based GIGN and the National Police, Research Assistance Intervention Dissuasion (RAID) acted to resolve two simultaneous hostage incidents.[420] Similarly, during the 2008 Mumbai terrorist attack, India deployed its Marine Commando (MARCO) counterterrorism force, the NSG, the Maharashtra State Anti-Terrorism Squad (ATS), as well as local police to counter the attack of ten Lashkar-e-Tabia (LeT) terrorists.[421] Military MARCO squads with an NSG team entered the Taj Mahal hotel to rescue hotel guests and defeat the terrorists, although it is reported that the MARCO units were limited in the tasks they could be assigned as they were not trained or equipped for the complex operation that was required to clear the hotel.[422] Similarly, MARCO and NSG teams assaulted the Oberoi Trident hotel, while police and firefighters rescued hotel guests.[423] At a third site, the Jewish center at Nariman House, the LeT terrorists seized hostages and engaged in negotiations. Police rescued a number of hostages, however, fearing that the remaining ones would be killed, the NSG assaulted the building. The terrorists were killed in a room-to-room gun battle. Unfortunately, all of the hostages held there "were found dead: they had been horrifically tortured before being murdered."[424]

The employment of military Special Forces units to conduct what are fundamentally law enforcement missions highlights their ability to operate within the human rights paradigm. This is not to say there is no difference between military and police units, and even within the military units themselves. The "Tier 1" special mission units of the United States military (e.g. SEAL Team 6, the Army Delta Force) provide different capabilities than non-covert special units (e.g. Rangers),[425] and "Special Forces" units (the "Green Berets").[426] The United States Marines has also developed a hostage rescue capability.[427] Generally military Special Forces units are normally better equipped than specialized police units to execute

---

[419] *Id.*

[420] *France's Rescue Operations, supra* note 164.

[421] David Kilcullen, Out of the Mountains: The Coming of Age of the Urban Guerrilla 57–60 (2013).

[422] Scott-Clark & Levy, *supra* note 415, at 207.

[423] *Id.* at 59.

[424] *Id.* at 60.

[425] Naylor, supra note 141, at 288–89 (for reference to the various "tiers" of special forces capability). *See also* Bernard J. Brister, *Canadian Special Operations Forces: A Blueprint For The Future*, Canadian Military Journal 29 (2004), http://www.journal.forces.gc.ca/vo5/no3/special-02-eng.asp (for a discussion of different allied Special Forces structures, and Tier 1 models).

[426] Kevin Maurer, Gentlemen Bastards: On the Ground with America's Elite Special Forces 15–16 (2012).

[427] Hope Hodge Seck, *MARSOC remains a growing, changing force after 9 years*, Marine Corps Times (Feb. 21, 2015, 11:38 EST), http://www.marinecorpstimes.com/story/military/2015/02/21/marsoc-at-9-anniversary/23607665/, and Dan Lamothe, *Marine Corps realigns its Special Operations, sends elite troops to Middle East*, Wash. Post, Jan. 20, 2015, https://www.washingtonpost.com/news/checkpoint/wp/2015/01/20/marine-corps-realigns-its-special-operations-sends-elite-troops-to-middle-east/.

hostage missions internationally that involve hostilities (e.g., Entebbe and Sierra Leone). They can also participate in a wide range of other direct action missions across the armed conflict spectrum. In contrast, police units can offer a different approach towards the use of force, and normally provide specialized investigatory skills.[428] This was seen in the contribution of an FBI HRT member for the purposes of protecting the integrity of the crime scene, collecting evidence, and presenting that evidence in court after the unsuccessful hostage rescue mission against pirates who had seized the *SV Quest* off Somalia in 2011.[429] It is indicated that an FBI HRT member was on the scene during the rescue of Jessica Buchanan by U.S. Navy SEALs."[430]

Specialized police units may also have a wartime role. This can be seen in the presence of FBI HRT members on Special Operations Forces raids against Al Qaeda in Afghanistan and Iraq. The FBI expertise extended to exploiting digital media and other media to locate terrorists, interrogations, sensitive site exploitation, as well as preserving evidence and maintain a chain of custody "should any suspect be transferred to the United States for trial."[431] The support to military forces can extend to other hostage rescue tasks. A HRT member also appears to have participated in the unsuccessful October 2013 Navy SEAL operation that stormed a compound in Somalia in pursuit of a suspect following the Westgate Mall attack.[432] Two FBI agents were also part of the U.S. Army Delta Force team that seized Abu Anas al-Libi, in Tripoli, Libya, that same weekend.[433] This police involvement highlights the degree to which international criminal law enforcement and conduct of hostilities have begun to converge.

At times controversy has arisen regarding differences between military and police approaches toward the use of force. That controversy has often centered on allegations of a more aggressive "shoot to kill" approach being adopted by military forces than police units (see Chapter 13, section 13.2.2, *Shooting to Kill*). While there may be operational pressures placed on the law enforcement normative regime because of the violence involved, that does not mean its fundamental tenets linked to the policing mission must be changed. It is reported that the FBI HRT members in Iraq contemplated but did not change their rules of engagement or deadly force policy because they were not the "door kickers."[434] However, it is clear that in terms of certain counterterrorism operations, such as hostage rescue, there has been a significant convergence of roles, tactics, and force applied when employed in either an armed conflict or policing role.

---

[428] AROSTEGUI, *supra* note 408, at 204 (for a discussion of the different shooting and training approaches for the GIGN and GSG-9 "police" units in the 1990s compared to other Special Forces units).

[429] *The Hostage Rescue Team Part 6: Mission in the Gulf of Aden*, THE FEDERAL BUREAU OF INVESTIGATION (Mar. 27, 2013), http://www.fbi.gov/news/stories/2013/february/the-hostage-rescue-team-30-years-of-service.

[430] BUCHANAN ET AL., *supra* note 373, at 267.

[431] Adam Goldman & Julie Tate, *Inside the FBI's Secret Relationship with the Military's Special Operations*, WASH. POST, Apr. 10, 2014, http://www.washingtonpost.com/world/national-security/inside-the-fbis-secret-relationship-with-the-militarys-special-operations/2014/04/10/dcca3460-be84-11e3-b195-dd0c1174052c_story.html. *See also* Schmidle, *supra* note 170 (where it is noted that for a July, 2014 hostage rescue mission in Syria the FBI not only provided the initial intelligence, but was given captured cell phones and seized blankets for forensic exploitation).

[432] Goldman & Tate, *supra* note 431.

[433] *Id.*

[434] *Id.*

The degree to which there has been such convergence is reflected in Israeli hostage rescue units. In the aftermath of the 1974 Ma'alot terrorist attack, the Israeli police hostage rescue unit, Yamam, was established.[435] This appears to have placed that unit in direct competition with the Sayeret Matkal, the elite Israeli military Special Forces unit. It is reported that the tension that initially existed between those two units abated after 2002. The mass hostage takings by Chechen terrorists at the Moscow theatre (2002) and at a school in Beslan (2004) in Russia prompted joint training sessions between the Israeli units.[436] At that time Israel was also confronted with the second intifada, which with its insurgency levels of violence likely prompted further cooperation. Apparently, the rivalry between the military and police units were resolved by assigning Yamam as the main hostage rescue unit. The Sayeret Matkal's "primary assignment is the field of special operations, and that at the same time they will serve as the secondary unit and assist the Yamam in hostage rescue situations when required . . . ."[437] Like a number of specialist police counterterrorism units worldwide, the activities of the Yamam do not appear to be limited to Israel or the terrorist threat on its borders. It is reported that "a team from Israel's elite counter-terrorism unit had assisted the Kenyan authorities in handling the hostage crisis" arising from the September 2013 Westgate Mall attack.[438]

Yamam's role appears to not be limited to hostage rescue. In 2007 it was disclosed, "during the second intifada the Yamam managed to kill some 50 terrorists en route to suicide bombings and also killed an additional 129 wanted terrorists."[439] The unit also arrested 550 terrorist suspects for questioning.[440] The capture of terrorists was claimed to be the goal of an Israeli military Special Forces unit operating in the Occupied Territories, the Duvdevan, which often works undercover disguised as Arabs.[441] The performance of that role included not using "lethal force unless it is necessary,"[442] avoiding forced entries into homes[443] and training with shooting drills to "make every effort to avoid collateral casualties."[444] These restrictions on the use of force resonate with a human rights–based law enforcement approach, whether applied as a matter of policy or law. The goal is to achieve a peaceful

---

435 Rebecca Anna Stoil, *Yamam Hostage Rescue Team's Tactics Revealed*, JERUSALEM POST (Jan. 16, 2007) [hereinafter *Yaman tactics*], http://www.jpost.com/Israel/Yamam-hostage-rescue-teams-tactics-revealed. *See also* Bar-Zohar & Mishal, *supra* note 202, at 257-58 (for reference to Yamam carrying out a 1988 hostage rescue on a bus seized by Arab terrorists).

436 Stoil, *supra* note 435.

437 Or Heller, *Rescuing Hostages*, ISRAEL DEFENCE (Aug. 24, 2013), http://www.israeldefense.co.il/en/content/rescuing-hostages.

438 Richard Norton-Taylor & Vikram Dodd, *Nairobi Attack: Israel Takes Lead Role in Advising Kenyan Forces*, THE GUARDIAN, Sept. 23, 2013, http://www.theguardian.com/world/2013/sep/23/nairobi-attack-israel-advising-kenyan-forces and Yaniv Kubovich, *Police Counterterrorism Unit Gets Green Light to Operate Outside of Israel*, HAARETZ (Sep. 15, 2014, 9:53 PM), http://www.haaretz.com/news/diplomacy-defense/.premium-1.615955. *See also* Daniel Howden, *Terror in Westgate Mall: The Full Story of the Attacks that Devastated Kenya*, THE GUARDIAN (Oct. 4, 2013, 3:12 BST), http://www.theguardian.com/world/interactive/2013/oct/04/westgate-mall-attacks-kenya-terror (where it is noted personnel from London's Metropolitan Police antiterrorism unit were also present).

439 Stoil, *supra* note 435.

440 *Id.*

441 AARON COHEN, BROTHERHOOD OF WARRIORS 198–99 (2008).

442 *Id.*

443 *Id.* ("We would always prefer to use the former method [knocking on a door, announcing they were the IDF and asking the occupants to come out]; explosives and gunplay increase the odds of fatality on both sides.").

444 *Id.* at 126.

surrender, however, it is also clear the philosophy was to react with overwhelming force against any terrorists refusing to surrender and attempting to kill members of the unit.[445]

It is the unique capabilities of Special Forces that provide States options extending to operating internationally under a law enforcement framework even in an armed conflict. That capability is not limited to hostage rescue missions. It can also seen in respect of the option that is frequently presented to capture rather than kill a terrorist suspect. While the May 2011 bin Laden raid by U.S. Navy SEALs was not planned specifically as a capture mission, it provides the most graphic demonstration of how far this counterterrorism capability has evolved. It was a raid that can claim approbation similar to that of the Entebbe and Sierra Leone missions as a remarkable feat of arms. The bin Laden mission highlights the extent to which, under the right circumstances, outcomes can be provided during armed conflict that closely resemble those mandated under a law enforcement paradigm. It is a mission that also indicates that it is the ability to control the immediate area of operations which makes capturing rather than killing a terrorist a feasible option (see Chapter 7, section 7.4.2, *The "Targeted Killing" Case and Control*).

Like the Entebbe and Sierra Leone raids, the 2011 bin Laden mission was a bold operation. It involved the air insertion of U.S. special operations forces and supporting personnel in an assault on a compound in Abbottabad, Pakistan. The operation was designed to neutralize Osama bin Laden as the senior leader of an opposing organized armed force (i.e., Al Qaeda), and included the possibility of capture.[446] In contrast to the Entebbe operation, the assault was not against an objective where hostages were being closely protected by terrorists or by State armed forces. This decreased the likelihood there would be fighting in order to gain entry to the objective. However, the similarity of capture and rescue missions has led to one analyst of such U.S. activity since World War II to describe them collectively as "liberation and seizure operations."[447] Covert access is invariably a desired goal of both hostage rescue and "capture" operations. As was demonstrated in the bin Laden operation, covert access might also be an essential part of direct action mission designed to kill a terrorist leader. The lack of any hostages in that situation probably removed some pressure concerning the maintenance of the element of surprise. However, the likelihood of having to confront armed members of Al Qaeda at the bin Laden compound and the chance of alerting Pakistani authorities who had not been warned of the attack no doubt enhanced the risk of an armed confrontation once that element of surprise was lost. Further, although there were no hostages, the anticipated presence of women and children in the bin Laden compound still mandated the application of targeting precautions regarding the use of force.

What is perhaps most significant in terms of tactics employed during the raid is that, after a decade of war in Iraq and Afghanistan, the covert entry of compounds had become such a routine part of U.S. Special Forces and other Coalition forces operations.[448] Those operations were often numerous and repetitive. It is reported that in Iraq operations by

---

[445] *Id.* at 198–99.

[446] Mark Lander & Helene Cooper, *New U.S. Account Says Bin Laden Was Unarmed During Raid*, N.Y. Times (May 3, 2011), http://www.nytimes.com/2011/05/04/world/asia/04raid.html.

[447] Isby, *supra* note 112, at 13–20.

[448] Eric Schmitt & Thom Shanker, Counterstrike: The Untold Story of America's Secret Campaign Against Al Qaeda 259 (2011).

Special Forces Task Force 714 to capture or kill suspected Al Qaeda personnel had gone from 18 raids in August 2004 to the point where "in August 2006, we were up to 300 raids a month—ten a night."[449] These missions included capturing rather than killing fighters.[450] It is also reported a 2008 raid reached into Syria to kill a senior Al Qaeda militant located at a "safe house." The target, Abu Ghadiya, was responsible for running a foreign fighter network into Iraq.[451] One participant in the bin Laden raid noted "[t]his target wasn't any more complicated than hundreds of others we'd assaulted over the years."[452] The questions that often arise during contemporary operations even during armed conflict are: Does the tactical situation permits that capture as an option, and is such an outcome a desired goal? The decision to use covert entry, thereby establishing a significant level of control over the compound, rather than a kinetic option such as aerial strikes with drones or aircraft, no doubt contributed to the post-raid discussion about whether Osama bin Laden was armed, and if he could have been captured rather than killed.[453]

Placing "boots on the ground," this time literally in bin Laden's living space, can create different expectations than the use of standoff weapons. To be clear, under humanitarian law the head of Al Qaeda, an organized armed group engaged in armed conflict, was a lawful target.[454] As the operation unfolded Osama bin Laden was shot and killed. However, there would have been a requirement to detain him, if it was established he was in fact surrendering or was otherwise *hors de combat*. It has been reported that, "if bin Laden were to conspicuously surrender, the rules of engagement of the SEALs adhered to meant he would have to be taken into custody."[455] A lawyer is also reported to have indicated during a pre-operation briefing to the assault team, "if he does not pose a threat, you will detain him."[456] This is consistent with other reports in the aftermath of the operation that the question of capturing or killing bin Laden was threat based. As the president's chief counterterrorism advisor is reported to have stated, "[i]f we had the opportunity to take Bin Laden alive, if he didn't present any threat, the individuals involved were able and prepared to do that."[457]

The tactics used by the SEALs took on even more of a human rights–based law enforcement flavor with targeting rules being based on the threat presented, rather than being status based. Discussions among State legal advisors and certainly military lawyers regarding the killing of bin Laden will undoubtedly focus on the shooting on sight being authorized under humanitarian law, with any decision to capture being a policy decision. That

---

449 *Generation Kill: A Conversation with Stanley McChrystal*, 92 FOREIGN AFF. 2, 4 (Mar./Apr., 2013) [hereinafter *Generation Kill*]. *See also* NAYLOR, supra note 141, at 288.

450 *Generation Kill, supra* note 449. *See also* NAYLOR, supra note 141, at 260 (where it is noted that "[a]s happened with most task force missions, the operators usually snatched the wanted individual without a fight.").

451 Seán D. Naylor, *Killing Abu Ghadiya*, FOREIGN POL'Y (Aug. 31, 2015), https://foreignpolicy.com/2015/08/31/deltaforce-seals-syria-iraq-terrorism/?utm_source=Sailthru&utm_medium=email&utm_campaign=New%20Campaign&utm_term=%2AEditors%20Picks.

452 OWEN, *supra* note 367, at 174.

453 Lander & Cooper, *supra* note 446 (for the discussion regarding the feasibility of capture).

454 CHARLIE SAVAGE, POWER WARS: INSIDE OBAMA'S POST 9/11 PRESIDENCY 266-67 (2015)(indicating the same conclusion was reached by United States administration lawyers).

455 PETER L. BERGEN, MANHUNT: THE TEN YEAR SEARCH FOR BIN LADEN FROM 9/11 TO ABBOTTABAD 186 (2012). *See also* SAVAGE, *supra* note 454, at 266–68.

456 OWEN, *supra* note 367, at 177.

457 Lander & Cooper, *supra* note 446.

determination is correct at law. However, regardless of how it is legally framed, this operation graphically demonstrates the evolving nature of many counterterrorist missions. In some circumstances they have become increasingly like law enforcement operations concerning the manner in which they are conducted. Sufficient control is often established at the scene of the operation, even if only for a limited period of time, so that the capture of an "unprivileged belligerent" is not only possible but routinely carried out.

Nor is threat-based targeting limited to the U.S. military or the bin Laden operation. There appears to have been a subtle shift away from status-based targeting to a threat/no threat decision-making process on the part of other personnel conducting raids.[458] Such an approach has been reported to be the one adopted by Canadian Special Forces operating in Afghanistan. Their commander is said to have explained, an "operator will apply the same standards he would in Toronto for a domestic event. This means that CANSOFCOM operators meet, and frequently exceed, the standards for laws of armed conflict."[459] This shift is reflective of the type of conflict (i.e., insurgency/counterinsurgency), as well as the influence that has been exerted by greater attention being paid overall to human rights–based norms.

It is important to note the change in operational approach toward a police-type response has not been not limited to Special Forces operations (see Chapter 12, section 12.2.1, *The Military Task: Applying Force in the First and Last Resort*). In Afghanistan, due to pressures associated with counterinsurgency, on-site tactics for some operations conducted by Special Forces developed to the point of surrounding a building and calling for occupants to surrender, "similar to what police did in the United States."[460] The reality is that many contemporary operations exist at the intersection of the humanitarian and human rights law. At the heart of this discussion are often questions regarding the effectiveness of the more restrictive law enforcement approach, the management of risk to the lives of the soldiers and civilians, and the ability of security forces to accomplish their assigned mission. While there is a requirement to apply a law enforcement approach when conducting policing missions, its application as a matter of policy during armed conflict ultimately can depend upon an assessment of the advantages and limitations of the human rights–based framework.

---

[458] Eric Pouw, International Human Rights Law and the Law of Armed Conflict in the Context of Counterinsurgency 317 (Nov. 14, 2013) (unpublished Phd. Dissertation, University of Amsterdam) (on file with the author).

[459] Mercedes Stephenson, *Canada's Elite Commandos Aim to Capture, Not Kill*, TORONTO SUN, Dec. 12, 2010, http://www.torontosun.com/comment/columnists/mercedes_stephenson/2010/12/10/16509801.html. *See also* Pouw, *supra* note 459, at 317.

[460] OWEN, *supra* note 368, at 141. *But see* NAYLOR, *supra* note 141, at 369–70 (indicating that the "call-out" approach was "tremendously unpopular with strike forces" because it provided an opportunity to destroy valuable intelligence).

[S]ometimes the violence offered to us will admit of no other remedy than the exertion of open force; and under such circumstances we may lawfully have recourse to it.[1]

11 Law Enforcement and "Self-Defense"

11.1. Introduction

It is evident in a number of contemporary security operations that States will act internationally in a law enforcement context to protect their citizens abroad. During armed conflict, particularly when conducting counterinsurgency or counterterrorism operations, States will also apply a human rights–based law enforcement approach to govern some uses of force. This can be required as a matter of law, and in other cases a policy choice may be made to privilege a policing response. A defensive law enforcement response may also be applied in reaction to a transnational armed attack by a non-State actor. The result can be that human rights–based law enforcement is frequently applied in the context of an "overarching" or "limited" application of self-defense principles (see Chapter 3, section 3.3, *An "Overarching" Application*, and section 3.4, *A More "Limited" Theory*). This raises the question of how the normative framework governing State self-defense interacts with the human rights–based paradigm.

Given the strategic-level application of the law governing State self-defense, it could be expected the interface between that body of law and human rights–based law enforcement would operate in a manner similar to its interaction with humanitarian law (see Chapter 3, section 3.6.3, *Applying Self-Defense Principles Across the Armed Conflict Spectrum*). The principles governing the State exercise of self-defense (e.g., proportionality, necessity, imminence, and immediacy[2]) will broadly influence the scope of State action is taken, however,

[1] Emer de Vattel, The Law of Nations 676 (Knud Haakonssen ed., 2008) (1797).

[2] Tallin Manual on the International Law Applicable to Cyber Warfare, at 61–66 (Michael N. Schmitt ed., 2013) [hereinafter Tallin Manual] (for a discussion of State self-defense principles).

it is the law enforcement framework that controls how force is actually used at the tactical level.[3] Further, the relatively low levels of violence traditionally associated with law enforcement means it is logical to view that paradigm as operating comfortably within an overarching legal framework governing the exercise of State self-defense.

While fairly straightforward in theory, an assessment of the interface between the State self-defense and human rights–based bodies of law requires consideration of a number of complicating factors. These factors include both paradigms being "self-defense" oriented and the confusion that can arise because of the use of common terminology (i.e., necessity, proportionality, and imminence). Particularly challenging in assessing how force is controlled at the tactical level is the dominant position that the right to act in self-defense in its recourse to war form has attained in the international law dialogue regarding the use of force. How broadly or narrowly that right is assessed can have a significant impact on how either humanitarian law or human rights law are interpreted to apply. Unfortunately, the international legal community has paid relatively little attention to the application of law enforcement–based norms beyond State borders or the interface between that human rights–based paradigm and national self-defense. The nature of twenty-first-century transnational threats (terrorist groups, transnational criminal organizations, and ordinary criminal gangs), the maturity of the human rights normative framework, and the enhanced capability of State police and other nonmilitary security forces to operation internationally is forcing greater consideration of how the policing framework interacts with State self-defense.

The chapter looks at this twenty-first-century challenge by first exploring the theoretical interface between State self-defense and the human rights law framework. This will include an overview of the common roots of the law enforcement and State self-defense paradigms, followed by an assessment of their different application as these bodies of law have developed. Next, the historical internal security role for human rights–based law enforcement will be assessed in comparison to its more internationally focused humanitarian law counterpart. What becomes evident is that as the law enforcement paradigm developed it became associated with a relatively narrow authorization to use force. That authority is based on a restrictive interpretation of when a threat is sufficiently imminent to justify action in personal or individual self-defense. It will be demonstrated that looking at law enforcement solely through this narrower self-defense lens can be misleading. Although the law enforcement paradigm is focused primarily on the defensive use of force, the additional authority to use force required in order to enforce the law will be explored. It is the authority to use force for what is often termed "mission accomplishment" purposes that can separate the law enforcement paradigm from the narrower human rights–based notion of individual self-defense.

The analysis then turns to one of the most significant challenges that arise in reconciling the interaction between the State self-defense and law enforcement paradigms. Strategically focused self-defense paradigms can exert a dominant influence over the assessment of the

---

[3] Gary P. Corn, *Should the Best Offense Ever Be a Good Defense*, Strategy Research Project 26–27 (Carlisle Barracks, PA: U.S. Army War College, Apr. 15, 2014) ("In contradistinction to the *jus ad bellum*, however, the LOAC and IHRL frameworks regulate the force states may use at the micro level through their designated agents, under color of state authority, against individual human beings.").

exercise of self-defense at the tactical level. This can include an interpretation of State self-defense according to narrow principles that seem to fit best within a human rights law paradigm. The confusion that potentially arises from the simultaneous application of the State and the human rights–based self-defense paradigms will be explored in the context of rules of engagement (ROE) doctrine, UN peacekeeping, and the U.S. Standing Rules of Engagement (SROE) framework.[4] What becomes apparent is that the focus on State self-defense can mask consideration of how human rights–based law enforcement norms should be applied during international operations. This becomes evident in two very different examples: the development of "robust" peacekeeping and the application of ROE during the U.S.-led fight for Fallujah in November 2004.

Interpretive approaches that permit an overly narrow or broad interpretation of self-defense principles can be problematic. In a peacekeeping context this can potentially lead to overly restrictive interpretations being applied in terms of what force military forces can apply during UN operations. Conversely, the Fallujah example points to the opposite result where broad interpretations of State self-defense principles may be permitted to influence what should be more individualized exercises of self-defense. Finally, the application of law enforcement rules in the international security environment will be discussed in order to highlight some of the contemporary pressures being exerted on a body of law that was primarily designed to regulate internal State order during times of relative peace.

11.2. Distinguishing Law Enforcement

11.2.1. THE TERMINOLOGY CHALLENGE

One of the challenges in assessing the scope of human rights–based law enforcement applied on international operations or in response to transnational attacks is the use of terminology that is common to both the law governing the exercise of State self-defense and that controlling the conduct of hostilities. For example, both the State self-defense framework and human rights–based law enforcement apply principles of proportionality, necessity, and imminence.[5] Of course this is not the only source of terminological confusion since proportionality is also a key concept applied under humanitarian law to regulate targeting.[6] This use of common terminology presents a number of challenges for international law. Due to their general application, these terms often lack precision.[7]

---

[4] CHAIRMAN OF THE JOINT CHIEFS OF STAFF INSTR. 3121.01B, STANDING RULES OF ENGAGEMENT (SROE)/ STANDING RULES FOR THE USE OF FORCE (SRUF) FOR U.S. FORCES (13 JUNE 2005), *in* OPERATIONAL LAW HANDBOOK, Chapter 5, Rules of Engagement Appendix A, at 90-104 (Lieutenant-Commander David Lee, ed., 2015) [hereinafter also CJCSI], http://www.loc.gov/rr/frd/Military_Law/pdf/operational-law-handbook_2015.pdf. *See also id.* at 81 (where the CJCSI includes the following notice: "NOTE: The current SROE are under revision and JAs [Judge Advocates] should obtain the latest version.").

[5] TALLIN MANUAL, *supra* note 2, at 61–66, and the *UN Basic Principles on the Use of Force*, at ¶ 9.

[6] AP I, art. 57(2)(a)(iii), and Public Committee Against Torture in Israel v. Israel, Israel Supreme Court [16 December 2006], 46 ILM 375, ¶¶ 42–46, at 395–96 (2007).

[7] Thomas M. Franck, *On Proportionality of Countermeasures in International Law*, 102 AM. J. INT'L L. 715, 716 (2008) ("the principle of proportionality has mostly eluded definition in any but the most general terms.").

While on one level, principles of broad application such as proportionality offer flexible analytical tools for the resolution of complex issues,[8] one of the frustrations for the practitioner is the lack of a clear definition. This can undermine the ultimate usefulness of the principle. In a humanitarian law context the assessment of proportionality has been equated to dealing with an "apples and oranges" phenomenon[9] of attempting to "compare tanks destroyed to the number of serious civilian injuries or deaths. . . ."[10] In respect of exercising the right of State self-defense, the proportionality assessment can involve a comparison of disparate factors such as the numbers of dead and destruction of property against the protection of sovereignty. This is said to be a task that requires "the comparison of incommensurables."[11] Further complicating the situation is that while the same term, "proportionality" is used in different contexts, there remains an element of commonality to what is considered: the relevant importance of the death and destruction resulting from the use of force. In application the term proportionality often becomes a form of fig leaf masking deeper ethical and moral issues associated with the use of force in entirely different contexts.

The use of the same terms hints at the common genesis of these bodies of law but also highlights the degree to which these normative frameworks have come to be seen as being separated by almost inviolable legal boundaries. Each body of law developed the application of the same "foundational" principle in a different way. The problem that arises in the contemporary security context, which requires a holistic application of the law, is that little guidance is provided on how to apply common terminology when these traditionally separated bodies of law intersect and overlap. Using the same terminology is less of a problem if it is considered there is no overlap or perhaps even interface in the application of the various legal domestic and international legal disciplines. However, such a separation simply breaks down in the reality of contemporary operations. In this respect one is reminded of Field Marshal Helmuth von Moltke's adage that "[n]o plan survives first contact with the enemy."[12] This is exactly what occurs when these legal theories meet the operational reality of contemporary conflict. At the points of interface of these bodies of law, their underlying principles inevitably have to be reconciled with one another. Common terminology complicates that challenge.

The theory that these various bodies of law operate entirely separately does not survive the practical challenges of fighting insurgencies and terrorism. One particularly challenging area is where human rights–based norms are suggested to be the appropriate framework to govern the use of force by a State in defense of its interests or nationals. Given the frequency with which these operations are now occurring and the gradual, if reluctant, acceptance by the international community that States will act in self-defense against non-State actors, it becomes important to discern how these principles actually work together. To understand the interface and overlap of the policing and State self-defense paradigms it is

---

[8] Aharon Barak, Proportionality: Constitutional Rights and Their Limitations 467 (2012).

[9] Franck, *supra* note 7, at 729.

[10] Michael N. Schmitt, *Faultlines in the Law of Attack*, *in* Testing the Boundaries of International Humanitarian Law 277, 293 (Susan Breau & Agnieszka Jachec-Neale eds., 2006).

[11] David Rodin, War & Self-Defense 115 (2002).

[12] Robert Debs Heinl, Jr., Dictionary of Military and Naval Quotations 239 (1966).

important to first consider both their common roots and how they have developed to meet different goals.

11.2.2. SELF-DEFENSE: COMMON ROOTS AND DIFFERENT GOALS

The challenge of de-conflicting the operation of law enforcement norms and State self-defense principles is not limited to simply assessing the meaning of necessity, proportionality, and imminence. There is a deeper relationship between the concept of individual self-defense, which underpins the use of force in the human rights context and the use of defensive force by States. This is based on a close historical connection between the principles applied under both bodies of law. In this regard it has been noted "the structure of international law suggests a strong analogy between individual self-defence and national-defence . . . ."[13] As Terry Gill has noted, self-defense under international law "shares a common legal origin with other manifestations of self-defence, such as the right of personal self-defence under national criminal law, both of which trace their origins to the natural law doctrine of just war tradition."[14] In ethical terms much of the debate regarding the State right of self-defense has been framed in the context of personal self-defense, from which "it gains enormous intuitive legitimacy."[15] It is exactly that, a debate by analogy, which "has had a profound influence on just war thinking and is explicitly invoked by both international lawyers and philosophers."[16]

However, the two legal regimes have developed along different paths. It is suggested that national self-defense is a "separate legal regime which is relevant only to States and is governed solely by international law criteria and conditions contained in the UN Charter and in customary international law."[17] Where it is viewed that the State also provides the authority for the exercise of unit and personal self-defense, a distinction is made between the recourse to war basis for national self-defense and that applicable to the other defensive force applied in the course of official duties.[18] Under a theory suggested by Gary Corn, that authority derives from the "public authority doctrine," which provides that acts such as the taking of life, detaining someone, or destroying property are not crimes if done with the proper public authority.[19] That doctrine can include limitations based in domestic and international human rights law.[20] One example of the different legal approaches is that under domestic law "the test of necessity is applied throughout the period of defensive action."[21] The result is that if a threat can be removed with a lesser level of force "it would

---

[13] RODIN, *supra* note 11, at 110.

[14] Terry D. Gill, *Legal Basis of the Right of Self-Defence under the UN Charter and under Customary International Law*, *in* THE HANDBOOK OF THE INTERNATIONAL LAW OF MILITARY OPERATIONS 187, 187–188 (Terry Gill & Dieter Fleck eds., 2010) [hereinafter HANDBOOK OF INTERNATIONAL OPERATIONAL LAW].

[15] RODIN, *supra* note 11, at 107.

[16] *Id. See* YORAM DINSTEIN, WAR AGGRESSION AND SELF-DEFENCE ¶ 497, at 188 (5th ed. 2011), and MICHAEL WALZER, JUST AND UNJUST WARS 58 (3rd ed. 2000) (for reference to the "domestic analogy").

[17] Gill, *supra* note 14, at 188.

[18] Corn, *supra* note 3, at 13 ("The point is that the *jus ad bellum* [law governing the recourse to war] is not the normative framework from which the authority stems for sub-national self-defense.").

[19] *Id.* at 20–21.

[20] *Id.* at 22.

[21] RODIN, *supra* note 11, at 111.

not be permissible to proceed with the infliction of a fatal blow."[22] That is not necessarily the approach applied to the use of force on an international level. For proponents of a more limited application of self-defense principles (see Chapter 3, section 3.4, *A More "Limited" Theory*), the test of necessity is only applied at the start of the conflict rather than throughout the war. As a result a State "may prosecute its war to final victory even after the point at which this is no longer necessary to reverse or frustrate the initial use of force which provided the justification for the war."[23]

In the same manner as with its relationship with humanitarian law, the law governing State self-defense operates at a strategic level influencing when and the degree to which force under the law enforcement framework is used (see Chapter 3, section 3.6.3, *Applying Self-Defense Principles Across the Armed Conflict Spectrum*). The principles governing the use of force during law enforcement operations can be viewed as being separate from, but nestled within, the overall State self-defense normative framework. In comparing these bodies of law it has been suggested that "[t]he purposes of *jus ad bellum* [law governing the recourse to war] are different from those of *jus in bello* [humanitarian law] and human rights, and the threshold for use of force under human rights law (outside of armed conflict) is higher than the threshold for use of force in the *jus in bello*."[24] A key distinguishing feature is that the use of force under a human rights law paradigm is "conduct" or "threat" based, rather than "status" (i.e., membership in an organized armed group) based.[25] Further, human rights law allows for the use of force outside the context of an armed conflict. The use of lethal force is only permitted when "the threat to life is 'imminent' and cannot reasonably be ameliorated by other means, such as detention."[26] Reflecting its operation at the tactical level, human rights–based law enforcement norms are more narrowly focused on the use of force by individual security personnel. This can be seen in the UN *Basic Principles on the Use of Force and Firearms by Law Enforcement Officials*, which indicates they shall not use firearms against persons except in self-defense or defense of others when faced with imminent threats.[27]

The relationship between the law governing State self-defense and law enforcement does not fundamentally change if security officials apply domestic law principles to enforce the law during international operations. In this respect, the International Tribunal for the Former Yugoslavia ruled in *Prosecutor v. Kordić* that the self-defense principle enshrined in Article 31(1)(c) of the Rome Statute "reflects provisions found in most national criminal codes and may be regarded as constituting a rule of customary international law." [28] The international and domestic human rights legal principles applied during law enforcement are substantially same, even taking into account variations in domestic law.[29] For example, after reviewing whether the ROE for an operation enforcing the blockade of the Gaza strip by Israeli naval forces reflected international requirements for law enforcement, the

---

22 *Id.* at 111–12.
23 *Id.* at 112.
24 Gabor Rona & Raha Wala, *No Thank You to a Radical Rewrite of the Jus ad Bellum*, 107 AM. J. INT'L L. 386, 387 (2013).
25 Corn, *supra* note 3, at 28.
26 Rona & Wala, *supra* note 24, at 387.
27 *UN Basic Principles on the Use of Force*, at ¶ 9.
28 Prosecutor v. Kordić, Case No. IT-95-14/2-T, Judgment, ¶ 451 (Int'l Crim. Trib. for the Former Yugoslavia, 2001), http://www.icty.org/x/cases/kordic_cerkez/tjug/en/kor-tj010226e.pdf.
29 Corn, *supra* note 3, at 32.

Public Commission to Examine the Maritime Incident of 31 May 2010 noted that "[t]he same principles are prevalent in Israeli domestic law governing the use of force in self-defense."[30] Regardless of whether international or domestic human rights norms are applied, a key characteristic separating these bodies of law from State self-defense or conduct of hostilities norms is the unique role performed by law enforcement.

11.2.3. LAW ENFORCEMENT AND ITS UNIQUE ROLE

In determining the degree to which a law enforcement framework can offer an effective option to control the use of force on international operations, it is helpful to look at how it developed in comparison to humanitarian law. An important issue is whether the human rights–based constraints on the use of force permit security forces to successfully address the wide range of threats that can arise during contemporary operations. This discussion inevitably raises an issue that has traditionally represented the fault line between the human rights– and conduct of hostilities–based frameworks. That is the greater acceptance of collateral damage in a humanitarian law context. The traditional law enforcement paradigm is much more constrained in terms of the levels of force authorized and the acceptance of collateral effects.[31] It provides that the "intentional lethal use of firearms may only be made when strictly unavoidable in order to protect life."[32] It is focused on self-defense and defense of others, as well as the arrest of a suspect. Force is to be used "only when less extreme means are insufficient to achieve these objectives."[33] In contrast, the conduct of hostilities norms under humanitarian law acknowledges there is a requirement during armed conflict to authorize a more liberal use force with potentially more significant collateral effects.

Historically, the emergence of these fundamental distinguishing factors for the two bodies of law was intimately tied to the authority wielded by the State to govern society (i.e., the proper authority) (see Chapter 4, section 4.4.1, *Public, Private, and Mixed War*). This did not mean that the use of deadly force was not a part of maintaining internal order. An initial pacifist desire to constrain the use of force, based on religious beliefs, was challenged by the realities of governance. This was evident when the Christian church assumed a governance role.[34] The acceptance of the use of deadly force, and its connection to governance is associated with just war theory. This just war concept "is elaborated and given authority by St. Augustine (A.D. 354–430), at a time when Christianity was linked with the secular power of the Empire."[35] The requirement to govern inevitably raised the issue of killing for the purposes of maintaining law, order and security. Intertwined in the evolving just war

---

30 1 PUBLIC COMMISSION TO EXAMINE THE MARITIME INCIDENT OF 31 MAY 2010, ¶ 210, at 246 (2011).

31 INTERNATIONAL HUMANITARIAN LAW AND THE CHALLENGES OF CONTEMPORARY ARMED CONFLICT, Report 32nd International Conference of the Red Cross and Red Crescent EN 32IC/15/11, at 16 (2015), *available at* https://www.justsecurity.org/wp-content/uploads/2015/11/2015-ICRC-Report-IHL-and-Challenges-of-Armed-Conflicts.pdf.

32 *UN Basic Principles on the Use of Force*, at ¶9.

33 *Id.*

34 IAN BROWNLIE, INTERNATIONAL LAW AND THE USE OF FORCE BY STATES 5 (1963) ("The early Christian Church refused to accept war as moral in any circumstances.... This period of extreme pacifism lasted for three centuries after Christ....").

35 *Id.*

debate were issues such as killing in individual self-defense, the death penalty for criminals and the application of the doctrine of double effect (e.g., targeting proportionality) in wartime.[36]

It is the development of the right to act in self-defense in the just war tradition and its application to a range of violence associated with governance that highlights how the internally focused law enforcement paradigm developed along a narrower path. As articulated by St. Thomas Aquinas, the justification for the use of force in individual self-defense was seen as the natural reaction to protect oneself (the good), which was not unlawful if the assailant was killed (the evil).[37] However, in the context of external threats this was subsequently expanded to "the killing of civilians in war when civilians are killed as part of the attempted achievement of some military goal."[38] In effect, the humanitarian law version of the proportionality principle represents an expanded version of individual self-defense. It arose in the context of the greater violence associated with intersocietal (i.e., primarily inter-State) conflict. This linkage to just war theory and the connection between individual self-defense and armed conflict in the development of the principle of proportionality again reflects the intertwined genesis of human rights and humanitarian law concepts that are now generally treated with the passage of time in a more exclusionary fashion.

Warfare between States is different. As Clausewitz noted, combat is "not a contest between individuals."[39] The dividing line between self-defense connected to governance and the authority to use force that actively contemplates collaterally killing uninvolved civilians, developed in relation to a fundamental distinction in the roles performed by the proper authority in relation to "private" and "public" war (see Chapter 4, section 4.4.1, *Public, Private, and Mixed War*). Internally, concerns over the power wielded by the State, including the use of force by its agents, led to the development of domestic human rights that restricted the actions of that authority when acting in respect of its own citizens. This created a more constrained environment within which security is maintained, and one that directly led to greater restrictions on the authority of State officials to use force internally. These restrictions on the use of force have become entrenched in what are now often referred to as law enforcement norms. They are reflected in international human rights law, which developed from its domestic counterpart (see Chapter 5, section 5.3.1, *Prior to the Mid-Twentieth Century*).

Externally, the State, as the proper authority under international law, was confronted by different threats. First, other "sovereigns" or States usually threatened a substantively increased level of violence due to their ability to harness the resources of the society they represented. As has been noted, "modern means of death and destruction would never have been possible without the state, its ministry of defense . . . and its regular, uniformed, bureaucratically managed armed forces."[40] Second, the threat from other States emanated

---

36 Colm McKeogh, Innocent Civilians: The Morality of Killing in War 62–65 (2002).

37 *Id.* at 64 ["which he saw as the natural reaction of a living organism (and which therefore could not be bad)."].

38 *Id.* at 65.

39 Carl Von Clausewitz, On War 95 (Michael Howard & Peter Paret eds. & trans., 1989) (Rather, "[w]arfare comprises everything related to the fighting forces—everything to do with their creation, maintenance and use.").

40 Martin Van Creveld, The Rise and Decline of the State 249 (1999). *See also* Ward Thomas, The Ethics of Destruction 61–62 (2001) (for a discussion of the rise of the mass army).

primarily from an external territory. In other words, from territory over which the potential target State exercised no control. The violence was mainly "cross border." The options for a threatened State were to submit or to act to reduce the military threat facing it including if necessary completely conquering the enemy territory. The armed forces of a State acted at the level of national self-defense. However, it also meant normally having to use greater levels of violence to defeat the army, navy, and later the air forces of the opposing State. These greater levels of violence, which were enhanced by the development of increasingly advanced weaponry, resulted from the development of warfare on an industrial scale. It is clear that inter-State violence normally occurred at a significantly different level than that normally associated with internal law enforcement.

Although humanitarian law was developed to restrain the impact of inter-State warfare, there was incorporated within its provisions an explicit acknowledgment of the likelihood of collateral damage. More permissive rules for the use of force, as well as the impact of such use, matched the reality of greater levels of violence being applied when States fought each other. This did not mean there was no room for the application of the more restricted law enforcement norms. Defeating the enemy would often mean taking over governance (e.g., as an occupying power), thereby putting a State in a position of being required to apply levels of force more closely associated with policing.[41] Further, specific humanitarian law treaty provisions have recognized the authority to use force in individual self-defense.[42] As was discussed in Chapter 4, conflict with groups other than States was largely left unaddressed in terms of treaty law until the end of World War II when Common Article 3 of the Geneva Conventions was applied to conflicts not of an international character.[43] It was not until 1977 that the international legal community developed the two Additional Protocols designed to regulate aspects of conflict with such groups. The default position has been for States to apply a human rights–based law enforcement approach to deal with non-State actors operating within their borders. While in an operational law context human rights law began to take on a significantly higher profile in the last two decades of the twentieth century (see Chapter 4, section 4.4.3, *A Reluctant Application of Humanitarian Law*), it is in the post-9/11 period that greatest attention has been paid to the application of both humanitarian law and human rights law on transnational operations involving non-State actors (see Chapter 5, section 5.3.2.3, *Divergence and Growing Reconciliation*).

The development of human rights–based law enforcement framework along different lines and for different reasons created a distinction that presents some of the greatest challenges for those seeking to apply human rights–based norms during international operations. Whether assessed on a domestic law basis or under international human rights law, law enforcement levels of force have been designed to control "internal" violence. If it is to be applied to conflicts rising above the level of "situations of internal disturbances and tensions, such as riots, isolated, and sporadic acts of violence and other acts of a similar nature,"[44] then it must be prepared to deal with the potential for a significant level of

---

[41] 1907 Land Warfare Regulations, art. 43, and GC IV art. 64.

[42] GC I, Arts. 22(1) and 22(2) (arming medical personnel and sentries, etc.); AP I, Arts. 65(3) and 67(1)(d) (arming persons for self-defense relating to civil defense). *See also* The 1998 Rome Statute, art. 31(1)(c).

[43] *See, e.g.*, GC IV, art. 3.

[44] AP II, art. 1(2).

collateral casualties and damage. It is not that a human rights–based approach may not in some circumstances tolerate at least some level of collateral death or injury of bystanders.[45] However, under that body of law "the use of force must avoid as far as possible deaths or injuries of bystanders, while the IHL [international humanitarian law] principle of proportionality prohibits only excessive incidental civilian losses."[46]

There is some indication that the contemporary terrorist threat is placing the traditional human rights approach of avoiding collateral death and injury under considerable stress. In a 2011 decision, *Finogenov v. Russia*, the European Court of Human Rights accepted that the death of 129 hostages at a Moscow theater, resulting primarily from the use of an incapacitating gas by security forces while attempting a rescue, "was not in the circumstances a disproportionate measure."[47] An estimated 730 hostages were rescued.[48] The court sought to distinguish between the use of such gas and situations involving the use of "bombs or air missiles" in Russian antiterrorist operations.[49] However, the view that collateral deaths of such magnitude might be justified under a human rights–based normative framework represents a marked departure from many common interpretations of how these norms restrain the use of force. It also raises the issue of the point at which applying human rights–based norms to counter violent threats posed by organized armed groups fundamentally changes how that body of law is viewed and applied (see Chapter 13, *Limits of Law Enforcement and Human Rights Law*).

While the unique role performed by law enforcement has its genesis in internal State policing, those norms are also applied on international operations in a number of contexts. For example, the requirement to deal with civilian unrest rather than engage in combat can arise in a myriad of international tasks such as protecting humanitarian relief supplies or providing security at embassies. For example, military forces may be used to protect embassies in foreign lands, as is evidenced by the U.S. employment of Marines to protect diplomatic and consular premises.[50] While local police and other security forces may provide an outer security layer, this does not completely remove the possibility that embassy security force personnel may have to interface with host nation civilians. The security threat can range from ordinary criminal activity and demonstrations to more aggressive threats.

---

45 Andronicou v. Cyprus, Application Nos. 86/1996/705/897, Eur. Ct. H. R. ¶ 194 (Oct. 9, 1997), http://hudoc.echr.coe.int/sites/eng/pages/search.aspx?i=001-58102, and Kerimova v. Russia, Application Nos. 17170/04, 20792/04, 22448/04, 23360/04, 5681/05 and 5684/05 Eur. Ct. H. R. ¶ 246 (May 3, 2011), http://hudoc.echr.coe.int/sites/eng/pages/search.aspx?i=001-104662 ("Those measures could presumably comprise the deployment of armed forces equipped with combat weapons, including military aircraft . . . and could entail, as a regrettable but unavoidable consequence, human casualties."). *See also* Gabrella Blum & Philip Heymann, *Law and Policy of Targeted Killing*, 1 HARV. NAT. SEC. J. 145, 148 (2010) ("In domestic law enforcement, the police must hold their fire if they believe that there is a danger to bystanders, except where using lethal force against a suspect if reasonably believed likely to reduce the number of innocent deaths.").

46 THE USE OF FORCE IN ARMED CONFLICTS INTERPLAY BETWEEN THE CONDUCT OF HOSTILITIES AND LAW ENFORCEMENT PARADIGMS, ICRC Report 9 (Nov. 2013), *available at* http://www.icrc.org/eng/assets/files/publications/icrc-002-4171.pdf.

47 Finogenov v. Russia, VI Eur. Ct. H. R. ¶ 236, at 409 (2011) (extracts). *See also, id.* at ¶ 8, at 374 (where it is indicated over 900 people were held hostage).

48 *Id.* ¶ 24, at 377 ("As a result of the operation the majority of hostages were released (over 730 people). The exact number is unknown . . . .").

49 *Id.* ¶ 232, at 408.

50 Kris Connor, *Marine Security Guard Battalion becomes Marine Embassy Security Command* (May 5, 2007), http://www.quantico.usmc.mil/sentry/StoryView.aspx?SID=938.

This reality combined with duties of an occupying power and the adoption of a police primacy approach in counterinsurgency operations means that military forces are increasingly engaging in law enforcement–style operations (see Chapter 12, section 12.2.1, *The Military Task: Applying Force of First and Last Resort*).

The rules governing the use of force must reflect the broad range of threats and tasks facing military units. As was noted regarding the 1979 takeover of the U.S. embassy by Iranian students, "[t]he Marine's Rules of Engagement did not permit shooting into a crowd of civilian demonstrators led by women who claimed they intended nothing more serious than a sit-in."[51] Such a human rights–based "law enforcement"–type restriction is understandable given that the situation was not one of armed conflict. However, the potential seriousness of the threat was also graphically demonstrated in 1983 when Hezbollah vehicle-borne improvised explosive device attacks were made on the American embassy[52] and the U.S. Marine and French paratroop barracks in Beirut, Lebanon.[53] This means foreign security forces may be required to conduct law enforcement–type crowd control activities, or in extreme circumstances even engage in hostilities with very little time or opportunity to distinguish the true nature of the threat.

The fine line between policing and combat can also arise in other situations of riots and demonstrations. Contemplating the use of deadly force when confronted with such demonstrations and disorder can be one of the most challenging operational situations confronting counterinsurgent forces. The 2003 decision by U.S. forces in Fallujah, Iraq, to fire into a crowd reported by the military to contain armed insurgents raised considerable questions as to whether the soldiers were faced with a law enforcement or an armed conflict situation.[54] The operational challenge is that such uses of force can adversely impact security by further inciting an insurgency.[55] The seizure of the U.S. embassy in Iran by militant students in 1979 demonstrates the fine line between civil unrest and more threatening activity. This latter point was even more graphically demonstrated in the tragic deaths of U.S. diplomatic personnel and CIA-employed security contractors at American facilities in Benghazi, Libya, in 2012. In that case an organized attack attributed to fundamentalist Islamist militia employing heavy machine guns, RPGs, and mortars clearly demonstrated levels of violence more closely associated with armed conflict than law enforcement.[56]

---

[51] Tony Geraghty, Black Ops: The Rise of Special Forces in the C.I.A., the S.A.S., and Mossad 40 (2010).

[52] David Crist, The Twilight War: The Secret History of America's Thirty-Year Conflict with Iran 131–32 (2012) (this attack took place on April 17, 1983. Sixty-three persons were killed including seventeen Americans, of which seven were CIA employees).

[53] *Id.* at 135–38 (the attacks took place on October 23, 1983, with 241 American service personnel and 58 French paratroopers being killed).

[54] Violent Response: The U.S. Army In Al-Falluja, 15 Human Rights Watch, Iraq, NO. 7 (E), June 2003, http://www.hrw.org/reports/2003/iraqfalluja/iraqfalluja.pdf.

[55] Dick Camp, Operation Phantom Fury: The Assault and Capture of Fallujah, Iraq 18 (2009) (where it is indicated "[t]he shooting inflamed the Sunni population, many of whom traced the beginning of the resistance to the incident."). *See also* Linda Robinson, Masters of Chaos: The Secret History of the Special Forces 337–38 (2004), and Thomas E. Ricks, Fiasco: The American Military Adventure in Iraq 138–42 (2007) (for reference to incidents having that effect in Mosul and Fallujah, Iraq, respectively).

[56] Paul Schemm & Maggie Michael, *Libyan Witnesses Recount Organized Benghazi Attack*, Assoc. Press (Oct. 27, 2012), http://bigstory.ap.org/article/libyan-witnesses-recount-organized-benghazi-attack. *See also* Mitchell Zuckoff, 13 Hours: The Inside Account of What Really Happened in Benghazi 211–20, 254–80 (2014) (outlining the violence of the attack on the compound including the use of mortars).

In the contemporary operating environment war is being fought among the people and, increasingly, in an urban environment. The issue is not whether military forces deployed on international operations will be required to apply human rights–based law enforcement norms, but rather under what circumstances such restrictions on the use of force must or should be applied. In some cases it will be obvious. Some international operations are purely law enforcement ones. However, there are also situations of overlap, not unlike during internal conflict, where a State or its military commanders will be able to make a policy choice as to whether to apply a law enforcement or conflict of hostilities paradigm to control the use of force. Ultimately, the decision to favor law enforcement over conduct of hostilities rules can depend on how effective the human rights–based regime is in dealing with the security threat. This is not a legal jurisdictional issue as to whether human rights law applies, but rather a functional question regarding the efficacy of the law enforcement approach. In this respect, there is only an overlap between the law enforcement and conduct of hostilities norms if both legal regimes "can get the job done." Ultimately, there are limits on what law enforcement can be asked to do if it is to be applied in a manner consistent with its historically unique role.

11.2.4. IMMINENCE AND THE NATURE OF THE THREAT

One of the challenges in applying a law enforcement paradigm to counter transnational security threats is an interpretive approach that views human rights law as only permitting the use of force in very restricted circumstances. This, at times, has led to the view that the use force is almost exclusively authorized within a narrowly proscribed zone of acting in self-defense or defense of others. Such an approach is evident in William Abresch's analysis favoring the use of human rights law for the governance of noninternational armed conflict. It is suggested that the same rules can be applied to "battles as for arrests" so that "[r]esort to lethal force is more likely to be lawful if the insurgent is actively participating in battle, because then he poses an actual or imminent threat to others and capturing him would more likely unreasonably endanger government soldiers. But there is no *per se* rule that insurgents may be targeted with lethal force."[57] In this interpretation the focus is on a reactive response to an actual or imminent threat.

In a State self-defense context, the discussion of when a threat is "imminent" often prompts reference to the *Caroline* incident test.[58] That standard requires that a State "show a necessity of self-defence, instant and overwhelming, leaving no choice of means, and no moment for deliberation."[59] However, the exercise of self-defense in the recourse to war context is different than for personal self-defense. It has increasingly been acknowledged that State self-defense incorporates some ability to act in a more anticipatory fashion (see Chapter 3, section 3.2, *The Continuing Application of Self-Defense Principles*). The authority to use force under human rights law is more restricted. For example, it has been suggested, in comparison to a State, which requires a reasonable period of time within which

---

[57] William Abresch, *A Human Rights Law of Internal Armed Conflict: The European Court of Human Rights in Chechnya*, 16 EUR. J. INT'L L. 741, 759 (2005).

[58] DINSTEIN, *supra* note 16, ¶ 726, at 274.

[59] *Id.* (quoting American Secretary of State Daniel Webster.).

to respond to an armed attack because of internal decision making, the "description of a human being under attack as having 'no moment for deliberation' would be accurate."[60] This is entirely consistent with the view that State self-defense operates at a strategic level requiring the consideration of vastly different threats than those posed to someone acting on an individual basis. The reaction as an individual to a threat is normally more narrowly prescribed than that available to States.

In a 2006 report, Philip Alston, the Special Rapporteur on extrajudicial, summary, or arbitrary executions, linked imminence to a strict necessity to use deadly force to prevent loss of life. That report stated:

> In most such situations, this necessity is the result of a threat's imminence. This too serves as a safeguard. When a criminal is already *in the process or visibly on the verge of using a weapon*, there can be little doubt regarding the inevitability of violence if immediate recourse to lethal force is not taken.[61]

Determining how imminence is assessed is key in establishing whether a law enforcement response can effectively address contemporary security threats. The problem is that a very narrow interpretation of imminence does not usually match the nature of the threat posed by organized armed groups. This is particularly evident when the threat presented by the leadership structure of a transnational terrorist group is considered. Much like the case for IED networks (see Chapter 7, section 7.5, *The IED and Suicide Bombing Threat: Arrest or Kill?*), it is the leadership, production, and logistical cells or elements that can present the greatest threat over the long term. Focusing on attacks in progress as suggested by the Special Rapporteur does not ultimately address the threat in terms of its "operating mind," the scope of the threat, or the capacity to repeatedly strike.

Given the group nature of terrorist and insurgent threats, as well as ready availability of computers, "Internet telephony," and cell and satellite phones with which to exercise command and control, the "operating mind" of an insurgent or terrorist organization may be located far from the point of attack.[62] This geographical separation can pose a barrier to applying a law enforcement approach if the assessment of the imminence of the threat is focused on seeing a criminal actually using a weapon. Addressing the threat posed by organized armed groups requires a more proactive and anticipatory approach than waiting and hoping to be in a position to see an attack unfold, possibly at multiple locations. At the heart of this issue is whether the threat is treated only as an individualized one at the point of attack. The challenge is that in terms of presenting the greatest danger it is the threat posed by an organized group as a whole that is operationally and strategically most significant.

It is no coincidence that group organization is one of the foundational criteria for an armed conflict, as well as a distinguishing feature for the application of humanitarian law

---

[60] *Id.* at ¶ 708, at 267.

[61] *Report of the Special Rapporteur on extrajudicial, summary or arbitrary executions*, UN Doc. E/CN.4/2006/53 (Mar. 8, 2006) (by Philip Alston) ¶ 49, at 19 [hereinafter *The 2006 Alston Report*] (emphasis added).

[62] Cathy Scott-Clark & Adrian Levy, The Siege: 68 Hours Inside the Taj Hotel 119, 150, 166, 236, 278 (2013) (for reference to the Lashkar's use of cheap Internet telephony network to communicate from Pakistan with the Mumbai attackers).

targeting norms against non-State actors. The increased potential for violence presented by organized armed groups, whether serving States or non-State actors, presents threats on a different order of magnitude than individual criminals or most criminal organizations. Although transnational criminal organizations can be extremely violent, once a group uses that violence for a political purpose the threat is moved even further down the conflict continuum (see Chapter 6, section 6.3.2, *Challenging the State*). The situation can evolve into an armed conflict where the authority to kill an opponent permits the targeting of members of the armed groups, as well as civilians taking a direct part in hostilities.[63] However, if the existence of an armed conflict has not been established, can the planning and deliberation stages be treated as an element of an "imminent" attack? It seems most analysts think not. Targeting the leadership, planners, and support personnel appears to more comfortably fit within the concept of "status"-based targeting, particularly where it is likely to be extremely difficult to determine exactly when operational preparation is being carried out. The imminence of an attack is less obvious when it is not in progress, or a person is not "visibly on the verge of using a weapon . . . ."[64]

Unless there is an armed conflict in existence there is no authority to conduct targeting under humanitarian law and therefore on the basis of membership.[65] It was the acknowledgment of being able to target members of organized armed groups that was the most significant aspect of the International Committee of the Red Cross (ICRC) *Interpretive Guidance*, although subsequent disagreements over how such membership is determined has often masked this point.[66] Once status-based targeting is permitted, the imminence of an attack is not relevant under humanitarian law unless the military forces are performing a law enforcement role regarding uninvolved civilians. The adoption of a narrowly defined concept of imminence for the law enforcement paradigm ultimately limits the ability of that human rights–based framework to counter real and significant armed conflict–related security threats. Further, in situations where human rights and humanitarian law are seen to be overlapping it is the requirement for a narrowly defined imminent threat that can restrict the utility of a law enforcement approach and privilege a conduct of hostilities one, whether applied as a matter of law or policy.

This does not mean there is no scope for a human rights–based approach to be applied in respect of some of significant security threats that can arise even in the context of an armed conflict. Not all of the force used involves conduct of hostilities–based targeting.

---

[63] Interpretive Guidance on the Notion of Direct Participation in Hostilities Under International Humanitarian Law, International Committee of the Red Cross 69 (2009) [hereinafter Interpretive Guidance].

[64] *The 2006 Alston Report*, *supra* note 61, ¶ 49, at 19. *See, e.g.*, Aerial Drone Deployment on 4 October 2010 in Mir Ali/Pakistan, 157 I.L.R. 722, 750 (2013) (where, applying a humanitarian law analysis, it was stated that the targeting proportionality of collateral damage to civilian property in a drone strike "holds all the more true in the case at hand, given that the aerial drone deployment may well have prevented the planned suicide attack that was to be carried out with the participation of B.E.").

[65] Prosecutor v. Tadić, Case No. IT-94-1-A, Decision On The Defence Motion For Interlocutory Appeal On Jurisdiction, ¶ 67 (Int'l Crim. Trib. for the Former Yugoslavia, 1995), http://www.icty.org/x/cases/tadic/acdec/en/51002.htm.

[66] ICRC Interpretive Guidance, *supra* note 63, at 34 (where the ICRC adopts a test called "continuous combat function," which applies narrow criteria for assessing membership focused on direct participation in hostilities). For a critique of this approach, *see* Kenneth Watkin, *Opportunity Lost: Organized Armed Groups and the ICRC "Direct Participation in Hostilities" Interpretive Guidance*, 42 N.Y.U. J. Int'l L. & Pol. 641, 683–686 (2010).

As has been noted, "[t]he law enforcement framework allows for the use of weapons and for the force necessary to seize ships or deal with hostage situations."[67] Further, since many insurgent and terrorist attacks occur "amongst the people," the paradigm that is applied to deal with potential threats is often initially human rights–based with the application of escalation of use of force rules. This is particularly evident at roadblocks, checkpoints, vehicle control points, and military objectives where the guards have to distinguish between ordinary civilians and the threats emanating from persons hiding among them.[68]

Human rights–based use of force procedures may also be applied because it is often law enforcement personnel who are located at the point of attack. Those attacks can occur in market places, mosques, churches, religious shrines, and other points of gathering where the terrorist action may be viewed as having the greatest impact. It is the presence of police at the local level that is one sign the counterinsurgency effort is working.[69] As a result, it may be the police who are most often targeted and therefore have to respond with force because they are the most visible local representatives of the government.[70] The police are likely to apply force consistent with their training and operational culture. The use of law enforcement means in these scenarios is largely reactive. This does leave open the question of whether human rights law should authorize a more anticipatory use of force in order to remove the insurgent or terrorist threat before it manifests itself, as it often does, in the killing of innocent civilians.

It may be helpful to look at twenty-first-century approaches to imminence in a recourse to war sense to see if there are any parallels to be drawn in respect of individual self-defense and law enforcement. The 9/11 attacks have prompted some reconsideration of what constitutes an imminent threat. There has been an acknowledgment in a State self-defense context that the *Caroline* incident approach to imminence may be too narrowly framed. It has been suggested there is a "need for the development of concept of what constitutes an 'imminent' armed attack to meet new circumstances and new threats."[71] In this vein the U.S. Army *Operational Law Handbook* relies on Michael Schmitt's analysis of imminence to suggest a response is justified "when (1) evidence shows that an aggressor has committed itself to an armed attack, and (2) delaying a response would hinder the defender's ability to mount a meaningful defense."[72] It is not clear that there is anything approaching an

---

[67] NOAM LUBELL, EXTRATERRITORIAL USE OF FORCE AGAINST NON-STATE ACTORS 226 (2010).

[68] OPERATIONAL LAW HANDBOOK, *supra* note 4, ¶ G 1., at 89 (where it is noted Escalation of Force measures are "simply the modern variant of what used to be called 'graduated force measures.' When time and circumstances permit, Soldiers should attempt to use lesser means of force."). *See also* Kenneth Watkin, *Use of Force during Occupation: Law Enforcement and Conduct of Hostilities*, 94 INT'L REV. RED CROSS 267, 308–09 (2012) (for a discussion of the challenges of conducting checkpoint duty and operating in markets, etc. and the use of human rights based law enforcement rules for the use of force).

[69] Jennifer Morrison Taw, *The Role of Police in the Colombian Insurgency*, *in* POLICING INSURGENCIES: COPS AS COUNTERINSURGENTS 136, 143 (C. Christine Fair & Sumit Ganguly eds., 2014) (regarding the expansion of a police presence in Columbia in 2002–03).

[70] DAVID H. BAYLEY & ROBERT M. PERITO, THE POLICE IN WAR: FIGHTING INSURGENCY, TERRORISM, AND VIOLENT CRIME 74 (2010).

[71] VAUGHN LOWE, INTERNATIONAL LAW 278 (2007). *See also* 75 BRIT. Y.B. INT'L L. 822, 823 (2004) (for the analysis of the right to self-defense by the UK Attorney-General, upon which Professor Lowe's conclusions are based).

[72] OPERATIONAL LAW HANDBOOK, *supra* note 4, ¶ III. d., at 7 (relying on Michael Schmitt, *Preemptive Strategies in International Law*, 24 MICH. J. INT'L L. 513, 535 (2003)).

international consensus as to whether such criteria should be applied as part of the State self-defense analysis. However, it is also evident that the very narrow pre-9/11 interpretation of imminence based on the *Caroline* incident approach is not viewed now as accurately reflecting the challenge facing States at the transnational level.

It is important to confirm that the broader application of the State self-defense imminence principle occurs at a national strategic level. This raises the issue of whether a similar approach is appropriate in respect to a more tactically focused use of law enforcement against terrorist and insurgent groups. Applying that reasoning by analogy to a law enforcement concept of self-defense would not make the use of force preemptive but rather more anticipatory. It would also look at the defense of others (i.e., society at large) in a broader context. Such an approach recognizes that the certainty of seeing a weapon in the hand of a suspect does not address the whole range of threats posed by non-State actors. This is particularly evident with the cyber-enhanced ability to exercise command over long distances.[73]

The challenge is that the expansion of individual self-defense to accept such anticipatory uses of force could fundamentally change well-accepted law enforcement approaches toward the use of force. It is likely there will be a resistance to such an expansion of human rights–based concepts of self-defense. Such resistance is not problematic as long as there is a corresponding acknowledgment that unless the law enforcement normative framework can address the full range of threats posed by insurgent and terrorist groups, which at some point does require the application of a broader understanding of what constitutes an imminent threat, its utility in addressing contemporary security threats will of necessity be more limited. Otherwise, overly restricted interpretations of the authority to use force can create virtual zones of immunity or safe haven within which organized armed groups can continually plan and execute attacks.

If the law enforcement paradigm is incapable of addressing these group threats, then the international community may have to assess whether an armed conflict is really in existence and turn to humanitarian law to consider if the threat posed by a group sitting in such a safe haven is more consistent with the existence of an armed conflict. In this respect, a reasonable argument can be made that the inability of the law enforcement paradigm to address the threat serves as a further indication such an armed conflict is in existence. This issue is one that arises in respect of "shoot to kill" allegations and the nature of the threat posed by non-State actor groups (see Chapter 13, section 13.2.2, *Shooting to Kill*). What is clear is there can be no normative void.

11.2.5. MISSION ACCOMPLISHMENT

Another issue that arises when considering the efficacy of a human rights–based law enforcement response is the overemphasis that is often placed on force being used in self-defense or the defense of others. It is not necessarily the law itself but interpretations of the law that have created this overemphasis. In this regard, the law enforcement paradigm allows the use of force on a much broader scale than is often acknowledged. The position

[73] TALLIN MANUAL, *supra* note 2, ¶ 5, at 65 (where reference is made to conducting an "anticipatory" attack against individuals gathered in one place who are believed will carry out an attack two weeks in the future).

that the law enforcement paradigm is primarily focused on self-defense may have arisen from a view that the right to life, as protected under human rights law, verges on the absolute. This perspective is perhaps best reflected in the statement, "[n]o derogation is permitted from the right to life, and none is needed."[74] The right to life does, and should, occupy a special position among rights. As has been noted, "[w]hen life is thought deprived, it is impossible to enjoy any fundamental freedom."[75] This is particularly important for security forces to keep in mind, whether it is assessed from a legal, moral, or operational effectiveness perspective. However, while it is essential to acknowledge the fundamental importance of the right to life, it would be misleading to suggest this leaves little scope for the use of deadly force, if necessary, to maintain order in society. This holds true whether that force is used domestically or internationally.

Under human rights law a person can be deprived of life in certain circumstances, including situations other than when using force in individualized self-defense. Ensuring the security of the State can require police and military forces acting in a law enforcement role to use force, not only in self-defense or defense of others but also to make others desist from illegal activity. The effectiveness of law enforcement on international operations primarily rests on its utility in forcing compliance with the law. The statement that no derogation is permitted from the right to life must be understood in the context that limits on that right are already built, not only into the treaty law but also commonly accepted principles and practices regarding the use of force by law enforcement personnel.

Perhaps the clearest evidence of the treaty limits is found in Article 2(2) of the European Convention on Human Rights, which states:

> Deprivation of life shall not be regarded as inflicted in contravention of this article when it results from the use of force which is no more than absolutely necessary:
> (a) in defense of any person from unlawful violence;
> (b) *in order to effect a lawful arrest or to prevent escape of a person lawfully detained*;
> (c) *in action lawfully taken for the purpose of quelling a riot or insurrection.*[76]

The additional circumstances articulated in the European Convention on Human Rights for when deadly force can be used beyond self-defense (e.g., effect an arrest and quelling a riot or insurrection) reflects the reality that governing entities must be able to use force to maintain order in society.

Similar justifications for depriving life can be found in the *U.N. Basic Principles on the Use of Force.* Officials are prohibited from using firearms "except in self-defence or defence of others against the imminent threat of death or serious injury, *to prevent the perpetration of a particularly serious crime involving grave threat to life, to arrest a person presenting such a danger and resisting may their authority, or to prevent his or her escape*, and only when less extreme means are insufficient to achieve these objectives."[77] It has been noted that these principles "in some aspects even refine the conditions and modalities established

---

[74] *The 2006 Alston Report, supra* note 61, ¶ 47, at 18.

[75] Yoram Dinstein, *Terrorism as an International Crime*, ISR. Y.B. HUM. RTS. 55, 63 (1987).

[76] European Convention on Human Rights, art. 2(2) (emphasis added).

[77] *UN Basic Principles on the Use of Force*, at ¶ 9 (emphasis added).

by conventional human rights law for the resort to lethal force in law enforcement operations" and as a result are "increasingly relied on in the judgments of the European Court of Human Rights."[78] Importantly, they also present situations in which the use of deadly force is authorized outside the context of acting in self-defense or the defense of others where the imminence of the threat is such a key factor.

The use of deadly force to prevent serious crimes, carry out an arrest, prevent an escape, or even quell riots and disturbances is linked to a deadly threat. However, the nature of that threat can take on a different character than when strictly acting in self-defense. This appears particularly relevant in terms of the imminence principle. The person being arrested or escaping only has to present a "danger" rather than actually be using or imminently about to use force in order for deadly force to justified. This does not mean that the authority to use force is as broad as that available under humanitarian law. However, the use of lethal force by law enforcement officials in these circumstances reflects the performance of a preventative role that is designed to protect broader society. In military terminology it might be described as the authority to use force for particular purposes outside of acting in self-defense. In other words, for "mission accomplishment."

The term "mission accomplishment" may be used in a broader context than law enforcement in a classic criminal sense. It can be applied to specific missions based on the Chapter VII UN Charter provision for taking "such action by air, sea, or land forces as may be necessary to maintain or restore international peace and security."[79] It is common in a UN context, where self-defense is ordinarily narrowly interpreted, to provide any required additional authority to use of force in the ROE in order to successfully complete Chapter VI–defined missions.[80] The term is also referred to in the U.S. SROE. Specifically, "[t]he purpose of the SROE is to provide implementation guidance on the application of force *for mission accomplishment* and the exercise of self-defense."[81] Those rules are set out in what are called "supplemental measures."[82] Those measures "are primarily used to define limits or grant authority for the use of force for mission accomplishment," however, they may also be used "to limit individual self-defense by members of their units."[83] The number and scope of mission accomplishment ROE may be directly related to how broadly the right to self-defense authority is interpreted to be in the governing legal framework. A narrowly defined right to self-defense is likely to result in a greater number of mission specific ROE, always subject to their legal basis being identified at law.

Given how narrowly the right to self-defense is defined under human rights law, it can be expected that mission accomplishment ROE would be required for the performance of many law enforcement duties. Mission accomplishment authority would not justify an

---

[78] Nils Melzer, Targeted Killing in International Law 203 (2008).

[79] UN Charter, art. 42.

[80] Hans F. R. Boddens Hosang, *Force Protection, Unit Self-Defence, and Extended Self-Defence, in* Handbook of International Operational Law, *supra* note 14, Rule 22.04, 413, at 417 ("United Nations Forces under United Nations command and control have the right to defend themselves against an (imminent) attack *as well as against armed attempts to interfere with the execution of the mandate.*") (emphasis added). *See also* Corn, *supra* note 3, at 44 ("The UN's model ROE clearly contemplate use of force authority beyond self-defense, that is, for mission accomplishment.").

[81] CJCSI, *supra* note 4, Enclosure A ¶ 1. a., A-1 at 93 (emphasis added).

[82] *Id.* Enclosure I., I-1 to I-2 at 99-100.

[83] *Id.* Enclosure I, ¶ 2. b., I-1 at 99.

attack against a target on the basis of membership in an organized armed group. However, it conceivably might be required if deadly force is to be used to remove a guard who is blocking access to a location where a hostage is being held (see Chapter 12, section 12.2.2, *Force and Hostage Rescue*). In such a case it might be reasonably believed that the guard could warn those holding the hostage, potentially leading to that person's death or injury. Assuming the use of deadly force would not be justified under a more expansive interpretation of imminence when acting to protect others, it might in exceptional circumstances fall under the authority "to prevent the perpetration of a particularly serious crime involving grave threat to life."[84] This may particularly be the case where it can be demonstrated there is a pattern of hostages being killed or injured when a rescue effort is compromised.

There is doctrinal recognition of the expanded authority to use force to enforce the law during international operations. The *San Remo Rules of Engagement Handbook*, which outlines a commonly understood approach toward the use of force by military forces, states, "[b]roadly speaking, during peacetime, the use of force is permitted in *self-defence*, in the *exercise of law enforcement authority*, and *to accomplish operations or missions* specifically authorised by a higher national authority or other governing body, such as the U.N. Security Council."[85] This is a reference to a broader concept of the use of force than acting in self-defense, although law enforcement is separated from the accomplishment of "missions" in this statement. The authority to use force is extended beyond self-defense to constabulary action on the high seas (such as against piracy), enforcing Security Council Resolutions, and other operations conducted outside the context of an armed conflict. It therefore outlines an authority to use force for law enforcement and mission accomplishment, as well as acting in self-defense.

Given the acknowledgment of such additional authority to use force for mission accomplishment, the question remains as to how it relates to self-defense. Geoffrey Corn and Eric Jensen highlight the traditional division of use force authority in their very useful explanation of the U.S. SROE. It is suggested that "[t]he genius of the SROE is in its bifurcation between rules governing self-defense and mission accomplishment."[86] It is further noted that this division of force employment authority "is indicative of both the nature of the mission as well as the nature of anticipated threats. . . ."[87] While this discussion of SROE is nation specific the overall approach of recognizing an expanded authority to use force beyond self-defense in the form of mission accomplishment–type rules is not unique to the United States.

A key issue to be addressed is how broadly self-defense in a law enforcement context is defined. This is because not all nations agree on how self-defense should be interpreted. That disagreement is most often represented in ROE. In looking at ROE it must be stressed they are not "law." Rules of engagement are fundamentally a command tool established to control the use of force. This use of force framework can reflect not only legal requirements

[84] *UN Basic Principles on the Use of Force*, at ¶ 9.

[85] Rules of Engagement Handbook, Institute of Humanitarian Law¶ 13, at 4 (2009), *available at* https://www.usnwc.edu/getattachment/7b0d0f70-bb07-48f2-af0a-7474e92d0bb0/San-Remo-ROE-Handbook [hereinafter ROE Handbook] (emphasis added).

[86] Geoffrey S. Corn & Eric Talbot Jensen, *Untying the Gordian Knot: A Proposal for Determining Applicability of the Laws of War to the War on Terror*, 81 Temple L. Rev. 787, 809 (2008).

[87] *Id.* at 810.

but also operational and policy considerations. However, such rules also indicate a national, and sometimes Coalition, understanding of what is legally permissible during international operations,[88] even though they may also "impose greater restrictions than those required by the law."[89] Often defined in terms of "personal,"[90] "unit,"[91] and "national"[92] self-defense, some nations have a very limited view of what constitutes self-defense. This can be seen in respect of the use of terms such as "hostile act"[93] and "hostile intent,"[94] which were first developed to provide guidance on the exercise of State self-defense.[95] For some States terms like hostile act and hostile intent "are related to mission-accomplishment, rather than self-defence."[96] However, other States incorporate those terms as criteria for when force may be used in self-defense.[97]

Therefore, the military forces of one State may apply a broad interpretation of self-defense that does not require a significant number of additional rules dealing with hostile intent or hostile act. Another State may incorporate those concepts into separate rules in order to accommodate a narrower view of self-defense. Regardless of the approach adopted by States, it is both lawful and prudent to develop ROE that are sufficiently comprehensive to ensure there is effectively a similar level of force authorized across a Coalition. This can include additional rules for mission accomplishment, which account for the most restrictive interpretation of self-defense by a Coalition partner even if it is not necessary for all States.[98]

---

[88] McCann and others v. United Kingdom, App. No. 18984/91, Eur. Ct. H.R., Judgment 156 (1995), http://hudoc.echr.coe.int/sites/eng/pages/search.aspx?i=001-57943 (where the European Court of Human Rights noted "that the rules of engagement issued to the soldiers and the police in the present case provide a series of rules governing the use of force which carefully reflect the national standard as well as the substance of the Convention standard."). *See also* Boddens Hosang, *supra* note 80, at 415–16 (where it is suggested there is a NATO unique authority (within a NATO-led or NATO commanded operation) for Alliance units to use force to assist one another in the event of an imminent attack as a form of "extended self-defence").

[89] Operational Law Handbook, *supra* note 4, ¶ B. 3, at 82.

[90] ROE Handbook, *supra* note 85, Annex D, at 83 (where individual self-defense is defined as "the right of an individual to defend himself or herself (and in some cases other individuals) from *hostile act* or *hostile intent*.").

[91] *Id.* Annex D, at 85 ("***unit self-defence*** the right of unit commanders to defend their unit, other units of their nation, and other specified units against *hostile act* or *hostile intent*.").

[92] *Id.* Annex D, at 83 ("***national self-defence*** the defence of a nation, a nation's forces, and a nation's persons and their property. Generally, unit commanders may exercise the right of *national self-defence* only when authorised by *higher authority*.").

[93] *Id.* Annex D, at 82 ("***hostile intent*** the threat of an imminent *hostile act*.").

[94] *Id.* Annex D, at 82 ("***hostile act*** an *attack* or other use of force against a nation, the *Force* or other designated persons or proper.").

[95] Corn, *supra* note 3, at 10–11 ("Although necessity is universally accepted as a predicate to exercising self-defense at any level, the hostile act/hostile intent (HA/HI) framework developed specifically to implement national self-defense authorities.").

[96] ROE Handbook, *supra* note 85, ¶ 8, at 3. *See* Boddens Hosang, *supra* note 80, at 424–25 (for a discussion of the different U.S. and European approaches toward hostile intent and hostile act).

[97] Corn, *supra* note 3, at 10 (from a U.S. SROE perspective, "[t]he common thread running through all types and levels of self-defense are the SROE concepts of hostile act and hostile intent."), and Boddens Hosang, *supra* note 80, ¶ 1 at 425.

[98] Corn, *supra* note 3, at 5 ("U.S. forces have either adopted the NATO understanding of non-imminent HA/HI or applied an aggressive view of SROE self-defense rules."). *See* Boddens Hosang, *supra* note 80, ¶ 3 at 425 ("The ROE regulating the use of force against hostile acts and hostile intent in the context of a NATO operation give those nations requiring such ROE to act: while they are superfluous, but thereby not detrimental, to those nations that already claim an inherent right to act.").

The requirement to authorize the use force for purposes beyond self-defense can arise in a diverse range of international operations. For example, Maritime Interception Operations (MIO) ranging from "querying the master of a vessel to stopping, boarding, inspecting, searching and even seizing the cargo or vessel" [99] may be ordered under different legal bases including Security Council Resolutions, consent, visit and search under the law of the sea, hot pursuit, and potentially as a matter of countermeasures.[100] While the use of deadly force outside the context of self-defense during MIO is authorized only in exceptional cases, it may, for example, be applied to enable incapacitating force, such as firing shots into the rudder of a ship in order to cause it to stop.[101] The requirement to use force beyond self-defense can also arise in other operations, including noncombatant evacuation missions, humanitarian intervention, or, more generally, operations that have come to be termed "stability operations."[102]

A classic example of when mission accomplishment ROE are applied is when a rescue force is conducting a noncombatant evacuation mission as part of the exercise of national self-defense. During such an operation an organized armed group manning a roadblock may present a significant obstacle to a relief force attempting to reach trapped personnel. That group also might not be overtly threatening any of the nationals being rescued. Further, the intervening force is not likely at that stage be involved in the armed conflict, although it will probably be operating in an atmosphere of considerable tension and disorder. With a goal of entering and egressing the territory as quickly as possible, the challenge for intervening military commanders is to reach the threatened personnel and, if necessary, clear an evacuation route. In doing so, a determination must be made as to what force is potentially required to successfully accomplish the mission. This will include consideration of the degree to which military forces have to expose personnel to risk for the purposes of confirming an armed force blocking the route poses a threat or will obstruct passage. In such a complex operating environment the requirement to identify the intentions of armed groups, or even individual armed civilians, will normally lead to the exposure of military personnel to some degree of risk.[103]

Depending upon the available intelligence or as a result of repeated contacts the situation can evolve into one where an elevated level of threat is posed by blocking organized armed groups. In such situations a more proactive use of force than simply acting in self-defense may be required to carry out the evacuation of the nationals. That authority would have to be found in mission accomplishment ROE. It also may not be clear at what point the violence will cross into the realm of an armed conflict to which conduct of hostilities rules would apply. A threat may be posed by a criminal gang or rioting civilians that will not

---

[99] Wolf Heintschel Von Heinegg, *Maritime Interception/Interdiction Operations*, *in* HANDBOOK OF INTERNATIONAL OPERATIONAL LAW, *supra* note 14, 375, at 375.

[100] *Id.* at 380–81.

[101] *Id.* at 393.

[102] Kenneth Watkin, *Stability Operations: A Guiding Framework for "Small Wars" and Other Conflicts of the Twenty-First Century?*, 85 INT'L L. STUD. 411 (2009).

[103] Michael R. Gordon, *Attack on U.S. Aircraft Foils Evacuation in South Sudan*, N.Y. Times, Dec. 21, 2013, http://www.nytimes.com/2013/12/22/world/africa/south-sudan.html?_r=0 (highlighting the risks inherent in non-combatant evacuation operations when three United States military aircraft were fired on resulting in a number of casualties to the rescue force).

rise to the level of an armed conflict. Until the threshold of armed conflict is crossed the use of force remains effectively governed by a human rights–based framework. This use of force will be based primarily on the right to self-defense, along with an additional authority to apply a level of force necessary to complete the mission. Such mission accomplishment authority is not unlike that available to law enforcement personnel that allows them to break down barriers and use force to gain entry to a scene of a hostage taking or to carry out an arrest.

11.3. Doctrinal Confusion: State Self-Defense and Law Enforcement

11.3.1. RULES OF ENGAGEMENT DOCTRINE

Regardless of whether the force used during a State exercise of self-defense is authorized under a law enforcement paradigm, a conflict of hostilities framework, or both, the State must keep the violence within the overarching framework of the State self-defense principles: necessity, proportionality, and imminence. This raises the question of how the law enforcement–based self-defense/mission accomplishment authority to use force is affected by those principles. Complicating this analysis is the simultaneous application of humanitarian law–based conduct of hostilities rules during armed conflict. It is how broadly or narrowly the concept of self-defense is interpreted at the strategic level that can determine what impact the law governing State self-defense has on the force used in a law enforcement context.

As has been noted, a generally restrictive interpretation of the principles applied to the use of force under a law enforcement paradigm should in theory be relatively easy to separate the application of State self-defense principles from that human rights–based body of law. However, this can turn out to be a far more difficult challenge in practice. A common approach for categorizing various forms of self-defense is reflected in the *San Remo Rules of Engagement Handbook* where "national self-defense"[104] is defined with reference to Article 51 of the UN Charter, while "unit self-defense" and "individual self-defense" refer to decisions to use force in language that is more personal in nature. For example, with the "unit self-defense" authorization commanders "have the right to defend ***their*** unit and other units from ***their*** nation in the face of an *attack* or imminent *attack*."[105] The term "individual self-defense" "refers to the right of an individual to defend himself or herself (and in some cases other individuals) from an *attack* or imminent *attack*."[106] Of note, some cases of individual self-defense may extend to the defense of others. Perhaps the best example of unit self-defense arising in an individualized as opposed to a national self-defense context is the May 1987 incident involving Iraqi Exocet missiles being fired at the *USS Stark*.[107] The American ship failed to respond adequately to an indiscriminate attack by an Iraqi airplane

[104] ROE HANDBOOK, *supra* note 85, ¶ 8 d., at 3 ("As recognised in Article 51 of the United Nations Charter, refers to the right of a nation to defend itself against armed *attack*, and for most nations, the threat of imminent armed *attack*.").

[105] *Id.* ¶ 8 b., at 3 (bold italics added).

[106] *Id.* ¶ 8 a., at 3

[107] CRIST, *supra* note 52, at 220–29.

attacking Iranian shipping.[108] Both missiles struck, disabling the warship and killing thirty-seven sailors.[109] While the *USS Stark* was clearly in the Gulf on a national mission, the circumstances of the attack points to a specific narrow exercise of unit self-defense.

One possible interpretation of the distinction between these forms of self-defense is that national self-defense involves the application of traditional State self-defense principles where the intention of the attack is to strike at a symbol of the State, while an individual- or unit-level defensive response can be viewed as more closely approximating a "personal" defensive act.[110] However, drawing the line between national, unit, or individual self-defense in order to distinguish the application of recourse to war versus human rights–based principles can be problematic. This can be seen in a unit self-defense context in the *Case Concerning Oil Platforms (Islamic Republic Of Iran V. United States Of America)*, where it was suggested by the International Court of Justice that an attack on a single warship could be considered to be an armed attack for the purposes of Article 51 of the UN Charter.[111] The ease with which national self-defense can be separated from individual acts also appears to depend upon whether a minimalist approach is taken toward the threshold for an armed attack. It has been suggested, "there is no reason to remove small-scale armed attacks (above a *de minimis* [minimal] threshold) from the Article 51 spectrum of armed attacks."[112] In this regard, if a single soldier firing across a border at a member of another State's armed forces is viewed as an armed attack, then there may be relatively little room left for an "individual" defensive response.[113]

The difficulty in making a distinction between national and unit self-defense is evident in another Persian Gulf example, the July 1988 shoot-down of an Iranian civilian passenger plane by the *USS Vincennes*. The sophisticated Aegis cruiser had been deployed to the Gulf as part of an expanded effort to protect vessels in the Gulf from Iranian attack. The *USS Vincennes* had aggressively entered Iranian waters and fired on Iranian boats it viewed as a threat, sinking two Revolutionary Guard boats and damaging another. It was in that context that the civilian airliner believed to be an attacking Iranian F-14 aircraft was then shot down.[114] While the warship was acting defensively, the overall situation makes that action appear to have been more of an exercise of national self-defense than the previous incident involving the *USS Stark*.

---

[108] *Id.* at 224 (an attempt was made to fire chaff to confuse the incoming missiles, however, general quarters was not sounded and the CWIS antimissile system was not engaged).

[109] *Id.* at 226.

[110] Mary Ellen O'Connell, *The True Meaning of Force*, AJIL Unbound (Aug. 4, 2014, 9:56 AM), http://www.asil.org/blogs/true-meaning-force (where Yoram Dinstein's concept of an "on-the sport-reaction" is said to be "akin to personal self-defense, which is found in all national criminal codes. It is part and parcel of the human right to life." However, it is stated that it is not part of self-defense under Article 51 of the UN Charter). *But see* Dinstein, *supra* note 16, ¶ 641, at 243 (where it is indicated when discussing "on-the-spot" reactions that "all self-defense is national self-defence. There is a quantative, but no qualitative difference between a single unit responding to an armed attack and the entire military structure doing so.").

[111] Case Concerning Oil Platforms (Islamic Republic Of Iran V. United States Of America), 2003 I.C.J. ¶ 72, at 195.

[112] Dinstein, *supra* note 16, ¶ 557, at 211.

[113] *Id.* ¶ 555, at 210 (the non-armed attack example of a soldier firing across a border and hitting a tree or cow leaves open the question of whether wounding or killing another State's soldier would meet the minimal armed attack threshold).

[114] Crist, *supra* note 52, at 363–70.

The legal challenge for both commanders and drafters of ROE is determining what degree of imminence is required for the respective types of self-defense. Under ROE doctrine distinctions between the self-defense-based paradigms (i.e., State self-defense and human rights law) are usually not addressed. One innovative approach already mentioned is the application of the common law–based "public order doctrine" to the use of force to unit and individual self-defense.[115] This situates the exercise of unit or individual self-defense separately from the State self-defense framework that applies broader concepts of imminence.[116] Instead, such action would be underpinned by "the body of law that regulates the conduct of non-State actors vis-à-vis individual human beings: that is IHRL [international human rights law] generally, and more specifically the prohibition against arbitrary killings as applied within and outside the context of armed conflict."[117] Based on the post-Westphalian view that the use of violence is monopolized in the hands of the State, this approach sits well with the just war–based proper authority principle (see Chapter 4, section 4.4.1, *Public, Private, and Mixed War*).[118] The public order doctrine would extend to the use of force outside of an armed conflict context to action such as law enforcement.[119] Members of the military act as State agents, and their authority to use force "need not necessarily be triggered by an actual threat of unlawful violence," rather there is only a need to be protecting or furthering a lawfully recognized interest. [120] Under this theory the use of force would be constrained by the requirement to comply with the principles of necessity and proportionality. In this respect, "the force used must not be excessive under the circumstances."[121]

The challenge to adopting this approach is that the authority to use force in self-defense in both a unit and individual context remains in many circumstances closely tied to the State exercise of self-defense. This is because of the existence of a close connection between the object being targeted and the State. For example, given the importance of a warship in terms of being a national symbol, as well as its cost of replacement, it is more likely that the State self-defense principles of necessity, proportionality, and imminence will govern the response by the unit commander rather than the narrower human rights–based principles. In this regard it has been suggested that a model where unit self-defense is viewed as "a tactical level representation of the sovereign right of nations to national self-defence" provides the most clear and coherent explanation of this concept.[122]

The distinction in the application of these self-defense principles is most likely seen in how individual self-defense is applied. However, as has been noted, even here the drawing

[115] Corn, *supra* note 3, at 20–24. *See also Memorandum for the Attorney General, Re: Applicability of Federal Criminal Laws and the Constitution to Contemplated Lethal Operations Against Shaykh Anwar al-Aulaqi*, Office of the Assistant Attorney General, July 16, 2010, at 14–19, https://www.aclu.org/sites/default/files/assets/2014-06-23_barron-memorandum.pdf (where an assessment of the public authority justification was also part of the U.S. opinion regarding the lawfulness of drone strikes in areas outside active hostilities).

[116] Corn, *supra* note 3, at 17.

[117] *Id.* at 18.

[118] *Id.*

[119] *Id.* at 22.

[120] *Id.* at 23.

[121] *Id.* at 23.

[122] Boddens Hosang, *supra* note 80, ¶ 7 at 422. *See also id.* Rule 22.07 ¶¶ 1–7, at 420–22 (where unit self-defense is assessed as a form of "collective personal self-defence," as a corollary to the right to life and as an inherent sovereign right). *See also* Dinstein, *supra* note 16, ¶ 641, at 243.

of a definitive line of separation can be problematic. An attack on an individual member of a nation's armed forces could raise questions about whether the incident was an attack on that person's State as well. Questions regarding the application of State self-defense can also arise when civilian persons and objects are targeted.[123] However, it also seems evident both soldiers and civilians can be subjected to attacks by persons who are unconnected to another State or a non-State actor. If a local criminal attacks a soldier in a souk, the response would resemble that of an individual acting in self-defense in a domestic setting. It would be difficult to argue there is a strong State connection in this instance, except perhaps that the soldier was deployed to the territory in question on official duties. Much depends upon the context of the attack.

Finally, the use of force as part of law enforcement does not always seem to fit comfortably with the broader concept of State self-defense. These situations can occur when armed forces conduct law enforcement operations during both international (e.g., occupation) and noninternational armed conflict (e.g., the "police primacy" approach to counterinsurgency), or simply when interfacing with ordinary civilians. A particular challenge in applying law enforcement norms during international operations is grappling with the dominant place that the exercise of State self-defense has attained in military doctrine. The impact of State self-defense principles is understandable to the extent the armed forces of a State are such a visible representation of its power. Individual service members and units normally act with their association with the State as the driving force and with a status visibly represented in their uniforms and markings. The dominance accorded to State self-defense is also entirely consistent with the traditional focus of international law on restricting, if not completely eliminating, inter-State uses of force. However, this creates a situation where consideration of law enforcement norms and even the application of humanitarian law may be improperly subsumed under overly restrictive or broad interpretations of the State self-defense principles of necessity, proportionality, imminence, and immediacy. The challenge is how to avoid undue influence of those principles in situations where force is used in individual self-defense or for law enforcement and mission accomplishment purposes conducted pursuant to a human rights law paradigm.

11.3.2. UNITED NATIONS OPERATIONS AND "ROBUST PEACEKEEPING"

The challenge of articulating the principles governing the use of force in a "law enforcement" context exists across a wide range of international operations. These include the UN peacekeeping operations. The history of UN peacekeeping is particularly instructive since the scope for acting in defense of the mission mandate has created a tension over the years between the strategic interpretation of self-defense and the authority to act more broadly. Such broader action can be required at the tactical level. The traditional narrowness of the self-defense framework, or conversely the expansiveness of the concept of defense of the mission associated with what has been called "robust peacekeeping," has made it challenging at

---

[123] DINSTEIN, *supra* note 16, ¶ 573, at 217 (where it is noted "the *jus ad bellum* [humanitarian law] does not recognize a difference between attacks against lawful targets (combatants or military objectives) and unlawful targets (civilians or civilian objects)" and therefore such an attack "validates a response irrespective of the military or civilian nature of the objects or persons struck.").

times to assess the degree to which the force used is properly governed by a human rights or humanitarian law framework.

The conduct of UN operations is one area where there has been controversy and confusion regarding what force is authorized by a mission mandate. This has been particularly evident in the context of UN Charter Chapter VI peacekeeping. The reliance on self-defense as the overarching framework for peacekeeping can be contrasted with Chapter VII mandates, which typically authorize "such action by air, sea or land forces as may be necessary to maintain or restore international peace and security."[124] Notwithstanding the traditional distinction between Chapter VI and Chapter VII operations, peacekeeping has from its inception had to address the need to use very broad applications of force. One of the classic cases was the UN Operation in the Congo (ONUC), which ran from July 1960 until June 1964.[125] While also tasked with assisting in the maintenance law and order,[126] that early peacekeeping mission ultimately witnessed 20,000 UN-mandated troops engaged in "fighting against those seeking succession."[127] In this context, the notion of self-defense was expanded far beyond that required for law enforcement to include the conduct of hostilities in the effort to carry out the assigned operational mandate.[128] In other words, force was expansively authorized for the defense of the mandate or the mission. In this context, "self-defense" at a mission level involved both humanitarian and human rights law–based uses of force.

However, the "defense of the mandate" approach did not remain a static paradigm. The idea of acting in self-defense under a peacekeeping mandate became considerably more constrained over the course of the 1970s and 1980s. This approach reflected the importation of a more restrictive interpretation of the authority of States to act in self-defense. It appears to have represented an integration of Cold War related recourse to war principles into the framework of what force is required to meet mission mandates. Importantly, it was an interpretation of self-defense that narrowly approximated that applied in a human rights law context. The result was that when United Nations authorized the deployment of military forces to the Balkans the 1990s in the midst of an armed conflict, the authority to use force in self-defense was often very narrowly interpreted. This meant because "they were bound by the parameters of a Chapter 6 mission, no one could fire unless fired upon, and the peacekeepers could return only equal fire."[129] This placed commanders in the position of not feeling they could even engage in a mission accomplishment use of force that is considered a common aspect of domestic law enforcement. A tragic example of this commonly applied restrictive approach occurred in the lead-up to the 1994 Rwandan genocide. A request by the UN Force Commander to proactively seize weapons caches that were established in

---

124 United Nations Charter, Chapter VII, art. 42.

125 The Blue Helmets: A Review of United Nations Peace-keeping 215–59 (2nd ed. 1990) (for an overview of the ONUC mission).

126 *Id.* at 226.

127 Christine Gray, International Law and the Use of Force 263 (3rd ed. 2008).

128 Bruce Oswald, Helen Durham, & Adrian Bates, Documents on the Law of UN Peace Operations 563 (2010).

129 Carol Off, The Ghosts of Medak Pocket: The Story of Canada's Secret War 172 (2004) (although it is also noted, "[a] force commander had a fair bit of latitude to change the rules—if he was willing to take the risk . . . .").

violation of the UN-brokered Kigali Secure Weapons Area agreement was not authorized because of the mission's narrowly defined Chapter VI peacekeeping mandate.[130]

These 1990s interpretations of the authority to use force during operations mandated to keep the peace were based on interpretations of international law that provided less authority to act than is found in the domestic jurisdictions of many countries. It became evident that such interpretations were neither morally nor operationally sustainable. In this regard, "the chaotic situation in Bosnia forced the leaders of contingents, and then the officials at the UN, to change that policy."[131] In the aftermath of a litany of controversial and criticized missions, more expansive interpretations of self-defense for peacekeeping missions began to be applied. This eventually resulted in the development of what was called "robust peacekeeping." That term is used to describe the situation where the Security Council, while not authorizing peace enforcement as envisaged under Chapter VII, has "given United Nations peacekeeping operations 'robust' mandates authorizing them to 'use all necessary means' to deter forceful attempts to disrupt the political process, protect civilians under imminent threat of physical attack, and/or assist the national authorities in maintaining law and order."[132]

From a doctrinal perspective, "robust peacekeeping" is distinguished from peace enforcement on the basis that the former still requires the consent of the main parties involved, while the latter does not.[133] The recognition of "robust peacekeeping" reflected the operational need for a broader authority to use force. This was because "[t]he environments into which United Nations peacekeeping operations are deployed are often characterized by the presence of militias, criminal gangs, and other spoilers who may actively seek to undermine the peace process or pose a threat to the civilian population."[134] Although the term "self-defense" has often been thought of as being narrowly prescribed in respect of peacekeeping, that concept was now interpreted to be "*applicable to the defence of the UN mission and its mandate* as well as the personal safety of those under the operational command and control of the UN."[135] In effect, this new doctrinal approach recognized the requirement to use force for mission accomplishment in addition to acting in self-defense or the defense of others. What was not as clear is what legal regime was to govern the use of force at the tactical level.

Notwithstanding the recognition of broader mission mandates, there remained a significant effort to limit the authority to use force during peacekeeping. The 2008 *United Nations Peacekeeping Operations: Principles and Guidelines* indicates the principle of the "*non-use of*

---

[130] Romeo Dallaire & Brent Beardsley, Shake Hands with the Devil 146 (2003).

[131] Jocelyn Coulon, Soldiers of Diplomacy: The United Nations, Peacekeeping, and the New World Order 177 (Phyllis Aronoff & Howard Scott trans., 1998).

[132] United Nations Peacekeeping Operations: Principles and Guidelines, Department of Peacekeeping Operations 34 (2008), *available at* http://www.dtic.mil/doctrine/doctrine/interagency/un_prin_guide.pdf [hereinafter Peacekeeping: Principles and Guidelines].

[133] *Id.*

[134] *Id.*

[135] Oswald et al., *supra* note 128, at 200 (emphasis added). *But see* Boddens Hosang, *supra* note 80, ¶ Rule 22.05, at 425 (where it is suggested the exercise of the mandate may be limited "when the agreement of the parties is a necessity for mission accomplishment").

force except in self-defense and defense of the mandate" is applied in such operations.[136] This wording reflects an overall restrictive approach toward the use of force that permeates the *Guidelines*. However, even within this restrictive approach the reference to "defense of the mandate" effectively acknowledges an authority use of force that on its face extends beyond the concept of individual self-defense. The question remains as to what the difference is between robust peacekeeping and peace enforcement. The *Peacekeeping: Principles and Guidelines* explains the difference in terms of robust peacekeeping being focused on the use of force at the tactical level.[137] In contrast, "peace enforcement" is said to involve "the use of military force at the strategic or international level, which is normally prohibited for Member States under Article 2(4) of the *Charter*, unless authorized by the Security Council."[138]

The suggested strategic versus tactical dichotomy loses much of its impact when other statements by UN officials are considered. The exercise of self-defense by the UN personnel in a peacekeeping context has been referred to as an "inherent right" and in terms of preserving a "collective and individual self-defence."[139] This is the unique wording found Article 51 of the UN Charter and unfortunately suggests an invocation of State self-defense principles.[140] There was also reference to the inherent right to act in self-defense being "consistently provided for in the rules of engagement established for each peacekeeping operation since its inception . . . ."[141] These comments suggest a blending of self-defense principles of a strategic nature with direction on the use of force at the tactical level.

As with State self-defense, the operation of self-defense principles such as the inherent right to act should apply at the strategic level (see Chapter 3, section 3.6.3, *Applying Self-Defense Principles Across the Armed Conflict Spectrum*). In contrast, the actual use of force is governed by humanitarian or human rights law as is appropriate to the circumstances. The force authorized should match the nature of the operational activity (i.e., enforcing criminal law, or engaging in combat). However, the *Peacekeeping: Principles and Guidelines* suggests an interpretive approach that creates uncertainty as to what principles are being applied (i.e., State or individual self-defense principles). It is an approach that is more consistent with overly restrictive interpretations of State self-defense and even human rights law principles.

The *Guidelines* indicate force is to be used as a measure of "last resort."[142] Further, it is stated:

> The use of force by a United Nations peacekeeping operation should always be calibrated in a precise, proportional and appropriate manner, within the principle of the minimum force necessary to achieve the desired effect, while sustaining consent for the mission and its mandate.[143]

---

[136] PEACEKEEPING: PRINCIPLES AND GUIDELINES, *supra* note 132, at 34.

[137] *Id.* (emphasis added).

[138] *Id.* at 34–35.

[139] OSWALD ET AL., *supra* note 128, at 563 (referring to an explanation provided by the UN Office of Legal Affairs (OLA) found in UN JURIDICAL YEARBOOK 371–72 (1993)).

[140] Corn, *supra* note 3, at 2–3.

[141] OSWALD ET AL., *supra* note 128, at 563 (referring to an explanation provided by the UN Office of Legal Affairs (OLA) found in UN JURIDICAL YEARBOOK 371–72 (1993)).

[142] PEACEKEEPING: PRINCIPLES AND GUIDELINES, *supra* note 132, at 35.

[143] *Id.*

These same principles are indicated to be applicable to both the rules of engagement (ROE) for the military component, and the Directives on the Use of Force (DUF) for the police component.[144] What is not clear is if this wording simply represents a policy-based restriction on the use of force, or a belief that even combat related uses of force must be applied in this way at law. The latter case seems more likely. However, it is also noted, "these ROE and DUF should be sufficiently robust to ensure that a United Nations peace-keeping operation retains its credibility and *freedom of action to implement its mandate*."[145]

Problematically, international humanitarian law is not designed for application in such an overly restricted legal framework. This is an issue that arises in another context where human rights tribunals choose to only apply humanitarian law "as far as possible" in interpreting the application of human rights treaty provisions during armed conflict.[146] This represents a dominant and, in effect, overarching role for human rights law that it has not historically performed in respect of humanitarian law. It is one thing to apply human rights law principles as a matter of policy in restricting how operations are conducted in a humanitarian law context (i.e., during armed conflict) and yet another to suggest this latter body of law is restricted by the former as a matter of law.

The ROE–Soldier's Pocket Card issued for the UN Mission in Sudan (UMIS) reflects this narrower approach with reference to the use of minimum force, the escalation of force, and even consideration that shots be fired "at non-vital parts of the body in order not to kill."[147] There is some indication of a broader authority to use force found in the direction to "[a]void or minimize collateral damage."[148] Further, the authorization to use deadly force to "resist attempts by any person or group that limits or intends to limit the freedom of movement of UN personnel, humanitarian workers or individuals designated by the Head of Mission" reflects a mission accomplishment goal.[149] The question that remains to be answered is whether the force used to accomplish that mission is actually governed by State self-defense principles, a law enforcement paradigm, or the conduct of hostilities framework.

Unfortunately, the *Peacekeeping: Principles and Guidelines* does not substantively address the reality that the UN forces may be engaged in an armed conflict during robust peacekeeping. Peacekeepers could very well be involved in combat. In such situations the *Principles and Guidelines* on the use of force are potentially problematic since the suggested authority to use force is far more restrictive than that available under humanitarian law. This ultimately raises the question of whether the operational pressures that lead to the

---

[144] *Id.*

[145] *Id.* (emphasis added).

[146] *See e.g.*, Hassan v. The United Kingdom, Application No. 29750/09, Eur. Ct. H. R. ¶ 104 (2014), http://hudoc.echr.coe.int/sites/eng/pages/search.aspx?i=001-146501 (the safeguards set out in the European Convention on Human Rights are to be accommodated "as *far as possible* with the taking of prisoners of war and the detention of civilians who pose a risk to security under the Third and Fourth Geneva Conventions.") (emphasis added). *But see* Advisory Opinion on the Legality of the Threat or Use of Nuclear Weapons 1996 I.C.J. 226, ¶ 25, at 240 (where the ICJ indicates that an arbitrary deprivation of life protected under the International Covenant on Civil and Political Rights "can only be decided by reference to the law applicable in armed conflict *and not deduced from the terms of the Covenant itself*.") (emphasis added).

[147] OSWALD ET AL., *supra* note 128, at 565, Annex J Rules of Engagement-Soldiers Pocket Card, Rule 7.

[148] *Id.* Rule 8, at 565.

[149] *Id.* Rule 5, at 565.

adoption of the concept of "robust" peacekeeping have been fully accepted or understood. The approach reflected in these guidelines needs to be reconciled with a 1999 *Secretary-General's Bulletin: Observance by United Nations Forces of International Humanitarian Law*.[150] While the *Peacekeeping: Principles and Guidelines* does make an initial reference to the *Secretary-General's Bulletin*, it is not mentioned again when the issue of the use of force is addressed.[151] There is no attempt to reconcile the two sets of guidance on the use of force, even though the *Secretary-General's Bulletin* indicates the fundamental principles and rules of international humanitarian law are applicable to UN forces[152] "when in situations of armed conflict they are actively engaged therein as combatants."[153] This includes "in *peacekeeping operations* when the use of force is permitted in self-defence."[154]

An acknowledgment that humanitarian law may be applicable to some aspects of peacekeeping operations does not mean that human rights–based law enforcement norms consistent with the *Peacekeeping: Principles and Guidelines* would not also apply to UN-mandated missions. The United Nations does employ police and periodically engages its military personnel in law-and-order functions. In those situations, including when assisting in maintaining order and security, the use of force would be governed by human rights–based principles. Those principles are outlined in the more narrowly proscribed *UN Basic Principles on the Use of Force*, as well as in policing standards articulated by both international and domestic courts and tribunals.[155] The conduct of such law enforcement operations would ordinarily include an authority to act in self-defense and defense of others, as well as an ability to use force to enforce the law (e.g., mission accomplishment). Such an authority to use force for mission accomplishment is entirely consistent with the defense of the mandate approach that underpins the development of robust peacekeeping.

Complicating the analysis of when humanitarian law is seen to apply to UN operations is the fact that the *Secretary-General's Bulletin* only applies to forces operating under that organization's command and control, and not "in UN-authorized peace operations under the command and control of another entity such as a state, regional arrangement or agency."[156] Non-UN-commanded operations are not "bound by administrative issuances of the Secretary-General such as the Bulletin."[157] Even when the *Secretary-General's Bulletin* covers State armed forces, they remain subject to obligations imposed by "their own national laws relating to the laws of war."[158] Given the overall lack of clarity regarding the exact application of the principles governing the use of force, the tendency is for States to rely heavily on their own national interpretations of ROE. The result is that individual State approaches on how broadly self-defense is to be interpreted will affect the manner in which troop-contributing nations fulfill their duties. However, in order to be effective, the

---

[150] *Id.* at 201–05.

[151] PEACEKEEPING: PRINCIPLES AND GUIDELINES, *supra* note 132, at 15–16.

[152] OSWALD ET AL., *supra* note 128, at 200.

[153] *Id.* at 202 (where Section 1.1 of the *Secretary-General's Bulletin* is reproduced) (emphasis added).

[154] *Id.*

[155] *UN Basic Principles on the Use of Force.*

[156] OSWALD ET AL., *supra* note 128, at 200.

[157] *Id.*

[158] *Id.*

mandate of military forces in most contemporary operations must include the ability to use force beyond a narrowly defined concept of self-defense.

Unfortunately, the development of "robust peacekeeping" has not ended either the confusion or controversy regarding UN-mandated missions. For a peace enforcement mission that authorizes measures "as may be necessary to maintain or restore international peace and security,"[159] such wording clearly provides sufficient scope to use force for mission accomplishment in addition to acting in self-defense. However, uncertainty can remain even in respect of Chapter VII peace enforcement missions. There have been operations, such as UNAMSIL in Sierra Leone, which were viewed as "pacific" in nature, since reference to "[t]he all important authority 'to use all necessary means' was not there."[160] While the mandate contemplated the use of force beyond self-defense in respect to aiding operations, "UNAMSIL chose not to interpret Resolution 1270 as was intended: they would resort to lethal forces only if attacked."[161] This raises the issue of whether Chapter VII missions launched without the traditional reference to using all necessary means are in effect a form of "robust peacekeeping."

Even when UN missions have a Chapter VII mandate this does not mean that the personnel involved will not be required to perform law enforcement–type functions. For example, the 1999 Australian led INTERFET (International Force in East Timor) mission, consisting of 10,000 troops from 22 nations,[162] was authorized to take necessary measures "to restore peace and security in East Timor, to protect and support UNAMET in carrying out its tasks and, within force capabilities, to facilitate humanitarian assistance operations . . . "[163] Although the entry into East Timor was largely nonviolent,[164] during its initial stages there were a series of combat engagements with militia forces.[165] However, even in this context a major influence on the success of the mission was a counter-war strategy that combined tight rules of engagement with an energetic approach to suppress "all forms of disorder, violence and lawfulness."[166] It is reported "[i]ndividuals found using or carrying a lethal weapon such as a firearm were subject to immediate disarmament, detention, and if they resisted to lethal force."[167] The INTERFET operation appears to have involved the use force in a variety of circumstances ranging from combat with organized groups to disarming individuals. The disarming of civilians who were not members of an organized armed group or taking a direct part of hostilities would appear to fall within a human rights–based law enforcement framework governing any use of force. This highlights that the application of human rights norms can be an integral part of any UN-mandated operations.

---

[159] UN Charter, art. 42.

[160] Richard Connaughton, Modern Warfare: The True Story of Conflict from the Falklands to Afghanistan 259 (2008).

[161] *Id.*

[162] David Kilcullen, The Accidental Guerrilla: Fighting Small Wars in the Midst of a Big One 189 (2009).

[163] S.C. Res. 1264, UN Doc. S/RES/1264, ¶ 3 (Sept. 15, 1999).

[164] Kilcullen, *supra* note 162, at 204.

[165] *Id.* at 194.

[166] *Id.* at 205

[167] *Id.*

In summary, during UN-mandated operations such as peacekeeping or even in respect of Chapter VII–based operations where there is a failure to mention the authority to use all necessary means, the use of force will not necessarily be restricted to acting in defense of self or others. Robust peacekeeping was developed precisely because the theoretical notion of peacekeepers acting solely in self-defense did not always match the mission mandate or the prevailing security situation on the ground. Some missions will even require the application of humanitarian law–based conduct of hostilities rules to control the use of force if UN peacekeepers become engaged in "combat." In the circumstances where a human rights–based law enforcement approach is applied, fulfillment of the mandate usually requires an authority to use force for the purposes of mission accomplishment. Because of the complexity of the security environments in which UN-mandated operations occur, it can be that both human rights and humanitarian law–based authorities to use force have to be applied in order to properly act in "defense of the mandate." It is in this context the suggestion found in the *Peacekeeping Principles and Guidelines* that the use of force is governed by restrictive concepts of minimum force, and the use of force in the last resort, may be most challenging to apply. It represents a tactical-level insertion of strategic narrowly interpreted self-defense-based criteria, when the use of force may be more appropriately governed by mission accomplishment or conduct of hostilities rules.

In this context the use of the term "self-defense" to describe the basis for peacekeeping can be misleading. A description of the overall peacekeeping mandate as acting in self-defense or in defense of the mandate may not fully reflect the force required at the tactical level to complete the mission. What is determinative of the required legal framework are the threats being faced (e.g., organized armed groups or criminal gangs), the mission that military forces are being asked to complete, and the overall context within which operations are being conducted (e.g., levels of violence). In a manner similar to the application of State self-defense principles, the peacekeeping mandate provides a strategic envelope within which human rights and humanitarian law are applied. In this regard the overarching principle of self-defense (e.g., defense of the mandate) referred to in a peacekeeping context influences but should not actually control the force applied by the peacekeepers. Humanitarian law governs the conduct of hostilities. Where the mission is one that occurs outside the context of an armed conflict, or a decision is made to apply a policing approach even when hostilities are taking place, the force used is governed by human rights–based principles. In doing so the performance of law enforcement duties permits the use of force beyond personal self-defense.

11.3.3. UNITED STATES RULES OF ENGAGEMENT AND "SELF-DEFENSE"

The use of the term "self-defense" as an overarching concept for providing guidance on the use of force is not limited to peacekeeping operations. It is also evident in the manner in which the U.S. ROE doctrine has been applied. The SROE is found in the Chairman of the Joint Chiefs of Staff Instruction[168] on ROE, which also includes Standing Rules for the Use of Force (SRUF).[169] The latter rules apply to law enforcement and security duties and

[168] CJCSI, *supra* note 4, at Annex A, A-1 to A-4, at 95–98.

[169] Corn & Jensen, *supra* note 86, at 808.

"are intended to apply in what are relatively clear peacetime/nonconflict situations."[170] The SROE apply not only during wartime but also to almost "everything the military does outside the continental United States."[171] This includes soldiers stationed outside the United States, aircrew on disaster assistance operations, Marines on shore leave, sailors cruising in the Mediterranean, and "members of the military patrolling neighborhoods on a United Nations peace enforcement mission or fighting in the streets against a counterinsurgency."[172] The SROE also applies to "air and maritime homeland defense missions conducted within U.S. territory, unless otherwise directed by the Secretary of Defense (SecDef)."[173] These rules control the use of force in situations ranging from individuals acting in personal self-defense to the participation of military units and formations in armed conflict. It is the scope of these operations that affect how the concept of self-defense is applied.

The SROE's bifurcated approach of relying on self-defense and mission accomplishment rules raises a key operational law issue. That is the degree to which the authority to use force can be found within the confines of self-defense or must provided through separate "authorizations to use force to accomplish the designated operational mission."[174] Ultimately, how broadly or narrowly the right to use force in self-defense is viewed affects what additional mission accomplishment ROE might be required for the successful conduct of an operation. In theory, self-defense is described in narrow terms. The applicable principles resemble those found in the human rights–based law enforcement framework. In this respect it is indicated, "the engagement authorization provided by the self-defense prong of the ROE essentially extends traditional criminal law enforcement self-defense and defense of others principles to the operational environment."[175]

However, what becomes evident is that the scope of self-defense is highly context based. What are described as criminal law–related self-defense principles are not necessarily applied in an insurgency context with the same restrictiveness that has come to be associated with domestic law enforcement. This appears to be because the threats posed by insurgent and terrorist groups are significantly greater than that offered by ordinary criminals. The result is that the term "self-defense" is often applied in an expansive manner that encompasses what are effectively armed conflict situations. Just as the United Nations has developed a "robust" interpretation of self-defense to address the force needed during more complex and violent operations (see section 11.3.2, *United Nations Operations and "Robust Peacekeeping,"* above), the SROE appear to permit such an approach as well.

The 2003 invasion of Iraq provides a recent example of pressures that the self-defense construct is placed under during contemporary armed conflict. It has been indicated that during invasion and subsequent engagement of Iraqi troops by the United States, military forces "were governed by mission accomplishment ROE, which provided robust authority to engage any Iraqi soldier upon contact."[176] In contrast, once control was established

---

170 *Id.*

171 *Id.*

172 *Id.* at 808–09.

173 CJCSI, *supra* note 4, ¶ 3 a., at 90. *But see* Operational Law Handbook, *supra* note 4, para. III. B., at 82 (where it is indicated the SRUF apply to "land-based homeland defense missions within U.S. territory").

174 Corn & Jensen, *supra* note 86, at 816.

175 *Id.* at 814.

176 *Id.* at 808.

throughout Iraq, the threat came from "civilians who opposed the U.S. presence in Iraq."[177] In those situations "it is self-defense principles that are implemented by the ROE authorizing U.S. forces to employ necessary force in response to an imminent threat directed to them or other innocent individuals."[178] A key distinguishing factor between the invasion and control phases was whether the enemy was declared to be a hostile force in the ROE. As Professors Geoffrey Corn and Eric Jensen note, in order to "destroy defeat or neutralize a designated enemy force or organization, such as the Iraqi Army in 2003, personnel associated with that force will be declared hostile pursuant to the ROE."[179]

Groups that can be declared hostile include "[a]ny civilian, paramilitary or military force or terrorist(s)."[180] Such a designation leads to "status"-based targeting, rather than relying on "conduct" or "threat" based activity, which normally involves a person exhibiting "hostile intent" or "hostile act." The application of status-based targeting rules indicates, "that the state views the nature of the military mission as sufficient to trigger the targeting authority of the laws of war."[181] It also "should trigger an analogous requirement to comply with fundamental regulatory obligations derived from the laws of war."[182] This has been described as the "[u]ltimate de facto indicator of armed conflict."[183] As a matter of law, targeting on the basis of status (i.e., membership in an organized armed group) may only occur when an armed conflict is in existence and humanitarian law can be applied.

The threat facing Coalition forces was viewed differently as control was established over Iraq. This is reflected in the ROE provided to U.S. Marines fighting in Fallujah in 2004. It can sometimes be difficult to separate which rules included in a tactical-level ROE instruction are self-defense-based from those that are mission accomplishment oriented. As the *Operational Law Handbook* notes, "the line between action for mission accomplishment and action is self-defense is not always clear."[184] However, the units involved in the Fallujah fight were provided an authority to use force that appears to have been primarily self-defense focused. This is reflected in the declaration in the ROE that "[n]o forces are declared hostile."[185] Instead, reliance was placed on individuals who were carrying arms being seen to demonstrate hostile intent and hostile act. Unless there was evidence to the contrary, they were viewed as posing "an imminent threat to Coalition Forces."[186] This is not the type of threshold criterion for the use of deadly force normally associated with a law enforcement approach. Although the lack of a declaration of a hostile force might suggest the violence was not occurring in the context of an armed conflict, that was not the case. It has been

[177] *Id.* at 811 ("In these situations, U.S. forces responded not against declared or known hostile forces, but against an otherwise protected civilian, who had decided to take up arms and act hostile to US forces.").

[178] *Id.*

[179] *Id.* at 816.

[180] CJCSI, *supra* note 4, Enclosure A, ¶ 3. d., A-3, at 97.

[181] Corn & Jensen, *supra* note 86, at 818.

[182] *Id.* at 819.

[183] *Id.* at 818 (see title to § V).

[184] Operational Law Handbook, *supra* note 4, ¶ E. 2. a. (4), at 83.

[185] Camp, *supra* note 55, at 150.

[186] *Id.* (while the ROE did not declare opposing forces as hostile, it indicated that individuals "who are carrying arms openly are demonstrating hostile act/intent unless there is evidence to the contrary; pose an imminent threat to Coalition Forces, and may be attacked....").

suggested "[i]n effect, the ROE authorized the Marines to attack a class of individuals on sight—a notion anathema to accepted notions of self-defense."[187]

The adoption of the "carrying arms openly" hostility threshold in the Fallujah ROE highlights a unique role performed by ROE in controlling the use of force. Notwithstanding the reliance on a self-defense framework in these ROE, the situation facing the American-dominated Coalition forces was one of urban combat. In the November 2004 battle for Fallujah, those insurgents were estimated to be 1,000 "hard core and two thousand part-timers."[188] Tactical intelligence indicated there were "several lines of obstacles and fortified lines of resistance containing over three hundred well-constructed defensive positions—road blocks, Hesco barriers, berms, and sniper and fighting pits."[189] These self-defense-based ROE were robust enough to sustain the use of tank main gunfire, mortar fire, artillery, and close air support in combat that resulted in the death of an estimated 1,200 insurgents,[190] but at a cost of 70 Americans killed and 609 wounded.[191] In addition, "[e]ighteen thousand of Fallujah's 39,000 buildings were damaged or destroyed."[192] What Fallujah ROE framework did accomplish was to place greater restrictions on the use of force in that tactical environment than might have been the case had the action been treated exclusively under a humanitarian law–based conduct of hostilities rules. However, the fact that the ROE adopted a self-defense approach, thereby requiring the security forces involved to adopt a more reactive posture, did not alter the reality the U.S. military was engaged in an armed conflict against determined insurgent forces.

The application of principles commonly associated with the human rights framework, such as the escalation of force, occurred in a context that was clearly an armed conflict and not simply a situation of peacetime law enforcement.[193] As Division Commander Major-General Richard Natonski noted, the intensity of the resistance was such that before risking the lives of soldiers or Marines by sending them into a building they would in some cases "drop the structure."[194] To do so they used "everything from tanks a close range to D9 armored bulldozers to 500-pound joint direct attack munitions (JDAMs)."[195] Another indication of the armed conflict context within which force was employed can be seen in the Army's use of automatic counter-radar, which unfortunately in one instance resulted in a Marine mortar position being struck in a "blue-on-blue" incident.[196] While the goal was then to "isolate that one specific building rather than damaging the entire neighborhood," this is not what is ordinarily considered to be a human rights law–related use of force. However, as was quite rightly noted in the context of the conflict, "[i]f you're taking

[187] Corn, *supra* note 3, at 5.

[188] Bing West, No True Glory: A Frontline Account of the Battle for Fallujah 256 (2005).

[189] Camp, *supra* note 55, at 143.

[190] *Id.* at 295.

[191] West, *supra* note 188, at 316.

[192] *Id.*

[193] Camp, *supra* note 55, at 151 ("When force is used, consider escalation of force procedures and limit force to that amount necessary to terminate the threat.").

[194] *Id.* at 152.

[195] *Id.*

[196] *Id.* at 212.

fire from a building, you're going to respond with whatever force is necessary to eliminate the threat."[197]

The existence of an armed conflict meant that it was permissible under humanitarian law to target the jihadist insurgents in Fallujah based on their membership in organized armed groups fighting the Iraqi government and Coalition forces. The reliance on hostile intent and hostile act in the 1st Marine Division ROE as the basis for engaging the insurgents reflects two challenges that arise when conducting operations "amongst the people." First, there is an obligation to identify potential opponents as members of organized armed groups in order for them to be targeted on that basis. This can be extremely difficult when insurgents wear civilian attire. Second, the conflict at that stage was an insurgency/counterinsurgency, and one occurring largely in an urban and therefore civilian environment. Although a large proportion of the civilian population had left Fallujah prior to the attack, an estimated 400 civilians remained.[198] There was a heightened operational requirement to exert greater control over the use of force at the tactical level in order to limit collateral civilian casualties and damage (see Chapter 7, section 7.6.2.2, *Proportionality and Counterinsurgency*).

The use of force in self-defense is highly context based. For example, if soldiers deployed on international operations are confronted with civilian rioters, then the application of self-defense-based ROE should approximate that of domestic law enforcement. However, the threat posed in Fallujah by well-armed and organized insurgents was different, as were the types of weapons, tactics, and levels of force necessary to neutralize that threat. In this instance the U.S. ROE permitted the use of force on a significantly broader scale than that normally associated with a human rights law–based framework. The concept of self-defense was applied in a mission context that primarily involved armed conflict and not law enforcement. In a sense it resembles the thinking behind the development of the theory of "robust" peacekeeping. This expansive interpretation of what constituted self-defense suggests a closer affinity with the exercise of national self-defense than human rights–based policing.

This does not mean the Fallujah ROE did not contemplate law enforcement–type activity. The rules also authorized the use of deadly force to protect "civilians from crimes that are likely to cause death or serious bodily harm, such as murder or rape."[199] There was also an authority to detain civilians "if they interfere with mission accomplishment or if requested for self-defense."[200] This wording reflects traditional human rights–based principles. What is not clear is how the principles governing the use of force in respect of civilians, which appear to be law enforcement in nature, were differentiated under the ROE from the manner in which the concepts of hostile intent and hostile act were applied to insurgents. As has been noted, escalation of force procedures and limitation of using only necessary force were required for all uses of force.

A particular challenge in assessing the Fallujah ROE is that the SROE doctrine applies a definition of imminence that is more easily associated with State self-defense than law enforcement. It is in this context that the blending of national self-defense with the authority to act in individual self-defense appears most evident. The authority to use

---

[197] *Id.* at 152 (quoting former chief of staff Brigadier-General Dunford).

[198] Ricks, *supra* note 55, at 398–99, and Camp, *supra* note 55, at 163–64.

[199] Camp, *supra* note 55, at 151.

[200] *Id.*

force in self-defense under SROE encompasses anticipatory action. In this regard the 2015 *Operational Law Handbook* notes:

> **anticipatory self-defense serves as a foundational element in the Chairman of the Joint Chiefs of Staff's Standing Rules of Engagement (CJCS SROE), as embodied in the concept of hostile intent,** which makes it clear to commanders that they do not, and should not, have to absorb the first hit before their right and obligation to exercise self-defense arises.[201]

There is no requirement to absorb the first hit under the human rights–based self-defense paradigm. However, the SROE also specifically states, "[i]mminent does not necessarily mean immediate or instantaneous."[202] This statement does not reflect the narrower meaning of imminence commonly associated with human rights–based law enforcement norms. When dealing with civilians who are not taking a direct part in the hostilities, the principle of imminence that is applied should clearly reflect the narrower meaning commonly associated with human rights law and criminal law enforcement.

The application of a broader self-defense framework, which relies on an expanded concept of imminence, sets the scene for a potentially more liberal use of force. Such an expanded interpretation of imminence could have the opposite effect of what the *Peacekeeping: Principles and Guidelines* seeks to do in respect of robust peacekeeping. The peacekeeping guidelines appear to be too restrictive, while the Fallujah ROE potentially created a situation where force might be too liberally applied in respect of civilians not participating in the fighting.[203] The challenge for those drafting ROE in either situation is to ensure that within the overarching self-defense framework the authority provided to use force is appropriate to the mission being accomplished and is adequately set out in the rules applied at the tactical level. In other words, ROE based not on the strategic level principles associated with State self-defense or the defense of a peacekeeping mandate, but rather grounded in human rights or humanitarian law.

11.4. The Way Ahead

There is considerable potential for confusion caused by the use of common terminology in the three main bodies of law governing the use of force: State self-defense, humanitarian law, and human rights law. What is clear is that human rights–based notions of self-defense have traditionally embraced a much narrower theoretical framework within which deadly force can be used. A key element in differentiating between human rights law and the other bodies of law is the application of a more restricted interpretation of how imminent an attack must be before deadly force is viewed as a justifiable response. However, the authority

---

[201] Operational Law Handbook, *supra* note 4, § III. Self-defense, ¶ C. 3. b., at 6.

[202] *Id.* ¶ E. 2. e., at 84.

[203] Corn, *supra* note 3, at 5. ("Without saying so specifically, however, the [Fallujah] ROE conflated offensive targeting and self-defense concepts, thereby creating unnecessary confusion and potentially questionable uses of force.").

to use force in a law enforcement context is not simply limited to acting in self-defense. As has been discussed, an overreliance on the authority to use force for individual self-defense or the defense of others is misleading. Whether viewed through a domestic or international law lens, the maintenance of law and order requires that security forces be authorized to use force for what doctrinally has come to be described in ROE as "mission accomplishment." It is this broader authority to use force that realistically provides the opportunity to apply human rights law–based action in order to address many contemporary security threats associated with insurgencies, terrorism, and transnational criminals.

Since transnational threats are increasingly being posed by transnational criminal organizations and criminal gangs, determining the effectiveness of human rights–based law enforcement will be a key aspect of maintaining international security. If there is no armed conflict, then human rights law is the only available legal framework for undertaking enforcement action. Further, as has been noted, in the context of an armed conflict there is a significant overlap between the human rights law and humanitarian law normative frameworks. It is the determination of how effectively law enforcement principles can be applied to address threats in conflicts which occur "amongst the people" that will establish whether this normative framework truly provides an attractive policy alternative to the potentially more violent conduct of hostilities rules provided for under humanitarian law (see Chapter 12, *The Narrow Operational and Normative Gap*). Linked to that discussion is the requirement to determine the limits of law enforcement to ensure that human rights law–based approach is not pledged to take on a task it was not designed for and cannot complete without altering its fundamental nature (see Chapter 13, *Limits of Law Enforcement and Human Rights Law*). As has been discussed in this chapter, the concept of imminence is one area of limitation.

Human rights law–based action does not only need to be differentiated from humanitarian law. With a greater awareness of the potential for law enforcement responses to be applied internationally, there is a requirement to properly situate "policing" within the legal framework governing defensive action undertaken by States. As a result of the use of common terminology there is a critical need to differentiate law enforcement from State self-defense and peacekeeping mandates. Again an important difference is the application of the principle of imminence. There is an increasing acceptance that imminence, as applied in a State self-defense context, includes a degree of anticipatory action. Traditionally, imminence has been more narrowly interpreted under human rights law. While a State may act defensively to protect its citizens and interests, including entering the sovereign territory of another State, this does not mean the use of deadly force will be more readily justified as part of a law enforcement response. The relatively more narrowly prescribed self-defense, defense of others, and mission accomplishment rules of the law enforcement paradigm would have to be applied to any use of force outside the context of the conduct of hostilities during armed conflict.

Since law enforcement is normally associated with lower levels of violence, it can be expected that the use of force by security officials, whether military, police, or intelligence personnel, would ordinarily fit comfortably within the broader framework of defensive action taken by States. However, a number of factors can alter this relationship. First, there is the tendency by human rights tribunals to apply human rights law to what are clearly armed conflict situations. This is evident in a series of cases in the European Court on

Human Rights where human rights law was used to assess airpower-based attacks during the Chechen conflict (see Chapter 13, section 13.4, *Human Rights Overreach*).[204] In some regions of the world (e.g., Europe, and Central and South America) human rights law is asked to perform the role for which humanitarian law was developed. Second, there is the use of what might ordinarily be viewed as war-fighting equipment and tactics to conduct law enforcement. Perhaps the clearest example of this phenomenon is the 2012 use of helicopter gunships to destroy pirate infrastructure in Somalia.[205] The purpose in that instance was law enforcement, however, the means were clearly designed for use in warfare. This should be kept in mind when consideration is given to the potential use of military forces to stop other criminal activity such as illegal migration into Europe.[206]

Third, the levels of violence necessary to counter terrorists, transnational criminal organizations, or criminal gangs all point to situations where the pressure will likely be to expand the interpretation of human rights law–based principles of necessity, proportionality, and imminence to the point where uses of force in the context of law enforcement could begin to resemble that permitted under humanitarian law.[207] This can be seen in the apparent acceptance of a significant number of collateral deaths and injuries resulting from the Russian hostage rescue operation highlighted in the European Court of Human Rights decision, *Finogenov v. Russia*.[208] Fourth, as is reflected in the discussion of the Fallujah ROE, care needs to be taken to differentiate law enforcement from the exercise of national self-defense. In particular, a failure to identify the requirement for a narrow application of human rights–based principles in a law enforcement context has the potential to erase the distinction between national self-defense and more individualized efforts to protect oneself and others. Finally, as narrow as the law enforcement paradigm may seem in comparison to its humanitarian law counterpart, its broader incorporation of a mission accomplishment authority to use force must be permitted to function within a potentially narrow interpretation of the overarching self-defense framework authorized for peacekeeping operations under Chapter VI of the UN Charter.

---

[204] For decisions of the European Court dealing with the use of airpower, *see, e.g.*, Isayeva v. Russia, Application Nos. 57947/00, 57948/00, 57949/00, Eur. Ct. H. R. (2005), http://hudoc.echr.coe.int/sites/eng/pages/search.aspx?i=001-68379, and Isayeva v. Russia Application No. 57959/00, Eur. Ct. H. R. (2005), http://hudoc.echr.coe.int/sites/eng/pages/search.aspx?i=001-68381, and Kerimova v. Russia, Application Nos. 17170/04, 20792/04, 22448/04, 23360/04, 5681/05 and 5684/05 Eur. Ct. H. R. (May 3, 2011), http://hudoc.echr.coe.int/sites/eng/pages/search.aspx?i=001-104662.

[205] Eliav Lieblich, *Quasi-Hostile Acts: The Limits on Forcible Disruption Operations under International Law*, 32 B.U. INT'L L.J. 101 (2014), http://papers.ssrn.com/sol3/papers.cfm?abstract_id=2324906 (for an analysis of the May 15, 2012, operation by European Union naval forces that involved the use of helicopter gunships to reportedly target pirate speed boats, fuel depots, and an armed store located in Somalia).

[206] Somini Senguta, *U.N. Wants to Let Europe Use Military Force to Stop Migrant Smuggling Boats*, N.Y. TIMES, May 6, 2015, http://www.nytimes.com/2015/05/07/world/europe/un-wants-to-let-europe-use-military-force-to-stop-migrant-smuggling-boats.html?_r=0.

[207] DAVID KILCULLEN, OUT OF THE MOUNTAINS: THE COMING OF AGE OF THE URBAN GUERRILLA 89–102 (2013) (for an outline of the background to the control exercised by the international crime syndicate, the "Shower Posse," over a "garrison district" in Kingston Jamaica and the fighting that ensued in an attempt to enforce an American extradition request against its leader).

[208] Finogenov v. Russia, IV Eur. Ct. H. R. Rep. 365 (2011) (extracts).

The FBI's role [in Iraq] raised thorny questions about the bureau's rules of engagement and whether its deadly-force policy should be modified for agents in war zones . . . . Ultimately, the FBI decided that no change was necessary.[1]

12 The Narrow Operational and Normative Gap

12.1. Introduction

The employment of military forces is commonly equated with traditional combat and a more proactive use of force, which has been described as the use of force in the "first resort."[2] This reference to a "measure of first resort" has been presented in contrast to force being applied in a State self-defense context as a matter of "last resort."[3] Last resort is also commonly used to describe human rights limitations on the application of force whether applied in peacetime or during armed conflict.[4] However, these general descriptors can mask the nuanced and demanding environments within which contemporary operations are conducted. This chapter will look at the frequently narrow operational and normative gap between law enforcement and conduct of hostilities missions. It explores the use of military forces in policing roles, as well as a unique aspect of hostage rescue where military forces often adopt tactics more consistent with a law enforcement approach, while security personnel operating under a law enforcement framework frequently apply force more closely associated with

---

[1] Adam Goldman & Julie Tate, *Inside the FBI's Secret Relationship with the Military's Special Operations*, WASH. POST, Apr. 10, 2014, http://www.washingtonpost.com/world/national-security/inside-the-fbis-secret-relationship-with-the-militarys-special-operations/2014/04/10/dcca3460-be84-11e3-b195-dd0c1174052c_story.html (quoting John Yacone, an assistant FBI Director who commanded the HRT).

[2] Geoffrey Corn, *Self-defense Targeting: Blurring the Line between the Jus ad Bellum and the Jus in Bello*, 88 INT'L L. STUD. 57, 74 (2012).

[3] *Id.* at 68.

[4] Laurie R. Blank, *Defining the Battlefield in Contemporary Conflict and Counterterrorism: Understanding the Parameters of the Zone of Combat*, GEO. J. INT'L & COMP. L. 1, 14 (2010). *See also* YORAM DINSTEIN, NON-INTERNATIONAL

the conduct of hostilities. A review of the 1997 Peruvian hostage rescue at the Japanese embassy in Lima, Peru, will highlight the risks, elevated levels of violence, and controversy that can accompany decisions to conduct such rescues.

Second, there is the challenge faced by security forces in respect of the unique threat posed by suicide bombers, improvised explosive devices (IEDs) and the use of other explosive devices where the dividing line between conduct of hostilities– and law enforcement–based uses of force are not as clear-cut as the traditional theoretical boundaries might ordinarily suggest. The reaction by security forces is often the same regardless of whether the threat arises in the context of law enforcement or conduct of hostilities tasks. In analyzing this issue, particular attention is paid to the legal test applicable to personnel using deadly force that requires they have a reasonable belief such force is necessary to resolve the incident.

12.2. Narrowing the Gap

12.2.1. THE MILITARY TASK: APPLYING FORCE OF FIRST AND LAST RESORT

The argument has been made that "when a State employs combat power in a manner that indicates it has implicitly invoked LOAC [law of armed conflict] principles (by employing deadly force as a *measure of first resort*), it is engaged in an armed conflict. As a result, it is bound to comply with core LOAC principles."[5] As has been noted, the use of force in the last resort is a crucial element of human rights–based law enforcement.[6] The last resort versus first resort dichotomy has unsurprisingly been suggested to distinguish these two normative frameworks when human rights law is compared to the humanitarian law authority to use force.[7] This terminology presents the respective frameworks governing the use of force as being dramatically different. Such references to force being applied in "first" or "last" resort once again raises the question of how the law enforcement and conduct of hostilities normative frameworks interact during international security operations.

The use of the terms "first" and "last" resort can be helpful as a form of shorthand to explain the difference in the application of conduct of hostilities rules and human rights–based law enforcement norms. However, there is also a danger that reliance on such generalizations can be potentially misleading. For example, the requirement to use force in a law enforcement situation may reflect little of the deliberative, sequential, measured response suggested by "last resort." The threat may present itself under circumstances leaving limited time for such deliberation (see Chapter 12, section 12.3.4, *Strictly Necessary, Proportionate, and Overwhelming*, below). Similarly, reference to measures of "first resort" can mask the strict legal requirements to apply precautionary steps when conducting attacks under humanitarian law, with the principle of distinction being a particularly key one where opponents hide among the population.[8] Force must be

ARMED CONFLICTS IN INTERNATIONAL LAW ¶ 114, at 38 (2014) (referring to human rights law permitting the use of force only as a last resort).

[5] Corn, *supra* note 2, at 74 (emphasis added).

[6] *UN Basic Principles on Use of Force*, at ¶ 4.

[7] Gary P. Corn, *Should the Best Offense Ever Be a Good Defense*, Strategy Research Project 28 (Carlisle Barracks, PA: U.S. Army War College, Apr. 15, 2014).

[8] AP I, art. 48.

applied during hostilities within what is by design a restrictive framework set out under humanitarian law.

A careful look at international humanitarian law targeting precautions is instructive. Among those steps is the obligation found in Article 57(2)(ii) of Additional Protocol I to "take all feasible precautions in the choice and means of attack with a *view to avoiding, and in any event minimizing*, incidental loss of civilian life, injury to civilians and damage to civilian objects."[9] While this represents a different form of "minimum force" than that associated with human rights–based law enforcement,[10] it nonetheless requires consideration of less destructive means and alternative operational methods. In respect of the rules of engagement (ROE) for the Marines operating in Fallujah in 2004, this requirement might be read into the simple rule: "[m]inimize collateral damage to innocent persons and property."[11] The "avoid or minimize" precaution serves to restrict the circumstances under which force can be lawfully applied. In this respect it is unfortunate that much of the dialogue about targeting has focused on the morally challenging and often emotionally charged final step of the targeting process, the assessment of excessive collateral civilian casualties and damage (i.e., proportionality).[12] It is through the application of the preceding "avoid and minimize" precaution that collateral effects are likely to be most effectively limited and a proportionate result achieved.

The obligation to avoid or at least minimize collateral effects highlights that the force used during the conduct of hostilities can actually be considerable more restrained than the colloquial term, "force of first resort" suggests. This requirement, combined with the pressures to limit collateral effects when fighting among the people during an insurgency (see Chapter 7, section 7.6.2.2 *Proportionality and Counterinsurgency*), can result in a more considered and measured application of force during contemporary operations than may be the popular perception. Adding the requirement for military personnel to become involved in policing establishes an operating environment that, at times, can appear even further removed from traditional conventional inter-State conflict. It is an environment where the normative gap between human rights–based law enforcement and conduct of hostilities norms can be significantly narrowed.

From an analytical and practical point of view, the approach of using the deployment of military forces as a factual indication of an armed conflict has merit (see Chapter 9, section 9.4.3, *The "Totality of the Circumstances" and Other Criteria*).[13] On its face this would suggest the deployment of forces trained and intended to use of force as a "first resort." However, as with many principles of general application such an interpretation requires closer analysis, particularly in respect of the complex security environment involving insurgencies, terrorism, and transnational crime. One example suggesting that not all uses of military force occur as a force of first resort can be seen in the restrictions identified under humanitarian law for the shoot-down of hijacked aircraft (see Chapter 13, section 13.5, *The*

[9] AP I, art. 57(2)(ii) (emphasis added).

[10] *UN Basic Principles on Use of Force*, at ¶¶ 9, 13 and 14.

[11] DICK CAMP, OPERATION PHANTOM FURY: THE ASSAULT AND CAPTURE OF FALLUJAH, IRAQ 150 (2009).

[12] AP I, art. 57(2)(a)(iii).

[13] THE GENEVA CONVENTIONS OF 12 AUGUST 1949: COMMENTARY, (IV) GENEVA CONVENTION RELATIVE TO THE PROTECTION OF CIVILIAN PERSONS IN TIME OF WAR, Art. 3, ¶ 1. A. 2., at 35 (Jean S. Pictet ed., 1958).

Shoot-Down of Hijacked Aircraft).[14] Further, the equating of armed conflict with military forces employing combat power does not fully address the scope of contemporary operations, since that same capability can be and indeed frequently is used internationally in a law enforcement role even during armed conflict.[15]

The engagement of military Special Forces units on policing tasks and the blurring of military and policing roles is evident in a variety of contemporary operations (*see* Chapter 10, section 10.4, *Converging Tactics: Special Forces, Law Enforcement, and Hostilities*). Conventional military forces have also found themselves similarly engaged during the post-9/11 period. Policing is a fundamental aspect of counterinsurgency. When working within a framework of institutions focused on maintaining order and security, it has been noted doctrinally "COIN [counterinsurgency] operations more closely resemble police work than combat operations."[16] The nature of counterinsurgency operations is such that military forces themselves will sometimes be required to perform police-like functions, such as "protecting key installations, controlling access to relatively secure areas, manning checkpoints, or detaining spoilers."[17] This is certainly the case before appropriate police forces can be effectively deployed.[18] The complex nature of contemporary conflict with insurgent armed groups relying heavily on IEDs has also forced military forces to adopt forensic policing methods to counter that threat. This has included the acquisition of air-portable forensic laboratories.[19] The resulting "introduction of scientific methods has reshaped counterinsurgency tactics, mixing police and military work, creating a seamless bridge between evidence collected on the battlefield and courtroom prosecutions years from now."[20]

Military forces have been deployed on policing operations during situations of occupation and while countering insurgencies generally. Iraq provides an obvious contemporary example where military forces were ultimately forced to engage in policing functions. In the aftermath of the 2003 invasion of Iraq, the police force (unlike the Iraqi army and other interior security forces[21]) was not formally disestablished; "they did dissolve in large part. Many though not all, abandoned their stations and went home."[22] Further, the "de-Ba'athification" program aimed at removing Saddam-era loyalists also resulted in the removal of high-level

---

14 COMMENTARY ON THE HPCR MANUAL ON INTERNATIONAL LAW APPLICABLE TO AIR AND MISSILE WARFARE, Program on Humanitarian Policy and Conflict Research at Harvard University Rule 68, at 166–67 (2010), http://ihlresearch.org/amw/Commentary%20on%20the%20HPCR%20Manual.pdf. *See also* WILLIAM BOOTHBY, THE LAW OF TARGETING 242-43 (2012).

15 Corn, *supra* note 2, at 74.

16 THE U.S. ARMY, MARINE CORPS, COUNTERINSURGENCY FIELD MANUAL ¶ 7-29, at 247 (2007) [hereinafter THE COUNTERINSURGENCY MANUAL].

17 DAVID H. BAYLEY & ROBERT M. PERITO, THE POLICE IN WAR: FIGHTING INSURGENCY, TERRORISM, AND VIOLENT CRIME 80 (2010).

18 *Id.* at 77.

19 Tu Thanh Ha, *CSI Afghanistan: Canadian Forces Bring Forensics to the Battlefield*, THE GLOBE AND MAIL (Sept. 6, 2012, 10:21 AM), http://www.theglobeandmail.com/news/national/canadian-forces-bring-forensics-to-the-battlefield/article4181923/?page=all.

20 *Id.*

21 THOMAS E. RICKS, FIASCO: THE AMERICAN MILITARY ADVENTURE IN IRAQ 162 (2007).

22 Matt Sherman & Josh Paul, *The Role of Police in Counterinsurgency Operations in Iraq, 2003–2006*, *in* POLICING INSURGENCIES: COPS AS COUNTERINSURGENTS 230 (C. Christine Fair & Sumit Ganguly eds., 2014) [hereinafter POLICING INSURGENCIES].

police officers.[23] Complicating the security situation was "Saddam's last-minute opening of prisons, criminal gangs were energized and found all the recruits they needed to launch large robberies of banks, public buildings and other sources of wealth."[24] The result was a security void that needed to be filled. It is inevitable that "[t]he higher level of insecurity confronted by foreign forces, the greater number of police roles the military may need to perform."[25]

Therefore, while there was little preconflict preparation or support for military forces to become involved in policing, it was inevitable the invading forces would be tasked to re-establish order. Unfortunately, "[r]esponsibility for law and order fell to coalition military forces that were neither trained nor equipped to perform police functions.... US soldiers complained they had not been trained to fight crime and should not be asked to arrest offenders."[26] Similarly, it is indicated that the British forces were not prepared to perform the necessary law enforcement duties in their area of responsibility.[27] The focus was primarily on conventional operations. This further highlights the degree to which State military forces generally have not been properly trained and equipped for the full spectrum of twenty-first-century warfare. A focus solely on conventional operations can have significant negative ramifications for mission success.

Inevitably, military forces became directly engaged in policing, and in assisting the reconstitution of an Iraqi policing capability. Early attempts to bring the situation under control involved joint patrols of U.S. soldiers and those Iraqi police who did continue to perform their duties.[28] However, even in situations where military forces, such as British units in Basra, made a conscious effort to change from a combat to a "hearts and minds" approach,[29] it was not a guarantee that stability could be easily restored.[30] Even when some level of stability is attained this does not mean that the policing role for military forces ends. This can be noted in a 2007 request made by the U.S. commander in Iraq for additional military police to be deployed since "the center of gravity was shifting from kinetic warfighting to police operations that would instill a law-and-order environment in the country."[31] Such specialist military police personnel and "third force" paramilitary police forces provide a valuable resource to commanders when they are available. However, there can be, due to their limited numbers, a requirement to employ conventional units in a policing role "in lieu of" military police. This happened in Iraq "during the times when MP units were hen's teeth scarce and the coalition was overwhelmed with route security, police training, and detainee holding operations."[32] It was a concept that was employed in Afghanistan as well.[33]

---

[23] *Id.* at 231.

[24] Gordon Cucullu & Chris Fontana, Warrior Police 63 (2011), and Bayley & Perito, *supra* note 17, at 7.

[25] Bayley & Perito, *supra* note 17, at 80.

[26] *Id.* at 8.

[27] Frank Ledwidge, Losing Small Wars: British Military Failure in Iraq and Afghanistan 30 (2011).

[28] Bayley & Perito, *supra* note 17, at 8.

[29] John Keegan, The Iraq War 182 (2004).

[30] Al-Skeini v. The United Kingdom, IV Eur. Ct. H. R. Rep. ¶¶ 22, at 119–20 (2011) (for reference to the 2008 Aitken Report and the challenges facing British soldiers in maintaining security in a situation considered too dangerous for the deployment of civilian police officers).

[31] Cucullu & Fontana, *supra* note 24, at 188.

[32] *Id.* at 267.

[33] *Id.*

Military involvement in a policing function can also result from the treatment of insurgents as criminals. Once detained, the criminal status assigned to captured personnel often means the application of a law enforcement approach regarding their treatment.[34] Further, not all threats posed to military forces emanate from insurgents or others who might be considered to be taking a direct part in hostilities. Criminal activity that accompanies the social discord associated with armed conflict can create an operational environment that includes well-armed criminal elements. It has been noted that although there can be considerable overlap between criminal and insurgent groups, the former "are generally locally focused and rarely move beyond a parochial environment with crime unless a significant power vacuum exists."[35] Even when drug trafficking and other crimes take on an international dimension, this does not necessarily mean that criminal activity is part of the insurgency, although as was seen in Afghanistan, Taliban reliance on the drug trade to finance operations raised challenging issues as to whether traffickers could be targeted (see Chapter, section 6.3, *Insurgency and Crime*).[36]

When tasked with maintaining law and order, regular conventional military forces must be prepared to only apply tactics and use force that is governed by human rights–based law enforcement principles. In other words, force that is used in the "last resort." The criminals would only be dealt with under a conduct of hostilities framework if they were taking a direct part in hostilities.[37] Although persons engaged in criminal activity can pose a significant threat, they remain civilians who have a protected status under international humanitarian law.[38] The resulting altered conceptual framework under which operations may be conducted in a counterinsurgency context was reflected in the postinvasion phase of the Iraq conflict when the ROE issued to U.S. forces changed from status- to conduct- or threat-based targeting rules (see Chapter 11, section 11.3.3, *United States Rules of Engagement and "Self-Defense"*).[39] It is because of the almost certain requirement for military forces to perform policing functions of one form or another during contemporary counterinsurgency operations that the incorporation of human rights norms within humanitarian law should be reinforced in military doctrine and during predeployment training (see Chapter 15, section 15.2, *Legal Advisors and Twenty-First-Century Warfare*).[40]

As was seen in both Afghanistan and Iraq, a key feature in contemporary counterinsurgency operations has been the development of police mentoring teams tasked with training and advising local police forces. United States military doctrine indicates that civilians should lead in the area of police training.[41] However, in Afghanistan, initial programs run by the German government and private contractors were viewed as not providing enough or sufficiently qualified police personnel. The training of the Afghan police ultimately became

---

[34] The Counterinsurgency Manual, *supra* note 16, ¶ 6-95, at 230.

[35] Randall Wilson, Blue Fish in a Dark Sea: Police Intelligence in a Counterinsurgency 15 (2013).

[36] Tony Geraghty, Black Ops: The Rise of Special Forces in the C.I.A., the S.A.S., and Mossad 348 (2010).

[37] AP I, art. 51(3).

[38] AP I, arts. 50 and 51.

[39] Geoffrey S. Corn & Eric Talbot Jensen, *Untying the Gordian Knot: A Proposal for Determining Applicability of the Laws of War to the War on Terror*, 81 Temple L. Rev. 787, 810–11 (2008).

[40] AP I, art. 75, and AP II, art. 4.

[41] The Counterinsurgency Manual, *supra* note 16, ¶ 6-98, at 231.

a military mission carried out under the auspices of the United States led Combined Security Transition Command–Afghanistan (CSTC-A).[42] This effort had its detractors. Highlighting the nature of the challenge, it has been suggested that "internal security forces built up by the United States have historically been poorly motivated, undisciplined, and often ruthless, owing largely to the cynical reasons people have joined."[43] Similar issues arose regarding the quality and type of recruits involved in British efforts to establish a new police force in Basra following the 2003 invasion.[44] That effort included the support and training provided to the Iraqi Serious Crimes Unit, which "became synonymous with terror for Basrawis, and more than a nuisance for the British."[45] It has also been suggested that the association of British forces in Helmand Province, Afghanistan, with corrupt and abusive Afghan police seriously undermined its credibility with the local population.[46]

The engagement of military personnel on police training and mentoring duties highlights another key aspect of contemporary operations. It is not enough to simply train military personnel on ROE that reflect law enforcement principles. There is a requirement for military planners, staff officers, and commanders to be knowledgeable and indeed comfortable with human rights–based policing. They must understand what the policing role entails, the type of forces necessary, and the training required to effectively carry out those duties. This has increasingly become a professionalism issue. Proficiency in the military profession demands knowledge and vision that is broader than the conduct of conventional operations. Given the significant engagement of U.S. military personnel in governance and ultimately policing issues in Iraq and Afghanistan, it therefore comes as no surprise that the 2015 U.S. Army *Operational Law Handbook* contains a prominent chapter on human rights.[47]

Afghanistan police initiatives included training provided by the Italian Carabinieri to Afghan Civil Order Police (ANCOP) units.[48] Coalition teams tasked with mentoring Afghan police were also staffed with military personnel. For example, in March 2009 an ISAF operation in Arghandab and lower Shah Wali Kot region of Afghanistan to pin down insurgents and target IED cells included "searches by the ANP, who were supported by Canadian police mentor teams, known as POMLETs."[49] The Afghan police often conducted searches and other activities with support from Coalition military units including the provision of close air support.[50] Such cooperation with indigenous police was not unique

---

[42] Austin Long, *The Police in Afghanistan, 2002–11*, *in* POLICING INSURGENCIES, *supra* note 22, 203, at 206–07. *See also id.* at 206.

[43] Jeremy Kuzmarov, *Afghanistan, Part I—"You Have to Not Mind Killing Innocents": American COIN Operations in Afghanistan and the Violence of Empire*, *in* HEARTS AND MINDS: A PEOPLE'S HISTORY OF COUNTERINSURGENCY 181, 188 (Hannah Gurman ed., 2013).

[44] LEDWIDGE, *supra* note 27, at 32–33.

[45] *Id.* at 33.

[46] *Id.* at 71, 81.

[47] OPERATIONAL LAW HANDBOOK, Chap. 3, at 47–57 (Lieutenant-Commander David Lee, ed., 2015), http://www.loc.gov/rr/frd/Military_Law/pdf/operational-law-handbook_2015.pdf.

[48] ROBERT M. PERITO, AFGHANISTAN'S CIVIL ORDER POLICE: VICTIM OF ITS OWN SUCCESS, Special Report 307, United States Institute for Peace 9 (May 2012), http://www.usip.org/publications/afghanistans-civil-order-police.

[49] CARL FORSBERG, THE TALIBAN'S CAMPAIGN FOR KHANDAHAR, Afghanistan Report 3, Institute for the Study of War 50 (Dec. 2009), http://www.understandingwar.org/sites/default/files/The_Talibans_Campaign_For_Kandahar.pdf.

[50] CUCULLU & FONTANA, *supra* note 24, at 245–56 (for an outline of an Afghan police led search supported by U.S. military units).

to Afghanistan. In both Iraq and Afghanistan special indigenous police units operated with foreign military Special Forces units including Special Weapons and Tactics (SWAT) units to carry out arrests of insurgent leaders.[51] The deployment of a SWAT capability with U.S. Military Police units in Iraq further highlights the overlap of roles, functions, and capabilities that can occur between military and civil police units.[52]

There also existed indigenous paramilitary police, such as the Public Order Battalions in Iraq. These were national crowd and riot control units, which did not receive training in criminal justice or have formal powers of arrest.[53] For example, once Fallujah had been cleared by U.S. Marines in November 2004, the Public Order Battalions were brought in to perform a role that "was dictated in large part by the Marines and was essentially limited to the management of secondary roadblocks at the entrances to town, the protection of key infrastructure such as the water tower and electricity generating sites, and to a visible presence at intersections around the town."[54] A similar capacity was developed in Afghanistan in the form of the ANCOP.[55] However, in terms of their interface with the civilian population, the governing legal framework for these paramilitary units would have remained human rights based.

Military forces may also be required to operate closely with their own national police forces. In situations where local police forces are nonexistent or not up to the task, it has been suggested that foreign police, "stability police units, operational police, advisers, trainers," should be deployed with the authority to "use force primarily to enforce criminal law or to protect themselves or others from attack."[56] One of the significant challenges of this approach relates to the ability of foreign police and military forces to operate among a population, which has a significantly different linguistic and cultural background. As has been noted, "[l]ocal forces have inherent advantages over outsiders in a counterinsurgency campaign."[57] They have better access to the public for intelligence purposes, generally do not need translators, and understand tribal structure and loyalties.[58] Further, successes gained in a more domestic setting are not necessarily easily transferred elsewhere. It has been observed with respect to British operations in Basra, Iraq, following the 2003 invasion that "[t]his was emphatically not Northern Ireland, although the reflex of senior officers was understandably to fall back on what they knew."[59]

While the traditional view may be that it is extraordinary for States to use law enforcement personnel outside of their borders,[60] that perspective is quickly changing. Police bring unique skills and capabilities to the "fight." Law enforcement officials have been increasingly integrated into contemporary military operations related to counterinsurgency,

---

[51] Sherman & Paul, *supra* note 22, at 236.

[52] Cucullu & Fontana, *supra* note 24, at 65.

[53] Sherman & Paul, *supra* note 22, at 236 (these units were initially formed as the Civil Intervention Force). *See also* Bayley & Perito, *supra* note 17, at 76–77 (criticizing the use of police forces as counterinsurgent forces).

[54] Sherman & Paul, *supra* note 22, at 248.

[55] Perito, *supra* note 48, at 2.

[56] Bayley & Perito, *supra* note 17, at 80.

[57] John A. Nagl, Learning to Eat Soup with a Knife: Counterinsurgency Lessons from Malaya and Vietnam xiv (2005).

[58] *Id.*

[59] Ledwidge, *supra* note 27, at 25.

[60] Corn & Jensen, *supra* note 39, at 821.

counterterrorism, and the countering of international criminal groups. The involvement of a police intelligence capability can be a key factor in the successful dismantling of an insurgent or terrorist force.[61] In this regard, police intelligence provides a particularly important capability, because "[i]ncreasingly, experts are noting . . . similarities between organized crime and insurgent networks."[62] An assessment by a French criminologist saw little difference between Somali pirates, smugglers in the Gulf of Aden, militias in Mogadishu, and Al Shabaab. They are gangs and "are all just kids."[63] The challenge is that some of these gangs are organized participants in an armed conflict (i.e., the militias and Al Shabaab), directly challenge State governance, and are armed with a capability of inflicting considerable death and destruction.

The increasingly integrated role performed by civilian police and military forces is reflected in the involvement of FBI personnel on post-9/11 operations. That involvement was considerably broader than the deployment of members of the hostage rescue unit (see Chapter 10, section 10.4, *Converging Tactics: Special Forces, Law Enforcement, and Hostilities*). It was noted in 2007 that "hundreds of FBI employees have been embedded with the military in Iraq and Afghanistan on a rotating basis."[64] Law enforcement personnel provide expertise in forensic investigation, detainee handling, the investigation of IEDs, and intelligence gathering. The integration of these personnel into military operations has been extended to the point they "may travel with the military to conduct nighttime raids, collect intelligence, and secure terrorist safe houses."[65] An enhanced FBI role can be seen in the post-torture controversy formation of an FBI-led "High Value Detainee Interrogation Team," which interrogates detainees for intelligence prior to handing them over for possible prosecution.[66] As a result, the military forces involved must be prepared to support and assist these police assets. As counterinsurgency and counterintelligence operations have matured, the military and policing roles have sometimes become increasingly blended. Finally, as is reflected in numerous operations, such as those countering piracy off the coast of Somalia and in a myriad other international operations (e.g., hostage rescue, counter drugs), many States deploy military forces internationally in what might be termed as purely law enforcement situations such as intercepting the importation of drugs.[67]

When considering how military forces are employed, it is important to look at these activities across a broad range of States. This is because some States like the United States may prefer to see military forces employed as "war fighters." For example, it has been noted

---

[61] Wilson, *supra* note 35, at 152.

[62] *Id.* at 6.

[63] James Fergusson, The World's Most Dangerous Place: Inside the Outlaw State of Somalia 124 (2013) (quoting Daniel LaDouceur, an expert in youth gang culture).

[64] *Speech, Robert S. Mueller III*, Federal Bureau of Investigation (2007), http://www.fbi.gov/news/speeches/the-fbi-and-the-military-combining-forces-to-keep-america-safe [hereinafter the *Mueller Speech*]. *See also* Sean Naylor, Relentless Strike: The Secret History of Joint Special Operations Command 276 (2015) ("At the height of the campaign against Al Qaeda in Iraq, there were nearly . . . eighty FBI personnel ("Fox Bravos," in JSOC speak) in Balad.").

[65] *Mueller Speech*, *supra* note 64.

[66] Jamie Dettmer, *After Syria Raid, FBI Interrogates ISIS Wife and Yazidi Slave*, The Daily Beast (May 17, 2015, 06:46 PM), http://www.thedailybeast.com/articles/2015/05/17/after-syria-raid-fbi-questions-isis-wife-and-yazidi-slave.html.

[67] Paul W. Forget, *Law Enforcement Detachments and the Canadian Navy: A New Counter-Drug Capability*, 7 Cdn. Nav. Rev. (2011), http://www.navalreview.ca/wp-content/uploads/public/vol7num2/vol7num2art2.pdf.

while the maintenance of maritime security has been important, "the US Navy has placed greater emphasis on war-fighting than most other naval services have done."[68] This approach must be contrasted with the reality that law enforcement conducted as part of a constabulary role on the high seas has been a long-standing role for naval forces.[69] Focusing on higher end conflict can wrongly influence perceptions of what ROE are required for international operations. Many States regularly deploy military forces on international law enforcement operations, which do not involve combat. The result is that the international deployment of military forces is not necessarily viewed as a form of combat power solely subject to humanitarian law. This is the case even for a country such as Canada that has, like the United States, traditionally applied the default humanitarian law policy approach in respect of international operations.[70] Military personnel must still apply the force of "last resort" approach during law enforcement operations.

Given the complexity of counterinsurgency and counterterrorism operations, there are a significant number of operational situations where military personnel may be called upon to use force as both a "first" and "last" resort. The application of human rights–based law enforcement and humanitarian law conduct of hostilities norms may have to be considered almost simultaneously. The threat may come from enemy fighters in one instance and from civilians not taking part in hostilities in another. For military personnel the result has been that "[s]oldiers traditionally trained to identify and destroy the enemy must adopt the policeman's paradigm of self-defense and the gradual escalation of force."[71]

12.2.2. FORCE AND HOSTAGE RESCUE

12.2.2.1. Unique Challenges and Responses

The narrowing of the normative gap between the conduct of hostilities and law enforcement paradigms is particularly evident in the context of hostage rescue. Such operations provide useful examples with which to assess the interaction between these normative regimes because hostage rescues span the conflict spectrum from purely criminal activity to inter-State armed conflict. Since hostages rescues are conducted by both military and police units, albeit ones with specialized capabilities, they serve to further highlight the narrowing differences in how such operations are conducted and force is used (see Chapter 10, section 10.4, *Converging Tactics: Special Forces, Law Enforcement, and Hostilities*). The taking of hostages is a phenomenon that has increased in frequency in the twenty-first century, in part, because of the willingness of significant portions of the international community to pay ransoms regardless of whether the actors involved are jihadists, other terrorist groups, international criminal organizations, or pirates.[72]

---

68 Martin N. Murphy, Somalia: The New Barbary?: Piracy and Islam in the Horn of Africa 165 (2011).

69 D. P. O'Connell, The Influence of Law on Sea Power 1 (1975).

70 Law of Armed Conflict at the Operational and Tactical Levels, B-GJ-005-104/FP-021 ¶ 1702. 2., at 17–1 (2001-08-13).

71 Matthew R. Modarelli, *Military Police Operations and Counterinsurgency*, *in* Small Wars Journal (2008), http://smallwarsjournal.com/blog/journal/docs-temp/151-modarelli.pdf?q=mag/docs-temp/151-modarelli.pdf.

72 Rukmini Callimachi, *Paying Ransoms, Europe Bankrolls Qaeda Terror*, N.Y. Times, July 29, 2014, http://www.nytimes.com/2014/07/30/world/africa/ransoming-citizens-europe-becomes-al-qaedas-patron.html?_r=0,

Hostage rescue operations frequently involve situations where the requirement to apply deadly force arises. There have been numerous incidents in the past decade in Afghanistan, Iraq, Nigeria, Sierra Leone, Columbia, Somalia, and elsewhere where the life of hostages were threatened and it was determined an assault was the appropriate means to effect release.

Whether conducted as part of the conduct of hostilities or thought of as an exercise of law enforcement authority, the force applied often involves the use of deadly force against the hostage takers. Operations to save hostages can be not only dangerous but can also logistically complex. In the aftermath of an era of plane hijacking by terrorists in the 1970s and 1980s, there was a worldwide development by States of a hostage rescue capability.[73] However, the existence of specialized "Tier 1" units and the capacity to project those forces internationally is restricted to a relatively small number of countries. There are several factors that make hostage rescues unique operations, whether looked at from a military or policing perspective. A key issue complicating the planning and decision-making process is the limited time period within which an operation often has to be conducted. There is even a basic fundamental challenge of first locating and then maintaining situation awareness concerning the circumstances under which hostages are being held. Tragically, two hostages were killed in a drone strike in Yemen in January 2015 because U.S. authorities did even not know they were in that location.[74]

In some instances detained persons have been moved prior to a rescue taking place. Perhaps the most famous incident was the 1970 Son Tay Raid in North Vietnam,[75] however, the challenge of mounting a rescue operation and finding the hostage in place remains an issue as was evidenced in 2014 in Syria[76] and Yemen.[77]A very real concern is that groups of hostages may be split up, thereby making it difficult to conduct a rescue operation. This happened in Lebanon in 1985 after an aircraft hijacking. Hezbollah terrorists seized nine Americans who were "spirited into hiding and scattered throughout the city to thwart any attempted rescue operations by U.S. military forces."[78] Another example was *Operation Barras* conducted Sierra Leone in 2000, where the concern was that the British hostages might be transferred farther into the jungle. This turned out to be a key consideration for launching the raid to rescue them (see Chapter 10, section 10.3.3, *Modern Counterterrorism: Sierra Leone*).[79] In 2010, hostages held in a largely inaccessible area of Afghanistan by a criminal gang were

---

and Adjoa Anymadu, *With Somali Pirates, Pay the Ransom Until There's Global Consensus*, N.Y. TIMES,Aug. 7, 2013, http://www.nytimes.com/roomfordebate/2013/08/07/when-ransoms-pay-for-terrorism/with-somali-pirates-pay-the-ransom-until-theres-global-consensus.

[73] LEROY THOMPSON, THE RESCUERS: THE WORLD'S TOP ANTI-TERRORIST UNITS 55–156 (1986).

[74] CHARLIE SAVAGE, POWER WARS: INSIDE OBAMA'S POST 9/11 PRESIDENCY 287-88 (2015). .

[75] BENJAMIN F. SCHEMMER, THE RAID 209 (1976)(prisoners taken in conflict are technically not "hostages", however their rescue often involves the same use of force issues).

[76] Michael D. Shear & Eric Schmitt, *In Raid to Save Foley and Other Hostages, U.S. Found None*, N.Y. TIMES, Aug. 20, 2014, http://www.nytimes.com/2014/08/21/world/middleeast/us-commandos-tried-to-rescue-foley-and-other-hostages.html.

[77] *Yemen Raid: Rescuers Missed Western Hostages*, BBC NEWS (Nov. 27, 2014), http://www.bbc.com/news/world-middle-east-30225146.

[78] BRUCE HOFFMAN, INSIDE TERRORISM 174 (2006), and MATTHEW LEVITT, HEZBOLLAH: THE GLOBAL FOOTPRINT OF LEBANON'S PART OF GOD 39 (2013).

[79] WILL FOWLER, CERTAIN DEATH IN SIERRA LEONE: THE SAS AND OPERATION *BARRAS* 2000, at 32 (2010).

under a grave threat, if not from the immediate captors then by others to whom the persons may be transferred. The transfer of hostages across the porous border to Pakistan would have made it significantly more difficult to locate them once moved.[80]

There was also a concern in Iraq following the 2003 invasion that the seizure of foreigners had become a criminal enterprise. In this respect, "if your prize was American or British, the greatest return could come from selling them to extremists."[81] The same issue arose regarding the 2011 abduction of an American and a Danish aid worker in Somalia. During unsuccessful negotiations for their release the kidnappers threatened to transfer the hostages to Al Shabaab.[82] Another factor weighing heavily on any decision to proceed with a rescue is the conditions of captivity and the risk to the health of the hostages. [83] However, these are not the only factors that need to be addressed. Such operations are often conducted under difficult and complex geographic conditions, with an incomplete intelligence picture and normally under significant time constraints. As was noted in respect of one hostage rescue operation in Afghanistan, there was a danger that launching too early would result in the hostages being executed, while waiting too long could mean the hostages would disappear and "resurface weeks later facing execution on an al-Qaeda video."[84] The intelligence being relied on, whether human or signals related, may also be temporally limited in terms of usefulness. There can be significant limitations on what intelligence can be collected. Technical means, such as drones, satellites, and interception of communications, can provide certain information. However, there also can be significant advantages in having human intelligence and even visual observation of where a hostage is being held.

Not every decision to act can be based on observations of an actual immediate threat to the hostages such as occurred in the 2009 *Maersk Alabama* incident.[85] There will often be a requirement to interpret the best available information in order to assess how imminent a threat may be without having direct evidence that it will be carried out. In one instance in Afghanistan there was an intercepted phone call where the Taliban urged the kidnappers to "make a declaration of intent."[86] This was interpreted to mean killing a hostage, which was viewed as a "game changer," resulting in the decision to make an assault.[87] A real concern may also exist that a hostage may be killed once the captors know a rescue operation has been launched. This has occurred in situations of armed conflict,[88] with kidnappings by

---

80 Julian Borger, *Linda Norgrove: US Navy Seal Faces Disciplinary Action over Grenade Death*, THE GUARDIAN, Oct. 13, 2010, http://www.guardian.co.uk/world/2010/oct/13/linda-norgrove-us-commando-disciplinary.

81 MARK URBAN, TASK FORCE BLACK 126 (2010).

82 JESSICA BUCHANAN, ERIK LANDEMALM, & ANTHONY FLACCO, IMPOSSIBLE ODDS: THE KIDNAPPING OF JESSICA BUCHANAN AND HER DRAMATIC RESCUE BY SEAL TEAM SIX 209 (2013).

83 *Id.* at 233–34.

84 Sean Rayment, *How the British Hostages Were Rescued in Afghanistan*, THE TELEGRAPH, June 3, 2012, http://www.telegraph.co.uk/news/worldnews/asia/afghanistan/9307833/How-the-British-hostages-were-rescued-in-Afghanistan.html.

85 Robert D. McFadden & Scott Shane, *In Rescue of Captain, Navy Kills 3 Pirates*, N.Y. TIMES, Apr. 13, 2009, http://www.nytimes.com/2009/04/13/world/africa/13pirates.html?pagewanted=all.

86 Rayment, *supra* note 84.

87 *Id.*

88 Chris Kraul, *Rebels Executed Four Longtime Captives, Colombia Officials Say*, L.A. TIMES, Nov. 27, 2011, http://articles.latimes.com/2011/nov/27/world/la-fg-colombia-farc-20111127.

groups associated with transnational terrorists,[89] and those carried out by other organized groups.[90]

The situation can also be such that any sighting of the security forces attempting the rescue will result in a firefight. With any group holding hostages there may be personnel on guard who are blocking access to where hostages are being held. Those guards may not themselves be directly or immediately threatening the lives of the hostages in terms of their being in very close physical proximity to where those persons being rescued are being held. However, the giving of a warning may not only compromise the mission but also result in the harming of the hostages. The challenge is how the guards can be neutralized without alerting their companions in the immediate vicinity of the hostages who may then kill the detainees. This is less of an issue if the guards are members of an organized armed group participating in an armed conflict with the intervening State. They can be killed outright because of their status as members of that group and therefore as direct participants in hostilities. However, the hostage takers may not be participants in an armed conflict. In those circumstances they can only subjected to a law enforcement response. The presence of those guards can directly impact on how an operation is conducted and significantly limit the ability to carry out a rescue without a more proactive use of deadly force. This raises the difficult question of whether a warning is required under a human rights–based law enforcement approach, and if the guards can be proactively engaged with deadly force (see Chapter 11, section 11.2.4, *Imminence and the Nature of the Threat*).

While exceptional, the *UN Principles on the Use of Force* contemplates the use of deadly force without warning in situations where "to do so *would unduly place the law enforcement officials at risk or would create a risk of death or serious harm to other persons, or would be clearly inappropriate or pointless in the circumstances of the incident*."[91] Clearly, the potential risk to law enforcement officials and the hostages would be valid considerations in deciding to forgo a warning. This could be of particular relevance when working in a hostile environment where prior warning could result in rescue force being attacked or otherwise obstructed. Concern that a hostage might be moved prior to or even during an operation would also appear to fall within the exception set out in these principles. The result is that hostage rescue operations often proceed without any warning in order to maximize the element of surprise.

Attempting to subdue even small groups of armed hostage takers without resorting to deadly force can present a significant challenge. The human rights–based legal regime has its genesis in the less violent tasks of suppressing ordinary crime, and "situations of internal disturbances and tensions, such as riots, isolated and sporadic acts of violence and other acts of a similar nature."[92] However, the use of deadly force, even during law enforcement–based hostage rescue operations, is quite common. The likelihood of deadly force being used is often increased by the willingness many hostage takers to employ explosives (see section

---

[89] Greg Mullany, *French Soldier Killed in Somalia Commando Raid*, N.Y. Times, Jan. 12, 2013, http://www.nytimes.com/2013/01/13/world/africa/france-somalia-hostage-raid.html.

[90] Elisabetta Povoledo & Johan F. Burns, *Failed Raid to Rescue Hostages in Nigeria Stirs Italy's Anger*, N.Y. Times, Mar. 9, 2012, http://www.nytimes.com/2012/03/10/world/africa/britain-italy-in-spat-over-failed-rescue-try-in-nigeria.html.

[91] *UN Principles on the Use of Force* ¶ 10 (emphasis added).

[92] AP II, art. 1(2).

12.3.4, *Strictly Necessary, Proportionate, and Overwhelming*, below). Hostage situations present a category of incidents where once a decision is made that an assault is necessary, the use of deadly force by security personnel against the hostage takers is highly likely.

The deadly outcome that can accompany hostage rescue operations is reflected in the reported freeing of four aid workers in Afghanistan held by a group that had been in contact with the Taliban. The status of the kidnappers is not clear, although it is reported they were demanding a £6 million ransom and the release of a colleague, and "the kidnappers were in communication with the Taliban."[93] They "were armed with the ubiquitous but deadly AK47 assault rifle, rocket propelled grenades and a PKM, a Russian belt-fed machine-gun, which is capable of shooting down a helicopter."[94] In conducting the rescue "[t]he soldiers [British SAS] moved into the darkness shooting dead the kidnappers with silenced weapons. Several were dispatched with a 'double tap' the preferred method of killing—two bullets in the center of the forehead. The US special forces cleared and secured their target, killing seven kidnappers in the process...."[95] After initial confusion regarding whether all the hostages were located, their rescue was reported along with the indication that four more kidnappers had been killed.[96]

To the extent that the kidnappers may have been members of an organized armed group engaged in the insurgency in Afghanistan, the use of force fits comfortably within conduct of hostilities rules. If reliance was being placed on a human rights–based law enforcement framework, the use of force would have to be justified on the basis of unique security situation facing the rescue force, the lethal risks they faced personally, and in particular the deadly threat to the hostages posed by the kidnappers. Similarly, in a 2011 hostage rescue in Somalia it appears that the American SEAL liberators shot the criminal abductors as soon as the alarm began to be given as the operation unfolded.[97] Further, as was demonstrated in the shooting of three Somali pirates in order free the captain of the *Maersk Alabama*, deadly force may need to be used directly against those in close vicinity of the hostages.[98] While this does not mean that the hostage takers will necessarily all be killed during an operation, this form of assault inherently carries with it significant risks and deadly consequences.[99]

Hostage rescue situations do not mean there is no room for negotiations or even warnings where appropriate or feasible. Hostage situations have been resolved through negotiations. However, the seizure of hostages in remote areas where it is difficult to establish and maintain control over the scene can make communications difficult, negotiations complicated, and ultimately fraught with uncertainty. A former Canadian diplomat, Robert Fowler, and his Canadian government assistant, Louis Guay, were seized in Niger in 2010 by AQIM, and held for 130 days. He has noted "[o]ur release was in so many respects 'a near run thing'...."[100] Many hostages are simply executed if the negotiations are not

---

[93] Rayment, *supra* note 84.

[94] *Id.*

[95] *Id.*

[96] *Id.*

[97] Buchanan et al., *supra* note 82, at 250.

[98] Mark Owen, No Easy Day: The Autobiography of a Navy SEAL 89–98 (2012).

[99] *Frenchman Dies in Somali Rescue*, BBC News (Apr. 11, 2009), *available at* http://news.bbc.co.uk/2/hi/in_depth/7994201.stm.

[100] Robert R. Fowler, A Season in Hell 312 (2011).

successful.[101] In a law enforcement context and in certain hostilities situations, capture is preferred. However, when the decision is made to launch an assault, the risk to both the hostages and the hostage rescuers is often such that the use of deadly force is required to successfully complete the mission. The likelihood of death rather than capture of the kidnappers is not unique to situations of armed conflict or international operations. Domestic hostage rescue situations can also result in the death of a hostage taker.[102]

Generally, courts are sensitive to the unique demands of hostage rescue and the risks that are an integral part of such activity even in what can be viewed as more domestic applications of force. The European Court of Human Rights dealt with a hostage rescue–related assault in *Andronicou v. Cyprus*.[103] In that case a special police unit was tasked with ending an incident where a distraught man armed with a shotgun was holding his fiancée hostage. Negotiations failed, and it was believed the man would harm his hostage and then commit suicide by midnight. An assault was ordered during which the hostage taker fired on the police officers and the hostage. Tragically both the hostage and the hostage taker were killed when the police responded to that threat. In discussing the force used, a badly split court found under the circumstances it was not excessive, although twenty-five bullets struck the hostage taker[104] and the woman was hit by two police bullets, which caused her death.[105]

The court noted it was regrettable that so much force was used. However, the decision went on to state:

> the Court cannot with detached reflection substitute its own assessment of the situation for that of the officers who were required to react in the heat of the moment in what was for them a unique and unprecedented operation to save life. The officers were entitled to open fire for this purpose and to take all measures which they honestly and reasonably believed were necessary to eliminate any risk either to the young woman's life or to their own lives.[106]

This reasoning is reflective of the approach taken by the U.S. Supreme Court that "[d]etached reflection cannot be demanded in the presence of an upturned knife,"[107] and that "[p]olice officers are often forced to make split-second judgments—in circumstances that are tense, uncertain, and rapidly evolving—about the amount of force that is necessary in a particular situation."[108] In effect, the unique circumstances of these situations result in the justification of the use of deadly force even under the restrictive law enforcement regime.

---

[101] *Id.* at 312–13.

[102] *See, e.g.*, *Toronto Hostage-taker Had History of Domestic Violence: Reports*, CBC News (Aug. 26, 2004), http://www.cbc.ca/news/canada/story/2004/08/25/hostage_union040825.html.

[103] Andronicou v. Cyprus, Application Nos. 86/1996/705/897, Eur. Ct. H. R. (Oct. 9, 1997), http://hudoc.echr.coe.int/sites/eng/pages/search.aspx?i=001-58102.

[104] *Id.* ¶ 107.

[105] *Id.* ¶¶ 111–14.

[106] *Id.* ¶ 192 and ¶¶ 193–94. *See also* Bubbins v. The United Kingdom, App. No. 50196/99, Eur. Ct. H.R., ¶ 139 (2005), http://hudoc.echr.coe.int/sites/eng/pages/search.aspx?i=001-68548.

[107] Brown v. United States 41 Sup. Ct. 502 (1921).

[108] Graham v. Connor 109 S. Ct. 1872 (1989).

Assaults against terrorists hijacking aircraft,[109] buses,[110] and trains,[111] or seizing embassies[112] and other buildings[113] have frequently resulted in the deaths of most if not all of the hostage takers. The hostages also face considerable risk of death.[114] They are not only in danger of being killed by the hostage takers prior to a rescue, the hostages may also be killed or injured during the assault. This can be unintentional by the rescuers.[115] One such incident was a 1985 Egyptian Special Forces assault on an airliner in Malta where two Egyptian sky marshals and five hostages were killed by the terrorists prior to the rescue operation, and three of the terrorists and fifty-seven hostages died during the assault.[116] In 1974, terrorists holding hostages in a school in Ma'alot, Israel, killed twenty-one children with gunfire and grenades, and wounded fifty more as Israeli Special Forces personnel attempted a rescue.[117] This incident and a subsequent 1975 hostage taking at the Savoy Hotel in Tel Aviv where three soldiers and eight hostages died[118] prompted "a stringent reappraisal of Sayeret Matkal [Israeli Special Forces unit] doctrine and procedures in counter-terrorism."[119]

Other incidents involving significant casualties among the hostages occurred in Saudi Arabia and Russia. In 1979 Sunni fundamentalists seized the Grand Mosque in Mecca. The rescue operation was carried out by Saudi security services with advice being provided by the French Groupe d'Intervention de la Gendarmerie Nationale (GIGN). The rescue involved the use of a potent chemical said to be a gas similar to that used by Russian troops in the 2002 Moscow theater incident.[120] The Mecca operation is reported to have resulted in 26 pilgrims being killed and 100 wounded. Saudi government forces casualties were indicated to be 127 killed and 450 wounded. The rebel deaths were reported as 117,[121] along with 170 militants arrested and 67 subsequently beheaded.[122] While most of those held captive were released prior to the assault, a number of pilgrims

---

109 Chris McNabb, Storming Flight 181: GSG 9 and the Mogadishu Hijack 1977 (2011).

110 Michael Bar-Zohar & Nissim Mishal, No Mission is Impossible: The Death Defying Missions of the Israeli Special Forces 257–58 (2015).

111 Peter Harclerode, Secret Soldiers: Special Forces in the War Against Terrorism 280–82 (2000).

112 Luis Giampietri, 41 Seconds to Freedom: An Insider's Account of the Lima Hostage Crisis, 1996–97 (2007).

113 Harclerode, *supra* note 111, at 285–86 (where in 1981 the Spanish *Grupo Especial de Operaciones* (GEO) of the Polica Nacional conducted a rescue operation in respect of 200 hostages seized in a bank by a 27 man gang although it was subsequently determined the hostage takers were members of a criminal gang rather than a terrorist group).

114 Bar-Zohar & Mishal, *supra* note 110, at 257–58 (for an account of the "Mother's Bus" incident in which terrorists threatened to kill a hostage every ½ hour. The terrorists killed three hostages including two during the assault. All the hostage takers were killed).

115 Saul David, Operation Thunderbolt: Flight 139 and the Raid on Entebbe Airport, The Most Audacious Hostage Rescue Mission in History 299–300 (2015)(outlining where one hostage was mistaken as a terrorist and killed by rescuers during the assault). *See also* Naylor, *supra* note 64, at 371–75 (outlining the circumstances surrounding the death of British hostage Linda Norgrove caused by the use of a grenade during a SEAL Team 6 rescue operation in Afghanistan in 2010).

116 J. Paul de B. Taillon, Hijacking and Hostages: Government Responses to Terrorism 150–56 (2002).

117 Muki Betser, Secret Soldier 275 (1996).

118 Iddo Netanyahu, Yoni's Last Battle: The Rescue at Entebbe, 1976, at 69 n. * (2002).

119 Simon Dunstan, Israel's Lightning Strike: The Raid on Entebbe 1976, at 12 (2009).

120 Yaroslav Trofimov, The Siege of Mecca: The Forgotten Uprising in Islam's Holiest Shrine and the Birth of Al Qaeda 191–92 (2007).

121 *Id.* at 224–25.

122 Dilip Hiro, War Without End: The Rise of Islamist Terrorism and Global Response 140 (2002), and Trofimov, *supra* note 120, at 222, 239 (where it is indicated 63 prisoners were beheaded).

and captured police and military personnel had remained hostage. The number of hostages killed during the assault is reported to have been around 26, with 110 wounded.[123] However, the casualties are believed to have been much higher than stated.[124] It is obvious that the retaking of the mosque and rescuing the hostages involved significant levels of violence that is inconsistent with the incident being viewed as a uniquely law enforcement operation.

In 2002, approximately 900 hostages were seized in a Moscow theater by more than 40 Chechen terrorists, of whom 18 were suicide bombers.[125] Although over 730 hostages were released,[126] 129 hostages died when Russian Special Forces Russian used fentanyl gas to incapacitate the terrorists prior to an assault.[127] It was commandos from the Federal Security Service (FSB) Alpha counterterrorism unit that led the operation.[128] A subsequent siege of a School in Beslan, Ossetia, in 2004 stands out as the bloodiest of hostage rescues. That incident saw 1,100 school children, parents, and teachers taken hostage by Chechen terrorists[129] seeking an end to the Chechen conflict and a withdrawal of Russian forces.[130] On the first day of the siege the terrorists were reported to have executed approximately 20 males in order to intimidate other hostages, let Russian authorities know their threats were credible, and "get rid of physically strong male hostages who could have caused the terrorists difficulties . . . ."[131]

After three days of negotiations, explosions led to a poorly coordinated assault by secret service, police, and military formations. The weapons used by the security forces during the assault included 125 mm tank cannon and RPO-A incendiary (thermo baric) rockets.[132] It is said that official Russian statistics indicate 330 individuals were killed, including 317 hostages. Of the hostages killed, 186 were children. Seven hundred and twenty-eight persons were wounded. Two bombs in the gymnasium were detonated, killing some of the hostages, and the terrorists also fired on a number of them.[133] The majority of the deceased hostages perished when the roof of the school collapsed, reportedly caused by the attack carried out by Russian Special Forces. Of the 34 terrorists only one was captured.[134] The disastrous operation is said to have prompted significant changes in Russian counterterrorist training and the development of a better hostage negotiation capability.[135]

Even with the most precise and careful rescue operations the threat to the hostages and rescuers can be significant. Compounding that risk is the use of explosive devices by hostage takers. Examples of when such devices have been deployed by terrorists include the

---

[123] Trofimov, *supra* note 120, at 224.

[124] *Id.* at 225.

[125] Finogenov v. Russia, VI Eur. Ct. H. R. ¶ 8, at 374 (2011) (extracts) (where it is indicated over 900 people were held hostage).

[126] *Id.* ¶ 24, at 377.

[127] *Id.*

[128] Mark Galeotti, Russian Security Forces and Paramilitary Forces since 1991, at 38 (2013).

[129] *Id.* at 18.

[130] John B. Dunlop, The 2002 Dubrovka and 2004 Beslan Hostage Crises: A Critique of Russian Counter-Terrorism 75 (2006).

[131] *Id.* at 53–54.

[132] Galeotti, *supra* note 128, at 19.

[133] *Id.* at 18.

[134] *Id.* at 19 (where it is reported 334 hostages, 12 policemen and soldiers, and all but one of the over 30 hostage takers died)

[135] *Id.*

1972 Sabena airlines hostage rescue in Tel Aviv,[136] the 1977 Mogadishu hijacked aircraft rescue,[137] the 1997 Japanese embassy seizure in Lima, Peru,[138] the 1994 seizure of an Air France passenger jet,[139] the 2002 the Moscow theater incident,[140] and the 2004 Beslan School siege.[141] Explosive devices, whether they are IEDs, suicide vests, landmines, or some other form of explosive device, can increase the risks to members of hostage rescue units and the hostages.[142] It is even more likely that assaulting personnel will use deadly force if there is reason to believe the hostage takers may be wearing suicide vests or have rigged explosives to detonate during a rescue operation (see section 12.3.4, *Strictly Necessary, Proportionate, and Overwhelming*, below).

One particular intelligence challenge is determining if the existence of explosive devices can be definitively established. In this respect during the Israeli 1976 Entebbe rescue operation, even though there were indications it was unlikely that the terminal building where hostages were held was rigged with demolitions,[143] there was a lingering uncertainty as to whether explosive devices would be ignited.[144] One participant in the assault stated that during the operation "[a]t any second, I feared, the terrorists would ignite the explosives they planted in the hallway."[145] The threat does not need to be as elaborate as a large improvised explosive. The potential use of grenades or homemade explosives by the hostage takers significantly changes the threat posed to the hostages and the rescuers.[146] At one point during the 1976 Entebbe raid, Israeli commandos did not know if individuals they encountered were terrorists or hostages. A grenade belt was noticed around one man's waist, and upon shooting the terrorist, a grenade was dropped and exploded.[147]

During the 1977 Mogadishu hostage rescue operation by Germany's GSG-9, one of the terrorists on the hijacked plane managed to release two primed hand grenades prior to being shot. However, the grenades were of poor quality and only two passengers were wounded.[148] A similar discovery of a grenade being carried by a terrorist occurred during the 1980 London embassy siege.[149] In 1994 Algerian terrorists of the Armed Islamic Group (AIG) used a grenade in an attempt to repel a GIGN assault on a seized commercial airliner at Marseille airport.[150] During the 2008 Mumbai attack, each of the Lashkar-e-Tabia (LeT)

---

136 Bar-Zohar & Mishal, *supra* note 110, at 140–53.

137 McNabb, *supra* note 109, at 38.

138 Giampietri, *supra* note 112, at 10–11, 31, 36–37.

139 Marten C. Arostegui, Twilight Warriors: Inside the World's Special Forces 441–42 (1996).

140 Finogenov v. Russia, VI Eur. Ct. H. R. ¶ 8, at 374 (2011) (extracts).

141 Galeotti, *supra* note 128, at 18.

142 Bar-Zohar & Mishal, *supra* note 110, at 149 (where it is noted the presence of explosives on the Sabena passenger plane in Tel Aviv in 1970 made finding and neutralizing the charges the main concern of the rescuers).

143 Netanyahu, *supra* note 118, at 79, William H. McRaven, Spec Ops: Case Studies in Special Operations Warfare: Theory and Practice 372 (1995) and Bar-Zohar & Mishal, *supra* note 110, at 13.

144 McRaven, *supra* note 143, at 347–48.

145 Betser, *supra* note 117, at 330.

146 McNabb, *supra* note 109, at 17. *See* Bar-Zohar & Mishal, *supra* note 110, at 141, 146 and 151 (for reference to hijackers being armed with grenades during the seizure of a Sabena airlines flight in 1970).

147 McRaven, *supra* note 143, at 362 and David, *supra* note 115, at 302–03.

148 McNabb, *supra* note 109, at 64–65. *See also* David, *supra* note 115, at 9–10 and 28–29 (for reference to the Entebbe hijackers carrying grenades and planting explosive devices during that 1976 hijacking).

149 Geraghty, *supra* note 36, at 210.

150 Arostegui, *supra* note 139, at 446.

terrorists were equipped with AK 47s, 8 kilos of RDX explosives, and hand grenades.[151] The terrorists who carried out the Japanese embassy seizure in Lima, Peru, were, in addition to carrying automatic weapons and rifles, armed with grenades and a bag of *quesos rusos*[152] (an improved explosive consisting of a metal container such as a beer or fruit can packed with plastic explosives and nails or scrap metal).[153]

That resort must be made to deadly force in circumstances where the captors put their hostages at such risk highlights the determination (often fanaticism) and deadly threat posed by many hostage takers. The result is that notwithstanding the critical importance of the right to life, difficult decisions will often have to be made requiring the balancing of that right in respect of the hostages, the hostage takers, and the security personnel. Ultimately, these are situations where, even under a law enforcement paradigm, deadly force frequently has to be resorted to in circumstances that come close to approximating its use in armed conflict. However, the unique circumstances of hostage rescue do not only affect force used when applied as a law enforcement matter. Even when the hostage takers can be targeted under conduct of hostilities rules because they are direct participants in hostilities, there still remains the requirement to consider the risk to both State security personnel and the hostages themselves in deciding to carry out an assault. This has included during international armed conflict (see Chapter 10, section 10.3.1, *Rescues During Armed Conflict*). Such considerations can dramatically change the tactics and levels of force that might otherwise be used against these non-State actors in ordinary combat.

12.2.2.2. The Japanese Embassy, Lima, Peru, 1997

The use of deadly force during hostage rescue operations can be controversial. One such example is the seizure of the Japanese embassy in Lima, Peru, on December 17, 1996, by members of the Túpac Amaru Revolutionary Movement (MRTA), and the subsequent rescue operation carried out by Peruvian Special Forces. Both the factual situation and the analysis of the legal approaches applied by the Inter-American Commission on Human Rigths and Inter-American Court of Human Rights highlight the complexities of hostage rescues occurring at the boundaries of human rights and humanitarian law.[154] The April 1997 assault to rescue hostages at the Japanese embassy took place in a context a long-running insurgency. The MRTA was a terrorist group (they were admirers of Che Guevara) that started operations in 1984, and at their peak in the late 1980s numbered 3,000 members. By late 1996 the group had been largely decimated by an amnesty program and arrests.[155] However, fourteen members of the MRTA seized the embassy during a celebration for the

[151] Cathy Scott-Clark & Adrian Levy, The Siege: 68 Hours Inside the Taj Hotel 123 (2013).

[152] Giampietri, *supra* note 112, at 12.

[153] *Id.* at 10.

[154] Eduardo Nicolás Cruz Sánchez, Et Al Peru, Case 12.444, Merits, Inter-Am. Comm'n H.R., Report No. 66/10 (2011), https://www.oas.org/en/iachr/decisions/court/PEME12.444ENG.pdf and Case of Cruz Sánchez et al. v. Peru, Series C no. 292, Preliminary Objections, Merits, Reparations and Costs, Inter-Am. Ct. Hum. Rts (2015), http://www.corteidh.or.cr/docs/casos/articulos/seriec_292_esp.pdf.

[155] James Brooke, *The Rebels and the Cause: 12 Years of Peru's Turmoil*, N.Y. Times, Dec. 19, 1996, http://www.nytimes.com/1996/12/19/world/the-rebels-and-the-cause-12-years-of-peru-s-turmoil.html.

Japanese emperor's birthday.[156] Among the approximately 600 hostages was the Peruvian foreign minister, members of the president's family, senior Peruvian military and police officers, members of the judiciary, other government officials, and members of the foreign diplomatic corps.[157] The demands of the terrorists included changes in economic policy, as well as the release and safe conduct of imprisoned MRTA members.[158]

The terrorists were armed with rifles, automatic weapons, rocket launchers, pistols and grenades. They also planted improvised explosives and landmines to block any assault on the embassy. Handheld devices (i.e., claymore "clackers") generating an electric current were connected to daisy chains of improvised explosive devises placed around doors and windows. The explosives were wired to ignite without a delay.[159] At an early point during the siege one terrorist was reported to be wearing a suicide vest along with a ring attached to a fuse lighter.[160] The hostage takers also practiced quick reaction drills where one part of their group would plan to repel any assault, while others would actively seek out and kill hostages.[161] Negotiations took place over a four-month period in an attempt to try and resolve the dispute.[162] Most of the hostages were released by the beginning of January 1997, with seventy-two being left in the residence.[163] During the negotiations the security forces used the time to dig tunnels under the embassy to facilitate an assault should it be required.[164]

On April 22, 1997, after a siege of 126 days,[165] the Peruvian government gave direction to assault the embassy even though their tunneling operation had been discovered.[166] This direction came after reaching the conclusion negotiations were fruitless and out of concern for the remaining hostages.[167] The operational plan identified the goals of seizing control of the property, rescuing the hostages, as well as capturing or eliminating the MRTA hostage takers. It also indicated they "(should) not commit excesses of any sort, maintaining absolute respect for HR [human rights]."[168] The layout of the residence, location of hostages, morale,

---

156 Case of Cruz Sánchez et al. v. Peru, Series C no. 292, Preliminary Objections, Merits, Reparations and Costs, Inter-Am. Ct. Hum. Rts ¶ 144-46, at 43-44 (2015), http://www.corteidh.or.cr/docs/casos/articulos/seriec_292_esp.pdf and Eduardo Nicolás Cruz Sánchez, Et Al Peru, Case 12.444, Merits, Inter-Am. Comm'n H.R., Report No. 66/10 ¶ 50-52 (2011), https://www.oas.org/en/iachr/decisions/court/PEME12.444ENG.pdf.

157 Case of Cruz Sánchez et al. v. Peru, Series C no. 292, Preliminary Objections, Merits, Reparations and Costs, Inter-Am. Ct. Hum. Rts ¶ 144, at 43 (2015), http://www.corteidh.or.cr/docs/casos/articulos/seriec_292_esp.pdf.

158 *Id.* ¶ 148, at 44–45.

159 Giampietri, *supra* note 112, at 10–11 and 31. *See also* Case of Cruz Sánchez et al. v. Peru, Series C no. 292, Preliminary Objections, Merits, Reparations and Costs, Inter-Am. Ct. Hum. Rts ¶ 146, at 44 (2015), http://www.corteidh.or.cr/docs/casos/articulos/seriec_292_esp.pdf.

160 Giampietri, *supra* note 112, at 34.

161 *Id.* at 98.

162 Eduardo Nicolás Cruz Sánchez, Et Al Peru, Case 12.444, Merits, Inter-Am. Comm'n H.R., Report No. 66/10 ¶ 57 (2011), https://www.oas.org/en/iachr/decisions/court/PEME12.444ENG.pdf.

163 *Id.* ¶ 55.

164 *Id.* ¶ 65 ("After detonating several underground explosions, approximately 143 commanders (sic) [commandos] formed into different groups entered the Embassy residence through tunnels . . . .").

165 Harclerode, *supra* note 111, at 522.

166 Giampietri, *supra* note 112, at 137.

167 Eduardo Nicolás Cruz Sánchez, Et Al Peru, Case 12.444, Merits, Inter-Am. Comm'n H.R., Report No. 66/10 ¶ 16 (2011), https://www.oas.org/en/iachr/decisions/court/PEME12.444ENG.pdf. *See also* Clifford Krauss, *Peru Troops Rescue Hostages; Rebels Slain as Standoff Ends*, N.Y. Times, Apr. 23, 1997, http://www.nytimes.com/1997/04/23/world/peru-troops-rescue-hostages-rebels-slain-as-standoff-ends.html?pagewanted=all&src=pm.

168 Eduardo Nicolás Cruz Sánchez, Et Al Peru, Case 12.444, Merits, Inter-Am. Comm'n H.R., Report No. 66/10 ¶ 62 (2011), https://www.oas.org/en/iachr/decisions/court/PEME12.444ENG.pdf. *See also* Case of Cruz Sánchez et al.

terrorist positions and routine, and their munitions and armaments had been provided by some of the hostages, including retired Peruvian Special Forces Admiral Luis Giampierti, by means of a hidden microphone smuggled into the residence.[169] Approximately 140 Peruvian Special Forces and a contingent of Marines (in a sniper and demolitions role) carried out the assault.[170]

The operation involved explosives being detonated below the ballroom, dining room, and kitchen where the terrorists normally congregated,[171] as well as simultaneous assaults from tunnels and an explosive entry through the embassy front door.[172] A Peruvian assault commander was killed by rifle fire, a landmine wounded one commando, and another was killed when several booby traps were set off on a stairway. The resistance included the detonation of explosives and the use of grenades and automatic weapons. The assaulters killed terrorists as they began to hunt down hostages. The fighting was brought to an end when explosive charges were used to bring a ceiling down on a room used as a terrorist armory. This act set off grenades, Semtex explosives, and rocket-propelled grenade (RPG) rockets, killing the terrorist inside. [173] The use of explosives in this manner to end hostage situation is not unique to the Japanese embassy operation. Indian "Black Cat" NSG personnel used explosives against the LeT terrorists holed up on a room at the Taj Hotel during the 2008 Mumbai attack. This brought about an end to that rampage.[174]

The assault portion of the Peruvian operation lasted 16 minutes, with the total time required to complete the mission, including the evacuation of the hostages, being 33 minutes.[175] Two commandos, one hostage, and all fourteen terrorists were killed, with fourteen State agents and eight hostages being wounded.[176] What was to become the subject of further investigation was whether three of the terrorists were killed after they no longer posed a danger to the hostages or state agents.[177] In killing the terrorists Admiral Giampierti indicates that the commandos either aimed for hostage takers' heads[178] or, in

---

v. Peru, Series C no. 292, Preliminary Objections, Merits, Reparations and Costs, Inter-Am. Ct. Hum. Rts ¶ 153, at 48 (2015), http://www.corteidh.or.cr/docs/casos/articulos/seriec_292_esp.pdf.

169 Giampietri, *supra* note 112, at 109–15 and Case of Cruz Sánchez et al. v. Peru, Series C no. 292, Preliminary Objections, Merits, Reparations and Costs, Inter-Am. Ct. Hum. Rts ¶¶ 159 and 161, at 50 (2015), http://www.corteidh.or.cr/docs/casos/articulos/seriec_292_esp.pdf.

170 Giampietri, *supra* note 112, at 91. *See also* Eduardo Nicolás Cruz Sánchez, Et Al Peru, Case 12.444, Merits, Inter-Am. Comm'n H.R., Report No. 66/10 ¶ 66 (2011), https://www.oas.org/en/iachr/decisions/court/PEME12.444ENG.pdf (where the rescue force is indicated to have numbered 143).

171 Giampietri, *supra* note 112, at 89 and 168. *See id.* at 171 (where it is indicated six terrorists were killed by the initial explosions). *But see* Harclerode, *supra* note 111, at 522 ("Five of the terrorists playing football were killed instantly as the ground blew up beneath their feet. . . .").

172 Giampietri, *supra* note 112, at 171.

173 *Id.* at 166–175.

174 Scott-Clark & Levy, *supra* note 151, at 271–72.

175 Case of Cruz Sánchez et al. v. Peru, Series C no. 292, Preliminary Objections, Merits, Reparations and Costs, Inter-Am. Ct. Hum. Rts ¶ 166, at 53 (2015), http://www.corteidh.or.cr/docs/casos/articulos/seriec_292_esp.pdf.

176 Eduardo Nicolás Cruz Sánchez, Et Al Peru, Case 12.444, Merits, Inter-Am. Comm'n H.R., Report No. 66/10 ¶ 65 (2011), https://www.oas.org/en/iachr/decisions/court/PEME12.444ENG.pdf.

177 Id. ¶¶ 142 and 158 and Case of Cruz Sánchez et al. v. Peru, Series C no. 292, Preliminary Objections, Merits, Reparations and Costs, Inter-Am. Ct. Hum. Rts ¶ 287, at 88 (2015), http://www.corteidh.or.cr/docs/casos/articulos/seriec_292_esp.pdf.

178 Giampietri, *supra* note 112, at 172 and 174.

at least one instance, their center of mass.[179] One reason for not firing at the center of mass was reported to be that the terrorist might be wearing a bulletproof vest (what is referred to as a "bullet bouncer").[180] It was also indicated that the rescuers were trained to use a double tap method of firing where "[e]very fallen enemy would get two rounds to the head."[181] The double tap shots to the head were also said to be "[s]tandard military procedure."[182] The senior officer on the ground was reported to have indicated during training that the goal was to quickly rescue the hostages and that "[t]here would be no time to take prisoners, which would only give the terrorists the opportunity to kill hostages and detonate their explosives."[183]

A 2011 *Report of the Inter-American Commission on Human Rights* dealing with this incident referred to the shooting tactic employed during the operation as "selective instinctive shooting."[184] It was said to consist of "firing three shots in two seconds to the vital parts of the adversary, followed by a shot to the head of the enemy; 'if the criminal was still alive, he was given the *coup de grâce*.' "[185] Reference was also made to this action being the "security shot."[186] The Inter-American Court of Human Rights also referred to "selective instinctive shooting". The judgment made note of a Peruvian Army doctrine manual describing this shooting technique, but the analysis by the Court of the killing of the hostages was ultimately to concentrate more on the issue of whether the three terrorists alleged to have been executed were taking a direct part in hostilities, or had been killed after the point they became *hors de combat*.[187]

Despite the initial positive public reaction to the operation, considerable controversy arose from allegations that a number of the hostage takers survived the assault but were subsequently executed.[188] The Inter-American Commission on Human Rights investigation[189] determined that three hostages takers were extrajudicially executed, thereby denying them of their right to life under Article 4(1) of the American Convention on Human

---

179 *Id.* at 171.

180 *Id.* at 141.

181 *Id.* at 108. *See also id.* at 173 (for reference to the "double tap" method of shooting being used).

182 *Id.* at 174 ("*Tucos* [terrorists] on the stairs had taken so many double taps, their heads were little more than shredded chunks of bloody meat.").

183 *Id.* at 108 ("Years of fighting taught these soldiers that to leave a fallen terrorist alive was to risk a bullet in the back."). *See also* David, *supra* note 115, at 298 (where reference is made to terrorists being shot during the Entebbe operation "to make sure they were out of action.").

184 Eduardo Nicolás Cruz Sánchez, Et Al Peru, Case 12.444, Merits, Inter-Am. Comm'n H.R., Report No. 66/10 ¶ 66 (2011), https://www.oas.org/en/iachr/decisions/court/PEME12.444ENG.pdf.

185 *Id.*

186 *Id.* ¶ 156.

187 Case of Cruz Sánchez et al. v. Peru, Series C no. 292, Preliminary Objections, Merits, Reparations and Costs, Inter-Am. Ct. Hum. Rts ¶ 163 n. 154, at 51 and ¶ 287, at 88 (2015), http://www.corteidh.or.cr/docs/casos/articulos/seriec_292_esp.pdf.

188 Frank Bajak & Franklin Briceno, *Peru's Famed Hostage Raid Investigated*, Assoc. Press (May 18, 2012), http://www.foxnews.com/world/2012/05/18/peru-famed-hostage-raid-investigated/.

189 *IACHR Takes Case Involving Peru to Inter-American Court*, Press Release Inter-American Court of Human Rights (Dec. 11, 2011)[hereinafter IACHR Press Release], http://www.oas.org/en/iachr/media_center/PReleases/2011/137.asp (The Inter-American Commission on Human Rights (IACHR) filed an application with the Inter-American Court of Human Rights (IA Court HR) in Case No. 12.444, Eduardo Nicolás Cruz Sánchez et al., (Operation Chavín de Huántar), Peru.).

Rights.[190] One terrorist was determined to have been apprehended and therefore was not posing any threat to the hostages or security personnel. That individual "was summarily executed by a *coup de grâce* to the neck fired by state agents while he was immobilized."[191] The report also indicates two other terrorists "were neutralized by military agents, begged for their lives, and nonetheless were extrajudicially killed, receiving multiple bullet wounds to vital parts of their bodies that were intended to eliminate them."[192] What the report did not address was whether the "selective instinctive shooting" tactic was itself legitimate, or constituted part of a "shoot to kill" policy (see Chapter 13, section 13.2.2, *Shooting to Kill*). Instead, the report highlighted its use in that instance to "eliminate, rather than neutralize the enemy."[193] In this context the term "neutralize" appears to be associated with the exercise of control over, and ultimately capture of the terrorists.[194]

An important factor was the nature of the violence. Unlike its European counterpart, the Inter-American human rights system has historically demonstrated some willingness to interpret its human rights mandate through the lens of humanitarian law even in respect of internal armed conflicts. Determining the existence of an armed conflict is essential in terms of identifying the rules governing the use of force.[195] A Peruvian military court had looked upon the incident as "a struggle with the characteristics of a military battle,"[196] although the Inter-American Commission dismissed such courts as not being independent or impartial.[197] The Commission did appear to accept that the hostage rescue took place in the context of a long running internal armed conflict, of which the MRTA was one of the active participants.[198] However, notwithstanding the existence of an armed conflict the Commission relied most heavily on human rights legal principles with reference being made to standards such as the requirement to use minimum force and only using such force as is absolutely necessary.[199] Further, the report stated lethal force could only be used where strictly unavoidable in order to protect themselves or other persons from imminent threat of death or serious injury. However, it could also be used "to otherwise maintain law and order where strictly necessary and proportionate."[200] In this context there was an acknowledgment of the requirement to use force to protect the lives of the hostages. Interestingly, the Commission did state it was "well aware that the kidnapping of diplomats and civilians *violates the basic principles of International Humanitarian Law* and it is likewise cognizant

[190] Eduardo Nicolás Cruz Sánchez, Et Al Peru, Case 12.444, Merits, Inter-Am. Comm'n H.R., Report No. 66/10 ¶ 228 a) (2011), https://www.oas.org/en/iachr/decisions/court/PEME12.444ENG.pdf.

[191] *Id.* ¶ 142.

[192] *Id.* ¶ 160.

[193] *Id.* ¶ 158.

[194] *Id.* ¶¶ 22, 141 and 158.

[195] Lindsay Moir, *Law and the Inter-American Human Rights System*, 25 HUM. RTS. Q. 182, 212 (2003).

[196] Eduardo Nicolás Cruz Sánchez, Et Al Peru, Case 12.444, Merits, Inter-Am. Comm'n H.R., Report No. 66/10 ¶ 194 (2011), https://www.oas.org/en/iachr/decisions/court/PEME12.444ENG.pdf.

[197] *Id.* ¶ 195.

[198] *IACHR Press Release, supra* note 189, and Eduardo Nicolás Cruz Sánchez, Et Al Peru, Case 12.444, Merits, Inter-Am. Comm'n H.R., Report No. 66/10 ¶¶ 44-49 (2011), https://www.oas.org/en/iachr/decisions/court/PEME12.444ENG.pdf.

[199] Eduardo Nicolás Cruz Sánchez, Et Al Peru, Case 12.444, Merits, Inter-Am. Comm'n H.R., Report No. 66/10 ¶ 120 (2011), https://www.oas.org/en/iachr/decisions/court/PEME12.444ENG.pdf.

[200] *Id.* ¶ 121.

that the individuals in the MRTA's power were at constant risk of their lives and personal integrity."[201] This reference to humanitarian law suggests the principles ultimately applied by the Commission could have been broader than traditional interpretations of human rights law.

Significantly, the use of force issues in this case focused on extrajudicial killing after capture as opposed to tactics applied during the assault.[202] Indeed, the use of tunnels and explosives at the commencement of the assault to kill many of the hostage takers places the selective shooting issue on a different and seemingly less violent plane regarding the force required to end the siege. Traditionally, the presence of hostages renders it less likely that force other than small arms, or perhaps a limited amount of explosives to gain entry, will be used.[203] However, the task of assaulting a building protected by landmines and improvised explosive charges, as well as the involvement of terrorists committed to killing the hostages armed with automatic weapons, rifles, grenades, and possibly in one instance a suicide vest, presented an exceedingly high-risk situation for both the hostages and their rescuers. Whether viewed through a human rights–based law enforcement lens, or a law of armed conflict lens this was clearly an exceptionally dangerous situation for a hostage rescue operation.

The key issue dealt with by the Commission was whether the force was used "against individuals who no longer present a threat [. . .], such as individuals who have been apprehended by authorities, have surrendered, or who are wounded and abstain from hostile acts. . . ."[204] On one level this issue appears to be no different from what would arise under humanitarian law regarding the protection that must be provided to persons who are *hors de combat*.[205] Under the law governing the conduct of hostilities, it is prohibited to deny quarter to an enemy.[206] Allegations regarding the summary execution of detainees following hostilities had also been the subject of inquiry in the 1997 *Abella v. Argentina* decision.[207] The Commission's previous handling of this issue was in the context of an incident where it was considered that an armed conflict was in existence and humanitarian law did apply.[208] What

---

201 *Id.* ¶ 132 (emphasis added).

202 Case of Cruz Sánchez et al. v. Peru, Series C no. 292, Preliminary Objections, Merits, Reparations and Costs, Inter-Am. Ct. Hum. Rts ¶ 287, at 88 (2015), http://www.corteidh.or.cr/docs/casos/articulos/seriec_292_esp.pdf

203 *But see* NAYLOR, *supra* note 64, at 371–75 (where it is indicated that the use of a grenade by a member of a rescue force led to the death of a hostage).

204 Eduardo Nicolás Cruz Sánchez, Et Al Peru, Case 12.444, Merits, Inter-Am. Comm'n H.R., Report No. 66/10 ¶ 126 (2011), https://www.oas.org/en/iachr/decisions/court/PEME12.444ENG.pdf.*Id.* (quoting from IACHR, Report on Terrorism and Human Rights, OEA/Ser.L/V/ll.116, Doc. 5 rev. 1 corr., October 22, 2002, para. 91.).

205 CUSTOMARY INTERNATIONAL HUMANITARIAN LAW STUDY 164 (Jean-Marie Henckaerts & Louise Doswald-Beck eds., 2005) (where the customary international law rule is set out as:

> Rule 47. Attacking persons who are recognised as *hors de combat* is prohibited. A person *hors de combat* is:
> (a) anyone who is in the power of an adverse party;
> (b) anyone who is defenceless because of unconsciousness, shipwreck, wounds ¶ or sickness; or
> (c) anyone who clearly expresses an intention to surrender; ¶ provided he or she abstains from any hostile act and does not attempt to escape.)

See also AP I, art. 41, and the 1907 Land Warfare Regulations, art. 23(c).

206 A.P.V. ROGERS, LAW ON THE BATTLEFIELD 48 (2004).

207 Case 11.137, Abella v. Argentina, Inter-Am. Y.B. on H.R. ¶ 218 (1997) (Commission report).

208 *Id.* ¶ 156.

is important to note is that there is significant symmetry between human rights law and humanitarian law on this issue. However, in making a distinction between the one hostage who was clearly apprehended, and the two who were surrounded (and "neutralized"), the Commission Report appears to have introduced a Least Restrictive Means (LRM) type test, without actually directly referencing it as such [see Chapter 7, section 7.4.2., *The "Targeted Killing Case" and Control*]. In those circumstances such an approach would appear to require the capture rather than killing of the two terrorists.[209]

In contrast to the Commission approach Inter-American Court of Human Rights clearly indicated that a non-international armed conflict was in existence between the State of Peru and the MRTA.[210] The Court also approached the application of humanitarian law in a more overt fashion. The legal analysis did initially concentrate on the principles governing the use of force under human rights law,[211] even to the extent of making specific reference to the *United Nations Basic Principles on the Use of Force and Firearms by Law Enforcement Officials*. This included an acknowledgement of the application of overarching principles of legality, absolute necessity and proportionality, which were identified as having their source in those human rights law based United Nations principles on the use of firearms.[212]

While these human rights principles are associated with the use of force in the "last" resort the Court also referred particularly to three factors: the existence of a non-international armed conflict; the use of force was carried out in the context of a hostage rescue; and the alleged victims were not civilians, but rather members of the MRTA.[213] In indicating an armed conflict with the MRTA existed reliance was put on the use of force being governed specifically by humanitarian law (e.g. Common Article 3 of the *1949 Geneva Conventions*). That law was applied to interpret the prohibition against the arbitrary deprivation of life protected under Article 4(1) of the American Convention on Human Rights.[214] This included the application of the humanitarian law principles of distinction, targeting precautions, and proportionality.[215]

The reliance by the Court on humanitarian law was not unfettered. Reference was still made to the human rights law based precaution of minimizing the use of lethal force.[216]

---

209 Ryan Goodman, *The Power to Kill or Capture Enemy Combatants*, 24 EUR. J. INT'L. L. 819 (2013)[hereinafter *The Power to Kill*]. For a critique of this theory *see* Geoffrey S. Corn, Laurie R. Blank, Chris Jenks & Eric Talbot Jensen, *Belligerent Targeting and the Invalidity of a Least Harmful Means Rule*, 89 INT'L. L. STUD. 536 (2013) and Michael N. Schmitt, *Wound, Capture, or Kill: A Reply to Ryan Goodman's 'The Power to Kill or Capture Enemy Combatants'*, 24 EUR. J. INT'L. L. 855 (2013)[hereinafter *A Reply to Ryan Goodman*]. *See also* Ryan Goodman, *The Power to Kill or Capture Enemy Combatants: A Rejoinder to Michael N. Schmitt*, 24 EUR. J. INT'L. L. 863 (2014).

210 Case of Cruz Sánchez et al. v. Peru, Series C no. 292, Preliminary Objections, Merits, Reparations and Costs, Inter-Am. Ct. Hum. Rts ¶¶ 140-42, at 42, ¶ 268, at 82 (2015), http://www.corteidh.or.cr/docs/casos/articulos/seriec_292_esp.pdf.

211 *Id.* ¶¶ 257–65, at 79–81.

212 Case of Cruz Sánchez et al. v. Peru, Series C no. 292, Preliminary Objections, Merits, Reparations and Costs, Inter-Am. Ct. Hum. Rts ¶ 264 at 80 and ¶ 265, notes 325–28, at 81 (2015), http://www.corteidh.or.cr/docs/casos/articulos/seriec_292_esp.pdf.

213 *Id.* ¶ 266, at 81.

214 *Id.* ¶ 272, at 84.

215 *Id.* ¶ 273, at 84.

216 *Id.* ¶ 283 n. 361, at 87. (where specific reference is made to Ergi v. Turkey, Application no. 66/1997/850/1057, Eur. Ct. H. R. ¶ 79 (1998), http://hudoc.echr.coe.int/eng?i=001-58200, and Finogenov v. Russia, VI Eur. Ct. H. R. ¶ 208, at 402 (2011)(extracts), which reinforced the requirement to minimize the use of force and placed particular emphasis on operational planning).

In determining the legal requirements were met regarding the conduct of the operation,[217] factors such as the mission specific training, the choice of qualified personnel, the use of a replica structure, and a plan for taking detainees were considered.[218] The focus of the mission was clearly the safe rescue of the hostages.[219] The interface between human rights and humanitarian law appears to have been reconciled in terms of while the mission was planned to minimize civilian casualties, etc. it ultimately was the humanitarian law concept of active or direct participation in hostilities that governed the use of force against the hostages.[220] These human rights law based principles appear to have been applied in an overarching manner similar to how recourse to war legal principles apply to self-defence based uses of force. The Court concluded that, while the apprehended hostage taker had been executed,[221] the evidence had not been established the other two terrorists were no longer participating in hostilities at the time of their death (i.e., *hors de combat*).[222] In reaching this conclusion the Court did not address the "neutralize" argument (i.e., LRM issue) as the killings were viewed as having taken place in the context of an ongoing operation.[223] The reliance on participation in hostilities regarding the status of the MRTA members also made the discussion about the method of using force (i.e., "selective instinctive shooting") unnecessary. Overall, the analysis more accurately reflects the speed, confusion, and need for decisive action inherent in hostage rescue operations.

It is significant that the Inter-American Court of Human Rights applied humanitarian law principles. What is not known is the approach that would have been taken by the Court had there not been an ongoing armed conflict, and instead the MRTA was viewed solely as a criminal terrorist organization. Operationally it is difficult to see that in the circumstances a successful rescue might have been substantively planned any differently. The result likely would have been either a more expansive interpretation of human rights law principles by a review body permitting a broader use of force under a law enforcement framework, or a determination that levels of violence required to overcome that organized armed group (operating with a political purpose) created as situation of a short duration armed conflict.

Adopting the armed conflict approach more accurately portrays the nature of the violence, and avoids the potential longer-term negative effects of permitting an expansive interpretation of human rights norms. The situation would not be unlike that which occurred in the *Abella v. Argentina* example.[224] In this respect the requirement for State authorities to adopt a more aggressive use of force in the face of the type of armaments used by the terrorists, and to overcome the threat to the hostages and rescuers posed by the MRTA (an organized armed group) would likely become a key factor in assessing if the armed conflict threshold had been crossed (see Chapter 9, section 9.3.4., *The Totality of the Circumstances and Other Criteria*, and Chapter 13, section 13.2.2., *Shooting to Kill*). Regardless of the

---

217 *Id.* ¶ 287, at 88.
218 *Id.* ¶¶ 284–86, at 87–88.
219 *Id.* ¶ 286, at 88.
220 *Id.* ¶ 287, at 88.
221 *Id.* ¶¶ 316–19, at 98.
222 *Id.* ¶¶ 339–43, at 107.
223 *Id.* ¶¶ 339–40, at 107.
224 Case 11.137, Abella v. Argentina, Inter-Am. Y.B. on H.R. ¶ 218 (1997) (Commission report).

normative framework applied the question of when the threat has been removed, or a target is hors de combat is likely to remain an important accountability issue.

The Japanese embassy incident, and the issues raised in its aftermath, once again highlight the significant risks and high levels of violence that are often associated with counterterrorist hostage rescues. These are operations that often occur at the intersection of the human rights and humanitarian law frameworks. Regardless of whether such operations arise in the context of an armed conflict or as part of law enforcement, they usually take place after certain criteria are met (e.g., grave threats to hostages, negotiations have proved fruitless or impracticable, there is an inability to warn without escalating the situation, the inappropriateness or ineffectiveness of alternative courses of action). However, even when conducted under the human rights–based law enforcement paradigm, it is likely the hostage takers will be subjected to similar, if not exactly, the same levels of force as if the assault was carried out during an armed conflict scenario. The normative gap narrows significantly when groups armed organized groups must be forcibly relieved of their hostages.

Further, it would be incorrect to assume all elements of a human rights–based response are less violent or destructive than one grounded in humanitarian law. For example, some ammunition used in a law enforcement–based operation can result in the hostage taker being at greater risk of injury or death, although it also offers an advantage in limiting risk to the hostages. This can occur because of the ability to use hollow point or "frangible" ammunition against the hostage taker in a policing context. Such ammunition helps avoid overpenetration and presents less of a risk of collateral damage to hostages.[225] The use of this ammunition, while not prohibited for all international operations (since not all hostage rescues are conducted in the context of an armed conflict), could be problematic in a conduct of hostilities scenario where expanding ammunition is prohibited.[226] Its use in a domestic law enforcement setting may also attract controversy.[227] Similarly, the authority to use chemical agents in a law enforcement context can also mean more less lethal options are available to assist in those situations than can be used during armed conflict.[228] However, as the 2002 Moscow theater incident[229] and the 1979 siege of Mecca tragically demonstrate, the use of some chemical agents is not without considerable risk.[230]

---

[225] For a discussion of the differences between law enforcement and armed conflict approaches to the use of hollow point ammunition, *see* Kenneth Watkin, *Chemical Agents and Expanding Bullets: Limited Law Enforcement Exceptions or Unwarranted Handcuffs?*, 82 INT'L L. STUD. 193 (2006), and Robin Geiß, *Poison, Gas and Expanding Bullets: The Extension of the List of Prohibited Weapons at the Review Conference at the International Criminal Court in Kampala*, 13 Y.B. INT'L L. 337 (2010). *See also* The 1998 Rome Statute, arts. 8(2)(b)(xix) and 8(2)(v)(xv), and *Frangible Ammunition*, GLOBALSECURITY.ORG, http://www.globalsecurity.org/military/systems/munitions/frangible.htm (last visited June 23, 2015).

[226] Watkin, *supra* note 225, at 197–99, and Geiß, *supra* note 225, at 343–45.

[227] David Leppard & Jonathan Calvert, *The Web of Terror*, THE SUNDAY TIMES, July 17, 2005, at 12 (in respect of the London Specialist Firearms Officers counterterrorist teams operating in 2005 it is noted "[t]he rifles . . . use 'frangible ammunition that releases all its energy in the target's body, instead of passing through it and endangering nearby civilians.'"), and Sandra Laville, *Met Police to Use Hollow Point Bullets that Killed Jean Charles de Menezes*, THE GUARDIAN, May 11, 2011, http://www.guardian.co.uk/uk/2011/may/11/met-police-hollow-bullets-menezes.

[228] Watkin, *supra* note 225, at 196–97.

[229] Finogenov v. Russia, VI Eur. Ct. H. R. 365 (2011) (extracts).

[230] TROFIMOV, *supra* note 120, at 192 (setting out the potential lethality of the chemical agent).

One issue for which there is no gap between the two normative frameworks is in respect of the obligation to protect those who have surrendered, are incapacitated, or otherwise fall under the control of the hostage rescuers. Once sufficient control is established, and the hostage takers no longer pose a threat to the hostages or the security personnel, they must be detained and treated humanely. Not only are they not to be killed, the detainees must be protected from torture or any form of abuse. To permit that to happen is as illegal under humanitarian law (i.e., the obligation to protect those who are *hors de combat*) as it would be during law enforcement–based operations. What is evident is that one factor having a significant impact on whether deadly force will be used to remove the threat posed by hostage takers is the potential that they have explosive devices, whether in the form of IEDs, suicide vests, rocket-propelled grenades, or grenades. However, this is an issue that is not unique to hostage rescue or a threat posed to the relative small number of military or police specialized units that normally conduct such operations. Explosive weapons in their various forms have become the predominant threat on many contemporary operations. It is to that threat and the unique challenges it presents that the analysis will now turn.

12.3. Explosives: The Game Changer

12.3.1. A UBIQUITOUS WEAPON

The pressure on security forces to use deadly force approximating a "first resort" is not unique to hostage rescue. It also arises in the context of protecting both the population and security forces from the effects of explosive attacks, whether in the form of suicide bombers, vehicle borne or other IEDs, the use of RPGs, booby traps, suicide vests, or simply grenades. The use of explosives is a game changer because of their destructive power and the frequently clandestine manner in which they are employed. Non-State actors have long used explosives as a means of attack. It has been noted that the development of "destructive and portable weaponry," such as dynamite, which was invented in 1866, was "to prove the terrorists' weapon of choice."[231] Similarly, "[a]ll modern insurgencies including aspects of the Spanish Civil War in Europe and T.E. Lawrence's Middle East campaigns, have included as part of their central strategy the use of explosives and mines by guerrillas to disrupt mechanized transport and supply lines."[232] There was heavy reliance of explosives in Vietnam following World War II, where French forces fighting the Viet Minh "trudged along embankments [of paddy fields], which were often mined with devices improvised from dud French shells and bombs . . . ."[233] The *Global Terrorism Index 2014* indicates that from 2000 to 2013 explosives were used in 60 percent of terrorist

---

[231] Max Boot, Invisible Armies: An Epic History of Guerrilla Warfare from Ancient Times to the Present 209 (2013).

[232] James Pettifer, The Kosova Liberation Army: Underground War to Balkan Insurgency, 1948–2001, at 70 (2013).

[233] Michael Burleigh, Small Wars, Far Away Places: Global Insurrection and the Making of the Modern World 220 (2013).

attacks, although in some regions such as sub-Saharan Africa the use of firearms was more prevalent.[234]

In contemporary conflict the insurgent use of bombs and IEDs is viewed as an integral and very dangerous part of asymmetric warfare. The ability to surreptitiously place and detonate bombs (including by means of a suicide attack) has long been viewed as an asymmetric alternative to States having the ability to use airpower and precision technology.[235] Improvised explosive devices became the emblematic weapon of the 2003 Iraq War, with the conflict being referred to by one commentator as the "war of the roadside bomb."[236] One third of the U.S. troops "killed in the first year of the insurgency were victims of these bombs, as were about two thirds of those wounded severely enough to require medical evacuation out of Iraq."[237] The threat from the use of IEDs has continued as evident in 2014 in the conflict between the Islamic State and Kurdish, Persmerga, forces in Iraq.[238]

A hallmark weapon during the postinvasion Iraq conflict was "what Americans called 'V-bids' or VBIEDs—Vehicle-Borne Improvised Explosive devices, driven by suicide attackers against highly symbolic targets, which cause large numbers of casualties."[239] The use of IEDs "combined the two qualities sought by all those who use political violence: capability, in that they were effective, and resilience, in that they did not expose those using them to excessive risk."[240] The use of such explosives has also been an integral part of other conflicts. In 1999, Hezbollah killed the commander of Israeli soldiers in southern Lebanon by detonating "an IED as his car rounded a bend . . . sending it over the cliff in flames."[241] During the 2000–05 Al-Aqsa intifada in the Occupied Territories, "the placement of improvised explosive devices (IEDs) on roads leading to settlements was an especially lethal tactic that accounted for numerous military and civilian casualties."[242] In 2015 the combination of IED and suicide attack was been identified as a particularly effective and low-cost form of attack employed by the Islamic State.[243]

Such attacks are not limited to the better-known conflict areas such as Iraq, Afghanistan, and the Occupied Territories. Another example of the improvised hand-thrown explosives is the use of what was believed to be a small "jelly" or "gelatina" bomb and satchel charges

---

[234] Global Terrorism Index 2014: Measuring and Understanding the Impact of Terrorism, Institute for Economics and Peace 30 (2014), http://www.visionofhumanity.org/sites/default/files/Global%20Terrorism%20Index%20Report%202014_0.pdf [hereinafter Global Terrorism Index 2014].

[235] Roger Triquinier, Modern Warfare: A French View of Counterinsurgency 21, n.1 (2006), Daniel Byman, A High Price: The Triumphs & Failures of Israeli Counterterrorism 100, 131 (2011), and Assaf Moghadam, The Globalization of Martyrdom 32 (2008), and Aaron Mehta, *General: ISIL Using IEDs as Guided Munitions*, Defense News (June 19, 2015, 2:51 PM EDT), http://www.defensenews.com/story/defense/land/weapons/2015/06/19/isis-isil-ied-iraq-syria-coalition-pgm-suicide-truck-bomb/28984469/.

[236] Ricks, *supra* note 21, at 217.

[237] *Id.*

[238] Alice Fordham, *Ambushes, Mines and Booby Traps: ISIS Militants Change Tack*, NPR (Oct. 21, 2014), http://www.npr.org/blogs/parallels/2014/10/21/357791239/ambushes-mines-and-booby-traps-isis-militants-change-tack.

[239] Urban, *supra* note 81, at 25.

[240] Jason Burke, The 9/11 Wars 129 (2011).

[241] Byman, *supra* note 235, at 226.

[242] Sergio Catignani, *The Israel Defense Forces and the Al-Aqsa Intifada: When Tactical Virtuosity Meets Strategic Disappointment*, *in* Counterinsurgency in Modern Warfare 233, 235 (Daniel Marston & Carter Malkasian eds., 2010).

[243] Mehta, *supra* note 235.

by attackers during the 2012 assault on the CIA annex in Benghazi, Libya.[244] In September 2012 the head of the U.S. Joint IED Defeat Organization indicated that "[s]ince January 2011, there have been more than 10,000 global IED events occurring in 112 countries that were executed by more than 40 regional and transnational threat networks."[245] This would seem to indicate that not all IED devices were detonated in the context of an armed conflict. The manner in which these devices are detonated has become increasingly more sophisticated, prompting a technological and tactical struggle of cat and mouse between insurgents and the State security forces seeking to stop them.

Particularly effective explosive weapons are those that can penetrate armored vehicles. One such weapon used in Iraq was the RKG-3 HEAT hand grenade, taken from stockpiles of weapons located around country following the invasion.[246] However, one of the most dangerous weapons in Iraq was Enhanced Formed Projectiles (EFPs) provided by Iran to Iraqi Shi'a insurgents.[247] These munitions, which consisted of a metal or PVC pipe with the front end encased in concave or steel disk, created a high-velocity molten-shaped slug when it exploded that easily penetrated armored vehicles.[248] The effect was that "[a]rmoured vehicles previously considered invulnerable where being smashed by a new kind of roadside bomb . . . ."[249] A variety of terrorist, insurgent, and other organized armed groups have also made use of RPGs. This includes against aircraft with incidents occurring in Somalia, Afghanistan, and Iraq.[250] The IED "war" in Afghanistan became a struggle of "cat and mouse" between Coalition and the Taliban with the latter group developing sophisticated tactics, various high and low technical means of igniting the bomb,[251] and "a 'low metal content' IED" to thwart counter IED measures.[252] Similar challenges were faced in Iraq.[253]

A challenge for security forces is that the use of explosives is not limited to warfare or acknowledged armed conflicts. As was evidenced in the 1995 criminal bombing by Timothy McVeigh of a U.S. government building in Oklahoma, domestic "lone wolf" terrorists can make such attacks.[254] During the 1970s and 1980s it was terrorist groups

---

244 Mitchell Zuckoff, 13 Hours: The Inside Account of What Really Happened in Benghazi 212 (2014).

245 A. J. Bosker, *IEDs Will Remain "Weapon of Choice" for Decades*, Joint IED Defeat Organization News Service (Sept. 12, 2012), https://www.jieddo.mil/news_story.aspx?ID=1488.

246 John Ismay, *Insight into How Insurgents Fought in Iraq*, N.Y. Times, Oct. 17, 2013, *available at* http://atwar.blogs.nytimes.com/2013/10/17/insight-into-how-insurgents-fought-in-iraq/.

247 Naylor, *supra* note 64, at 297.

248 David Crist, The Twilight War: The Secret History of America's Thirty-Year Conflict with Iran 520 (2012).

249 Ledwidge, *supra* note 27, at 34.

250 Mark Bowden, Blackhawk Down: A Story of Modern War 79 (2010), Ray Rivera, Alissa J. Rubin, & Thom Shanker, *Copter Downed by Taliban Fire; Elite U.S. Unit among Dead*, N.Y. Times, Aug. 6, 2011, http://www.nytimes.com/2011/08/07/world/asia/07afghanistan.html?pagewanted=all&_r=0, John F. Burns, *The Struggle for Iraq: The Occupation; Troops Hold Fire for Negotiations at 3 Iraqi Cities*, N.Y. Times, Apr. 12, 2004, http://www.nytimes.com/2004/04/12/world/struggle-for-iraq-occupation-troops-hold-fire-for-negotiations-3-iraqi-cities.html?pagewanted=all&src=pm.

251 Toby Harnden, Deadmen Risen: The Welsh Guards and the Real Story of Britain's War in Afghanistan 211–13 (2011).

252 *Id.* at 206.

253 Ricks, *supra* note 21, at 217 (discussing the evolution of IED bombs and countermeasures).

254 Jessica Stern, Terror in the Name of God: Why Religious Militants Kill 141 (2003).

operating in mainland Europe, which were dealt with as a criminal threat that made regular use of bombs.[255] So did the IRA in Northern Ireland, which as an organization was also dealt with under a law enforcement paradigm. In this regard, in that conflict "[t]he statistics for 1971 still make grim reading. Security Forces recorded 6,948 violent incidents of all types during the year, including more than 1,000 bombs that caused fifty deaths."[256] Over the years the use of RPGs has also not been limited to armed conflict situations. For example, during the 1970s there were a number of attempts made to attack Israeli civilian aircraft with RPGs and missiles.[257] In 1981 the terrorist group, the Red Army Faction, fired an RPG at the armored Mercedes of the commander of U.S. land forces in Europe while it was stopped at a red light in Heidelberg, Germany. That attack was unsuccessful.[258]

The bombings in Bali, 2002;[259] Casablanca[260] and Istanbul, 2003;[261] Madrid, 2004;[262] London, 2005;[263] and Kampala, 2010[264] stand out as examples of the use of explosives by terrorists in the twenty-first century under the circumstances (i.e., occurring within peaceful States) that were amenable to a law enforcement response. The use of explosive devices (e.g., IEDs, suicide vests, and grenades) is also a tactic employed by many terrorist groups to deter hostage rescues (see Chapter 11, section 11.2.2.1, *Unique Challenges and Responses*). Grenades may also be used in other circumstances as was demonstrated during the 2015 raid by French police on an Islamic hideout in the St. Denis area of Paris. Both the terrorists and the police used grenades during a shootout that reflected a level of violence not associated with ordinary crime or a law enforcement response.[265] As has been noted, the hostage takers involved in the 1997 seizure of the Japanese embassy in Lima, Peru, were armed with improvised grenades.[266]

This challenge is not unique to armed conflict. In respect of transnational criminal organizations Mexico saw its first car bomb attack on July 15, 2010.[267] This means of attack "with their attendant casualty generating capacity—poses a threat similar to the darkest days of narco-violence in Colombia or the insurgencies in Iraq and Afghanistan."[268] It also serves as a further indication of the narrowing gap between criminal activity and insurgency. Importantly, the threat from explosives is not limited to bombs. It is reported there were

---

[255] HARCLERODE, *supra* note 111, at 101 (two examples include 1984 car bomb attacks by the Red Army Faction on U.S. Air Forces bases in Germany).

[256] TONY GERAGHTY, THE IRISH WAR: THE MILITARY HISTORY OF A DOMESTIC CONFLICT 41 (1998).

[257] MICHAEL BURLEIGH, BLOOD AND RAGE: A CULTURAL HISTORY OF TERRORISM 172–73 (2009).

[258] *Id.* at 258

[259] BURKE, *supra* note 240, at 156–57.

[260] *Id.* at 157.

[261] *Id.* at 158–59.

[262] *Id.* at 159.

[263] *Id.* at 209.

[264] FERGUSSON, *supra* note 63, at 38.

[265] *Paris Attacks: How the Saint Denis Raid Unfolded*, BBC NEWS (Nov. 19, 2015), http://www.bbc.com/news/world-34867205 (the terrorists threw grenades at the police during a fierce gun battle, and the police threw "about 20 grenades inside the apartment.").

[266] GIAMPIETRI, *supra* note 111, at 10.

[267] John P. Sullivan, *Explosive Escalation? Reflections on the Car Bombing in Cuidad Juarez, in* MEXICO'S CRIMINAL INSURGENCY 81, 81 (John P. Sullivan & Robert J. Bunker eds., 2013).

[268] *Id.* at 82.

101 grenade attacks in Mexico against government buildings in the three and a half years prior to 2010.[269]

The bomb-making skill set is also one that has over the years been the subject of inter-group training on an international scale. It is reported that Palestinian groups pioneered "the 'networking' dimension of international terrorism."[270] The international transfer of bomb- and IED-making skills has included Irish terrorists training Columbian rebels;[271] Hezbollah and the Iranian Quds Force operating in Iraq training Shi'a militant forces;[272] and ex-members of Saddam Hussein's Republican Guard passing on IED-making skills to Al Qaeda, the Taliban, and other jihadist organizations.[273] Such training is shared amongst geographically separated jihadist groups.[274] Hezbollah formed a new Unit 3800 to help train Iraqi insurgents in Iran. In 2008 there were "between forty and sixty Lebanese running regular classes on EFPs and small-unit tactics in four small training camps scattered about Iran: in Tehran, Ahvaz, Elam, and Qom."[275] There is also an indication "of technology migration from Iraq to Syria via the Lebanese militia group Hezbollah."[276] Similarly, drug cartel bomb makers "are experimenting and importing TTP from other conflicts to exploit on their own battleground."[277]

The spread of knowledge regarding the making and employment of explosive devices is now significantly enhanced through the use of the Internet.[278] This is the kind of information "sought out not just by sophisticated terrorist organizations but also by disaffected individuals prepared to use terrorist tactics to advance their own idiosyncratic agendas."[279] The threat this entails and the use of explosives cut across the spectrum of violence from domestic and international terrorist groups to insurgencies and armed conflict. A dominating feature of twenty-first-century conflict from a State perspective is not one of strategic bombing or other similarly destructive aspects of inter-State war (e.g., artillery strikes), but rather the presence of explosive devices in the hands of non-State actors. The threat from explosive devises is ubiquitous, raising the question of under what circumstances their use crosses the dividing line between law enforcement into armed conflict in terms of how the intensity of violence threshold is assessed. The fact that explosives are often used by terrorists or other groups in circumstances that are solely subject to a law enforcement response or are preferred to be dealt with under a policing approach points to human rights–based law needing to provide an effective means of addressing many of these highly destructive attacks. Once again the potential exists for both an overlap of hostilities and law

---

269 Sullivan, *supra* note 267, at 83.

270 HOFFMAN, *supra* note 78, at 77

271 BURLEIGH, *supra* note 257, at 344.

272 ROBERT M. CASSIDY, COUNTERINSURGENCY AND THE GLOBAL WAR ON TERROR: MILITARY CULTURE AND IRREGULAR WAR 2 (2008).

273 ABDEL BARI ATWAN, AFTER BIN LADEN: AL QAEDA, THE NEXT GENERATION 140–41 (2012).

274 SAVAGE, *supra* note 74, at 340 (where it is noted that in 2011 al Shabaab member, Ahmed Warsame, was captured in international waters returning from Yemen where AQAP "had been teaching him to build bombs and he was returning to Somalia to share that knowledge with the group.").

275 CRIST, *supra* note 248, at 522.

276 Ismay, *supra* note 246.

277 Sullivan, *supra* note 267, at 82.

278 GABRIEL WIEMANN, TERROR ON THE INTERNET 123–29 (2006).

279 *Id.* at 124.

enforcement paradigms, along with a convergence of norms and ultimately tactics employed by security forces.

12.3.2. SUICIDE MISSIONS

A hallmark of contemporary conflict, the roadside bomb, is rivaled by the threat of suicide attacks, although statistically such bombings only constituted 6 percent of all terrorist attacks in 2013.[280] Suicide bombing has become a particularly dangerous mode of attack in the post-9/11 world. They are statistically calculated to be four and one half times more deadly than other forms of terrorist attack.[281] The tactic of using suicide missions predates this century, extending back to biblical times.[282] In contemporary armed conflict the results are deadly. It has been reported that worldwide, 1,269 suicide missions between 1981 and April 2007 caused more than 50,000 casualties.[283] Hezbollah was an early practitioner of this form of attack. However, the modern root of suicide attacks, "and particularly of religious inspired SMs [suicide missions], are found in post-1979 Iran, from which the tactic had spread to Lebanon in the first place."[284] Car bombs were a common weapon in the Lebanese civil war, which was made all the more deadly when Hezbollah "put a unique spin on this Lebanese tradition by putting a human behind the wheel."[285] A devastating vehicle-borne attack on a major Israeli military headquarters on November 11, 1982, marked the beginning of this form of attack.[286] It was followed by a detonation at the American embassy in Beirut in April 1983,[287] and in October 1983, the destruction of U.S. Marine and French military headquarters, respectively.[288] In those October attacks, the United States lost 241 Marines and the French 58 paratroopers.[289]

Hezbollah was not the only proponent of suicide bombing. The Liberation Tigers of Tamil Eelam (LTTE) were impressed by Hezbollah's successful use of this form of asymmetric warfare.[290] During the Sri Lanka conflict, "suicide bombing became the LTTE's signature mode of attack."[291] The high-profile assassinations by suicide bomber of former Indian Prime Minister Rajiv Gandhi in 1991, and then Sri Lankan President Ranasinghe Premadasa two years later, highlight the impact and reach of this mode of violence.[292] As was demonstrated in Israel, attacks by terrorists can be directed solely against civilians. Suicide missions were a prominent part of the Israel conflict from 1993 to 1998, and again from 2000 until 2005. During the second intifada (September 2000 until December 2005), "156 male and 8 female suicide bombers conducted a total of 147 SMs in Israel, the West

---

[280] GLOBAL TERRORISM INDEX 2014, *supra* note 234, at 30.

[281] *Id.* at 32.

[282] MOGHADAM, *supra* note 235, at 9–26.

[283] *Id.* at 38–39 (statistics based University of Haifa database).

[284] *Id.* at 17.

[285] CRIST, *supra* note 248, at 129.

[286] *Id.* at 128.

[287] *Id.* at 131–32.

[288] *Id.* at 135–38.

[289] *Id.* at 138.

[290] MOGHADAM, *supra* note 235, at 22.

[291] HOFFMAN, *supra* note 78, at 143.

[292] *Id.* at 144 (the LTTE denied responsibility for the attacks).

Bank, and Gaza."[293] Five hundred and twenty-seven people were killed and 3,350 were wounded. A total of 450 attacks were reported as having been prevented.[294] The Al-Aqsa intifada suicide attacks against Israeli civilian targets were "carried out in crowded public transportation buses, street markets, cafes, night clubs and restaurants in order to kill as many Israeli civilians as possible."[295]

The *Global Terrorism Index 2014* reports that "of the 2,500 suicide attacks since 2000, over 90 per cent occurred in MENA [Middle East and North Africa] and South Asia."[296] The Taliban, Al Qaeda and its affiliates, and the Islamic State have used this tactic the most since 2000.[297] Theories regarding the basis for contemporary suicide attacks include them being "mainly a response to foreign occupation,"[298] part of "outbidding" to compete among other groups "for the support of the local population, thereby increasing their 'market share' among that community,"[299] or as a result of the rise of Al Qaeda and "its guiding ideology the Salafi Jihad."[300] Significantly, Al Qaeda is viewed as having "all but perfected the tactic . . . " and "has institutionalized SMs to an extent not seen in other groups by instilling the spirit of self-sacrifice in the collective psyche of virtually all its fighters . . . ."[301] It is a "tactic that the group has managed to successfully employ by air, land, and sea."[302] Suicide bombings have expanded since 2001 to attacks on a global scale.[303] The attacks of 9/11 and the bombings in London (2007), and Ankara, Beirut and Paris (2015) are prime examples of such attacks. Consistent with the transnational nature of the conflict with the jihadist movement, these attacks occur not only in areas involving ongoing insurgencies but also within States far from the "battlefield."[304]

It is reported that suicide attacks "have occurred in at least 31 countries . . . . Of these, 628 attacks—nearly half of all suicide attacks between 1981 and April 5, 2007 (49.1 percent)—occurred in Iraq."[305] This form of attack has been increasingly integrated into contemporary insurgent operating procedures. In Afghanistan in 2010 "suicide attacks occurred at a rate of about three per week and the use of IEDs nearly doubled."[306] As was reported by General Stanley McChrystal, the American commander of the Special Forces organization tasked with dismantling AQI, by 2005 "leaders, even midlevel ones, began wearing suicide vests

---

293 Moghadam, *supra* note 235, at 23.

294 *Id.* (for statistics relies on *Suicide Bombing Terrorism during the Current Israeli-Palestinian Confrontation* 2, 5 (Sept. 2000–Dec. 2005), http://www.terrorism-info.org.il/data/pdf/PDF_19279_2.pdf).

295 Sergio Catignani, Israeli Counter-Insurgency and the Intifadas 106–07 (2008).

296 Global Terrorism Index 2014, *supra* note 234, at 32.

297 *Id.* at 34.

298 Robert A. Pape, Dying to Win: The Strategic Logic of Suicide Terrorism 23 (2005).

299 Moghadam, *supra* note 235, at 36 (referring to the outbidding theory developed by Mia Bloom, in Dying to Kill: The Allure of Suicide Terrorism (2005)).

300 *Id.* at 251.

301 *Id.* at 78.

302 *Id.*

303 *Id.* at 59.

304 S.C. Res. 2249, UN Doc. S/RES/2249 *Preamble* (Nov. 20, 2015) ("its recruitment and training of foreign terrorist fighters whose threat affects all regions and Member States, *even those far from conflict zones*, the Islamic State in Iraq and the Levant (ISIL, also known as Da'esh), constitutes a global and unprecedented threat to international peace and security".) (emphasis added).

305 Moghadam, *supra* note 235, at 41(statistics based on a University of Haifa suicide terrorism database).

306 Burke, *supra* note 240, at 476.

constantly, usually sleeping in them so that if our men breached their doors and headed toward their cots, they could light themselves off in the darkness."[307] The detonation of an explosives belt by a terrorist during the previously mentioned "St. Denis" raid in November 2015 to capture Islamic State terrorists in Paris highlights that such use is not limited to war zones in the Middle East, or Africa.[308] Clearly, a committed adversary prepared to react in this fashion can make any attempt at capture an exceedingly dangerous enterprise.

Not all martyrdom operations exclusively involve the use of explosive belts or vehicle turned into bombs. What are described as "fidayeen attacks" involve complex attacks against government positions or civilian targets by small groups of militants armed with automatic weapons and grenades, as well as suicide devices.[309] While essentially suicide missions, the goal is "not to be martyred right away, but to inflict as much damage on the enemy in order to inspire fear in others."[310] Such attacks have been carried out by the LeT in the Kashmir[311] and by militant groups in Afghanistan,[312] Pakistan[313] and Somalia,[314] and by the Islamic State in Paris.[315] These attacks are seen by some as a way of circumventing the prohibition on suicide under Islam.[316] As a result of this expanding suicide bombing phenomenon, civilian and military security forces have been handed essentially the same, and extremely challenging, role of countering that threat regardless of whether they are operating domestically, in occupied territory, or internationally in the midst of an ongoing insurgency.

Inevitably, the threat of such attacks puts military personnel, police, and even independent security contractors in an unenviable and exceedingly difficult position. They have to assess whether persons on foot, on bicycle, or in vehicles who attempt to blend seamlessly into the civilian population are about to ignite a bomb killing themselves, any nearby civilians, and likely the security personnel as well. Suicide bombing has prompted exceptional and, at times, controversial security measures such as the Israeli security fence-wall, which seeks to put a barrier between the terrorists and their likely intended target. Although the

---

307 Stanley McChrystal, My Share of the Task: A Memoire 184 (2013). *See also* Naylor, *supra* note 64, at 284 (for the description of a raid on a farmhouse in Iraq by the SAS where three of the five insurgents wore suicide vests. Two were killed before igniting their vests and the third insurgent detonated, but caused no assault force casualties).

308 Katrin Bennhold, *For Woman Dead in French Police Raid, Unlikely Path to Terror*, N.Y. Times (Nov. 20, 2015), http://www.nytimes.com/2015/11/21/world/europe/hasna-aitboulahcen-paris-attacks.html?hp&action=click&pgtype=Homepage&clickSource=story-heading&module=first-column-region&region=top-news&WT.nav=top-news.

309 *Afghanistan Mid-Year Report 2015 Protection Of Civilians In Armed Conflict*, United Nations Assistance Mission in Afghanistan 2 n.4 (August, 2015)("UNAMA definition of 'complex attack' refers to a deliberate and coordinated attack which includes a suicide device (i.e., BBIED, VBIED), more than one attacker and more than one type of device (i.e., BBIED and mortars). All three elements must be present for an attack to be considered complex.").

310 Stephen Tankel, Storming the World Stage: The Story of Lashkar-e-Taiba 65 (2013).

311 *Id.* at 64–65.

312 Alissa J. Rubin, Ray Rivera, & Jack Healy, *U.S. Embassy and NATO Headquarters Attacked in Kabul*, N.Y. Times, Sept. 13, 2014, http://www.nytimes.com/2011/09/14/world/asia/14afghanistan.html?pagewanted=all&_r=0.

313 Jon Boone & Ewen MacAskill, *More Than 100 children killed in Taliban Attack on Pakistan School*, The Guardian, Dec. 16, 2014, http://www.theguardian.com/world/2014/dec/16/taliban-attack-army-public-school-pakistan-peshawar.

314 Robyn Kriel & Omar Nor, *15 Die When Militants Lay Siege to Hotel With Bombs and Guns in Somalia*, CNN (Nov. 1, 2015, 5:53 AM ET), http://www.cnn.com/2015/11/01/africa/somalia-hotel-attack/

315 Adam Nossiter, Aurelien Breeden & Katrin Bennold, *Three Teams of Coordinated Attackers Carried Out Assault on Paris, Officials Say; Hollande Blames ISIS*, Int'l N.Y. Times, Nov. 14, 2015, http://www.nytimes.com/2015/11/15/world/europe/paris-terrorist-attacks.html.

316 Tankel, *supra* note 310, at 64–65.

International Court of Justice ruled, "the construction of the wall constitutes action not in conformity with various international legal obligations incumbent upon Israel,"[317] the security fence-wall is reported to have reduced terrorist attacks by 30 percent and civilian casualties by 50 percent during 2002 to 2003. Attacks were reduced by 75 percent from mid-2003 to mid-2004.[318]

Humanitarian-focused groups, such as the United Nations and the International Committee of the Red Cross (ICRC),[319] which often eschew exceptional security measures in order to maintain contact with the local population, may themselves present particularly vulnerable targets. As a result they have had to reassess their security arrangements and operations. Like other uses of explosive devices, suicide bombing is ultimately a threat that transcends the dividing line between law enforcement and armed conflict. It is also a threat that requires action be taken under very time-constrained circumstances when there is often limited information available to the person seeking to disrupt an attack. As will be discussed, the nature of the threat is such that a response available under a law enforcement paradigm is, under many circumstances, virtually the same as that available to military personnel operating under conduct of hostilities rules. Steps can be taken to dismantle the organizations that carry out suicide bombing and plant IEDs (see Chapter 7, section 7.5, *The IED and Suicide Bombing Threat: Arrest or Kill?*). However security forces such as those manning checkpoints and performing patrols are often placed in position where they must immediately react to the sudden threat posed by these weapons.

12.3.3. CONFRONTING THE EXPLOSIVE THREAT

The threat posed by IEDs, other explosive devices, and suicide attacks has to be addressed by all security personnel whether they are civilian police, part of a military force, or private security contractors. Drivers of transport vehicles as well as soldiers on the "front lines" face death and injury from improvised explosive devices. These attacks can occur at checkpoints, while driving in convoys, patrolling in civilian market places, and even while eating at dining facilities located behind supposedly secure compound walls.[320] A distinguishing feature about the threat posed by IEDs and suicide bombers is that both forms of attack are primarily group-based activities. It is the participation of organized armed groups in such violent activity that forms one of the key bases for separating ordinary criminal acts from participation in a noninternational armed conflict (see Chapter 9, section 9.4.2, *Group Organization*).[321] Improvised explosive devices may be used in non-armed conflict situations. However, the vast majority of "explosive"-based attacks occur in the context of ongoing hostilities.

When an armed conflict is in existence, more proactive kinetic means may be applied to counter the IED threat under the conduct of hostilities framework. This includes targeting

---

317 Legal Consequences Of The Construction Of A Wall In The Occupied Palestinian Territory, 2004 I.C.J. ¶¶ 138, at 194.

318 Catignani, *supra* note 295, at 167–68.

319 McChrystal, *supra* note 307, at 107.

320 *Id.* at 167–68.

321 Prosecutor v. Tadić, Case No. IT-94-1-A, Decision On The Defence Motion For Interlocutory Appeal On Jurisdiction, ¶ 70 (Int'l Crim. Trib. for the Former Yugoslavia, 1995), http://www.icty.org/x/cases/tadic/acdec/en/51002.htm.

by drones, other aircraft, or artillery assets that provide a means to strike at not only the bombers and IED placers but also the commanders, planners, and support personnel who are members of the organized armed group. However, even during an armed conflict, the application of a law enforcement approach favoring capture may be preferred when seeking to dismantle a bombing cell or other organization posing such a threat (see Chapter 7, section 7.5, *The IED and Suicide Bombing Threat: Arrest or Kill?*).[322] Importantly, due to the fact that counterinsurgency and counterterrorist operations occur among the civilian population, it is the human rights–based law enforcement response that may form, at least initially, the main security force reaction to these weapons.

The significant danger posed by explosive devices to the security personnel and the civilians they are tasked to protect raises questions regarding how different a law enforcement–based response is from that permitted by the rules governing the conduct of hostilities. On its face the more restrictive human rights law–based policing paradigm would suggest that there is considerably more leeway under humanitarian law to use force to counter IEDs or suicide bombers. Certainly, this is true to the extent that the conduct of hostilities paradigm permits the targeting not only of the persons laying and setting off the explosives but also other members of the organized armed group engaged in commanding, planning, and directly supporting those activities. However, the unique tactical environments of counterinsurgency and counterterrorism operations often present a very restrictive environment, even in a humanitarian law context, within which to react to explosive threats. Depending upon the circumstances this can result in a response in a hostilities context that is substantively the same as if human rights norms were to be applied.

Perhaps the best example of when such a limited response can occur is during the manning of a checkpoint in occupied territory or in a State where an insurgency is taking place. It might be asked whether checkpoint duty should considered to be a policing task rather than an aspect of military operations carried out during hostilities. It has been suggested that the protection of checkpoints and roadblocks is a law enforcement matter, while "the response by armed forces to an attack against one of their convoys within territory under their control would be governed by the paradigm of hostilities." [323] While those scenarios can be linked to the identified normative frameworks, such distinctions cannot be applied generally. This is because the identified frameworks do not fully represent the circumstances under which security forces may be required to use force while employed in those situations.

For example, military guards at the gates of defense installations or serving at checkpoints put in place to restrict civilian access to areas of military operations would primarily be engaged in interacting with the civilian population. However, they must also be prepared to respond to insurgent threats. Military personnel may also set up a checkpoint in an urban area specifically to search for IED builders and suicide bombers (e.g., persons taking a direct part in hostilities) rather than be engaged in traditional law enforcement duties. In each of those situations force may be applied against insurgents under the humanitarian law–based

---

322 Catignani, *supra* note 295, at 168, and Wilson, *supra* note 35, at 147–51(for a discussion of the use of policing techniques to dismantle insurgent organizations).

323 Nils Melzer, *Conceptual Distinction and Overlap Between Law Enforcement and the Conduct of Hostilities*, *in* The Handbook of The International Law of Military Operations 33, 44 (Terry Gill & Dieter Fleck eds., 2010).

conduct of hostilities paradigm. Conversely, angry civilians or even criminal elements may attack a military convoy for a variety of reasons not directly related to the hostilities (e.g., to obtain food, water, or other valuable supplies). The response by military guards in that situation must be consistent with human rights–based principles governing the use of force (e.g., necessity, proportionality, and imminence).

It is not the security activity being carried out or the object being protected that should be the main consideration in determining what law applies. Rather, it is the nature of the threat (e.g., the group or person involved, levels of violence) that has a particular impact on what legal paradigm governs the use of force by security personnel. These are similar to the criteria applied in determining if an armed conflict is in existence.[324] Consistent with the complexity of many contemporary security situations, the law enforcement and conduct of hostilities normative frameworks must often be applied simultaneously. However, since the threat to the security personnel can emanate from a wide range of disgruntled civilians, criminals, or participants in hostilities hiding among the civilian population, the default position for the use of force will normally be the human rights–based law enforcement framework. In this respect, regardless of whether the security personnel apply human rights law or the human rights norms contained in humanitarian law (e.g., AP I, art. 75, AP II, art. 4, or customary international law), the response would remain the same. Until the threat can be positively identified as emanating from a person taking a direct part in hostilities, it is the law enforcement paradigm that must be applied.

The response by police and military personnel to threats generated from "amongst the people" frequently involves the application of escalation of force procedures (verbal and visual warnings, warning shots, etc.). Such a response, which is associated with the human rights law framework, can provide a useful means by which to assess the nature of the threat. As the escalation procedures are applied it may become evident whether the person involved is a member of an organized armed group participating in hostilities or not. Such procedures also help avoid an overuse of force that endangers the civilian population. The increasing reliance placed on escalation procedures is reflected in the development and deployment of technology such as "laser dazzler" weapons that is intended to warn civilians they are approaching military checkpoints and convoys.[325] Nor will all suicide missions end up with the attackers detonating their charges. There will be some bombers who are not willing or are unable to complete their attack. Both police and military security personnel need to have procedures in place that permit them to identify and then neutralize the threat in order to take a bomber into custody who has established a willingness to surrender.[326]

When force is used at a checkpoint or on similar duties, it will in many circumstances be in self-defense or defense of others in a classic narrow sense, even if the threat is subsequently

---

[324] Kenneth Watkin, *Use of Force during Occupation: Law Enforcement and Conduct of Hostilities*, 94 Int'l Rev. Red Cross 267, 310–14 (2012).

[325] Jeff Hecht, *Nonlethal Lasers: Nonlethal Lasers Deter Attacks and Warn Away Noncombatants*, Photonic Frontiers (Apr. 1, 2013), http://www.laserfocusworld.com/articles/print/volume-49/issue-04/features/photonics-frontiers--nonlethal-lasers--nonlethal-lasers-deter-at.html.

[326] Peter Beaumont, *Israeli Forces Prevent Suspected Suicide Bomb Attempt*, The Guardian, May 30, 2014, http://www.theguardian.com/world/2014/may/30/israeli-forces-prevent-suspected-suicide-bomb-attempt-west-bank, and Byman, *supra* note 235, at 159–60.

determined to be insurgency or terrorist related. It may only become apparent at the last minute there is a threat from a suicide bomber or an IED. This can leave security personnel little time for deliberation or reflection. The effect is that regardless of whether such attacks are addressed as part of law enforcement or under the conduct of hostilities–based paradigm the reaction by the security personnel will often fundamentally be the same. The challenge in these scenarios will be determining when there is sufficient information to justify the use deadly force, decide whether an escalatory approach is feasible in the circumstances, and decide whether a defensive response can be taken in time to protect the persons or object being targeted.

Not all reactions to suicide attacks or IEDs will be governed by human rights norms or involve a graduated response. One approach for addressing the potential explosive threat is the use of well-positioned sniper teams. For example, in Iraq "[t]he most effective counterbomb tactic turned out to be the low-tech sniper. U.S. troops learned to hide and spy on spots, such as traffic circles, where bombs were likely to be placed."[327] When a soldier determines the attacker is a member of an organized armed group involved in the hostilities, he or she would not have to wait until an attack is imminent to use force when operating under the conduct of hostilities paradigm. Perhaps informed by other means such as aerial surveillance from drones, visual observation from sniper teams and other observation posts, and by recognizing certain TTPs being employed by the insurgents, a belief can be reasonably established that certain individuals can be targeted well before they strike.[328] Once it becomes evident that the threat is caused by a person directly participating in hostilities, then the more permissive rules governing the use of force in armed conflict would be available to address the threat.

In many circumstances the more reactive and narrowly self-defense-oriented human rights law concept of the use force may not be adequate to defeat organized armed groups threatening significant levels of violence. However, until a positive determination of enemy action is established, the default position is one of applying the law enforcement framework. This reality highlights the unique nature of contemporary conflict. The operating environment is one where military personnel may frequently be deployed in situations where they will apply human rights–based principles, if not always operate uniquely under the paradigm of human rights law (i.e., apply human rights norms found in humanitarian law). This can mean that military and police personnel are often similarly situated in responding to an explosive threat, which can present itself with little warning and be exceedingly deadly.

12.3.4. STRICTLY NECESSARY, PROPORTIONATE, AND OVERWHELMING

While under the many circumstances human rights provides the initial normative framework for both a law enforcement and military forces response to a potential explosive threat, questions remain as to what level of force is ultimately justified. There is also the issue of how aggressively that force can be applied to ensure an explosive attack does not succeed. In other words, once the threat is recognized as an explosive one, does the human rights framework provide an effective means to deal with the challenge. Given the potential destructive

[327] Ricks, *supra* note 21, at 218.

[328] Watkin, *supra* note 289, at 312.

power of these threats, the response by security forces understandably will often involve a very robust use of deadly force. However, questions remain regarding the point at which the line is crossed from applying force justified under a law enforcement paradigm to what is effectively the application of humanitarian law–based conduct of hostilities rules.

This is an issue that has been looked at in terms of whether States have directed a "shoot to kill" policy be put into effect by security personnel operating under a law enforcement mandate (see Chapter 13, section 13.2.2, *Shooting to Kill*). Shoot to kill policies have been equated to "targeted killing,"[329] a term that is normally associated with the conduct of hostilities. However, the introduction of generalized and emotive terms such as "shoot to kill" or "targeted killing" to describe law enforcement–based uses of force can be problematic for much the same reasons that too much reliance may be put on "first" or "last" resort terminology. These shorthand phrases can mask the requirement for nuanced and situation-specific assessments of the point at which the normative gap between human rights and conduct of hostilities paradigms may be narrowed virtually to the point of disappearance. This is particularly the case when an armed conflict–linked term like "targeted killing" is incorporated into assessments of what is clearly a law enforcement–based use of force.[330] It is not immediately evident what is substantively gained by the use of terminology such as "shoot to kill" or "targeted killing" in this context.

There is considerable merit to the observation made by Philip Alston that "[t]he rhetoric of shoot-to-kill serves only to displace clear legal standards with a vaguely defined licence to kill, risking confusion among law enforcement officers, endangering innocent persons, and rationalizing mistakes, while avoiding the genuinely difficult challenges that are posed by the relevant threat."[331] It is noteworthy is that Nils Melzer ultimately reaches the conclusion that the law enforcement paradigm does not categorically prohibit the use of deadly force in "very exceptional circumstances," although it does impose extensive restraints on such action.[332] It is only by assessing the capacity of the law enforcement paradigm in its own right to address the very exceptional circumstances arising from the type of explosive threats faced by military and police forces, as well as the civilians they protect, that its utility can be established for use across a wide range of contemporary operations. As has been noted, "[t]his may require the use of lethal force against a suspect, but only when doing so is proportionate and strictly unavoidable to prevent the loss of life."[333]

The challenge in contemporary operations is that the threat from explosives is exceptional, but also, as has been established, a ubiquitous one that makes consideration of the use of deadly force necessary across a wide range of circumstances. In assessing the permissible action for thwarting such attacks, it is first useful to consider what is at stake. A 2006 *Report of the Special Rapporteur on Extrajudicial, Summary or Arbitrary Executions* indicates in a very direct way there are two human rights engaged: the right not to be blown up

329 NILS MELZER, TARGETED KILLING IN INTERNATIONAL LAW 9 (2008).

330 *Id.* 239–39 (for an assessment of the 2000 Swiss *Ewald K.* case where a police sniper killed a man holding a rifle who had already seriously wounded a police officer during a failed attempt at storming the apartment).

331 *Report of the Special Rapporteur on extrajudicial, summary or arbitrary executions*, UN Doc. E/CN.4/2006/53 (Mar. 8, 2006) (by Philip Alston) ¶ 45, at 17 [hereinafter *The 2006 Alston Report*].

332 MELZER, *supra* note 329, at 239.

333 *The 2006 Alston Report*, *supra* note 331, ¶ 47, at 18.

by a bomb and the right not to be arbitrarily shot by security authorities.[334] However, the calculation is not one that factually weighs equally in the final result. A high proportion of the casualties of suicide attacks are civilians. As has been noted, in many instances insurgent and terrorist groups specifically target civilians. These casualties are caused overwhelmingly by those groups and not by security forces using force in the attempt to stop them.

For example, the first six months of 2012 witnessed 637 civilian casualties in Afghanistan as a result of suicide attacks and what are termed "complex attacks" that included suicide devices.[335] A July 2012 report by the UN Assistance Mission in Afghanistan (UNAMA) indicated that during this period, "[t]aking into consideration tactic [*sic*] which use IEDs, such as suicide and complex attacks, IEDs overall caused 53 percent of all civilian deaths and injuries in the first six months of 2012."[336] Overall "[a]nti-Government Elements were responsible for *80 percent of all civilian casualties*, while 10 percent were attributed to Pro-Government Forces. UNAMA was unable to attribute responsibility to any party to the conflict in 10 percent of cases."[337] Statistically, a significantly greater number of civilians can be at risk from terrorist attacks using explosives than from an overreaction by government forces. However, concerns can legitimately be expressed about the use of protective force by State security forces. It is a use of force that must be restrained. As the Commander of ISAF noted, in the latter part of 2009 the security forces were responsible for fewer civilian casualties, "but were still killing far too many Afghans, particularly at checkpoints, and needed to better shield them."[338]

It is the range and complexity of the attacks that can make countering them so challenging. In Afghanistan, "[s]uicide attacks ranged in type from those carried out by single individuals either wearing vests or driving vehicles charged with explosives, to multiple suicide bombers that initiated complex attacks involving large numbers of fighters."[339] Nor is the challenge simply one of complexity. Depending upon the situation there may be literally only a split second for security forces to respond. This compressed decision-making period can occur as a vehicle ignoring warning signals speeds toward a checkpoint or convoy; a suspicious person is in a position of observation who may be preparing to initiate a command-detonated or cell-phone-activated IED; or when an insurgent posing as a civilian makes an arm motion that may be the throwing an RKG-3 HEAT hand grenade at a passing vehicle. The consequences of error can be tragic for civilians in the surrounding area if, for example, the bomber decides to ignite the weapon in a crowded market. Even more challenging is trying to understand and react to situations where the enemy constantly changes tactics to thwart attempts at intercepting an attack. For example, changing from cell-phone ignition of IEDs to the use of garage door openers avoided busy or misplaced calls, permitting the explosives to be detonated, in the words of one insurgent, from "20 meters away, with your

[334] *Id.*

[335] United Nations Assistance Mission in Afghanistan, *Afghanistan Mid-Year Report 2012 Protection Of Civilians In Armed Conflict* 15 (July 2012).

[336] *Id.* at 3.

[337] *Id.* at 2 (emphasis added).

[338] McChrystal, *supra* note 307, at 386.

[339] *Id.*

hands in your pockets and your finger on the button."[340] This reality impacts directly on the significance and the urgency of the "shoot, do not shoot" decision facing security authorities who may be eighteen-year-old soldiers or poorly educated and trained local police forces.

Notwithstanding this threat, it has been suggested in the context of the suicide bombing campaign during the Israeli-Palestinian conflict that security personnel should first challenge a suspect to show they are not carrying explosives, and only if they refuse could they be fired upon.[341] Confronting a suspect in this way is no doubt intended to ensure adequate grounds are established to justify the use of deadly force. However, the likelihood the bomb would be ignited at that point also makes mandating this course of action potentially problematic. In certain circumstances the simple act of confronting the suicide bomber may result in the explosives being detonated with considerable loss of life to the security personnel involved, as well as other soldiers and civilians in the area. Suggestions such as this reflect a tendency on the part of some lawyers to seek a full and at times seemingly exhaustive application of precautions, warnings, and less lethal reactions to a perceived threat.

However, the situation simply may not permit the application of any precautionary measures, and the law does not require such measures to be applied where, for example, nonviolent means are ineffective or do not offer "any promise of achieving the intended result."[342] As has been noted, "the use of lethal force may prove strictly unavoidable when such tactics would unduly risk death or serious harm to law enforcement officers or other persons."[343] It has even been suggested that the risk of an imminent devastating attack on the lives of hundreds of civilians could under stringent circumstances (i.e., no other less lethal option, it would be impossible to find the attackers again, and a detention option was not feasible) justify a jeep containing terrorists being targeted by a missile. In other words, "it might theoretically be possible to claim that an action of this type could be legitimate under the rules of human rights law . . . ."[344]

A key issue is the legal test for when deadly force can be used. The Special Rapporteur on Extrajudicial, Summary or Arbitrary Executions correctly indicated that security personnel could not act on mere suspicion.[345] However, his 2006 report dealing with the use of force against suicide bombers goes on to suggest, "there is no legal basis for shooting to kill for any reason other than *near certainty* that to do otherwise will lead to loss of life."[346] While this may be a statement of policy, or what the Rapporteur believes the law should be, it must be emphasized this is not and should not become the test at law. That standard is far too high for most tactical situations, although it has been applied by the United States since

---

340 Ahmed S. Hashim, Insurgency and Counter-Insurgency in Iraq 162 (2006) (quote from an insurgent in Greg Grant, *Anatomy of an IED*, Army Times, Aug. 15, 2005, at 14).

341 Antonio Cassese, *Expert Opinion on Whether Israel's Targeted Killings of Palestinian Terrorists Is Consonant with International Humanitarian Law*, ¶ 16, at 9, http://www.stoptorture.org.il/files/cassese.pdf.

342 *UN Principles on the Use of Firearms* ¶ 4.

343 *The 2006 Alston Report, supra* note 331, ¶ 48.

344 Noam Lubell, Extraterritorial Use of Force Against Non-State Actors 176 (2010). *But see id.* at 176–77 (he also notes the actual strike on Al Qaeda terrorists in 2002 upon which his hypothetical is based did not meet this factual profile and therefore that killing was not lawful).

345 *The 2006 Alston Report, supra* note 331, ¶ 50.

346 *Id.* ¶ 59 (emphasis added).

2013 as a matter of policy for drone strikes occurring outside areas of ongoing conflict.[347] However, this latter form of strategic-level targeting takes place in a unique operational environment where there is often considerable intelligence, as well as the time to make a more fully informed decision.

As lawyers who have provided targeting advice know, it is a common axiom there is never enough intelligence or time to make a perfectly informed decision. More importantly, ordinary soldiers and policemen tasked with reacting on the spot to explosive threats must frequently make an analysis of the available facts within an exceedingly compressed time period. Split-second judgments about the amount of force required for a particular situation have to be made by police officers in tense, uncertain, and rapidly evolving circumstances.[348] Further, security personnel must also remain very cognizant of the consequences of error. Either acting or not acting could mean people will die and the mission could fail. In such circumstances there is a danger that requiring a "near certainty" standard could end up contributing to the loss of innocent life to the extent it may cause security personnel to hesitate too long while they seek to gather sufficient facts.

The legal test for using force is properly one of "reasonable belief." That standard must be based on objective criteria, however, it also allows for an honest but mistaken belief. The reasonable belief test provides for a margin of error that accurately reflects the nature of the decision, as well as the difficult circumstances under which it often has to be made. It must be remembered that given the stakes involved, detached reflection is not required of a decision maker.[349] The reasonable belief standard also highlights yet another area of congruence between human rights and humanitarian law since the same test applies to the decision to use of force whether operating in a law enforcement capacity[350] or making a targeting decision during hostilities.[351] Given these factors, the decision to shoot or not to shoot may end up being the same regardless of which paradigm is applied to regulate the use of force.

It should be of particular concern to the military and police leadership if mistakes are made. This was starkly evidenced in the shooting of John Charles de Menezes in London, England, on July 22, 2005. Metropolitan Police officers, acting in the wake of successful July 7, 2005 suicide bombings and unsuccessful July 21, 2005 attacks in London, killed an innocent Brazilian citizen trying to enter the subway system.[352] Just five days before the de Menezes shooting, an officer from the police special firearms unit had been quoted in a London paper as saying in the aftermath of the July 7 bombing: "[t]he most effective way of dealing with someone with explosives is to shoot them in the head."[353] That was exactly what was to pass as a result of a series of mistakes made by the security team tailing de

---

[347] *Obama's Speech on Drone Policy*, N.Y. TIMES, May 23, 2013, http://www.nytimes.com/2013/05/24/us/politics/transcript-of-obamas-speech-on-drone-policy.html.

[348] Graham v. Connor 109 S. Ct. 1872 (1989).

[349] Brown v. United States 41 Sup. Ct. 502 (1921).

[350] McCann v. United Kingdom, Application No. 18984/91, Eur. Ct H. R. ¶ 200, at 173 and ¶ 5, at 179 (1995), http://hudoc.echr.coe.int/sites/eng/pages/search.aspx?i=001-57943.

[351] Prosecutor v. Galić, ICTY, IT-98-29-T, Judgement and Opinion, ¶¶ 50, 51, and 55 (Int'l Crim. Trib. for the Former Yugoslavia, 2003), http://www.icty.org/x/cases/galic/tjug/en/gal-tj031205e.pdf (where the test is applied in respect of humanitarian law).

[352] MELZER, *supra* note 329, at 230.

[353] Leppard & Calvert, *supra* note 227, at 12.

Menezes as a potential bombing suspect.[354] Tragically, the consequence of those errors was the death of this innocent civilian.

The after-incident review included criticism by a coroner's jury,[355] but the civilian prosecution authority decided to not lay charges against the individual security personnel involved.[356] In making that decision it was noted, "[a]ny successful prosecution would have to prove that the officers did not 'honestly and genuinely' hold those beliefs [that Mr. de Menezes was a suicide bomber who would blow up a train]."[357] An assessment of the incident criticizes the general application of precautions. This included not making an attempt to arrest before the suspect reached the underground station, as well as questioning whether the operating officers' "mistaken belief could also be regarded as objectively reasonable."[358] In this instance the overall circumstances of this tragedy were properly brought under additional scrutiny. However, the prosecution determination of an honest and mistaken belief on the part of the officers involved appears to be one that could be properly made at law.

There was a prosecution and conviction of the Office of Commissioner of Police for an offence of the Health and Safety at Work Act 1974 of failing to provide for the health, safety, and welfare of Jean Charles de Menezes on July 22, 2005.[359] This action is reflective of a general trend, also found in the European Court of Human Rights and elsewhere, to concentrate on systemic issues, safeguards, and the after-the-fact accountability process, rather than second-guess the decisions made by security personnel. What is of note is that it appears concerns over the de Menezes shooting have not removed the possibility that such tactics will be used in the future.[360] This is because the reality is that there remains a requirement to act decisively in the face of determined terrorist willing to use explosives to cause mass casualties.[361]

---

354 Mark Hughes, *Seven Mistakes that Cost De Menezes His Life*, THE INDEPENDENT, Dec. 13, 2008, http://www.independent.co.uk/news/uk/crime/seven-mistakes-that-cost-de-menezes-his-life-1064466.html.

355 Sandra Laville, *Jean Charles de Menezes Inquest Records Open Verdict*, THE GUARDIAN, Dec. 12, 2008, http://www.guardian.co.uk/uk/2008/dec/12/de-menezes-verdict.

356 *CPS statement: Charging decision on the fatal shooting of Jean Charles de Menezes*, The Crown Prosecution Service (July 17, 2006), http://www.cps.gov.uk/news/latest_news/146_06/, and *CPS Statement—Shooting of Jean Charles de Menezes*, The Crown Prosecution Service (Feb. 13, 2009), http://www.cps.gov.uk/news/latest_news/103_09/index.html. *See also Police Officers in Jean Charles de Menezes Shooting Escape Punishment*, THE TELEGRAPH, Oct. 2, 2009, http://www.telegraph.co.uk/news/uknews/law-and-order/6253971/Police-officers-in-Jean-Charles-de-Menezes-shooting-escape-punishment.html (where it is reported that the Independent Police Complaints Commission also determined that it was not going to recommend disciplinary action against the officers involved in the shooting).

357 Peter Walker, *No Criminal Charges in De Menezes Shooting*, THE GUARDIAN, July 17, 2006, http://www.guardian.co.uk/uk/2006/jul/17/menezes.uksecurity.

358 MELZER, *supra* note 329, at 237.

359 *CPS statement: Charging decision on the fatal shooting of Jean Charles de Menezes*, *supra* note 356 (outlining the charges against the Office of Commissioner of Police), and James Sturke, *Met Police Guilty over De Menezes Shooting*, THE GUARDIAN, Nov. 1, 2007, http://www.guardian.co.uk/uk/2007/nov/01/menezes.jamessturcke2. ("The force was fined £175,000 and ordered to pay £385,000 costs after an Old Bailey jury found it had breached health and safety rules and failed in its duty to protect members of the public in the killing of the innocent Brazilian electrician at Stockwell station, south London, on July 22 2005.").

360 John Steele, *Shoot-to-Kill Tactic Still Viable, Say Police Chiefs*, THE TELEGRAPH, Mar. 8, 2006, http://www.telegraph.co.uk/news/uknews/1512406/Shoot-to-kill-tactic-still-viable-say-police-chiefs.html.

361 Vikram Dodd, *Scotland Yard Creates SAS-style Unit to Counter Threat of Terrorist Gun Attack*, THE GUARDIAN (June 29, 2015, 19:23 BST), http://www.theguardian.com/uk-news/2015/jun/29/scotland-yard-creates-sas-style-unit-to-counter-threat-of-terrorist-gun-attack.

As has been highlighted previously in this chapter, the challenge of having to use deadly force under such difficult, time-constrained circumstances is not limited to suicide attacks and roadside bombs. It is also an inherent part of assaults conducted during hostage rescue operations. The context within such law enforcement or counterterrorist action is taken is a key issue. Even a hostage taking, in what might be described as a purely domestic criminal context, still carries with it the increased likelihood that the hostage taker will be killed once an assault decision has been taken. While each specific case must be considered individually, any assessment must be situated within the broader reality of the nature of the threat posed to the hostages. In particular, the potential use of explosive devices by the hostage takers means there is a strong likelihood they will be subjected to lethal force by rescuers.

In practical terms, given the speed of such assaults and the numbers of armed hostage takers that have to be dealt with, there usually is a very small window of opportunity within which it can be established the kidnappers are not posing such a lethal risk to those being rescued. It has been suggested in respect of the 1980 Libyan embassy rescue in London that because of earlier claims the building was wired with explosives by persons who claimed to be seeking martyrdom, "[t]he only certain way of saving the hostages' lives, therefore, was to kill the terrorists unless it was unequivocally clear that they were identifying themselves for what they were and were very plainly surrendering."[362] Under these circumstances it is the actions of the terrorists that may be determinative. If they do not take steps to clearly surrender, thereby establishing such a threat is not being posed, then lethal force may have to be considered. The terrorists simply may not be willing to surrender. However, at some point the use of force edges increasingly toward a status-based form of targeting based on the group threat posed by the terrorists.

During the 1979 siege in Mecca it is reported that none of the gunmen willingly gave up when called with bullhorns to surrender. The operational commander indicated no one surrendered on their own, they were captured.[363] Further, the ease with which a cell phone or button capable of igniting an explosive device can be hidden from view significantly raises the likelihood that deadly force will be used by security forces to address the threat. The use of deadly force is also very likely to result in situations when the attack is part of a fidayeen mission involving attackers who are determined to die. Police did capture one of the ten LeT terrorists during the 2008 Mumbai attack while he was a passenger in a disabled car. Even then the terrorist killed one of the policemen as he struggled to get out of the car.[364] However, it is also reported the assessment of the officer in charge of the Indian NSG personnel assigned to assault the LeT members in Chabad House and the Trident-Oberoi and Taj hotels was that there "would be no negotiations, and 'It was kill or be killed.'"[365]

The idea of a "double tap" shot to remove the threat posed by a terrorist or hostage taker previously referred to in the Afghanistan[366] and the Japanese embassy[367] examples

---

[362] Tony Geraghty, Inside the Special Air Service 180 (1980).

[363] Trofimov, *supra* note 120, at 212.

[364] Scott-Clark & Levy, *supra* note 151, at 147.

[365] *Id.* at 250.

[366] Rayment, *supra* note 84.

[367] Giampietri, *supra* note 112, at 173.

is consistent with the approach of ensuring explosives or other armaments cannot be used against the hostages. In this respect, "[w]hen targeting a bomber preparing to activate a radio controlled improvised explosive device, for example, a shot to the head resulting in instant death may be necessary to ensure non-activation."[368] This approach appears to have long been part of counterterrorism training,[369] although units applying a police philosophy continue to seek to avoid the use of deadly force.[370] The use of "double taps" is reported to be part of the training for India's NSG counterterrorism force[371] and was used by U.S. Special Forces in the operation against Saddam Hussein's sons in July 2003.[372] Of note, in addition to firing two-round bursts, the training applied by the SAS was reported in 1988 to emphasize "sustained and accurate firepower; a continuous burst of fire where the impact of the bullets keeps the terrorist's hands away from his body so that there is no opportunity to press a button."[373]

It is the extremely dangerous circumstances of hostage rescue and the common use of explosive devices by hostage takers that makes it a common and prudent practice to use overwhelming force until the threat is removed or neutralized. Such action may constitute the minimum force necessary to remove the explosive threat.[374] Such incidents are on a different point on the reaction scale than traditional law enforcement–based concepts. In this regard the risks presented during the Japanese embassy siege in Lima, Peru, consisting of improvised explosive devices, suicide vests, landmines, automatic weapons, high-powered riles, and a terrorist group with a plan to kill hostages lent that situation to overwhelming force being used by the military force during the assault (see Chapter 12, section 12.2.2.2, *The Japanese Embassy, Lima, Peru, 1997*).

However, in terms of law enforcement there is a difference between using deadly force in this fashion and having a generalized "shoot to kill" policy. Only minimum force that is absolutely necessary can be justified under a human rights normative framework. While many hostage rescues will often result in most if not all of the hostage takers being killed if an assault is ordered, such outcomes must be placed under close scrutiny. Any investigation should look at the intelligence available prior to the assault, the training of the security personnel, and the ROE applied during the operation. The objective factors relied on by security personnel and the reasonableness of any belief formed that not only deadly but also overwhelming force was the proportionate force necessary to resolve the incident must be carefully scrutinized. When feasible, forensic evidence must be obtained. In many

---

368 Steven Haines, *Northern Ireland 1968–1998*, *in* INTERNATIONAL LAW AND THE CLASSIFICATION OF CONFLICTS 117, 138 (Elizabeth Wilmshurst ed., 2012).

369 CHARLES A. BECKWITH & DONALD KNOX, DELTA FORCE: THE U.S. COUNTER-TERRORISM UNIT AND THE IRANIAN HOSTAGE RESCUE MISSION 145 (1983).

370 THOMPSON, *supra* note 73, at 74

371 SCOTT-CLARK & LEVY, *supra* note 151, at 248.

372 DAVID C. ISBY, LEAVE NO MAN BEHIND: LIBERATION AND CAPTURE MISSIONS 347 (2004).

373 JAMES ADAMS, ROBIN MORGAN, & ANTHONY BAMBRIDGE, AMBUSH: THE WAR BETWEEN THE SAS AND THE IRA 64 (1988). *See also* CHRISTOPHER WHITCOMB, COLD ZERO: INSIDE THE HOSTAGE RESCUE TEAM 47 (2001) (where it is noted that eight of ten rounds in law enforcement gunfights never hit anyone, and "the highest probability for success comes with aiming at the largest part of the offending person and pulling the trigger until he goes down or you run out of rounds.").

374 Haines, *supra* note 368, at 138.

circumstances this will include quarantining the weapons used and ensuring the assault members do not discuss the incident with one another prior to obtaining their statements.

Close scrutiny must also be made of what criteria are relied on in determining that a suspect is no longer a threat. In other words, what indicia of surrender did the members of the assaulting force rely on? The existence of the standard operating procedures for the provision of medical assistance to captured persons; the proper handling of detainees; training on detainee handling; the designation of personnel, including potentially those assigned from other units for that purpose; and the carrying of equipment such as flex cuffs to secure detainees are all indicators that detentions were anticipated. There is a difference between directing personnel to use the force to the extent necessary to remove the deadly threat and ordering that prisoners should not be taken. Any suggestion that no prisoners are to be taken, or a statement that could reasonably be interpreted as such, is exceedingly problematic. What cannot be lost sight of is that for the personnel seeking to remove a deadly threat to hostages this often means not just using lethal force but also intentionally aiming to kill rather than seeking to wound the suspect. Shooting with the intent to kill in circumstances where the facts exist to make it a reasonable and justifiable action under human rights law is different from having a generalized "shoot to kill" policy (see Chapter 13, section 13.2.2, *Shooting to Kill*).[375]

In the aftermath of an incident it may be attractive to argue that an incident should have been resolved by shooting to wound.[376] For example, it has been suggested one option available to air marshals reacting to an incident on an airplane was "a single shot fired in the shoulder or arm which was holding the bag [containing a potential bomb]."[377] However, this argument does not seem to adequately take into account the stress, time constraints, the need to protect civilians, and the risk being assumed by security forces seeking to remove the terrorist threat. The outcome in Ottawa in 2014 where security personnel engaged a terrorist who had killed a Canadian Forces member and stormed the Parliament buildings highlights the challenges. In that case the security forces were found to have fired fifty-six shots, hitting the terrorist thirty-one times. It is reported that a police investigation found that the use of force was justified.[378]

It has been suggested that a shoot to wound approach must be applied under humanitarian law during hostilities.[379] However, this suggestion has been challenged on a number of bases, including that it would be problematic to apply "in the heat and confusion of the battlefield"[380] and the "situations presenting a viable possibility of wounding instead of

---

[375] *The 2006 Alston Report, supra* note 331, ¶ 51 (where it is acknowledged States may have to use "shoot to kill" policies when dealing with suicide bombers, albeit within strictly defined parameters).

[376] MELZER, *supra* note 329, at 238.

[377] *Id.*

[378] Evan Soloman, *Ottawa Gunman Michael Zehaf-Bibeau Was Shot 31 Times, Police Report to Reveal,* CBC NEWS (June 2, 2015, 10:24 AM ET), http://www.cbc.ca/news/politics/ottawa-gunman-michael-zehaf-bibeau-was-shot-31-times-police-report-to-reveal-1.3096073.

[379] INTERPRETIVE GUIDANCE ON THE NOTION OF DIRECT PARTICIPATION IN HOSTILITIES UNDER INTERNATIONAL HUMANITARIAN LAW, International Committee of the Red Cross 82 n.221 (February 2009) (quoting J. PICTET, DEVELOPMENT AND PRINCIPLES OF INTERNATIONAL HUMANITARIAN LAW 75 (1985)). *See also* Goodman, *The Power to Kill, supra* note 209 at, 819 (arguing for a least restrictive means approach).

[380] Schmitt, *A Reply to Ryan Goodman,* supra note 209 at 856.

killing are so rare that it is counter-intuitive to conclude that States intended the 'method' language to extend to such circumstances."[381] Further, the option of using nonlethal weapons "are seldom viable when involved in classic combat or when enemy forces are themselves employing lethal force."[382] As a matter of law the shoot to wound argument seems better positioned as one perspective of what the law "should" be rather than what it presently is (i.e., the *lex ferenda* versus *lex lata* debate).

Courts may continue to look for security forces to attempt to wound. For example, in the 1995 *McCann v. United Kingdom* case, the European Court of Human Rights specifically ruled, when considering the control and organization of the operation leading to the 1987 use of force by British SAS personnel that killed three IRA members in Gibraltar, that it was not clear they had been trained in shooting to wound. Further, the soldier's reflex action lacked the necessary degree of caution or duty of care reflected in instructions on the use of force by the police. That failure "suggests a lack of appropriate care in the control and organisation of the arrest operation."[383] This conclusion was reached in a context where one of the soldiers involved had testified "that it had been discussed with the soldiers that there was an increased chance that they would have to shoot to kill since there would be less time where there was a 'button' device . . . ."[384]

There are circumstances where shooting at the extremities of a person, rather than firing at the center of the body, or specifically to shoot to kill are used. For example, it appears to be applied by the Israel Defence Forces in their escalation of force procedures.[385] This has included the controversial use of the Ruger suppressed .22 caliber sniper rifle, with fire intended to be directed at legs of protesters throwing rocks during riots and demonstrations.[386] However, even such use of force can result in death[387] and may not be applied because of the wounding of targets can incite further violence.[388] Similarly, ROE issued by the United Nations suggested firing at "non-vital parts of the body."[389] The use of such tactics can be very much situation and training dependent.

The weapons used may also reflect the primary roles assigned to the Special Forces units. In the 1990s it was reported that the French GIGN preferred a .357 magnum revolver because "[b]eing a police unit called on to intervene in both terrorist and purely criminal cases,

---

381 *Id.* at 857.

382 *Id.* at 857 n.9.

383 McCann v. United Kingdom, Application No. 18984/91, Eur. Ct H. R. ¶ 212, at 52 (1995), http://hudoc.echr.coe.int/sites/eng/pages/search.aspx?i=001-57943 (emphasis added).

384 *Id.* ¶ 211.

385 1 Public Commission to Examine the Maritime Incident of 31 May 2010, ¶ 206, at 244 (2011).

386 *B'Tselem Demands That Judge Advocate General Prohibit Use of Ruger Rifle to Disperse Demonstrators*, B'Tselem - The Israeli Information Center for Human Rights in the Occupied Territories (Mar. 1, 2009), http://www.btselem.org/firearms/20090301_use_of_ruger_rifle_in_demonstrations_prohibit, and *B'Ttselem Calls on IDF to Stop Using Sniper Rifle, .22 Caliber Bullets*, Haaretz (Oct. 7, 2015, 5:02 AM), http://www.haaretz.com/news/diplomacy-defense/1.679081.

387 Tamar Pileggi, *Palestinian Suspect Killed during West Bank Arrest Operation*, The Times of Israel (July 27, 2015, 9:00 AM), http://www.timesofisrael.com/palestinian-suspect-killed-in-arrest-attempt/.

388 Noam Amir & Maariv Hashavua, *Israeli Combat Troops React Angrily to New Rules of Engagement in West Bank*, The Jerusalem Post (Aug. 12, 2015, 9:02), http://www.jpost.com/Arab-Israeli-Conflict/Israel-tightens-rules-of-engagement-for-combat-troops-in-West-Bank-411868.

389 Bruce Oswald, Helen Durham, & Adrian Bates, Documents on the Law of UN Peace Operations 565 (2010), Annex J Rules of Engagement-Soldiers Pocket Card, Rule 7.

the French group requires every round to guarantee strong stopping power while avoiding unnecessary casualties."[390] In contrast, the British SAS approach using 9 mm pistols and close quarter battle training "reflects the extreme environment of guerrilla war in Northern Ireland, where deadly force is required, along with speed and flexibility, in circumstances of close proximity during ambushes against multiple groups of heavily armed well-trained dangerous criminals." (see Chapter 13, section 13.2.1, *The "Troubles"*).[391] Each approach has its operational application. The training and tactics used by the security forces involved are important considerations when carrying out operational planning and making decisions regarding the use of a particular unit.[392]

A significant challenge is that the emphasis placed on shooting to wound as outlined above does not address the difficulty experienced by the even best marksmen or members of highly trained specialized units in making a wounding shot. This is particularly true during the stress and confusion of high-risk situations.[393] Such a capability is significantly less likely to exist among regularly trained soldiers and police who usually have to confront explosive threats. In this respect, "[a]ll soldiers are trained to fire at the main part of the human torso as the most effective means of ensuring that a round hits it target and has the desired 'stopping' effect."[394] Further, as an FBI Hostage Rescue Team (HRT) marksman has stated, "[t]rying to shoot someone in the leg or the arm when he is threatening your life is just about impossible for most people."[395]

In reality, the functional role and training of most security personnel and the operational situation may simply not match the approach suggested by the court in the *McCann Case*. Just as verbally challenging suicide bombers in some cases may be counterproductive, so may seeking to shoot to wound. The idea that an attempt should be made to shoot to wound was resisted in the context of the "shoot to kill" debate during the Northern Ireland conflict (see Chapter 13, section 13.2.2, *Shooting to Kill*). It was viewed as an idea that was based on a myth evoking a Hollywood fantasy.[396] It is also exceedingly difficult to use firearms to stop speeding cars.[397] There is considerable potential risk to hostages or other civilians in the vicinity of an attack (e.g., a crowded marketplace) should a shot fired at a target's extremities either miss or pass through an arm or leg.

It is the concern about the stopping power of rounds, the potential for over penetration, and the risk posed to hostages that has prompted the use of hollow point or frangible ammunition in a domestic criminal law context. As is evident in the aftermath of the de Menezes shooting, police forces still consider the use of overwhelming lethal force, including hollow point bullets, to counter explosive threats a necessary option. The result is that

---

[390] AROSTEGUI, *supra* note 139, at 204.

[391] *Id.*

[392] NAYLOR, *supra* note 64, at 374 (One of the critiques of a 2010 hostage rescue operation in which a hostage was killed by a rescue force grenade was that the unit involved did not do hostage rescue training during its pre-deployment workup and instead 'focused on "combat clearance" techniques.').

[393] MARK URBAN, BIG BOYS' RULES: THE SECRET STRUGGLE AGAINST THE IRA 6 (1992).

[394] Haines, *supra* note 368, at 139.

[395] WHITCOMB, *supra* note 373, at 46.

[396] ADAMS ET AL., *supra* note 373, at 33.

[397] *See also* RICHARD DOHERTY, THE THIN GREEN LINE: THE HISTORY OF THE ROYAL ULSTER CONSTABULARY GC 1922–2001, at 171 (2012).

under specific situations such as hostage rescue, and even when seeking to remove an explosive threat, any attempt to consider wounding a suspect may have to be quickly discounted depending upon the tactical situation confronting the soldiers or police personnel. Neither the threat nor the reaction falls easily into the paradigm of everyday policing.

The result is that efforts to stop insurgents and terrorists employing IEDs or conducting suicide attacks carry with them the increased likelihood that it will be necessary to kill the perpetrators in order to protect both civilians and security personnel. This is particularly the case in hostage rescue situations when an assault is required to resolve the incident. That reality must be made clear by legal advisors in the advice given to decision makers before such assaults or actions are put into place. Such advice must avoid suggesting the incident can be resolved by shots to wound unless those trained and actually conducting the assault view such action as feasible. Above all, the lethal consequences of the action and the likely requirement for an after-action inquiry must be clear to all decision makers, including politicians who often become engaged in high-profile incidents. A requirement to justify why such lethal force was used in the circumstances will undoubtedly be part of the postincident accountability process.

12.4. A Unique Challenge

The deployment of military forces by a State will continue to provide a useful indicator of when an armed conflict is in existence. Such a deployment can mean there is a greater potential for violence to occur because of the application of the conduct of hostilities framework to govern the use of force. Unfortunately, contributing to the increased levels of violence is a frequently more unrestrained approach that foreign military forces adopt toward the use of force when operating in other countries. In this regard the traditional restrained application of force required in a domestic setting is not always transferred to international operations.[398] Of course, this can be avoided to a large degree by treating the local population as much as possible as if it was your own. Significantly, the nature of contemporary security threats (i.e., insurgency, terrorism, and crime) requires a much more nuanced assessment of the roles being performed by security forces on contemporary operations than is often overtly acknowledged within military circles. This is because military forces are not simply deployed as "war fighters." They also contribute to the maintenance of law and order in the territories in which they are operating. On occasion the operation may be a purely law enforcement one. "Policing" in one form or another can be in many circumstances an integral part of military operations (e.g., occupation, counterinsurgency, counterterrorism, hostage rescue, noncombatant evacuation, counterpiracy).

Further, the deployment of military forces does not necessarily mean "combat power" will be applied even during armed conflict. The armed forces of a State do not always use force as a first resort. Indeed, any insistence that State armed forces only be deployed on conventional operations involving combat operations represents a clear and, sometimes in terms

[398] LEDWIDGE, *supra* note 27, at 185–86 (for a discussion of the lack of investigation and accountability accorded to civilian deaths in Afghanistan in comparison to British operations in Northern Ireland).

of outcome, tragic form of denial of the type of action required to maintain international security. The nature of the "enemy,", and the largely civilian environment within which they operate, inevitably results in military forces having to engage in law enforcement–type operations. As was demonstrated in Iraq and Afghanistan this can include directly policing a civilian population, supporting foreign and national police, and the training and mentoring of local police forces. The result is that the professional skills required to command, plan, and execute operations in twenty-first-century operations must include a clear understanding on the part of military personal of human rights law–based restrictions on the use of force.

However, the same operating environment that privileges the application of a policing approach also frequently demands significantly more of the law enforcement framework than is traditionally associated with domestic policing. This can be seen in unique operations such as hostage rescue, as well as with more generalized tasks where the risk posed by explosives are a ubiquitous and deadly threat. While a restrictive approach to the use of force must be applied when operating under a human rights legal framework, there are circumstances in which the use of deadly force is very likely to be required. In applying that lethal force, the legal test is not that there must be "near certainty" that doing otherwise will lead to a loss of life. While that test may be adopted for policy reasons, the legal standard for the use of force under these circumstances is the same as for targeting under humanitarian law: a reasonable belief the use of such force is necessary. This standard not only provides a workable and fair standard, it serves as further evidence of the narrowness of the normative gap that can exist between the law enforcement and conduct of hostilities paradigms.

It is evident that in many circumstances where force is used it will take on a lethal nature regardless of whether it is considered under a law enforcement or conduct of hostilities paradigm. However, what must be clearly understood is that even when lethal force can be applied, there is a need to recognize there are limits to what a law enforcement approach can operationally achieve. Counterinsurgency and, frequently, counterterrorism operations occur in the context of an armed conflict. At its heart such conflict involves the conduct of warfare against organized armed groups and a requirement to respond to threats of "explosive" violence that far exceed what is normally seen in a domestic peacetime setting. As international lawyers increasingly embrace the idea that human rights law, with its traditionally more restrained authority to use force, applies in these violent situations, it will become important to identify the limits of what that normative framework can do. It is only by identifying the advantages and the limits of both the human rights and humanitarian law authority to use force that appropriate policy and command choices can be made regarding such use.

[O]ne cannot kill one's way to victory; however as Sir Robert Thompson has pointed out, one cannot defeat an established insurgency without killing.[1]

13 Limits of Law Enforcement and Human Rights Law

13.1. Introduction

It is a simple reality that human rights–based uses of force comprises a significant part of contemporary military operations. Human rights law may be applied as a matter of law or policy to control the use of force during security operations, both at home and overseas. However, given the different rules governing the conduct of hostilities and the traditionally more restrictive approach toward the use of force under human rights law, it is inevitable questions will arise regarding the limits of this latter body of law and the function associated with it: law enforcement. This chapter explores some of those limits from three perspectives: practical limitations, overreach in application, and limits caused by interpretation.

Practical limitations are assessed in the context of the long-running insurgency in Northern Ireland, which the UK government chose to address as a domestic law enforcement matter. Reviewing the scope of the violence during the "Troubles" (the conflict in Northern Ireland), the government response and unique factors impacting on the success of the chosen approach are particularly helpful due to the tendency of some legal analysts to rely on that conflict to underpin an argument that "terrorism" should be dealt with as a law enforcement matter. Interestingly, within the military there can also be an overreliance on this example in arguing how counterinsurgency should be conducted. As will be noted, applying a uniquely criminal law approach cannot necessarily be easily transferred to most conflict in the world's ungoverned spaces. The analysis will then explore the "shoot to kill"

[1] John T. Fishel, *Forward*, *in* Max G. Manwaring, The Complexity of Modern Asymmetric Warfare x (2012).

controversy that arose during the "Troubles," where allegations were made that the human rights–based authority to use force was exceeded by the actions of the security forces such that they appeared to be applying hostilities-related force. This controversy raises questions regarding the point at which a law enforcement approach may no longer be effective and whether that assessment can assist in determining when a conduct of hostilities response may be permitted at law and is required in fact. Finally, the impact of developing militarized police forces to extend the application of a law enforcement role during higher end security operations will be discussed.

The problem of overreach in the application of human rights–based restrictions on the use of force will be explored in the context of the jurisprudence of the European Court of Human Rights. This analysis will focus on that court's application of restrictive human rights principles to aerial bombing and the use of other armaments. This places that body of law in direct competition with the humanitarian–based rules intended to govern the conduct of hostilities during armed conflict. Finally, interpretive limits on the use of force will be looked at in the context of various post-9/11 State responses to the threat posed by hijacked aircraft. The analysis establishes that narrow interpretations of human rights law will face challenges when required to weigh the value of the lives of aircraft passengers as a decision is made about whether a civilian airliner should be shot down. The approach to this shoot-down question applied by Canada and the United States will be contrasted with that of Germany and Australia.

13.2. The Shoot to Kill Controversy

13.2.1. THE "TROUBLES"

The conflict in Northern Ireland, known as the "Troubles," ran from 1969 until 1998. The "Troubles" has become the classic example of a low-level insurgency that ultimately proved to be amenable to a law enforcement response. However, it is a conflict that has, in many respects, taken on mythical and at times misleading proportions that impact directly on contemporary operations. For the British military it appears to have created a false sense of having a counter-insurgency expertise that was transferable to all forms of insurgent conflict and other forms of small wars. One of the lessons to be learned from their experience in Iraq and Afghanistan is that "[s]uccess and expertise in fighting an insurgency in north-west Europe does not imply such prowess elsewhere."[2] For others this conflict is relied on to suggest contemporary terrorism is fundamentally a law enforcement matter. Such a view has been supported by reference to the UK reservation to the 1977 Additional Protocols, made in the midst of that conflict, that indicates armed conflict is a situation "not constituted by the commission of ordinary crimes *including acts of terrorism* whether concerted or in isolation."[3]

---

[2] Frank Ledwidge, Losing Small Wars: British Military Failure in Iraq and Afghanistan 164 (2011).

[3] AP I, UK Reservation/Declaration (July 2, 2002), http://www.icrc.org/applic/ihl/ihl.nsf/Notification.xsp?action=openDocument&documentId=0A9E03F0F2EE757CC1256402003FB6D2 (emphasis added). *See also* Mary Ellen O'Connell, *The Choice of Law Against Terrorism*, 4 Nat'l Sec. L. & Pol'y 343, 348 (2010) (where reference is made to the British, Germans, Italians, and Indians having dealt with terrorist challenges using law enforcement methods, but then go on to refer specifically to the UK reservation regarding armed conflict, ordinary crime, and terrorism).

The "Troubles" involved a complex legal and factual situation, which makes it difficult to conclude it did not reach the level of an armed conflict. However, it also occurred in a political and military context that highlights the effective application of a uniquely policing approach can be dependent on a number of factors including good governance, an established and responsive justice system, a capable functioning police force, the ability to exercise control in an area of operations, and an operating environment where cultural similarities facilitate rather than hinder operations. In other words, it obviously did not occur in a failed State, in a "foreign" land, or in areas where the government had lost complete control.

In discussing the application of a law enforcement approach, and to set the scene for a discussion of the shoot to kill controversy that arose during operations in Northern Ireland, it is helpful to review in some detail the nature of that conflict. The following outline addresses paramilitary group organization; the levels and intensity of violence; the procurement, production, and reliance on explosive devices; and the means used to confront the terrorist threat. As will be established, the "Troubles" did not fall into the categories of ordinary criminal activity or more limited acts of terrorism. Importantly, this conflict highlights that States can choose to adopt a law enforcement approach, even when the violence may qualify as an armed conflict. However, the nature of the violence was such that at crucial points questions began to be asked whether security forces, both military and police, had moved beyond the constraints of human rights–based law enforcement regarding the manner in which force was being used. The "Troubles" involved a near thirty-year conflict that highlights some of the operational limits of a law enforcement approach. Neither a law enforcement nor conduct of hostilities approach may provide a quick path to victory: a return to normalcy.

In terms of participants in the violence, there was what has been described as a "complicated cast of characters" involving:

> the British government (including the army, locally recruited regiments, and the police force), republican paramilitaries (especially the PIRA [Provisional Irish Republican Army], the Irish National Liberation Army, or ILNA, and others and loyalist paramilitaries (including the Ulster Volunteer Force, or UVF, Ulster Defence Association/Ulster Freedom Fighters or UDA/UFF, the Loyalist Volunteer Force, or LVF, and others).[4]

Both the Republican and Protestant paramilitary groups were loosely organized along military lines.[5] In a development that is not uncommon with insurgences, the IRA organization changed over the course of the conflict from a traditional military structure to a more cellular one. This was brought on by the operational challenges those groups faced as the security services became increasingly effective and focused on the insurgent threat.[6] In this respect,

---

[4] Audrey Kurth Cronin, How Terrorism Ends: Understanding the Decline and Demise of Terrorist Campaigns 43 (2011).

[5] Michael Burleigh, Blood & Rage: A Cultural History of Terrorism 297 (2009).

[6] Richard Iron, *Britain's Longest War: Northern Ireland 1967–2007*, in Counterinsurgency in Modern Warfare 157, 160 (Daniel Marston & Carter Malkasian eds., 2008).

"[b]y 1984, the IRA guerrilla activists were reduced to an estimated 250 diehards, but this was an experienced, highly disciplined, and ideologically devout group supported by a large number of sympathizers."[7] In contrast, the Protestant groups were identified to largely be criminals operating "more for bloodthirsty revenge and monetary gain than for political reasons . . . ."[8] This aspect of the Protestant groups highlights once again the thin line that often exists between insurgents, terrorists, and criminals. What is evident is that the loose organization and diversity of these groups was not unlike that seen in more contemporary insurgencies, such as in Afghanistan and Iraq.

The IRA and paramilitary Loyalist groups conducted bombings that caused significant casualties.[9] However, particular attention is usually paid to the Provisional IRA because by the mid-1990s they had become responsible for almost half of those casualties.[10] That group has been assessed as "the most dedicated and effective of them all, and the only one that threatened the British government."[11] As journalist and author, Tony Geraghty, notes a British Army analysis of the over three-decade campaign in Ulster describes "the IRA as 'a professional, dedicated, highly skilled and resilient force' that had not been defeated."[12] While the use of the word "professional" appears terribly misplaced given the IRA involvement in indiscriminate bombings and other terrorist activity, there is no doubt regarding the significant challenge that group presented to the UK security forces. Further, such references to professionalism cannot in any way be confused with legitimacy. The IRA committed murder against members of the security forces and civilians by means of both direct and indiscriminate attacks. As has been noted, like many terrorist and insurgent groups, it also relied on a "bedrock of crime" to fund its activities.[13]

As for the duration and intensity of violence, the "Troubles" were not simply isolated terrorist or criminal acts. From 1968 until 1998 there were 3,251 fatalities with milestones including "bomb attacks on pubs in Belfast, Birmingham and the streets of Dublin and Monaghan."[14] The casualties included 650 British soldiers and 301 Royal Ulster Constabulary police officers.[15] While the number of attacks varied considerably over the years, in 1972 it is reported "there were 12,481 incidents, including 10,628 shootings and 1,853 bombings."[16] The IRA bombing campaign attacked security forces, both inside and outside Northern Ireland.[17] Those security forces were confronted with violence ranging from hooligans, often young children, making insults, and throwing bottles and bricks[18]

---

[7] Robert B. Asprey, War in the Shadows: The Guerrilla in History 1134 (1994).

[8] *Id.* at 1130.

[9] Richard Doherty, The Thin Green Line: The History of the Royal Ulster Constabulary GC 1922–2001, at 120 (2012).

[10] Cronin, *supra* note 4, at 43.

[11] Iron, *supra* note 6, at 157.

[12] Tony Geraghty, Black Ops: The Rise of Special Forces in the C.I.A., the S.A.S., and Mossad 227 (2010).

[13] James Adams, Robin Morgan, & Anthony Bambridge, Ambush: the War Between the SAS and the IRA 39 (1988).

[14] Tony Geraghty, The Irish War: The Military History of a Domestic Conflict v (1998).

[15] *Id.* at xvii.

[16] Adams et al., *supra* note 13, at 50. *See also* Geraghty, *supra* note 14, at 41 (for reference to 6,948 violent incidents (1,500 shootings and 1,000 bombs).

[17] *Id.* at 41–46.

[18] Michael Dewar, The British Army in Northern Ireland 112 (1985).

to roadside bombs, mortars, and coordinated assaults.[19] It was a violent struggle with extensive use made of explosives. Even the term, "improvised explosive device" (IED), "came from Northern Ireland, normally referring to IRA bombs made from fertilizer and more commonly used in rural areas."[20] Importantly, the "Troubles" also challenged the notion that in the right circumstances violence involving the use of explosives such as IEDs, mortars, and rocket-propelled grenades is not amenable to law enforcement–based resolution.

The conflict in Northern Ireland has proven challenging to categorize. It is not uncommon to consider the Northern Ireland conflict in terms of three distinct phases: "militarization (between 1969 and 1974); normalization (between 1975 and 1980); and the active counter-insurgency phase (1981–1994)."[21] It has also been suggested counterinsurgency activity was a feature of the police response throughout the conflict.[22] In contrast, one analysis has suggested the insurgency phase lasted from 1972 until 1974, and at some point after that dropped to a level of "civil disturbances."[23] However, the difficulty in assessing the civil disturbance threshold is evident in the statement that "point might be the end of the insurgency in 1974, or alternatively 1994 and the PIRA ceasefire of that year, notwithstanding PIRA's subsequent return to violence in 1997."[24] Although there were periodic lulls in the violence it is difficult to see the "Troubles" as simply constituting a civil disturbance after 1974. Some of the violence clearly presaged that experienced in later insurgencies in Iraq and Afghanistan. For example, on August 27, 1979, at Warrenpoint, Northern Ireland, a roadside bomb of an estimated 500 lbs of explosive was set off as a three-vehicle convoy of the 2nd Battalion of the Parachute Regiment passed by. IRA gunmen then engaged the surviving soldiers with gunfire. When a reconnaissance element of an airborne reaction force landed, a second 1,000 lb bomb was detonated killing twelve men. In all, eighteen soldiers died.[25]

Another factor elevating the violence above the realm of ordinary crime was that for substantial periods certain land areas could not be entered without significant risk. A correspondingly major military effort was required to mitigate the threat. The "Troubles" also involved both urban and rural operations. In the early 1970s there was a "shift of the major guerrilla effort to the South Armagh countryside, which in view of a three hundred-mile-long border and limited number of army and police brought an awkward challenge . . . ."[26] In 1979 "[r]emotely-detonated bombs caused nearly thirty per cent of Security Force deaths in an IRA campaign against the Army which concentrated on rural and border areas. In

---

[19] Mark Urban, Big Boys' Rules: The Secret Struggle Against the IRA 221–22 (1992) (outlining a series of attacks against police stations in Tyrone in 1985).

[20] Richard Connaughton, Modern Warfare: The True Story of Conflict from the Falklands to Afghanistan 300 (2008).

[21] Mark Cochrane, *The Role of the Royal Ulster Constabulary in Northern Ireland*, *in* Policing Insurgencies: Cops as Counterinsurgents 107, 108 (C. Christine Fair & Sumit Ganguly eds., 2014) [hereinafter Policing Insurgencies].

[22] *Id.*

[23] Steven Haines, *Northern Ireland 1968–1998*, *in* International Law and the Classification of Conflicts 117, 135 (Elizabeth Wilmshurst ed., 2012).

[24] *Id.*

[25] Dewar, supra note 18, at 158–60, and Urban, *supra* note 19, at 85–86.

[26] Asprey, *supra* note 7, at 1131.

Belfast and other towns the victims were mostly civilians. . . ."[27] Bases in Armagh, particularly Crossmaglen, were very difficult and dangerous to reach by road.[28] A Royal Ulster Constabulary (RUC) Special Branch officer indicated that near the border with Ireland "if you dial 999 the police won't come, not until the soldiers have secured the area first, and the soldiers won't come until their engineers have cleared the route for bombs."[29] For example, "[i]f a large convoy is required to resupply Crossmaglen the route has to be cleared and picqueted."[30] In such circumstances "the simplest method of routine supply and liaison is by helicopter and this is indeed the method normally used."[31] This situation has a resonance with some contemporary insurgencies. The loss of operational mobility and relatively benign access to areas by security forces is a key factor in assessing whether a law enforcement response is feasible (see Chapter 7, section 7.4.2, *The "Targeted Killing" Case and Control*).

Bombings also occurred outside Northern Ireland, including British Army targets, Harrods department store in London, a mortar attack on 10 Downing Street, the bombing of the Baltic Exchange and Canary Wharf, and an attempt to murder Prime Minister Margaret Thatcher and the cabinet at the Grand Hotel in Brighton.[32] The violence extended to the European continent with attacks in the Netherlands, Belgium, and Germany.[33] Like contemporary IED operations, the development of a bomb-making capability within the IRA demanded organization, skill, and experience.[34] In 1998 it was noted that "[a]fter almost three decades, Irish Republicans were the world's leaders in the design of Improvised Explosive Devices"[35] although the claim for that title would now likely be made by members of Islamic jihadist groups. The attacks were often indiscriminate. A bombing that attracted particular revulsion was a November 1987 car bomb at a Remembrance Day ceremony in Enniskillen.[36]

Considerable efforts were undertaken by the IRA to smuggle military weapons and explosives into Northern Ireland, however, the IRA also "developed a home-made, do-it-yourself capability for arms manufacture which makes any effort to achieve weapons-control a doomed enterprise."[37] Liberal use was made of mortars and other weapons of war, including locally manufactured anti-armor grenades.[38] A mortar round hit one aircraft as it tried to land at a base at Crossmaglen, South Armagh.[39]

Further the fire from .50 caliber guns hit twelve aircraft brining down four of them between 1977 and 1993.[40] Efforts were even made to obtain antiaircraft missiles in the

---

27 DEWAR, *supra* note 18, at 158.

28 *Id.* at 205.

29 JAMES RENNIE, THE OPERATORS: ON THE STREETS WITH 158 COMPANY 222 (1996).

30 DEWAR, *supra* note 18, at 205

31 *Id.*

32 BURLEIGH, *supra* note 5, at 330, and Iron, *supra* note 6, at 164.

33 URBAN, *supra* note 19, at 125.

34 GERAGHTY, *supra* note 14, at 204–06.

35 *Id.* at 204.

36 BURLEIGH, *supra* note 5, at 331.

37 GERAGHTY, *supra* note 14, at 188.

38 URBAN, *supra* note 19, at 208–09.

39 GERAGHTY, *supra* note 14, at 196.

40 *Id.* at 186, and URBAN, *supra* note 19, at 209–10.

United States. However, the group involved was ultimately arrested.[41] These were simply not the weapons of ordinary criminals. The use of such weapons by armed groups is indicative of the existence of an armed conflict. Their use and the sophistication of the organization that developed and deployed them, highlight the challenge of attempting to control the threat through law enforcement means.

From a security force perspective the conduct of the initial counterterrorist campaign was marred by an overuse of force, such as the conduct of large-scale military operations and the killing of thirteen Catholics on Bloody Sunday.[42] Reforms introduced by Brigadier Frank Kitson in the early 1970s changed the focus toward intelligence-led operations.[43] However, tactics applied in colonial conflicts "were found to be wholly inappropriate for use on citizens of the United Kingdom."[44] The transition to an intelligence-led operation was followed by other initiatives "when a policy change placed emphasis on 'police primacy' in all operations."[45] This was encouraged by the police and occurred in spite of the desire of a number of senior Army officers to retain that responsibility.[46] It has been suggested that "[t]hrough time, the evolution and increasingly specialized capabilities of the RUC Special Branch were to prove to be the single greatest organizational weapon in countering planned insurgent activity."[47] This highlights the essential requirement of having an effective local police force. This is a situation that is virtually nonexistent in the "ungoverned spaces" of the word, and frequently law enforcement capabilities are inadequate in many States being challenged by insurgents. In Northern Ireland the effort also remained heavily supported by military forces. The security response included "framework operations" involving military and police forces "patrolling, base security, searching for terrorist munitions, over surveillance, and control of movement," and covert operations consisting of covert surveillance, agent handling, and occasionally direct action "to ambush, kill, or capture terrorists engaged in operations."[48] However, one of the most effective techniques, "satellite patrolling," proved to not be transferrable to British operations in Iraq.[49]

It has also been argued that "[t]he single most effective weapon against terrorism in Northern Ireland was unquestionably intelligence."[50] The continuing effort to concentrate on intelligence-led operations was to lead to an exceptionally broad array of British civilian and military intelligence agencies operating in Northern Ireland.[51] Those agencies were involved not only in finding but also, it has been suggested, "licensed to take the war to the enemy."[52] The RUC developed its own intelligence, surveillance, criminal forensic, and enforcement specialists.[53] As in other situations of insurgency, the police and military forces deployed

---

41 URBAN, *supra* note 19, at 128–29.

42 ASPREY, *supra* note 7, at 1127–28.

43 URBAN, *supra* note 19, at 36–38.

44 Cochrane, *supra* note 21, at 112.

45 ASPREY, *supra* note 7, at 1130.

46 DOHERTY, supra note 9, at 131.

47 Cochrane, *supra* note 21, at 107.

48 Iron, *supra* note 6, at 167.

49 LEDWIDGE, *supra* note 2, at 31–32.

50 Cochrane, *supra* note 21, at 124.

51 GERAGHTY, *supra* note 14, at 130–31.

52 *Id.* at 131.

53 *Id.*

units that often had complimentary capabilities (i.e., 14 Intelligence Company/RUC E4A, and the SAS/RUC Headquarters Mobile Support Unit-HMSU).[54] A key factor in the ability of British forces to conduct covert surveillance was that they were operating within what was basically their own territory, could make ready use of local government records,[55] and blended in both physically and culturally with the population.[56] Once again this was to prove much more challenging in Iraq where British soldiers carried out surveillance in local clothes and drove civilian vehicles,[57] but found it much more difficult to blend into the surrounding population.[58] While there were successes in Iraq, "[i]t was very difficult to go out into the streets...."[59]

Throughout the "Troubles" the British government consistently took the position that the terrorist activity was a criminal matter and "at no point formally acknowledged that the violence generated by PIRA or any other non-state armed group reached the Common Article 3 threshold...."[60] While the security response was generally reflective of counter-insurgency doctrine, there was a focus on maintaining an exclusive application of the law enforcement paradigm for dealing with terrorism. The ability to carry out effective policing and other security operations and apply a law enforcement approach in Northern Ireland was very much a result of the use of "indigenous" security personnel and robustness of the existing law enforcement establishment (i.e., police, lawyers, courts, and prisons). Of course these are the key indicia of effective governance. Other counterinsurgency environments, such as in Afghanistan and Iraq, have had to rely on foreign security forces and often on military units until the national law enforcement infrastructure can be reconstituted or adequately developed.

The UK decision to apply a law enforcement framework put its security forces in conflict with a terrorist organization willing to use levels of violence that was far beyond what was commonly associated with criminal activity or even terrorism at the time. However, it was not law enforcement in a classic domestic peacetime environment. It has been noted "Britain's laws on terrorism were rooted in traditional military notions of counter-insurgency, particularly through the use of 'temporary' emergency laws suspending the usual rights of prompt access to lawyers and court protection."[61] However, another commentator has also suggested "[t]he result is an unprecedented legal framework whose oddity makes it hard to place in the British tradition."[62] It was an approach that felt "terrorism could be more effectively managed by the ordinary if somewhat altered criminal justice and prison

---

54 *Id.* at 130–31, 135.

55 Rennie, *supra* note 29, at 97.

56 *Id.* at 161.

57 Mark Urban, Task Force Black 168 (2010).

58 *Id.* at 109

59 *Id.* at 198. *See also* Ledwidge, *supra* note 2, at 220–23.

60 Haines, *supra* note 23, at 130 (the author indicates he was informed various legal advisors within the British government, in particular Army military lawyers, were of the view the Common Article 3 threshold was crossed in the 1970s). *See also* Kieran McEvoy, Paramilitary Imprisonment in Northern Ireland: Resistance, Management and Release 15 (2001) ("One central tenet... which pervaded British policy in Northern Ireland has been the disavowal of the notion of the state as a combatant in the Northern Ireland conflict.").

61 Geraghty, *supra* note 14, at 96.

62 Charles Townshend, Britain's Civil Wars: Counterinsurgency in the Twentieth Century 69 (1986).

systems."[63] Special legislation included the suspension of the right to jury trial through the creation of the Diplock courts in order to deal with a security situation that was stretching the bounds of the law enforcement normative framework.[64] In a similar vein there was the 1974 Prevention of Terrorism (Temporary Provisions) Act, which "was a means of gathering Intelligence rather than punishing the guilty."[65] In 1982, of the 5,500 people detained only 2 percent were charged.[66] Further, a "supergrass" system in operation between 1981 and 1983, which relied on Loyalist and Republican terrorists to put 600 suspects behind bars, proved to be unworkable when their evidence was excluded as being unreliable.[67]

This was not the only area where the law enforcement framework found itself under stress as from 1974 onward emphasis was placed on dealing with terrorist threat as a criminal matter. It extended to the prison system in the form of a "struggle between the prisoners' assertion of their status as *political* prisoners and attempts by the prison authorities to deny, undermine and manage that assertion."[68] New use of force rules in the form of a "yellow card" were also introduced that required "soldiers not to shoot unless they, or their comrades, or civilians were at imminent risk of death."[69] These constraints reflected the principle that a "soldier quite rightly operates within the constraints of the common law."[70] However, it has also been suggested that the "yellow card" principles were instituted to ensure "all operations were subject to police primacy so as to maintain the fiction the Irish problem was not an armed conflict but an internal security issue."[71] The restrictions on the use of force resulted in the prosecution, but ultimately an acquittal, of a soldier for shooting at a vehicle causing the death of the driver and a passenger speeding through a vehicle checkpoint.[72] This case highlights that the State can require compliance with domestic legal obligations even if the violence qualifies as an armed conflict.

The use of a law enforcement framework to govern the use of force was particularly challenging because of the nature and extent of the violence. The security forces were required to change their approach over the course of the conflict. Tony Geraghty notes that up until 1976 "the SAS [Special Air Service] role had been to provide skilled Intelligence Officers to run the undercover surveillance teams," however, "the regiment's role increasingly was to ambush IRA terrorists known to be on their way to a target and unlikely to surrender without a fight."[73] The late 1980s saw an escalation of violence, in part brought on by the importation of arms from Libya.[74] That violence included a May 8, 1987, IRA attack involving an attempt to destroy the RUC Laughgall police station with a mechanical "digger carrying a

---

[63] McEvoy, *supra* note 60, at 223. *See also* Oren Gross & Fionnuala Ní Aoláin, Law in Time of Crisis: Emergency Powers in Theory and Practice 275–82 (2006).

[64] Geraghty, *supra* note 14, at 94–96, and Asprey, *supra* note 7, at 1129.

[65] Geraghty, *supra* note 14, at 96.

[66] *Id.*

[67] Cochrane, *supra* note 21, at 116.

[68] McEvoy, *supra* note 60, at 1.

[69] Geraghty, *supra* note 12, at 218.

[70] Dewar, supra note 18, at 112.

[71] Geraghty, *supra* note 12, at 218–19.

[72] Geraghty, *supra* note 14, at 103–10.

[73] *Id.* at 123.

[74] *Id.* at 124.

bomb in its excavator, covered by a screen of bricks."[75] The result was eight members of the IRA being killed in an ambush.[76] This was followed by the badly executed reprisal bombing by the IRA in November 1987 at Enniskillen. There was also a 1988 security force operation in Gibraltar that saw the SAS kill three IRA terrorists who were believed to be about to detonate a car bomb positioned to kill British military personnel.[77] On August 20, 1988, a remote controlled IED was used by the IRA to kill thirty-five members of a British regiment on the main road from Belfast.[78]

The action taken by the security forces in the late 1980s to counter the increased violence led to allegations of improper uses of force.[79] This included not only the SAS but also the RUC's special firearms, the HMSU.[80] The use of force by security personnel was brought before the European Court of Human Rights, raising the question of whether a "shoot to kill" policy was being employed.[81] Allegations such as these, when aligned against the British claim to be exclusively applying a law enforcement approach for countering the terrorist threat, highlights the narrow boundary that can exist between the policing and conduct of hostilities paradigms when countering insurgent violence. Overall, the insurgency and government response were both affected by the unique situation in Northern Ireland. While representative of a situation where significant security threats could be contained and defeated by means of law enforcement, it was also unique in that the violence occurred in an extremely well-functioning State containing robust governance structures.

13.2.2. SHOOTING TO KILL

A shoot to kill policy is an extremely controversial concept in a human rights context.[82] The types of shootings that attracted scrutiny in Northern Ireland varied from what were viewed as ambushes,[83] stopping fleeing terrorist suspects,[84] and actions being taken to prevent the

---

75 *Id.* at 126.

76 ASPREY, *supra* note 7, at 1135. *See also* PETER HARCLERODE, SECRET SOLDIERS: SPECIAL FORCES IN THE WAR AGAINST TERRORISM 545–48 (2000) (for an outline of the Laughgall operation).

77 ASPREY, *supra* note 7, at 1135. *See also* HARCLERODE, *supra* note 76, at 549–56.

78 ADAMS ET AL., *supra* note 13, at 8–9.

79 McCann v. United Kingdom, App. No. 18984/91, Eur. Ct. H.R., Judgment (1995), http://hudoc.echr.coe.int/sites/eng/pages/search.aspx?i=001-57943, and Kelly v. United Kingdom, App. No. 30054/96, Eur. Ct. H.R., Judgment (2001), http://hudoc.echr.coe.int/sites/eng-press/pages/search.aspx?i=001-59453.

80 GERAGHTY, *supra* note 14, at 135. *See also* DOHERTY, *supra* note 9, at 170–71.

81 McCann v. United Kingdom, App. No. 18984/91, Eur. Ct. H.R., Judgment, ¶ 118 (1995), http://hudoc.echr.coe.int/sites/eng/pages/search.aspx?i=001-57943, McKerr v. United Kingdom, Application No. 28883/95, Eur. Ct. H.R., Judgment ¶ 160 (2001), http://hudoc.echr.coe.int/sites/eng/pages/search.aspx?i=001-59451, and Kelly v. United Kingdom, App. No. 30054/96, Eur. Ct. H.R., Judgment ¶ 83 (2001), http://hudoc.echr.coe.int/sites/eng-press/pages/search.aspx?i=001-59453.

82 NILS MELZER, TARGETED KILLING IN INTERNATIONAL LAW 22–24 (2008) (where the concept of a shoot to kill policy is dealt with as a form of targeted killing).

83 Kelly v. United Kingdom, App. No. 30054/96, Eur. Ct. H.R., Judgment ¶¶ 12–25 (2001), http://hudoc.echr.coe.int/sites/eng-press/pages/search.aspx?i=001-59453. *See also* GERAGHTY, *supra* note 14, at 124–27 (for a summary of the Loughgall attack), and *id.* at 123–24 (outlining a series of engagements in 1977 where IRA terrorists were killed by the SAS at or near explosives and firearms caches. Two innocent civilians were killed.).

84 McKerr v. United Kingdom, Application No. 28883/95, Eur. Ct. H.R., Judgment, ¶ 11, and ¶ 19 (2001), http://hudoc.echr.coe.int/sites/eng/pages/search.aspx?i=001-59451 (for outline of the facts of the case including an excerpt from the trial judgment).

ignition of car bombs.[85] Allegations of the existence of a shoot to kill policy arose, in part, because of the frequency with which suspects were killed.[86] However, the general attitude of members of the security services also came to be questioned because of comments such as those reportedly made by a senior British officer who referred to the killing of three IRA suspects as "the Gibraltar cull."[87] "Shoot to kill" carries with it the aura of status-based targeting, which is a fundamental distinguishing feature between the application of conduct of hostilities norms and a policing approach. In other words, shooting someone simply because they are a member of a "terrorist" group.

The question of whether security forces are no longer properly applying the restrictions of a law enforcement paradigm, and therefore have adopted a shoot to kill policy, must be considered in a broader context by looking at a full range of security operations being undertaken by the State. It has been suggested that proof the security services were not applying a shoot to kill policy (i.e., status-based targeting) was evident in the fact that the top IRA men knew they were easily identifiable and "accessible to the security forces should they wish to kill them. Their freedom is testimony to the nonsense of a shoot-to-kill policy."[88] This statement suggests that the shoot to kill issue does not fully introduce the idea of status-based targeting, as it usually is traditionally understood in a humanitarian law–based targeting context.

That the uses of force in question did not involve widespread targeting of identifiable IRA members based on their status provides some indication that the shoot to kill controversy may have been less about a conduct of hostilities approach being adopted and more about whether the human rights–based normative regime was breached. However, allegations of a shoot to kill policy can be based on a more subtle expansion of the authority to use force. This can be seen in the 1992 FBI Hostage Rescue Team shooting of a criminal suspect at Ruby Ridge, Idaho. The rules of engagement (ROE) in that incident extended to "deadly force can and should be employed" where any adult male is observed with a weapon either before or after a surrender announcement is made.[89] This direction raised questions about whether an improper shoot to kill order had been given.[90] The carrying of arms openly is reflective of a humanitarian law–style criterion related to combatant status,[91] or evidence of direct participation in hostilities (i.e., combat function).[92] The close connection that these FBI ROE have to a hostilities-based approach can be seen when they are compared to the

---

85 McCann v. United Kingdom, App. No. 18984/91, Eur. Ct. H.R., Judgment, ¶¶ 23–31 (1995), http://hudoc.echr.coe.int/sites/eng/pages/search.aspx?i=001-57943 (for an outline of the perceived threat from a car bomb). *See also* Geraghty, *supra* note 12, at 220–21.

86 Geraghty, *supra* note 14, at 119–32 (for an outline of British Special Air Service operations in Northern Ireland), and Geraghty, *supra* note 12, at 221.

87 Geraghty, *supra* note 14, at 75.

88 Adams et al., *supra* note 13, at 89.

89 Christopher Whitcomb, Cold Zero: Inside the Hostage Rescue Team 216 (2001).

90 *Id.* at 219 ("I did wonder at the time, whether the new rules of engagement were a shoot-to-kill order.")

91 GC III, art. 4. A. (2)(c) & (6). *See also*, Idaho v. Horiuchi, 253 F.3d 359, 377, *vacated as moot*, 266 F.3d 979 (9th Cir. 2001), *quoted in* William C. Banks & Peter Raven-Hansen, *Targeted Killing and Assassination: The U.S. Legal Framework*, 37 U. Rich. L. Rev. 667, 677 (2003) (where in respect of the Ruby Ridge shooting it was found that "wartime [shoot-to-kill] rules are patently unconstitutional for a police action.").

92 Interpretive Guidance on the Notion of Direct Participation in Hostilities Under International Humanitarian Law, International Committee of the Red Cross 35 (Feb. 2009).

direction provided to U.S. Marines in Fallujah in 2004. The Fallujah ROE indicated individuals carrying arms openly were demonstrating hostile act/intent, posed an imminent threat, and could, subject to certain limitations, be attacked.[93] It is recalled that the observation has been made that attacking a class of individuals is an "anathema to accepted notions of self-defense."[94]

This raises the issue of what the human rights normative framework demands in terms of its application. Much depends on the context. In most cases focusing simply on the "shoot/don't shoot" decision is likely to result in an absolution of the security force member who pulls the trigger. This is because the confrontation with the terrorists often places the soldier or member of the police force in a self-defense situation. This situation was evident in the November 1982 incident where three members of the RUC HMSU fired on a car carrying three IRA members fleeing a checkpoint. As the trial judge, Lord Justice Gibson, indicated, he "would wish most emphatically to repudiate any idea that I would approve or that the law would countenance what has been described as a shoot-to-kill policy on the part of the police."[95] However, as often happens in such situations, the judge demonstrated considerable deference to the security personnel having to make difficult and time-sensitive decisions under considerable threat of death or injury. This resulted in a determination in that the accused policemen were not guilty.[96] Ironically, that judge was himself to subsequently die on the border in April 1987 as a result of an IRA bomb.[97] Similarly, in *McCann v. United Kingdom*, involving the 1987 SAS shooting of three IRA members in Gibraltar, the European Court of Human Rights accepted "that the soldiers honestly believed, in the light of the information that they had been given . . . that it was necessary to shoot the suspects in order to prevent them from detonating a bomb and causing serious loss of life . . . . The actions which they took, in obedience to superior orders, were thus perceived by them as absolutely necessary in order to safeguard innocent lives."[98]

However, in the *McCann v. United Kingdom* decision, the European Court also determined that there was insufficient control and organization of the arrest operation. Further there was inadequate planning and erroneous assumptions were made about the terrorists' intentions.[99] This portion of the judgment reflects a human rights accountability focus on an evaluation of State action leading up to the point of confrontation, including operational planning and the various options available to security force personnel. What is noteworthy is that the court did not assess whether a shoot to kill policy was in place, but rather if the circumstances of that individual incident led to an unjustified shooting. This decision highlights what is likely a key point regarding the existence of a general shoot to kill policy. That

---

[93] Dick Camp, Operation Phantom Fury: The Assault and Capture of Fallujah, Iraq 150 (2009).

[94] Gary P. Corn, *Should the Best Offense Ever Be a Good Defense*, Strategy Research Project 5 (Carlisle Barracks, PA: U.S. Army War College, Apr. 15, 2014).

[95] McKerr v. United Kingdom, Application No. 28883/95, Judgment, ¶ 20 (2001), http://hudoc.echr.coe.int/sites/eng/pages/search.aspx?i=001-59451 (excerpt of trial judge's ruling).

[96] *Id.* ¶ 20.

[97] Geraghty, *supra* note 14, at 102.

[98] McCann v. United Kingdom, App. No. 18984/91, Eur. Ct. H.R., Judgment, ¶ 200 (1995), http://hudoc.echr.coe.int/sites/eng/pages/search.aspx?i=001-57943.

[99] *Id.* ¶ 212.

is whether there is a systemic handling of security operations by the security establishment in a way that makes the use of deadly force almost inevitable and, importantly, unjustified.

One 1988 analysis of whether a shoot to kill policy existed in Northern Ireland looked at the various choices available to security forces in stopping a possible IRA active service unit assassination of a target. The choices included warning the target and removing him to a place of safety; springing a trap on the terrorists as they collected their guns; or drawing them into an ambush at a place of choosing by the security forces.[100] It was suggested the "only answer was ambush" by laying a trap and drawing the terrorists into a killing ground, as "[t]he chances of success in any counter-terrorist operation increase in direct proportion to the *level of control* that can be exercised over the action."[101] The reference to control in this 1988 analysis is noteworthy. It was also an important aspect of the capture or kill analysis in the Israeli *Targeted Killing Case* (see Chapter 7, section 7.4.2, *The "Targeted Killing" Case and Control*).

An approach that has members of the security forces lying in wait rather than acting earlier to avoid an attack is obviously a controversial one. It carries with it a significant potential for risk for uninvolved civilians as well as for the security forces and the terrorists themselves. For example, the widow of Gervaise McKerr questioned why the police did not simply go to their home and "lift him."[102] The May 1987 thwarting of an IRA attack on the Laughgall police station has been criticized as permitting unstable bombs to being driven, while under surveillance, into "an SAS killing ground."[103] It "foreshadowed the strategy adopted less than a year later at Gibraltar. . . ."[104] The question of whether it would have been preferable during the Gibraltar operation to have arrested the suspects at the border rather than allow them to enter the territory endangering the population, was a key issue addressed by the European Court of Human Rights in *McCann v. United Kingdom*. It was viewed by the majority of the judges, in a badly divided court (10 to 9 judge split) as being preferable to avoid the risk to the population, even if it meant there was insufficient evidence to warrant their detention and trial.[105] The impact of the security official's decision to allow the terrorists to enter the territory was that "the scene was set in which the fatal shooting, given the intelligence assessments which had been made, was a foreseeable possibility if not a likelihood."[106] In contrast, the dissenting judges were of the view that "[t]o release them, after having alerted them to the state of readiness of the authorities, would be to increase the risk that they or other IRA members could successfully mount a renewed terrorist attack on Gibraltar."[107]

The significant judicial disagreement in the *McCann Case* about whether to preemptively stop terrorists thereby potentially leaving them free to strike again (perhaps because there is

---

[100] Adams et al., *supra* note 13, at 5–6 (the third option was seen as the best, but problematic due to the challenges of placing twenty-four-hour surveillance on a possible target).

[101] *Id.* at 6 (emphasis added).

[102] Urban, *supra* note 19, at 152.

[103] Geraghty, *supra* note 14, at 127.

[104] *Id.*

[105] McCann v. United Kingdom, App. No. 18984/91, Eur. Ct. H.R., Judgment, ¶¶ 203–205 (1995), http://hudoc.echr.coe.int/sites/eng/pages/search.aspx?i=001-57943.

[106] *Id.* ¶ 205.

[107] *Id.* ¶ 11 (dissenting judgment).

insufficient evidence to hold them) is reflective of the conclusion reached in the 1988 "shoot to kill" analysis referred to above. In that analysis it was suggested warning a person targeted for IRA assassination would mean he would remain on the hit list, and setting a trap at a weapons cache could be exploited by IRA, who would claim innocent persons were involved. It was viewed as better to catch the terrorist not only with the weapons but also in the act of using them.[108] This conclusion is open to challenge, as it seems that much depends upon the manner in which an operation is planned and conducted as to whether an arrest can be carried out. The operation should preferably be executed in a manner that ensures there is adequate proof of the suspect's involvement in a crime (e.g., having possession of weapons from a cache) and with detention foremost in mind.[109] However, it is preferable if the arrest takes place before an attack unfolds in order to limit the risk to the public. This issue has a resonance with the rationale for targeting IED cells as a group, rather than attempting to stop attacks immediately before a device is detonated (see Chapter 7, section 7.5, *The IED and Suicide Bombing Threat: Arrest or Kill?*). It should be remembered though that the nonviolent capture of terrorist suspects cannot always be guaranteed.

If the manner in which shootings occur are looked at from a systemic perspective, they must be considered in the overall context of security operations and the significant threat being faced. As was noted, "the 10,000 or so soldiers walking the streets of Northern Ireland have generally behaved with restraint when they have met people to be known to them as IRA members."[110] Further, the number of killings by covert security personnel varied throughout the conflict, depending upon the threat and how those forces were deployed.[111] For five years following December 1978, "the SAS did not kill anybody,"[112] and during service in Northern Ireland did capture IRA suspects.[113] The violence and the threats posed to the security personnel were exceptional in law enforcement terms. High-profile incidents where soldiers were killed attempting to arrest IRA members or seize arms undoubtedly contributed to the overall sense that the use of deadly force would likely be required.[114] Further, there appears to have been no discovery of an official shoot to kill policy,[115] although the attitude developed of that being the expected action of specialized units.[116] It has further been suggested that the adoption of such a policy would have legitimized the IRA as they "would have looked like bonafide soldiers in the eyes of the world."[117]

It is evident the ambush style of operation constituted an exceptional means of conducting law enforcement. The journalist Mark Uban has rightly noted "[t]he apparent conflict between the law of minimum force and the behaviour of some soldiers and police officers is

---

108 ADAMS ET AL., *supra* note 13, at 5–6.

109 URBAN, *supra* note 19, at 176 (where it is suggested a soldier could have waited until the weapons were placed in a car before a challenge was issued leading to the terrorist being killed).

110 *Id.* at 246.

111 *Id.* at 239–41.

112 *Id.* at 240

113 GERAGHTY, *supra* note 14, at 129.

114 ADAMS ET AL., *supra* note 13, at 95–98 (for a discussion of the case of Lance-Corporal Jones who was killed in 1978 challenging two IRA terrorists), and URBAN, *supra* note 19, at 71–72.

115 MELZER, *supra* note 82, at 24.

116 URBAN, *supra* note 19, at 70 and 164.

117 MICHAEL ASHER, SHOOT TO KILL: A SOLDIER'S JOURNEY THROUGH VIOLENCE 256 (1990).

at the root of a 'shoot to kill' policy in Northern Ireland."[118] He identifies three dimensions to the question of whether a shoot to kill policy has been put in place: are there other ways to stop the terrorists other than with deadly force; were the terrorists armed and warned about the potential use of force; and were the soldiers told to fire in a manner to permanently remove the threat?[119] After acknowledging that the shooting tactics of both police and Army personnel provides for the firing at vital organs, it is suggested that inquiries have normally centered on the first two factors. Importantly, with respect to a number of SAS shootings of the late 1980s, it has been stated, "there has been no official suggestion that the SAS ever challenged the terrorists—inviting them to give themselves up—before opening fire."[120]

It has also been indicated that there will be unease about shooting of terrorists unless "four quite specific conditions are met: those shot dead should be known terrorists, they should be on a murder mission, they should be armed and finally they should be seen to fire first, or be seen to respond aggressively to an order to surrender."[121] However, in these situations the lack of a challenge before firing does not necessarily establish law enforcement norms have been breached, since international law recognizes warnings may not have to be given in all law enforcement circumstances.[122] Clearly, questions such as these should be asked, and accountability demanded in respect of the use of force by security forces. However, in such cases this does not mean the limits of law enforcement will be clearly established as having been systemically breached. Tony Geraghty perceptively identified the challenge facing security personnel and the courts overseeing the actions of the SAS when he stated, "[t]hese were cases where minimum force, reasonable force, legal and illegal lethal force were so close as to be indistinguishable."[123] With the Special Forces arresting or killing IRA terrorists, "policing crossed over into military action, confirming the hybrid nature of the conflict that now existed."[124]

The controversy surrounding the security force reaction to the threats posed during the "Troubles" serves as a clear indication there are limits on the effectiveness of a law enforcement approach, although the debates in that instance centered on whether the aggressive tactics applied by the security forces were necessary. Questions regarding the limits of the law enforcement paradigm are particularly evident when it is believed that the setting up of what are effectively ambush operations, and shooting without warning, which might be equated to participation in armed conflict, is required to maintain order. That the exceptional reaction by the security establishment to the threats posed by the IRA and other groups were viewed as necessary in the "ideal" situation of Northern Ireland

---

[118] URBAN, *supra* note 19, at 70.

[119] *Id.*

[120] *Id.* at 246.

[121] ADAMS ET AL., *supra* note 13, at 171.

[122] *UN Basic Principles on Use of Force*, at ¶ 10. *See also* McCann v. United Kingdom, App. No. 18984/91, Eur. Ct. H.R., Judgment, ¶ 136 (1995), http://hudoc.echr.coe.int/sites/eng/pages/search.aspx?i=001-57943 (where the orders to open fire included direction that warnings were not required "where the giving of a warning or any delay in firing could lead to death or serious injury to a person whom it is your duty to protect, or to yourself, or to another member in your operation."

[123] GERAGHTY, *supra* note 14, at 123.

[124] *Id.* at 75.

(i.e., the institutions of law enforcement were well established) should cause considerable pause for advocates who seek to automatically apply a law enforcement approach to other insurgencies.

The conduct of physical surveillance, tracking of suspects, and the carrying out of arrests is often directly tied to the ability to control the territory where such operations are to be conducted. If such control does not exist or the risks involved in conducting operations preclude an ordinary policing response, then a reassessment of the approach applied by security forces is warranted. Where it has not been acknowledged by the State there is an armed conflict in existence, the requirement to systemically apply such exceptional measures can serve as an indicator that such a conflict exists both in fact and at law. If it has been acknowledged such a conflict exists, and the law enforcement option has been applied as a policy choice, then it also serves as an indication the operational situation may have reached the point from a practical perspective that a more robust response is required. In effect, the shoot to kill issue can act as a form of threshold as to whether the point has been reached that an armed conflict is not only in existence but that the application of a conduct of hostilities approach is required.

13.3. The Militarization of the Police

There are also questions regarding the impact on police forces should they be asked to perform what is fundamentally a military role. This issue arose with respect to the RUC HMSU, which attracted considerable criticism as a result of the November 1982 shooting of Gervaise McKerr and his accomplices, as well as two other terrorists at Lurgan. Further there was the killing of two INLA members in Armagh in December of that year.[125] In defense of the HMSU it has been noted that other European countries, such as Germany and France, employ specialized police units to deal with terrorists (i.e., the GSG-9 and the GIGN, respectively). However, the actions of the HMSU ultimately attracted too much controversy. Within a few months of this last incident "the will within the RUC to deploy its special units on missions to confront terrorists had collapsed," and "[t]he RUC's special units were given different tasks, often acting as back-up to the SAS . . . ."[126]

It has been suggested the political controversy resulting from claims the HMSU had adopted a shoot to kill policy "demonstrated that a civilian police service will always be subject to greater scrutiny than a similar military unit and convinced the RUC that actions of this kind were to be avoided whenever possible . . . ."[127] At a minimum, such units must be employed in carefully controlled circumstances and be subjected to a significant level of oversight. The issue is not whether their members will be called to account for the use of force, but rather what type of inquiry will be held. As a result, such units must develop a culture of accountability that permeates all aspects of operational planning and command.

---

[125] URBAN, *supra* note 19, at 151–52.

[126] *Id.* at 160.

[127] Cochrane, *supra* note 21, at 121.

They must be prepared to justify the force used and in most circumstances be prepared to account for every round fired.

While the controversy regarding the HMSU does not take away from the essential role the RUC played in maintaining order in Northern Ireland, it does highlight that human rights–based policing may not always be the appropriate response for some of the most serious threats faced during an insurgency. Otherwise, in order to confront the terrorist/insurgent threat, the understanding of what constitutes policing may ultimately have to change. Police forces maintaining order may have to become increasingly like their military counterparts. Regarding the human rights normative framework, the pressure to change in order to address more violent threats posed by organized armed groups is reflected in the possibility of having to accept that significant collateral damage may result from the force used and that force may have to be applied more proactively.[128] This is a trend that is the opposite of one faced by soldiers during insurgencies and counterterrorism. While military forces versed in humanitarian law are increasingly employed in situations more akin to law enforcement, police units used to a restrictive human rights normative and accountability framework are being asked to confront higher levels of violence.

Efforts to apply policing norms in situations involving the elevated violence associated with armed conflict should cause a moment of pause and considerable analysis for human rights advocates. It could mean the increased militarization for police forces, as happened with the RUC HMSU, if they are being asked to engage in combat-like situations. Questions have been raised generally in the United States regarding the broader impact of the perceived militarization of the police, the proliferation of SWAT teams, the levels of force used, and the follow-on effect on civil liberties.[129] Issues regarding the role of police and the degree to which they should be integrated into the community [130] or, conversely, fight like soldiers have also arisen in a counterinsurgency context with integration being seen as far more effective.[131] This is an issue where ordinary community-based policing and traditional human rights–based limits on the use of force are intimately intertwined. Some anti-State violence is simply not readily amenable to a law enforcement response, and "this leads inevitably to an increase in lethality with which insurgency is met and consequently greater civilian casualties."[132] Specialized paramilitary police forces can perform an important counterinsurgency role.[133] However, it has also been suggested that police forces should not be used to carry out what is fundamentally a military role, as "they often become a force within a force, operating according to their own priorities and stretching the limits of their legal authorization . . . ."[134]

---

128 Finogenov v. Russia, VI Eur. Ct. H. R. ¶ 236, at 409 (2011) (extracts).

129 Radley Balko, Rise of the Warrior Cop: The Militarization of America's Police Forces 333–36 (2013). *See also* War Comes Home: The Excessive Militarization of American Policing, American Civil Liberties Union (2014), https://www.aclu.org/sites/default/files/field_document/jus14-warcomeshome-report-web-rel1_1.pdf.

130 Randall Wilson, Blue Fish in a Dark Sea: Police Intelligence in a Counterinsurgency 6 (2013).

131 David P. Fidler, *Police in Counterinsurgency: The Challenge of Comprehensive Reformation*, *in* Policing Insurgencies, *supra* note 21, 319 at 338–39.

132 David H. Bayley & Robert M. Perito, The Police in War: Fighting Insurgency, Terrorism, and Violent Crime 76 (2010).

133 Fidler, *supra* note 131, at 336.

134 Bayley & Perito, *supra* note 132, at 76.

The observation has been made concerning the difficulties caused by "early stage and/or continued militarization of COIN [counterinsurgency] responses (for example, Philippines, Kenya, Northern Ireland, Punjab, Columbia, Pakistan, Iraq, and Afghanistan)—an approach that proved counterproductive and made the legitimacy mountain harder for counterinsurgents to scale."[135] A lesson learned from the experience in post-2003 invasion Iraq is that "the militarization of police—whether intentional or as a result of the absence of other security counterparts—will detract from their ability to play a roles in an integrated criminal justice sector, and contribute to their politicization."[136] It cannot be forgotten that in a counterinsurgency "the idea of the police becomes part of the normative battle for the 'hearts and minds' of the population waged between insurgents and counterinsurgents."[137] There is much to be gained by separating law enforcement from the conduct of hostilities by ensuring human rights–based limitations are applied in circumstances for which they were developed and where they can have the greatest effect. Stretched too far, the watering down of the human rights–based law enforcement norms and the use of police for "military" missions may actually undercut the counterinsurgency and counterterrorism effort and increase the overall level of insecurity for the civilian population.

13.4. Human Rights Overreach

A second area of limitation concerns situations where human rights law is applied to control what is fundamentally combat activity. This has arisen in the context of the approach adopted by the European Court of Human Rights, which historically has looked at use of force issues exclusively through a human rights lens regardless of the level of violence or the nature of the conflict.[138] It is important to note that what had traditionally been an almost total avoidance of dealing with international humanitarian law by the European Court changed substantively in the 2014 case *Hassan v. The United Kingdom*. The Court ruled that interpreting the European Convention on Human Rights pursuant to Article 31(3)(c) of the Vienna Convention in harmony with other rules of international law included applying humanitarian law to international armed conflicts.[139] However, by leaving the onus on a State to claim the applicability of humanitarian law, it is conceivable that human rights law may continue to be applied to such conflicts should no claim be made.

---

135 Fidler, *supra* note 131, at 326.

136 Matt Sherman & Josh Paul, *The Role of Police in Counterinsurgency Operations in Iraq*, 2003–6, *in* POLICING INSURGENCIES, *supra* note 21, 246 at 254.

137 Fidler, *supra* note 131, at 324.

138 *See, e.g.*, Isayeva v. Russia, Application Nos. 57947/00, 57948/00, 57949/00, Eur. Ct. H. R. (2005), http://hudoc.echr.coe.int/sites/eng/pages/search.aspx?i=001-68379 [hereinafter Isayeva v. Russia I], and Isayeva Application No. 57959/00, Eur. Ct. H. R. (2005), http://hudoc.echr.coe.int/sites/eng/pages/search.aspx?i=001-68381 [hereinafter Isayeva v. Russia II].

139 Hassan v. The United Kingdom, Application No. 29750/09, Eur. Ct. H. R. ¶ 102 (2014), http://hudoc.echr.coe.int/sites/eng/pages/search.aspx?i=001-146501. *See also* Varnava v. Turkey (Application Nos. 16064/90, 16065/90, 16066/90, 16068/90, 16069/90, 16070/90, 16071/90, 16072/90 and 16073/90), Eur. Ct. H. R. ¶ 185 (2009), http://hudoc.echr.coe.int/sites/eng/pages/search.aspx?i=001-94162 (which had previously ruled Article 2 of the European Convention on Human Rights (i.e., the "Right to Life") must be interpreted as far a possible in light of humanitarian law provisions applicable during international armed conflict).

Further, there remains the question of the willingness of the European Court of Human Rights to recognize the applicability of humanitarian law in noninternational armed conflict. The *Hassan v. United Kingdom* decision dealt exclusively with international armed conflict. [140] Noninternational armed conflicts are simply not addressed in that decision.[141] The issue of the application of human rights law rather than humanitarian law can still arise in the context of an armed conflict occurring within a member State, or as a result of the deployment of military forces to another State to assist them in dealing with an insurgency. The importance of this latter issue is enhanced by the predominance of such conflicts in the twenty-first century (see Chapter 2, section 2.2, *A Focus on the State*).

It is not evident that the European Court of Human Rights will readily embrace the use of humanitarian law to interpret human rights law in noninternational armed conflict, even though the Inter-American Court of Human Rights has done so.[142] The jurisprudence of the American court was not relied on in *Hassan v. The United Kingdom*. To a certain degree this highlights the regional focus of the European court. For conflict taking place within member States, the clear territorial claim for the application of the European Convention on Human Rights, derogation being available in respect of certain human rights during internal conflicts (i.e., in times of emergency),[143] and the more limited humanitarian treaty law applicable to such armed conflicts may be influential in causing the European court to rely uniquely on the provisions of its own convention. In this respect, the paucity of humanitarian law treaties applicable to noninternational armed conflicts may be particularly problematic. In order for many rules of humanitarian law to be applied to internal conflicts, the European Court of Human Rights would have to accept that customary humanitarian law can be used to interpret the European Convention on Human Rights. This might prove a challenge for a very traditional European court approach steeped in the predominantly civil code–based European legal system in which "civil law codes and statutes are the primary source of law. . . ."[144]

In addition, the European Court has a history of applying human rights law to assess even aerial bombing in the context of what was acknowledged to be an "illegal armed insurgency."[145] This was done despite the fact that the use of air attack is normally viewed as a quintessential armed conflict activity. In *Isayeva v. Russia I*, the court held there was a violation of Article 2 of the European Convention on Human Rights regarding an attack

---

[140] Hassan v. The United Kingdom, Application No. 29750/09, Eur. Ct. H. R. ¶ 101 (2014), http://hudoc.echr.coe.int/sites/eng/pages/search.aspx?i=001-146501.

[141] Hassan v. The United Kingdom, Application No. 29750/09, Eur. Ct. H. R. ¶ 104 (2014), http://hudoc.echr.coe.int/sites/eng/pages/search.aspx?i=001-146501 (the court only making reference to peacetime and, in contrast, international armed conflict).

[142] *See, e.g.*, Bámaca-Velásquez v. Guatemala, Inter-American Ct. H. R. ¶¶ 208–09 (2000), http://www.corteidh.or.cr/docs/casos/articulos/seriec_70_ing.pdf and Case of Cruz Sánchez et al. v. Peru, Series C no. 292, Preliminary Objections, Merits, Reparations and Costs, Inter-Am. Ct. Hum. (2015), http://www.corteidh.or.cr/docs/casos/articulos/seriec_292_esp.pdf. *See also* SANDESH SIVAKUMARAN, THE LAW OF NON-INTERNATIONAL ARMED CONFLICT 501–03 (2012).

[143] European Convention on Human Rights, art. 15.

[144] William Tetley, *Mixed Jurisdictions: Common Law vs Civil Law (Codified and Uncodified)* (pt. 1) (1999), http://www.cisg.law.pace.edu/cisg/biblio/tetley.html.

[145] Isayeva v. Russia I, Application Nos. 57947/00, 57948/00, 57949/00, Eur. Ct. H. R. ¶ 178 (2005), http://hudoc.echr.coe.int/sites/eng/pages/search.aspx?i=001-68379.

by Russian aircraft in Chechnya on what turned out to be a civilian convoy. Under Article 2 a deprivation of life is not regarded as having contravened the article when no more force than is absolutely necessary is used for the purposes of defending any person from unlawful violence; effecting a lawful arrest or preventing the escape of a person lawfully detained; or quelling a riot or insurrection.[146] That attack resulted in numerous civilian casualties and considerable damage.[147]

The case involved the type of conflict where it is likely the European Court on Human Rights will continue to favor the application of human rights law. It occurred on the territory of a Convention State, and the Russian government, consistent with the approach of many States, did not acknowledge the existence of an armed conflict in its own territory.[148] The court ruled that there was a breach of the victims' rights, in part; because of the use of military aviation and air-to-ground S-24 missiles was not undertaken with adequate care for civilian lives. The effects of the attack were disproportionate to the aim trying to be achieved.[149] This legal approach differs in a number of important respects from a humanitarian law–based one. The types of weapons used are relevant under humanitarian law to an assessment of means and methods of an attack considered under targeting precautionary measures.[150] However, the humanitarian law proportionality test involves a weighing of the military advantage to be gained from an attack against its likely excessive collateral effects.[151] That body of law does not deal with the proportionality of a response in relation to the threat unless objectively the attacks may cause unnecessary suffering or superfluous injury.[152] During armed conflict, the security forces may use an overwhelming, indeed disproportionate response against a lawful target. The European Court's consideration of proportionality in this instance appears to be a human rights–based application of the principle.

At times the European Court of Human Rights has incorporated conduct of hostilities wording into its human rights analysis.[153] The humanitarian law version of the proportionality test is reflected in *Isayeva v. Russia II*, which also involved aerial attacks as well as other armaments including tanks and multiple-rocket launchers. Reference was made to wording that seems to be taken from Additional Protocol I, Article 57(2)(a)(ii) regarding avoiding or limiting collateral casualties.[154] However, ultimately the legal test applied by the court was whether the force used was more than was absolutely necessary to achieve one of the purposes

[146] European Convention on Human Rights, art. 2(a), (b) and (c).

[147] Isayeva v. Russia I, Application Nos. 57947/00, 57948/00, 57949/00, Eur. Ct. H. R. ¶ 29 (2005), http://hudoc.echr.coe.int/sites/eng/pages/search.aspx?i=001-68379 (the government conceded sixteen civilians were killed and eleven wounded).

[148] William Abresch, *A Human Rights Law of Internal Armed Conflict: The European Court of Human Rights in Chechnya*, 16 EUR. J. INT'L L. 741, 764 (2005).

[149] Isayeva v. Russia I, Application Nos. 57947/00, 57948/00, 57949/00, Eur. Ct. H. R. ¶¶ 196–99 (2005), http://hudoc.echr.coe.int/sites/eng/pages/search.aspx?i=001-68379.

[150] AP I, art. 57(2)(a)(ii).

[151] AP I, art. 57(2)(a)(iii).

[152] W. Hays Parks, *Conventional Weapons and Weapons Reviews*, 8 INT'L Y.B. HUM. L. 55, 104 (2005) (where it is noted "the threshold for superfluous injury to combatants is high.").

[153] SIVAKUMARAN, *supra* note 142, at 504.

[154] Isayeva v. Russian II, Application No. 57950/00, Eur. Ct. H. R. ¶ 176 (2005), http://hudoc.echr.coe.int/sites/eng/pages/search.aspx?i=001-68381.

set out in Article 2 and strictly proportionate to the achievement of the permitted aims.[155] In effect, the court sought to avoid any recourse to force at all.[156] While a humanitarian law assessment of this case might also have reached the conclusion that this particular attack was not justifiable, the adherence to the human rights standards suggests a much more restrictive approach toward controlling State action during what is clearly an armed conflict. This conclusion is supported by the decision of the European Court of Human Rights in *Finogenov v. Russia*, where a distinction was made between the use of a gas, which was not intended to kill in a counterterrorism operation, and *Isayeva v. Russia II*, with its "use of airborne bombs to destroy a rebel group which was hiding in a village full of civilians."[157]

Perhaps the case that best highlights the restrictive human rights law approach adopted by the European Court is *Kerimova v. Russia*.[158] That 2011 case did acknowledge that the security situation may require "the deployment of armed forces equipped with combat weapons, including military aircraft," and may involve "regrettable but unavoidable consequence, human casualties."[159] However, the court clearly applied human rights law standards of using no more force than absolutely necessary, a strict and compelling test of necessity, and the requirement that the force used is strictly proportionate to the permitted aims.[160] Further, in determining there was a breach of the right to life of civilians who were collaterally killed and injured, it was stated regarding the use of aerial bombing:

> The Court has already held that using this kind of weapon in a populated area is impossible to reconcile with the degree of caution *expected from a law-enforcement body in a democratic society*. As was rightly pointed out by the applicants, no martial law or state of emergency had ever been declared in the Chechen Republic, and no derogation had been made under Article 15 of the Convention. The attacks in question therefore have to be examined *against a normal legal background*. Even when faced with a situation where the population of the town was held hostage by a large group of well-equipped and well-trained fighters, the authorities' primary aim should have been to protect lives from unlawful violence.[161]

Of note, the analysis was not completely divorced from the nature of the violence. The court did find "that the use of missiles, as opposed to the use of high-explosive aerial bombs, *might have been* more appropriate in the circumstances."[162]

---

[155] Isayeva v. Russia II, Application No. 57959/00, Eur. Ct. H. R. ¶ 173 (2005), http://hudoc.echr.coe.int/sites/eng/pages/search.aspx?i=001-68381.

[156] *Id.* ¶ 175 ("The authorities must take appropriate care to ensure that *any risk to life* is minimised.") (emphasis added).

[157] Finogenov v. Russia, VI Eur. Ct. H. R. ¶ 232, at 408 (2011) (extracts). *See also, id.* ¶ 216 ("the choice of means (gas) by the authorities was less dangerous than in *Isayeva* (bombs)"), and Kerimova v. Russia, Application Nos. 17170/04, 20792/04, 22448/04, 23360/04, 5681/05 and 5684/05 Eur. Ct. H. R. ¶ 253-54 (May 3, 2011), http://hudoc.echr.coe.int/sites/eng/pages/search.aspx?i=001-104662 (where the view was expressed the use of missiles rather than aerial bombs might have been more appropriate).

[158] Kerimova v. Russia, Application Nos. 17170/04, 20792/04, 22448/04, 23360/04, 5681/05 and 5684/05 Eur. Ct. H. R. (May 3, 2011), http://hudoc.echr.coe.int/sites/eng/pages/search.aspx?i=001-104662.

[159] *Id.* ¶ 246.

[160] *Id.* ¶ 238.

[161] *Id.* ¶ 253 (emphasis added).

[162] *Id.* ¶ 254 (emphasis added).

A critical problem that arises is that a human rights framework is not designed to assess all the factors that need to be considered when using force during armed conflict or dealing with the levels of violence that can be involved. Nor is it intended to operate as an overarching restraint on the use of force authorized under humanitarian law. The legal framework applied to control State and non-State actor participation in armed conflict must be prepared not only to consider the use of airpower or artillery in civilian areas but also a large number of attacks and potentially elevated levels of collateral civilian casualties and damage that is common in warfare. Human rights analyses tend to individualize and isolate attacks, whereas humanitarian law was developed in the context of group-based warfare involving industrialized levels of violence associated with inter-State conflict. Historically, these levels of violence are completely foreign to human rights law. As has been repeatedly argued in this book, this does not mean there is no role for the application of human rights norms during armed conflict. Just the opposite is true. As a matter of law and policy, law enforcement–type uses of force are part of contemporary security operations. The problem is when human rights law is applied to what is an armed conflict scenario involving elevated levels of violence with a clear potential for collateral civilian casualties and damage. As will be discussed in the next section, a court applying a human rights approach may be very challenged in making the types of decisions involving collateral impacts that are an ordinary and inherent part of the conduct of warfare (see section 13.5, *The Shoot-Down of Hijacked Aircraft*, below). In order to apply human rights law to govern the use of force in armed conflict, the court will become increasingly driven to adopt humanitarian law standards.

In contrast to human rights law, humanitarian law overtly acknowledges the possibility of greater collateral civilian casualties and damage precisely because it is designed to control the use of force in an environment where groups participate in conflict. Noninternational armed conflict can include combat between conventionally organized military forces. Even in conflict against non-State actors the use of airpower alone can run into the tens of thousands of strikes[163] and begin to take on the attributes of a strategic air campaign (e.g., targeting Islamic State oil production facilities).[164] It has been noted that the use of force in certain operations among the civilian population can change the conduct of hostilities proportionality assessment to a degree that it begins to reflect a human rights base approach (see Chapter 7, section 7.6.2.2, *Proportionality and Counterinsurgency*). However, that change is context specific, as is reflected in the high levels of violence that occurred in 2004 in Fallujah (see Chapter 11, section 11.3.3, *United States Rules of Engagement and "Self-Defense"*). A counterinsurgency is still a "war."

It is when the human rights accountability framework is applied to factual situations involving the levels of violence associated with the use of airpower, artillery, and other

---

[163] Benjamin S. Lambeth, Air Operations in Israel's War Against Hezbollah: Learning from Lebanon and Getting It Right in Gaza, Rand Project Air Force 251 (2011), http://www.rand.org/content/dam/rand/pubs/monographs/2011/RAND_MG835.pdf (outlining the large number of aerial strikes in the 2006 Israeli-Hezbollah conflict, and the 2008 conflict with Hamas), and Ben Hartman, *50 days of Israel's Gaza Operation, Protective Edge—By the Numbers*, The Jerusalem Post (Aug. 28, 2014, 0425 AM), http://www.jpost.com/Operation-Protective-Edge/50-days-of-Israels-Gaza-operation-Protective-Edge-by-the-numbers-372574 (outlining the aerial sorties during the 2014 conflict with Hamas).

[164] Kenneth Watkin, *Targeting "Islamic State" Oil Production Facilities*, 90 Int'l L. Stud. 499 (2014).

means of warfare that problems will inevitably arise. A key question is at what point does human rights law reach the limits of its effectiveness? This approach can have a direct negative impact on operational effectiveness and the ability of State armed forces to protect civilians. This latter point arises because, with the understandable focus of both humanitarian law and human rights law on collateral casualties and the death of innocent civilians, what is sometimes lost in the discussion is the mission that security forces have to protect civilians. Those civilians are the ones being defended from indiscriminate suicide bombing attacks, rescued as hostages, or otherwise protected from harm posed by terrorist groups. It is in this context that the military advantage being sought in an attack becomes a key part of assessment of its collateral effects. As a 2013 report analyzing the potential harm to civilians arising from an possible intervention by States on the side of rebels fighting the Syrian government noted, "[t]here is a complicated calculus to be done in comparing the threat the fighting poses to civilians and the potential harm that could be created through a military intervention, as well as not intervening at all."[165] It is this very weighing of the relative "good" and "evil" (i.e., the "double effect" doctrine) that resulted in the development of the conduct of hostilities–based proportionality standard. A human rights approach does not appear to comfortably address such calculations.

In practical terms, the approach adopted by the European Court of Human Rights can also have direct operational effect in terms of the ability of the States subject to its jurisdiction being able to work with other nations in a Coalition setting. The United States, Canada, Australia, and the majority of the world's nations are not bound exclusively by this human rights–based approach, or subject to the oversight of a human rights tribunal system. More importantly, as the likely Coalition lead on many international operations, the substantive rejection by the United States of the extraterritoriality of international human rights treaty law would appear to put it directly at odds with European States. The United States will primarily apply humanitarian law to regulate hostilities during international and noninternational armed conflicts, even though it has increasingly recognized the application of human rights norms (i.e., customary human rights law and within humanitarian law) (see Chapter 5, section 5.3.2.3, *Divergence and Growing Reconciliation*). In terms of operational planning, relying exclusively on a human rights paradigm based on the use of minimum force as last resort, a resistance to accepting collateral casualties and damage, and only using force where absolutely necessary does not necessarily sit comfortably with targeting precautions found under humanitarian law.

That European States have continued to be essential and effective Coalition partners in the past decade appears to be the result of a number of factors. First, the nature of many contemporary counterinsurgency and counterterrorism operations is such that there is a narrowing of the normative gap between human rights law and humanitarian law making the different approaches required to meet legal obligations less problematic. Second, because of the jurisdictional limits applied by the European Court there will continue to be operations that fall outside the scope of the European Convention on Human Rights. For example, in

---

[165] Syria: Minimizing Civilian Harm During Military Intervention Roundtable Outcome Document, Centre for Civilians in Conflict 2 (2013), http://civiliansinconflict.org/uploads/files/publications/Syria_Roundtable_brief_Feb_2013.pdf.

2012 the NATO legal advisor indicated that the countries engaged in a bombing campaign in Libya applied humanitarian law even though a significant number of participants were from European countries.[166] It may be that reliance was placed on the *Banković v. Belgium* decision, which limited the applicability of the jurisdiction of the European Court of Human Rights in respect of aerial bombing, conducted in a non-member country (i.e., to the legal space of the Contracting States).[167] This situation could also conceivably arise with regard to drone strikes, depending upon where the operations were conducted (e.g., outside member countries or on the high seas), and the jurisdiction of the court has not otherwise become engaged because of the "effective control" or "state agency control" tests.[168] Third, in other situations where the European treaty applies, the affected States may apply broad interpretations of the human rights law standards. These interpretations may approximate the humanitarian law rules sufficiently in order to allow continuing interoperability.

Finally, Coalition military commanders have proven adept at employing military forces in roles that accommodate unique national limitations. However, there are ultimately limits to how much accommodation can be made because of the significantly different theoretical approaches of human rights law and humanitarian law toward the use of force. For example, the human rights model reflected in the Chechnya cases[169] is not well structured for application to higher intensity combat. Even with the European Court having recognized the application of humanitarian law norms in *Hassan v. The United Kingdom* this does mean that the more restrictive principles of human rights will not heavily influence how force will be used. This can be seen in the reference to the Third and Fourth Geneva Conventions only being accommodated as far as possible.[170] This wording may not mean a completely unfettered application of the law governing armed conflict.

As has been discussed, such fettering is even more likely to occur in a noninternational conflict, assuming humanitarian law principles are even applied. Such influence can been seen in the reference to the principle of using minimum force, and only in situations of absolutely necessary in the Japanese embassy case.[171] It is not clear that the European Court of Human Rights will embrace the application of humanitarian law as readily as the Inter-American court appears to have done. Unfortunately, such references to human

---

[166] *Letter Peter Olson to Judge Kirsch*, OLA (2012)006, 4–5 (23 Jan. 2012) http://www.nato.int/nato_static/assets/pdf/pdf_2012_05/20120514_120514-NATO_1st_ICIL_response.pdf.

[167] Banković v. Belgium, XII Eur. Ct. H. R. Rep. 333, ¶ 80, at 358–59 (2001).

[168] Hassan v. The United Kingdom, Application No. 29750/09, Eur. Ct. H. R. ¶¶ 74–77 (2014), http://hudoc.echr.coe.int/sites/eng/pages/search.aspx?i=001-146501. *See also* Jaloud v. The Netherlands, Eur. Ct. H. R., Application No. 47708/08, ¶ 152 (2014), http://hudoc.echr.coe.int/sites/eng/pages/search.aspx?i=001-148367 (an incremental extension of human rights law jurisdiction is indicated where the court determined the degree of control necessary to engage the jurisdiction of the court was set at persons "passing through" a checkpoint).

[169] Isayeva v. Russia I, Application Nos. 57947/00, 57948/00, 57949/00, Eur. Ct. H. R. (2005), http://hudoc.echr.coe.int/sites/eng/pages/search.aspx?i=001-68379, Isayeva v. Russian II, Application No. 57950/00, Eur. Ct. H. R. (2005), http://hudoc.echr.coe.int/sites/eng/pages/search.aspx?i=001-68381, and Kerimova v. Russia, Application Nos. 17170/04, 20792/04, 22448/04, 23360/04, 5681/05 and 5684/05 Eur. Ct. H. R. (May 3, 2011), http://hudoc.echr.coe.int/sites/eng/pages/search.aspx?i=001-104662.

[170] Hassan v. The United Kingdom, Application No. 29750/09, Eur. Ct. H. R. ¶ 104 (2014), http://hudoc.echr.coe.int/sites/eng/pages/search.aspx?i=001-146501.

[171] Case of Cruz Sánchez et al. v. Peru, Series C no. 292, Preliminary Objections, Merits, Reparations and Costs, Inter-Am. Ct. Hum. Rts ¶ 283, at 87 (2015), http://www.corteidh.or.cr/docs/casos/articulos/seriec_292_esp.pdf.

rights norms can mask the reality that these incidents often have the factual attributes of an armed conflict and ultimately require the use of levels of force not those ordinarily associated with human rights law (see Chapter 12, section 12.2.2.2, *Japanese Embassy, Lima, Peru, 1997*).[172] At some point the application of a narrow uniquely human rights–based approach will impact negatively on the missions or roles that can be assigned to European States and ultimately the protection they can provide to civilians the Coalition is seeking to protect.

13.5. The Shoot-Down of Hijacked Aircraft

Another way in which the limits of a human rights approach may become evident is in terms of its utility in weighing the value of life, not only in relation to the different persons or groups who may be killed by violence but also against the importance of protecting objects of commercial and national defense or security value. As was reflected in the 9/11 attacks, these are the very issues at stake when a State considers whether it will shoot down a hijacked aircraft.[173] This weighing of the value of life is an inherent part of the decision-making process in targeting under humanitarian law where the military advantage to be gained from an attack is assessed against its likely collateral affects.[174] The question is whether the human rights normative framework is capable of acting in place of humanitarian law in protecting a State from such attacks.

The different approaches adopted by the United States and Canada, Germany and Australia toward the shoot-down of hijacked aircraft highlight the limitations of a human rights–based approach. Notably, the North American countries have addressed the threat of attacks from aircraft hijacked by non-State actors with a military response.[175] The attacks of 9/11 were of sufficient gravity that they justified an exercise of State self-defense. In this respect they met the threshold of an "armed attack" under Article 51 of the UN Charter. The hijacked aircraft had become a weapon of war.[176] That attack also marked the beginning of operations that led to the use of military force against Al Qaeda in Afghanistan and the commencement of an international armed conflict against the Taliban. To respond to the threat of hijacked civilian aircraft the Canadian and United States invoked the North American Air Defense (NORAD) Agreement, which among other tasks provides for the "surveillance and exercising operational control of the airspace of the United States and

[172] Eduardo Nicolás Cruz Sánchez, Et Al Peru, Case 12.444 Merits, Report No. 66/10 ¶ 120 (March 31, 2011), http://www.oas.org/en/iachr/decisions/cases.asp and Case of Cruz Sánchez et al. v. Peru, Series C no. 292, Preliminary Objections, Merits, Reparations and Costs, Inter-Am. Ct. Hum. Rts (2015), http://www.corteidh.or.cr/docs/casos/articulos/seriec_292_esp.pdf.

[173] The 9/11 Commission Report: Final Report of the National Commission on Terrorist Attacks Upon the United States 7–14 (2004) (for reference to attacks on the World Trade Center and the Pentagon and an intended strike on the Capitol Building or White House in Washington, D.C.).

[174] AP I, art. 57(2)(a)(iii).

[175] Elinor C. Sloan, Security and Defence in the Terrorist Era 85–89 (2005).

[176] Commentary on the HPCR Manual on International Law Applicable to Air and Missile Warfare, Program on Humanitarian Policy and Conflict Research at Harvard University Rule 1(t), ¶ 2, at 41 (2010), http://ihlresearch.org/amw/Commentary%20on%20the%20HPCR%20Manual.pdf [hereinafter Air and Missile Warfare Manual Commentary].

Canada."[177] The NORAD Agreement now specifically recognizes the global threat posed by terrorists and non-State actors.[178] The NORAD Agreement also acknowledges that criminal threats may be presented when it notes, "the non-military air and maritime activities associated with drug trafficking and other illegal transnational activities are a threat to their national security...."[179] Therefore the airspace of North America is patrolled for both conflict and law enforcement reasons. Importantly, as was highlighted with the 1973 shoot-down of a Libyan airliner by the Israeli air force, the decision to shoot down a civilian passenger aircraft will inevitably result in a significant loss of civilian life.[180]

The challenge for the North American approach, which is based on the exercise of national self-defense, is that in an everyday context most suspicious aircraft are likely to be engaged in criminal activity or simply be operating in a certain airspace without authority.[181] Ultimately, the NORAD response has to be capable of distinguishing between inattentive pilots, criminals, and transnational terrorist threats. The unique status of a civilian airliner, the threat they pose once hijacked, and additional limits on a State response are reflected in the *Air and Missile Warfare Manual Commentary*.[182] It has been stated "there is no doubt that the law expects and requires great care, it may be argued particular care, be employed when taking precautions prescribed by article 57 of AP I in relation to civilian airliners."[183] The *Commentary* provides for the shoot-down of hijacked aircraft under humanitarian law only if diversion for landing, inspection, and possible capture is not feasible;[184] all feasible means to exercise control of the aircraft are exhausted;[185] the circumstances are sufficiently grave to justify an attack;[186] and "expected collateral damage will not be excessive in relation to the military advantage anticipated and all feasible precautions have been taken...."[187] In this regard the lawful use of such

177 *Agreement Between the Government of Canada and the Government of the United States of America on the North American Aerospace Defense Command*, E105060, Article I, 2(b) (Apr. 28, 2006) (the treaty was first entered into on May 12, 1958), http://www.treaty-accord.gc.ca/text-texte.aspx?id=105060 [hereinafter *NORAD Agreement*].

178 *Id.* (see the Preamble).

179 *NORAD Agreement, supra* note 176 (Preamble). *See also* Sam Tabory, *Peru Revives Controversial Drug Plane Shoot-Down Law*, INSIGHT CRIME (Aug. 21, 2015), http://www.insightcrime.org/news-briefs/peru-revives-controversial-shoot-down-law-for-narco-planes (Several South American countries provide for the shoot-down of drug smuggling aircraft. The United States began backing away from supporting such programs after a 2001 incident where Peruvian authorities mistakenly shot down a small civilian plane carrying a U.S. missionary and a small child. Peru had relied on intelligence from U.S. counternarcotics "operatives." It is reported that "[l]ast year, the US government stopped sharing intelligence from anti-drug radars with authorities in Honduras following the passage of a shoot-down law in that country.").

180 DAN RAVIV & YOSSI MELMAN, SPIES AGAINST ARMAGEDDON: INSIDE ISRAEL'S SECRET WARS 137–38 (2012) (In February 1973 a Libyan civilian airliner was tragically shot down by the Israeli air force. The airliner refused to respond to radio calls or internationally recognized signals such as the flapping of wings by the fighters. Out of 122 passengers and crew, 105 were killed.).

181 Craig Mellow, *Would a Fighter Pilot Shoot Down a Private Airplane?*, AIR & SPACE MAG. (Mar. 2010), http://www.airspacemag.com/flight-today/dont-cross-that-line-5841988/?no-ist.

182 AIR AND MISSILE WARFARE MANUAL COMMENTARY, *supra* note 176.

183 WILLIAM BOOTHBY, THE LAW OF TARGETING 243 (2012).

184 AIR AND MISSILE WARFARE MANUAL COMMENTARY, *supra* note 176, Rule 68 (a), at 166.

185 *Id.* Rule 68(b), ¶ 2, at 167.

186 *Id.* Rule 68(c), at 167.

187 *Id.* Rule 68(d), at 167. *See also*, BOOTHBY, *supra* note 183, ¶ 12.6.1, at 242-43 (for reference to restrictive criteria found in the SAN REMO MANUAL ON INTERNATIONAL LAW APPLICABLE TO ARMED CONFLICTS AT SEA, Rules 57 and 58, 143-44 (1995) that are similar to those found in the Air and Missile Warfare Manual Commentary, supra note 176.).

force under this humanitarian law framework regarding hijacked aircraft resonates with the concept of the use of force as a last resort, which is more traditionally associated with a human rights–based law enforcement response (see Chapter 12, section 12.2.1, *The Military Task: Applying Force of First and Last Resort*). However, this humanitarian law framework is ultimately about shooting down airfcraft and is designed to carry out the very weighing of life that is integral to armed conflict decision making. It is not as clear that human rights law comfortably performs that role.

In contrast, the German government relied on a policing approach where the military provided assistance in confronting hijacked aircraft.[188] This resulted in litigation where the German Constitutional Court struggled with some of the type of life and death issues military commanders must regularly confront, often under severe time constraints and with less than perfect information available to them. In *Dr. H. et al. v. s.14.3 of the Aviation Security Act of 11 January 2005*, the court identified a number of challenges regarding the decision to shoot down a hijacked aircraft, including the reality that the motivation of the hijackers is often difficult to establish,[189] the possibility there could be limited communications with the aircraft,[190] people may be killed on the ground by aircraft wreckage,[191] the existence of a multitiered decision-making process and a small air space could lead to enormous time pressures and premature decisions,[192] and the factual basis for a shoot-down decision may not be determined with certainty.[193] Perhaps most significantly from a normative perspective, the court considered the rights and human dignity of the passengers as innocent victims and took the view "the weighing up of lives against lives according to the standard of how many people are possibly affected on the one side and how many on the other side is impermissible. The state may not kill people because they are fewer in number than the ones whom the state hopes to save by their being killed."[194] Thus the court avoided dealing with what is arguably one of the most important and difficult assessments that must be performed by a decision authority in making a shoot-down decision. It is an assessment that has real world consequences in terms of limiting the overall potential loss of life.[195]

The German court also did not address another difficult question, the weighing of the value of life against the protection of objects, such as installations of national security, or even commercial interest (e.g., the Pentagon, the White House, the World Trade Center, the New York Stock Exchange).[196] It is this type of assessment that is at heart of the

---

[188] Nina Naske & Georg Nolte, *"Aerial Security Law." Case No. 1 BvR 357/05. 115 BVerfGE 118*, 101 AM. J. INT'L L. 466 (2007) (for an overview of *Dr. H. et al. v. s.14.3 of the Aviation Security Act of 11 January 2005* (2006)).

[189] Dr. H. et al. v. s.14.3 of the Aviation Security Act of 11 January 2005, ¶ 123 (2006) http://www.bundesverfassungsgericht.de/SharedDocs/Entscheidungen/EN/2006/02/rs20060215_1bvr035705en.html.

[190] *Id.* ¶ 126.

[191] *Id.* ¶ 151.

[192] *Id.* ¶ 127.

[193] *Id.* ¶ 123.

[194] *Id.* ¶ 35.

[195] BENJAMIN WITTES & GABRIELLA BLUM, THE FUTURE OF VIOLENCE: ROBOTS AND GERMS, HACKERS AND DRONES 227 (2015) (where it is noted the attempted take over of United Flight 93, which resulted in the death of forty innocent people, "was a tragedy, but one with a body count of a different order of magnitude that if the plane had hit the Capitol or White House.").

[196] Kenneth Watkin, *Military Advantage: A Matter of "Value", Strategy and Tactics*, 17 INT'L Y.B. HUM. L. 315-16 (2014)(for further analysis of the 9/11 attacks and the factors to be applied in assessing "military advantage" in terms of protecting property of commercial, military, or political importance).

proportionality test under humanitarian law as force is used to protect not only citizens but also property.[197] This is not to say that all human rights–focused tribunals will be unable to make an assessment regarding the shoot-down of aircraft or accept the likelihood of collateral civilian casualties. As has been noted, in 2011 the European Court of Human Rights in *Finogenov v. Russia* demonstrated some willingness to accept the infliction of collateral casualties (e.g., 129 hostages killed) in what it viewed as a terrorist incident at the Moscow theater.[198] However, that decision has limits in that it dealt with civilians, who as hostages already had their lives placed in jeopardy. This is not the same issue as civilians who might be collaterally killed as a result of a use of force.[199] In the shoot-down scenario it is not just the civilians on the plane but also any persons at the target site and where the aircraft might crash who are at risk. Concern over the ability of a human rights law framework to respond to the threat of hijacked aircraft has prompted a suggestion that body of law should change to account for the likelihood of collateral casualties.[200]

From a domestic constitutional perspective the German court may have "emphasized the importance of a strictly text-based legal interpretation (*strikte Texttreue*) and, hence, objected to functional interpretation of the constitution with the purpose to empower the Armed Forces to fulfil domestic tasks apart from defense against an enemy's military attack."[201] However, it has not provided a practical solution to a real-world operational challenge. It can only be hoped that the effects of this form-trumps-function approach does not have to be experienced. This interpretative approach avoids engaging in the types of value judgments that are an inherent part of targeting decisions carried out by the lowest ranking soldier standing on checkpoints to commanders conducting strategic air campaigns. In adopting this position, strong support is provided for arguments doubting the ability of human rights law to broadly regulate these types of threats during hostilities.

Like Germany, Australia has also adopted a support to police or "call out" approach. However, in doing so it developed a third option for dealing with the threat of hijacked aircraft. Australia's defense legislation was amended in 2006 to authorize the military to potentially use force against a vessel or aircraft as part of the powers the Australian military performs in aid of civil defense (a traditional form of assistance to law enforcement in common law systems).[202] Those amendments included the authority to use force against aircraft, extending to destroying such aircraft; the power to order such measures; introduction of the defense of superior orders; and the authorization to do what is necessary to protect life or prevent serious injury and protect property designated as critical infrastructure.[203] The type of property included as critical infrastructure could include "nuclear

---

197 Simon Bronitt & Dale Stephens, *"Flying Under The Radar"—The Use of Lethal Force Against Hijacked Aircraft: Recent Australian Developments*, 7 Oxford U. Comm. L.J. 265, 276–77 (2007).

198 Finogenov v. Russia, VI Eur. Ct. H. R. ¶ 236, at 409 (2011) (extracts).

199 Eliav Lieblich, *Quasi-Hostile Acts: The Limits on Forcible Disruption Operations under International Law*, 32 B.U. Int'l L.J. 101, 113–14 (2014).

200 Robin F. Holman, *Law Enforcement, the Rogue Civil Airliner and Proportionality of Effects: An Analysis of International Human Rights Law*, A thesis submitted to McGill University in partial fulfillment of the requirements of the degree of LL.M. 109 (2010), http://www.collectionscanada.gc.ca/obj/thesescanada/vol2/QMM/TC-QMM-97268.pdf.

201 Oliver Lepsius, *Human Dignity and the Downing of Aircraft: The German Federal Constitutional Court Strikes Down a Prominent Anti-terrorism Provision in the New Air-transport Security Act*, 7 German L.J. 761, 776 (2007).

202 *Defence Act 1903* (Cth), ss. 51SE and 51ST.

203 Bronitt & Stephens, *supra* note 197, at 271–75.

and bio-hazard facilities," "gas pipelines, power lines and plants that supply hospitals."[204] While acknowledging that using deadly force to protect property is "morally controversial," it has been noted, "this way signifies a move beyond the traditional criminal law paradigm of self-defence (which includes defence of others) towards a much broader national security necessity paradigm."[205]

The Australian approach prompts yet another hybrid analogy where "terrorism threats inhabit a legal space that spans both domestic criminal law and international armed conflict."[206] This approach offers a more realistic framework than that offered by the German case for dealing with the security threat posed by hijacked aircraft.[207] However, an issue in the Australian context may be whether that State has been too subtle in relying on concepts such as the defense of superior orders, and "Australia's international obligations" as the basis for arguing humanitarian law may be invoked as a matter of law.[208] It has been suggested the reference to "international obligations" points to the UN Charter and the right of national self-defense.[209] However, a different interpretation of the legislation reaches the conclusion that "these provisions provide a legislative framework to respond to non-military threats that are beyond the scope of normal law enforcement and yet, although possibly prompting measures in national self-defence, fall short of armed conflict."[210]

With national self-defense potentially encompassing both law enforcement and armed conflict responses the challenge for Australian authorities justifying the shoot-down of a hijacked aircraft could be in establishing an armed conflict is in existence (see Chapter 14, section 14.3.2, *Commencement of Armed Conflict*). Recognition that the attacks of 9/11 were an "armed attack" prompting the exercise of self-defense and the triggering point for an international armed conflict suggests such a conclusion based on those facts. If such incidents are not viewed as an armed conflict, the question is raised whether a law enforcement response is demanded or some form of "self-defense" option is considered to be available. As has been noted, that latter approach is subject to considerable criticism (see Chapter 8, section 8.5.3, *The "Self-Defense" Option*, and section 8.6. *Seeking a Solution*). In any event, since the potential threats posed by civilian aircraft usually involve criminal activity or simply an unobservant pilot, this ultimately makes the application of a humanitarian law–based paradigm unlikely to occur in most situations.

What the three different approaches (NORAD, Germany, and Australia) indicate is that increasingly humanitarian law and both domestic and international human rights law may have to be considered simultaneously in addressing these unique transnational terrorist threats. It also highlights the holistic response that some States have adopted to counter these threats. This issue is relevant not only to hijacked aircraft but also to other "one-off"

---

[204] *Id.* at 274–75.

[205] *Id.* at 275.

[206] *Id.* at 277.

[207] Dr. H. et al. v. s.14.3 of the Aviation Security Act of 11 January 2005 (2006), http://www.bundesverfassungsgericht.de/SharedDocs/Entscheidungen/EN/2006/02/rs20060215_1bvr035705en.html.

[208] *Defence Act 1903* (Cth), s. 51ST.

[209] Bronitt & Stephens, *supra* note 197, at 273.

[210] Donald R. Rothwell & Cameron Moore, *Australia's Traditional Maritime Security Concerns and Post 9/11 Perspectives*, in MARITIME SECURITY: INTERNATIONAL LAW AND POLICY PERSPECTIVES FROM AUSTRALIA AND NEW ZEALAND 37, 49 (Natalie Klein, Joanna Mossop, & Donald R. Rothwell eds., 2010).

attacks such as those experienced in Mumbai (2008), Benghazi (2012), the Westgate Mall in Kenya (2013), or Peshawar (2014). Such incidents also serve as a warning that the use of a law enforcement framework may mask the true nature of what members of the military are being asked to do. While domestic or international law enforcement terminology may sometimes be used to describe the action taken by the State, the incident will often be more comfortably situated at the armed conflict end of the spectrum in terms of the decisions that have to be made. There is much to be gained, in terms of effectiveness and accountability, from being clear about the limits of a law enforcement response in order to ensure that the State and its agents are properly prepared and react appropriately to the threat.

13.6. Limits and the Need for Clarity

The increased recognition of the utility of using law enforcement operations to counter the often integrated threat posed by insurgents, terrorists, and transnational criminal organizations presents both an opportunity and a challenge for military commanders. It is evident in many situations that a human rights–based law enforcement approach has to be applied as a matter of law. In other situations, military commanders or the States they serve often adopt the more restrictive human rights normative framework as a matter of policy because of the operational advantages that can be gained. Ultimately, the reliance on law enforcement is closely linked to governance and the role that State security forces perform in maintaining order both internally and increasingly on international operations.

In an operational environment where the issue is not just whether the two normative regimes overlap but increasingly how they will be applied concurrently, the articulation of the limits of the human rights normative regime is crucial in explaining why certain levels of force are used or certain tactics are applied. The challenge for State security officials is to clearly identify the limits that exist in application of the human rights–based approach, particularly in the context of an armed conflict. Although the ICJ case, *Legal Consequences Of The Construction Of A Wall In The Occupied Palestinian Territory*, highlighted there is a significant overlap between the human rights and humanitarian normative regimes, neither body of law was intended to fully displace the other or necessarily cover exactly the same matters.[211] In this chapter, the limits of law enforcement have been suggested to be practical, arise because of "overreach," and be interpretive. These limits highlight both the positive and negative aspects of applying human rights–based norms to contemporary armed conflict.

In positive terms a number of the limits can be easily identified. The adoption of ambush-style tactics, the application of "shoot to kill" policies, or the need to use indirect fire weapons (e.g., military tactics) can serve as indicators of when the human rights–based paradigm is being breached or the threshold into armed conflict has been crossed. As was seen in the context of the shoot-down of hijacked aircraft, the adoption of an interpretation of human rights law that reaches the conclusion weighing the value of life is impermissible would

[211] Legal Consequences Of The Construction Of A Wall In The Occupied Palestinian Territory, 2004 I.C.J. Reports 136, ¶ 106, at 178.

effectively mean that a considerable portion of armed conflict–related violence might only be amenable to a humanitarian law normative paradigm.

There also remains the challenge of ensuring that the application of a human rights approach and the use of human rights terminology does not mask when force is ultimately being used in the context of an armed conflict. This issue could arise in the context of the United States having incorporated human rights norms into its conduct of hostilities based targeting assessment for drone strikes.[212] Analogies to the historic practice of "air policing"[213] have been made "in an age when many military tacticians argue that the terrorist hotspots of the world can be policed by unmanned aerial drones in place of ground troops."[214] There are important differences between the 1920s and the 21st Century use of airpower. At a minimum, contemporary operations conducted in the context of an armed conflict means that the rules of humanitarian law apply. Obviously human rights norms may also be applied. However, it remains that clarity regarding the governing legal framework is particularly important for security personnel whose actions will ultimately be judged pursuant to one of the human rights or humanitarian law normative regimes. In any domestic accountability proceedings it is challenging enough to explain the details of humanitarian law to a court that is likely most comfortable with national criminal law concepts. A failure by the State to clearly identify the applicable normative regime can be problematic for a member of the security services who has been accused of wrongdoing. It is only by clearly identifying the limits of law enforcement and transparently acknowledging the type of conflict, the normative regime being applied, and the reason why one body of law is being applied over the other that the overlap between the human rights and humanitarian law paradigms can be rationalized and the added protection their joint application provides be properly understood.

---

212 *Obama's Speech on Drone Policy*, NY Times, May 23, 2013), http://www.nytimes.com/2013/05/24/us/politics/transcript-of-obamas-speech-on-drone-policy.html

213 Philip Anthony, Pilots and Rebels: The Use of Aircraft in Unconventional Warfare 1918-1988, at 12-45 (1989) (In the aftermath of World War I a number of colonial powers: the British, French, Spanish and Italians used aerial bombing to maintain order in areas largely governed by tribal societies. The British applied the tactic to the greatest extent in territory such as Aden, Iraq, Somalia, Afghanistan and the Northwest frontier of India (Pakistan). Of note they mirror the geography of air strikes in the contemporary drone campaign).

214 James Fergusson, The World's Most Dangerous Place: Inside the Outlaw State of Somalia 245 (2013). *See also* Daniel P. Bolger, Why We Lost: A General's Inside Account of the Iraq and Afghanistan Wars 105-06 (2014).

V The Way Ahead

Those skilled in war cultivate the Tao [the way of humanity and justice] and preserve the laws and therefore are able to formulate victorious policies.[1]

14 A Holistic Solution

14.1. Introduction

As was indicated in Chapter 1, the international law governing armed conflict is at a crossroads. In the twenty-first century other States are no longer uniquely viewed as the most significant security threat. Instead, that threat is presented in the form of an exceptionally diverse set of non-State actors, which challenge not only the authority of States but also the very basis of the Westphalia system of governance. This does not mean that inter-State war, or what is categorized in formal legal terms as international armed conflict, has disappeared. However, it is conflict involving non-State groups operating as "pseudo States"; Salafi jihadists seeking to establish a "global" Caliphate by fighting both a perceived "near" and "far" enemy; and criminal gangs operating transnationally but also seeking their own unregulated domestic space that have dominated the twenty-first-century security dialogue.

Complicating the assessment of these conflicts is the post-9/11 shift of viewing terrorism as no longer being just a criminal act. Further, many of these security threats involve a progression from criminal activity/terrorism, to insurgency, and finally an effort to seize territory and govern in the place of States. The jihadists, while applying terrorist tactics, are also

---

[1] Sun Tzu, The Art of War 88 (Samuel B. Griffith trans., 1971) (where it is indicated a ninth-century commentator, Tu Mu, explained:

> The Tao is the way of humanity and justice; "laws" are regulations and institutions. Those who excel in war first cultivate their own humanity and justice and maintain their laws and institutions. By these means they make their governments invincible.)

often insurgents focused on displacing incumbent governments.[2] It is no coincidence that despite the fundamentalist religious motivation of the Salafi jihadists, their military strategy reflects revolutionary warfare doctrine, often applied with a vanguard/*focoist* approach.[3]

Added to this diverse mix of actors are criminal gangs that thrive in the same situations of instability in which the non-State groups also flourish. This situation can be a by-product of the operations of insurgent, terrorist, or transnational criminal organizations, or simply occur because the criminal actors operate in parts of the world that are spaces ungoverned by States. The security environment is complicated by the intertwined relationship between organized armed groups and their criminal counterparts. Smuggling, kidnapping for ransom, the drug trade, and other criminal activity fuel the operational capability of insurgent/terrorist groups and line the pockets of transnational and domestic criminal gangs. This "criminal" connection is further enhanced by the long-standing and universal categorization by States of these non-State actors as being illegitimate. That States are now forced to address threats from such a diverse group of non-State actors places a spotlight on their role in maintaining order in the face of a wide range of security threats that can occur internally, internationally, or concurrently in both arenas.

The focus of this concluding chapter is the application of a holistic operational law approach designed to address the evolving, diverse, and unique security threats facing the international community. Divided into four parts, the analysis will first place this legal challenge in context by considering it in relation to the doctrinal struggle being experienced by State military forces in addressing "hybrid" warfare. What becomes evident is that both the security community and its legal counterpart will need to situate the solution for countering contemporary non-State actor threats in an analytical framework that more broadly encompasses conventional conflict, irregular warfare, and criminal activity regardless of how an armed conflict is categorized legally.

The second part deals with two key issues that need to be addressed when carrying out operational law–based decision making. The first issue involves consideration of how the law governing the recourse to warfare, humanitarian law, and human rights law interacts in responding to transnational attacks and threats to nationals from non-State actors. This includes an analysis of the interface or "boundaries" between these areas of law; an assessment of when armed conflict commences in the context of such armed attacks; and identification of criteria that can be applied to separate an armed conflict response from situations requiring law enforcement. The second issue focuses on applying a human rights–based response. This includes when such a response must be applied as a matter of law, as a matter of State policy, or as a discretionary decision made by the chain of command. The ways in which law enforcement is privileged as a response to attacks at home and abroad are

---

2 Jessica S. Stern & J. M. Berger, ISIS: The State of Terror 59–63 (2015) (for an analysis of Al Qaeda's shift through its affiliates from a focus on global terrorism to local insurgencies). *See also* Samuel B. Griffith II, *Introduction*, *in* Mao Tse-tung, On Guerrilla Warfare 27 (Samuel B. Griffith trans., 2000) (1961) (where it is noted the displacement of local governments is an integral part of revolutionary warfare).

3 Michael W. S. Ryan, Decoding Al-Qaeda's Strategy: The Deep Battle Against America 230 (2013) (discussing the adoption by jihadist theorists of a "classic three-stage guerrilla war strategy"). *See also* David Kilcullen, Counterinsurgency 177 (2010) (for a discussion of jihadist reliance on the vanguard/*focoist* approach.).

explored, as is the inevitable concurrent application of law and policy in addressing contemporary security threats.

The third part provides an outline of a decision-making process that will assist in reconciling the application of the relevant bodies of international and domestic law. The recommended approach is summarized in a "decision tree" analysis set out in the Appendix. Particular attention is also paid to the role that responding States should play in acting as a proper authority in the functional absence of a territorial State when responding to threats in ungoverned spaces.

The final area of analysis highlights the link between policing and "normalcy" in reaching the conclusion that the ability to apply such a human rights–based approach provides an important indicator of success in respect of conflict with non-State actors.

14.2. From "Conventional" to "Hybrid" to "Holistic"

What is important for international lawyers to note is that the changes in approach demanded by twenty-first-century security threats have not just been occurring in the discipline of law. The operational landscape began to shift immediately after the end of the Cold War, forcing a change in military doctrine. The 1990s were a period defined by an increased acknowledgment of the participation of State military forces in conflicts that were not viewed as traditional warfare. The complexity of the operational environment is probably best represented by the UN interventions in the former Yugoslavia, although there were also other operations, such as a difficult intervention in Somalia and a horrific genocide in Rwanda. This period was the beginning of a significant period of change, which forced a reassessment of the way that State military forces viewed their role. These complex missions also prompted alterations in how UN peacekeeping was traditionally viewed and the degree to which humanitarian law applied to these "lesser" conflicts (see Chapter 11, section 11.3.2, *United Nations Operations and "Robust Peacekeeping"*). However, during this period those writing military doctrine were careful to categorize these types of activities as "military operations other than war" (MOOTW), presumably in order to maintain the sanctity of "real" war (i.e., conventional inter-State conflict).[4]

One of the challenges was that separation of "real" war from the other types of armed conflict was often difficult to fully articulate. It was acknowledged in doctrine that "MOOTW and war may often seem similar in action."[5] The conceptual differentiation between large-scale operations and MOOTW permitted States and their armed forces to remain largely focused on inter-State conventional conflict (their comfort zone), even though the likelihood of such conflict appeared to recede in the face of growing transnational threats posed by non-State actors. Representative of the international military community, "the [United States] army continued to prepare itself to fight wars as it wanted to fight them."[6] With the post-9/11 requirement to shift to counterterrorism and counterinsurgency operations,

---

[4] Joint Doctrine for Military Operations Other Than War, Joint Pub 3-07, ¶ 2, at I-1, (June 16, 1995).

[5] *Id.* at vii.

[6] John A. Nagl, Learning to Eat Soup with a Knife: Counterinsurgency Lessons from Malaya and Vietnam 207 (2005).

the result was often a mismatch between military capability and the type of doctrine, equipment, skills, and approach required to successful counter non-State foes. This was a mismatch that also ultimately had legal implications. The lessons learned by military commanders during this period were often painful ones both in terms of operational effectiveness and political impact (see Chapter 7, section 7.2, *Counterinsurgency in the Twenty-First Century*).

That specific new doctrine, training, and equipment acquisition initiatives were required to address the counterinsurgency challenge is representative of the larger problem created by the traditional overemphasis by States on inter-State conflict, a focus on conventional warfare, and a culture which often has demonstrated difficulty in adapting to a nonlinear battlespace.[7] The doctrinal struggle to define roles outside the context of conventional combat continued with the development of the concept of "stability operations," which was to become a "catch-all" phrase that sought to put other operational tasks on an equal footing with offensive and defensive action.[8] In effect, this doctrine raised the profile of activities related to governance and the protection of the civilian population.[9]

The crosscutting nature of the doctrinal shift can be seen in the statement that "[w]hether an operation is a peace operation preventing conflict or a large-scale combat operation, forces will always integrate offensive, defensive, *and stability tasks*."[10] For the U.S. Army, the primary "stability tasks" encompass the establishment of civil security and civil control, restoration of essential services, support to governance, and support to economic and infrastructure development.[11] An example of the role performed by the military in a achieving a safe and secure environment for the local population is "by ending hostilities, isolating belligerents and criminal elements, demobilizing armed groups, eliminating explosives and other hazards, and providing public order and safety."[12] Inherent in this role is engagement in a law enforcement function, either directly or in support of other security partners (see Chapter 12, section 12.2.1, *The Military Task: Applying Force of First and Last Resort*).

With U.S.-led counterinsurgency operations having ended in Iraq in 2011, and the withdrawal of most international forces from Afghanistan having commenced by that time as well, there began to surface critiques concerning the integration of counterinsurgency doctrine into the American "way of war."[13] To a certain extent, the recalibration of military

---

[7] ROBERT M. CASSIDY, COUNTERINSURGENCY AND THE GLOBAL WAR ON TERROR: MILITARY CULTURE AND IRREGULAR WAR 99–126 (2008) (for an outline of the cultural challenges the U.S. military faces in adapting to contemporary conflict that includes counterinsurgency operations).

[8] Kenneth Watkin, *Stability Operations: A Guiding Framework for "Small Wars" and Other Conflicts of the Twenty-First Century?*, 85 INT'L LEG. STUD. 411, 414–16 (2009). *But see* FRANK LEDWIDGE, LOSING SMALL WARS: BRITISH MILITARY FAILURE IN IRAQ AND AFGHANISTAN 203 (2011) (where it is noted that "[s]tabilization during armed conflict is a Sisyphean task, akin to attempting to put up a tent in a storm.").

[9] STABILITY, FM 3-07, Headquarters Department of the Army, at iv (June 2, 2014) [hereinafter STABILITY OPERATIONS MANUAL] ("Stability ultimately aims to establish conditions the local populace regards as legitimate, acceptable, and predictable. Stabilization is a process in which personnel identify and mitigate underlying sources of instability to establish the conditions for long-term stability.").

[10] *Id.* ¶ 1-1, at 1-1.

[11] *Id.* ¶ 1-6, at 1-2.

[12] *Id.* ¶ 1-2, at 1-1.

[13] RUSSEL F. WEIGHLEY, THE AMERICAN WAY OF WAR: A HISTORY OF UNITED STATES MILITARY STRATEGY AND POLICY xxii (1973) (where it is noted that American strategy has through most of the country's history been one of annihilation or seeking to overthrow the enemy's military power, rather than one of attrition, exhaustion, or

effort at the end of those operations needed to be undertaken, since military forces have to be fully prepared to fight conventional as well as unconventional wars (see Chapter 7, section 7.2, *Counterinsurgency in the Twenty-First Century*). However, at the heart of the military challenge is that even conventional wars can involve guerrilla warfare, and at some point can require the policing of a civilian population (e.g., occupation). Conversely, countering insurgencies usually involves participation in law enforcement operations and can extend to conventional conflict involving a more traditional use of force.[14]

As was noted in Chapter 1, the reality is that the fundamental nature of war does not change whether it is viewed as a conventional conflict or as an insurgency/counterinsurgency. In effect, "war is war."[15] To varying degrees all armed conflicts involve the full range of threats, and as a result require responses tailored to those threats. An early attempt to capture such diversity was made in the 1990s with the development of the doctrinal term "three block" wars.[16] Reflecting the broader range of conflict that is an inherent part of warfare, Figure 14.1 represents what the past decade of both international and noninternational armed conflict has demonstrated. Both legal categories—international and noninternational armed conflict (i.e., State and non-State armed conflict)—have the potential to include conventional and guerrilla warfare, as well as a requirement for State armed forces at some point to perform a law enforcement function.[17]

The difference between these two major forms of conflict is one of emphasis, with inter-State war normally involving a significantly greater degree of hostilities between conventional armed forces. Noninternational armed conflict is more likely to involve counterinsurgency, counterterrorism, and law enforcement operations. However, neither form of conflict has an exclusive application of any one of these types of operations.[18]

What would be problematic for States, and not likely to occur given the events of the past decade, is to have their military forces completely reject counterinsurgency doctrine and return to a unique focus on the inter-State conventional warfare. The twenty-first-century approach to warfare must remain "holistic" in nature. Consistent with this doctrinal

---

erosion that involves an indirect approach). *See* GIAN GENTILE, WRONG TURN: AMERICA'S DEADLY EMBRACE OF COUNTER-INSURGENCY (2013), and DOUGLAS PORCH, COUNTERINSURGENCY: EXPOSING THE MYTHS OF THE NEW WAY OF WAR (2013) (for critiques of American adoption of counterinsurgency warfare).

14 *US Airstrikes Target IS-held Oil Refineries in Syria*, FRANCE24 (Sept. 25, 2014), http://www.france24.com/en/20140925-syria-airstrikes-target-held-oil-refineries/. *See also* Kenneth Watkin, *Targeting "Islamic State" Oil Facilities*, 90 INT'L L. STUD. 499 (2014), and Kenneth Watkin, *Targeting in Air Warfare*, 41 ISR. Y.B. HUM. RTS. 1, 35–39 (2014) (for a discussion of the aerial targeting of oil producing facilities, economic targets, and the issue "war sustaining" targeting).

15 HEW STRACHAN, THE DIRECTION OF WAR 20-7-08 (2013).

16 Charles C. Krulak, *The Strategic Corporal: Leadership in the Three Block War*, MARINES MAG. (Jan. 1999), http://www.au.af.mil/au/awc/awcgate/usmc/strategic_corporal.htm ("The Corps has described such conflicts as the three block war, contingencies in which Marines may be confronted with the entire spectrum of tactical challenges in the span of a few hours within the space of three adjacent city blocks.").

17 STRACHAN, *supra* note 15, at 207 (for reference to both types of wars looking like one another particularly at the tactical level). *See id.* at 208 ("A binary notion of war creates a ready-made characterisation of wars based on theory but often insufficiently flexible to reflect reality.").

18 STABILITY OPERATIONS MANUAL, *supra* note 9, at ¶ 1-1. at 1-1 (which explains the application of stability tasks in a similar fashion. In this doctrine during "a peace operation, forces may still perform offensive tasks such as raids to capture war criminals. Conversely, in large-scale combat operations, forces perform stability tasks to control captured areas or provide emergency essential services.").

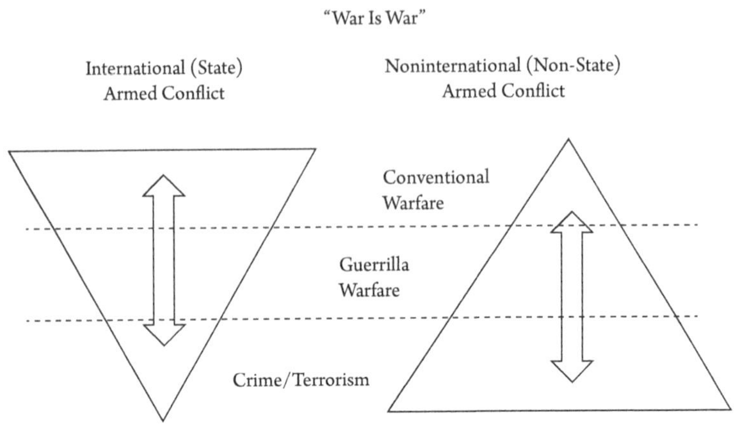

FIGURE 14.1 War is War

struggle, the altered security environment of this century has witnessed a definite move away from looking at conflict itself as being uniquely conventional or unconventional. This is seen in the common use of the term "hybrid" to explain various aspects of contemporary conflict.[19] This hybridization approach is also seen in the legal sphere, particularly in respect of the international community coming to terms with the simultaneous application of humanitarian and human rights law.[20]

It is in this more complicated security environment that the concept of "operational law" was developed. Given the analysis undertaken in this book, it might be described as the international and domestic laws applicable to the conduct of military operations, including the law governing the recourse to war, international humanitarian law, human rights law (i.e., international and domestic), and international criminal law. The concept of operational law has developed because of the need to resolve the simultaneous application, interaction, and overlap of the various bodies of law. The impact of these bodies of law on operations has become a very real practical concern. The theory of "exclusion," where each body of law is treated in isolation from the other, is simply inconsistent with the types of operational challenges faced by military commanders and the questions being asked of State legal advisors.

Operational law issues can be both diverse and complex. Legal questions can arise regarding the status of members of improvised explosive device (IED) and suicide bombing cells; the threat posed by hijacked civilian airliners; the permissible response to "one-off" attacks; the use of drones in territories that are not policed by the sovereign State; the conduct of air strikes on pirate infrastructure; and the boarding of dhows or other vessels that may be smuggling members of terrorist groups on the high seas. Questions may also

---

19 Peter R. Mansoor, *Introduction, in* HYBRID WARFARE: FIGHTING COMPLEX OPPONENTS FROM THE ANCIENT WORLD TO THE PRESENT 1–11 (Williamson Murray & Peter R. Mansoor eds., 2012) (for a discussion of hybrid warfare involving regular or irregular conflict), ANTHONY H. CORDESMAN, GEORGE SULLIVAN, & WILLIAM D. SULLIVAN, LESSONS OF THE 2006 ISRAELI-HEZBOLLAH WAR 80 (2007) (discussing Hezbollah's regular/irregular military structure), and DAVID KILCULLEN, THE ACCIDENTAL GUERRILLA: FIGHTING SMALL WARS IN THE MIDST OF A BIG ONE 38 (2009) (referring to organized armed groups that blend foreign and local fighters.).

20 CHARLIE SAVAGE, POWER WARS: INSIDE OBAMA'S POST 9/11 PRESIDENCY 344-47 (2015)(outlining the hybrid military-criminal approach to intelligence led interrogations followed by civilian trials).

arise concerning the arrest of a civilian supporting a terrorist group; the scope of justifiable collateral civilian casualties and damage during counterinsurgency operations; and the conduct of rescue operations to save foreign nationals kidnapped by criminal gangs, who then may plan to sell them to enemy armed groups. Issues involving actions to be taken against non-State actors and criminal gangs have prompted some of the most challenging operational law questions of this century. The challenges these issues present are compounded by their largely having arisen outside the context of inter-State conflict, which dominated the legal dialogue of the previous century. Of necessity, the solutions to these pressing legal issues must be firmly grounded in a holistic approach that seeks to apply all relevant bodies of law that can affect operations.

14.3. Responding to Transnational Attacks and Threats to Nationals

14.3.1. ARMED CONFLICT OR LAW ENFORCEMENT?

One aspect of contemporary conflict that has forced greater consideration of the interface between the law governing the recourse to war, humanitarian law, and human rights law has been the requirement for States to respond domestically to "one-off" cross-border assaults by non-State actors or to take action to defend nationals against isolated attacks in distant lands. In considering this issue it is important to first recall the roles performed by each of these bodies of law. For example, the law governing the recourse to war acts at the strategic level influencing the use of force through the application of its necessity, proportionality, imminence, and immediacy principles. However, within this overarching recourse to war framework, it is either humanitarian law or human rights law that directly regulates use of force decisions, particularly at the tactical level (see Chapter 3, section 3.6.3, *Applying Self-Defense Principles Across the Armed Conflict Spectrum*, and Chapter 11, section 11.2.2, *Self-Defense: Common Roots and Different Goals*). From a practitioner's perspective this makes the question of whether there is an armed conflict in existence a crucial issue. That determination impacts directly on what rules of engagement (ROE) are issued and the tactics used to counter a non-State actor threat. Without the existence of an armed conflict, human rights law must govern the State response.[21]

A particular challenge is that while there has been a concerted theoretical focus in the post-9/11 period on the exercise of State self-defense against transnational non-State actors, debates concerning the categorization of such conflict, and considerable discussion about thresholds for noninternational armed conflict, the practical impact of adopting a particular theory has generally not been incorporated into the analysis.[22] For example, what is the effect of setting a high threshold for a noninternational armed conflict? Does a resulting legal requirement to apply a human rights–based response to "real-world" threats provide an adequate framework within which to deal with the types and scale of violence being

[21] Prosecutor v. Tadić, Case No. IT-94-1-A, Decision On The Defence Motion For Interlocutory Appeal On Jurisdiction, ¶ 67 (Int'l Crim. Trib. for the Former Yugoslavia, 1995), http://www.icty.org/x/cases/tadic/acdec/en/51002.htm.

[22] Daniel Bethlehem, *Principles Relevant to the Scope of a State's Right as Self-Defense Against an Imminent or Actual Armed Attack by Nonstate Actors*, 106 AM. J. INT'L L. 770, 773 (2012) (commenting on the divide between theory and "operational realities").

faced by the State? As has been noted, "it is important that principle is sensitive to the practical realities of the circumstances that it addresses, even as it endeavors to prohibit excess and the egregious pursuit of national interest."[23]

A crucial task for State legal advisors has been to consider the nature of the threats and the practical responses available to counter these threats. However, one question is the degree to which these practical assessments have been properly grounded in legal theory. The boundaries inherent in the theoretical models serve to constrain and ultimately better direct the State response.[24] At the same time the theoretical models are of little practical use if the full impact of their implementation is not considered during their development. Consideration of the practical effect of a particular theoretical choice can serve to test its suitability to address the security challenge presented to a State. In a sense, both practitioners and theoreticians have often seemed to be talking past one another. Ultimately, neither theory nor practice can be divorced from one another.

The analysis by both practitioners and theorists must be more comprehensive. It is not enough to simply indicate that a State has the authority to act in self-defense. In terms of effectiveness, mission accomplishment, and the protection of the civilian population, it matters whether humanitarian law governs the response or such State action is limited to human rights–based law enforcement. For State legal advisors this assessment must also include consideration of the strategic, operational, and tactical effect of a particular theoretical legal approach when determining which normative framework should apply. Adopting a more holistic analysis that fully considers the implications of adopting one theoretical model over another is particularly important given the stakes involved and the unsettled nature of these areas of law in the context of transnational non-State actor threats. It is the uncertainty surrounding the status and the application of numerous legal theories claiming to regulate the same issue that has undoubtedly influenced some States to adopt a policy of applying humanitarian law on international operations (see Chapter 4, section 4.4.4, *A Policy Solution for Legal Uncertainty*), or suggestions there are unique "self-defense" rules for governing the use of force (see Chapter 8, section 8.5.3, *The "Self-Defense" Option*).

What is evident is the importance of determining if an armed conflict is in existence. In doing so, placing inordinate reliance on a narrow strict reading of the *Prosecutor v. Tadić* criteria, with its reference to "protracted armed violence,"[25] could lead to an interpretation that any "one-off" attack carried out solely by a non-State actor (i.e., one not controlled, harbored, or supported by a State) must be dealt with as a law enforcement matter. This is because a single attack would not likely be of sufficient duration to be considered "protracted."[26] The challenge such an interpretation presents is that many of these attacks by

---

[23] *Id.*

[24] Quincy Wright, *The Outlawry of War and the Law of War*, 47 AM. J. INT'L L. 365, 368 (1953).

[25] Prosecutor v. Tadić, Case No. IT-94-1-A, Decision On The Defence Motion For Interlocutory Appeal On Jurisdiction, ¶ 70 (Int'l Crim. Trib. for the Former Yugoslavia, 1995), http://www.icty.org/x/cases/tadic/acdec/en/51002.htm.

[26] *See, e.g., Leiden Policy Recommendations on Counter-Terrorism and International Law, in* COUNTER-TERRORISM STRATEGIES IN A FRAGMENTED INTERNATIONAL LEGAL ORDER: MEETING THE CHALLENGES, Annex, ¶ 62, at 721 (Larissa Van Den Herik & Nico Schrijver eds., 2013) ("it would include protracted violence meaning that acts are not perpetrated in isolation but as part of an ongoing campaign."), and YORAM DINSTEIN, NON-INTERNATIONAL ARMED CONFLICTS IN INTERNATIONAL LAW ¶ 101, at 33 (2014). *But see* NOAM LUBELL, EXTRATERRITORIAL USE OF FORCE AGAINST NON-STATE ACTORS 105 (2010) (where it is noted the demand for protracted violence is problematic in addressing what are short-term military hostilities).

non-State actors do not involve ordinary criminals acts, and there are significant limitations inherent in responding with force that is constrained by human rights–based law enforcement rules. Many non-State actor threats cannot be meaningfully addressed solely as a law enforcement matter (see Chapter 13, *Limits of Law Enforcement and Human Rights Law*).

The considerable threat posed by organized armed groups and the scale of military response that may be required to address such threats was evident in the 2000 rescue of British military hostages taken by the West Side Boys in Sierra Leone. While it may be attractive to try and mask that incident under the cloak of "law enforcement," any operation that involves the air insertion of a Special Forces squadron, a Parachute Regiment company, support from helicopter gunships, and indirect mortar fire, and which anticipates and results in an extensive firefight against an organized armed group armed with weapons of war, undermines the credibility of such a claim (see Chapter 10, section 10.3.3, *Modern Counterterrorism: Sierra Leone*). A similar issue arises in respect of the attacks of 9/11 carried out by means of hijacked aircraft. Setting aside the question of whether those hijackings were a continuation of prior attacks against U.S. interests,[27] had Al Qaeda not been considered to have been harbored or supported by the Taliban government in Afghanistan, the question would arise whether the attacks were of sufficient intensity to be considered to constitute an armed conflict. If the intensity requirement is tied to a duration of weeks, then in theory the 9/11 attacks would not occur in the context of an armed conflict and the response would be limited to a law enforcement one. This presumably could include any response that reached into "ungoverned spaces" from which Al Qaeda may have planned and initiated those attacks.

In the context of a 9/11-type hijacking, there are a number of factors that militate against a conclusion that a response would have to be limited to law enforcement. First, there is a general recognition that the attacks constituted an "armed attack" for which a response could be justified under Article 51 of the UN Charter.[28] As such, this form of attack would have met the gravity standards set out in *Military and Paramilitary Activities in and against Nicaragua Merits*.[29] The human and property losses certainly support the conclusion the violence was similar to that inflicted by States during armed conflict.[30] Second, there is the treatment of such a threat as one warranting a military response by countries such as the United States, Canada, and apparently Australia (see Chapter 13, section 13.5, *The Shoot-Down of Hijacked Aircraft*).

---

[27] *See, e.g.*, THE 9/11 COMMISSION REPORT: FINAL REPORT OF THE NATIONAL COMMISSION ON TERRORIST ATTACKS UPON THE UNITED STATES 190–91 (2004) [hereinafter 9/11 COMMISSION REPORT] (for a discussion of the 2000 attack on the *USS Cole* in Yemen). *See also id.* at 46 (where the 9/11 Commission took the view "the conflict did not begin on 9/11. It had been declared years earlier, most notably in a declaration faxed early in 1998 to an Arabic-language newspaper in London.").

[28] Bethlehem, *supra* note 22, at 774, and David Kretzmer, *The Inherent Right to Self-Defence and Proportionality in Jus Ad Bellum*, 24 EUR. J. INT'L L. 235, 273 (2013).

[29] Military and Paramilitary Activities in and against Nicaragua Merits, 1986 I.C.J. 13, ¶ 195, at 103, ¶ 247, at 127 (1986) (for reference to degrees of "gravity" and "armed attack").

[30] *September 11 Anniversary Fast Facts*, CNN (Sept. 11, 2013), http://www.cnn.com/2013/07/27/us/september-11-anniversary-fast-facts/ (where it is indicated 2,753 people died in the World Trade Center incident, 184 at the Pentagon, and 40 when the hijacked plane crashed near Shanksville, Pennsylvania. The estimated economic impact of these terrorist incidents is indicated to be well over $247 billion).

Finally, as is set out in the 2010 *Air and Missile Warfare Manual Commentary*, [31] the events of 9/11 have been tied to a humanitarian law response. As that *Commentary* notes:

> There is also an urgent need to confront new methods of warfare (however gruesome), introduced by international terrorism. At least since 11 September 2001, the law of armed conflict has been forced to consider, e.g., the use of a hijacked civilian airliner as a weapon . . . .[32]

The *Air and Missile Warfare Manual Commentary* does not specifically identify the 9/11 attacks as an armed conflict. However, it is suggested that "[a]s the 2001 terrorist attacks against the Twin Towers in New York City show, hijacked civilian airliners may also be used as a means of warfare in flagrant violation of the law of international armed conflict."[33] In the *Commentary* the term "law of international armed conflict" is used in the context of conflict between states.[34] It is clear the 9/11 attacks motivated the incorporation of hijacked aircraft as a means of warfare into that manual, which is exclusively devoted to humanitarian law.[35]

That the 9/11 attacks have been "technically" identified with an international armed conflict is not surprising given the reliance that has been placed on the argument that the Taliban was harboring and actively supporting that terrorist group (see Chapter 9, section 9.3.2, *Harboring and Active Support*). However, as was demonstrated in the German case, *Dr. H. et al. v. s.14.3 of the Aviation Security Act of 11 January 2005* (2006),[36] States may also choose to deal with the threat of hijacked aircraft by applying a human rights law–based law enforcement approach. The question remains whether the intensity of violence inflicted by Al Qaeda as an organized armed group would permit that attack to be considered an "armed conflict" had the nominal government in Afghanistan not been the harboring and active support provided to Al Qaeda. From a factual perspective it is difficult to see how it would not or how a law enforcement response could provide an effective response without significant changes to the type and scale of force permitted under that normative regime (see Chapter 13, section 13.5, *The Shoot-Down of Hijacked Aircraft*).

What the hijacked aircraft example does demonstrate is the inherent complexity and, to a significant degree, the existing formalism and artificiality of conflict categorization involving attacks by non-State actors. In theory, at least, it is relatively straightforward to

---

[31] Commentary on the HPCR Manual on International Law Applicable to Air and Missile Warfare, Program on Humanitarian Policy and Conflict Research at Harvard University, (2010), http://ihlresearch.org/amw/Commentary%20on%20the%20HPCR%20Manual.pdf.

[32] *Id. Introduction*, ¶ A at 1. *See also id.* § J, Rule 58, ¶ 10 at 156 ("Given the events of 9/11, it is impossible to ignore the danger that civilian airliners can be hijacked (and then used as means of attack) or otherwise be employed in ways harmful to the enemy.").

[33] *Id. Definitions*, Rule 1(t), ¶ 2, at 41.

[34] *Id. Definitions*, Rule 1(s), at 41.

[35] *Id.* ¶ D (d), at 6 ("There was discussion of human rights law without agreement. Most members of the Group of Experts believe that it has only minimal bearing on air and missile warfare in international armed conflicts because the law of armed conflict is *lex specialis*.").

[36] Dr. H. et al. v. s.14.3 of the Aviation Security Act of 11 January 2005 (2006), http://www.bundesverfassungsgericht.de/SharedDocs/Entscheidungen/EN/2006/02/rs20060215_1bvr035705en.html.

distinguish between international and noninternational armed conflict. This is true regardless of which threshold criteria are favored for assessing the existence of a noninternational armed conflict. Common Article 2 of the 1949 Geneva Conventions is viewed as outlining a broader set of circumstances for the beginning of an international armed conflict,[37] which may not even involve any force being used.[38] In contrast, conflicts not of an international character are based on more restrictive criteria, whether they are found in Common Article 3 of the 1949 Geneva Conventions, the *Prosecutor v. Tadić* test,[39] the Additional Protocol II reference to "situations of internal disturbances and tensions, such as riots, isolated and sporadic acts of violence and other acts of a similar nature, as not being armed conflicts,"[40] the *Abella v. Argentina* case,[41] or a "totality of circumstances" approach (see Chapter 9, section 9.4.3, *The "Totality of the Circumstances" and Other Criteria*). Each provides for a different and higher threshold of armed conflict than is applicable to inter-State violence. These bases for establishing the existence of a noninternational armed conflict reflects the fact that threats presented by non-State actors are usually viewed differently in terms of the scale and the types of violence than those posed by State actors.

As with many of the legal issues arising at the boundaries of these bodies of law (i.e., law governing the recourse to war, humanitarian law, and human rights law), the application of theory makes their distinctions sometimes appear less clear. That is because some attacks that are factually carried out solely by non-State actors (e.g., 9/11 attacks) can qualify as an international armed conflict (see Chapter 9, section 9.3.1, *Non-State Actors Controlled by a State*, section 9.3.2, *Harboring and Active Support*, and section 9.3.3, *Entering Territory of Another State: International Armed Conflict*). This means the broader criteria of Common Article 2 of the 1949 Geneva Conventions would apply. Theoretically, the actions of the non-State actor would qualify as an international armed conflict even though the threatened State and the controlling or harboring State never fire a shot at one another. Seemingly, the non-State actor would not even have to use any force or may only use limited violence that would not meet the "conflict not of an international character" threshold. However, in practice these lower level threats are likely to be met, at least initially, with a law enforcement response, even though it is the non-State actor's link to a State that governs how their involvement is ultimately categorized and not the intensity of the violence.

A practical challenge is likely to be that in many instances the requirement to respond to attacks made by non-State actors will occur under circumstances where it is initially not known if there is other State involvement or, if so, the full nature of that support. While

---

[37] Yoram Dinstein, The Conduct of Hostilities under the Law of International Armed Conflict ¶ 70, at 28–29 (2nd ed. 2010) ("LOAIC [Law of International Armed Conflict] is brought to bear upon the conduct of hostilities between sovereign States, even if these hostilities fall short of war, namely, constitute a mere incident.").

[38] Yoram Dinstein, War, Aggression and Self-Defence ¶ 22, at 10 (5th ed. 2011), and Leslie Green, The Contemporary Law of Armed Conflict 71–72 (3rd ed. 2008).

[39] Prosecutor v. Tadić, Case No. IT-94-1-A, Decision On The Defence Motion For Interlocutory Appeal On Jurisdiction, ¶ 70 (Int'l Crim. Trib. for the Former Yugoslavia, 1995), http://www.icty.org/x/cases/tadic/acdec/en/51002.htm. *See also* Anthony Cullen, The Concept of Non-International Armed Conflict in International Law 131–33 (2010) (for a discussion of the different standards for the commencement of international armed conflicts and noninternational armed conflicts, respectively).

[40] AP II, art. 1(2).

[41] Case 11.137, Juan Carlos Abella v. Argentina, Inter-Am. Y.B. on H.R. ¶ 1 (1997) (Commission report), http://www1.umn.edu/humanrts/cases/1997/argentina55-97a.html.

this was not the case during the 1976 Entebbe raid,[42] it was for the 9/11 attacks. Although the determination that the Taliban government was harboring the Al Qaeda attackers was made fairly quickly, it was not until after the attack was over.[43] In the meantime decisions had to be made by political leaders and military commanders regarding the nature of the attack and what force could lawfully be used to defend against the threat. Another example of delay in attributing responsibility for an attack is the October 2000 incident involving the *USS Cole* in Yemen. Even by December of that year there was only a "preliminary judgment" that Al Qaeda had "supported the attack." It is reported, "[t]his . . . . was not the conclusion they [the president and his officials] needed in order to go to war or deliver an ultimatum to the Taliban threatening war."[44] Without that linkage, the attack on the warship would have to be dealt with as some form of noninternational armed conflict or as a law enforcement matter. Of note, the immediate defensive reaction by the crew of the *USS Cole* may actually have been the same regardless of the categorization of the conflict. Much like when serving at a checkpoint, personnel on guard duty normally apply some form of escalation procedure to separate criminal acts from military threats. However, as with checkpoint scenarios, guards are not locked into a law enforcement–style response if an armed conflict exists and it is reasonably believed the attackers are direct participants in hostilities (see Chapter 12, section 12.3.3, *Confronting the Explosive Threat*).

Therefore, unlike a situation where conventionally organized, uniformed, and equipped regular armed forces are involved, it may not be immediately clear that another State is actually controlling the non-State actor or otherwise legally responsible for that cross-border attack. In responding to most of these attacks by non-State actors, the State authorities will normally lack the necessary advance information and certainly not have the advantage of hindsight in making a decision as to whether the attack constitutes an international or noninternational armed conflict. As a practical matter, non-State actors are most readily associated with noninternational armed conflict. The result is that a threatened State is likely to default to an approach that views all non-State actors potentially the same way. Even if invoked on a policy basis, the approach is likely to be one that applies the noninternational armed conflict threshold when dealing with a non-State actor "armed attack."

This approach is consistent with how these groups are often dealt with by practitioners. The multiplicity of often confusing and highly technical theoretical approaches for identifying when an armed conflict involving non-State actors is occurring, and what type of conflict it might be, can lead to a simple question being posed. That is whether an "armed conflict" is in existence. It is not that the categorization of the various types of armed conflict as international or noninternational does not matter. Such categorization is important, both in theory and practice. For example, it is clear that only limited treaty law and not all customary humanitarian law rules apply in a noninternational conflict.[45] However, it

---

42 J. Paul de B. Taillon, Hijacking and Hostages: Government Responses to Terrorism 110 (2002), and Iddo Netanyahu, Yoni's Last Battle: The Rescue at Entebbe, 1976, at 39 (2002).

43 9/11 Commission Report, *supra* note 27, at 330 (It was on the evening of September 11, 2001, that the U.S. decision was made to "punish not just the perpetrators of the attacks, but also those who harbored them.").

44 *Id.* at 195.

45 Prosecutor v. Tadić, Case No. IT-94-1-A, Decision On The Defence Motion For Interlocutory Appeal On Jurisdiction, ¶¶ 96–127 (Int'l Crim. Trib. for the Former Yugoslavia, 2 October 1995), http://www.icty.org/x/cases/tadic/acdec/en/51002.htm.

cannot be forgotten there are also significant similarities in how humanitarian law regulates hostilities regardless of how conflict is characterized. This can be seen in the areas of targeting, the prohibition of abuse or torture of detainees, and the obligation to humanely threat those in the custody of security officials (see Chapter 9, section 9.3.4.4, *The Practical Impact*, and section 9.5, *Addressing the Challenge*).[46] It is questions regarding targeting and what force is authorized at law that are likely to be foremost in the minds of commanders as they respond to an attack.

Nor are what appear at times to be artificial legal distinctions limited to the initial attack. They can also arise in terms of the State response. For example, during the "iconic" 1976 Entebbe counterterrorism raid, the engagement of Ugandan troops by Israeli forces made that conflict an international one. The weapons used and some of fighting (including suppressing Ugandan fire from the control tower) was entirely consistent with an armed conflict (see Chapter 10, section 10.3.2, *"Iconic" Counterterrorism: A Thunderbolt Strikes Entebbe*).[47] However, if no Ugandan troops had been present it is not clear that by applying a narrow interpretation of the *Prosecutor v. Tadić* test (i.e., duration of the violence) an armed conflict of any form would be considered to be in existence since the entire raid lasted only ninety minutes. That would be the theoretical result even though factually the threat of violence posed by a terrorist group defending the terminal could have been fundamentally the same as that presented by the Ugandan guards who were assisting the terrorists.

During such a raid, humanitarian law rules would likely need to be considered, although some form of "robust" law enforcement rules might also be successfully applied, depending upon the circumstances and the willingness of courts to stretch the theoretically restrictive principles governing the use of force under human rights law. However, human rights–based rules cannot always meet the operational need. At some point the theory has to match the facts on the ground if the mission is to succeed. As a result, it is likely a more liberal interpretation of what constitutes a noninternational armed conflict (i.e., a lower threshold) will continue to find greater support within the international legal community. This will better allow for a realistic assessment of the violence, with States still having the option of tailoring (i.e., restricting) their use of force to match the threat.

The broader acceptance of the existence of a noninternational armed conflict arising from non-State actor attacks may appear to threaten the historically restrictive recourse to war framework. However, in assessing most non-State actor attacks it is still important to conceptually separate them from State-on-State violence. This should help limit concerns that recognizing those attacks fall under Article 51 of the UN Charter will undermine the legal standards governing the recourse to war between States. The restrictive State-based standards has contributed significantly to limiting international armed conflict since World War II. There seems to be no reason why the approach developed during the last century to regulate inter-State violence needs to change. Indeed, it should be vigorously maintained. In

---

46 DINSTEIN, *supra* note 26, at ¶¶ 677–700, at 213–19 (for a discussion of some of the similarities of how humanitarian law applies to international and noninternational armed conflict centered on the principles of distinction and unnecessary suffering).

47 MUKI BETSER, SECRET SOLDIER 334 (1996) (where it is indicated that several rocket-propelled grenades and heavy machine-gun fire from an armored car was directed at the control tower to suppress fire from its occupants).

contrast, the unique nature of transnational non-State actors means criteria specific to that threat should be applied to help establish when States can legitimately respond to attacks from organized armed groups by applying humanitarian law rules. Logically, the closer the violence factually resembles a generic situation of armed conflict, the easier it is to also make an "armed attack" determination. A recognition of a right to respond to non-State attacks with military action grounded in armed conflict only represents an incremental change in the law governing the recourse to war, which seeks to apply international norms to a unique transnational threat. It is not a rejection or repudiation of the existing law governing the recourse to war applicable to inter-State conflict.

As a result, an issue to be addressed is whether the violence is sufficiently elevated to be equated to an armed attack as is contemplated by Article 51 of the UN Charter. In this regard, the requirement for a certain intensity of violence in order to have a noninternational armed conflict tracks well with the "scale and effects" standard set out in the *Military and Paramilitary Activities in and against Nicaragua Merits* decision in order to determine an armed attack has occurred rather than a mere frontier incident.[48] In this regard, a series of suicide bomber attacks;[49] hijacked planes being used as substitute cruise missiles (e.g., the 9/11 attacks);[50] or the violent seizure of an embassy by an organized armed group using mortars, explosive devices, and military tactics[51] all appear to involve uses of force that is significant enough to meet both the armed attack and armed conflict thresholds. That will be the case particularly if the *Abella v. Argentina* test,[52] the "situations of internal disturbances and tensions" threshold for law enforcement,[53] or the broader "totality of the circumstances" approach are also considered along with the *Prosecutor v. Tadić* criteria.[54] This does not mean that defensive responses outside the context of an armed conflict (i.e., law enforcement–based actions) such as extraterritorial action to protect nationals may never be justified under either Article 51 of the UN Charter or as a matter of customary law (see Chapter 10, section 10.2.1, *A State Responsibility*). However, it seems reasonable that if an armed conflict is established as a result of non-State actor violence, then the right for a State to act in self-defense under Article 51 of the UN Charter would be established as well.

---

[48] Military and Paramilitary Activities in and against Nicaragua Merits, ICJ Reports 103, ¶ 194 (1986). *See also* Christine Gray, International Law and the Use of Force 178–81 (3rd ed. 2008). *But see* Dinstein, *supra* note 38, ¶ 555, at 210 ("The assumption that a 'mere frontier incident' can have no 'scale and effects' is quite bothersome.").

[49] Dapo Akande & Thomas Liefländer, *Clarifying Necessity, Imminence, and Proportionality in the Law of Self-Defense*, 107 Am. J. Int'l L. 563, 567 (2013).

[50] 9/11 Commission Report, *supra* note 27, at 46 (where it is indicated the mission crew commander at NORAD on September 11, 2001, is reported to have remarked "that 'This is a new type of war.'" The Report notes, "He was, and is, right.").

[51] Mitchell Zuckoff, 13 Hours: The Inside Account of What Really Happened in Benghazi 211–20, 254–80 (2014) (outlining the violent attack on the embassy annex involving mortars, rocket propelled grenades, and coordinated assaults).

[52] Case 11.137, Juan Carlos Abella v. Argentina, Inter-Am. Y.B. on H.R. ¶ 1 (1997) (Commission report), http://www1.umn.edu/humanrts/cases/1997/argentina55-97a.html.

[53] AP II, art. 1(2).

[54] Prosecutor v. Tadić, Case No. IT-94-1-A, Decision On The Defence Motion For Interlocutory Appeal On Jurisdiction, ¶ 70 (Int'l Crim. Trib. for the Former Yugoslavia, 1995), http://www.icty.org/x/cases/tadic/acdec/en/51002.htm.

It is evident that a particular challenge, given the unsettled nature of this area of law, is reaching consensus on the threshold to be applied for determining when a noninternational armed conflict is in existence. However, if the legal theory cannot adequately address these threats, then the pressure will build to adopt new theories or at least interpretations of the law (see Chapter 8, section 8.5.3, *The "Self-Defense" Option*). From a practitioner's perspective, it is normally much less complicated to decide what the response needs to be. The criteria applied when assessing these non-State actor threats to determine if an armed conflict is in existence should include the degree of organization of the group (hierarchical, horizontal, cellular, or "hybrid"); how that group conducts its operations (e.g., the tactics used); the weapons used in an attack, including explosives such as grenades, IEDs, rocket-propelled grenades, suicide belts; and the type of State security forces, weapons, tactics, and so forth reasonably required to defeat the threat (e.g., missiles, mortars, heavy weapons, airpower—meaning that State armed forces have to engage in "combat"). Added to the criteria for conflict categorization are contextually important factors such as the nature of the target (embassy, warship, a representative of State power, or symbols of that State's commercial strength) and other indicia such as the kidnapping is for a political purpose.

This approach means that the State response to many contemporary non-State actor attacks will occur in the context of an armed conflict, while those not rising to the legal and factual level of such a conflict must be addressed as a human rights–based law enforcement matter. It is essential to emphasize that the categorization of the situation as an armed conflict does not demand a conduct of hostilities approach. The manner of the response remains in the discretion of the State, unless it has to be dealt with as a law enforcement matter (i.e., no armed conflict exists). In any event, States will frequently be subject to considerable domestic and international pressure to favor a law enforcement response either in whole or in part even when an armed conflict is considered to be in existence. This is particularly likely to occur in respect of incidents occurring within the State or on territory physically under its control. This can result in a bifurcated response, as was evident in the French reaction to the November 2015 Islamic State directed attacks in Paris. While declaring the attacks as an act of war the response within France (and more broadly in Europe) was a police and intelligence led domestic law enforcement one, albeit with the declaration of a state of emergency and a significant mobilization of police and military resources.[55] Externally, France immediately stepped up its aerial bombing against Islamic State campaign in Syria.[56] While this conclusion is not likely to be popular with either armed conflict or human rights exclusionists, it does more accurately reflect the practical requirement of having to apply both the law and policy to decisions about what security responses are appropriate to the situation.

---

[55] Aurelien Breeden, Kimiko de Freytas-Tamura & Katrin Bennhold, *Call to Arms in France Amid Hunt for Belgian Suspect in Paris Attacks*, INT'L N.Y. TIMES, Nov. 16, 2015, http://mobile.nytimes.com/2015/11/17/world/europe/paris-terror-attack.html?hp&action=click&pgtype=Homepage&clickSource=story-heading&module=span-ab-top-region*ion=top-news&WT.nav=top-news&referer= and *Paris Attacks: France Mobilises 115,000 Security Personnel*, BBC NEWS (Nov. 17, 2015), http://www.bbc.com/news/world-europe-34840699?utm_source=Sailthru&utm_medium=email&utm_campaign=New%20Campaign&utm_term=%2AMorning%20Brief.

[56] Breeden, de Freytas-Tamura & Bennhold, *supra* note 55.

14.3.2. COMMENCEMENT OF THE ARMED CONFLICT

A key issue with respect of international armed conflicts is determining when they actually begin. This must also be resolved in respect of non-State actor attacks since it is only at that point that the security forces can apply conduct of hostilities rules to defend against the threat. Again it is helpful to first look at this question from the perspective of the 9/11 hijackings. The attack by Al Qaeda is noteworthy in a number of respects, but in this context particularly because it eventually led to a Coalition of States invading Afghanistan and ousting the Taliban government. Did the international armed conflict, and with it the justification for using conduct of hostilities rules, start from the moment the Al Qaeda operatives embarked on their operation, or did there actually have to be violence caused by the terrorist group? Did the 9/11 attacks start when the terrorists boarded the aircraft, with the hijacking of the first aircraft, the attempt by passengers to take back control of one aircraft, or the crashing of the last plane into its target?[57] Does the point at which the conflict commences depend upon what the responding State does? If the North American Air Defense (NORAD) could have responded in time and shot down the planes, would the armed conflict have occurred at that point?[58] Another potential point of commencement for the international armed conflict also might be when the Coalition forces entered Afghanistan.

To a certain extent this issue is not dissimilar from the debate surrounding when an "armed attack" crystalizes such that a defensive response under Article 51 of the UN Charter is justified. Start points that rely on a State response appear problematic. There does seem to be something intuitively wrong if responsibility is placed on the responding State for bringing about an armed conflict by lawfully exercising its right to self-defense.[59] It should be the actions of the terrorist group that mark the commencement of the conflict, rather than have the defensive response by a State become a determinative factor. Waiting until one State invades another territory leaves unaddressed the issue of how the threatened State can respond to the non-State actor attack as it unfolds. This is an approach that sits most comfortably with the twentieth-century focus on inter-State armed conflict, but seems out of place given the nature of the violence leading up to such action.

For a 9/11-style attack, a conflict might possibly be considered to commence when the plane is hijacked. It is from that point that national leaders and security officials are put into a position of having to make decisions about how to respond to the security threats. More likely, with the hijacking of passenger jets historically having been considered a crime, the determination that a commercial airliner has become a weapon of war employed against

---

[57] 9/11 COMMISSION REPORT, *supra* note 27, at 32–33 (setting out the timings for the takeover of the planes by the hijackers, the defensive action by passengers on United Airlines Flight 93 causing it to crash in Pennsylvania, and the crash of the other hijacked aircraft into their targets).

[58] *Id.* at 40–46 (outlining the decision-making process concerning the issuance of a shoot-down order on 9/11).

[59] *But see* Mary Ellen O'Connell, *The Choice of Law Against Terrorism*, 4 NAT'L SEC. L. & POL'Y 343, 355–56 n.65 (2010) (where the US/UK response is suggested as the start of the post-9/11 conflict in Afghanistan), and INTERNATIONAL LAW ASSOCIATION, FINAL REPORT OF THE USE OF FORCE COMMITTEE: THE MEANING OF ARMED CONFLICT 31 (Aug. 2010), http://www.ila-hq.org/en/committees/index.cfm/cid/1022 (where the opinion adopted by this group of international lawyers was that an attack by North Korea that sank a South Korean ship did not create an armed conflict because there was no response by the attacked State). *See also* LUBELL, *supra* note 26, at 90 (setting out opposing views on this issue).

the State would likely be delayed until the intentions of the hijackers and true nature of the threat can be determined. This will not necessarily be an easy conclusion to reach, however, given the attacks of 9/11 there is an obvious precedent upon which to base the analysis.[60] In the case of hijackings that could be when it reasonably appears a plane seized by, or on behalf of an organized armed group might be crashed into objects of national or commercial importance. As a practical matter, it seems that such terrorist action would constitute not only an "armed attack" but also mark the simultaneous commencement of an armed conflict for which State security forces can, if necessary, apply force governed by humanitarian law.

As previously discussed, there is the question of what the status of the conflict would have been if the non-State actors involved in the 9/11 hijacking were not controlled or harbored and supported by another State. Having accepted that non-State actors can carry out "armed attacks" international law must then deal with the impact of recognizing that reality. Factually, nothing would change beyond not having a link to a State. The nature of the threat and the decisions confronting the security officials remain the same. In responding to such a threat an assessment would have to be made whether an armed conflict is in existence or if a law enforcement response is required. If a conduct of hostilities approach is needed to counter a non-State actor linked to a State, it would logically also be necessary where the terrorist group is acting on its own. In the context of "one-off" attacks by terrorist groups not linked to a State, a requirement that the violence take place over a certain period of time (i.e., "protracted armed violence[61]) before an armed conflict is considered to exists is problematic. It does not fully reflect the nature of the security challenge facing a State or the type of response required to defeat it. Considering that potentially high levels of civilian death and destruction can result from an armed attack, which may only last minutes or at most hours, a duration requirement in those circumstances seems out of place.

It is not that the duration requirement is not a useful criterion. However, as has been noted, the criteria articulated in the *Prosecutor v. Tadić* decision and other ICTY jurisprudence for establishing the existence of a noninternational armed conflict were developed by a court in the context of reviewing a situation of violence that was already well established (see Chapter 9, section 9.4.1, *Conflict Criteria and the Intensity of the Violence*). A criterion of duration appears to have little relevance to an unfolding "armed attack" where State security forces are required to make an immediate decision regarding acting in self-defense against a non-State actor. An approach toward conflict categorization developed from the unique perspective of one type of noninternational armed conflict cannot necessarily be readily or completely transferable to all contemporary security threats.

The seizure of foreign hostages in another State by organized armed groups (acting with political motives) raises similar questions regarding the nature of the violence and if categorized as such when an armed conflict begins. The decision on conflict categorization may not appear to be straightforward as the incident unfolds. For example, during the 1976

---

[60] Olivier Corten, The Law Against War: The Prohibition on the Use of Force in Contemporary International Law 60-61 (Christopher Sutcliffe trans., 2010) (where the author suggests police measures are more likely to be applied).

[61] Prosecutor v. Tadić, Case No. IT-94-1-A, Decision On The Defence Motion For Interlocutory Appeal On Jurisdiction, ¶ 70 (Int'l Crim. Trib. for the Former Yugoslavia, 1995), http://www.icty.org/x/cases/tadic/acdec/en/51002.htm.

Entebbe operation the involvement of the Ugandan government seems not to have taken place until after the hijacked plane entered that country. What was the status of that incident in terms of conflict categorization prior to the Ugandan engagement? It is not clear that the hijackers would have been considered to be members of an organized armed group. At that time the international legal community largely saw hijacking as a criminal act. At what point did the hijacking become an armed conflict? When Ugandan soldiers were tasked with guarding the air terminal, Israeli planes entered Ugandan airspace, or at the point the Israeli commandos fired on those guards as the rescue commenced on the ground? In effect, when would the rescue force be authorized to apply conduct of hostilities rules in seeking to free the hostages?

One approach is to consider that an armed conflict is in existence when it is reasonably likely military action associated with combat will be required to resolve the situation. The seizure of a plane by a terrorist group does not usually involve levels of violence associated with an armed conflict. The intensity of violence threshold may not actually become a factor until the State armed forces begin to plan for the rescue of its nationals. What may be required is a prospective application of the threshold criteria for an armed conflict with a non-State actor, including making an assessment of the threat of violence posed by the armed group, its degree of organization, and the nature of the State response required to free the hostages. Such an analysis places particular weight on the "means" test for determining an armed conflict is in existence,[62] the deployment of "regular military forces" criteria found in the ICRC *Commentary*,[63] and ultimately a "totality of the circumstances" approach (see Chapter 9, section 9.4.3, *The "Totality of the Circumstances" and Other Criteria*).

As was observed in *Operation Barras* in 2000, UK military forces planned for the use of significant military force in Sierra Leone because of the level of organization of the West Side Boys and the armaments they possessed. These two factors significantly elevated the potential for violence and the likelihood the operation would be carried out in the context of an armed conflict. The anticipated requirement to use elevated levels of violence was reflected in operational planning, which provided for British forces using helicopter gunships and indirect fire weapons as part of the assault on the armed group compounds (see Chapter 10, section 10.3.3, *Modern Counterterrorism: Sierra Leone*). Military force, weapons of war, and violence associated with combat were required to successfully carry out the release of the hostages. In this circumstance it would be reasonable to conclude the armed conflict actually commenced with the initial hostage taking carried out by an organized armed group capable of engaging in armed conflict levels of violence, but it was not evident until the totality of the circumstances were assessed.

Arguing that the armed conflict commences only when the State armed forces or the non-State actor actually start using weapons does provide a greater degree of certainty. Problematically, it also makes the determination of an armed conflict dependent upon the choice of tactics adopted by the parties to the hostilities. One issue with this approach is

---

[62] Geoffrey Corn, *Hamdan, Lebanon, and the Regulation of Hostilities: The Need to Recognize a Hybrid Category of Armed Conflict*, 40 VAND. J. TRANSNAT'L L. 295, 346 (2007).

[63] THE GENEVA CONVENTIONS OF 12 AUGUST 1949: COMMENTARY, GENEVA CONVENTION (I) FOR THE AMELIORATION OF THE CONDITION OF THE WOUNDED AND SICK IN ARMED FORCES IN THE FIELD Art. 3, ¶ 1A(2), at 49 (Jean S. Pictet ed., 1958).

that if the State initially limits its use of force or chooses to adopt less robust tactics in order to protect the hostages, there may not be a clear picture as the operation commences whether it is being undertaken in the context of an armed conflict or not. In contrast, using other criteria, such as the wrongful act of the well armed and politically motivated non-State actor (e.g., hostage taking or hijacking of an aircraft), as a factor to be considered in determining if an armed conflict exists provides a useful point at which to begin planning to ensure an appropriate response. The actual determination that there is an armed conflict would still have to be based on an assessment of all the available evidence.

As early a determination as possible will help ensure that appropriate rules of engagement are issued. Much depends upon the nature of the threat and the means reasonably required to deal with that threat. For reasons of accountability (including on the part of politicians) and to ensure the military personnel involved are certain about their authority to use force, it is important that the status of the violence and its point of commencement not be in doubt at the time force has to be used. Therefore, from an operational law perspective the decision regarding conflict characterization cannot wait for a postincident assessment once all the facts are established and there is time to consider factors such as the duration of the violence.

14.3.3. SEPARATING ARMED CONFLICT FROM LAW ENFORCEMENT

A different standard for establishing when a noninternational armed conflict is in existence remains necessary because of the uniqueness of non-State actor violence. Unlike State-based attacks, not all violent acts caused by non-State actors that prompt a defensive response by a State will take place in the context of an armed conflict. Given the connection between non-State groups and crime, it is particularly important that there is a way to distinguish between armed conflict and law enforcement situations. From a practical perspective the Additional Protocol II criterion highlighting that "situations of internal disturbance and tensions, such as riots, isolated and sporadic acts of violence" do not constitute an armed conflict can be a particularly useful distinguishing characteristic.[64] However, this is just one of a number of criteria that have been suggested as international law struggles with this issue (see Chapter 9, section 9.4, *Determining the Existence of a Noninternational Armed Conflict*).

Non-armed-conflict-related attacks could take place in various forms. As was indicated in *Prosecutor v. Boškoski*, "isolated acts of terrorism may not reach the threshold of armed conflict."[65] Therefore, even if a particular group has political motivation, it may not be either sufficiently organized or carry out violent acts of sufficient intensity to engage the armed conflict threshold. For example, if a non-State actor operating in one State crosses the border to carry out bank robberies in another State in order to fund its operations, those thefts would appear to be purely criminal matters and subject solely to a law enforcement–based response. Further, as has been demonstrated, States will react across the conflict spectrum to defend their nationals from criminal acts such as kidnapping. Such operations have

---

[64] AP II, art. 1(2).

[65] Prosecutor v. Boškoski, Case No. IT-04-82-T 90, Judgement ¶190 (Int'l Crim. Trib. for the Former Yugoslavia, July 10, 2008), http://www.icty.org/x/cases/boskoski_tarculovski/tjug/en/080710.pdf.

included intervening in failed or failing States and on the high seas to protect both lives and property (see Chapter 10, section 10.2.2, *Taking Action Across the Conflict Spectrum*).

In protecting persons on their own territory from cross-border criminal threats, the defensive response can extend to conducting operations within another State. Such threats can also lead to requests for assistance from a neighboring State. This can occur because of the porousness of many international borders and the existence of "ungoverned" spaces from where such gangs may operate. One 2012 example is when a Somali gang operating close to the Kenyan border entered that country to seize foreign hostages at a refugee camp. The group were possibly Al Shabaab sympathizers but were not identified as members of that organized armed group.[66] A subsequent joint Kenyan/Somalia operation rescued the hostages in Somalia.[67] Had the Somali government been unable or unwilling to police that region of the country, it would have been difficult to fault Kenyan authorities for entering Somalia unilaterally with a law enforcement response. Attacks on government authorities by transnational criminal organizations, such as those operating in Mexico, Columbia, and other regions of Central and South America, are another example of criminal activity that results in human rights–based defensive responses by States.

Given the degree of organization and armaments available to criminal gangs in many areas of the world, the human rights–based normative paradigm may face its greatest test in terms of its permitting the use of force necessary to counter these threats (see Chapter 12, section 12.3.4, *Strictly Necessary, Proportionate, and Overwhelming*). What this means is that notwithstanding the focus that has been placed in the post-9/11 period on the exercise of State self-defense through the use of drones against attacks or threatened attacks by terrorist groups, it is law enforcement that can govern many defensive reactions by States. The result is that, with many existing threats posed by non-State actors never rising to the level of an armed conflict, the potential of having to apply a law enforcement approach must be considered in military planning and operations. This includes when dealing with non-State actor violence within the borders of the threatened State or when occurring against interests and nationals of that State in another territory.

14.4. Applying a Human Rights–Based Law Enforcement Approach

14.4.1. AS A MATTER OF LAW

For a variety of reasons law enforcement, and the human rights normative paradigm upon which it is based, have become a defining feature of many contemporary security operations. The challenge for military commanders and their legal advisors is to determine when a policing response is required by law or is the preferred State response for meeting the goal of maintaining order. Simply stated, when an incident does not meet the armed conflict threshold, then human rights law is the applicable normative regime. As has been noted, it is also clear historically that the application of human rights–based law enforcement

---

[66] James Macharia, *Gunmen Kidnap Aid Workers from Kenya Camp, Driver Killed*, REUTERS (June 29, 2012), http://www.reuters.com/article/2012/06/29/us-kenya-kidnap-idUSBRE85S0KN20120629.

[67] Jessica Murphy, *Kenyan, Somali Troops Rescue Kidnapped Aid Workers, Including Two Canadians*, THE TORONTO SUN (July 2, 2012, 05:27 PM), http://www.torontosun.com/2012/07/02/canadian-aid-worker-rescued-in-somalia-army.

represents the default approach for States domestically even when there is an armed conflict (see Chapter 4, section 4.4.3, *A Reluctant Application of Humanitarian Law*). Further, even a cursory look at State practice in dealing with non-State actors reveals a bias toward law enforcement in a wide variety of international operational scenarios. As the following analysis demonstrates, assessed strategically, a strong argument can be made that whether at law or in practice the human rights–based law enforcement approach constitutes a dominate normative approach governing the use of force applied to counter contemporary non-State actor threats.

Examples of where a law enforcement approach is applied domestically as a legal requirement include States dealing with international criminal organizations capable of threatening considerable violence, such as the Mexican drug cartels, Columbian *bandas criminales*, Brazilian gangs in the Flavelas, the Jamaican "Shower Posse," and the Central American Maras. Although these international criminal organizations are organized and extremely violent, they remain profit motivated. As a result, the unstated but logical requirement that these groups operate for a political purpose before an armed conflict is recognized serves as a significant brake on the State response to these threats in terms of the potential levels of force and means used. This in turn is intended to provide a greater level of protection to the civilian population. The key issue to be addressed is whether, at some point, the activities of the criminal gang could rise to the level of an insurgency because the group starts to interfere with State governance (see Chapter 6, section 6.3.2, *Challenging the State*). However, even where they may begin to resemble an insurgency and take on the attributes of an armed conflict, the State reaction is likely to remain law enforcement focused.

The requirement for some form of "political" criterion to be met can be contrasted with an approach that seeks to extend the humanitarian protection of humanitarian law by claiming that an armed conflict is in existence based solely on the intensity of violence and the involvement of sufficiently organized armed groups. In effect, there are two competing humanitarian goals. The obligation placed on the State to apply the more limiting law enforcement framework, which is an integral part of the claim there is no armed conflict, is pitted against the opportunity to insert what are seen as more protective provisions for civilians associated with international humanitarian law.[68] This humanitarian protection is not the only advantage to be gained in admitting an armed conflict is occurring. It has been suggested that the application of international humanitarian law also facilitates the holding of "both state and non-state actors accountable for their violations of the laws and customs of war."[69] It has also been rightly pointed out that it can be problematic if a threat posed by a non-State actor is treated over the long term as a permanent emergency. This can allow that emergency to become a cloaking device for low-intensity conflict.[70]

Notwithstanding the positive impact of extending humanitarian protection to such violence, insufficient discussion has occurred regarding the advantages to be gained by privileging the more restrictive human rights law framework for controlling the State use

---

[68] Felicity Szesnat & Annie R. Bird, *Columbia*, *in* ELIZABETH WILMHURST ED., INTERNATIONAL LAW AND THE CLASSIFICATION OF CONFLICTS 203, 226 (2012).

[69] OREN GROSS & FIONNUALA NÍ AOLÁIN, LAW IN TIME OF CRISIS: EMERGENCY POWERS IN THEORY AND PRACTICE 362 (2006).

[70] *Id.* at 360.

of force. This is particularly evident against groups that are fundamentally criminal gangs rather than political actors. Nor is it clear that permitting a prolonged period of emergency and civil strife to be dealt with under a human rights law framework is necessarily a more problematic outcome than applying an armed conflict paradigm. As is evident in the situation of Columbia, some armed conflicts, like emergencies, can also last decades.[71] The same can be said for the Northern Ireland "Troubles," which lasted nearly thirty years (see Chapter 13, section 13.2.1, *The "Troubles"*). Further, seeking a declaration there is an armed conflict also appears to place inordinate reliance on the effectiveness of international systems of criminal and human rights accountability over that provided by domestic courts.[72] It is not obvious that these international systems of accountability are always more effective than their domestic counterparts. States and the international community should always be reticent to cross the threshold into armed conflict in respect of what is essentially economically driven violence.

The fact that counterinsurgency doctrine prefers a "police primacy" approach is another powerful indicator that policing should be favored over a purely military response in situations involving transnational criminal gangs. The adoption of a "war" paradigm to deal with what is primarily criminally motivated activity should only be considered when it is clear that what is being threatened is the ability of the State to govern and more than a policing response is required for the State to succeed. In other words, the situation is one of an armed conflict that cannot be won without such a response. The human rights law paradigm can often be interpreted sufficiently broadly to handle many of the elevated forms of violence threatened by transnational criminal organizations. There is much to be gained in reinforcing the role of policing in terms of avoiding the death and destruction that can result from armed conflict.

As has been seen with numerous contemporary piracy and kidnapping incidents, a law enforcement approach may also have to be applied outside of a State as a matter of law when criminal gangs threaten that State's nationals or interests (see Chapter 10, section 10.3.5, *Rescues and Law Enforcement*). In many circumstances the acting State will be exercising its right to act defensively to protect nationals. When what it occurring is only criminal activity, it must be tackled with a policing approach regardless of whether the territorial State or foreign authorities carry out the enforcement action. It also does not matter if a responding State gains entry to territory with the consent of that sovereign State or such intervention is made unilaterally. The basis for entry does not alter the nature of the threat or the legal paradigm under which it must be dealt with. Given the degree of organization of these criminal gangs, their armaments, and the geographical challenges presented in reaching and then operating in "ungoverned spaces," conducting these operations can make the application of a human rights law–based approach exceedingly challenging. In addition, operations to rescue the hostages can require a use of force that significantly narrows the normative gap between the conduct of hostilities and law enforcement (see Chapter 11, section 12.2.2.1, *Unique Challenges and Responses*). However, the key point is that the matter

[71] Szesnat & Bird, *supra* note 68, at 204–14 (outlining the history of the conflict in Columbia from 1960s to present day).

[72] *But see id.* at 226 (where it is suggested "there is no reason why members of these groups cannot be held to account for any alleged crimes they may have committed during that conflict, within the domestic system of the affected State.").

remains a policing one whether carried out by either military forces or law enforcement personnel.

The performance of law enforcement duties by military forces is not confined to territory controlled by States. Naval forces regularly perform constabulary duties on the high seas (i.e., outside the territorial waters of States). As the term "constabulary" suggests, these enforcement matters such as countering piracy, drug smuggling, and trafficking in persons; immigration control; fisheries enforcement; and environmental protection involve law enforcement. Canada's 1995 fisheries dispute with Spain saw a Spanish fishing trawler impounded and its captain charged in a Canadian court. This incident fundamentally remained a law enforcement matter, although the potential existed for the two countries to become involved in the defense of national interests, with both Spain and Canada dispatching armed vessels to the area of the dispute.[73] This particular clash of State interests provides an example of when enforcing domestic fisheries laws could have theoretically led to a situation of "armed conflict" between two States. One of those States might have been put in a position of applying self-defense principles regulating the recourse to war in order to limit any outbreak or expansion of a conflict should the warships had confronted one another (see Chapter 3, section 3.2, *The Continuing Application of Self-Defense Principles*). This scenario highlights once again that these bodies of law (both domestic and international) do not necessarily operate in isolation. As a result, legal advisors may have to be prepared to consider a possible application of the whole range of international law even when initially operating within a law enforcement paradigm.

The requirement to apply a policing approach has also been seen to frequently arise during armed conflict simply because of the instability and crime that accompanies politically motivated insurgency and terrorism. That criminal activity can seriously undermine efforts to promote security and can be "a funding mechanism for a variety of violent nonstate armed groups. Indeed, the rise of criminal organizations is part of a much broader global phenomenon in which violent nonstate armed groups are challenging the Westphalian state."[74] Operations to maintain law and order can take place in a variety of situations, including combating crime as an occupying power, dealing with local civilians while conducting counterinsurgency missions, or carrying out rescue operations to release hostages held by criminal gangs in "ungoverned spaces" being fought over by States and their insurgent challengers. The important issue for military forces is not the need to address criminal activity but rather how best to do it. At the forefront of suggested action is the requirement for States to develop a broad strategy toward combatting organized crime,[75] including integrating law enforcement specific intelligence-gathering capability.[76]

Further, notwithstanding the focus on drones and targeted killing, during hostilities most supporters of armed groups must be dealt with through law enforcement. Those supporters can entail a significant portion of the population. In this regard, British officials

---

[73] Mark Clayton, *War over Fish Sticks? Canada, Spain Spar*, CHRISTIAN SCIENCE MONITOR, Mar. 13, 1995, http://www.csmonitor.com/1995/0313/13061.html.

[74] PAUL WILLIAMS, CRIMINALS, MILITIA, AND INSURGENTS: ORGANIZED CRIME IN IRAQ 269 (2009).

[75] *Id.* at 263.

[76] *Id.* at 263–65. *See also* RANDALL WILSON, BLUE FISH IN A DARK SEA: POLICE INTELLIGENCE IN A COUNTER-INSURGENCY 147–51 (2013).

in Northern Ireland identified the criteria for early success by an insurgent movement as having "the active support of five to ten per cent of the population and the passive support of at least twenty-five per cent."[77] A key issue is whether the persons providing support to armed conflict are members of an organized armed group or, more narrowly, persons taking a direct part in hostilities. This raises the issue of their legal status as possible targets under humanitarian law. However, overall only a small portion of the support provided to an armed group in an insurgency qualifies as direct participation in hostilities. The overwhelming majority of those supporters have to be dealt with by both military and police security authorities by means of a law enforcement response.

Finally, given the interconnected nature of today's world, particularly because of the Internet, a substantial amount of that support, financial or otherwise, can come from countries where there are no hostilities being conducted. There does remain the potential for some persons outside the warring State to be direct participants in hostilities. This can raise difficult questions concerning the geography of war and the authority of States to take military action. Such direct participation can occur in the realm of "real time" intelligence gathering over social media networks[78] or, as was seen in the 2008 Mumbai attack, by means of the exercise of command and control from Pakistan by telephone.[79] However, in most cases this external support would not constitute direct participation, and as a result a policing response would also be required to deal with those supporters. As each of these examples has reinforced, the legal requirement to apply human rights–based law enforcement is an integral part of contemporary security operations.

14.4.2. A POLICY CHOICE BY STATES

Even where an attack by a non-State actor meets the threshold of an armed conflict or the levels of ongoing internal violence are such that an insurgency is occurring, the State may still choose as a matter of policy to apply a human rights–based law enforcement approach. The idea that a State may make a choice in the normative framework under which it may be held accountable, and by extension members of its security forces as well, is reflected in *Hassan v. The United Kingdom*.[80] A State has some control over what body of law is to be applied in assessing its actions, although there would still be a requirement for an armed conflict to be established before humanitarian law can be applied as a matter of law. In placing the onus on the State to claim humanitarian law applies, the *Hassan* decision also highlights that human rights law remains the default normative regime governing the use of

---

77 James Pettifer, The Kosova Liberation Army: Underground War to Balkan Insurgency, 1948–2001, at 137 (2013).

78 Graeme Smith, *How Social Media Users Are Helping NATO Fight Gadhafi in Libya*, Globe and Mail, June 14, 2011, http://www.theglobeandmail.com/news/world/how-social-media-users-are-helping-nato-fight-gadhafi-in-libya/article583325/ (for an example where passing of the location of potential targets to a NATO Twitter account during the 2011 Libya campaign could be viewed as direct participation in hostilities).

79 Cathy Scott-Clark & Adrian Levy, The Siege: 68 Hours Inside the Taj Hotel 119–20, 153–57, 266–67 (2013).

80 Hassan v. The United Kingdom, Application No. 29750/09, Eur. Ct. H. R. ¶ 107 (2014), http://hudoc.echr.coe.int/sites/eng/pages/search.aspx?i=001-146501 (The court could not assume a State "intends to modify the commitments which it has undertaken by ratifying the Convention in the absence of a clear indication to that effect.").

force. What is evident in respect of either internal or international conflict is that a decision by the State to apply human rights law as opposed to humanitarian law to govern the use of force is not likely to questioned by a court or tribunal.

The classic example where a State chose domestically to apply a human rights–based policing approach was in Northern Ireland during the "Troubles." Although the indicia of an insurgency were present, the United Kingdom consistently treated the activities of Irish Republican Army and the other "terrorist" groups as a law enforcement matter (see Chapter 13, section 13.2.1, *The "Troubles"*). In such a situation, a military commander must view that decision as direction provided by the government. As a result, that commander has no discretion to apply a hostilities-based humanitarian law approach toward the use of force, unless the government subsequently acknowledges the existence of an armed conflict against established criteria and then gives direction to apply that normative framework to its operations.

Adopting a policy approach favoring law enforcement may not be effective in all insurgencies. Northern Ireland was unique in that a majority of its population supported the government. There was also an effective police force, which adapted itself to the changing security environment as the conflict unfolded, as well as military forces that relinquished their initial leading role. In the context of this counterinsurgency effort there was clearly an emphasis on "police primacy." Importantly, there was also a robust and effective justice system to back up the policing and security efforts. With law enforcement capacity ultimately being linked to governance, it is clear that the security forces of a number of States are not capable of performing in as effective fashion. As has been evident in Afghanistan and Iraq during the post-9/11 period, a key weakness in national and Coalition counterinsurgency efforts has often been ineffective and corrupt national police forces.[81] Weak policing is inevitably reflective of broader problems in a justice system and the overall governance exhibited by a State. It also has a direct impact on the efficacy of a law enforcement–based approach.

The degree to which countries default to a policing response is demonstrated in the German approach toward dealing with hijacked aircraft, where the military was tasked to support the police (see Chapter 13, section 13.5, *The Shoot-Down of Hijacked Aircraft*). The Australian "call out" framework for military forces provides a similar basis for reacting to this threat. Although the latter national approach has been interpreted to permit the application of humanitarian law rules, the reliance on an aid in the civil defense authority highlights that action within or at the approaches to national territory presents a conceptually different situation. The closer the connection to national territory the more likely it is that the deployment of military units may be subordinated to domestic policing. Even in terms of the American and Canadian adoption of a self-defense-based reaction to the threat of hijacked aircraft (i.e., NORAD), the prevalence of criminal or inattentively piloted aircraft inevitably results in a bias toward viewing threats, at least initially, as ones being amenable

---

[81] Austin Long, *The Police in Afghanistan, 2002–11*, *in* POLICING INSURGENCIES: COPS AS COUNTERINSURGENTS 203, 205–207 (C. Christine Fair & Sumit Ganguly eds., 2014), and Matt Sherman & Josh Paul, *The Role of Police in Counterinsurgency Operations in Iraq, 2003–6*, *in id.* at 229–40. *See also* LEDWIDGE, *supra* note 8, at 32–33 and 71 (for a discussion of the Iraqi and Afghan police forces).

to a law enforcement approach. Further, it cannot be forgotten that the attackers of 9/11 boarded the aircraft at U.S. airports after having obtained entry into that country. The suggestion that "[t]he failure of September 11 was one of law enforcement, not of criminal law" can be extended to the position it was not a failure of humanitarian or other international law either.[82] The challenge following that attack has been how to reconcile the interface between domestic law enforcement and international humanitarian law in order to properly address this type of threat.

A government decision to apply human rights–based law enforcement to counter violence occurring in situations that qualify as armed conflicts can be controversial. It potentially exposes military personnel to greater personal risk and causes significant operational and tactical challenges if what the security forces are facing is an organized armed group that is clearly engaged in warfare. This challenge is particularly acute where, in order to defeat the terrorist or insurgent threat, military tactics and the use of heavier weaponry (i.e., mortars, artillery, aerial bombing) have to be considered. Their use does not fit comfortably within the ordinary constraints of a legal framework that is human rights based. It is in this context that a realistic professional assessment needs to be made as to whether the restrictions arising from the adoption of a law enforcement approach are too operationally limiting. If the conclusion is that is the case, then a government acknowledgment of the operational context within which force issued should be sought (see Chapter 13, section 13.2.2, *Shooting to Kill*). However, given the sensitivity with which many governments view any admission there is an armed conflict occurring on their territory, or that criminal activity has reached the level of such a conflict, the likely result is that a more restricted use of force will continue to be required of security forces than might otherwise be permitted under humanitarian law.

The determination by a State as to whether an armed conflict will be treated as such is a political decision that has significant operational repercussions. This was graphically illustrated in Columbia where, during the government of President Uribe, the conflict with the Revolutionary Armed Forces of Colombia (FARC) was viewed as a fight against terrorism rather than an armed conflict.[83] A similar classification challenge has also occurred in Nigeria with respect to its battle with Boko Haram.[84] The potential complexity of the task facing security forces in seeking to reconcile the application of human rights law and humanitarian law in an operating environment where the political leadership has struggled to classify the conflict has been particularly evident in Columbia. In that country a special Advisory Board was eventually established to classify the situation with respect to each non-State armed group and to decide "whether Red [humanitarian law–based, with a mixture of human rights rules], or Blue [international human rights law–based] cards are to be

---

[82] Gross & Ní Aoláin, *supra* note 69, at 378 (quoting Kent Roach in September 11: Consequences for Canada 23 (2003) in the context of 9/11 not being a failure of criminal law but rather enforcement).

[83] Szesnat & Bird, *supra* note 68, at 214–15. *But see also id.* at 215 (where it is also noted that the Colombian Supreme Court ruled in *Gian Carlo Suarez*, Supreme Court of Columbia, Radicado No. 32.22, 59 (21 September 2009) that the conflict had been underway for forty years).

[84] *Nigeria: Read Full Text of Jonathan's Declaration of Emergency Rule*, Vanguard, May 14, 2013, http://www.vanguardngr.com/2013/05/read-full-text-of-jonathans-declaration-of-emergency-rule/ (where the president of Nigeria refers to Boko Haram as participating in a rebellion and insurgency, but labels them as "terrorists" in declaring an Emergency).

applied when engaging with each non-state armed group."[85] The utility of such a process is logically linked to timely determinations, and it has been noted, "the Board runs the risk of being out of touch with current developments, which may potentially lead to the wrong rules being applied."[86]

In applying the various bodies of law there is also a possibility that the interpretation of their governing principles will become blended. This is an area where care must be exercised. While they must be holistically applied, each body of law brings with it unique features that are tailored to the operational threats being faced by security forces and the risks presented to a civilian population. In terms of both State and individual accountability, it is essential that there be clarity as to what normative regimes are governing the use of force and how they relate to one another even when a human rights–based approach is incorporated as a matter of policy. Lawyers must be clear about what each body of law states prior to advising on how its various parts should be holistically applied. Such clarity is even more important in a security environment where "the line that separates policy and politics from law has grown blurry."[87] Lawyers often engage in legal policy decision-making,[88] but policy decisions on what body of law is to be applied should be made more broadly by the executive of government or the chain of command. Not the least of these unique features of international law is that armed conflict normally permits a broader use of force than restrictive interpretations of State self-defense or human rights law ordinarily allow. The optimum approach is for there to be clear direction from the government as to whether it considers an armed conflict is in existence, an indication if a law enforcement approach is to be uniquely applied as a matter of policy, and how any conduct of hostilities action is to be reconciled with State self-defense or human rights law–based limitations (e.g., police primacy, privileging capture).

14.4.3. DISCRETION EXERCISED BY THE "CHAIN OF COMMAND"

The degree to which a requirement to restrictively apply force is mandated by law or by the chain of command is also an important consideration. In a military context this "policy" application is often referred to as the "exercise of command discretion." In applying an exclusionary interpretive model, military legal advisors will often refer to humanitarian law providing the legal basis for the use of force, but that military commanders retain considerable discretion to apply law enforcement–type restrictions (e.g., capture rather than kill). A problem with this law/military discretion model, which is often discussed in the context of lower levels of commanders having almost unfettered discretion, is that it does not fully account for the impact of strategic or operational level restrictions on the use of force. For example, under ROE doctrine, commanders may have a discretionary authority to further restrict the authority to use force that they are provided. In hierarchical organizations like the military, the ability to exercise discretion at lower levels of command may be

[85] Szesnat & Bird, *supra* note 68, at 230.

[86] *Id.*

[87] SAVAGE, *supra* note 20, at 38.

[88] *Id.* at 37-38 (for a discussion of legal policy decision making where government lawyers "decide which interpretations of the law are reasonably available and which are not.").

significantly proscribed by higher level commanders. These restrictions can have considerable impact on the tactics and types of force used at the tactical level.

For military personnel, policy restrictions made at a political level or at a higher command level can have the force of law given the requirement to obey lawful commands. For example, in the 2013 *Drone Speech* by President Obama, who is both a political leader and the commander in chief of the U.S. armed forces, restrictions on the use of force were issued from the strategic level. One requirement was that there be a "near certainty" that no civilian casualties will result prior to conducting an attack.[89] This near certainty standard sets a much higher standard than the legal test of having a "reasonable belief" (also called an "honest belief")[90] that such casualties might occur.[91] Such command-directed restrictions are not limited to presidential directives. As was seen in the context of the ROE issued for the November 2004 Fallujah battle, the exercise of discretion regarding what legal principles will be applied to control force is often made at higher levels of military command.[92] Consistent with the legal/political interface that occurs at the strategic level, it might not be immediately evident to subordinate commanders whether the restrictions on the use of force have been imposed because of legal or policy reasons. In this context, if nationally directed ROE reflect the application of State self-defense principles, then the discretion available at the tactical level to apply the humanitarian law authority to use force to its full effect may be limited for what are ultimately treated as "legal reasons" (e.g., lawful commands).

A nationally directed restriction may also insert human rights–based obligations directly into the ROE. This can be seen in Rule 2 of the Columbian "Red Card," which directs that "[i]f the circumstances permit, demobilization and capture are favored over deaths in combat."[93] As one analysis of this Rule notes, "rule 2 reflects a policy decision, which is probably due to the influence of international human rights law."[94] These ROE-based restrictions effectively mandate at least a portion of the human rights–based law enforcement paradigm be applied in armed conflict situations. Where ROE restrictions are applied at the operational level of war, the freedom by a subordinate to apply what might ordinarily be action permitted by humanitarian law will be similarly affected. An example of this is the decision by ISAF commanders to restrict the use of airpower in a counterinsurgency context.[95]

---

89 *Obama's Speech on Drone Policy*, N.Y. Times, May 23, 2013 [hereinafter the *Drone Policy*], http://www.nytimes.com/2013/05/24/us/politics/transcript-of-obamas-speech-on-drone-policy.html.

90 Prosecutor v. Galić, ICTY, IT-98-29-T, Judgement and Opinion, ¶¶ 50, 51, and 55 (Int'l Crim. Trib. for the Former Yugoslavia, 2003), http://www.icty.org/x/cases/galic/tjug/en/gal-tj031205e.pdf.

91 McCann v. United Kingdom, Application No. 18984/91, Eur. Ct. H. R. ¶ 200 (1995), http://hudoc.echr.coe.int/sites/eng/pages/search.aspx?i=001-57943 ("where it is based on an honest belief which is perceived, for good reasons, to be valid at the time but which subsequently turns out to be mistaken. To hold otherwise would be to impose an unrealistic burden on the State and its law-enforcement personnel in the execution of their duty, perhaps to the detriment of their lives and those of others.").

92 Dick Camp, Operation Phantom Fury: The Assault and Capture of Fallujah, Iraq 150–51 (2009)(for ROE issued at the Divisional level).

93 Szesnat & Bird, *supra* note 68, at 229.

94 *Id.*

95 ISAF Commander's Tactical Directive (July 6, 2009), ISAF Headquarters [hereinafter the *Tactical Directive*], http://www.nato.int/isaf/docu/official_texts/Tactical_Directive_090706.pdf.

14.4.4. A LAW ENFORCEMENT RESPONSE TO ATTACKS AT HOME AND ABROAD

The privileging of a law enforcement approach is evident when States are confronted with what are frequently treated as individual "one-off" attacks by non-State actors. While at some point these assaults may be viewed as part of an overall armed conflict or constitute an armed conflict on its own, their geographical and temporal separation often results in their being considered, at least initially, on an individual basis. Further, at the time an attack begins to unfold it may not be possible with any certainty to attribute responsibility to a particular non-State actor or a controlling or harboring State. Such attacks can take the form of hijacked aircraft, fidayeen-style attacks from external groups, or "homegrown" small cell and individual jihadist motivated attacks.

It is significant that attacks can occur from the sea, in the air, or on land. The complexity of these attacks is evident in their sometimes being physically generated externally but with the final phase taking place from locations within the State (e.g., the 9/11 attack, and the Mumbai, Westgate Mall incidents). In this regard, the externally generated attacks may look the same as those undertaken by "homegrown" cells. They can involve a mixed group of external and internal participants as occurred in Paris in 2015.[96] The State response may be viewed as entirely an internal policing matter or as being subject to regulation by self-defense principles with an appropriate application of the humanitarian law or human rights law normative frameworks. For most States joining the international coalition in 2014 against the Islamic State, it was attacks such as these that were seen as the greatest direct security threat.[97] There had been mixed signals as to how concrete that threat might be.[98] However, attacks occurring in Tunisia, France, and Kuwait in June 2015;[99] the destruction of a Russian airliner over the Sinai widely attributed to an Islamic State linked group[100] and a suicide attack in Turkey in October;[101] as well as attacks Lebanon[102] and France[103]

---

[96] Breeden, de Freytas-Tamura & Bennhold, *supra* note 55 (in respect of the November 2015 Islamic State directed attacks in Paris the French President, Francois Hollande, "said the attacks had been "planned in Syria, organized in Belgium, perpetrated on our soil with French complicity.").

[97] *Iraq Crisis: Cameron Warns of IS Threat to UK*, BBC NEWS (Aug. 17, 2014), http://www.bbc.com/news/uk-28823281, and *Written testimony of DHS Secretary Jeh Johnson for a House Committee on Homeland Security hearing titled "Worldwide Threats to the Homeland,"* Homeland Security (Sept. 17, 2014), http://www.dhs.gov/news/2014/09/17/written-testimony-dhs-secretary-jeh-johnson-house-committee-homeland-security.

[98] Mark Mazetti & Helene Cooper, *U.S. Officials and Experts at Odds on Threat Posed by ISIS*, N.Y. TIMES, Aug. 24, 2014, http://www.nytimes.com/2014/08/23/us/politics/us-isnt-sure-just-how-much-to-fear-isis.html?_r=0, Richard Norton-Taylor, *Islamist Terror Threat to West Blown Out of Proportion—Former MI6 Chief*, THE GUARDIAN, July 7, 2014, http://www.theguardian.com/uk-news/2014/jul/07/islamist-terror-threat-out-proportion-former-mi6-chief-richard-dearlove.

[99] Greg Miller, *Three Attacks Deepen Fears About Islamic State's Global Reach*, WASH. POST., June 27, 2015, http://www.washingtonpost.com/world/national-security/three-attacks-deepen-fears-about-islamic-states-reach/2015/06/27/04b82dd2-1c1d-11e5-ab92-c75ae6ab94b5_story.html.

[100] *Russia Plane Crash: 'Terror Act' Downed A321 Over Egypt's Sinai*, BBC News (Nov. 17, 2015), http://www.bbc.com/news/world-europe-34840943.

[101] *Ankara Blasts: Bombers 'Linked to Islamic State'*, BBC NEWS (Oct. 15, 2015), http://www.bbc.com/news/world-europe-34533522.

[102] Kareen Shaheen, *ISIS Claims Responsibility as Suicide Bombers Kill Dozens in Beirut*, THE GUARDIAN (Nov. 12, 2015, 20:02 GMT), http://www.theguardian.com/world/2015/nov/12/beirut-bombings-kill-at-least-20-lebanon.

[103] *Three Hours of Terror in Paris, Moment by Moment*, N.Y. TIMES, Nov. 15, 2015, http://www.nytimes.com/interactive/2015/11/13/world/europe/100000004036728.app.html?nytapp=ipad&_r=0.

in November 2015 demonstrate the global security impact of the Islamic State jihadist ideology.[104]

Even where such attacks are viewed as rising to the level of an armed conflict, they will normally happen so quickly and unexpectedly that the first responders are likely to be police units. As a matter of training, these law enforcement personnel will first attempt to apply force based on human rights law to attempt to contain the violence. This is particularly likely to occur with locally planned and unaffiliated but still jihadist-motivated attacks by individuals or small groups. It is what happened in October 2014 with attacks on Canadian military personnel and the forced entry into the Parliament buildings in Ottawa by a lone gunman.[105] This is not to suggest that resolution of the incidents involving attacks by small groups armed with military weapons, grenades, or other explosives or by even suicide bombers acting individually will necessarily fall within the category of "ordinary" policing. However, a significant change since the late 1970s has been the improved capability of police antiterrorism units in a number of States, which represents the positive aspect of the militarization of police. Assistance to policing provided by military units, such as Special Forces, and paramilitary units in a number of countries facilitates many of these more isolated incidents being treated as law enforcement matters. When acting in this capacity those military and paramilitary units may operate using law enforcement (i.e., human rights law) based ROE.

The challenge of dealing with groups capable of using considerable violence is not limited to jihadists or other terrorist groups. As has been experienced in Central and South America, drug cartels can field well-armed and tactically capable groups organized along military lines (see Chapter 6, section 6.3.1, *The Nature of the Threat*). The issue of whether a law enforcement response will be effective can depend on the capacity of State authorities to contain the incident and ultimately affect an arrest or remove the threat through the use of force. The ability to control territory is intimately linked to effective governance (i.e., a political issue). As a result, even the exceptionally violent 2009 assault on the safe house of Arturo Beltrán Leyva by Mexican Marines can still be viewed as being subject to a human rights–based normative framework.[106] Similarly, the ability of Brazilian police to operate within the Rio de Janeiro Flavelas, albeit often with military backing, points to the maintenance of a policing response to the criminal activities and threats posed by drug gangs.[107]

What is problematic is when criminal gangs are in a position to deny access to certain areas by State security personnel seeking to conduct law enforcement. The issue of area

---

104 S.C. Res. 2249, UN Doc. S/RES/2249 *Preamble* (Nov. 20, 2015)(where the global nature of the Islamic State threat is recognized).

105 Josh Wingrove, Steven Chase, Bill Curry, & Jill Mahoney, *Attack on Parliament Hill: PM Cites Terrorist Motive*, The Globe and Mail, Oct. 24, 2014, http://www.theglobeandmail.com/news/national/parliament-shooting/article21217602/?page=all, and Steven Chase & Tu Thanh Ha, *Suspect in Quebec Attack Was Arrested by RCMP This Summer*, The Globe and Mail, Oct. 21, 2014, http://www.theglobeandmail.com/news/national/quebec-hit-and-run/article21187200/.

106 Ioan Grillo, El Narco: Inside Mexico's Criminal Insurgency 126 (2011) (Beltrán Leyva and five of his aides were killed in the assault).

107 *Brazil Forces Occupy Favela ahead of World Cup*, BBC News (Mar. 30, 2014), http://www.bbc.com/news/world-latin-america-26809732 ("when more than 1,000 police backed by soldiers and armoured vehicles took up position in Mare, one of Rio largest shantytowns.").

denial is inevitably intertwined with the levels and frequency of the violence and the capabilities of the armed groups. One example which may point more to a situation resembling armed conflict than one of simple law enforcement was the 2010 fighting between Jamaican government forces and a drug gang known as the "Shower Posse" and its allies in the heavily defended Tivoli Gardens "Garrison." In that incident it is reported security forces used mortars, an indirect fire weapon, against the organized group.[108] Yet there must be a rational connection between the security situation (i.e., an armed conflict involving an organized armed group and a sufficient intensity of violence, as well as a political purpose) and the methods used before force can be justified as an armed conflict response. In these situations the uncertainty regarding the thresholds to be applied in categorizing noninternational armed conflict, particularly with a criminal organization, and the historic State inclination to avoid legitimizing non-State actors, will likely result in a bias toward an incident being viewed as a law enforcement matter.

When military units are called out to assist the police authorities, such as occurred in 2008 in Mumbai, it may never be acknowledged or fully resolved whether the violence reaches the level of an armed conflict. Such a determination may occur some time after the incident or not at all. Until a determination is made, an incident will ordinarily be dealt with under a law enforcement paradigm, even if it begins to take on the attributes of armed conflict. As one commentator has noted in respect of the 2010 Jamaican violence, "[t]he distinction between war and crime, and between domestic and international affairs, had effectively disappeared in Kingston, just as it had in Mumbai."[109] It is the exceptional nature of these threats, which often involve the use of explosive devices, and the elevated risk to the civilian population that lead to the acceptance by the courts of a more permissive use of lethal force under the human rights law paradigm to resolve these incidents (see Chapter 12, section 12.3.4, *Strictly Necessary, Proportionate, and Overwhelming*).

A State may not equivocate at all regarding the nature of the violence, as was demonstrated by the French President, Francois Hollande in the aftermath of the 2015 ISIS attacks in Paris. He clearly stated the attacks were an act of war committed by a "terrorist army".[110] Not all responses to even isolated attacks will be law enforcement–based. For some operations it may be determined ahead of an attack that a conduct of hostilities approach could be necessary to deal with a particular type of non-State actor threat (e.g., hijacked aircraft), or that conclusion may be quickly reached as the incident unfolds. For example, it was noted in respect of the 1979 efforts by Saudi security forces to regain control of the Grand Mosque at Mecca seized by extremists, "[t]he bloody battles of recent days made it clear: this was

---

108 David Kilcullen, Out of the Mountains: The Coming of Age of the Urban Guerrilla 89 (2013) (referred to in cable sent from U.S. Chargé d'affairs to Washington).

109 *Id.* at 99.

110 Adam Nossiter, Aurelien Breeden & Katrin Bennold, *Three Teams of Coordinated Attackers Carried Out Assault on Paris, Officials Say; Hollande Blames ISIS*, Int'l N.Y. Times, Nov. 14, 2015, http://www.nytimes.com/2015/11/15/world/europe/paris-terrorist-attacks.html. *But see* Noah Barkin & Sabine Siebold, *Germany to Support Military Campaign Against IS After French Appeal*, Reuters, Nov. 26, 2015, 8:11 PM GMT, http://uk.reuters.com/article/uk-mideast-crisis-germany-france-idUKKBN0TF0ZU20151126 (the German Defense Minister "told German television later on Thursday that Germany was not at war because it is not fighting a sovereign country but a "murderous gang" referring to Islamic State.").

open warfare, not a policing action. The regular army had to be in command."[111] This was evident in the rebel use of Molotov cocktails against Saudi armored personnel carriers[112] and ambushes carried out against security forces as they attempted to enter the narrow confines of the mosque.[113] It is difficult not to see that violence in practical terms and at law (e.g., applying the *Abella v. Argentina* case, "totality of the circumstances" criteria) as being armed conflict–related.

An extraterritorial defensive use of force by security personnel can also occur in situations where an attack begins with what factually looks like mob violence and then escalates to the level of an armed conflict.[114] This appears to be the case in the 2012 defense of the U.S. diplomatic facilities in Benghazi, Libya, by largely contractor security personnel in response to an attack by a mixed force of Al Qaeda–affiliated groups.[115] Until the armed conflict threshold is crossed, the use of force is governed by human rights norms. In this situation, even when the violence began to take on the attributes of an armed conflict there continued to be limitations applied to the use of force. During the course of an intense firefight some of the attackers used a house near the compound as a refuge. However, the operators "never fired at the building because they didn't know whether women and children were inside."[116] This protective approach toward uninvolved civilians suggests either the ROE or the training of the security contractors involved mirrored the U.S. Standing Rules of Engagement. In terms of exercising self-defense, it has been noted that these rules extend criminal law enforcement self-defense principles to the operational environment.[117] Similarly, had a military quick reaction force been able to respond to this incident, absent any specific mission accomplishment ROE or a declaration of hostile forces, it appears likely they too would have applied similar defensive principles. Again these "one-off" attacks and State action to defend nationals prompt the application of a human rights law approach, often as a matter of policy.

14.4.5. APPLYING LAW AND POLICY

While law enforcement has a significantly greater role to play in contemporary conflict, it would be inaccurate to suggest many of the non-State actor insurgent and terrorist threats can be defeated by means of a policy decision to solely apply a human rights–based

---

111 Yaroslav Trofimov, The Siege of Mecca: The Forgotten Uprising in Islam's Holiest Shrine and the Birth of Al Qaeda 151 (2007).

112 *Id.* at 155–56.

113 *Id.* at 128–31.

114 Michael Morell & Bill Harlow, The Great War of Our Time: The CIA's Fight Against Terrorism from Al Qa'ida to ISIS 205–10 (2015) (for a sophisticated analysis of the changing nature of the three attacks made on the diplomatic facilities and Annex in Benghazi).

115 Perm. S. Com. on Intelligence, 113th Cong., Investigative Rep. on the Terrorist Attacks on U.S. Facilities in Benghazi, Libya, September 11–12, 2012, at 17 (Nov. 21, 2014), https://intelligence.house.gov/sites/intelligence.house.gov/files/documents/Benghazi%20Report.pdf (members from AQIM, AQI, the Mohammed Jamal Network (MJN), Ansar al-Shari'a (AAS), and Abu Abdayda Ibn Jarah Battalion (UJB), and Qadafi loyalists were identified as participants in the attack on the U.S. facilities).

116 Zuckoff, *supra* note 51, at 235.

117 Geoffrey S. Corn & Eric Talbot Jensen, *Untying the Gordian Knot: A Proposal for Determining Applicability of the Laws of War to the War on Terror*, 81 Temple L. Rev. 787, 814 (2008).

approach. What must be acknowledged is that counterinsurgency and many counterterrorism operations will involve participation in armed conflict. A decision to apply a human rights law–based approach in whole, or in part, as a matter of policy does not change the fundamental character of an insurgency/counterinsurgency conflict as a form of warfare. The requirement to overmatch insurgent and terrorist forces frequently means that State security forces have to engage in hostilities.[118] As a result, military operations may often primarily involve the conduct of hostilities, with the policing function being mostly carried out by police units. Such operations can involve conventional military forces, including the use of drones and other aircraft.

However, the conduct of these military operations will be impacted by the overall context within which they are conducted. The result is often a requirement to incorporate law enforcement norms and with it human rights law into operational planning and direction, such as privileging arrest over killing in order to enable subsequent prosecution of the terrorist leader of an organized armed group.[119] The interaction of law and policy can be seen in Coalition operations against the Islamic State. What must be remembered is that this is a Coalition where the territorial State, Iraq, will do the heaviest lifting. The United States formed a coalition of States to counter the 2014 territorial gains of the Islamic State in Iraq. However, as of late 2015 Western States have avoided putting significant numbers of "boots on the ground" beyond isolated raids to rescue hostages[120] or seeking to capture an Islamic State leader.[121] Included in these operations has been the targeting of oil-production facilities controlled by the Islamic State. Such targeting has long been associated with strategic air campaigns.[122] One potential problem that arises is that this may wrongly give the impression that the foreign States are not involved in a counterinsurgency.

Large-scale intervention on the ground with foreign forces against the Islamic State armed groups has likely not occurred because of donor State fatigue with the post-9/11 wars; lingering controversy over the 2003 Iraq invasion; and a concern that those forces will be viewed as "occupiers," which will further stoke jihadist recruiting and resistance. Instead, Iraqi military, Shi'a militia, and Kurdish security forces provide the ground element. However, the conflict fundamentally is an insurgency/counterinsurgency. In reclaiming territory from the Islamic State, particular attention will have to be paid by Iraqi security forces and their Coalition allies to a counterinsurgency approach that is consistent with Iraq's governance role and one that protects the population over which it wants to re-establish control. This means applying a doctrine that places emphasis on human rights, policing, and a more restrained use of force.

---

[118] Christopher Paul et al., Paths to Victory, Rand xxviii (2013).

[119] Helene Cooper & Eric Schmitt, *ISIS Official Killed in U.S. Raid in Syria, Pentagon Says*, N.Y. Times, May 16, 2015, http://www.nytimes.com/2015/05/17/world/middleeast/abu-sayyaf-isis-commander-killed-by-us-forces-pentagon-says.html?hp&action=click&pgtype=Homepage&module=first-column-region&region=top-news&WT.nav=top-news&_r=0 (where reference is made to a plan to interrogate and prosecute the Islamic State financier, Abu Sayyaf).

[120] Oliver Holmes & Jason Szep, *U.S. Hostage Rescuers Dropped from Night Sky: Syria Activist*, Reuters (Aug. 22, 2014), http://www.reuters.com/article/2014/08/23/us-syria-usa-raid-idUSKBN0GM26620140823.

[121] Missy Ryan, *In Intensified Islamic State Effort, U.S. to Send Elite Targeting Force to Iraq*, Wash. Post. (Dec. 1, 2015), https://www.washingtonpost.com/news/checkpoint/wp/2015/12/01/in-intensified-islamic-state-effort-u-s-to-send-elite-targeting-force-to-iraq/ and Cooper & Schmitt, *supra* note 119.

[122] Watkin, *Targeting "Islamic State" Oil Facilities*, *supra* note 14, at 501–02. *See also* Watkin, *Targeting in Air Warfare*, *supra* note 14, at 5–16 (discussing airpower theory and the conduct of strategic air campaigns).

While some Western States may view their role as being a conventional one, limited uniquely by humanitarian law, the overall military effort will require a blending of the conduct of hostilities and human rights law–based "policing." Given the predominately Shi'a Muslim makeup of the Iraqi forces, the effort to defeat the Islamic State will likely prove challenging. One challenge that has to be faced in gaining effective control over the Sunni areas dominated by the Islamic State is that the Iraqi police force is relatively ineffective. Like the military, its membership also reflects the intersectarian divide since they are primarily Shi'a Moslem.[123] It is the political accommodation of the Sunni and Kurdish populations by the Shi'a-dominated Iraqi government and a secure environment being provide to those groups that are the most important aspect of the task facing the government. These are also potentially the weakest link in the apparent plan to defeat of the Islamic State.[124]

Inadequate or discriminatory policing is not the only way that the civilian population can become alienated. What was apparent early in the operations against the Islamic State is that the pressure to avoid collateral civilian casualties is not limited to counterinsurgency operations taking place on the ground. Inevitably, restrictions will also be placed on States supporting that effort with the use of airpower, even though this approach is not always popular with airpower advocates.[125] The negative impact on overall mission goals arising from the operational effect of potentially alienating the civilian population is as evident in respect of strikes in areas the Iraqi government does not control, as it is in areas it does control.[126] Generally, the collateral civilian casualties resulting from the use of airpower has become a particularly sensitive public policy issue as a result of the perception created in Afghanistan,[127] Libya,[128] Pakistan, Somalia, and Yemen that such casualties could or at least should be reduced to almost zero. In other words, according to a human rights law standard. This can be seen in August 2015 media reports that Canadian air strikes against ISIS in Iraq caused civilian casualties, while the government position remained that "Canada is quite confident that with all the strikes that we've executed, there is absolutely no evidence of civilian casualties associated with our strikes."[129] Impacting on the position taken by each

---

123 Sherman & Paul, *supra* note 81, at 249 (where it is noted that by 2007 the Iraqi police force had become increasingly Shi'a-dominated.).

124 *Proud Sunni Neighborhood Writhes under Iraq's Shiite Security Forces*, L.A. TIMES, July 20, 2014, http://www.latimes.com/world/middleeast/la-fg-iraq-adhamiya-20140720-story.html (where Iraqi security forces are viewed as an occupying army by Sunni residents).

125 Kristina Wong, *US Aim for "Zero Civilian Casualties" Draws Criticism*, THE HILL (June 24, 2015, 6:00 AM), http://thehill.com/policy/defense/policy-strategy/245932-us-aims-for-zero-civilian-casualties-in-war-vs-isis.

126 Nabih Bulos & Patrick J. Mcdonnell, *Iraq News Reports Say 18 Civilians Killed in Strike Aimed at Militants*, L.A. TIMES, Oct. 6, 2014, http://www.latimes.com/world/middleeast/la-fg-iraq-civilian-casualties-20141006-story.html ("Responding to the outcry, Prime Minister Haider Abadi announced last month that the country's armed forces had suspended aerial bombardments in areas where civilians are present.").

127 *Tactical Directive*, *supra* note 95.

128 *Letter Peter Olson to Judge Kirsch*, OLA (2012)006, at 3 (Jan. 23, 2012), http://www.nato.int/nato_static/assets/pdf/pdf_2012_05/20120514_120514-NATO_1st_ICIL_response.pdf ("Not one of the targets struck, involving over 7700 weapons, was approved for attack, or in fact attacked, if either those designating and approving the target or the pilot executing it *had any evidence or other reason to believe that civilians would be injured or killed by a strike*.") (emphasis added).

129 Steven Chase, *Canadian Armed Forces Accused of Killing Civilians in Iraq Air Strike*, THE GLOBE AND MAIL (Aug. 28, 2015, 3:03PM EDT), http://www.theglobeandmail.com/news/politics/canadian-armed-forces-accused-of-killing-civilians-in-iraq-air-strike/article26138351/.

party is the limited ability to investigate the allegations.[130] In any event, causing zero civilian collateral casualties and damage will likely be an impossible standard to maintain given the nature of such warfare.

The indication that Coalition forces would be applying a humanitarian law standard rather than the *Drone Policy* test of "near certainty" is unlikely to remove the public (both local and international) sensitivity to civilian death and injury.[131] Nor is a broad interpretation of a humanitarian-based targeting authority likely to lead to strategic success. As has been noted, when applying a humanitarian law analysis, the military advantage to be gained from not alienating the civilian population is likely to result in a narrower acceptable "zone of proportionality" (see Chapter 7, section 7.6.2.2, *Proportionality and Counterinsurgency*). This does not mean that the legal test for targeting proportionality equates to that traditionally permitted under human rights law. However, concern over potential collateral effects may in certain circumstances result in narrower policy-directed restrictions on the use of airpower that move toward creating a "human rights law like" effect. Those planning and conducting the raids will have their ability to exercise discretion within the already restrictive humanitarian law–based zone of proportionality even more narrowly proscribed.

The discretion exercised by military commanders can be influenced by human rights–based norms on matters other than the use of airpower. It is in the context of contemporary counterinsurgency and many counterterrorism operations where human rights–based norms will most likely be applied. Restrictions on the use of force can lead to greater reliance on capture rather than kill missions. In part, the privileging of a law enforcement approach is significantly impacted by the view consistently adopted by States, as the proper authority, that insurgents and terrorists are criminals (see Chapter 4, section 4.3, *Impact of the Proper Authority Principle*). Further, the criminalization of these non-State participants is directly reflected in the adoption of an "arrest or kill" methodology, which has been applied even to members of an opposing armed group. Where an arrest for prosecution rather than capture for interrogation purposes is the goal of a mission, then human rights law, including the requirements for due process, automatically becomes part of the operational law decision making. Both humanitarian law and human rights law (including domestic law) have to be considered.

Importantly, assessments regarding risk to soldiers civilians and whether there is adequate control exercised over the area of operations would almost inevitably have to be undertaken when national- or operational-level direction is provided to military commanders to favor capture over killing an opponent who is taking a direct part in hostilities. Such direction inherently gives the military commander discretion to only conduct capture missions if they meet acceptable operational parameters and to restrict subordinate action if they do not. Even low-level commanders may have to apply parameters set by human rights law–based considerations when determining if an arrest operation can be conducted in lieu of a targeted killing (see Chapter 7, section 7.5, *The IED and Suicide Bombing Threat: Arrest or Kill?*). Of note, this outcome is entirely consistent with the approach of the Israel Supreme

---

130 *Id.* ("The Forces say they didn't consider it necessary to make public this particular allegation because there was insufficient evident to mount a formal investigation.").

131 *Drone Policy, supra* note 89.

Court in the *Targeted Killing Case*, which indicated this was a legal obligation for Israeli forces operating in the Occupied Territories. That case is widely viewed as incorporating human rights–based factors into humanitarian law decision making (see Chapter 7, section 7.4.2, *The "Targeted Killing" Case and Control*). While framed in the terms of a legal obligation, the exercise of such discretion also reflects practical limitations potentially facing all military commanders.

14.5. Confronting Transnational Violence: A Holistic Approach

Armed conflict between States and non-State actors presents itself in two basic forms: conflicts that develop internal to a State and transnational violence generated from or occurring outside that State. This latter form of conflict has posed a particular challenge for State security forces as the non-State actor threat originates from outside the affected State. This type of threat has forced consideration of the complex interface between the various bodies of law applicable to operations. It has also resulted in a fundamental shift in how non-State actor conflict is viewed. Although State military forces and the United Nations had already begun to struggle with how to categorize such conflict in the 1990s (e.g., "MOOTW," "robust peacekeeping"), it was Al Qaeda's attacks on Western States and interests (e.g., the "far enemy") that spurred a broader requirement for the international community to deal systemically with this "external" security threat. As can be seen in the U.S. Supreme Court decision, *Hamdan v. Rumsfeld*, non-State actor conflict is no longer viewed by many international lawyers to be strictly confined by borders. Rather, its transnational character is increasingly defining that threat.[132] The threat posed by non-State actors has proven unique in that those groups have demonstrated a significant capability to strike transnationally within a target State, as well as attack that State's "interests" in areas outside its borders.

A unique feature of this transnational threat is that nations have been forced to deal with threats emanating from areas of the world where another sovereign government is not willing or able to control the actions of non-State actors operating on or from its territory. At law it is the territorial State that is responsible for controlling its sovereign territory and removing such non-State actor threats. In reality, some States are too weak or are otherwise ill equipped to do so. In this sense the failure to control the activities of these transnational actors appears to be more one of domestic governance and of law enforcement on the part of the territorial State rather than a purely international law issue. However, it is a failure with significant international security ramifications, as States targeted by these non-State actors must still address its consequences.

In an ideal situation use could be made of the UN Charter provisions providing for collective action to address transnational threats to international peace and security. In reality, as with many inter-State disputes, it is often the second prong of the bifurcated framework of international security that is applied to respond to the actions of transnational non-State actors: the exercise of the right to self-defense. It is the pitting of States against non-State actors, which has moved the legal analysis beyond a twentieth-century legal approach, that

[132] Hamdan v. Rumsfeld, 45 I.L.M. 1130, 1154 (2006) (U.S. Supreme Court 2006).

was uniquely focused on regulating inter-State conflict. As has been demonstrated, the security threats posed by non-State actors able to operate from ungoverned spaces spans the conflict spectrum from simple criminal acts to "State-like" groups fielding significant hybrid or even conventionally organized military forces. Importantly, the violence often occurs at the intersection between various bodies of law: the exercise of State self-defense, humanitarian law, and international and domestic human rights law.

The practical solution adopted by a number of States has increasingly been to holistically apply these bodies of law under the rubric of "operational law." This approach offers a number of advantages, including the opportunity to tailor the response to the nature of the threat, as well as lower the overall levels and types of force used. In applying this approach there must also be an emphasis on ensuring "respect for and protection of the civilian population."[133] While this duty is outlined specifically in humanitarian law it has a more universal appeal extending to whatever legal framework is applied to regulate the violence. Adopting an operational law approach has not been without its challenges since it requires military commanders and their legal advisors to apply a different and ultimately more strategic perspective when responding to the non-State actor threat.

It has not always been easy to gain the consensus necessary to apply the various bodies of law in a coherent and operationally relevant manner. This is because of the existence of legal theories and interpretations that emphasize a separation, notwithstanding their overlap in application; a lack of consensus on the threshold for noninternational armed conflict, and therefore the application of humanitarian law; a resistance on the part of some States toward acknowledging human rights law applies to military forces conducting international operations; the diverse nature of the transnational security threats; and the only relatively recent acknowledgment by a significant portion of the international legal community that the law governing State self-defense even applies to non-State actors.

This does not mean that a greater consensus on some legal issues has not been forming. More than a decade after the 9/11 attacks a number of these issues now seem less controversial (e.g., self-defense against non-State actors, the application of human rights norms on operations). Further, there is also a greater understanding of the nature of the threat and the advantages of looking beyond the use of a conduct of hostilities approach in order to address the full range of contemporary security threats. In applying an operational law approach there are key factors relevant to the application of these bodies of law This is an approach that favors law enforcement. At the same time there must be recognition that there are limits to the effectiveness of the human rights law and policing (see Chapter 13, *Limits of Law Enforcement and Human Rights Law*).

An operational law–based decision process, which can be holistically applied to transnational conflict in order to facilitate the response to non-State actor threats, is set out in the Appendix to this chapter. This process establishes key decision points, with a particular emphasis being placed on defaulting to the application of human rights–based law enforcement when it provides an effective means of dealing with the threat. Such a default should be the normative standard to which all States are expected to act. This favoring of law enforcement occurs whether operating inside or outside the targeted State. However,

[133] AP I, art. 48.

there are factors that make it more likely a policing response to be adopted when operating domestically. It is an approach that looks at whether the violence can be categorized as an armed conflict but also focuses on the effectiveness thresholds of human rights law regarding the use of force when such conflict is in existence. This holistic application of the law rejects the traditional exclusionary approaches, except in the sole situation where an armed conflict is not occurring. This one exclusionary threshold factor requires that a law enforcement response be applied.

Conceptually "transnational" conflict can be divided into two types. The first type involves an externally originated attack directed at the threatened State and which is taking place on the territory of that State, its airspace, or its territorial waters. To a certain degree this approach can mirror traditional State versus non-State actor conflict where the final stages of the violence can be internally generated with the difference being that the threat is not necessarily homegrown (e.g., the 9/11 attacks). Consistent with the historical approach adopted by States regarding internal violence, such attacks are ordinarily dealt with, at least initially, as a law enforcement matter (see section 14.4.4, *A Law Enforcement Response to Attacks at Home and Abroad*, above). The exception would be if an armed conflict is occurring and it becomes evident that the defensive use of military forces, tactics, and armament associated with the conduct of hostilities is required. To the extent the violence reaches the level of an insurgency, this form of transnational conflict would still favor a "police primacy" approach.

The second type of attack is also externally generated, but is one directed by non-State actors at targets such as citizens, military forces, embassies, or commercial symbols of one State located in another State, or on the high seas. These non-State actor attacks may originate from within the State where those armed groups ordinarily operate (e.g., its ungoverned spaces) or on the territory of a third State. These attacks are transnational in terms of the threatened State being required to react to threats outside its borders (i.e., the other States are unable or unwilling to deal with the threat), as well as the cross-border effect that non-State actors hope to have in carrying out the attacks. Again, the application of a conduct of hostilities–based response by the threatened State is dependent upon an "armed conflict" being established. A threatened State's response is also subject to there being legal authority to enter the territory of another State to conduct operations. Such authority includes the exercise of State self-defense or otherwise acting in defense of nationals.

As has been noted throughout this book, the utility of a law enforcement approach and the ability of States to operate ground forces in "ungoverned spaces" is leading to a greater use of the "arrest" option. In a conflict with transnational terrorists, which is often discussed as a global insurgency crossing international borders, it is becoming increasingly difficult to argue that the standard doctrinal response applied to internal insurgencies (e.g., police primacy) should not also be applied on a cross-border basis.

In what is likely to be a viewed as a controversial suggestion, the response by the targeted State should become expected to be human rights–based when feasible, and it can be applied to effectively deal with the threat. It is always available to the State to explain why that option could not be applied in the circumstances. The ability to apply a law enforcement response to such "external" violence is dependent upon a number of factors, including the capabilities of State security forces; geography, including distances involved; the organization and potential scale of violence threatened by the terrorist or insurgent group;

the degree of control that can be exercised by the affected State security forces over the immediate area of operations; and the risks to the soldiers involved and to civilians living in proximity to such action.

In seeking to carry out such an approach, the capabilities of the security forces will be a particularly important factor. The reality is that not all States can project the necessary military or police forces across the globe to respond to these threats in that fashion. The use of this approach will likely be much easier to carry out in adjacent States, or at least those limited to a regional context. The ability to extraterritorially apply either a conduct of hostilities or human rights–based law enforcement approach is enhanced by means of a capable blue-water navy, cooperative allies, and the forward basing of military forces. However, a privileging of the "law enforcement" option will not provide a complete solution to the security challenge facing States from transnational actors in the context of an armed conflict. When power projection can be carried out, it is often in the form of airpower (i.e., drones or other aircraft), missile strikes, or Special Forces raids where capture may be an unlikely outcome.[134] This leaves a kinetic option as a likely response in many, if not most, armed conflict situations.

As can be seen in the 2012 European Union use of helicopters to conduct missiles strikes in an antipiracy operation in Somalia[135] and the proposed use of European military forces to destroy ships involved in illegal migration from Libyan territory, the use of such military force in a law enforcement context, although exceedingly rare, is not beyond consideration in international operations.[136] As can be seen in the Libyan example, the limited willingness or ability of the local authorities (e.g., the "Masked Brigade" militia) to stop the smuggling almost demands some action be taken by affected States to "police" the situation, particularly because of the tragic humanitarian consequences that can result.[137] What is left unresolved is what levels and kinds of force will human rights law tolerate being used to dismantle the criminal organizations engaged in this trafficking. Even more problematic from a human rights perspective is the shoot-down of drug trafficking planes in South America.[138]

The impact of the greater use of the arrest option is that national borders are being viewed less as an absolute barrier to defensive action (see Chapter 9, section 9.3.7.3, *Borders*

---

[134] Seán D. Naylor, *Killing Abu Ghadiya*, FOREIGN POL'Y (Aug. 31, 2015), https://foreignpolicy.com/2015/08/31/deltaforce-seals-syria-iraq-terrorism/?utm_source=Sailthru&utm_medium=email&utm_campaign=New%20Campaign&utm_term=%2AEditors%20Picks (for an outline of a 2008 Special Forces raid from Iraq into Syria to kill an Al Qaeda militant).

[135] Eliav Lieblich, *Quasi-Hostile Acts: The Limits on Forcible Disruption Operations under International Law*, 32 B.U. INT'L L.J. 101 (2014).

[136] *Stop the Boats*, THE ECONOMIST, May 19, 2015, http://www.economist.com/news/europe/21651693-european-union-ministers-are-turning-their-navies-mediterranean-migrant-traffickers-stop.

[137] Patrick Kingsley, *Libyan Militia Captures Smugglers Allegedly Responsible for Migrant Deaths*, THE GUARDIAN (Aug. 29, 2015, 12:54 BST), http://www.theguardian.com/world/2015/aug/29/libyan-militia-captures-smugglers-allegedly-responsible-for-migrant-deaths.

[138] Sam Tabory, *Peru Revives Controversial Drug Plane Shoot-Down Law*, INSIGHT CRIME (Aug. 21, 2015), http://www.insightcrime.org/news-briefs/peru-revives-controversial-shoot-down-law-for-narco-planes ("Peru was one of the last major drug producing and trafficking nations in Latin America not to have a contemporary shoot-down law on the books. Bolivia, Brazil, Columbia, Honduras, Paraguay, and Venezuela all have laws in place that authorize the downing of drug planes.").

or Barriers?). In external operations against transnational non-State actor groups engaged in armed conflict, the application of human rights norms to govern State action may been seen to occur as a matter of law (e.g., the *Targeted Killing Case*[139]), or more likely as a matter of policy (i.e., the 2013 *Drone Policy*[140]). The approach applied can depend upon national legal requirements regarding the use of military force (i.e., the decision of a national court or constitutional interpretation). Conceptually the threatened State could be viewed as operating in the place of the territorial State within "ungoverned space." Certainly, the full obligations of governance would not be invoked. However, as with any State dealing with "criminal" elements, there is a resonance with the exercise of its traditional law enforcement function, albeit on an international basis. This move toward extraterritorial interventions raises legally, politically, and morally complex questions regarding what obligations there may be toward the foreign population within which the non-State actor may be operating. However, as was discussed in Chapter 9, the vertical relationship between an intervening State and an enemy population as well as the fundamental nature of the right to life argues against treating those civilians any differently than that State's own citizens (see Chapter 9, section 9.3.7.4, *Acting in Place of the Territorial State*).[141]

Consideration of the doable should be part of any accountability assessment. In a very fundamental way what often are presented as intractable disagreements over the application of different normative frameworks must give way to an assessment of how these complex issues can be resolved in a manner that meets operational needs and protects the civilian population. In this respect, a legal advisor must not only set out the available legal frameworks but also assist the decision maker in determining how those various bodies of law can be applied and, if required, reconciled in order to meet operational goals. In applying a holistic response, consideration must be given not only to the boundaries of the various bodies of law but also the reasonable policy choices that can be applied to resolve the inevitable interface and overlap. This places legal advisors in a space of advising on various options at the intersection of policy and law. This is simply a reality of contemporary operational law decision making.

In situations where the issue is one involving an exercise of discretion by a military commander, that decision will ultimately have to be justified by explaining how it fits within the legal framework governing the military operation. As was seen in the *Targeted Killing Case*, there is no decision regarding the use of force, even one made in the exercise of professional military judgment, which is completely divorced from the law. In that context, this issue was framed in terms of whether that exercise of professional discretion is one that falls within the zone of reasonable activity.[142] Ultimately, an assessment of reasonableness

---

139 *See* Aerial Drone Deployment on 4 October 2010 in Mir Ali/Pakistan, 157 I.L.R. 722, 749 (2013) (where the human rights norms identified in the Public Committee Against Torture in Israel v. Israel, Israel Supreme Court [16 December 2006], 46 ILM 375 (2007) were applied to drone strikes in Pakistan).

140 *Drone Policy*, *supra* note 89.

141 *See* Eyal Benvenisti, *The Legal Battle to Define the Law on Transnational Asymmetric Warfare*, 20 DUKE J. COMP. & INT'L L. 339, 351 (2010).

142 Public Committee Against Torture in Israel v. Israel, Israel Supreme Court [16 December 2006], 46 ILM 375, ¶ 57, at 399 (2007). *See also* A. P. V. ROGERS, LAW ON THE BATTLEFIELD 110–11 (2nd ed. 2004) (for a discussion of the reasonableness of decision making by a commander).

requires an objective assessment against an established standard. Explaining and justifying why operational choices are made, rather than simply saying the law generally sets the boundaries, and a commander is free to decide as he or she wishes, contributes to the maintenance of normative controls over decision making. It also reflects an essential requirement of successful military operations. Where such accountability is not established, the use of force can quickly become counterproductive.[143]

The analysis of operational courses of action and the use of force should be influenced by an approach that takes the view the value attributed to human life must be the same regardless of whether civilians live inside or outside a threatened State. As the 2013 *Drone speech* demonstrates, the universal nature of rights and the dignity of humans is increasingly impacting on the international security dialogue.[144] Civilians in "ungoverned spaces" deserve no less protection than those located within a State that exercises effective governance. The result is that the intervening State should perform a proper authority role regardless of where civilians are located or what force is used. The failure of the territorial State to fulfill its obligations should not empower the responding State to ignore the role it is capable of performing. It might be argued that another State should not be permitted to interfere with what is fundamentally a sovereign responsibility of the territorial State. However, that is not what is being suggested. The situation is analogous to the role performed by the occupying power, which cannot usurp the legal status of the territorial State. An occupier still has "temporary managerial powers, for the period until a peaceful solution is reached."[145]

It is not evident beyond a sense of formalism attached to national borders why the State conducting operations should not be required to carry out a role involving a special trust toward uninvolved civilians of the enemy or any other State during cross-border deployments against non-State actors. It is also an outcome that on its face mirrors the lessons learned by British forces as they took control of territory as they expanded their empire in the nineteenth century.[146] While that more humane policy was initially directed toward persons under their direct control,[147] it also became evident attitudes were changing. For example, the more bellicose use of "air policing" by colonial powers to keep indigenous tribes in line[148] had by the 1930s ceased because of international opinion against such action.[149] A normative framework has to be applied to deal with civilians, and one based

---

143 LEDWIDGE, *supra* note 8, at 190 (where it is noted the accountability for the use of force on international operations is not always evenly applied. He states, "[c]ivilians killed in their own country by foreign forces are accorded the briefest of glances by comparison [to friendly fire, 'blue on blue' incidents].").

144 *Id.*

145 EYAL BENVENISTI, THE INTERNATIONAL LAW OF OCCUPATION 6 (1993) (where in respect of occupation it is noted, "[d]uring that limited period the occupant administers the territory on behalf of the sovereign. Thus the occupant's status is conceived to be that of a trustee.").

146 CHARLES W. GWYNN, IMPERIAL POLICING 14–15 (1934).

147 *Id.* at 14.

148 PHILIP ANTHONY TOWLE, PILOTS AND REBELS: THE USE OF AIRCRAFT IN UNCONVENTIONAL WARFARE 1918–1988, at 9–45 (1989) (where it is noted the British, French, Spanish, and Italians made use of aircraft to control territory. The British applied the tactic to the greatest extent. This occurred in territory such as Aden, Iraq, Somalia, Afghanistan, and the Northwest frontier of India), and John Pilmott, *The British Experience*, *in* THE ROOTS OF COUNTER-INSURGENCY: ARMIES AND GUERRILLA WARFARE, 1900–1945, at 17, 24–25 (1988).

149 TOWLE, *supra* note 148, at 52–55.

on minimum force more accurately reflects the nature of the required interface with even hostile foreign populations. Based on human rights norms, it matters little if the authority is found in customary humanitarian law or human rights law. It would be an odd outcome that lessons learned in previous centuries could be so easily abandoned on the basis that the creation of State boundaries means the policing role should to be left solely in the hands of a territorial State, regardless of whether it can carry out that function.

A holistic approach places emphasis on the function performed by the law. A key factor to consider in this application of the law is the strategic purposes for which force is being applied. In other words, what action most realistically contributes to victory in the "fight" against non-State actors? This includes questions regarding how the situation can be returned to one of "peace" and the obligations that rise for States in their role as a "proper authority." Ultimately, the goal of these security operations is to return the situation to one of "normalcy." As will be discussed in the next section, the exercise of State authority in this context is most often associated with human rights and the application of a law enforcement approach.

14.6. Law Enforcement and the Search for "Normalcy"

In order to succeed, the operational experience of the first decade of the twenty-first century has demonstrated that the various bodies of law impacting on operations have to be applied holistically. The application of these legal regimes cannot be done in a contextual vacuum. In terms of "grand strategy," the focus must remain looking beyond war toward eventual peace.[150] A key aspect of government decision making is the desire to attain strategic goals, defeat enemies in armed conflict, and maintain law and order within the areas of State responsibility. One challenge for States in the twenty-first century is that non-State actors (e.g., Salafi jihadists) who want to do away with the present international governance structure pose a significant portion of the global security threat. Not only do they not recognize State territorial boundaries, they seek to eliminate them altogether. In addition, other more criminally oriented transnational challengers to State authority also often ignore international borders and seek their own zones of impunity from which they can carry on their profit-motivated activity.

Faced with threats such as these, which can be integrated within the social fabric of society, it seems unlikely that there will be a dramatic moment of victory or a celebratory parade so commonly associated with traditional armed conflict. Part of the reason for this is that the non-State actors themselves are such diverse and complex participants in the conflicts. Armed conflicts with non-State actors are also "often of a fluctuating nature, typified by temporary lulls in the armed violence or instability in the level of organize of the non-State party to the conflict."[151] The methods identified for defeating them are also

[150] B. H. LIDDELL HART, STRATEGY 322 (2nd ed. 1991).

[151] INTERNATIONAL HUMANITARIAN LAW AND THE CHALLENGES OF CONTEMPORARY ARMED CONFLICT, Report 32nd International Conference of the Red Cross and Red Crescent EN 32IC/15/11, at 10 (2015), *available at* https://www.justsecurity.org/wp-content/uploads/2015/11/2015-ICRC-Report-IHL-and-Challenges-of-Armed-Conflicts.pdf [hereinafter the ICRC CHALLENGES REPORT].

inevitably varied. Paul Staniland has suggested a counterinsurgency strategy "will work best if it is guided by the organizational structure of the insurgent group being targeted."[152] As he notes, groups with urban-rural divisions may be weakened with a strategy that focuses on co-opting rural areas, a divide-and-conquer strategy may work against groups with feuding elites, and an integrated group might be approached with "a mix of negotiations and directed force. . . ."[153] Vanguard groups are susceptible to decapitation strategies, however, "[t]he state needs to quickly break apart leadership networks before local alliances can be formed."[154] While attempts at decapitating the enemy leadership can in some circumstances have a positive effect, it can also fail as targeting mistakes can turn civilians against the government "and lay the basis for local alliances."[155]

Another analysis views the conflict between the United States and Al Qaeda as being similar to "the one between a government and its domestic challengers."[156] This is because of "[t]he shift of the global configuration of power to a strictly unipolar one, and the ability of both the hegemon and the traditional challengers to work across state borders."[157] The suggested reaction is for the State to centralize and restructure agencies in its response to the transnational challengers.[158] The same technologies that favor the challengers "allow the centralization of large institutions and complex organizations."[159] In this respect, "[e]thnic groups, social classes, civilizations, religions, and nations do not engage in conflict or strategic interaction—organizations do."[160] A centralized approach for dealing with decentralized challengers ultimately reflects a domestic model that has always privileged a law enforcement approach. This theory has a certain resonance with the view such conflicts are international in character.

It has been suggested elsewhere that a movement such as AQAM, which consists of alliances with other armed groups and applies a cellular structure, presents a radically different challenge since "they combine the power of the organizations by uniting economic, political or military power."[161] Therefore, "[w]eakening the group comes not from attacking specific cells or individuals . . . . but rather through breaking the bonds of its alliances."[162] More recently, it has been argued that the United States should tailor its response with a strategic framework that combines direct engagement with terrorists relying heavily on Special Forces; forward partnering with local governments; and off-shore balancing relying on air, naval, and rapidly deployable ground forces.[163] This seems to describe the approach that

---

152 Paul Staniland, Networks of Rebellion: Explaining Insurgent Cohesion and Collapse 229 (2014).

153 *Id.*

154 *Id.* at 47.

155 *Id.*

156 Abdulkader H. Sinno, Organizations at War in Afghanistan and Beyond 301 (2008).

157 *Id.*

158 *Id.*

159 *Id.*

160 *Id.* at 3.

161 Anthony Vinci, Armed Groups and the Balance of Power: The International Relations of Terrorists, Warlords and Insurgents 125 (2009).

162 *Id.* at 126.

163 Seth Jones, A Persistent Threat: The Evolution of al Qa'ida and Other Salafi Jihadists, Rand 49–62 (2014), http://www.rand.org/content/dam/rand/pubs/research_reports/RR600/RR637/RAND_RR637.pdf.

has been adopted by the United States after its difficult counterinsurgency experiences in Afghanistan and Iraq. Notably, this approach does not advocate taking on a primacy role, which "might involve utilizing conventional forces, as the United States did in Iraq and Afghanistan."[164] It is the local national governments that will have to act. Counterterrorism operations are not limited to drone strikes or the use of other airpower. The Special Forces raids have involved, at times to a significant extent, arrest rather than kill missions. None of these suggested approaches for dealing with the transnational non-State actor threat rely on a classic conventional warfare strategy or suggest a moment of clear-cut victory.

Overall, Audrey Kurth Cronin argues there are "six patterns in the decline and ending of campaigns emerge from the history of terrorism: (1) capture or killing the group's leader, (2) entry of the group into a legitimate political process, (3) achievement of the group's aims, (4) implosion or loss of the group's public support, (5) defeat and elimination by brute force, and (6) transition from terrorism to other forms of violence."[165] Ultimately, terrorism is fundamentally "always a struggle for legitimacy"[166] and "is only a fundamental threat to the nation-state to the extent that it successfully mobilizes the hopes, dreams, energies, and resources of the people—while the government does not."[167] Once again, defeating terrorism is linked to good governance, which itself is connected to human rights and an effective justice system and police force. Groups that transition from terrorism to an insurgency "gain legitimacy and increase their ability to take control of states themselves."[168] This is why "on the whole, arresting a leader damages a campaign more than does killing him, probably because the jailing of a leader demystifies him and demonstrates the power of the legal edifice of the state."[169] As one U.S. terrorism analyst has indicated in 2011, "[t]he task . . . . is to push the threat to a lower level, and manage—and accept—a degree of risk."[170] In 2012, Jeh Johnson, the U.S. Department of Defense legal advisor, indicated that the end of the conflict with Al Qaeda would occur when the group had been organizationally destroyed. One indicator would be when that group could no longer strategically attack the United States. At that point the security effort would become primarily a law enforcement and intelligence agency effort against its scatter remnants, or unaffiliated groups.[171] As Harold Koh subsequently highlighted "the hard part . . . would be drawing a line between al-Qaeda and new Islamic groups that spring up."[172]

In this battle for legitimacy, the ability of a State to manage the violence with a law enforcement response is a key indicator of success. Being able to effectively respond with

---

[164] *Id.* at 52–3.

[165] Audry Kurth Cronin, How Terrorism Ends: Understanding the Decline and Demise of Terrorist Campaigns 8 (2011).

[166] *Id.* at 205.

[167] *Id.* at 205–6.

[168] *Id.* at 205.

[169] *Id.* at 201.

[170] Eric Schmitt & Thom Shanker, Counter Strike: The Untold Story of America's Secret Campaign Against Al Qaeda 277 (2011) (referencing statements made by John Tyson, a Defense Intelligence Agency terrorism expert.).

[171] *Jeh Johnson Speech at the Oxford Union*, Lawfare (Nov. 30, 2012, 12:01 PM), https://www.lawfareblog.com/jeh-johnson-speech-oxford-union. *See also* Savage *supra* note 20, at 683-85.

[172] Savage *supra* note 20, at 684, and Harold Hongju Koh, *"How to End the Forever War"*, N.Y. Times (May 7, 2013), http://graphics8.nytimes.com/packages/pdf/world/2013/KOHSPEECH.pdf.

a human rights–based framework controlling the use of force is a clear sign of "normalcy" in the context of how a State maintains order within society.[173] The ability to shift to an exclusive law enforcement response may occur long before an armed conflict is viewed as having technically ceased.[174] The result can be decisions taken at the highest strategic levels to privilege a law enforcement response or to apply human rights–based norms during the conduct of hostilities against non-State actors regardless of whether the conflicts occur internally or internationally. If the State can treat the "terrorists" as criminals, a strong message is sent to the international community. In terms of the threat posed by jihadist groups challenging the State's role as the proper authority, the ability to deal with an emerging threat with this human rights–based paradigm highlights that the matter is fundamentally positioned at the lowest stage of the three-tiered revolutionary war–style doctrine that has been apparently embraced by these groups.[175] The first stage, which Abu Mus'Ab al-Suri calls "the stage of exhaustion," is one where the conflict has not transitioned to situations that the guerrilla force controls territory or commenced more conventional operations.[176] It is a stage involving "methods of assassination, small raids and ambushes, and selective bomb attacks . . . ."[177] The 2005 analysis by al-Suri in *The Military Theory of the Global Islamic Resistance Call*[178] identified that the jihadist movement was, after the defeats following 9/11, in that first stage.[179] As is seen with the Islamic State and some insurgent groups associated with AQAM (e.g., Al Shabaab), there are jihadist groups that have since moved well beyond this initial level of violence. In this context, victory still remains elusive.

Al-Suri also noted that in the first stage there is a particular role for individual terrorism and "Global Islamic Resistance Units."[180] Patterned after selected historical incidents, where assassinations, bombing attacks, or other acts of terrorism were carried out,[181] he suggests the "spontaneous operations performed by individuals and cells here and there over the whole world without connection between them, have put the local and international

---

[173] Benjamin Wittes & Gabriella Blum, The Future of Violence: Robots and Germs, Hackers and Drones 166 (2015)("Law enforcement belongs to the everyday, the commonplace. It is the normal state of affairs; all states experience some level of crime . . . ."). *See also Analytical Report, Fragmentation Of International Law: Difficulties Arising From The Diversification And Expansion of International Law*, Report of the Study Group of the International Law Commission Fifty-eighth session Geneva, 1 May–9 June and 3 July–11 August 2006, UN Doc. A/CN.4/L.682 13 April 2006, ¶ 104, at 57 (where war is identified as being the exception to the "to the normality of peace").

[174] ICRC Challenges Report, *supra* note 151, at 10-11 (where it is noted there are two views for assessing the end of an armed conflict: hostilities fall below the threshold of protracted armed violence, or one of the parties ceases to exist or no longer meets the organization requirements. The ICRC approach is to wait for a complete cessation of hostilities before assessing the whether there is a real risk of a resumption of hostilities).

[175] Norman Cigar, Al-Qa'ida's Doctrine for Insurgency: 'Adb Al-'Aziz Al-Muqrin's A Practical Course for Guerrilla War 20–22 (2009) (outlining the three "phases" of warfare), and Brynjar Lia, Architect of Global Jihad: The Life of Al-Qaeda Strategist Abu Mus'ab al-Suri 471–73 (2008) (setting out the three "stages" of conflict). *See also* Ryan, *supra* note 3, at 230, and Sinno, *supra* note 156, at 302 ("revolutionaries adopt organizational strategies across ideological divides.").

[176] Lia, *supra* note 175, at 471.

[177] *Id.*

[178] *Id.* at 347–483 (for excerpts of this work).

[179] *Id.* at 373.

[180] *Id.* at 382–407.

[181] *Id.* at 364–65 (examples include the killing of Rabbi Meir Kahane in the United States; the 1993 attempt by Ramzy Yusuf to blow up the World Trade Center in New York; the killing of female Israeli students by a Jordanian Border Guard; attacks on tourists and foreigners in Morocco, Jordan, etc.).

intelligence apparatus in a state of confusion. . . ."[182] The goal is to change spontaneous action into a "strategic operational method."[183] Under this theory, that approach was to be assisted through the formation of "Secret Agitation Units," consisting to one to three persons, who were to "spread the Call and redistribute its literature, its research studies, and its various programs by clandestine means, especially over the Internet."[184] This jihadist doctrine has not been fully realized, although there is no doubt that self-radicalization has become a significant security threat.

It has been argued that the most "effective jihadist terrorists are generally graduates of training camps or war zones, rather than the passive consumers of jihadist propaganda on the Web."[185] Notwithstanding this view, the 2012 Boston bombers appear to have gained knowledge on how to prepare their bombs over the Internet.[186] The activities of groups, such as the "Toronto 18,"[187] and even "lone wolf" radicalized individuals who murdered a British soldier in the streets of London in 2013[188] or carried out the murders of soldiers in Canada, reflect this more decentralized approach.[189] Importantly, this activity also represents an element of the jihadist threat that is most amenable to a law enforcement response, particularly in States with effective policing and robust domestic justice systems. Looked at strategically, international lawyers should encourage the default to a traditional law enforcement reaction for these types of security threats.

That human rights law maintains a pre-eminent position in some legal analyses is reflected in the *Advisory Opinion on the Legality of the Threat or Use of Nuclear Weapons* decision where the interaction between human rights law and humanitarian law is resolved by the use of the latter body of law to interpret whether there has been an arbitrary deprivation of the right to life as a human right.[190] From a legal perspective, the preference accorded to human rights law can also be seen in legal instruments that provide certain rights cannot be derogated from even in times of war or other public emergency threatening the nation.[191]

---

182 *Id.* at 366.

183 *Id.* at 391.

184 *Id.* at 449.

185 Peter L. Bergen, The Longest War: The Enduring Conflict Between American and Al-Qaeda, 205 (2011).

186 Michael Cooper, Michael S. Schmidt, & Eric Schmitt, *Boston Suspects Are Seen as Self-Taught and Fueled by Web*, N.Y. Times, Apr. 23, 2013, http://www.nytimes.com/2013/04/24/us/boston-marathon-bombing-developments.html?pagewanted=all, and Azmat Khan, *The Magazine that "Inspired" the Boston Bombers*, PBS Frontline (Apr. 30, 2013), http://www.pbs.org/wgbh/pages/frontline/iraq-war-on-terror/topsecretamerica/the-magazine-that-inspired-the-boston-bombers/.

187 Isabel Teotonio, *Toronto 18*, The Star, http://www3.thestar.com/static/toronto18/ (last visited Mar. 23, 2015).

188 Chris Greenwood, Rebecca Evans, & Martin Robinson, *Fiancée and Estranged Wife of Soldier Lee Rigby Flee Murder Trial in Tears as Jury Is Shown CCTV Footage of Moment "Muslim Converts Ran Him Down Before Almost Decapitating Him with Meat Cleaver and Knives,"* Mail Online (Nov. 29, 2013), http://www.dailymail.co.uk/news/article-2515493/Soldier-Lee-Rigby-murdered-mutilated-decapitated-Woolwich-attack-court-hears.html.

189 David D. Kirkpatrick, *Attacks in West Raise New Fears over ISIS' Influence*, N.Y. Times, Oct. 24, 2014, http://www.nytimes.com/2014/10/25/world/middleeast/string-of-attacks-in-west-raise-new-fears-over-isis-influence.html?smid=nytcore-ipad-share&smprod=nytcore-ipad&_r=0.

190 Advisory Opinion on the Legality of the Threat or Use of Nuclear Weapons 1996 I.C.J. 226, ¶ 25, at 240. *See also* Hassan v. The United Kingdom, Application No. 29750/09, Eur. Ct. H. R. ¶ 104 (2014), http://hudoc.echr.coe.int/sites/eng/pages/search.aspx?i=001-146501.

191 Malcolm N. Shaw, International Law 274–75 (6th ed. 2008).

Derogations can occur for some rights in times of emergency, with that being suggested to be an indication of the flexibility of human rights treaties.[192] However, such derogations exist within a framework where there are procedural requirements of notification and proclamation. This makes taking such a step an exceptional one.[193] There is also a requirement that any derogation be proportionate and consistent with other State obligations under international law.[194] The European Convention on Human Rights and the American Convention on Human Rights do provide for derogation in "time of war," which extends to the right to life.[195] However, the "war" derogation provisions have not been applied in either internal or international armed conflicts or interpreted by the European Court of Human Rights.[196] In addition, as is highlighted in *Hassan v. The United Kingdom*, the application of derogations in respect of "emergencies" for conflict external to a State is not the normal practice.[197]

While the formal system of derogations found in human rights treaties may not always provide a practical solution to many security challenges, it does still point to a preference for a default application of human rights law. It highlights the degree to which human rights law, and with it law enforcement, are connected to a sense of normalcy. It is certainly the preferred means by which States address internal violence. Making a policy decision to privilege a law enforcement response, whether acting within that State or on operations within another State, is entirely consistent with the existing international law framework and the strategic goals of many security operations. It also offers a principled approach toward reconciling the overlap that can occur between human rights law and international humanitarian law during armed conflict. In other words, default to law enforcement as much as possible. As was demonstrated by the United Kingdom in its fight with the IRA this might be able to be done by exclusively adopting a law enforcement framework. However, it can also be selectively applied even while engaged in hostilities whenever a policy of applying the law enforcement model or some of its principles offers an effective option for dealing with the security threat.

Applying human rights norms to regulate the use of force and ultimately soften the impact of the conduct of hostilities approach provides the State a method of responding to non-State actor threats that has a more universal appeal than the blunt use of military force. This is likely why the U.S. drone policy, with its reliance on higher standards of "near certainty" for collateral damage and seeking first to capture rather than kill an opponent, has largely been viewed in a positive light.[198] As has been suggested in respect of targeted killing, "[t]he government's interest should be to tame violence, not exacerbate it. Where

---

192 Marko Milanovic, Extraterritorial Application of Human Rights Treaties: Law, Principles, and Policy 111 (2011).

193 Gross & Ní Aoláin, *supra* note 69, at 259–61 (for a discussion of the procedural requirements for derogations).

194 *Id.* at 261.

195 European Convention on Human Rights, art. 15(1), and the American Convention on Human Rights, OAS Treaty Series No. 36; 1144 UNTS 123; 9 ILM 99 (1969) art. 27(1).

196 Marco Milanovic, *Extraterritorial Derogations from Human Rights Treaties in Armed Conflict*, *in* Collected Courses of the Academy of European Law 8 (Nehal Bhuta ed., Oxford University Press) (Forthcoming), *available at* http://papers.ssrn.com/sol3/papers.cfm?abstract_id=2447183 (last visited July 21, 2015).

197 Hassan v. The United Kingdom, Application No. 29750/09, Eur. Ct. H. R. ¶ 40 (2014), http://hudoc.echr.coe.int/sites/eng/pages/search.aspx?i=001-146501.

198 *Drone Policy*, *supra* note 89.

alternatives exist they should be pursued not simply as a matter of law, but also as a matter of sound policy."[199] This type of response has also been evident in the consideration of arrest before killing as applied by military commanders in Afghanistan, Iraq, and elsewhere. In this type of conflict the application of human rights–based norms should be seen as a force multiplier in terms of attaining the strategic goal of peace, rather than having that body of law invoked only after the peace has been hopefully "won."

Given the nature of the threat, the ultimate goal should be to reduce the challenges to State authority to a level where they can be dealt with by ordinary law enforcement means. Even during situations of armed conflict, a policy option of defaulting to a human rights–based approach whenever possible highlights not only that goal but also an ability to operate at the level of a normal peacetime environment. The ability to do so provides tangible and quite positive evidence that success is being achieved. Arresting insurgents or terrorists who otherwise can be targeted as direct participants in conflict confirms their "criminal" nature and reinforces the status of the State as the proper authority. However, care must be exercised. Here the courts have an important role to play. Too narrow an interpretation of the human rights law authority to use force will force the violence to be dealt with as an armed conflict. At the same time there is also risk in too liberal an approach that empowers overreach on the part of human rights law. Courts have to remain very sensitive to the effect of a loosening of the restrictions provided for in human rights law regarding the impact it can have on the broader and longer term role that body of law performs in a peacetime environment.

It will be an exceptional situation where there will not be a requirement to simultaneously apply some or all of the bodies of law impacting on operations in counterinsurgency and counterterrorism operations. The strength of a holistic approach, which contemplates the availability of the law enforcement option, is that it provides States and their security personnel with the full range of potential responses. It also demands that the practical effect of theoretical options be considered for the purposes of assessing their viability. This enables States not only to counter the threat across the full spectrum of violence but also permits security officials to make intelligent operational choices that are uniquely tailored to the nature of that threat.

---

[199] Gabriella Blum & Philip Heymann, *Law and Policy of Targeted Killing*, 1 HARV. NAT. SEC. J. 145, 170 (2010).

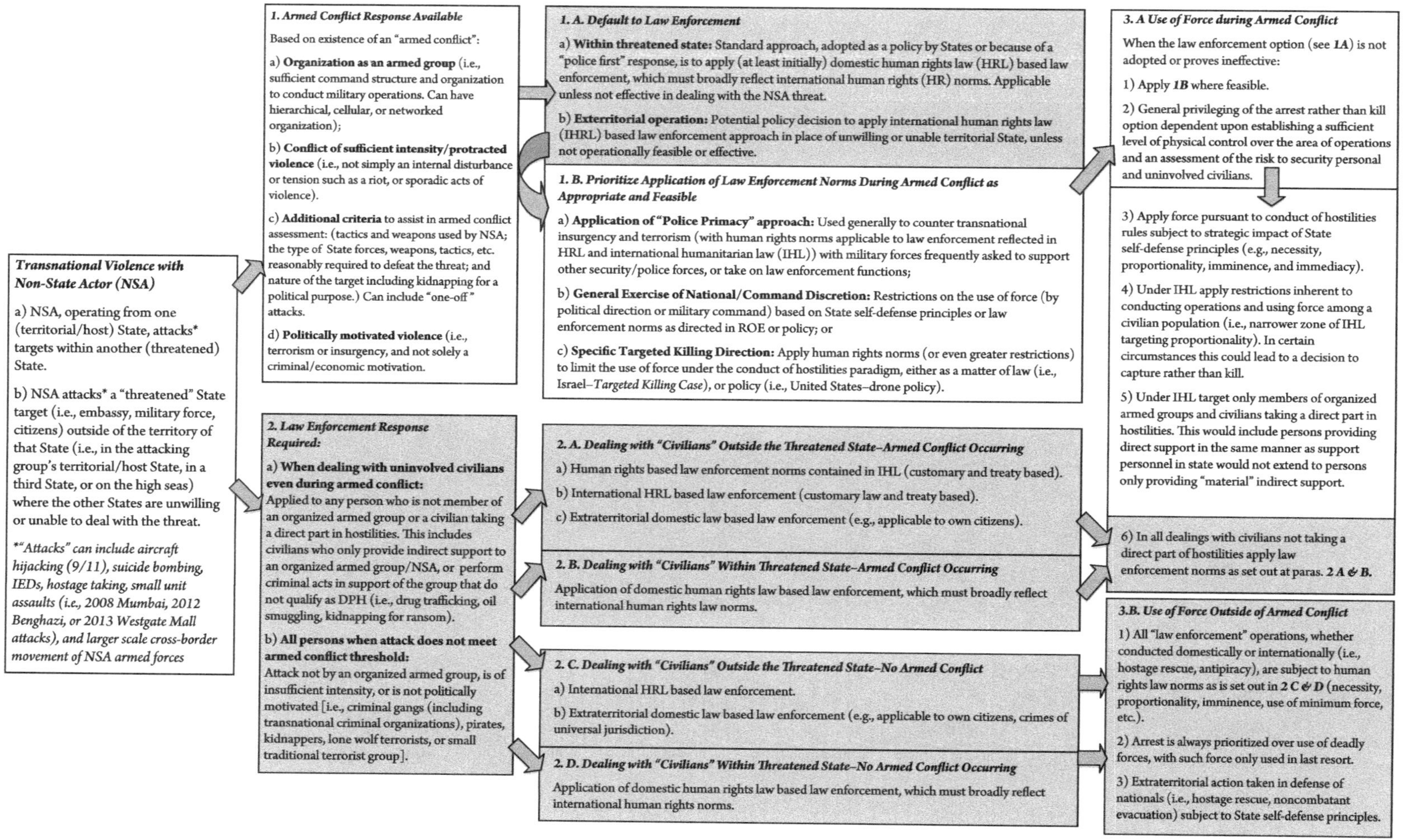
Transnational Violence with Non-State Actor (NSA)
a) NSA, operating from one (territorial/host) State, attacks* targets within another (threatened) State.
b) NSA attacks* a "threatened" State target (i.e., embassy, military force, citizens) outside of the territory of that State (i.e., in the attacking group's territorial/host State, in a third State, or on the high seas) where the other States are unwilling or unable to deal with the threat.
*"Attacks" can include aircraft hijacking (9/11), suicide bombing, IEDs, hostage taking, small unit assaults (i.e., 2008 Mumbai, 2012 Benghazi, or 2013 Westgate Mall attacks), and larger scale cross-border movement of NSA armed forces
1. Armed Conflict Response Available
Based on existence of an "armed conflict":
a) Organization as an armed group (i.e., sufficient command structure and organization to conduct military operations. Can have hierarchical, cellular, or networked organization);
b) Conflict of sufficient intensity/protracted violence (i.e., not simply an internal disturbance or tension such as a riot, or sporadic acts of violence).
c) Additional criteria to assist in armed conflict assessment: (tactics and weapons used by NSA; the type of State forces, weapons, tactics, etc. reasonably required to defeat the threat; and nature of the target including kidnapping for a political purpose.) Can include "one-off" attacks.
d) Politically motivated violence (i.e., terrorism or insurgency, and not solely a criminal/economic motivation.
2. Law Enforcement Response Required:
a) When dealing with uninvolved civilians even during armed conflict:
Applied to any person who is not member of an organized armed group or a civilian taking a direct part in hostilities. This includes civilians who only provide indirect support to an organized armed group/NSA, or perform criminal acts in support of the group that do not qualify as DPH (i.e., drug trafficking, oil smuggling, kidnapping for ransom).
b) All persons when attack does not meet armed conflict threshold:
Attack not by an organized armed group, is of insufficient intensity, or is not politically motivated [i.e., criminal gangs (including transnational criminal organizations), pirates, kidnappers, lone wolf terrorists, or small traditional terrorist group].
1. A. Default to Law Enforcement
a) Within threatened state: Standard approach, adopted as a policy by States or because of a "police first" response, is to apply (at least initially) domestic human rights law (HRL) based law enforcement, which must broadly reflect international human rights (HR) norms. Applicable unless not effective in dealing with the NSA threat.
b) Exterritorial operation: Potential policy decision to apply international human rights law (IHRL) based law enforcement approach in place of unwilling or unable territorial State, unless not operationally feasible or effective.
1. B. Prioritize Application of Law Enforcement Norms During Armed Conflict as Appropriate and Feasible
a) Application of "Police Primacy" approach: Used generally to counter transnational insurgency and terrorism (with human rights norms applicable to law enforcement reflected in HRL and international humanitarian law (IHL)) with military forces frequently asked to support other security/police forces, or take on law enforcement functions;
b) General Exercise of National/Command Discretion: Restrictions on the use of force (by political direction or military command) based on State self-defense principles or law enforcement norms as directed in ROE or policy; or
c) Specific Targeted Killing Direction: Apply human rights norms (or even greater restrictions) to limit the use of force under the conduct of hostilities paradigm, either as a matter of law (i.e., Israel–Targeted Killing Case), or policy (i.e., United States–drone policy).
2. A. Dealing with "Civilians" Outside the Threatened State–Armed Conflict Occurring
a) Human rights based law enforcement norms contained in IHL (customary and treaty based).
b) International HRL based law enforcement (customary law and treaty based).
c) Extraterritorial domestic law based law enforcement (e.g., applicable to own citizens).
2. B. Dealing with "Civilians" Within Threatened State–Armed Conflict Occurring
Application of domestic human rights law based law enforcement, which must broadly reflect international human rights law norms.
2. C. Dealing with "Civilians" Outside the Threatened State–No Armed Conflict
a) International HRL based law enforcement.
b) Extraterritorial domestic law based law enforcement (e.g., applicable to own citizens, crimes of universal jurisdiction).
2. D. Dealing with "Civilians" Within Threatened State–No Armed Conflict Occurring
Application of domestic human rights law based law enforcement, which must broadly reflect international human rights norms.
3. A Use of Force during Armed Conflict
When the law enforcement option (see 1A) is not adopted or proves ineffective:
1) Apply 1B where feasible.
2) General privileging of the arrest rather than kill option dependent upon establishing a sufficient level of physical control over the area of operations and an assessment of the risk to security personal and uninvolved civilians.
3) Apply force pursuant to conduct of hostilities rules subject to strategic impact of State self-defense principles (e.g., necessity, proportionality, imminence, and immediacy).
4) Under IHL apply restrictions inherent to conducting operations and using force among a civilian population (i.e., narrower zone of IHL targeting proportionality). In certain circumstances this could lead to a decision to capture rather than kill.
5) Under IHL target only members of organized armed groups and civilians taking a direct part in hostilities. This would include persons providing direct support in the same manner as support personnel in state would not extend to persons only providing "material" indirect support.
6) In all dealings with civilians not taking a direct part of hostilities apply law enforcement norms as set out at paras. 2 A & B.
3.B. Use of Force Outside of Armed Conflict
1) All "law enforcement" operations, whether conducted domestically or internationally (i.e., hostage rescue, antipiracy), are subject to human rights law norms as is set out in 2 C & D (necessity, proportionality, imminence, use of minimum force, etc.).
2) Arrest is always prioritized over use of deadly forces, with such force only used in last resort.
3) Extraterritorial action taken in defense of nationals (i.e., hostage rescue, noncombatant evacuation) subject to State self-defense principles.

The humanitarian law of war is a form of human rights law operating for the protection of those who cannot be properly safeguarded by the country to which they belong.[1]

15 Preparing for Twenty-First-Century Conflict

15.1. Introduction

The application of an operational law approach in contemporary operations is not without challenges. By definition it requires a broad knowledge of international and domestic law. It also demands a nuanced and sophisticated assessment of the threats and available responses. A key task is how to get legal advisors ready to provide wide-ranging operational law advice in a complex operating environment extending from law enforcement on the high seas (e.g., counterpiracy), to arresting member of organized armed groups hiding among the population, to more conventional conflict against insurgents involving house-to-house fighting (e.g., the 2004 Fallujah assault).

For State legal advisors this means being educated and trained, not only in international humanitarian law but also human rights law (international and domestic), the law governing the recourse to war, and international criminal law. In addition, there is an increasing requirement to know aspects of the legal systems of Coalition partners, as well as that of the territory in which you are operating (i.e., common law or civil law). This final chapter outlines the type of education, training, and other preparation that could be considered in order to produce legal advisors who can provide the required operational law advice in the complex security environment of the twenty-first century. The analysis discusses the need to incorporate the lessons learned from the post-9/11 period, view contemporary conflict as occurring on a much broader continuum of violence, encourage the academic community

[1] Richard Baxter, *Human Rights in War*, *in* HUMANIZING THE LAWS OF WAR: SELECTED WRITINGS OF RICHARD BAXTER 341, 346 (Detlev F. Vagts et al. eds., 2013).

to ground their theoretical work in the world of practice, and seek to adapt to, rather than resist, the changing legal and security landscape.

15.2. Legal Advisors and Twenty-First-Century Warfare

The scope of the challenge facing State legal advisors cannot be overstated in terms of the range of legal knowledge required of young military legal officers and civilian lawyers who enter government service primarily educated in the domestic laws of their own country. A further challenge in most State military cultures is the concentration made by military forces, and by extension their legal advisors, on preparation for the conduct of hostilities in inter-State warfare. This in turn causes legal advisors to emphasize knowledge of humanitarian law. This familiarity with conventional hostilities, while essential, has often occurred to the exclusion of other critical aspects of operational roles that are required to be performed by State military forces. This includes a lack of emphasis on the obligations relating to maintaining order among the civilian population. A focus on conventional warfare and the law governing hostilities is often evident in organizations where the "law of armed conflict," or the "law of war," is referred to in an exclusionary fashion. In a military context any demand by senior lawyers that these latter terms be used to the exclusion of "humanitarian law," or any indication that reference to human rights law is to be discouraged, says much about the limited scope of the strategic vision that can exist within a legal organization. Conversely, advocates of an exclusive application of human rights law and law enforcement, who largely reside within academia and some human rights organizations, provide an equally restricted conceptual framework with which to attempt to address exceptionally wide-ranging contemporary security threats.

A lack of readiness for other than conventional warfare became evident in the aftermath of the 2003 invasion of Iraq when U.S. military lawyers found themselves as unprepared for the occupation phase of the conflict as the rest of the military forces. The doctrinal and legal focus was not on what would come to be viewed as stability operations.[2] It has been observed that the U.S. military generally found itself "untrained and unfamiliar with tasks that they needed to accomplish to secure victory, such as restoring civil order, creating an interim government, establishing essential services, and ensuring that the judicial system was operational."[3] Similarly, the legal advisors were "responsible for cobbling the Iraqi judicial system back together . . . yet they had not been trained on the civil law system or on Iraqi law, procedure, or custom."[4] What is particularly noteworthy about this situation is that occupation law is part of the law governing international armed conflict and an integral aspect of conventional warfare.[5]

---

[2] Kenneth Watkin, *Stability Operations: A Guiding Framework for "Small Wars" and Other Conflicts of the Twenty-First Century?*, 85 INT'L LEG. STUD. 411, 412–14 (2009) (for a more detailed discussion of stability operations).

[3] Matthew R. Hover, *The Occupation of Iraq: A Military Perspective on Lessons Learned*, 94 INT'L REV. RED CROSS 339, 342 (2012).

[4] *Id.* at 343.

[5] YORAM DINSTEIN, THE INTERNATIONAL LAW OF BELLIGERENT OCCUPATION 1–2 (2009).

It has been suggested that in Iraq the military effort, including that of its legal advisors, appears to have been more focused on winning battles, rather than winning the war.[6] In legal terms this means concentration was placed on the conduct of hostilities provisions of humanitarian law, rather than on the full spectrum of legal obligations that arise even during international armed conflict. Notably, those obligations do not simply relate to the re-establishment and operation of a foreign legal system[7] but more broadly having expertise regarding a century-old obligation to apply international human rights norms to "restore, and ensure, as far as possible public order and safety, while respecting, unless absolutely prevented, the law in force in the country."[8] The negative operational consequences of failing to ensure adequate training in areas such as detainee handling were graphically demonstrated in the 2004 Abu Graib incident, which is widely viewed as having stoked the flames of the insurgency in Iraq[9] and marred the overall perception of the counterinsurgency effort.[10] This, like the torture of detainees, represented a strategic failure.[11]

The lack of attention to legal issues outside the context of the actual hostilities is not unique to the U.S. military. Deficiencies in the handling of detainees, interrogation, and tactical questioning were investigated regarding the British armed forces in Iraq.[12] The problems experienced by the Canadian Forces in Somalia in 1993, which included the torture and death of a young Somali detainee, were in part the result of a lack of doctrine and training on the detainee handling, particularly in other than conventional armed conflict situations.[13] While considerable effort was made following the Somalia incidents to prepare legal doctrine and conduct training regarding the treatment of detainees, a comprehensive Canadian Forces doctrine manual for the handling of detainees was only put in place just prior to the main deployment of the Canadian Forces to Afghanistan in 2002. In that case it was based significantly on the UK manual.

States such as the United Kingdom, the United States, and Canada are committed to having their military forces comply with the rule of law and their humanitarian law obligations.

---

[6] Hover, *supra* note 3, at 344.

[7] GC IV, art. 64.

[8] 1907 Hague Land Warfare Regulations, art. 43.

[9] THOMAS E. RICKS, FIASCO: THE AMERICAN MILITARY ADVENTURE IN IRAQ 290–91 (2007).

[10] JASON BURKE, THE 9/11 WARS 135 (2011) ("Abuse was of course far from universal . . . but it had become an integral part of the American military and counter-terrorist effort.").

[11] CHRIS MACKEY & GREG MILLER, THE INTERROGATORS: INSIDE THE SECRET WAR AGAINST AL QAEDA 477 (2004) (Where an interrogator concludes the reason why torture should not be used is "simply because it is wrong. It dehumanizes us, undermines our cause, and, over the long run breeds more enemies of the United States than coercive interrogation methods will ever allow us to capture."), and ERIK SAAR & VIVECA NOVAK, INSIDE THE WIRE 248 (2005) ("Guantánamo is a rallying cry throughout the Arab and Muslim world, even our closest allies oppose us in this venture. The bottom line is this: the minimal intelligence we are gathering from those held in Cuba is not worth the harm we are doing to our international reputation.").

[12] *See, e.g.*, III THE REPORT OF THE BAHA MOUSA INQUIRY, The Rt Hon Sir William Gage (Chairman) 1267–86 (2011), https://www.gov.uk/government/uploads/system/uploads/attachment_data/file/279192/1452_iii.pdf (for a listing of the seventy-three recommendations arising from the Public Inquiry into the death of an Iraqi citizen detained by British forces in Iraq in 2003).

[13] 5 DISHONOURED LEGACY: LESSONS OF THE SOMALIA AFFAIR, Report of the Commission of Inquiry into the Deployment of Canadian Forces to Somalia 1477–81 (1997) [hereinafter SOMALIA REPORT] (outlining eighteen recommendations to improve training and doctrine relating to peace support operations and the incorporation of non-traditional military training.).

However, even those forces have struggled to fully incorporate that law into the wide range of operational activities that occur in international armed conflict, let alone conflicts "not of an international character." This demonstrates the scope of the educational, training, and doctrinal challenge. Added to that challenge is the increasing realization there is a need for a strategic vision regarding the application of the law that extends beyond the traditional focus on a narrow range of conventional military operations. Integrating an operational law approach to the wider range of conflict involving non-State actors requires a vision that incorporates the inevitability of irregular war and the need to interface with the civilian population. This vision can only be realized by ensuring an adequate allocation of staff resources to prepare legal doctrine and conduct the appropriate training regarding these contemporary threats.

The knowledge required to prepare relevant doctrine extends far beyond a familiarity with the law. The lawyers involved must understand how the military functions and fights at all levels of command (i.e., strategic, operational, and tactical). At a minimum, this includes knowledge of contemporary military operations that involve what is increasingly called "hybrid warfare" (both regular and irregular conflict). They also must be comfortable with human rights law and its application in a policing context. This is a type of warfare that is very likely to include some form of law enforcement. Obviously, military forces have been engaged in policing-related duties (see Chapter 12, section 12.2.1, *The Military Task: Applying Force of First and Last Resort*). This is a role that is not going to disappear with the withdrawal of Western States from counterinsurgency operations in Iraq and Afghanistan. In one form or another law enforcement with human rights law–related restrictions on the use of force is an enduring part of counterterrorism, counterinsurgency, and regular military taskings (e.g., naval constabulary duties). The knowledge of the State legal advisors and the legal preparations for operations must match those tasks.

Challenges remain. It has been suggested that for the U.S. military the lessons of Iraq and Afghanistan have been learned. This is represented in government directives and in military doctrine placing "stability operations on an equal footing with conventional combat operations."[14] Similarly, as has been noted previously, the International and Operational Law Department of the Judge Advocate Generals Legal Center and School (United States Army) has developed an operational law manual, the *Operational Law Handbook*, which incorporates the hard-learned lessons of the post-9/11 period. The *Handbook* acknowledges the potential application of customary human rights law to overseas operations[15] and includes an entire chapter on the application of human rights law.[16] This document is exceptionally forward thinking. It demonstrates the leadership that the U.S. military has long displayed in integrating law into their operations. However, the *Handbook* is only a single service document (i.e., U.S. Army) and was generated in the context of an overall strategic atmosphere within which the United States basically maintains its position that there is

---

[14] Hover, *supra* note 3, at 445.

[15] Operational Law Handbook, §IV, 53–54 (Lieutenant-Commander David Lee, ed., 2015) [hereinafter Operational Law Handbook], http://www.loc.gov/rr/frd/Military_Law/pdf/operational-law-handbook_2015.pdf

[16] *Id.* Chapter 3, at 47–57.

no extraterritorial application of human rights treaty law.[17] This raises the question of how deeply imbedded its approach has become within that military, or the degree to which it reflects broader government thinking on this issue.

Further, there is likely to remain a legal interpretation within some parts of the broader international military legal community that humanitarian law operates entirely to exclude the general operation of human rights law during armed conflict (i.e., as a form of *lex specialis*, or specialized law). A residual reference can be found in the 2014 *Operational Law Handbook*, which indicates "[i]n situations of armed conflict, where the LOAC [law of armed conflict] provides specific guidance, LOAC will likely set the rules, not IHRL [international human rights law], and provide authoritative guidance for military action."[18] In addition, rather than make reference to the need to apply human rights–based law enforcement norms on operations, there remains the policy applied by some States of requiring compliance with the law of armed conflict in all military operations outside the context of armed conflict.[19] A unitary focus on humanitarian law lingers even though there appears to have been some adjustment in approach so human rights law is applied to supplement that body of law.[20] It also seems that a previous reluctance to recognize the application of human rights norms, at least in terms of their integration into humanitarian law in the context of the treatment of detainees, is in the process of being changed (see Chapter 5, section 5.3.2.3, *Divergence and Growing Reconciliation*). However, overall, it looks as though the acceptance of a wider application of the applicability of human rights law is in a state of flux with the strategic dialogue only now catching up to the operational realities of the past decade.

It should be noted that the United States is not alone in presenting an unclear picture on whether or how human rights law is applied to contemporary operations. Canada's position is not clearly stated, with reference having to be made to an International Committee of the Red Cross (ICRC) document, *The Use Of Force In Armed Conflicts: Interplay Between The Conduct Of Hostilities And Law Enforcement Paradigms*, to get an idea of its limited acceptance of the extraterritoriality of human rights law.[21] Mitigating this lack of clarity somewhat is the Canadian position presented in litigation related to its detainee policy in Afghanistan that "where an individual detained in the context of armed conflict is charged with criminal offences, Common Article 3 and Article 75 of Protocol I Additional to the Geneva Conventions apply, either directly or as a matter of customary international law, to guarantee a number of due process rights."[22] This approach is consistent with

---

[17] H.R. Committee, Concluding observations on the fourth periodic report of the United States of America, Human Rights Committee, UN Doc. CCPR/C/USA/CO/4, ¶ 4. (Apr. 23, 2014).

[18] Operational Law Handbook, *supra* note 15, at § V A. 3., at 55.

[19] *DoD Law of War Program*, Department of Defense, Directive Number 2311.01E, May 9, 2006 (Incorporating Change 1, November 15, 2010, Certified Current as of February 22, 2011), http://www.dtic.mil/whs/directives/corres/pdf/231101e.pdf, ¶ 4.1, at 2. *See also* Department of Defense Law of War Manual, Office of the General Counsel ¶ 3.1.1.2, at 71–72 (June 2015), http://www.dod.mil/dodgc/images/law_war_manual15.pdf.

[20] Operational Law Handbook, *supra* note 15, at § V A. 3., at 55.

[21] The Use Of Force In Armed Conflicts: Interplay Between The Conduct Of Hostilities And Law Enforcement Paradigms, ICRC Expert Meeting 5 n.21 (Nov. 2013).

[22] Amnesty International Canada and British Columbia Civil Liberties Union v. Chief of Defence Staff for the Canadian Forces, et al., Case File No. T-324-07, Respondent's Factum, ¶ 83, at 26 (18 January 2008), http://bccla.org/wp-content/uploads/2012/06/20010118-Afghan-Detainees-Factum-Crown.pdf (Afghanistan was considered to be a non-international armed conflict for the purposes of the litigation).

the recognition by the U.S. Supreme Court in *Hamdan v. Rumsfeld* that Article 75 of Additional Protocol I reflects customary international law provisions that applied in the noninternational armed conflict with Al Qaeda (see Chapter 5, section 5.3.2.3, *Divergence and Growing Reconciliation*).[23]

The uncertainty regarding the scope of the various bodies of law impacting on operations is not limited to the North American continent. With the European Court of Human Rights only recently agreeing in *Hassan v. The United Kingdom* to apply humanitarian law in order to interpret human rights treaty law, the full scope of how European military forces and their legal advisors will apply these bodies of law also remains very uncertain.[24] Unfortunately, this overall lack of certainty is compounded by the various and often diverse theories posited within the academic community on important legal issues. Such issues include the right to respond to non-State actor threats, the defense of nationals, the interface between the law governing the recourse to war and humanitarian law, the interaction between human rights law and humanitarian law, the use of drones, the thresholds to be applied for categorization of noninternational armed conflict, and what constitutes direct participation in hostilities. There is also insufficient analysis of the impact on human rights law should it be asked to regulate the use of force in the context of hostilities, rather than confronting ordinary criminal activity. Given the time that has passed since 9/11, the lack of consensus on these fundamental issues should raise concerns.

More has to be done at a strategic level to acknowledge that a broad range of laws, including customary international human rights law, apply to contemporary military operations. Changes in the traditional narrow focus of military doctrine would benefit from top-down government direction. However, until such clarity is provided and the discussion moves beyond almost ideologically driven arguments over the extraterritoriality of treaty law and the "special law" debate, military forces must prepare for the reality of the type of wars they have to fight. Not just the ones they would like to engage in. By far the dominant form of twenty-first-century armed conflict involves irregular warfare (e.g., guerrillas, insurgents, and terrorists). While it can have conventional aspects, this is a type of conflict firmly grounded in violence involving insurgency, terrorism, and criminal activity. It is also not new. Small wars have always been the predominate form of warfare.

Even in a conventional conflict, the requirement to act as an occupier or otherwise interface with uninvolved civilians means that human rights norms will govern aspects of the force used by State security forces. As has been discussed, State armed forces are regularly tasked to carry out policing duties in a wide range of operations (see Chapter 12, section 12.2.1, *The Military Task: Applying Force of First and Last Resort*). Broader recognition of that fact and actually changing the culture of State military forces will take a commitment beyond simply developing counterinsurgency doctrine to ensure troops do not stand by and permit wide-scale looting to occur. It requires leadership, including within the legal community, and a commitment to not repeat past mistakes. It has been suggested that the necessary cultural change within State military forces may have to wait until the next generation

23 Hamdan v. Rumsfeld, 45 I.L.M. 1130, 1155 (2006) (U.S. Supreme Court 2006).

24 Hassan v. The United Kingdom, Application No. 29750/09, Eur. Ct. H. R. ¶ 103 (2014), http://hudoc.echr.coe.int/sites/eng/pages/search.aspx?i=001-146501.

of leaders who have experienced the challenges of contemporary conflict reach more than senior levels of command.[25] This sentiment could be equally applied to the legal advisors. That is in the process of happening now, over a decade after 9/11.

One issue hanging over how military forces will prepare for future conflict is whether there will be an attempt to completely abandon counterinsurgency and counterterrorism doctrine and return to the traditional emphasis on "winning battles" (see Chapter 7, section 7.2, *Counterinsurgency in the Twenty-First Century*). The debate about prioritizing conventional inter-State conflict over that with non-State actors can be seen in the views expressed by senior U.S. military officers in late summer 2015 that Russia represents the most important existential threat to the United States,[26] while politicians and senior security officials indicate the most important threat remains Salafi jihadists.[27] It is evident this struggle for prioritization of the threat will continue. However, what is unlikely to occur to the extent it has following previous conflicts, such as with the U.S. military during the post-Vietnam period, is to have a complete abandonment of counterinsurgency or counterterrorism. In part, this is because a significant portion of many State armed forces—in particular air force and Special Forces—will continue to engage in or advise on both counterterrorism and counterinsurgency operations.

For example, in 2014 a number of States deployed military advisors to assist the Iraqi armed forces in countering the Islamic State push toward Baghdad and to ultimately "degrade and destroy" ISIS.[28] Given the quasi-conventional nature of Islamic State forces, it is inevitable that some combat resembled inter-State warfare. However, reclaiming the portions of Iraq seized by that group is ultimately a counterinsurgency task where avoiding the alienation of the population will be an overriding strategic goal. The foreign military forces deployed to advise and support the Iraqi military will have to be prepared to apply counterinsurgency doctrine, including championing a "police primacy" approach. This involves the application of more restrictive proportionality assessments in respect of the aerial targeting being carried out in support of various Iraqi forces (see Chapter 7, section 7.6.2.2, *Proportionality and Counterinsurgency*). The lawyers assisting those deployed advisors must be prepared to consider the full range of laws impacting on operations, including those of the territorial State in the course of that campaign. In order to do that, both the military doctrine and training received must fully reflect the operational law requirements inherent in such a complex operating environment.

---

[25] Hover, *supra* note 3, at 346 ("It is now up to current and future leaders to ensure that history does not repeat itself."). *See also* FRANK LEDWIDGE, LOSING SMALL WARS: BRITISH MILITARY FAILURE IN IRAQ AND AFGHANISTAN 267 (2011) (where a similar point is made that in respect to veterans of the Afghanistan and Iraq conflicts, "[t]hese promising leaders need to be cherished *now*. They must be given a chance . . . .").

[26] Paul Shinkman, *The Greatest Threat to America*, U.S. NEWS, Aug. 7, 2015, http://www.usnews.com/news/the-report/articles/2015/08/07/the-greatest-threat-to-america.

[27] *John Kerry: Daesh Biggest Threat to US, Not Russia*, MIDDLE EAST MONITOR (July 11, 2015, 11:40), https://www.middleeastmonitor.com/news/americas/19782-john-kerry-daesh-biggest-threat-to-us-not-russia, and Shinkman, *supra* note 26.

[28] *Oct. 10: U.S. Military Conducts Airstrikes Against ISIL in Syria and Iraq*, U.S. CENTRAL COMMAND NEWS RELEASE (Oct. 10, 2014), http://www.centcom.mil/en/news/articles/oct.-10-u.s.-military-conducts-airstrikes-against-isil-in-syria-and-iraq ("The strikes were conducted as part of the President's comprehensive strategy to degrade and destroy ISIL.").

An interesting bellwether of change regarding the approach that State armed forces take toward contemporary operations may be the extent to which military doctrine and training concentrates solely on engaging in combat, rather than the full spectrum of tasks required even during inter-State conflict. An alternative approach would be for standing doctrine and readiness training to incorporate the requirement to be prepared to conduct conventional warfare, counterinsurgency, counterterrorism, and stability operations; participate in UN-led operations, and carry out law enforcement as a normal aspect of operations. Even if this were not done, mission-specific training would need to address any additionally required skill sets.[29] What must be avoided in the future is the situation where military forces do not know what action to take should they be placed in a position of occupier or are confronted with a burgeoning insurgency as happened in Iraq in 2003. The retraining of conventionally trained military forces by the U.S. military in counterinsurgency operations in theater was impressive, but unfortunately there was a cost in losing the operational initiative until that process was put in place.[30]

For legal advisors this means moving away from a purely "law of armed conflict" regime to one that fully embraces the requirement to apply international human rights law, the domestic law of both the sending and territorial States, the law governing the recourse to war, and international criminal law. There should also be a recognition that the law of armed conflict is not solely or, in many situations, primarily about the conduct of hostilities. The focus on criminal law arises because of the increasing emphasis on accountability and the broader phenomenon of lawfare where States, human rights advocates, and non-State actors battle over public perceptions of legitimacy.

There should be a concerted effort to provide legal officers with an understanding of both the civil and common law systems. The issue does not only arise in the context of being able to address "host nation" problems, such as occurred for State military forces in Iraq where the major participants in the occupying forces inculcated in the common law were confronted with dealing with a civil law system. It also extends to understanding how each body of law may impact on approaches taken by Coalition partners toward the law governing operations. Coalition operations inevitably involve nations with interpretative approaches based on one of these two legal systems. Two reasons have been suggested for differences in legal interpretations:

> Firstly, common law statutes have to be read against a case law background, while civil law codes and statutes are the primary source of under Montesquieu's theory. Secondly, civil law judges are influenced by Rousseau's theory that the State is the source of all rights under the social contract, while English judges favour Hobbes' theory that the individual agreed to forfeit to the State only certain rights.[31]

---

29 SOMALIA REPORT, *supra* note 13, at 1479 (the problem of relying solely on general purpose combat training to prepare for the complex military operations of the 1990s involving non-State actors was recognized in a recommendation to devote the entire pre-deployment training period to "primarily to mission specific training.").

30 Carter Malkasian, *Counterinsurgency in Iraq: May 2003–January 2010*, *in* COUNTERINSURGENCY IN MODERN WARFARE 287, 296–301 (Daniel Marston & Carter Malkasian eds., 2008) (where it is noted that the counterinsurgency reforms did not start until late 2004 and was hindered in part by the highly centralized conventional warfare command structure).

31 William Tetley, *Mixed Jurisdictions: Common Law vs Civil Law (Codified and Uncodified)* (pt. 1) (Nov. 29, 1999), *available at* http://www.cisg.law.pace.edu/cisg/biblio/tetley.html.

The result can be a greater reliance on treaty texts by lawyers adopting a civil law approach and, conversely, more of a willingness on the part of common law lawyers to embrace customary international law. These different sources of legal interpretation can have operational effects that legal officers need to be sensitive to.

The requirement for State legal advisors to understand international human rights law extends beyond advising commanders on matters related to counterinsurgency, counterterrorism, law enforcement, and the maintenance of order. Legal advisors must be increasingly ready to address the critiques of human rights groups and deal with accountability processes that for some coalition States are framed uniquely in terms of human rights violations (e.g., European Court of Human Rights, American Court of Human Rights). Conversely, lawyers from those States must be prepared to work with military forces from the larger number of countries that are not subject to the jurisdiction of such tribunals and that do not operate with an overriding human rights law focus. In other words, there is a reliance on international humanitarian law that may be fettered by a policy application of human rights norms, but not as a matter of law.

In addition, different views adopted by human rights and humanitarian law communities, respectively, have been expressed on numerous important legal issues. There is also a debate regarding what form of investigation is required (e.g., civilian, military police, or the military chain of command). The human rights law approach is suggested to require any civilian death to be investigated,[32] while under humanitarian law the requirement to investigate only arises if there is a reasonable suspicion that such excessive collateral injury, death, or destruction had occurred.[33] State legal advisors must be ready to engage on issues such as these at the time of an "incident" in order to ensure a proper investigation is carried out. They also may have to counter calls for investigations that may not be required at law. This can only be done properly and convincingly if those advisors understand both areas of law.

Law is perhaps the most strategic of disciplines in that it primarily finds its genesis in the political actions of States. It is often said, although usually overstated, that States "make" international law. Setting aside the role of courts (both international and domestic) and the effect other bodies such as human rights organizations may have on the advancement of international law, States do make law in terms of creating treaties and establishing customary international law. It is essential that lawyers providing advice to State military forces become fully engaged at the strategic level in order to ensure their very practical "world" is reflected in government decisions, legal theory, arguments before international and domestic courts, and as part of the broader academic dialogue. Avoiding the use of Latin and instead using plain language to explain the legal theories will enhance this effort at communication outside the relatively small international legal community.

At the same time the role performed by academics, human rights advocates, and the media would benefit considerably by becoming more sensitive to the reality of

---

32 McKerr v. United Kingdom, Application No. 28883/95, Eur. Ct. H.R., Judgment ¶ 111 (2001), http://hudoc.echr.coe.int/sites/eng/pages/search.aspx?i=001-59451.

33 Kenneth Watkin, *Use of Force during Occupation: Law Enforcement and Conduct of Hostilities*, 94 Int'l Rev. Red Cross 267, 297–98 (2012). *See also Report of the Special Rapporteur on the promotion and protection of human rights and fundamental freedoms while countering terrorism, Ben Emmerson*, UN Doc. A/68/389 (Sept. 18, 2013) ¶¶ 41–45, at 11–13 (for a discussion of transparency and investigations).

contemporary conflict and the practical effects of the various theories and interpretations of the law presented for public discourse. This must include an accurate assessment of the true nature of the threat based on established facts and a realistic appraisal of how the suggested legal framework can effectively deal with it. The theory must be capable of enabling success at the tactical level. The analysis could also be made more multidisciplinary in nature by incorporating history, political science, and other relevant perspectives. In other words, facilitate the stepping out of an insular legal world that, like many State militaries, has to a great significant degree remained overly focused on twentieth-century inter-State conflict.

On one level it can be argued that the existing multiplicity of views on a number of international law issues, such as the categorization of non-State actor conflict (see Chapter 9, section 9.3, *Categorizing Conflict with Non-State Actors*), reflects the strength and robustness of the academic dialogue. Such debate is essential. However, robust or not, in order to be of any use a theory must be capable of being applied in an effective manner, lead to success, and ultimately prioritize the protection of the population. The tendency by many practitioners to look for generic indicia of an armed conflict rather than engage in seemingly endless and highly technical debates about conflict characterization highlights the limits and the unsettled nature of international law in this area.

In Chapter 1 it was noted that international law is at a crossroads. British General David Richards described the challenges facing that military in 2009 as "arguably at least our generation's horse and tank moment."[34] It could be argued the post-9/11 period also represents a watershed moment for international law. Although the post–Cold War period saw the beginning of a change from the focus on inter-State armed conflict that so heavily dominated international law, it really took the shock of the 2001 attacks on the twin towers in New York and on the Pentagon to force the international law community to begin to confront the changing nature of conflict. This does not mean there is no room for views that rely on traditional approaches such as the antiwar sentiment that fueled much of the significant advances in international law last century both in terms of the law governing the recourse to war and humanitarian law (see Chapter 2, section 2.3, *The Destructive Power of the State*).[35] The State use of force must be even more constrained in noninternational armed conflict than it is in inter-State warfare, with a fundamental difference being the extent it involves fighting "amongst the people."[36] However, non-State actor conflict is even less likely to be successfully "outlawed" than its inter-State counterpart, and labeling such a threat solely as a law enforcement matter does not fundamentally address the true nature of the threat or what methods and means may be required to defeat it.

The threats, and the international security environment within which they are presented, have fundamentally changed since 9/11. To a degree, the legal analysis of contemporary conflict has shifted. There is greater acceptance that States may act in self-defense against non-State actors, and that in some cases at least borders can be crossed to protect nationals and

---

[34] LEDWIDGE, *supra* note 25, at 257 (quoting from a speech to the Royal United Services Institute).

[35] Kenneth Watkin, *Assessing Proportionality: Moral Complexity and Legal Rules*, 8 Y.B. HUM. L. 1, 32–34 (2005) (for a discussion of the impact of pacifism and pacificism on proportionality and targeting).

[36] RUPERT SMITH, THE UTILITY OF FORCE: THE ART OF WAR IN THE MODERN WORLD 3 (2005).

national interest when territorial States are unwilling or unable to carry out their responsibilities as the "proper authority." That said, only limited emphasis outside the military has been made of the operational law approach for analyzing conflict.[37] What is necessary are a much greater acceptance and a deeper acknowledgment of the need for a holistic response to confront these unique twenty-first-century threats operating at the boundaries between humanitarian law, human rights law, the law governing the recourse to war, and international criminal law.

Given the breadth of knowledge required of an operational legal advisor, the education and training challenge is a significant one. Instructing lawyers to be able to provide advice in the area of humanitarian law must remain a major focus of the initial education and training of military legal advisors. However, significant steps can be taken by increasing international human rights law training. As one U.S. judge advocate noted in 2010, "the time is coming, if it has not already arrived, when judge advocates will require a more sophisticated knowledge of human rights law, not merely in an academic sense, but also as a practical aspect of operations."[38] In reality, the time is now long past when lawyers can deploy or serve in senior operational law advisory positions without an established knowledge in this area of the law. These are issues commanders and individual soldiers will have to address. Their legal advisors must be prepared to as well, or they will do a disservice to the "client." With that in mind, consideration should be given to having international human rights law form at least one-quarter of the initial instruction of military legal advisors. Making this happen will require a culture change not unlike that facing the broader military community that has demonstrated itself to be so focused on conventional operations.

Addressing human rights obligations in the context of humanitarian law training on occupation would provide an additional avenue through which both areas of the law and their interaction could be discussed simultaneously. To a certain extent the task of providing instruction on law enforcement issues is made easier by the lawyers initially undertaking their duties with a good understanding of domestic law (through their formal education) that is, of course, human rights based. However, the course of instruction should also introduce those lawyers to the international human rights case law and enforcement mechanisms (i.e., human rights tribunals), the law governing the recourse to war, a basic comparison between the civil and common law systems, and international criminal law. Where possible, the legal positions regarding the operation of human rights law by the major potential Coalition partners should be addressed. Importantly, national military legal doctrine (e.g., the U.S. *Operational Law Handbook*[39]) addressing these areas of law should be developed.

Armed with this broader based instruction, more detailed aspects of how these various bodies of law impact on operations can be incorporated into subsequent specialized training is areas such as targeting, the handling of detainees, the investigation of alleged war crimes and criminal acts, and Special Forces operations. Mission-specific training can also be designed to incorporate information regarding host nation laws and any unique operational law issues that may arise. For example, one participating State might be tasked with

---

[37] *For an exception, see* THE HANDBOOK OF THE INTERNATIONAL LAW OF MILITARY OPERATIONS (Terry Gill & Dieter Fleck eds., 2010).

[38] Brian J. Brill, *Human Rights: Time for Greater Judge Advocate Understanding*, THE ARMY LAWYER 54, 54 (2010).

[39] OPERATIONAL LAW HANDBOOK, *supra* note 15.

operating the detainee handling facility for a Coalition operation. In that context, specific instruction regarding the humane treatment of detainees, standards of interrogation, and the law governing the transfer of detainees should address a range of legal issues based in humanitarian and human rights law, as well as international criminal law. There should also be a focus on incorporating "lessons learned" from ongoing operations to ensure the education and training program is kept current. This more advanced training can be augmented by a directed self-study program, as well as by postgraduate education. Academic instruction should be incorporated into military courses and education. Participation in postgraduate programs has become an essential part of the equipping of legal advisors with the knowledge and analytical skills necessary to engage at the strategic level both inside and outside of government. Military legal officers selected for advanced studies should be required to study areas of the law other than humanitarian law in order to broaden their knowledge base.

Ultimately, State legal advisors must be comfortable in providing legal advice across the wide range of operations that military forces have been and will continue to be asked to perform to maintain order in its broadest sense in our increasingly globalized and interconnected society. The security landscape and the type of threats directed at States and their citizens come from an exceedingly diverse group of non-State actors. The relative certainty associated with inter-State warfare has given way to a dangerous and often murky interaction between terrorists, insurgents, and criminal organizations that transcends the protection ordinarily offered by international borders. State military forces must be capable not only of fighting conventional wars but also acting to maintain order and offer protection to the civilian population across the whole spectrum of conflict. The small wars of this century require a response that matches the diversity of the threat and the transnational scope within which it operates.

The various bodies of law applicable to contemporary operations have developed along specific lines to address particular challenges: restrain State recourse to war, regulate combat, police societies, and hold wrongdoers accountable. For a variety of historical and other reasons, and particularly because of the overwhelming focus of international law on inter-State conflict in the twentieth century, these areas of law have developed in a fashion that has separated them, often artificially, from their common roots. However, contemporary military operations increasingly necessitate the simultaneous operation of these various bodies of law. Humanitarian law or human rights law, and frequently both areas of law, will govern the use of force operations against non-State actors. Even when humanitarian law is applied, there has been an increasing reliance on human rights legal norms to restrict the use of force. However, when States act, those bodies of law often have to be applied within the overarching strategic of the law governing the recourse to war. In addition, decisions to use force will be frequently be subjected to an accountability framework governed by both international and domestic criminal law.

The boundaries developed to define the laws applicable to the use of force during the past century have less relevance in a quickly changing security environment dominated by a jihadist movement bent on eliminating international borders and the State system of order upon which international law and our way of life is based. The result is that in the twenty-first century, security forces are often required to operate and use force at the limits or boundaries of the settled law. In doing so they increasingly need to reconcile and integrate the application of these legal disciplines. Maintaining distinctions in the roles performed by these bodies of law, such as the human rights focus on maintaining order within a State while protecting its citizens, remains an important goal. Each area of law has a role to play. What is crucial is that the boundaries of these bodies of law are not allowed to become barriers to maintaining law and order and protecting civilians regardless of where they might live. It is in this context that an "operational law" approach to security operations must be further developed.

Bibliography

ABBAS, HASSAN, THE TALIBAN REVIVAL: VIOLENCE AND EXTREMISM ON THE PAKISTAN-AFGHANISTAN FRONTIER (2014).

Abbott, Kirby, *"Terrorists: Criminals, Combatants or . . . .": The Question of Combatancy*, *in* THE MEASURES OF INTERNATIONAL LAW: EFFECTIVENESS, FAIRNESS AND VALIDITY 366 (2004).

ABIZAID, JOHN P. & ROSA BROOKS, RECOMMENDATIONS AND REPORT OF THE TASK FORCE ON US DRONE POLICY, The Stimson Center (2014), http://www.stimson.org/images/uploads/task_force_report_final_web_062414.pdf.

Abresch, William, *A Human Rights Law of Internal Armed Conflict: The European Court of Human Rights in Chechnya*, 16 EUR. J. INT'L L. 741 (2005).

ADAMS, JAMES ET AL., AMBUSH: THE WAR BETWEEN THE SAS AND THE IRA (1988).

Akande, Dapo, *Classification of Armed Conflicts: Relevant Legal Concepts*, *in* INTERNATIONAL LAW AND THE CLASSIFICATION OF CONFLICTS 32 (Elizabeth Wilmshurst ed., 2012).

Akande, Dapo & Milanović , Marco, *The Constructive Ambiguity of the Security Council's ISIS Resolution*, EJIL: *Talk!* (Nov. 21, 2015), http://www.ejiltalk.org/the-constructive-ambiguity-of-the-security-councils-isis-resolution/.

Akande, Dapo & Thomas Liefländer, *Clarifying Necessity, Imminence, and Proportionality in the Law of Self-Defense*, 107 AM. J. INT'L L. 563 (2013).

Aldrich, George H., *Prospects for United States Ratification of Additional Protocol I to the 1949 Geneva Conventions*, 85 AM. J. INT'L L. 1 (1991).

Alston, Phillip, *The CIA and Targeted Killings Beyond Borders*, 2 HARV. NAT'L. SEC. J. 283 (2011).

Anderson, Kenneth, *Targeted Killing and Drone Warfare: How We Came to Debate Whether There Is a "Legal Geography of War," in* FUTURE CHALLENGES IN NATIONAL SECURITY AND LAW (Peter Berkowitz ed., 2011), http://www.futurechallengesessays.com.

______, *Efficiency in Bello and ad Bellum: Targeted Killing Through Drone Warfare* (2011), *available at* http://papers.ssrn.com/sol3/papers.cfm?abstract_id=1812124.

______, *Readings: Civilian Intelligence Agencies and the Use of Armed Drones by Ian Henderson*, Lawfare (June 27, 2014, 3:00 PM), http://www.lawfareblog.com/2014/06/readings-civilian-intelligence-agencies-and-the-use-of-armed-drones-by-ian-henderson/.

Archick, Kristen, *U.S.-EU Cooperation Against Terrorism, Congressional Research Service* (2013), http://www.fas.org/sgp/crs/row/RS22030.pdf.

Arimatsu, Louise & Michael N. Schmitt, *Attacking "Islamic State" and the Khorasan Group: Surveying the International Law Landscape*, Colum. J. Transnat'l L. Bulletin 1 (2014), http://jtl.columbia.edu/attacking-islamic-state-and-the-khorasan-group-surveying-the-international-law-landscape//.

Arostegui, Marten C., Twilight Warriors: Inside the World's Special Forces (1996).

Arquilla, John, Insurgents, Raiders, and Bandits: How Masters of Irregular Warfare Have Shaped Our World (2011).

Arquilla, John & David Ronfeldt, The Advent of Netwar (1996).

Asher, Michael, Shoot to Kill: A Soldier's Journey Through Violence (1990).

Asprey, Robert B., 1 War in the Shadows: The Guerrilla in History (1975).

______, War in the Shadows: The Guerrilla in History (1994).

Atwan, Abdel Bari, After Bin Laden: Al Qaeda, The Next Generation (2012).

Aussaresses, Paul, The Battle of the Casbah: Counterterrorism and Torture in Algeria (Robert L. Miller trans., 2006).

Balko, Radley, Rise of the Warrior Cop: The Militarization of America's Police Forces (2013).

Barak, Aharon, Proportionality: Constitutional Rights and Their Limitations (2012).

Barfield, Thomas, Afghanistan: A Cultural and Political History (2010).

Bar-Zohar, Michael & Nissim Mishal, Mossad: The Greatest Missions of the Israeli Secret Service (2014).

______, No Mission is Impossible: The Death Defying Missions of the Israeli Special Forces (2015).

Baxter, Richard R., So-called *"Unprivileged Belligerency": Spies, Guerrillas, and Saboteurs*, *in* Humanizing the Laws of War: Selected Writings of Richard Baxter 37 (Detlev F. Vagts et al. eds., 2013).

______, *A Skeptical Look at the Concept of Terrorism*, *in* Detlev F. Vagts et al. eds., Humanizing the Laws of War, 211.

______, *Human Rights in War*, *in* Detlev F. Vagts et al. eds., Humanizing the Laws of War, 244.

______, *Humanitarian Law or Humanitarian Politics?*, *in* Detlev F. Vagts et al. eds., Humanizing the Laws of War, 287.

Bayley, David H. & Robert M. Perito, The Police in War: Fighting Insurgency, Terrorism, and Violent Crime (2010).

Beckett, Ian F. W., Modern Insurgencies and Counter-insurgencies: Guerrillas and their Opponents since 1750 (2001).

Beckwith, Charles A. & Donald Knox, Delta Force: The U.S. Counter-Terrorism Unit and the Iranian Hostage Rescue Mission (1983).

Beer, Yishai, *Humanity Considerations Cannot Reduce War's Hazards Alone: Revitalizing the Concept of Military Necessity*, 26 EUR. J. INT'L L. 801 (2015)

Beittel, S., *Mexico's Drug Trafficking Organizations: Source and Scope of the Rising Violence*, Congressional Research Service (Apr. 15, 2013), http://fpc.state.gov/documents/organization/155587.pdf.

Bellinger, John B. & William J. Haynes II, *A US Government Response to the International Committee of the Red Cross Study Customary International Humanitarian Law*, 89 INT'L REV. RED CROSS 443 (2007).

Bellinger, John B., & Vijay M. Padmanabhan, *Detention Operations in Contemporary Conflicts: Four Challenges for The Geneva Conventions and Other Existing Law*, 105 AM. J. INT'L L. 201 (2011).

BEN-NAFTALI, ORNA ED., INTERNATIONAL HUMANITARIAN LAW AND INTERNATIONAL HUMAN RIGHTS LAW (2011).

Bennett, Matthew, *The German Experience*, *in* THE ROOTS OF COUNTERINSURGENCY: ARMIES AND GUERRILLA WARFARE 1900–1945 (Ian F. W. Beckett ed., 1988).

BENVENISTI, EYAL, THE INTERNATIONAL LAW OF OCCUPATION (1993).

______, *Rethinking the Divide Between Jus ad Bellum and Jus in Bello in Warfare Against Nonstate Actors*, 34 YALE J. INT'L L. 541 (2009).

______, *The Legal Battle to Define the Law on Transnational Asymmetric Warfare*, 20 DUKE J. COMP. & INT'L L. 339 (2010).

Beran, Matthew, *How Counterinsurgency Changes "Military Advantage"*, THE ARMY LAWYER (2010), *available at* http://www.loc.gov/rr/frd/Military_Law/pdf/08-2010.pdf.

BERGEN, PETER L., THE LONGEST WAR: THE ENDURING CONFLICT BETWEEN AMERICA AND AL-QAEDA (2011).

______, MANHUNT: THE TEN-YEAR SEARCH FOR BIN LADEN FROM 9/11 TO ABBOTTABAD (2012).

BEST, GEOFFREY, WAR AND LAW SINCE 1945 (2002) (1994).

Bethlehem, Daniel, *Principles Relevant to the Scope of a State's Right as Self-Defense Against an Imminent or Actual Armed Attack by Nonstate Actors*, 106 AM. J. INT'L L. 770 (2012).

BETSER, MUKI, SECRET SOLDIER (1996).

BIDDLE, STEPHEN & JEFFREY A. FRIEDMAN, THE 2006 LEBANON CAMPAIGN AND THE FUTURE OF WARFARE: IMPLICATIONS FOR ARMY AND DEFENSE POLICY (2008), http://www.strategic-studiesinstitute.army.mil/pdffiles/pub882.pdf.

BLANFORD, NICHOLAS, WARRIORS OF GOD: INSIDE HEZBOLLAH'S THIRTY-YEAR STRUGGLE AGAINST ISRAEL (2011).

Blank, Laurie R., *Defining the Battlefield in Contemporary Conflict and Counterterrorism: Understanding the Parameters of the Zone of Combat*, GEO. J. INT'L & COMP. L. 1 (2010).

Blank, Laurie R. & Geoffrey S., *Corn, Syria Must be Held to the Law of War*, CNN (Apr. 4, 2012), http://www.cnn.com/2012/04/04/opinion/blank-corn-syria-war/index.html.

THE BLUE HELMETS: A REVIEW OF UNITED NATIONS PEACE-KEEPING (2nd ed. 1990).

Blum, Gabriella & Philip Heymann, *Law and Policy of Targeted Killing*, 1 HARV. NAT. SEC. J. 145 (2010).

Boddens Hosang, Hans F. R., *Force Protection, Unit Self-Defence, and Extended Self-Defence*, *in* THE HANDBOOK OF THE INTERNATIONAL LAW OF MILITARY OPERATIONS 413 (Terry Gill & Dieter Fleck eds., 2010).

BOLGER, DANIEL P., WHY WE LOST: A GENERAL'S INSIDE ACCOUNT OF THE IRAQ AND AFGHANISTAN WARS (2014).

BOOT, MAX, THE SAVAGE WARS OF PEACE: SMALL WARS AND THE RISE OF AMERICAN POWER (2002).

______, INVISIBLE ARMIES: AN EPIC HISTORY OF GUERRILLA WARFARE FROM ANCIENT TIMES TO THE PRESENT (2013).

BOOTHBY, WILLIAM H., THE LAW OF TARGETING (2012).

Bothe, Michael, *The Law of Neutrality*, *in* THE HANDBOOK OF INTERNATIONAL HUMANITARIAN LAW 571 (Dieter ed., 2nd ed. 2008).

BOTHE, MICHAEL, KARL JOSEF PARTSCH, & WALDEMAR A. SOLF, NEW RULES FOR VICTIMS OF ARMED CONFLICTS: COMMENTARY ON THE TWO 1977 PROTOCOLS ADDITIONAL TO THE GENEVA CONVENTIONS OF 1949 (1982).
BOWDEN, MARK, BLACKHAWK DOWN: A STORY OF MODERN WAR (2010).
BOWETT, D. W., SELF-DEFENSE IN INTERNATIONAL LAW (The Lawbook Exchange Ltd., 2009) (1958).
BOYNE, WALTER J., THE INFLUENCE OF AIRPOWER UPON HISTORY (2005).
BRABAZON, JAMES, MY FRIEND THE MERCENARY (2010).
Brands, Hal, *Third-Generation Gangs and Criminal Insurgency in Latin America*, *in* JOHN P. SULLIVAN & ROBERT J. BUNKER, MEXICO'S CRIMINAL INSURGENCY 43 (2012).
BRAUDY, LEO, FROM CHIVALRY TO TERRORISM: WAR AND THE CHANGING NATURE OF MASCULINITY (2003).
BREAU, SUSAN, MARIE ARONSSON, & RACHEL JOYCE, DISCUSSION PAPER 2: DRONE ATTACKS, INTERNATIONAL LAW, AND THE RECORDING OF CIVILIAN CASUALTIES OR ARMED CONFLICT, Oxford Research Group (June 2011), http://www.oxfordresearchgroup.org.uk/sites/default/files/ORG%20Drone%20Attacks%20and%20International%20Law%20Report.pdf.
Brill, Brian J., *Human Rights: Time for Greater Judge Advocate Understanding*, THE ARMY LAWYER 54 (2010).
Bronitt, Simon & Dale Stephens, *"Flying Under the Radar"—The Use of Lethal Force Against Hijacked Aircraft: Recent Australian Developments*, 7 OXFORD U. COMM. L.J. 265 (2007).
BROWNLIE, IAN, INTERNATIONAL LAW AND THE USE OF FORCE BY STATES (1963).
______, PRINCIPLES OF PUBLIC INTERNATIONAL LAW (7th ed. 2008).
BRUCE, VICTORIA, KARIN HAYES, & JORGE ENRIQUE BOTERO, HOSTAGE NATION: COLUMBIA'S GUERRILLA ARMY AND THE FAILED WAR ON DRUGS (2010).
BUCHANAN, JESSICA, ERIK LANDEMALM, & ANTHONY FLACCO, IMPOSSIBLE ODDS: THE KIDNAPPING OF JESSICA BUCHANAN AND HER DRAMATIC RESCUE BY SEAL TEAM SIX (2013).
Bunker, Robert J., *The Mexican Cartel Debate: As Viewed Through Five Divergent Fields of Security Studies*, *in* JOHN P. SULLIVAN & ROBERT J. BUNKER, MEXICO'S CRIMINAL INSURGENCY 107 (2012).
BURKE, JASON, AL QAEDA: THE TRUE STORY OF RADICAL ISLAM (2004).
______, THE 9/11WARS (2011).
BURLEIGH, MICHAEL, BLOOD & RAGE: A CULTURAL HISTORY OF TERRORISM (2009).
______, SMALL WARS, FAR AWAY PLACES: GLOBAL INSURRECTION AND THE MAKING OF THE MODERN WORLD (2013).
BYMAN, DANIEL, A HIGH PRICE: THE TRIUMPHS & FAILURES OF ISRAELI COUNTERTERRORISM (2011).
______, *Why Drones Work: The Case for Washington's Weapon of Choice*, 92 FOREIGN AFF. 33 (July/August 2013).
CABLE, LARRY E., CONFLICT OF MYTHS: THE DEVELOPMENT OF AMERICAN COUNTERINSURGENCY DOCTRINE AND THE VIETNAM WAR (1986).
CALLWELL, C. E., SMALL WARS: THEIR PRINCIPLES AND PRACTICE (3rd ed. 1996) (1906).
CAMP, DICK, OPERATION PHANTOM FURY: THE ASSAULT AND CAPTURE OF FALLUJAH, IRAQ (2009).
Cannizzaro, Enzo, *Contextualizing Proportionality: Jus ad Bellum and Jus in Bello in the Lebanese War*, 88 INT'L REV. RED CROSS 779 (2006).
CASSESE, ANTONIO, INTERNATIONAL LAW (2nd ed. 2005).
______, INTERNATIONAL CRIMINAL LAW (2008).
______, *Expert Opinion on Whether Israel's Targeted Killings of Palestinian Terrorists Is Consonant with International Humanitarian Law*, http://www.stoptorture.org.il/files/cassese.pdf.

CASSIDY, ROBERT M., COUNTERINSURGENCY AND THE GLOBAL WAR ON TERROR 117 (2008).

CATIGNANI, SERGIO, ISRAELI COUNTER-INSURGENCY AND THE INTIFADAS (2008).

______, *The Israel Defense Forces and the Al-Aqsa Intifada: When Tactical Virtuosity Meets Strategic Disappointment, in* COUNTERINSURGENCY IN MODERN WARFARE 233 (Daniel Marston & Carter Malkasian eds., 2010).

CHEHAB, ZAKI, INSIDE THE RESISTANCE: THE IRAQI INSURGENCY AND THE FUTURE OF THE MIDDLE EAST (2005).

______, INSIDE HAMAS: THE UNTOLD STORY OF THE MILITANT ISLAMIC MOVEMENT (2007).

Chima, Jugdep S., *The Punjab Police and Counterinsurgency against Sikh Militants in India: The Successful Convergence of Interests, Identities and Institutions, in* POLICING INSURGENCIES: COPS AS COUNTERINSURGENTS 258 (C. Christine Fair & Sumit Ganguly eds., 2014).

CHUN, CLAYTON K. S., THE LAST BOARDING PARTY: THE USMC AND THE SS MAYAGUEZ 1975 (2011).

CHURCHILL, WINSTON, THE STORY OF THE MALAKIND FIELD FORCE: AN EPISODE OF FRONTIER WAR (Octopus Publishing Group 1989) (1898).

CIGAR, NORMAN, AL-QA'DA'S DOCTRINE FOR INSURGENCY (2009).

THE CIVILIAN IMPACT OF DRONES: UNEXAMINED COSTS, UNANSWERED QUESTIONS, Center for Civilians in Conflict/Columbia Law School (2012), *available at* http://web.law.columbia.edu/human-rights-institute/counterterrorism/drone-strikes/civilian-impact-drone-strikes-unexamined-costs-unanswered-questions.

CLANCY, TOM, CARL STINER, & TONY KOLTZ, SHADOW WARRIORS: INSIDE THE SPECIAL FORCES (2002).

Cochrane, Mark *The Role of the Royal Ulster Constabulary in Northern Ireland, in* POLICING INSURGENCIES: COPS AS COUNTERINSURGENTS 107 (C. Christine Fair & Sumit Ganguly eds., 2014).

COCKBURN, ANDREW, KILL CHAIN: THE RISE OF THE HIGH-TECH ASSASSINS (2015).

COCKBURN, PATRICK, MUQTADA AL-SADR AND THE SHIA INSURGENCY IN IRAQ (2008).

______, THE JIHADIS RETURN: ISIS AND THE NEW SUNNI UPRISING (2014).

COHEN, AARON, BROTHERHOOD OF WARRIORS (2008).

COMMENTARY ON THE HPCR MANUAL ON INTERNATIONAL LAW APPLICABLE TO AIR AND MISSILE WARFARE, Program on Humanitarian Policy and Conflict Research at Harvard University (2010), http://ihlresearch.org/amw/Commentary%20on%20the%20HPCR%20Manual.pdf.

CONNAUGHTON, RICHARD, MODERN WARFARE: THE TRUE STORY OF CONFLICT FROM THE FALKLANDS TO AFGHANISTAN (2008).

COOLEY, JOHN K., UNHOLY WARS: AFGHANISTAN, AMERICA AND INTERNATIONAL TERRORISM (2000).

CORDESMAN, ANTHONY H., GEORGE SULLIVAN & WILLIAM D. SULLIVAN, LESSONS OF THE 2006 ISRAELI-HEZBOLLAH WAR (2007).

Corn, Gary P., *Should the Best Offense Ever Be a Good Defense*, Strategy Research Project (Carlisle Barracks, PA: U.S. Army War College, Apr. 15, 2014).

Corn, Geoffrey, *Hamdan, Lebanon, and the Regulation of Hostilities: The Need to Recognize a Hybrid Category of Armed Conflict*, 40 VAND. J. TRANSNAT'L. L. 295 (2007).

______, *Mixing Apples and Hand Grenades: The Logical Limit of Applying Human Rights Norms to Armed Conflict*, 1 J. INT'L HUM. LEG. STUD. 52 (2010).

______, *Self-defense Targeting: Blurring the Line between the Jus ad Bellum and the Jus in Bello*, 88 INT'L L. STUD. 57 (2012).

Corn, Geoffrey S. & Eric Talbot Jensen, *Untying the Gordian Knot: A Proposal for Determining Applicability of the Laws of War to the War on Terror*, 81 TEMPLE L. REV. 787 (2008).

______, *Transnational Armed Conflict: A "Principled" Approach to the Regulation of Counter-Terror Combat Operations*, 42 ISR. L. REV. 1 (2009).

Corn, Geoffrey S., Laurie R. Blank, Chris Jenks, & Eric Talbot Jensen, *Belligerent Targeting and the Invalidity of a Least Harmful Means Rule*, 89 INT'L L. STUD. 536 (2013).

Corn, Geoffrey S. & Gary P. Corn, *The Law of Operational Targeting: Viewing the LOAC Through an Operational Lens*, 47 TEX. INT'L L.J. 337 (2012).

CORTEN, OLIVIER, THE LAW AGAINST WAR: THE PROHIBITION ON THE USE OF FORCE IN CONTEMPORARY INTERNATIONAL LAW (2012).

CORUM, JAMES S. & WRAY R. JOHNSON, AIRPOWER IN SMALL WARS: FIGHTING INSURGENTS AND TERRORISTS (2003).

COULON, JOCELYN, SOLDIERS OF DIPLOMACY: THE UNITED NATIONS, PEACEKEEPING, AND THE NEW WORLD ORDER (Phyllis Aronoff & Howard Scott trans., 1998).

CRAWFORD, EMILY, THE TREATMENT OF COMBATANTS AND INSURGENTS UNDER THE LAW OF ARMED CONFLICT (2010).

CRAWFORD, JAMES, THE CREATION OF STATES IN INTERNATIONAL LAW (2nd ed. 2006).

CRAWFORD, NETA, ACCOUNTABILITY FOR KILLING (2013).

CRIST, DAVID, THE TWILIGHT WAR: THE SECRET HISTORY OF AMERICA'S THIRTY-YEAR CONFLICT WITH IRAN (2012).

CRONIN, AUDRY KURTH, HOW TERRORISM ENDS: UNDERSTANDING THE DECLINE AND DEMISE OF TERRORIST CAMPAIGNS (2009).

______, *Why Drones Fail: When Tactics Drive Strategy*, 92 FOREIGN AFF. 44 (May/June 2013)

______, *ISIS IS NOT A TERRORIST GROUP: WHY COUNTERRORISM WON'T STOP THE LATEST JIHADIST THREAT*, FOR. AFF. 87 (MAR./APR. 2015).

CRYER, ROBERT ET AL., AN INTRODUCTION TO INTERNATIONAL CRIMINAL LAW AND PROCEDURE (2nd ed. 2010).

CUCULLU, GORDON & CHRIS FONTANA, WARRIOR POLICE (2011).

CULLEN, ANTHONY, THE CONCEPT OF NON-INTERNATIONAL ARMED CONFLICT IN INTERNATIONAL HUMANITARIAN LAW (2010).

CUTLER, THOMAS J., BROWN WATER, BLACK BERETS (1988).

Dabone, Zakaria, *International Law; Armed Groups in a State-centric System*, 93 INT'L REV. RED CROSS 395 (2011).

DALLAIRE, ROMEO & BRENT BEARDSLEY, SHAKE HANDS WITH THE DEVIL (2003).

David, Paul A., *Path Dependence—A Foundational Concept for Historical Social Science*, CLIOMETRICA—1 J. HIST. ECONOMICS & ECONOMETRIC HIST. 91 (2007).

DAVID, SAUL, OPERATION THUNDERBOLT: FLIGHT 139 AND THE RAID ON ENTEBBE AIRPORT, THE MOST AUDACIOUS HOSTAGE RESCUE MISSION IN HISTORY (2015).

David, Steven R., *Fatal Choices: Israel's Policy of Targeted Killing*, *in* DEMOCRACIES AND SMALL WARS 138 (Efraim Inbar ed., 2003).

DAVIDSON SMITH, G., COMBATING TERRORISM (1990).

DEBRAY, REGIS, REVOLUTION IN THE REVOLUTION? (Bobye Ortiz trans., 1967)

Deeks, Ashley S., *Pakistan's Sovereignty and the Killing of Osama Bin Laden*, 15 AM. SOC. INT'L L. INSIGHTS (May 5, 2011), http://www.asil.org/insights/volume/15/issue/11/pakistans-sovereignty-and-killing-osama-bin-laden.

______, *"Unwilling or Unable": Toward a Normative Framework for Extra-Territorial Self-Defense*, VA. J. OF INT'L L. 483 (2012).

______, *U.S. Airstrikes Against ISIS in Syria? Possible International Legal Theories*, LAWFARE (Aug. 23, 2014, 3:04 PM), http://www.lawfareblog.com/us-airstrikes-against-isis-syria-possible-international-legal-theories.

Dennis, Michael J., *Application of Human Rights Treaties Extraterritorially in Times of Armed Conflict and Military Occupation*, 99 AM. J. INT'L .L. 119 (2005).

Department of Defense Dictionary of Military and Associated Terms, Nov. 8, 2012 (as amended through Mar. 15, 2015), http://www.dtic.mil/doctrine/dod_dictionary/index.html.

Dewar, Michael, The British Army in Northern Ireland (1985).

Dilegge, David, *Foreword: Mexico's Criminal Insurgency and SWJ El Centro*, *in* John P. Sullivan & Robert J. Bunker, Mexico's Criminal Insurgency xi (2012).

Dinstein, Yoram, *Terrorism and Afghanistan*, 85 Int'l L. Stud. 43 (2008).

______, The International Law of Belligerent Occupation (2009).

______, The Conduct of Hostilities Under the Law of International Armed Conflict (2nd ed. 2010).

______, War Aggression and Self-Defence (5th ed. 2011).

______, Non-International Armed Conflicts in International Law (2014).

Doherty, Richard, The Thin Green Line: The History of the Royal Ulster Constabulary GC 1922–2001 (2012).

Doswald-Beck, Louise ed., San Remo Manual on International Law Applicable to Armed Conflicts at Sea (1995).

______, *The Right to Life in Armed Conflict: Does International Humanitarian Law Provide All the Answers?*, 88 Int'l Rev. Red Cross 881 (2006).

Draper, G. I. A. D., The Red Cross Conventions (1958).

______, *The Status of Combatants and the Question of Guerrilla Warfare*, 24 Brit. Y.B. Int'l L. 172 (1971).

______, *Combatant Status: A Historical Perspective*, 11 Mil. L. and Law of War Rev. 135 (1972).

______, *The Legal Classification of Belligerent Individuals*, *in* Reflections on the Law and Armed Conflict 196 (Michael A. Meyer & Hilaire McCoubrey eds., 1998).

Droege, Cordula, *Elective Affinities? Human Rights and Humanitarian Law*, 90 Int'l Rev. Red Cross 501 (2008).

Dunlop, John B., The 2002 Dubrovka and 2004 Beslan Hostage Crises: A Critique of Russian Counter-Terrorism (2006).

Dunstan, Simon, Israel's Lightning Strike: The Raid on Entebbe 1976 (2009).

Ellis, John, From the Barrel of a Gun: A History of Guerrilla, Revolutionary and Counter-Insurgency Warfare, from the Romans to the Present (1995).

Fadok, David S., *John Boyd and John Warden: Airpower's Quest for Strategic Paralysis*, *in* The Paths of Heaven: The Evolution of Airpower Theory 357 (Phillip S. Meilinger ed., 1997).

Fair, C. Christine & Sumit Ganguly, *Introduction: The Police in Counterinsurgency Operations*, *in* Policing Insurgencies: Cops as Counterinsurgents 1 (C. Christine Fair & Sumit Ganguly eds., 2014).

Fall, Bernard, *The Theory and Practice of Insurgency and Counterinsurgency*, 17 Nav. Col. Rev., 21 (1965).

Fergusson, James, A Million Bullets: The Real Story of the British Army in Afghanistan (2008)

______,Taliban: The True Story of the World's Most Feared Guerrilla Fighters (2010).

______, The World's Most Dangerous Place: Inside the Outlaw State of Somalia (2013).

Ferraro, Tristan, *Determining the Beginning and End of an Occupation under International Humanitarian Law*, 94 Int'l Rev. Red Cross 133 (2012).

Fidler, David P., *Police in Counterinsurgency: The Challenge of Comprehensive Reformation*, *in* Policing Insurgencies: Cops as Counterinsurgents 317 (C. Christine Fair & Sumit Ganguly eds., 2014).

Fishel, John T., *Forward*, *in* MANWARING, MODERN ASYMMETRIC WARFARE, vii.
FISHER, LOUIS, NAZI SABOTEURS ON TRIAL (2nd ed. 2005).
FLORY, WILLIAM E. S., PRISONERS OF WAR (1942).
Forget, Paul W., *Law Enforcement Detachments and the Canadian Navy: A New Counter-Drug Capability*, 7 CDN. NAV. REV. 4 (2011), http://www.navalreview.ca/wp-content/uploads/public/vol7num2/vol7num2art2.pdf.
FORSBERG, KARL, THE TALIBAN'S CAMPAIGN FOR KANDAHAR, AFGHANISTAN REPORT (2009), *available at* http://www.understandingwar.org/sites/default/files/The_Talibans_Campaign_For_Kandahar.pdf.
FOWLER, ROBERT R., A SEASON IN HELL: MY 130 DAYS IN THE SAHARA WITH AL QAEDA (2011).
FOWLER, WILLIAM, OPERATION BARRAS THE SAS RESCUE MISSION: SIERRA LEONE 2000 (2005).
______, CERTAIN DEATH IN SIERRA LEONE: THE SAS AND OPERATION *BARRAS* 2000 (2010).
FOX, GREGORY H., HUMANITARIAN OCCUPATION (2008).
FRANCK, THOMAS M., RECOURSE TO FORCE: STATE ACTION AGAINST THREATS AND ARMED ATTACKS (2002).
______, *On Proportionality of Countermeasures in International Law*, 102 AM. J. INT'L L. 715 (2008).
FREEDMAN, LAWRENCE, STRATEGY: A HISTORY (2013).
Frowein, Jochen Abr. & Nico Krisch, *Article 42*, *in* THE CHARTER OF THE UNITED NATIONS: A COMMENTARY 749 (Bruno Simma et al. eds., 2002).
GALEOTTI, MARK, RUSSIAN SECURITY FORCES AND PARAMILITARY FORCES SINCE 1991 (2013).
GALULA, DAVID, COUNTERINSURGENCY WARFARE: THEORY AND PRACTICE (1964).
Gardam, Judith, *Proportionality and Force in International Law*, 87 AM. J. INT'L L. 391 (1993).
______, NECESSITY, PROPORTIONALITY AND THE USE OF FORCE BY STATES (2004).
GARNER, BRYAN A. ED., BLACK'S LAW DICTIONARY 519 (10th ed. 2009).
GARNER, BRYAN A., GARNER'S DICTIONARY OF LEGAL USAGE, http://www.oxforddictionaries.com.
GAT, AZAR, WAR IN HUMAN CIVILIZATION (2006).
Geiß, Robin, *Armed Violence in Fragile States*, 91 INT'L REV. RED CROSS 127 (2009).
______, *Poison, Gas and Expanding Bullets: The Extension of the List of Prohibited Weapons at the Review Conference at the International Criminal Court in Kampala*, 13 Y.B. INT'L L. 337 (2010).
Generation Kill: A Conversation with Stanley McChrystal, 92 FOREIGN AFF. 2 (Mar./Apr., 2013).
GENTILE, GIAN, WRONG TURN: AMERICA'S DEADLY EMBRACE OF COUNTER-INSURGENCY (2013).
GERAGHTY, TONY, INSIDE THE SPECIAL AIR SERVICE (1980).
______, THE IRISH WAR: THE MILITARY HISTORY OF A DOMESTIC CONFLICT (1998).
______, BLACK OPS: THE RISE OF SPECIAL FORCES IN THE C.I.A., THE S.A.S., AND MOSSAD (2010).
GERGES, FAWAZ, THE FAR ENEMY: WHY JIHAD WENT GLOBAL (2005).
______, THE RISE AND FALL OF AL-QAEDA 190 (2011).
GIAMPIETRI, LUIS, 41 SECONDS TO FREEDOM: AN INSIDER'S ACCOUNT OF THE LIMA HOSTAGE CRISIS, 1996–97 (2007).
Gill, Terry D., *Legal Basis of the Right of Self-Defence under the UN Charter and under Customary International Law*, *in* THE HANDBOOK OF THE INTERNATIONAL LAW OF MILITARY OPERATIONS, 187 (Terry Gill & Dieter Fleck, 2010).
Gill, Terry & Dieter Fleck, *Concept and Sources of the International Law of Military Operations*, *in* THE HANDBOOK OF THE INTERNATIONAL LAW OF MILITARY OPERATIONS, 3 (Terry Gill & Dieter Fleck, 2010).
Gill, Terry D. & Paul A. L. Ducheine, *Rescue of Nationals*, *in* THE HANDBOOK OF THE INTERNATIONAL LAW OF MILITARY OPERATIONS, 217 (Terry Gill & Dieter Fleck, 2010).

Gioa, Andrea, *The Role of the European Court of Human Rights in Monitoring Compliance with Humanitarian Law in Armed Conflict*, *in* Ben-Naftali ed., INTERNATIONAL HUMANITARIAN LAW AND INTERNATIONAL HUMAN RIGHTS LAW, 201.

GIUSTOZZI, ANTONIO, KORAN, KALASHNIKOV AND LAPTOP: THE NEO-TALIBAN INSURGENCY IN AFGHANISTAN (2008).

______, EMPIRES OF MUD: WARS AND WARLORDS IN AFGHANISTAN (2009).

Glasser, Hans-Peter, *Internationalized Non-International Armed Conflicts: Case Studies of Afghanistan, Kampuchea, and Lebanon*, 33 AM. U. L. REV. 145 (1983).

______, *An Appeal for Ratification by the United States*, 81 AM. J. INT'L L. 912 (1987).

GLOBAL TERRORISM INDEX 2014: MEASURING AND UNDERSTANDING THE IMPACT OF TERRORISM, Institute for Economics and Peace (2014), http://www.visionofhumanity.org/sites/default/files/Global%20Terrorism%20Index%20Report%202014_0.pdf.

GOLDSTEIN, JOSHUA S., WINNING THE WAR ON WAR (2011).

Gomez, Juan Carlos, *Twenty-First-Century Challenges: The Use of Military Forces to Combat Criminal Threats*, 88 INT'L L. STUD. 279 (2012).

Goodman, Ryan, *The Detention of Civilians in Armed Conflict*, 103 AM. J. INT'L L. 48 (2009).

______, *The Power to Kill or Capture Enemy Combatants*, 24 EUR. J. INT'L L. 819 (2013).

______, *The Power to Kill or Capture Enemy Combatants: A Rejoinder to Michael N. Schmitt*, 24 EUR. J. INT'L L. 863 (2014).

______, *Australia, France, Netherlands Express Legal Reservations about Airstrikes in Syria [Updated]*, JUST SECURITY (Sept. 25, 2014, 8:01 AM), http://justsecurity.org/15545/australia-france-netherlands-express-legal-reservations-airstrikes-syria/.

Goodman, Ryan & Sara Knuckey, *Remarkable Statement by the UN Secretary General on US Airstrikes in Syria*, JUST SECURITY (Sept. 23, 2014, 5:21 PM), http://justsecurity.org/15456/remarkable-statement-secretary-general-airstrikes-syria/.

GOULD, TERRY, WORTH DYING FOR: CANADA'S MISSION TO TRAIN POLICE IN THE WORLD'S FAILING STATES (2014).

Grant, Rebecca, *The Redefinition of Airpower*, 86 AIR FORCE: J. OF THE AIR FORCE ASSOC. 32 (2003).

GRAY, CHRISTINE, INTERNATIONAL LAW AND THE USE OF FORCE (3rd ed. 2008).

GRAY, COLIN S., ANOTHER BLOODY CENTURY: FUTURE WAR (2005).

GREY, STEPHEN, GHOST PLANE: THE TRUE STORY OF THE CIA TORTURE PROGRAM (2006).

Green, Leslie C., *The Relations between Human Rights Law and International Humanitarian Law: A Historical Overview*, *in* TESTING THE BOUNDARIES OF INTERNATIONAL HUMANITARIAN LAW 49 (Susan C. Breau & Agnieszka Jachec-Neale eds., 2006).

______, THE CONTEMPORARY LAW OF ARMED CONFLICT (3rd ed. 2008).

GREENTREE, TODD, CROSSROADS OF INTERVENTION: INSURGENCY AND COUNTERINSURGENCY LESSONS FROM CENTRAL AMERICA (2008).

Greenwood, Christopher, *Self-Defense and the Conduct of International Armed Conflict*, *in* INTERNATIONAL LAW AT A TIME OF PERPLEXITY: ESSAYS IN HONOUR OF SHABTAI ROSENNE 273 (Yoram Dinstein ed., 1989).

______, *Terrorism and Humanitarian Law—The Debate over Additional Protocol I*, 19 ISR. Y.B. HUM. RTS. 187 (1989).

______, *Scope of Application of Humanitarian Law*, *in* THE HANDBOOK OF INTERNATIONAL HUMANITARIAN LAW, 45 (Dieter Fleck ed., 2nd ed. 2008).

______, *The Definition of the Term "Humanitarian Law"*, *in* THE HANDBOOK OF INTERNATIONAL HUMANITARIAN LAW, 1 (Dieter Fleck ed., 2nd ed. 2008).

Griffith, Samuel B., *Introduction*, *in* MAO TSE-TUNG, ON GUERRILLA WARFARE (Samuel B. Griffith trans., 2000) (1961).

Grillo, Ioan, El Narco: Inside Mexico's Criminal Insurgency (2011).

Gross, Oren & Fionnuala Ní Aoláin, Law in Times of Crisis: Emergency Powers in Theory and Practice (2006).

Grotius, Hugo, 1 & 3 The Rights of War and Peace (Richard Tuck ed., Liberty Fund Inc. (2005) (1646).

Gwynn, Charles W., Imperial Policing (1934).

Haines, Steven, *Northern Ireland 1968–1998*, *in* International Law and the Classification of Conflicts 117 (Elizabeth Wilmshurst ed., 2012).

Hakimi, Monica, *A Functional Approach to Targeting and Detention*, 110 Mich. L. Rev. 1365 (2012).

Hammer, Ellen & Marina Salvin, *The Taking of Hostages in Theory and Practice*, 38 Am. J. Int'l L. 20 (1944).

Hampson, Françoise J., *The Relationship between International Humanitarian Law and Human Rights from the Perspective of a Human Rights Treaty Body*, 90 Int'l Rev. of the Red Cross 549 (2008).

______, *Afghanistan 2001–2010*, *in* International Law and the Classification of Conflicts 242 (Elizabeth Wilmshurst ed., 2012).

Harclerode, Peter, Secret Soldiers: Special Forces in the War Against Terrorism (2000).

Harel, Amos & Avi Issacharoff, 34 Days: Israel, Hezbollah and the War in Lebanon (2008).

Harnden, Toby, Deadmen Risen: The Welsh Guards and the Real Story of Britain's War in Afghanistan (2011).

Hashim, Ahmed S., Insurgency and Counter-insurgency in Iraq (2006).

Hastings, Max & Simon Jenkins, The Battle for the Falklands (1983).

Hathaway, Oona A. et al., *The Relationship Between International Humanitarian Law and Human Rights Law in Armed Conflict*, 96 Minn. L. Rev. 1883 (2012).

Hecht, Jeff, *Nonlethal Lasers: Nonlethal Lasers Deter Attacks and Warn Away Noncombatants*, Photonic Frontiers (Apr. 1, 2013), http://www.laserfocusworld.com/articles/print/volume-49/issue-04/features/photonics-frontiers--nonlethal-lasers--nonlethal-lasers-deter-at.html.

Heinl, Robert Debs, Dictionary of Military and Naval Quotations (1966).

Helfstein, Scott Nassir Abdullah & Muhammad al-Obaidi, Deadly Vanguards: A Study of al-Qa'ida's Violence Against Muslims, Combatting Terrorism Center at West Point, Occasional Paper Series 6 (Dec. 2009), http://www.ctc.usma.edu/wp-content/uploads/2010/10/deadly-vanguards_complete_l.pdf.

Henckaerts, Jean-Marie & Louise Doswald-Beck eds., Customary International Humanitarian Law Study (2005).

Henderson, Ian, *Civilian Intelligence Agencies and the Use of Armed Drones*, 13 Y.B. Int'l Hum. L. 133 (2010).

Herridge, Catherine, The Next Wave: On the Hunt for Al Qaeda's American Recruits (2011).

Heymann, Philip B., Terrorism and America (1998).

Hiro, Dilip, War Without End: The Rise of Islamist Terrorism and Global Response (2002).

Hmoud, Mahmoud, *Are New Principles Really Needed? The Potential of the Established Distinction Between Responsibility for Attacks by Nonstate Actors and the Law of Self Defense*, 107 Am. J. Int'l Law 576 (2013).

Hobsbawm, Eric, Bandits (reprint 2012) (2011).

Hoffman, Bruce, Inside Terrorism (2006).

______, *The Myth of Grass-Roots Terrorism Subtitle: Why Osama bin Laden Still Matters*, FOREIGN AFF. 133 (May 2008–June 2008).

Holley, I. B., *Reflections on the Search for Airpower Theory*, *in* THE PATHS OF HEAVEN: THE EVOLUTION OF AIRPOWER THEORY 357 (Phillip S. Meilinger ed., 1997).

Holman, Robin F., *Law Enforcement, the Rogue Civil Airliner and Proportionality of Effects: An Analysis of International Human Rights Law*, A thesis submitted to McGill University in partial fulfillment of the requirements of the degree of LL.M. (2010), http://www.collectionscanada.gc.ca/obj/thesescanada/vol2/QMM/TC-QMM-97268.pdf.

Horn, Bernd, *Defining Terrorism*, *in* THE DIFFICULT WAR: PERSPECTIVES ON INSURGENCY AND SPECIAL OPERATIONS FORCES 111 (Emily Spencer ed., 2009).

Horowitz, Jonathon, *A Legal Map of Air Strikes in Syria (Part 1)*, JUST SECURITY (Dec. 7, 2015), 1:20 PM, https://www.justsecurity.org/28167/legal-map-airstrikes-syria-part-1/.

______, *A Legal Map of Airstrikes in Syria (Part 2)*, JUST SECURITY (Dec. 8, 2015, 1:55 PM), https://www.justsecurity.org/28199/legal-map-airstrikes-syria-part-2/

HORNE, ALISTAIR, A SAVAGE WAR OF PEACE: ALGERIA 1954–1962 (2006) (1977).

Hover, Matthew R., *The Occupation of Iraq: A Military Perspective on Lessons Learned*, 94 INT'L REV. RED CROSS 339 (2012).

HUBBELL, JOHN G., P.O.W.: A DEFINITIVE HISTORY OF THE AMERICAN PRISONER-OF-WAR EXPERIENCE IN VIETNAM, 1964–1973 (1976).

HULL, EDMUND J., HIGH-VALUE TARGET: COUNTERING AL QAEDA IN YEMEN (2011).

HUMAN RIGHTS ORGANIZATIONS' AND UNITED NATIONS' RECOMMENDATIONS FOR TARGETED KILLINGS, Stimson: Managing Across Borders Initiative (May 2014), http://www.stimson.org/images/uploads/human_rights_groups_and_un_special_rapporteur_recommendations.pdf.

Huneeus, Alexandra, *International Criminal Law by Other Means: The Quasi-Criminal Jurisdiction of the Human Rights Courts*, 107 AM. J. INT'L L. 1 (2013).

HUNT, LYNN, INVENTING HUMAN RIGHTS (2007).

HYDE, CHARLES CHENEY, 1 INTERNATIONAL LAW: CHIEFLY AS INTERPRETED AND APPLIED BY THE UNITED STATES (1945).

Iron, Richard, *Britain's Longest War: Northern Ireland 1967–2007*, *in* COUNTERINSURGENCY IN MODERN WARFARE 157 (Daniel Marston & Carter Malkasian eds., 2010).

ISBY, DAVID C., LEAVE NO MAN BEHIND: LIBERATION AND CAPTURE MISSIONS (2004).

ISHAY, MICHELINE R., THE HISTORY OF HUMAN RIGHTS: FROM ANCIENT TIMES TO THE GLOBALIZATION ERA (2004).

Issacharoff, Samuel & Richard H. Pildes, *Targeted Warfare: Individuating Enemy Responsibility*, 88 N.Y.U. L. REV. 1521 (2013).

Jackson, Richard B., *Perfidy in Non-International Armed Conflict*, 88 INT'L L. STUD. 237 (2012).

Jensen, Eric T., *Targeting of Persons and Property*, *in* MICHAEL W. LEWIS, THE WAR ON TERROR AND THE LAWS OF WAR 37 (2009).

JOES, ANTHONY JAMES, URBAN GUERRILLA WARFARE (2007).

JOHNSON, JAMES TURNER, MORALITY AND CONTEMPORARY WARFARE (1999).

Jeh Johnson Speech at the Oxford Union, LAWFARE (Nov. 30, 2012, 12:01 PM), https://www.lawfareblog.com/jeh-johnson-speech-oxford-union.

Jones, Marcus, *Fighting "this nation of liars to the very end", The German Army in the Franco-Prussian War, 1870–1871*, *in* HYBRID WARFARE: FIGHTING COMPLEX OPPONENTS FROM THE ANCIENT WORLD TO THE PRESENT 171 (Williamson Murray & Peter R. Mansoor eds., 2012).

JONES, SETH G., HUNTING IN THE SHADOWS: THE PURSUIT OF AL QA'IDA SINCE 9/11 (2012).

______, A PERSISTENT THREAT: THE EVOLUTION OF AL QA'IDA AND OTHER SALAFI JIHADISTS, RAND (2014), http://www.rand.org/content/dam/rand/pubs/research_reports/RR600/RR637/RAND_RR637.pdf.

Joyner, Charles C., *Reflections on the Lawfulness of Invasion*, 78 AM. J. INT'L L. 133 (1984).
KATZ, YAAKOV & YOAZ HENDEL, ISRAEL VS. IRAN: THE SHADOW WAR (2012).
KEEGAN, JOHN, A HISTORY OF WARFARE (1993).
______, THE IRAQ WAR (2004).
KEELEY, LAWRENCE H., WAR BEFORE CIVILIZATION: THE MYTH OF THE PEACEFUL SAVAGE (1996).
KENNEDY, ROBERT M., HOLD THE BALKANS: GERMAN ANTIGUERRILLA OPERATIONS IN THE BALKANS, 1941–1944 (2000).
KILCULLEN, DAVID, THE ACCIDENTAL GUERRILLA: FIGHTING SMALL WARS IN THE MIDST OF A BIG ONE (2009).
______, COUNTERINSURGENCY (2010).
______, OUT OF THE MOUNTAINS: THE COMING OF AGE OF THE URBAN GUERRILLA (2013).
Killebrew, Robert, *A Volatile Brew*, *in* JOHN P. SULLIVAN & ROBERT J. BUNKER, MEXICO'S CRIMINAL INSURGENCY 53 (2012).
KINDRED, HUGH M. ET. AL., INTERNATIONAL LAW: CHIEFLY AS INTERPRETED AND APPLIED IN CANADA (6th ed. 2000).
KITSON, FRANK, LOW INTENSITY OPERATIONS: SUBVERSION INSURGENCY & PEACEKEEPING (1971).
KLAIDMAN, DANIEL, KILL OR CAPTURE (2012).
Kolb, Robert, *Origin of the Twin Terms Jus Ad Bellum/Jus in Bello*, 320 INT'L REV. RED CROSS (1997), https://www.icrc.org/eng/resources/documents/misc/57jnuu.htm.
______, *The Relationship Between International Humanitarian Law and Human Rights Law: A Brief History of the 1948 Universal Declaration of Human Rights and the 1949 Geneva Conventions*, 38 INT'L REV. RED CROSS 409 (1998).
KOLENDA, CHRISTOPHER, THE COUNTERINSURGENCY CHALLENGE: A PARABLE OF LEADERSHIP AND DECISION MAKING IN MODERN CONFLICT (2012).
KRASKA, JAMES, MARITIME POWER AND THE LAW OF THE SEA (2011).
Kretzmer, David, *The Inherent Right to Self-Defence and Proportionality in Jus Ad Bellum*, 24 EUR. J. INT'L L. 235 (2013).
Krulak, Charles C., *The Strategic Corporal: Leadership in the Three Block War*, MARINES MAGAZINE (Jan. 1999), *available at* http://www.au.af.mil/au/awc/awcgate/usmc/strategic_corporal.htm.
Kunz, Josef L., *Individual and Collective Self-Defense in Article 51 of the Charter of the United Nations*, 41 AM. J. INT'L L. 872 (1947).
______, *The Chaotic Status of the Laws of War and the Urgent Necessity for Their Revision*, 45 AM. J. INT'L L. 37 (1951).
______, *Bellum Justum and Bellum Legale*, 45 AM. J. INT'L L. 528 (1951).
Kuzmarov, Jeremy, *Afghanistan, Part I—"You Have to Not Mind Killing Innocents": American COIN Operations in Afghanistan and the Violence of Empire*, *in* HEARTS AND MINDS: A PEOPLE'S HISTORY OF COUNTERINSURGENCY 181 (Hannah Gurman ed., 2013).
LACEY, JIM ED., A TERRORIST'S CALL TO GLOBAL JIHAD: DECIPHERING ABU MUSAB AL-SURI'S ISLAMIC JIHAD MANIFESTO (2008).
LAMB, CHRISTOPHER J. & EVAN MUNSING, SECRET WEAPON: HIGH-VALUE TARGET TEAMS AS AN ORGANIZATIONAL INNOVATION, Strategic Perspectives 4, Institute for National Strategic Studies (2011).
LAMBETH, BENJAMIN S., AIRPOWER AGAINST TERROR: AMERICA'S CONDUCT OF OPERATION ENDURING FREEDOM (2005).
LAQUEUR, WALTER, GUERRILLA WARFARE: A HISTORICAL & CRITICAL STUDY (2006) (1976).
Lauterpacht, H., *The Problem of the Revision of the Law of War*, 29 BRIT. Y.B. INT'L L. 360 (1952).
LEDWIDGE, FRANK, LOSING SMALL WARS: BRITISH MILITARY FAILURE IN IRAQ AND AFGHANISTAN (2011).

LEEBAERT, DEREK, TO DARE & CONQUER: SPECIAL OPERATIONS AND THE DESTINY OF NATIONS, FROM ACHILLES TO AL QAEDA (2006).

LEECH, GARRY, THE FARC: THE LONGEST INSURGENCY (2011).

Leiden Policy Recommendations on Counter-Terrorism and International Law, *in* COUNTER-TERRORISM STRATEGIES IN A FRAGMENTED INTERNATIONAL LEGAL ORDER: MEETING THE CHALLENGES, Annex, 706 (Larissa Van Den Herik & Nico Schrijver eds., 2013).

Lepsius, Oliver, *Human Dignity and the Downing of Aircraft: The German Federal Constitutional Court Strikes Down a Prominent Anti-terrorism Provision in the New Air-transport Security Act*, 7 GERMAN. L.J. 761 (2007).

LEVIE, HOWARD, PRISONERS OF WAR IN INTERNATIONAL ARMED CONFLICT, 59 INT'L L. STUD. (1977).

LEVITT, MATTHEW, HAMAS: POLITICS, CHARITY, AND TERRORISM IN THE SERVICE OF JIHAD (2006).

______, HEZBOLLAH: THE GLOBAL FOOTPRINT OF LEBANON'S PART OF GOD (2013).

LEVY, JACK S. & WILLIAM R. THOMPSON, CAUSES OF WAR (2010).

LEWIS, BERNARD, THE ASSASSINS: A RADICAL SECT IN ISLAM (1967).

LEWIS, DAMIEN, OPERATION CERTAIN DEATH (2004).

Lewis, Michael W., *The Law of Aerial Bombardment in the 1991 Gulf War*, 97 AM. J. INT'L L. 481 (2003).

LIA, BRYNJAR, ARCHITECT OF GLOBAL JIHAD: THE LIFE OF AL-QAIDA STRATEGIST ABU MUS'AD AL-SURI (2008).

LIDDELL HART, B. H., STRATEGY (2nd ed. 1991).

Lieber, Francis, *Guerrilla Parties Considered with Reference to the Law and Usages of War*, *in* RICHARD SHELLY HARTIGAN, LIEBER'S CODE AND THE LAW OF WAR 31 (1983).

Lieblich, Eliav, *Quasi-Hostile Acts: The Limits on Forcible Disruption Operations under International Law*, 32 B.U. INT'L L.J. 101 (2014).

LINDHOUT, AMANDA & SARA CORBETT, A HOUSE IN THE SKY (2013).

LIVING UNDER DRONES: DEATH, INJURY, AND TRAUMA TO CIVILIANS FROM US DRONE PRACTICES IN PAKISTAN, Stanford Law School/NYU School of Law (Sept. 2012), *available at* http://www.livingunderdrones.org/report/.

Long, Austin, *The Police in Afghanistan, 2002–11*, *in* POLICING INSURGENCIES: COPS AS COUNTERINSURGENTS 203 (C. Christine Fair & Sumit Ganguly eds., 2014).

LOWE, VAUGHN, INTERNATIONAL LAW (2007).

LUBELL, NOAM, EXTRATERRITORIAL USE OF FORCE AGAINST NON-STATE ACTORS (2010).

MacKenzie, Jean, *Afghanistan—Part II, Counterinsurgency in Afghanistan: Myth or Reality?*, *in* HEARTS AND MINDS: A PEOPLE'S HISTORY OF COUNTERINSURGENCY 200 (Hannah Gurman ed., 2013).

MACKEY, CHRIS & GREG MILLER, THE INTERROGATORS: INSIDE THE SECRET WAR AGAINST AL QAEDA (2004).

MAHAN, ALFRED THAYER, THE INFLUENCE OF SEA POWER UPON HISTORY 1660–1783 (1889) (2009).

Malkasian, Carter, *Counterinsurgency in Iraq: May 2003–January 2010*, *in* COUNTERINSURGENCY IN MODERN WARFARE, 287 (Daniel Marston & Carter Malkasian eds., 2010).

MALLISON, SALLY V. & W. THOMAS MALLISON, ARMED CONFLICT IN LEBANON, 1982: HUMANITARIAN LAW IN A REAL WORLD SETTING (1985).

MANDSAGER, DENNIS ED., RULES OF ENGAGEMENT HANDBOOK, Institute of Humanitarian Law (2009), *available at* https://www.usnwc.edu/getattachment/7b0d0f70-bb07-48f2-af0a-7474c92d0bb0/San-Remo-ROE-Handbook.

Mansoor, Peter R., *Introduction*, *in* HYBRID WARFARE: FIGHTING COMPLEX OPPONENTS FROM THE ANCIENT WORLD TO THE PRESENT 1 (Williamson Murray & Peter R. Mansoor eds., 2012).

Manwaring, Max G., Gangs Pseudo-Militaries and Other Modern Mercenaries (2010).

_____, The Complexity of Modern Asymmetric Warfare (2012).

Marighella, Carlos, Minimanual of the Urban Guerrilla (2008) (1972).

Marston, Daniel & Carter Malkasian, *Introduction*, *in* Counterinsurgency in Modern Warfare, 13 (Daniel Marston & Carter Malkasian eds., 2010).

Martin, Matt J. & Charles W. Sasser, Predator Remote-Control Air War Over Iraq and Afghanistan: A Pilot's Story (2010).

Maurer, Kevin, Gentlemen Bastards: On the Ground with America's Elite Special Forces (2012).

Mazzetti, Mark, The Way of the Knife: The CIA, A Secret Army, and the War at the Ends of the Earth (2013).

Metz, Steven, Rethinking Insurgency, Strategic Studies Institute, U.S. Army War College (2007).

McCoy, Alfred W., A Question of Torture: CIA Interrogation from the Cold War to the War on Terror (2006).

McChrystal, Stanley, My Share of the Task: A Memoire (2013).

McDermott, Terry & Josh Meyer, The Hunt for KSM (2012).

McElwee, William, The Art of War: Waterloo to Mons (1974).

McEvoy, Kieran, Paramilitary Imprisonment in Northern Ireland: Resistance, Management and Release (2001).

McKeogh, Colm, Innocent Civilians: The Morality of Killing in War (2002).

McLaughlin, Rob, *Legal Policy Considerations and Conflict Characterizations at the Threshold between Law Enforcement and Non-International Armed Conflict*, 13 Melb. J. Int'l L. 1 (2012).

McNabb, Chris, Storming Flight 181: GSG 9 and the Mogadishu Hijack 1977 (2011).

McNeal, George S., *Targeted Killing and Accountability*, 102 Geo. L.J. 681 (2014).

McRaven, William H., Spec Ops: Case Studies in Special Operations Warfare: Theory and Practice (1995).

Mellow, Craig, *Would a Fighter Pilot Shoot Down a Private Airplane?*, Air & Space Magazine (Mar. 2010), http://www.airspacemag.com/flight-today/dont-cross-that-line-5841988/?no-ist.

Melzer, Nils, Targeted Killing in International Law (2008).

Melzer, Nils, *Keeping the Balance Between Military Necessity and Humanity: A Response to Four Critiques of the ICRC's Interpretive Guidance on the Notion of Direct Participation in Hostilities*, 42 N.Y.U. J. of Int'l L. and Pol. 831 (2010).

_____, *Conceptual Distinction and Overlap Between Law Enforcement and the Conduct of Hostilities*, *in* The Handbook of the International Law of Military Operations, 33 (Terry Gill & Dieter Fleck, 2010).

Merom, Gil, Why Democracies Lose Small Wars (2003).

Meron, Theodor, *The Time Has Come for the United States to Ratify Geneva Protocol I*, 88 Am. J. Int'l L. 678 (1994).

_____, *The Humanization of Humanitarian Law*, 94 Am. J. Int'l L. 239 (2000).

Metz, Steven, Rethinking Insurgency, Strategic Studies Institute, U.S. Army War College (2007), http://www.strategicstudiesinstitute.army.mil/pdffiles/PUB790.pdf.

Milanović, Marko, *Lessons for Human Rights and Humanitarian Law in the War on Terror: Comparing Hamdan and the Israeli Targeted Killings Case*, 89 Int'l Rev. Red Cross 373 (2007).

_____, *Norm Conflicts, International Humanitarian Law and Human Rights Law*, *in* International Humanitarian Law and International Human Rights Law 95 (Orna Ben-Naftali ed., 2011).

______, EXTRATERRITORIAL APPLICATION OF HUMAN RIGHTS TREATIES: LAW, PRINCIPLES, AND POLICY (2011).

______, *Extraterritorial Derogations from Human Rights Treaties in Armed Conflict*, *in* COLLECTED COURSES OF THE ACADEMY OF EUROPEAN LAW, Nehal Bhuta ed., Oxford University Press (Forthcoming), *available at* http://papers.ssrn.com/sol3/papers.cfm?abstract_id=2447183.

Modarelli, Matthew R., *Military Police Operations and Counterinsurgency*, *in* SMALL WARS JOURNAL (2008), http://smallwarsjournal.com/blog/journal/docs-temp/151-modarelli.pdf?q=mag/docs-temp/151-modarelli.pdf.

MOGHADAM, ASSAF, THE GLOBALIZATION OF MARTYRDOM (2008).

Moir, Lindsay, *The Historical Development of the Application of Humanitarian Law in Non-International Armed Conflicts to 1949*, 47 INT'L & COMP L.Q. 337 (1998).

______, THE LAW OF INTERNAL ARMED CONFLICT 60 (2002).

______, *Law and the Inter-American Human Rights System*, 25 HUM. RTS. Q. 182 (2003).

______, *Towards the Unification of International Humanitarian Law*, *in* INTERNATIONAL CONFLICT AND SECURITY LAW (Richard Burchill, Nigel D. White, & Justin Morris eds., 2005).

______, REAPPRAISING THE RESORT TO FORCE: INTERNATIONAL LAW, *JUS AD BELLUM* AND THE WAR ON TERROR (2010).

MOLZAHN, CORY OCTAVIO RODRIGUEZ FERREIRA & DAVID A. SHIRK, DRUG VIOLENCE IN MEXICO: DATA AND ANALYSIS THROUGH 2012, TRANS-BORDER INSTITUTE, Feb. 15, 2013, http://justiceinmexico.files.wordpress.com/2013/02/130206-dvm-2013-final.pdf.

MORELL, MICHAEL & BILL HARLOW, THE GREAT WAR OF OUR TIME: THE CIA'S FIGHT AGAINST TERRORISM FROM AL QA'IDA TO ISIS (2015).

Moussa, Jasmine, *Can Jus Ad Bellum Override Jus In Bello? Reaffirming the Separation of the Two Bodies of Law*, 90 INT'L REV. RED CROSS 963 (2008).

MURPHY, MARTIN N., SOMALIA: THE NEW BARBARY?: PIRACY AND ISLAM IN THE HORN OF AFRICA (2011).

Murphy, Sean D., *U.S. Reaction to ICJ Judgment in Iranian Oil Platforms Case*, 98 AM. J. INT'L L. 597 (2004).

MURRAY, WILLIAMSON & ROBERT H. SCALES, THE IRAQ WAR: A MILITARY HISTORY (2003).

NAGL, JOHN A., LEARNING TO EAT SOUP WITH A KNIFE: COUNTERINSURGENCY LESSONS FROM MALAYA AND VIETNAM (2005).

NAJI, ABU BAKR, THE MANAGEMENT OF SAVAGERY: THE MOST CRITICAL STAGE THROUGH WHICH THE UMMA WILL PASS (William McCants trans., 2006).

NAYLOR, SEAN, RELENTLESS STRIKE: THE SECRET HISTORY OF JOINT SPECIAL OPERATIONS COMMAND (2015).

Naseemullah, Adnan, *Police Capacity and Insurgency in Pakistan*, *in* POLICING INSURGENCIES: COPS AS COUNTERINSURGENTS 177 (C. Christine Fair & Sumit Ganguly eds., 2014).

Naske, Nina & Georg Nolte, *"Aerial Security Law." Case No. 1 BvR 357/05. 115 BVerfGE 118*, 101 AM. J. INT'L L. 466 (2007).

NETANYAHU, IDDO, YONI'S LAST BATTLE: THE RESCUE AT ENTEBBE, 1976 (2002).

THE 9/11 COMMISSION REPORT: FINAL REPORT OF THE NATIONAL COMMISSION ON TERRORIST ATTACKS UPON THE UNITED STATES (2004).

Nolte, Georg, *Article 2(7)*, *in* THE CHARTER OF THE UNITED NATIONS: A COMMENTARY 148 (Bruno Simma et al. eds., 2002).

NORTON, AUGUSTUS RICHARD, HEZBOLLAH: A SHORT HISTORY (2007).

Nurick, Lester & Roger W. Barrett, *Legality of Guerrilla Forces Under the Laws of War*, 40 AM. J. INT'L L. 563 (1946).

O'CONNELL, D. P., THE INFLUENCE OF LAW ON SEA POWER (1975).

O'Connell, Mary Ellen, *Combatants and Combat Zones*, 43 U. Rich. L. Rev. 845 (2009).

______, *Lawful Use of Drones*, Congress of the United States, Hearing: Rise of the Drones II: Examining the Legality of Unmanned Targeting (Apr. 28, 2010), https://www.fas.org/irp/congress/2010_hr/042810oconnell.pdf.

______, *The Choice of Law Against Terrorism*, 4 Nat'l Sec. L. & Pol'y 343 (2010).

______, *Drones Under International Law*, International Debate Series 4 (Oct. 8, 2010), *available at* http://law.wustl.edu/harris/documents/OConnellFullRemarksNov23.pdf.

______, *Unlawful Killing with Combat Drones: A Case Study of Pakistan, 2004–2009* (Notre Dame Law School Legal Studies Research Paper No. 09-43, Final Draft, July 2010), https://www.law.upenn.edu/institutes/cerl/conferences/targetedkilling/papers/OConnellDrones.pdf.

______, *Dangerous Departures*, 107 Am. J. Int'l L. 380 (2013).

______, *The True Meaning of Force*, AJIL Unbound (Aug. 4, 2014, 9:56 AM), http://www.asil.org/blogs/true-meaning-force.

______, *The True Meaning of Force: A Further Response to Tom Ruys in the Interest of Peace*, AJIL Unbound (Sept. 4, 2014, 10:04 AM), http://www.asil.org/blogs/true-meaning-force-further-response-tom-ruys-interest-peace.

Odom, Thomas P., Dragon Operations: Hostage Rescues in the Congo, 1964–1965 (1988).

Off, Carol, The Ghosts of Medak Pocket: The Story of Canada's Secret War (2004).

O'Neill, Bard E., Insurgency & Terrorism: From Revolution to Apocalypse (2005).

Orend, Brian, The Morality of War (2006).

Oswald, Bruce, Helen Durham, & Adrian Bates, Documents on the Law of UN Peace Operations (2010).

Otto, Roland, *Neighbours as Human Shields? The Israel Defense Forces' "Early Warning Procedure" and International Humanitarian Law*, 86 Int'l Rev. Red Cross 771 (2004).

Ousby, Ian, Occupation: The Ordeal of France 1940–1944 (1997).

Owen, Mark, No Easy Day: The Autobiography of a Navy Seal (2012).

Oxford Dictionaries, http://www.oxforddictionaries.com.

Pape, Robert A., Dying to Win: The Strategic Logic of Suicide Terrorism (2005).

Paret, Peter, *Introduction*, *in* Makers of Modern Strategy: from Machiavelli to the Nuclear Age 3 (Peter Paret ed., 1986).

Parks, W. Hays, *Memorandum of Law: Executive Order 12333 and Assassination*, The Army Lawyer 4 (1989), http://www.loc.gov/rr/frd/Military_Law/pdf/12-1989.pdf.

______, *Conventional Weapons and Weapons Reviews*, 8 Int'l Y.B. Hum. L. 55 (2005).

Paul, Christopher et al., Paths to Victory, Rand (2013).

Paulus, Andreas & Mindia Vashakmadze, *Asymmetrical War and the Notion of Armed Conflict—A Tentative Conceptualization*, 91 Int'l Rev. Red Cross 95 (2009).

Paust, Jordan J., *Self-Defense Targetings of Non-State Actors and Permissibility of U.S. Use of Drones in Pakistan*, 19 J. Transnat'l L. & Pol'y 237 (2010).

Pejic, Jelena, *The Protective Scope of Common Article 3: More than Meets the Eye*, 93 Int'l Rev. Red Cross 189 (2011).

Perito, Robert M., Afghanistan's Civil Order Police: Victim of Its Own Success, Special Report 307, United States Institute for Peace (May 2012), http://www.usip.org/publications/afghanistans-civil-order-police.

Peters, Gretchen, How Opium Profits the Taliban (2009), *available at* http://www.usip.org/files/resources/taliban_opium_1.pdf.

Pettifer, James, The Kosova Liberation Army: Underground War to Balkan Insurgency, 1948–2001 (2013).

PICTET, JEAN S. ED., THE GENEVA CONVENTIONS OF 12 AUGUST 1949, COMMENTARY: (I) GENEVA CONVENTION FOR THE AMELIORATION OF THE CONDITION OF THE WOUNDED AND SICK IN ARMED FORCES IN THE FIELD (1952).

______, THE GENEVA CONVENTIONS OF 12 AUGUST 1949, COMMENTARY: (III) THE GENEVA CONVENTION RELATIVE TO THE TREATMENT OF PRISONERS OF WAR (1960).

______, THE GENEVA CONVENTIONS OF 12 AUGUST 1949: COMMENTARY, (IV) GENEVA CONVENTION RELATIVE TO THE PROTECTION OF CIVILIAN PERSONS IN TIME OF WAR (1958).

Pilmott, John, *The British Experience*, *in* THE ROOTS OF COUNTER-INSURGENCY: ARMIES AND GUERRILLA WARFARE 1900–1945, at 17 (Ian F. W. Beckett ed., 1988).

PINKER, STEVEN, THE BETTER ANGELS OF OUR NATURE: WHY VIOLENCE HAS DECLINED (2011).

Pomper, Stephen, *Toward a Limited Consensus on the Loss of Civilian Immunity in Non-International Armed Conflict: Making Progress through Practice*, 88 INT'L LEG. STUD. 181 (2012).

PORCH, DOUGLAS, COUNTERINSURGENCY: EXPOSING THE MYTHS OF THE NEW WAY OF WAR (2013).

Pouw, Eric, International Human Rights Law and the Law of Armed Conflict in the Context of Counterinsurgency (Nov. 14, 2013) (unpublished Phd. Dissertation, University of Amsterdam).

PROVOST, RENÉ, INTERNATIONAL HUMAN RIGHTS AND HUMANITARIAN LAW (2002).

Prud'homme, Nancie, *Lex Specialis: Oversimplifying a More Complex and Multi-faceted Relationship?*, 40 ISR. L. REV. 356 (2007).

PRUGH, GEORGE S., LAW AT WAR 1964–1973 (1975).

RABASA, ANGEL & PETER CHALK, COLOMBIAN LABYRINTH (2001).

Randelzhofer, Albrecht, *Article 2(4)*, *in* THE CHARTER OF THE UNITED NATIONS: A COMMENTARY 112 (Bruno Simma et al. eds., 2nd ed. 2002).

______, *Article 51*, *in* THE CHARTER OF THE UNITED NATIONS: A COMMENTARY 788 (Bruno Simma et al. eds., 2002).

RAVIV, DAN & YOSSI MELMAN, SPIES AGAINST ARMAGEDDON (2012).

Reagan, Ronald, *Letter of Transmittal*, 81 AM. J. INT'L L. 910 (1987).

Reinhold, Theresa, *State Weakness, Irregular Warfare, and the Right to Self-Defense Post 9/11*, 105 AM. J. INT'L L. 244 (2011).

REISMAN, W. MICHAEL & JAMES E. BAKER, REGULATING COVERT ACTION (1992).

RENNIE, JAMES, THE OPERATORS: ON THE STREETS WITH 158 COMPANY (1996).

THE REPORT OF THE CONSTITUTION PROJECT'S TASK FORCE ON DETAINEE TREATMENT, The Constitution Project (2013), *available at* http://detaineetaskforce.org/pdf/Full-Report.pdf.

2 RESTATEMENT OF THE LAW: THE FOREIGN RELATIONS LAW OF THE UNITED STATES (1987).

RICE, DESMOND & ARTHUR GAVSHON, THE SINKING OF THE BELGRANO (1984).

RICKS, THOMAS E., FIASCO: THE AMERICAN MILITARY ADVENTURE IN IRAQ (2007).

ROACH, KENT, THE 9/11 EFFECT: COMPARATIVE COUNTER-TERRORISM (2011).

Roberts, Adam, *Prolonged Military Occupation: The Israeli-Occupied Territories Since 1967*, 84 AM. J. INT'L L. 44 (1990).

______, *Against War*, *in* THE OXFORD HISTORY OF MODERN WAR 317 (Charles Townshend ed., 2005).

______, *Transformative Military Occupation: Applying the Laws of War and Human Rights*, 100 AM. J. INT'L L. 580 (2006).

______, *The Equal Application of the Laws of War: A Principle Under Pressure*, 90 INT'L REV. RED CROSS 931 (2008).

ROBERTS, ADAM & RICHARD GUELFF EDS., DOCUMENTS ON THE LAWS OF WAR (3rd ed. 2003)

ROBINSON, LINDA, MASTERS OF CHAOS: THE SECRET HISTORY OF THE SPECIAL FORCES (2004).

______, *The Future of Special Operations: Beyond Kill and Capture*, FOREIGN AFF. 110 (2012).

Rodin, David, War & Self-Defense (2002).

Rodley, Nigel S. & Matt Pollard, The Treatment of Prisoners Under International Law (3rd ed. 2009).

Rogers, A. P. V., Law on the Battlefield (2004).

Rollins, John & Liana Sun Wyler, Terrorism and Transnational Crime: Foreign Policy Issues for Congress, Congressional Research Service (June 11, 2013), https://www.fas.org/sgp/crs/terror/R41004.pdf.

Römer, Jan, Killing in a Gray Area Between Humanitarian Law and Human Rights (2010).

Rona, Gabor & Raha Wala, *No Thank You to a Radical Rewrite of the Jus ad Bellum*, 107 Am. J. Int'l L. 386 (2013).

Root, Elihu, *The Basis of Protection to Citizens Residing Abroad*, 4 Proc. of the Am. Soc. of Int'l L. at Its Annual Meeting 16 (1910).

Rosas, Allan, The Legal Status of Prisoners of War: A Study in International Humanitarian Law Applicable in Armed Conflict (1976).

Rothwell, Donald R. & Cameron Moore, *Australia's Traditional Maritime Security Concerns and Post 9/11 Perspectives*, *in* Maritime Security: International Law and Policy Perspectives from Australia and New Zealand 37 (Natalie Klein, Joanna Mossop, & Donald R. Rothwell eds., 2010).

Rottman, Gordon L., The Los Baños Prison Camp Raid: The Philippines 1945 (2010).

Ruys, Tom, "Armed Attack" and Article 51 of the UN Charter: Evolutions in Customary Law and Practice (2010).

______, *The Meaning of "Force" and the Boundaries of the Jus Ad Bellum: Are "Minimal" Uses of Force Excluded from UN Charter Article 2(4)?*, 108 Am. J. Int'l L. 159 (2014).

______, *The True Meaning of Force—A Reply to Mary Ellen O'Connell* AJIL Unbound (Sept. 3, 2014, 9:30 AM), http://www.asil.org/blogs/true-meaning-force—-reply-mary-ellen-o'connell.

Ryan, Michael W. S., Decoding Al-Qaeda's Strategy: The Deep Battle Against America (2013).

Saar, Erik & Viveca Novak, Inside the Wire (2005).

Sageman, Marc, Leaderless Jihad: Terror Networks in the Twenty-First Century (2008).

Sageman, Marc & Bruce Hoffman, *Does Osama Still Call the Shots? Subtitle: Debating the Containment of al Qaeda's Leadership*, 87 Foreign Aff. 163 (July/Aug. 2008).

Sandoz, Yves et al. eds., Commentary on the Additional Protocols of 8 June 1977 to the Geneva Conventions of 12 August 1949 (1987), *available at* http://www.cicr.org/ihl.nsf/COM/470-750073?OpenDocument.

Sanger, David E., Confront and Conceal: Obama's Secret Wars and Surprising Use of American Power (2012).

Sassoli, Marco, *Ius ad Bellum and Jus in Bello—The Separation between the Legality of the Use of Force and Humanitarian Rules to Be Respected in Warfare: Crucial or Outdated?*, *in* International Law and Armed Conflict: Exploring the Faultlines 241 (Michael Schmitt & Jelena Pejic eds., 2007).

______, *The Role of Human Rights and International Humanitarian Law in New Types of Armed Conflicts*, *in* International Humanitarian Law and International Human Rights Law 34 (Orna Ben-Naftali ed., 2011).

Saul, Ben, Defining Terrorism in International Law (2006).

Savage, Charlie, Power Wars: Inside Obama's Post 9/11 Presidency (2015).

Scahill, Jeremy, Dirty Wars: The World Is a Battlefield (2013).

Scobbie, Iain, *Gaza*, *in* International Law and the Classification of Conflict 280 (Elizabeth Wilmshurst ed., 2012).

SCHEMMER, BENJAMIN F., THE RAID (1976).

SCHEUER, MICHAEL, THROUGH OUR ENEMIES' EYES: OBAMA BIN LADEN, RADICAL ISLAM, AND THE FUTURE OF AMERICA (2006).

SCHMITT, ERIC & THOM SHANKER, COUNTERSTRIKE: THE UNTOLD STORY OF AMERICA'S SECRET CAMPAIGN AGAINST AL QAEDA (2011).

Schmitt, Michael N., *Effects-Based Operations and the Law of Aerial Warfare*, 5 WASH. GLOB. STUD. L. REV. 265 (2006).

______, *Faultlines in the Law of Attack*, *in* TESTING THE BOUNDARIES OF INTERNATIONAL HUMANITARIAN LAW 277 (Susan Breau & Agnieszka Jachec-Neale eds., 2006).

______, *Responding to Transnational Terrorism under the Jus ad Bellum: A Normative Framework*, *in* INTERNATIONAL LAW AND ARMED CONFLICT: EXPLORING THE FAULTLINES 157 (Michael Schmitt & Jelena Pejic eds., 2007).

______, *"Change Direction" 2006: Israeli Operations in Lebanon and the International Law of Self-Defense*, 84 INT'L L. STUD. 265 (2008).

______, *Targeting and International Humanitarian Law in Afghanistan*, 85 INT'L L. STUD. 307 (2009).

______, *Drone Attacks Under the Jus Ad Bellum and Jus In Bello: Clearing the "Fog of Law"*, 13 Y.B. INT'L HUM. L. 311 (2010).

______, *Unmanned Combat Aircraft Systems and International Humanitarian Law: Simplifying the Oft Benighted Debate*, 30 B.U. INT'L L.J. 595 (2012).

______, *Wound, Capture, or Kill: A Reply to Ryan Goodman's "The Power to Kill or Capture Enemy Combatants,"* 24 EUR. J. INT'L L. 855 (2013).

______, *Charting the Legal Geography of Non-International Armed Conflict*, 90 INT'L L. STUD. 1 (2014).

______, *Narrowing the International Law Divide: The Drone Debate Matures*, YALE J. INT'L L. ONLINE (2014), http://www.yjil.org/docs/pub/o-39-schmitt-narrowing-the-international-law-divide.pdf.

SCHMITT, MICHAEL ED., TALLIN MANUAL ON THE INTERNATIONAL LAW APPLICABLE TO (2013).

SCOTT-CLARK, CATHY & ADRIAN LEVY, THE SIEGE: 68 HOURS INSIDE THE TAJ HOTEL (2014).

Sewall, Sarah, *Introduction to the University of Chicago Press Edition*, *in* THE U.S. ARMY, MARINE CORPS, COUNTERINSURGENCY FIELD MANUAL xxi (2007).

SHAHZAD, SYED SALEEM, INSIDE AL-QAEDA AND THE TALIBAN: BEYOND BIN LADEN AND 9/11 (2011).

SHAPIRO, JACOB N., THE TERRORIST'S DILEMMA: MANAGING VIOLENT COVERT ORGANIZATIONS (2013).

SHAW, MALCOLM N., INTERNATIONAL LAW (6th ed. 2008).

Sherman, Matt & Josh Paul, *The Role of Police in Counterinsurgency Operations in Iraq, 2003–2006*, *in* POLICING INSURGENCIES: COPS AS COUNTERINSURGENTS 227 (C. Christine Fair & Sumit Ganguly eds., 2014).

SHULTZ, RICHARD H. & ANDREA J. DEW, INSURGENTS, TERRORISTS AND MILITIAS: WARRIORS OF CONTEMPORARY COMBAT (2006).

Shy, John & Thomas W. Collier, *Revolutionary War*, *in* MAKERS OF MODERN STRATEGY: FROM MACHIAVELLI TO THE NUCLEAR AGE 815 (Peter Paret ed., 1986).

SINNO, ABDULKADER H., ORGANIZATIONS AT WAR IN AFGHANISTAN AND BEYOND (2008).

Sivakumaran, Sandesh, *Re-evisaging the International Law of Internal Armed Conflict*, 22 EUR. J. INT'L L. 219 (2011).

______, THE LAW OF NON-INTERNATIONAL ARMED CONFLICT (2012).

SLIM, HUGO, KILLING CIVILIANS: METHOD, MADNESS, AND MORALITY IN WAR (2008).

SLOAN, ELINOR C., SECURITY AND DEFENCE IN THE TERRORIST ERA (2005).

Smith, E. D., Counter-Insurgency Operations: 1 Malaya and Borneo (1985).
Smith, Jamie, Gray Work: Confessions of an American Paramilitary Spy (2015).
Smith, Paul J., The Terrorism Ahead: Confronting Transnational Violence in the Twenty-First Century (2008).
Smith, Rupert, The Utility of Force: The Art of War in the Modern World (2005).
Soafer, Abraham D., *The Sixth Annual Waldemar A. Solf Lecture in International Law: Terrorism, the Law, and the National Defense*, 126 Mil. L. Rev. 89 (1989).
Solis, Gary D., The Law of Armed Conflict: International Humanitarian Law in War (2010).
Spaight, J. M., War Rights on Land, 1911, (1911).
______, Air Power and War Rights (3rd ed. 1947).
Spencer, Emily, *The People Puzzle*, *in* The Difficult War: Perspectives on Insurgency and Special Operations Forces 79 (Emily Spencer ed., 2009).
Staniland, Paul, Networks of Rebellion: Explaining Insurgent Cohesion and Collapse (2014).
Stanley, George F. G., Canada's Soldiers: The Military History of an Unmilitary People (3rd ed. 1974).
Stern, Jessica, Terror in the Name of God: Why Religious Militants Kill (2003).
Stern, Jessica S. & J. M. Berger, ISIS: The State of Terror (2015).
Stevens, Anthony, The Roots of War & Terror (2004).
Steyn, Johan, *Guantánamo Bay: The Legal Black Hole* (the 27th FA Mann Lecture, Nov. 25, 2003), *reprinted in* 53 Int'l & Comp. L. Q. 1 (2004).
Stiff, Peter, The Silent War: South African Recce Operations 1969–1994 (2009).
Stout, Mark E. et al., Terrorist Perspectives Project: Strategic and Operational Views of Al Qaida and Associated Movements (2008).
Strachan, Hew, The Direction of War (2013).
Stürchler, Nicholas, The Threat of Force in International Law (2007).
Sullivan, John P., *Future Conflict: Criminal Insurgencies, Gangs and Intelligence*, *in* John P. Sullivan & Robert J. Bunker, Mexico's Criminal Insurgency 29 (2012).
______, *Criminal Insurgencies in the Americas*, *in* John P. Sullivan & Robert J. Bunker, Mexico's Criminal Insurgency 71 (2012).
______, *Explosive Escalation? Reflections on the Car Bombing in Cuidad Juarez*, *in* John P. Sullivan & Robert J. Bunker, Mexico's Criminal Insurgency 81 (2012).
Sullivan, John P. & Robert J. Bunker, *Iraq & The Americas: 3 Gen Gangs Lessons and Prospects*, *in* John P. Sullivan & Robert J. Bunker, Mexico's Criminal Insurgency 1 (2012).
Sullivan, John P. & Adam Elkus, *State of Siege: Mexico's Criminal Insurgency*, *in* John P. Sullivan & Robert J. Bunker, Mexico's Criminal Insurgency 7 (2012).
Sullivan, John P. & Adam Elkus, *Cartel v. Cartel: Mexico's Criminal Insurgency*, *in* John P. Sullivan & Robert J. Bunker, Mexico's Criminal Insurgency 59 (2012).
Suter, Keith, An International Law of Guerrilla Warfare (1984).
Sutherland, Daniel E., A Savage Conflict: The Decisive Role of Guerrillas in the American Civil War (2009).
Sweets, John F., The Politics of Resistance in France, 1940–1944 (1976).
Syria: Minimizing Civilian Harm During Military Intervention Roundtable Outcome Document, Centre for Civilians in Conflict (2013), http://civiliansinconflict.org/uploads/files/publications/Syria_Roundtable_brief_Feb_2013.pdf.
Szesnat, Felicity & Annie R. Bird, *Columbia*, *in* International Law and the Classification of Conflicts 203 (Elizabeth Wilmshurst ed., 2012).

DE B. TAILLON, J. PAUL, HIJACKING AND HOSTAGES: GOVERNMENT RESPONSES TO TERRORISM (2002).

TANKEL, STEPHEN, STORMING THE WORLD STAGE: THE STORY OF LASHKAR-E-TAIBA (2013).

Taw, Jennifer Morrison, *The Role of Police in the Colombian Insurgency*, *in* POLICING INSURGENCIES: COPS AS COUNTERINSURGENTS 136 (C. Christine Fair & Sumit Ganguly eds., 2014).

Tetley, William, *Mixed Jurisdictions: Common Law vs Civil Law (Codified and Uncodified)* (pt. 1) (1999), http://www.cisg.law.pace.edu/cisg/biblio/tetley.html.

THOMAS, WARD, THE ETHICS OF DESTRUCTION (2001).

THOMPSON, LEROY, RAGGED WAR: THE STORY OF UNCONVENTIONAL AND COUNTER-REVOLUTIONARY WARFARE (1994).

THOMPSON, ROBERT, DEFEATING COMMUNIST INSURGENCY (1966).

Tladi, Dire, *The Nonconsenting Innocent State: The Problem with Bethlehem's Principle 12*, 107 AM. J. INT'L L. 570 (2013).

TOWLE, PHILIP ANTHONY, PILOTS AND REBELS: THE USE OF AIRCRAFT IN UNCONVENTIONAL WARFARE 1918–1988 (1989).

TOWNSHEND, CHARLES, BRITAIN'S CIVIL WARS: COUNTERINSURGENCY IN THE TWENTIETH CENTURY (1986).

______, TERRORISM: A VERY SHORT INTRODUCTION (2002).

TRINQUIER, ROGER, MODERN WARFARE: A FRENCH VIEW OF COUNTERINSURGENCY (2006) (1964).

TROFIMOV, YAROSLAV, THE SIEGE OF MECCA: THE FORGOTTEN UPRISING IN ISLAM'S HOLIEST SHRINE AND THE BIRTH OF AL QAEDA (2007).

TSE-TUNG, MAO, ON GUERRILLA WARFARE (Samuel B. Griffith trans., 2000) (1961).

Turner, L. S. et al., *Optimizing Deadly Persistence in Kandahar: Armed UAV Integration in the Joint Tactical Fight*, 13 CAN. ARMY J. 117 (2010).

TZU, SUN, THE ART OF WAR (Samuel B. Griffith trans., 1971) (1963).

UCKO, DAVID H., THE NEW COUNTERINSURGENCY ERA (2009).

URBAN, MARK, BIG BOYS' RULES: THE SECRET STRUGGLE AGAINST THE IRA (1992).

______, TASK FORCE BLACK (2010).

VAN CREVELD, MARTIN, THE RISE AND DECLINE OF THE STATE (1999).

DE VATTEL, EMER, THE LAW OF NATIONS (Knud Haakonssen ed., 2008) (1797).

VENTER, A. J., GUNSHIP ACE: THE WARS OF NEALL ELLIS, HELICOPTER PILOT AND MERCENARY (2011).

Verma, Arvind, *The Police and India's Maoist Insurgency*, *in* POLICING INSURGENCIES: COPS AS COUNTERINSURGENTS 291 (C. Christine Fair & Sumit Ganguly eds., 2014).

VINCI, ANTHONY, ARMED GROUPS AND THE BALANCE OF POWER: THE INTERNATIONAL RELATIONS OF TERRORISTS, WARLORDS AND INSURGENTS (2009).

VON CLAUSEWITZ, CARL, ON WAR (Michael Howard & Peter Paret eds. & trans., 1989).

VON DACH, H., TOTAL RESISTANCE: SWISS ARMY GUIDE TO GUERRILLA WARFARE AND UNDERGROUND OPERATIONS (1965).

von Elbe, Joachim, *The Evolution of the Concept of the Just War in International Law*, 33 AM. J. INT'L 1 (1939).

VON GLAHN, GERHARD, THE OCCUPATION OF ENEMY TERRITORY (1957).

von Heinegg, Wolff Heintschel, *Current Legal Issues in Maritime Operations: Maritime Interception Operations in the Global War on Terrorism, Exclusion Zones, Hospital Ships and Maritime Neutrality*, 80 INT'L L. STUD. 207 (2006).

______, *Maritime Interception/Interdiction Operations*, *in* THE HANDBOOK OF THE INTERNATIONAL LAW OF MILITARY OPERATIONS, 375 (Terry Gill & Dieter Fleck, 2010).

WALKER, ANDREW, WHAT IS BOKO HARAM?, Special Report, United States Institute for Peace (2012), *available at* http://www.usip.org/files/resources/SR308.pdf.

WALZER, MICHAEL, JUST AND UNJUST WARS (3rd ed. 1977).

______, ARGUING ABOUT WAR (2004).

WAR COMES HOME: THE EXCESSIVE MILITARIZATION OF AMERICAN POLICING, American Civil Liberties Union (2014), https://www.aclu.org/sites/default/files/field_document/jus14-warcomeshome-report-web-rel1_1.pdf.

WARDEN, JOHN A., THE AIR CAMPAIGN: PLANNING FOR COMBAT (1988), *available at* http://www.au.af.mil/au/awc/awcgate/warden/ward-toc.htm.

WARDLAW, GRANT, POLITICAL TERRORISM (2nd ed. 1990).

WARRICK, JOBY, THE TRIPLE AGENT (2011).

Watkin, Kenneth, *Warriors, Obedience and the Rule of Law*, 3 THE ARMY DOCTRINE AND TRAINING BULLETIN (Winter 2000/Spring 2001).

______, *Controlling the Use of Force: A Role for Human Rights Norms in Contemporary Armed Conflict* 1 (2004).

______, *Assessing Proportionality: Moral Complexity and Legal Rules*, 8 Y.B. OF INT'L HUM. L. 3 (2005)

______, *Warriors Without Rights? Combatants, Unprivileged Belligerents, and the Struggle over Legitimacy*, 2 Occasional Paper Series (Program on Humanitarian Policy and Conflict Research Harvard University, Winter 2005), http://www.hpcrresearch.org/sites/default/files/publications/OccasionalPaper2.pdf.

______, *Chemical Agents and Expanding Bullets: Limited Law Enforcement Exceptions or Unwarranted Handcuffs?*, 82 INT'L L. STUD. 193 (2006).

______, *Stability Operations: A Guiding Framework for "Small Wars" and Other Conflicts of the Twenty-First Century?*, 85 INT'L L. STUD. 411 (2009).

______, *Opportunity Lost: Organized Armed Groups and the ICRC "Direct Participation in Hostilities" Interpretive Guidance*, 42 N.Y.U. J. INT'L L. & POL. 641 (2010).

______, *Use of Force during Occupation: Law Enforcement and Conduct of Hostilities*, 94 INT'L REV. RED CROSS 267 (2012).

______, *The Cyber Road Ahead: Merging Lanes and Legal Challenges*, in 89 INT'L L. STUD. 472 (2013).

______, *Targeting in Air Warfare*, 44 ISR. Y.B. ON HUM. RTS. 1 (2014).

______, *Targeting "Islamic State" Oil Production Facilities*, 90 INT'L L. STUD. 499 (2014).

______, *Military Advantage: A Matter of "Value", Strategy and Tactics*, 17 INT'L Y.B. HUM. L. 277 (2014).

WAYNE, MARTIN, CHINA'S WAR ON TERRORISM: COUNTER-INSURGENCY, POLITICS AND INTERNAL SECURITY (2008).

WEIGHLEY, RUSSEL F., THE AMERICAN WAY OF WAR: A HISTORY OF UNITED STATES MILITARY STRATEGY AND POLICY (1973).

WEST, BING, NO TRUE GLORY: A FRONTLINE ACCOUNT OF THE BATTLE FOR FALLUJAH (2005).

______, THE STRONGEST TRIBE: WAR, POLITICS AND THE ENDGAME IN IRAQ (2009).

Whitaker, Richard M., *Civilian Protection Law in Military Operations: An Essay*, Department of the Army Pamphlet 27-50-288 (Nov. 1996), http://www.loc.gov/rr/frd/Military_Law/pdf/11-1996.pdf.

WHITCOMB, CHRISTOPHER, COLD ZERO: INSIDE THE HOSTAGE RESCUE TEAM (2001).

WIEMANN, GABRIEL, TERROR ON THE INTERNET (2006).

WILLIAMS, PAUL, CRIMINALS, MILITIA, AND INSURGENTS: ORGANIZED CRIME IN IRAQ (2009).

WILLIAMSON, TONY, COUNTER STRIKE ENTEBBE (1976).

WILMSHURST, ELIZABETH, PRINCIPLES OF INTERNATIONAL LAW ON THE USE OF FORCE BY STATES IN SELF-DEFENCE (Oct. 2005), http://www.chathamhouse.org/publications/papers/view/108106.

Wilson, Randall, Blue Fish in a Dark Sea: Police Intelligence in a Counter-insurgency (2013).
Woods, Elliot D., *The Soft Knock*, Va. Q. Rev. (Oct. 9, 2009), http://www.vqronline.org/news/soft-knock.
Wright, Lord, *The Killing of Hostages as a War Crime*, 25 Brit. Y.B. Int'l L. (1948).
Wright, Quincy, *Changes in the Conception of War* 18 Am. J. Int'l L. 755 (1924).
______, *The Outlawry of War and the Law of War*, 47 Am. J. Int'l L. 365 (1953).
Zarate, Juan C., Treasury's War: Unleashing of a New Era of Financial Warfare (2013).
Zegveld, Liesbeth, The Accountability of Armed Opposition Groups in International Law (2002).
Zuckoff, Mitchell, 13 Hours: The Inside Account of What Really Happened in Benghazi (2014).

REPORTS/STATEMENTS OF GOVERNMENTS

Canada

1 & 5 Dishonoured Legacy: Lessons of the Somalia Affair, Report of the Commission of Inquiry into the Deployment of Canadian Forces to Somalia (1997).
Special Committee on the Canadian Mission in Afghanistan, Meeting No. 14, 40th Parliament 2nd Session, Evidence No. 14 (Nov. 4, 2009), http://www.parl.gc.ca/content/hoc/Committee/402/AFGH/Evidence/EV4209808/AFGHEV14-E.PDF.
Fighting International Terrorism and Crime, Permanent Mission of Canada to the United Nations, http://www.canadainternational.gc.ca/prmny-mponu/canada_un-canada_onu/positions-orientations/peace-paix/terrorism-terrorisme.aspx?lang=eng.

Israel

The Operation in Gaza 21 December, 2008–18 January 2009, Legal and Factual Aspects, The State of Israel (2009).
1 Public Commission to Examine the Maritime Incident of 31 May 2010 (2011).
Salah Shehadeh-Special Investigatory Commission, Israel Ministry of Foreign Affairs (Feb. 21, 2011), http://mfa.gov.il/MFA/AboutIsrael/State/Law/Pages/Salah_Shehadeh-Special_Investigatory_Commission_27-Feb-2011.aspx.
2 Public Commission to Examine the Maritime Incident of 31 May 2010 (2013).

United Kingdom

AP I, UK Reservation/Declaration (July 2, 2002), http://www.icrc.org/applic/ihl/ihl.nsf/Notification.xsp?action=openDocument&documentId=0A9E03F0F2EE757CC1256402003FB6D2.
Testimony of Attorney-General Lord Goldsmith, 660 *Hansard H.L.* (Apr. 21, 2004) 370 (U.K.), *available at* http://hansard.millbanksystems.com/lords/2004/apr/21/international-self-defence.
The Manual of The Law of Armed Conflict, United Kingdom Ministry of Defence (2004).
CPS statement: Charging decision on the fatal shooting of Jean Charles de Menezes, The Crown Prosecution Service (July 17, 2006), http://www.cps.gov.uk/news/latest_news/146_06/.
CPS Statement—Shooting of Jean Charles de Menezes, The Crown Prosecution Service (Feb. 13, 2009), http://www.cps.gov.uk/news/latest_news/103_09/index.html.
III The Report of the Baha Mousa Inquiry, The Rt Hon Sir William Gage (Chairman) (2011), https://www.gov.uk/government/uploads/system/uploads/attachment_data/file/279192/1452_iii.pdf.

United States of America

THE SEIZURE OF THE MAYAGUEZ: A CASE STUDY OF CRISIS MANAGEMENT, U.S. Government Accountability Office (May 11, 1976), http://www.gao.gov/assets/240/233910.pdf.

Speech, Robert S. Mueller III, Federal Bureau of Investigation, http://www.fbi.gov/news/speeches/the-fbi-and-the-military-combining-forces-to-keep-america-safe (2007).

Clinton, Hillary Rodham, *Reaffirming America's Commitment to Humane Treatment of Detainees* Press Statement, (Mar. 7, 2011), http://m.state.gov/md157827.htm.

Koh, Harold Hongju, *The Obama Administration and International Law* (Mar. 25, 2010), http://www.state.gov/s/l/releases/remarks/139119.htm.

______, Legal Adviser, Department of State, *Responses to Questions Submitted by Senator Richard G. Lugar, Libya and War Powers: Hearing Before the Committee on Foreign Relations, U.S. Senate*, 112th Congress, First Session (June 28, 2011).

______, Legal Advisor, U.S. Department of State, Remarks at USCYBER-COM Inter-Agency Legal Conference (Sept. 18, 2012), http://www.state.gov/s/l/releases/remarks/197924.htm.

NATIONAL STRATEGY FOR COUNTERTERRORISM, The White House (June 2011), http://www.whitehouse.gov/sites/default/files/counterterrorism_strategy.pdf.

Letter Attorney General Eric Holder to The Honorable Patrick Leahy, 22 May 2013, http://www.justice.gov/slideshow/AG-letter-5-22-13.pdf.

Opening Statement Mary E. McLeod Acting Legal Adviser U.S. Department of State Committee Against Torture, Mission of the United States, Geneva, Nov. 12–13, 2014, https://geneva.usmission.gov/2014/11/12/acting-legal-adviser-mcleod-u-s-affirms-torture-is-prohibited-at-all-times-in-all-places/.

PERM. S. COM. ON INTELLIGENCE, 113th CONG., INVESTIGATIVE REP. ON THE TERRORIST ATTACKS ON U.S. FACILITIES IN BENGHAZI, LIBYA, SEPTEMBER 11–12, 2012 (Nov. 21, 2014), https://intelligence.house.gov/sites/intelligence.house.gov/files/documents/Benghazi%20Report.pdf.

THE NATIONAL MILITARY STRATEGY OF THE UNITED STATES OF AMERICA 2015, (June 2015), http://www.jcs.mil/Portals/36/Documents/Publications/2015_National_Military_Strategy.pdf.

Department of Justice White Paper, 4 (undated), http://msnbcmedia.msn.com/i/msnbc/sections/news/020413_DOJ_White_Paper.pdf.

Memorandum for the Attorney General, Re: Applicability of Federal Criminal Laws and the Constitution to Contemplated Lethal Operations Against Shaykh Anwar al-Aulaqi, Office of the Assistant Attorney General, July 16, 2010, at 27, https://www.aclu.org/sites/default/files/assets/2014-06-23_barron-memorandum.pdf.

Letter from the United States Ambassador to the United Nations to The President of the United Nations Security Council (June 17, 2014), https://www.documentcloud.org/documents/1201215-power-letter-to-un-about-khattala.html.

Letter to United Nations Secretary General from United States Ambassador Samantha J. Power, dated 23 September, 2014, reproduced in Marty Lederman, *The War Powers Resolution and Article 51 Letters Concerning Use of Force in Syria Against ISIL and the Khorasan Group [UPDATED to add statement of the U.N. Secretary-General]*, Just Security (Sept. 23, 2014, 3:21 PM), http://justsecurity.org/15436/war-powers-resolution-article-51-letters-force-syria-isil-khorasan-group/.

Preston, Stephen W., *The Legal Framework for the United States' Use of Military Force Since 9/11*, Remarks by the General Counsel of the Department of National Defense on the Legal Framework for the United States' Use of Force Since 9/11 (Apr. 10, 2015), http://www.defense.gov/Speeches/Speech.aspx?SpeechID=1931.

INTERNATIONAL ORGANIZATIONS

Human Rights Watch

Israel, The Occupied West Bank And Gaza Strip, And The Palestinian Authority Territories, Jenin: IDF Military Operations, Human Rights Watch, Vol. 14, No. 3 (E) (May, 2002), http://www.hrw.org/reports/2002/israel3/israel0502.pdf.

Violent Response: The U.S. Army In Al-Falluja, 15 Human Rights Watch, Iraq, NO. 7 (E), June 2003, http://www.hrw.org/reports/2003/iraqfalluja/iraqfalluja.pdf.

International Commission on Intervention and State Sovereignty

The Responsibility to Protect: Report of the International Commission on Intervention and State Sovereignty, ICSS, Dec., 2001, http://responsibilitytoprotect.org/ICISS%20Report.pdf.

International Committee of the Red Cross and Humanitarian Law

XV Official Records of the Diplomatic Conference on the Reaffirmation and Development of International Humanitarian Law Applicable in Armed Conflicts, Geneva (1974–1977) CDDH/III/SR. 43 (1978).

Interpretive Guidance on the Notion of Direct Participation in Hostilities Under International Humanitarian Law, International Committee of the Red Cross (2009).

International Humanitarian Law and the challenges of contemporary armed conflicts, Report 31st International Conference Of The Red Cross And Red Crescent, EN 31IC/11/5.1.2 (28 November–1 December 2011), http://www.icrc.org/eng/assets/files/red-cross-crescent-movement/31st-international-conference/31-int-conference-ihl-challenges-report-11-5-1-2-en.pdf.

Third Meeting Of Experts: The Use Of Force In Occupied Territory 29–30 October 2009, Geneva, *in* Occupation and Other Forms of Administration of Foreign Territory, ICRC (Tristan Ferraro ed., 2012).

The Use of Force in Armed Conflicts: Interplay Between the Conduct of Hostilities and Law Enforcement Paradigms, ICRC (Nov. 2013), http://www.icrc.org/eng/assets/files/publications/icrc-002-4171.pdf.

INTERNATIONAL HUMANITARIAN LAW AND THE CHALLENGES OF CONTEMPORARY ARMED CONFLICT, Report 32nd International Conference of the Red Cross and Red Crescent EN 32IC/15/11 (2015), *available at* https://www.justsecurity.org/wp-content/uploads/2015/11/2015-ICRC-Report-IHL-and-Challenges-of-Armed-Conflicts.pdf

International Crisis Group

Dismantling Colombia's New Illegal Armed Groups: Lessons From A Surrender, International Crisis Group Latin America Report N°41 (June 8, 2012), http://www.crisisgroup.org/~/media/Files/latin-america/colombia/41-dismantling-colombias-new-illegal-armed-groups-lessons-from-a-surrender.pdf.

International Law Commission

Fragmentation Of International Law: Difficulties Arising From The Diversification And Expansion Of International Law, Rep. of the Int'l L. Comm'n., 58th session, May 1–June 9, July 3–Aug. 11 2006, UN Doc. A/CN.4/L.702 (July 18, 2006), http://legal.un.org/ilc/documentation/english/a_cn4_l682.pdf.

International Law Association

INTERNATIONAL LAW ASSOCIATION, FINAL REPORT OF THE USE OF FORCE COMMITTEE: THE MEANING OF ARMED CONFLICT (Aug. 2010), http://www.ila-hq.org/en/committees/index.cfm/cid/1022.

NATO

ISAF Commander's Tactical Directive (July 6, 2009) ISAF Headquarters, http://www.nato.int/isaf/docu/official_texts/Tactical_Directive_090706.pdf.

COMISAF NIGHT OPERATIONS TACTICAL DIRECTIVE, Headquarters International Security Assistance Force (Dec. 2011), http://www.isaf.nato.int/images/docs/20111105%20nuc%20night%20operations%20tactical%20directive%20(releaseable%20version)%20r.pdf.

Letter Peter Olson to Judge Kirsch, OLA (2012) 006, 4, Jan. 23, 2012, http://www.nato.int/nato_static/assets/pdf/pdf_2012_05/20120514_120514-NATO_1st_ICIL_response.pdf.

United Nations

A MORE SECURE WORLD: OUR SHARED RESPONSIBILITY, REPORT OF THE HIGH-LEVEL PANEL ON THREATS, CHALLENGES AND CHANGE (2004), https://www.un.org/en/peacebuilding/pdf/historical/hlp_more_secure_world.pdf.

Report of the International Commission of Inquiry on Darfur, Pursuant to Security Council Resolution 1564 of 18 September 2004, Report to the Secretary General, Geneva (Jan. 25, 2005), http://www.un.org/news/dh/sudan/com_inq_darfur.pdf.

Extrajudicial, summary or arbitrary executions, *Civil And Political Rights, Including The Questions Of Disappearances And Summary Executions*, Comm'n. on Human Rights, UN Doc. E/CN.4/2006/53 (Mar. 8, 2006) (by Philip Alston), http://www.unhcr.org/refworld/pdfid/45377b100.pdf.

UNITED NATIONS PEACEKEEPING OPERATIONS: PRINCIPLES AND GUIDELINES, Department of Peacekeeping Operations (2008), *available at* http://www.dtic.mil/doctrine/doctrine/interagency/un_prin_guide.pdf.

Report of the Special Rapporteur on extrajudicial, summary or arbitrary executions, Philip Alston, UN Doc. A/HRC/14/24/Add.6 (May 28, 2010), http://www2.ohchr.org/english/bodies/hrcouncil/docs/14session/A.HRC.14.24.Add6.pdf.

Human Rights Council, *Report of the independent international commission of inquiry on the Syrian Arab Republic*, UN Doc. A/HRC/19/69 (Feb. 22, 2012), http://daccess-dds-ny.un.org/doc/UNDOC/GEN/G12/106/13/PDF/G1210613.pdf?OpenElement.

Report of the Monitoring Group on Somalia and Eritrea pursuant to Security Council resolution 2060 (2012): Somalia, *Letter to the Security Council*, UN Doc. S/2013/413 (July 12, 2013), http://www.securitycouncilreport.org/atf/cf/%7B65BFCF9B-6D27-4E9C-8CD3-CF6E4FF96FF9%7D/s_2013_413.pdf.

United Nations Assistance Mission in Afghanistan, *Afghanistan Mid-Year Report 2012 Protection Of Civilians In Armed Conflict* (July 2012).

Report of the Special Rapporteur on extrajudicial, summary or arbitrary executions, UN Doc. AA/68/382 (Sept. 13, 2013) (by Christof Heyns), http://www.un.org/en/ga/search/view_doc.asp?symbol=A/68/382.

Report of the Special Rapporteur on the promotion and protection of human rights and fundamental freedoms while countering terrorism, Ben Emmerson, UN Doc. A/68/389 (Sept. 18, 2013).

H.R. Committee, Concluding observations on the fourth periodic report of the United States of America, Human Rights Committee, UN Doc. CCPR/C/USA/CO/4 (Apr. 23, 2014).

Performance Peacekeeping: Final Report of the Expert Panel on Technology and Innovation in UN Peacekeeping, Under-Secretary Department of Peacekeeping Operations & Under-Secretary Field Operations (Dec. 22, 2014).

MILITARY MANUALS AND DOCTRINE PUBLICATIONS

Canada

Law of Armed Conflict at the Operational and Tactical Levels, B-GJ-005-1 04/FP-021 (2001-08-13).

CF Code of Conduct CF Personnel, Office of the Judge Advocate General B-GG-005-027/AF-023 2-9 (undated).

United States of America

U.S. War Department, Instructions for the Government of Armies of the United States in the Field, General Orders No. 100, Arts. (April 24, 1863), *in* Lieber's Code, 45–71.

Joint Doctrine for Military Operations Other Than War, Joint Pub 3-07 (16 June 1995).

Small Wars Manual, United States Marine Corps (2009) (1940).

Joint Doctrine for Military Operations Other Than War, Joint Pub 3-07 (June 16, 1995), http://www.bits.de/NRANEU/others/jp-doctrine/jp3_07.pdf.

The U.S. Army, Marine Corps, Counterinsurgency Field Manual (2007).

DoD Law of War Program, Department of Defense, Directive Number 2311.01E, May 9, 2006 (Incorporating Change 1, November 15, 2010, Certified Current as of February 22, 2011), http://www.dtic.mil/whs/directives/corres/pdf/231101e.pdf.

Counterinsurgency, Joint Publication (Nov. 22, 2013), http://www.dtic.mil/doctrine/new_pubs/jp3_24.pdf.

Legal Support to the Operational Army, FM-04, Headquarters Department of the Army (Jan. 2012), http://armypubs.army.mil/doctrine/DR_pubs/dr_a/pdf/fm1_04.pdf.

Joint Targeting, United States Joint Publication (Jan. 31, 2013).

Stability, FM 3-07, Headquarters Department of the Army (June 2, 2014).

Counterterrorism, JP 3-26 (Oct. 24, 2014), http://www.dtic.mil/doctrine/new_pubs/jp3_26.pdf.

Detainee Operations, JP3-63 (Nov. 14, 2014), http://www.dtic.mil/doctrine/new_pubs/jp3_63.pdf.

Department of Defense Law of War Manual, Office of the General Counsel (June 2015), http://www.dod.mil/dodgc/images/law_war_manual15.pdf.

Operational Law Handbook (Lieutenant-Commander David Lee, ed., 2015), http://www.loc.gov/rr/frd/Military_Law/pdf/operational-law-handbook_2015.pdf

United Kingdom

Manual of Military Law (His Majesty's Stationary Office, 1916) (1914).

The Manual of The Law of Armed Conflict, United Kingdom Ministry of Defence (2004).

Table of Cases

International Courts and Tribunals
- European Court of Human Rights
- International Criminal Court
- International Court of Justice
- International Criminal Tribunal for Rwanda
- International Criminal Tribunal for the former Yugoslavia
- Inter-American Commission on Human Rights
- Inter-American Court on Human Rights
- Nuremburg Military Tribunals

National Courts
- Canada
- Germany
- Israel
- The Netherlands
- United Kingdom
- United States

INTERNATIONAL COURTS AND TRIBUNALS

European Court of Human Rights

Al-Skeini v. The United Kingdom, IV Eur. Ct. H. R. Rep. 99 (2011): 103*n*88, 120*n*216, 149*n*181, 152*n*200, 230*n*86, 234*n*114, 402*n*152, 487*n*30

Andronicou v. Cyprus, Application Nos. 86/1996/705/897, Eur. Ct. H. R. (Oct. 9, 1997), http://hudoc.echr.coe.int/sites/eng/pages/search.aspx?i=001-58102: 139*n*122, 452*n*45, 497

Banković v. Belgium, XII Eur. Ct. H. R. Rep. 333 (2001): 558

Bubbins v. The United Kingdom, App. No. 50196/99, Eur. Ct. H.R. (2005), http://hudoc.echr.coe.int/sites/eng/pages/search.aspx?i=001-68548: 497*n*106
Ergi v. Turkey, Application no. 66/1997/850/1057, Eur. Ct. H. R. (1998), http://hudoc.echr.coe.int/eng?i=001-58200: 507*n*216
Finogenov v. Russia, VI Eur. Ct. H. R. 365 (2011) (extracts): 139*n*122, 154, 154*n*208 260*n*292, 452, 481, 499*nn*125–127, 500*n*140, 507*n*216, 508*nn*217–223, 509*n*229, 551*n*128, 555, 562
Hassan v. The United Kingdom, Application No. 29750/09 Eur. Ct. H. R. (2014), http://hudoc.echr.coe.int/sites/eng/pages/search.aspx?i=001-146501: 103*n*88, 124–125, 127–128, 152*n*200, 471*n*146, 552–553, 558, 592–593, 614*n*190, 615, 624
Isayeva v. Russia, Application Nos. 57947/00, 57948/00, 57949/00 Eur. Ct. H. R. (2005): 125*n*22, 419*n*305, 481*n*204, 552*n*138, 553–554, 558*n*169
Isayeva v. Russia Application No. 57959/00 Eur. Ct. H. R. (2005), http://hudoc.echr.coe.int/sites/eng/pages/search.aspx?i=001-68381: 125*n*22, 419*n*305, 481*n*204, 552*n*138, 554–555, 558*n*169
Jaloud v. The Netherlands, Eur. Ct. H. R., Application No. 47708/08 (2014), http://hudoc.echr.coe.int/sites/eng/pages/search.aspx?i=001-148367: 103*n*88, 558*n*168
Kelly v. United Kingdom, App. No. 30054/96, Eur. Ct. H.R. Judgment (2001), http://hudoc.echr.coe.int/sites/eng-press/pages/search.aspx?i=001-59453: 544*n*79, 544*n*81, 544*n*83
Kerimova v. Russia, Application Nos. 17170/04, 20792/04, 22448/04, 23360/04, 5681/05 and 5684/05 Eur. Ct. H. R. (2011), http://hudoc.echr.coe.int/sites/eng/pages/search.aspx?i=001-104662: 125*n*22, 139*n*122, 452*n*45, 481*n*204, 555, 558*n*169
McCann and others v. United Kingdom, App. No. 18984/91, Eur. Ct. H.R., Judgment (1995), http://hudoc.echr.coe.int/sites/eng/pages/search.aspx?i=001-57943: 21*n*105, 131*n*57, 286*n*118, 462*n*88, 525*n*350, 530, 531, 544*n*79, 544*n*81, 545*n*85, 546–548, 549*n*122, 596*n*91
McKerr v. United Kingdom, Application No. 28883/95, Eur. Ct. H.R., Judgment (2001), http://hudoc.echr.coe.int/sites/eng/pages/search.aspx?i=001-59451: 131*n*62, 544*n*81, 544*n*84, 546*nn*95–96, 547, 550, 627*n*32
Sargsyan v. Azerbaijan, Eur. Ct. H. R., Application No. 40167/06 (2015), http://hudoc.echr.coe.int/eng?i=001-155662: 103*n*86
Varnava v. Turkey, Application Nos. 16064/90, 16065/90, 16066/90, 16068/90, 16069/90, 16070/90, 16071/90, 16072/90 and 16073/90, Eur. Ct. H. R. (2009), http://hudoc.echr.coe.int/sites/eng/pages/search.aspx?i=001-94162: 552*n*139

International Criminal Court

Prosecutor v. Gombo, Case No. ICC-01/05-01/08, Decision Pursuant to Article 61(7)(a) and (b) of the Rome Statute (International Criminal Court, 15 June 2009), http://www.icc-cpi.int/iccdocs/doc/doc699541.pdf: 367*n*195

International Court of Justice

Advisory Opinion on the Legality of the Threat or Use of Nuclear Weapons, 1996 I.C.J. 226: 56, 73, 82, 85, 122*n*3, 124, 125, 127, 128, 302*n*225, 303*n*234, 471*n*146, 614
Case Concerning United States Diplomatic and Consular Staff in Tehran (United States v. Iran), 1980 I.C.J. 3: 338, 409
Case Concerning Oil Platforms (Islamic Republic Of Iran V. United States Of America), 2003 I.C.J. 158: 73, 78, 87–88, 331–332, 390, 392–393, 465
Legal Consequences Of The Construction Of A Wall In The Occupied Palestinian Territory, 2004 I.C.J. 133: 35*n*10, 125, 127, 153, 237*n*141, 262, 517–518, 564
Military and Paramilitary Activities in and against Nicaragua Merits, 1986 I.C.J. 13: 71, 331, 335*n*30, 389–390, 577, 582

International Criminal Tribunal for Rwanda

Prosecutor v. Akayesu, Case No. ICTR-96-4-T, Judgement (Int'l Crim. Trib. for Rwanda, Sept. 2, 1998), http://www.unictr.org/sites/unictr.org/files/case-documents/ictr-96-4/trial-judgements/en/980902.pdf: 375*n*238, 377*n*247

International Criminal Tribunal for the former Yugoslavia

Prosecutor v. Boškoski, Case No. IT-04-82-T 90, Judgement (Int'l Crim. Trib. for the Former Yugoslavia, July 10, 2008), http://www.icty.org/x/cases/boskoski_tarculovski/tjug/en/080710.pdf: 290*n*146, 317*n*321, 376, 587

Prosecutor v. Galić, ICTY, IT-98-29-T, Judgement and Opinion (Int'l Crim. Trib. for the Former Yugoslavia, 2003), http://www.icty.org/x/cases/galic/tjug/en/gal-tj031205e.pdf: 286*n*117, 525*n*351, 596*n*90

Prosecutor v. Haradinaj, Case No. IT-04-84-T, Judgement, (Int'l Crim. Trib. for the Former Yugoslavia, Apr. 3, 2008), http://www.icty.org/x/cases/haradinaj/tjug/en/080403.pdf: 161*n*15, 293*n*168, 367*n*195, 372*n*222, 374*n*231, 375, 417–418, 425

Prosecutor v. Kordić, Case No. IT-95-14/2-A 90, Judgement (Int'l Crim. Trib. for the Former Yugoslavia, Dec. 17, 2004), http://www.icty.org/x/cases/kordic_cerkez/acjug/en/cer-aj041217e.pdf: 138*n*118, 347*n*232, 448

Prosecutor v. Kunarac, No. IT-96-23-T & IT-96-23/1-T, Judgement (Int'l Crim. Trib. for the Former Yugoslavia, Feb. 22, 2001), http://www.icty.org/x/cases/kunarac/tjug/en/kun-tj010222e.pdf: 134*n*78

Prosecutor v. Limaj, Case No. IT-03-66-T 37, Judgement (Int'l Crim. Trib. for the Former Yugoslavia, Nov. 30, 2005), http://www.icty.org/x/cases/limaj/tjug/en/lim-tj051130-e.pdf: 161*n*15, 167, 372*n*222, 374–375, 417*n*301

Prosecutor v. Milŏsević, Case No. IT-02-54-T, Decision on Motion for Judgment of Acquittal (Int'l Crim. Trib. for the Former Yugoslavia, June 16, 2004), http://www.icty.org/x/cases/slobodan_milosevic/tdec/en/040616.pdf: 372*n*222, 373–374

Prosecutor v. Tadić, Case No. IT-94-1-A, Decision On The Defence Motion For Interlocutory Appeal On Jurisdiction (Int'l Crim. Trib. for the Former Yugoslavia, 1995), http://www.icty.org/x/cases/tadic/acdec/en/51002.htm: 14*n*68, 34*n*2, 107, 109, 161*n*14, 175, 262*n*298, 292, 293, 315*n*314, 316, 329*n*4, 331*n*11, 348*n*96, 354, 363, 365–367, 370, 373, 417, 419, 456*n*65, 518*n*321, 575*n*21, 576, 579, 580*n*45, 581, 582, 585

Prosecutor v. Tadić, Case No. IT-94-1-T, Opinion and Judgement (Int'l Crim. Trib. for the Former Yugoslavia, May 7, 1997), http://www.icty.org/x/cases/tadic/tjug/en/tad-tsj70507JT2-e.pdf: 366, 367

Prosecutor v. Tadić, No. IT–94–1–A, Appeals Judgment (Int'l Crim. Trib. for the Former Yugoslavia, July 15, 1999), http://www.icty.org/x/cases/tadic/acjug/en/tad-aj990715e.pdf: 35*n*10, 335*nn*30–31

Inter-American Commission on Human Rights

Case 11.137, Abella v. Argentina, Inter-Am. Y.B. on H.R. (1997) (Commission report), http://www1.umn.edu/humanrts/cases/1997/argentina55-97a.html: 293, 354, 355, 366, 367, 375*n*237, 376*n*246, 377, 506, 508, 579, 582, 600

Eduardo Nicolás Cruz Sánchez, et al Peru (*"Japanese Embassy Case"*), Case 12.444, Merits, Inter-Am. Comm'n H.R., Report No. 66/10 (2011), https://www.oas.org/en/iachr/decisions/court/PEME12.444ENG.pdf: 501*n*154, 502*n*156, 502*nn*162–164, 502*nn*167–168, 503*nn*176–177, 504*nn*184–186, 504*n*189, 505*nn*190–194, 505*nn*196–200, 506*n*204, 558*n*171, 559*n*172

Inter-American Court on Human Rights

Bámaca-Velásquez v. Guatemala, Inter-American Ct. H. R. (2000), http://www.corteidh.or.cr/docs/casos/articulos/seriec_70_ing.pdf: 553*n*142
Case of Cruz Sánchez et al. v. Peru (*"Japanese Embassy Case"*), Series C no. 292, Preliminary Objections, Merits, Reparations and Costs, Inter-Am. Ct. Hum. Rts (2015), http://www.corteidh.or.cr/docs/casos/articulos/seriec_292_esp.pdf: 125*n*21, 501*n*154, 502*nn*156–159, 503*nn*168–169, 503*n*175, 503*n*177, 504*n*187, 506*n*202, 507*nn*210–216, 553*n*142, 558*n*171, 559*n*172

Nuremburg Military Tribunals

The Hostage Case, Trials of War Criminals, Washington: Government Printing Office (1950): 94, 399*n*116

NATIONAL COURTS

Canada

Amnesty International Canada v. Canada (Chief of the Defence Staff), 2008 FCA 401, [2009] 4 F.C.R. 149, http://reports.fja-cmf.gc.ca/eng/2008/2008fc336.html: 149*n*182, 149*n*184
Amnesty International Canada v. Canada (Chief of the Defence Staff) (F.C.), 2008 FC 336, [2008] 4 F.C.R. 546, http://reports.fja-cmf.gc.ca/eng/2008/2008fc336.html: 47*n*113, 149*n*182, 271*n*18

Germany

Aerial Drone Deployment on 4 October 2010 in Mir Ali/Pakistan, 157 I.L.R. 722 (2013): 231*n*94, 236*n*140, 238*n*152, 239*n*157, 288*n*135, 319*n*328, 370*n*211, 376*n*246, 456*n*64, 608*n*139
Dr. H. et al. v. s.14.3 of the Aviation Security Act of 11 January 2005 (2006), File no: 1 BvR 357/05. (German Federal Constitutional Court), http://www.bundesverfassungsgericht.de/entscheidungen/rs20060215_1bvr035705en.html: 261, 561–563, 578

Israel

Al-Bassiouni v. Prime Minister, HCJ 9132/07 (unpublished, Jan. 30, 2008): 343
Public Committee Against Torture in Israel v. Israel (*"Targeted Killing Case"*), Israel Supreme Court [16 December 2006], 46 ILM 375 (2007): 21*n*106, 131*n*60, 230, 231–237, 243, 245, 247, 258*n*280, 261, 262–263, 265, 286, 320–321, 341–349, 361, 376*n*245, 377*n*249, 445*n*6, 604, 608

The Netherlands

Nuhanović v. The State of the Netherlands [Court of Appeal in The Hague] Case No. 200.020.174/01 (July 2011) (Neth.) http://uitspraken.rechtspraak.nl/inziendocument?id=ECLI:NL:GHSGR:2011:BR5388: 21*n*107
The State of the Netherlands v. Nuhanović [Supreme Court of the Netherlands] Case No. 12/03324 (Sept. 16, 2013), http://www.asser.nl/upload/documents/20130909T125927-Supreme%20Court%20Nuhanovic%20ENG.pdf: 21*n*107

United Kingdom

Al-Bassiouni v. Prime Minister, HCJ 9132/07 (unpublished, Jan. 30, 2008): 343

United States

Al-Bihani v. Obama, 590 F.3d 866, 872 (2010), en banc rehearing denied, 619 F.3d 1 (D.C. Cir. 2010): 309*n*278, 310*n*282
Brown v. United States, 41 Sup. Ct. 502 (1921): 497*n*107, 525*n*349
Ex Parte Quirin, 317 U.S. 1 (1942): 94, 264
Graham v. Connor, 109 S. Ct. 1872 (1989): 497*n*108, 525*n*348
Hamdan v. Rumsfeld, 45 I.L.M. 1130 (2006) (U.S. Supreme Court 2006): 27, 48, 67*n*86, 98*n*48, 99, 148, 345–347, 351, 353, 355, 356–358, 364, 604, 624
Hamdan v. United States of America, 696 F. 3d 1250 (2012): 310
Holder v. Humanitarian Law Project, 130 S. Ct. 2705 (2010): 307*n*259, 309
Idaho v. Horiuchi, 253 F.3d 359, 377, vacated as moot, 266 F.3d 979 (9th Cir. 2001): 545

Index

Abaslon, Danish warship, 430
Abdikadir, Abdikadir Mohamed (known as Ikrimah), 276
Abella v. Argentina, Inter-Am. Y.B. on H.R. (1997) (Commission report), 293, 354–55, 366–67, 375*n*237, 376*n*246, 377, 506, 508, 579, 582, 600
Abu Sayyaf, 170, 250, 277, 309, 325, 601
Additional Protocol I (AP I), articles, 96–97, 145–46
  Article 1(2), 161*n*16
  Article 1(4), 45, 96*n*30, 97, 358
  Article 35(2), 232*n*104
  Article 41, 506*n*205
  Article 41(2)(a), 232*n*103
  Article 43, 246*n*202
  Article 44, 96*n*29
  Article 48, 92*n*5, 334*n*29, 426*n*352, 484*n*8, 605*n*133
  Article 50, 92*nn*5–6, 488*n*38
  Article 51, 86, 92*nn*5–6, 488*n*38
  Article 51(1), 363*n*176
  Article 51(2), 80*n*182
  Article 51(3), 246*n*202, 405*n*174, 488*n*37
  Article 51(5)(b), 132*n*69
  Article 52, 58, 246
  Article 52(2), 79*n*166, 87*n*223, 92*n*5, 133*n*70, 246*n*201
  Article 57, 79–80, 84*n*200, 92*n*5, 132*n*69, 273*n*33, 281, 560
  Article 57(2), 244*n*190
  Article 57(2)(a)(ii), 248*n*213, 249*n*222, 485*n*9, 554
  Article 57(2)(a)(iii), 57*n*79, 75*n*137, 79*n*166, 154*n*211, 245–46, 258, 445*n*6, 485*n*12, 559*n*174
  Article 65(1), 138*n*118
  Article 65(3), 451*n*42
  Article 67, 138*n*118
  Article 67(1)(d), 451*n*42
  Article 75, 133*n*75, 142–43, 147–49, 150*n*189, 156*n*220, 262, 488*n*40
  Article 75(1), 137*n*108, 143*n*150
  Article 75(2), 143*n*150
  Article 75(2)(a), 138*n*109
  Article 75(2)(c), 399*n*117
  Article 75(2)(ii), 137*n*107
  Article 75(3), 138*n*111
  Article 75(4), 138*n*111
  Article 75(7), 138*n*111
  Article 87, 133
  disinterest in amending, 145–46
Additional Protocol II (AP II), articles, 145–46
  Article 1(1), 167
  Article 1(2), 175*n*149, 377*n*247, 392*n*64, 451*n*44, 495*n*92, 579*n*40, 582*n*53, 587*n*64
  Article 3, 44, 97, 99

Additional Protocol II (AP II), articles (*Cont.*)
Article 4, 133*n*75, 142–43, 148–49, 156*n*220, 488*n*40
Article 4(1), 137*n*108
Article 4(2)(a), 137*n*107
Article 4(2)(c), 399*n*117
Article 5, 148
Article 5(3), 137*n*108
Article 6, 142*n*144, 148
Article 6(2), 138*n*111
Article 7(2), 137*n*108
disinterest in amending, 145–46
humanitarian law, reluctant application of, 109
Advisory Opinion on the Legality of the Threat or Use of Nuclear Weapons, 56, 73, 82, 85, 122*n*3, 124–25, 127–28, 302*n*225, 303*n*234, 471*n*146, 614
Aerial Drone Deployment on 4 October 2010 (Mir Ali/Pakistan), 231*n*94, 236*n*140, 238*n*152, 239*n*157, 288*n*135, 319*n*328, 370*n*211, 376*n*246, 456*n*64, 608*n*139
Afghanistan, 46–49, 115–16, 209–13, 215–24, 271, 280–83, 288, 336–40, 359–61, 486–91, 493–94, 511–13, 621–23
International Security Assistance Forces (ISAF) and post 9/11 Afghanistan conflict, 47–48, 63, 217, 220, 248, 489, 523
military forces engagement in policing, 484–90
night raids and house searches, 111–12
ungoverned spaces and, 116
Afghan Taliban, 116, 193–95, 251, 360
Air and Missile Warfare Program on Humanitarian Policy and Conflict Research (2009), Manual Commentary on, 11, 560, 578
Akayesu; Prosecutor v., 375*n*238, 377*n*247
Al Aqsa (second Intifada, 2000-2005), 235, 343, 438, 511, 515–16
Al-Baghdadi, Abu Bakr, 200–202, 206
Al-Bassiouni v. Prime Minister, 343
Al-Bihani v. Obama, 309*n*278, 310*n*282
al-Libi, Abu Anas (also known as Nazih Abdul-Hamed al-Ruqai), 276, 437
Al-Muqrin, Abd Al-'Aziz, 8*n*41
A Practical Course for Guerrilla Warfare, 115*n*172, 206, 613*n*175
Al Nusra Front (ANF), 200–203, 298*n*208
Al Qaeda, 69, 114–15, 168–69, 179–81, 239–42, 250–53, 272–74, 294–98, 307–9, 335–39, 368, 425–27, 439–40, 577–78, 600
affiliates and associates, 4*n*11, 186, 193, 197*nn*344–45, 198–99, 202–4, 252, 294, 296, 323, 570*n*2
Al Qaeda Central, 169, 198–99, 208, 294–95
alternate spellings of, 4*n*4
and an Islamic state, 197–208
bin Laden, Osama, 200, 202–4
FBI HRT, Special Forces raids, Afghanistan and Iraq, 437
global battlefield, 4
groups, 198 (*See also specific Al Qaeda-related groups*)
hybrid warfare, 7
jihadist movement, 205
jihadist organization, 199
leaderless jihad theory, 294–95
Levant (*al-Sham* in Arabic), 4
Libya, 197–98
Mali, Bamako Hotel attack (2015), 403, 425
Management of Savagery: The Most Critical Stage Through Which the Umma Will Pass (Abu Bakr Naji), 115, 229*n*83, 298*n*205, 403
network concept, 191, 198–99
Pakistan, 115–16
Special Forces Task Force 714, 440
strategy of, 204
structure and organization of, 197–98
suicide bombing, 200
Yemen, 6, 67, 114–15, 193, 198, 205, 212, 252
Al Qaeda and Associated Movements (AQAM), 199, 202–5, 611, 613
Al Qaeda in Iraq (AQI), 8*n*46, 115, 168, 198–200, 204, 206–7, 250, 250*n*225, 253, 272, 399, 402–3, 600*n*115
al-Zarqawi, Abu Musab, 176, 199–200, 206, 236, 250*n*235, 303*n*235, 402
counterinsurgency, 236
counterinsurgency and counterterrorism, 209
narrow operational and normative gap, 491, 516
non-international armed conflict, determining existence of, 371
suicide bombing threat, capture or kill, 242
Al Qaeda in the Arabian Peninsula (AQAP), 169, 198, 202, 206, 212, 275, 303, 514*n*274
Al Qaeda in the Indian Subcontinent (AQIS), 194
Al Qaeda in the Islamic Maghreb (AQIM), 169, 173, 198–99, 298, 425, 496, 600*n*115
Al Qaida. *See* Al Qaeda
Al Qa'ida. *See* Al Qaeda
al-Ruqai, Nazih Abdul-Hamed (also known as Abu Anas al-Libi), 276, 437
Al-Shabaab, 165, 168–69, 189, 198–99, 202, 273, 275–76, 282, 305, 368, 426–27, 429–30, 491, 494, 514*n*274, 588
al Sham (Levant), 4
Al-Skeini v. The United Kingdom, 103*n*88, 120*n*216, 149*n*181, 152*n*200, 230*n*86, 234*n*114, 402*n*152, 487*n*30

Al-Suri, Abu Mus'ab, 4, 8*n*41, 168, 294*n*176, 295
The Military Theory of the Global Islamic Resistance Call, 205–6, 613
Al-Zarqawi, Abu Musab, 199–200, 206
bombing of UN Headquarters, Baghdad (2003), 200
death of, 200, 236, 250*n*235, 303*n*235, 402
drone strikes against, 176
guerrilla operations of, 206
ambush style operations, 541, 543–44, 547–49, 564
American Captain Phillips, rescue from Somali pirates *(2009)*, 328, 397, 427–28, 430, 491, 494, 496
American Convention on Human Rights, Article 4(1), 504, 507
Amin, Idi, 397*n*102, 406, 408
Amnesty International Canada v. Canada (Chief of the Defence Staff) (F.C.), 47*n*113, 149*n*182, 149*n*184, 271*n*18
Andronicou v. Cyprus, 139*n*122, 452*n*45, 497
Aulaqi, Anwar, 68*n*94, 98*n*42, 300, 300*n*216, 302, 466*n*115
Australia
INTERFET (International Force in East Timor, 1999), 473
shoot-down of hijacked aircraft, 562–63, 593
Autonomous Zone of Algiers, 165
avengers, 179–80

Balkans (1990), 468
Bámaca-Velásquez v. Guatemala, 553*n*142
banda criminales (BACRIM), 171, 176, 196–97
Bandit Queen Phoolan Devi, 180
banditry, 179–80
Bankovic v. Belgium, 558
Basic Principles on the Use of Force and Firearms by Law Enforcement Official (United Nations), 128, 131*nn*57–59, 227*n*74, 333*n*28, 445*n*5, 448, 449*n*32, 459, 461*n*84, 472, 484*n*6, 485*n*10, 495, 507, 524*n*342, 549*n*122
Baxter, Richard, 94, 97, 121, 141, 181
belligerents
belligerency, definition of, 161
doctrine of belligerent equality, 13, 39
lawful status of, 37
post-WW II status, extension to illegal aggressions, 13
POW status of belligerents, 92, 94–95, 98
recognition of, 34, 44, 46, 96, 104, 106–7, 161
unlawful belligerent, 94–95
unprivileged belligerents
capture of, 441
drone strikes, "conduct of hostilities" approach, 309–10
enemy unprivileged belligerent, 95
inter-state conflict, status of unprivileged beligerent, 334
proper authority principle, 93–95, 97–99
bellum justum. See just war *(bellum justum)*
bellum legale (lawful war), 41
Beltrán-Leyva, Arturo, 171, 175, 598
Benghazi attack (Libya, 2012), 276, 293, 368, 391–92, 453, 512, 600
Beslan School siege (2004), 438, 499–500
Bethlehem principles, theory *vs.* operational reality, 50–51, 53
bin Laden, Osama, 226, 228, 251, 291, 294–97, 358
al Qaeda, 200, 202–4
harboring and active support, 338
jihadist movement, 203–4
raid by Navy SEALs, Team 6 (Pakistan, 2011), 247–48, 272, 284, 305, 385, 439–41
takfiri, 117
Black Cats (India), 435, 503
Boko Haram (Nigeria), 98, 118, 199, 205, 222, 424–26, 594
Border Protection Group 9 (GSG-9) (Germany), 423–24, 434–37, 500, 550
Boškoski; Prosecutor v., 290*n*146, 317*n*321, 376, 587
Brown v. United States, 497*n*107, 525*n*349
brown water warfare operations (South Vietnam, 1966–1968), 155
Bubbins v. The United Kingdom, 497*n*106
Bush administration and global war on terror, 312–13
Bush Doctrine (2002 National Security Strategy), 56

caliphate, 4, 193–94, 569
Cambodia, seizure of *SS Mayaguez* (1975), 397, 400–402
campaign against terrorism (CAT), 20, 181
Canada
Amnesty International Canada v. Canada (Chief of the Defence Staff) (F.C.), 47*n*113, 149*n*182, 149*n*184, 271*n*18
responses to terrorism, 185–86
Captain Phillips, rescue from Somali pirates (2009), 328, 397, 427–28, 430, 491, 494, 496
capture or kill
Abu Sayyaf, 170, 250, 277, 309, 325, 601
suicide bombing threat, 215, 237–43
Targeted Killing Case, 231–32
Caroline incident, 49, 66, 302, 396, 454, 457–58
cartels, 170–71, 173–75, 177, 183

Case Concerning Oil Platforms (Islamic Republic Of Iran V. United States Of America), 73, 78, 87–88, 331–32, 390, 392–93, 465
Case Concerning United States Diplomatic and Consular Staff in Tehran (United States v. Iran), 338, 409
cellular organizations, 165
 decentralized and cellular groups, 372
 insurgent cell, distinguished from criminal gang, 168
 non-state actors and armed conflicrt, adoption of cellular structure by, 371
Charter of the United Nations. *See* United Nations Charter
Chatham House *Principles of International Law on the Use of Force by States in Self-Defence (2005)*, 51, 88
The Civilian Impact of Drones: Unexamined Costs, Unanswered Questions, 270
civil wars
 Lebanese Civil War, 515
 Spanish Civil War, 510
 U.S. Civil War, 37–38, 105
classic international law, stages of violence against established authorities, 160–61
Clausewitz, Carl von, 7*n*37, 23, 33*n*1, 34, 75–76, 80*n*180, 114*n*166, 166, 187*n*254, 419, 450
Clinton, Hillary, 147–48, 177, 358
collateral casualties, 139, 428, 433, 438
 drone strikes, 319, 321, 326
 human rights-based law application restrictions, overreach, 554, 556–57, 562
 insurgents and target "value," 247, 252
 proportionality and counterinsurgency, 259
 self-defense principles, strategic approach *vs.* tactical approach, 82, 85
 shoot-down of hijacked aircraft, 562
 targeting, strategy and tactics, 82
collateral damage, 256, 410, 418, 420, 551
 drone strikes, 281
 improvised explosive devices (IEDs) suicide bombing threat, capture or kill, 242–43
 insurgents and target "value," 246–48, 250
 law enforcement and self-defense, 449, 451, 456
 proportionality and counterinsurgency, 258–59, 261
 shoot-down of hijacked aircraft, 560
 suicide bombing threat, capture or kill, 242–43
Colombia
 banda criminales (BACRIM), 171, 176, 196–97
 insurgency and crime, challenging the state, 171–73, 175–76
 Revolutionary Armed Forces of Colombia (FARC), 66, 171–73, 176, 196–97, 212, 222, 260, 307, 349, 403, 405, 422, 594
 state-conducted hostage rescue, 421–23
 Unified Action Groups for Personal Liberty (GAULA), 422–23, 435
Common Article 2. *See* Geneva Convention, Common Article 2
Common Article 3. *See* Geneva Convention, Common Article 3
conduct of hostilities approach, drone stikes. *See* drone strikes, "conduct of hostilities" approach
conflict law, development of, 33–34
conflict not of an international character
 criteria, 579
 drone strikes, 314
 Israeli Supreme Court and cross-border terrorism, 345
 proper authority principle, 99
 transnational armed conflicts, 356
Congo
 Stanleyville rescue operation (1964), 403
 United Nations Operations in the Congo, 468
conventional warfare, 9, 21, 46, 140, 225, 253, 257, 353, 572–73, 574*f*, 612, 620, 626
 state-centric focus, 34–35
 state *vs.* non-state actor conflicts, 46
counterinsurgency, 215–65. *See also* insurgency; military advantage
 control and operational choices, 228–37
  geography, 228
  Interpretive Guidance, least restrictive means (LRM) analysis, 231–32
  jurisdiction, 230
  mosaic war, 230
  nature of control, 228–30
  state and non-State actors, exercise of exclusive control functions, 229–30
  suicide bombing, 237–39
  targeted killing, 236–37
  "Targeted Killing" Case and control, 231–37
  The "Troubles" (Northern Ireland), 236
 and counterterrorism, integrated response, 208–13
 direct participation in hostilities, 239, 241–43
 drone strikes, 267, 311
 existence of armed-conflict, threshold, 265
 governance
  human rights and humanitarian laws, counterinsurgency frameworks, 262–64

law enforcement, counterinsurgency and primacy of, 228–30
twenty-first century counterinsurgency, 220–21
humanitarian law
human rights and, overlapping normative frameworks, 261–65
law enforcement, counterinsurgency and primacy of, 223, 225–29
proportionality and counterinsurgency, 258–64
human rights law
and humanitarian laws, overlapping normative frameworks, 261–65
law enforcement, counterinsurgency and primacy of, 223, 225–29
proportionality and counterinsurgency, 260–63
improvised explosive devices (IEDs)
and counterterrorism, integrated response, 154
overlapping normative frameworks, 265
and suicide bombing threat, capture or kill, 215, 237–43
law enforcement
counterinsurgency and primacy, 223–30
and self-defense, police primacy, 453, 467
levels of war, 254–55
military advantage. *See* military advantage
one-off attacks by non-state actors, 219, 265
overview, 215–16
police primacy, 215, 224, 227
counterinsurgency doctrine, 224, 227
human rights and humanitarian laws, counterinsurgency frameworks, 262, 265
insurgents and target "value," 249
law enforcement, 224, 227
law enforcement and self-defense, 453, 467
non-international armed conflicts, 360–61, 370, 378
targeted killing, 249
proper authority, police primacy, non-international armed conflicts, 360–61
proportionality and, 258–61
suicide bombing, 215, 237–39, 241–43
targeted killing
control and operational choices, 236–37
improvised explosive devices (IEDs) and suicide bombing threat, capture or kill, 243
police primacy, 249
twenty-first century counterinsurgency, 216–23
counterinsurgency and counterterrorism, integrated response, 208–13
counterinsurgency doctrine, 209
improvised explosive devices (IEDs), 154
jihadist, 212–13
The Military Theory of the Global Islamic Resistance Call, 205–6, 613
minimum force, 209, 213
national self-defense, 213
Obama, Barack, 211, 213
Rome Statute, Article 8(2)(f), 366
Salafi jihadist, 4, 8, 197, 212, 228
small wars, 212
Syrian Conflict, 108–9
ungoverned spaces, 212
counterinsurgency math, 242, 257
counterterrorism
counterinsurgency and, integrated response, 208–13
Leiden Policy Recommendations on Counter-Terrorism and International Law (2010), 51
modern counterterrorism, 412–15
operations, 8–9
drone strikes. *See headings under* drone strikes
extraterritorial protection of nationals, 276
hostages. *See* hostage rescue
one-off attacks by non-state actor, 293, 311, 317
proper authority, 361
self-defense, limited application theory and, 66–69
Special Operations Forces, counterterrorism and direct action, 271–77
suicide bombing, 268, 283, 285, 307
U.S. Counterterrorism Manual (2014), 255
Covenant on Civil and Political Rights
Article 14, ¶7, 142*n*149
Article 17, ¶1, 142*n*149
Crimea, Russia's annexation of (2014), 17, 35
criminal gangs, 3, 5–6, 9. *See also* transnational criminal organizations
insurgent cell, distinguished from criminal gang, 168
Mexican gangs and Calderón administration, 173–75, 177
third-generation transnational gangs, 172–73
ungoverned spaces, control over territory, 118–19
criminal insurgency, 26, 118, 167, 170, 172–75, 177–78, 422, 513, 598
cross-border attacks by non-state actors, 48–52
authority for engagement, 48
Bethlehem principles, theory *vs.* operational reality, 50–51, 53
exercise of self-defense, effect of 9/11 on, 49
Israeli Supreme Court and, 341–49
non-international armed conflict, 48–52
parallelism, 52
state attribution requirement, 48–52
UN Charter, Article 51, 48–52

Cruz-Sánchez, et al v. Peru. See Japanese Embassy Case (Lima, Peru, 1997)
Customary International Humanitarian Law Study (2005, International Review of the Red Cross), 14, 81, 109, 126, 146
  Rule 47, 506*n*265

dacoit, 180
Da'esh (Islamic State in Iraq and the Levant (ISIL)), 298*n*207, 324*nn*353–54, 516*n*304, 625
deadly force, mission accomplishment, 460
Department of Defense Law of War Manual (2015), 22*n*110, 37, 93*n*8, 94*n*22, 95*n*23, 98*n*44, 110*n*136, 119*n*210, 122*n*3, 124*n*14, 147*n*172, 148, 239*n*154, 259*n*282, 309*n*273, 387*n*30, 623*n*19
Department of Justice Drone Opinion, 300*n*216, 302–5
destructive power of state, 37–42
  antiwar movement, impact upon, 40–42
  development of destructive capabilities, 37–38
  inter-state conflict, limiting the recourse to war, 40–42
  just war *(bellum justum)*, 37–39, 41
  lawful war *(bellum legale)*, 41
  pacificism, 39, 41, 628*n*35
  pacifism, 39, 41, 75, 449*n*34, 628*n*35
  UN Charter, limiting recourse to war, 40–42
  unregulated states, destructive capabilities, 40
de Vattel, Emer, 441
direct action
  insurgents and target "value," 252
  Special Operations Forces, counterterrorism and direct action, 271–77
direct participation in hostilities
  controlling state involvement in non-international armed conflict, 53
  drone strikes
    "conduct of hostilities" approach, 303, 306, 308–10
    financing, as direct participation in hostilities, 308–10
    lack of consensus, 270
  financing as direct participation in hostilities, 308–10
  force and hostage rescue, 508
  holistic operational law, application of human rights-based approach, 595
  humanitarian law and human rights law interface, 132
  improvised explosive devices (IEDs) and suicide bombing threat, 239, 241–43
  insurgency and crime, 170
  Interpretive Guidance on the Notion of Direct Participation in Hostilities Under International Humanitarian Law (International Committee of the Red Cross), 238–42, 303, 307, 321, 456
  Japanese Embassy Case (Lima, Peru; 1997), 508
  law enforcement
    counterinsurgency and primacy of, 226, 232, 239, 241–43
    and self-defense, 456
  narrow operational and normative gap, 456, 508
  shoot to kill policy, 545
  suicide bombing threat, capture or kill, 239, 241–43
  "Targeted Killing" Case, 232
domestic law
  legal challenges, 3, 9, 14*f*, 15–16, 19
  transnational armed conflicts, 353
"double tap" shooting method, 496, 504, 527–28
double effect doctrine, 61, 247, 450, 557
Dr. H. et al. v. s.14.3 of the Aviation Security Act of 11 January 2005, 261, 561–63, 578
Drone Speech (2013) (President Obama), 3–4, 20–21, 181, 191, 298–99, 322, 363, 596, 609
drone strikes, 277–87
  Additional Protocol I, Article 57, 281
  Aerial Drone Deployment on 4 October 2010 (Mir Ali/Pakistan), 231*n*94, 236*n*140, 238*n*152, 239*n*157, 288*n*135, 319*n*328, 370*n*211, 376*n*246, 456*n*64, 608*n*139
  antiwar influence, 318
  approaches
    "conduct of hostilities" approach, 300–311, 318–19
    "law enforcement" theory, 287–300, 319–22
    "self-defense" option, 311–19
  The Civilian Impact of Drones: Unexamined Costs, Unanswered Questions, 270
  collateral casualties, 319, 321, 326
  collateral damage, 281
  conflict not of an international character, 314
  controlling use, 284–87
  counterinsurgency doctrine, 267, 311
  direct participation in hostilities
    "conduct of hostilities" approach, 303, 306, 308–10
    lack of consensus, 270
  general use of, 280–85
  geography of war, 292, 297, 313
  high-profile attacks, 278–80

Holder v. Humanitarian Law Project, 307*n*259, 309
humanitarian law
"conduct of hostilities" approach, 307–8, 311
Interpretive Guidance, 303, 307–8
"law enforcement" theory, thresholds, 292
"self-defense" option, 311–19
valid target, 307–8
international criminal law, liability for war crimes under, 286
lack of consensus, 269–71
"law enforcement" theory, 287–300, 319–22
law of armed conflict (LOAC), 301–2, 316
Living Under Drones: Death, Injury, and Trauma to Civilians From US Drone Practices in Pakistan Study (Stanford Law School & NYU Law School), 269–70
means and authority test, 316
military activity, use across broad spectrum, 281–84
mission accomplishment and, 460–61
narrow scope of covert effort, 281–82
national self-defense, 312, 315*n*312
neutrality law, 302–7, 310, 321, 323, 326
non-international armed conflict, 288, 292–93
non-State actor threats, security challenges, 5
one-off attacks by non-state actor
"conduct of hostilities" approach, 311
"law enforcement" theory, 293
threshold for existence of armed-conflict, 317
"ordinariness" of drones, 280–85
organized armed groups, 292–93
overview, 267–69
"pattern of life," 281
personality strikes, 278–82
police primacy, 311, 322
"positive identification" (PID), 281
proper authority principle
"law enforcement" theory, 320
"self-defense" option, 318
transnational terrorism, 290–91
protracted armed violence, 293, 317
self-defense and protection of nationals, 277–78
"self-defense" option, 311–19
short of war, 313, 316
signature strikes, 278–82
small wars, 319
Special Operations Forces, 277–78, 281–85
special terms, 278–80
state self-defense, 289–90, 313–16
suicide bombing, 268, 283, 285
targeted killing, 286, 312, 319–21
terminology, drone-specific, 281
totality of the circumstances, 317
transnational criminal organizations, 283–84
transnational terrorism, proper authority principle, 290–91
UN Charter, Article 51, 318–19
"conduct of hostilities" approach, 301
"law enforcement" theory, 290, 296
unwilling or unable justification for drone strikes, 325
ungoverned spaces, 268, 271, 298, 300, 318, 320, 322
uniqueness of, 277–80
UN Security Council Resolutions
Resolution 1373 (2001), 304*n*240
Resolution 1973 (2011), 304
Resolution 2249 (2015), 298, 323–25
unwilling or unable test, 322–26
valid target, humanitarian law, 307–8
drone strikes, "conduct of hostilities" approach, 300–311, 318–19
applicable law, 310–11
basis for strikes, 303
counterinsurgency doctrine, police primacy, 311
Department of Justice Drone Opinion, 300*n*216, 302–5
direct participation in hostilities, 303, 306, 308–10
criteria, 308–9
financing, as direct participation in hostilities, 308–10
financing as direct participation in hostilities, 308–10
focoist approach, 308
functional participation in armed conflict, 303
Global War on Terrorism (GWOT), 304
guerrilla warfare, 307–9
Holder v. Humanitarian Law Project, 307*n*259, 309
humanitarian law, 311
imminence, 303
international armed conflict, 301–2, 305–6
Interpretive Guidance, 303, 307–8
law enforcement theory, distinguished from, 301
lawful use of, 300–301
law of armed conflict (LOAC), 301–2
neutrality law, 302–7, 310, 321, 323, 326
one-off attacks by non-state actors, 311
police primacy, 311
San Remo Manual on International Law Applicable to Armed Conflicts at Sea, 305–6, 311
self-defense determination, 304–5
State-centric approach, 301

drone strikes, "conduct of hostilities" approach (*Cont.*)
  suicide bombing, 307
  targeting, status-based, 303
  UN Charter, Article 51, 301
  ungoverned spaces, 300
  unlawful combatant, 309
  unprivileged belligerents, 309–10
  unresolved issues, 310–11
  UN Security Council Resolutions
    Resolution 1373 (2001), 304*n*240
    Resolution 1973 (2001), 304
  valid target, humanitarian law, 307–8
  vanguard approach, 208, 308, 570
drone strikes, "law enforcement" theory, 287–300, 319–22
  Al Qaeda and leaderless jihad theory, 294–95
  application challenges, 290, 298–99
  basis of, 289–90
  conduct of hostilities theory, distinguished from, 301
  humanitarian law thresholds, 292
  intensity concept, 293
  isolated acts of terrorism theory, 294–96
  jihadists threats, 296–97
  one-off attacks, 293
  Oxford Research Group, 288–90, 299, 321
  proper authority, role of state, 320
  protracted armed violence, 293
  reaction beyond territorial borders, comparison to threats on high seas, 300
  scenarios, 289–90
  situations addressed by, 293
  State self-defense, interpretation of, 290
  UN Charter, Article 51, 290, 296
  unresolved issues, 290, 293
  use under theory, requirements, 297–98
  variations of, 288–90
  war concept of self-defense, 291–92
drone strikes, "self-defense" option, 311–19
  armed conflict, establishment of
    combat power test, robust self-defense approach, 316–17
    threshold, naked self-defense approach, 314
  Bush administration and global war on terror, 312–13
  gap filling, 314–15, 317–18
  humanitarian law, 311–19
  jus cogens, 314
  law governing the recourse to war, 311–13
  law of armed conflict (LOAC), 316
  naked self-defense, 311–17, 320, 355
  proper authority principle, 318
  robust self-defense approach, 314–17
  special law, 312
  ungoverned spaces, 318
Duvdevan, 243n120, 438

East Timor (1999), 473
England
  John Charles de Menezes shooting (2005), 509, 525–26, 531
Entebbe, Israeli hostage rescue (*Operation Thunderbolt* (*Yonatan*, 1976)), 28, 185, 328, 397, 405–12, 418, 420–21, 434, 437, 439, 498, 500, 504, 580–81, 586
equal application principle, 12–13
  insurgency, 167
  just war theory and, 39
  limited application theory, 65
  proper authority principle, humanitarian law and, 92
  self-defense principles, application during armed conflict, 65, 79
  strategic approach *vs.* tactical approach, 79
  targeting, strategy and tactics, 79
Ergi v. Turkey, 507*n*218
European Convention on Human Rights, 124–25, 138, 557, 615
  Article 2, 552*n*139, 553–54
  Article 2(2), 459–60, 481
  Article 15(1), 615*n*195
  Article 31(3)(c), 552
explosives. *See also* improvised explosive devices (IEDs)
  confronting explosive threats, 518–21
  criminal insurgency, 513
  development and deployment of technology, 520
  hostage rescue, 528–29
  human rights law, 520–21
  Intifadas, 516
  jihadist, 514, 516
  knowledge in making and employing explosives, 514–15
  lex ferenda, 530
  lex lata, 530
  lone wolf attacks, 198, 371, 512, 614
  narrow operational and normative gap, 510–32
  rules of engagement (ROE), 530
  shoot to kill policy, 522, 528–31
  shoot to wound approach, 529–31
  strictly necessary, proportionate, and overwhelming, 521–32
  suicide missions, 515–19, 521, 523–25
  targeted killing, 522, 525, 527–28, 533
  ubiquitous weapon, 510–15

use of force, reasonable belief standard, 525–26
Vehicle-Borne Improvised Explosive devices (VBIEDs), 511
extrajudicial killing, 506
extraterritorial law enforcement, 52, 66, 349–52, 356–57, 394
categorizing State *vs.* non-state actors conflicts, 349–52, 378–80
autonomy of terrorist groups, 352
guerrilla war, 350
humanitarian law, 350–53
non-international armed conflict, 350–51
protracted armed violence, 352
short of war, 350
terminology confusion, 351–52
transnational armed conflict, 356–57
limited application theory, application during armed conflict, 66–67
non-international armed conflict, 52
non-state actors and armed conflicts, 349–52, 378–80
protection of nationals. *See* extraterritorial protection of nationals
extraterritorial protection of nationals, 385–98
antiwar bias, 398
counterterrorism operations, 276
non-combatant evacuation operations (NEO), 328, 393–95, 463
state responsibility, 385–95
taking action across conflict spectrum, 395–98
totality of the circumstances, 391
UN Charter, Article 51, 384, 387, 389, 391–92, 395

failed states, 52, 104, 114*n*161, 431, 537
Falklands/Malvinas conflict (1980) *(General Belgrano warship)*, 60, 77, 292, 419
Fallujah assault (2004), 28, 201, 206, 485, 490, 596
carrying arms openly threshold, 477
existence of armed conflict, 478
human rights-based law application restrictions, overreach, 546, 556
imminence principles, application difficulties, 478–79
law enforcement and self-defense, 445, 453, 476–79, 481
proportionality and counterinsurgency, 259
rules of engagement (ROE) framework, 476–79
terrorism, 189
ungoverned spaces, 116
FBI Hostage Rescue Team (HRT), 435, 437, 483, 531, 545
fidayeen attack, 192, 219, 517, 527, 597
financing, as direct participation in hostilities, 308–10
Finogenov v. Russia, 139*n*122, 154, 260*n*292, 452, 481, 499*nn*125–27, 500*n*140, 507*n*216, 508*nn*217–23, 509*n*229, 551*n*128, 555, 562
focoist approach, 208, 308, 570
urban focoist theory, 187–88
force of first and last resort, 484–92
Additional Protocol II, Article 4, 488*n*40
Afghanistan, military forces engagement in policing, 484–90
improvised explosive devices (IEDs), 484, 486, 491
Iraq, 2003 U.S. invasion of; military forces engagement in policing, 486–87, 490
law of armed conflict (LOAC), 484
military forces engagement in policing, 486–92
occupations and counter insurgencies, 486–87
piracy, 491–92
war fighters, 491
force of last resort
hostage rescue, 507
Japanese Embassy Case (Lima, Peru; 1997), 507
narrowing the gap, 492
UN mandated operations, robust peacekeeping, 470
Fragmentation Of International Law: Difficulties Arising From The Diversification And Expansion Of International Law (2006 Study), 11*n*57, 123, 127–28
France, 185, 203, 294, 296, 323–24, 368, 395, 403, 406, 411, 429–30, 434–36, 573, 583, 597
Groupe d'Intervention de la Gendarmerie Nationale (GIGN), 434–37, 498, 500, 530, 550
piracy, rescue operations within Somalia, 429–30
shooting of Charlie Hebdo journalists (2015), 436

Galić; Prosecutor v., 286*n*117, 525*n*351, 596*n*90
gangs. *See* criminal gangs
Gaza, 89, 105, 117, 155, 163, 202, 229, 235, 237, 282, 287, 341, 343–44, 346–47, 556
General Belgrano warship (Falklands/Malvinas conflict, 1980), 60, 77, 292, 419
Geneva Convention, 143–44
articles
Additional Protocols. *See* Additional Protocol I (AP I), articles; Additional Protocol II (AP II), articles
generally. *See* Geneva Convention, Articles
ratification of, 11

Geneva Convention, Articles. *See also* Geneva Convention, Common Article 2; Geneva Convention, Common Article 3
Article 4, 13*n*65
Article 12, 137*n*107–37*n*108
Article 22(1), 138*n*118, 451*n*42
Article 22(2), 138*n*118, 451*n*42
Second Geneva Convention (GC II)
Article 12, 137*n*107–37*n*108
Third Geneva Convention (GC III)
Article 4. A. (2)(c), 545*n*91
Article 4. A. (6), 545*n*91
Article 4A(2), 358*n*157
Article 4.A.(2), 358*n*157
Article 13, 137*n*108
Article 17, 137*n*107
Article 87, 137*n*107
Articles 99-108, 138*n*111
Fourth Geneva Convention (GC IV), 95
Article 5, 95*n*26, 138*n*111
Article 27, 133*n*75, 137*n*108
Article 32, 137*n*107
Article 34, 399*n*117
Article 37, 137*n*108
Article 54, 102*n*79
Article 55, 102*n*81
Article 56, 102*n*80
Article 64, 97, 102, 138*n*112, 217*n*10, 331*n*10, 402*n*149, 451*n*41, 621*n*7
Article 66, 97
Articles 71-75, 138*n*111
Article 76, 95*n*26
Article 147, 399*n*117
post-World War II, humanitarian law and human rights law interface, historical background, 143–44
Geneva Convention, Common Article 2, 340*n*57, 579
humanitarian law and human rights law interface, prior to mid-twentieth century, 138
non-state actors and armed conflict
ICRC Commentary to Common Article 2, Geneva Convention, 330
inter-state conflict, 330–32
transnational armed conflict, 352
proper authority principle and humanitarian law, 97
Geneva Convention, Common Article 3, 3*n*2, 34, 399*n*117, 507, 542, 579
hostage rescue, 507
hostage rescue, narrow operational and normative gap, 507
humanitarian law and human rights law interface
Article 3(1), 137*n*108
Article 3(1)(a), 137*n*107
general principles recognized by civilized nations, 133
historical background, prior to mid-twentieth century, 134*n*79, 138
post-World War II, divergence and growing reconciliation, 139, 148, 150*n*189
ICRC Commentary to Common Article 3, Geneva Convention, 377
insurgency and crime, 161, 176
law enforcement and self-defense, 451
non-international armed conflict, proper authority, 357
non-state actors, internal conflicts with international character, 44
non-state actors and armed conflict, 357
proper authority principle
non-international armed conflict, proper authority, 357
state and governance, reluctant application of humanitarian law, 106–7
shoot to kill controversy, 542
transnational armed conflict, 352–55
geography of war, 21, 27
drones, 292, 297, 313
external support and direct participation, 592
law enforcement, counterinsurgency and primacy of, 228
non-international armed conflict, proper authority, 359
Germany
Border Protection Group 9 (GSG-9), 423–24, 434–37, 500, 550
piracy, planned rescue operations by GSG-9 (2009), 435–36
shoot-down of hijacked aircraft, 561–63
Global Terrorism Index, 510–11, 516
Global War on Terrorism (GWOT), 181–82, 304, 312–13
Godane, Abi, 165*n*54, 276–77
Gombo; Prosecutor v., 367*n*195
governance, 25–26, 320, 359, 431, 608–10
human rights and humanitarian laws, counterinsurgency frameworks, 262–64
insurgency, 167
insurgency and crime, 177–79
law enforcement, counterinsurgency and primacy of, 228–30
law enforcement and self-defense, 449–51
proper authority principle, humanitarian law and definition of, 101–2
reluctant application of humanitarian law, 104–9
in respect of sovereign territory, 102
responsibilities, 102–3

terrorism, 190–91
twenty-first century counterinsurgency, 220–21
Graham v. Connor, 497*n*108, 525*n*348
grand strategy, 76, 81, 610
Grotius, Hugo, 17, 99
Groupe d'Intervention de la Gendarmerie Nationale (GIGN), 434–37, 498, 500, 530, 550
guerrilla, 93–94, 105–6, 112–16, 159–60, 162–66, 186–89, 191, 203, 206–9, 244, 307–9, 538–39, 573–74
guerrilla warfare, 6–9, 34, 187, 206, 218, 350, 360, 531, 570, 613
classic three stage warfare strategy, 8*n*41, 160–61, 570*n*3
drone strikes, "conduct of hostilities" approach, 307–9
focoist/vanguard approach, 208, 308, 570
strategy, 8*n*41, 570*n*3
twenty-first century counterinsurgency, 218
urban focoist theory, 187–88
urban guerrilla warfare, 183, 187–88, 235, 371

Hague Land Warfare Regulations
Article 2, 93*n*12
Article 23(c), 506*n*205
Article 42, 103, 230*n*87
Article 43, 102, 132*n*67, 133*n*74, 138*n*113, 402*n*149, 451*n*41, 621*n*8
Article 46, 133*n*75
Hamas, 105, 117–18, 192, 195, 229, 231, 237, 282, 284, 286–87, 341, 556
counterinsurgency and counterterrorism, 212
drone strikes, "conduct of hostilities" approach, 308
Israeli Supreme Court and cross-border terrorism, 344–47
Hamdan v. Rumsfeld, 27, 48, 67*n*86, 98*n*48, 99, 148, 345–47, 351, 353, 355–58, 364, 604, 624
Hamdan v. United States of America, 310
The Handbook of the International Law of Military Operations (Gill and Fleck), 22
Haraakat-I Inquilab Islam-I, 204
Haradinaj; Prosecutor v., 161*n*15, 293*n*168, 367*n*195, 372*n*222, 374*n*231, 375, 417–18, 425
harboring, 337–39
UN Charter, Article 51, 301
hard knock approach, 248–49, 361
Hassan v. The United Kingdom, 103*n*88, 124–25, 127–28, 152*n*200, 471*n*146, 552–553, 558, 592–93, 614*n*190, 615, 624
Hezbollah, 8, 169, 192, 195, 284, 514–15
counterinsurgency and counterterrorism, 212
drone strikes, "conduct of hostilities" approach, 308
extraterritorial law enforcement, 351–52
Israeli Supreme Court and cross-border terrorism, 344–46
Israel-Lebanon Conflict (2006), 66–67, 86–87, 117–18, 163, 182, 229, 344–45
high-profile attacks, drone strikes, 278–80
Hizbollah. *See* Hezbollah
Holder, Eric, 20, 237*n*142
Holder v. Humanitarian Law Project, 307*n*259, 309
holistic operational law, 569–617
conventional warfare, 571–73
direct participation in hostilities concept, application of human rights-based approach, 595
evolution of approach, 571–75
law enforcement
human rights-based approach, application of, 588–604
"normalcy" and, 610–16
overview, 569–71
transnational attacks and threats to nationals, responding to, 575–88
transnational violence
confronting, holistic approach, 604–10
decision tree analysis (chart), 617
hors de combat
ICRC Rule 47, 506*n*205
Japanese embassy case (Lima, Peru; 1997), 125*n*21, 484, 500–510, 513, 527–28, 558–59
Targeted Killing Case, 231–33
hostage rescue, 399–433
armed conflict, rescues during, 399–405
Colombia, state-conducted hostage rescue, 421–23
commonality of tactical approach, 384–85
Common Article 3, 507
direct participation in hostilities, 508
explosives, 528–29
FBI Hostage Rescue Team (HRT), 435, 437, 483, 531, 545
and force, 492–510
force of last resort, 507
foreign involvement in, 423–26
humanitarian law and human rights law interface, 132
human rights-based law application restrictions, overreach, 558
India, counterpiracy operations, 430–31
Japanese Embassy Case (Lima, Peru; 1997), 125*n*21, 484, 500–510, 513, 527–28, 558–59
law enforcement and protection of nationals, 431–33
Lynch, Jessica, rescue of, 401

hostage rescue (*Cont.*)
Mayaguez rescue operation (1975), 397, 400–402
narrow operational and normative gap, 510, 527–28, 531–33
Nigeria (2012), 424–25
Operation Barras, hostage rescue by British Special Forces (Sierra Leone, 2000), 412–20, 434, 493, 586
Operation Thunderbolt (Yonatan, 1976), Entebbe, Israeli hostage rescue, 28, 185, 328, 397, 405–12, 418, 420–21, 434, 437, 439, 498, 500, 504, 580–81, 586
Panama, rescue operations, 401
piracy in Somalia, 426–31
rescue of American Captain Phillips *(2009),* 328, 397, 427–28, 430, 491, 494, 496
states acting unilaterally to rescue their own nationals, 426–31
rescue operations, personnel and tactics, 281–85, 433–41
rescues and law enforcement, 421–31
Russia, rescue operation, 430
short-term conflicts, 415–21
Somailia, piracy, 426–31
Special Boat Squadron, Nigeria hostage rescue (2012), 424–25
Special Operations Forces
counterterrorism and direct action, 273
personnel and tactics, 281–85
Special Boat Squadron, Nigeria hostage rescue (2012), 424–25
U.S. Navy SEALs, 260, 273, 275, 428, 437, 439
Stanleyville rescue operation (Congo, 1964), 403
state-conducted hostage rescue, Colombia, 421–23
states acting unilaterally to rescue their own nationals, Somalia piracy, 426–31
state's right to intervene, 432
status-based targeting, 303, 441, 456, 476, 527, 545
transnational armed conflict, 354–55
UN Security Council Resolution 1851 (2008), 427–28
U.S. Navy SEALs, 260, 273, 275, 428, 437, 439
Vietnam War, prisoners of war rescue, 195–96, 400, 493
hostage rescue and force, 492–510
challenges and responses, 492–501
Common Article 3, 507
direct participation in hostilities, 508
"double tap" shooting method, 496, 504, 527–28
explosives, establishing existence of, 500–501
force of last resort, 507
hostages, risk of death, 498–99
imminence of threat, 494–95
increased frequency in, 492–93
Inter-American Commission on Human Rights, 501, 504–5, 507
Inter-American Court of Human Rights, 501, 504, 507–8, 553, 558
negotiations and warnings, 496–97
UN Basic Principles on the Use of Force and Firearms by Law Enforcement Official, 495
use of deadly force, 495–97, 500–501
use of force approach, differences between military and police, 437–38
hostage rescues during armed conflict, 399–405
actors involved, 399–400
gain access to hostage location, limitations, 401
humanitarian law based conduct, status of insurgent group in armed conflict, 405
occupied territory, defense of nationals in, 402
post-9/11 hostage taking, 403–4
riot control agents, 401–2
safety of hostages, 400–401
transnational armed conflict, 354–55
The Hostage Case, 94, 399*n*116
HPCR Manual on International Law Applicable to Air and Missile Warfare (2010 Commentary), 123
humanitarian law and human rights law interface, 121–56
accountability, 133
counterinsurgency, overlapping normative frameworks, 261–65
exclusionary approach arguments and operational reality, 122–34
direct participation in hostilities, 132
human rights norms not recognized, 132
law enforcement framework, 131–32
non-international armed conflict, 134
rescue operations, personnel and tactics, 132
shoot to kill, avoidance of, 131–32
similarities and overlap, 122–34
special law, 134
"Special law" and exclusion debate, 122–30
Universal Declaration of Human Rights (1948), 134
extraterritorial application of human rights treaty law, U.S. and Israeli position on, 149–50
general principles recognized by civilized nations, 132–33

historical background. *See* humanitarian law and human rights law interface, historical background
human rights law and humanitarian law, contrasted, 36–37, 556
international criminal law, 133
international humanitarian law (IHL), 110–12, 122, 124, 135, 139*n*122, 152*n*202, 156
international human rights law (IHRL), 124, 134*n*79, 135–36, 152*n*202
overlap and convergence, 152–56
improvised explosive devices (IEDs), 154
international or non-international conflict, 155
military forces application of human rights-based concepts, 154–56
suicide bombing, 154
U.S. Operational Law Handbook, 153*n*204
use of force, 155–56
overview, 121–22
proper authority principle, 145, 147
targeting, 132–33
uninvolved citizens, 133
Universal Declaration of Human Rights, 128
use of force, principles of distinction and proportionality, 132–33
humanitarian law and human rights law interface, historical background, 134–52
abuses of World War II, 134
European Convention on Human Rights, 138
Geneva Convention, Articles, 143–44
Common Article 2, 138
Common Article 3, 133, 134*n*79, 138
Article 3(1), 137*n*108
Article 3(1)(a), 137*n*107
general principles recognized by civilized nations, 133
historical background, prior to mid-twentieth century, 134*n*79, 138
post-World War II, divergence and growing reconciliation, 139, 148, 150*n*189
globalization, impact of, 137
Hague Land Warfare Regulations, Article 43, 138*n*113
hierarchy of rights, 138
international humanitarian law (IHL), 110–12, 122, 124, 135, 139*n*122, 152*n*202, 156
international human rights law (IHRL), 124, 134*n*79, 135–36, 152*n*202
prior to mid-twentieth century, 134–39
religious humanism, 135
right to life, 138
UN Charter, Article 2(7), 141
Universal Declaration of Human Rights (1948), 136, 139
World War II, aftermath of, 15, 139–52
humanitarian law *(jus in bello)*. *See also* humanitarian law and human rights law interface
application, complexity of situations involving hostilities with non-state actors, 34
The Civilian Impact of Drones: Unexamined Costs, Unanswered Questions, 270
counterinsurgency
human rights and, overlapping normative frameworks, 261–65
law enforcement, counterinsurgency and primacy of, 223, 225–29
proportionality and counterinsurgency, 258–64
Customary International Humanitarian Law Study (2005, International Review of the Red Cross), 14, 81, 109, 126, 146
Rule 47, 506*n*265
definition of, 10
drone strikes, 292, 303, 307–8, 311–19
human rights law and humanitarian law
contrasted, 36–37, 556
counterinsurgency, overlapping normative frameworks, 261–65
interface. *See* humanitarian law and human rights law interface
legal challenges, 11–16, 14*f*
military advantage, proportionality test, 57, 245, 247, 254, 560–61, 603
narrowing the gap, 505–10
proper authority principle and. *See* proper authority principle, humanitarian law and
proportionality
and counterinsurgency, 258–64
military advantage, proportionality test, 57, 245, 247, 254, 560–61, 603
targeting proportionality, 57, 74, 77–79, 81–82, 254, 360, 456*n*64, 603
Secretary-General's Bulletin: Observance by United Nations Forces of International Humanitarian Law (1999), 110*n*139, 120*n*214, 365, 379, 472
targeting proportionality, 57, 74, 77–79, 81–82, 254, 360, 456*n*64, 603
UN mandated operations, robust peacekeeping and, 472–74
UN Security Council, authorization of military intervention for human protection purposes, 10–11

human rights law, 14*f*, 18–19, 23–26, 28–30, 535–65, 592–603, 614–16, 622–24, 629–30. *See also* humanitarian law and human rights law interface
ambush style operations, 564
application restrictions, overreach, 552–59
coalition military commanders, accommodation of unique national limitations, 558
collateral civilian casualties, 554, 556–57, 562
double effect doctrine, 557
European Convention on Human Rights, 552–54, 557
hostage rescue, 558
humanitarian law and human rights law, contrasted, 36–37, 556
human rights framework, assessment of factors when using force, 556
international armed conflict, 552–53
levels of violence, 556
limits of effectiveness of human right law, 557
minimum force, 557–58
proportionality test, 554
suicide bombing, 557
UN Charter, Article 51, 559
use of force, minimum requirements, 555
counterinsurgency
and humanitarian laws, overlapping normative frameworks, 261–65
law enforcement, counterinsurgency and primacy of, 223, 225–29
proportionality and counterinsurgency, 260–63
European Convention on Human Rights
Article 2, 552*n*139, 553–54
Article 31(3)(c), 552
explosives, 520–21
extrajudicial killing, 21, 506
hostage rescue, application restrictions, 558
humanitarian law. *See* humanitarian law and human rights law interface
humanitarian law and human rights law
contrasted, 36–37, 556
counterinsurgency, overlapping normative frameworks, 261–65
law of armed conflict (LOAC), 150
legal challenges, 14*f*, 18–19
limitations and need for clarity, 564–65
militarization of police, 550–52
mission accomplishment and, 458–61
overview, 535–36
proportionality and counterinsurgency, 260–63
shoot-down of hijacked aircraft, 559–65
shoot to kill controversy, 536–50
state self defense. *See* state self-defense and human rights-based law enforcement
targeting, hostilities based, 565
transnational criminal organizations, 564
Universal Declaration of Human Rights (1948), 134, 136, 139–40
Hussein, Saddam, 17, 207, 236, 514, 528
hybrid insurgent organizations, 163
hybrid warfare, 7–8

ICRC. *See* International Committee of the Red Cross (ICRC)
Idaho v. Horiuchi, 545
Ikrimah (Abdikadir Mohamed Abdikadir), 276
immediacy, self-defense legal principle, 55–57, 65, 68
terminology issue, 57*n*11
imminence, self-defense legal principle
Caroline incident, 49, 66, 302, 454
humanitarian law, 303, 494–95
human rights-based law enforcement, 454–58
law enforcement and self-defense, 478–80
recourse to war, 56–57, 466, 478–79
improvised explosive devices (IEDs), 484, 486, 491, 500, 510–13, 516, 518–19, 521, 523, 528, 532
collateral damage, 242–43
counterinsurgency
and counterterrorism, integrated response, 154
overlapping normative frameworks, 265
direct participation in hostilities, 239, 241–43
force of first and last resort, 484, 486, 491
group nature of IED attacks, 238
human rights and humanitarian laws, counterinsurgency frameworks, 265
Interpretive Guidance, 238–42
narrowing the gap, 484, 486, 491, 500, 510–13, 516, 518–19, 521, 523, 528, 532
non-international armed conflicts, determining existence of, 371–72
non-state actors and armed conflict, group organization, 371–72
suicide bombing threat, capture or kill, 215, 237–43
Targeted Killing Case, 234, 237
terrorism, 154, 194
The "Troubles" (Northern Ireland), 539–40
Vehicle-Borne Improvised Explosive devices (VBIEDs), 511
India
Al Qaeda in the Indian Subcontinent (AQIS), 194
Black Cats, 435, 503
Lashkar-e-Tabia (LeT), 192–94, 265, 436, 500, 503, 517, 527

Maharashtra State Anti-Terrorism Squad (ATS), 436
Marine Commando (MARCO), 436
Mumbai terrorist attack (2008), 192–93, 293, 367–68, 372, 435–36, 500, 503, 527, 592, 599
National Security Guard (NSG), 435–36, 503, 527–28
individual self-defense, 10, 47, 138, 416, 444
doctrinal confusion, 464, 466, 470, 478, 480
state self-defense, distinguished, 447, 450–51, 457–58, 460, 462, 464
insurgency, 160–80. *See also headings under* counterinsurgency; insurgency and crime
Additional Protocol II, APII, Article 1(1), 167
Al Qaeda and an Islamic state, 197–208
classic international law, stages of violence against established authorities, 160–61
Common Article 3, 161, 176
criteria for, 160–61
definition of, 161
equal application principle, 167
governance, 167
guerrilla, 159–60, 162–66
insurgents. *See* insurgents
Islamic State in Iraq and al Sham (ISIS), 160, 168
just war, 167
kidnapping, 159–60
legal challenges, 167
organizations. *See* insurgent organizations
overview, 159–60
political dimension of insurgency, 167
political purpose test, 167–68, 178
protracted armed violence, 161
pseudo state, ISIS as, 160*n*4
purpose of conflict, 166–68
recognition of belligerency doctrine, 161
Rome Statute
Article 8(2)(d), 161*n*16
Article 8(2)(f), 161*n*16, 366
Taliban, 164
terrorism and. *See* terrorism
totality of the circumstances, 161, 175
type of conflict directed against established authority, 161
UN Security Council, Resolution 2060 (2012), 165*n*52
violence and organization criteria, 167
insurgency and crime, 6, 168–80
avengers, 179–80
banda criminales (BACRIM), 171, 176, 196–97
banditry, 179–80
capture or kill operation, Abu Sayyaf, 170, 250, 277, 309, 325, 601
cartels, 170–71, 173–75, 177, 183
challenging the state, 173–79
Common Article 3, 161, 176
direct participation in hostilities, 170
governance, political impact of criminal insurgency on State, 177–79
insurgents. *See* insurgents
kidnapping, 168–69, 171, 176
nature of threat, 168–73
organizations
cartels, 170–71, 173–75, 177, 183
distinction between criminal and ordinary insurgency group, political purpose basis, 178
transnational criminal organizations, 160, 170, 172–73, 176–77, 179
vigilante based organizations, 171–72, 174
other criminal activity, 179–80
pseudo paramilitaries, 171
third-generation transnational gangs, 172–73
totality of the circumstances, 175
transnational criminal organizations, 160, 170, 172–73, 176–77, 179
ungoverned spaces, 168, 178–79
vigilante based organizations, 171–72, 174
insurgent organizations, 162–65
cellular guerrilla organizations, 165
insurgent cell, distinguished from criminal gang, 168
conventional military organization by non-State actors, 163
criminal organizations
cartels, 170–71, 173–75, 177, 183
distinction between criminal and ordinary insurgency group, political purpose basis, 178
transnational criminal organizations, 160, 170, 172–73, 176–77, 179
vigilante based organizations, 171–72, 174
gangs
insurgent cell, distinguished from criminal gang, 168
third-generation transnational gangs, 172–73
hybrid insurgent organizations, 163
organized armed groups, 161–62
success or failure of, 164–65
tactical changes, adaptations, 165
target "value," 244, 251–52
terrorism, insurgent groups using, 190–93
terrorists and insurgents, distinctions and relationship, 190–97, 360
third-generation transnational gangs, 172–73
transnational criminal organizations, 160, 170, 172–73, 176–77, 179

insurgent organizations (*Cont.*)
  types of, 163–65
  ungoverned spaces, 164
  vigilante based organizations, 171–72, 174
  violence and organization criteria, 167
insurgents, 3–7
  counterinsurgency and counterterrorism operations, 8–9
  criminal insurgency. *See* insurgency and crime
  hostage rescues during armed conflict, status of insurgent group in armed conflict, 405
  jihadist-based conflict, global insurgency, 4–5
  non-international armed conflict, proper authority, 360
  target "value," 244–53
    collateral casualties, 247, 252
    collateral damage, 246–48, 250
    conventional warfare, 253
    direct action, 252
    double effect doctrine, 247
    guerrilla, 244
    hard knock approach, 248–49, 361
    levels of war, 250
    organized armed groups, 244, 251–52
    police primacy, 249
    small wars, 244
    soft knock approach, 248–49, 361
    targeted killing, 245–47, 250–52
  terrorists and insurgents, distinctions and relationships, 190–97, 360
Inter-American Commission on Human Rights, 501, 504–5
INTERFET (International Force in East Timor), 473
internal conflicts with "international character"
  Common Article 3, 44
  humanitarian law, extension of law to conflicts with non-state actors, 44–45
  non-state actors, internal conflicts with international character, 44–46
"internal disturbances and tensions, such as riots, isolated and sporadic acts of violence and other acts of a similiar nature", 35, 154, 161, 175, 354, 377, 392, 451, 495, 579, 582, 587
international armed conflict, 35–36, 44–46, 94–99, 330–42, 355–58, 366–67, 396–97, 578–81, 620–23
  categorizing State *vs.* non-state actors conflicts
    entering territory of another state, international armed conflict, 339–41
    transnational armed conflicts, 355–56
  definition of, 35
  drone strikes, "conduct of hostilities" approach, 301–2, 305–6
  human rights-based law, application restrictions, overreach, 552–53
  Israeli Supreme Court and cross-border terrorism, 346, 348–52
  post-World War II humanitarian law and human rights law interface, historical background, 147–48
  proper authority principle, humanitarian law and, 95–97
International Committee of the Red Cross (ICRC)
  Customary International Humanitarian Law Study (2005), 14, 81, 109, 126, 146
    Rule 47, 506*n*265
  ICRC Commentary to Common Article 2, Geneva Convention, 330
  ICRC Commentary to Common Article 3, Geneva Convention, 377
  Interpretive Guidance on the Notion of Direct Participation in Hostilities Under International Humanitarian Law, 238–42, 303, 307, 321, 456
  occupied territory existence, test for establishing, 103
  post 9/11 Afghanistan conflict, 46
  Rule 47, *hors de combat*, 506*n*205
  World War II, aftermath of, 140–41
international criminal law, 97, 133
  drone strikes, liability for war crimes, 286
  genesis of, 15
  humanitarian rights law based, 432
  legal challenges, 14*f*, 15, 19, 21
  liability for war crimes under, 286
  1990s through 2000s, 144
  operational law and, 574
  police, wartime role of, 437
  terrorism definition for purposes of, 181
international humanitarian law (IHL), 110–12, 122, 124, 135, 139*n*122, 152*n*202, 156
international human rights law (IHRL), 124, 134*n*79, 135–36, 152*n*202
international law, traditional interpretation, 19–20
International Security Assistance Forces (ISAF)
  air power, restriction on use of, 596
  establishment by UN Security Council, 47
  post 9/11 Afghanistan conflict, 47–48, 63, 217, 220, 248, 489, 523
Interpretive Guidance on the Notion of Direct Participation in Hostilities Under International Humanitarian Law (International Committee of the Red Cross, 2009), 238–42, 303, 307, 321, 456
  drone strikes, 303, 307
  least restrictive means (LRM) analysis, 231–32

suicide bombing threat, capture or kill, 238–42
inter-state conflict, 34–37, 330–35
Additional Protocol I, Article 48, 334*n*29
belligerents, status of unprivileged, 334
broad categorization of, 331–33
controlling state involvement in armed conflict, 34–37
decline in, 16–17, 35–36
degree of organization of armed group, no assessment criteria, 331
forceful response, no requirement, 332
Geneva Convention, Articles
Common Article 2, 330–32
Fourth Geneva Convention
Article 2, 330*n*9
Article 64, 331*n*10
humanitarian law, 329–35
international armed conflict, definition, 35
international (inter-state) or non-international (intra-state) conflicts, 37
limited levels of force, 332–33
limiting the recourse to war, 40–42
antiwar movement, impact upon, 40–42
just war, 41
lawful war *(bellum legale)*, 41
pacificism, 39, 41, 628*n*35
pacifism, 39, 41, 75, 449*n*34, 628*n*35
post-World War I effects, 39–40
UN Charter, limiting recourse to war, 40–42
unregulated states, destructive capabilities, 40
main bodies of international law, 334–35
non-international armed conflict, 35
non-state actor conflicts *vs.*, 9
non-state participants, bias, 334
organized armed groups, 334
party to a conflict, 331
post-World War II effects, 16–17
recognition of belligerency doctrine, 34
regulation of state-related conflict, legal categories, 35–37
risk of inter-state armed conflicts, examples, 35
scholarly research, 34
situations of internal disturbances and tensions, 35
treatment of war under international law, 35
UN Charter, Article 51, 331
UN Charter, limiting recourse to war, 40–42
unprivileged belligerents, status of, 334
warfare, universal characteristic of human existence, 36
Intifadas
definition of, 341
first Intifada (1987-1993), 234*n*120, 343
second Intifada, Al Aqsa (2000-2005), 235, 343, 438, 511, 515–16
Iran, 35, 86, 96, 107, 159, 453, 512, 514–15
Case Concerning Oil Platforms (Islamic Republic Of Iran V. United States Of America), 73, 78, 87–88, 331–32, 390, 392–93, 465
Case Concerning United States Diplomatic and Consular Staff in Tehran (United States v. Iran), 338, 409
shoot down of Iranian civilian passenger plane by *USS Vincennes* (1988), 465
Iraq, 3–6, 168–69, 199–204, 206–13, 215–26, 235–36, 280–81, 401–4, 453, 485–91, 511–14, 620–22, 625–26. *See also* Al Qaeda in Iraq (AQI); Fallujah assault (2004)
attack on *USS Stark*, 464–65
Islamic State in Iraq and al Sham (ISIS), 4, 115, 160, 168, 170, 200–202, 221–22, 277, 296, 308, 323–25, 368, 404, 597, 599–602
Islamic State of Iraq (ISOI), 200, 206
kidnappings, post-2003 invasion, 402–3
U.S. 2003 invasion of, 475–76
Irish National Liberation Army (INLA), 112, 131–32, 185, 371, 419–20, 513, 528, 530–31, 537–40, 543–50, 615
Irish Republican Army (IRA), 130–31, 537, 593
irregular warfare, 7, 180, 216*n*4
Bandit Queen Phoolan Devi, 180
inevitability of, 622
regular *vs.* irregular conflicts, 9
Isayeva v. Russia, 125*n*22, 419*n*305, 481*n*204, 552*n*138, 553–55, 558*n*169
Islamic state, Al Qaeda and
Al-Baghdadi, abu Bakr, 200–202, 206
Al-Muqrin, Abd Al-'Aziz, 206
Al Nusra Front (ANF), 200–203, 298*n*209
Al Qaeda and Associated Movements (AQAM), 199, 202–5
Al Qaeda Central, 169, 198–99, 208, 294–95
Al Qaeda in Iraq (AQI), 198–200, 204, 206–7
Al Qaeda in the Arabian Peninsula (AQAP), 198, 202, 206
Al Qaeda in the Islamic Maghreb (AQIM), 198–99
Al Shabaab, 198–99, 202
bin Laden, Osama, 200, 202–4
Boko Haram, 199, 205
caliphate, 4, 193–94, 569
focoist/vanguard approach, 208, 308, 570
guerrilla warfare, 203, 206–9
Haraakat-I Inquilab Islam-I, 204
Islamic State in Iraq and al Sham (ISIS), 200–202

Islamic state, Al Qaeda and (*Cont.*)
Islamic State of Iraq (ISOI), 200, 206
jihadist movement, 205
jihadist organization, 199
The Military Theory of the Global Islamic Resistance Call, 205–6
Salafi jihadist, 4, 8, 197, 212, 228
Syrian Conflict, 200–202, 287, 295, 323–26
Taliban, 198–99
Islamic State in Iraq and al Sham (ISIS), 4, 168, 170, 200–202, 221–22, 277, 296, 323–25, 368, 404, 597, 599–602
drone strikes, 308
insurgency, 160, 168
non-state actors and armed conflicts, 368
as pseudo state, 160*n*4
twenty-first century counterinsurgency, 221–22
ungoverned spaces, 115
wilayat concept, 201
Islamic State in Iraq and the Levant (ISIL, also known as Da'esh), 298*n*207, 324*nn*353–54, 516*n*304, 625
Islamic State of Iraq (ISOI), 200, 206
isolated acts
"internal disturbances and tensions, such as riots, isolated and sporadic acts of violence and other acts of a similiar nature," 35, 154, 161, 175, 354, 377, 392, 451, 495, 579, 582, 587
isolated acts of terrorism theory, 294–96
isolated attacks, 85
isolated suspects, 259
Israel
Entebbe, Israeli hostage rescue (*Operation Thunderbolt (Yonatan*, 1976)), 28, 185, 328, 397, 405–12, 418, 420–21, 434, 437, 439, 498, 500, 504, 580–81, 586
extraterritorial law enforcement, Israeli position on application of human rights treaty law, 149–50
Operation Defensive Shield (2002), 235
Public Committee Against Torture in Israel v. Israel. See Targeted Killing Case
Sayeret Matkal, military Special Forces unit, 434, 438, 498
strike on Syrian nuclear reactor (2007), 57
Yamam, tactical hostage rescue team, 435, 438–39
Israel Defense Forces (IDF), 407
Sayeret Matkal, Special Forces unit, 434, 438, 498
Duvdevan, 243n120, 438
Israeli Supreme Court and cross-border terrorism, 341–49
American approach contrasted, 345–47
conflict not of an international character, 345
Hamas, 344–47
Hezbollah, 344–46
humanitarian law, 347–49
international armed conflict, 346, 348–52
non-international armed conflict, 346
Operation Cast Lead, 347
Operation Protective Edge, 344
practical impact, 347–48
suicide bombing, 343
Targeted Killing case, 341–43
unique security threat, 343–45
unlawful combatants, 348–49
Israel-Lebanon Conflict (1982), 66, 78, 205, 350, 352, 360, 493
Israel-Lebanon Conflict (2006), 62–63, 66–67, 69, 78, 82*n*193, 87, 117, 163, 182, 189, 229, 344–45
Israel/Palestinian Conflict
counterinsurgency, 234–36
Jenin resistance, 235
Legal Consequences Of The Construction Of A Wall In The Occupied Palestinian Territory, 35*n*10, 125, 127, 153, 237*n*141, 262, 517–18, 564
Occupied Territories, 234–36
Operation Defensive Shield (2002), 235
ungoverned spaces, 117

Jaloud v. The Netherlands, 103*n*88, 558*n*168
Jamaican Shower Posse, 481*n*207, 589, 599
Japanese Embassy Case (Lima, Peru; 1997), 125*n*21, 484, 500–510, 513, 527–28, 558–59
American Convention on Human Rights, Article 4(1), 504, 507
Common Article 3, 507
direct participation in hostilities, 508
explosives, 503
extrajudicial killing, 506
Giampierti, Peruvian Special Forces Admiral Luis Giampierti, 503
Hague Land Warfare Regulations Article 23(c), 506*n*205
hors de combat, 125*n*21, 484, 500–510, 513, 527–28, 558–59
humanitarian law, 505–10
human rights law, extrajudicial killing, 506
ICRC Rule 47, 506*n*205
Inter-American Commission on Human Rights, 501, 504–5, 507
Inter-American Court of Human Rights, 501, 504, 507–8, 553, 558
last resort, 507
minimum force, 485, 505
totality of the circumstances, 508
Túpac Amaru Revolutionary Movement (MRTA), 501–2, 505–8

use of force, individuals who no longer present a threat, 506–7, 510
Jenin resistance, 235
jihadist, 190, 193–94, 197–208, 267–68, 294–98, 597–98, 601, 613–14
counterinsurgency and counterterrorism, integrated response, 212–13
drone strikes, "law enforcement" theory, 296–98
explosives, 514, 516
leaderless jihad, 294–95
Salafi jihadist, 4, 8, 160, 197–98, 199*n*364, 202, 207, 212, 228, 516, 569–70, 625
ungoverned spaces, 114–16
jihadist-based conflict, global insurgency, 4–5
John Charles de Menezes shooting (2005), 509, 525–26, 531
Joint Task Force 2 (JTF 2), 434*n*405
jus ad bellum. *See* law governing the recourse to war (*jus ad bellum*)
jus cogens
drone strikes, "self-defense" option, 314
jus in bello. *See* humanitarian law (*jus in bello*)
just cause principle, 13
just war (*bellum justum*), 25, 37–39, 44, 46, 240–41
application of self-defense principles, 55–56
bellum justum, 41
destructive capabilities of the state, 37–38
development of rules governing, 37–39
equal application principle and, 39
insurgency, 167
law enforcement and self-defense, 447, 449–50
legal challenges, 10–13
nineteenth century developments, 37–38
pacificism, growth of, 39
pacifism, growth of, 39
proper authority principle, humanitarian law and, 91–93
reintroduction of principles, 38–39
The Responsibility to Protect Report (International Commission on Intervention and State Sovereignty, 2001), just war principles, 10–13
self-defense principles
application during armed conflict, 70–72
application of, 55–56
strategic approach *vs.* tactical approach, 75, 77
UN Charter, influence of just war theory on, 41
UN Security Council decision regarding self-defense, 70–72

Kallay, Foday, 412, 416
Kellogg-Briand Pact (1928), 39
Kelly v. United Kingdom, 544*n*79, 544*n*81, 544*n*83
Kenya, Westgate Mall attack (2013), 293, 368, 437–38, 564, 597
Kerimova v. Russia, 125*n*22, 139*n*122, 452*n*45, 481*n*204, 555, 558*n*169
Khattala, Ahmed Abu, 276, 391–92
kidnapping, 9, 422–23, 429, 432, 583, 587, 590. *See also headings under* hostage rescue
insurgency, 159–60
insurgency and crime, 168–69, 171, 176
internal conflicts, 403
Iraq, post-2003 invasion, 402–3
Mali, kidnapping and murder of French journalists (2013), 426
non-international armed conflicts, 403–4
Revolutionary Armed Forces of Colombia (FARC), 403, 405, 422
terrorism, 184, 188, 190
transnational armed conflicts, 354–55
Koh, Harold, 20–21, 147, 213, 311, 390*n*47, 612
Kordić; Prosecutor v., 138*n*118, 347*n*232, 448
Kosovo Liberation Army (KLA), 372–75, 376*n*244
Kunarac; Prosecutor v., 134*n*78
Kuwait, 17, 70–71, 296, 597

La Familia, 170*n*94, 171
Lashkar-e-Tabia (LeT), 192–94, 265, 436, 500, 503, 517, 527
Latin terminology, 10–11
law enforcement
counterinsurgency and primacy of law enforcement. *See* law enforcement, counterinsurgency and primacy of
drone strikes, "law enforcement" theory, 287–300, 319–22
extraterritorial law enforcement, 52, 66, 349–52, 356–57, 394
and human rights law. *See* human rights law
self-defense
and human rights. *See* state self-defense and human rights-based law enforcement
and law enforcement. *See* law enforcement and self-defense
law enforcement, counterinsurgency and primacy of, 223–28
blended rules of engagement (ROE) cards, 226–27
conventional warfare, 225
direct participation in hostilities, 226, 232, 239, 241–43
geography of war, 228
governance, 228–30
human rights and humanitarian law, 223, 225–29

law enforcement, counterinsurgency and primacy of (*Cont.*)
maintenance of security through policing, 224–26
minimum force, 227
police primacy, 224, 227
Salafi jihadist, 4, 8, 197, 212, 228
small wars, 225–26, 228
soft law, 227
Westphalian system of government, state structure, 228
law enforcement and self-defense, 443–81. *See also* state self-defense and human rights-based law enforcement
collateral damage, 449, 451, 456
Common Article 3, 451
counterinsurgency doctrine, 453, 467
direct participation in hostilities, 456
European Convention on Human Rights, Article 2(2), 459–60, 481
Fallujah assault (2004), 445, 453, 476–79, 481
governance, 449–51
Hague Land Warfare Regulations, Article 43, 451*n*41
humanitarian law, 441, 448
targeting, 445, 455–56
human rights-based law enforcement. *See* state self-defense and human rights-based law enforcement
human rights law, 477–81
insurgency/counterinsurgency, 478
just war, 447, 449–50
law governing the recourse to war, 447–48
levels of violence, 480–81
minimum force, 470–71, 474
mission accomplishment, 458–64
non-combatant evacuation operations (NEO), 328, 393–95, 463
operational law, 457, 462–63, 475–76, 479
organized armed groups, 455–56, 478
overview, 443–45
police primacy, 453, 467
proper authority principle, 449–50
robust peacekeeping, 467
rules of engagement (ROE), 445, 461–62, 464, 470–71, 473–74, 483, 485, 488
shoot to kill, 458
standing rules of engagement (SROE), U.S., 445, 460–62, 474–75, 478–79
state self-defense and human rights-based law enforcement, 449–50
targeting
hostilities based targeting, 456, 565
humanitarian law and, 445, 455–56
proportionality, 450, 507
status-based targeting rules, 303, 441, 456, 476, 527, 545
strategic level targeting, 525
transnational criminal organizations, 480–81
UN Charter
Article 2(4), 470
Article 42, 460*n*79, 468*n*124, 473*n*159
Article 51, national self-defense, defined, 464
lawful war *(bellum legale)*, 41
law governing the recourse to war *(jus ad bellum)*, 10–12, 19, 21, 24–25, 39, 41–43, 45–47, 49–57, 61–62, 64, 73–75, 78–79, 84–85, 88–89, 311–13, 315, 383, 389–91, 393–94, 447–48, 574–75, 582, 628–30
antiwar movement, 628
definition of, 10
drone strikes, "self-defense" option, 311–13
law of armed conflict (LOAC), 18*n*87, 256, 320, 444*n*3, 623
drone strikes, 301–2, 316
force of first and last resort, 484
humanitarian law
and human rights law interface post-World War II, historical background, 150
proper authority principle, 110
The Manual of the Law of Armed Conflict (United Kingdom), 13*n*65, 95*n*27, 310*n*279, 387
non-international armed conflict, proper authority, 362–63
post-World War II, humanitarian law and human rights law interface, historical background, 150
proper authority
humanitarian law and, 110
non-international armed conflict, 362–63
"law of 9/11", 20–24
Law of War Manual (Department of Defense, 2015), 22*n*110, 37, 93*n*8, 94*n*22, 95*n*23, 98*n*44, 110*n*136, 119*n*210, 122*n*3, 124*n*14, 147*n*172, 148, 239*n*154, 259*n*282, 309*n*273, 387*n*30, 623*n*19
The Manual of the Law of Armed Conflict (United Kingdom), 13*n*65, 95*n*27, 310*n*279, 387
leaderless jihad, 294–95
League of Nations, 39–40, 75
least restrictive means (LRM), 231–33, 507–8
Lebanon-Israel Conflict (1982), 66, 78, 205, 350, 352, 360, 493
Lebanon-Israel Conflict (2006), 62–63, 66–67, 69, 78, 82*n*193, 87, 117, 163, 182, 189, 229, 344–45

legal advisors and twenty-first-century warfare, 620
legal challenges, 10–20. *See also* humanitarian law; human rights law
ability of contemporary law to regulate current conflicts, 10–20
application of law, legal boundaries and, 15–18
bodies of law that impact armed conflict, 14*f*
domestic law, 3, 9, 14*f*, 15–16, 19
equal application principle, 12–13, 39, 65, 79, 92, 167
historical focus on State, 12
humanitarian law, 11–16, 14*f*
human rights law, 14*f*, 18–19
international criminal law, 14*f*, 15, 19, 21
international law, traditional interpretation of, 19–20
interstate conflicts, 12
issues overlooked, 10–14
just cause principle, 13
just war principles, 10–13
Latin terminology, 10–11
legal boundaries and interpretational limits, 18–20
overview, 10–20
role of State, proper or lawful authority to maintain peace, 13
theory and practice, schism between, 17–18
twentieth-century law, 16–17, 23
Legal Consequences Of The Construction Of A Wall In The Occupied Palestinian Territory, 35*n*10, 125, 127, 153, 237*n*141, 262, 517–18, 564
Leiden Policy Recommendations on Counter-Terrorism and International Law (2010), 51
Levant (*al-Sham* in Arabic), 4
level of threat, mission accomplishment, 463–64
levels of violence, 183
types of, distinction, 161
levels of war, 72–78
operational level, 22–23, 76–78, 228–37, 569–617
strategic level, 74–76, 78–83
tactical level, 74, 78–83
Levya, Arturo Beltrán, 171, 175, 598
lex ferenda, 10–11, 393, 530
lex lata, 10–11, 237, 393, 530
lex specialis. *See* special law *(lex specialis)*
Liberation Tigers of Tamil Eelam (LTTE), 192, 515
Libya, 278, 296, 298, 592, 602
Al Qaeda and, 197–98
Benghazi attack (2012), 276, 293, 368, 391–92, 453, 512, 600
Entebbe, Israeli hostage rescue (*Operation Thunderbolt (Yonatan*, 1976)), 406, 408
2011 Conflict, 83–84, 281–82, 398
Lieber Code, 38, 50
Limaj; Prosecutor v., 161*n*15, 167, 372*n*222, 374–75, 417*n*301
Living Under Drones: Death, Injury, and Trauma to Civilians From US Drone Practices in Pakistan Study (Stanford Law School & NYU Law School), 269–70
London Specialist Firearms Command (SCO), 435
lone wolf attacks, 198, 371, 512, 614
Lord Salisbury, 267
Los Baños raid (Philippines, 1945), 399*n*112, 400–401
Loyalist Volunteer Force (LVF), 537
Lynch, Jessica, rescue of, 401

Mahan, Alfred Thayer, 34
Maharashtra State Anti-Terrorism Squad (ATS), 436
Mali, 4
Bamako, Hotel attack (2015), 403, 425
kidnapping and murder of French journalists (2013), 426
Management of Savagery: The Most Critical Stage Through Which the Umma Will Pass (Abu Bakr Naji), 115, 229*n*83, 298*n*205, 403
Marighella, Carlos *(Minimanual of the Urban Guerrilla)*, 164*n*46, 187–88
Marine Commando (MARCO), 436
Mayaguez rescue operation (1975), 397, 400–402
McCann and others v. United Kingdom, 21*n*105, 131*n*57, 286*n*118, 462*n*88, 525*n*350, 530–31, 544*n*79, 544*n*81, 545*n*85, 546–48, 549*n*122, 596*n*91
McChrystal, General Stanley, 5, 8*n*46, 108*n*125, 204*n*412, 217*n*15, 236*n*134, 238*n*147, 242*n*175, 285*n*109, 371*n*215, 440*n*449, 516–18, 523*n*338
McKerr v. United Kingdom, 131*n*62, 544*n*81, 544*n*84, 546*nn*95–96, 547, 550, 627*n*32
means and authority test, 353, 355, 376, 586
drone strikes, 316
Mecca siege (1979), 434*n*408, 498–99, 509, 527, 599
Menezes shooting (2005), 509, 525–26, 531
Mexico
cartels, 24, 170–71, 177
gangs and Calderón administration, 173–75, 177
militarization of police, 550–52

military advantage
 and counterinsurgency, 244–61
  insurgents and target "value," 244–53
  proportionality and counterinsurgency, 258–61
  strategic military advantage and civilian casualties, 254–58
 military necessity, 245
 proportionality
  assessments, 79
  and counterinsurgency, 258–61
  humanitarian law proportionality test, 245, 247, 254, 560–61, 603
  targeting proportionality, 79, 81–82, 254
Military and Paramilitary Activities in and against Nicaragua, 71, 331, 335*n*30, 389–90, 577, 582
military necessity, 245
military operations other than war (MOOTW), 364, 571, 604
The Military Theory of the Global Islamic Resistance Call (Abu Mus'ab Al-Suri), 205–6, 613
Milŏsević; Prosecutor v., 372*n*222, 373–74
Minimanual of the Urban Guerrilla (Carlos Marighella), 164*n*46, 187–88
minimum force, 435, 470–71, 474, 528, 610
 counterinsurgency and counterterrorism, integrated response, 209, 213
 human rights-based law application restrictions, overreach, 557–58
 Japanese Embassy Case (Lima, Peru; 1997), 485, 505
 law enforcement, counterinsurgency and primacy of, 227
 narrow operational and normative gap, 485, 505
 proper authority principle, humanitarian law and, 106
 shoot to kill policy, 548–49
mission accomplishment, 458–64
 deadly force, 460
 definition of self-defense in law enforcement context, 461–63
 drone strikes and, 460–61
 European Convention on Human Rights, Article 2(2), 459–60, 481
 human rights law and, 458–61
 level of threat, 463–64
 rules of engagement (ROE), disagreements regarding, 461–62, 464
 standing rules of engagement (SROE), 460–62
 state self-defense and human rights-based law enforcement, distinguished, 458–64
 UN Charter, Chapter VII, 458–61
modern counterterrorism, 412–15
modern terrorism, 185
mosaic war, 119, 230
Moscow Theatre incident (2002), 154, 260, 452, 498–500, 509, 562
Mumbai terrorist attack (India, 2008), 192–93, 293, 367–68, 372, 436, 500, 503, 527, 592, 599

Nabhan, Saleh Ali Saleh, 273, 275
Naji, Abu Bakr *(Management of Savagery: The Most Critical Stage Through Which the Umma Will Pass)*, 115, 229*n*83, 298*n*205, 403
naked self-defense, 311–17, 320, 355
 and robust self-defense approach, distinguished, 317–18
 threshold, 314
Nariman House Jewish Center, 436
National Liberation Army (ELN), 165, 171, 196, 422, 426
National Security Guard (NSG) (India), 435–36, 503, 527–28
 Black Cats, 435, 503
national self-defense
 counterinsurgency and counterterrorism, integrated response, 213
 defined, UN Charter, Article 51, 464
 drone strikes, 312, 315*n*312
 and human rights-based law enforcement, 444, 447, 451, 462*n*92, 462*n*95, 463–66, 478, 481
The National Military Strategy of the United States of America 2015, 36, 221
necessity, self-defense legal principle of
 Advisory Opinion on the Legality of the Threat or Use of Nuclear Weapons, 56, 73, 82, 85
 Caroline incident, 49, 66, 302, 454
 human rights law, 447–48, 454–55, 458
 military necessity, 245
 recourse to war, 56–58, 61, 63, 260–61
Neo-Taliban, 193, 199
networks, terrorists, 191, 198–99
neutrality concept, humanitarian law and, 59
neutrality law, 27, 326
 drone strikes, 302–7, 310, 321, 323, 326
 "Special law" and exclusion debate, 122
 state and governance, reluctant application of humanitarian law, 105
Nicaragua
 Military and Paramilitary Activities in and against Nicaragua, 71, 331, 335*n*30, 389–90, 577, 582
Nigeria
 Boko Haram, 98, 118–19, 199, 205, 222, 424–26, 594

hostage rescue (2012), 424–25
Night Operations Tactical Directive (2011), 248–49, 256
9/11 attacks, impact of
cross-border attacks by non-state actors, effect of attacks on exercise of self-defense, 49
operational challenges, "law of 9/11," 20–24
post-World War II humanitarian law and human rights law interface, 147–48
non-combatant evacuation operations (NEO), 328, 393–95, 463
non-international armed conflict, 43–54, 507. *See also* non-international armed conflict, determining existence of
antiwar movement, 53
application of law governing, non-state actor conflicts, 53
authority to engage, domestic source theory, 43–44, 48, 54
Common Article 3, 44, 357
conflict evolution, 46–48
cross-border attacks by non-state actors, 48–52
direct participation in hostilities concept, 53
drone strikes, 288, 292–93
extraterritorial law enforcement, 52
humanitarian law
extension of law to conflicts with non-state actors, 44–45
proper authority, 360–61
insurgencies, 46–48
internal conflicts with "international character," 44–46
international law, 43–44
non-state actors, 43–54
application of Geneva Convention, 45
cross-border attacks by, 48–52
internal conflicts with international character, 44–46
interpretive struggle, application of law governing conflicts, 53
one-off attacks by, threshold for determining existence of conflict, 366, 369, 372
proper authority, 42–44, 356–63, 379
acting in place of the territorial state, 361–63
Additional Protocol I, Article 51(1), 363*n*176
bin Laden, Osama, 358
controlling state involvement in armed conflict, 42–44
counterinsurgency doctrine, 360
counterinsurgency operations, 361
counterterrorism operations, 361
Geneva Convention, Articles
Common Article 3, 357
Third Geneva Convention, Article 4A(2), 358*n*157
geographical parameters, 359–61
geography of war, 359
governance, 359
guerrilla war, 360
Hamdan v. Rumsfeld, 356–58
humanitarian law, 360–61
human rights-based law enforcement, 359–61
insurgents and/or terrorist, 360
law of armed conflict (LOAC), 362–63
Obama, Barack, 358
police primacy, 360–61
police primacy approach, 361–62
power and responsibility, concentration of, 358
protection of civilians, 362–63
protracted armed violence, 363
state as, 43
targeted killing, 358–59, 361
ungoverned spaces, 358–60
recognition of belligerency doctrine, 44, 46
state self-defense concept, 43
transnational threats rising to armed attack by non-state actors, 52–53
UN Charter, Article 51, cross-border attacks by non-state actors, 48–52
violence threshold of armed attack, 53
wars of national liberation, 45–46
non-international armed conflict, determining existence of, 363–78
accountability, 376–77
active support, 378
Additional Protocol II, Article 1(2), 377*n*247
Common Article 3, 377
ICRC Commentary to Common Article 3, Geneva Convention, 377
conflict criteria, 363–70
counterinsurgency doctrine, 372
group organization, 370–78
adoption of cellular structure, 371
decentralized and cellular groups, 372
destructive activity, 371–72
improvised explosive devices (IEDs), 371–72
nature of organized armed group, 371
one-off attacks, 372
varied organizational types, 370–72
human rights law, 363–65
and humanitarian-based law, overlap of, 377–78
ICRC Commentary to Common Article 3, Geneva Convention, 377
intensity of violence, 365–70, 378
internal disturbances and tensions, 377

non-international armed conflict, determining existence of (*Cont.*)
international law governing recourse to war, existence questioned, 43
limitations of law enforcement, 367–69
one-off attacks by non-state actors, threshold, 366, 369, 372
police primacy, 370, 378
retrospective view of violence, analysis and court decisions, 369
Rome Statute, Article 8(2)(f), 366
threshold, 365–66
"totality of the circumstances," 329, 375–76
non-state actors and armed conflicts, 3–9, 327–80. *See also* non-international armed conflict
active support, 337–39
non-international armed conflicts, determining existence of, 378
non-state actors controlled by a state, 337
Additional Protocol II, Article 1(2), 377*n*247
belligerents. *See* belligerents
categorizing State *vs.* non-state actors conflicts, 335–63, 378–80
entering territory of another state, international armed conflict, 339–41
extraterritorial law enforcement, 349–52, 378–80
harboring, 337–39
non-international armed conflict, proper authority, 42–44, 356–63, 379
non-state actors controlled by a state, 335–37
protracted armed violence, 352, 363
transnational armed conflict, 352–56
Common Article 2
ICRC Commentary to Common Article 2, Geneva Convention, 330
inter-state conflict, 330–32
transnational armed conflict, 352
Common Article 3, 44, 357
ICRC Commentary to Common Article 3, Geneva Convention, 377
non-international armed conflict, proper authority, 357
non-international armed conflicts, determining existence of, 377
transnational armed conflict, 352–55
complexities of, 34
cross-border attacks by non-state actors, 48–52
cross-border terrorism, Israeli Supreme Court and, 341–49
drone strikes, 5
extraterritorial law enforcement, 349–52, 378–80
Fourth Geneva Convention
Article 2, 330*n*9, 340*n*57
Article 64, 331*n*10
group organization, improvised explosive devices (IEDs), 371–72
guerrilla warfare, 6–9
harboring and active support, 337–39
human rights law, application of, 36–37
insurgents. *See* insurgents
international borders, lack of acceptance and respect, 4
inter-state conflict, 330–35
inter-state *vs.* non-state actor conflicts, 9
lone wolf attacks, 371
military operations other than war (MOOTW), 364, 571, 604
non-combatant evacuation operations (NEO), 328, 393–95, 463
non-international armed conflict, 43–54
non-state actors controlled by a state, 335–38
one-off attacks by non-state actors, threshold for existence of armed-conflict, 366, 369, 372
organized armed groups, 328–29, 366, 371
overview, 3–9, 327–30
police primacy, 370
privileging state participants over non-state protagonists, 92–97
proper authority, state role as, 36–37
protracted armed violence, 329, 352, 363, 366–67
recognition of belligerency doctrine, 378
robust peacekeeping, 365
Rome Statute, Article 8(2)(f), 366
Secretary-General's Bulletin: Observance by UN Forces of International Humanitarian Law (1999), 110*n*139, 120*n*214, 365, 379, 472
small wars, threshold for existence of armed-conflict, 366, 369, 372
state, role as proper authority, 36–37
state as primary actor in conflict, applied norms, 36–37
Third Geneva Convention, Article 4A(2), 358*n*157
totality of the circumstances, 329, 375–76
transnational armed conflict, 352–56, 365, 367, 378
transnational criminal organizations, 6–7
criminal gangs, 3, 5–6, 9, 168, 172–73
types of criminal activity, 9
Noriega, Manuel, 401
North American Air Defense (NORAD), 582*n*50, 584
North American Air Defense (NORAD) Agreement, 559–60, 563

Northern Ireland. *See The "Troubles"* (Northern Ireland)
Nuhanović v. The State of the Netherlands, 21*n*107

Obama, Barack, 20, 279–80, 309–11, 313, 320–21, 324, 358, 429
counterinsurgency, control and operational choices, 237
counterinsurgency and counterterrorism, integrated response, 211, 213
Drone Speech (2013), 3–4, 20–21, 181, 191, 298–99, 322, 363, 596, 609
Occupied Territories
Israel/Palestinian conflict, 234–36
Legal Consequences Of The Construction Of A Wall In The Occupied Palestinian Territory, 35*n*10, 125, 127, 153, 237*n*141, 262, 517–18, 564
occupied territory
definition of, 103
establishing existence of, test favored by ICRC, 103
Oil Platforms Case (Islamic Republic Of Iran V. United States Of America), 73, 78, 87–88, 331–32, 390, 392–93, 465
Oklahoma City bombing (1995), 372*n*216, 512
Oman, 115, 420, 428
one-off attacks by non-state actors, 53, 108, 219, 311. *See also* small wars
counterinsurgency, 219, 265
counterterrorism operations, 293, 311, 317
duration criterion, 366, 369
holistic response to, 563–64, 575–76, 597, 600
proper authority principle, humanitarian law and, 108
threshold for determining existence of armed-conflict, 265, 317, 366, 369, 372, 419–20, 585, 597
operational challenges, 506
extrajudicial killing, 21, 506
"law of 9/11," 20–24
operational law concept, 22–23
traditionalists *vs.* integrators, 23
operational law. *See* holistic operational law
definition of, 22*n*10, 574
Operational Law Handbook, U.S., 18*n*87, 19, 124, 150–51, 153*n*204, 457, 475*n*173, 476, 479, 489, 622–23, 629
operational level of war, 22–23, 76–78
counterinsurgency, 228–37
Operation Barras (British operation), 412–20, 493, 586
Operation Cast Lead, 347
Operation Defensive Shield (Israel, 2002), 235
Operation Dragon Rouge, 403
Operation Enduring Freedom (OEF), 47, 86, 337
Operation Just Cause (1989), 401
Operation Palliser, 412
Operation Praying Mantis, 88
Operation Protective Edge, 229*n*81, 344
operations other than war
East Timor, 473
Rwandan genocide (1994), 144*n*154, 373*n*223, 468–69, 571
Somalia, rescue operations, 426–31
(former) Yugoslavia, 34–35, 107, 161, 167, 292–93, 315–17, 335, 348, 366–67, 372, 374–76, 417–19, 448, 575–76, 579–80
Operation Thunderbolt (Yonatan, 1976), Entebbe, Israeli hostage rescue, 28, 185, 328, 397, 405–12, 418, 420–21, 434, 437, 439, 498, 500, 504, 580–81, 586
organized armed groups
drone strikes, 292–93
insurgent organizations, 161–62
insurgents and target "value," 244, 251–52
inter-state conflict, 334
law enforcement and self-defense, 455–56, 478
non-international armed conflict, determining existence of, 371
non-state actors and armed conflicts, 328–29, 366, 371
proper authority principle, humanitarian law and, 107–8
self-defense, limited application theory, 65, 67
suicide bombing threat, capture or kill, 239, 241–42
Oxford Research Group
drone strikes, 288–90, 299, 321

pacificism, 39, 41, 628*n*35
Pacific Rim, excessive maritime claims, 35
pacifism, 75, 449*n*34, 628*n*35
growth of, 39
inter-state conflict, limiting the recourse to war, 41
Padilla, José, 264–65
Pakistan, 219, 231, 236, 238–39, 251, 253, 255, 280, 282–84, 287–91, 298, 358–61
Aerial Drone Deployment on 4 October 2010 (Mir Ali/Pakistan), 231*n*94, 236*n*140, 238*n*152, 239*n*157, 288*n*135, 319*n*328, 370*n*211, 376*n*246, 456*n*64, 608*n*139
Al Qaeda and, 115–16
Living Under Drones: Death, Injury, and Trauma to Civilians From US Drone Practices in Pakistan Study (Stanford Law School & NYU Law School), 269–70
Tehrek-e-Taliban Pakistan (TTP), 115, 193–94, 514

Palestine/Israel conflict
counterinsurgency, 234–36
Jenin resistance, 235
Legal Consequences Of The Construction Of A Wall In The Occupied Palestinian Territory, 35*n*10, 125, 127, 153, 237*n*141, 262, 517–18, 564
Occupied Territories, 234–36
cost-benefit analysis, 234–35
Operation Defensive Shield (2002), 235
ungoverned spaces, 117
Palestine Liberation Organization (PLO), 66, 196*n*332, 205, 251, 251*n*226, 268*n*4, 350, 408
Panama, rescue operations, 401
parallelism, 52, 142
"pattern of life", 281
personality strikes, 278–82
Philippines
Los Baños raid (1945), 399*n*112, 400–401
Santos Tomás raid (1945), 400–401
Philippines National Police Special Action Force, 435
piracy, 169, 171, 397, 419, 435, 461, 491–92, 590–91
Somalia piracy, rescue operations, 328, 397, 426–31 491, 494, 496
police primacy, 26, 54, 590, 593, 595, 606, 625
counterinsurgency, 215, 224, 227
human rights and humanitarian laws, counterinsurgency frameworks, 262, 265
insurgents and target "value," 249
law enforcement and self-defense, 453, 467
non-international armed conflicts, 360–61, 370, 378
targeted killing, 249
drone strikes, 311, 322
human rights and humanitarian laws, counterinsurgency frameworks, 262, 265
insurgents and target "value," 249
law enforcement, counterinsurgency and primacy of, 224, 227
law enforcement and self-defense, 453, 467
non-international armed conflicts
determining existence of, 370, 378
proper authority, 360–62
non-state actors and armed conflicts, 370
proper authority
counterinsurgency, non-international armed conflicts, 360–61
humanitarian law and, 112
non-international armed conflicts, 360–62
targeted killing, 249
The "Troubles" (Northern Ireland), 541, 543
ungoverned spaces, 112–13
political purpose test, 167–68, 178
Popular Front for the Liberation of Palestine (PFLP), 185, 406
"positive identification" (PID), 281
A Practical Course for Guerrilla Warfare (Abd Al-'Aziz Al-Muqrin), 115*n*172, 206, 613*n*175
Principles of International Law on the Use of Force by States in Self-Defence (Chatham House, 2005), 51, 88
proper authority principle, 91–120
Boko Haram, 98
Common Article 2, privileging state participants over non-state protagonists, 97
Common Article 3, 106–7, 357
conflict not of an international character, 99
controlling state involvement in armed conflict
non-international armed conflict, 42–44
non-state actors, armed conflicts involving state, role as, 36–37
counterterrorism operations, 361
criminals, persons without proper authority categorized as, 98–99
drone strikes, 290–91, 318, 320
impact of, 97–99
law of armed conflict (LOAC)
humanitarian law and, 110
non-international armed conflict, 362–63
non-international armed conflict, 42–44, 98, 356–63, 379. *See* non-international armed conflict *main heading for more specific entries*
geography of war, 359
insurgents, 360
non-state actors, armed conflicts involving, 36–37
overview, 91–92
police primacy, non-international armed conflicts, 360–61
state and governance, 99–120
failed states, 104
governance, 101–4
inter-State conflict, 99–100
occupied territory, 103–4
public, private, and mixed war, 99–100
recognition of belligerency doctrine, 104–6
State's exercise of governance, 101–4
wars of national liberation, 109
state self-defense and human rights-based law enforcement, 449–50
ungoverned spaces, 112–19
unprivileged belligerent, 93–95, 97–99
proper authority principle, humanitarian law and equal application principle and, 92

humanitarian law, policy solution for legal uncertainty, 109–12
humanitarian law and human rights law interface, 145, 147
international armed conflict and, 95–97
international humanitarian law (IHL) application, 110–12, 122, 124, 135, 139*n*122, 152*n*202, 156
law of armed conflict (LOAC), 110
military advantage, proportionality test, 57, 245, 247, 254, 560–61, 603
nonterritorial State, 104
police primacy, 112
privileging state participants over non-state protagonists, 92–97
affected groups, 93
background, 92–94
belligerents
enemy unprivileged belligerent, 95
POW status of, 92, 94–95, 98
recognition of belligerency doctrine, 96
unlawful belligerent, 94–95
unprivileged belligerent, 93–95
Common Article 2, 97
equal application principle, 92
guerrilla, 93–94
Hague Land Warfare Regulations, Article 2, 93*n*12
humanitarian law, 91–96
international criminal law, 97
just war, 91–93
legal challenges, 92
legitimizing non-state participant in international armed conflicts, resistance to, 95–97
non-international armed conflict, 96
POW status of belligerents, 92, 94–95, 98
uniform application of law, impossibility of, 92–93
wars of national liberation, 96–97
proportionality and counterinsurgency, 258–64
state and governance, reluctant application of humanitarian law, 104–9
proportionality
application during armed conflict, 57, 70
and counterinsurgency, 258–61
humanitarian law
and human rights law interface, 132–33
military advantage, proportionality test, 57, 245, 247, 254, 560–61, 603
targeting proportionality, 57, 74, 77–79, 81–82, 254, 360, 456*n*64, 603
human rights-based law
application restrictions, overreach, 554
and humanitarian law interface, 132–33
limited application theory, 64
military advantage
assessments, 79
and counterinsurgency, 258–61
humanitarian law proportionality test, 245, 247, 254, 560–61, 603
targeting proportionality, 79, 81–82, 254
"overarching" application theory, 58, 61–62
recourse to war, 83–88
conduct of hostilities, 83–84
defensive responses, 85–88
isolated attacks, 85
law governing the recourse to war, 84–85
levels of war, 85, 87
national self-defense, 87
overarching effect of, 85
Principles of International Law on the Use of Force by States in Self-Defence (Chatham House, 2005), 88
state self-defense, 83–84
targeting decision, 84
UN Charter, Article 51, 86
UN Security Council, Resolution 1973 (2011), 83
strategic self-defense approach *vs.* tactical self-defense approach, 72–78
targeting, 78–83, 450, 507
humanitarian law, 57, 74, 77–79, 81–82, 254, 360, 456*n*64, 603
law enforcement and self-defense, 450, 507
military advantage, 79, 81–82, 254
strategic proportionality, 84
strategic self-defense approach *vs.* tactical self-defense approach, 78–83
UN Charter, Article 51, 86
proportionality and counterinsurgency, 258–61
Additional Protocol I, Article 57(2)(a)(iii), 258
collateral casualties, 259
collateral damage, 258–59, 261
humanitarian law, 258–64
human rights law, 260–63
insurgency/counterinsurgency, 259
isolated suspects, 259
rules of engagement (ROE), 259
proportionality assessment
state self-defense, armed conflict situations, 83–88
Prosecutor v. See name of defendant
protection of nationals, 383–441
Additional Protocol II, Article 1(2), 392*n*64
categorization of threats, challenges, 383–84
contemporary security environment, defining characteristic, 384
drone strikes, 277–78
extraterritorial protection of nationals, 385–98

protection of nationals (*Cont.*)
hostage rescue, 399–433 (*See also headings under* hostage rescue for more specific entries)
operations, personnel and tactics, 433–41
rescues and law enforcement, 421–31
rescues during armed conflict, 399–405
short-term conflicts, 415–21
kidnapping. *See* kidnapping
overview, 383–85
rescue operations, personnel and tactics, 433–41
responding to transnational attacks and threats to nationals, holistic operational law, 575–88
UN Charter
Article 2(4), 390, 392–93
Article 51, 384, 387, 389, 391–92, 395, 577, 581–82, 584
ungoverned spaces, 383, 389, 396, 398, 432
unwilling or unable test, 387, 389, 394, 396, 408
protracted armed violence, 417, 576, 585, 613
drone strikes, 293, 317
extraterritorial law enforcement, 352
insurgency, 161
non-state actors and armed conflict, 329, 352, 363, 366–67
proper authority principle, 107, 363
transnational armed conflicts, 354
Provisional Irish Republican Army (PIRA), 537, 539, 542
pseudo paramilitaries, 171
pseudo state, 569
ISIS as, 160*n*4
public authority principle, 447, 466*n*115
Public Committee Against Torture in Israel v. Israel. See Targeted Killing Case

Quirin, Richard, 94, 264–65

rebellion, defined, 161
Red Cross. *See* International Committee of the Red Cross (ICRC)
religious humanism, 135
rescue operations, personnel and tactics, 433–41. *See also* hostage rescue
commonality of tactical approach, 384–85
concentration of skills, 433–34
deployment of military and police counterterrorist units, 436
military or police forces, decision to use, 435–37
Research Assistance Intervention Dissuasion (RAID), 435–36
Special Air Service (SAS), 132, 274, 412–17, 420–21, 423, 434–35, 493, 496, 517, 526–28, 530–31, 538, 542–50
Special Boat Service (SBS), 413–15, 434
Specialist Firearms Command (SCO), 435
specialized police units, wartime role, 437
Special Operations Forces, 281–85
Vietnam, American POW rescue, 195–96, 400, 493
Special Operations Forces, Task Force 714, 440
status-based targeting, 303, 441, 456, 476, 527, 545
tactical developments, 439–41
tactics applied during armed conflict, 384
Unified Action Groups for Personal Liberty (GAULA), 422–23, 435
U.S. Army Delta Force, 8*n*42, 276, 404, 412, 434*n*405, 436–37
U.S. Navy SEALs, 260, 273, 275, 428, 437, 439
use of force approach, differences between military and police, 437–38
Vietnam, American POW rescue by Special Operations Forces, 195–96, 400, 493
Research Assistance Intervention Dissuasion (RAID), 435–36
The Responsibility to Protect Report (International Commission on Intervention and State Sovereignty, 2001), 10–11, 71
just war principles, 10–13
Revolutionary Armed Forces of Colombia (FARC), 66, 171–73, 175–76, 196–97, 212, 222, 260, 307, 349, 403, 405, 422, 594
Revolutionary Guard (Iran), 465
right to life, 138
robust peacekeeping, 153, 365, 467, 469–74, 479, 571, 604
law enforcement and self-defense, 467
non-state actors and armed conflicts, 365
UN mandated operations, 467–74
robust self-defense approach, drone strikes, 314–17
Rome Statute
Article 8(2)(a)(viii), 399*n*117
Article 8(2)(b)(iv), 75*n*137, 257*n*273
Article 8(2)(c)(iii), 399*n*117
Article 8(2)(d), 161*n*16
Article 8(2)(f), 161*n*16, 366
Royal Ulster Constabulary Headquarters Mobile Support Unit (HMSU), 542, 544, 546, 550–51
Royal Ulster Constabulary (RUC), 531, 538–44, 546, 550–51
Ruby Ridge (Idaho, 1992), 545

rules of engagement (ROE), 28, 437, 440, 470–71, 473–74, 483, 485, 488
blended rules of engagement (ROE) cards, 226–27
explosives, 530
Fallujah assault (2004), 476–79
imminence, 466
mission accomplishment
disagreements regarding, 461–62, 464
standing rules of engagement (SROE), 460–62
proportionality and counterinsurgency, 259
public order doctrine, 466
shoot to kill policy, 545
Soldier's Pocket Card issued for the UN Mission in Sudan (UMIS), 471
standing rules of engagement (SROE), 445, 460–62, 474–75, 478–79, 600
mission accomplishment, 460–62
state self-defense and human rights-based law enforcement
doctrinal confusion, 464–67
United States rules of engagement (ROE) and "self-defense," 474–79
types of self-defense, distinctions, 464–67
Russia
annexation of Crimea (2014), 17, 35
Beslan School siege (2004), 438, 499–500
Kerimova v. Russia, 125*n*22, 139*n*122, 452*n*45, 481*n*204, 555, 558*n*169
rescue operations, piracy, 429
Turkish shootdown of Russian war plane, Turkey/Syria border (2015), 35
Rwandan genocide (1994), 144*n*154, 373*n*223, 468–69, 571

Salafi jihadist, 4, 8, 160, 197–98, 199*n*364, 202, 207, 212, 228, 516, 569–70, 625
San Remo Manual on International Law Applicable to Armed Conflicts at Sea, 62, 74–75, 77, 305–6, 461, 464
Paragraph 22, 305, 311, 394*n*77
Santos Tomás raid (Philippines, 1945), 400–401
Sargsyan v. Azerbaijan, 103*n*86
Sayeret Matkal, Israeli military Special Forces unit, 434, 438, 498
Secretary-General's Bulletin: Observance by United Nations Forces of International Humanitarian Law (1999), 110*n*139, 120*n*214, 365, 379, 472
self-defense
humanitarian law, 55–62
human rights-based law enforcement, 445–79
just war, 55–56
legal principles
immediacy, 55–57, 65, 68
imminence. *See* imminence, self-defense legal principle
necessity. *See* necessity, self-defense legal principle of
proportionality. *See* proportionality
less certain boundaries, 88–89
overview, 55
protection of nationals. *See* protection of nationals
recourse to war, 55–89
limited application theory, 63–70
"overarching" application theory, 58–63
strategic approach *vs.* tactical approach, 72–88
UN Charter, Article 51, 56, 63*n*50, 65, 67, 70–72
self-defense limited application theory, 63–70
application challenges, 64–65
attacks viewed in isolation, 68
constraining influences, exemption from, 65
counterterrorism operations and, 66–69
distinguished from overarching application theory, 65–66, 69
equal application principle, 65
extraterritorial law enforcement, 66–67
humanitarian law, 69
law governing the recourse to war, 64
legal challenges, 65
level of conflict, 64
non-state actors, 63, 66–67
organized armed groups, 65, 67
proportionality assessment, 64
protection of nationals, 63–64
short of war, 64–65, 69
UN Charter, Article 51, 65, 67, 70
war *vs.* situation short of war, 64–65, 69
self-defense "overarching" application theory, 58–63
Additional Protocol I, Article 52, 58
antiwar movement, 61
distance warfare, 61
distinguished from limited application theory, 65–66, 69
double effect doctrine, 61
example of application theory, 59–60
geographic limitations-62, 55–62
humanitarian law, 55–62
hybrid warfare, 61
indirect application of overarching theory, post-9/11 Afghanistan conflict (2002), 63
issues affected by overarching approach, 58–59
law governing the recourse to war, 61–62
Lebanon, occupation by Israel (1982), 66

self-defense "overarching" application theory (*Cont.*)
- necessity principles, 58–63
- neutrality, 59
- non-state actors, 62
- proportionality, 58, 61–62
- *San Remo Manual on International Law Applicable to Armed Conflicts at Sea*, 62
- temporal and geographic scope of conflict, 58
- UN Charter, Article 51, 63*n*50
- where warfare is conducted, 59

self-defense option, drone strikes. *See* drone strikes, "self-defense" option

Shehada, Saleh, 231, 271*n*22, 282, 286

shoot-down of hijacked aircraft, 260–61, 559–65
- Additional Protocol I
  - Article 57, 560
  - Article 57(2)(a)(iii), 559*n*174
- *Air and Missile Warfare Program on Humanitarian Policy and Conflict Research (2009), Manual Commentary on*, 11, 560, 578
- call-out approach, Australia, 562–63, 593
- collateral casualties, 562
- collateral damage, 560
- differing approaches, 559–63
- German approach, 561–63
- humanitarian law, 551–63
- national self-defense, 560, 563
- North American Air Defense (NORAD) Agreement, 559–60, 563
- one-off attacks by non-state actors, holistic response to, 563–64
- transnational criminal organizations, 564
- UN Charter, Article 51, 559
- weighing the value of life, 559

shoot to kill policy
- ambush style operations, 541, 543–44, 547–49
- avoidance of shoot to kill, 131–32
- Common Article 3, 542
- controversy, 536–50
- direct participation in hostilities, 545
- explosives, 522, 528–31
- human rights normative framework, application demands, 546–47
- law enforcement and self-defense, 458
- minimum force, 548–49
- rules of engagement (ROE), 545
- shooting to kill policy, 544–50
- suicide bombing, 548
- targeted killing, 544–45
- *The "Troubles"* (Northern Ireland), 535–44, 549–50
- shoot to wound, 529–31

short of war, 410, 579
- drones, 313, 316
- extraterritorial law enforcement, 350
- war *vs.* situation short of war, 64–65, 69

Shower Posse (Jamaica), 481*n*207, 589, 599

Sierra Leone
- *Operation Barras,* hostage rescue by British Special Forces(2000), 412–20, 434, 493, 586
- Revolutionary United Front (RUF), 173, 412
- UNAMSIL, 473
- West Side Boys, 28, 412–17, 577, 586

Sierra Leone, Revolutionary United Front (RUF), 173

signature strikes, 278–82

Sinaloa cartel, 170*n*94, 171, 173

small wars, 112, 114, 140–41, 149, 225–26, 228, 536, 572, 624–25. *See also* one-off attacks by non-state actors
- counterinsurgency and counterterrorism, 212
- definition, 36
- drone strikes, 319
- guerrilla warfare, suppression of, 36
- insurgents and target "value," 244
- one-off attacks by non-state actors, 53, 108, 219, 311
- statistics, 36
- twenty-first century counterinsurgency, 218–19, 223

social banditry, 179–80

soft knock approach, 248–49, 361

soft law, 128–29, 227

Somalia piracy, hostage rescue, 426–31
- Planned German rescue operations by GSG-9 (2009), 435–36
- rescue of American Captain Phillips *(2009)*, 328, 397, 427–28, 430, 491, 494, 496

South West Africa People's Organization (SWAPO), 162

Special Air Service (SAS), 132, 274, 412–17, 420–21, 423, 434–35, 493, 496, 517, 526–28, 530–31, 538, 542–50

Special Boat Service (SBS), 413–15, 434

Special Boat Squadron (Nigeria hostage rescue, 2012), 424–25

Special Operations Forces, counterterrorism and direct action, 271–77. *See also* drones
- capture rather than kill, 273, 275–77
- direct action mission, definition of, 272
- direct participation in hostilities, 270
- distinguishing between U.S. Special Forces and special operations units, 8*n*42
- generic application of the term, 8*n*42
- hostage rescue operations, 273
- lack of consensus, 269–71

operations carried out at invitation of territorial State, 275
primary counterterrorist means, definition, 272–73
rescue operations, personnel and tactics, 281–85
Special Air Service (SAS), 274
specially trained personnel, 273
totality of the circumstances, 276
transnational criminal organizations, 274
types of operations, 273
ungoverned spaces, 271
U.S. Army Delta Force, 8*n*42, 276, 404, 412, 434*n*405, 436–37
U.S. Navy SEALs, Team 6, 273–76
Western military forces, change in focus and expectations, 274–75
Specialist Firearms Command (SCO), 435
special law *(lex specialis)*, 10, 18, 134, 338, 578, 623–24
definition of, 11
drone strikes, "self-defense" option, 312
and exclusion debate, 122–30
human rights and humanitarian laws, counterinsurgency frameworks, 262
Special Mission Units, 8*n*42
Special operations forces
special mission units, 8*n*42
Tier I, 8*n*42, 274, 436, 493
Tier II, 274
U.S. Army Delta Force, 8*n*42, 276, 404, 412, 434*n*405, 436–37
Sri Lanka, Liberation Tigers of Tamil Eelam (LTTE), 192, 515
SS Maersk Alabama (2009), 397, 427–28, 430, 494, 496
SS Mayaguez (1975), 397, 400–402
standing rules of engagement (SROE), 445, 474–75, 478–79, 600
mission accomplishment, 460–62
Stanleyville Congo rescue operation (1964), 403
state involvement in armed conflict, controlling, 33–54
Common Article 3, 44
destructive power of state, 37–42
development of conflict law, 33–34
inter-state conflict, 34–37
non-international armed conflict, 43–54
non-state actors, armed conflicts involving
Common Article 3, internal conflicts with international character, 44
complexities of, 34
human rights law, application of, 36–37
non-international armed conflict, 43–44
state, role as proper authority, 36–37
state as primary actor in conflict, applied norms, 36–37
UN Charter, Article 51, 48–52
overview, 33–34
proper authority principle
non-international armed conflict, 42–44
non-state actors, armed conflicts involving state, role as, 36–37
state self-defense
drone strikes, 289–90, 313–16
human rights-based law enforcement. *See* state self-defense and human rights-based law enforcement
proportionality assessment, armed conflict situations, 83–88
state self-defense and human rights-based law enforcement
distinguished, 445–64
armed conflict, existence of, 455–56
Common Article 3, 451
differing goals, 447–49
double effect doctrine, 450
governing principles, state self-defense, 443
imminence of threat, assessment, 454–58
individual self-defense concept, 447, 450–51, 457–58, 460, 462, 464
legal approaches, 447–49
mission accomplishment, 458–64
nature of threat, 454–58
policing and combat, 452–53
proper authority principle, 449–50
public authority principle, 447
terminology challenge, 445–47
unique role of law enforcement, 449–54
urban environments, 454
doctrinal confusion, 464–79
individual self-defense concept, 464, 466, 470, 478, 480
public authority principle, 466*n*115
rules of engagement (ROE) doctrine, 464–67
UN Charter, Article 51, 464–65, 470
United States rules of engagement (ROE) and "self-defense," 474–79
UN mandated operations, robust peacekeeping, 467–74
national self-defense, 444, 447, 451, 462*n*92, 462*n*95, 463–66, 478, 481
defined, UN Charter, Article 51, 464
and unit defense, distinction, 464–67
principles of state self-defense, 467
UN Charter, Article 51
application restrictions, 559
doctrinal confusion, 464–65, 470
national self-defense, 464
unit self-defense, 460*n*80, 462*n*91, 464–66
and national defense, distinction, 464–67

The State of the Netherlands v. Nuhanović, 21*n*107
state *vs.* non-state actors conflicts, categorizing, 335–63, 378–80
status-based targeting, 303
strategic self-defense approach *vs.* tactical self-defense approach, 72–88
antiwar movement, 75
collateral casualties, 82, 85
equal application principle, 79
grand strategy, 76, 81, 610
Hezbollah, 86–87
humanitarian law, 72–78, 84–89
just war, 75, 77
law governing the recourse to war, 73–75, 78–79
League of Nations, 75
legal challenges, 79
levels of war, 72–78
proportionality, 72–78
state self-defense proportionality assessment, armed conflict situations, 83–88
targeting proportionality, 78–83
Rome Statute, Article 8(2)(b)(iv), 75*n*137
San Remo Manual on International Law Applicable to Armed Conflicts at Sea, 74–75, 77
state self-defense proportionality assessment, armed conflict situations, 83–88
strategic concept, generally, 74–76
tactical concept, generally, 74, 79–80
targeting, strategy and tactics, 78–83
suicide bombing, 26, 215
capture or kill, 239, 241–43
control and operational choices in counterinsurgency, 237–39
counterterrorism operations, 268, 283, 285, 307
drone strikes, 268, 283, 285, 307
humanitarian law and human rights law interface, overlap and convergence, 154
human rights-based law application restrictions, overreach, 557
improvised explosive devices (IEDs) and suicide bombing threat, capture or kill, 215, 237–43
Israeli Supreme Court and cross-border terrorism, 343
narrow operational and normative gap, explosives, 515–19, 521, 524–25
shoot to kill policy, 548
statistics, 516, 523
terrorism and insurgency, Al Qaeda, 200
threat of attacks, 517–18
Sun Tzu, 295, 569
Susanese People's Liberation Army (SPLA), 162
SV Quest, 428
Syria, Israeli strike on nuclear reactor (2007), 57
Syrian Conflict, 108–9
air strikes against Islamic State in Syria (2015), 51, 325, 583
Al Qaeda and an Islamic state, 200–202, 287, 295, 323–26
shootdown of Russian war plane, Turkey/Syria border (2015), 35

tactical proportionality, 78, 82, 85
Tadić; Prosecutor v., 14*n*68, 34*n*2, 35*n*10, 107, 109, 161*n*14, 175, 262*n*298, 292–93, 315*n*314, 316, 329*n*4, 331*n*11, 335*nn*30–31, 348*n*96, 354, 363, 365–67, 370, 373, 417, 419, 456*n*65, 518*n*321, 575*n*21, 576, 579, 580*n*45, 581–82, 585
Taliban, 47, 111–13, 115–16, 168–69, 193–95, 198–99, 298, 337–38, 496, 512
Afghan Taliban, 116, 193–95, 251, 360
insurgency, 164
insurgents and target "value," 251–53
Neo-Taliban, 193, 199
suicide bombing threat, capture or kill, 242
twenty-first century counterinsurgency, 219, 221–22
Tamil Tigers, 192, 515
The Tao, defined, 569
targeted killing, 279–80, 603–4. *See also headings under* drone strikes; *Targeted Killing Case*
control and operational choices, 236–37
counterinsurgency, 236–37
equal application principle, 79
explosives, 522, 525, 527–28, 533
hard knock approach, 248–49, 361
hostilities based targeting, 456, 565
humanitarian law, 445, 455–56
and human rights law interface, general principles recognized by civilized nations, 132–33
improvised explosive devices (IEDs), 243
insurgents and target "value," 244–53
Israeli Supreme Court and cross-border terrorism, 341–43
non-international armed conflict, proper authority, 358–59, 361
police primacy, 249
proportionality, 450, 507
shoot to kill controversy, 544–45
status-based targeting rules, 303, 441, 456, 476, 527, 545
strategic level targeting, 525

strategic proportionality, 84
strategy and tactics, 78–83
suicide bombing threat, capture or kill, 243
targeting proportionality, humanitarian law, 57, 74, 77–79, 81–82, 254, 360, 456*n*64, 603
Targeted Killing Case, 21*n*106, 131*n*60, 230–37, 243, 245, 247, 258*n*280, 261–63, 265, 286, 320–21, 341–49, 361, 376*n*245, 377*n*249, 445*n*6, 604, 608
Additional Protocol I, Article 35(2), 232*n*104
Additional Protocol I, Article 41(2)(a), 232*n*103
capture, consideration as a matter of law, 236
capture or kill, 231–32
direct participation in hostilities, 232
Hamas, 231, 237
hors de combat, 231–33
human rights law, application of, 233
improvised explosive devices (IEDs), 234, 237
lex lata, 237
Team 6. *See* U.S. Navy SEALs
Tehrek-e-Taliban Pakistan (TTP), 115, 193–94, 514
terrorism, 180–90. *See also* counterterrorism; suicide bombing; transnational criminal organizations
Al Qaeda and an Islamic state, 197–208
caliphate, 4, 193–94, 569
campaign against terrorism (CAT), 20, 181
Canadian, U.S., and UK responses to terrorism, 185–86
counterterrorism. *See* counterinsurgency and counterterrorism, integrated response
dacoit, 180
defining, difficulties, 180–82, 184
Drone Speech (2013) (President Obama), 3–4, 20–21, 181, 191, 298–99, 322, 363, 596, 609
Entebbe, Israeli hostage rescue (*Operation Thunderbolt (Yonatan*, 1976)), 185
Fallujah assault (2004), 189
focoist approach, 208, 308, 570
genesis of, 183
Global War on Terrorism (GWOT), 181–82, 304, 312–13
governance, 190–91
guerrilla, 186–89
guerrilla warfare, 187–88
improvised explosive devices (IEDs), 194
and insurgency
Al Qaeda and an Islamic state, 197–208
counterinsurgency. *See* counterinsurgency and counterterrorism, integrated response
insurgent groups using, 190–93
intelligence as effective weapon against terrorism, 541–42
international criminal law, terrorism definition for purposes of, 181
irregular warfare, 180
isolated acts of terrorism theory, 294–96
Israeli Supreme Court and cross-border terrorism, 341–49
jihadist, 190
kidnapping, 184, 188, 190
levels of violence, 183
modern terrorism, 185
1960s-1980s, 186–87
overview, 159–60
patterns of decline and ending campaigns, 612
post-World War II political violence, 187
struggle for legitimacy, 612–13
terrorist activity, 184–85
twentieth century changes in, 183, 186
typology of, 183
UN Charter, Article 51, 183, 187*n*248
UN Security Council Resolution 2249 (2015), 203*n*400
urban focoist theory, 187–88
urban guerrilla warfare, 183, 187–88, 235, 371
vanguard approach, 208, 308, 570
terrorists, 190–97
autonomy of terrorist groups, extraterritorial law enforcement, 352
groups. *See specific group name*
groups, autonomy of, 352
guerrilla, 191
insurgent groups using terrorism, 190–93
insurgents and terrorists, distinction and relationship, 190–97, 360
jihadist, 193–94
networks, 191, 198–99
non-international armed conflict, proper authority, 360
terrorist activity, 184–85
transnational criminal organizations, 196
third-generation transnational gangs, 172–73
Thompson, Sir Robert, 113, 535
totality of the circumstances, 27, 418, 425, 485, 579, 582, 586
drones, 317
extraterritorial protection of nationals, 391
insurgency, 161, 175
Japanese Embassy Case (Lima, Peru; 1997), 508
narrow operational and normative gap, 485, 508
non-state actors and armed conflict, 329, 375–76
Special Operations Forces, counterterrorism and direct action, 276

transnational armed conflicts, 27, 365, 367, 378, 419
categorizing State *vs.* non-state actors conflicts, 352–56
Common Article 2, 352
Common Article 3, 352–55
conflict not of an international character, 356
conventional warfare, 353
counterinsurgency doctrine, 353
domestic law application, 353
extraterritorial law enforcement, 356–57
geography, threat not limited by, 353
international armed conflict, 355–56
kidnapping and hostage rescue, 354–55
means and authority test, 353, 355, 376, 586
means necessary to remove or neutralize threat, 354
naked self-defense, 355
non-international armed conflict, 355
protracted armed violence, 354
transnational armed conflicts, distinction from traditional international conflict and noninternational conflict, criteria, 353
triggers for existence of conflict, 353, 355
holistic operational law approach
decision tree analysis (chart), 617
responding to attacks and threats to nationals, 575–88
transnational violence, confronting, 604–10
hostage rescues during armed conflict, 354–55
non-international armed conflict, transnational threats rising to armed attack by non-state actors, 52–53
non-state actors and armed conflicts, 352–56, 365, 367, 378
responding to attacks and threats to nationals, holistic approach, 575–88
armed conflict or law enforcement, 575–83
commencement of armed conflict, 584–87
separating armed conflict from law enforcement, 587–88
UN Charter, Article 51, 577, 581–82, 584
transnational attacks and threats to nationals, responding
holistic operational law, 575–88
armed conflict or law enforcement, 575–83
commencement of armed conflict, 584–87
separating armed conflict from law enforcement, 587–88
UN Charter, Article 51, 577, 581–82, 584
transnational criminal organizations, 6–7, 26, 196, 564, 570, 588, 590
criminal gangs, 3, 5–6, 9, 168, 172–73
drones, 283–84
drone strikes, 283–84
generally, 6–7
insurgency, 160
insurgency and crime, 170–73, 176–77, 179
law enforcement and human rights law, limitations, 564
law enforcement and self-defense, 480–81
limitations and need for clarity, 564
shoot-down of hijacked aircraft, 564
Special Operations Forces, counterterrorism and direct action, 274
terrorists, 196
third-generation transnational gangs, 172–73
types of criminal activity, 9
transnational terrorism, proper authority principle, 290–91
transnational violence, holistic operational law
confronting, holistic approach, 604–10
decision tree analysis (chart), 617
Trinquier, Roger, 159
The "Troubles" (Northern Ireland), 29, 52, 76, 175, 179, 236, 262, 328, 590, 593
categorization of conflict, 539
Common Article 3, 542
conflict participants, 537
counterinsurgency doctrine, 542
failed states, 537
guerrilla, 538–39
human rights and humanitarian laws, counterinsurgency frameworks, 262
improvised explosive devices (IEDs), 539–40
intelligence as effective weapon against terrorism, 541–42
Irish National Liberation Army (INLA), 537–40, 543–50
Irish Republican Army (IRA), 537
Loyalist Volunteer Force (LVF), 537
period of conflict, 536
phases of conflict, 539
police primacy, 541, 543
Provisional Irish Republican Army (PIRA), 537, 539, 542
Royal Ulster Constabulary Headquarters Mobile Support Unit (HMSU), 542, 544, 546
Royal Ulster Constabulary (RUC), 531, 538–44, 546
shoot to kill policy, 535–44, 549–50
small wars, 536
smuggling of military weapons, 540–41
Special Air Service (SAS), 538, 542–50
"special law" and exclusion debate, 130
targeting, 544–45
Ulster Freedom Fighters (UFF), 537

Ulster Volunteer Force (UVF), 537
United Kingdom, law enforcement framework, 542–44
use of force rules, 543
yellow cards, 543
Túpac Amaru Revolutionary Movement (MRTA), 501–2, 505–8
Turkey
Ergi v. Turkey, 507*n*218
shootdown of Russian war plane, Turkey/Syria border (2015), 35
Varnava v. Turkey, 552*n*139
twentieth-century law, legal challenges, 16–17, 23
twenty-first-century conflict preparation, 619–31
twenty-first century counterinsurgency, 216–23
Afghanistan
International Security Assistance Forces (ISAF) and post 9/11 Afghanistan conflict, 217, 220
Boko Haram, 222
British context, 219
broad perspective of, 219
counterinsurgency doctrine, 218–20, 223
fundamental aspect of, 219
governance, 220–21
guerrilla war, 218
insurgency/counterinsurgency, 220–21
International Security Assistance Forces (ISAF), post 9/11 Afghanistan conflict, 217, 220
Iraq, 217–18
irregular warfare, 216*n*4
Islamic State in Iraq and al Sham (ISIS), 221–22
The National Military Strategy of the United States of America 2015, 221
one-off attacks by non-state actors, 219
small wars, 218–19, 223
Special Operations Forces, 218
Taliban, 219, 221–22

Uganda
Entebbe, Israeli hostage rescue (*Operation Thunderbolt (Yonatan*, 1976)), 405–12
Ukraine, Russia's annexation of Crimea (2014), 17, 35
Ulster Freedom Fighters (UFF), 537
Ulster Volunteer Force (UVF), 537
ungoverned spaces, 25–27, 29–30, 358–60, 590–91, 605–6
Afghanistan, Taliban and, 116
Al Qaeda in Iraq (AQI), 115–16
bin Laden, Osama, 117
Boko Haram, 118–19
counterinsurgency and counterterrorism, 212
counterinsurgency campaigns, conduct and aims of, 119
criminal insurgency, 118
drone strikes, 268, 271, 298, 318, 320, 322
"conduct of hostilities" approach, 300
"self-defense" option, 318
failed states, 114*n*161
Fallujah assault (2004), 116
gang control over territory, 118–19
guerrilla, 112–16
Hamas, seizure of Gaza (2007), 117
insurgency and crime, 168, 178–79
insurgent organizations, 164
Islamic State in Iraq and al Sham (ISIS), 115
Israel-Hezbollah war (2006), 117–18
jihadist, 114–16
lack of State control, historically, 114–15
mosaic war, 119
Pakistan, Al Qaeda and, 115–16
Palestinian state, 117
police primacy, 112–13
savagery, 115–17
State exercise of control, 113–14
takfiri extremists, 117
Yemen, Al Qaeda and, 114–15
Unified Action Groups for Personal Liberty (GAULA), 422–23, 435
United Kingdom
Al-Skeini v. The United Kingdom, 103*n*88, 120*n*216, 149*n*181, 152*n*200, 230*n*86, 234*n*114, 402*n*152, 487*n*30
Bubbins v. The United Kingdom, 497*n*106
counterterrorism operations, use of drones, 283
drones, use of, 283
drone strikes, 283
Hassan v. The United Kingdom, 103*n*88, 124–25, 127–28, 152*n*200, 471*n*146, 552– 553, 558, 592–93, 614*n*190, 615, 624
Kelly v. United Kingdom, 544*n*79, 544*n*81, 544*n*83
The Manual of the Law of Armed Conflict, 13*n*65, 95*n*27, 310*n*279, 387
McCann and others v. United Kingdom, 21*n*105, 131*n*57, 286*n*118, 462*n*88, 525*n*350, 530–31, 544*n*79, 544*n*81, 545*n*85, 546–48, 549*n*122, 596*n*91
McKerr v. United Kingdom, 131*n*62, 544*n*81, 544*n*84, 546*nn*95–96, 547, 550, 627*n*32
Northern Ireland, *"The "Troubles." See* "Troubles" (Northern Ireland)
responses to terrorism, 185–86
Special Boat Squadron, Nigeria hostage rescue (2012), 424–25

United Nations
Basic Principles on the Use of Force and Firearms by Law Enforcement Official, 128, 131*nn*57–59, 227*n*74, 333*n*28, 445*n*5, 448, 449*n*32, 459, 461*n*84, 472, 484*n*6, 485*n*10, 495, 507, 524*n*342, 549*n*122
mandated operations, robust peacekeeping, 467–74
Secretary-General's Bulletin: Observance by United Nations Forces of International Humanitarian Law (1999), 110*n*139, 120*n*214, 365, 379, 472
United Nations Charter. *See also* United Nations Charter, Article 51
Preamble, 41*n*67
Article 2(4), 40*n*63–40*n*65, 41, 43, 52, 390, 392–93, 470
Article 2(7), 41, 141
Article 4, 102*n*74
Article 42, 460*n*79, 468*n*124, 473*n*159
Chapter VII, 458–61
customary law requirement limitations, 70
creation of, 33
just war theory, influence of theory on Charter, 41
limitation on states recourse to war, framework, 40–42
United Nations Charter, Article 51
cross-border attacks by non-state actors and, 48–52
drones, 318–19
"conduct of hostilities" approach, 301
restricted "law enforcement" theory, 290, 296
unwilling or unable justification for drone strikes, 325
extraterritorial protection of nationals, state responsibility, 384, 387, 389, 391–92, 395
holistic operational law, responding to transnational attacks and threats to nationals, 577, 581–82, 584
human rights-based law
application restrictions, overreach, 559
enforcement, doctrinal confusion, 464–65, 470
inter-state conflict, 331
law enforcement and self-defense
national self-defense, defined, 464
state self-defense and human rights-based law enforcement, 464–65, 470
non-international armed conflict, cross-border attacks by non-state actors, 48–52
non-state actors and inter-state armed conflict, 331
self-defense principles, application during armed conflict
application of self-defense principles, 56
limited application theory, 65, 67, 70
"overarching" application theory, 63*n*50
reporting requirements, 71
strategic proportionality, state self-defense assessment, 86
UN Security Council decision regarding self-defense, 70–72
shoot-down of hijacked aircraft, 559
state self-defense and human rights-based law enforcement, doctrinal confusion, 464–65, 470
terrorism, 183, 187*n*248
United Nations mandated operations, robust peacekeeping, 467–74
Chapter VI mandated operations, 468–69, 481
Chapter VII mandated operations, 468–69, 473–74
conduct of operations, 468
defense of mandate approach, 468, 474
humanitarian law and, 472–74
"last resort," use of force as measure of, 470
mission mandates, 468–74
non-use of force in self-defense, 469–70
peace enforcement, distinguished from robust peacekeeping, 469–70
"self-defense" term to describe basis for peacekeeping, 474
UN Charter, Article 2(4), 470
UN Peacekeeping Operations: Principles and Guidelines, 469–72, 474, 479
UN Security Council Resolutions, 469, 473
United Nations Mission in Sierra Leone (UNMASIL), 412–13
United Nations Mission in Sudan (UMIS), 471
United Nations Operations in the Congo, 468
United Nations Security Council. *See also* United Nations Security Council Resolutions
humanitarian law, authorization of military intervention for human protection purposes, 10–11
International Security Assistance Forces (ISAF), establishment by Security Council, 47
mandated operations, robust peacekeeping, 469
UN Charter, Article 51 and UN Security Council decision regarding self-defense, 70–72
United Nations Security Council Resolutions, 469
Resolution 1264 (1999), 473*n*163
Resolution 1270 (1999), 473
Resolution 1368 (2001), 50
Resolution 1373 (2001), 50, 304*n*240, 307*n*257

Resolution 1386 (2001), 47*n*109
Resolution 1564 (2004), 109*n*134
Resolution 1851 (2008), 427–28
Resolution 1973 (2011), 83, 304
Resolution 2060 (2012), 165*n*52
Resolution 2249 (2015), 47*n*112, 203*n*400, 298, 323–25
United Self-Defence Groups of Colombia (AUC), 171–72
United States
carrying arms openly hostility threshold, 477
existence of armed conflict, 478
groups declared as hostile, 476
objections, Additional Protocol I would legitimize terrorist groups, 146
responses to terrorism, 185–86
rules of engagement (ROE) and "self-defense," 474–79
standing rules of engagement (SROE), 445, 460–62, 474–75, 478–79
United States Diplomatic and Consular Staff in Tehran Case (United States v. Iran), 338, 409
United States Operational Law Handbook, 18*n*87, 19, 124, 150–51, 153*n*204, 457, 475*n*173, 476, 479, 489, 622–23, 629
unit self-defense, 460*n*80, 462*n*91, 464–66
Universal Declaration of Human Rights, 136, 139–40
Articles 3 and 28, 128
unlawful belligerent, 94–95
unlawful combatant, 94
drone strikes, 309
Targeted Killing case, 348–49
unprivileged belligerents. *See also* belligerents
capture of, 441
drone strikes, "conduct of hostilities" approach, 309–10
enemy unprivileged belligerent, 95
inter-state conflict, status of unprivileged beligerent, 334
post-WW II status, extension to illegal aggressions, 13
proper authority principle, 93–95, 97–99
unwilling or unable test
drone strikes, 322–26
protection of nationals, 387, 389, 394, 396, 408
urban focoist theory, 187–88
urban guerrilla warfare, 183, 187–88, 235, 371
U.S. Army Delta Force, 8*n*42, 276, 404, 412, 434*n*405, 436–37
U.S. Congressional Authorization to Use Military Force (2001), 48
U.S. Counterterrorism Manual (2014), 255
U.S. Navy SEALs
Abdulka-dir Warsame, operation to seize, 275
Team 6, 8*n*42, 434*n*405, 436, 498*n*115
bin Laden raid (Pakistan, 2011), 247–48, 272–76, 284, 305, 385, 439–41
rescue of Jessica Buchanan, 260, 437
USS Cole, 336*n*37, 577*n*27, 580
USS Stark, 464–65
USS Vincennes, 465

valid target, humanitarian law, 307–8
vanguard/focoist approach, 208, 308, 570
urban focoist theory, 187–88
Varnava v. Turkey, 552*n*139
Vehicle-Borne Improvised Explosive devices (VBIEDs), 511
Vietnam War, American POW rescue, 195–96, 400, 493
vigilante based organizations, 171–72, 174

Waldock, Sir Humphrey, 393, 396, 410
war concept of self-defense, 291–92
warfare as characteristic of human existence, 36
war on terrorism, 67*n*86, 351
Global War on Terrorism (GWOT), 181–82, 304, 312–13
Warsame, Abdulka-dir, 275–76, 277*n*57, 514*n*274
Westgate Mall attack (Kenya, 2013), 293, 368, 437–38, 564, 597
Westphalian system of government, 569
challenges to, 591
post-Westphalian period, 38, 43, 99, 466
state structure, 228
West Side Boys (Sierra Leone), 28, 412–17, 577, 586
wilayat concept, 201
World War I
deaths, statistics, 34*n*4
post war effects on inter-state conflict, limiting recourse to war, 39–40
World War II
deaths, statistics, 34*n*4
post war effects
belligerent status, extension to illegal aggressions, 13
concept of state, 18
deaths, statistics, 34*n*4
human rights development, 15
inter-state conflicts, decline in, 16–17
post-war humanitarian law and human rights law interface, historical background, 15, 139–52
accountability, 151
application of human rights norms, renewed interest, 149

World War II (*Cont.*)
post-war humanitarian law and human rights law interface, historical background (*Cont.*)
Civilian Protection Law (CPL), 148
Common Article 3, 139, 148, 150*n*189
conventional warfare, 140
criticisms, 144–47
customary human rights law, 19, 150*n*189, 151, 153, 622
Customary International Humanitarian Law Study (2005, International Review of the Red Cross), 146
Department of Defense Detainee Program, 148
divergence and growing reconciliation, 143–52
extraterritorial application of human rights treaty law, U.S. and Israeli position on, 149–50
forced integration and treaty law, 141–43
disinterest in enacting new treaty law, 145–46
guerrilla wars, 141, 145
human rights community, disinterest in amending existing treaty law, 144–46
International Committee of the Red Cross (ICRC), 141
parallelism, 142
small wars, 140–41, 145
gaps in application of human rights norms, 152
Geneva Convention, Articles, 143–44
impact of 9/11, 147–48
international armed conflict, 147–48
International Committee of the Red Cross (ICRC), 140–41
international criminal law, 1990s through 2000s, 144
International Law Commission, 140
law of armed conflict (LOAC), 150
neglect, 139–40
nonbinding recommendations from human rights instruments, 147
objecting states, 145–46
operational law, 149–50
parallelism, 142
post-World War II treaty law, 134–35, 141–43
proper authority principle, 145, 147
small wars, 149
UN Convention Against Torture, 149–50, 153
Universal Declaration of Human Rights (1948), 136, 139–40
U.S. objections, Additional Protocol would legitimize terrorist groups, 146
U.S. Operational Law Handbook, 150–51
wars of national liberation, 145

Yamam, Israeli tactical hostage rescue team, 435, 438–39
yellow cards, 543
Yemen
Al Qaeda, 6, 67, 114–15, 193, 198, 212, 252
Al Qaeda and Associated Movements (AQAM), 205
Al Qaeda in the Arabian Peninsula (AQAP), 198, 212
Department of Justice Drone Opinion, 302
drone strikes, 255, 275, 280, 282–84, 287–89, 321, 493
"ungoverned spaces," 114–15
USS Cole, 336*n*37, 577*n*27, 580
(former) Yugoslavia, 34–35, 107, 161, 167, 292–93, 315–17, 335, 348, 366–67, 372, 374–76, 417–19, 448, 575–76, 579–80

Zetas, 170–71

Printed in the USA/Agawam, MA
November 15, 2017

662637.006